D0855706

VICKERS INDUSTRIAL HYDRAULICS MANUAL

(Sixth Printing)

Library of Congress Catalog Card Number 91-121667

Vickers Industrial Hydraulics Manual

ISBN 0-9634162-0-0

Training Center
2730 Research Drive
Rochester Hills, Michigan 48309-3570

PREFACE

"To cherish traditions, old buildings, ancient cultures, and graceful lifestyles is a worthy thing — but in the world of technology, to cling to outmoded methods of manufacture, old product lines, old markets, or old attitudes among management and workers is a prescription for suicide."

(author unknown)

Meeting the workplace challenge of increasing skills, productivity, and knowledge while remaining competitive requires continuous on-the job training. The rapidly expanding availability of global products, increased complexity of supporting systems, and the continual changing technologies makes it necessary to update this manual periodically.

This version now includes metric conversions for all units of measure, extensive changes to the Contamination Control chapter, and an added appendix for Noise Control in Hydraulic systems.

We know you will find this manual to be a very helpful document and will recommend it to your associates.

Vickers Training Center
2730 Research Drive
Rochester Hills, Michigan 48309-3570
Tel: 313-853-1100
Fax: 313-853-1061

The standards for graphic symbols and color-coding of flows and pressures that were established by the Joint Industry Conference (J.I.C.) and the American National Standards Institute (A.N.S.I.) have been used throughout this manual. Significance of the symbols can be found in Chapter 2 and Appendix B. The color key is marked below.

RED	Operating or System Pressure
BLUE	Exhaust Flow
GREEN	Intake or Drain
YELLOW	Measured (Metered) Flow
ORANGE	Reduced Pressure, Pilot Pressure, or Changing Pressure
VIOLET	Intensified Pressure
BLANK	Inactive Fluid

TABLE OF CONTENTS

CHAPTER 1

AN INTRODUCTION TO HYDRAULICS

The study of hydraulics deals with the use and characteristics of liquids. Since the beginning of time, man has used fluids to ease his burden. It is not hard to imagine a caveman floating down a river, astride a log with his wife and towing his children and other belongings aboard a second log with a rope made of twisted vines.

Earliest recorded history shows that devices such as pumps and water wheels were known in very ancient times. It was not, however, until the 17th century that the branch of hydraulics with which we are to be concerned first came into use. Based upon a principle discovered by the French scientist Pascal, it relates to the use of confined fluids in transmitting power, multiplying force and modifying motions.

Pascal's Law, simply stated, says this:

Pressure applied on a confined fluid is transmitted undiminished in all directions, and acts with equal force on equal areas, and at right angles to them.

This precept explains why a full glass bottle will break if a stopper is forced into the already full chamber. The liquid is practically noncompressible and transmits the force applied at the stopper throughout the container (Figure 1-1). The result is an exceedingly higher force on a larger area than the stopper. Thus it is possible to break out the bottom by pushing on the stopper with a moderate force.

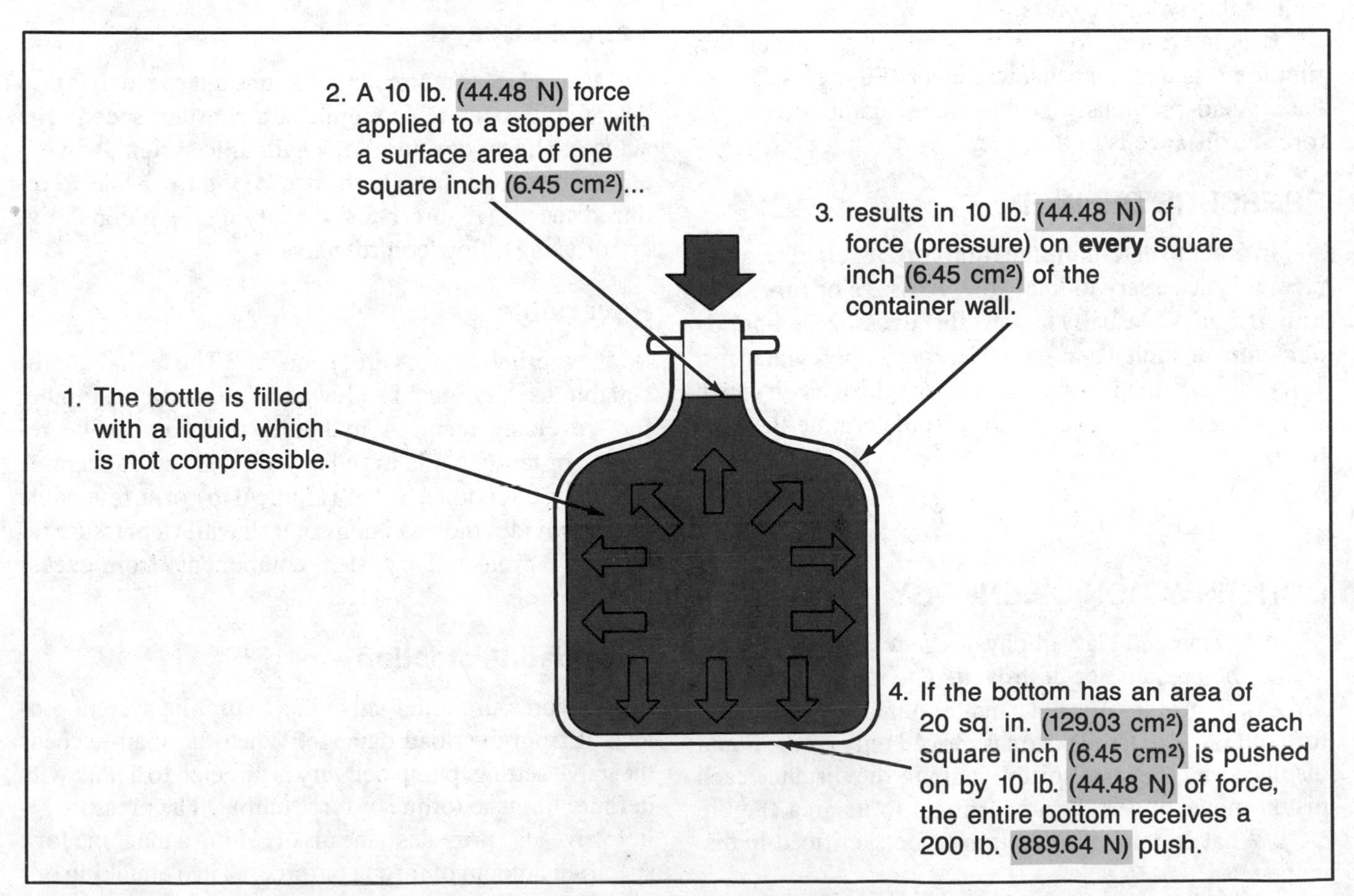

2. A 10 lb. (44.48 N) force applied to a stopper with a surface area of one square inch (6.45 cm²)...
3. results in 10 lb. (44.48 N) of force (pressure) on **every** square inch (6.45 cm²) of the container wall.
1. The bottle is filled with a liquid, which is not compressible.
4. If the bottom has an area of 20 sq. in. (129.03 cm²) and each square inch (6.45 cm²) is pushed on by 10 lb. (44.48 N) of force, the entire bottom receives a 200 lb. (889.64 N) push.

Figure 1-1. Pressure (force per unit area) is transmitted throughout a confined fluid.

Perhaps it was the very simplicity of Pascal's Law that prevented people from realizing its tremendous potential for some two centuries. Then, in the early stages of the industrial revolution, a British mechanic named Joseph Bramah utilized Pascal's discovery in developing a hydraulic press.

Bramah decided that, if a small force on a small area would create a proportionally larger force on a larger area, the only limit to the force a machine can exert is the area to which the pressure is applied.

Figure 1-2A shows how Bramah applied Pascal's principle to the hydraulic press. The applied force is the same as on the stopper in Figure 1-1 and the small piston has the same one square inch (6.45 square centimeters) area. The larger piston though, has an area of 10 in^2 (64.52 cm^2). The larger piston is pushed on with 10 lb of force per square inch (0.69 bar) (68.94 kPa), so that it can support a total weight or force of 100 lb (444.82 N).

The forces or weights which will balance using this apparatus are proportional to the piston areas. Thus, if the output piston area is 200 in^2 (1290.32 cm^2) the output force will be 2000 lb (8896.44 N) (assuming the same 10 lb of push on each square inch (0.69 bar) (68.94 kPa)). This is the operating principle of the hydraulic jack, as well as the hydraulic press.

It is interesting to note the similarity between this simple press and a mechanical lever (Figure 1-2B). As Pascal had previously stated—here again force is to force as distance is to distance.

PRESSURE DEFINED

In order to determine the **total** force exerted on a surface, it is necessary to know the **pressure** or **force** on a **unit** of area. We usually express this **pressure** in **pounds per square inch (bar) (kilopascals)**, abbreviated psi (bar) (kPa). Knowing the pressure and the area on which it is being exerted, one can readily determine the total force.

$$Force = Pressure \times Area$$

CONSERVATION OF ENERGY

A fundamental law of physics states that energy can neither be created nor destroyed. The multiplication of force in Figure 1-2A is not a matter of getting something for nothing. The large piston is moved only by the liquid displaced by the small piston making the distance each piston moves inversely proportional to its area (Figure 1-3). What is gained in force must be sacrificed in distance.

HYDRAULIC POWER TRANSMISSION

Hydraulics now could be defined as a means of transmitting power by pushing on a confined liquid. The input component of the system is called a **pump**; the output is called an **actuator**.

While for the sake of simplicity we have shown a single small piston, most power driven pumps incorporate multiple pistons, vanes or gears as their pumping elements. Actuators are linear, such as the cylinder shown in Figure 1-4A; or rotary, such as the hydraulic motor shown in Figure 1-4B.

The hydraulic system is not a **source** of power. The power source is a **prime mover** such as an electric motor or an engine which drives the pump. The reader might ask, therefore, why not forget about hydraulics and couple the mechanical equipment directly to the prime mover? The answer is in the versatility of the hydraulic system, which gives it advantages over other methods of transmitting power.

ADVANTAGES OF HYDRAULICS

As described below, there are several important advantages of hydraulic systems.

Variable Speed

Most electric motors run at a constant speed. It is also desirable to operate an engine at a constant speed. The actuator (linear or rotary) of a hydraulic system, however, can be driven from high speeds (Figure 1-5A) to reduced speeds (Figure 1-5B) by varying the pump delivery or using a flow control valve.

Reversible

Few prime movers are reversible. Those that are reversible usually must be slowed to a complete stop before reversing them. A hydraulic actuator can be reversed instantly while in full motion without damage. A four-way directional valve (Figure 1-6) or a reversible pump provides the reversing control, while a pressure relief valve protects the system components from excess pressure.

Overload Protection

The pressure relief valve in a hydraulic system protects it from overload damage. When the load exceeds the valve setting, pump delivery is directed to a tank with definite limits to torque or force output. The pressure relief valve also provides a means of setting a machine for a specified amount of torque or force, as in a chucking or a clamping operation.

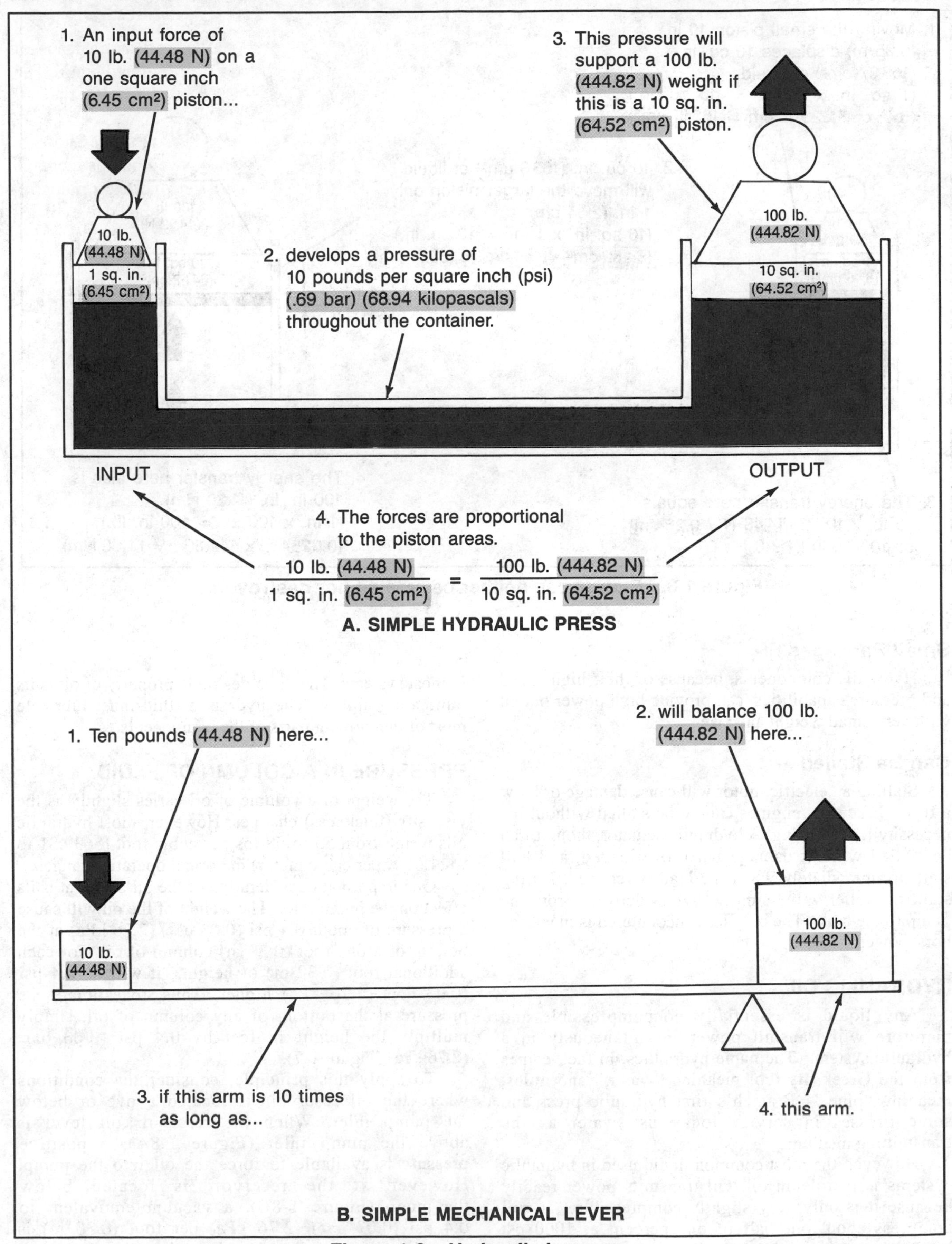
1. An input force of 10 lb. (44.48 N) on a one square inch (6.45 cm²) piston...
10 lb. (44.48 N)
1 sq. in. (6.45 cm²)
2. develops a pressure of 10 pounds per square inch (psi) (.69 bar) (68.94 kilopascals) throughout the container.
3. This pressure will support a 100 lb. (444.82 N) weight if this is a 10 sq. in. (64.52 cm²) piston.
100 lb. (444.82 N)
10 sq. in. (64.52 cm²)
INPUT
OUTPUT
4. The forces are proportional to the piston areas.
10 lb. (44.48 N) / 1 sq. in. (6.45 cm²) = 100 lb. (444.82 N) / 10 sq. in. (64.52 cm²)
A. SIMPLE HYDRAULIC PRESS
1. Ten pounds (44.48 N) here...
2. will balance 100 lb. (444.82 N) here...
10 lb. (44.48 N)
100 lb. (444.82 N)
3. if this arm is 10 times as long as...
4. this arm.
B. SIMPLE MECHANICAL LEVER

Figure 1-2. Hydraulic leverage.

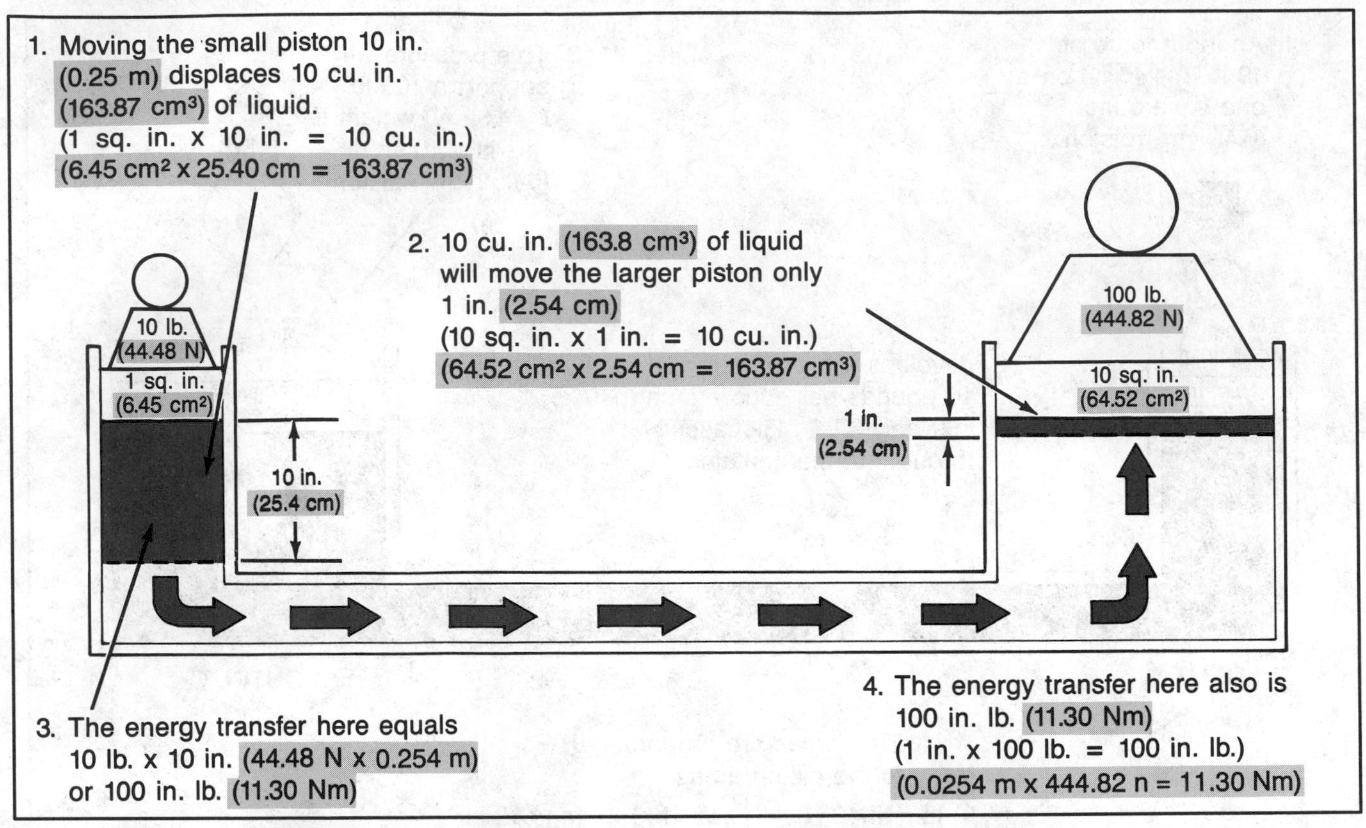

1. Moving the small piston 10 in. (0.25 m) displaces 10 cu. in. (163.87 cm³) of liquid.
(1 sq. in. x 10 in. = 10 cu. in.)
(6.45 cm² x 25.40 cm = 163.87 cm³)
2. 10 cu. in. (163.8 cm³) of liquid will move the larger piston only 1 in. (2.54 cm)
(10 sq. in. x 1 in. = 10 cu. in.)
(64.52 cm² x 2.54 cm = 163.87 cm³)
10 lb.
(44.48 N)
1 sq. in.
(6.45 cm²)
10 in.
(25.4 cm)
1 in.
(2.54 cm)
100 lb.
(444.82 N)
10 sq. in.
(64.52 cm²)
3. The energy transfer here equals 10 lb. x 10 in. (44.48 N x 0.254 m) or 100 in. lb. (11.30 Nm)
4. The energy transfer here also is 100 in. lb. (11.30 Nm)
(1 in. x 100 lb. = 100 in. lb.)
(0.0254 m x 444.82 n = 11.30 Nm)

Figure 1-3. Energy can neither be created nor destroyed.

Small Packages

Hydraulic components, because of their high speed and pressure capabilities, can provide high power output with very small weight and size.

Can Be Stalled

Stalling an electric motor will cause damage or blow a fuse. Likewise, engines cannot be stalled without the necessity for restarting. A hydraulic actuator, though, can be stalled without damage when overloaded, and will start up immediately when the load is reduced. During stall, the relief valve simply diverts delivery from the pump to the tank. The only loss encountered is in wasted horsepower.

HYDRAULIC OIL

Any liquid is essentially noncompressible and therefore will transmit power instantaneously in a hydraulic system. The name hydraulics, in fact, comes from the Greek, **hydro**, meaning "water," and **aulos**, meaning "pipe." Bramah's first hydraulic press and some presses in service today use water as the transmitting medium.

However, the most common liquid used in hydraulic systems is petroleum oil. Oil transmits power readily because it is only very slightly compressible. It will compress about one-half of one percent at 1000 psi (68.94 bar) (6894.00 kPa) pressure, a negligible amount in most systems. The most desirable property of oil is its lubricating ability. The hydraulic fluid must lubricate most of the moving parts of the components.

PRESSURE IN A COLUMN OF FLUID

The weight of a volume of oil varies slightly as the viscosity (thickness) changes. However, most hydraulic oils weigh from 55 to 58 lbs per cubic foot (8639.81 to 9111.07 N per cubic meter) in normal operating ranges.

One important consideration of the oil's weight is its effect on the pump inlet. The weight of the oil will cause a pressure of about 0.4 psi (0.03 bar) (2.76 kPa) at the bottom of a one-foot (0.30 m) column of oil. For each additional foot (0.30 m) of height, it will be 0.4 psi (0.03 bar) (2.76 kPa) higher. Thus, to estimate the pressure at the bottom of any column of oil, simply multiply the height in feet by 0.4 psi (0.03 bar) (2.76 kPa) (Figure 1-7).

To apply this principle, consider the conditions where the oil reservoir is located above or below the pump inlet. When the reservoir oil level is above the pump inlet (Figure 1-8A), a positive pressure is available to force the oil into the pump. However, if the reservoir is located below the pump (Figure 1-8B), a vacuum equivalent to 0.4 psi (0.03 bar) (2.76 kPa) per foot (0.30 m) is needed to "lift" the oil to the pump inlet. Actually the oil

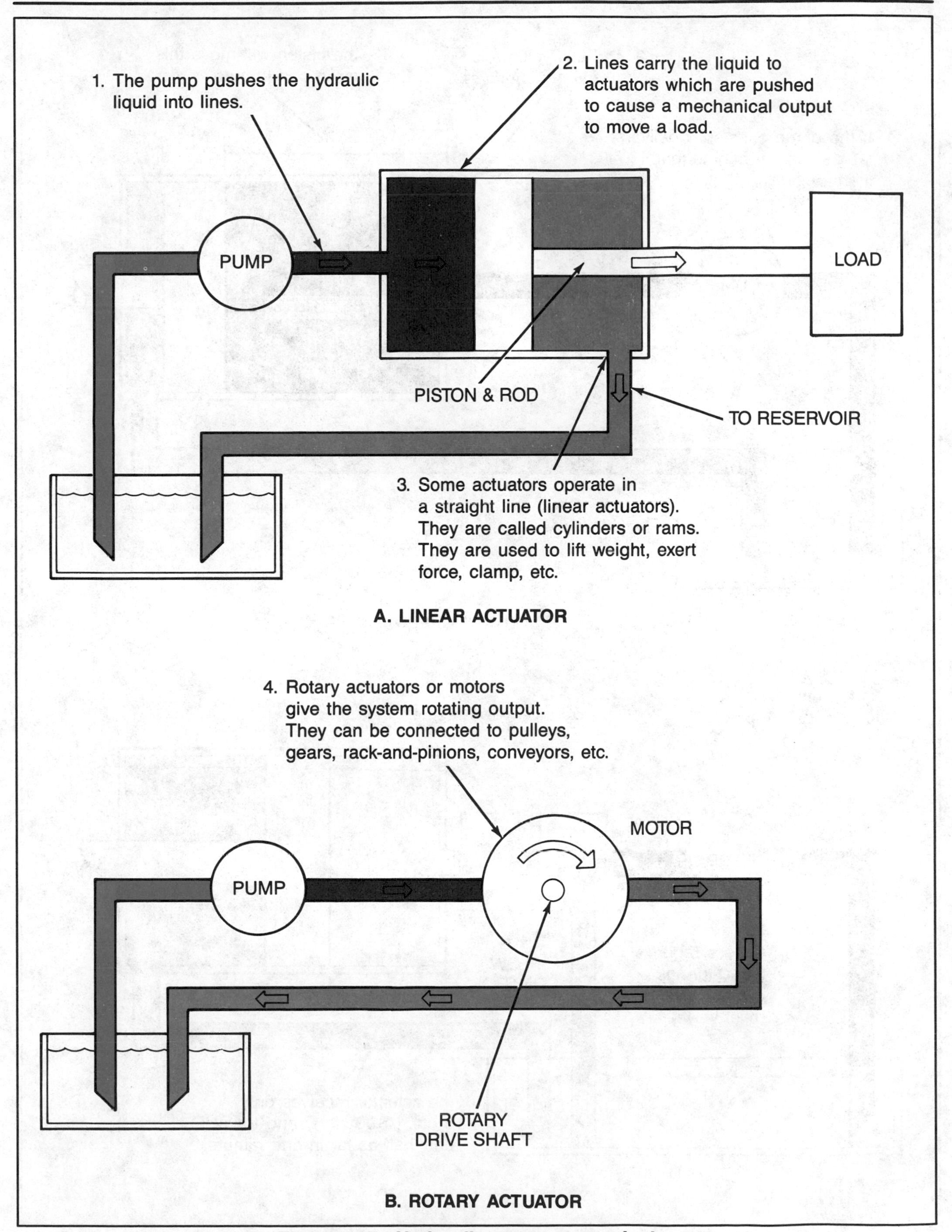

1. The pump pushes the hydraulic liquid into lines.
2. Lines carry the liquid to actuators which are pushed to cause a mechanical output to move a load.
PUMP
LOAD
PISTON & ROD
TO RESERVOIR
3. Some actuators operate in a straight line (linear actuators). They are called cylinders or rams. They are used to lift weight, exert force, clamp, etc.
A. LINEAR ACTUATOR
4. Rotary actuators or motors give the system rotating output. They can be connected to pulleys, gears, rack-and-pinions, conveyors, etc.
MOTOR
PUMP
ROTARY
DRIVE SHAFT
B. ROTARY ACTUATOR

Figure 1-4. Hydraulic power transmission

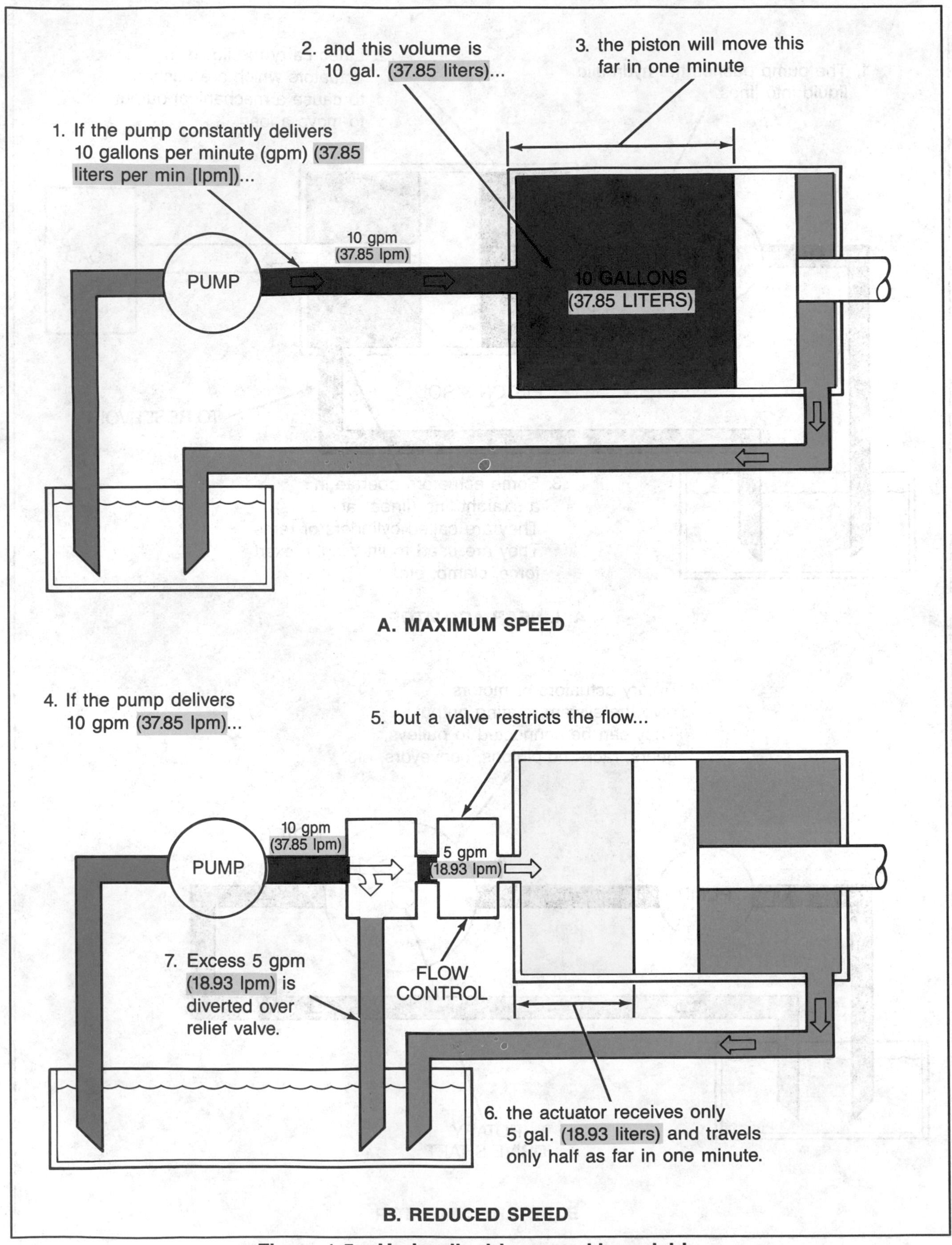

Figure 1-5. Hydraulic drive speed is variable.

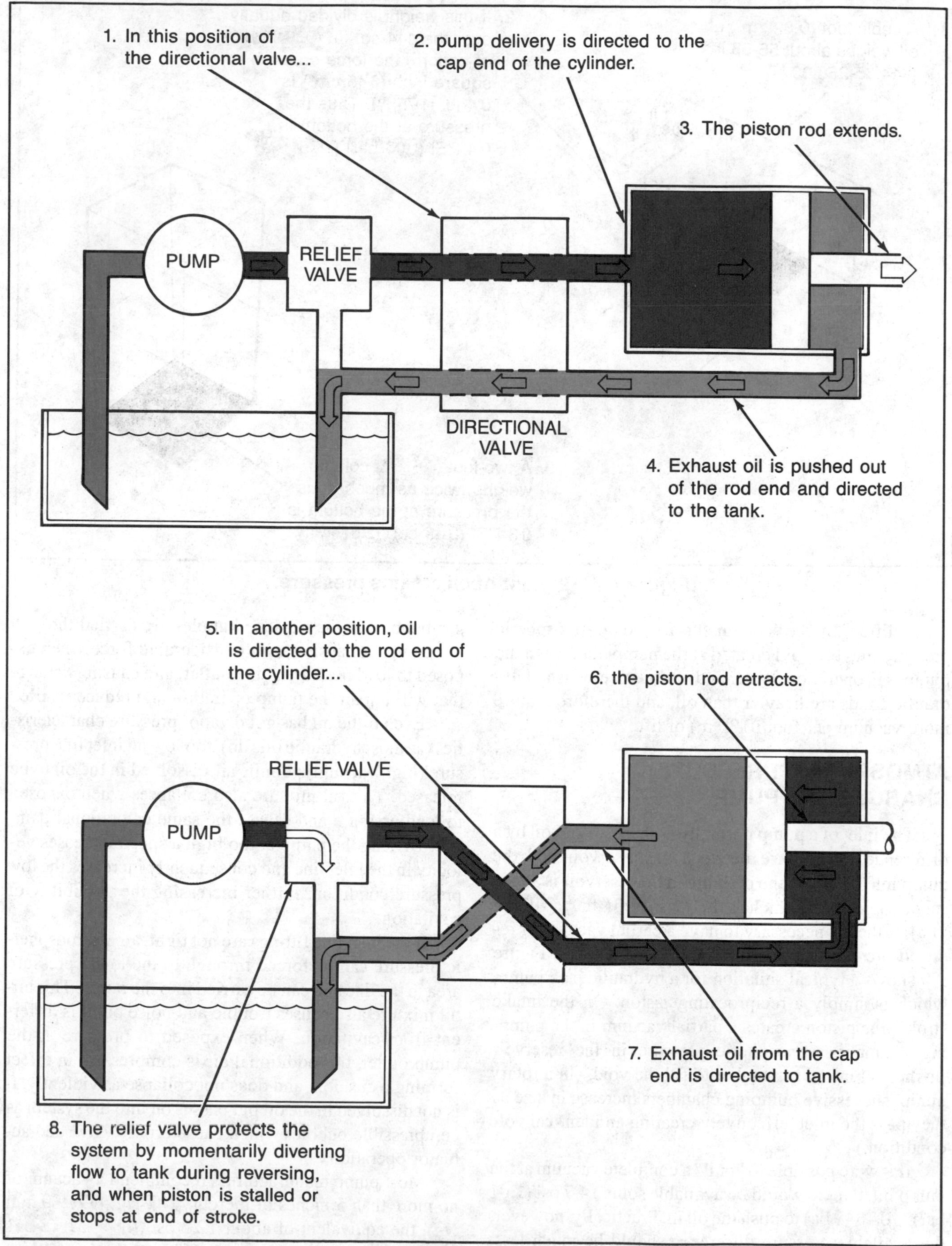

1. In this position of the directional valve...
2. pump delivery is directed to the cap end of the cylinder.
3. The piston rod extends.
PUMP
RELIEF VALVE
DIRECTIONAL VALVE
4. Exhaust oil is pushed out of the rod end and directed to the tank.
5. In another position, oil is directed to the rod end of the cylinder...
6. the piston rod retracts.
RELIEF VALVE
PUMP
7. Exhaust oil from the cap end is directed to tank.
8. The relief valve protects the system by momentarily diverting flow to tank during reversing, and when piston is stalled or stops at end of stroke.

Figure 1-6. Hydraulic drives are reversible.

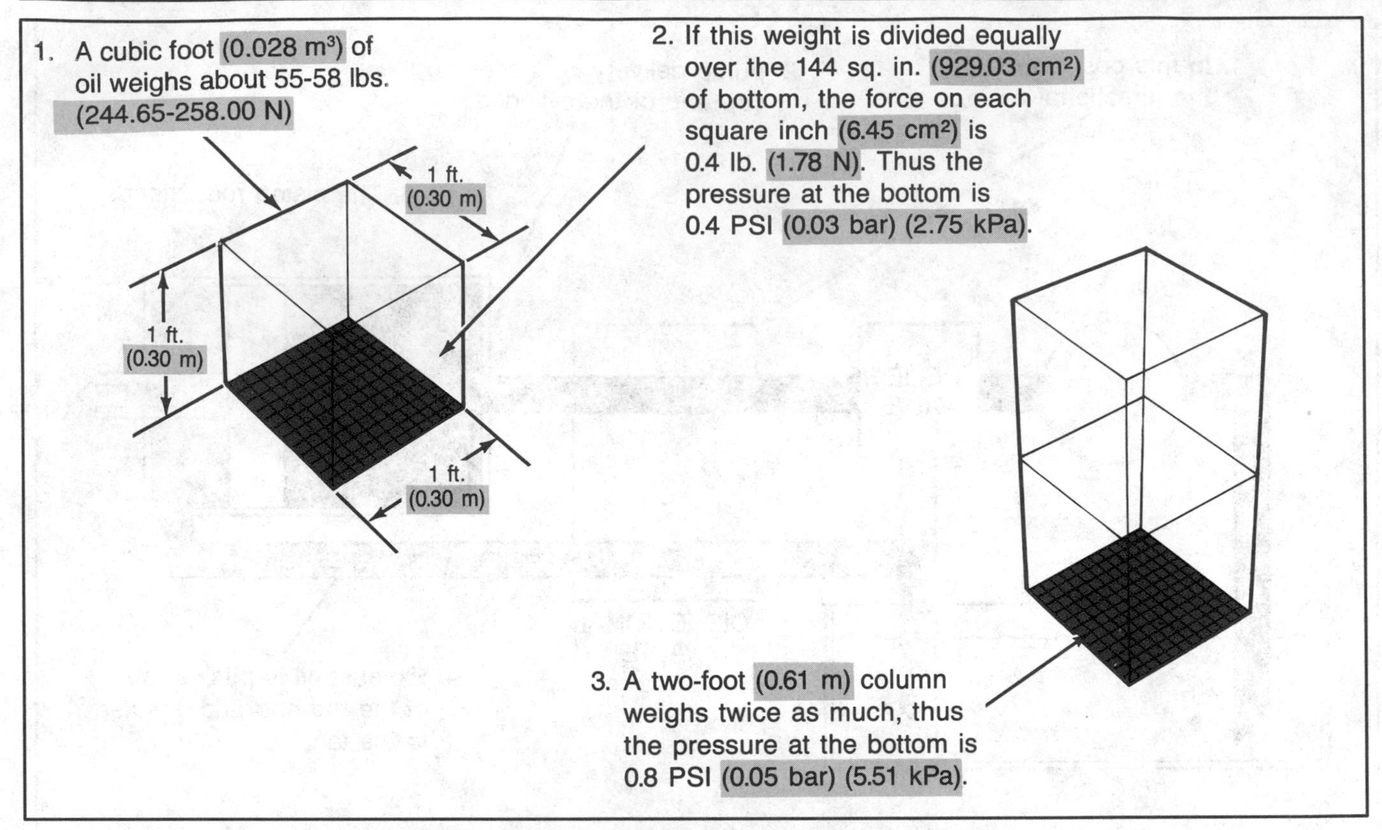

1. A cubic foot (0.028 m³) of oil weighs about 55-58 lbs. (244.65-258.00 N)
1 ft. (0.30 m)
1 ft. (0.30 m)
1 ft. (0.30 m)
2. If this weight is divided equally over the 144 sq. in. (929.03 cm²) of bottom, the force on each square inch (6.45 cm²) is 0.4 lb. (1.78 N). Thus the pressure at the bottom is 0.4 PSI (0.03 bar) (2.75 kPa).
3. A two-foot (0.61 m) column weighs twice as much, thus the pressure at the bottom is 0.8 PSI (0.05 bar) (5.51 kPa).

Figure 1-7. Weight of oil creates pressure.

is not "lifted" by the vacuum, it is forced by atmospheric pressure into the void created at the pump inlet when the pump is in operation. Water and various fire-resistant hydraulic fluids are heavier than oil, and therefore require more vacuum per foot (0.30 m) of lift.

ATMOSPHERIC PRESSURE CHARGES THE PUMP

The inlet of a pump normally is charged with oil by a difference in pressure between the reservoir and the pump inlet. Usually the pressure in the reservoir is atmospheric pressure, which is 14.7 psi (1.01 bar) (101.34 kPa). It then is necessary to have a partial vacuum or reduced pressure at the pump inlet to create flow. Figure 1-9 shows a typical situation for a hydraulic jack pump, which is simply a reciprocating piston. On the intake stroke, the piston creates a partial vacuum in the pumping chamber. Atmospheric pressure in the reservoir **pushes** oil into the chamber to fill the void. (In a rotary pump, successive pumping chambers increase in size as they pass the inlet, effectively creating an identical void condition.)

If it were possible to "pull" a complete vacuum at the pump inlet, there would be available some 14.7 psi (1.01 bar) (101.34 kPa) to push the oil in. Practically, however, the available pressure difference should be much less. For one thing, liquids vaporize in a vacuum. This puts gas bubbles in the oil. The bubbles are carried through the pump; collapsing with considerable force when exposed to load pressure at the outlet, and causing damage that will impair the pump operation and reduce its life.

Even if the oil has good vapor pressure characteristics (as most hydraulic oils do), too low an inlet line pressure (high vacuum) permits air dissolved in the oil to be released. This oil mixture also collapses when exposed to load pressure and causes the same cavitational damage. Driving the pump at too high a speed increases velocity in the inlet line and consequently increases the low pressure condition, further increasing the possibility of cavitation.

If the inlet line fittings are not tight, air at atmospheric pressure can be forced through to the lower pressure area in the line and can be carried into the pump. This air-oil mixture also causes trouble and noise but it is different from cavitation. When exposed to pressure at the pump outlet, this additional air is compressed, in effect forming a cushion, and does not collapse as violently. It is not dissolved in the oil but passes on into the system as compressible bubbles which cause erratic valve and actuator operation.

Most pump manufacturers recommend a vacuum of no more than 5 inches of mercury (in. Hg) (127.00 mm Hg), the equivalent of about 12.2 psi (0.84 bar) (84.11 kPa) absolute at the pump inlet. With 14.7 psi (1.01 bar)

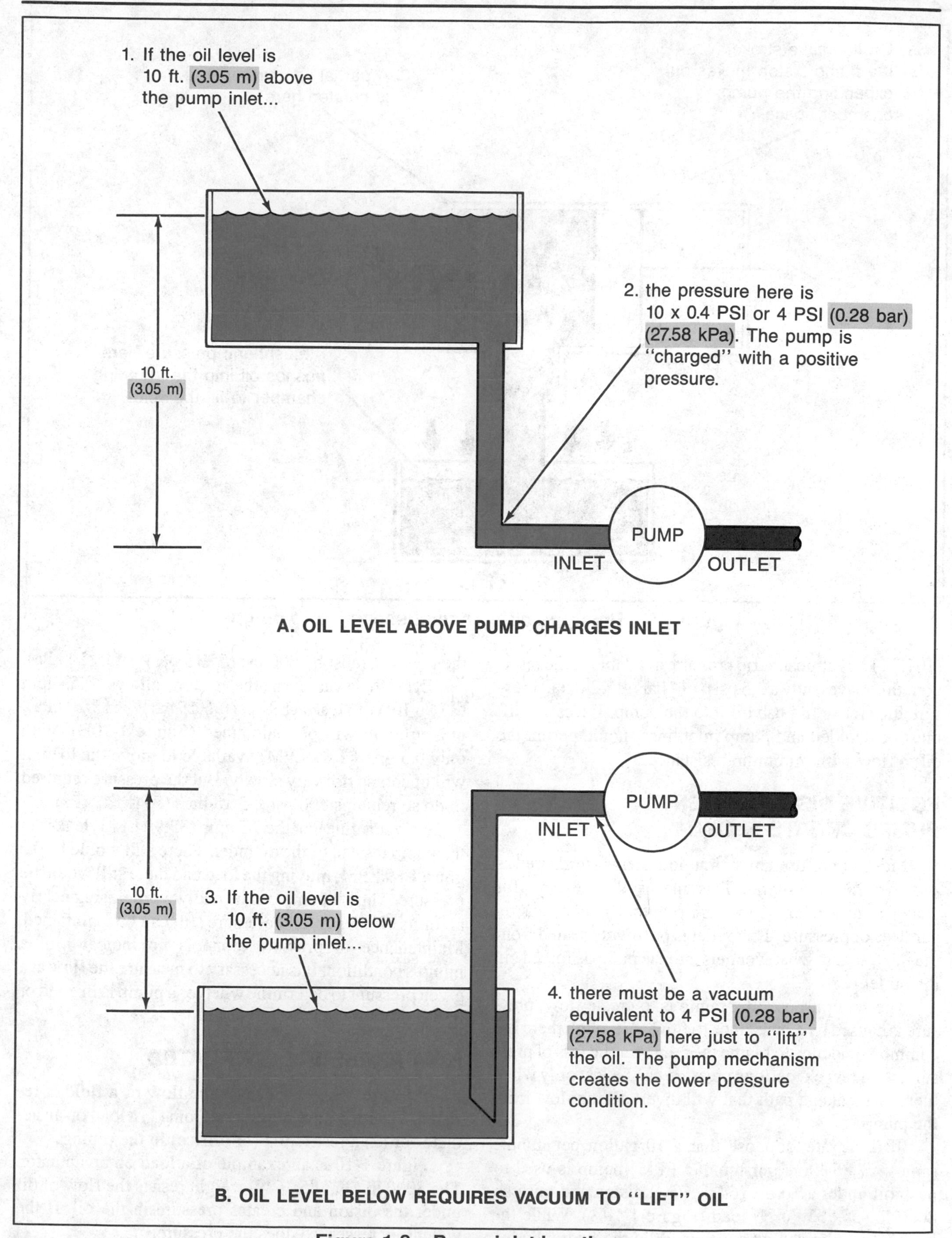

1. If the oil level is
10 ft. (3.05 m) above
the pump inlet...
10 ft.
(3.05 m)
2. the pressure here is
10 x 0.4 PSI or 4 PSI (0.28 bar)
(27.58 kPa). The pump is
"charged" with a positive
pressure.
PUMP
INLET
OUTLET
A. OIL LEVEL ABOVE PUMP CHARGES INLET
PUMP
INLET
OUTLET
10 ft.
(3.05 m)
3. If the oil level is
10 ft. (3.05 m) below
the pump inlet...
4. there must be a vacuum
equivalent to 4 PSI (0.28 bar)
(27.58 kPa) here just to "lift"
the oil. The pump mechanism
creates the lower pressure
condition.
B. OIL LEVEL BELOW REQUIRES VACUUM TO "LIFT" OIL

Figure 1-8. Pump inlet locations.

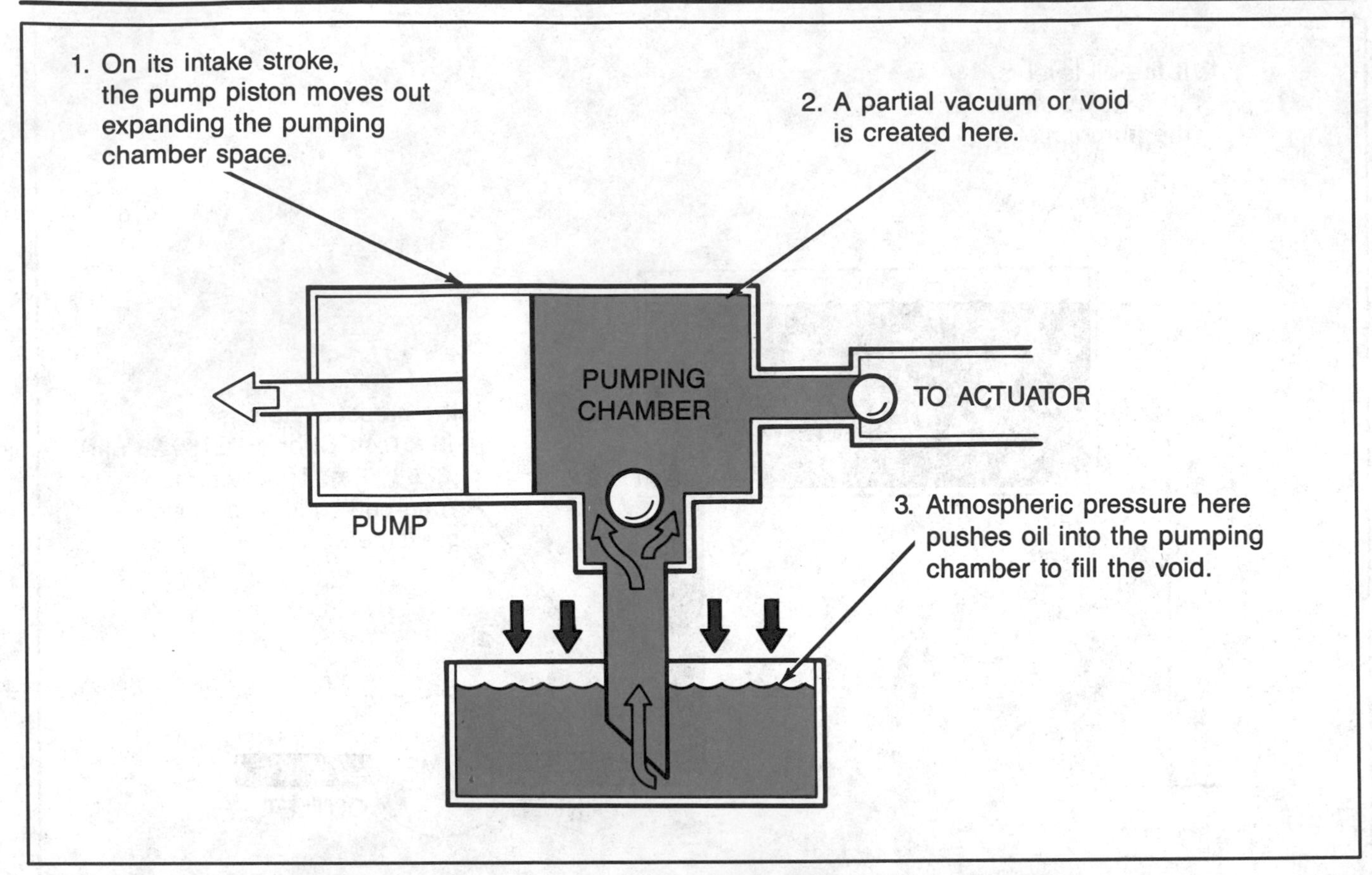

1. On its intake stroke, the pump piston moves out expanding the pumping chamber space.
2. A partial vacuum or void is created here.
PUMPING CHAMBER
TO ACTUATOR
PUMP
3. Atmospheric pressure here pushes oil into the pumping chamber to fill the void.

Figure 1-9. Pressure difference pushes oil into pump.

(101.34 kPa) atmospheric pressure available at the reservoir, this leaves only a 2.5 psi (0.17 bar) (17.24 kPa) pressure difference to push oil into the pump. Excessive lift must be avoided and pump inlet lines should permit the oil to flow with minimum resistance.

POSITIVE DISPLACEMENT PUMPS CREATE FLOW

Most pumps used in hydraulic systems are classed as **positive displacement**. This means that, except for changes in efficiency, the pump output is constant regardless of pressure. The outlet is positively sealed from the inlet, so that whatever gets into the pump is forced out the outlet port.

The sole purpose of a pump is to create flow; pressure is caused by a resistance to flow. Although there is a common tendency to blame the pump for the loss of pressure, with few exceptions pressure can be lost only when there is a leakage path that will divert **all** the flow from the pump.

To illustrate, suppose that a 10 gallon per minute (gpm) (37.85 liters per minute, LPM) pump is used to push oil under a 10 in^2 (64.52 cm^2) piston and raise an 8000 lb (35,585.76 N) load (Figure 1-10A). While the load is being raised **or supported** by the hydraulic oil, the pressure must be 800 psi (55.15 bar) (5515.20 kPa).

Even if a hole in the piston allows 9.5 gpm (35.96 LPM)to leak at 800 psi (55.15 bar) (5515.20 kPa), pressure still will be maintained (Figure 1-10B). With only 0.5 gpm (1.89 LPM) available to move the load, it will of course rise very slowly. But the pressure required to do so remains 800 psi (55.15 bar) (5515.20 kPa).

Now imagine that the 9.5 gpm (35.96 LPM) leak is in the pump instead of the cylinder. There still would be 0.5 gpm (1.89 LPM) moving the load and there still would be pressure. Thus, a pump can be badly worn, losing nearly all of its efficiency, and pressure still can be maintained. Maintenance of pressure alone is no indicator of a pump's condition. It is necessary to measure the **flow** at a given pressure to determine whether a pump is in good or bad condition.

HOW PRESSURE IS CREATED

Pressure results whenever the flow of a fluid is resisted. The resistance may come from (1) a load on an actuator or (2) a restriction (or orifice) in the piping.

Figure 1-10 is an example of a load on an actuator. The 8000 lb (35,585.76 N) weight resists the flow of oil under the piston and creates pressure in the oil. If the weight increases, so does the pressure.

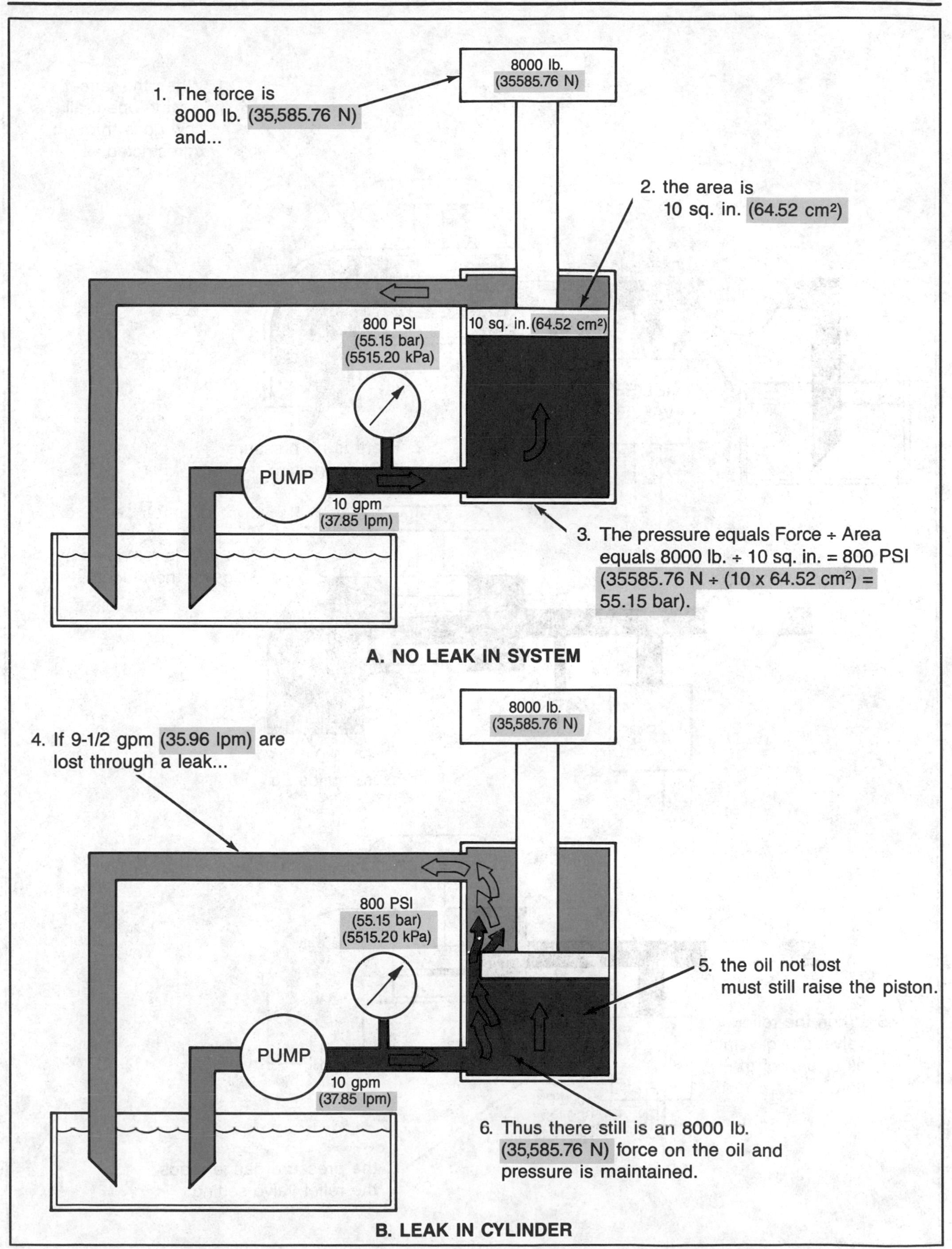

Figure 1-10. Pressure loss requires full loss of pump output.

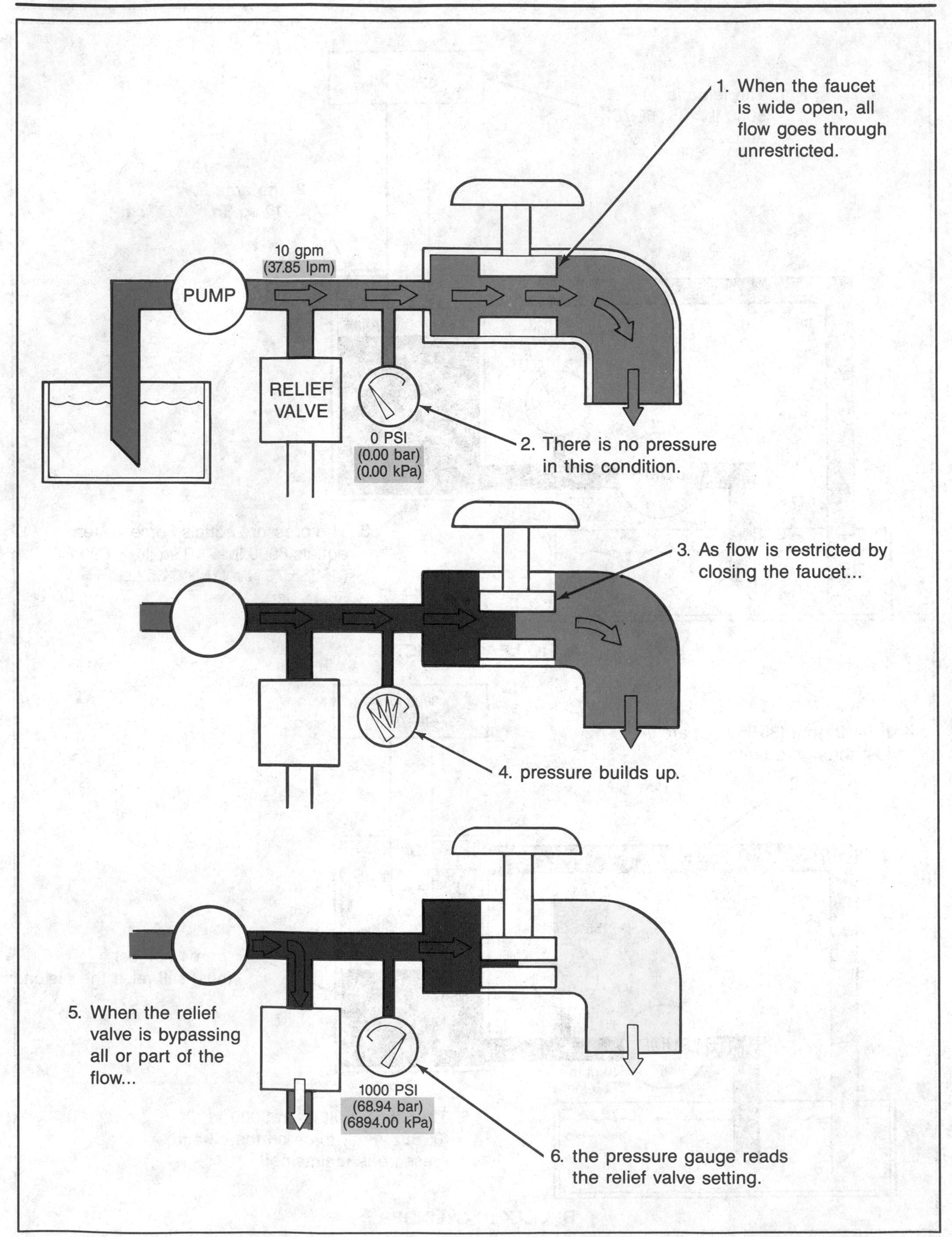

1. When the faucet is wide open, all flow goes through unrestricted.
10 gpm
(37.85 lpm)
PUMP
RELIEF
VALVE
0 PSI
(0.00 bar)
(0.00 kPa)
2. There is no pressure in this condition.
3. As flow is restricted by closing the faucet...
4. pressure builds up.
5. When the relief valve is bypassing all or part of the flow...
1000 PSI
(68.94 bar)
(6894.00 kPa)
6. the pressure gauge reads the relief valve setting.

Figure 1-11. Pressure caused by restriction and limited by pressure control valve.

In Figure 1-11, a 10 gpm (37.85 LPM) pump has its outlet connected to a pressure relief valve set at 1000 psi (68.94 bar) (6894.00 kPa) and to an ordinary water faucet. If the faucet is wide open, the pump delivery flows out unrestricted and there is no reading on the pressure gauge.

Now suppose that the faucet is gradually closed. It will resist flow and cause pressure to build up on the upstream side. As the opening is restricted, it will take increasingly more pressure to push the 10 gpm (37.85 LMP) through the restriction. Without the relief valve, there would theoretically be no limit to the pressure build-up. In reality, either something would break or the pump would stall the prime mover.

In our example, at the point where it takes 1000 psi (68.94 bar) (6894.00 kPa) to push the oil through the opening, the relief valve will begin to open. Pressure then will remain at 1000 psi (68.94 bar) (6894.00 kPa). Further closing of the faucet will simply result in less oil going through it and more going over the relief valve. With the faucet completely closed, all 10 gpm will go over the relief valve at 1000 psi (68.94 bar) (6894 kPa).

It can be seen from the above that a relief valve or some other pressure limiting device should be used in all systems using a positive displacement pump.

PARALLEL FLOW PATHS

An inherent characteristic of liquids is that they will always take the path of least resistance. Thus, when two parallel flow paths offer different resistances, the pressure will increase only to the amount required to take the easier path.

In Figure 1-12A, the oil has three possible flow paths. Since valve A opens at 100 psi (6.89 bar) (689.40 kPa), the oil will go that way and pressure will build up to only 100 psi (6.89 bar) (689.40 kPa). Should flow be blocked beyond valve A (Figure 1-12B), pressure would build up to 200 psi (13.78 bar) (1378.80 kPa); then oil would flow through valve B. There would be no flow through valve C unless the path through valve B should also become blocked.

Similarly, when the pump outlet is directed to two actuators, the actuator which needs the lower pressure will be first to move. Since it is difficult to balance loads exactly, cylinders which must move together are often connected to each other mechanically.

SERIES FLOW PATH

When resistances to flow are connected in series, the pressures add up. In Figure 1-13 are shown the same valves as in Figure 1-12 but connected in series. Pressure gauges placed in the lines indicate the pressure normally required to open each valve plus back pressure from the valve downstream. The pressure at the pump is the sum of the pressures required to open individual valves.

PRESSURE DROP THROUGH AN ORIFICE

An orifice is a restricted passage in a hydraulic line or component, used to control flow or create a pressure difference (pressure drop).

In order for oil to flow through an orifice, there must be a pressure difference or **pressure drop** through the orifice. (The term “drop” comes from the fact that the lower pressure is always downstream.) Conversely, if there is no flow, there is no difference in pressure across the orifice.

Consider the condition surrounding the orifice in Figure 1-14A. The pressure is equal on both sides; therefore, the oil is being pushed equally both ways and there is no flow.

In Figure 1-14B, the higher pressure pushes harder to the right and oil does flow through the orifice. In Figure 1-14C, there is also a pressure drop; however, the flow is less than in Figure 1-14B because the pressure difference is lower.

An increase in pressure drop across an orifice will always be accompanied by an increase in flow.

If flow is blocked beyond an orifice (Figure 1-14D), the pressure will immediately equalize on both sides of the orifice in accordance with Pascal's Law. This principle is essential to the operation of many compound pressure and flow control valves.

PRESSURE INDICATES WORK LOAD

Figure 1-10 illustrated how pressure is generated by resistance of a load. It was noted that the pressure equals the force of the load divided by the piston area. We can express this relationship by the general formula:

$$P = \frac{F}{A} \qquad P = \frac{F}{10 \times A}$$

In this relationship:
P is pressure (pounds per square inch (bar))
F is force (pounds (newtons))
A is area (square inches (square centimeters))

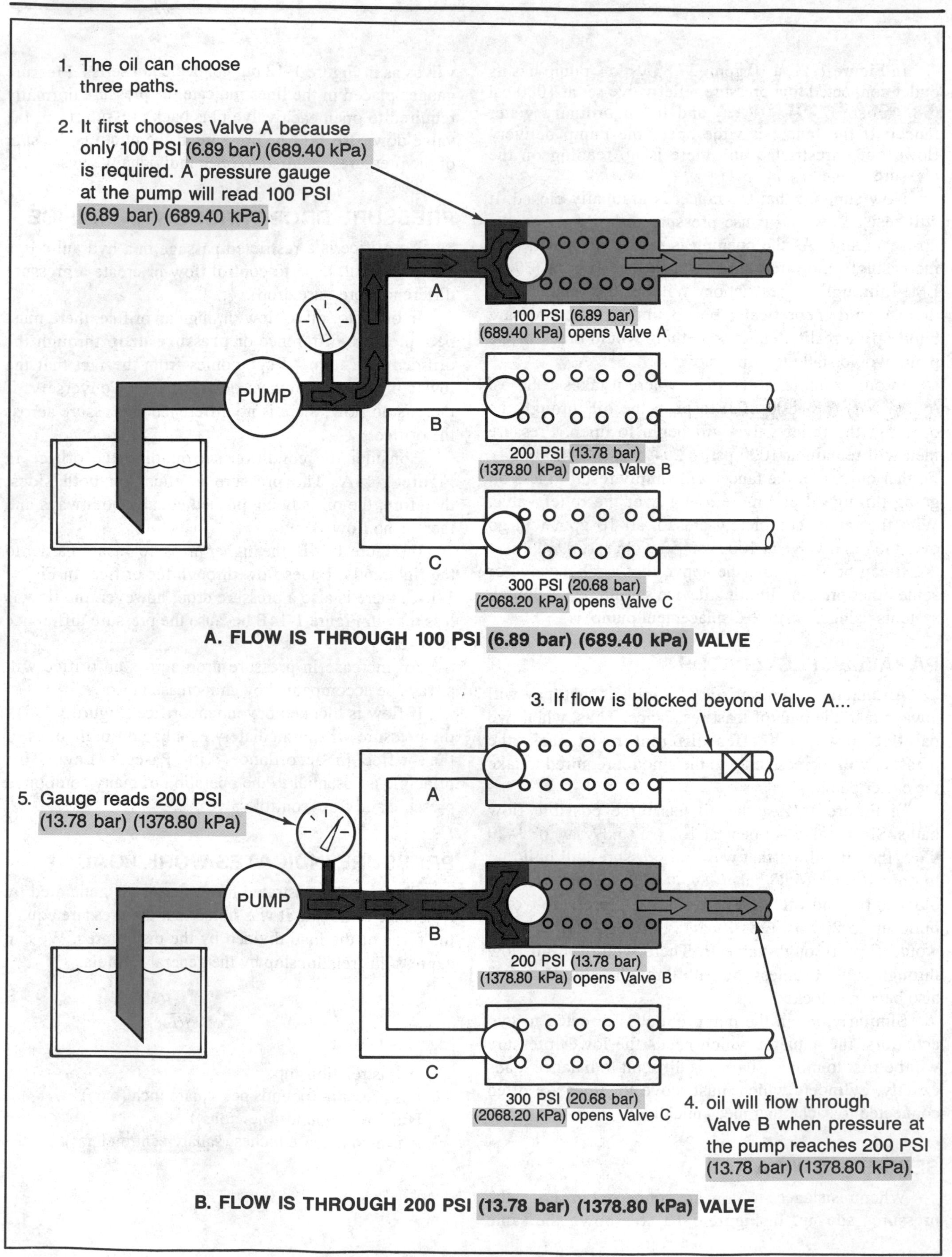
1. The oil can choose three paths.
2. It first chooses Valve A because only 100 PSI (6.89 bar) (689.40 kPa) is required. A pressure gauge at the pump will read 100 PSI (6.89 bar) (689.40 kPa).
PUMP
A
B
C
100 PSI (6.89 bar) (689.40 kPa) opens Valve A
200 PSI (13.78 bar) (1378.80 kPa) opens Valve B
300 PSI (20.68 bar) (2068.20 kPa) opens Valve C
A. FLOW IS THROUGH 100 PSI (6.89 bar) (689.40 kPa) VALVE
3. If flow is blocked beyond Valve A...
5. Gauge reads 200 PSI (13.78 bar) (1378.80 kPa)
PUMP
A
B
C
200 PSI (13.78 bar) (1378.80 kPa) opens Valve B
300 PSI (20.68 bar) (2068.20 kPa) opens Valve C
4. oil will flow through Valve B when pressure at the pump reaches 200 PSI (13.78 bar) (1378.80 kPa).
B. FLOW IS THROUGH 200 PSI (13.78 bar) (1378.80 kPa) VALVE

Figure 1-12. Parallel flow paths.

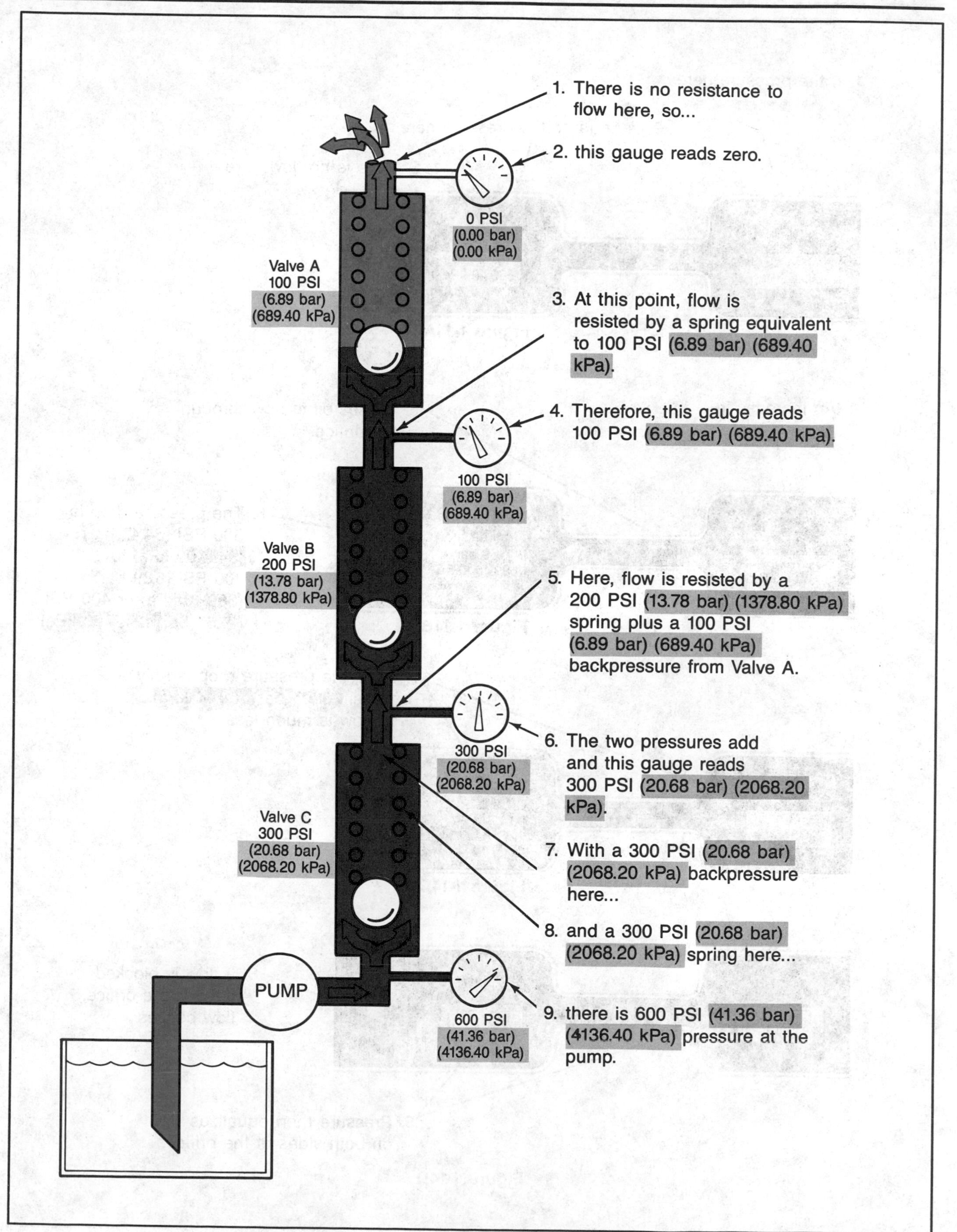

1. There is no resistance to flow here, so...
2. this gauge reads zero.
0 PSI
(0.00 bar)
(0.00 kPa)
Valve A
100 PSI
(6.89 bar)
(689.40 kPa)
3. At this point, flow is resisted by a spring equivalent to 100 PSI (6.89 bar) (689.40 kPa).
4. Therefore, this gauge reads 100 PSI (6.89 bar) (689.40 kPa).
100 PSI
(6.89 bar)
(689.40 kPa)
Valve B
200 PSI
(13.78 bar)
(1378.80 kPa)
5. Here, flow is resisted by a 200 PSI (13.78 bar) (1378.80 kPa) spring plus a 100 PSI (6.89 bar) (689.40 kPa) backpressure from Valve A.
6. The two pressures add and this gauge reads 300 PSI (20.68 bar) (2068.20 kPa).
300 PSI
(20.68 bar)
(2068.20 kPa)
Valve C
300 PSI
(20.68 bar)
(2068.20 kPa)
7. With a 300 PSI (20.68 bar) (2068.20 kPa) backpressure here...
8. and a 300 PSI (20.68 bar) (2068.20 kPa) spring here...
PUMP
600 PSI
(41.36 bar)
(4136.40 kPa)
9. there is 600 PSI (41.36 bar) (4136.40 kPa) pressure at the pump.

Figure 1-13. Series resistances add pressure.

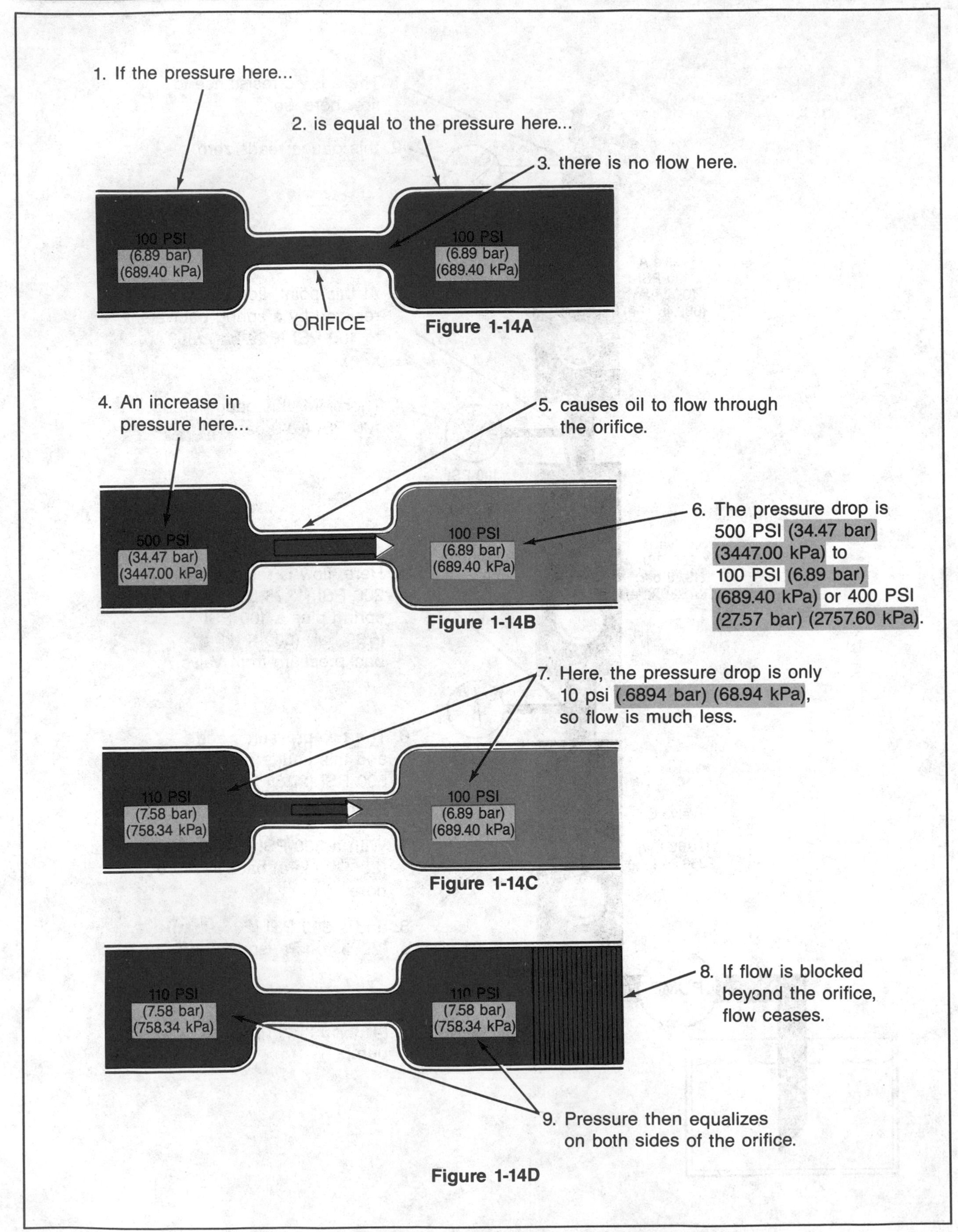
1. If the pressure here...
2. is equal to the pressure here...
3. there is no flow here.
100 PSI
(6.89 bar)
(689.40 kPa)
100 PSI
(6.89 bar)
(689.40 kPa)
ORIFICE
Figure 1-14A
4. An increase in pressure here...
5. causes oil to flow through the orifice.
500 PSI
(34.47 bar)
(3447.00 kPa)
100 PSI
(6.89 bar)
(689.40 kPa)
6. The pressure drop is 500 PSI (34.47 bar) (3447.00 kPa) to 100 PSI (6.89 bar) (689.40 kPa) or 400 PSI (27.57 bar) (2757.60 kPa).
Figure 1-14B
7. Here, the pressure drop is only 10 psi (.6894 bar) (68.94 kPa), so flow is much less.
110 PSI
(7.58 bar)
(758.34 kPa)
100 PSI
(6.89 bar)
(689.40 kPa)
Figure 1-14C
110 PSI
(7.58 bar)
(758.34 kPa)
110 PSI
(7.58 bar)
(758.34 kPa)
8. If flow is blocked beyond the orifice, flow ceases.
9. Pressure then equalizes on both sides of the orifice.
Figure 1-14D

Figure 1-14. Pressure drop and flow through an orifice.

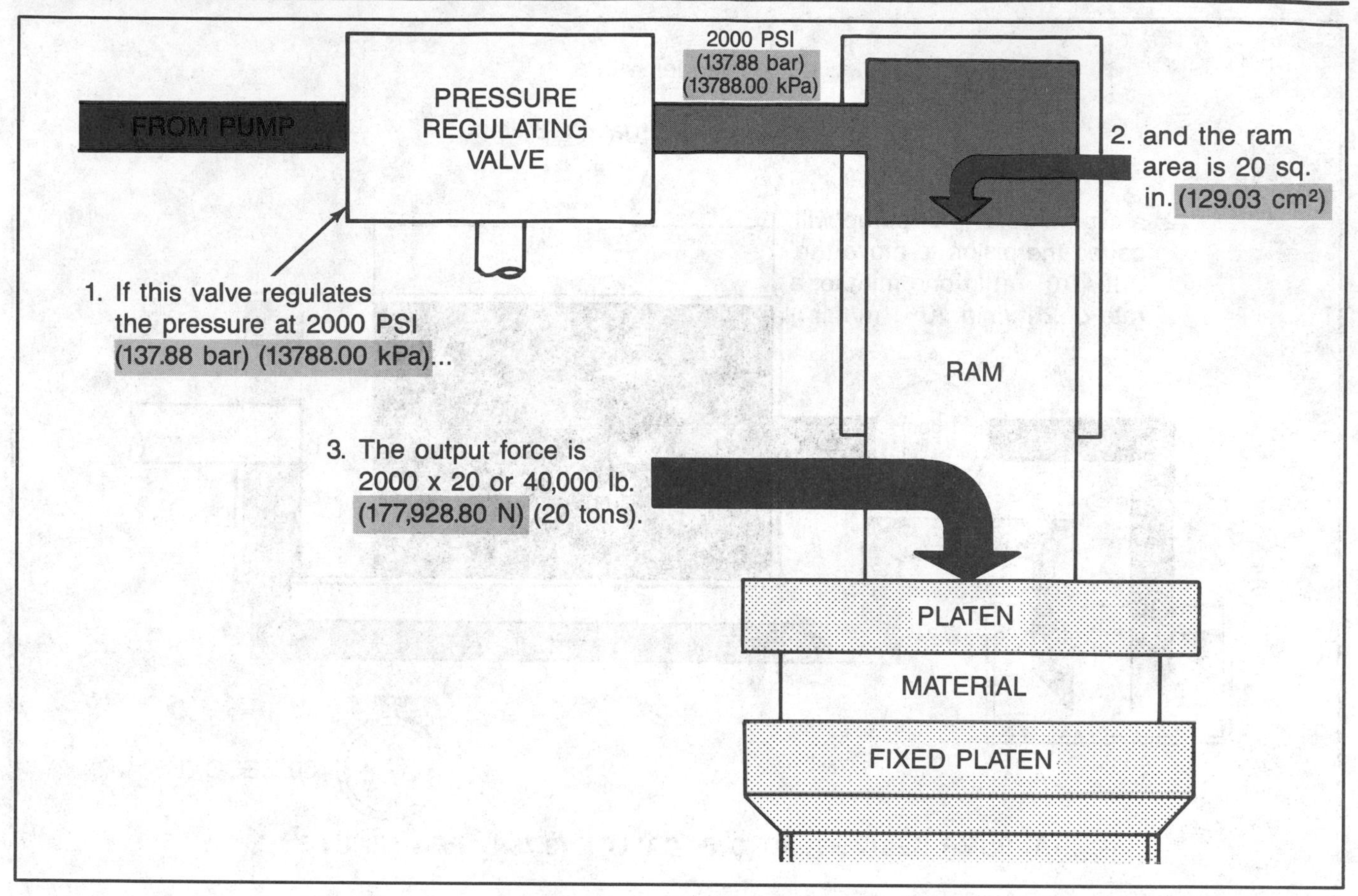

Figure 1-15. Force equals pressure multiplied by area.

From this can be seen that an increase or decrease in the load will result in a like increase or decrease in the operating pressure. In other words, **pressure is proportional to the load**, and a pressure gauge reading indicates the work load at any given moment.

Pressure gauge readings normally ignore atmospheric pressure. That is, a standard gauge reads zero at atmospheric pressure. **An absolute** gauge reads 14.7 psi (1.01 bar) (101.34 kPa) sea level atmospheric pressure. Absolute pressure is usually designated "psia."

FORCE IS PROPORTIONAL TO PRESSURE AND AREA

When a hydraulic cylinder is used to clamp or press, its output force can be computed as follows:

$$F = P \times A \qquad F = P \times A \times 10$$

Again:
P is pressure (psi (bar))
F is force (pounds (newtons))
A is area (square inches (square centimeters))

As an example, suppose that a hydraulic press has its pressure regulated at 2000 psi (137.88 bar) (13,788.00 kPa) (Figure 1-15) and this pressure is applied to a ram area of 20 in^2 (129.03 cm^2). The output force will then be 40,000 lbs or 20 tons (177,928.8 N).

COMPUTING PISTON AREA

The area of a piston or ram can be computed by this formula:

$$A = \frac{\pi}{4} \times d^2$$

A is area (square inches or square centimeters)
d is diameter of the piston (inches or centimeters)
π/4 is 0.7854

These pressure, force and area relationships are sometimes illustrated as shown below to aid in remembering the equations.

$$F = P \times A$$

$$P = \frac{F}{A}$$

$$A = \frac{F}{P}$$

F
P | A

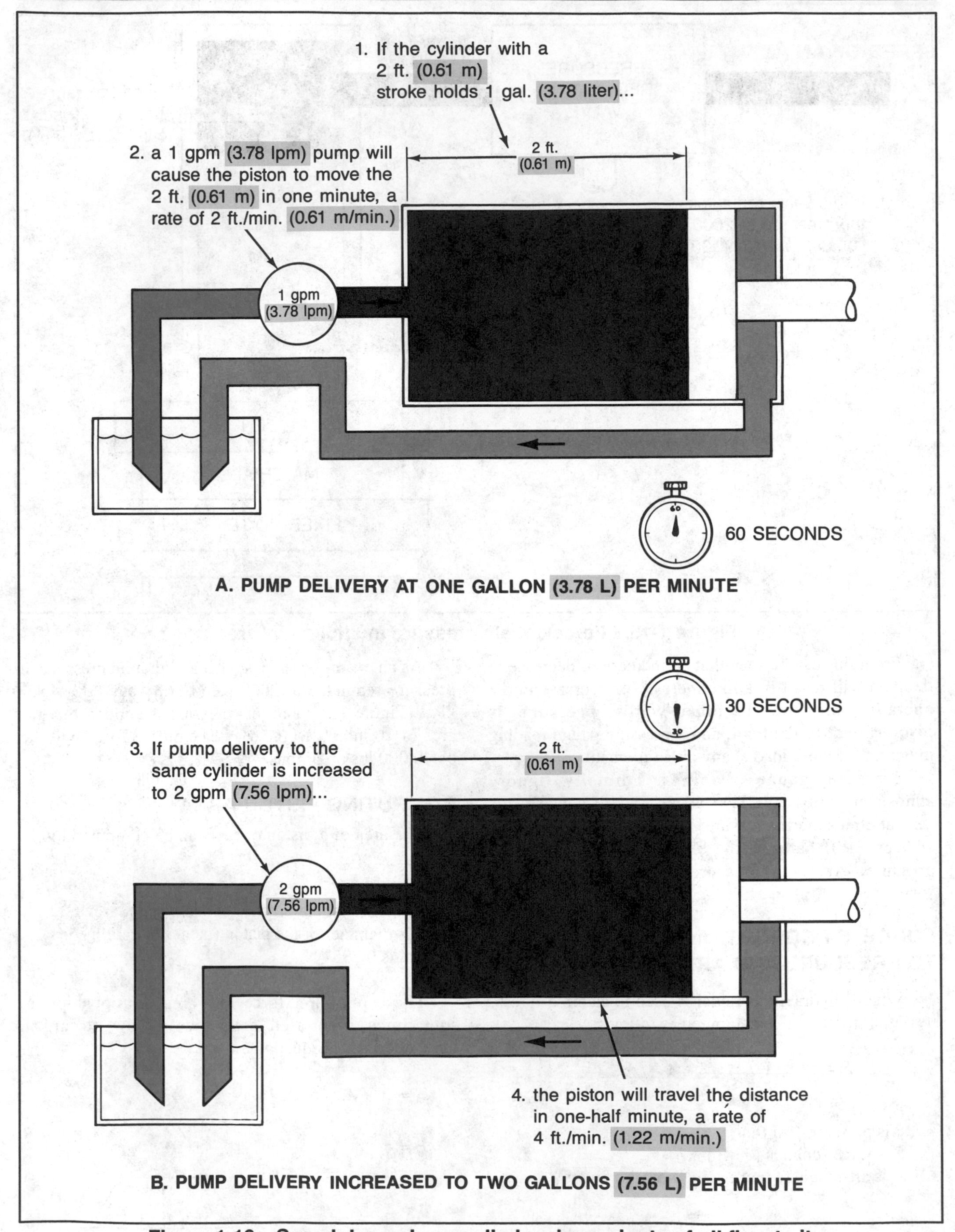

1. If the cylinder with a
2 ft. (0.61 m)
stroke holds 1 gal. (3.78 liter)...
2. a 1 gpm (3.78 lpm) pump will
cause the piston to move the
2 ft. (0.61 m) in one minute, a
rate of 2 ft./min. (0.61 m/min.)
2 ft.
(0.61 m)
1 gpm
(3.78 lpm)
60 SECONDS
A. PUMP DELIVERY AT ONE GALLON (3.78 L) PER MINUTE
30 SECONDS
3. If pump delivery to the
same cylinder is increased
to 2 gpm (7.56 lpm)...
2 ft.
(0.61 m)
2 gpm
(7.56 lpm)
4. the piston will travel the distance
in one-half minute, a rate of
4 ft./min. (1.22 m/min.)
B. PUMP DELIVERY INCREASED TO TWO GALLONS (7.56 L) PER MINUTE

Figure 1-16. Speed depends on cylinder size and rate of oil flow to it.

SPEED OF AN ACTUATOR

How fast a piston travels or a motor rotates depends on its size and the rate of oil flow into it. To relate flow rate to speed, consider the volume that must be filled in the actuator to cause a given amount of travel.

In Figure 1-16, note that both cylinders have the same volume. Yet, the piston shown in Figure 1-16B will travel twice as fast as the one shown in Figure 1-16A because the rate of oil flow from the pump has been doubled. If either cylinder had a smaller diameter, its rate would be faster. Or if its diameter were larger, its rate would be less, assuming of course the pump delivery remained constant.

The relationship may be expressed as follows:

$$speed = \frac{vol.\ /\ time}{area}$$

$$vol.\ /\ time = speed \times area$$

$$area = \frac{vol.\ /\ time}{speed}$$

V/t
s
a

v/t = in^3/minute or cm^3/minute
a = in^2 or cm^2
s = in/minute or cm/minute

From this we can conclude: (1) that the force or torque of an actuator is directly proportional to the pressure and independent of the flow, (2) that its speed or rate of travel will depend upon the amount of fluid flow without regard to pressure.

VELOCITY IN PIPES

The velocity at which the hydraulic fluid flows through the lines is an important design consideration because of the effect of velocity on friction.

Generally, the recommended velocity ranges are:
Pump Inlet Line = 2-4 feet per second
(.61-1.22 meters per second)
Working Lines = 7-20 feet per second
(2.13-6.10 meters per second)

In this regard, it should be noted that:

1. The velocity of the fluid varies **inversely** as the **square** of the inside diameter.
2. Usually, friction of a liquid flowing through a line is proportional to the **velocity**. However, should the flow become turbulent, friction varies as the **square** of the velocity.

Figure 1-17 illustrates that **doubling** the inside diameter of a line **quadruples** the cross-sectional area; thus the velocity is only one-fourth as fast in the large line. Conversely, **halving** the diameter **decreases** the area of one-fourth and **quadruples** the oil velocity.

Friction creates turbulence in the oil stream and of course resists flow, resulting in an increased pressure drop through the line. Very low velocity is recommended for the pump inlet line because very little pressure drop can be tolerated there.

DETERMINING PIPE SIZE REQUIREMENTS

Two formulas are available for sizing hydraulic lines.

If the gpm (LPM) and desired velocity are known, use this relationship to find the inside cross-sectional area:

$$AREA = \frac{gpm \times 0.3208}{velocity\ (in\ feet\ per\ second)}$$

$$AREA\ (mm^2) = \frac{LPM \times 16667}{velocity\ (in\ mm\ per\ second)}$$

When the gpm and size of pipe are given, use this formula to find what the velocity will be:

$$VELOCITY\ (feet\ per\ second) = \frac{gpm}{3.117 \times area}$$

$$VELOCITY\ (mm\ per\ second) = \frac{LPM \times 16667}{area\ (mm^2)}$$

Area = square inches (square millimeters)

In Chapter 4, you will find a nomographic chart which permits making these computations by laying a straight edge across printed scales.

SIZE RATINGS OF LINES

The nominal ratings in inches for pipes, tubes, etc., are not accurate indicators of the inside diameter.

In standard pipes, the actual inside diameter is larger than the nominal size quoted. To select pipe, you'll need a standard chart which shows actual inside diameters (see Chapter 4).

For steel and copper tubing, the quoted size is the outside diameter. To find the inside diameter, subtract twice the wall thickness (Figure 1-18).

WORK AND POWER

Whenever a force or push is exerted through a distance, **work** is done.

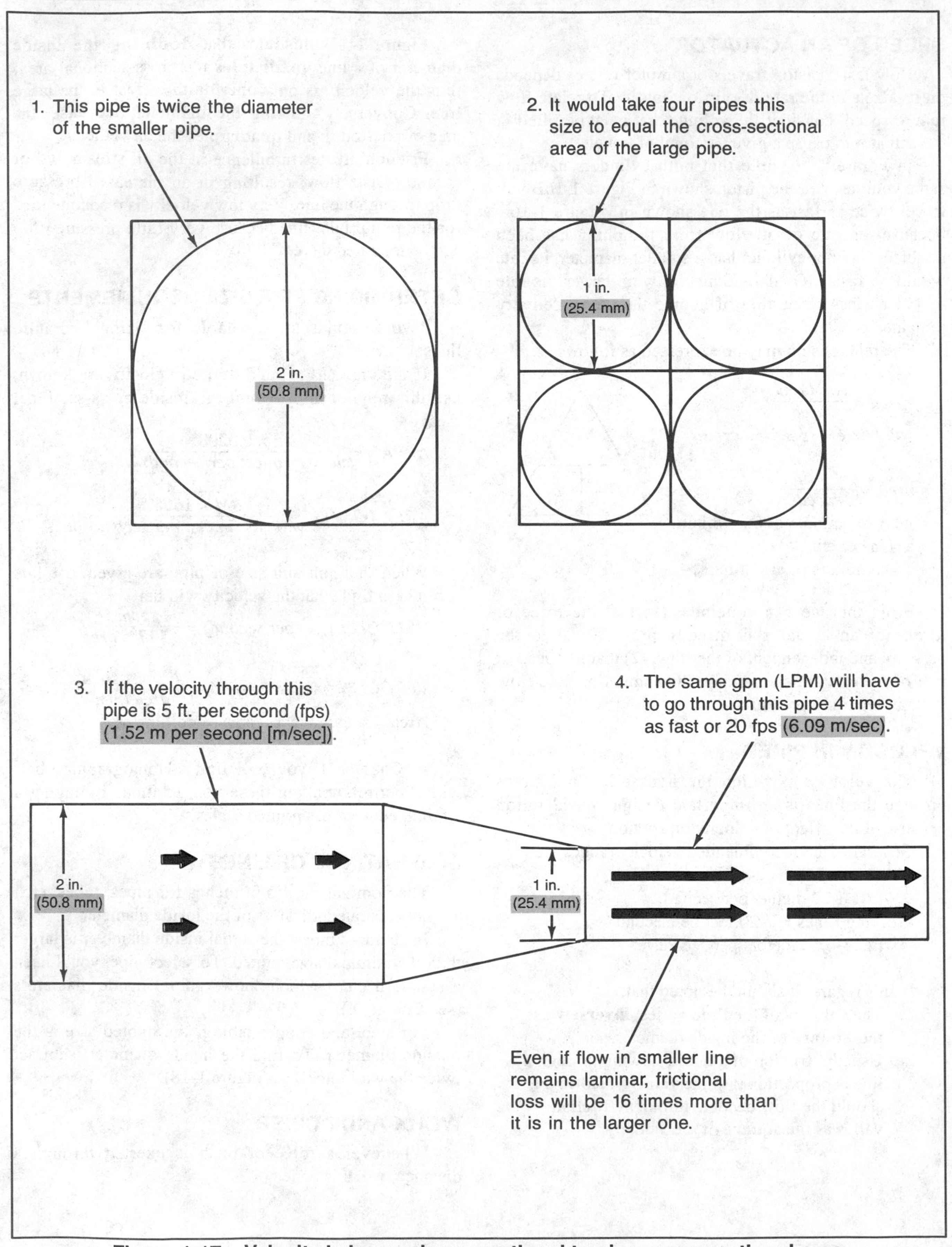
1. This pipe is twice the diameter of the smaller pipe.
2 in.
(50.8 mm)
2. It would take four pipes this size to equal the cross-sectional area of the large pipe.
1 in.
(25.4 mm)
3. If the velocity through this pipe is 5 ft. per second (fps) (1.52 m per second [m/sec]).
2 in.
(50.8 mm)
4. The same gpm (LPM) will have to go through this pipe 4 times as fast or 20 fps (6.09 m/sec).
1 in.
(25.4 mm)
Even if flow in smaller line remains laminar, frictional loss will be 16 times more than it is in the larger one.

Figure 1-17. Velocity is inversely proportional to pipe cross-sectional area.

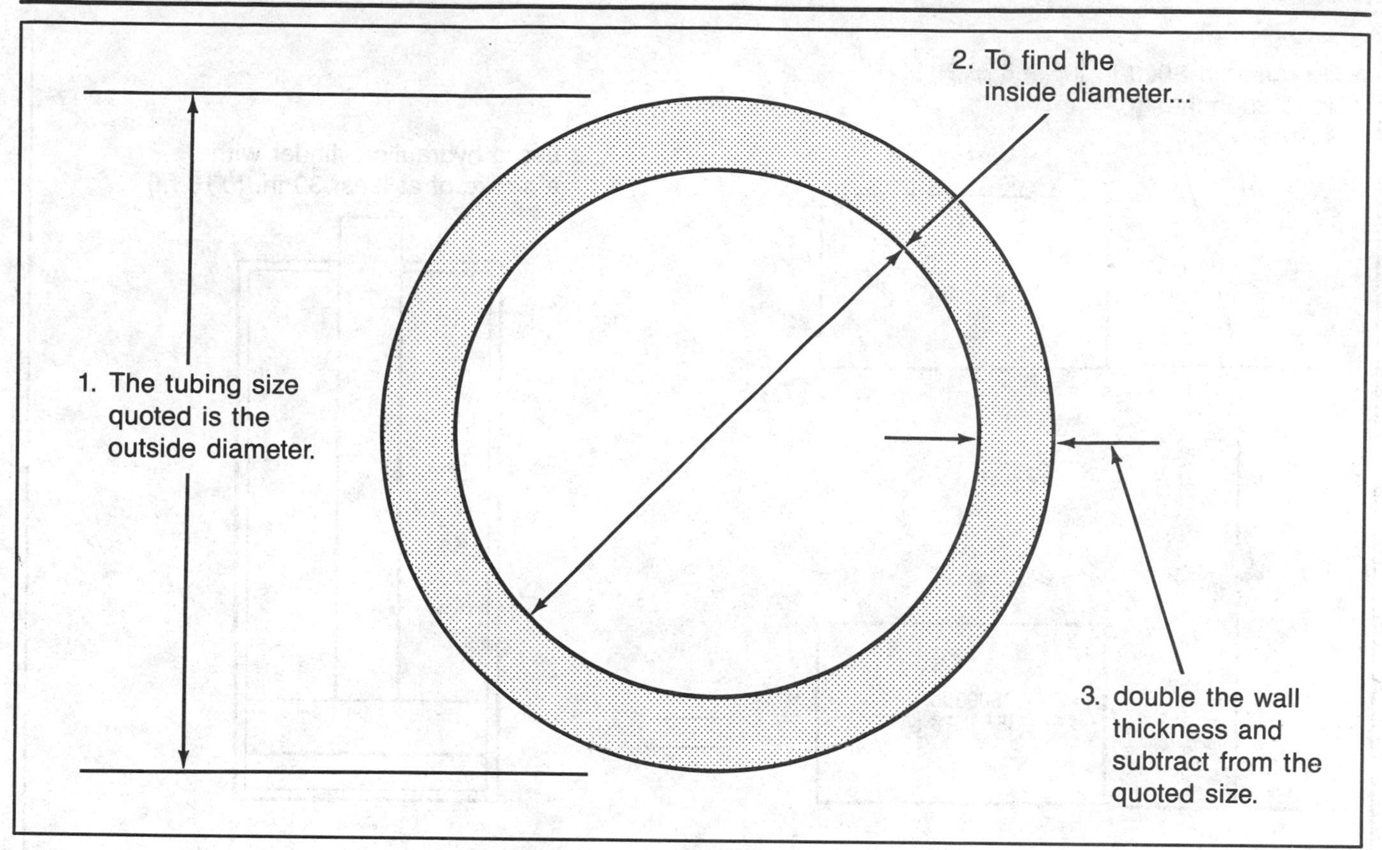

2. To find the inside diameter...
1. The tubing size quoted is the outside diameter.
3. double the wall thickness and subtract from the quoted size.

Figure 1-18. Tubing inside diameter.

$$WORK = force \times distance$$

Work is usually expressed in foot pounds (joules). For example, if a 10 lb (44.48 N) weight is lifted 10 ft (3.05 m), the work is 10 lb (44.48 N) x 10 ft (3.05 m) or 100ft lb (135.66 J).

The formula above for work does not take into consideration how fast the work is done. The **rate** of doing work is called **power.**

To visualize power, think of climbing a flight of stairs. The work done is the body's weight multiplied by the height of the stairs. But it is more difficult to run up the stairs than to walk. When you run, you do the same work at a faster rate.

$$POWER = \frac{force \times distance}{time} \; or \; \frac{work}{time}$$

The usual unit of power is the **horsepower (watt)**, abbreviated hp (W). It is equivalent to 33,000 lb lifted one foot in one minute. (One watt is equal to 1 newton lifted one meter in one second.) It also has equivalents in electrical power and heat.

$$1\ hp = \frac{33,000\ ft\ lb}{min.} \; or \; \frac{550\ ft\ lb}{sec.}$$

$$1W = \frac{1N\,m}{sec.}$$

1 hp = 746 watts (electrical power)
1 hp = 42.4 btu/minute (heat power)

Obviously, it is desirable to be able to convert hydraulic power to horsepower (watt) so that the mechanical, electrical, and heat power equivalents will be known.

HORSEPOWER IN A HYDRAULIC SYSTEM

In the hydraulic system, speed and distance are indicated by the gpm (LPM) flow and force is indicated by pressure. Thus, we might express hydraulic power this way:

$$POWER = \frac{gallons}{minutes} \times \frac{pounds}{square\ inches}$$

$$POWER = \frac{liters}{minutes} \times \frac{newtons}{square\ meter}$$

To change the relationship to mechanical units, we can use these equivalents:

1 gallon = 231 cubic inches (in^3)
12 inches = 1 foot

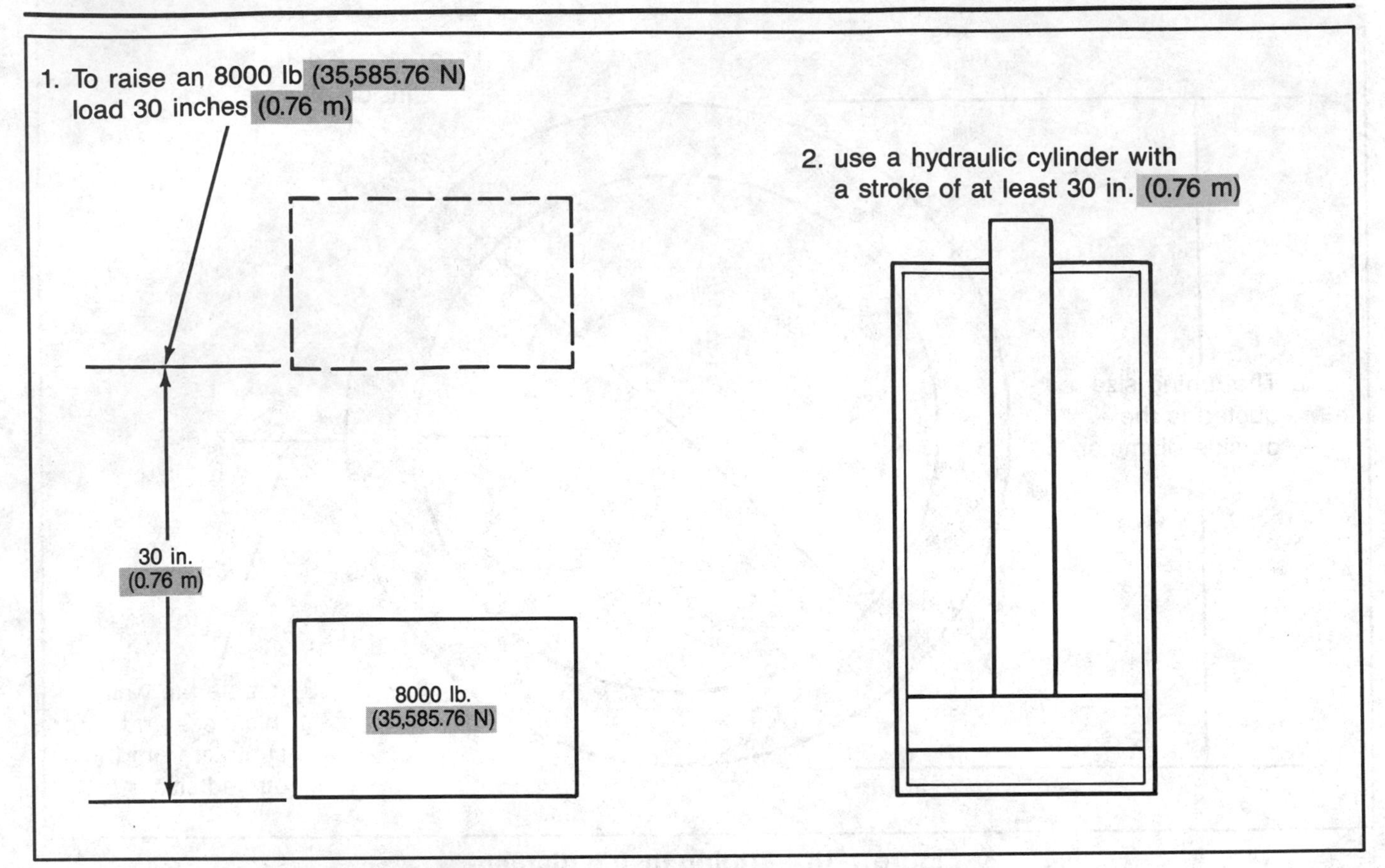
1. To raise an 8000 lb (35,585.76 N)
load 30 inches (0.76 m)
2. use a hydraulic cylinder with
a stroke of at least 30 in. (0.76 m)
30 in.
(0.76 m)
8000 lb.
(35,585.76 N)

Figure 1-19. Use a cylinder to raise a load.

Thus:

$$\text{POWER} = \frac{\text{gallon}}{\text{min.}} \text{ x } \left(\frac{231 \text{ in}^3}{\text{gallon}}\right) \text{x } \frac{\text{pound}}{\text{in}^2} \text{ x } \left(\frac{1 \text{ foot}}{12 \text{ in.}}\right)$$

$$= \frac{231 \textit{ foot pounds}}{12 \textit{ minutes}}$$

This gives us the equivalent mechanical power of one gallon per minute flow at one psi of pressure. To express it as horsepower, divide by 33,000 pounds/minute:

$$\frac{\frac{231 \textit{ ft lb}}{12 \textit{ min.}}}{\frac{33\,000 \textit{ ft lb}}{1 \textit{ min}}} = 0.000583$$

Thus, one gallon per minute flow at one psi equals 0.000583 hp ($4.347 \text{x} 10^{-4}$kW). The total horsepower for any flow condition is:

$hp = gpm \text{ x } psi \text{ x } 0.000583$

or

$$hp = \frac{gpm \text{ x } psi \text{ x } 0.583}{1000}$$

or

$$hp = \frac{gpm \text{ x } psi}{1714}$$

$$kW = \frac{LPM \textit{ x } bar}{600}$$

The third formula is derived by dividing 1000 by 0.583.

These horsepower formulas tell the exact power being used in the system. The horsepower required to drive the pump will be somewhat higher than this since the system is not 100 percent efficient.

If we assume an average efficiency of 83 percent, this relationship can be used to estimate power input requirements:

$$hp = gpm \text{ x } psi \text{ x } 0.0007$$

$$kW = LPM \textit{ x } bar \text{ x } 0.002$$

HORSEPOWER AND TORQUE

It also is often desirable to convert back and forth from horsepower to torque without computing pressure and flow.

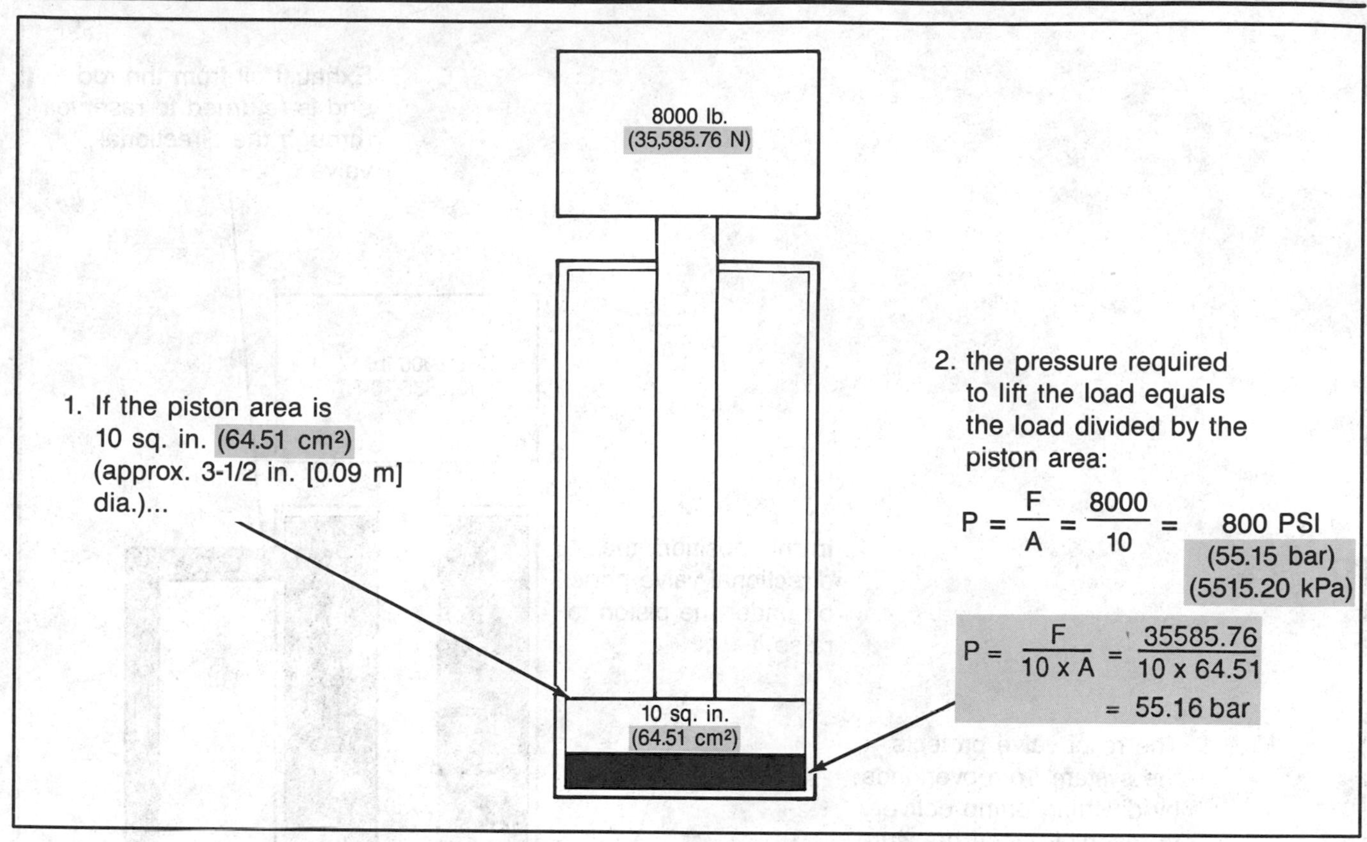

Figure 1-20. Choosing cylinder size.

These are general torque-power formulas for any rotating equipment:

$$torque = \frac{63025 \times hp}{rpm}$$

$$torque = \frac{9550 \times kW}{rpm}$$

$$hp = \frac{torque \times rpm}{63025}$$

$$kW = \frac{torque \times rpm}{9550}$$

Torque in these formulae must be in pound-inches or newton-meters.

DESIGNING A SIMPLE HYDRAULIC SYSTEM

From the information given in this chapter, it is possible to design a simple hydraulic circuit. Following is a description of how the job might proceed. See Figures 1-19 through 1-21.

A Job To Be Done

All circuit design must start with the job to be done. There is a weight to be lifted, a tool head to be rotated, or a piece of work that must be clamped.

The job determines the type of actuator that will be used.

Perhaps the first step should be the selection of an actuator.

If the requirement were simply to raise a load, a hydraulic cylinder placed under it would do the job. The stroke length of the cylinder would be at least equal to the distance the load must be moved (Figure 1-19). Its area would be determined by the force required to raise the load and the desired operating pressure. Let's assume an 8000 lb (35,585.76 N) weight is to be raised a distance of 30 in. (0.76 m) and the maximum operating pressure must be limited to 1000 psi (68.94 bar) (6894.00 kPa). The cylinder selected would have to have a stroke length of at least 30 in. (0.76 m) and with an 8 in^2 (51.61 cm^2) area piston it would provide a maximum force of 8000 lb (35,585.76 N). This, however, would not provide any margin for error. A better selection would be a 10 in^2 (64.52 cm^2) cylinder permitting the load to be raised at 800 psi (55.15 bar) (5515.20 kPa) and providing the capability of lifting up to 10,000 lb (44,482.20 N) (Figure 1-20).

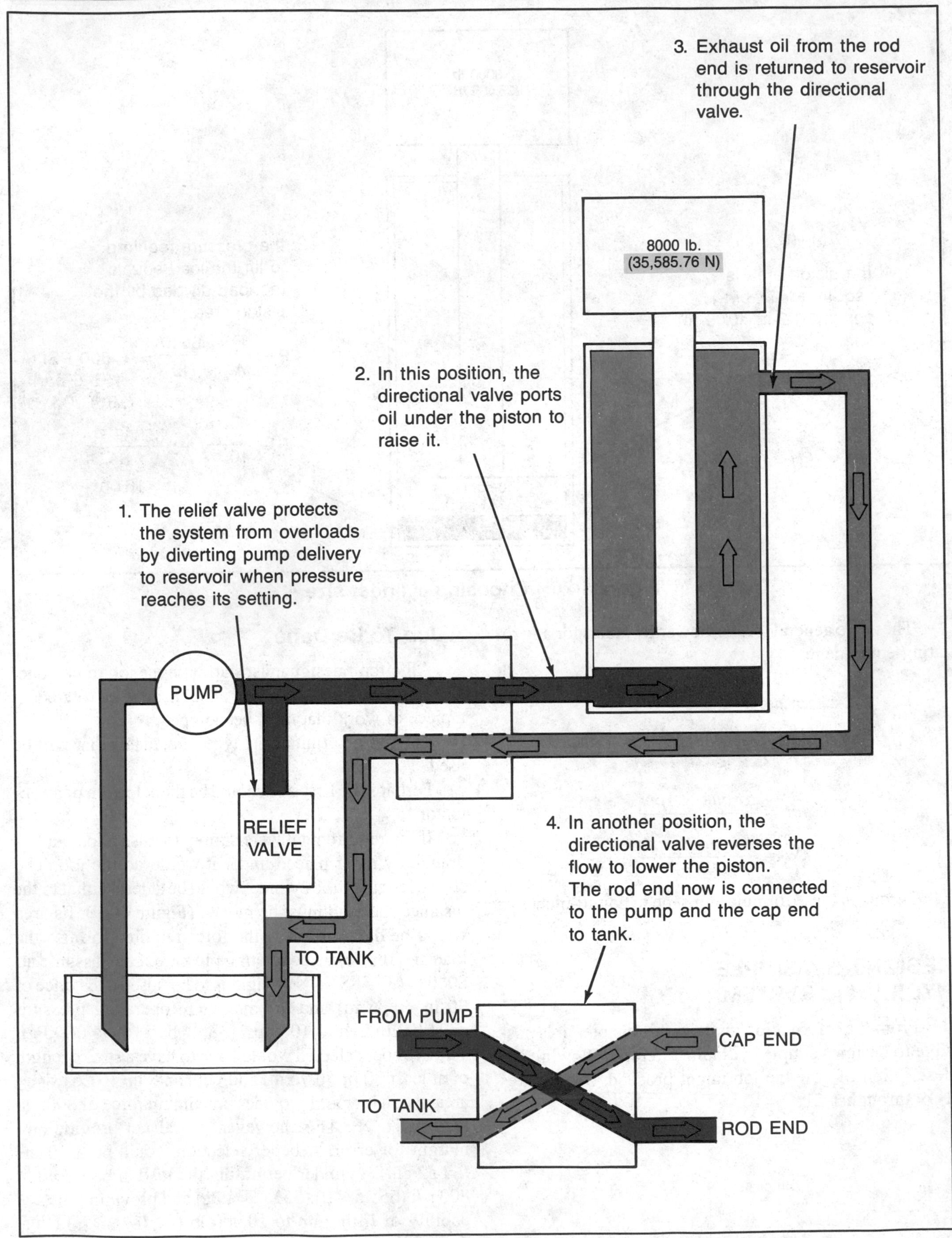
3. Exhaust oil from the rod end is returned to reservoir through the directional valve.
8000 lb.
(35,585.76 N)
2. In this position, the directional valve ports oil under the piston to raise it.
1. The relief valve protects the system from overloads by diverting pump delivery to reservoir when pressure reaches its setting.
PUMP
RELIEF VALVE
TO TANK
4. In another position, the directional valve reverses the flow to lower the piston. The rod end now is connected to the pump and the cap end to tank.
FROM PUMP
CAP END
TO TANK
ROD END

Figure 1-21. Valving to protect and control the system.

The upward and downward travel of the piston rod would be controlled by a directional valve as shown in Figure 1-21. If the load is to be stopped at intermediate points in its travel, the directional valve should have a neutral position in which oil flow from the underside of the piston is blocked to support the weight on the cylinder. The rate at which the load must travel will determine the pump size. The 10 in^2 (64.52 cm^2) piston will displace 10 in^3 (163.87 cm^3) for every inch (0.03 m) it lifts. Extending the piston rod 30 in (0.76m) will require 300 in^3 (4916.12 cm^3) of fluid. If it is to move at the rate of 10 in per second (0.25 m per second), it will require 100 in^3 (1638.71 cm^3) of fluid per second or 6000 in^3 (98,322 cm^3) per minute. Since pumps are usually rated in gallons per minute (liters per minute), it will be necessary to divide 6000 (98,322) by 231 (cubic inches per gallon) (1000 cm^3 per liter) to convert the requirements into gallons per minute (liters per minute).

$$\frac{6000}{231} = 26\ gpm$$

$$\frac{98322}{1000} = 98.32\ LPM$$

The hp (wattage) needed to drive the pump is a function of its delivery and the maximum pressure at which it will operate. The following formula will determine the size of the electric motor required:

$$hp = gpm \times psi \times 0.0007$$
$$hp = 26 \times 1000 \times 0.0007 = 18.2$$

$$kW = LPM \times bar \times 0.002$$
$$kW = 98.32 \times 68.94 \times 0.002 = 13.6$$

To prevent overloading of the electric motor and to protect the pump and other components from excessive pressure due to overloads or stalling, a relief valve set to limit the maximum system pressure should be installed in the line between the pump outlet and the inlet port to the directional valve (Figure 1-21).

A reservoir sized to hold approximately two to three times the pump capacity in gallons per minute (liters per minute), filters, and adequate interconnecting piping would complete the system.

CONCLUSION

This chapter has presented a brief introductory overview of hydraulics to demonstrate the basic principles involved in hydraulic system operation. There are, of course, countless variations of the system presented. Many of these will be developed with a more detailed study of operating principles and components in future chapters.

QUESTIONS

1. State Pascal's Law.
2. Define pressure.
3. If a force of 1000 lb (4448.22 N) is applied over an area of 20 in^2 (129.03 cm^2), what is the pressure?
4. What is meant by "conservation of energy?"
5. What is the output component of a hydraulic system named? The input component?
6. What is the prime mover?
7. Name several advantages of a hydraulic system.
8. What is the origin of the term "hydraulics?"
9. What makes petroleum oil suitable as a hydraulic fluid?
10. What is the pressure at the bottom of a 20 ft (6.10 m) column of oil?
11. What can you say definitely about the pressures on opposite sides of an orifice when oil is flowing through it?
12. What pressure is usually available to push liquid into the pump inlet?
13. Why should the pump inlet vacuum be minimized?
14. What is the function of the pump?
15. Why is loss of pressure usually not a symptom of pump malfunction?
16. How is pressure created?
17. If three 200 psi (13.78 bar) (1378.80 kPa) check valves are connected in series, how much pressure is required at the pump to push oil through all three?
18. What is the formula for pressure developed when moving a load with a cylinder?
19. What is the formula for the maximum force output of a cylinder?
20. What determines the speed of an actuator?
21. What is the relationship between fluid velocity and friction in a pipe?
22. What is work? Power?
23. How do you find horsepower (wattage) in a hydraulic system?
24. With which component does the design of a hydraulic circuit begin?
25. What determines the size of the pump needed in a hydraulic circuit?
26. What is the piston area of a 5 in. (127 mm) cylinder?
27. What does the relief valve do?
28. What does a directional valve do?

CHAPTER 2

PRINCIPLES OF POWER HYDRAULICS

This chapter is divided into three sections:

- Principles of Pressure
- Principles of Flow
- Hydraulic Graphical Symbols

The first two sections will further develop the fundamentals of the physical phenomena that combine to transfer power in the hydraulic circuit. The third section, illustrating graphical symbols for circuit diagrams, will deal with the classes and functions of lines and components. All this material will serve as a background for following chapters on the equipment that makes up a hydraulic system.

PRINCIPLES OF PRESSURE

A Precise Definition

It has been noted that the term hydraulics is derived from a Greek word for water. Therefore, it might be assumed correctly that the science of hydraulics encompasses any device operated by water. A water wheel or turbine (Figure 2-1) for instance, is a hydraulic device.

However, a distinction must be made between devices which utilize the impact or momentum of a moving liquid and those which are operated by pushing on a confined fluid; that is, by pressure.

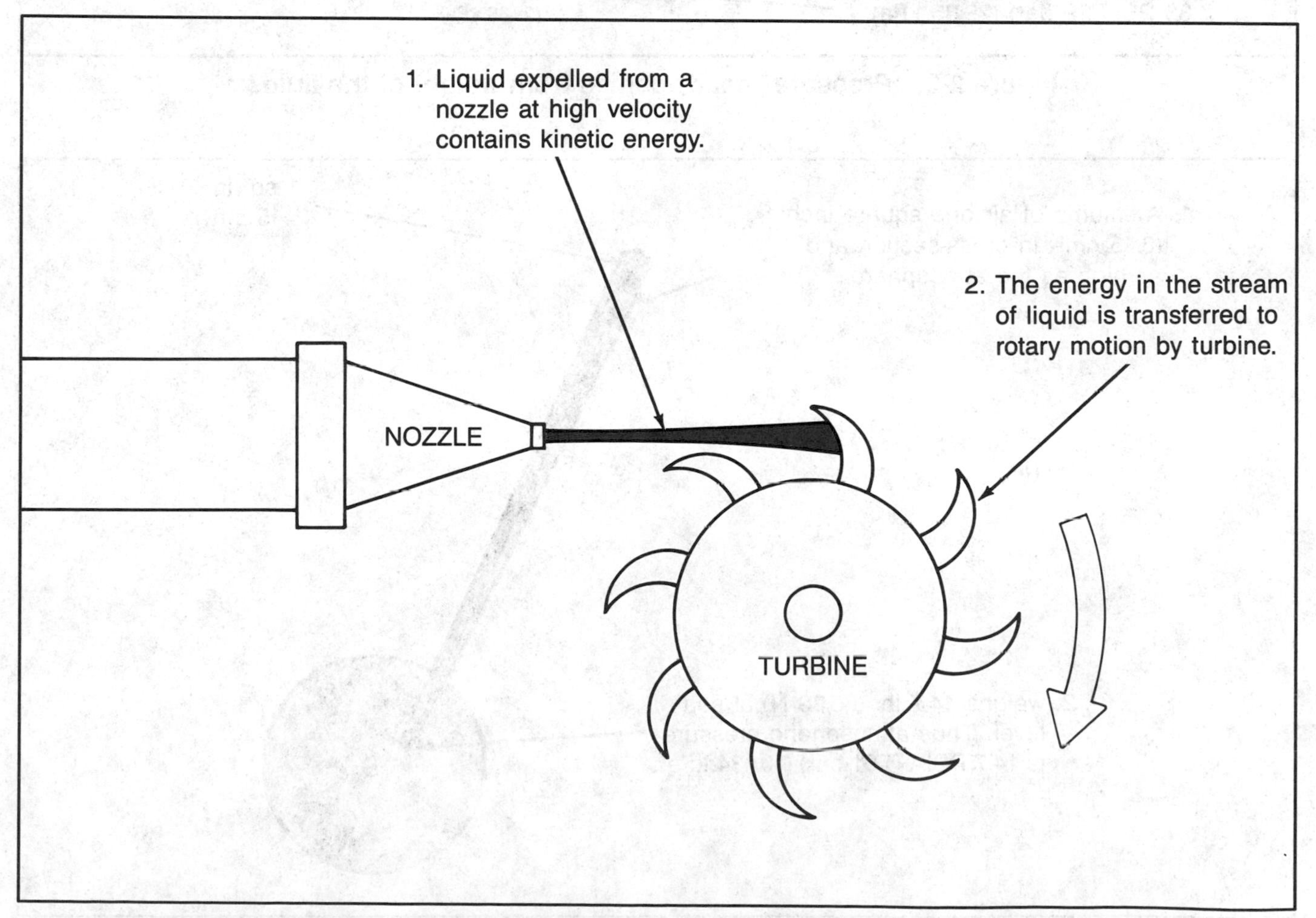

1. Liquid expelled from a nozzle at high velocity contains kinetic energy.
2. The energy in the stream of liquid is transferred to rotary motion by turbine.
NOZZLE
TURBINE

Figure 2-1. Hydrodynamic device uses kinetic energy rather than pressure.

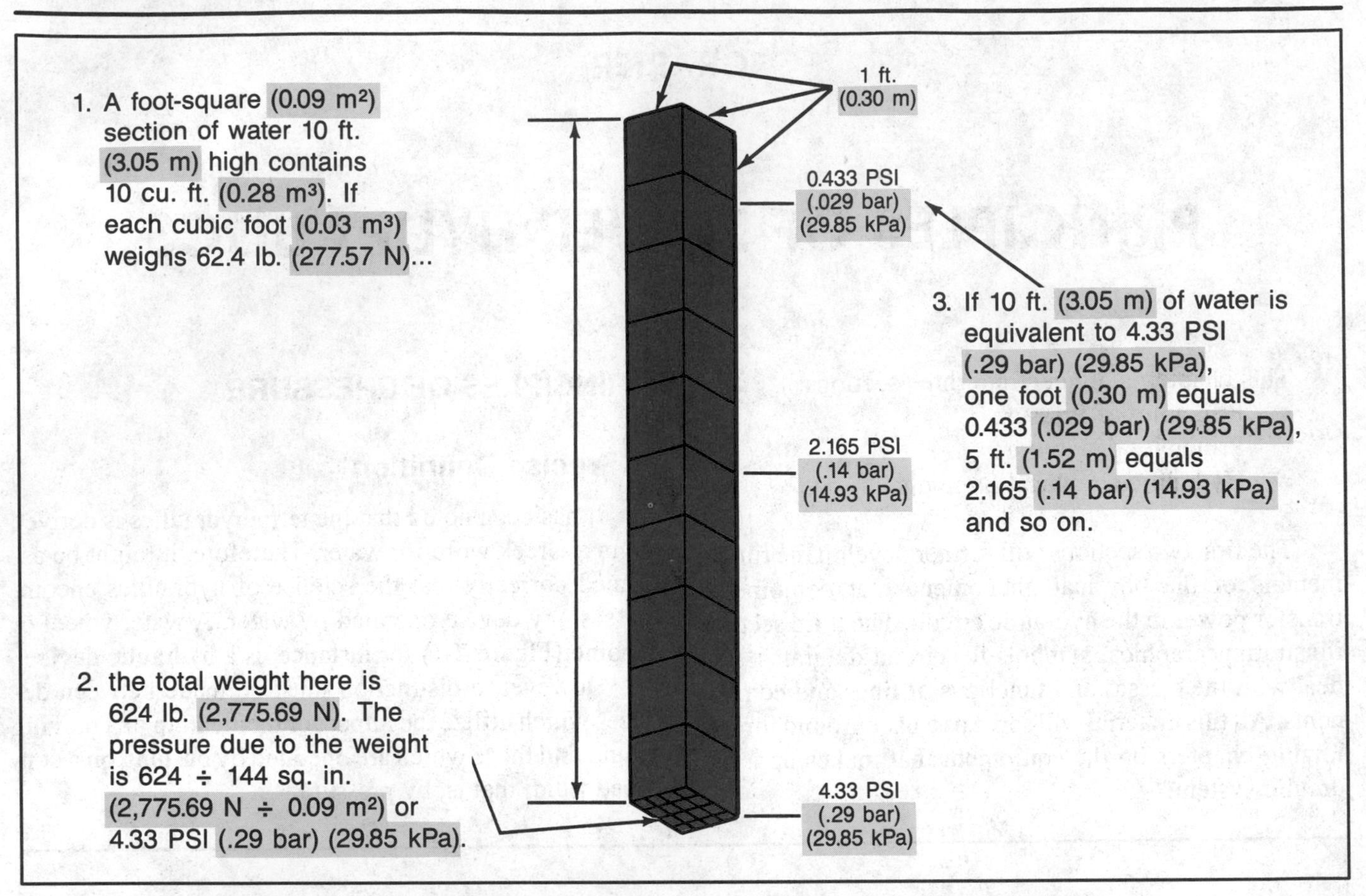

1. A foot-square (0.09 m²) section of water 10 ft. (3.05 m) high contains 10 cu. ft. (0.28 m³). If each cubic foot (0.03 m³) weighs 62.4 lb. (277.57 N)...
1 ft.
(0.30 m)
0.433 PSI
(.029 bar)
(29.85 kPa)
3. If 10 ft. (3.05 m) of water is equivalent to 4.33 PSI (.29 bar) (29.85 kPa), one foot (0.30 m) equals 0.433 (.029 bar) (29.85 kPa), 5 ft. (1.52 m) equals 2.165 (.14 bar) (14.93 kPa) and so on.
2.165 PSI
(.14 bar)
(14.93 kPa)
2. the total weight here is 624 lb. (2,775.69 N). The pressure due to the weight is 624 ÷ 144 sq. in. (2,775.69 N ÷ 0.09 m²) or 4.33 PSI (.29 bar) (29.85 kPa).
4.33 PSI
(.29 bar)
(29.85 kPa)

Figure 2-2. Pressure "head" comes from weight of the fluid.

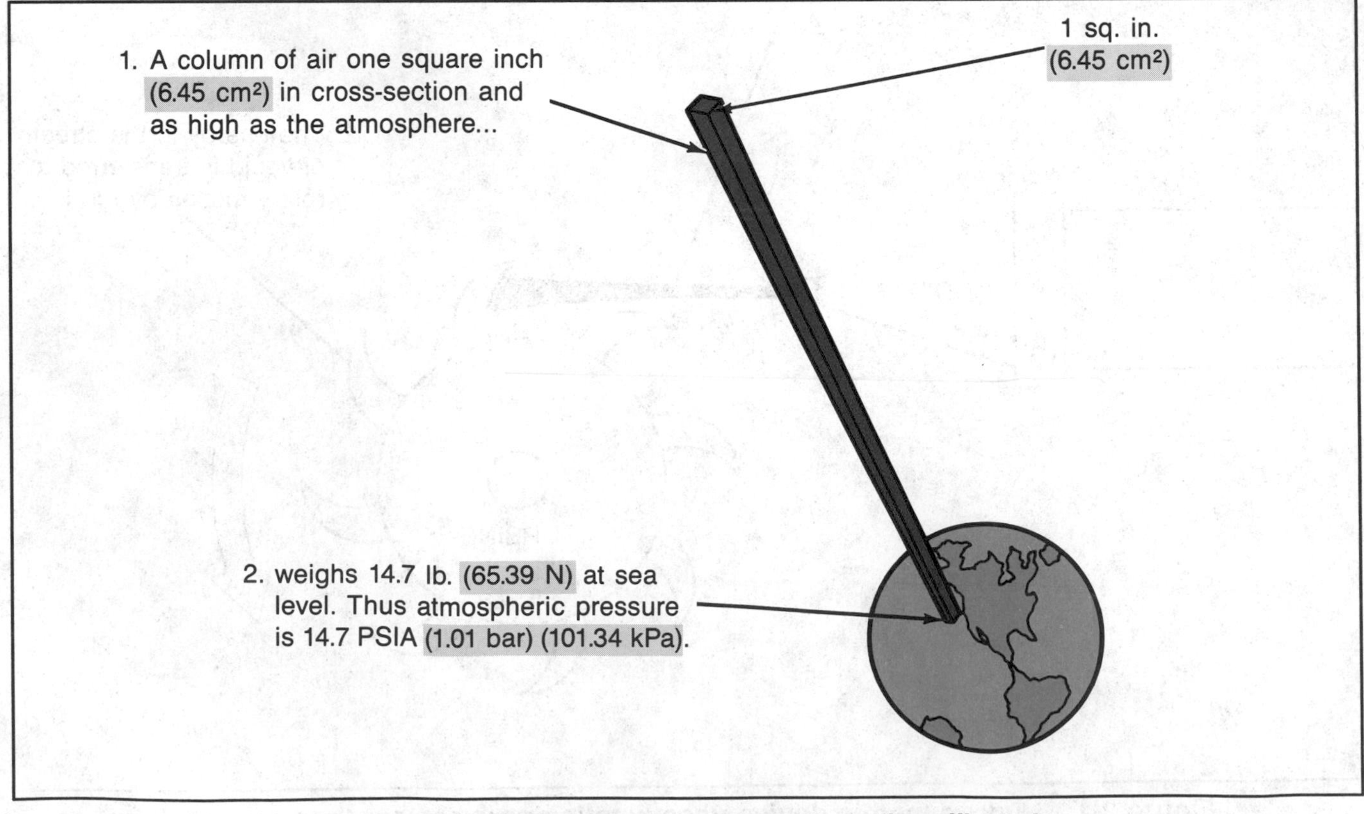

Figure 2-3. Atmospheric pressure is a "head" of air.

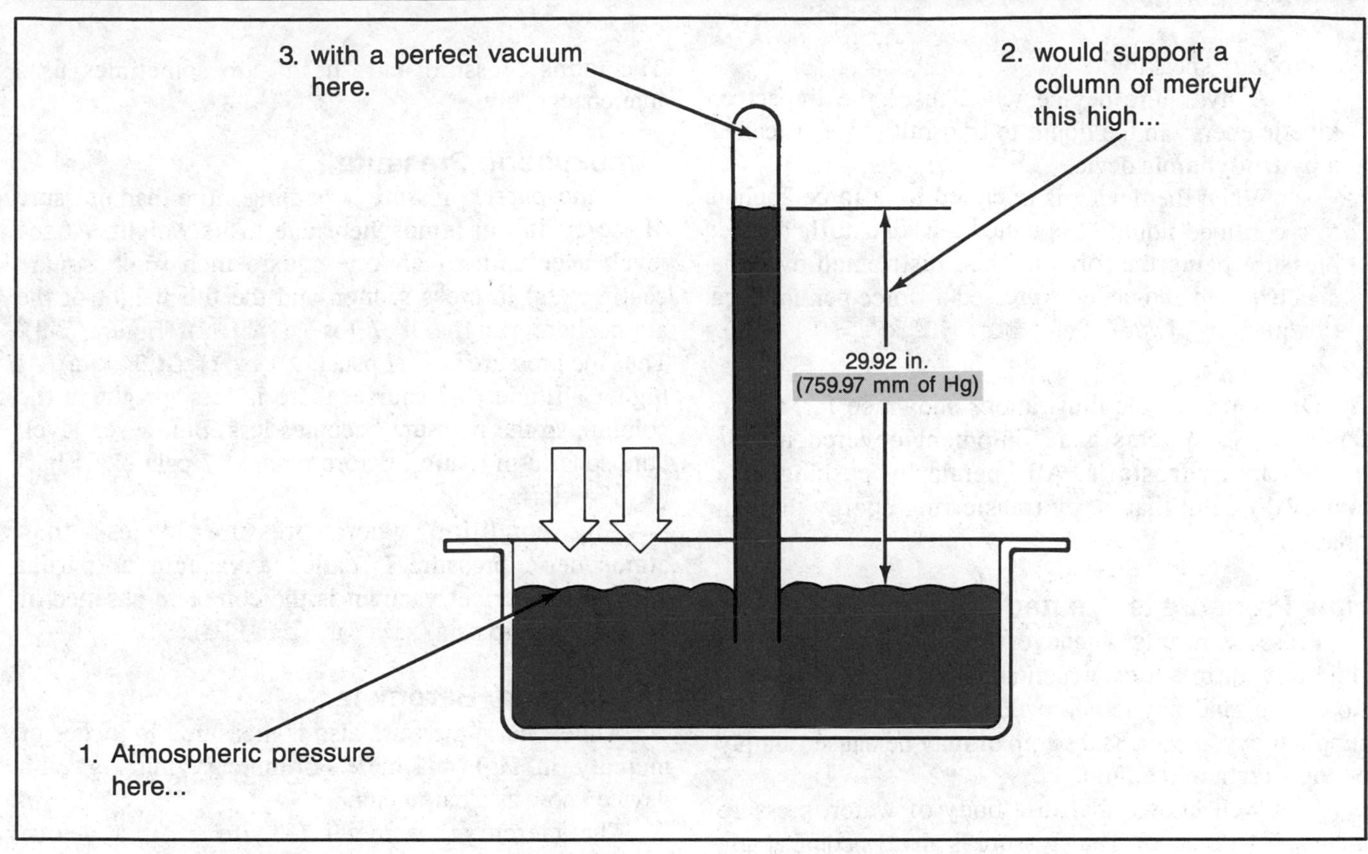
3. with a perfect vacuum here.
2. would support a column of mercury this high...
29.92 in.
(759.97 mm of Hg)
1. Atmospheric pressure here...

Figure 2-4. The mercury barometer measures atmospheric pressure.

	PSIA (POUNDS PER SQUARE INCH ABSOLUTE) (bar) (kPa)	PSI (POUNDS PER SQUARE INCH GAUGE) GAUGE SCALE (bar) (kPa)	IN. HG ABS. (INCHES OF MERCURY ABSOLUTE) BAROMETER SCALE (mbar)	IN. HG (INCHES OF MERCURY) VACUUM SCALE (mbar)	FEET OF OIL ABSOLUTE (meters)	FEET OF WATER ABSOLUTE (bar)
3 ATMOSPHERES ABSOLUTE 2 ATMOSPHERES GAUGE	44.1 (3.04) (304.06)	29.4 (2.02) (202.68)	(90) (3047.75)		111 (33.83)	102 (3.04)
2 ATMOSPHERES ABSOLUTE 1 ATMOSPHERE GAUGE	29.4 (2.02) (202.68)	14.7 (1.01) (101.34)	(60) (2031.83)		74 (22.55)	68 (2.03)
1 ATMOSPHERE ABSOLUTE (ATMOSPHERIC PRESSURE)	14.7 (1.01) (101.34)	0 (0.00) (0.00)	29.92 (30) (1013.21)(1015.92)	0 (0.00)	37 (11.28)	34 (1.01)
	10 (0.68) (68.94)	−5 (−0.34) (−34.47)	20 (677.29)	10 (338.64)	24 (7.31)	22⅔ (0.67)
	5 (0.34) (34.47)	−10 (−0.68) (−68.94)	10 (338.64)	20 (677.29)	12 (3.65)	11½ (0.34)
PERFECT VACUUM	0 (0.00) (0.00)	−15 (−1.03) (−103.41)	0 (0.00)	29.92 (1013.21)	0 (0.00)	0 (0.00)

============= Indicates that the scale is not used in this range. Values are shown for comparison only.

Figure 2-5. Pressure and vacuum scale comparison.

Properly speaking:

- A hydraulic device which uses the impact or kinetic energy in the liquid to transmit power is called a **hydrodynamic** device.
- When the device is operated by a force applied to a confined liquid, it is called a **hydrostatic** device: pressure being the force applied distributed over the area exposed and being expressed as force per unit area (lbs/in^2 = psi, N/m^2 = Pa, 1 bar = 100 kPa = 0.1 MPa).

Of course, all the illustrations shown so far, and in fact, all the systems and equipment covered in this manual are **hydrostatic**. All operate by pushing on a confined liquid; that is, by transferring energy through pressure.

How Pressure Is Created

Pressure results whenever there is a resistance to fluid flow or to a force which attempts to make the fluid flow. The tendency to cause flow (or the push) may be supplied by a mechanical pump or may be caused simply by the weight of the fluid.

It is well known that in a body of water, pressure increases with depth. The pressure is always equal at any particular depth due to the weight of the water above it. Around Pascal's time, an Italian scientist named Torricelli proved that if a hole is made in the bottom of a tank of water, the water runs out fastest when the tank is full and the flow rate decreases as the water level lowers. In other words, as the "head" of water above the opening lessens, so does the pressure.

Torricelli could express the pressure at the bottom of the tank only as "feet of head," or the height in feet (meters) of the column of water. Today, with the pound per square inch (psi) (newton per square meter, Pa) as a unit pressure, we can express pressure anywhere in any liquid or gas in more convenient terms. All that is required is knowing how much a cubic foot (cubic meter) of the fluid weighs.

As shown in Figure 2-2, a "head" of one foot (.30 m) of water is equivalent to 0.433 psi (0.03 bar) (2.99 kPa); a five-foot head (1.52 m) of water equals 2.17 psi (0.15 bar) (14.96 kPa), and so on. And as shown earlier, a head of oil is equivalent to about 0.4 psi per foot (0.09 bar) (9.05 kPa per meter).

In many places, the term "head" is used to describe pressure, no matter how it is created. For instance, a boiler is said to "work up a head of steam" when pressure is created by vaporizing water in confinement. The terms pressure and "head" are sometimes used interchangeably.

Atmospheric Pressure

Atmospheric pressure is nothing more than pressure of the air in our atmosphere due to its weight. At sea level, a column of air one square inch (6.45 square centimeters) in cross section and the full height of the atmosphere weighs 14.7 lbs (65.39 N) (Figure 2-3). Thus the pressure is 14.7 psia (1.01 bar) (101.34 kPa). At higher altitudes, of course, there is less weight in the column, so the pressure becomes less. Below sea level, atmospheric pressure is more than 14.7 psia (1.01 bar) (101.34 kPa).

Any condition where pressure is less than atmospheric pressure is called a vacuum or partial vacuum. A perfect vacuum is the complete absence of pressure or zero psia (zero bar) (zero kPa).

The Mercury Barometer

Atmospheric pressure also is measured in inches of mercury (in. Hg) (millimeters of mercury, mm Hg) on a device know as a barometer.

The mercury barometer (Figure 2-4), a device invented by Torricelli, is usually credited as the inspiration for Pascal's studies of pressure. Torricelli discovered that when a tube full of mercury is inverted in a pan of the liquid, the column in the tube will fall only a certain distance. He reasoned that atmospheric pressure on the surface of the liquid was supporting the weight of the column of mercury with a perfect vacuum at the top of the tube.

In a normal atmosphere, the column will always be 29.92 in.(759.97 mm) high. Thus, 29.92 (usually rounded off to 30) in Hg (760.00 mm Hg) becomes another equivalent of the pressure of one atmosphere.

Measuring Vacuum

Since vacuum is pressure below atmospheric, vacuum can be measured in the same units. Thus, vacuum can be expressed as psia or psi (bar) (kPa) (in negative units) as well as in inches of mercury (millimeters of mercury).

Most vacuum gauges, however, are calibrated in inches of mercury (millimeters of mercury). A perfect vacuum, which will support a column of mercury 29.92 in. (760.00 mm) high is 29.92 in. Hg (760.00 mm Hg). Zero vacuum (atmospheric pressure) reads zero on a vacuum gauge.

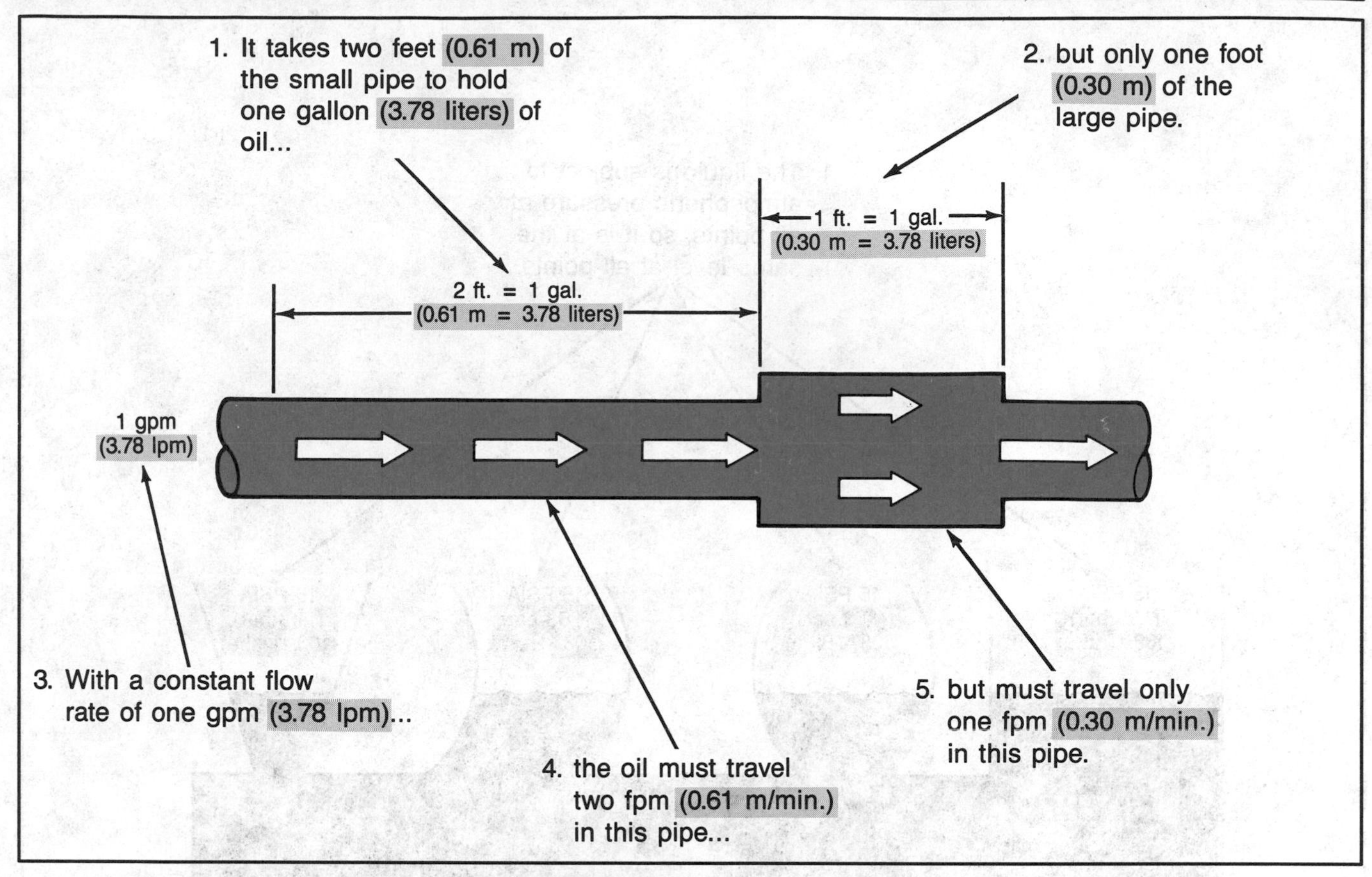

1. It takes two feet (0.61 m) of the small pipe to hold one gallon (3.78 liters) of oil...
2. but only one foot (0.30 m) of the large pipe.

(0.30 m = 3.78 liters)

(0.61 m = 3.78 liters)

4. the oil must travel two fpm (0.61 m/min.) in this pipe...
5. but must travel only one fpm (0.30 m/min.) in this pipe.

Figure 2-6. Flow is volume per unit of time; velocity is distance per unit of time.

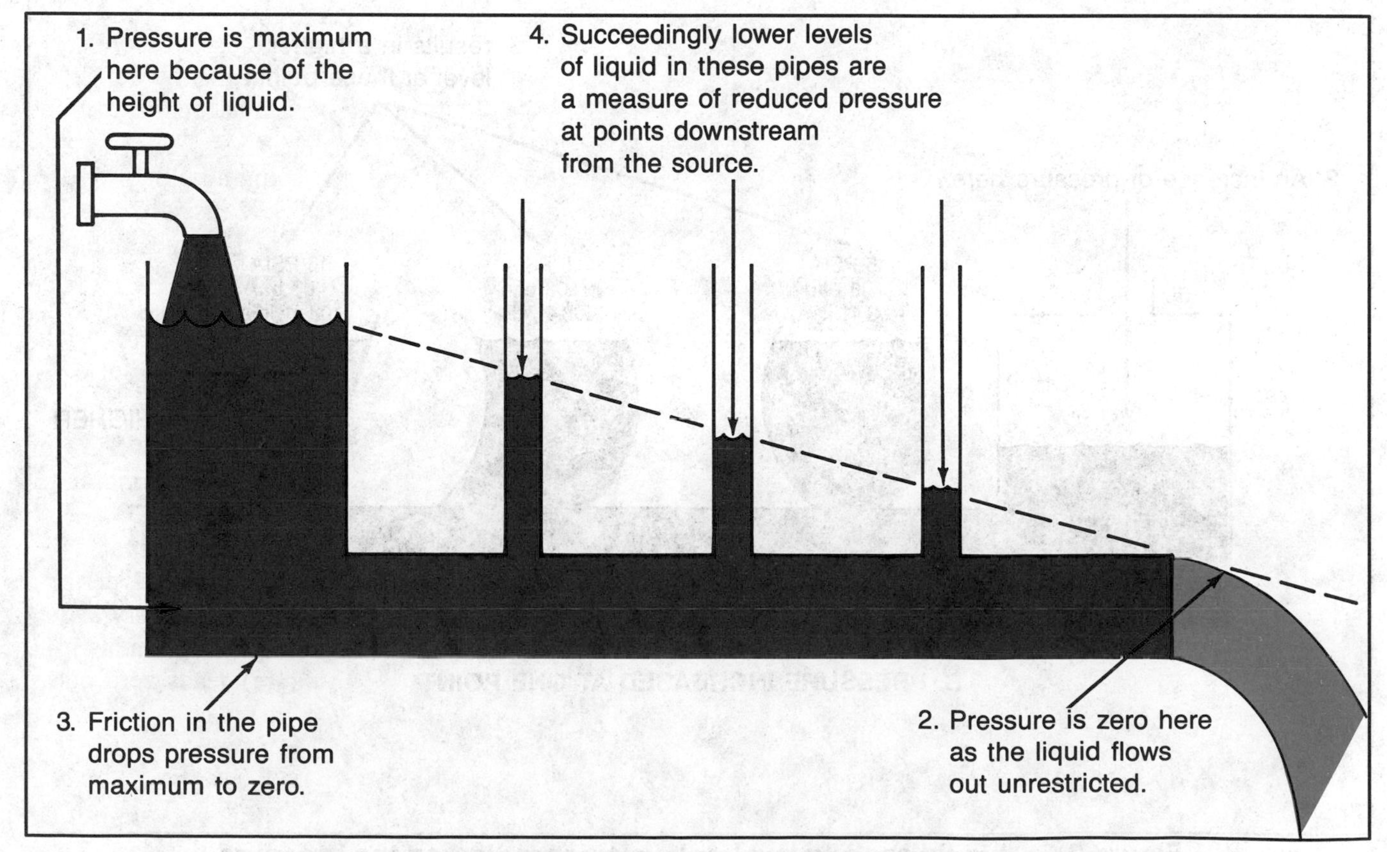

1. Pressure is maximum here because of the height of liquid.
4. Succeedingly lower levels of liquid in these pipes are a measure of reduced pressure at points downstream from the source.
3. Friction in the pipe drops pressure from maximum to zero.
2. Pressure is zero here as the liquid flows out unrestricted.

Figure 2-7. Friction in pipes results in a pressure drop.

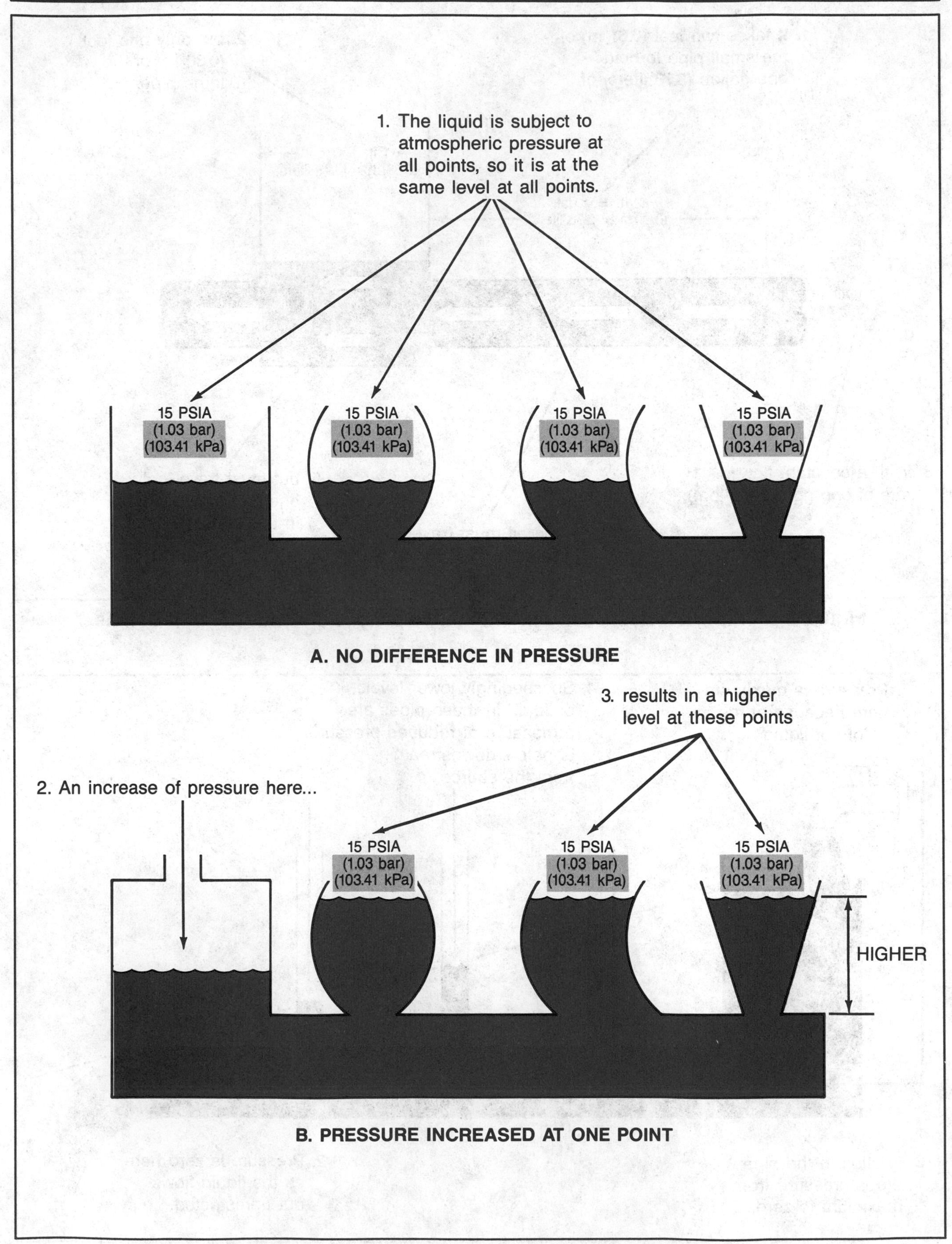

Figure 2-8. Liquid seeks a level or levels depending on the pressure.

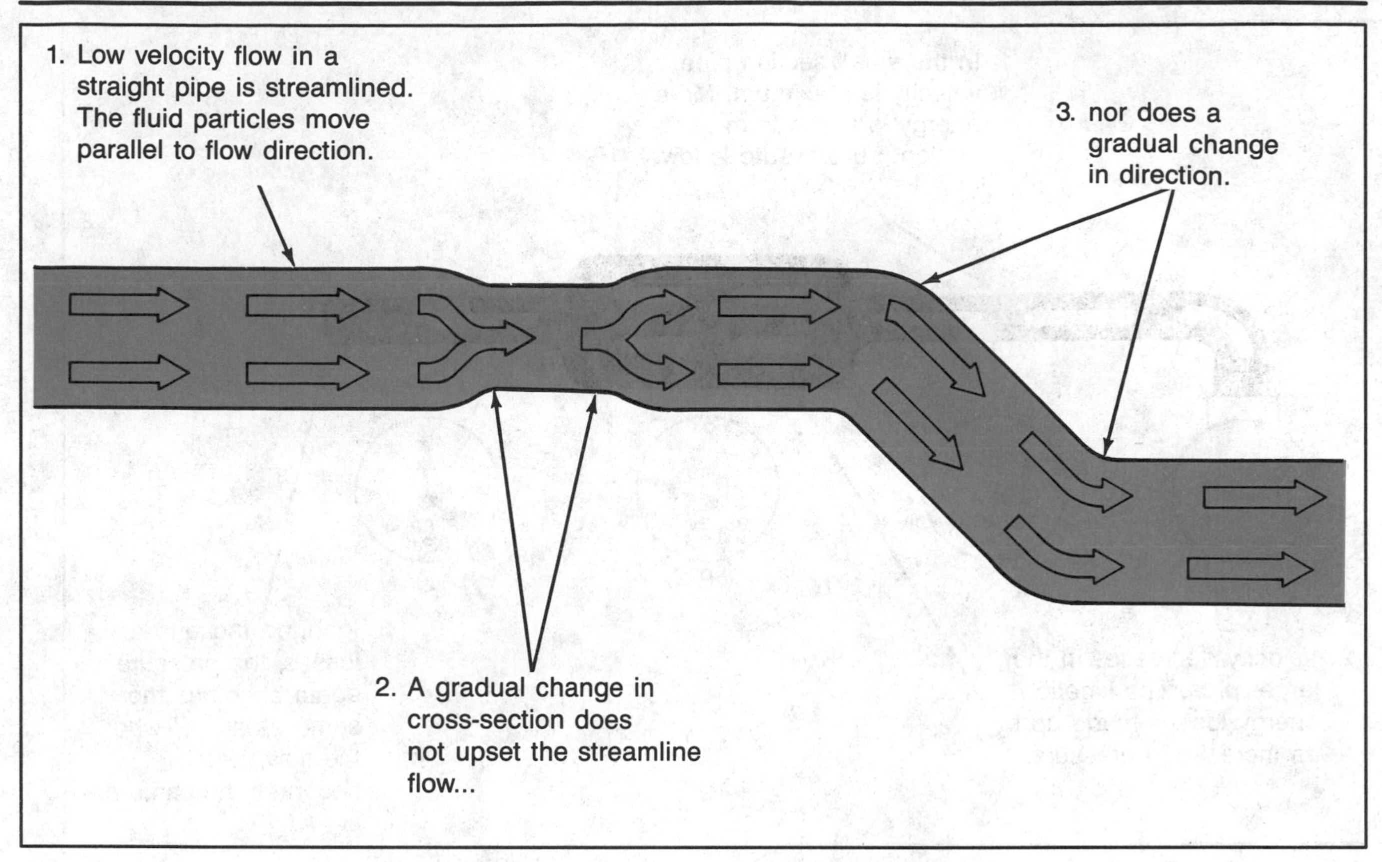

1. Low velocity flow in a straight pipe is streamlined. The fluid particles move parallel to flow direction.
3. nor does a gradual change in direction.
2. A gradual change in cross-section does not upset the streamline flow...

Figure 2-9. Laminar flow is in parallel paths.

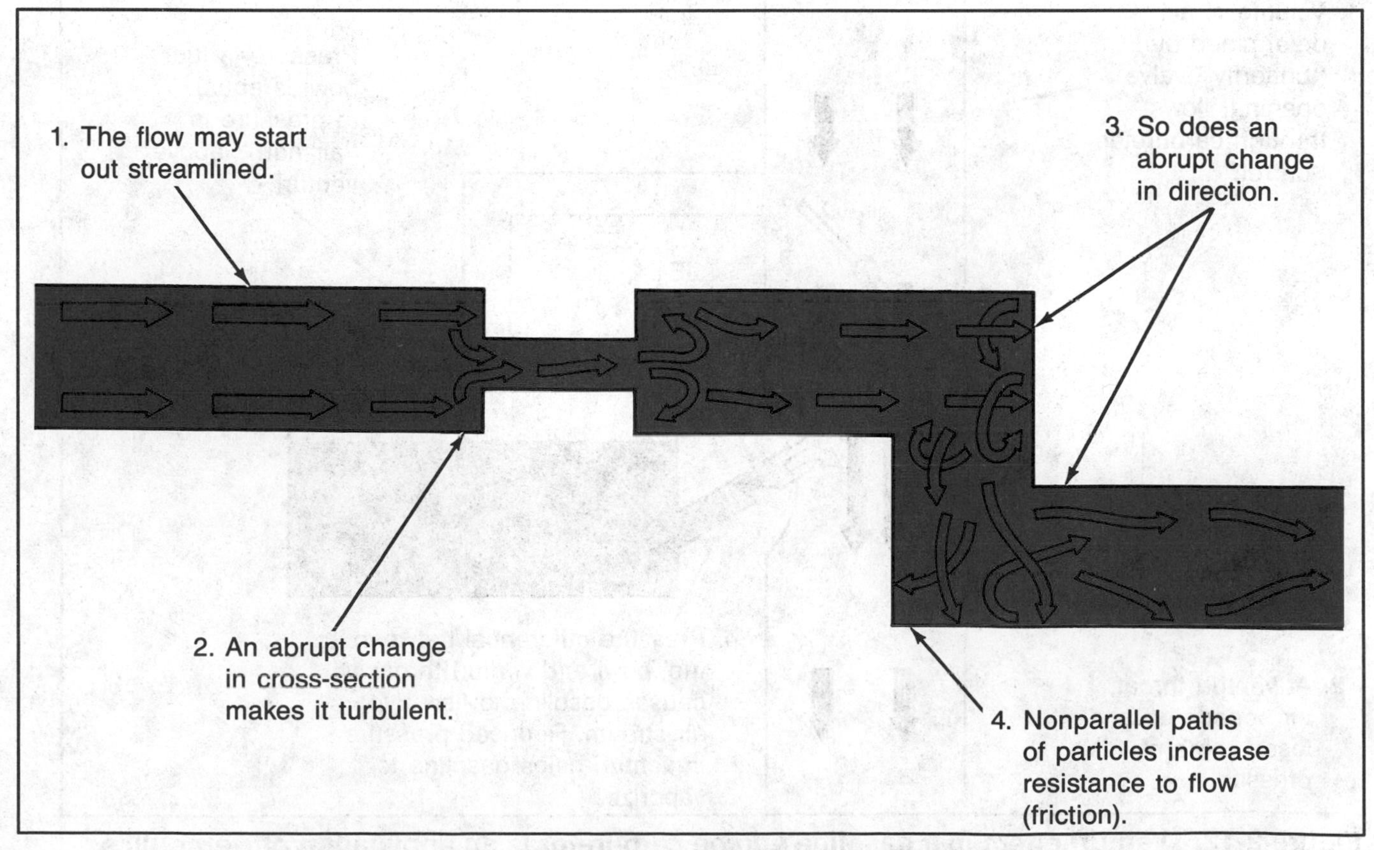

1. The flow may start out streamlined.
3. So does an abrupt change in direction.
2. An abrupt change in cross-section makes it turbulent.
4. Nonparallel paths of particles increase resistance to flow (friction).

Figure 2-10. Turbulence results in flow resistance.

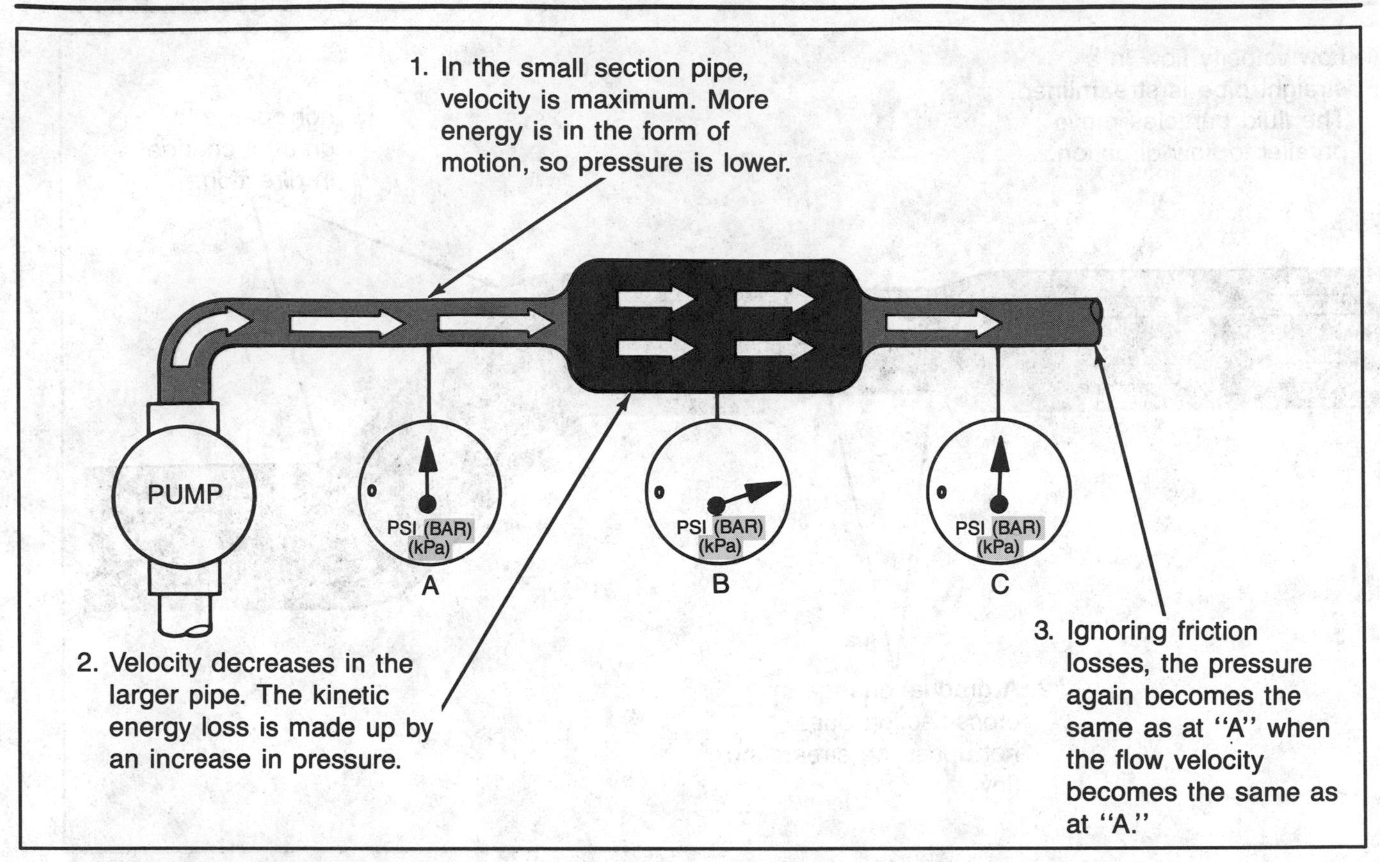

1. In the small section pipe, velocity is maximum. More energy is in the form of motion, so pressure is lower.

3. Ignoring friction losses, the pressure again becomes the same as at "A" when the flow velocity becomes the same as at "A."

Figure 2-11. The sum of pressure and kinetic energy is constant with a constant flow rate.

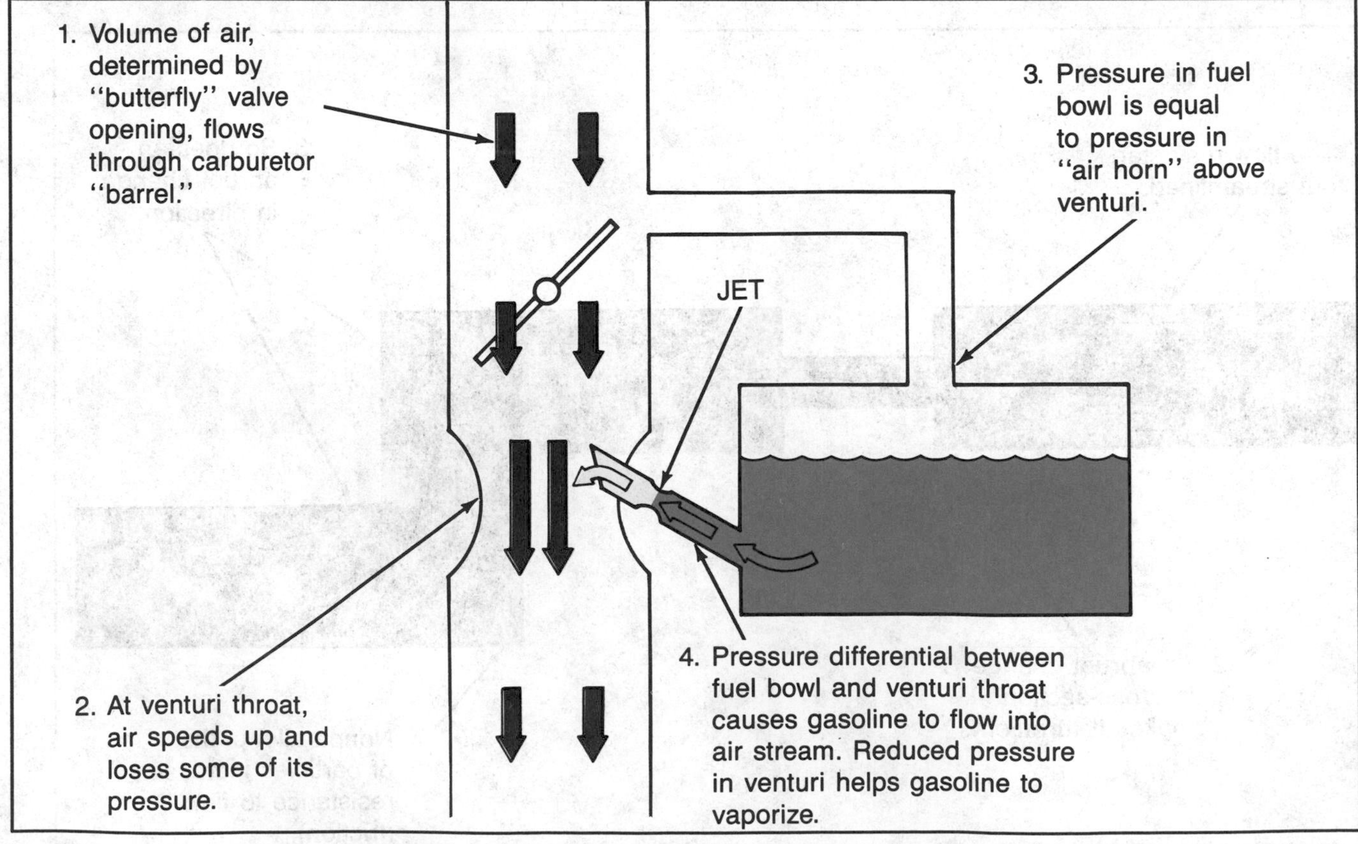

1. Volume of air, determined by "butterfly" valve opening, flows through carburetor "barrel."
3. Pressure in fuel bowl is equal to pressure in "air horn" above venturi.
JET
2. At venturi throat, air speeds up and loses some of its pressure.
4. Pressure differential between fuel bowl and venturi throat causes gasoline to flow into air stream. Reduced pressure in venturi helps gasoline to vaporize.

Figure 2-12. Venturi effect in a gasoline engine carburetor is an application of Bernoulli's principle.

Summary of Pressure and Vacuum Scales

Since a number of ways of measuring pressure and vacuum have been discussed, it would be well to place them all together for comparison. Figure 2-5 presents a summary of pressure and vacuum measurements:

1. An **atmosphere** is a pressure unit equal to 14.7 psi (1.01 bar) (101.34 kPa).
2. Psia (pounds per square inch absolute) (kilopascals) is a scale which starts at a perfect vacuum (0 psia (0.00 bar) (0.00 kPa)). Atmospheric pressure at sea level is 14.7 (1.01 bar) (101.34 kPa) on this scale.
3. Psi (pounds per square inch gauge) (bar) (kilopascals) is calibrated in the same units as psia but ignores atmospheric pressure. Gauge pressure may be abbreviated psig (bar) (kPa).
4. To convert from psia to psig:
   Gauge Pressure + 14.7 = Absolute Pressure
   Absolute Pressure - 14.7 = Gauge Pressure.
5. Atmospheric pressure on the barometer scale is 29.92 in. Hg (760.00 mm Hg). Comparing this to the psia scale, it is evident that:
   1 psi (0.07 bar) (6.89 kPa) = 2 in. Hg (50.80 mm Hg) (approximately).
   1 in. Hg (25.40 mm Hg) = .5 psi (0.03 bar) (3.45 kPa) (approximately).
6. An atmosphere is equivalent to approximately 34 ft. (10.36 m) of water or 37 ft. (11.28 m) of oil.

PRINCIPLES OF FLOW

Flow is the action in the hydraulic system that gives the actuator its motion. Pressure gives the actuator its force, but flow is essential to cause movement. Flow in the hydraulic system is created by the pump.

How Flow Is Measured

There are two ways to measure the flow of a fluid: **Velocity** is the average speed of the fluid's particles past a given point or the average distance the particles travel per unit of time. It is usually measured in feet per second (fps) (meters per second, m/s), feet per minute (fpm) (meters per minute, m/min) or inches per second (ips) (centimeters per second, cm/s).

Flow rate is a measure of the **volume** of fluid passing a point in a given time. Large volumes are measured in gallons per minute (gpm) (liters per minute, LPM). Small volumes may be expressed in cubic inches per minute (cubic centimeters per minute).

Figure 2-6 illustrates the distinction between velocity and flow rate. A constant flow of one gallon per minute (3.78 LPM) either increases or decreases in velocity when the cross section of the pipe changes size.

Flow Rate and Speed

The speed of a hydraulic actuator, as was illustrated in Chapter 1 (Figure 1-16), always depends on the actuator's size and the rate of flow into it. Since the size of the actuator will generally be expressed in cubic inches (cubic centimeters), use these conversion factors:

$$1\ gpm = 231\ in^3/minute$$

$$gpm = \frac{in^3/minute}{231}$$

$$in^3\ /\ minute = gpm \times 231$$

$$1\ LPM = 1000 cm^3/minute$$

$$LPM = \frac{cm^3/minute}{1000}$$

$$cm^3\ /\ minute = LPM \times 1000$$

Flow and Pressure Drop

Whenever a liquid is flowing, there must be a condition of unbalanced force to cause motion. Therefore, when a fluid flows through a constant-diameter pipe, the pressure will always be slightly lower downstream with reference to any point upstream. This difference in pressure or **pressure drop** is required to overcome friction in the line.

Figure 2-7 illustrates pressure drop due to friction. The succeeding pressure drops (from maximum pressure to zero pressure) are shown as differences in head in succeeding vertical pipes.

Fluid Seeks a Level

Conversely, when there is no pressure difference on a liquid, it simply seeks a level as shown in Figure 2-8A. If the pressure changes at one point (Figure 2-8B) the liquid levels at the other points rise only until their weight is sufficient to make up the difference in pressure. The difference in height (head) in the case of oil is one foot (.30 m) per 0.4 psi (2.76 kPa). Thus it can be seen that additional pressure difference will be required to cause a liquid to flow up a pipe or to **lift** the fluid, since the force due to the weight of the liquid must be overcome. In circuit design, naturally, the pressure required to move the oil mass and to overcome friction must be added to the pressure needed to move the load. In most applications, good design minimizes these pressure "drops" to the point where they become almost negligible.

Laminar and Turbulent Flow

Ideally, when the particles of a fluid move through a pipe, they will move in straight, parallel flow paths (Figure 2-9). This condition is called **laminar** flow and occurs at low velocity in straight piping. With laminar flow, friction is minimized.

Turbulence is the condition where the particles do not move smoothly parallel to the flow direction (Figure 2-10). Turbulent flow is caused by abrupt changes in direction or cross section, or by too high velocity. The result is greatly increased friction, which generates heat, increases operating pressure and wastes power.

Bernoulli's Principle

Hydraulic fluid in a working system contains energy in two forms: Kinetic energy by virtue of the fluid's weight and velocity, and potential energy in the form of pressure.

Daniel Bernoulli, a Swiss scientist, demonstrated that in a system with a constant flow rate, energy is transformed from one form to the other each time the pipe cross-section size changes.

As shown in Figure 2-11, when the cross-sectional area of a flow path increases, the velocity (kinetic energy) of the fluid decreases. Bernoulli's principle says that if the flow rate is constant, the sums of the kinetic energy and the pressure energy at various points in a system must be constant. Therefore, if the kinetic energy decreases, it results in an increase in the pressure energy. This transformation of energy from one kind to the other keeps the sum of the two energies constant. Likewise, when the cross-sectional area of a flow path decreases, the increase of kinetic energy (velocity) produces a corresponding decrease in the pressure energy.

The use of a venturi in an automobile engine carburetor (Figure 2-12) is a familiar example of Bernoulli's principle. Air flowing through the carburetor barrel is reduced in pressure as it passes through a reduced cross section of the throat. The decrease in pressure permits gasoline to flow, vaporize and mix with the air stream.

Bernoulli's principle is an important factor in the design of spool-type hydraulic valves. In such valves, changes in fluid velocity are common. If the maximum flow rate of the valve is exceeded, the pressure changes as a result of Bernoulli's principle can produce unbalanced axial forces within the valve. These forces may become great enough to overpower the valve's actuator and cause the valve to malfunction.

Figure 2-13 shows the combined effects of friction and velocity changes on the pressure in a line.

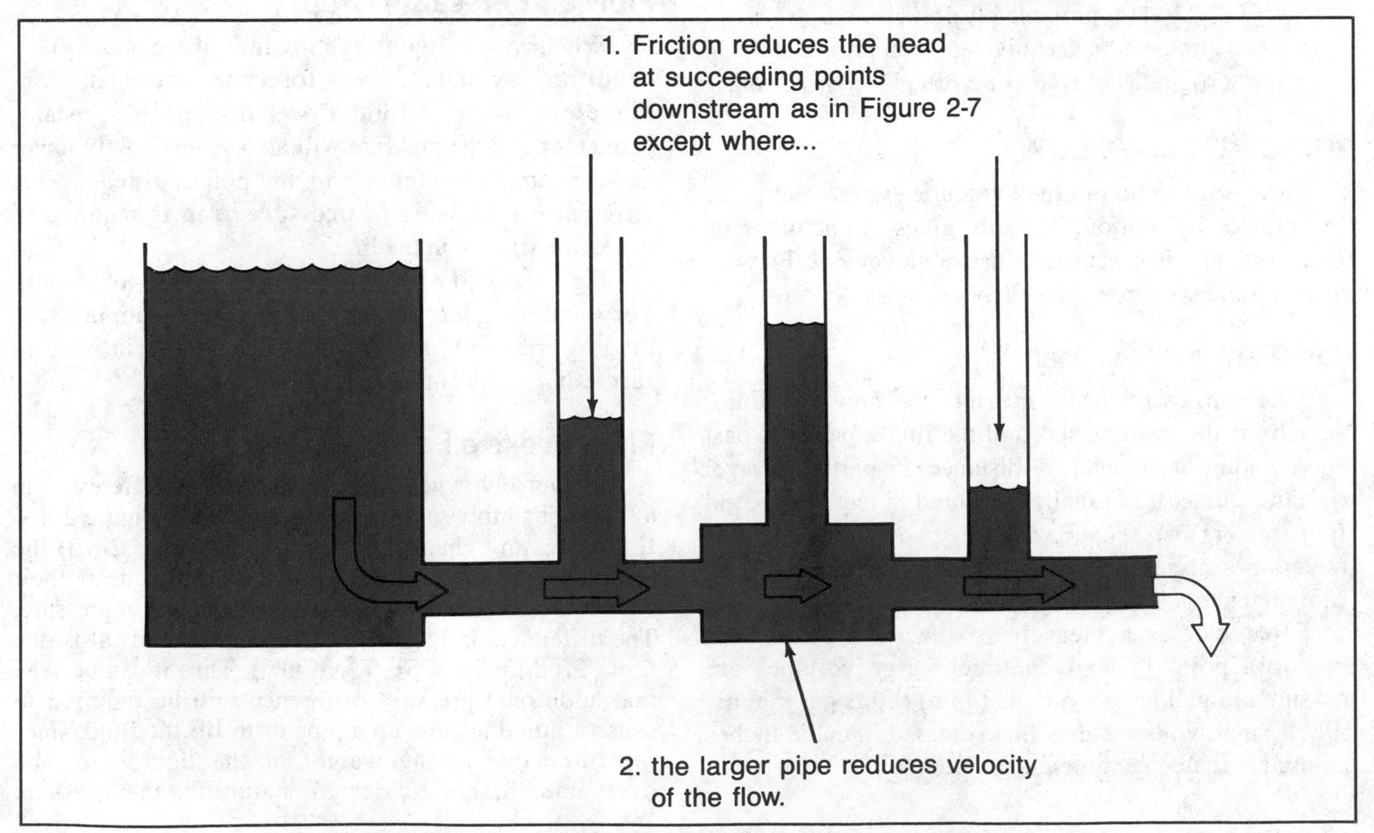

1. Friction reduces the head at succeeding points downstream as in Figure 2-7 except where...
2. the larger pipe reduces velocity of the flow.

Figure 2-13. Friction and velocity affect pressure.

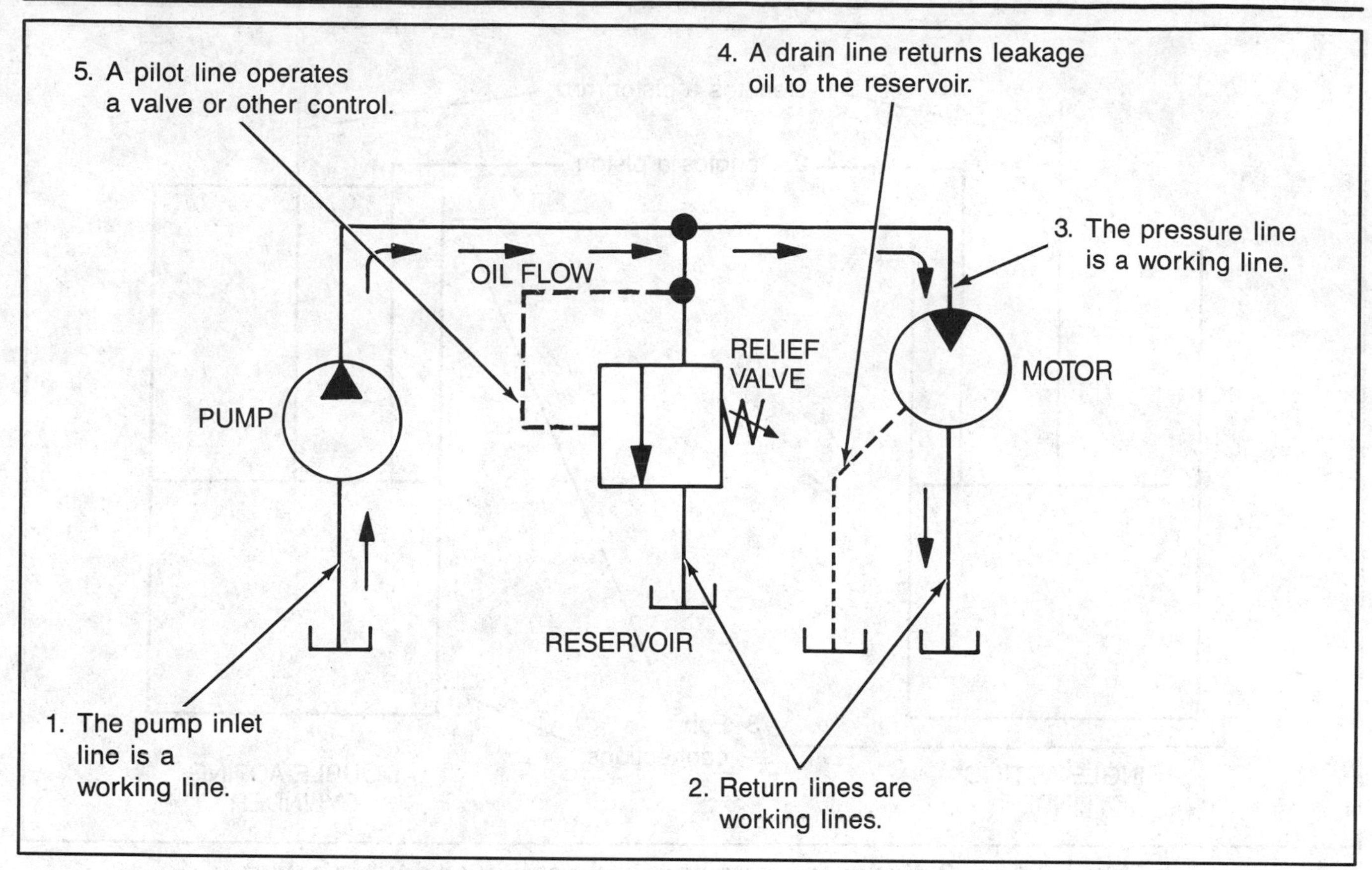

Figure 2-14. Three classifications of lines.

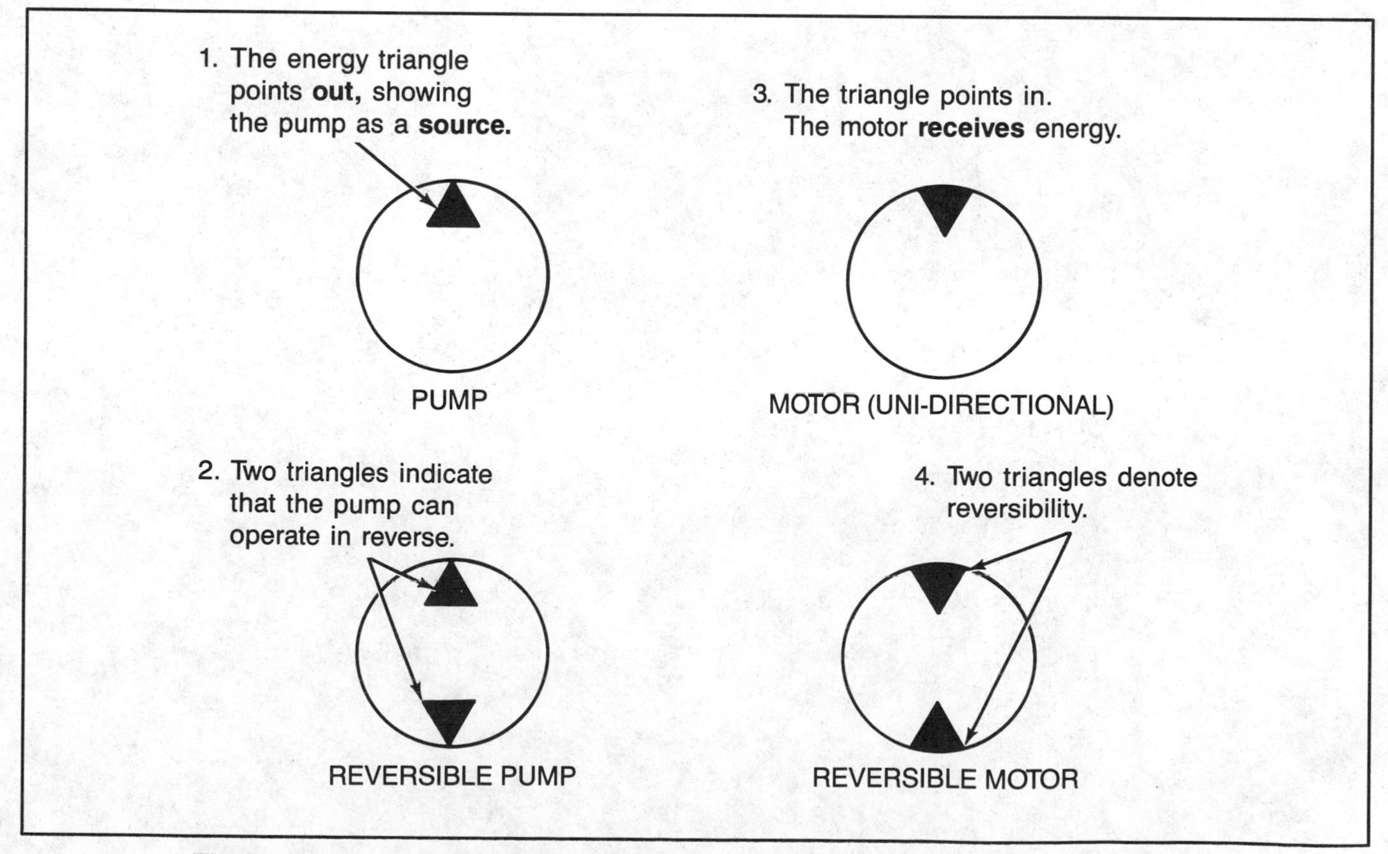

Figure 2-15. A circle with energy triangles symbolizes a pump or motor.

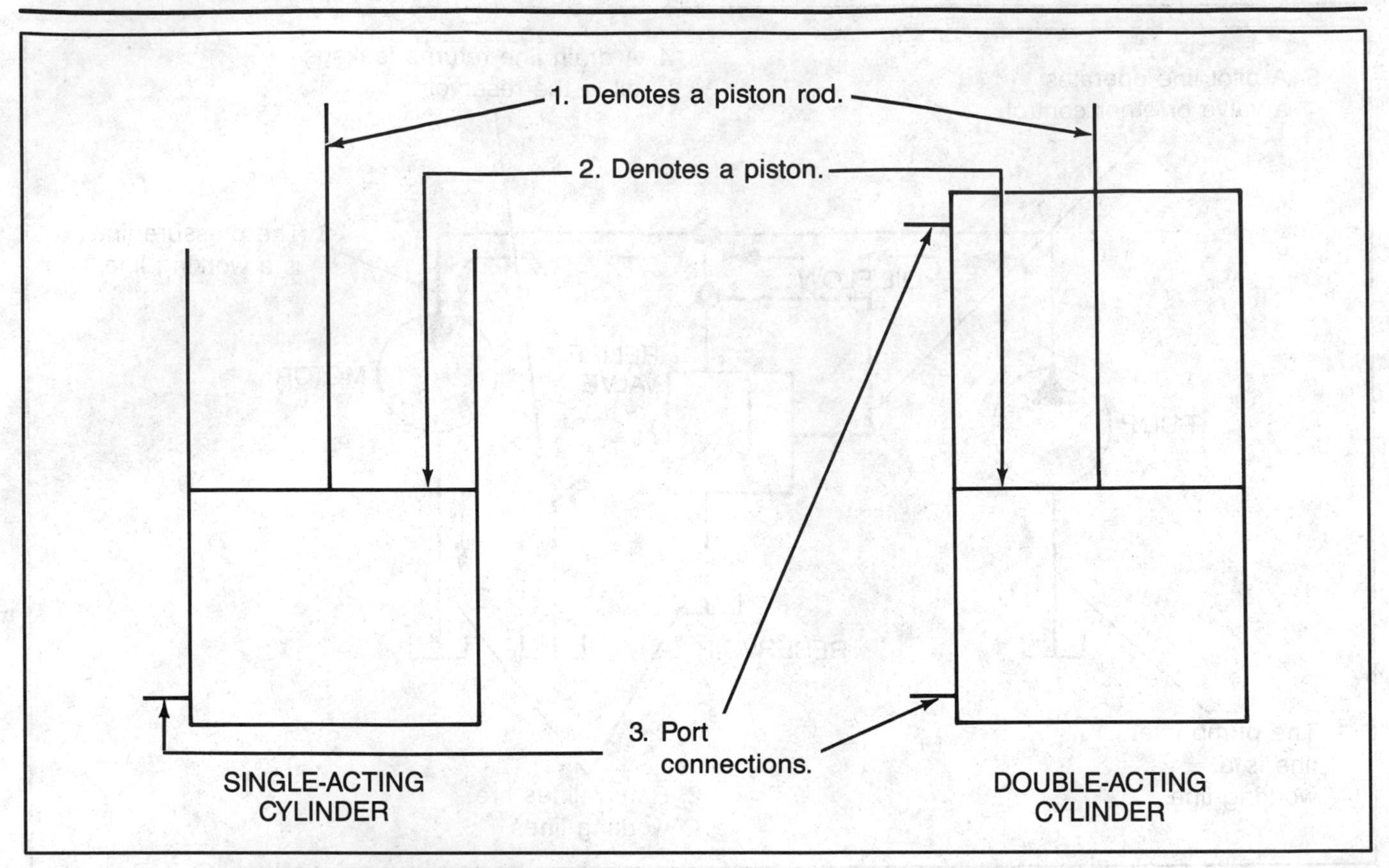

Figure 2-16. Cylinder symbols are single acting or double acting.

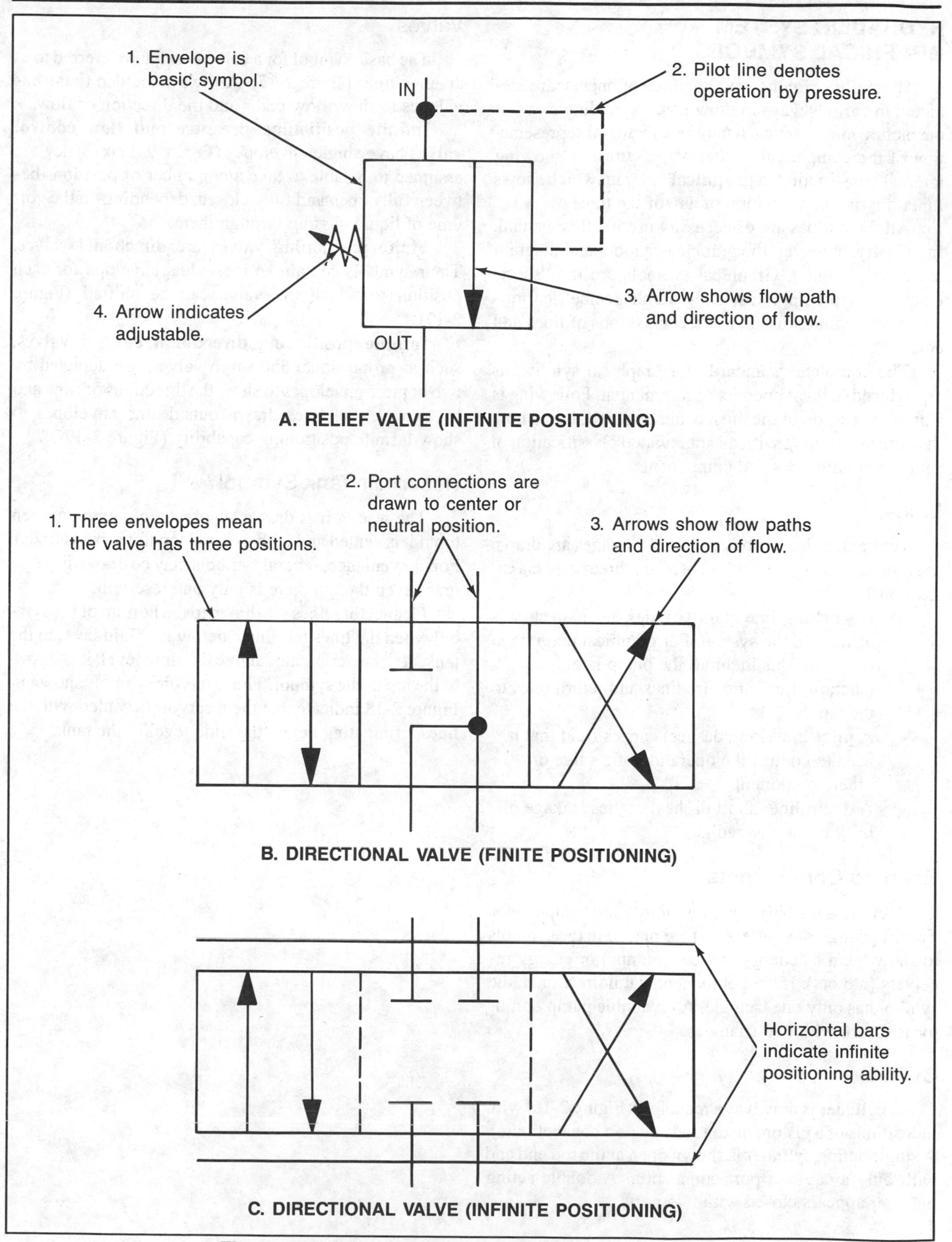
1. Envelope is basic symbol.
IN
2. Pilot line denotes operation by pressure.
4. Arrow indicates adjustable.
3. Arrow shows flow path and direction of flow.
OUT
A. RELIEF VALVE (INFINITE POSITIONING)
2. Port connections are drawn to center or neutral position.
1. Three envelopes mean the valve has three positions.
3. Arrows show flow paths and direction of flow.
B. DIRECTIONAL VALVE (FINITE POSITIONING)
Horizontal bars indicate infinite positioning ability.
C. DIRECTIONAL VALVE (INFINITE POSITIONING)

Figure 2-17. An envelope is the basic valve symbol.

HYDRAULIC SYSTEM GRAPHICAL SYMBOLS

Hydraulic circuits and their components are depicted in various ways in drawings. Depending on what the picture must convey, it may be a **pictorial** representation of the components' exteriors; a **cutaway** showing internal construction; a **graphical** diagram which shows function; or a combination of any of the three.

All three types are of necessity used in this manual. In industry, however, the graphical symbol and diagram are most common. Graphical symbols are the "shorthand" of circuit diagrams, using simple geometric forms which show functions and interconnections of lines and components.

The complete "standard" for graphical symbols is reproduced in the Appendix of this manual. Following is a brief exposition of the most common symbols and how they are used, along with an abbreviated classification of some hydraulic lines and components.

Lines

Hydraulic pipes, tubes and fluid passages are drawn as single lines (Figure 2-14). There are three basic classifications:

- A **working** line (solid) carries the main stream of flow in the system. For graphical diagram purposes, this includes the pump inlet (suction) line, pressure lines and return lines to the tank.
- A **pilot** line (long dashes) carries fluid that is used to control the operation of a valve or other component.
- A **drain** line (short dashes) carries leakage oil back to the reservoir.

Rotating Components

A circle is the basic symbol for rotating components. Energy triangles (Figure 2-15) are placed in the symbols to show them as energy sources (pumps) or energy receivers (motors). If the component is unidirectional, the symbol has only one triangle. A reversible pump or motor is drawn with two triangles.

Cylinders

A cylinder is drawn as a rectangle (Figure 2-16) with indications of a piston, piston rod and port connection(s). A single acting cylinder is shown open at the rod end and with only a cap-end port connection. A double-acting cylinder appears closed with two ports.

Valves

The basic symbol for a valve is a square referred to as an **envelope** (Figure 2-17). Arrows are added to the envelopes to show flow paths and the direction of flow.

Infinite positioning pressure and flow control valves have single envelopes (Figure 2-17A). They are assumed to be able to take any number of positions between fully open and fully closed, depending on the volume of liquid passing through them.

Finite positioning valves are directional valves. Their symbols contain an individual envelope for each position to which the valve can be shifted (Figure 2-17B).

Infinite positioning directional control valves, such as proportional and servo valves, are depicted by two or more envelopes to show the directions of flow and by two parallel lines drawn outside the envelopes to show infinite positioning capability (Figure 2-17C).

Reservoir Tank Symbol

The reservoir is drawn as a rectangle with an open top if it is vented and with a closed top if it is pressurized. For convenience, several symbols may be drawn in a diagram even though there is only one reservoir.

Connecting lines are drawn to the bottom of the symbol when the lines terminate below the fluid level in the tank. If a line terminates above the fluid level, it is drawn to the top of the symbol. The reservoir symbols shown in Figure 2-18 indicate that the reservoir is vented with the lines terminating below the fluid level in the tank.

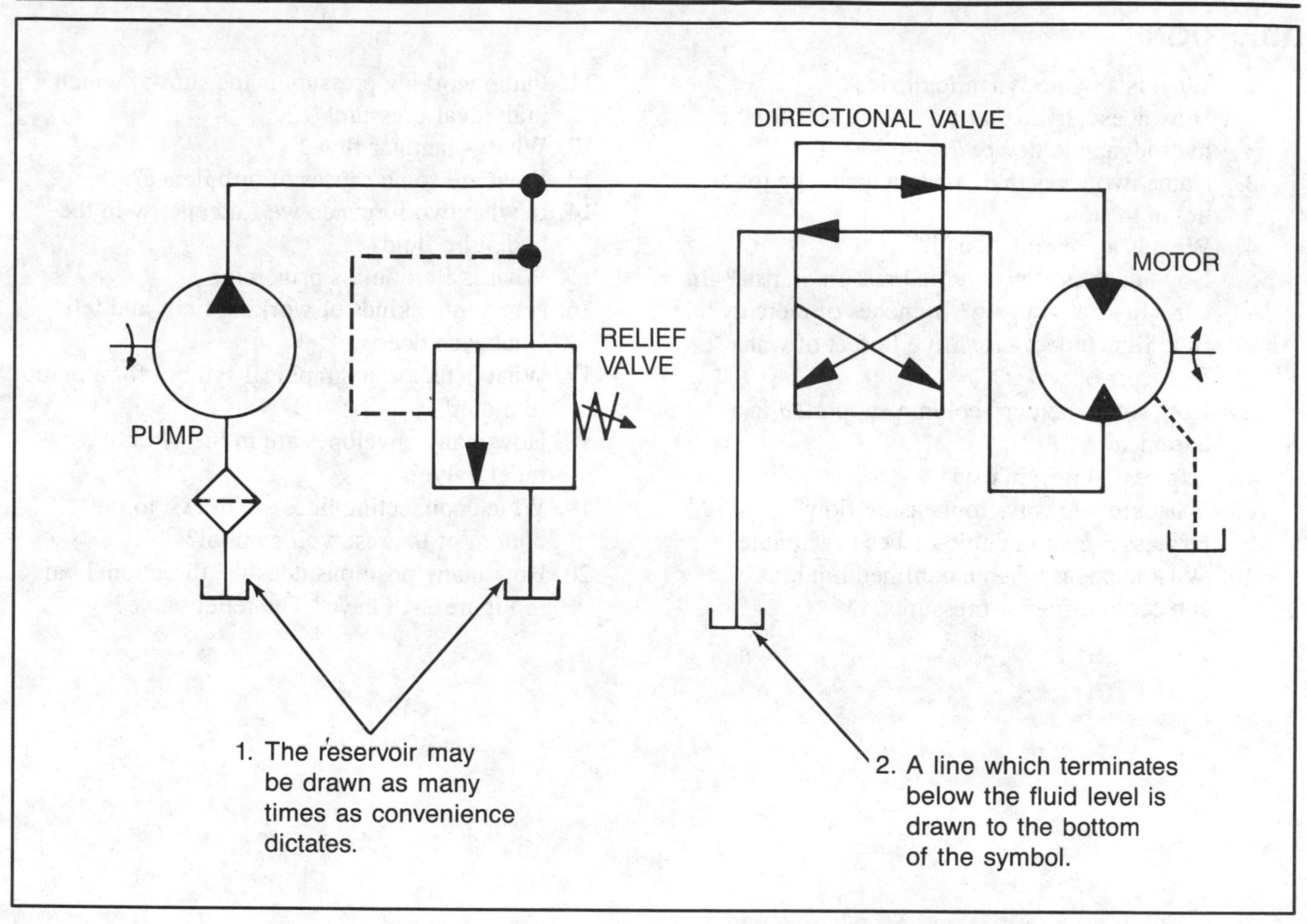

Figure 2-18. Graphical diagram of motor-reversing circuit.

CONCLUSION

Figure 2-18 shows a graphical diagram of an entire hydraulic circuit. Note that there is no attempt to show the actual size, shape, location or construction of any component. The diagram does show function and connections, which suffice for most purposes.

Variations and refinements of these basic symbols will be dealt with in the chapters on components and systems.

QUESTIONS

1. What is a hydrodynamic device?
2. How does a hydrostatic device differ from a hydrodynamic device?
3. Name two ways that create a tendency for a liquid to flow.
4. What is a pressure "head?"
5. How much is atmospheric pressure in psia? In bar? In kPa? In psig? In inches of mercury? In millimeters of mercury? In feet of water? In meters of water?
6. How is the mercury column supported in a barometer?
7. Express 30 psig in psia.
8. What are two ways to measure flow?
9. Express 5 gpm in cubic inches per minute.
10. What happens when a confined liquid is subject to different pressures?
11. Pump working pressure is the sum of which individual pressures?
12. What is laminar flow?
13. What are some causes of turbulence?
14. In what two forms do we find energy in the hydraulic fluid?
15. What is Bernoulli's principle?
16. Name three kinds of working lines and tell what each does.
17. What is the basic graphical symbol for a pump or motor?
18. How many envelopes are in the symbol for a relief valve?
19. Which connecting lines are drawn to the bottom of the reservoir symbol?
20. How many positions does the directional valve in Figure 2-18 have? The relief valve?

CHAPTER 3

HYDRAULIC FLUIDS

Proper selection and care of hydraulic fluid for a machine will have an important effect on how the machine performs and on the life of the hydraulic components. The formulation and application of hydraulic fluids are sciences in themselves, far beyond the scope of this manual. In this chapter, you will find the basic factors involved in the choice of a fluid and its proper use.

A fluid has been defined in Chapter 1 as any liquid or gas. However, the term fluid has come into general use in hydraulics to refer to the liquid used as the power-transmitting medium. In this chapter, fluid will mean the hydraulic fluid, whether a specially compounded petroleum oil or one of the special fire-resistant fluids, which may be a synthetic compound.

PURPOSES OF THE FLUID

The hydraulic fluid has four primary purposes: to transmit power, to lubricate moving parts, to seal clearances between parts, and to cool or dissipate heat.

Power Transmission

As a power transmitting medium, the fluid must flow easily through lines and component passages. Too much resistance to flow creates considerable power loss. The fluid also must be as incompressible as possible so that action is instantaneous when the pump starts or a valve shifts.

Lubrication

In most hydraulic components, internal lubrication is provided by the fluid. Pump elements and other wearing parts slide against each other on a film of fluid (Figure 3-1). For long component life the oil must contain the necessary additives to ensure high antiwear characteristics. Not all hydraulic oils contain these additives.

Vickers recommends the use of good quality hydraulic oils containing adequate quantities of antiwear additives. For general hydraulic service, these oils offer superior protection against component wear and the advantage of long service life. In addition, they provide good demulsibility as well as protection against rust and oxidation. These oils are generally known as antiwear-type hydraulic oils.

Experience has shown that the 10W and 20W-20 SAE viscosity automotive crankcase oils, having letter designations SC through SF, are excellent for severe hydraulic service where there is little or no water present. The only adverse effect is that their detergent additives tend to hold water in a tight emulsion, preventing the separation of water and oil, even when left standing for long periods of time. It should be noted that water problems are infrequently experienced in the use of these crankcase oils in machinery hydraulic systems. Condensation has not been a problem in normal operation.

Both automotive crankcase oils and antiwear-type hydraulic oils are highly recommended for mobile equipment hydraulic systems.

Sealing

In many instances, the fluid is the only seal against pressure inside a hydraulic component. In Figure 3-1, there is no seal ring between the valve spool and body to minimize leakage from the high-pressure passage to the low-pressure passages. The close mechanical fit and the viscosity of the oil determine leakage rate.

Cooling

Circulation of the fluid through lines and around the walls of the reservoir (Figure 3-2) allow the fluid to give up the heat that has been generated in the system.

QUALITY REQUIREMENTS

In addition to these primary functions, the hydraulic fluid may have a number of other quality requirements. Some of these are to:

- Prevent rust
- Prevent formation of sludge, gum, and varnish
- Depress foaming
- Maintain its own stability and thereby reduce fluid replacement cost
- Maintain relatively stable body over a wide temperature range

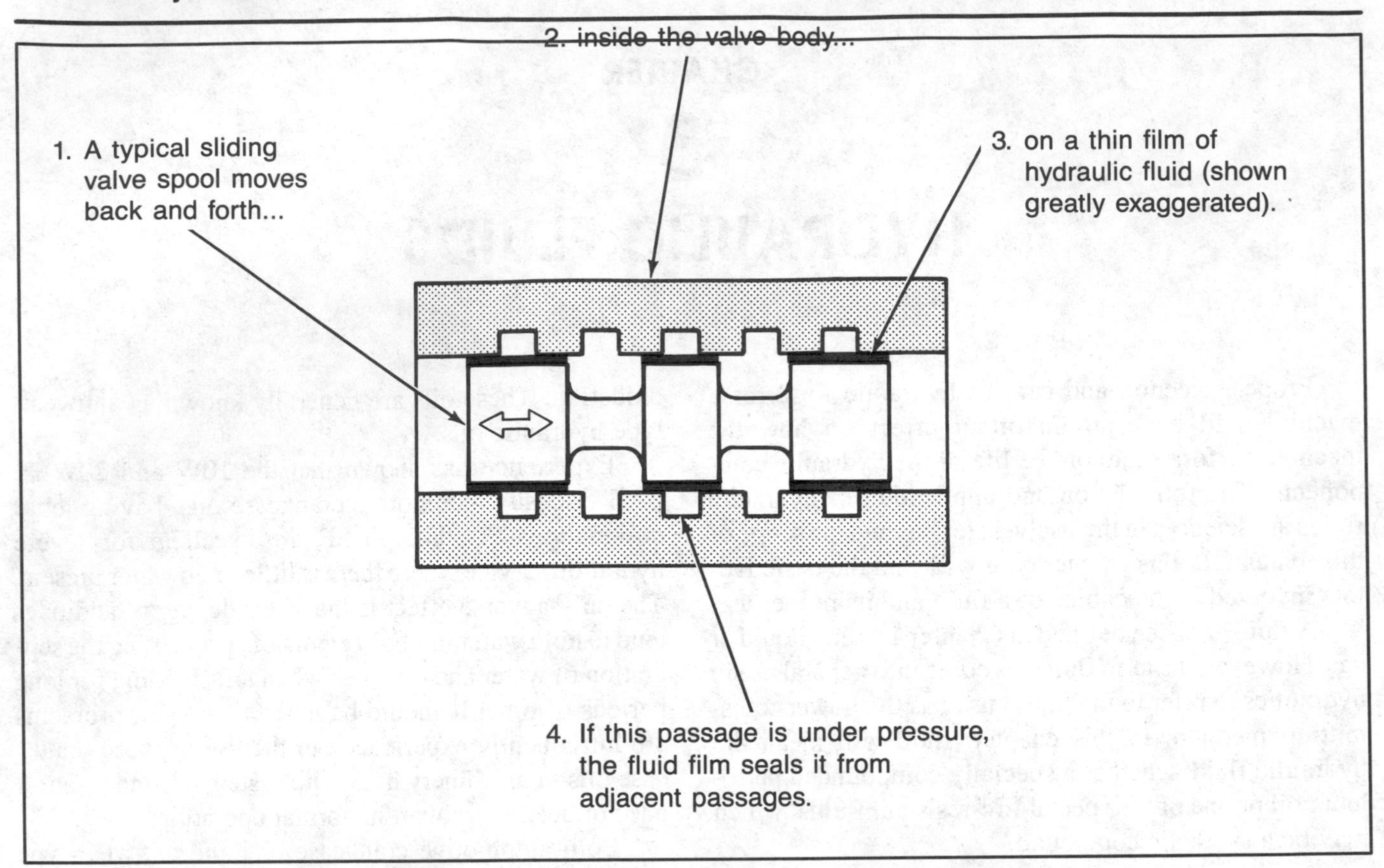

2. inside the valve body...
1. A typical sliding valve spool moves back and forth...
3. on a thin film of hydraulic fluid (shown greatly exaggerated).
4. If this passage is under pressure, the fluid film seals it from adjacent passages.

Figure 3-1. Fluid lubricates working parts.

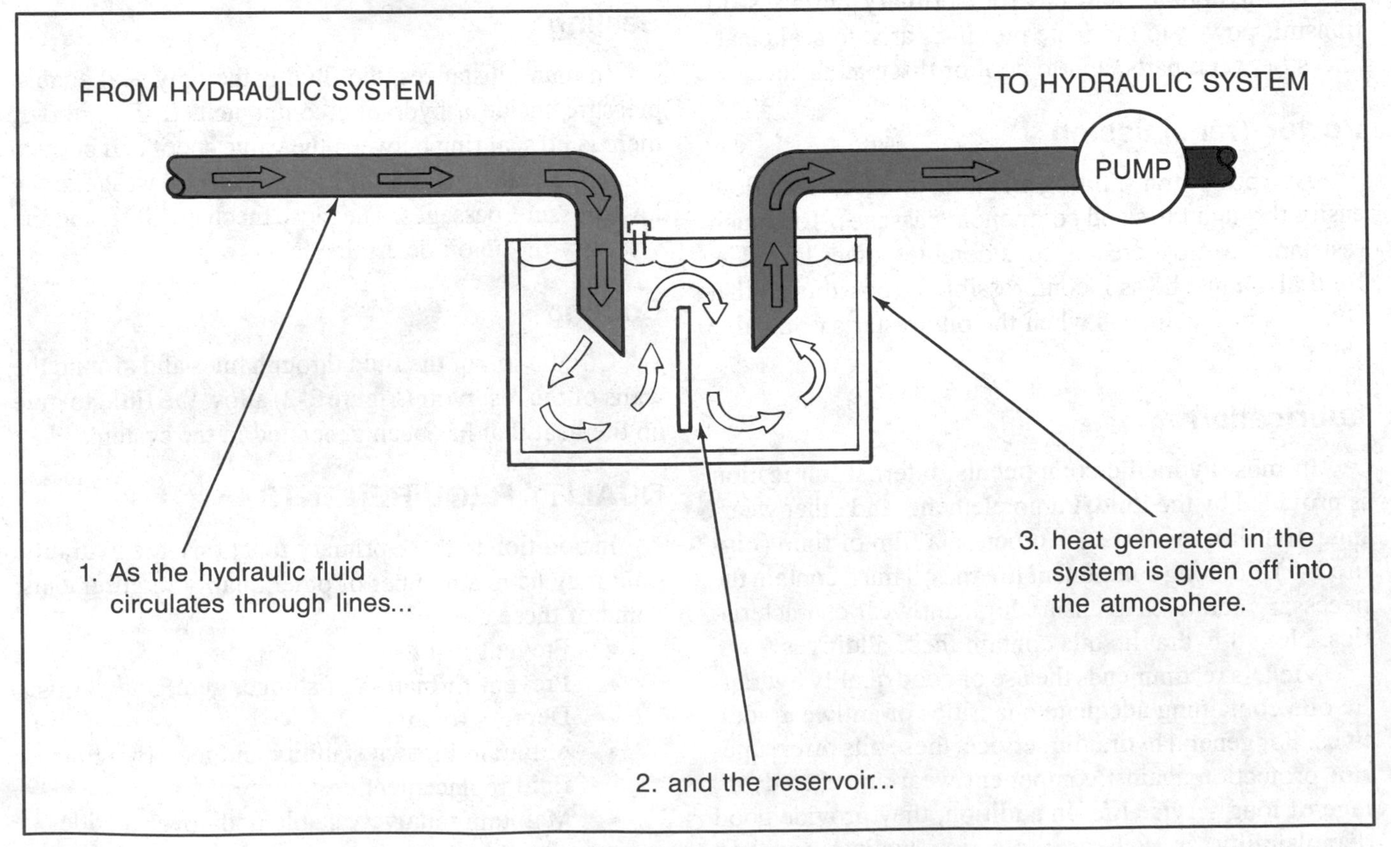

FROM HYDRAULIC SYSTEM
TO HYDRAULIC SYSTEM
PUMP
1. As the hydraulic fluid circulates through lines...
2. and the reservoir...
3. heat generated in the system is given off into the atmosphere.

Figure 3-2. Circulation cools the system.

- Prevent corrosion and pitting
- Separate out water
- Be compatible with seals and gaskets

These quality requirements often are the result of special compounding and may not be present in every fluid.

FLUID PROPERTIES

Let us now consider the properties of hydraulic fluid which enable it to carry out its primary functions and fulfill some or all of its quality requirements:

- Viscosity
- Pour Point
- Lubricating Ability
- Oxidation Resistance
- Rust and Corrosion Protection
- Demulsibility

Viscosity

Viscosity is the measure of the fluid's resistance to flow; or an inverse measure of fluidity. If a fluid flows easily, its viscosity is low. You also can say that the fluid is thin or has low body. A fluid that flows with difficulty has a high viscosity. It is thick or high in body.

Viscosity Is a Compromise. For any hydraulic machine, the actual fluid viscosity must be a compromise. A high viscosity is desirable for maintaining sealing between mating surfaces.

However, too high a viscosity increases friction, resulting in:

- High resistance to flow
- Increased power consumption due to frictional loss
- High temperature caused by friction
- Increased pressure drop due to resistance
- Possibility of sluggish or slow operation
- Difficulty in separating air from oil in reservoir

And should the viscosity be too low:

- Internal leakage increases
- Excessive wear and even seizure may occur under heavy load due to breakdown of the oil film between moving parts
- Pump efficiency may decrease, causing slower operation of the actuator
- Leakage losses may result in increased temperatures

Defining Viscosity. Some methods of defining viscosity are: absolute (dynamic) viscosity; kinematic viscosity in centistokes (cSt); relative viscosity in Saybolt Universal Seconds (SUS); and SAE numbers (for automotive oils). In the United States, hydraulic fluid viscosity requirements are specified in either centistokes or SUS or both.

Absolute Viscosity. Absolute viscosity is defined as the resistance encountered when moving one layer of liquid over another. Absolute (dynamic) viscosity is defined as the force per unit of area required to move one parallel surface at a given speed past another parallel surface separated by a given fluid film thickness. In the SI system, force is expressed in newtons (N); area in square meters (m^2).

The common SI unit of measurement for absolute viscosity is the centipoise (cP), which is one-hundredth of a poise (P). A corresponding, numerically equal SI unit is the millipascal second (mPa s), which is one-thousandth of a pascal second (Pa s).

Following are conversions between the two units of measure for absolute viscosity in the SI system of measurement:

$$1\ Poise\ (P) = 0.1\ Pascal\ Second\ (Pa\ s)$$

$$1\ Pascal\ (Pa) = \frac{N}{m^2}$$

$$1\ Pa\ s = \frac{Ns}{m^2}$$

$$1\ Poise = \frac{0.1Ns}{m^2} = 0.1\ Pa\ s$$

$$1\ Centipoise\ (cP) = 0.001\ Pa\ s = 1\ mPa\ s$$

Kinematic Viscosity. Kinematic viscosity is the most common way of measuring viscosity. It is measured by the amount of time needed for a fixed volume of oil to flow through a capillary tube. The coefficient of absolute viscosity, when divided by the density of the liquid is called the kinematic viscosity.

The official SI unit for kinematic viscosity is m^2/s (meters squared per second), but the centistoke (cSt), which is mm^2/s (millimeters squared per second), is commonly adopted in the petroleum industry.

The absolute viscosity (cP) of a fluid at any temperature is equivalent to its kinematic viscosity (cSt) at that temperature times its density $[(kg/m^3) \times 10^{-3}]$ at the same temperature.

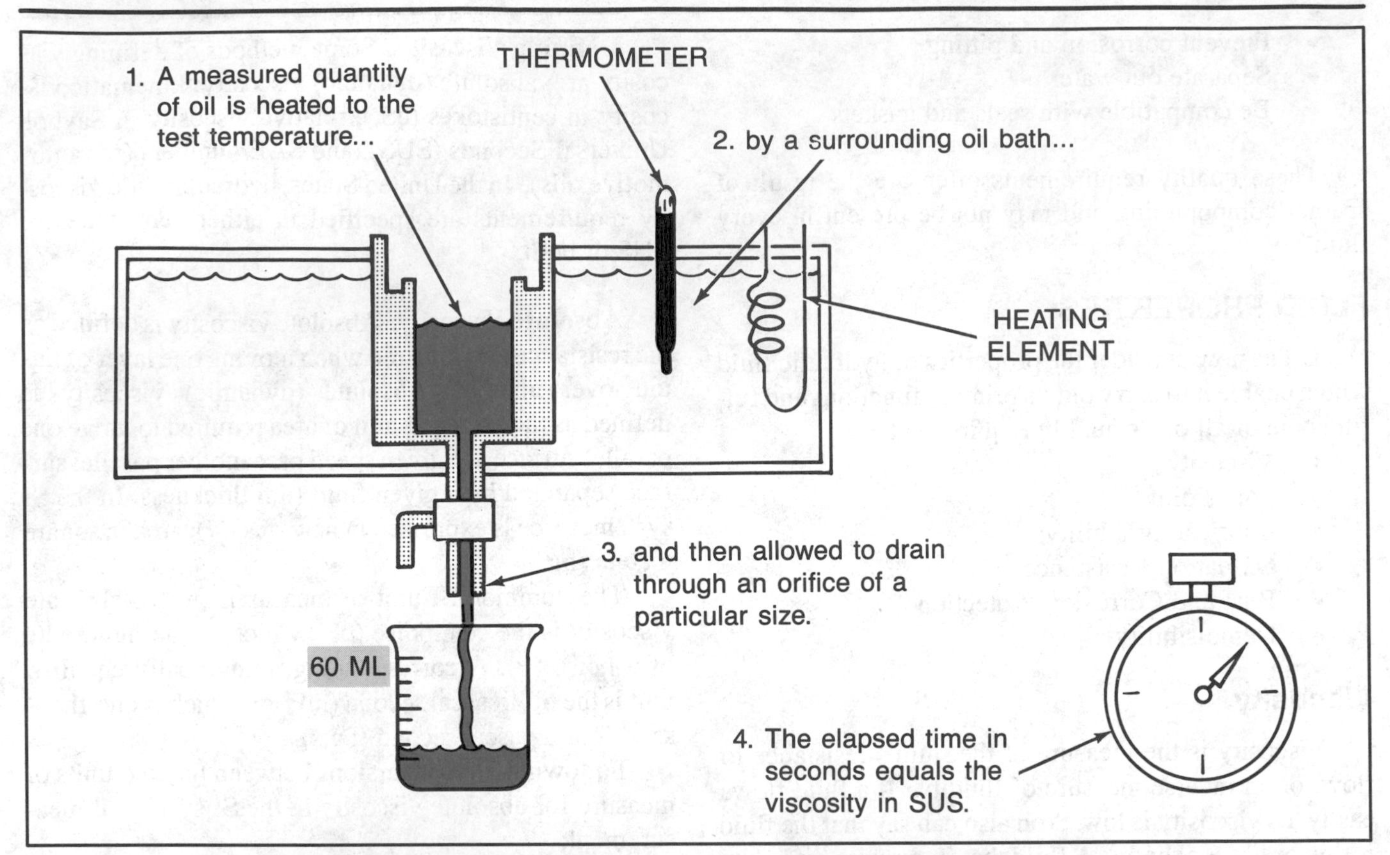
1. A measured quantity of oil is heated to the test temperature...
THERMOMETER
2. by a surrounding oil bath...
HEATING ELEMENT
3. and then allowed to drain through an orifice of a particular size.
60 ML
4. The elapsed time in seconds equals the viscosity in SUS.

Figure 3-3. Saybolt viscosimeter measures relative viscosity.

Following are conversions between absolute and kinematic viscosity:

$$\textit{Centipoise} = \textit{Centistoke} \times \textit{Density}$$

OR

$$cP = cSt \times Kg/m^3 \times 10^{-3}$$

$$\textit{Centistoke} = \frac{\textit{Centipoise}}{\textit{Density}}$$

OR

$$cSt = \frac{cP}{\frac{kg}{m^3} \times 10^{-3}}$$

SUS Viscosity. For practical purposes, it may serve to know the relative viscosity of the fluid. Relative viscosity is determined by timing the flow of a given quantity of the fluid through a standard orifice at a given temperature.

There are several measurement methods in use. A very common method in the United States is the Saybolt Viscosimeter (Figure 3-3). The time it takes for the measured quantity of liquid to flow through the orifice is measured with a stopwatch. The viscosity in Saybolt Universal Seconds (SUS) equals the elapsed time.

Obviously, a thick liquid will flow slowly, and the SUS viscosity will be higher than that of a thin liquid which flows faster. Since oil becomes thicker at a low temperature and thins when warmed, the viscosity must be expressed as so many SUS at a given temperature. The tests are usually made at 40°C and 100°C.

For industrial applications, hydraulic oil viscosities usually are in the vicinity of 150 SUS (32 cSt) at 40°C. It is a general rule that the viscosity should never go below 45 SUS (5.8 cSt) or above 4000 SUS (860 cSt), regardless of temperature. Where temperature extremes are encountered, multiviscosity grade fluids may be specified (e.g., 5W-30, 15W-40, etc., engine oils or multigrade hydraulic oils with relatively stable viscosity characteristics).

ISO Viscosity Grades. The International Standard Organization (ISO) has established a table of ISO grades for industrial oils based on viscosity ranges at different temperatures. These grades are provided in Figure 3-4. Many fluid suppliers now incorporate the ISO viscosity grade in their brand designation (e.g., Nuto H32, Tellus 68, Rando HD46, etc.).

ISO Viscosity Grade	Midpoint Kinematic Viscosity cSt at 40°C (104°F)	Kinematic Viscosity Limits cSt at 40°C (104°F) Minimum	Maximum
ISOVG2	2.2	1.98	2.42
ISOVG3	3.2	2.88	3.52
ISOVG5	4.6	4.14	5.06
ISOVG7	6.8	6.12	7.48
ISOVG10	10	9.00	11.0
ISOVG15	15	13.5	16.5
ISOVG22	22	19.8	24.2
ISOVG32	32	28.8	35.2
ISOVG46	46	41.4	50.6
ISOVG68	68	61.2	74.8
ISOVG100	100	90.0	110
ISOVG150	150	135	165
ISOVG220	220	198	242
ISOVG320	320	288	352
ISOVG460	460	414	506
ISOVG680	680	612	748
ISOVG1000	1000	900	1100
ISOVG1500	1500	1350	1650

Figure 3-4. ISO viscosity classification for industrial oils.

SAE VISCOSITY GRADES FOR ENGINE OILS

SAE Viscosity Grade	Viscosity (cP) at Temperature °C (°F) Max	Viscosity (cSt) at 100°C (212°F) Min	Max
0W	3250 at −30 (−22)	3.8	—
5W	3500 at −25 (−13)	3.8	—
10W	3500 at −20 (−4)	4.1	—
15W	3500 at −15 (5)	5.6	—
20W	4500 at −10 (14)	5.6	—
25W	6000 at −5 (23)	9.3	—
20	—	5.6	Less than 9.3
30	—	9.3	Less than 12.5
40	—	12.5	Less than 16.3
50	—	16.3	Less than 21.9

NOTE: 1cP = 1mPa·s; 1 cSt = 1mm²/s

Figure 3-5. SAE viscosity grades for engine oils.

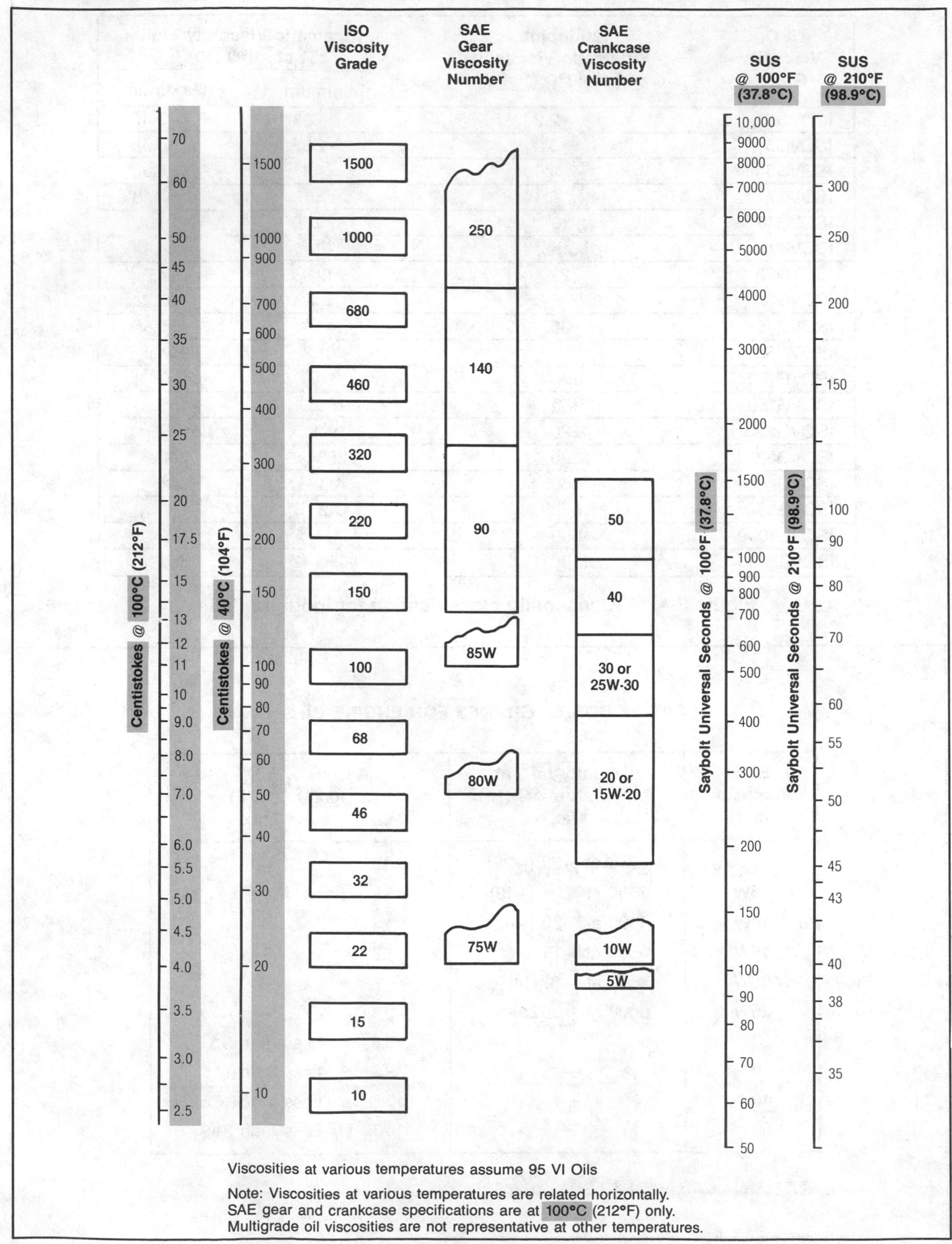

Figure 3-6. Comparative viscosity classifications.

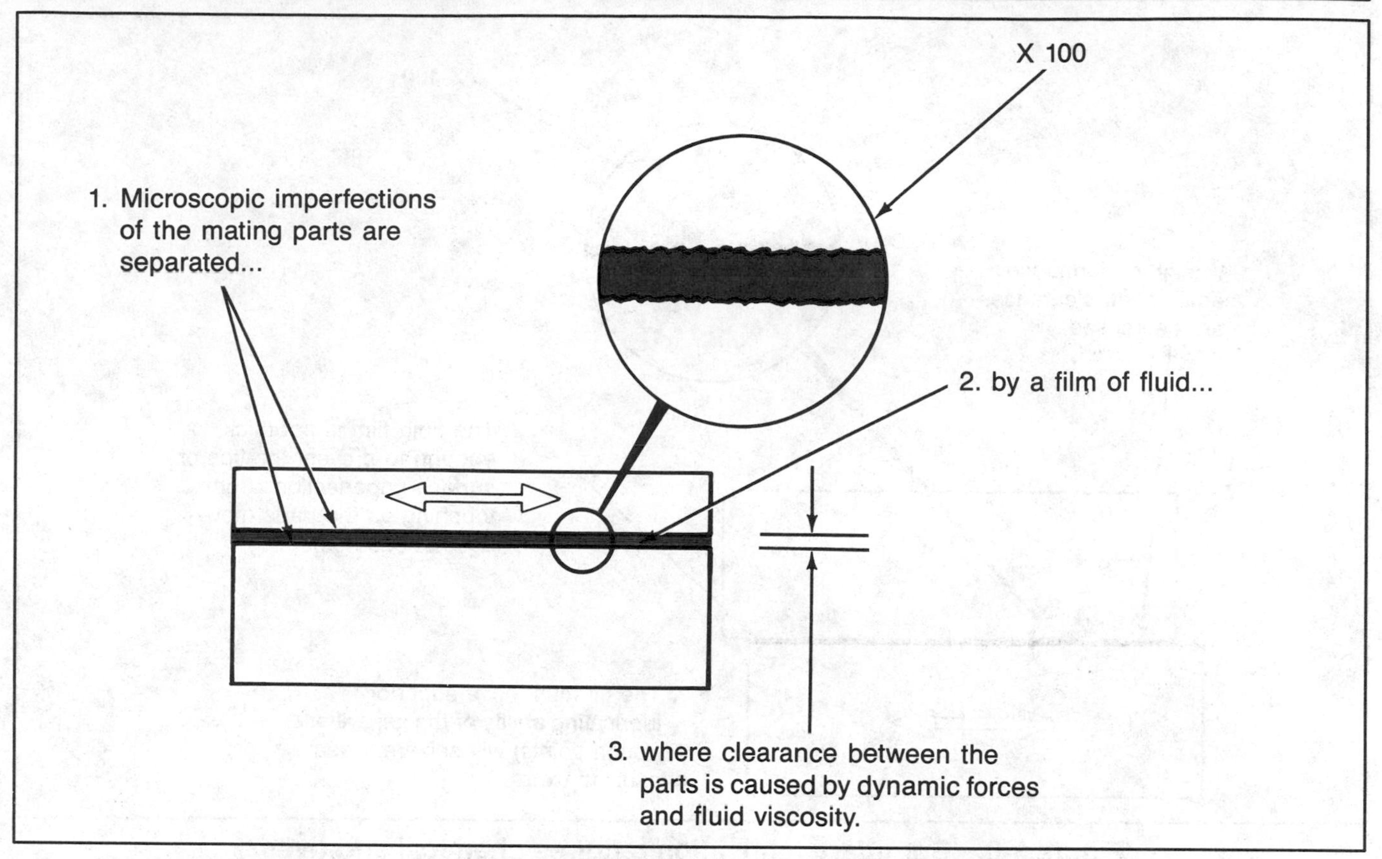

X 100
1. Microscopic imperfections of the mating parts are separated...
2. by a film of fluid...
3. where clearance between the parts is caused by dynamic forces and fluid viscosity.

Figure 3-7. Full film lubrication prevents metal-to-metal contact.

SAE Numbers. The Society of Automotive Engineers (SAE) has established numbers to specify ranges of viscosities of engine oils at specific test temperatures. Winter numbers (0W, 5W, 10W, 15W, etc.) are determined by tests at cold temperatures. Summer oil numbers (20, 30, 40, 50, etc.) designate the SUS range at 212°F (100°C). Figure 3-5 is a chart of the SAE viscosity ranges.

There has been a worldwide trend to adopt the SI system of measuring viscosity and temperature. In the interim period of conversion there may be some confusion. Figure 3-6 is meant to serve as a useful guide.

Viscosity Index. Viscosity index is an arbitrary measure of a fluid's resistance to viscosity change with temperature changes. A fluid that has a relatively stable viscosity at temperature extremes has a high viscosity index (VI). A fluid that is very thick when cold and very thin when hot has a low VI.

The following chart is a comparison between a 50 VI and a 90 VI oil. Compare these actual viscosities at three temperatures:

VI	0°F (-18°C)	104°F (40°C)	212°F (100°C)
50	2600 cSt	32 cSt	4.5 cSt
90	1730 cSt	32 cSt	5.1 cSt

Note that the 90 VI oil is thinner at 0°F (-18°C) and thicker at 212°F (100°C), while both have the same viscosity at 104°F (40°C).

The original VI scale was from 0 to 100, representing the poorest to best VI characteristics then known. Today, chemical additives and refining techniques have increased the VI of some oils to well over 100. In a machine that runs at relatively constant temperatures the viscosity index of the fluid is less critical.

Pour Point

Pour point is the lowest temperature at which a fluid will flow. It is a very important specification if the hydraulic system will be exposed to extremely low temperatures. Generally, the pour point of the selected fluid should be 20°F (-6.67°C) below the lowest temperature to be encountered.

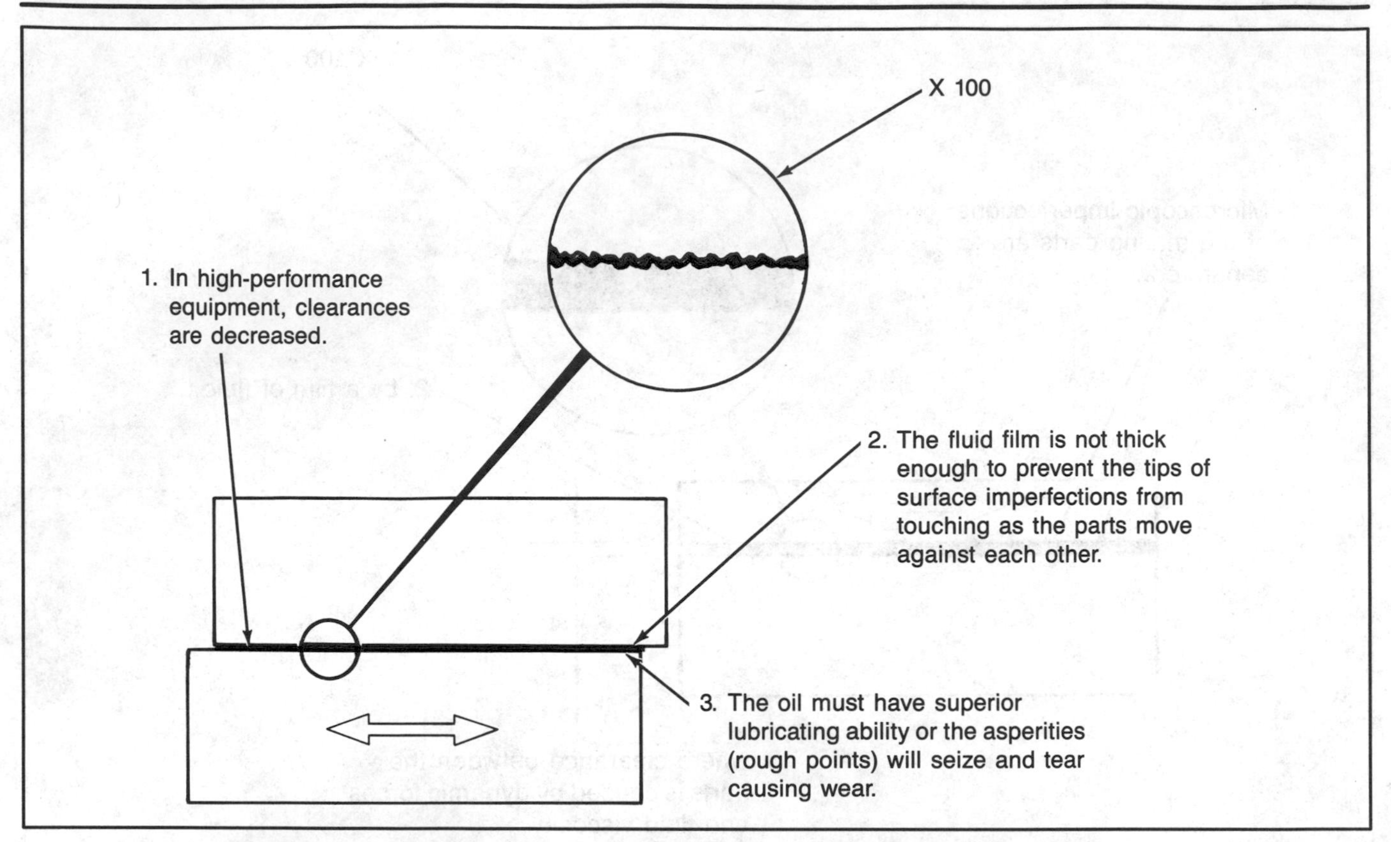

X 100
1. In high-performance equipment, clearances are decreased.
2. The fluid film is not thick enough to prevent the tips of surface imperfections from touching as the parts move against each other.
3. The oil must have superior lubricating ability or the asperities (rough points) will seize and tear causing wear.

Figure 3-8. Boundary lubrication requires chemical additives.

Lubricating Ability

It is desirable for hydraulic system moving parts to have enough clearance to run together on a substantial film of fluid (Figure 3-7). This condition is called full-film lubrication. As long as the fluid has adequate viscosity, the minute imperfections in the surfaces of the parts do not touch.

However, in certain high performance equipment, increased speeds and pressure, coupled with lower clearances, cause the film of fluid to be squeezed very thin (Figure 3-8) and a condition called boundary lubrication occurs. Here, there may be metal-to-metal contact between the tips of the two mating part surfaces and some enhancement of lubricating ability with special extreme pressure (EP) and antiwear additives is needed.

Oxidation Resistance

Oxidation, or chemical union with oxygen, is a serious reducer of the service life of a fluid. Petroleum oils are particularly susceptible to oxidation, since oxygen readily combines with both carbon and hydrogen in the oil's makeup.

Most of the oxidation products are soluble in the oil, and additional reactions take place in the products to form gum, sludge, and varnish. The first stage products which stay in the oil are acid in nature and can cause corrosion throughout the system, in addition to increasing the viscosity of the oil. The insoluble gums, sludge, and varnish plug orifices, increase wear, and cause valves to stick.

Catalysts. There are always a number of oxidation catalysts, or helpers, in a hydraulic system. Heat, pressure, contaminants, water, metal surfaces, and agitation all accelerate oxidation once it has started. Temperature is particularly important. Tests have shown that below 135°F(57°C), oil oxidizes very slowly. But the rate of oxidation (or any other chemical reaction) approximately doubles for every 18°F (-8°C) increase in temperature.

Oil refiners incorporate additives in hydraulic oils to resist oxidation since many systems operate at considerably higher temperature. Their additives either:

- Stop oxidation from continuing immediately after it starts (chain breaker type)

  OR
- Reduce the effect of oxidation catalysts (metal deactivator type)

Rust and Corrosion Prevention

Rust (Figure 3-9) is the chemical union of iron (or steel) with oxygen. Corrosion is a chemical reaction between a metal and a chemical—usually an acid. Acids

Figure 3-9. Rust caused by moisture in the oil.

result from the chemical union of water with certain elements.

Since it is usually not possible to keep air and atmosphere-borne moisture out of the hydraulic system, there will always be opportunities for rust and corrosion to occur. During corrosion, particles of metal are dissolved and washed away (Figure 3-10). Both rust and corrosion contaminate the system and promote wear. They also allow excessive leakage past the affected parts and may cause components to seize.

Rust and corrosion can be inhibited by incorporating additives that "plate" on the metal surfaces to prevent their being attacked chemically.

Demulsibility

Demulsibility is the ability of a fluid to separate out water. Small quantities of water can be tolerated in most systems. In fact, some antirust compounds promote a degree of emulsification, or mixture with any water that gets into the system. This prevents the water from settling and breaking through the antirust film. However, too much water in the oil will promote the collection of contaminants and can cause sticky valves and accelerated wear. With proper refining, a hydraulic oil can have a high degree of demulsibility.

Use of Additives

Since most of the desirable properties of a fluid are at least partly traceable to additives, it might be supposed that commercial additives could be incorporated in any oil, to make it more suitable for a hydraulic system.

Refiners, however, warn against this, saying that additives must be compatible with the base fluid and with each other and further that this compatibility cannot be determined in the field. Unless one has laboratory facilities for ascertaining their compatibility it is best to leave the use of additives to the discretion of the fluid manufacturer.

PETROLEUM OIL AS A HYDRAULIC FLUID

Petroleum oil is still by far the most highly used base for hydraulic fluids. The characteristics or properties of petroleum oil fluids depend on three factors:

- The type of crude oil used
- The degree and method of refining
- The additives used

In general, petroleum oil has excellent lubricity. Some crude oils have better than average lubricating or

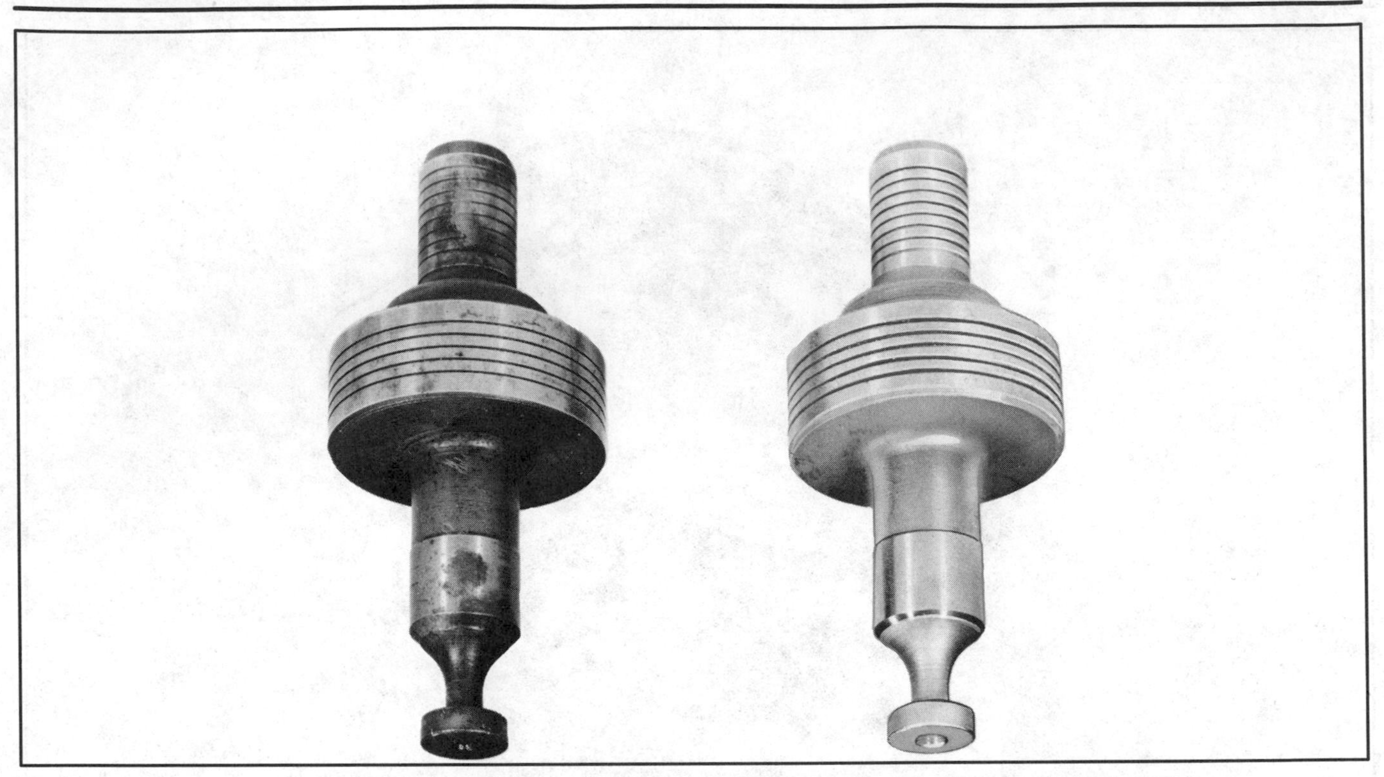

Figure 3-10. Corrosion caused by acid formation in the hydraulic oil.

antiwear properties. Depending on their makeup, some crude oils may display higher demulsibility, more oxidation resistance at higher temperatures, or higher viscosity index than others. Oil naturally protects against rust, seals well, dissipates heat easily, and is easy to keep clean by filtration or gravity separation of contaminants. Most of the desirable properties of a fluid, if not already present in the crude oil, can be incorporated through refining or additives.

A principal disadvantage of petroleum oil is that it will burn. For applications where fire could be a hazard, such as heat treating, welding, die casting, hydroelectric welding, and many others, there are several kinds of fire-resistant fluids available.

FIRE-RESISTANT FLUIDS

There are four basic types of fire-resistant hydraulic fluids:

	ISO Designation
High Water Fluids	HFA
Water in Oil (Invert Emulsions)	HFB
Water-Glycols	HFC
Synthetics	HFD

High Water Fluids

High water fluids are also known as high water base fluids (HWBF), high water content fluids (HWCF), 5-95 fluids, or ISO HFA fluids. The original fluids were 5 percent "soluble oil" emulsions (oil-in-water) having a milky appearance, or 5 percent solutions (synthetic chemical additives in water) which are clear in appearance. Oil-in-water emulsions contain tiny droplets of specially refined oil dispersed in water. We say that water is the continuous phase and the fluid's characteristics are more like water than oil. Around 1980, microemulsions with considerably smaller droplet size were introduced. These emulsions have a translucent, almost transparent, appearance. The dispersed internal phase in microemulsions is synthetic in nature as opposed to mineral oil in "soluble oil" emulsions.

High water fluids are highly fire resistant and have excellent cooling characteristics. When properly formulated, these fluids offer strong protection against rust and some degree of steel-on-steel lubricity. However, because the viscosities approach that of water, pumps are generally derated in order to obtain adequate service life.

A recent development in this field has been the introduction of thickened fluids. They are commonly 90 percent water and 10 percent concentrate. These fluids have viscosities ranging from 30 to 80 cSt (140 to 370 SUS) at 40 °C.

General Considerations. The operating temperatures should be limited to a maximum of 120 °F (49°C) in in order to minimize evaporation and deterioration of the fluid. Often, this is not a problem as high water systems tend to run cooler than oil systems because they have higher specific heat and thermal conductivity characteristics. Temperatures below freezing (32°F or 0°C), however, may cause separation of the phases or otherwise affect the fluid additives.

Because of the relatively high density and low viscosity of the fluid, inlet conditions and fluid conductor sizing should be carefully controlled to keep the fluid at a relatively low velocity. Excessive turbulence produces a release of air and fluid vapor in the form of bubbles which cause cavitation, noise, erosion and wear.

Compatibility with Seals, Metals, and Protective Coatings. Because of the great diversity of fluids, the supplier of each specific fluid must be consulted for compatibility information. In general, the seals normally used with petroleum oils are satisfactory for these fluids. Asbestos, leather, paper or cork materials should not be used, because they tend to deteriorate in water.

Although high water fluids are formulated with corrosion inhibitors, galvanic action can occur with certain combinations of metal (such as copper-to-steel or unanodized aluminum-to-steel).

Petroleum compatible paints cannot be used because of the solvent effects of the fluids. Special epoxy-type paints have been found to be compatible with some of the fluids.

Water-In-Oil Emulsions

These fluids (ISO HFB) are commonly called "invert emulsions," and like the HFA fluids, depend on water content for their fire-resistant properties. Tiny droplets of water are dispersed in a continuous oil phase. Like oil, these fluids have good lubricity and body. Additionally, the dispersed water gives the fluid a better cooling ability. Rust inhibitors are incorporated for both the water and oil phases. Antifoam additives also are used without difficulty.

These emulsions usually contain about 40 percent water. However, some manufacturers furnish a fluid concentrate to which the customer adds water when the fluid is installed. It is necessary to replenish the water to maintain proper viscosity.

Other Characteristics. Operating temperatures must be kept low to avoid evaporation and oxidation. The fluid must circulate and should not be repeatedly thawed and frozen or the two phases may separate. Inlet conditions should be carefully chosen because of the higher density of the fluid and its inherent high viscosity.

Emulsions seem to have a greater affinity for contamination and require extra attention to filtration, including magnetic plugs to attract iron particles.

Compatibility with Seals and Metals. Emulsion fluids are generally compatible with all metals and seals found in petroleum hydraulic systems.

Changeover to Emulsions. When a hydraulic system is changed over to water-in-oil emulsion fluid, it should be completely drained, cleaned, and flushed. It is essential to get out any contamination (such as water-glycol fluids) which might cause the new fluid to break down. Most seals can be left undisturbed. Butyl dynamic (moving) seals should be replaced, however. In changing from synthetic fluids, seals must be changed to those rated for petroleum oil use.

Water-Glycol Type Fluids

Water-glycol fluids (ISO HFC) are compounded of:

- 35 percent to 45 percent water to provide resistance to burning
- A glycol (a synthetic chemical of the same family as permanent anti-freeze ethylene or other glycols)
- A water-soluble thickener to improve viscosity.

Water-glycol fluids also contain additives to prevent foaming, rust, corrosion and to improve lubrication.

Characteristics. Water-glycol fluids generally have good wear resistance characteristics, provided that high speeds and loads are avoided. The fluid has a high specific gravity (it is heavier than oil), which can create a higher vacuum at pump inlets. Certain metals such as zinc, cadmium, and magnesium react adversely with water-glycol fluids, generating gummy residues which plug orifices and filters and cause valve spools to stick. It is recommended that parts which are alloyed or plated with these metals not be used with water-glycol. Examples of such parts might be galvanized pipe, and zinc or cadmium plated strainers, fittings and reservoir accessories.

Many of the synthetic seal materials used with petroleum oils are also compatible with water-glycol fluid. Asbestos, leather, and cork-impregnated materials should be avoided in rotating seals, since they tend to absorb water.

Some disadvantages of these fluids are as follows:

- They must be measured continually for water content and evaporation must be made up continually to maintain required viscosity.

		WATER-BASE FLUIDS		NONWATER-BASE FLUIDS		
MATERIALS UNDER CONSIDERATION	PETROLEUM OILS	OIL AND WATER EMULSION	WATER-GLYCOL MIXTURE		PHOSPHATE ESTERS	
ACCEPTABLE SEAL AND PACKING MATERIALS	NEOPRENE, BUNA N	NEOPRENE, BUNA N, (NO CORK)	NEOPRENE, BUNA N, (NO CORK)		BUTYL, VITON*, VYRAM, SILICONE, TEFLON FBA	
ACCEPTABLE PAINTS	CONVENTIONAL	CONVENTIONAL	AS RECOMMENDED BY SUPPLIER		"AIR CURE" EPOXY AS RECOMMENDED	
ACCEPTABLE PIPE DOPES	CONVENTIONAL	CONVENTIONAL	PIPE DOPES AS RECOMMENDED, TEFLON TAPE			
ACCEPTABLE SUCTION STRAINERS	100 MESH WIRE 1-1/2 TIMES PUMP CAPACITY	40 MESH WIRE 4 TIMES PUMP CAPACITY	50 MESH WIRE, 4 TIMES PUMP CAPACITY			
ACCEPTABLE FILTERS	CELLULOSE FIBER, 200–300 MESH WIRE, KNIFE EDGE OR PLATE TYPE	GLASS FIBER, 200–300 WIRE, KNIFE EDGE OR PLATE	CELLULOSE FIBER, 200–300 MESH WIRE, KNIFE EDGE OR PLATE	CELLULOSE FIBER, 200–300 MESH WIRE, KNIFE EDGE OR PLATE TYPE (FULLER'S EARTH OR MICRONIC TYPE MAY BE USED ON NONADDITIVE FLUIDS.)		
ACCEPTABLE METALS OF CONSTRUCTION	CONVENTIONAL	CONVENTIONAL	AVOID GALVANIZED METAL AND CADMIUM PLATING		CONVENTIONAL	

* SOME LOW VISCOSITY PHOSPHATE ESTERS, INCLUDING THE ALKYL PHOSPHATE ESTERS (AERO) SUCH AS SKYDROL, ARE NOT COMPATIBLE WITH VITON SEALS.

Figure 3-11. Compatibility of hydraulic fluids and sealing materials.

- Evaporation may also cause loss of certain additives, thereby reducing the life of the fluid and of the hydraulic components.
- Operating temperatures must be kept low.
- The cost of water-glycol fluids (at the present time) is greater than that of conventional oils.

Changing to Water-Glycol. When a system is changed from petroleum oil to water-glycol, it must be thoroughly cleansed and flushed. Recommendations include removing original paint from inside the reservoir, changing zinc or cadmium plated parts, and replacing certain die cast fittings. It may also be necessary to replace aluminum parts unless properly treated, as well as any instrumentation equipment which is not compatible with the fluid.

Synthetic Fire-Resistant Fluids

Synthetic fire-resistant fluids (ISO HFD) are laboratory synthesized chemicals which are less flammable than petroleum oils. Typical of these are:

- Phosphate esters
- Polyol esters
- Halogenated (fluorinated and/or chlorinated) hydrocarbons
- Mixtures of phosphate esters or polyol esters and petroleum oil

Characteristics. Since the synthetics do not contain any water or other volatile material, they operate well at higher temperatures than water containing fluids. They also are suitable for higher pressure systems than the water containing fluids.

Synthetic fire resistant fluids do not operate best in low- temperature systems. Auxiliary heating may be required in cold environments.

Synthetic phosphate esters usually have high specific gravities (weight) and pump inlet conditions require special care when they are used. Some vane pumps are built with special bodies to provide the improved inlet conditions needed to prevent pump cavitation when a synthetic fluid is used.

The viscosity index of phosphate ester fluids is generally low ranging from 80 to as low as minus 40. Thus, they should not be used except where the operating temperature is relatively constant. Synthetic fluids are probably the most costly hydraulic fluids being used at this time.

Seal Compatibility. Synthetic fluids, because of their great diversity, will vary in their compatibility with seals. The phosphate esters are not compatible with the commonly used nitrile (Buna) and neoprene seals. Therefore, a changeover from petroleum, water-glycol, or water-oil fluids requires dismantling all the components to replace the seals.

Special seals made of compatible materials are available for replacement on all Vickers components. They can be purchased singly or in kits or can be built into new units ordered specifically for this type of fluid.

It is crucial to use materials that are compatible with synthetic fluids. Figure 3-11 is a chart showing the types of materials that are compatible with various hydraulic fluids.

QUESTIONS

1. Name the four primary purposes of the hydraulic fluid.
2. Name four quality requirements of a hydraulic fluid.
3. Define viscosity. What is the common unit of measuring viscosity?
4. How is viscosity affected by cold? By heat?
5. If viscosity is too high, what can happen to the system?
6. What is the viscosity index? When is the viscosity index important?
7. Which type of hydraulic fluid has the best natural lubricity?
8. Name several catalysts to oxidation of hydraulic oil.
9. How are rust and corrosion prevented?
10. What is demulsibility?
11. What are the three factors that determine the properties of a hydraulic oil?
12. What are the four basic types of fire-resistant hydraulic fluid?
13. Which type of hydraulic fluid is not compatible with Buna or neoprene seals?
14. Which type of fire-resistant hydraulic fluid is best for high temperature operation?
15. How does the specific gravity of the synthetic phosphate esters affect the pump inlet conditions?

CHAPTER 4

HYDRAULIC FLUID CONDUCTORS AND SEALS

This chapter is comprised of two parts. First is a description of the hydraulic system "plumbing"—the types of connecting lines and fittings used to carry fluid between the pumps, valves, actuators, etc. The second part deals with the prevention of leakage and the types of seals and seal materials required for hydraulic applications.

FLUID CONDUCTORS

Fluid conductors is a general term which embraces the various kinds of conducting lines that carry hydraulic fluid between components plus the fittings or connectors used between the conductors. Hydraulic systems today use principally three types of conducting lines: steel pipe, steel tubing, and flexible hose. Pipe may be the cheapest but its use often presents serious leakage problems, especially at higher operation pressures. Pipe is still in use in many installations (which accounts for its inclusion in this chapter) but it is gradually being replaced with tubing and hose. The future may see plastic plumbing which is slowly coming into use for certain applications.

Steel Pipe

Iron and steel pipes were the first conductors used in industrial hydraulic systems and are still used because of their low cost. Seamless steel pipe is recommended for hydraulic systems with the pipe interior free of rust, scale and dirt.

Sizing. Pipe and pipe fittings are classified by nominal size and wall thickness. Originally, a given size pipe had only one wall thickness and the stated size was the actual **inside** diameter.

Later, pipes were manufactured with varying wall thicknesses: standard, extra heavy, and double extra heavy. However, the outside diameter did not change. To increase wall thickness, the inside diameter was changed. Thus, the nominal pipe size alone indicates only the thread size for connections.

Schedule. Currently, wall thickness is being expressed as a **schedule** number. Schedule numbers are specified by the American National Standards Institute (ANSI).

For comparison, schedule 40 corresponds closely to standard. Schedule 80 essentially is extra heavy. Schedule 160 covers pipes with the greatest wall thickness under this system. The old double extra heavy classification is slightly thicker than schedule 160. Figure 4-1 shows pipe sizes up to 12 inches (305 mm) (nominal).

Sealing. Pipe threads are tapered (Figure 4-2) as opposed to tube and some hose fittings which have straight threads. Joints are sealed by an interference fit between the male and female threads as the pipe is tightened.

This creates one of the major disadvantages of pipe. When a joint is broken, the pipe must be tightened further to reseal. Often this necessitates replacing some of the pipe with slightly longer sections. However, the difficulty has been overcome somewhat by using modern compounds to reseal pipe joints. The sealant should always be applied to the male thread, never the female. Avoid the first two male threads from the end to keep sealer from contaminating the system. TFE-fluorocarbon tape is not recommended for use on hydraulic fittings because pieces may get into the system and jam close-fitting parts.

Special taps and dies are required for threading hydraulic system pipes and fittings. The threads are the "dryseal" (NPTF—National Pipe Tapered Fuel) type. They differ from standard pipe threads by engaging the roots and crests before the flanks; thus avoiding spiral clearance (Figure 4-2).

Fittings. Since pipe can have only male threads and does not bend, various types of fittings are used to make connections and change direction. The many fittings necessary in a pipe circuit present multiple opportunities for leakage, particularly as pressure increases. Threaded connections are used up to 1-1/4 inch (32 mm). Where larger pipes are needed, flanges are welded to the pipe

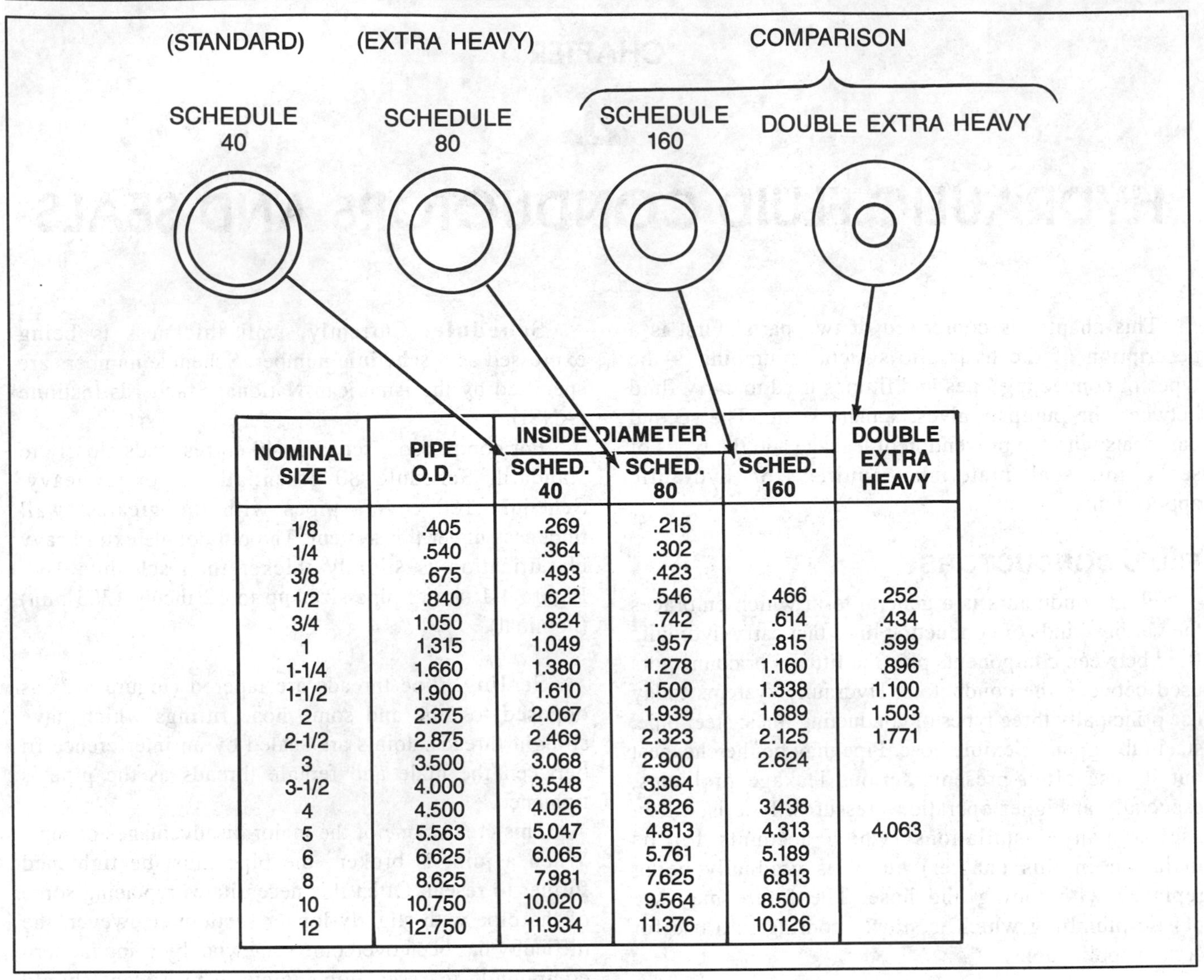

NOMINAL SIZE	PIPE O.D.	INSIDE DIAMETER			DOUBLE EXTRA HEAVY
		SCHED. 40	SCHED. 80	SCHED. 160	
1/8	.405	.269	.215		
1/4	.540	.364	.302		
3/8	.675	.493	.423		
1/2	.840	.622	.546	.466	.252
3/4	1.050	.824	.742	.614	.434
1	1.315	1.049	.957	.815	.599
1-1/4	1.660	1.380	1.278	1.160	.896
1-1/2	1.900	1.610	1.500	1.338	1.100
2	2.375	2.067	1.939	1.689	1.503
2-1/2	2.875	2.469	2.323	2.125	1.771
3	3.500	3.068	2.900	2.624	
3-1/2	4.000	3.548	3.364		
4	4.500	4.026	3.826	3.438	
5	5.563	5.047	4.813	4.313	4.063
6	6.625	6.065	5.761	5.189	
8	8.625	7.981	7.625	6.813	
10	10.750	10.020	9.564	8.500	
12	12.750	11.934	11.376	10.126	

Figure 4-1. Pipes currently are sized by schedule number.

(Figure 4-3). Flat gaskets or O-rings are used to seal flanged fittings.

Steel Tubing

Seamless steel tubing offers significant advantages over pipe for hydraulic plumbing. Tubing can be bent into any shape, is easier to work with, and can be used over and over without any sealing problems. Usually the number of joints is reduced. Tubing will handle higher pressure and flow with less bulk and weight. However, it may cost more, as do the fittings required to make tube connections.

Sizes. A tubing size specification refers to the **outside** diameter. Tubing is available in 1/16 inch (0.16 cm) increments from 1/8 inch (0.32 cm) to one inch (2.54 cm) O.D. and in 1/4 inch (0.64 cm) increments beyond one inch (2.54 cm). Various wall thicknesses are available for each size. The inside diameter, as previously noted, equals the outside diameter less twice the wall thickness.

Fittings. Tubing is not sealed by threads, but by various kinds of fittings (Figure 4-4). Some of these fittings seal by metal-to-metal contact. They are known as **compression** fittings and may be either the flared or flareless type. Others use O-rings or comparable seals. In addition to threaded fittings, flanged fittings also are available to be welded to larger sized tubing.

Flared Fitting. The 37-degree flare fitting is the most common fitting for tubing that can be flared. The fitting shown in Figure 4-4A seals by squeezing the flared end of the tube against a seal as the compression nut is tightened. A sleeve or extension of the nut supports the tube to damp out vibration.

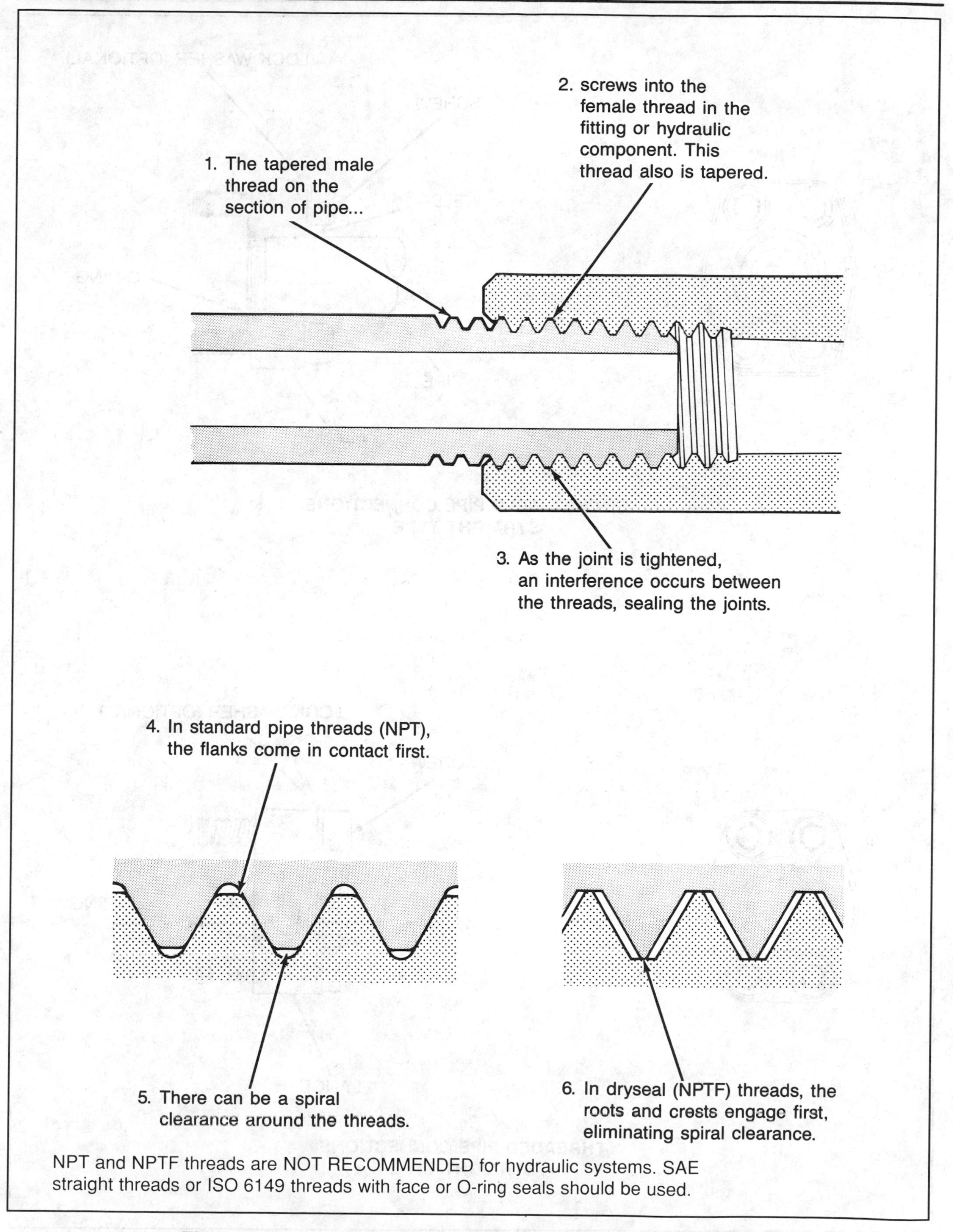

1. The tapered male thread on the section of pipe...
2. screws into the female thread in the fitting or hydraulic component. This thread also is tapered.
3. As the joint is tightened, an interference occurs between the threads, sealing the joints.
4. In standard pipe threads (NPT), the flanks come in contact first.
5. There can be a spiral clearance around the threads.
6. In dryseal (NPTF) threads, the roots and crests engage first, eliminating spiral clearance.
NPT and NPTF threads are NOT RECOMMENDED for hydraulic systems. SAE straight threads or ISO 6149 threads with face or O-ring seals should be used.

Figure 4-2. Hydraulic pipe threads are dry seal tapered type.

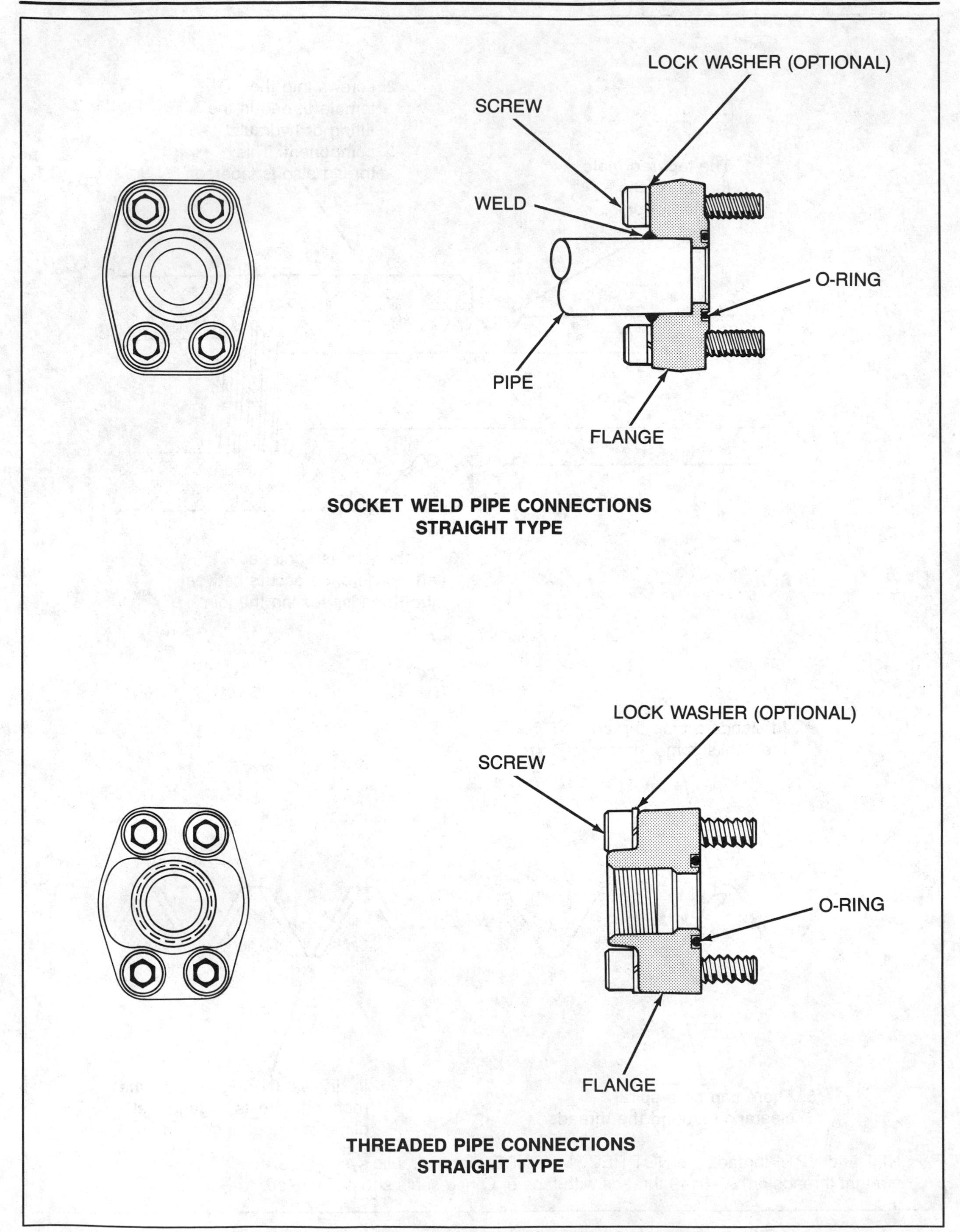

Figure 4-3. Flanged connections for large pipe.

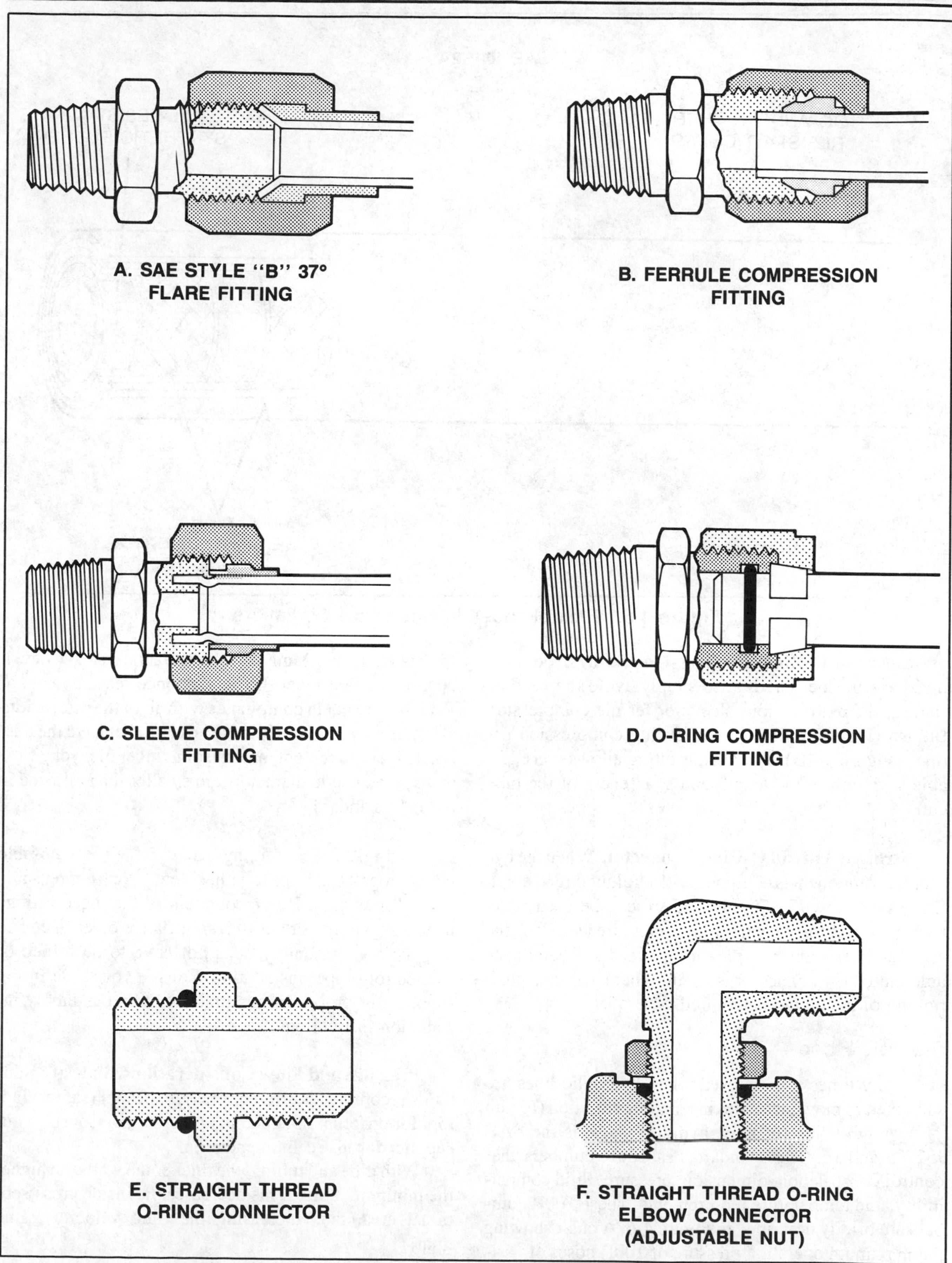

Figure 4-4. Threaded fittings and connectors used with tubing.

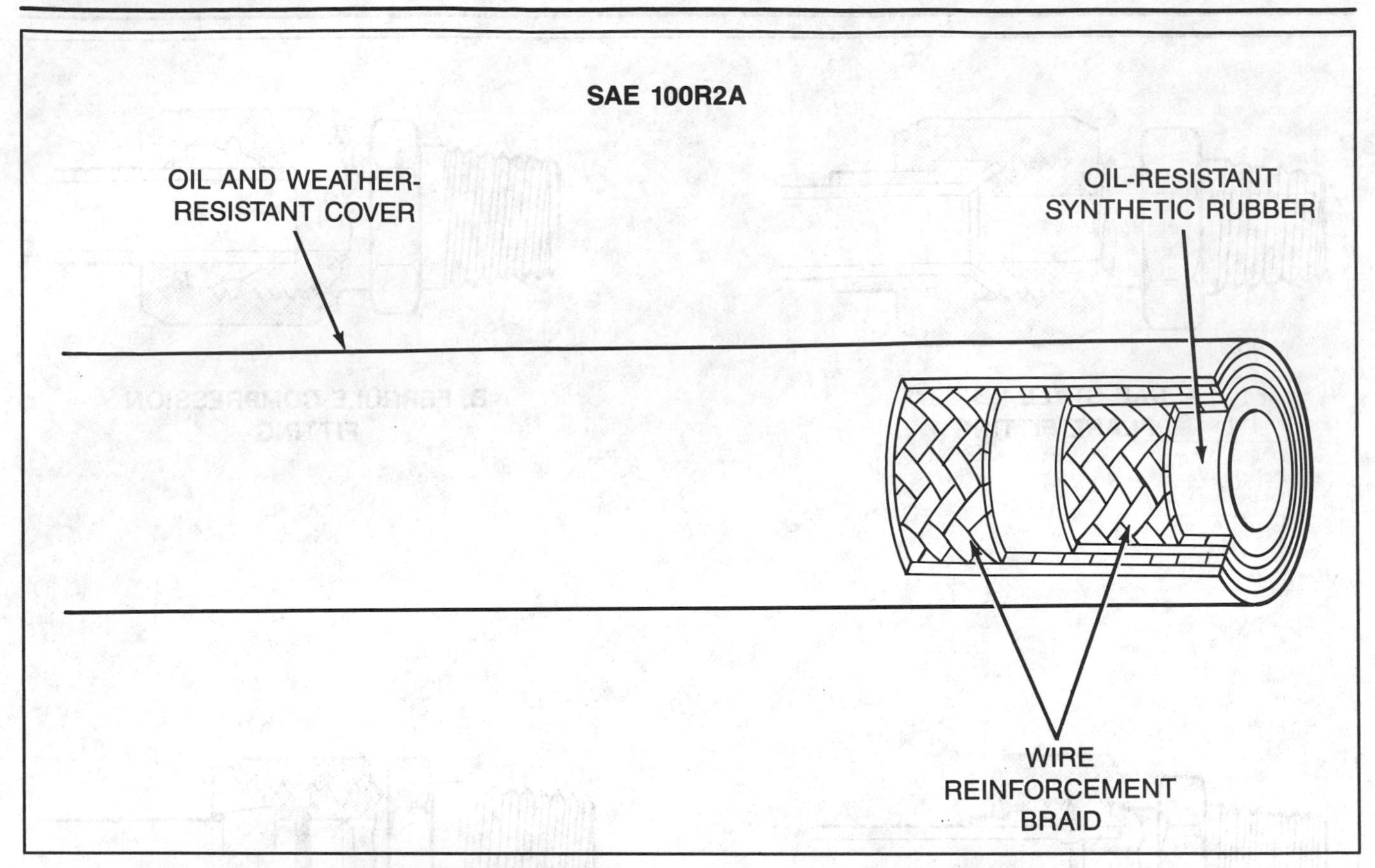

Figure 4-5. Flexible hose is constructed in layers.

Sleeve or O-Ring Compression Fittings. For tubing that can't be flared, or to simply avoid the need of flaring, there are various sleeve or ferrule compression fittings (Figure 4-4B&C), and O-ring compression fittings (Figure 4-4D). The O-ring fitting allows considerable variation in the length and squareness of the tube cut.

Straight Thread O-Ring Connector. When the hydraulic component is equipped with straight thread ports, fittings as shown in Figure 4-4E&F can be used. This type of connector is ideal for high-pressure use, since the seal becomes tighter as pressure increases. The adjustable nut elbow connector is useful where accurate positioning of the tubing is required.

Flexible Hose

Many times hose is used when hydraulic lines are subjected to movement, flexing, and/or vibration (Figure 4-5). A guideline for most hydraulic hose is the SAE J517 Standard. This standard has 100R numbers that control construction, dimension, pressure, fluid compatibility, and temperature requirements. These 100R numbers are briefly described in Figure 4-6. A chart showing the maximum operating pressure of 100R hoses of various sizes appears in Appendix C. Compliance with SAE J517 is voluntary. Many hose manufacturers have developed hoses that meet the performance requirements of J517, but are not in compliance with all of the dimensional requirements. Usually these hoses do not have the designated reinforcement material or number of layers. Normally, the outside diameters are smaller than allowed in the J517 standard.

Fittings. Hose fittings may be either reusable (screw-together, bolt-together, etc.) or nonreusable (crimp or swage). It is recommended that hose fittings have a swivel nut or an SAE split flange on each end so that the hose assembly does not have to be turned or twisted for proper installation. Consult your hose/fitting supplier for proper fitting attachment and assembly installation instructions.

Pressure and Flow Considerations. Industry standards recommend different pressure design (safety) factors for different conductors and systems. Consult your supplier for his recommendation.

With pipe and tubing, the thicker the walls the higher the minimum burst. This decreases the inside cross-sectional area, thus increasing the fluid velocity (same gpm) (LPM).

SAE 100R1
Type A — This hose shall consist of an inner tube of oil resistant synthetic rubber, a single wire braid reinforcement and an oil and weather resistant synthetic rubber cover. A ply or braid of suitable material may be used over the inner tube and/or over the wire reinforcement to anchor the synthetic rubber to the wire.

Type AT — This hose shall be of the same construction as Type A, except having a cover designed to assemble with fittings which do not require removal of the cover or a portion thereof.

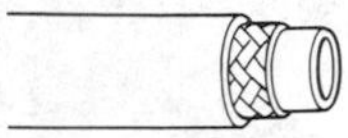

SAE 100R2
The hose shall consist of an inner tube of oil resistant synthetic rubber, steel wire reinforcement according to hose type as detailed below and an oil and weather resistant synthetic rubber cover. A ply or braid of suitable material may be used over the inner tube and/or over the wire reinforcement to anchor the synthetic rubber to the wire.

Type A — This hose shall have two braids of wire reinforcement.

Type AT — This hose shall be of the same construction as Type A, except having a cover designed to assemble with fittings which do not require removal of the cover or a portion thereof.

Type B — This hose shall have two spiral plies and one braid of wire reinforcement.

Type BT — This hose shall be of the same construction as Type B, except having a cover designed to assemble with fittings which do not require removal of the cover or a portion thereof.

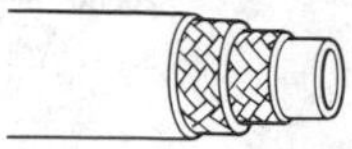

Type A

Type B

SAE 100R3
The hose shall consist of an inner tube of oil resistant synthetic rubber, two braids of suitable textile yarn and an oil and weather resistant synthetic rubber cover.

SAE 100R4
The hose shall consist of an inner tube of oil resistant synthetic rubber, a reinforcement consisting of a ply or plies of woven or braided textile fibers with a suitable spiral of body wire and an oil and weather resistant synthetic rubber cover.

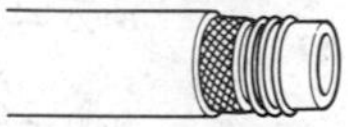

SAE 100R5
The hose shall consist of an inner tube of oil resistant synthetic rubber and two textile braids separated by a high tensile steel wire braid. All braids are to be impregnated with an oil and mildew resistant synthetic rubber compound.

SAE 100R6
The hose shall consist of an inner tube of oil resistant synthetic rubber, one braided ply of suitable textile yarn and an oil and weather resistant synthetic rubber cover.

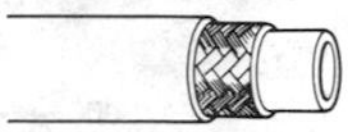

SAE 100R7
The hose shall consist of a thermoplastic inner tube resistant to hydraulic fluids with suitable synthetic fiber reinforcement and a hydraulic fluid and weather resistant thermoplastic cover.

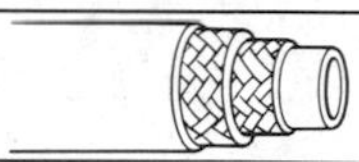

SAE 100R8
The hose shall consist of a thermoplastic inner tube resistant to hydraulic fluids with suitable synthetic fiber reinforcement and a hydraulic fluid and weather resistant thermoplastic cover.

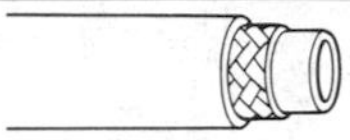

SAE 100R9
Type A — This hose shall consist of an inner tube of oil resistant synthetic rubber, 4-spiral plies of wire wrapped in alternating directions and an oil and weather resistant synthetic rubber cover. A ply or braid of suitable material may be used over the inner tube and/or over the wire reinforcement to anchor the synthetic.

Type AT — This hose shall be of the same construction as Type A, except having a cover designed to assemble with fittings which do not require removal of the cover or a Portion thereof.

SAE 100R10
Type A — This hose shall consist of an inner tube of oil resistant synthetic rubber, 4-spiral plies of heavy wire wrapped in alternating directions and an oil and weather resistant synthetic rubber cover. A ply or braid of suitable material may be used over the inner tube and/or over the wire reinforcement to anchor the synthetic rubber to the wire.

Type AT — This hose shall be of the same construction as Type A, except having a cover designed to assemble with fittings which do not require removal of the cover or a portion thereof.

SAE 100R11
This hose shall consist of an inner tube of oil resistant synthetic rubber, 6-spiral plies of heavy wire wrapped in alternating directions and an oil and weather resistant synthetic rubber cover. A ply or braid of suitable material may be used over the inner tube and/or over the wire reinforcement to anchor the synthetic rubber to the wire.

SAE100R12
This hose shall consist of an inner tube of oil resistant synthetic rubber, 4-spiral plies of heavy wire wrapped in alternating directions and an oil and weather resistant synthetic rubber cover. A ply or braid of suitable material may be used over or within the inner tube and/or over the wire reinforcement to anchor the synthetic rubber to the wire.

SAE100R13
This hose shall consist of an inner tube of oil resistant synthetic rubber, multiple spiral plies of heavy wire wrapped in alternating directions and an oil and weather resistant synthetic rubber cover. A ply or braid of suitable material may be used over or within the inner tube and/or over the wire reinforcement to anchor the synthetic rubber to the wire.

SAE100R14
Type A — This hose shall consist of an inner tube of polytetrafluorethylene (PTFE), reinforced with a single braid of 303XX series stainless steel.

Type B — This hose shall be of the same construction as Type A, but shall have the additional feature of an electrically conductive inner surface so as to preclude buildup of an electrostatic charge.

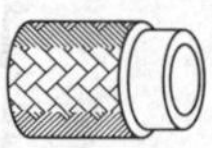

Type A 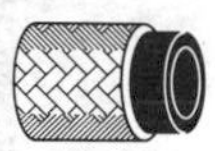 *Type B*

SAE 100R15 - Heavy-Duty, High Impulse Spiral Wire Reinforcement Hose
This hose shall consist of an inner tube of oil resistant synthetic rubber, multiple spiral plies of heavy wire wrapped in alternating directions and an oil-and-weather 9 resistant synthetic rubber cover.

SAE 100R16 - Compact Wire Braid Hoses of this type are smaller in diameter than two wire braid hoses with similar performance characteristics as 100R2, which gives them the ability to operate at smaller bend radii.
This hose shall consist of an inner tube of oil resistant synthetic rubber. wire braid reinforcement (one or two braids) and an oil and weather resistant synthetic rubber cover.

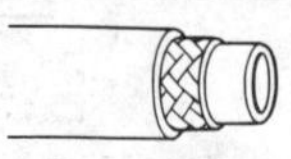

One Braid 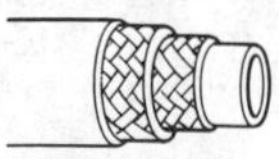 *Two Braids*

Figure 4-6. Description of 100R-type flexible hose.

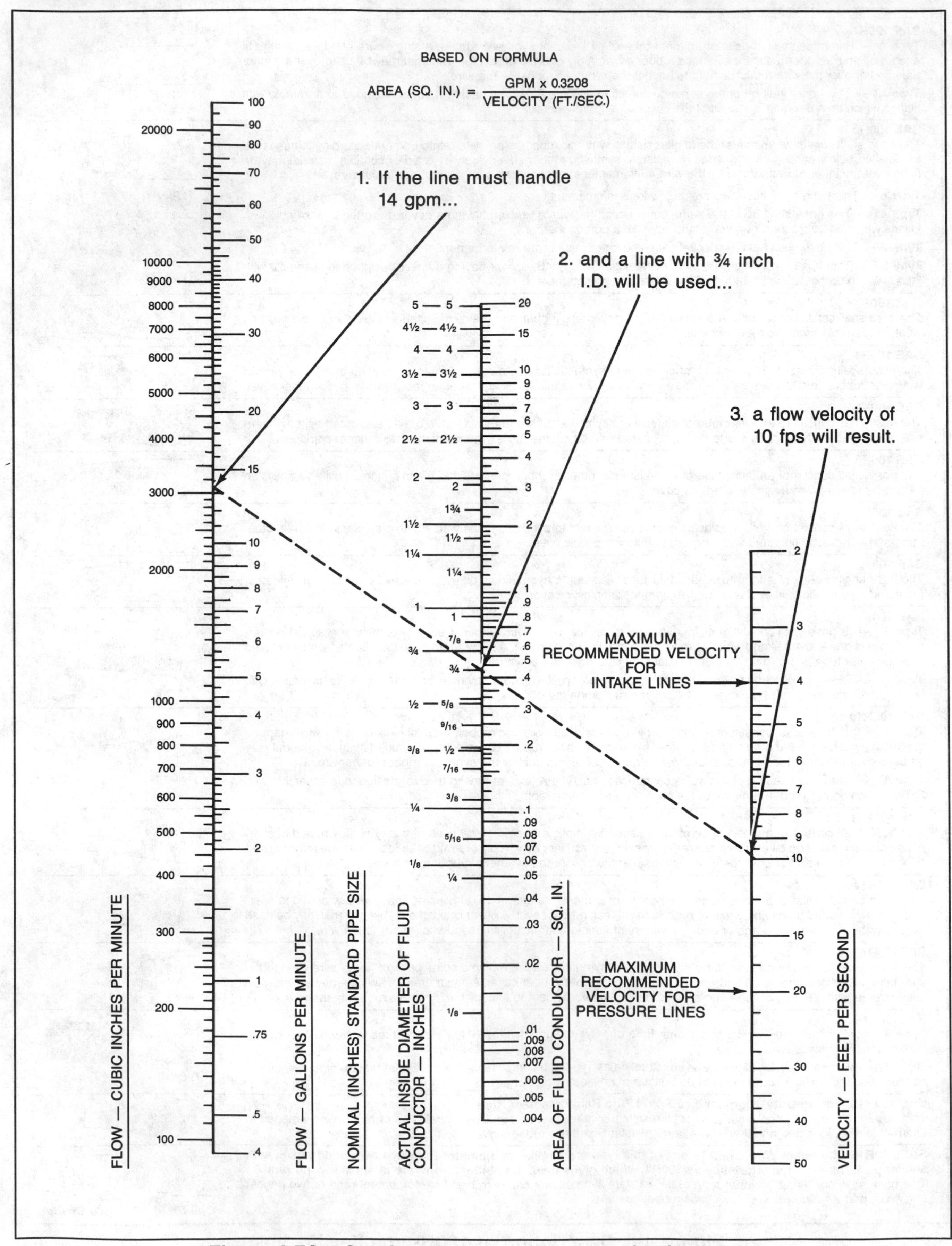

Figure 4-7A. Conductor inside diameter selection chart.

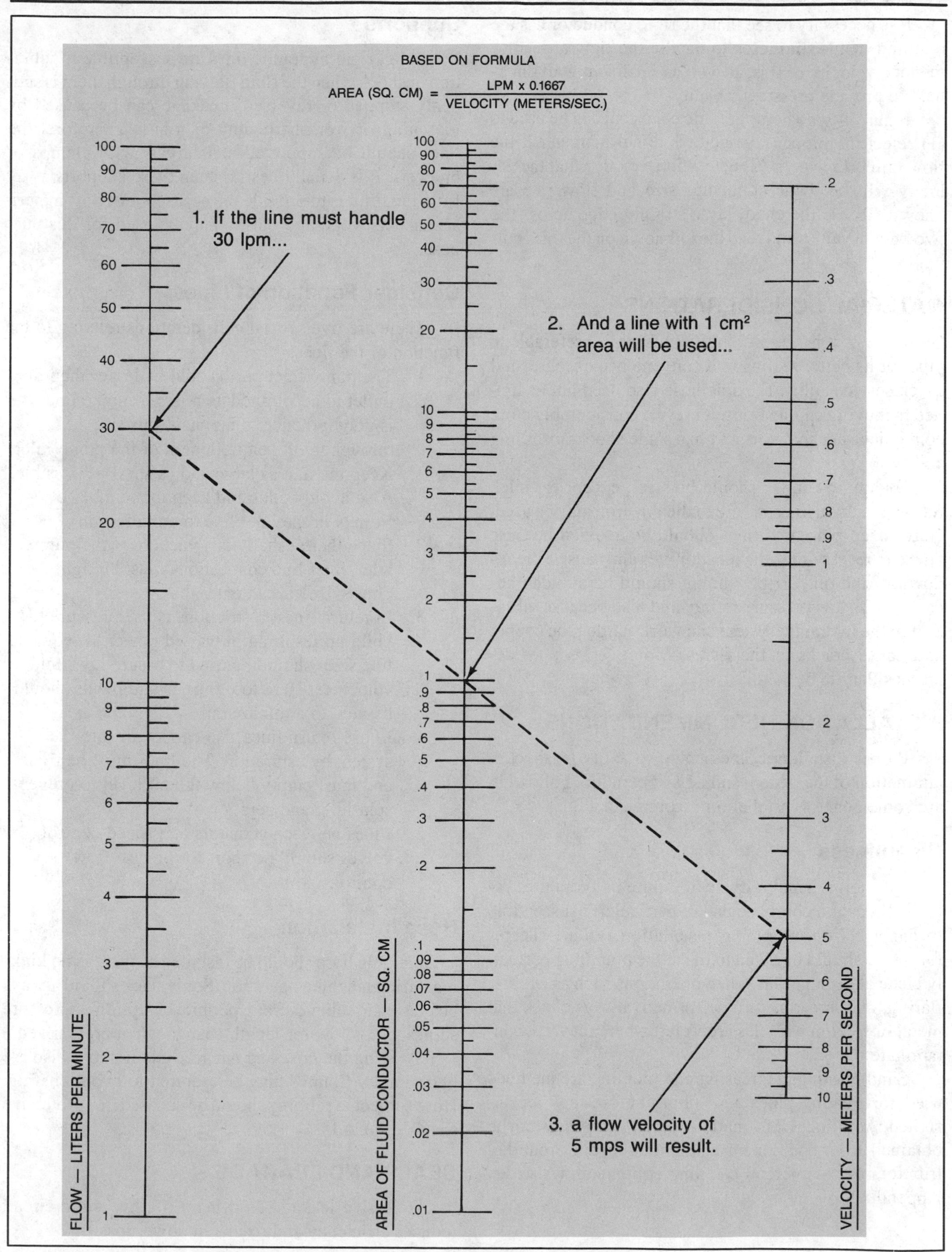

Figure 4-7B. Conductor inside diameter selection chart.

It is necessary to see that the rigid conductor has the required **inside diameter** to handle the flow at recommended velocity or less, as well as sufficient wall thickness to provide pressure capacity.

Figure 4-7 is a nomographic chart that can be used to (1) select the proper conductor internal diameter if the flow rate is known or (2) determine exactly what the velocity will be if the conductor size and flow rate are known. To use the chart, lay a straight edge across the two known values and read the unknown on the third column.

MATERIAL CONSIDERATIONS

If the cost is not prohibitive, tubing is preferable to pipe for its better sealing and convenience of reuse, and quick serviceability. Flexible hose also need not be limited to moving applications. It may be considerably more convenient in short runs and has some shock-absorbing ability.

Hydraulic fittings should be steel, except for inlet, return and drain lines, where malleable iron may be used. Galvanized pipe or fittings should be avoided because zinc can react with some oil additives and cause a breakdown of the oil. Copper tubing should be avoided because of its low pressure rating, and also because vibration in the hydraulic system can work-harden the copper and cause cracks at the flares. Moreover, copper decreases the life of the oil.

INSTALLATION RECOMMENDATIONS

Proper installation is essential to avoid leaks, contamination of the system and noisy operation. Following are some general installation recommendations.

Cleanliness

Dirty oil is a major cause of failure in hydraulic systems. Precision components are particularly susceptible to damage from plumbing installation residue. Therefore, care should be taken to make the plumbing perfectly clean at installation. When operations such as cutting, flaring, and threading are performed, always check that metal or foreign particles aren't left where they can contaminate the oil.

Sand blasting, degreasing and pickling are methods used for treating pipes and tubing before they are installed. Additional information on these processes can be obtained from component manufacturers and from distributors of commercial cleaning equipment. (Also see Appendix C.)

Supports

Long, rigid hydraulic lines are susceptible to vibration or shock when the fluid flowing through them is suddenly stopped or reversed. Leakage can be caused by loosening or work-hardening of joints. Therefore, the lines should be supported at intervals with clamps or brackets. It is usually best to keep these supports away from the fittings to provide the best support. Soft materials such as wood and plastic may be used for this purpose.

Consider Function of Lines

There are five special considerations relating to the function of the lines.

1. The pump inlet port should be larger than the outlet to accommodate a larger intake line. It is good practice to maintain this size throughout the entire length of the pump inlet. Keep the line as large and as short as possible. Also avoid bends and keep the number of fittings in the inlet line to a minimum.
2. Since there usually is a vacuum at the pump inlet, inlet line connections must be tight. Otherwise air can enter the system.
3. In return lines, restrictions cause pressure to build up resulting in wasted power. Adequate line sizes should be used to assure low fluid velocities. Here too, fittings and bends should be held to a minimum.
4. Loose return lines also can let air into the system by aspiration. The lines must be tight and must empty below the oil level to prevent splashing and aeration.
5. Lines between actuators and speed control valves should be short for precise flow control.

Hose Installation

Flexible hose should be installed so there is no kinking during machine operation. Some slack should always be present to relieve strain, permit absorption of pressure surges, and allow for length change when pressurized.

Twisting the hose and unusually long loops also are undesirable. Clamps may be required to avoid chafing. Hose subject to rubbing should be encased in a protective sleeve or guard.

SEALS AND LEAKAGE

Excessive leakage anywhere in a hydraulic circuit reduces efficiency and results in power loss.

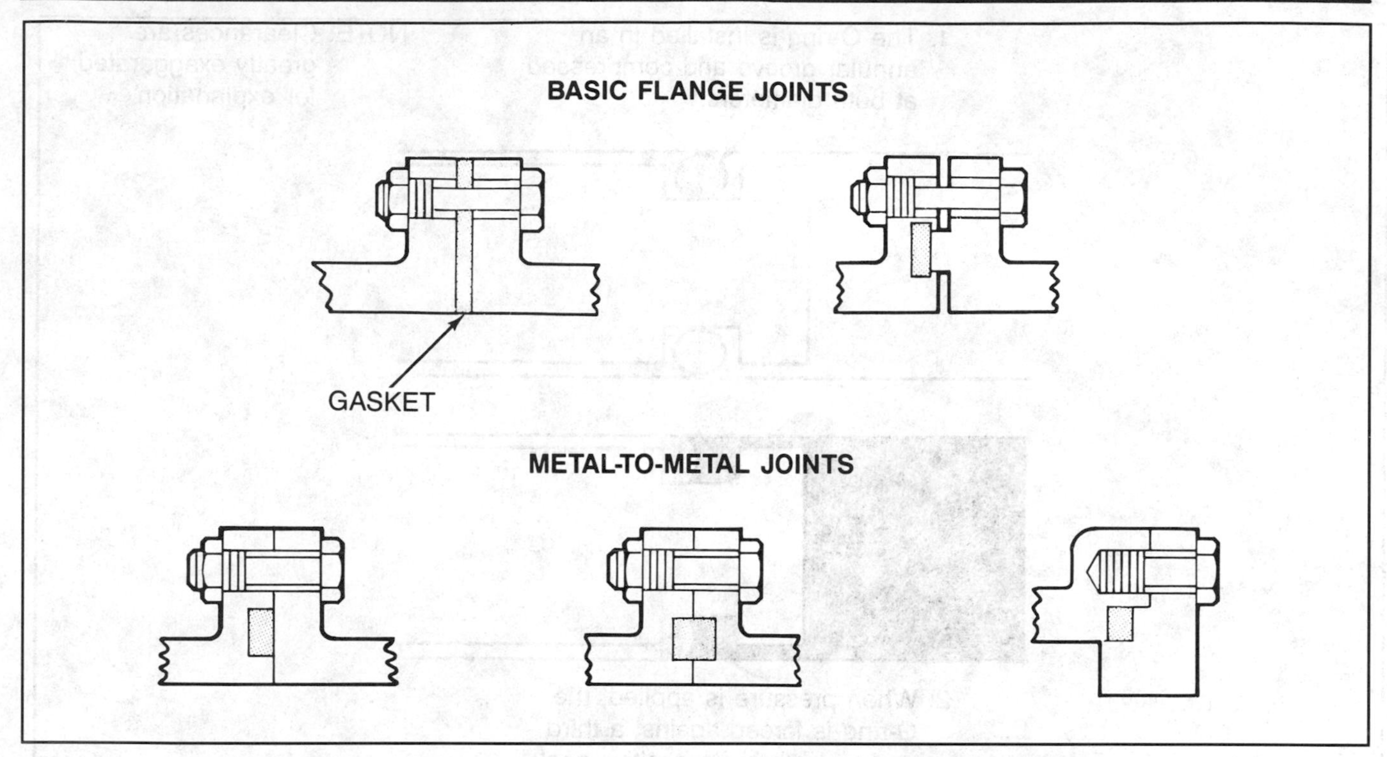

Figure 4-8. Flange gaskets and seals are typical static applications.

Internal Leakage

Most hydraulic system components are built with operating clearances which allow a certain amount of internal leakage. Moving parts, of course, must be lubricated and leakage paths may be designed in solely for this purpose. In addition, some hydraulic controls have internal leakage paths built in to prevent "hunting" or oscillation of valve spools and pistons.

Internal leakage, of course, is not loss of fluid. The fluid eventually is returned to the reservoir either through an external drain line or by way of an internal passage in the component.

Additional internal leakage occurs as a component begins to wear and clearances between parts increase. This increase in internal leakage can reduce the efficiency of a system by slowing down the work and generating heat.

Finally, if the internal leakage path becomes large enough, all the pump's output may be bypassed and the machine will not operate at all.

External Leakage

External leakage is unsightly and can be very hazardous. It is expensive because the oil that leaks out should not be returned to the system. The principal cause of external leakage is improper installation. Joints may leak because they weren't put together properly or because vibration or shock in the line loosened them. Failure to connect drain lines, excessive operating pressure, and contamination in the fluid all are common reasons for seals becoming damaged. The use of tapered pipe threads is the single largest contributor of external leakage in hydraulic systems and should be avoided in new designs and replaced in existing systems.

Sealing

Sealing is required to prevent fluid loss and to keep out contamination. There are various methods of sealing hydraulic components, depending on whether the seal must be **positive or nonpositive**, whether the sealing application is static or dynamic, how much pressure is on the system, and other factors.

- A **positive** seal prevents even a minute amount of fluid from getting past.
- A **nonpositive** seal allows a small amount of internal leakage; such as the clearance of a spool in its bore to provide a lubricating film.

Static Seals

A seal that is compressed between two rigidly connected parts is classified as a static seal. The seal itself may move somewhat as pressure is alternately applied and released, but the mating parts do not move in relation

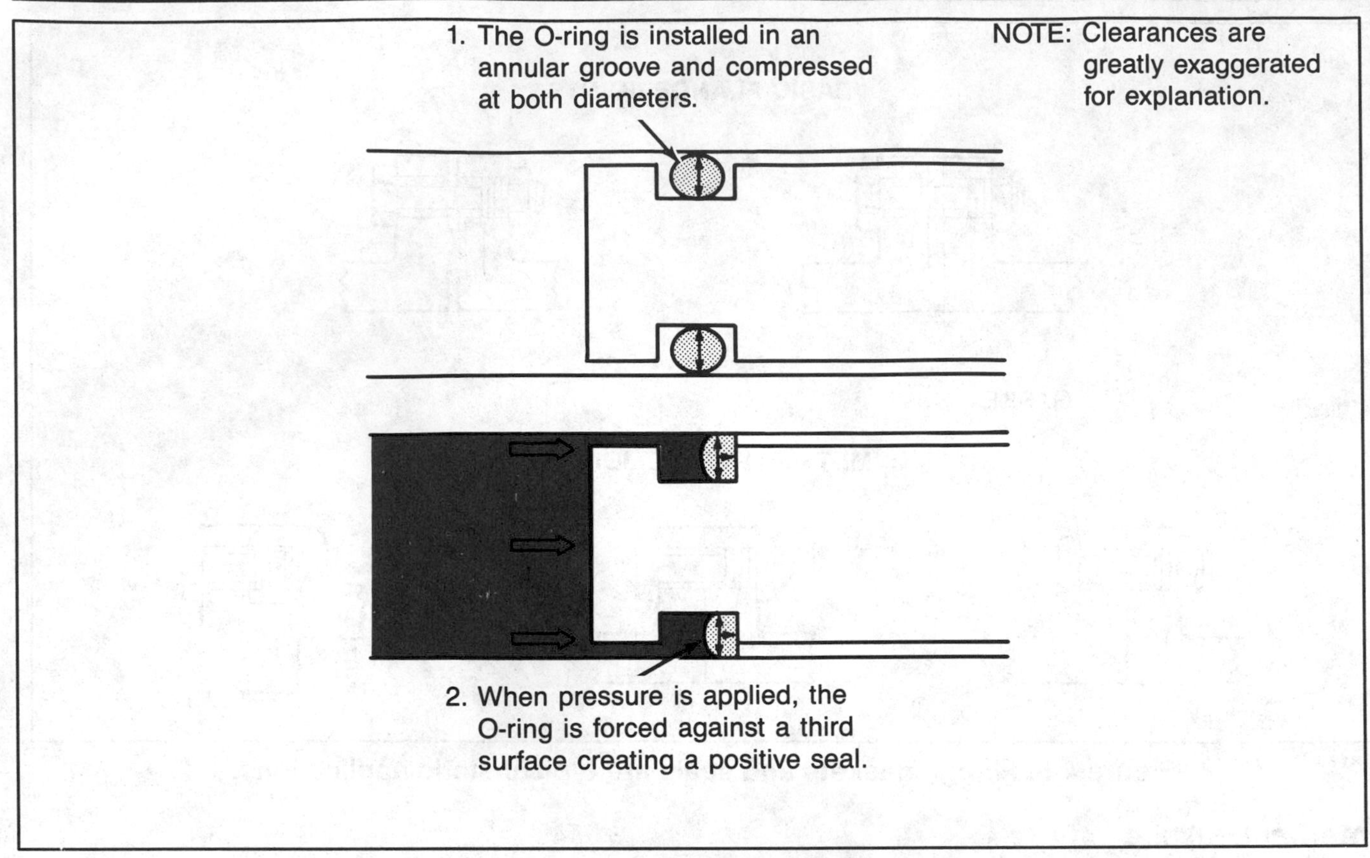

1. The O-ring is installed in an annular groove and compressed at both diameters.
NOTE: Clearances are greatly exaggerated for explanation.
2. When pressure is applied, the O-ring is forced against a third surface creating a positive seal.

Figure 4-9. An O-ring is a positive seal.

to each other.

Some examples of static seals are mounting gaskets, pipe thread connections, flange joint seals (Figure 4-8), compression fitting ferrules (Figure 4-4B), and O-rings. Static sealing applications are relatively simple. They are essentially "nonwearing" and usually are trouble-free if assembled properly.

Dynamic Seals

Dynamic seals are installed between parts which move relative to one another. Thus, at least one of the parts must rub against the seal, and therefore dynamic seals are subject to wear. This makes their design and application more difficult.

O-Ring Seals

Probably the most common seal in use in modern hydraulic equipment is the O-ring (Figure 4-9). An O-ring is a molded, synthetic rubber seal which has a round cross-section in the free state.

The O-ring is installed in an annular groove machined into one of the mating parts. At installation, it is compressed at both the inside and outside diameters. However, it is a pressure-actuated seal as well as a compression seal. Pressure forces the O-ring against one side of its groove and outward at both diameters. It thus seals positively against two annular surfaces and one flat surface. Increased pressure results in a higher force against the sealing surfaces. The O-ring, therefore, is capable of containing extremely high pressure.

O-rings are used principally in static applications. However, they are also found in dynamic applications where there is a short reciprocating motion between the parts. They are not generally suitable for sealing rotating parts or for applications where vibration is a problem.

Backup (Nonextrusion) Rings

At high pressure, the O-ring has a tendency to extrude into the clearance space between the mating parts (Figure 4-10). This may not be objectionable in a static application. But this extrusion can cause accelerated wear in a dynamic application. It is prevented by installing a stiff backup ring in the O-ring groove opposite the pressure source. If the pressure alternates, backup rings can be used on both sides of the O-ring.

T-Ring Seals

The T-ring seal (Figure 4-11) is used extensively to seal cylinder pistons, piston rods, and other reciprocating parts. It is constructed of synthetic rubber molded in

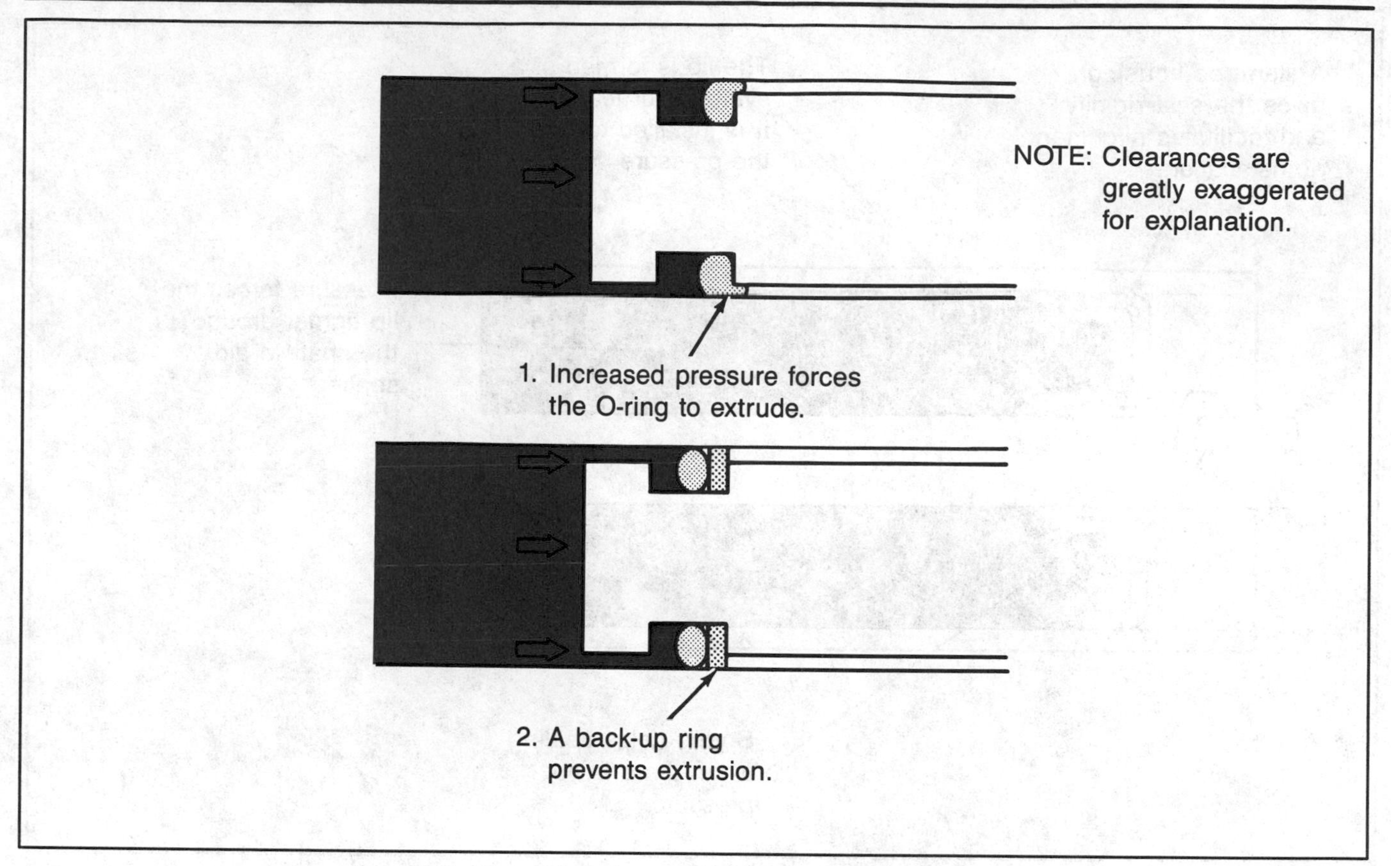
NOTE: Clearances are greatly exaggerated for explanation.
1. Increased pressure forces the O-ring to extrude.
2. A back-up ring prevents extrusion.

Figure 4-10. A backup ring is a nonextrusion ring.

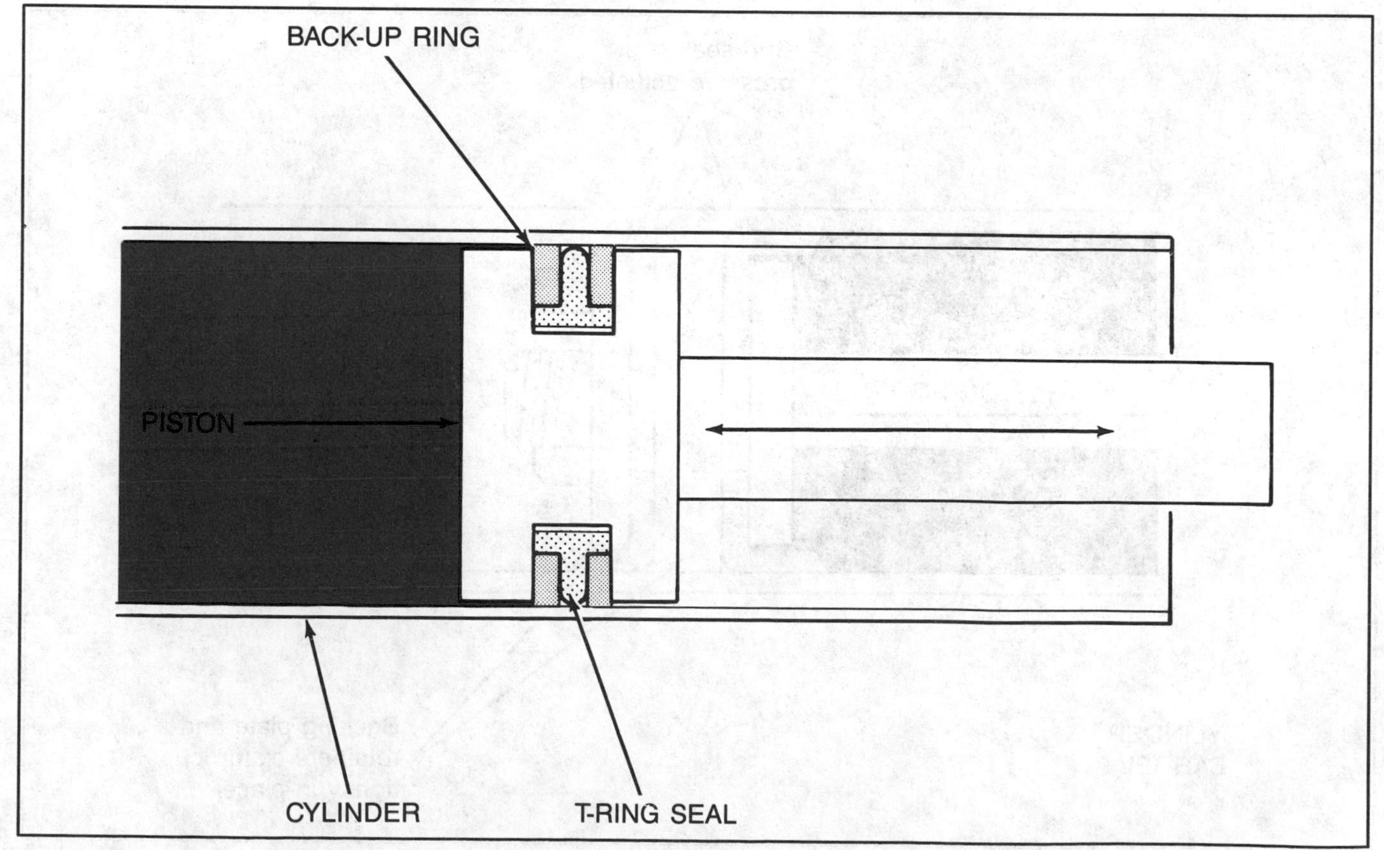
BACK-UP RING
PISTON
CYLINDER
T-RING SEAL

Figure 4-11. T-ring is a dynamic seal for reciprocating parts.

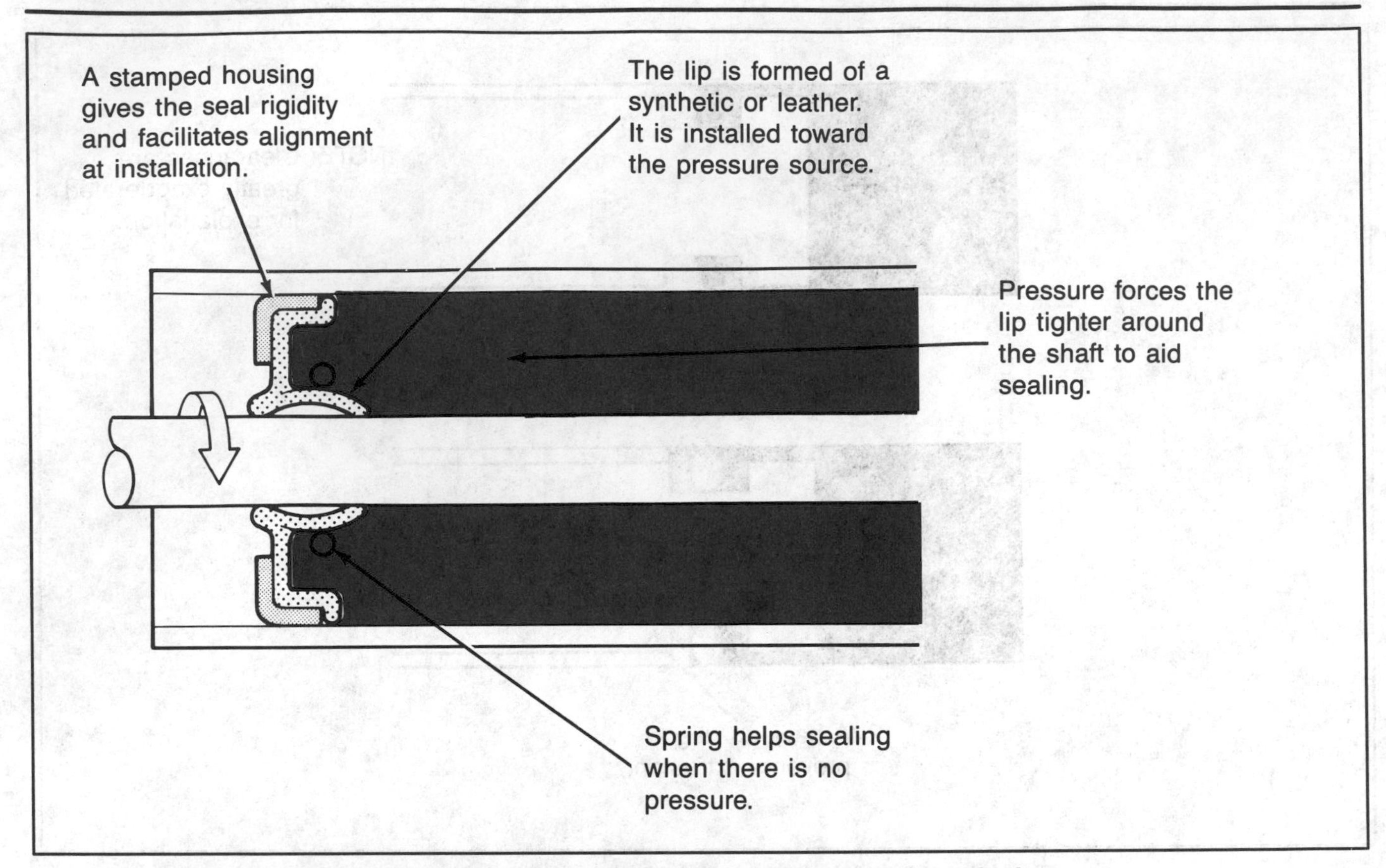

A stamped housing gives the seal rigidity and facilitates alignment at installation.
The lip is formed of a synthetic or leather. It is installed toward the pressure source.
Pressure forces the lip tighter around the shaft to aid sealing.
Spring helps sealing when there is no pressure.

Figure 4-12. Lip seals are used on rotating shafts.

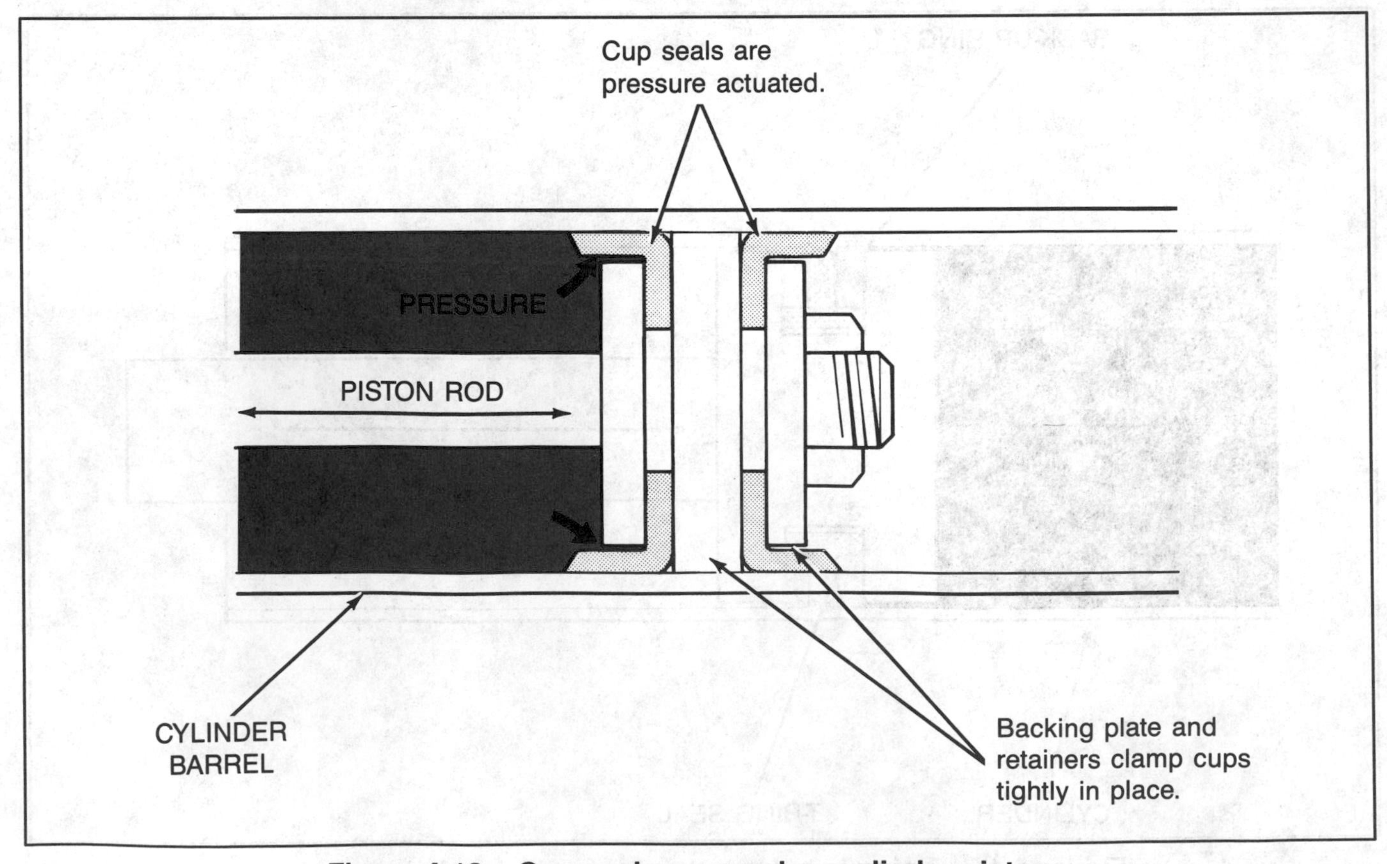

Cup seals are pressure actuated.
PRESSURE
PISTON ROD
CYLINDER BARREL
Backing plate and retainers clamp cups tightly in place.

Figure 4-13. Cup seals are used on cylinder pistons.

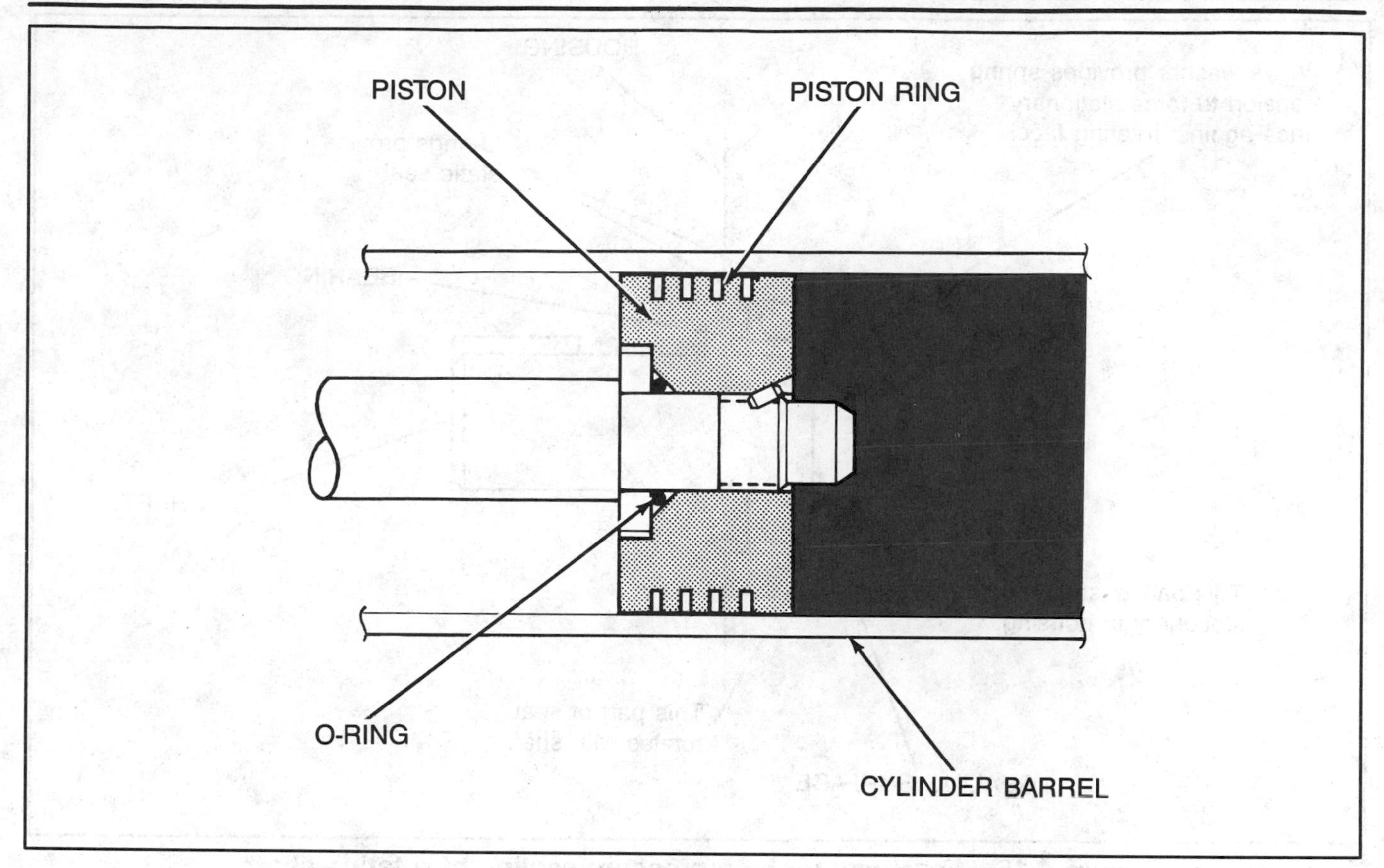

Figure 4-14. Piston rings are used for cylinder pistons.

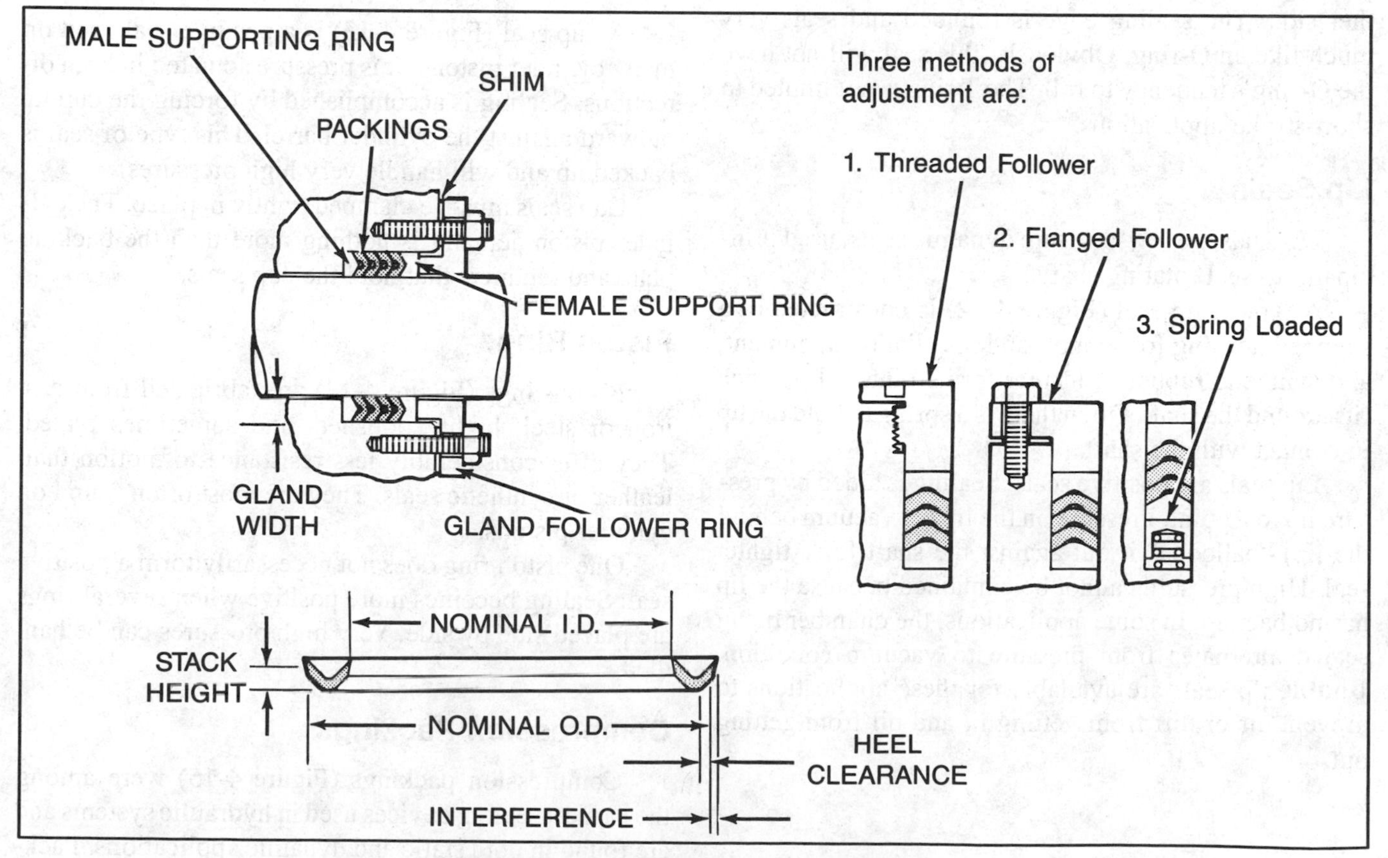

Figure 4-15. Compression packings.

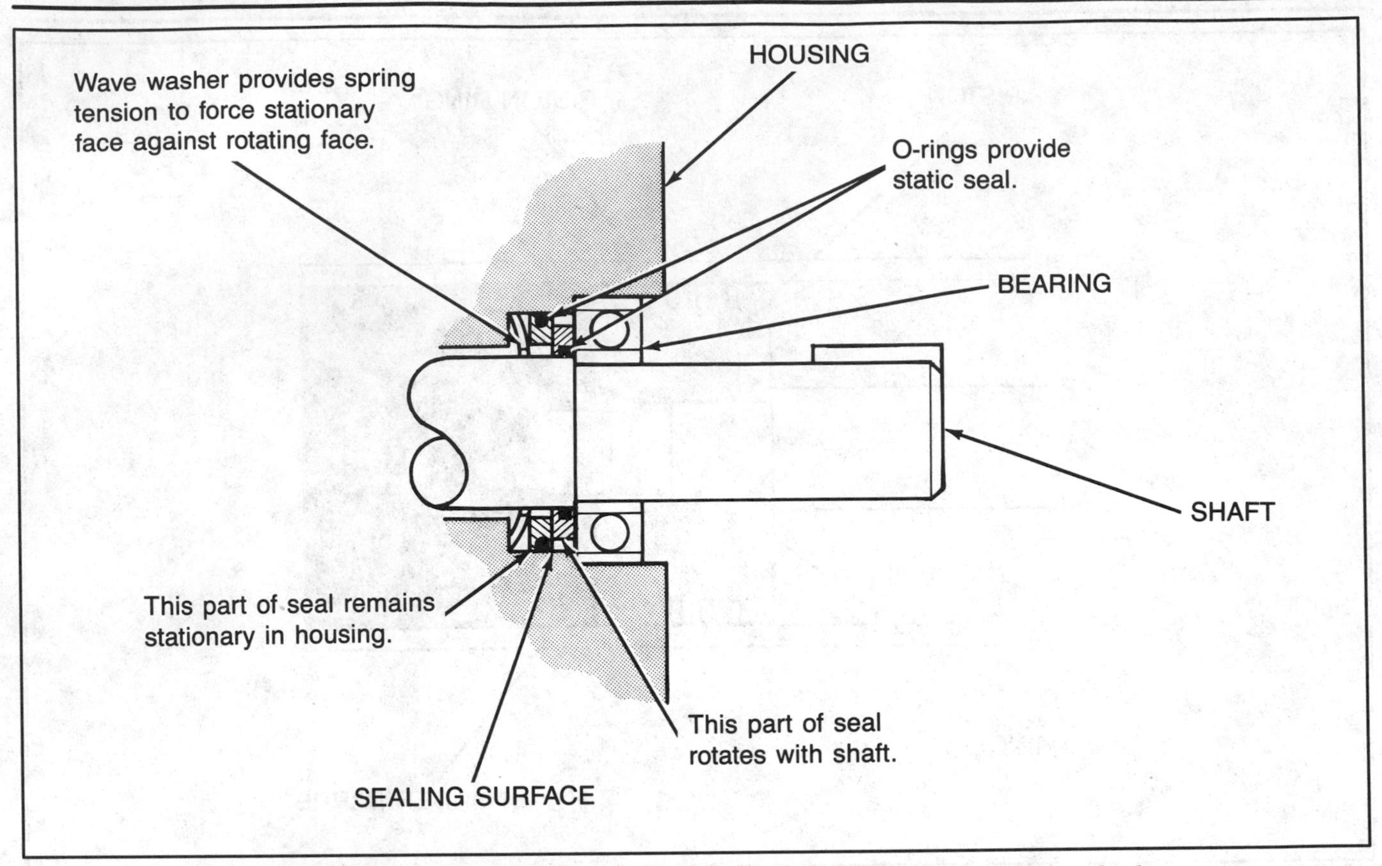

Figure 4-16. Face seal for high pressure sealing of rotating shaft.

the shape of a "T," and reinforced by backup rings on either side. The sealing edge is rounded and seals very much like an O-ring. Obviously, this seal will not have the O-ring's tendency to roll. The T-ring is not limited to short-stroke applications.

Lip Seals

Lip seals are low-pressure dynamic seals, used principally to seal rotating shafts.

A typical lip seal (Figure 4-12) is constructed of a stamped housing for support and installation alignment, and synthetic rubber or leather formed into a lip which fits around the shaft. Often there is a spring to hold the lip in contact with the shaft.

Lip seals are positive seals. Sealing is aided by pressure up to a point. Pressure on the lip (or vacuum behind the lip) "balloons" it out against the shaft for a tighter seal. High pressure cannot be contained because the lip has no backup. In some applications, the chamber being sealed alternates from pressure to vacuum condition. **Double** lip seals are available for these applications to prevent air or dirt from getting in and oil from getting out.

Cup Seals

A cup seal (Figure 4-13) is a positive seal used on many cylinder pistons. It is pressure actuated in both directions. Sealing is accomplished by forcing the cup lip outward against the cylinder barrel. This type of seal is backed up and will handle very high pressures.

Cup seals must be clamped tightly in place. The cylinder piston actually is nothing more than the backing plate and retainers that hold the cup seals.

Piston Rings

Piston rings (Figure 4-14) are fabricated from cast iron or steel, highly polished, and sometimes plated. They offer considerably less resistance to motion than leather or synthetic seals. They are most often found on cylinder pistons.

One piston ring does not necessarily form a positive seal. Sealing becomes more positive when several rings are placed side by side. Very high pressures can be handled.

Compression Packings

Compression packings (Figure 4-15) were among the earliest sealing devices used in hydraulic systems and are found in both static and dynamic applications. Pack-

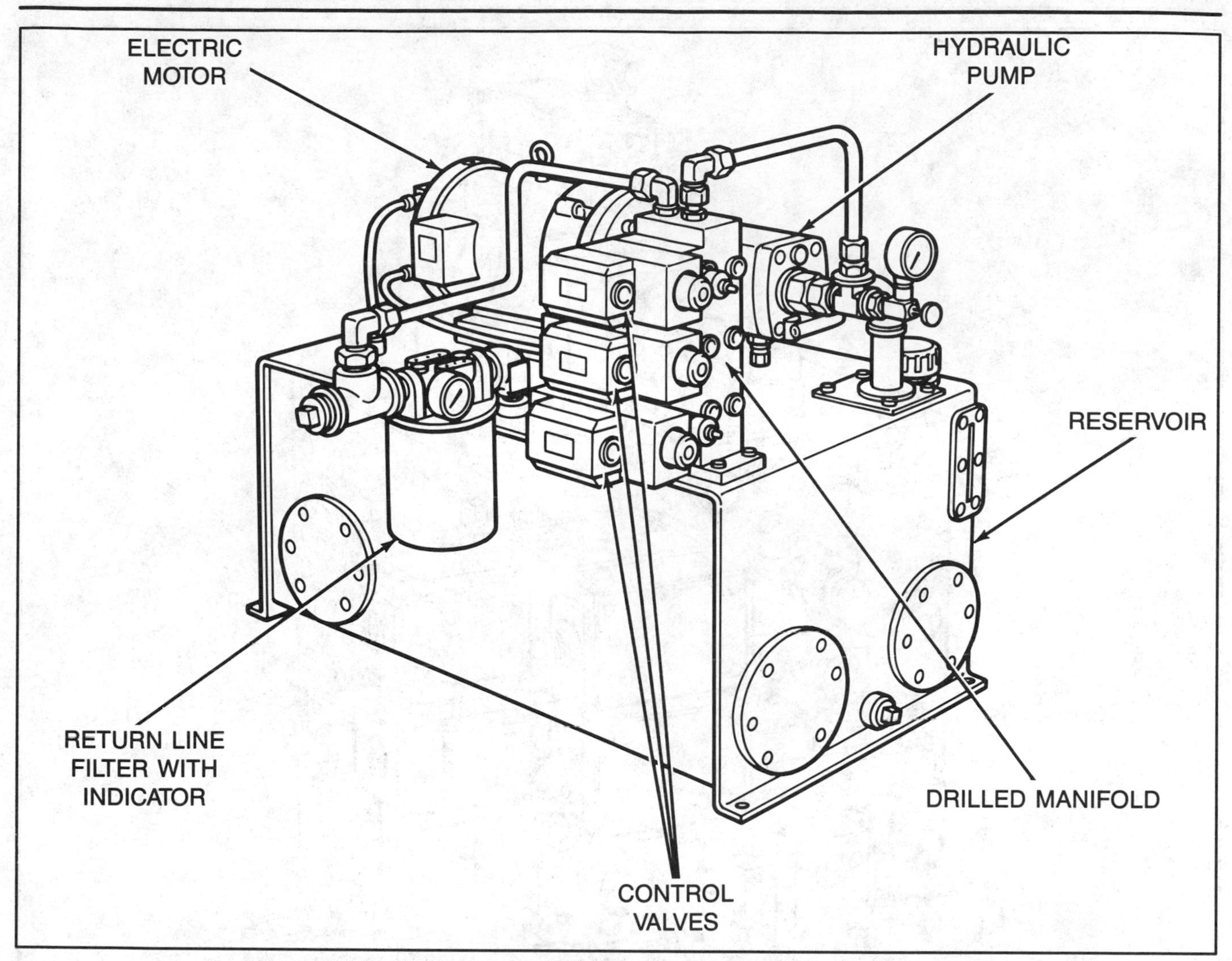

Figure 4-17. Back-mounting leaves connections undisturbed.

ings are being replaced in most static applications by O-rings.

Most packings in use today are molded or formed into "U" or "V" shapes, and multiple packings are used for more effective sealing. The packings are compressed by tightening a flanged follower ring against them. Proper adjustment is critical because excessive tightening accelerates wear. In some applications, the packing ring is spring-loaded to maintain the correct force and take up wear.

Face Seals

Face seals (Figure 4-16) are used in applications where a high-pressure seal is required around a rotating shaft. Sealing is accomplished by constant contact between two flat surfaces, often hard carbon and steel. The stationary sealing member is attached to the body of the component. The other is attached to the shaft and turns against the stationary member. One of the two parts is usually spring-loaded to improve contact initially and to take up wear. Pressure increases the contact force and tightens the seal.

Gaskets

Gaskets are flat sealing devices, usually fabricated in the shape of the flat mating surfaces to be sealed. Early designs of connection flanges and surface-mounted valves were sealed with gaskets. Today they have been largely replaced in hydraulic equipment by O-rings or formed packings.

SEAL MATERIALS

Leather, cork, and impregnated fibers were the earliest sealing materials for hydraulic equipment. They were used extensively until after the development of synthetic rubber during World War II. Natural rubber is seldom used as a sealing material because it swells and deteriorates in the presence of oil.

Synthetic rubbers (elastomers), however, are for the

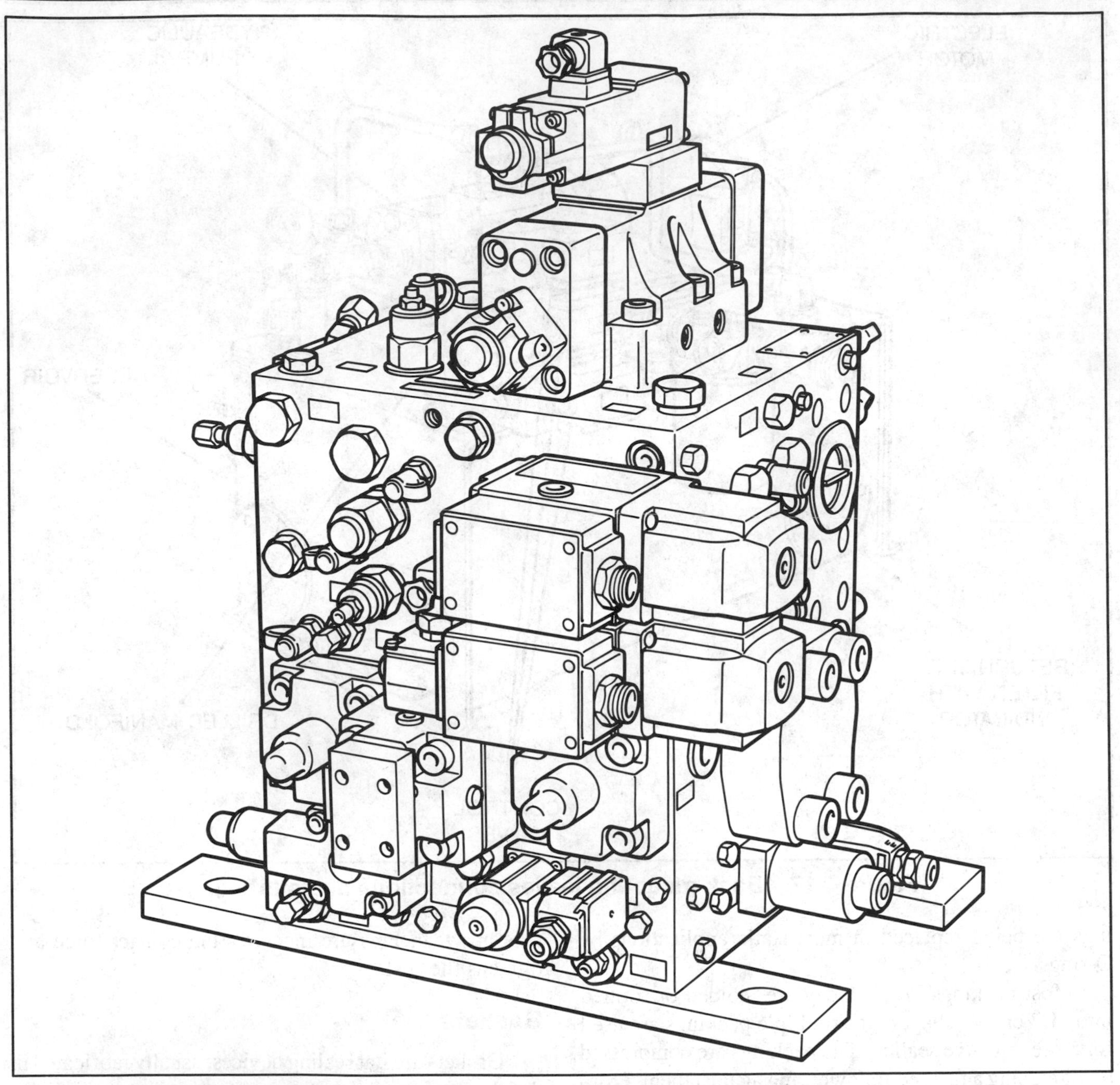

Figure 4-18. Drilled manifold contains interconnecting passages to eliminate piping between valves.

most part quite compatible with oil. Elastomers can be made in many compositions to meet various operating conditions. Most of the hydraulic equipment seals today are made of one of these elastomers: Nitrile (Buna-N), chloroprene (Neoprene), Teflon, EPR/EPDM (also known as EPM), or silicone.

Nitrile

The elastomer Nitrile (Buna-N) is a widely used sealing material in modern hydraulic systems. It is moderately tough, wears well, and is inexpensive. There are a number of compositions compatible with petroleum oil—most of them easily molded into any required seal shape.

Nitrile has a reasonably wide temperature range, retaining its sealing properties from -40°F to 250°F (-40°C to 121.11°C). At moderately high temperatures, it retains its shape in most petroleum oils where other materials tend to swell. It does swell, however, in some synthetic fluids.

Chloroprene

One of the earliest elastomers used in hydraulic system sealing was chloroprene. A tough material, it still is used in systems using petroleum fluids.

Plastics, Fluoro-Plastics and Fluoro-Elastomers

Several sealing materials are synthesized by combining fluorine with an elastomer or plastic. They include Kel-F, Viton A, and Teflon. Nylon is another synthetic material with similar properties. It is often used in combination with the elastomers to give them reinforcement. Both nylon and Teflon are used for backup rings as well as sealing materials. All have exceptionally high heat resistance (to 500°F) (260°C) and are compatible with most fluids.

Silicone

Silicone is an elastomer with a much wider temperature range than Nitrile, and is, therefore, a popular material for rotating shaft seals and static seals in systems that run from very cold to very hot. It retains its shape and sealing ability to -60°F (-51.11°C) and is generally satisfactory to 400 or 500°F (204.44 or 260°C).

At high temperatures, silicone tends to absorb oil and swell. This, however, is no particular disadvantage in static applications. Silicone is not used for reciprocating seals, because it tears and abrades too easily. Silicone seals are compatible with most fluids; even more so with fire-resistant fluids than petroleum.

PREVENTING LEAKAGE

The three general considerations in preventing leakage are:

1. Design to minimize the possibility (back, gasket or subplate mounting).
2. Proper installation.
3. Control of operating conditions.

Anti-Leakage Designs

It has already been noted that designs using straight thread connectors and welded flanges are less susceptible to leakage than pipe connections. **Back-mounting** of valves with all connections made permanently to a mounting plate has made a great difference in preventing leakage and in making it easier to service a valve (Figure 4-17). Many valves being built today are the back-mounted design. (The term gasket-mounted was originally applied to this design because gaskets were used on the first back-mounted valves. Gasket-mounted or subplate-mounted is still used to refer to back-mounted valves sealed by O-rings.)

A further advance from back-mounting is the use of manifolds (Figure 4-18). Some are drilled and some combine mounting plates with passage plates (sandwiched and brazed together), providing interconnections between valves and eliminating a good deal of external plumbing.

Proper Installation

Installation recommendations were covered earlier in this chapter. Careful installation, with attention to avoiding pinching or cocking a seal, usually assures a leak-proof connection. Manufacturers often recommend a special driver for inserting lip-type shaft seals to be certain they are installed correctly. Vibration and undue stress at joints, which are common causes of external leakage, also are avoided by good installation practice.

Operating Conditions

Control over operating conditions can be very important to seal life. A number of factors that can help prevent leakage are discussed below.

Contamination Prevention. An atmosphere contaminated with moisture, dirt, or any abrasive material shortens the life of shaft seals and piston rod seals exposed to the air. Protective devices should be used in contaminated atmospheres. Equally important are clean fluid and proper filtration to avoid damage to internal seals and surfaces.

Fluid Compatibility. Some fire-resistant fluids attack and disintegrate certain elastomer seals. Few seals, in fact, are compatible with all fluids. The fluid supplier should always be consulted when in doubt whether to change seals when a change is made in the type of fluid. (See Chapter 3) Fluid additives (added by the machine user) also may attack seals and should be used only at the recommendation of the fluid supplier. See Appendix C for a chart showing the compatibility of various materials with hydraulic fluids.

Temperature. At extremely low temperatures, a seal may become too brittle to be effective. At too high a temperature, a seal may harden, soften, or swell. The operating temperature should always be kept well within the temperature range of the seals being used.

Pressure. Excess fluid pressure puts an additional strain on oil seals and may "blow" a seal causing a leak.

Lubrication. No seal should ever be installed or operated dry. All must be lubricated prior to installation or the seal will wear quickly and leak.

Appendix E contains leakage troubleshooting analysis charts as well as procedures for correcting leaks once the cause has been determined.

QUESTIONS

1. How is a pipe size classified?
2. What is the schedule number of standard pipe?
3. How does a pipe thread seal?
4. What advantages does tubing have over pipe?
5. To what does the specified size of tubing refer?
6. How are tubing connections sealed?
7. How does a flexible hose contain pressure?
8. Give two reasons for fluid conductor supports.
9. What is a positive seal?
10. What is a static sealing application?

CHAPTER 5
RESERVOIRS

The main function of the reservoir in a hydraulic system is to store and supply hydraulic fluid for use by the system. The chapter discusses this and other reservoir functions such as heat exchange and deaeration. Reservoir components, including air breathers, baffle plates and filler caps are also described. A general discussion on reservoir sizing and reservoir designs is included along with explanations of various fluid temperature conditioning devices.

FUNCTIONS OF A RESERVOIR

Since, in addition to holding the system fluid supply, a reservoir can also serve several secondary functions, some system designers feel that the reservoir is the key to effective hydraulic system operation. Some examples of these functions are discussed below.

By transferring waste heat through its walls, the reservoir acts as a heat exchanger that cools the fluid within. As a deaerator, the reservoir allows entrained air to rise and escape while solid contaminants settle to the bottom of the tank, making it a fluid conditioner.

These are functions that can also be provided to the system by methods that do not involve the reservoir.

In some instances, the reservoir may be used as a platform to support the pump, motor, and other system components. This saves floor space and is a simple way to keep the pumps and valves at a good height for servicing.

RESERVOIR COMPONENTS

A typical industrial reservoir (Figure 5-1) is constructed of welded steel plate with end-plate extensions that support the unit. To reduce the chance of condensed moisture within the tank causing rust, the inside of the reservoir is painted with a sealer that is compatible to the fluid being used. Because the reservoir is designed for easy fluid maintenance, a plug is placed at a low point on the tank to allow complete drainage.

The various components that make up a reservoir are discussed in the following sections.

Oil Level Gage

To check the fluid level in the reservoir, either a sight glass or two small portholes are installed in the clean-out plates. This allows the upper and lower fluid limits to be checked without exposing the reservoir to the contamination that can occur when using a dipstick.

Breather Assembly

A vented breather cap is installed to accommodate the air exchange that results from the constant change of pressure and temperature within the tank. As the hydraulic cylinders extend and retract, air is taken in and expelled through this filter. Generally, the breather must be large enough to handle the airflow required to maintain atmospheric pressure, whether the tank is empty or full (the higher the flow rate, the larger the breather).

On a pressurized reservoir, the breather is replaced by an air valve that regulates the tank pressure between preset limits. An oil bath air filter is sometimes used in atmospheres that are exceptionally dirty.

Filler Opening

The filler opening is often a part of the breather assembly. The opening has a removable screen that keeps contaminants out of the tank when fluid is being added to the reservoir. A cap that will provide a tight seal should be chained to the reservoir.

Another type of filler opening is a quick-disconnect fitting, screwed into a pipe that extends to within a few inches of the bottom of the tank. A portable oil cart, equipped with a small pump and filter, supplies fluid to the tank. This keeps the new fluid clean and prevents contamination of the reservoir.

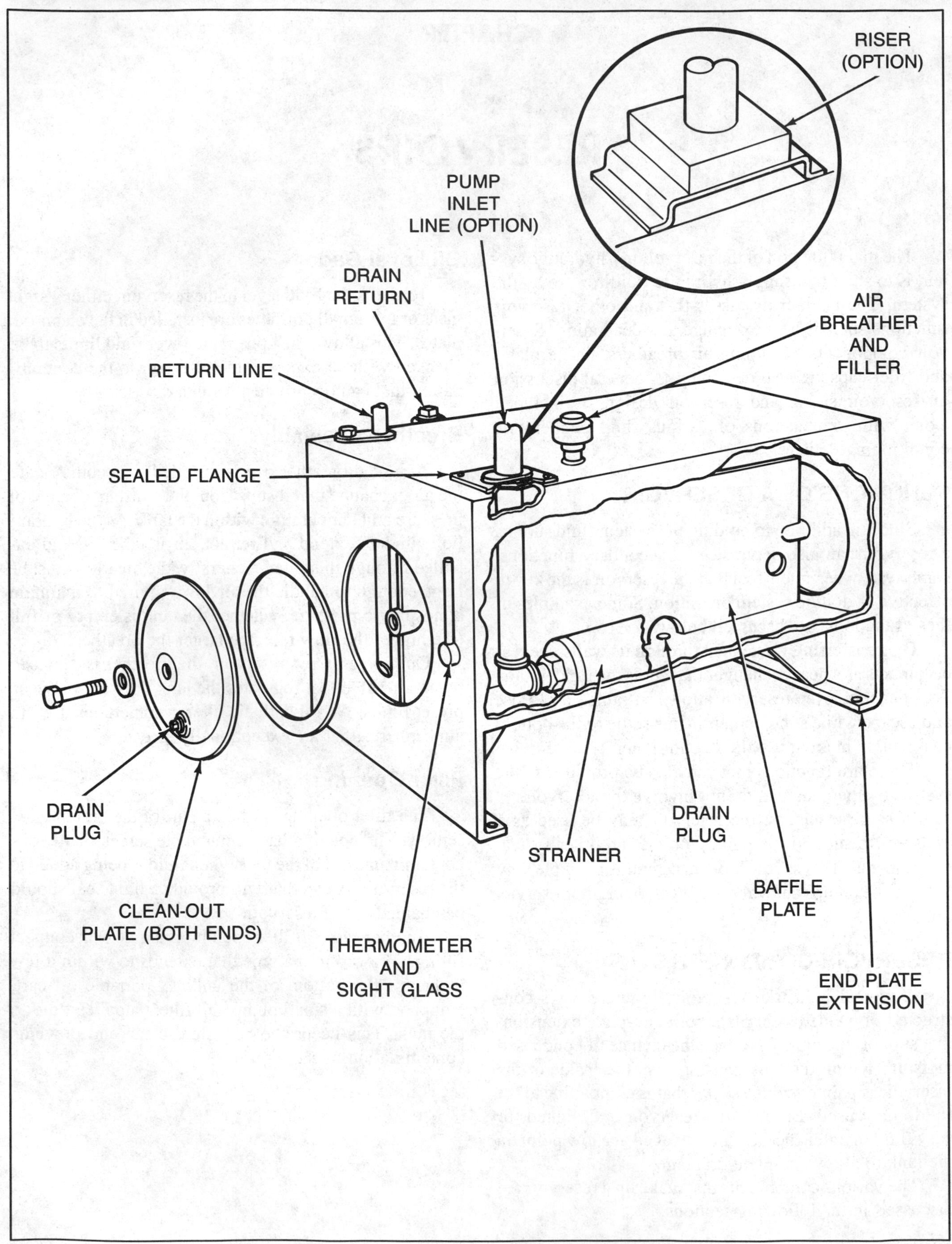

Figure 5-1. Typical industrial reservoir.

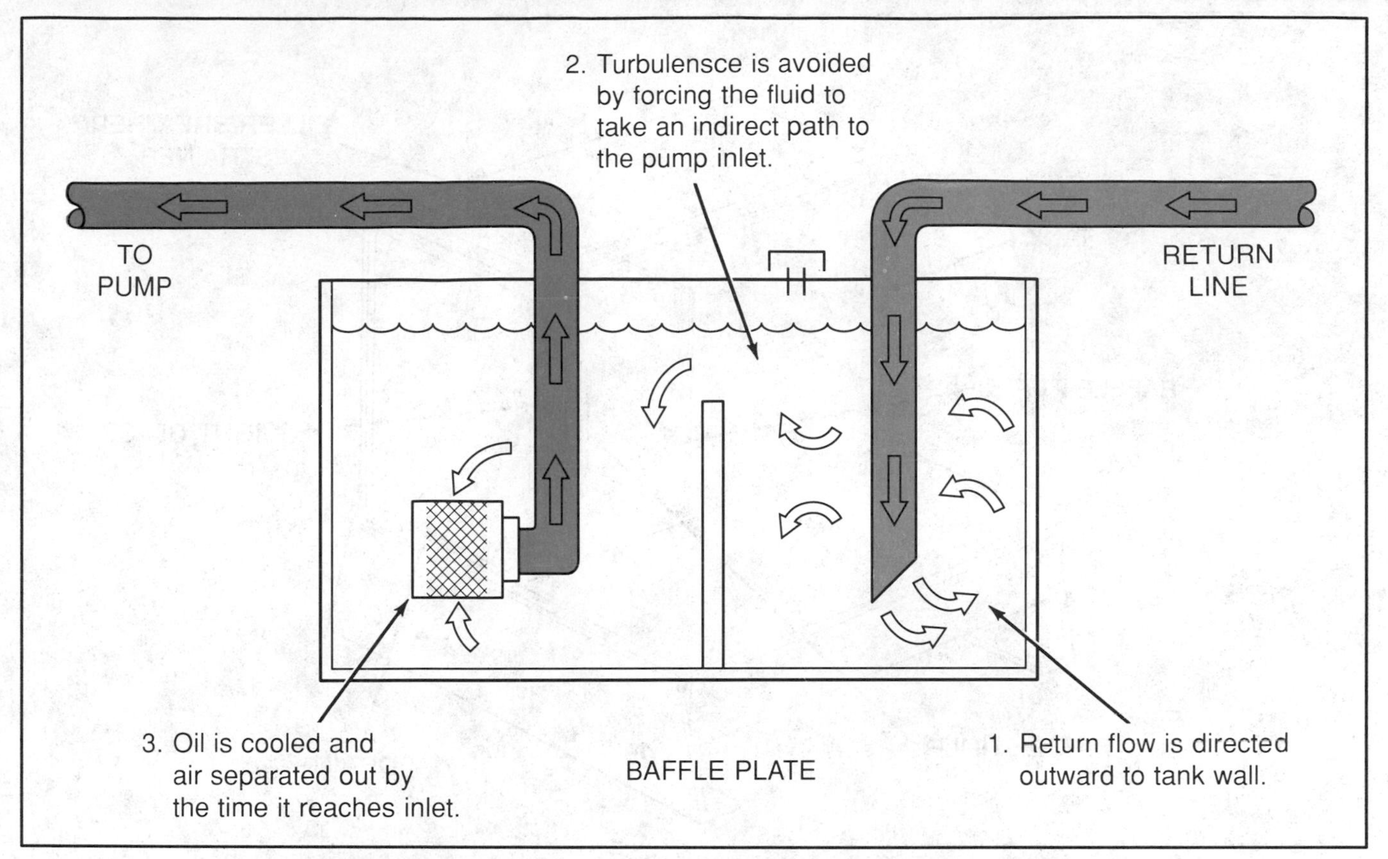
2. Turbulensce is avoided by forcing the fluid to take an indirect path to the pump inlet.
TO PUMP
RETURN LINE
3. Oil is cooled and air separated out by the time it reaches inlet.
BAFFLE PLATE
1. Return flow is directed outward to tank wall.

Figure 5-2. Baffle plate controls direction of flow in tank.

Clean-Out Plates

Clean-out plates are usually installed on both ends of the tank. This is especially true on reservoirs sized above ten gallons (37.85 liters). The plates are easily removed and large enough to provide complete access when the interior of the reservoir is being cleaned or painted.

Baffle Plate

Because fluid returning to the reservoir is usually warmer than the supply fluid and probably contains air bubbles, baffles are used to prevent the returning fluid from directly entering the pump inlet. A baffle plate (Figure 5-2) is installed lengthwise through the center of the tank, forcing the fluid to move along the reservoir walls, where much of the heat is dissipated to the outer surfaces of the reservoir. This long, low-velocity travel also allows the contaminants to settle at the bottom of the tank and provides an opportunity for the fluid to be cleared of any entrained air. The end result is less turbulence in the tank.

Line Connections and Fittings

Most lines leading to the reservoir terminate below the oil level. The line connections at the tank cover are often packed (sealed), slip-joint type flanges. This design prevents contaminants from entering through these openings and makes it easy to remove inlet line strainers for cleaning.

Connections made on the top of the reservoir are often set on risers to keep them above dirt and other contaminants that may collect on the reservoir.

To prevent the hydraulic fluid from foaming and becoming aerated, pump inlet lines must terminate below the fluid level, usually two inches (50 mm) from the bottom of the tank.

Valve drain lines may terminate above the fluid level, but it is generally better to extend them approximately two inches (50 mm) below the fluid level. In all cases, pump and motor drain lines must terminate below the lowest possible fluid level.

Lines that terminate near the tank bottom and are not equipped with strainers should be cut at a 45-degree angle. This prevents the line opening from bottoming in the tank and cutting off flow. On a return line, the angled opening is often positioned so that flow is directed at the tank walls and away from the pump inlet line.

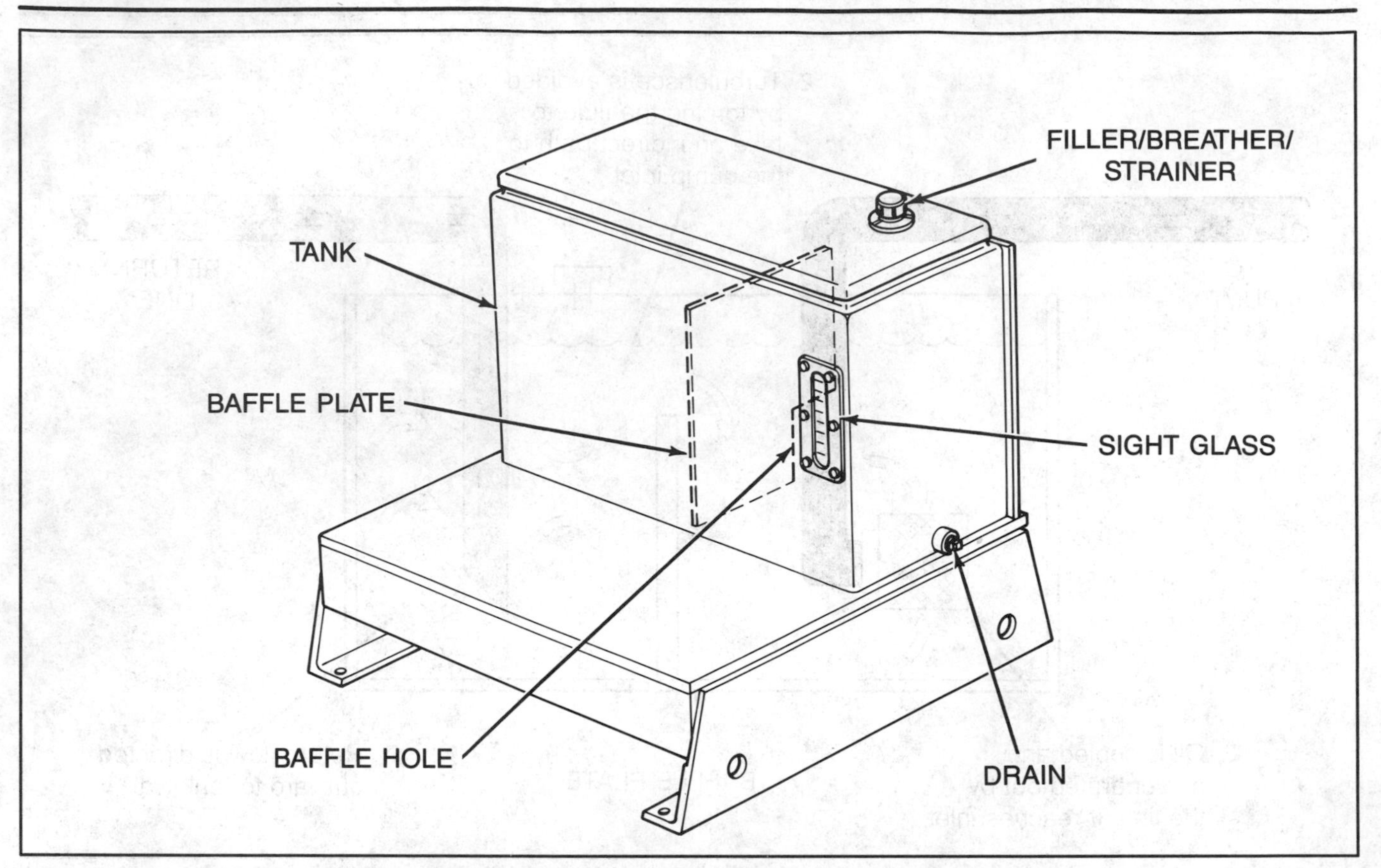

Figure 5-3. L-shaped reservoir design.

STANDARD RESERVOIR DESIGNS

In addition to the proprietary styles offered by manufacturers, three types of standard reservoir designs are commonly used today.

JIC Reservoirs

The Joint Industry Conference (JIC) reservoir design (Figure 5-1), sometimes referred to as Flat Top, is a horizontal tank with extensions that hold it several inches off the floor or the surface of a drip pan. This design permits increased air circulation and heat transfer from the bottom as well as from the walls of the tank.

JIC reservoirs are usually as deep as they are wide, with the length approximately twice the width. They are generally made from 9 or 11 gauge pickled and oiled steel. Single-bolt clean-out plates at each end provide access for cleaning. The bottom of the tank is concave and has a drain plug at its lowest point.

Filtered hydraulic fluid is pumped into the reservoir through a filler/breather cap assembly equipped with a fluid strainer. A sight-level glass that usually incorporates a thermometer is installed on one or two of the reservoir walls.

L-Shaped Reservoir

Another standard reservoir design is the L-shaped configuration (Figure 5-3), which consists of a vertical tank mounted to one side of a wide base. The other side of the base is used to mount the pump, motor, relief valve, and if needed, the heat exchanger. Because the fluid level in the tank is higher than the pump inlet, positive inlet pressure is maintained, minimizing the possibility of cavitation and loss of pumping action.

Pressure control and directional valves can be mounted to the side of the tank at the vertical section above the fluid level. If subplate valves are used, nearly all the piping will be inside the reservoir. This arrangement minimizes loss of fluid due to leaks.

The reservoir interior is accessed through a hinged top, which also permits a visual check of the returning fluid while the circuits are in operation. This can aid in system troubleshooting, when it is required.

The L-shaped design provides large surface areas for cooling. To promote air circulation, the base of the reservoir should be raised several inches off the floor.

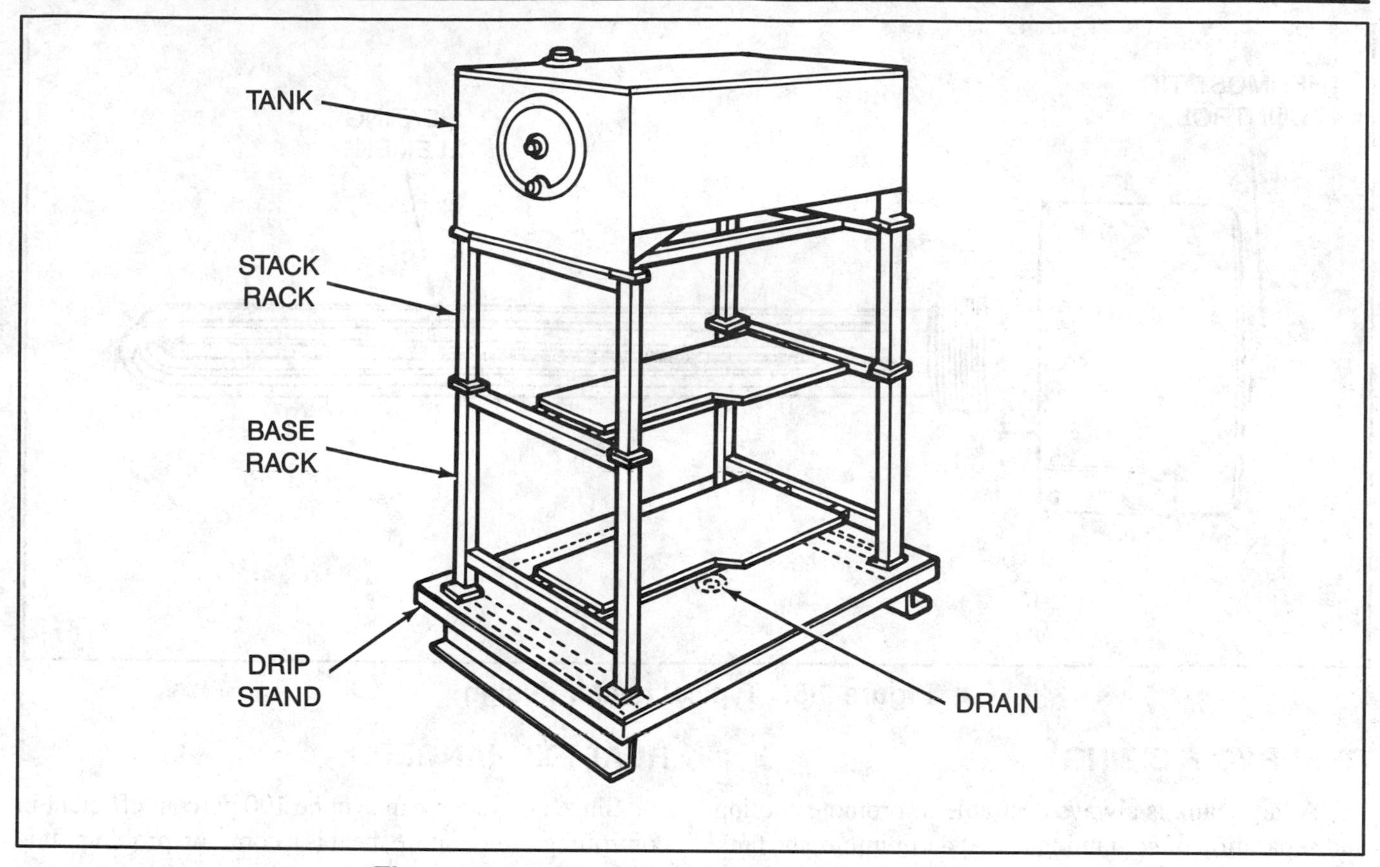

Figure 5-4. Overhead stack reservoir design.

Overhead Stack Reservoir

Another reservoir design concept is the overhead stack (Figure 5-4), which makes use of one or more modular rack-type frames stacked vertically with a standard horizontal tank at the top. Each frame is usually set up with one pump-motor assembly, with all pumps drawing fluid from the common reservoir at the top.

Because HWC fluids have relatively high density and low viscosity, they should be supplied under positive pressure to prevent pump cavitation. Most manufacturers supply the overhead reservoir as a matter of good hydraulic practice whenever HWCF is specified.

Although there is no limit to the number of stacking levels that can be used, safety and servicing problems may occur when the stack exceeds three levels (two frames plus the tank at the top).

Because the positive head condition developed is also beneficial to the operation of conventional oil units, the use of the stack configuration in non-HWCF applications is growing. The elevated design makes it possible to drain the reservoir easily, without having to pump the fluid out.

The vertical arrangement conserves floor space while providing an economy of scale. Two or more pump-motor assemblies with controls can be incorporated in the rack mounting with a common reservoir. The overhead reservoir also provides the capability for an HWCF changeover if necessary.

Reservoir Modifications

Here are some points to keep in mind for cases where these standard reservoir designs cannot be used or modified and a custom design must be developed.

- Make certain the reservoir has ample clean-out openings that will be accessible when the reservoir is in position.
- Be sure the walls and top of the reservoir are strong enough to support any equipment that may be mounted on it.
- Size the reservoir for at least 20 percent overcapacity to provide a reserve against unexpected demands on system capacity.
- Be sure the reservoir is provided with a means of filling, a level gauge, and a drain connection.

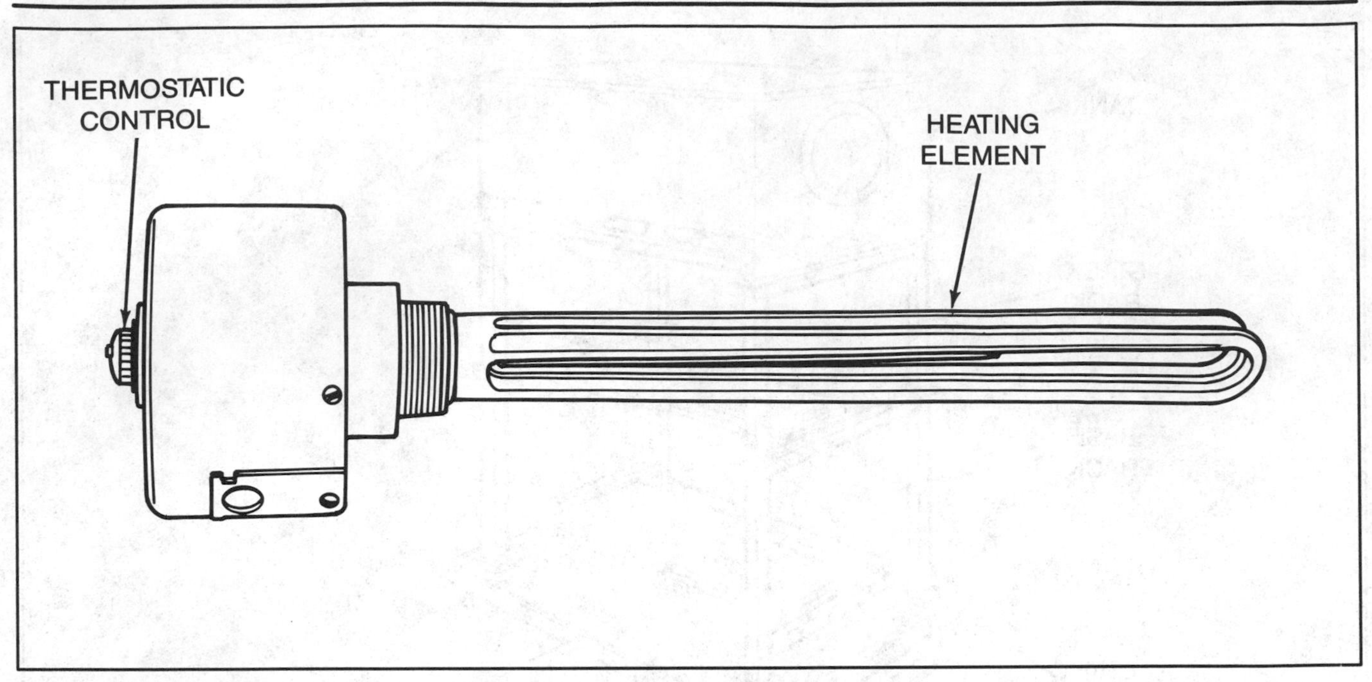

Figure 5-5. Typical heater design.

RESERVOIR SIZING

A large tank is always desirable to promote cooling and separation of contaminants. At a minimum, the tank must store all the fluid the system will require and maintain a high enough level to prevent a whirlpool effect at the pump inlet opening. If this occurs, air will be taken in with the fluid.

When determining reservoir size, it is important to consider the following factors:

- Fluid expansion caused by high temperatures.
- Changes in fluid level due to system operation.
- Exposure of the tank interior to excess condensation.
- The amount of heat generated in the system.

For industrial use, a general sizing rule is used:

Tank size (gallons) = pump gpm × 2
or pump gpm × 3

Tank size (litres) = pump LPM × 2
or pump LPM × 3

In mobile or aerospace systems, the benefits of a large reservoir may have to be sacrificed because of space limitations.

HEAT EXCHANGERS

Since no system can ever be 100 percent efficient in controlling temperature, heat is a common problem. It is customary to think of cooling when the fluid must be temperature conditioned, but there are some applications where the fluid must be heated. For example, fluids with a low viscosity index will not flow readily when cold and must be kept warm by heaters.

Although the design of the circuit has considerable effect upon the fluid temperature, heat exchangers are sometimes required when operating temperatures are critical or when the system cannot dissipate all the heat that is generated. The three types—heaters, air coolers, and water coolers—are discussed in the following sections.

Heaters

A heater (Figure 5-5) is installed in the reservoir, below the fluid level and close to the pump inlet. The low-heat density design (10 watts per square inch) (1.55 watts per square centimeter) prevents the fluid from burning and a thermostatic control is usually attached.

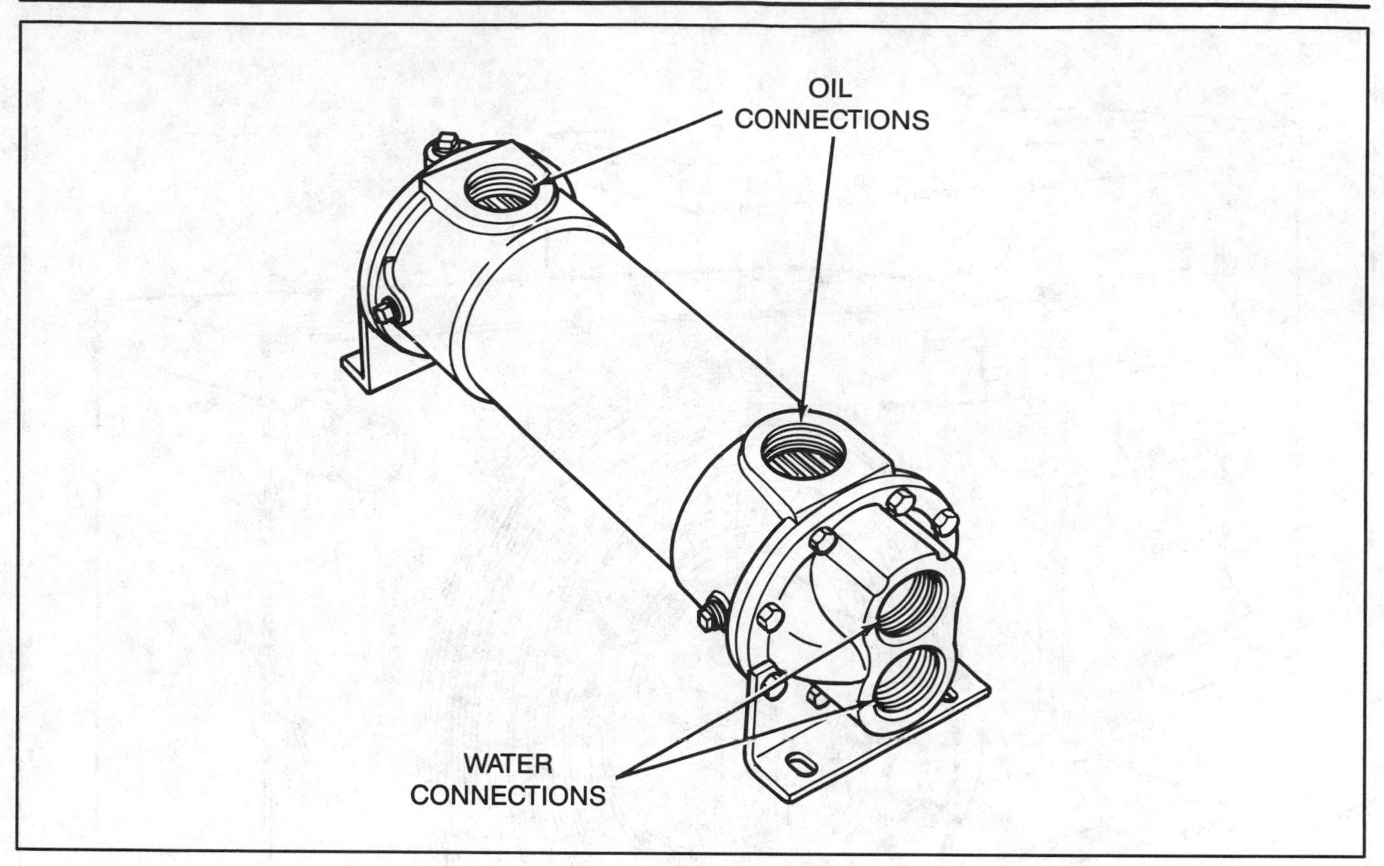

Figure 5-6. Typical water cooler design.

Water Coolers

In a typical water cooler (Figure 5-6), hydraulic fluid is circulated through the unit and around the tubes containing the water. The heat is removed from the hydraulic fluid by the water, which can be regulated thermostatically to maintain a desired temperature. The water is filtered to prevent the cooler from clogging. By circulating hot rather than cold water, this type of temperature control can also be used as a heater.

Water flow requirements are usually equal to between 1/4 and 1/3 of the system oil flow. The availability of cooling towers and water recycling reduces the cost, but water taken from these sources usually has a higher temperature than that of municipal systems.

Air Coolers

An air cooler (Figure 5-7) is used when water is not readily available for cooling. The fluid is pumped through tubes that are bonded to fins made of aluminum or some other metal that transfers heat to the outside air. The cooler usually has a blower to increase the heat transfer.

Air coolers are less efficient than water coolers and tend to be ineffective in areas of high ambient air temperatures. The initial installation cost is higher than that of water coolers, but the operational costs are usually less.

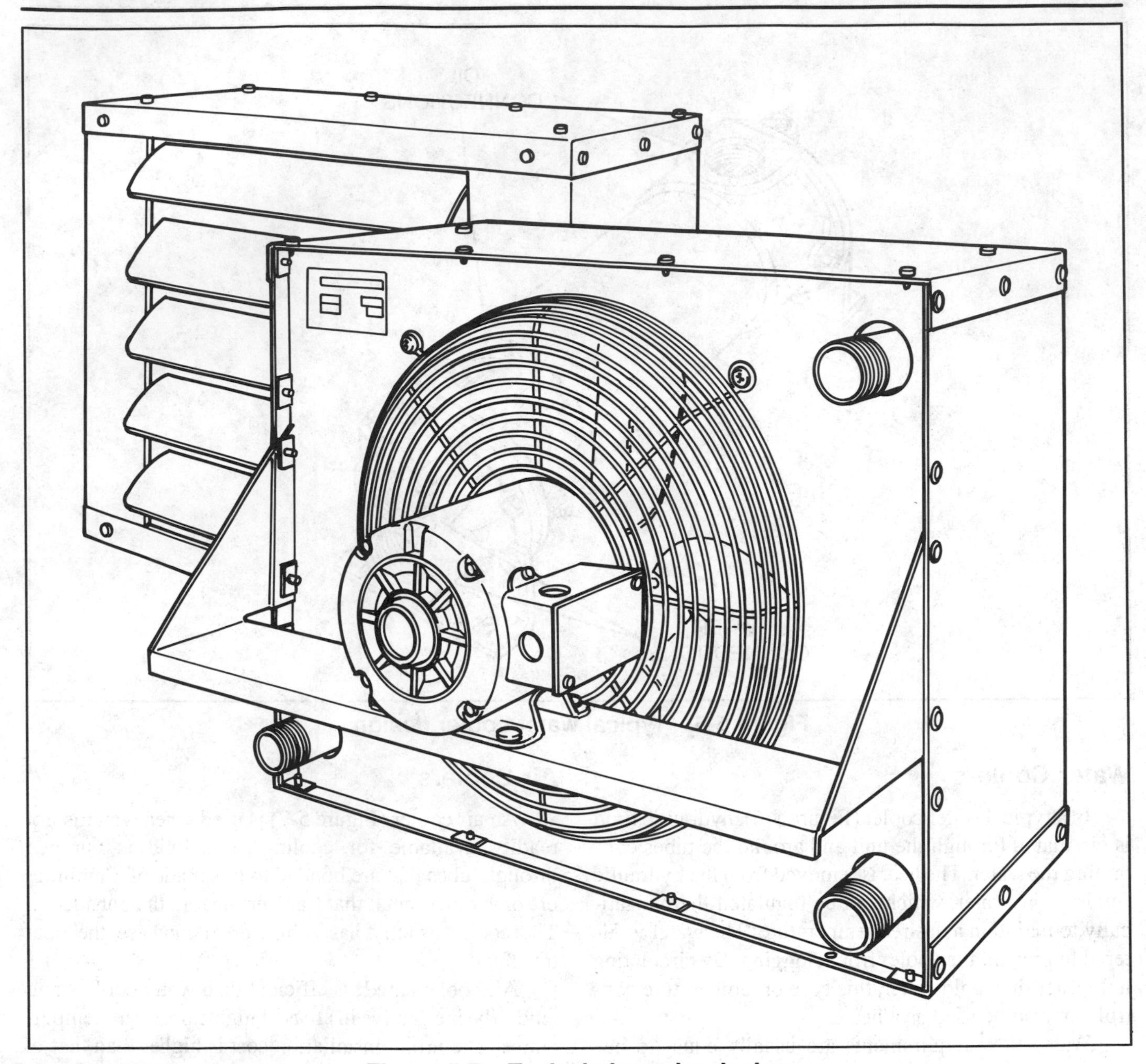

Figure 5-7. Typical air cooler design.

QUESTIONS

1. Name three possible functions of the reservoir.
2. Where should the reservoir drain plug be located?
3. What are the three most common standard reservoir designs?
4. What is the most desirable method of checking fluid level in the reservoir?
5. What is the purpose of the reservoir breather?
6. What is the purpose of using a riser with reservoir connections?
7. What is a reservoir baffle plate used for?
8. Why is a return line often cut at a 45-degree angle?
9. What would probably be an adequate size reservoir for a system with a 5 gpm pump?
10. What is the recommended heat density of a heater?

CHAPTER 6
CONTAMINATION CONTROL

Contamination control is both a relatively new engineering science and a well established, much practiced art among hydraulics personnel. Although a great deal is known about the prevention and control of contaminant buildup, it is estimated that over 80 percent of hydraulic system failures are due to poor fluid condition.

Obviously, the proper care of hydraulic fluid, both within a system and while stored and handled, has an important effect on machine performance and system component life. The purpose of this chapter is to provide you with information that leads to improved control of contamination in the systems you maintain and operate. In other words, the goal is to reduce the amount of harmful solid particles that are in hydraulic fluids. To accomplish this task, this chapter covers information on the effects of contamination on hydraulic machinery operation and life, how contamination is measured, component dirt tolerance ratings and how to use them, sources of contamination, how to prevent contamination during assembly and servicing, how to establish desired cleanliness levels, and how filters are specified and constructed. The primary focus of this chapter is the removal of particulate contamination from hydraulic systems.

THE EFFECT OF CONTAMINATION ON HYDRAULIC MACHINERY OPERATION AND LIFE

Several factors must be considered when examining the effect of contamination on the life and operation of hydraulic machinery. These factors include how contamination affects the function of hydraulic fluid, mechanical tolerances in components, defining the type of contamination, and how components fail when a system contains contamination.

The Functions of Hydraulic Fluid

Recall from an earlier chapter that the four primary functions of hydraulic fluid are to:

- Transmit power
- Cool or dissipate heat
- Lubricate moving parts
- Seal clearances between parts.

Solid contamination interferes with the first three of these functions. It interferes with power transmission by blocking or plugging small orifices in devices such as pressure valves and flow control valves. The action of a valve affected by solid contamination in this way is unpredictable and unsafe. A recent study conducted by M.I.T. found mechanical wear was responsible for 50% of surface degradation. The study also found surface degradation responsible for 70% of over-all loss of machine usefulness (See Figure 6-1).

Contamination can form a sludge on the reservoir walls and interfere with the cooling processes. This sludge impedes heat transfer from the fluid to the wall which will eventually result in higher system operating temperatures. Many times the combined affect of several types of contamination damage will cause the heat generating capacity of a normally operating unit to rise significantly.

The most serious effect that contamination can have on a system is when it affects the lubricating ability of hydraulic fluid. This can occur in several different ways. Very fine particles, smaller than a component's mechanical clearances, can collect in the clearance and eventually block the flow of lubricating fluid into the small space between moving parts and otherwise interfere with the component operation. This accumulation of very fine particulate contamination is known as silting.

The silting of valves prevents them from shifting. It can also interfere with the pump operation by causing the degradation of a pumping unit or rotating group which erodes efficiencies, and degrades pump performance. Contamination also affects actuators, and (as in the case of hydraulic motors) can cause degradation in performance efficiency or motor failure. In rotary actuators, the seals and mating dynamic surfaces can be damaged by the presence of contamination. The sealing mechanisms between the opposing sides of the pistons of cylinders can also be destroyed by contamination.

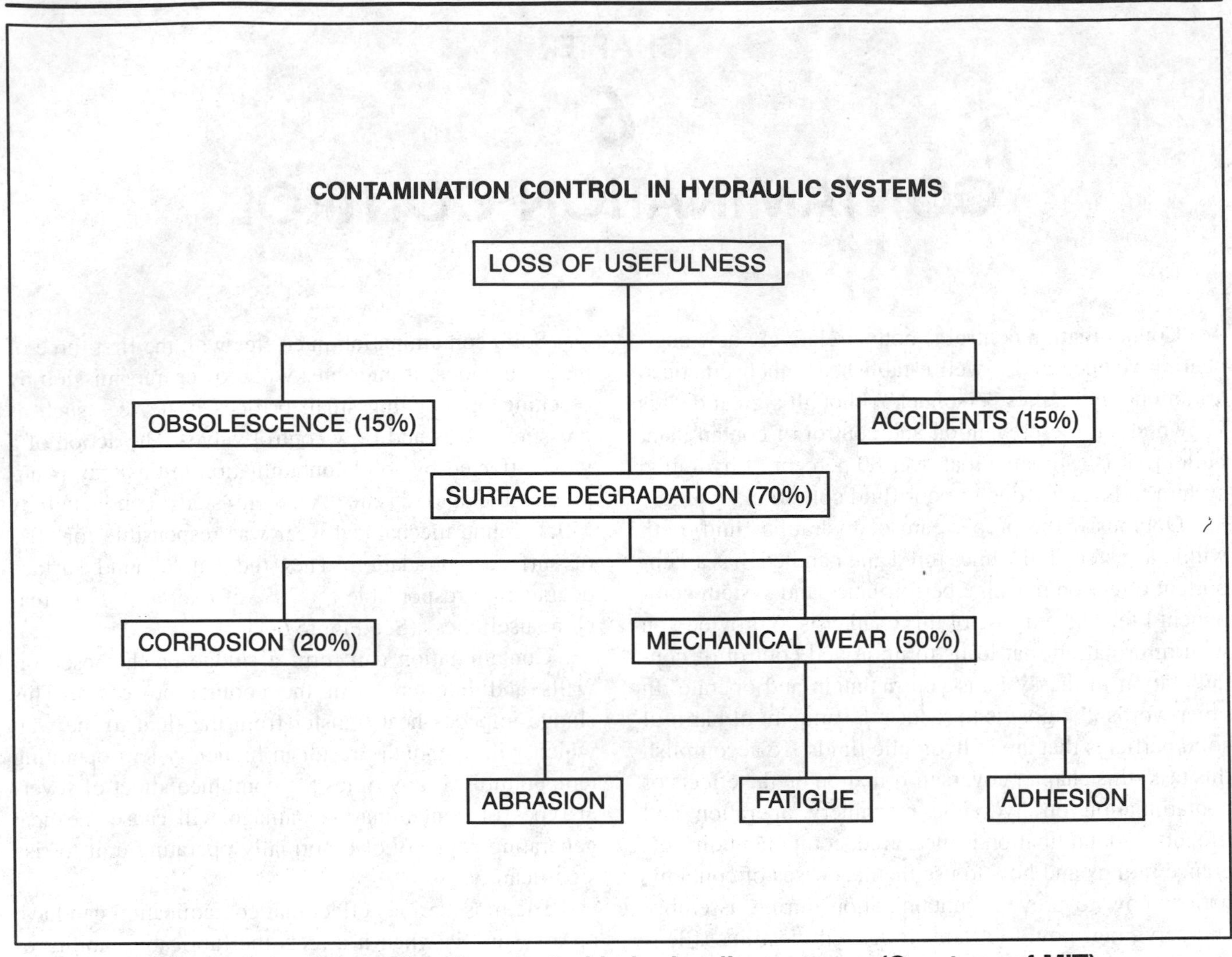

Figure 6-1. Contamination control in hydraulic systems. (Courtesy of MIT)

When the particles are about the same size as the clearance they must pass through, they rub against the moving parts, break down the film of lubricant, and cause the most wear and damage to the component surfaces (Figure 6-2). Wear generates more contaminants, increases leakage, lowers efficiency, and generates heat. The higher the pressure, the greater the problem. Wear takes many different forms.

Mechanisms of Wear

Internal mechanisms of wear include: abrasion, erosion, adhesion, fatigue, cavitation, corrosion, and aeration (Figure 6-3).

Abrasion. Solid particles in the hydraulic fluid grinding between moving surfaces is referred to as abrasion. Abrasion damages the surfaces and can create additional abrasive particles.

Abrasion is categorized into three classes (see Figure 6-4). This first abrasion class is referred to as one body abrasion. This type of abrasion is said to have occurred when particulate contamination damages a surface through contact between the fluid and the surface. A two body type abrasion occurs when an abrasive particle is embedded in a surface making contact with a second surface. If the hard asperity of one surface contacts a softer surface, a two body abrasion is also said to have occurred. Three body type abrasion refers to loose abrasive particles making contact with two surfaces.

Erosion. Erosion is similar to abrasion. Erosion occurs when high velocity particles strike surfaces, causing damage.

Adhesion. When metal contacts metal adhesion may occur..

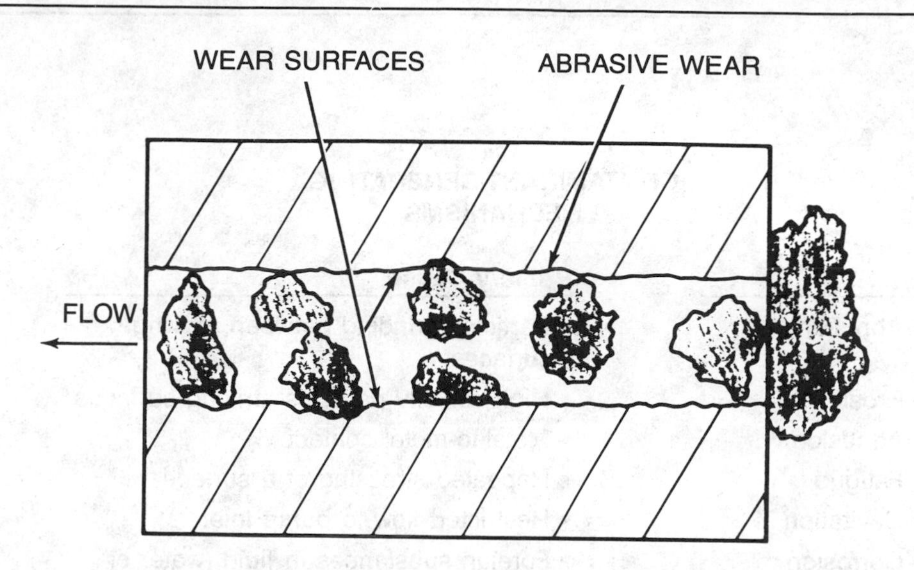

Particles, same size or slightly smaller than the clearance between moving surfaces, will interact with both surfaces to cause wear. Very large particles (right) do not normally get into critical clearance areas and thus cause little or no wear. Very small particles (less than one micrometer) usually flow through without abrading either surface.

Figure 6-2. The effect of different size particles on surface wear.

Fatigue. Fatigue refers to the repeated stressing of a surface. This stressing can eventually create particulate contamination (see Figure 6-5).

Cavitation. Cavitation is caused by a dynamic pressure reduction on hydraulic fluid, typically in a pump inlet, but also possible in actuators and other circuit elements under certain conditions. This reduction causes cavities to be formed in the fluid which may then collapse, as in the high pressure side of a pump, causing severe localized shock and heat damage to the metal pump surfaces. Pump cavitation is caused by exceeding recommended pump inlet vacuum. Excessive inlet vacuum can be caused by overspeeding the pump, restricted inlet lines or air breathers, plugged inlet strainers and oil viscosity being too high. Typical indications are pump noise and temperature rise in the system. There is little if any evidence of cavitation downstream of affected components as there is with aeration.

Corrosion. The deterioration of a surface due to corrosion (such as rust) is often caused by foreign substances in the fluid (e.g. water or chemicals).

Aeration. Aeration is the presence of air or gas bubbles in the fluid. The expansion and compression of these bubbles in the fluid causes the same type of damage as cavitation. However, downstream symptoms will often be very noticeable because of the compressibility of gas under pressure and the inability of gas to seal clearances. Sponginess in cylinder operation and erratic machine performance due to the presence of air will be observed. Poor reservoir design, leaky gas charged accumulators, low fluid level, and leaky fittings may all contribute to aeration.

Mechanical Clearances in Component

Manufacturing clearances within hydraulic components can be divided into two principle zones:

- up to 5 micrometers for high-pressure units
- 10 to 20 micrometers for low-pressure units

The actual clearance may vary widely, depending on the kind of component and component operating conditions. Refer to Figure 6-6 for typical clearance values in various components.

Large particles of solid contamination cannot pass into the clearances between moving parts and normally do not cause abrasion damage to moving surfaces. However, this kind of contaminant can collect at the entrance to a clearance and obstruct the flow of fluid between the moving parts. Large contaminants also jam pumps, valves, and motors.

Definition of a Micrometer. One micrometer is equal to one millionth of a meter or thirty-nine millionths of an inch. To put the size of a micrometer into perspective, consider that the average diameter of a grain of table salt is around .100 micrometers and the average size of a human hair is about 74 micrometers. A micrometer is also called a micron.

The naked eye can normally only resolve size down to 40 micrometers. A single micrometer is too small to be detected. Most of the harmful contamination in hydraulic fluid is below 40 micrometers in size. This means that contamination cannot be evaluated by a visual inspection.

CONTAMINANT-GENERATING MECHANISMS

Type	Primary Cause
• Abrasion	• Particles grinding between moving surfaces
• Erosion	• High velocity particles striking surfaces
• Adhesion	• Metal-to-metal contact
• Fatigue	• Repeated stressing of a surface
• Cavitation	• Restricted flow to pump inlet
• Corrosion	• Foreign substances in fluid (water or chemical)
• Aeration	• Gas bubbles in fluid

Figure 6-3. Contaminant-generating mechanisms.

Definition of Contamination

A contaminant is any material foreign to a hydraulic fluid that has a deleterious effect on the fluid's performance in a system. There are several ways to describe the contaminants associated with fluid systems, including their physical state, their originating activity, their properties and characteristics, and their effect on the system.

A contaminant in a fluid system can exist as a gas, liquid, or solid. A gas either can be in the dissolved state or entrained in the fluid as bubbles. A gas contaminant can be characterized by its density, absorption, adsorption, solubility, dew point, and reactiveness.

A liquid, depending on its compatibility with the hydraulic fluid, can be free, dissolved, or emulsified. Most precipitates such as gums and biological debris are also considered to be liquid in nature. A liquid contaminant can be characterized by its vapor pressure, freezing temperature, emulsifiability, filterability, and corrosivity.

Solid matter, also known as particulate matter, can be characterized in many ways, including density, hardness, compactability, settling ability, transportability, dispersibility, agglomeration characteristics, size, shape, and concentration.

Simply stated, contamination is any foreign substance that contributes to or causes harm to the hydraulic or lubrication fluid or the system or parts of the system in which the fluid flows. Contaminants can be gaseous, liquid, or solid.

Component Failure Modes

Particulate contaminants come in a variety of shapes and sizes. The majority are abrasive and cause wear-related failures in hydraulic components. There are three types of particulate contamination-induced failures: catastrophic failures, intermittent failures, and degradation failures.

Catastrophic Failures. Catastrophic failure is the rapid or sudden total disabling of a component or system. Catastrophic failures occur when a large particle gets stuck in a main component. For example, if a particle caused a vane to jam in a rotor slot, the pump or motor could seize. In spool valves, large-size particles could keep the valve from shifting. If the pilot orifice of a valve is blocked by particulate matter, a catastrophic failure could occur. Fine particles also can cause a catastrophic failure in a valve due to silt accumulating in the clearances.

Intermittent Failures. Intermittent failures are transitory in occurrence and are most often self-correcting or correctable without intrusive maintenance efforts.

ABRASION CLASSES

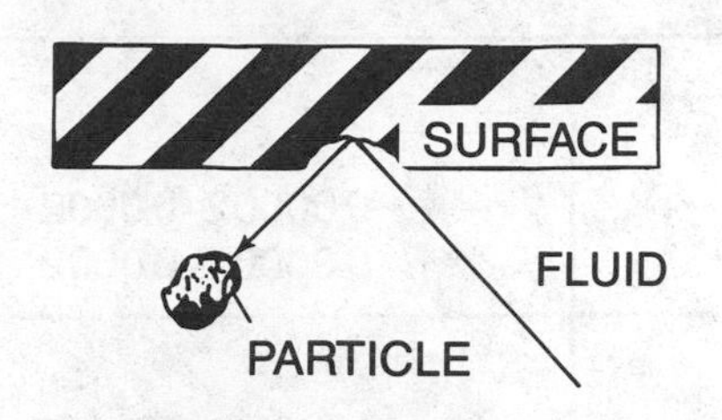

One Body

Contact between surface and the surrounding fluid.

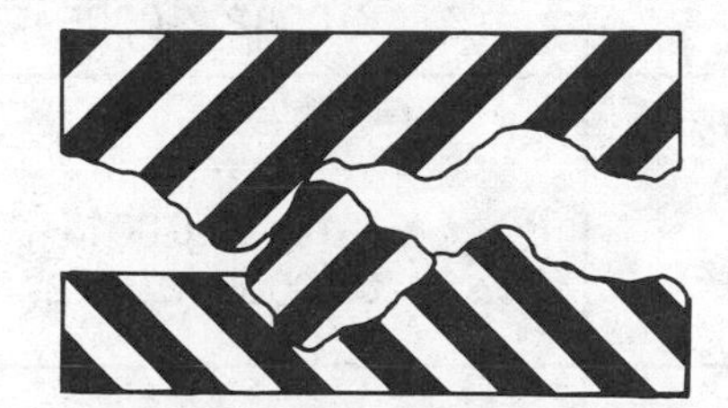

Two Body

- Abrasive particle embedded in a surface making contact with second solid surface.
- Hard asperity of the one surface contacting another softer surface.

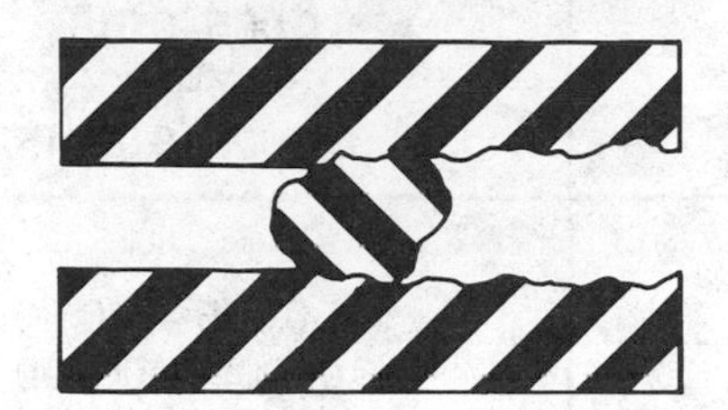

Three Body

Loose abrasive particle making contact with two surfaces.

Figure 6-4. Abrasion classes.

MECHANISM OF SLIDING WEAR WITH PARTICLES GREATER THAN THE FILM THICKNESS

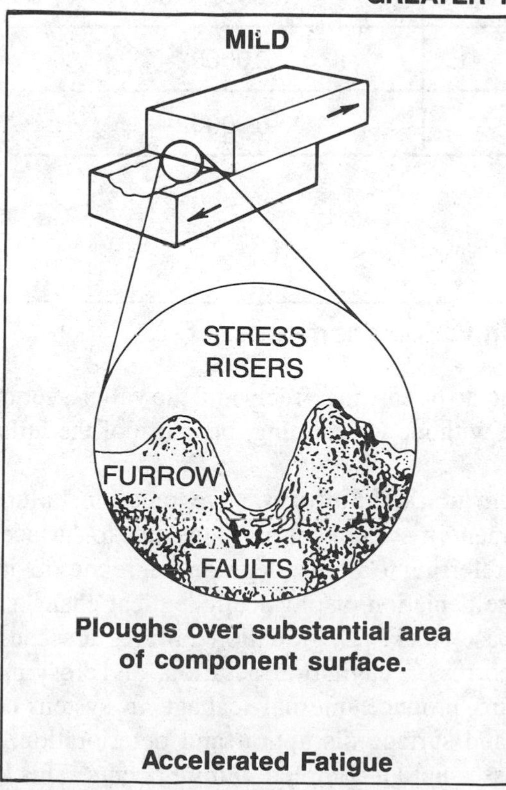

Ploughs over substantial area of component surface.

Accelerated Fatigue

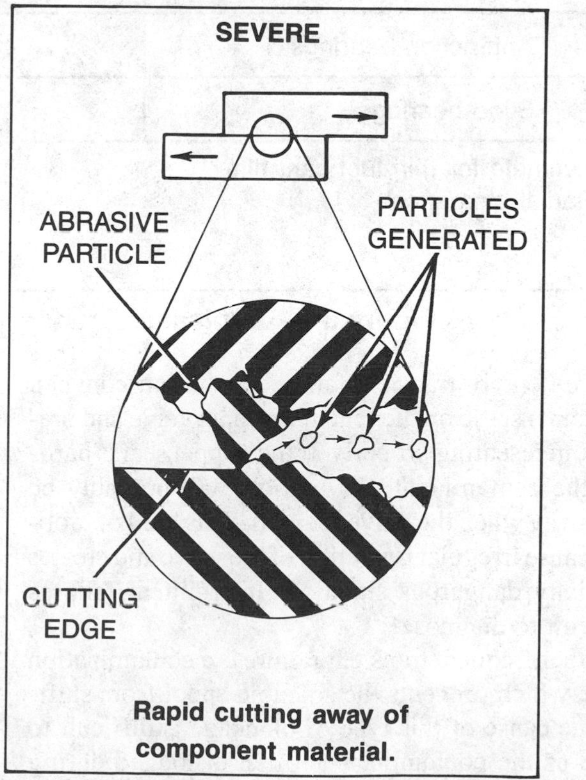

Rapid cutting away of component material.

Figure 6-5. Mechanism of sliding wear with particles greater than the film thickness.

	μm	in.
Gear pump (Pressure loaded) Gear to side plate Gear tip to case	 1/2-5 1/2-5	 0.00002–0.0002 0.00002–0.0002
Vane pump Tip of vane Sides of vane	 1/2-1* 5–13	 0.00002–0.00004 0.0002–0.0005
Piston pump Piston to bore (R)** Valve plate to cylinder	 5–40 1/2–5	 0.0002–0.0015 0.00002–0.0002
Servo valve Orifice Flapper wall Spool sleeve (R)**	 130–450 18–63 1–4	 0.005–0.018 0.0007–0.0025 0.00005–0.00015
Control valve Orifice Spool sleeve (R)** Disc type Poppet type	 130–10,000 1–23 1/2–1* 13–40	 0.005–0.40 0.00005–0.00090 0.00002–0.00004* 0.0005–0.0015
Actuators	50–250	0.002–0.010
Hydrostatic bearings	0–25	0.00005–0.001
Antifriction bearings	*1/2–	0.00002–
Slide bearings	*1/2–	0.00002–

* Estimate for thin lubricant film.
** Radial clearance.

Figure 6-6. Typical clearance values in various components.

An intermittent or transient failure can be caused when contaminants settle on the seat of a poppet valve and prevent it from reseating properly. If the poppet seat is harder than the contaminant, the particle will probably be washed away when the valve reopens. This kind of problem can cause irregular operation of the valve and creates a potentially dangerous situation. Intermittent failures are difficult to diagnose.

Another frequent transient failure is a contamination blockage which prevents the solenoid spool from shifting . The cause of this type of blockage is difficult to diagnose as the contaminant is often dislodged during the disassembly of the unit. Subsequent inspection finds the spool to be shifting freely and the valve is returned to service without determining the cause of the failure.

Degradation Failures. Degradation failure of a component or system simply translates to surface wear. When referring to fluids, degradation relates to additive package depletion or physical/chemical changes to the fluid base stock. Degradation failures are caused by wear, corrosion, cavitation, aeration, and erosion. These problems produce internal leakage in system components and surface disruption and deterioration, which progresses until a complete failure occurs. This kind of problem in a rotating group usually results in a cata-

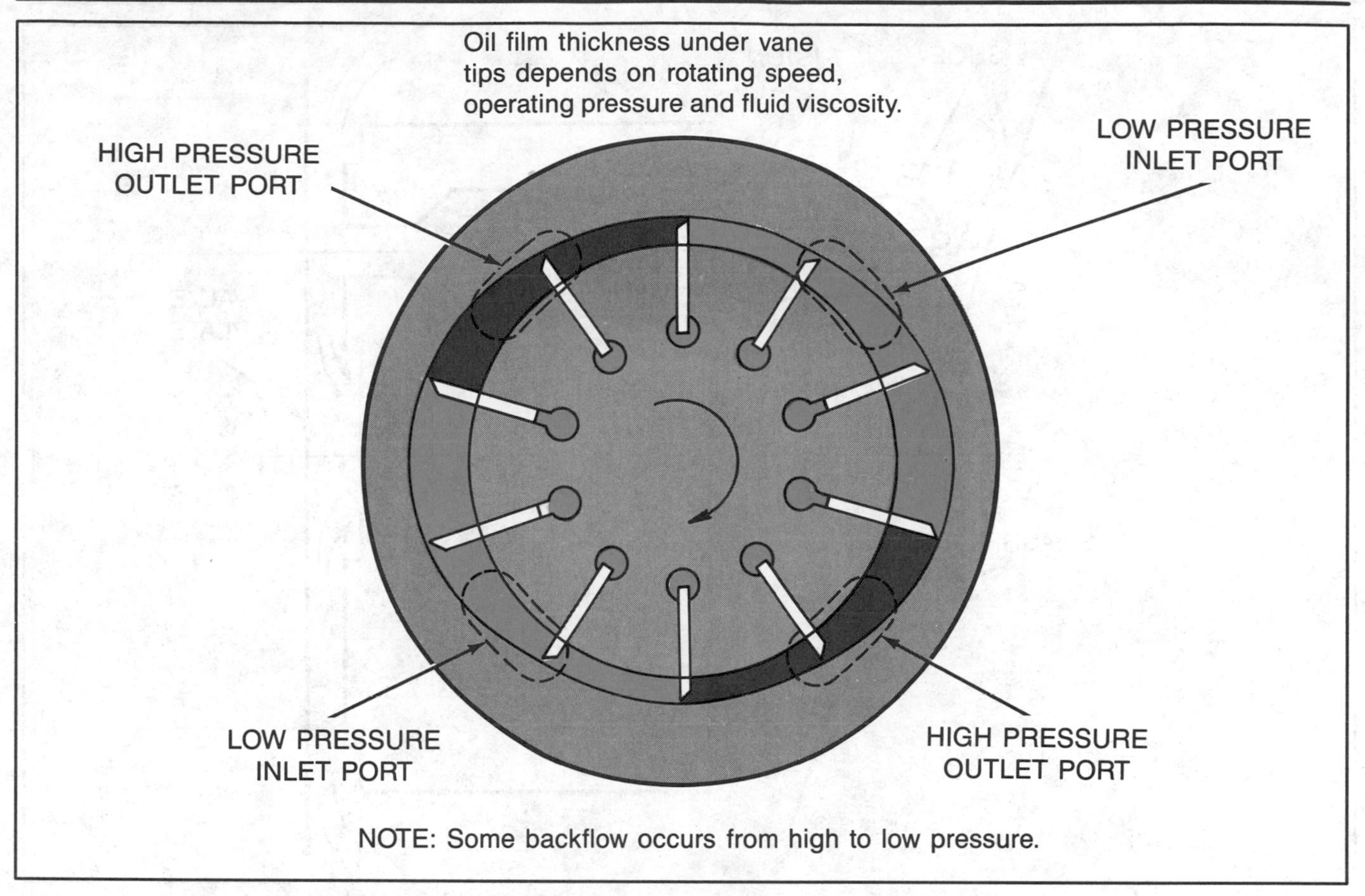
Oil film thickness under vane tips depends on rotating speed, operating pressure and fluid viscosity.
HIGH PRESSURE OUTLET PORT
LOW PRESSURE INLET PORT
LOW PRESSURE INLET PORT
HIGH PRESSURE OUTLET PORT
NOTE: Some backflow occurs from high to low pressure.

Figure 6-7. Critical clearance areas in a pressure balanced vane pump.

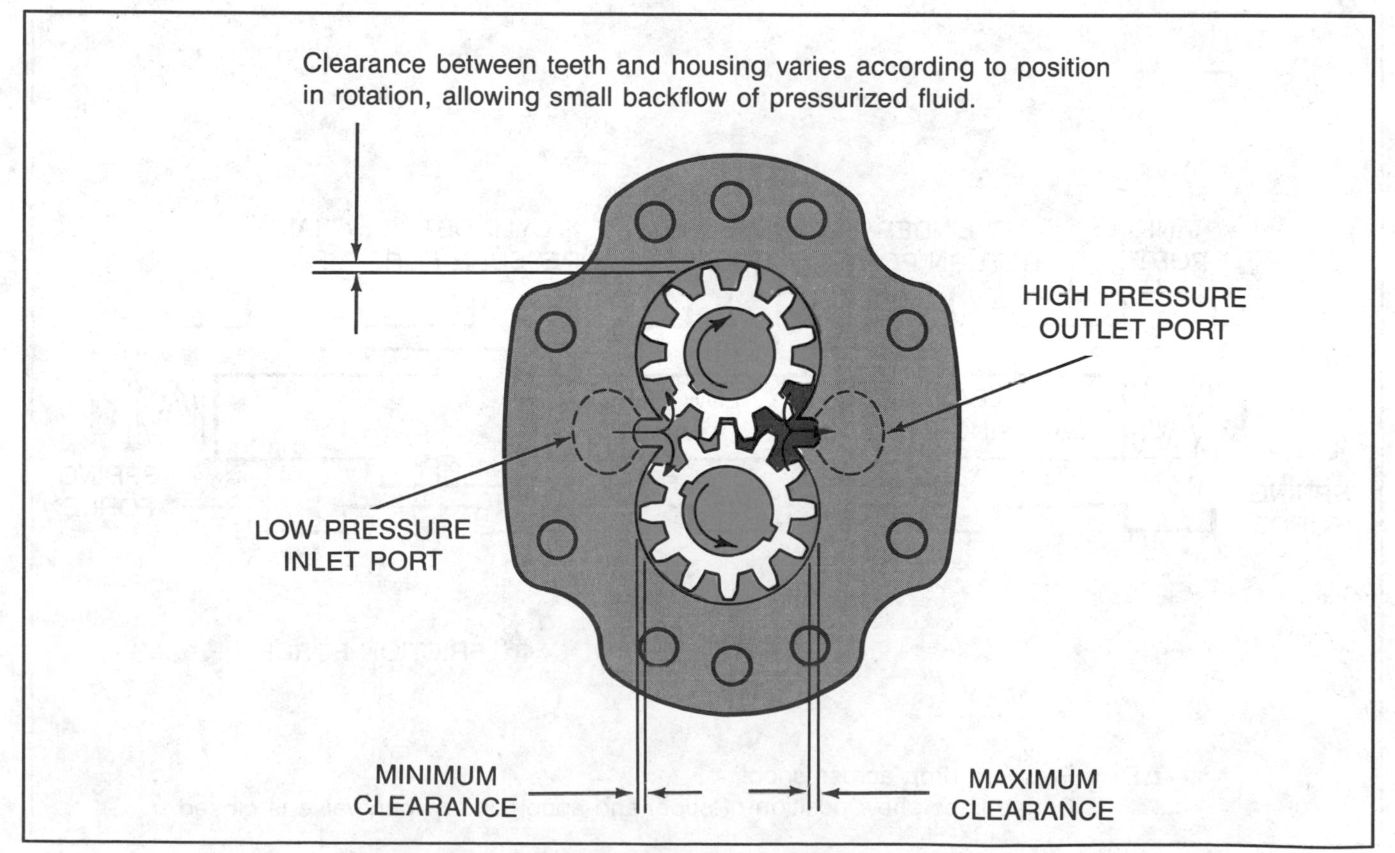
Clearance between teeth and housing varies according to position in rotation, allowing small backflow of pressurized fluid.
HIGH PRESSURE OUTLET PORT
LOW PRESSURE INLET PORT
MINIMUM CLEARANCE
MAXIMUM CLEARANCE

Figure 6-8. Critical clearance areas in a gear pump.

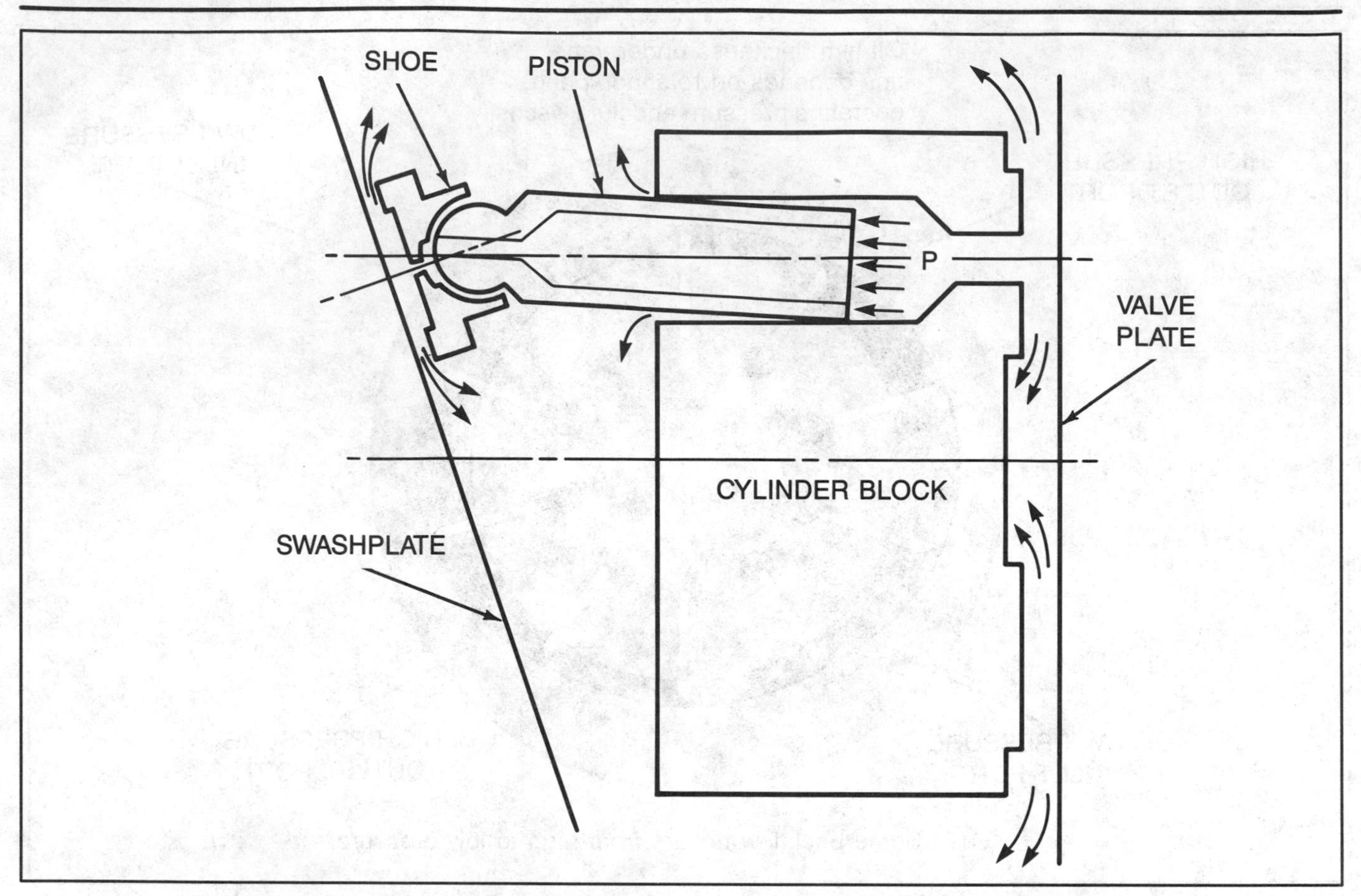

Figure 6-9. Critical clearance areas in an axial piston pump.

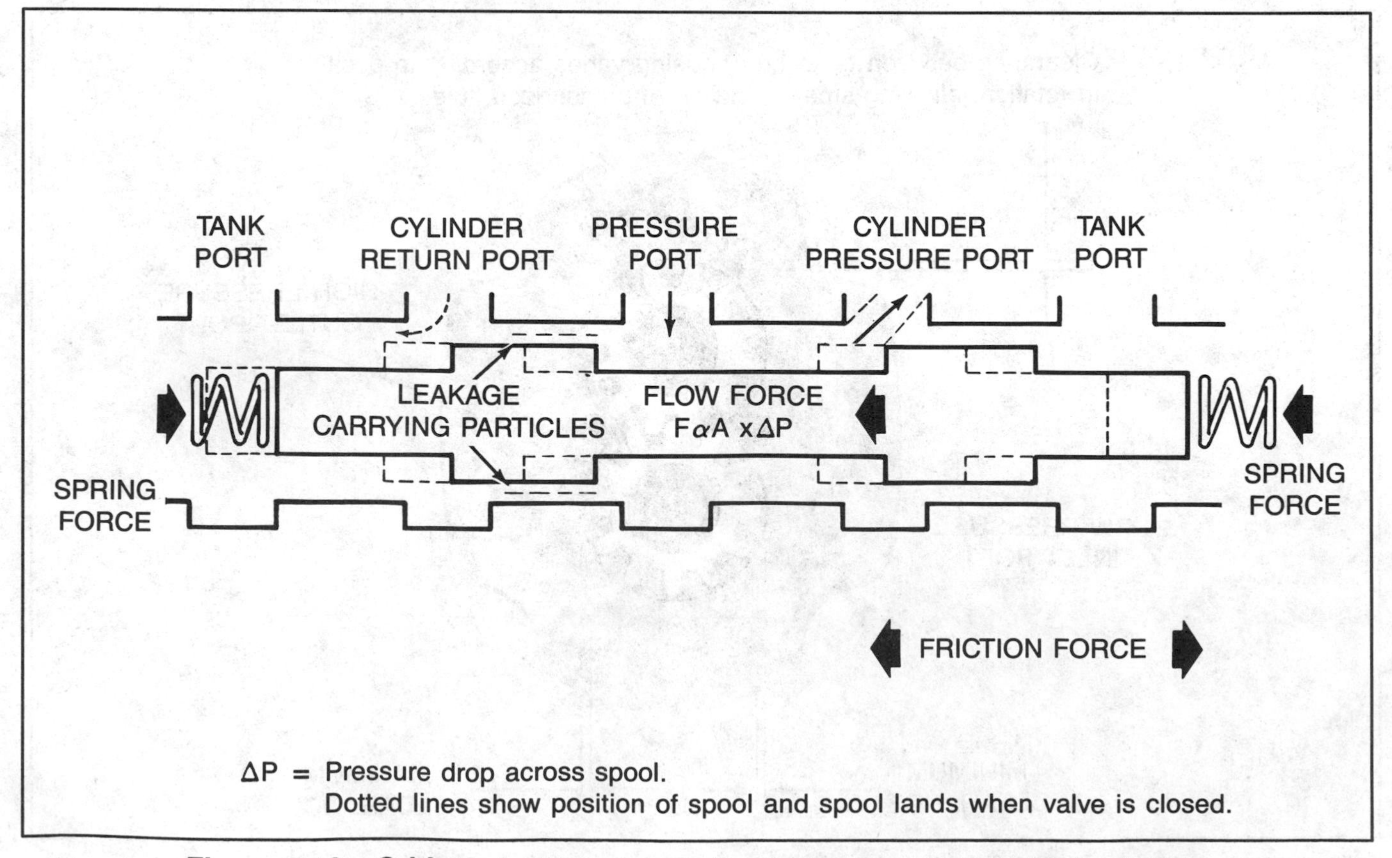

Figure 6-10. Critical clearance areas in a valve spool—longitudinal view.

strophic failure at some point. The contaminants most likely to cause wear are clearance-sized particles that barely pass between the moving parts, but bridge the clearance during their passage (Figure 6-7).

Contamination Failures in Components

Components are affected in different ways by contamination, depending on their design and the function they perform in the hydraulic circuit. Pressure, flow, component load, and critical clearance areas are a few of the factors that should be considered when analyzing why and how components fail as a result of contamination.

Pumps and Motors. Hydraulic pumps and motors have component parts that move relative to each other but are separated by a fluid-filled clearance. The components are usually loaded toward each other by pressure forces that also tend to force fluid through the clearances. If the fluid within these clearances is heavily contaminated, rapid degradation and probable seizure will occur. In low-pressure systems, pump design can tolerate relatively large clearances and contamination effects are somewhat lessened. At lower operating pressures, less force is available to drive particles into critical clearance areas. Higher pressures are of greater significance in determining the effect of contamination on a pump or motor because more tightly loaded clearances tend to be dangerous (see Figure 6-8).

Viscosity of the fluid also affects clearances. The film thickness should support loads hydrodynamically but the viscosity should be low enough to avoid cavitation within the pump or motor. A fluid with viscosity compatible with the pump inlet conditions should be used. Proper temperature control is also important for long life of the component and the fluid.

Pumps and motors are especially susceptible to clearance problems in the following areas:

- Vane pumps and motors—vane tip to cam ring; rotor to side plate; vane to vane slot. See Figure 6-7.
- Gear pumps and motors—tooth to housing; tooth to tooth; gear to side plate. See Figure 6-8.
- Axial piston pumps and motors—shoe to swashplate; cylinder block to valve plate; piston to cylinder block. See Figure 6-9.

Under normal operating conditions, the clearances within a pump or motor are self-adjusting, but with increasing pressure, the clearances become smaller. Under adverse operating conditions, particularly when shock loading is present, smaller clearances increase a pump's vulnerability to smaller contaminant particles. Even when clearances are nominally fixed, components under high loads may assume eccentric positions that make them more vulnerable to small particles. It is difficult to be precise about the magnitude of these clearances and the effect of different size particles in the clearances.

A pump or the rotating group should be replaced when it cannot deliver the required output at a given shaft speed, discharge pressure, and fluid temperature. Generally, a 10 percent loss of flow indicates that servicing is necessary. Quite often, degradation is undetected until catastrophic failure releases vast quantities of contaminant into the system. The system must be thoroughly cleaned and flushed after such a failure or the replacement pump and other parts will have a shorter-than-normal life expectancy. Where case drains are provided for pumps and motors, leakage flow from the drain can be measured as a guide to the component's condition. A change in case drain flow frequently indicates that the component is in a failure mode and has begun shedding particles that can ultimately cause catastrophic failures.

Directional Valves. The range for radial clearances between the bore and spool in most directional control valves is from 5 to 13 micrometers. In a practical sense, the production of perfectly round and straight bores is difficult, if not impossible. For this reason, it is not likely that any spool can be positioned in the exact center of the clearance band. For example, in a nominal one-eighth inch valve, a good spool fit is likely to have clearances of less than 2.5 micrometers.

The forces acting on the solenoid in an electrically operated valve are shown in Figure 6-10. These forces include flow, spring, friction, and inertia. Flow, spring, and inertia forces are built-in factors, but friction depends to a large extent on filtration. If the system is heavily contaminated, higher forces are required to move the spool.

Silting produces even more problems as contaminants are driven into clearances under pressure, eventually leading to fluid film breakdown and spool restriction (Figure 6-11). For example, consider a nominal one-eighth inch valve operating at 3000 psi (206.82 bar) (20,682.00 kPa). If this valve is infrequently operated, the area between the spool and bore can silt up. Test data has shown that 30 pounds of force (133.45 N) would be necessary to dislodge the spool. Because the spring and solenoid forces can exert only 10 pounds of force (44.48 N), the silting usually causes problems in the form of a solenoid failure.

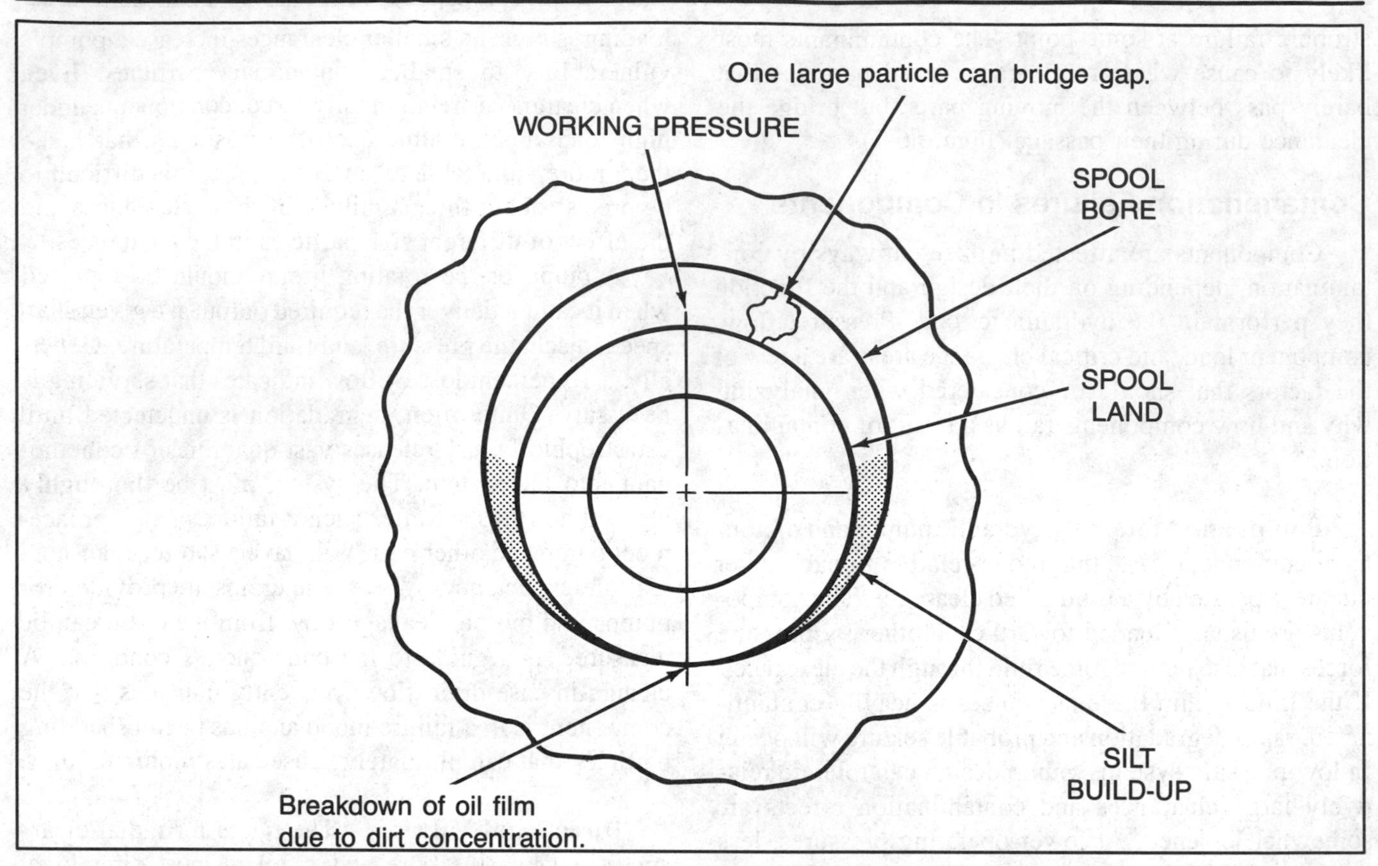

Figure 6-11. Critical clearance areas in a valve spool—end view.

Pressure Controls. Erosion of internal valve surfaces by highly abrasive particles suspended in high-velocity streams of fluid is a common occurrence in pressure controls. This is especially true of relief valves, which are subjected to maximum pressure drop and fluid velocities up to **90 ft/sec (27.43 m/s)**. Pilot control stages generally see low volumes at high velocities. Heavy fluid contamination affects both the stability and repeatability of valves.

Flow Controls. Orifice configuration is the most important factor in determining the contamination tolerance of flow control valves. Figure 6-12 shows two orifices having different shapes but equal areas. The groove-type orifice can tolerate a high contamination level, except when used at a low flow setting. The flat-cut orifice is much more susceptible to silting at all settings.

Contamination affects the performance of the pressure-reducing element in all types of pressure-compensated flow controls, regardless of the valve setting. Damage to the metering orifice can also occur and is particularly apparent at lower flow settings.

MEASURING THE AMOUNTS AND SIZES OF CONTAMINANTS

Whenever a particle contaminant analysis is performed on a hydraulic system, the validity of the results depends on several factors, including:

- Cleanliness control over the equipment used to obtain and analyze the fluid sample.
- Cleanliness control over the environment to which the sample may be exposed.
- Method used to obtain the sample.
- Method used to count particles in the sample.
- Accuracy of the equipment selected for the analysis and skill in its use.
- Accuracy in interpreting analysis results and in assessing the contamination in a fluid sample.

Contaminant analysis is a complex process involving many more factors than those listed above. However, knowledge of this process, coupled with an understanding of its importance to the performance of hydraulic systems, can be quite valuable to those charged with operating and maintaining these systems. Contamination control may be the single most important area in which the technician can directly contribute to operating efficiency and improved performance. For this reason, basic

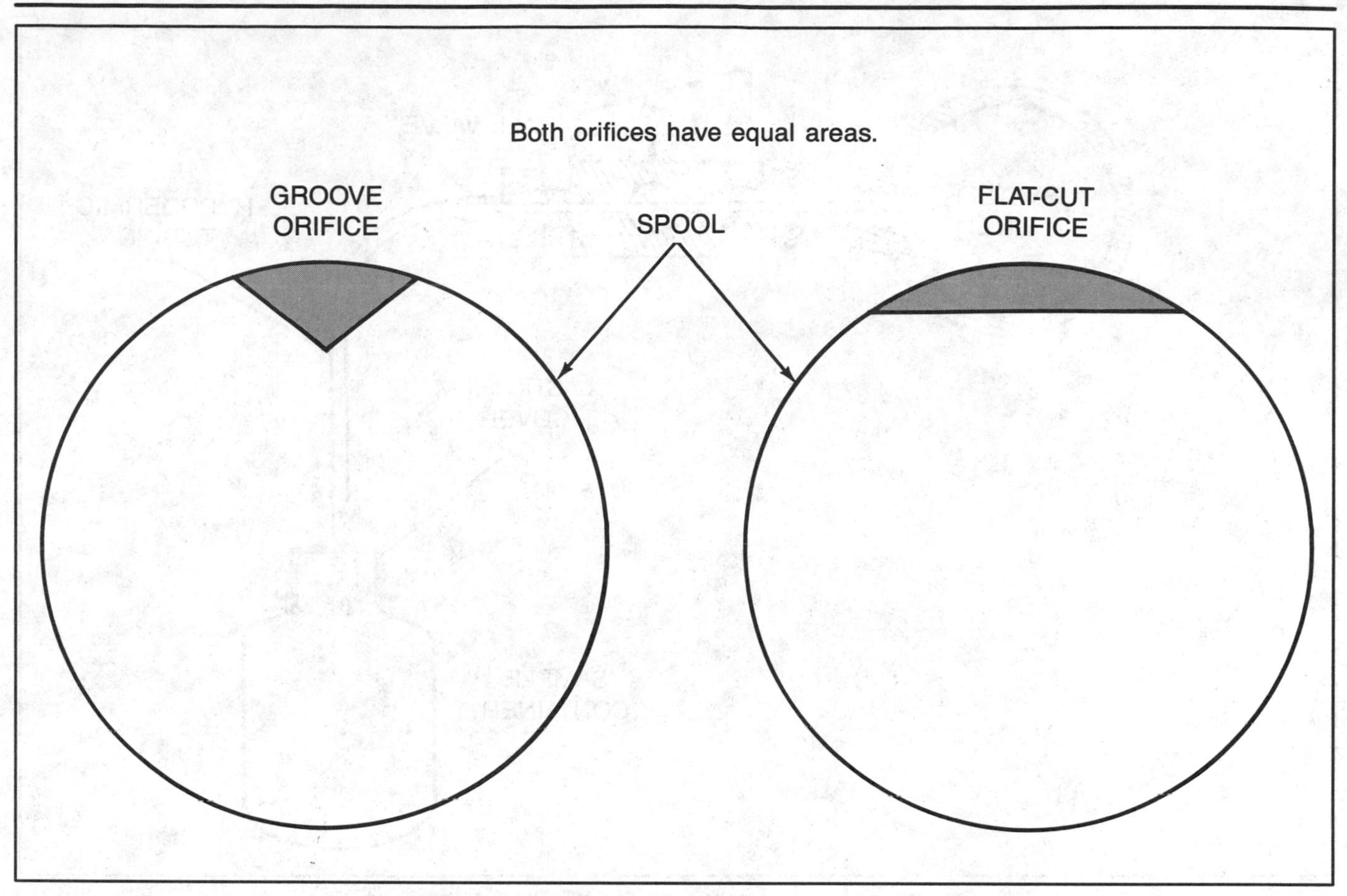

Figure 6-12. Flow control valve throttle sections.

information on how contamination data is collected and analyzed is presented in this section.

Cleanliness Control

Prior to obtaining a fluid sample for analysis, certain precautions must be taken to ensure that external contaminants from sampling equipment or from the environment are not introduced into the sample. Cleanliness control must be established for sampling appendages, sample containers, and exposed surfaces within the analyzer. In addition, the background level of all fluids used to flush, rinse, or dilute the sample must be maintained at a level that does affect the true contaminant makeup of the sample.

Methods of Taking Fluid Samples

The selection of the sampling method and its proper application are critical aspects of the overall analysis effort. The sampling method should be selected on either the basis of the type of sample needed or the purpose of the sample.

Static Sampling. If all that is required is a chemical or physical analysis of fluid precipitates and foreign particulates, a static sample is used. These samples are obtained from a fluid body at rest and are normally extracted at the center point of a fluid container. These samples do not reflect the contaminant conditions present in the system under operating conditions, nor are they useful in determining how much particulate contamination is distributed within the fluid. Consequently, dynamic sampling is the preferred method of sampling.

Dynamic Sampling. Many times, the most important factor in taking a fluid sample is to make sure that the sample is representative of the contamination level that exists within the system under actual operating conditions. In these instances, a dynamic sample must be taken from a fluid in motion. Both the location within the system from which the sample is taken and the period of time during the operation cycle in which the sample is taken are important in obtaining a representative sample. The preferred location for sampling is directly upstream of the return line filter element.

Laminar Flow Sampling. The basic types of dynamic sampling are laminar and turbulent flow sampling. In laminar flow sampling, the sample is extracted by a probe inserted into a region of the system where the

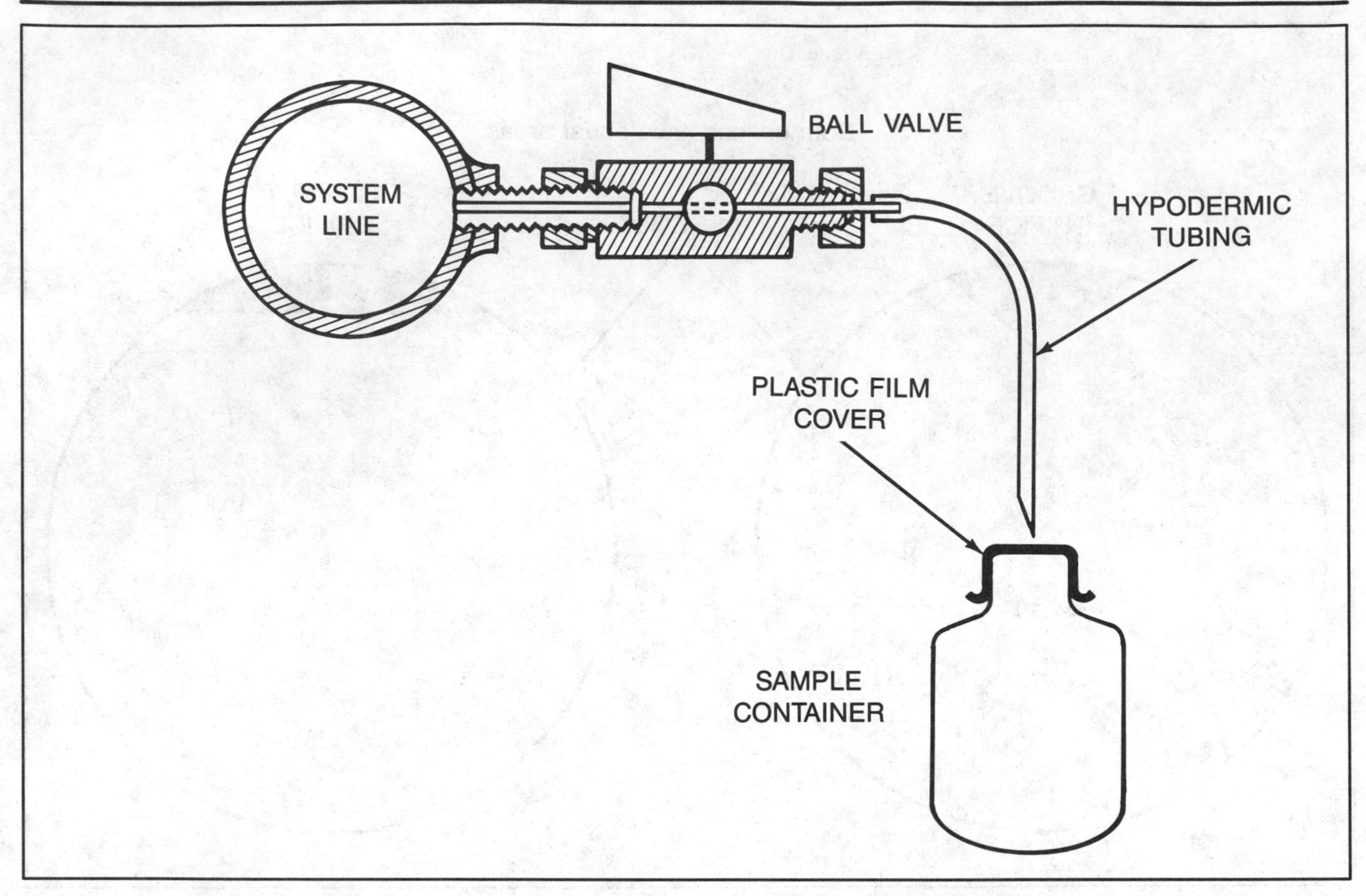

Figure 6-13. Inline sampling method.

flow of liquid is streamlined. This method can be extremely accurate but is quite difficult to implement. In laminar flow, the large contaminant particles are carried along in the boundary layer, while the small particles tend to flow in the center of the velocity profile. Also, in laminar flow, any dispersed material generally settles. Under these conditions, obtaining a representative sample is no simple matter.

Turbulent Flow Sampling. In most situations, turbulent flow is easier to establish and ensure than laminar flow. By its very nature, turbulent flow produces a violent mixing action and provides uniform distribution of particles in the sampling field. It can be shown that sample quality does not depend on the sampling flow or the probe configuration, as long as the sample is taken from the main stream in a turbulent region. In turbulent flow sampling, with dispersed particulate matter generally less than 100 micrometers in size, the quality of the sample is less affected by the sampling system geometry or the withdrawal rate. Therefore, a very simple system can be used to obtain a representative sample from a fluid conductor operating in turbulent flow.

Inline Sampling. The accepted procedure for extracting a dynamic sample is called inline sampling. In this procedure, fluid is removed from the mixing zone through fully opened and appropriately flushed ball valves and hypodermic tubing (Figure 6-13). This procedure minimizes the accidental introduction of trash and generated particles from the sampling device. The following guidelines should be followed when using the inline sampling procedure:

- Before the actual sample is obtained, a quantity of fluid (equal to at least five times the volume between the probe or port and the exiting point) should be allowed to flush from the assembly.
- The sample bottle should be filled to about 70 percent of its total volume to leave enough room to effectively agitate the sample before it is analyzed.
- Extract the sample fluid through fully opened ball valves connected to pressurized lines. To reduce the fluid velocity in the sample line to a desired value, the size and length of the hypodermic tubing can be varied.

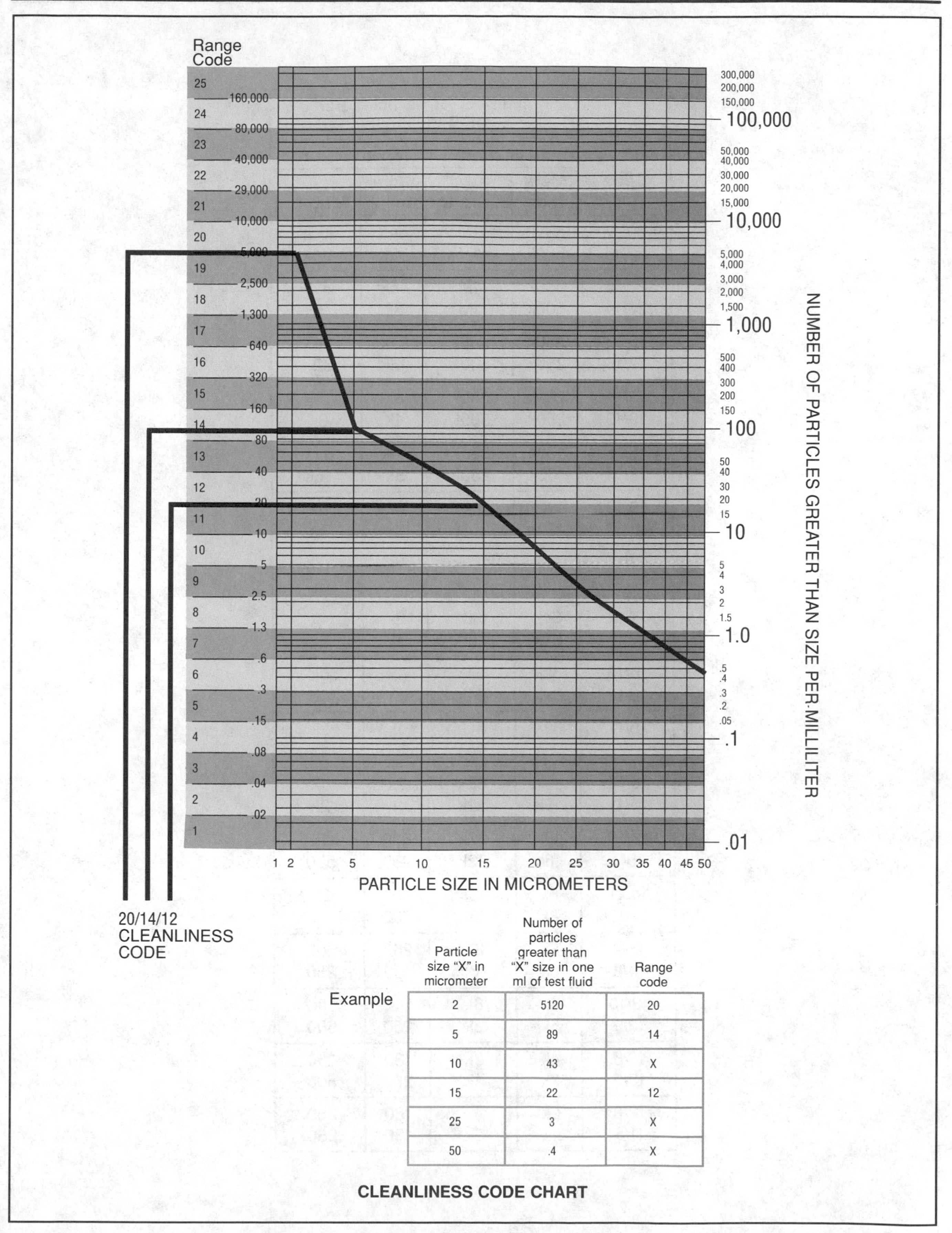

Particle size "X" in micrometer	Number of particles greater than "X" size in one ml of test fluid	Range code
2	5120	20
5	89	14
10	43	X
15	22	12
25	3	X
50	.4	X

Figure 6-14. ISO solid contaminant code.

Code	Number of particles per 100 milliliters			
	Over 5 μm		Over 15 μm	
	More than	& up to	More than	& up to
20/17	500k	1M	64k	130k
20/16	500k	1M	32k	64k
20/15	500k	1M	16k	32k
20/14	500k	1M	8k	16k
19/16	250k	500k	32k	64k
19/15	250k	500k	16k	32k
19/14	250k	500k	8k	16k
19/13	250k	500k	4k	8k
18/15	130k	250k	16k	32k
18/14	130k	250k	8k	16k
18/13	130k	250k	4k	8k
18/12	130k	250k	2k	4k
17/14	64k	130k	8k	16k
17/13	64k	130k	4k	8k
17/12	64k	130k	2k	4k
17/11	64k	130k	1k	2k
16/13	32k	64k	4k	8k
16/12	32k	64k	2k	4k
16/11	32k	64k	1k	2k
16/10	32k	64k	500	1k
15/12	16k	32k	2k	4k
15/11	16k	32k	1k	2k
15/10	16k	32k	500	1k
15/9	16k	32k	250	500
14/11	8k	16k	1k	2k
14/10	8k	16k	500	1k
14/9	8k	16k	250	500
14/8	8k	16k	130	250
13/10	4k	8k	500	1k
13/9	4k	8k	250	500
13/8	4k	8k	130	250
12/9	2k	4k	250	500
12/8	2k	4k	130	250
11/8	1k	2k	130	250

Figure 6-15. Table of ISO codes and corresponding contamination levels.

Particle Size Analysis Methods

The distribution of contaminant particle sizes in a fluid can be analyzed by many different methods. Each of these uses some particular property or combination of properties to distinguish one size particle from another. In some methods, the size dimension is measured directly, while in other methods, the dimension is derived from the measured physical behavior of the particles. Also, some methods detect and measure each particle, and other methods measure particles in bulk powder form. Brief explanations for some of the more common methods are given below.

Optical Methods. Several different optical methods are in use today or are being contemplated for use in the future. Imaging methods with reflected and transmitted light using microscopes have been employed to size particles for quite some time. The image-analyzing computer uses television equipment and a minicomputer along with a microscope to count and size particles. Both the conventional electron microscope and the scanning electron microscope are useful instruments in this field. Other optical methods include light extinction, light scattering, diffraction, laser and holographic techniques.

Automatic Particle Counters

The use of Automatic Particle Counters (APC's) is becoming increasingly popular. This method of counting and sizing particles directs a beam of light through an orifice. The variations in the disruption of the beam as the particles pass, allows a sensor to determine the number and size of the contamination particles.

The speed and accuracy of APC's have made this method very popular. The technology required for this type of particle analysis makes it possible to analyze a relatively small sample of the hydraulic fluid.

Using the ISO Solid Contaminant Code

To specify the cleanliness level of a fluid, the ISO Solid Contaminant Code (Figure 6-14) was established to express the degree of contamination. The code, which applies to all types of fluid systems, provides a simple, unmistakable, meaningful, and consistent means of worldwide communication between suppliers and users.

Description of the Procedure. The contaminant code is constructed by using the results of a particle count analysis to assign a pair of range numbers that represent the cleanliness level of the fluid. First, the contaminant particle size distribution results are used to determine the total number of particles above 5 micrometers in size per unit volume and the total number of particles above 15 micrometers in size per unit volume. These two sizes were selected because the concentration at the smaller size gives an accurate assessment of the amount of silt in the fluid, while the number of particles above 15 micrometers reflects the amount of wear-type particles in the fluid. The two values then are plotted on the graph and connected with a straight line. The range number for each value is read from the graph and the resultant pair of range numbers describes the cleanliness level of the fluid. The diagonal line connecting the range numbers represents the cleanliness profile of the fluid.

The ISO code range numbers are based on the fact that a step ratio of two for particle concentration is adequate both to differentiate between two significantly different systems and to allow for reasonable differences in measurement. Note that the numbers dividing the range numbers double (step ratio of two) from one range number to the next (Figure 6-14).

Example of the Procedure. For this example, the following particle count analysis results are used.

Particle Size Range	No. of Particles per 100 ml
5-15 micrometers	150,000
15-25 micrometers	5,000
25-50 micrometers	1,250
50-100 micrometers	250
Above 100 micrometers	50

The total number of particles above 5 micrometers in size per 100 ml of fluid is 156,550 and the total number of particles above 15 micrometers in size per 100 ml of fluid is 6,550. These two values are plotted on the graph shown in Figure 6-14.

Particle count levels are indicated on the log scales on the left and right sides of the graph. The left side scale represents the number of particles per milliliter greater than the indicated size. The scale on the right represents the number of particles per 100 milliliters greater than the indicated size. Either scale can be used depending on how the analysis results were reported. For this example, the right-hand scale is used. Particle size increments are indicated on the scale shown across the bottom of the graph. On the vertical line marked 5 micrometers, a point is plotted above the horizontal line marked 105 from the right-hand scale. This line represents the number 100,000. The small 2 just above this represents the number 200,000. 156,550 falls near the middle of these two horizontal lines.

Similarly, on the vertical line marked 15 micrometers, a point is plotted above the horizontal line marked 10^3 (1000) but below the line marked 10^4 (10,000).

6,550 falls a little above the horizontal line marked with the small 5 (5000). A line is then drawn between the two plotted points.

The range number corresponding to the point plotted on the 5 micrometer line is 18. The range number corresponding to the point plotted on the 15 micrometer line is 13. The cleanliness level profile for this fluid sample is 18/13.

The typical cleanliness levels of new fluid vary dramatically, but as a rule of thumb, the cleanliness level of new fluid typically falls between 18/15 and 21/18.

The International Standards Organization is considering adoption of code with a third range identified. Some manufacturers are already using three range codes which identify the number of particles larger than 2 microns as the first number in the code. It is likely that the three range cleanliness code will become a standard. When this happens the code will represent the particle count **ranges**.

Particles greater than 2 microns / Particles greater than 5 microns / Particles greater than 15 microns

USING COMPONENT CONTAMINATION TOLERANCE RATINGS

Specifications for hydraulic components include contamination tolerance ratings in the form of ISO code numbers that indicate the appropriate cleanliness level of the fluid required for the particular component. Figure 6-15 is a tabular representation of ISO codes and corresponding contamination levels. Code numbers are shown at the left of the table and the particle distributions for those codes are shown to the right of the code numbers. For example, a particular four-way directional valve might have a component contamination tolerance rating of 18/15. To interpret the meaning of this rating, a table like the one shown in Figure 6-15 should be consulted. In this example, a rating of 18/15 indicates that the fluid should have the following cleanliness level:

- Number of particles over 5 micrometers in size per 100 milliliters of fluid is more than 130,000 but not more than 250,000.
- Number of particles over 15 micrometers in size per 100 milliliters of fluid is more than 16,000 but not more than 32,000.

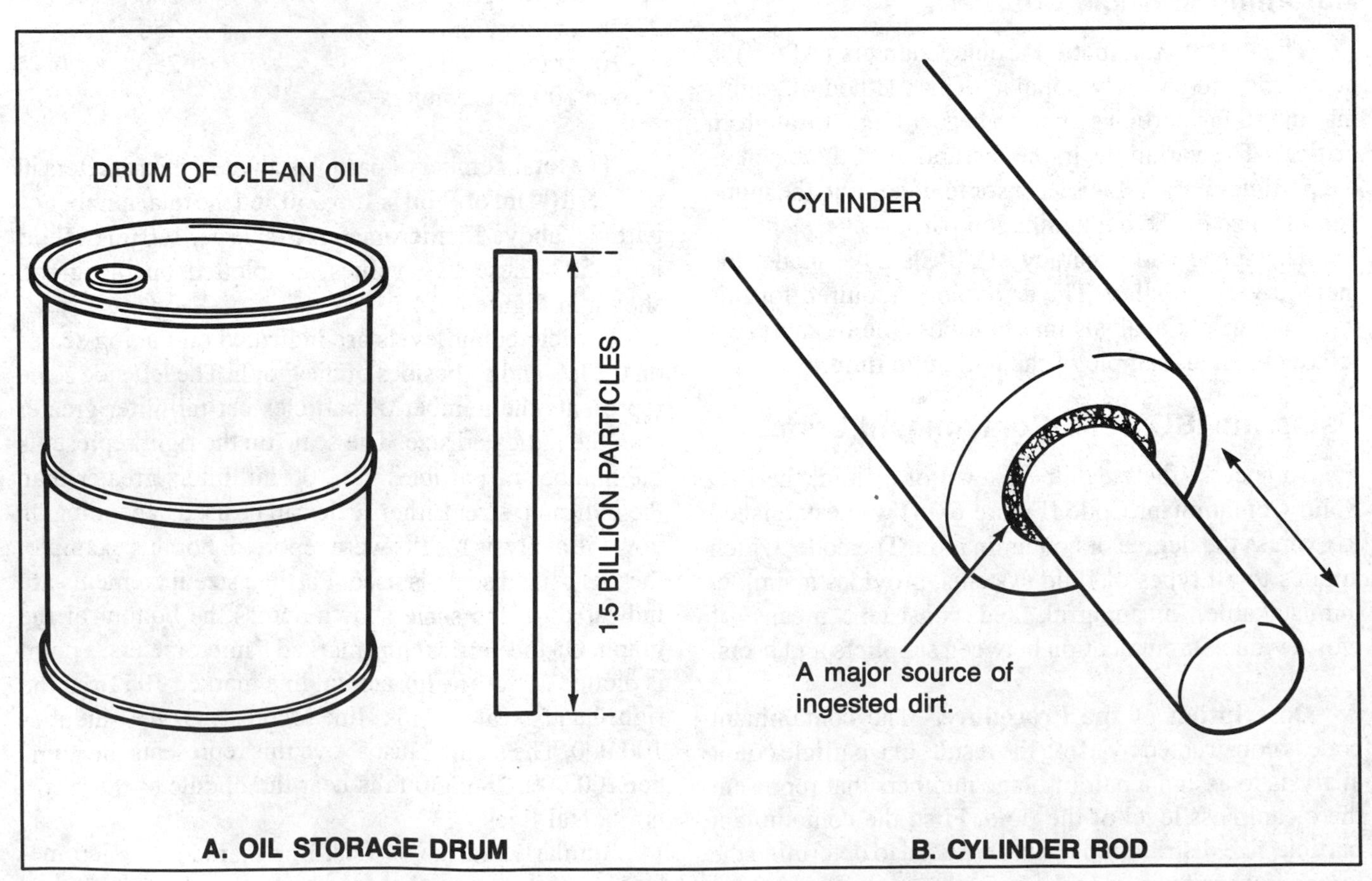

Figure 6-16. Ingressed contamination.

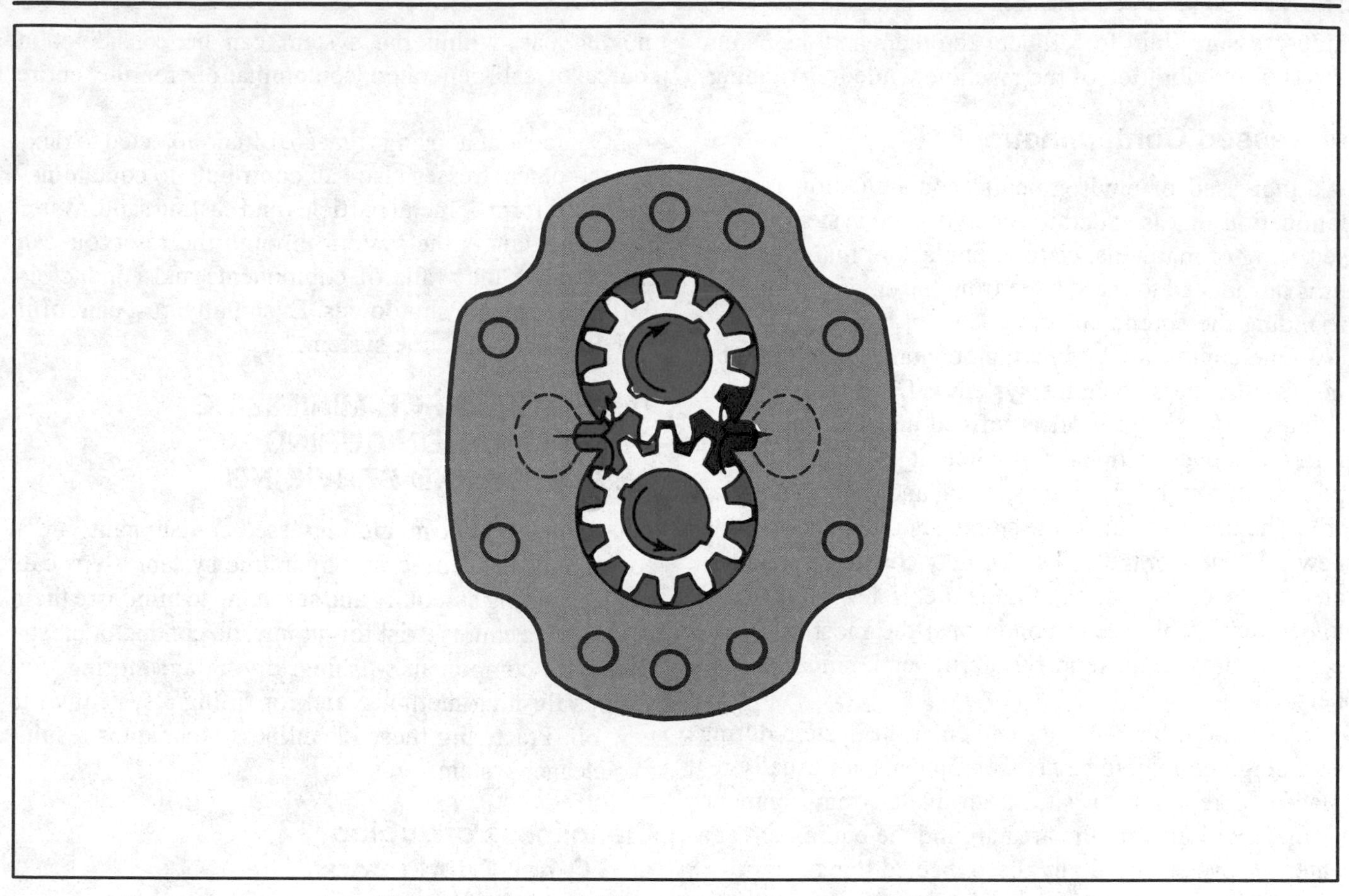

Figure 6-17. Self-generated contamination.

The lowest ISO code number should be used to select and place filters when using tolerance ratings to determine the cleanliness level for a system. Filters should be selected so the entire system is cleaned to a target level below the minimum cleanliness level. For example: if the recommended minimum cleanliness level for the most sensitive component in the system is 16/13 a 14/11 target cleanliness level should be selected. This standard should be applied to all samples regardless of the time or from where in the system it was taken.

SOURCES OF CONTAMINATION

The sources of contamination in hydraulic systems can be divided into three general categories:

- Built-in contamination
- Ingressed contamination
- Internally-generated contamination

Built-in Contamination

Hydraulic system manufacturers generally are careful to provide internally clean products but, in spite of these efforts, new equipment usually contains some built-in contamination. These contaminants might include burrs, chips, flash, dirt, dust, fibers, sand, moisture, pipe sealants, weld splatter, paints and flushing solutions. New components within a system may also become sources of contamination due to improper storage, handling, and installation practices. New directional valves, cylinders, relief valves, and pumps may contain contaminants that appear in the system fluid after a very short period of operation.

As the machine is assembled, the reservoir may accumulate rust, paint chips, or dust. Although the reservoir is cleaned prior to use, many contaminants are invisible to the human eye and are not removed by wiping with a rag or blowing off with an air hose.

Many people believe filters can perform a miraculous clean-up of sections of the system. Unfortunately, experience has shown that a single pass through a filter will not adequately protect a system component or zone.

A frequently overlooked source of contamination is through threading fittings together. For example: the making of one 5/8 inch pipe fitting introduces over 60,000 particles greater than 5 microns into the system. Installing several fittings introduces huge numbers of contaminates into the system during assembly.

Contaminants such as weld scales may not break off and enter the fluid stream until they are loosened by high-pressure fluid forced between them and the parent metal or by vibration of the machine while it is running.

Ingressed Contamination

Ingressed or environmental contamination is contamination that is added to the hydraulic system during servicing or maintenance (or from lack of maintenance) or is introduced to the system from the environment surrounding the equipment.

One common way in which contamination may be ingressed occurs when the system is filled with new oil (Figure 6-16A). New oil is refined and blended under fairly clean conditions but when it is delivered and pumped through filling lines, metal and rubber particles from the lines may enter the storage tanks along with the new oil. The storage tanks also may contain rust generated by the condensation of moisture. If new oil is stored under reasonably clean conditions, the most common contaminants in makeup fluid are metal, silica, and fibers.

Dirt and other particles can enter the system during servicing and maintenance. Components are usually replaced or repaired on site in an unclean environment. Contamination from the area around the equipment can enter the system from any disconnected line or port.

Another source of ingressed contamination is the air breather cap on the reservoir. Air enters the reservoir through the air breather cap every time a cylinder cycles or the oil level lowers for any reason such as thermal contraction. The breather is frequently a coarse screen that allows unfiltered dirt into the system. Many times, a breather cap that has become clogged due to lack of maintenance is removed and never replaced.

Contamination from the environment also can enter the system through power unit access plates that have been removed and not replaced. If access to strainers or other components depends on the removal of power unit covers, good resealing may not be possible.

Another major source of environmental contamination occurs when cylinder rods remain extended in a heavily contaminated atmosphere for long periods of time. Fine particles may settle on the rod and then be pulled into the system when the rods are retracted (Figure 6-16B). As seals and wipers on these rods wear, the contamination ingression rate can increase considerably.

Internally-Generated Contamination

This type of contamination is created internally within the system by the moving parts of hydraulic components. These contaminants are produced by wear, surface fatigue, cavitation, and decomposition and oxidation of the system fluid (Figure 6-17). Every internal moving part within the system can be considered a source of self-generated contamination for the entire system.

Component housings that are often subjected to flexing and other stresses also can contribute to contamination in the form of metal particles and casting sand. Water vapor that enters the system through the reservoir can condense on the walls of components and conductors during equipment shutdowns. Eventually, rust can form and be washed into the system.

TECHNIQUES FOR MINIMIZING CONTAMINATION DURING ASSEMBLY AND SERVICING

Although built-in and ingressed contaminants are a continuous problem in any hydraulic system, steps can be taken during assembly and servicing to minimize their effects. Procedures exist for tubing and conductor cleanliness, for component washing, drying and storing, for various flushing methods, and for filling a system with new oil. Practicing these cleanliness techniques results in a cleaner system.

Cleanliness of Tubing and Other Conductors

The following information on preparing pipes, tubing, hoses and fittings prior to installation in a hydraulic system should be considered a basic guideline only. Complete procedures are included in Appendix C of this manual.

Iron and steel pipes, tubing, and metal fittings must be absolutely clean, free from scale and other foreign matter, before they are installed. Metal conductors may be wire brushed, cleaned with a commercial pipe-cleaning apparatus, or pickled, depending on the amount of scale or rust initially present on the conductor. If pickling is used, the conductor must be degreased prior to the pickling procedure and thoroughly rinsed after it.

The inside edge of tubing and pipe should be deburred after cutting to remove any burrs that may be present. Tubing must not be welded, brazed, or soldered after system assembly as this makes proper cleaning impossible. Hose should be flexed several times to release any trapped dirt and then properly flushed.

Flanges must fit squarely on the mounting faces and be properly secured with bolts of the correct length. Threaded fittings should be inspected for metal slivers on the threads before installation. Cover all openings into the system to keep out contaminants during assembly or maintenance.

If conductors are stored for any length of time, they should be plugged to prevent the entrance of foreign mat-

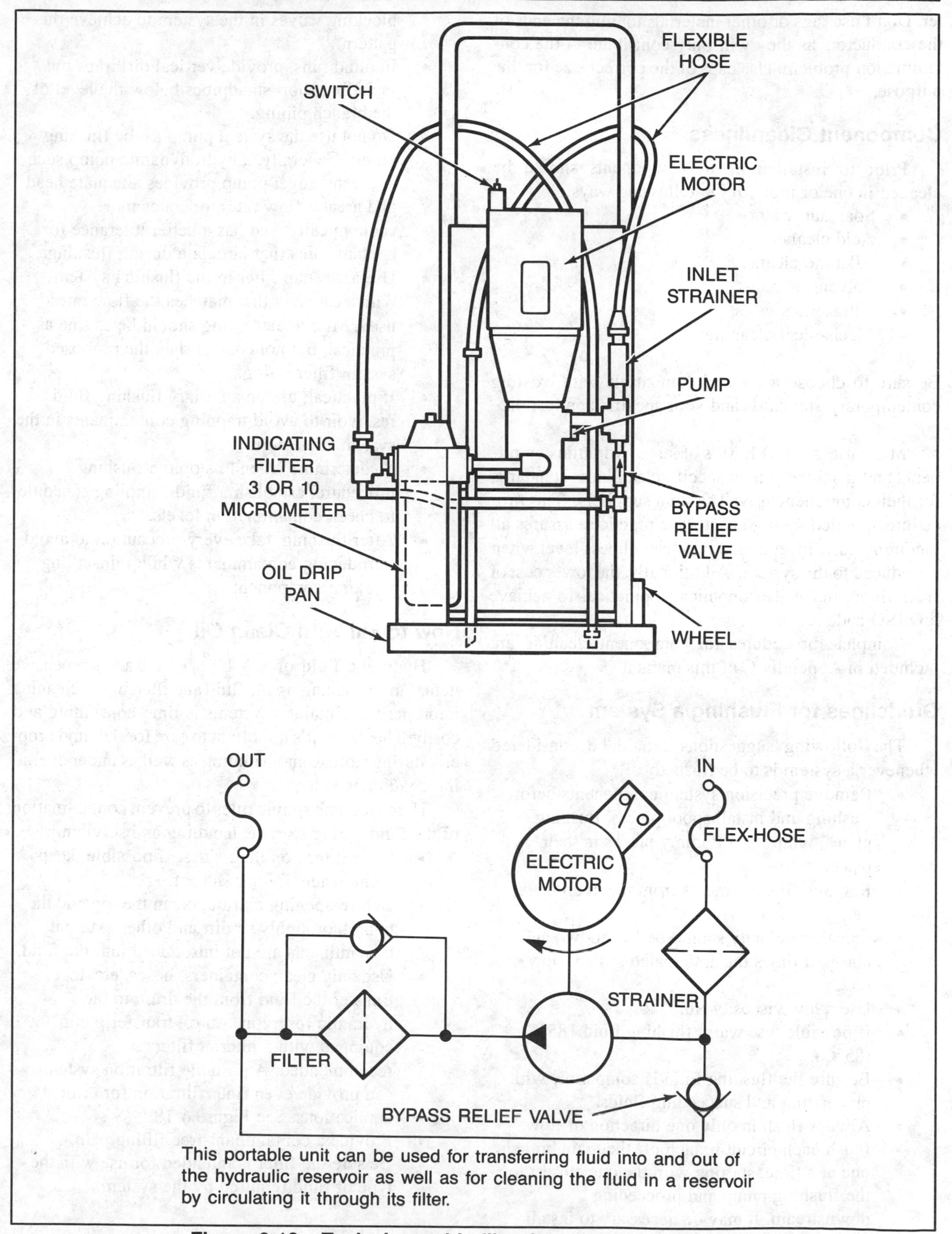

This portable unit can be used for transferring fluid into and out of the hydraulic reservoir as well as for cleaning the fluid in a reservoir by circulating it through its filter.

Figure 6-18. Typical portable filtration system and symbol.

ter. Don't use rags or other materials to plug the ends of the conductors as they will only contribute to the contamination problem. Use caps of the correct size for this purpose.

Component Cleanliness

Prior to installation, all components should be cleaned in one or more of the following ways:

- Soap and water
- Acid cleaner
- Alkaline cleaner
- Solvent
- Ultrasonics
- Mechanical cleaning

Be sure to choose a method compatible with existing contemporary standards and your specific application.

Many users and O.E.M.'s of large hydraulic components and machinery now specify cleanliness standards for their component as well as fluid suppliers. This procedure, coupled with good storage practices, assures all incoming parts meet a minimum cleanliness level when introduced to the system. Additionally, the lower cost of finer filters make it economically practical to achieve low ISO code levels.

Complete procedures for component cleaning are included in Appendix C of this manual.

Guidelines for Flushing a System

The following suggestions should be considered whenever a system is to be flushed:

- Remove precision system components before flushing and install spool pieces, flushing plates, jumpers or dummy pieces in their place.
- Remove filter elements from the main lines being flushed.
- Flushing velocity should be two to two and one-half times the anticipated system flow rate.
- Use a low viscosity fluid.
- If possible, use warm flushing fluid 185°F (85°C).
- Be sure the flushing fluid is compatible with preceeding and succeeding fluids.
- Always flush in only one direction of flow.
- Flush each circuit branch off the main branch one at a time, starting with the one closest to the flushing pump and proceeding downstream. It may be necessary to install blocking valves in the system to achieve this pattern.
- In blind runs, provide vertical dirt traps by including short standpipes below the level of the branch piping.
- Do not use the system pump as the flushing pump. Generally, a hydrodynamic pump such as a centrifugal pump provides adequate head and greater flow rates, operates more economically, and has a better tolerance for contaminants that circulate during flushing.
- Use a cleanup filter in the flushing system with a capacity that matches the flow rates used. Micrometer rating should be as fine as practical, but not coarser than the proposed system filter rating.
- If practical, use an auxiliary flushing fluid reservoir to avoid trapping contaminants in the system's reservoir.
- To determine when to stop the flushing procedure, establish a fluid-sampling schedule to check contamination levels.
- After flushing, take every precaution to avoid introducing contaminants while reinstalling working components.

How to Fill With Clean Oil

Hydraulic fluid of any kind is not an inexpensive item. Further, changing the fluid and flushing or cleaning improperly maintained systems is time-consuming and costly. Therefore, it's important to care for the fluid properly during storage and handling as well as under operating conditions.

Here are some simple rules to prevent contamination of the fluid during storage, handling and servicing:

- Store drums on their sides. If possible, keep them inside or under a roof.
- Before opening a drum, clean the top and the bung thoroughly so dirt and other external contaminants are not introduced into the fluid.
- Use only clean containers, hoses, etc. to transfer the fluid from the drum to the hydraulic reservoir. An oil transfer pump equipped with 5 micron filters is recommended. A portable filtration system can provide even finer filtration for critical applications. See Figure 6-18.
- Provide a contaminant-free filling path.
- Be sure the filter is intended for use with the type of fluid required by the system.

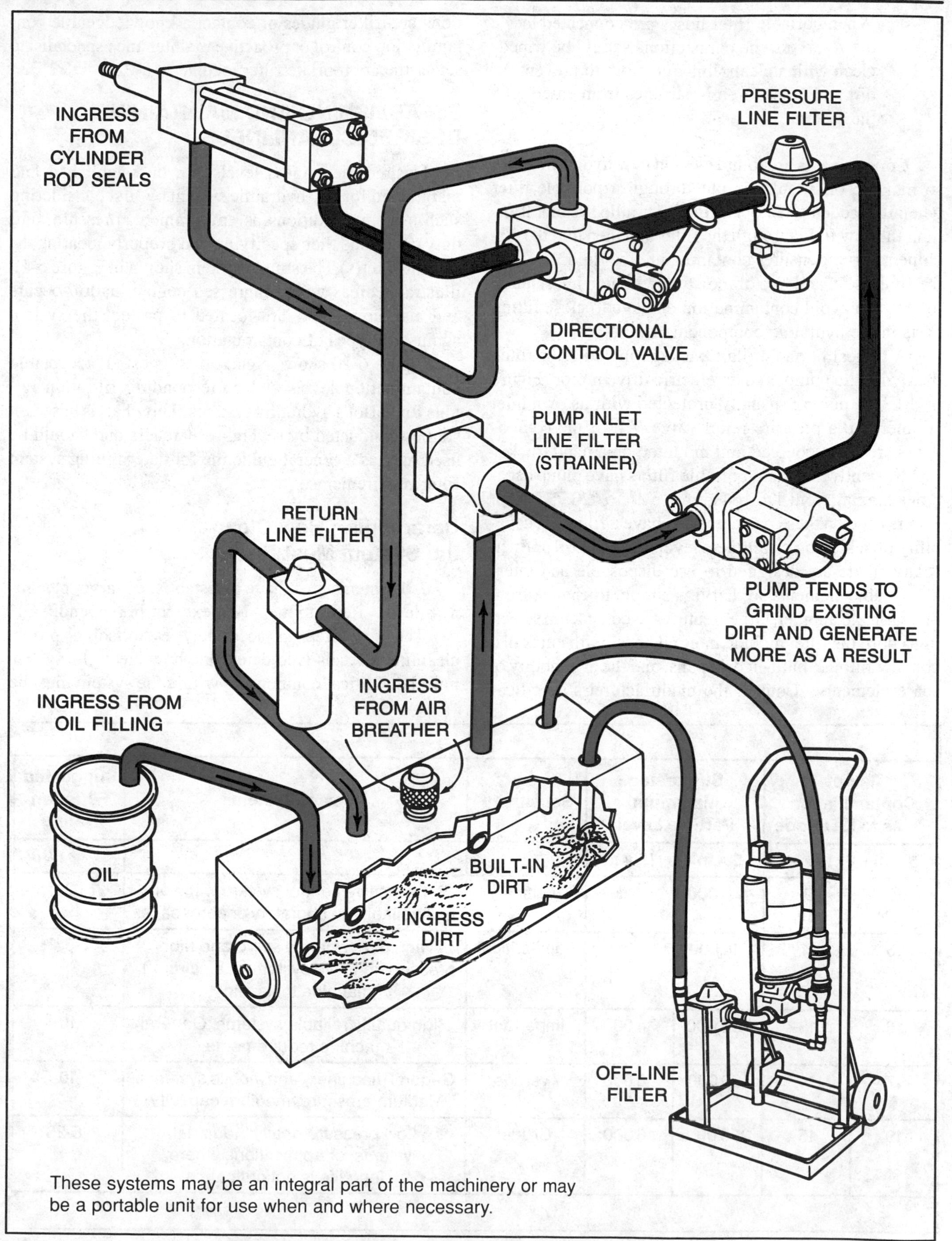

Figure 6-19. Basic filtration of a hydraulic system.

- When portable filter hoses are connected to the reservoir, the connections should be wiped clean with a clean, lint-free cloth to prevent dirt and other foreign particles from entering the system.

Contaminants are found even in new hydraulic fluid, so makeup oil should be put through a portable filter when it is added to an operating hydraulic system. Running the new fluid through the filter and then into the machine reservoir ensures contaminant-free fluid. Keeping the fluid clean and free of moisture will help it last much longer and avoid contamination damage to close fitting parts in the hydraulic components.

A typical portable filter package includes the filter, the hydraulic pump, and an electric drive motor (Figure 6-18). The pump is usually protected with its own inlet strainer and a pressure-relief valve. All of the components are close-coupled and mounted on a hand truck.

Currently available portable filters have pump capacities ranging from 4 to 164 gpm (15.14 to 620.79 LPM). Filters used on these packages can have a filtration capability of achieving cleanliness levels of ISO 12/9 or finer. Many filters are designed to use disposable and interchangeable elements so that it is simple to change from one rating to another. Other features to consider in selecting a portable filter include integral hoses with quick disconnect fittings, built-in drip pans, and the availability of spare elements. Consult the manufacturer's specifications and filter guides or contact a knowledgeable contamination control or proactive maintenance specialist to select the appropriate filter media.

ESTABLISHING AND MAINTAINING DESIRED CLEANLINESS

Desired cleanliness levels can be established and maintained for any hydraulic system by first considering such system conditions as environment, pressure, and duty cycle, and then specifying and properly locating the system filter(s). The simple system shown in Figure 6-19 illustrates areas where ingressed contamination occurs and also how filters are located to protect the system against all types of contamination.

Figure 6-20 shows a chart of suggested, acceptable contamination levels and corresponding filtration ratings for various hydraulic systems. This chart is based on studies conducted by several researchers and should be used only as a general guideline for determining system filter requirements.

Determining How Clean the System Must Be

All systems must have an established target cleanliness level. Refer to the ANSI excerpt in Appendix C.

The first step in accurately establishing proper cleanliness levels is to determine how clean the system must be. Critical components within the system must be

Target Contamination Class to ISO Code		Suggested Maximum Particle Level		Sensitivity	Type of System	Suggested Filtration Rating
5 µm	15 µm	5 µm	15 µm			$\beta_x >$ 100
13	9	4,000	250	Critical	Silt sensitive control system with very high reliability. Laboratory or aerospace.	3
15	11	16,000	1,000	Semi-critical	High performance servo and high pressure long life systems, i.e., aircraft, machine tools, etc.	5
16	13	32,000	4,000	Important	High quality reliable systems. General machine requirements.	10
17	14	130,000	8,000	Average	General machinery and mobile systems. Medium pressure, medium capacity.	10
19	15	250,000	16,000	Crude	Low pressure heavy industrial systems, or applications where long life is not critical.	15-25

Figure 6-20. Suggested acceptable contamination levels.

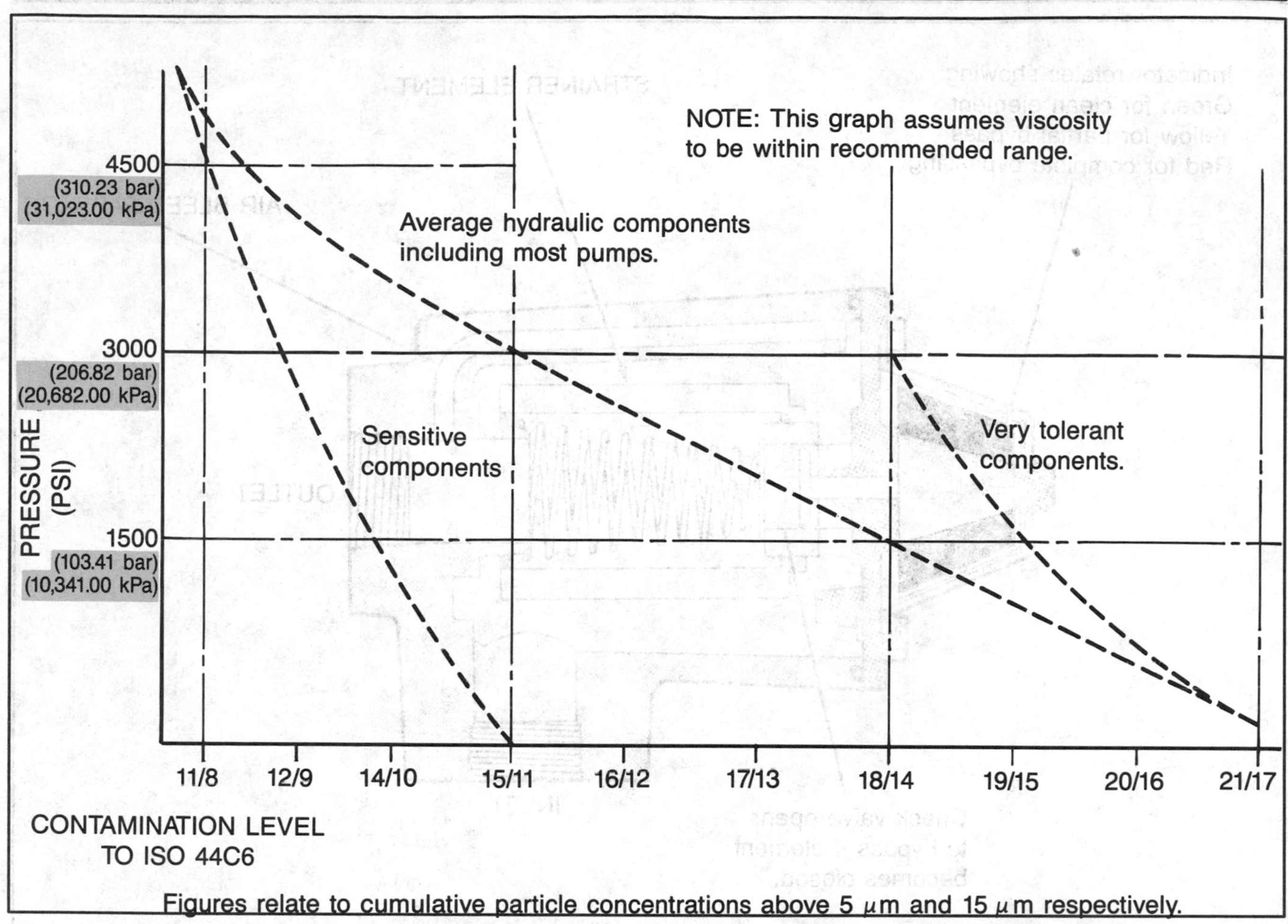

Figure 6-21. Suggested cleanliness levels for satisfactory component life.

identified and the ISO code for those components determined from component specifications or from charts developed for this purpose.

Although the most critical component in many systems is the pump, infrequently operated spool valves which are subjected to continuous pressure must be protected from high silt concentrations. Low setting flow controls, regardless of pressure, must also be protected from silt. Flow and pressure control valves also provide greater operating reliability and repeatability when silt content is under control.

After analyzing the dirt sensitivity levels of the components used in a system, a chart such as the one shown in Figure 6-21 can be used to determine acceptable recommended contamination ranges, if the system pressure is known. For example, if the system pressure is 1500 psi (103.41 bar) (10,341.00 kPa), a component identified as sensitive to contamination requires an ISO cleanliness range of 14/10. Therefore the target cleanliness level for this system should be no higher than 14/10.

Other factors to consider in filter selection, such as steady or pulsating flow conditions, flow rates, and filter pressure drop are covered later in this chapter.

Filter Location

There are 4 general areas in the system for locating a contamination control device: The inlet line (Figure 6-22), the pressure line (Figure 6-23), the return line (Figure 6-24), and an off loop or kidney loop type filtration/filling circuit (Figure 6-25).

Strainers are usually used for the inlet lines, while filters (or a last chance strainer type device) are used to protect downstream components from the catastrophic failures of upstream components.

Inlet Strainers. Figure 6-26 shows a typical strainer of the type installed on pump inlet lines inside the reservoir. Compared to a filter this type of strainer is relatively coarse, being constructed of fine mesh wire. A 100-mesh strainer protects the pump from particles above about 150 microns in size.

Two important requirements for any inlet strainer (and its associated conductors) are:

- The strainer must pass the full pump volume within the permitted inlet pressure drop for that pump.

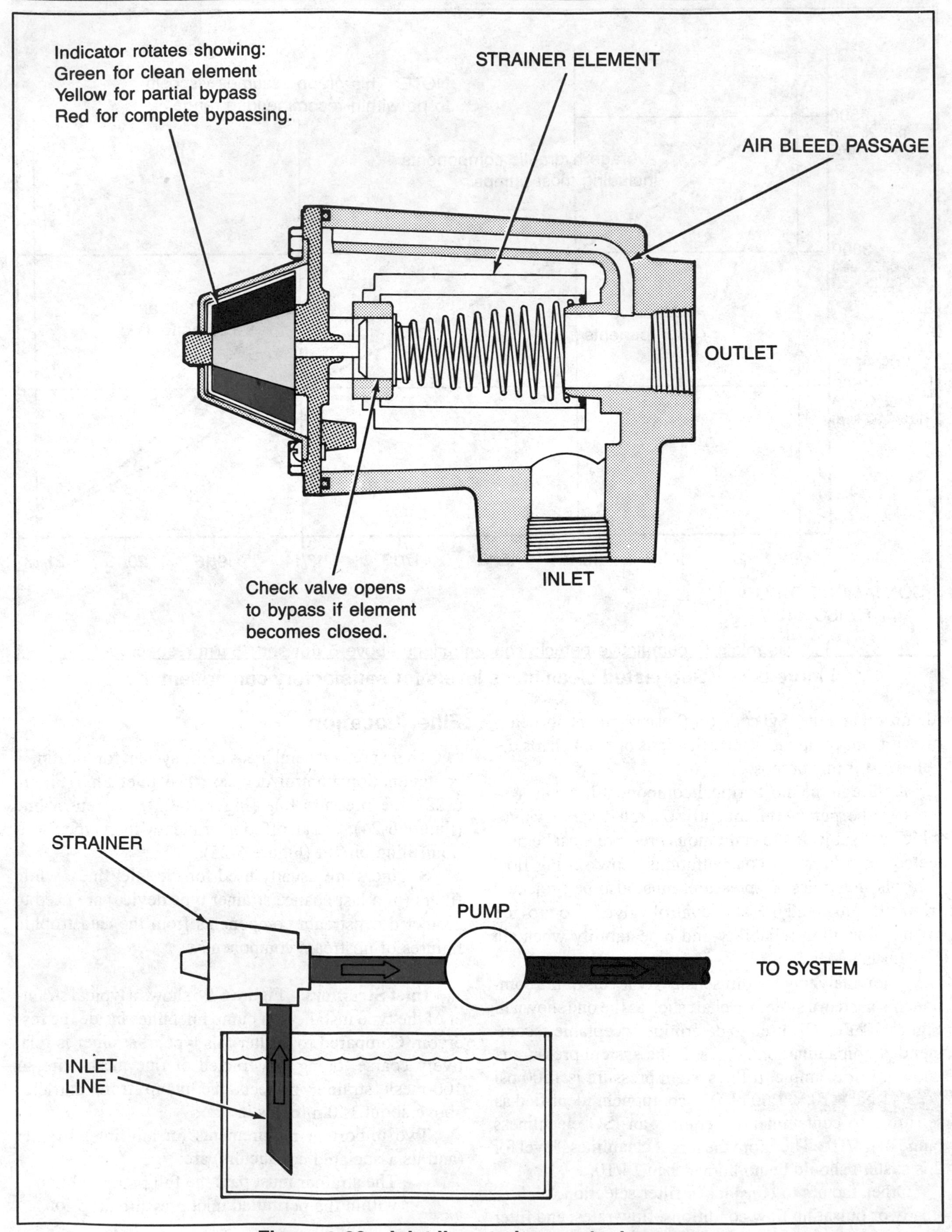
Indicator rotates showing:
Green for clean element
Yellow for partial bypass
Red for complete bypassing.
STRAINER ELEMENT
AIR BLEED PASSAGE
OUTLET
INLET
Check valve opens
to bypass if element
becomes closed.
STRAINER
PUMP
TO SYSTEM
INLET
LINE

Figure 6-22. Inlet line strainer protects pump.

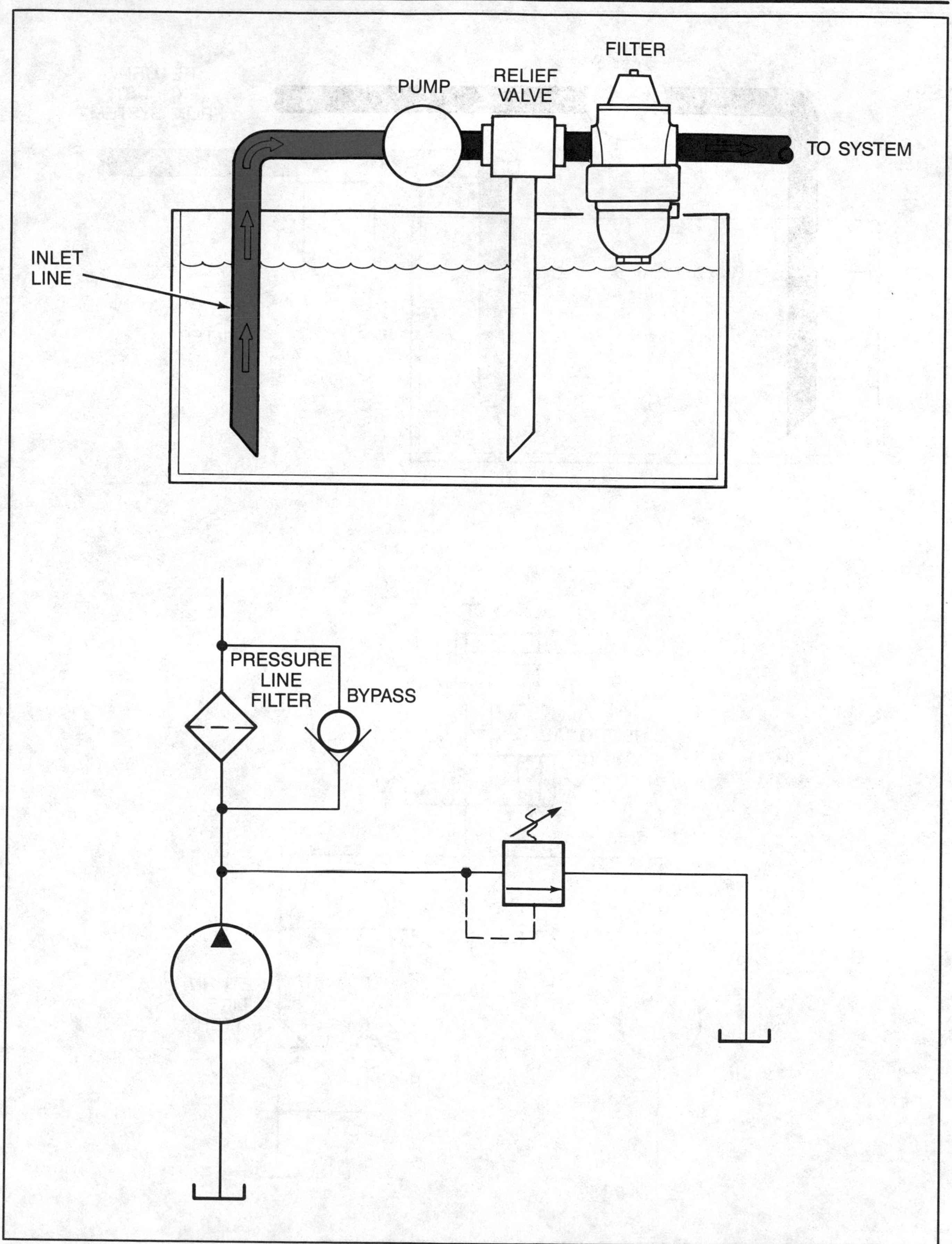

Figure 6-23. Pressure line filter is downstream from pump.

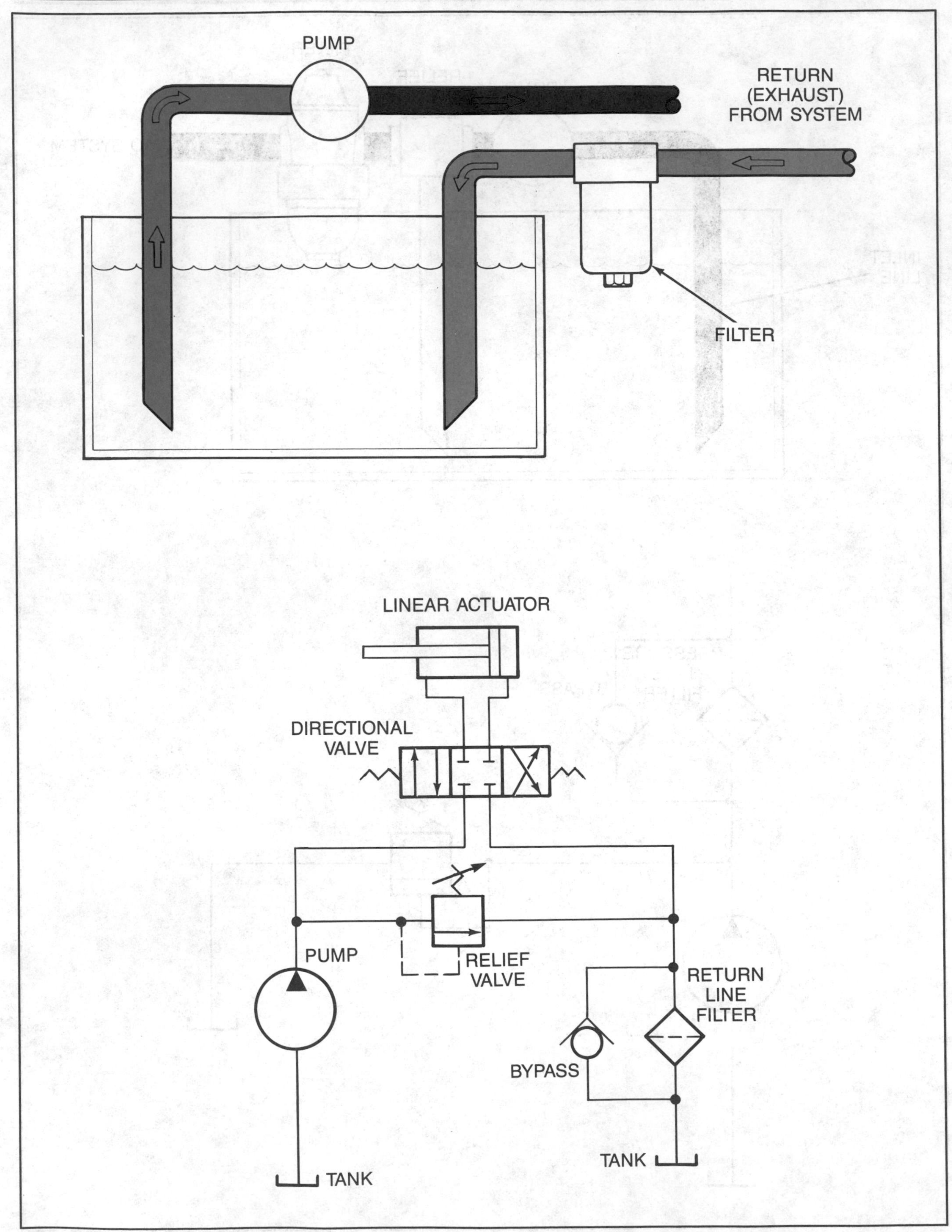

Figure 6-24. Return line filter keeps contamination from reservoir.

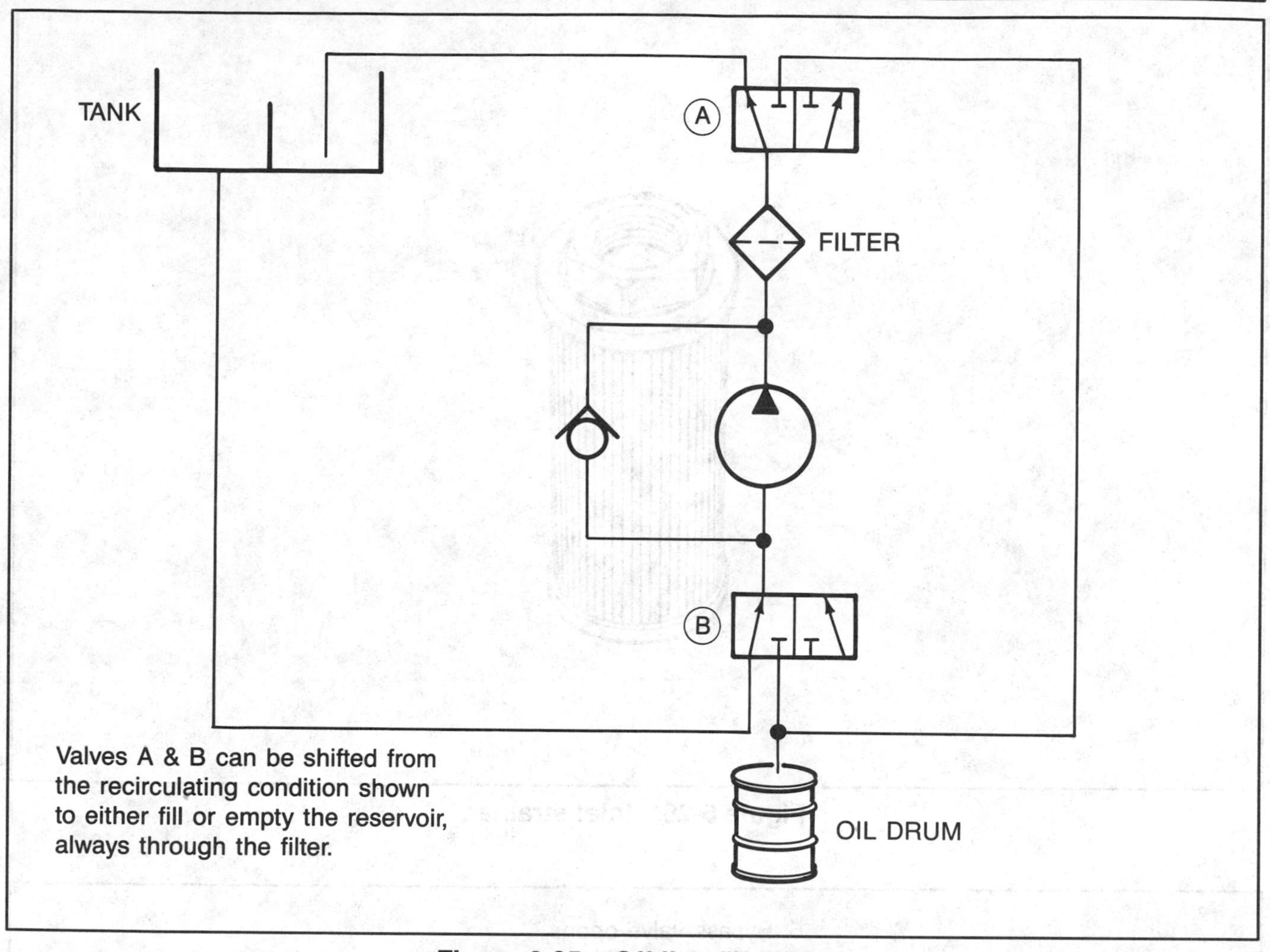

Figure 6-25. Off-line filtration.

- The strainer must provide bypass flow which is still within that limit when the strainer element is blocked.

The level of contamination entering the pump is a critical factor. Inlet strainers should only be used to prevent large particles from entering the pump and causing catastrophic failure.

Pressure Line Filters. A number of filters are designed for installation right in the pressure line (Figure 6-23) and can trap much smaller particles than inlet line strainers. Such a filter might be used where system components are less dirt-tolerant than the pump or to protect downstream components from pump deterioration. The filter traps fine contamination from the fluid as it leaves the pump.

Pressure line filters must be able to withstand the operating pressure of the system as well as any pump pulsations. Changing a pressure line filter element requires shutting down the hydraulic system, unless external bypass valves are provided or an a duplex filter is used.

Return Line Filters. Return line filters (Figure 6-24) also can trap very small particles before the fluid returns to the reservoir. A return line filter is nearly a must in a system with high-performance components which have very close clearances.

Full-flow return filters should have enough capacity to handle maximum return flow with minimal pressure drop. The performance of any return line filter depends on the magnitude of flow, pressure changes, and media selected.

The term "full flow" applied to a filter means that all the flow generated by the system passes through the filtering element. In most full-flow filters, however, there is a bypass valve preset to open at a given pressure drop to divert flow past the filter element. This prevents the element from being subjected to excessive pressures which could cause collapse.

Flow, as shown in Figure 6-27, is outside to inside; that is, from around the element through it to its center. The bypass opens when total flow can no longer pass

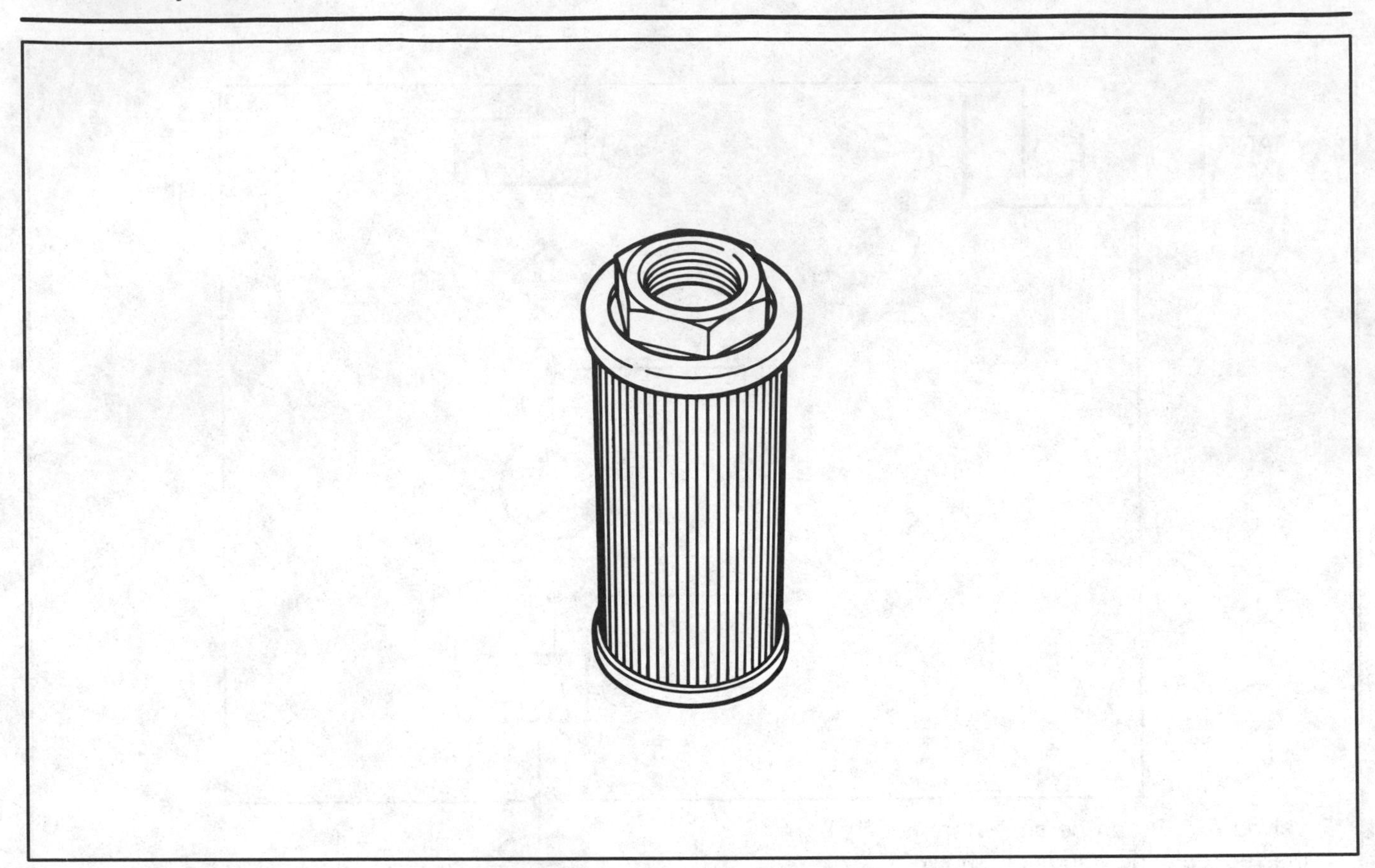

Figure 6-26. Inlet strainer.

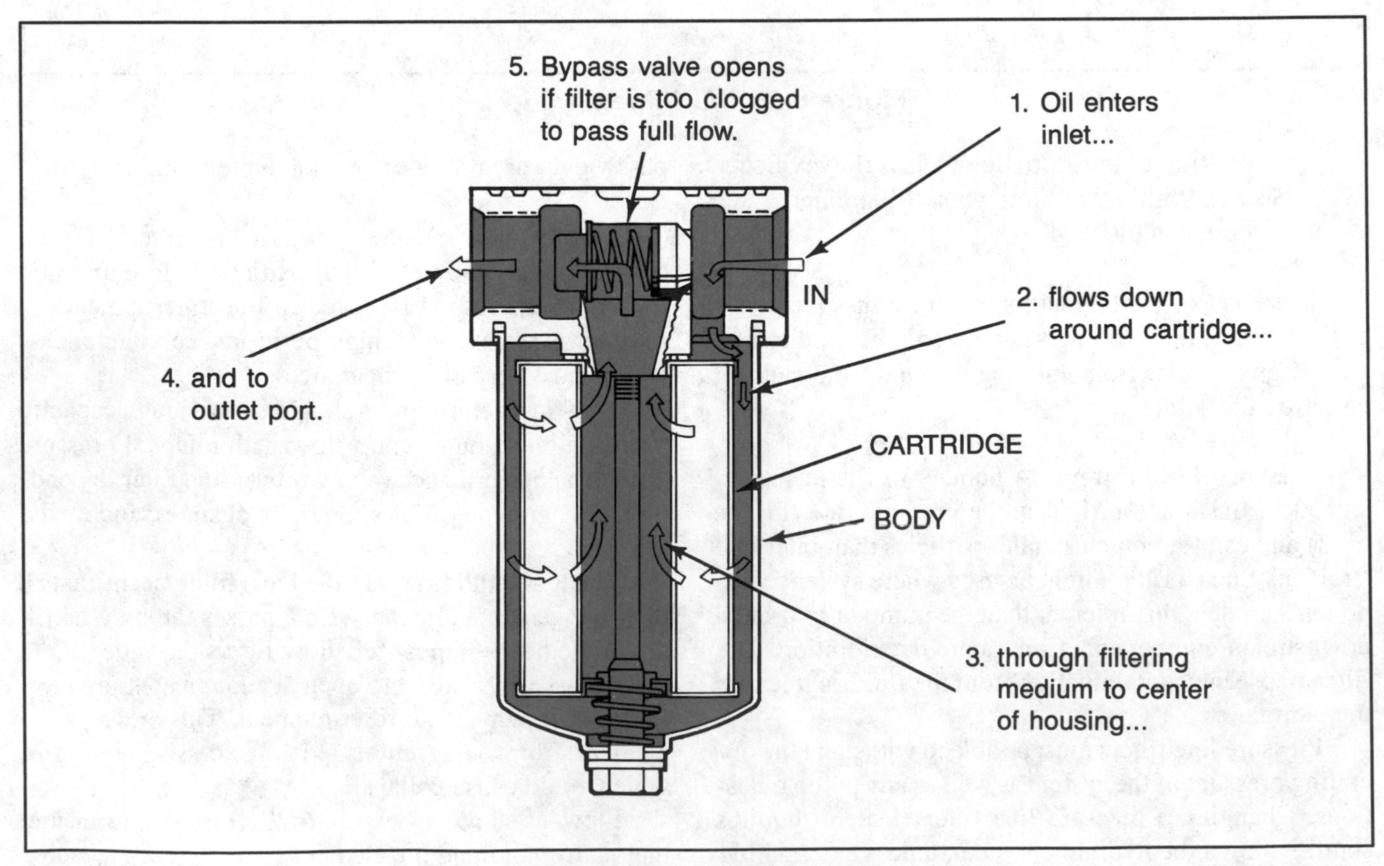
5. Bypass valve opens if filter is too clogged to pass full flow.
1. Oil enters inlet...
IN
2. flows down around cartridge...
4. and to outlet port.
CARTRIDGE
BODY
3. through filtering medium to center of housing...

Figure 6-27. Full-flow filter.

through the contaminate element without raising the pressure above the cracking pressure of the bypass valve.

Off-Line Filter Systems. The desirability and cost effectiveness of pressure and return line filters can be affected by shock, surges, pulsation, and vibration, depending on media types and how well they are supported. Steady flow, relatively free of pressure fluctuations, provides optimum filter performance. The simplest way to achieve this is to supplement the filter in the main system with an independently powered recirculating system where filter performance is subject to fewer variables. Off-line filter systems in which reservoir fluid is circulated through a filter at a constant rate are sometimes used when operating system conditions are severe and the needed quality of filtration is difficult to obtain within the operating system. See Figure 6-25.

With off-line filtration, flow rate or filter type can be altered readily without affecting the design of the main system. Furthermore, the off-line filter system can be run before starting the main system to clean the fluid in the reservoir and reduce the contamination level the pump is subjected to at start-up. A simple valve can be added to filter the initial charge of fluid and any subsequent make-up fluid. Off-line filtration should run continuously to provide clean fluid for every start-up.

Being independent of the main hydraulic system, off-line filters can be placed where they are most convenient to service. Element changes do not affect the main system as the change can be performed at any time without interrupting the operation of the main system.

HOW TO SPECIFY FILTERS

Specifying the correct filter or strainer for a given application requires consideration of several important factors, including: the minimum size of the particles to be controlled, the quantity or weight of the particles to be held, the flow rate capacity, the type of filter condition indicator provided, the pressure rating, the pressure drop through the filter element, and the filter's compatibility with the system fluid among others.

Filter or Strainer

There will probably always be controversy in the industry over the exact definitions of filters and strainers. In the past, many such devices were named filters, but technically classed as strainers. To minimize the controversy, We offer these definitions:

- **Filter** — a device whose primary function is the retention, by some porous medium, of insoluble contaminants from a fluid.
- **Strainer** — a coarse filter, usually metal, with pores larger than 50 microns.

To put it simply, whether the device is a filter or strainer, its function is to trap contaminants from fluid flowing through it. "Porous medium" simply refers to a screen or filtering material that allows fluid to flow through it, but stops other materials above a specific size.

Minimum Size of Particles to Be Trapped

A simple screen or a wire strainer is usually rated for filtering pore size by a **mesh** number or its near equivalent **standard sieve** number. The higher the mesh or sieve number, the finer the screen.

Filters used to be described by nominal and absolute ratings in microns. A filter nominally rated at 10 microns, for example, would trap some particles 10 microns in size or larger. The same filter's absolute rating, however, would be a somewhat higher size, perhaps 25 microns.

The absolute rating was, in effect, the size of the largest opening or pore size in the filter media. However, so many different standards for determining absolute ratings were used, that the term's value for comparative purposes was severely reduced.

Filter rating today is far more precise than in the days of the old nominal and absolute ratings. Some manufacturers still use the "10 or 25 micron filter" reference, but also use Beta ratios to more closely identify how efficiently a filter performs.

Beta Ratio. The Beta ratio or rating is used for most hydraulic or lubricating system filters and is determined under laboratory conditions using a procedure developed in the early 1970s. Although not a true measure of how well a filter will do in an operating system, the Beta rating is a good indicator of the filtration performance. The Multi-pass test used to determine Beta ratios has been incorporated into ISO and other standards organizations.

Simply stated, the Beta ratio of a filter during steady state flow test is the count upstream divided by the count downstream of fine test dust, based on any selected particle size:

$$Beta_{(x)} = \frac{Number\ of\ upstream\ particles > x}{Number\ of\ downstream\ particles > x}$$

$$x = particle\ size$$

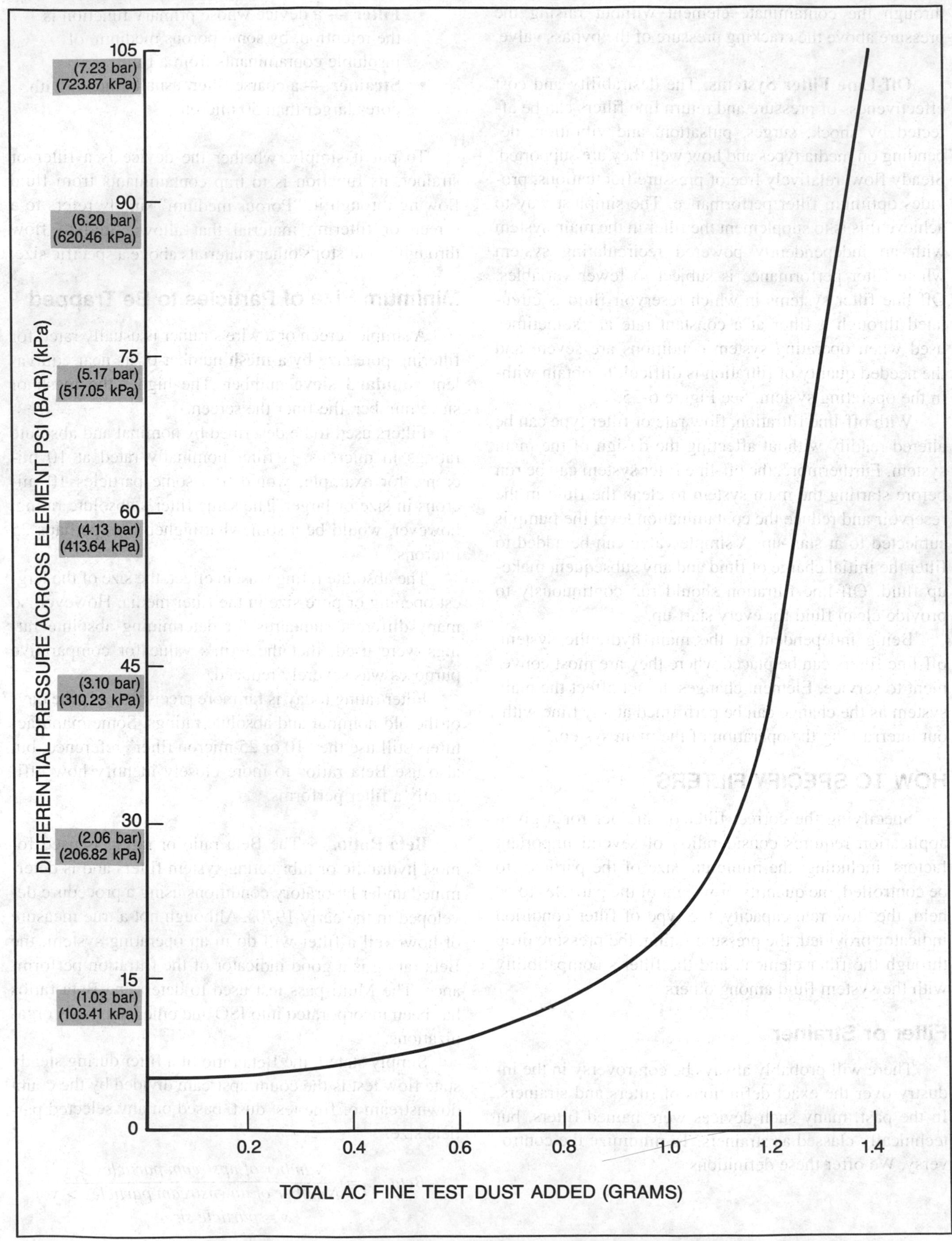

Figure 6-28. Typical dirt holding capacity curve for hydraulic filter element.

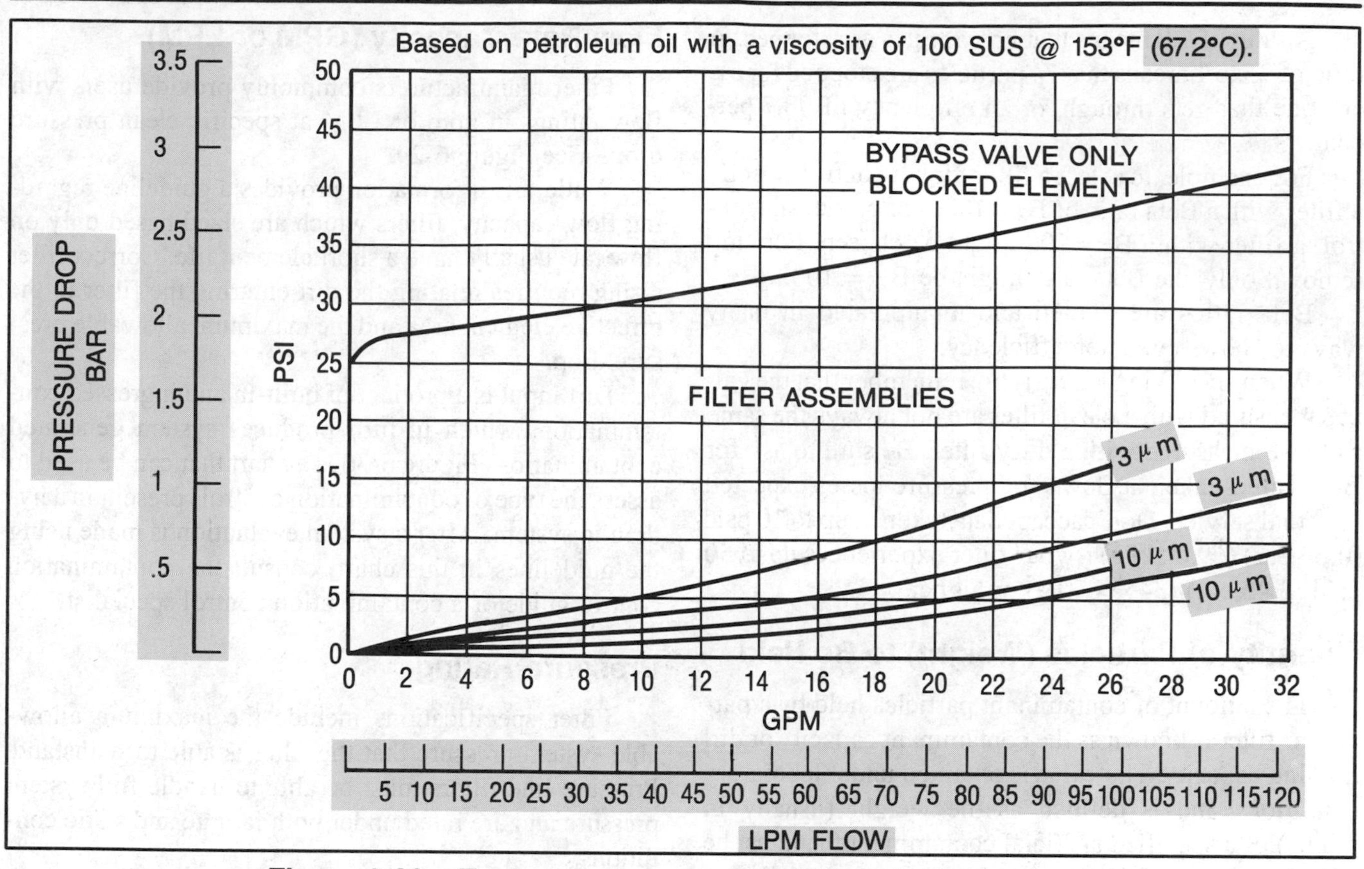

Figure 6-29. Typical performance and flow characteristics.

Contamination Source	Controller
Inbuilt in components, pipes, manifolds, etc.	Good flushing procedures, system not operated on load until acceptable contamination level obtained.
plus Present in initial charge of fluid	Integrity of supplier. Fluid stored under correct conditions (exclusion of dirt, condensation, etc.). Fluid filtered during filling.
plus Ingressed through air breather	An effective air breather with rating compatible with degree of fluid filtration.
plus Ingressed during fluid replenishment	Suitable filling points which ensure some filtration of fluid before entering reservoir.
plus Ingressed during maintenance	This task undertaken by responsible personnel. Design should minimize the effects.
plus Ingressed through cylinder rod seals	Effective wiper seals or, if airborne contamination, rods protected by suitable gaiters.
plus Further generated contamination produced as a result of the above and the severity of the duty cycle.	Correct fluid selection and properties (viscosity and additives) maintained. Good system design minimizing effects of contamination present on system components.

Figure 6-30. Guidelines for controlling contamination in hydraulic systems.

A ratio of 1.0 means that no particles are stopped. A ratio of 75 to 1 means that 74 particles are stopped for every one that gets through, or an efficiency of 98.7 percent.

For example, to select a filter for silt control, specify a filter with a Beta ratio of $B_3 = 100$. For partial silt control, a filter with a $B_5 = 100$ might be chosen. For chip removal only, the Beta ratio might be $B_{25} = 100$.

Beta ratios are plotted and manipulated in many ways to show separation efficiency.

When specifying a Beta ratio, remember that the values measured with a clean filter are not always the same as those measured with a dirty filter. Be sure to ask for Beta values taken at flows and pressure losses expected in actual service. Don't accept data taken at up to 10 psid (0.69 bar) (69.00 kPa) if your filter experiences up to 50 psid (3.45 bar) (345.00 kPa) or higher.

Quantity of Particles (Weight) to Be Held

The amount of contaminant particles held by a particular filter is known as the contaminant capacity or dirt holding capacity. The value is obtained under laboratory conditions and is defined as the weight (usually in grams) of a specified artificial contaminant that must be added upstream of the filter to produce a given differential pressure across the filter under specific conditions.

The artificial contaminant is added at a constant rate to a continuously recirculating oil system and the resultant increase in differential pressure is plotted against the weight of contaminant added, as shown in Figure 6-28. The resulting curve has a characteristic form which is constant for a given filter media.

Contaminant capacity is sometimes used as an indication of relative service life. This should only be done when all other variables are constant and equal among the units being compared. Filters must have enough dirt holding capacity to provide acceptable time intervals between element changes. While large capacity filters cost more initially, the differential costs are generally recovered quickly in reduced operating costs. Fewer element changes are required, reducing labor costs and downtime, resulting in higher productivity.

Flow Rate Capacity (GPM or LPM)

Filter manufacturers commonly provide users with flow ratings in gpm or LPM at specific clean pressure drops. See Figure 6-29.

While this information provides a guideline regarding flow capacity, filters which are sized based only on flow rate usually have a short element life. Correct filter sizing requires relating the dirt entering the filter to the effective element area and the maximum allowable pressure drop.

Dirt input is a product of built-in and ingressed contamination, which in turn produces system generated contamination. Figure 6-30 is a chart that can be used to assess the type of contamination controls present in a hydraulic system. After a system evaluation is made using the guidelines in this chart, consult the contamination control guide or a contamination control specialist.

Pressure Rating

Filter specifications include the maximum allowable system pressure that the filter is able to withstand. Pressure line filters must be able to handle full system pressure and are rated under both fatigue and static conditions.

Pressure Drop Through the Filter Assembly

The initial and terminal pressure drops (or pressure differentials) across a filter element are important. To minimize energy consumption, the average pressure drop across filters should be as low as possible. A pressure differential of 85 psid at 20 gpm wastes 1 hp at the pump. The maximum pressure drop under bypass conditions should be less than the collapse rating of the element. This maximum value, termed the terminal pressure drop (usually 45 psi (3 bar) (300 kPa)), is controlled by the filter manufacturer and depends on media strength and construction details.

Filter manufacturers provide charts, tables, and graphs to select filters on the basis of flow and pressure drop. These pressure-flow characteristics refer to a clean element and assembly. The pressure drop across a filter is a measure of the resistance to flow across the filter due to kinetic and viscous effects. In actual service, factors that affect the filter pressure drop include fluid viscosity, fluid specific gravity, flow, operating temperature, filter size, permeability of the filter medium, restrictions in the housing, and the degree to which the medium is loaded with contaminant.

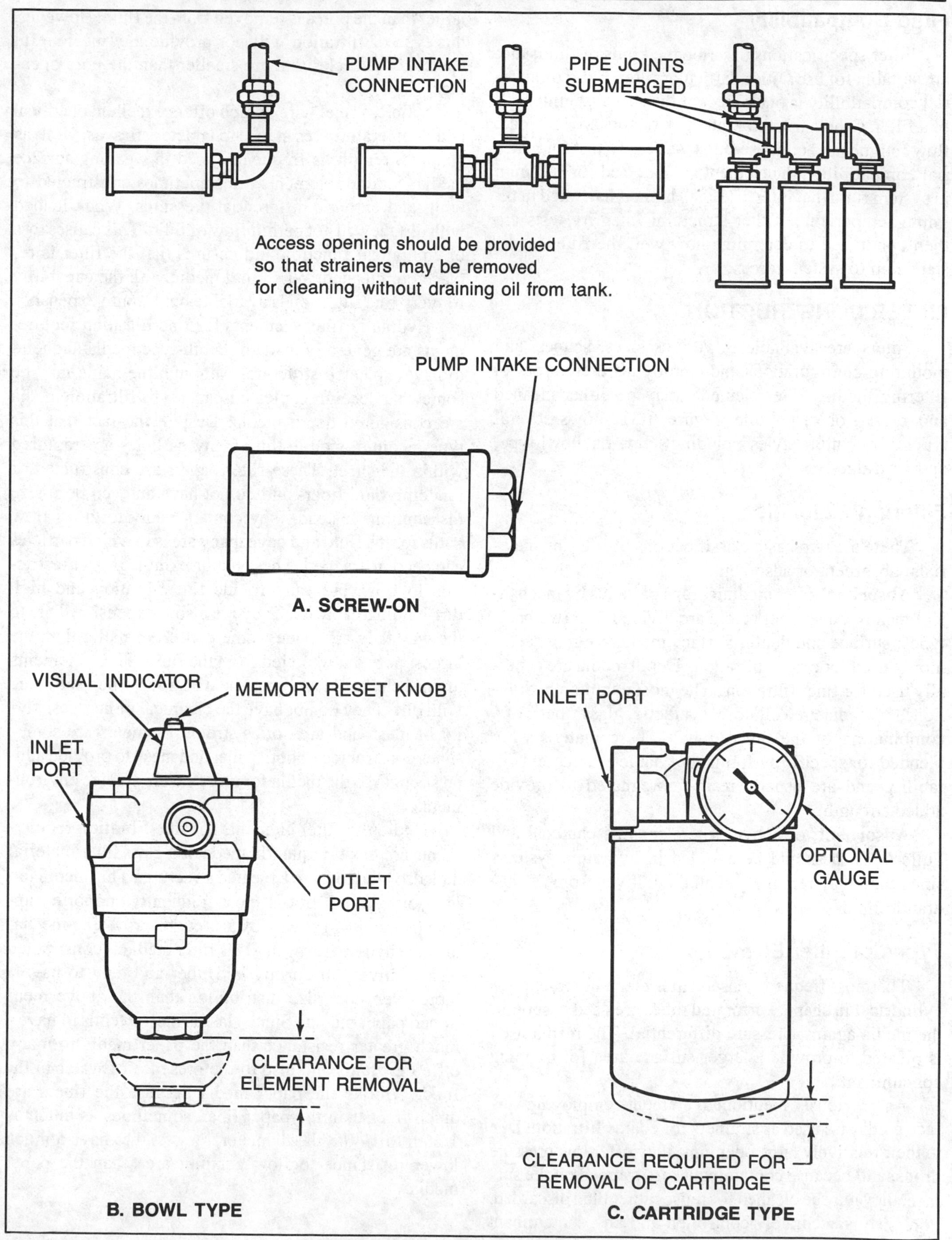

Figure 6-31. Filter mounting types.

Fluid Compatibility

Filter specifications describe the kinds of fluids that are suitable for use with a particular filter. Fluid or material compatibility is one aspect of the structural integrity of a filter. Other structural integrity factors are end load, flow fatigue, and collapse-burst values. To test for material compatibility, the element is immersed for 72 hours in system fluid that is 59°F (15°C) higher than rated maximum temperature. Other structural integrity tests are then conducted to determine how well the filter media stands up to system forces.

FILTER CONSTRUCTION

Filters are available in various sizes, shapes, and mounting configurations and employ different types of filtering media. Filter elements may be either cleaned and reused or disposable. Figure 6-31 shows several types of mounting styles including screw-on, bowl-type, and cartridge-type.

Filtering Materials

There are two basic classifications of filtering materials: absorbent or adsorbent.

Absorbent filter medium traps particles by mechanical means. Absorbent media are divided into two basic types: surface and depth. Surface media are most commonly used for coarse filtration. Depth media are generally used for finer filtration. The most common hydraulic filter media are cellulose, synthetic, glass fibers, or a combination of the preceeding. These materials are blended for specific performance characteristics and durability and are usually resin-impregnated to provide added strength.

Adsorbent, or active, filters such as charcoal and Fuller's earth should be avoided in hydraulic systems since they may remove essential additives from the hydraulic fluid.

Types of Filter Elements

The most frequently used filter element structure is cylindrical in shape. Perforated tubes are used to support the media against pressure differential. The media used is pleated to provide a larger surface area for trapping contaminant.

As previously mentioned, elements employing surface media are used as strainers for coarse filtration. Due to their relatively large openings, they offer low pressure drops, and because contaminants are trapped on the surface, they may be cleaned for reuse rather than discarded. Pore size is accurately controlled and any contaminant larger than the pores is removed from the fluid. However, this type of filtration will not provide any meaningful control of particles that are smaller than the pore openings.

Stainless steel wire, which offers excellent chemical and temperature resistance characteristics as well as greater strength, is frequently used in weaving surface media. Strainer elements are sometimes constructed to withstand extreme differential pressures. When loaded with dirt they do not permit flow of fluid. This causes loss of equipment function and ensures that the filter is replaced or cleaned. Wire cloth media with the capability for very fine filtration is available but is quite expensive.

Hydraulic filter elements for fine filtration requirements are generally made of depth-type media and tend to have a pleated structure. Although the name may be somewhat deceiving, pleated-paper type filtration media are considered depth media. Despite the fact that this type medium is quite thin, many particles are captured within the sheet. These filter papers are constructed of randomly laid fibers and do not have an even pore-size distribution. Instead, they contain many tortuous flow paths for the fluid and have many areas in which particles can become trapped. These media exhibit greater restriction to flow, but can provide fine filtration and high dirt-holding capacity. They are also inexpensive. Due to the way they are made, some particles smaller than the largest pore are removed from the fluid. These elements are not cleanable and must be discarded when saturated with dirt. They do not have the chemical or heat resistivity of mesh and are not as strong. Elements are sometimes constructed of both paper and mesh to provide added strength to the media for more severe service environments.

Hydraulic filter elements for fine filtration requirements are most frequently made of a glass fiber material and also tend to have a pleated structure. This media too is also very thin but is most frequently supported upstream and down stream by a strength-providing mechanism, usually wire mesh. This filter media is constructed of a relatively uniformly laid fiber and tend to have a more even pore size distribution than the above mentioncd paper media. Since glass is inert it tends to have a much greater resistance than the paper media to many different fluids and also to the presence of water in the fluid. Unlike the paper media because the fibers are much finer than the paper fiber, sometimes as much as 1/25th to 1/40th the diameter, they tend to have a much lower resistance to flow per unit area than the paper media.

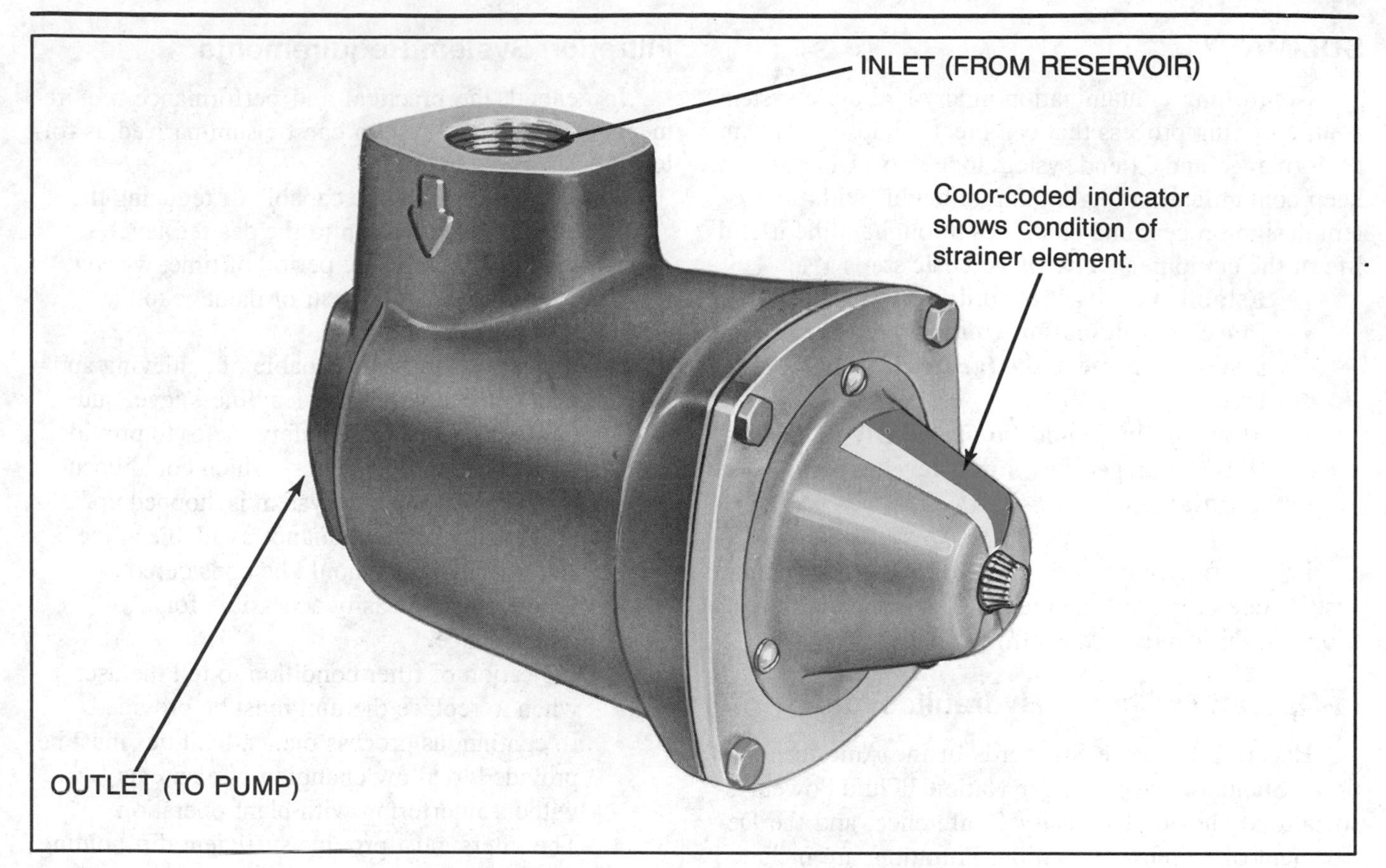
INLET (FROM RESERVOIR)
Color-coded indicator shows condition of strainer element.
OUTLET (TO PUMP)

Figure 6-32. Strainer condition indicator.

The glass media elements also tend to have relatively high dirt holding capacity in comparison to strainers and to paper media per unit volume. Advances in the manufacturing processes of glass media have made them extremely competitive with paper media. Because they control much finer particles than paper media, the result is greatly reduced operating expenses and greatly extended component life, filter element life and fluid life. Like the paper element these elements are not reusable and must be discarded when saturated with dirt. However since they are madc of glass they do tend to have a high resistance to chemicals and to heat and when the mesh is properly supported both upstream and down stream by stainless steel wire mesh they are quite strong.

Filter Bypass Valves

Generally, a bypass valve is incorporated into the housing to eliminate the possibility of element collapse or rupture when the element becomes clogged or when the fluid is very viscous. The bypass valve is almost a necessity for most filter assemblies operating under cold start-up conditions.

Filter Condition Indicators

Filter condition indicators are often used to tell when the filter element should be replaced (Figure 6-32). There are two basic types of indicators: those that are linked to the filter's bypass valve and those that react only to the differential pressure. Indicators that are linked directly to the bypass valve show the actual position of the bypass valve and indicate movement of the bypass valve. Differential pressure indicators show only that the preset pressure drop has been attained. This type is usually set at a pressure drop somewhat below that at which the bypass valve opens. Both types of indicators are configured in visual and electrical models or a combination of both. The electrical model can be used to light a warning lamp, activate a central buzzer, or even to shut a machine down until the element is replaced, tending to ensure that the system fluid is being filtered.

SUMMARY

Controlling contamination in any hydraulic system is an on-going process that can greatly improve system performance and extend system longevity. The effort to keep contamination to a minimum begins with the system design process and continues throughout the useful life of the equipment. **The three basic steps are:**

- **Establish a target cleanliness level.**
- **Place contamination control devices in the system to achieve the target cleanliness level.**
- **Monitor fluid condition regularly to ensure that the target cleanliness level is being maintained.**

By carefully considering the following key points, system operators and maintenance personnel can contribute significantly to this effort.

In-Operation Care of Hydraulic Fluid

Excerpts from the Standards of the American National Standards Institute, The National Fluid Power Association, The Joint Industry Conference, and the Department of Defense Regarding Filtration, are included in Appendix C.

Key areas for proper in-operation care of hydraulic fluid include:

- Preventing contamination by keeping the system tight and using proper air and fluid filtration devices and procedures.
- Establishing fluid change intervals so the fluid will be replaced before it breaks down. If necessary, the fluid supplier can test samples in the laboratory at specific intervals to help establish the frequency of change.
- Keeping the reservoir filled properly to take advantage of its heat dissipating characteristics and to prevent moisture from condensing on inside walls.
- Repairing all leaks immediately.

Filtration System Requirements

In general, the practical and performance requirements of a filtration system can be summarized as follows:

- The system must be capable of reducing the initial contamination to the desired level within an acceptable period of time, without causing premature wear or damage to the hydraulic components.
- The system must be capable of achieving and maintaining the target cleanliness level and must allow a suitable safety factor to provide for a concentrated ingress which could occur, for example, when a system is "topped up."
- The quality of maintenance available at the end user location should be considered.
- Filters must be easily accessible for maintenance.
- Indication of filter condition to tell the user when to replace the unit must be provided.
- In continuous process plants, facilities must be provided to allow changing of elements without interfering with plant operation.
- The filters must provide sufficient dirt holding capacity for an acceptable interval between element changes.
- The inclusion of a filter in the system must not produce undesirable effects on the operation of components, for example, high back pressures on seal drains.
- Sampling points must be provided to monitor initial and subsequent levels of contamination.

QUESTIONS

1. Name three factors that must be considered when examining the effect of contamination on the life and operation of hydraulic machinery.
2. Describe how contamination can interfere with power transmission.
3. Describe how contamination interferes with the cooling function of hydraulic fluid.
4. Describe how contamination interferes with the lubrication of moving parts.
5. How large is a micron?
6. Define contamination.
7. Name the three kinds of particulate contamination-induced failures in hydraulic components.
8. What areas within a vane pump or motor are the most susceptible to contamination-induced clearance problems?
9. What areas within a gear pump or motor are the most susceptible to contamination-induced clearance problems?
10. What areas within an axial piston pump or motor are the most susceptible to contamination-induced clearance problems?
11. How does pressure influence the effect of contamination on hydraulic components?
12. What type of contamination problem is most common with directional valves?
13. What is the most common type of contamination problem in pressure controls?
14. What is the most important factor determining contamination tolerance of flow control valves?
15. If only a chemical or physical analysis of fluid precipitants and foreign particulates is needed in assessing the contamination level of a system, what kind of sample would be taken?
16. If a fluid sample must be representative of the contamination level that exists within the system under actual operating conditions, what kind of sample would be taken?
17. What are the two types of dynamic sampling methods and which one is the easiest to use?
18. Name the particle size analysis methods.
19. A contaminant particle analysis is conducted with the following results (100 ml sample):
    - Number of particles above 5 micrometers equals 400,000
    - Number of particles above 15 micrometers equals 10,000.

    What is ISO code for this sample?
20. A certain in-line piston motor has a contamination tolerance rating of 16/13 or cleaner. To safely operate this motor, what are the maximum number of particulates greater than 5 micrometers and greater than 15 micrometers in size for a 100 milliliter sample?
21. What are three sources of particulate contamination in hydraulic systems?
22. Describe proper fluid storage and handling procedures.
23. Name two ways in which a portable filtration unit might be used.
24. Name three possible locations for a filter within a hydraulic system?
25. When might an off-line filtration system be used?
26. What does "full flow" mean?
27. Name five of the seven factors discussed in this chapter that should be considered when specifying a filter.
28. What is a filter? What is a strainer?
29. What is used today in place of nominal and absolute ratings for contamination control devices?
30. What is meant by a Beta ratio of $B_3 = 100$ and what is the efficiency?
31. What is the purpose of filter condition indicators?
32. Describe how an electric-type filter condition indicator might be used.
33. Describe the difference between surface and depth-type filtering material.

CHAPTER 7
HYDRAULIC ACTUATORS

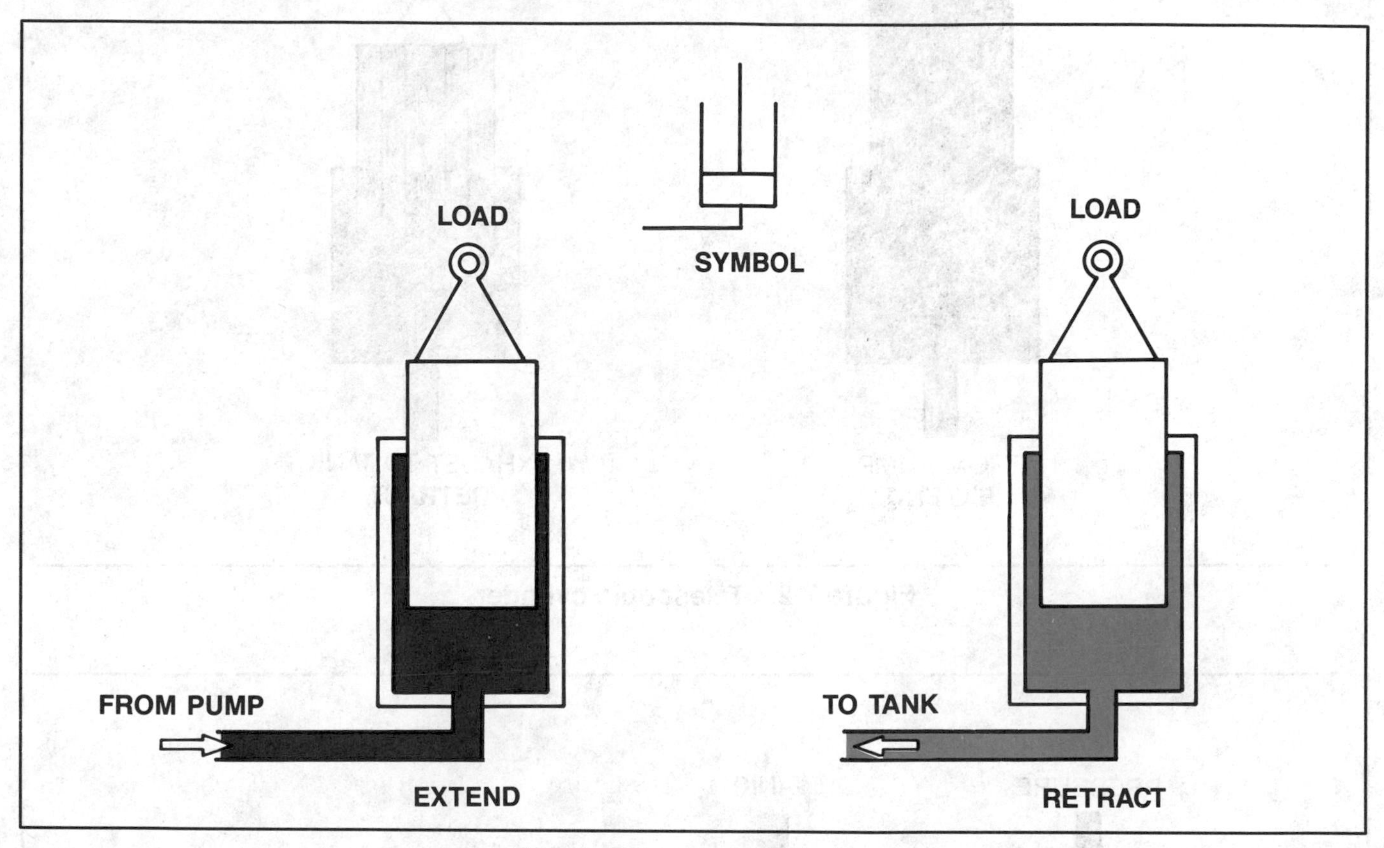

Figure 7-1. Ram cylinder.

The focus of this chapter is on the output member or actuator, a device for converting hydraulic energy into mechanical energy. Two types of hydraulic actuators are cylinders and motors. The type of job done and its power requirements determine the correct type and size motor or cylinder for an application.

CYLINDERS

Cylinders are **linear** actuators. This means that the output of a cylinder is a straight-line motion and/or force. The major function of a hydraulic cylinder is to convert hydraulic power into linear mechanical power.

Types of Cylinders

There are several types of cylinders including single-acting and double-acting cylinders. Design features of the most common types are explained below.

Single-Acting Cylinder. A single-acting cylinder is pressurized on one end only. The opposite end is vented to tank.

Ram. (Figure 7-1) Perhaps the simplest actuator is the ram type. It has only one fluid chamber and exerts force in only one direction. Most are mounted vertically and retract by the force of gravity on the load. Practical for long strokes, ram-type cylinders are used in elevators, jacks and automobile hoists.

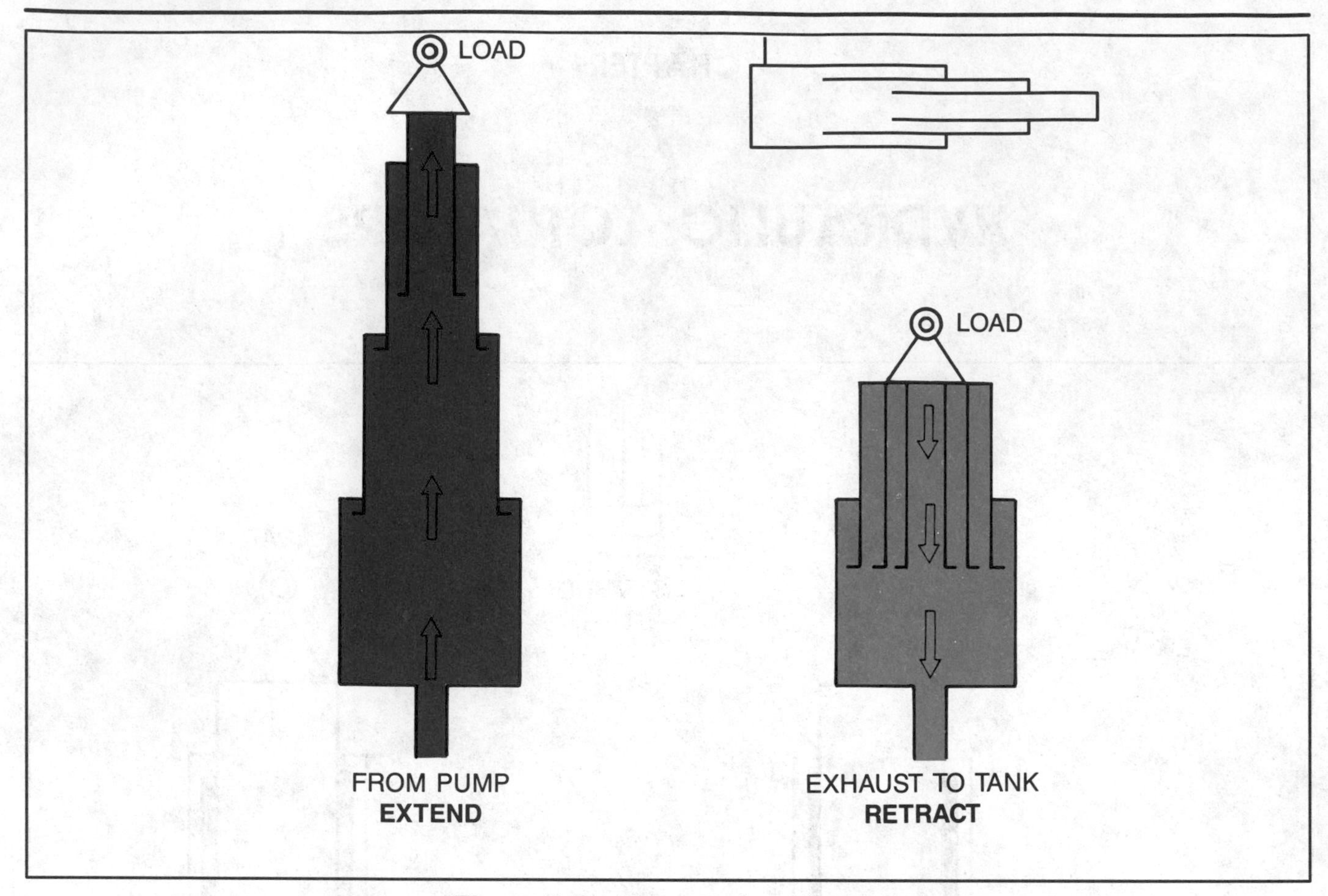

Figure 7-2. Telescopic cylinder.

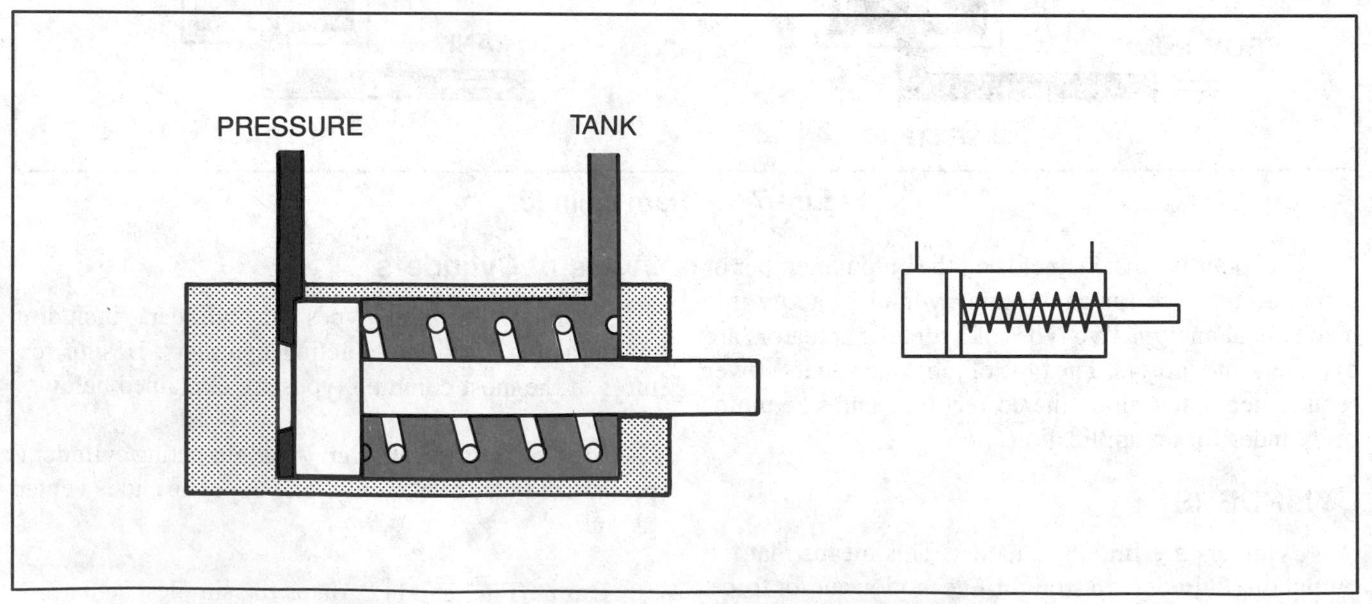

Figure 7-3. Spring return, single-acting cylinder.

Telescopic Cylinder. (Figure 7-2) Most telescoping cylinders are single-acting. The telescoping cylinder is equipped with a series of nested tubular rod segments called sleeves. These sleeves work together to provide a longer working stroke than is possible with a standard cylinder. Up to four or five sleeves can be used. The maximum force can be exerted when the cylinder is collapsed. In the extended position, the force is a function of the area of the smallest sleeve.

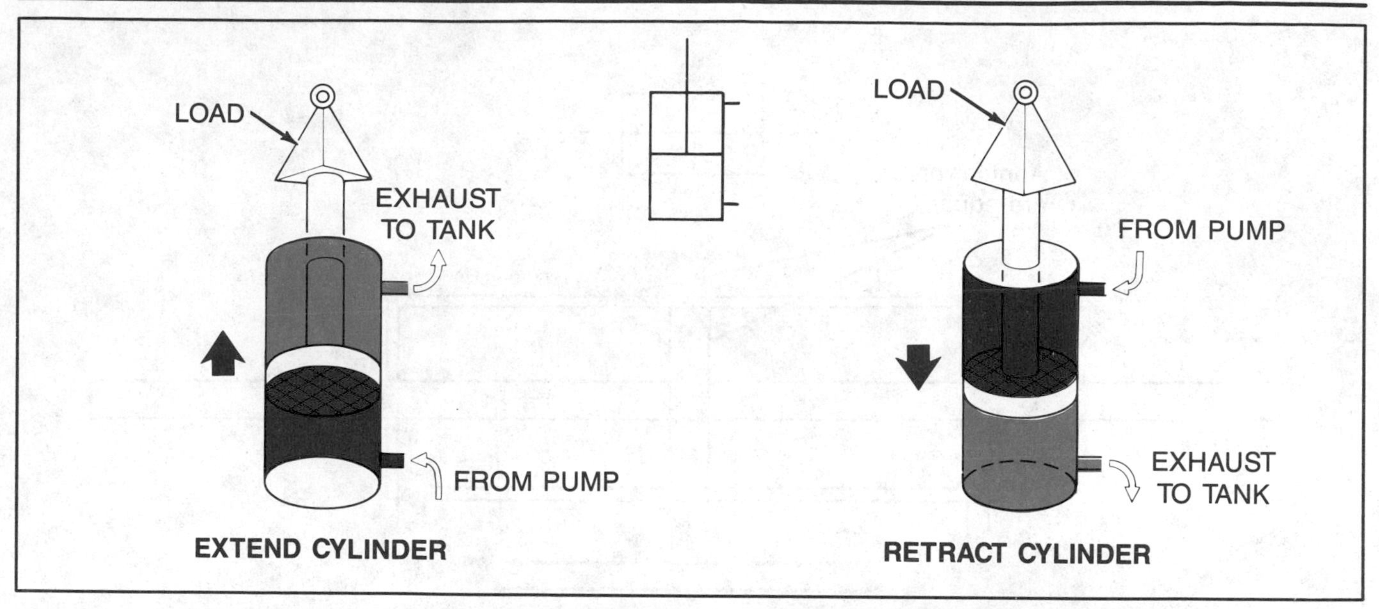

Figure 7-4. Basic double-acting cylinder.

Spring Return. (Figure 7-3) A spring return cylinder is also considered a single-acting cylinder. Pressure applied to the cap-end port compresses the spring as the rod extends. With pressure removed, the spring retracts the rod.

Double-Acting Cylinder. The double-acting cylinder is the most common type in industrial hydraulics. Hydraulic pressure is applied to either port, giving powered motion when extending or retracting.

Basic Double-Acting Cylinder. (Figure 7-4) The majority of cylinders in use are basic double-acting cylinders. These cylinders are classed as differential cylinders because there are unequal areas exposed to pressure during the extend and retract movements. The difference is caused by the cross-sectional area of the rod which reduces the area under pressure during retraction.

Extension is slower than retraction because more fluid is required to fill the swept volume of the piston. However, greater force is possible because the pressure operates on the full piston area.

When retracting, the same flow from a pump causes faster movement of the cylinder because the swept volume is less. With the same system pressure, the maximum force exerted by the cylinder is also less because of the smaller area under pressure.

Double-Rod Cylinder. (Figure 7-5) A double-rod cylinder is an example of a non-differential type cylinder. There are identical areas on each side of the piston, and they can provide equal forces in either direction. The double-rod cylinder is primarily used where it is advantageous to couple a load to each end, or where equal displacement is needed on each end. Any double-acting cylinder may be used as a single-acting unit by draining the inactive end to tank.

Tandem Cylinder. (Figure 7-6) By mounting two cylinders in line with a common rod, higher forces can be developed with a given pressure and bore size. This cylinder arrangement is mentioned as an example of design flexibility. It is one of many possible techniques that can be considered when applying basic cylinder principles.

Cylinder Construction and Operation

Figure 7-7 shows the cross section of a typical industrial hydraulic cylinder. The chrome plated steel piston rod and piston assembly is the moving part. A cushion collar and/or plunger features are added to the assembly when cushions are desired. The pressure-containing assembly is constructed of a steel cap or blind-end head, a steel body with honed finish, a rod-end head and the rod bearing.

Tie rods and nuts are used to hold the heads and body together. Static seals keep the joint pressure tight. The bearing is commonly retained with a plate and screws for easy removal. A rod wiper is provided to keep foreign material from entering into the bearing and seal area.

Sealing of moving surfaces is provided by the rod seal which prevents fluid from leaking along the rod, and by piston seals which prevent fluid from bypassing the piston.

Fluid is routed to and from the cylinder through ports in each of the heads.

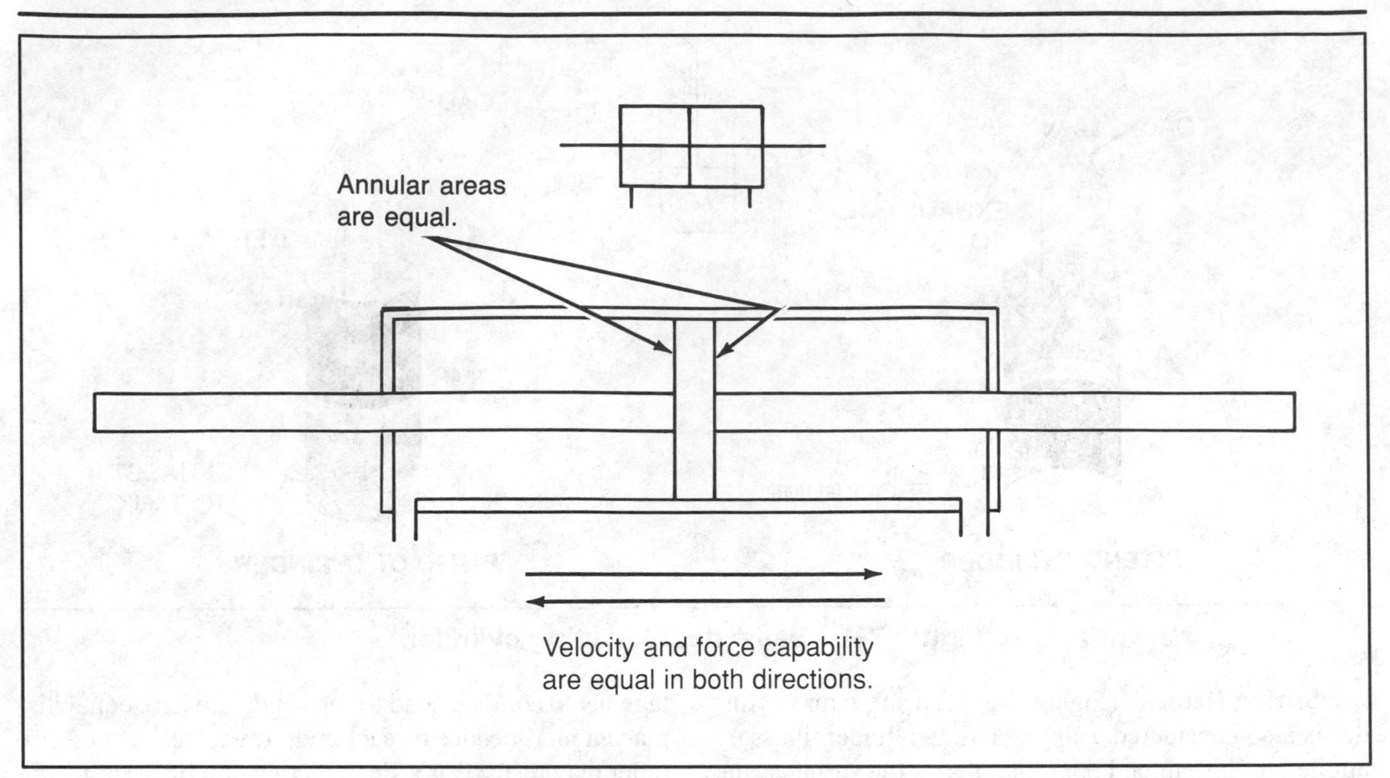

Figure 7-5. Double-rod cylinder.

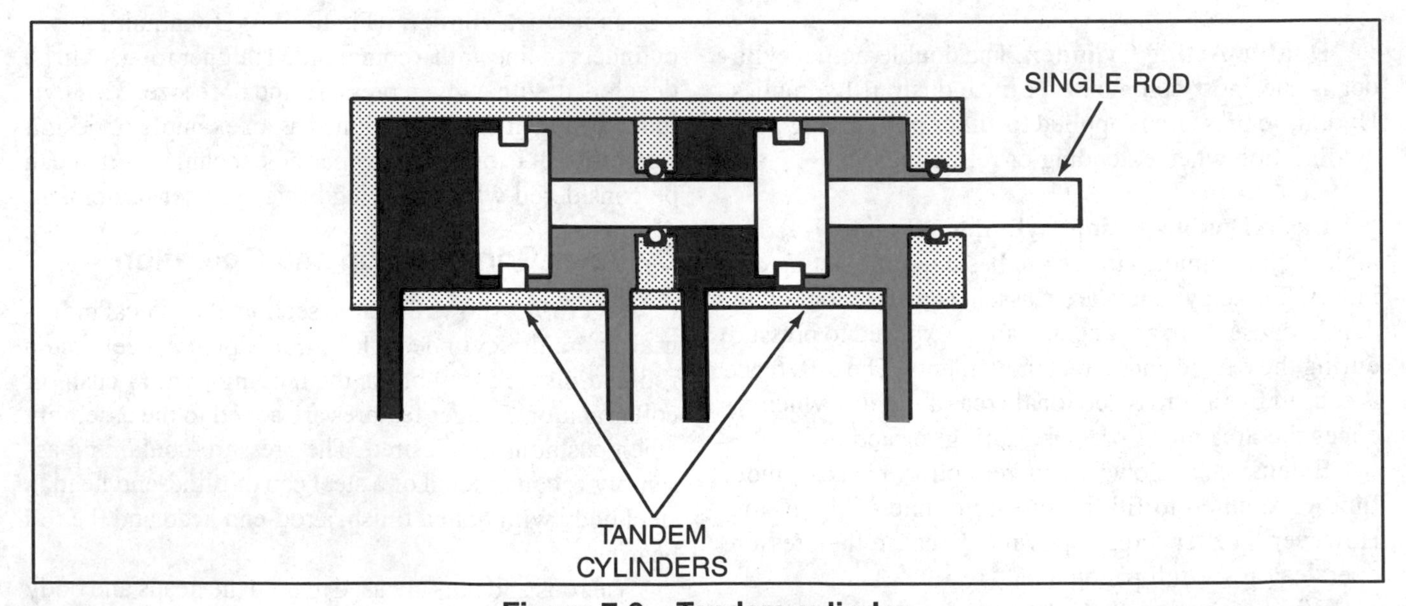

Figure 7-6. Tandem cylinder.

The cylinder rod will retract with pressure to the rod-end port and the other port connected to drain. To extend the rod, pressure the cap-end port and connect the rod-end port to drain.

Cylinder Mounting

The main function of a cylinder mount is to provide a means of anchoring the cylinder. There are a variety of ways to mount the cylinder including the tie rod, bolt mount, flange, trunnion, side lug and side tapped, and clevis. The tie rod is the most common type of industrial mount. A variety of cylinder mounts are shown in Figure 7-8.

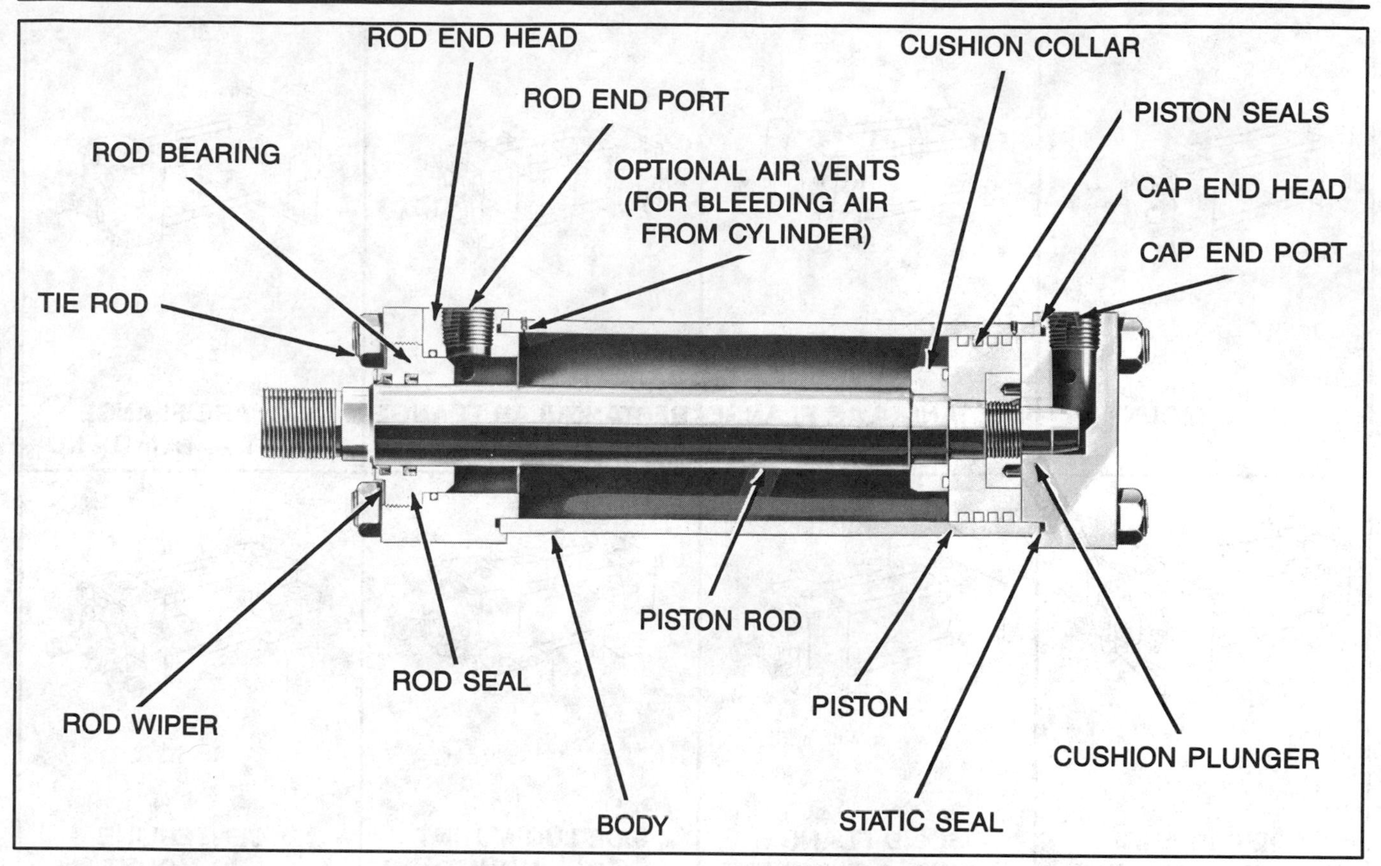

Figure 7-7. Typical cylinder construction.

Cylinder Ratings

The ratings of a cylinder include its size and pressure capability. Principal size features are:

- Piston diameter
- Piston rod diameter
- Stroke length

The pressure rating is established by the manufacturer. Refer to the cylinder nameplate or the manufacturer's catalog for this information.

Cylinder speed, the output force available, and the pressure required for a given load all depend on the piston area (0.785 multiplied by the diameter squared) when extending the rod. When retracting the rod, the area of the rod must be subtracted from the piston area.

Formulas for Cylinder Applications

The following data on cylinder application were developed in Chapter 1:

To find the speed of a cylinder when size and gpm (LPM) delivery are known, use the following formula:

$$Speed\ (in/min) = GPM \times \frac{231}{Effective\ Piston\ Area\ (in^2)}$$

$$Speed\ (mm/sec) = LPM \times \frac{16667}{Effective\ Piston\ Area\ (mm^2)}$$

To find the flow required for a given speed, use the following formula:

$$GPM = \frac{Effective\ Piston\ Area\ (in^2) \times Speed\ (in/min)}{231}$$

$$LPM = \frac{Effective\ Piston\ Area\ (mm^2) \times Speed\ (mm/sec)}{16667}$$

To find the force output for a given pressure, use the following formula:

$$Force\ (lbs.) = Pressure\ (psi) \times Effective\ Piston\ Area\ (in^2)$$

$$Force\ (N) = \frac{Pressure\ (bar) \times Effective\ Piston\ Area\ (mm^2)}{10}$$

To find the pressure required to exert a given force, use the following formula:

$$Pressure\ (psi) = \frac{Force\ (lbs.)}{Effective\ Piston\ Area\ (in^2)}$$

$$Pressure\ (bar) = \frac{Force\ (N) \times 10}{Effective\ Piston\ Area\ (mm^2)}$$

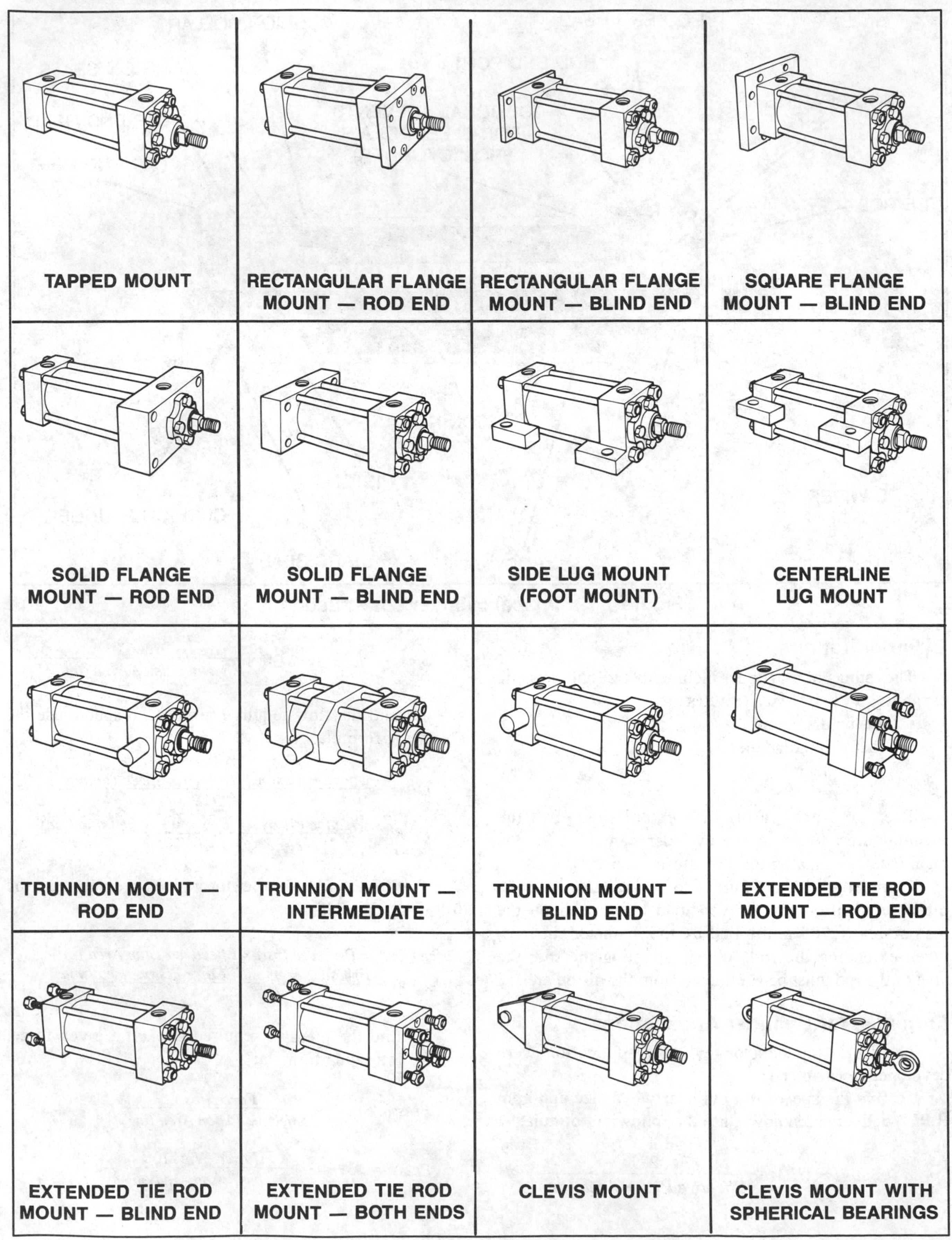

Figure 7-8. Cylinder mounting methods.

CHANGE	SPEED	EFFECT ON OPERATING PRESSURE	OUTPUT FORCE AVAILABLE
INCREASE PRESSURE SETTING	NO EFFECT	NO EFFECT	INCREASES
DECREASE PRESSURE SETTING	NO EFFECT	NO EFFECT	DECREASES
INCREASE GPM (LPM)	INCREASES	NO EFFECT	NO EFFECT
DECREASE GPM (LPM)	DECREASES	NO EFFECT	NO EFFECT
INCREASE CYLINDER DIAMETER	DECREASES	DECREASES	INCREASES
DECREASE CYLINDER DIAMETER	INCREASES	INCREASES	DECREASES

ABOVE TABLE ASSUMES A CONSTANT WORKLOAD.

Figure 7-9. Summary of effects of application changes on cylinder operation.

Figure 7-9 is a summary of the effects for changes in input flow, size, and pressure on cylinder applications.

Figure 7-10 lists piston areas, output forces and speeds for cylinders of various sizes. For example, assume that a cylinder has a 4-inch bore, a 2-inch rod and a pressure rating of 3000 psi. Looking under the column heading 3000, note that the maximum push force (extending) is 37,698 lb. The maximum pull force (retracting) is read from the same column but is opposite the 2-inch piston rod diameter row heading. In this case, the pull force is 28,272 lb. To travel one inch while extending requires 0.0544 gallons of fluid, but only 0.0408 gallons when retracting. The piston velocity with a 20.3 gpm flow rating is 6.2 inches per second extending and 8.3 inches per second when retracting. (Only a few cylinder sizes are shown in Figure 7-10. See Appendix C for a complete chart.)

Cylinder Features

In addition to the basic size and pressure rating, there are important options and features available for cylinders. The more important items are discussed below.

Seals. Cast iron piston rings are commonly used as the piston seal. Long service life is the important characteristic. However, when an external load acts on the cylinder, these rings also display a characteristic clearance flow that permits slow drifting with the control valve closed. Where such drifting cannot be tolerated, various other seal forms and materials are available. Rubber seals are common, but care must be taken to assure that the seal material is compatible with the fluid and the system temperatures.

Rod seals are most commonly made from rubber-like materials, though polymer (plastic) base materials such as Teflon, are gaining popularity. (Teflon is a DuPont trade name and there are other brands of the same material just as good as Teflon.)

With few exceptions, a rod wiper or scraper is provided. The importance of this device should not be overlooked. It keeps foreign material from entering the cylinder and the hydraulic system. The materials must be compatible with not only the fluid, but also with the environment that the rod is exposed to such as ice, dirt, dust, steam, water, etc. Maintenance of the scraper/wiper device is important but frequently forgotten.

Cylinder Cushions. Cylinder cushions (Figure 7-11) are often installed at either or both ends of a cylinder to slow down the movement of the piston near the end of its stroke to prevent the piston from hammering against the end cap. The figure shows the basic elements: plunger, adjustable cushion orifice and a check valve. This cushion configuration is used when the cylinder is retracting.

Figure 7-12 shows the construction when the rod is extending. In this situation, if the adjusting screw is closed it is possible to generate a pressure greater than the system relief setting. This happens for the following reason. Assume 3000 psi (206.82 bar / 20,682.00 kPa) is acting across the diameter of the 4-inch (102 mm) bore cylinder used as an example earlier in this section. To slow the motion, the resistive force must be greater than 37,698 lb (167,688.99 N). Because the area is smaller, due to the cushion plunger, the pressure will exceed 3000 psi (206.82 bar / 20,682.00 kPa). Assume the plunger diameter is 2-1/2 inches (64 mm). The pressure

Cyl. Bore Dia. Inch	Piston Rod Dia. Inch	Work Area Sq. in.	HYDRAULIC WORKING PRESSURE p.s.i.						Fluid Required Per In. Of Stroke		Port Size Dia. Inch	Fluid Velocity @ 15 Ft./Sec.	
			500	750	1000	1500	2000	3000	Gal.	Cu. In.		Flow GPM	Piston Vel. in/Sec.
1½	–	1.767	883	1325	1767	2651	3534	5301	.00765	1.767	½	11.0	24.0
	⅝	1.460	730	1095	1460	2190	2920	4380	.00632	1.460			29.0
	1	.982	491	736	982	1473	1964	2946	.00425	.982			43.1
2	–	3.141	1571	2356	3141	4711	6283	9423	.01360	3.141	½	11.0	13.5
	1	2.356	1178	1767	2356	3534	4712	7068	.01020	2.356			18.0
	1⅜	1.656	828	1242	1656	2484	3312	4968	.00717	1.656			25.6
2½	–	4.909	2454	3682	4909	7363	9818	14727	.02125	4.909	½	11.0	8.6
	1	4.124	2062	3093	4124	6186	8248	12372	.01785	4.124			10.3
	1⅜	3.424	1712	2568	3424	5136	6848	10272	.01482	3.424			12.4
	1¾	2.504	1252	1878	2504	3756	5008	7512	.01084	2.504			16.9
3¼	–	8.296	4148	6222	8296	12444	16592	24888	.0359	8.296	¾	20.3	9.4
	1⅜	6.811	3405	5108	6811	10216	13622	20433	.0295	6.811			11.5
	1¾	5.891	2945	4418	5891	8836	11782	17673	.0255	5.891			13.3
	2	5.154	2577	3865	5154	7731	10308	15462	.0223	5.154			15.2
4	–	12.566	6283	9425	12566	18849	25132	37698	.0544	12.566	¾	20.3	6.2
	1¾	10.161	5080	7621	10161	15241	20322	30483	.0440	10.161			7.7
	2	9.424	4712	7068	9424	14136	18848	28272	.0408	9.424			8.3
	2½	7.657	3828	5743	7657	11485	15314	22971	.0331	7.657			10.2

Figure 7-10. Data for various size cylinders.

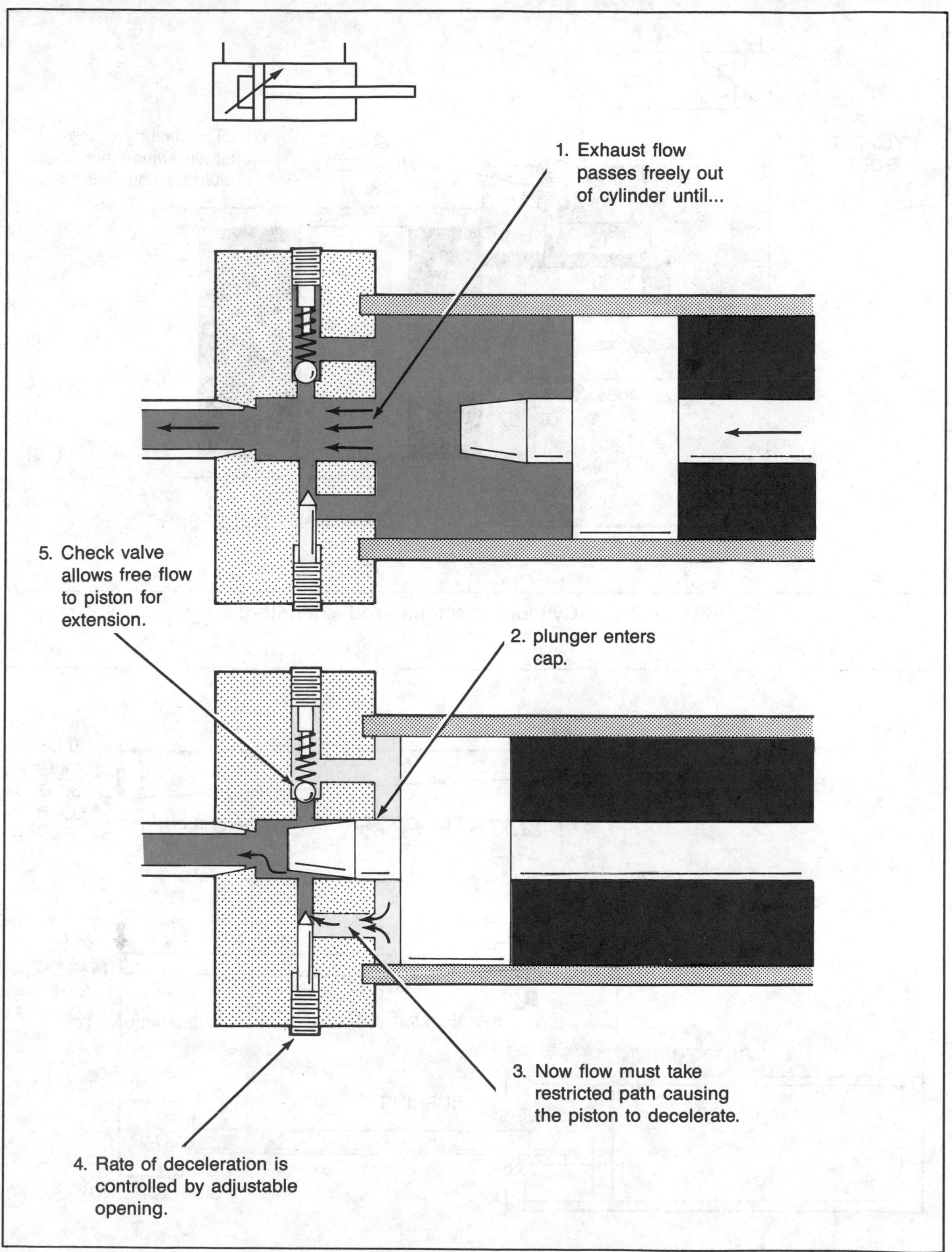

1. Exhaust flow passes freely out of cylinder until...
5. Check valve allows free flow to piston for extension.
2. plunger enters cap.
3. Now flow must take restricted path causing the piston to decelerate.
4. Rate of deceleration is controlled by adjustable opening.

Figure 7-11. Cylinder cushions-rod retracting.

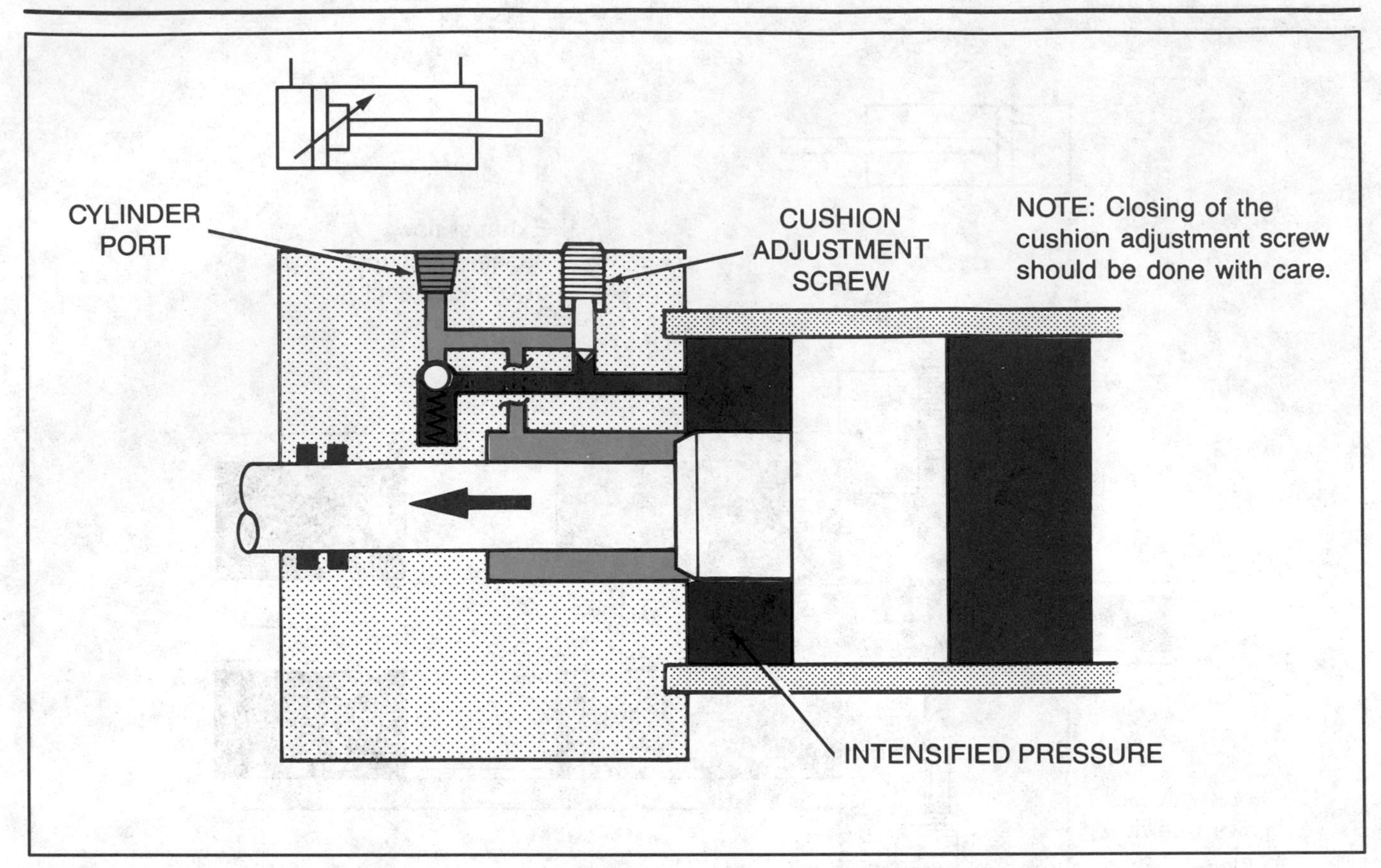

Figure 7-12. Cylinder cushions—rod extending.

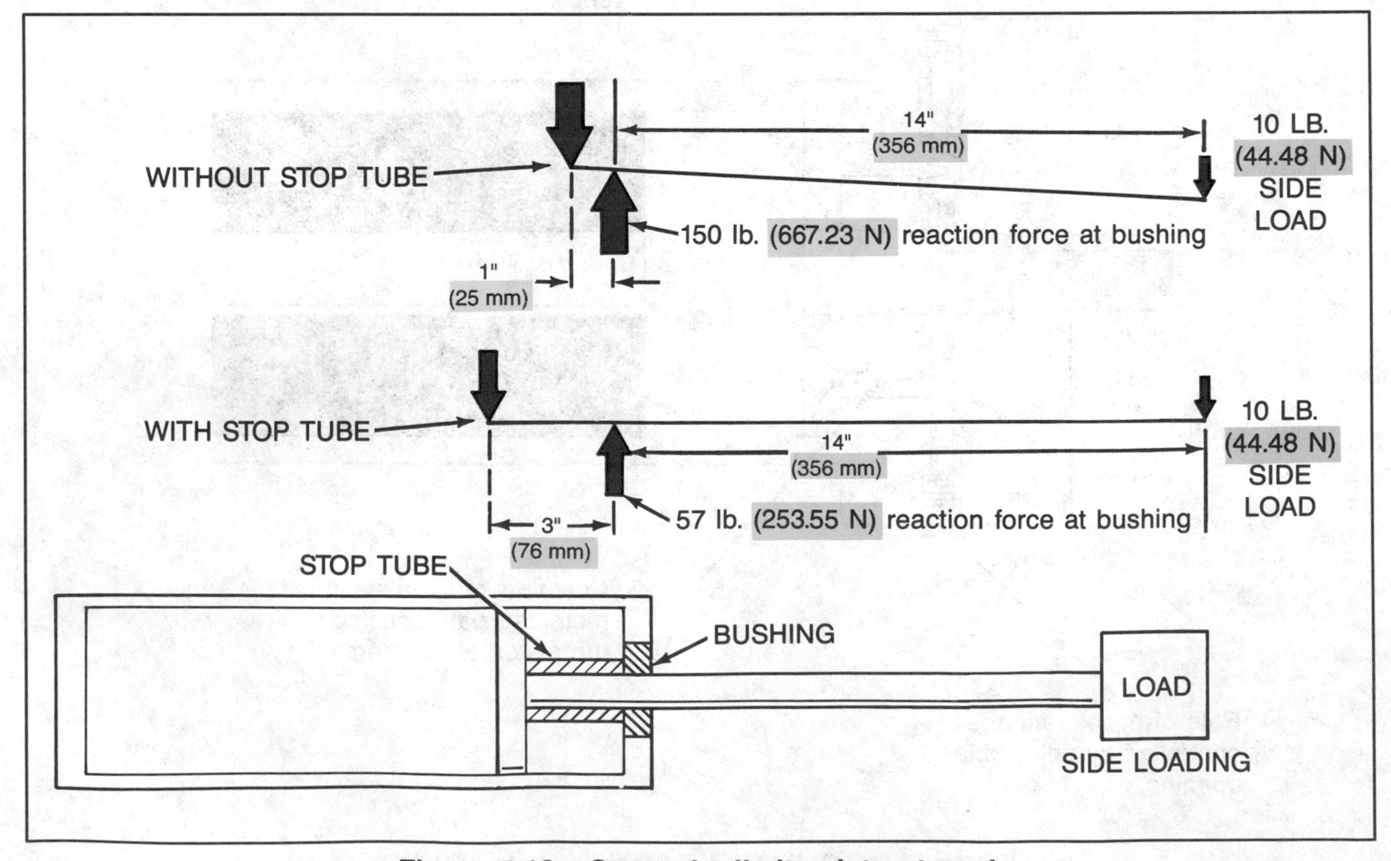

Figure 7-13. Stop tube limits piston travel.

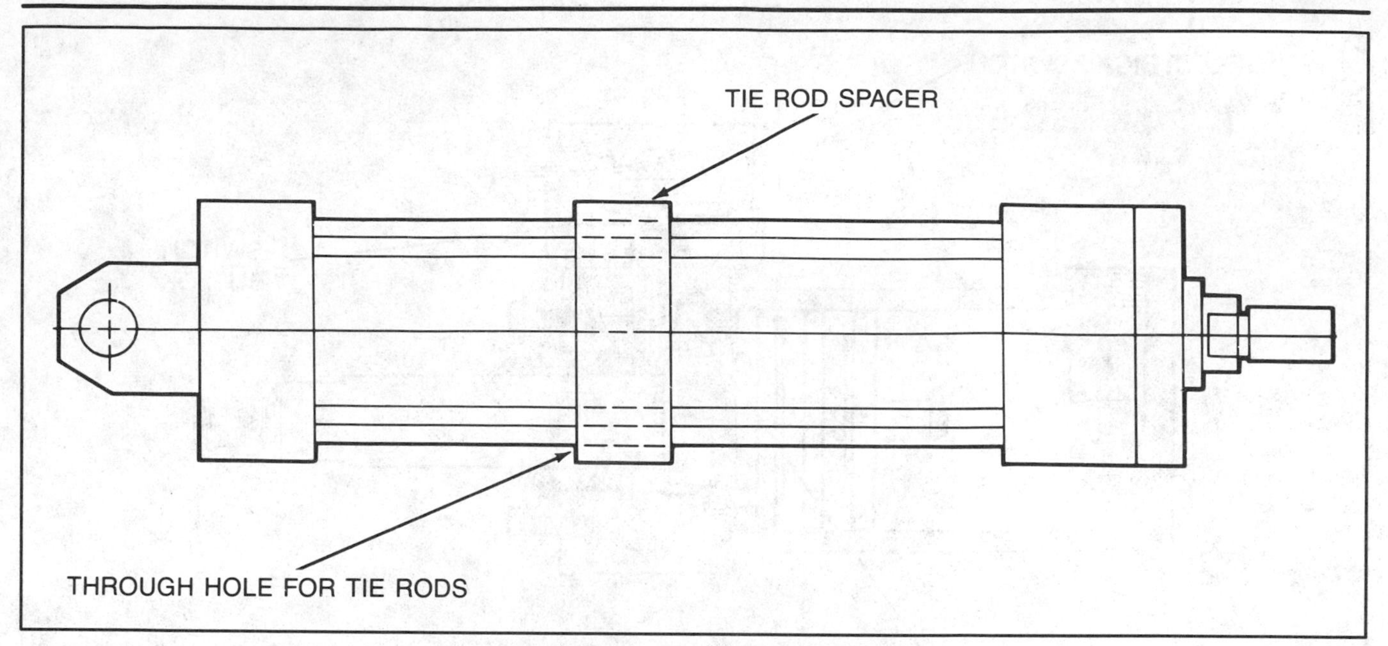

Figure 7-14. Tie rod spacer.

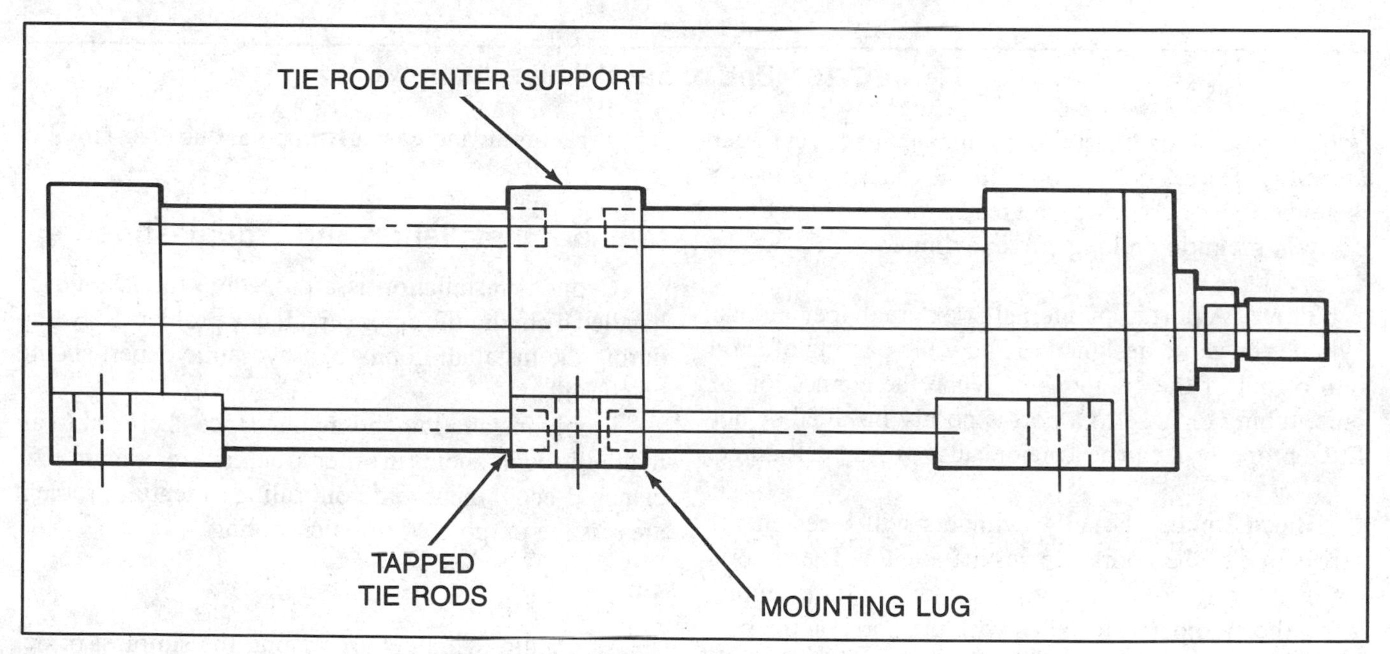

Figure 7-15. Tie rod center support.

developed would be 4920 psi (339.18 bar / 33,918.48 kPa). This value would be even larger with a heavy weight attached to the piston rod. Closing of the cushion adjustment screw should be done with care.

Stop Tubes. A stop tube (Figure 7-13) is usually a metal collar which fits over the piston rod next to the piston. It is used primarily on cylinders with a long stroke. The main function of the tube is to separate the piston and rod bushing when a long-stroke cylinder is fully extended. The majority of hydraulic applications do not require a stop tube. If a stop tube is necessary, it should be mentioned in the manufacturer's catalog.

Tie Rod Spacers. Tie rod spacers and center supports are used to improve the structural rigidity of long stroke tie rod cylinders. The spacer (Figure 7-14) keeps the tie rod in the proper position around the center line of the cylinder and acts much like a truss in preventing excessive deflection.

Figure 7-15 shows a tie rod center support. This support has side mounting lugs similar to side lug mount heads and serves as an additional mounting location. The tie rods are studded into the center support and the sup-

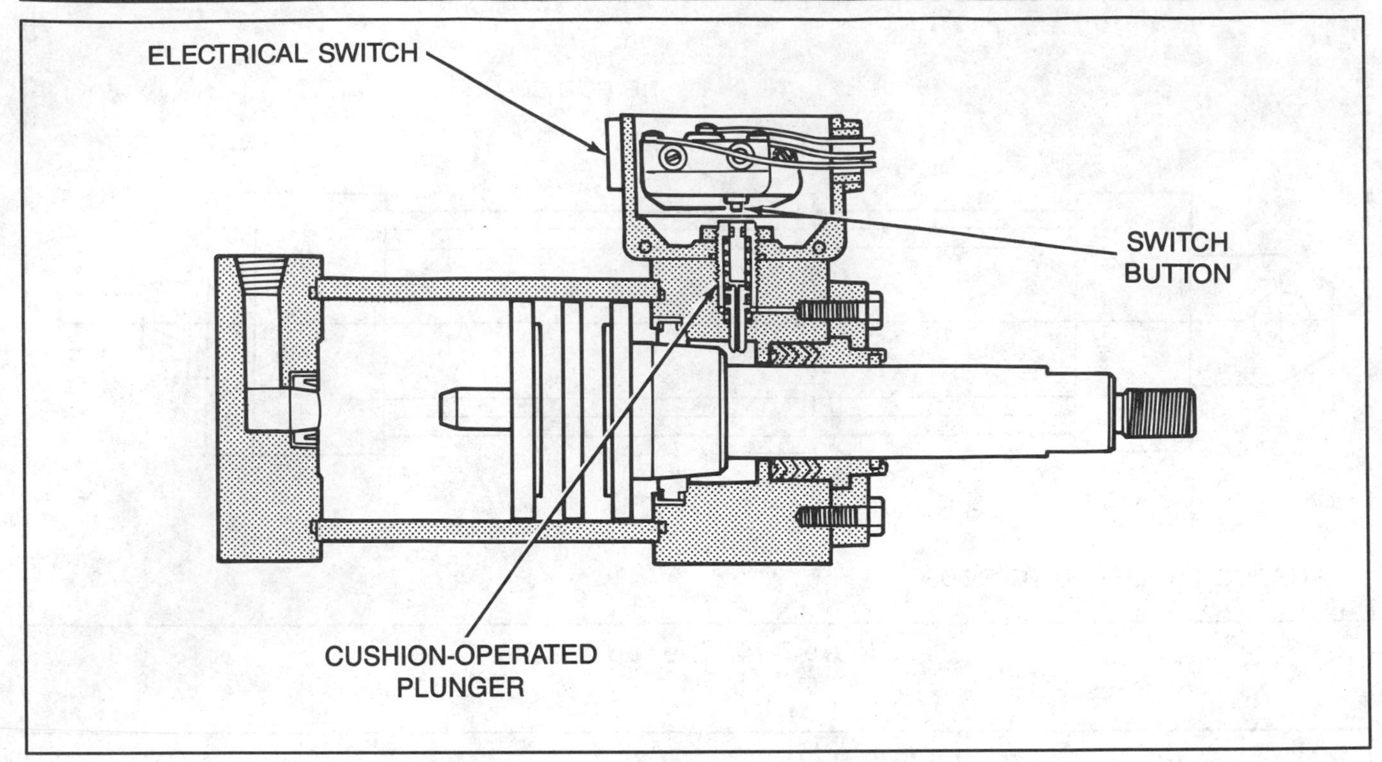

Figure 7-16. End of travel limit switch.

port becomes a load-carrying component of the cylinder assembly. The exact location of the tie rod center support is generally optional, which greatly increases the flexibility in mounting a long-stroke cylinder.

Ports. A port is an internal or external opening in a cylinder or valve designed to allow the passing of fluid into or out of the component. When the connection of pipe, tubing or hose to a port is poorly installed or not fitted properly, it can be a major cause of system leakage.

Bleed Ports. Usually cylinders will bleed themselves of air when ports are vertical, on top. Bleed ports are often desirable to remove entrained air, for example, when the cylinder is installed with the ports on the bottom. High performance and high speed heavy load applications are a few examples where air bleed ports are desirable.

Limit Switches. End of travel limit switches (Figure 7-16) are available that signal rod position to control and safety circuits. There are two common types: mechanical and proximity limit switches. Mechanical actuation of an electrical switch is accomplished when a button on the switch engages a plunger moved by the lead angle on a hardened cylinder cushion. Proximity switches are activated when a metal cushion plunger passes close to the magnetic pickup of the switch. This type is becoming increasingly popular due to its simplicity.

Cylinder Installation and Troubleshooting

Cylinder installation is a procedure that should be handled by hydraulic engineers. If any problems develop during the installation phase, a hydraulic expert should be consulted.

Proper installation and maintenance is crucial to all hydraulic components in order to achieve maximum efficiency. Recognizing and controlling potential problem areas is the purpose of troubleshooting.

Summary

Hydraulic cylinders are among the simplest of devices in fluid power having one moving part, the piston and rod assembly. However, forces generated by cylinders are also among the largest found in fluid power systems. Pressures in the cylinders can and often do exceed system relief settings. The life of the cylinder and the system can be highly dependent on proper specification and maintenance of a simple element, the rod wiper/scraper.

Like all system components, the cylinder can be damaged by contamination. Contaminants are not readily flushed out as is possible with rotary devices. General cleanliness practices, such as plugging ports until lines are connected, are important.

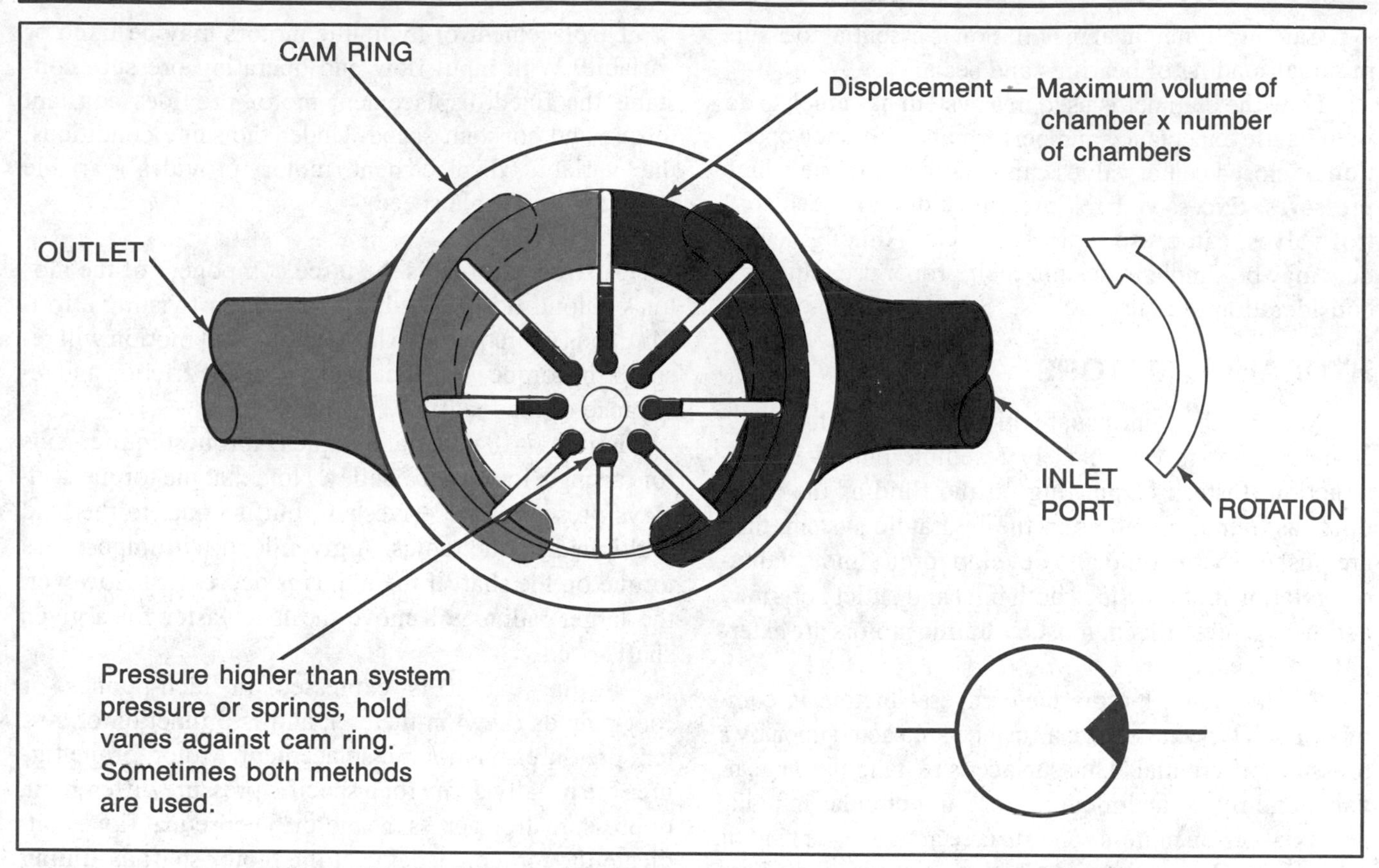

Figure 7-17. Motor displacement is capacity per revolution.

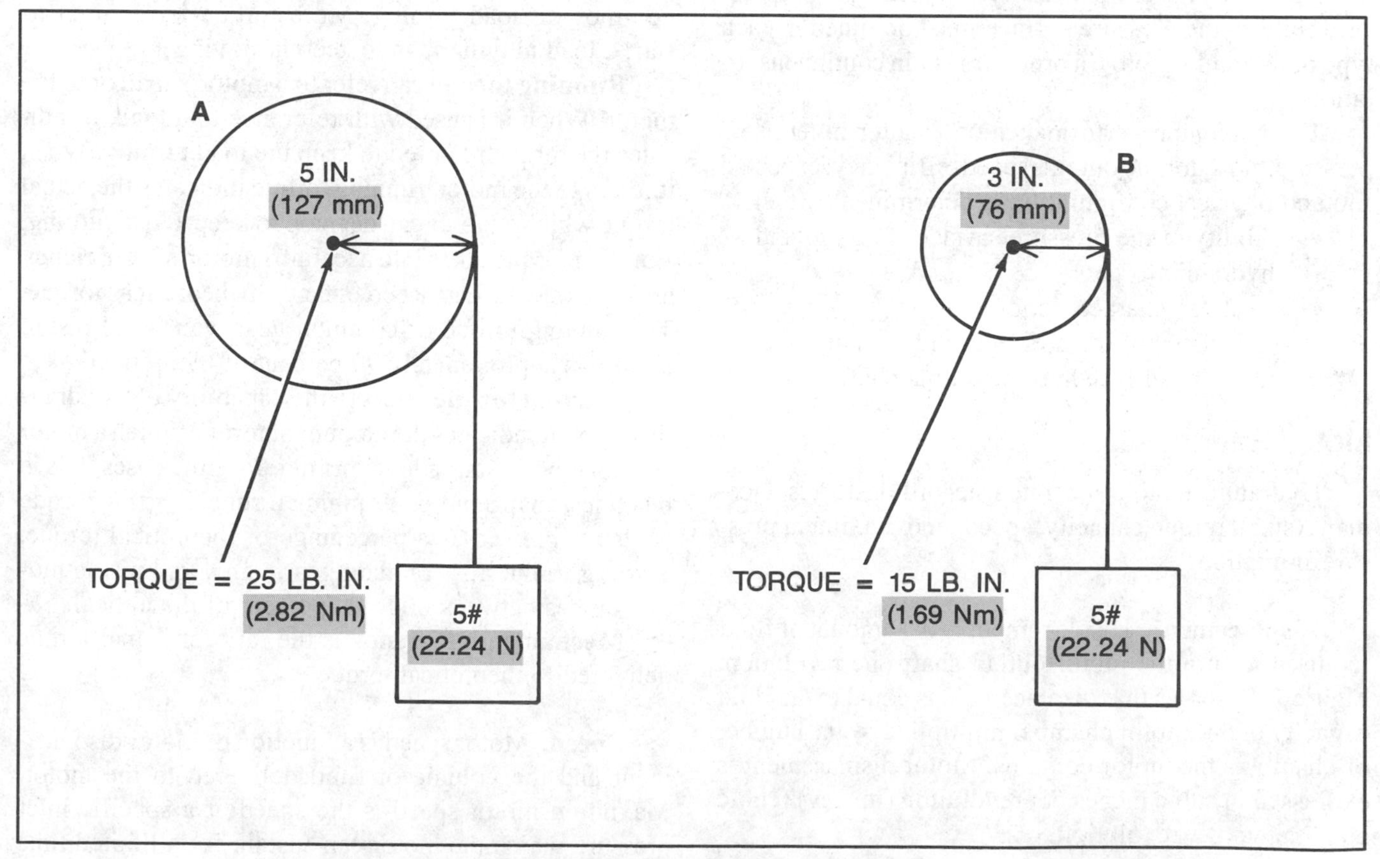

Figure 7-18. Torque equals load multiplied by radius.

Care in alignment at installation is essential to assure minimal loading of bearings and seals.

How the cylinder is used in a system has much to do with its life expectancy and performance. Sudden operation of closed center valves can generate extremely high pressures. Excessive back pressures, due to speed control valves, can cause rapid seal wear. Although cylinders may be simple in design, their proper use requires a consideration of many factors.

HYDRAULIC MOTORS

Motor is the name usually given to a rotary hydraulic actuator. Motors very closely resemble pumps in construction. Instead of pushing on the fluid as the pump does, as output members in the hydraulic system, they are pushed by the fluid and develop torque and continuous rotating motion. Since both inlet and outlet ports may at times be pressurized, most hydraulic motors are externally drained.

All hydraulic motors have several factors in common. Each type must have a surface area acted upon by a pressure differential. This surface is rectangular in gear, vane, and rotary abutment motors and circular in radial and axial piston motors. The surface area in each kind of motor is mechanically connected to an output shaft from which the mechanical energy is coupled to the equipment driven by the motor. Finally, the porting of the pressure fluid to the pressure surface must be timed in each type of hydraulic motor in order to sustain continuous rotation.

The maximum performance of a motor in terms of pressure, flow, torque output, speed, efficiency, expected life and physical configuration is determined by the:

- Ability of the pressure surfaces to withstand hydraulic force
- Leakage characteristics
- Efficiency of the means used to connect the pressure surface and the output shaft

Motor Ratings

Hydraulic motors are rated according to displacement (size), torque capacity, speed, and maximum pressure limitations.

Displacement. Displacement is the amount of fluid required to turn the motor output shaft one revolution. Figure 7-17 shows that displacement is equal to the fluid capacity of one motor chamber multiplied by the number of chambers the motor contains. Motor displacement is expressed in cubic inches per revolution (in^3/rev) (cubic centimeters per revolution, cm^3/rev).

Displacement of hydraulic motors may be fixed or variable. With input flow and operating pressure constant, the fixed-displacement motor provides constant torque and constant speed. Under the same conditions, the variable displacement motor provides variable torque and variable speed.

Torque. Torque is the force component of the motor's output. It is defined as a turning or twisting effort. Motion is not required to have torque, but motion will result if the torque is sufficient to overcome friction and resistance of the load.

Figure 7-18 illustrates typical torque requirements for raising a load with a pulley. Note that the torque is always present at the driveshaft, but is equal to the load multiplied by the radius. A given load will impose less torque on the shaft if the radius is decreased. However, the larger radius will move the load faster for a given shaft speed.

Torque output is expressed in inch-pounds or foot-pounds (newton-meters), and is a function of system pressure and motor displacement. Motor torque figures are usually given for a specific pressure differential, or pressure drop across the motor. Theoretical figures indicate the torque available at the motor shaft **assuming the motor is 100 percent efficient**.

Breakaway torque is the torque required to get a non-moving load turning. More torque is required to start a load moving than to keep it moving.

Running torque can refer to a motor's load or to the motor. When it is used with reference to a load, it indicates the torque required to keep the load turning. When it refers to the motor, running torque indicates the actual torque which a motor can develop to keep a load turning. Running torque takes into account a motor's inefficiency and is expressed as a percentage of theoretical torque. The running torque of common gear, vane, and piston motors is approximately 90 percent of theoretical.

Starting torque refers to the capability of a hydraulic motor. It indicates the amount of torque which a motor can develop to start a load turning. In some cases, this is much less than a motor's running torque. Starting torque is also expressed as a percentage of theoretical torque. Starting torque for common gear, vane, and piston motors ranges between 60 to 90 percent of theoretical.

Mechanical efficiency is the ratio of actual torque delivered to theoretical torque.

Speed. Motor speed is a function of motor displacement and the volume of fluid delivered to the motor. Maximum motor speed is the speed at a specific inlet pressure which the motor can sustain for a limited time

CHANGE	SPEED	EFFECT ON OPERATING PRESSURE	TORQUE AVAILABLE
INCREASE PRESSURE SETTING	NO EFFECT	NO EFFECT	INCREASES
DECREASE PRESSURE SETTING	NO EFFECT	NO EFFECT	DECREASES
INCREASE GPM (LPM)	INCREASES	NO EFFECT	NO EFFECT
DECREASE GPM (LPM)	DECREASES	NO EFFECT	NO EFFECT
INCREASE DISPLACEMENT (SIZE)	DECREASES	DECREASES	INCREASES
DECREASE DISPLACEMENT (SIZE)	INCREASES	INCREASES	DECREASES

ABOVE TABLE ASSUMES A CONSTANT LOAD.

Figure 7-19. Summary of effects of application changes on motor operations.

without damage. Minimum motor speed is the slowest, continuous, smooth rotational speed of the motor output shaft. Slippage is the leakage across the motor, or the fluid that moves through the motor without doing any work.

Pressure. Pressure required in a hydraulic motor depends on the torque load and the displacement. A large displacement motor will develop a given torque with less pressure than a smaller unit. The size and torque rating of a motor usually is expressed in pound-inches of torque per 100 psi of pressure (newton-meters per bar).

$$\frac{lb.\ in.}{100\ psi}$$

$$\frac{N\ m}{bar}$$

Formulas for Motor Applications

Listed below are the formulas used for applying hydraulic motors and determining flow and pressure requirements.

NOTE: All of the following formulas are for theoretical torque. An additional 10 percent to 35 percent torque capability may be needed to start a given load. Check the starting torque specifications on the installation drawings.

To find the **size of motor** required for a job, use the following formula:

$$Torque\ Rate\ (lb.\ in./100\ psi) = \frac{Torque\ Load\ (lb.\ in.)}{Desired\ Operating\ Pressure\ (psi) \times 0.01}$$

$$Torque\ Rate\ (N\ m/bar) = \frac{Torque\ Load\ (N\ m)}{Desired\ Operating\ Pressure\ (bar)}$$

For example, to handle a 500 pound-inch (56.49 newton-meter) load at 2000 psi (137.88 bar / 13,788.00 kPa), a 25 pound-inch/100 psi (0.41 newton-meter/bar) motor is required:

$$Size = \frac{500}{2000 \times 0.01} = \frac{500}{20} = 25\ lb.\ in./100\ psi$$

$$Size = \frac{56.49}{137.88} = .41\ newton\text{-}meters/bar$$

To find **working pressure** for a given size motor and load, use the following formula:

$$Operating\ Pressure\ (psi) = \frac{Torque\ Load\ (lb.\ in.) \times 100}{Motor\ Torque\ Rate\ (lb.\ in./100\ psi)}$$

$$Operating\ Pressure\ (bar) = \frac{Torque\ Load\ (N\ m)}{Motor\ Torque\ Rate\ (N\ m/bar)}$$

For example, a 50 pound-inch/100 psi (0.82 newton-meter/bar) motor develops 3000 psi (206.82 bar) with a load of 1500 pound-inches (169.47 newton-meters):

$$Pressure = \frac{1500 \times 100}{50} = 3000\ psi$$

$$Pressure = \frac{169.5}{0.82} = 206.82\ bar$$

To find the **maximum torque** for a given size motor, use the following formula:

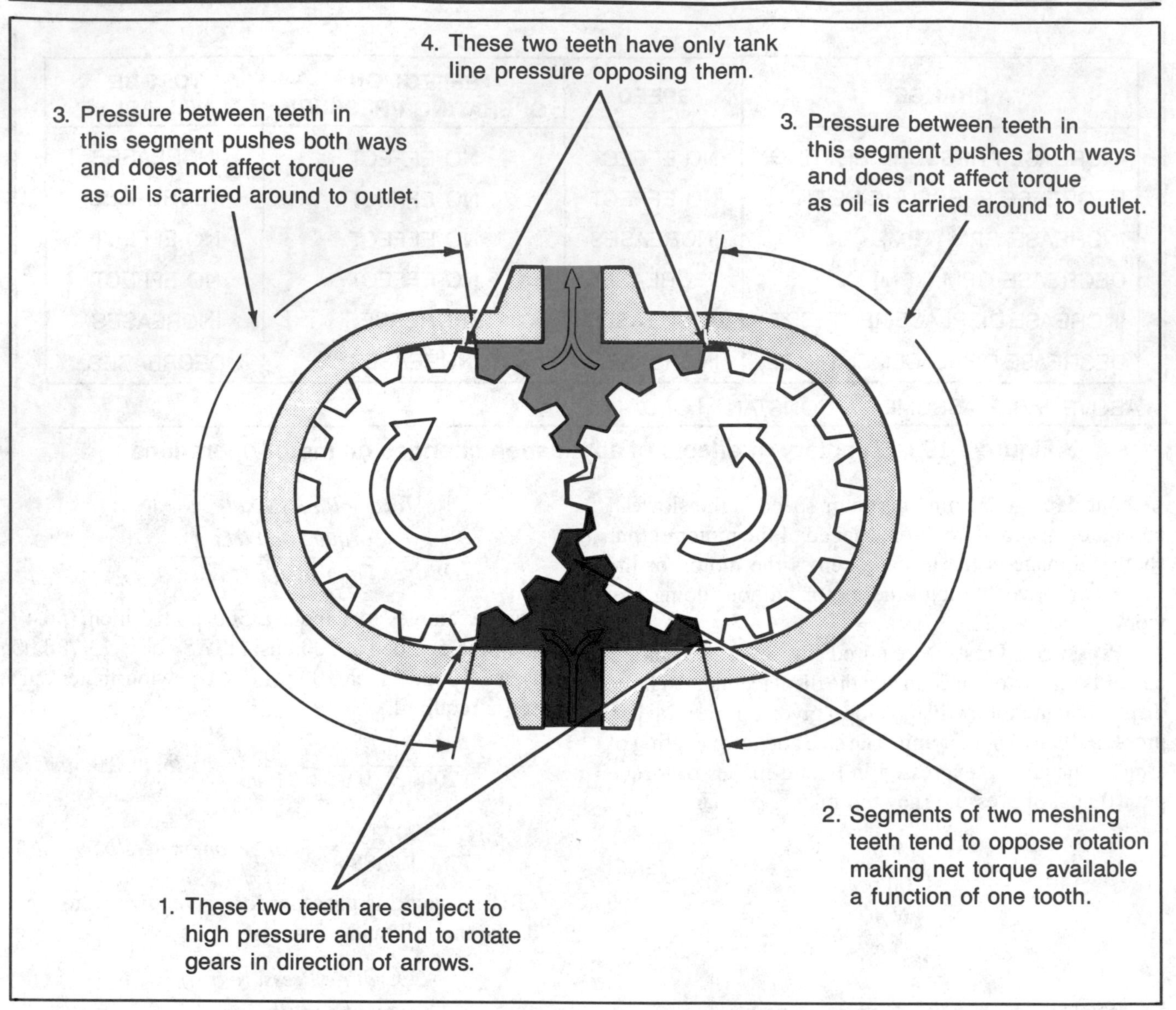

4. These two teeth have only tank line pressure opposing them.
3. Pressure between teeth in this segment pushes both ways and does not affect torque as oil is carried around to outlet.
3. Pressure between teeth in this segment pushes both ways and does not affect torque as oil is carried around to outlet.
2. Segments of two meshing teeth tend to oppose rotation making net torque available a function of one tooth.
1. These two teeth are subject to high pressure and tend to rotate gears in direction of arrows.

Figure 7-20. Torque development in an external gear motor.

$$\textit{Maximum Torque (lb. in.)} = \frac{\textit{Torque Rate (lb. in. /100 psi)} \times \textit{Max. psi}}{100}$$

$$\textit{Maximum Torque (N m)} = \textit{Torque Rate (N m/bar)} \times \textit{Max. Pressure (bar)}$$

Thus, a 10 pound-inch/100 psi (0.16 newton-meter/bar) motor rated at 2500 psi (172.35 bar/17,235 kPa) can handle a maximum load of 250 pound-inches (27.58 newton-meters):

$$\textit{Maximum Torque} = \frac{10 \times 2500}{100} = 250 \textit{ lb. in.}$$

$$\textit{Maximum Torque} = 0.16 \times 172.35 = 27.58 \textit{ N m}$$

To find **torque** when pressure and displacement are known, use the following formula:

$$\textit{Torque (lb. in.)} = \frac{\textit{Pressure (psi)} \times \textit{Displacement (in}^3\textit{/rev.)}}{2\pi\,(6.28)}$$

$$\textit{Torque (N/m)} = \frac{\textit{Pressure (bar)} \times \textit{Displacement (cm}^3\textit{/rev.)}}{20\pi}$$

To find **gpm requirements** for a given drive speed, use the following formula:

$$\textit{GPM} = \frac{\textit{Speed (rpm)} \times \textit{Displacement (in}^3\textit{/rev.)}}{231}$$

$$LPM = \frac{Speed\ (rpm) \times Displacement\ (cm^3/rev.)}{1000}$$

A motor with a displacement of 10 cubic inches per revolution (163.87 cubic centimeters per revolution) requires just over 43 gpm (162.77 LPM) to run at 1000 rpm:

$$GPM = \frac{1000 \times 10}{231} = 43.3\ gpm$$

$$LPM = \frac{1000 \times 163.87cm^3}{1000} = 163.87\ LPM$$

To find **drive speed** when displacement and gpm (LPM) are known, use the following formula:

$$RPM = \frac{GPM \times 231}{Displacement\ (in^3/rev.)}$$

$$RPM = \frac{LPM \times 1000}{Displacement\ (cm^3/rev.)}$$

Figure 7-19 summarizes the effects on speed, pressure and torque capacity for changes in motor applications. Note that the basic principles are identical to Figure 7-9 on cylinders.

Classes of Hydraulic Motors

Hydraulic motors can be classified by application into three categories:

- High Speed, Low Torque Motors (HSLT)
- Low Speed, High Torque Motors (LSHT)
- Limited Rotation Motors (Torque Actuators)

HSLT Motors. In many applications, the motor operates continuously at relatively high rpm. Examples are fan drives, generator drives, and compressor drives. While the speed is high and reasonably constant, the load may be either steady, as in fan drives, or quite variable, as in compressors or generators. HSLT motors are excellent for these kinds of applications. The four primary types of HSLT motors are in-line piston, bent-axis piston, vane and gear.

LSHT Motors. In some applications, the motor must move a relatively heavy load at lower speeds and fairly constant torque. A motor for a crane is one such application. LSHT motors are often used for performing this type of work. Some LSHT motors operate smoothly down to one or two rpm. LSHT motors are simple in design with a minimum of working parts and are quite reliable and generally less expensive than higher speed motors employing speed reducing devices.

Ideally, an LSHT motor should have high starting and stall torque efficiencies, and good volumetric and mechanical efficiencies. They should start smoothly under full load and provide full torque over their entire speed range. These motors should exhibit little or no torque ripple throughout their speed range, and velocity variation at an average speed at constant pressure should be minimal.

Basic LSHT motor designs include: internal gear, vane, rolling vane, radial piston and axial piston, and axial ball piston.

Limited Rotation Motors. Modifications of the rotary motor are sometimes found in industrial machinery where special motions are required. One is the limited rotation motor which will not permit continuous rotation in either direction. The vane-type version has a movable vane which forms two chambers in an annulus. Pressure exerted against either side of the vane causes it to rotate and turn the rotor and output shaft. Rotation is limited to less than 360 degrees by the width of the body segment containing inlet and outlet porting. Another version of the limited rotation motor is the piston type which converts the linear motion of a cylinder into rotary motion through a crank arm.

Types of Hydraulic Motors

There are a variety of hydraulic motors used in industrial applications. The type of motor that is used depends on the demands of each individual application. The following motors are reviewed in this chapter:

- Gear Motors—including external and internal (gerotor and rolator or orbital) motors
- Vane Motors—including unbalanced, balanced, fixed, variable and cartridge (high performance) types
- Piston Motors—including in-line, bent-axis, and radial motors (fixed, variable and cam type)
- Screw Motors
- Torque Generators—including vane and piston types

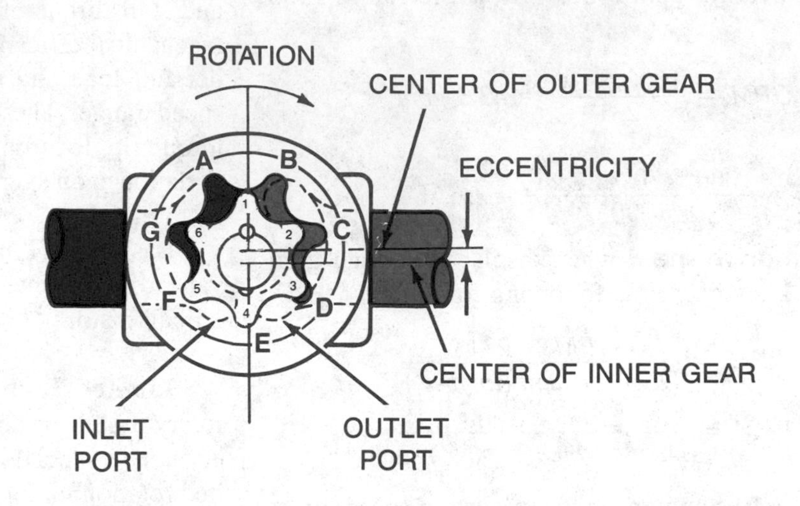

A

HIGH PRESSURE

LOW PRESSURE

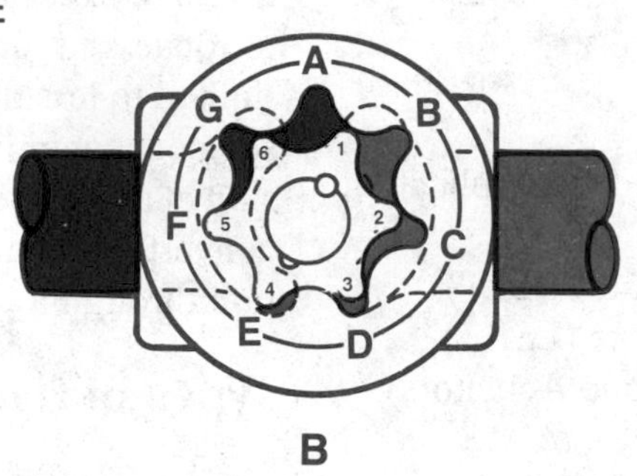

B

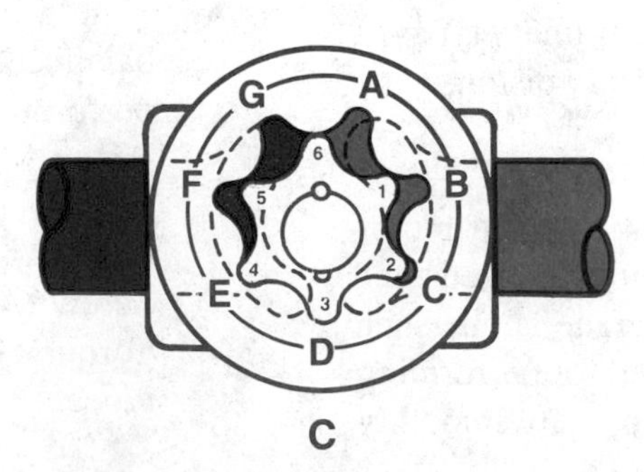

C

Direct-drive gerotor motor has internal and external gear set.
Both gears rotate during operation.

Figure 7-21. Direct-drive gerotor motor.

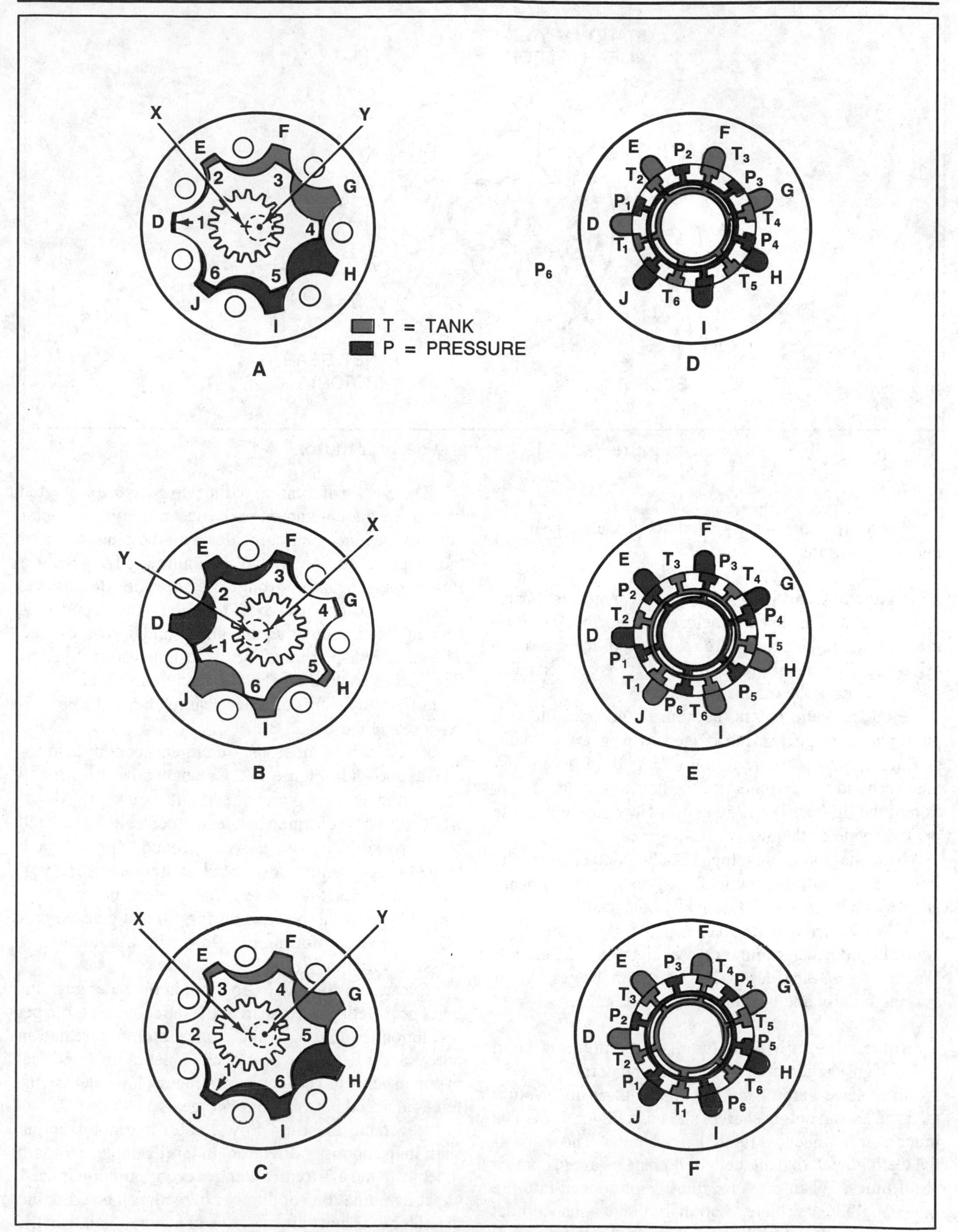

Figure 7-22. Orbiting gerotor motor.

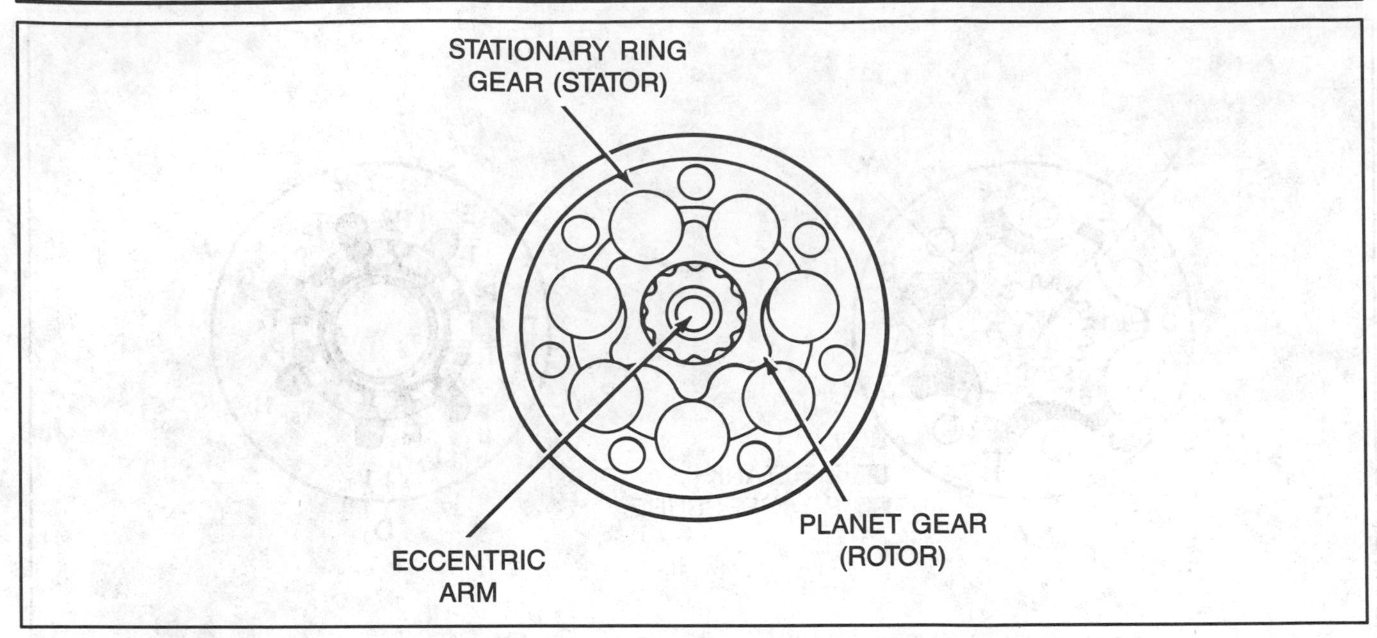

Figure 7-23. Roller-vane gerotor motor.

Gear Motors

There are two types of gear motors: external gear and internal gear motors.

External Gear Motor. External gear motors consist of a pair of matched gears enclosed in one housing (Figure 7-20). Both gears have the same tooth form and are driven by fluid under pressure. One gear is connected to an output shaft, the other is an idler.

Fluid pressure enters the housing on one side at a point where the gears mesh and forces the gears to rotate, as fluid at high pressure follows the path of least resistance around the periphery of the housing (colored portion of the figure). The fluid exits, at low pressure, at the opposite side of the motor.

Note that torque developed is a function of hydraulic imbalance of only one tooth of one gear at a time; the other gear and teeth are hydraulically balanced.

Close tolerances between gears and housing help control fluid leakage and increase volumetric efficiency. Wear plates on the sides of the gears keep the gears from moving axially and also help control leakage.

Internal Gear Motor. Internal gear motors fall into two categories: direct drive and orbiting gerotor motors. A direct-drive gerotor motor consists of an inner-outer gear set, and an output shaft as shown in Figure 7-21. The inner gear has one less tooth than the outer. The shape of the teeth is such that all teeth of both gears are in contact at all times. When pressure fluid is introduced into the motor, both gears rotate. Stationary kidney-shaped inlet and outlet ports are built into the motor housing.

The centers of rotation of the two gears are separated by a given amount known as the eccentricity. The center point of the inner gear coincides with the center point of the output shaft. As shown in Figure 7-21A, pressure fluid enters the motor through the inlet port. Because the inner gear has one less tooth than the outer, a pocket is formed between inner teeth 6 and 1, and outer socket A. The kidney-shaped inlet port is designed so that just as this pocket volume reaches its maximum, fluid flow is shut off, with the tips of inner gear teeth 6 and 1 providing a seal (Figure 7-21B).

As the pair of inner and outer gears continues to rotate, as shown in Figure 7-21C, a new pocket is formed between inner teeth 6 and 5, and outer socket G. Meanwhile, the pocket formed between inner teeth 6 and 1 and outer socket A has moved around opposite the kidney-shaped outlet port, steadily draining as the volume of the pocket decreases. The gradual, metered volume change of the pockets during fill and exhaust provides smooth, uniform fluid flow with a minimum of pressure variation (ripple).

Because of the one extra tooth in the outer gear, the inner gear teeth move ahead of the outer by one tooth per revolution. In Figure 7-21C, inner tooth 4 is seated in outer socket E. On the next cycle, inner tooth 4 will seat in outer socket F. This action produces a low relative differential speed between the two gears.

Reversing the fluid flow rotates the motor output shaft in the opposite direction. In this example, a 6-tooth inner gear and a 7-tooth outer gear configuration is used. Other combinations of number of teeth can be used, but the outer gear must always have one more tooth than the inner.

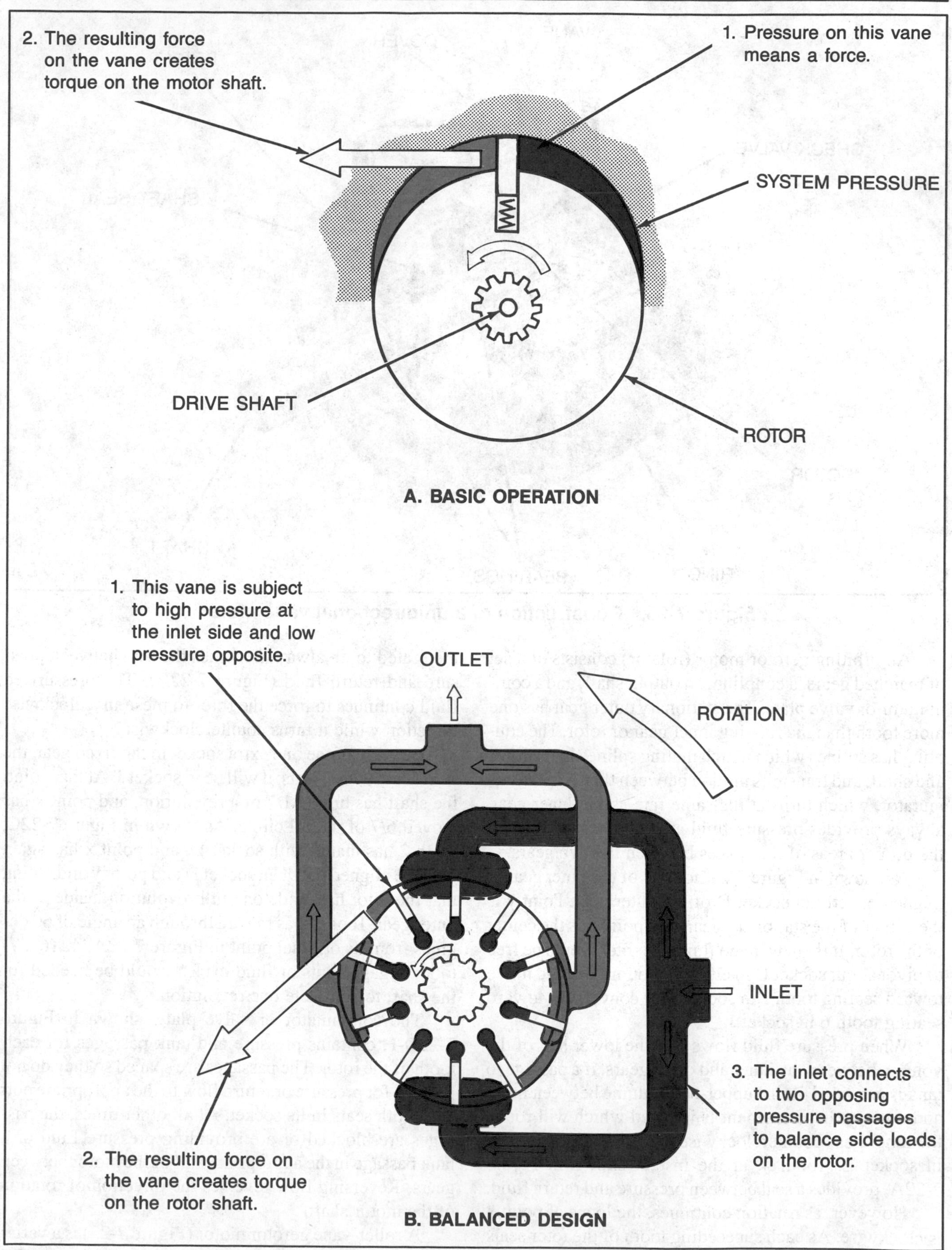

2. The resulting force on the vane creates torque on the motor shaft.
1. Pressure on this vane means a force.
SYSTEM PRESSURE
DRIVE SHAFT
ROTOR
A. BASIC OPERATION
1. This vane is subject to high pressure at the inlet side and low pressure opposite.
OUTLET
ROTATION
INLET
3. The inlet connects to two opposing pressure passages to balance side loads on the rotor.
2. The resulting force on the vane creates torque on the rotor shaft.
B. BALANCED DESIGN

Figure 7-24. Torque development in a balanced vane motor.

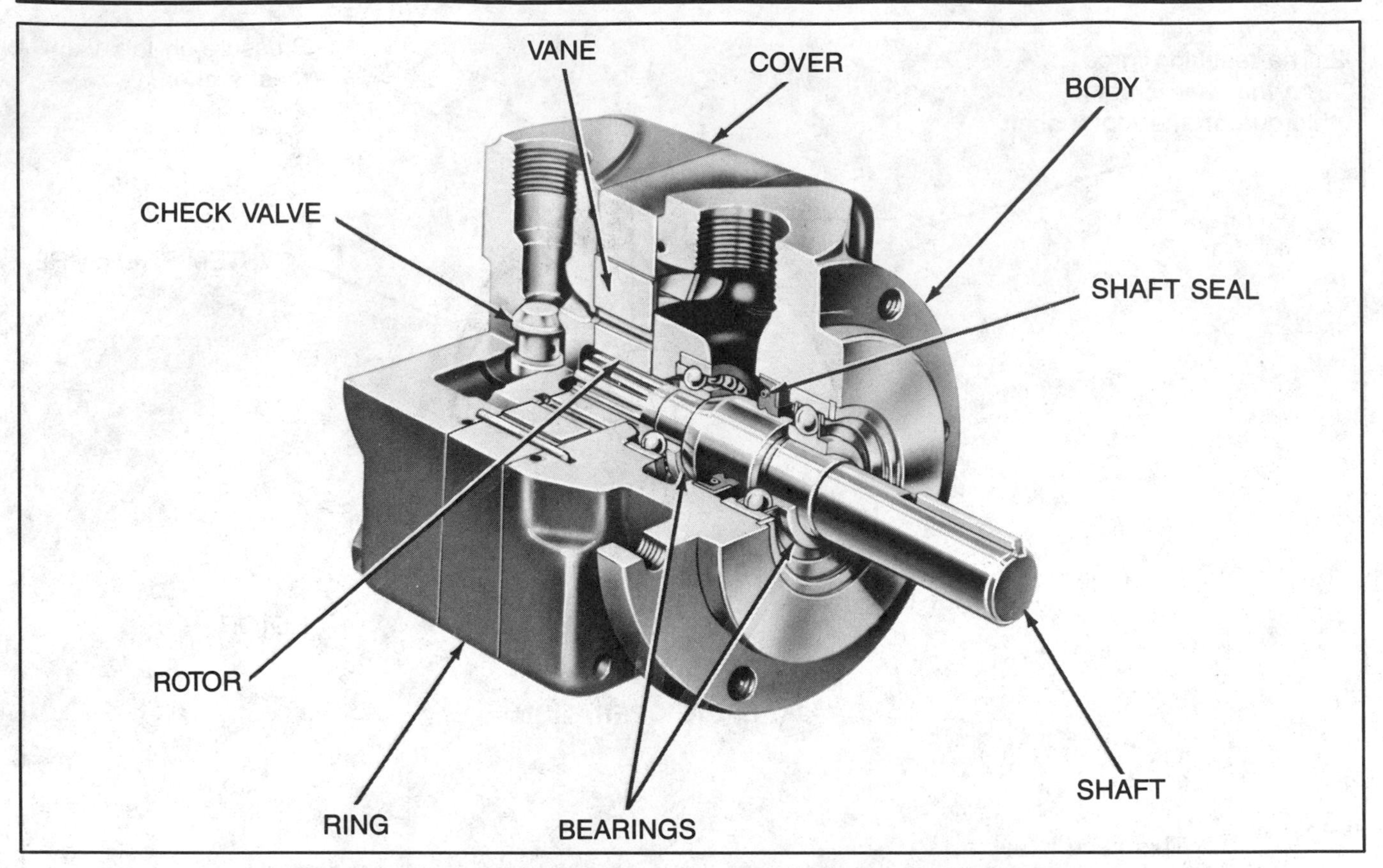

Figure 7-25. Construction of a unidirectional vane motor.

An orbiting gerotor motor (rolator) consists of a set of matched gears, a coupling, an output shaft, and a commutator or valve plate. The stationary outer gear has one more tooth than the rotating inner gear or rotor. The coupling has splines which match mating splines in the rotor and shaft, and transmits motion between them. The commutator, which turns at the same rate as the inner gear, always provides pressure fluid and a passage to tank to the proper areas of the spaces between the two gears.

As shown in Figure 7-22, tooth 1 of the inner gear is aligned exactly in socket D of the outer gear. Point y is the center of the stationary gear, and point x is the center of the rotor. If there were no fluid, the rotor would be free to pivot about socket D in either direction. It could move toward seating tooth 2 in socket E or conversely, toward seating tooth 6 in socket J.

When pressure fluid flows into the lower half of the volume between the inner and outer gears, if a passage to tank is provided for the upper-half volume between inner and outer gears, a moment is induced which will rotate the inner gear counterclockwise and start to seat tooth 2 in socket E. Tooth 4, at the instant shown in Figure 7-22A, provides a seal between pressure and return fluid.

However, as rotation continues, the locus of point x is clockwise. As each succeeding tooth of the rotor seats in its socket, the tooth directly opposite on the rotor from the seated tooth always becomes the seal between pressure and return fluid (Figure 7-22B). The pressurized fluid continues to force the rotor to mesh in a clockwise direction while it turns counterclockwise.

Because of the one extra socket in the fixed gear, the next time tooth 1 seats, it will be in socket J. At that point, the shaft has turned 1/7 of a revolution, and point x has moved 6/7 of its full circle. As shown in Figure 7-22C, tooth 2 has mated with socket D, and point x has again become aligned between socket D and point y, indicating that the rotor has made one full revolution inside of the outer gear. Tooth 1 has moved through an angle of 60 degrees from its original point in Figure 7-22A; 42 (6 x 7) tooth engagements or fluid cycles would be needed for the shaft to complete one revolution.

The commutator or valve plate, shown in Figure 7-22D-F, contains pressure and tank passages for each tooth of the rotor. The passages are spaced so they do not provide for pressure or return flow to the appropriate port as a tooth seats in its socket. At all other times, the passages are blocked or are providing pressure fluid or a tank passage in the appropriate half of the motor between gears. Reversing flow reverses the direction of rotation of the motor shaft.

A roller-vane gerotor motor (Figure 7-23) is a variation of the orbiting gerotor motor, with a stationary ring

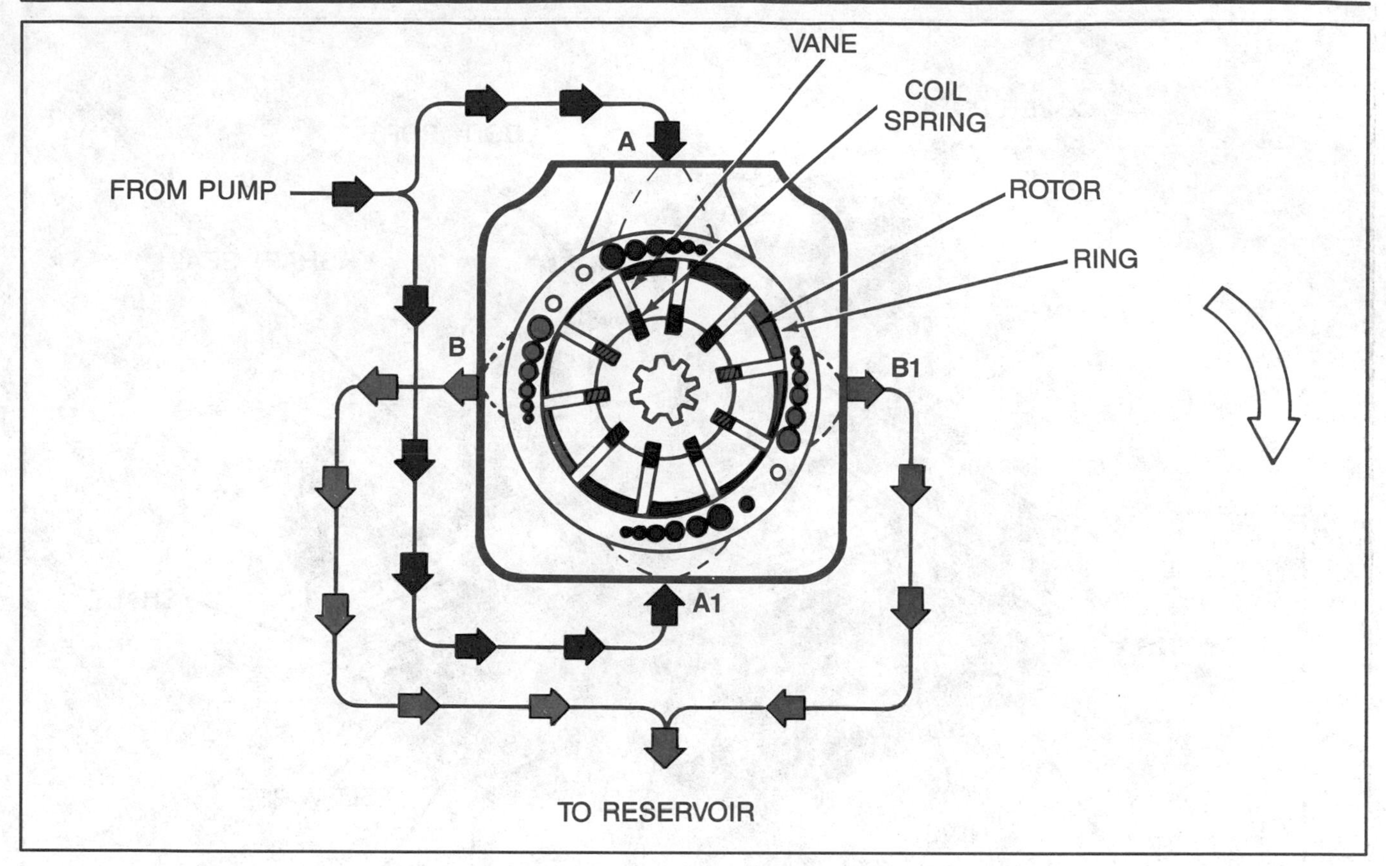

Figure 7-26. Operation of a high performance vane motor.

gear (or stator) and a moving planet gear (or rotor). Instead of being held by two journal bearings, the eccentric arm of the planetary is held by the intergear action of the 6-tooth rotor, and 7-socket stator. Instead of direct contact between the stator and rotor, roller vanes are incorporated to form the displacement chambers. The roller vanes reduce wear, enabling the motors to be used in a closed-loop, high-pressure hydrostatic circuit as direct-mounted wheel drives.

Vane Motors

In a vane motor, torque is developed by pressure on exposed surfaces of rectangular vanes which slide in and out of slots in a rotor splined to the driveshaft (Figure 7-24A). As the rotor turns, the vanes follow the surface of a cam ring, forming sealed chambers which carry the fluid from the inlet to the outlet.

In the **balanced design** shown in Figure 7-24B, the system working pressure at the inlet and the tank line pressure at the outlet are both forces in the system. These opposing pressures are directed to two interconnected chambers within the motor located 180 degrees apart. Any side loads which are generated oppose and cancel each other. The majority of vane motors used in industrial systems are the balanced design.

There are various design modifications that can be made to vane motors. A vane motor with a unidirectional or non-reversible design is shown in Figure 7-25. A check valve in the inlet port assures pressure to hold the vanes extended, eliminating the need for rocker arms, shuttle valves, or an external pressure source. One application might be a fan drive or similar device which would rotate in only one direction.

High Performance Vane Motors. The high performance vane motor (Figure 7-26) is a later design of balanced vane motor. It operates under the same principles to generate torque but has significant changes in construction.

In this design, the vanes are held out against the ring by coil springs. The entire assembly of ring, rotor, vanes and side plates is removable and replaceable as a unit (Figure 7-27). In fact, preassembled and tested "cartridges" are available for field replacement. These motors also are reversible by reversing flow to and from the ports. Both side plates function alternately as pressure plates (Figure 7-28), depending on the direction of flow.

Piston Motors

There are a variety of piston motor designs currently available. The demands of each industrial application

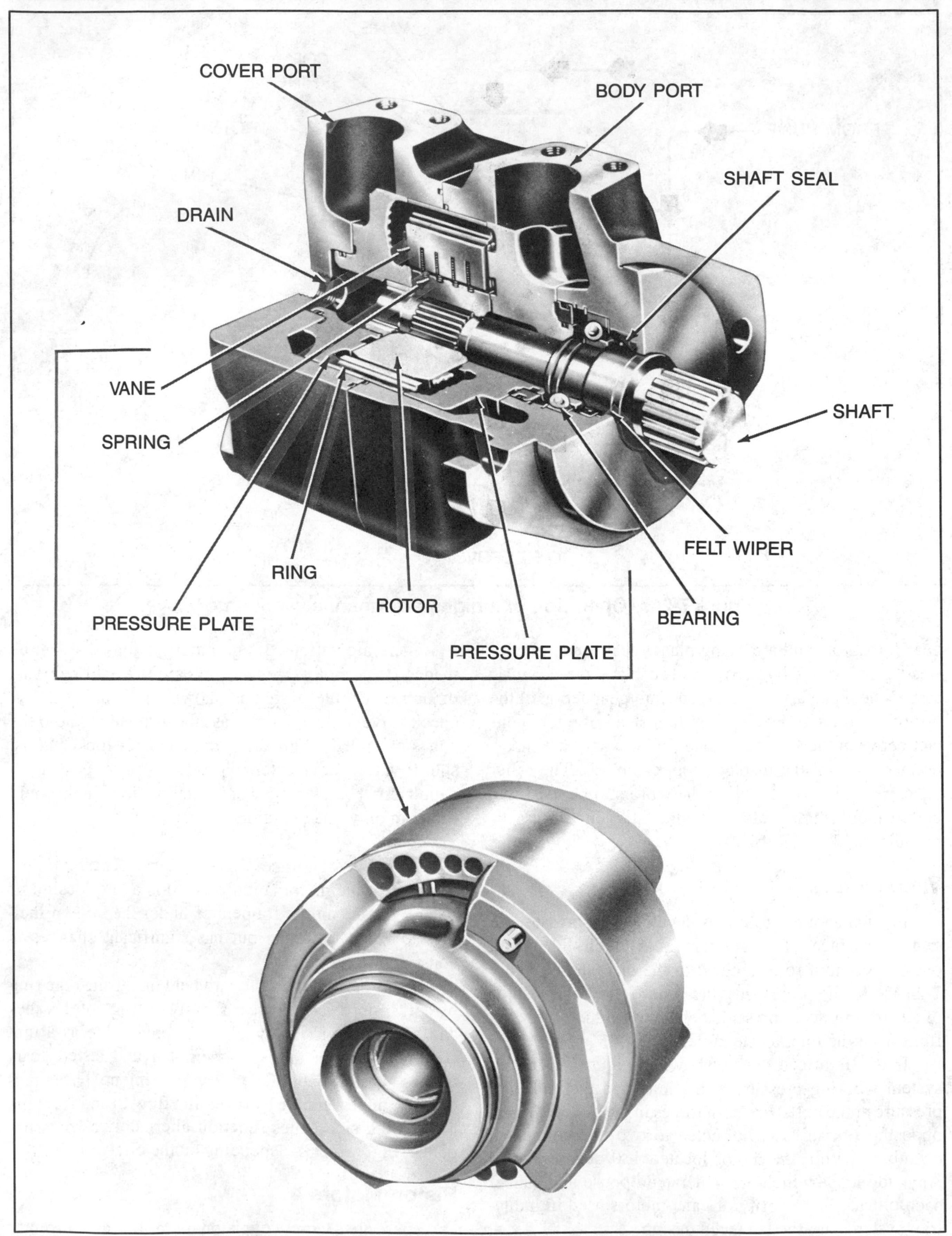

Figure 7-27. Construction of a high performance vane motor.

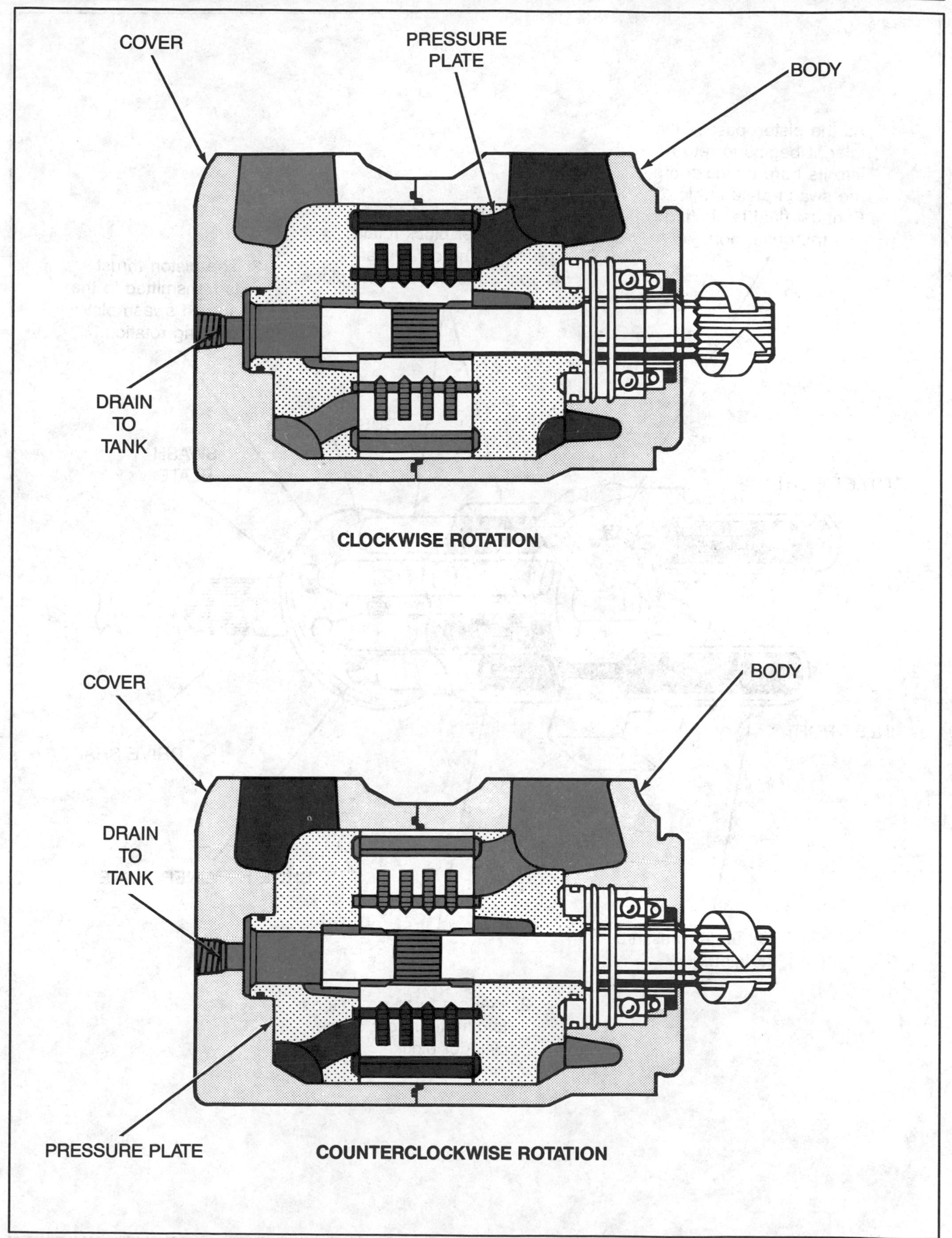

Figure 7-28. Side plates are pressure plates in the high performance design.

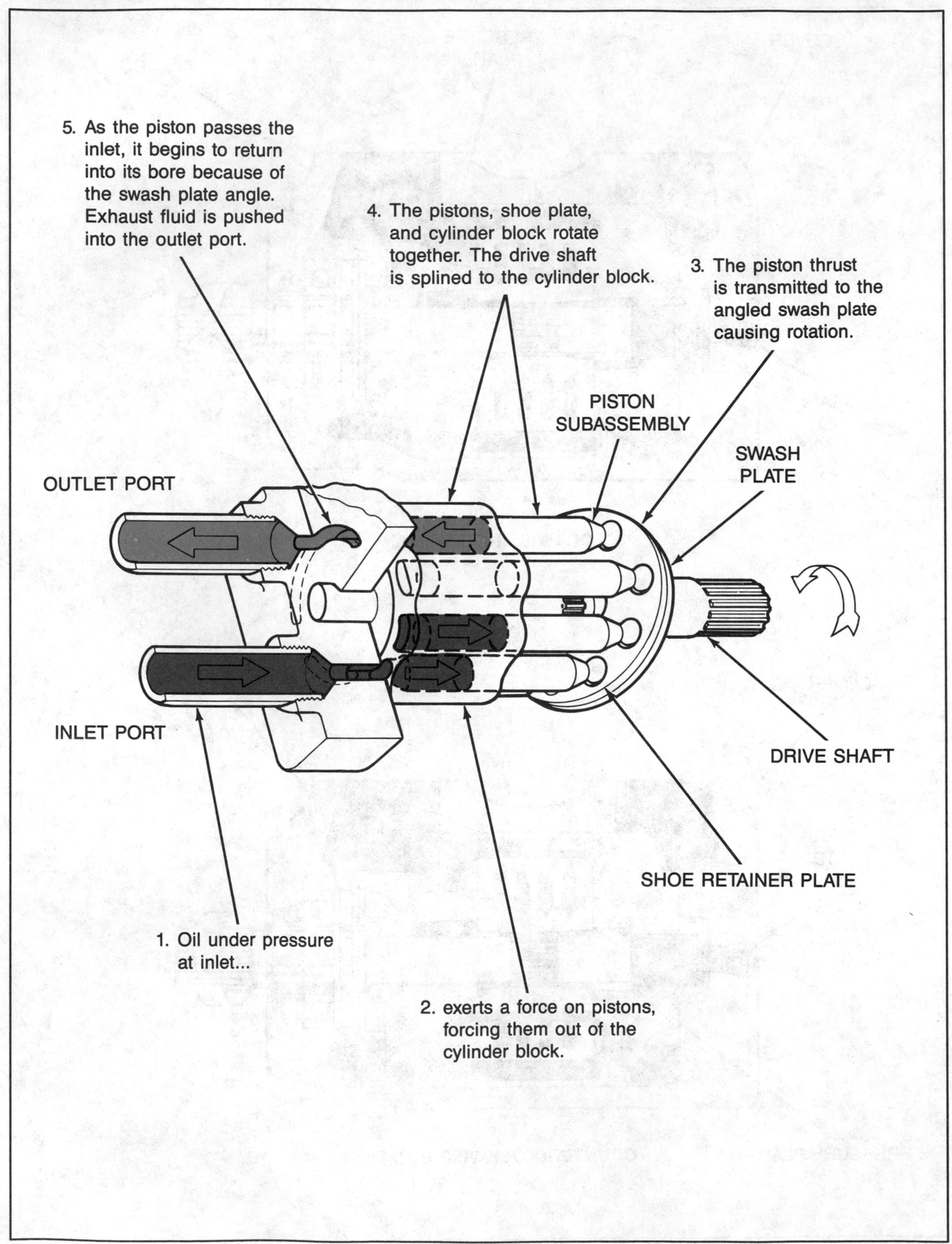

5. As the piston passes the inlet, it begins to return into its bore because of the swash plate angle. Exhaust fluid is pushed into the outlet port.
4. The pistons, shoe plate, and cylinder block rotate together. The drive shaft is splined to the cylinder block.
3. The piston thrust is transmitted to the angled swash plate causing rotation.
PISTON SUBASSEMBLY
SWASH PLATE
OUTLET PORT
INLET PORT
DRIVE SHAFT
SHOE RETAINER PLATE
1. Oil under pressure at inlet...
2. exerts a force on pistons, forcing them out of the cylinder block.

Figure 7-29. In-line piston motor operation.

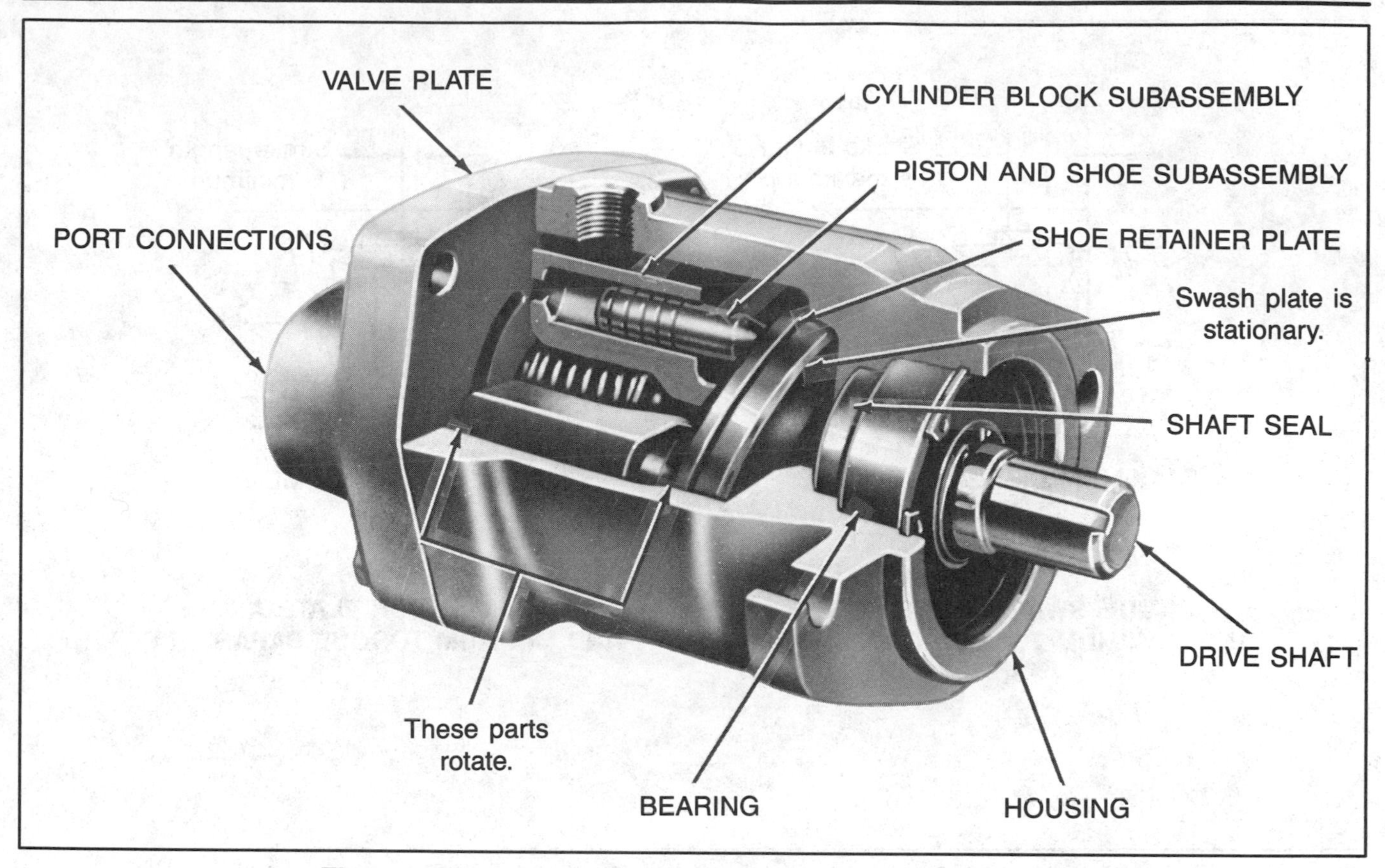

Figure 7-30. Fixed displacement in-line piston motor.

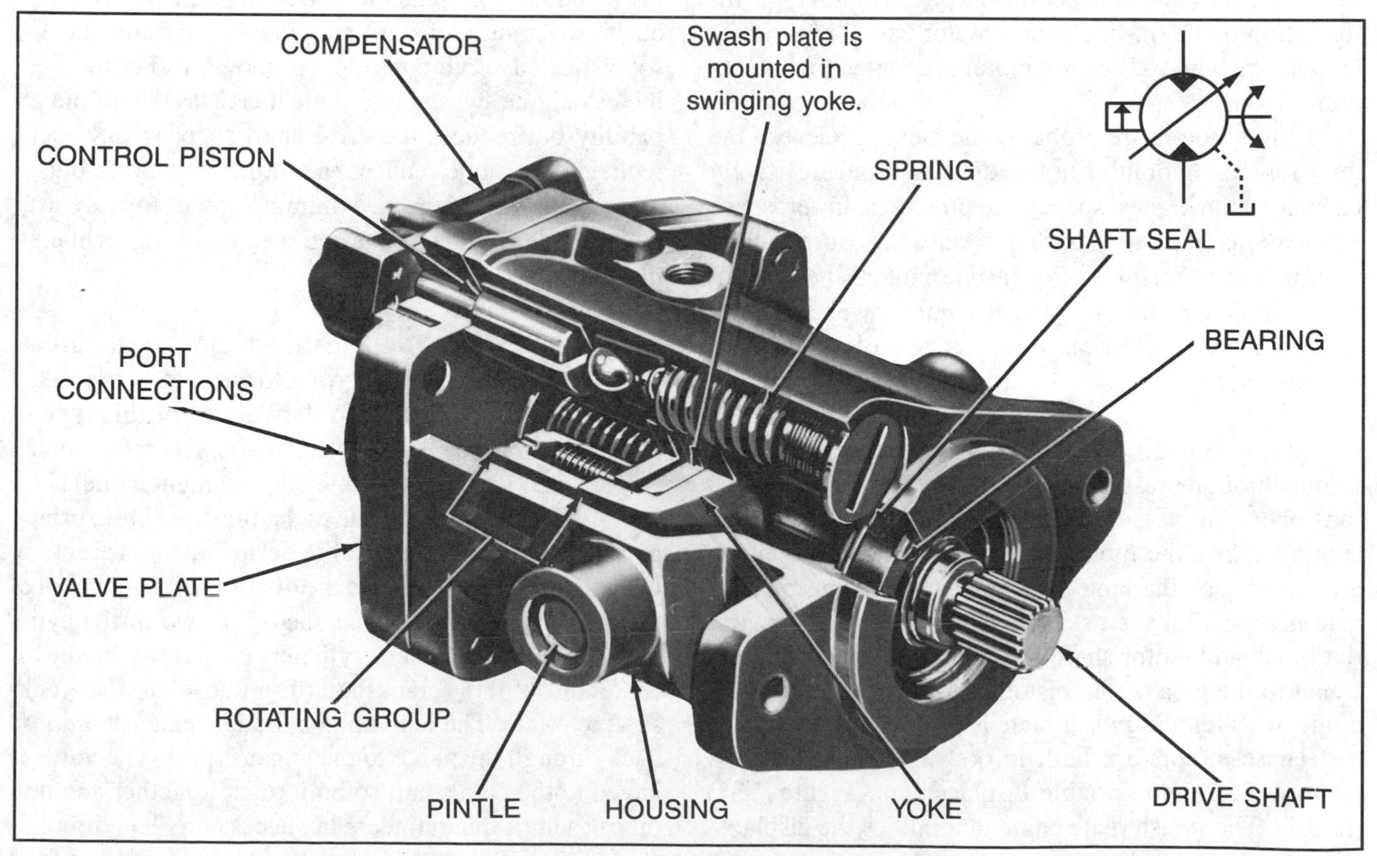

Figure 7-31. Variable displacement in-line piston motor.

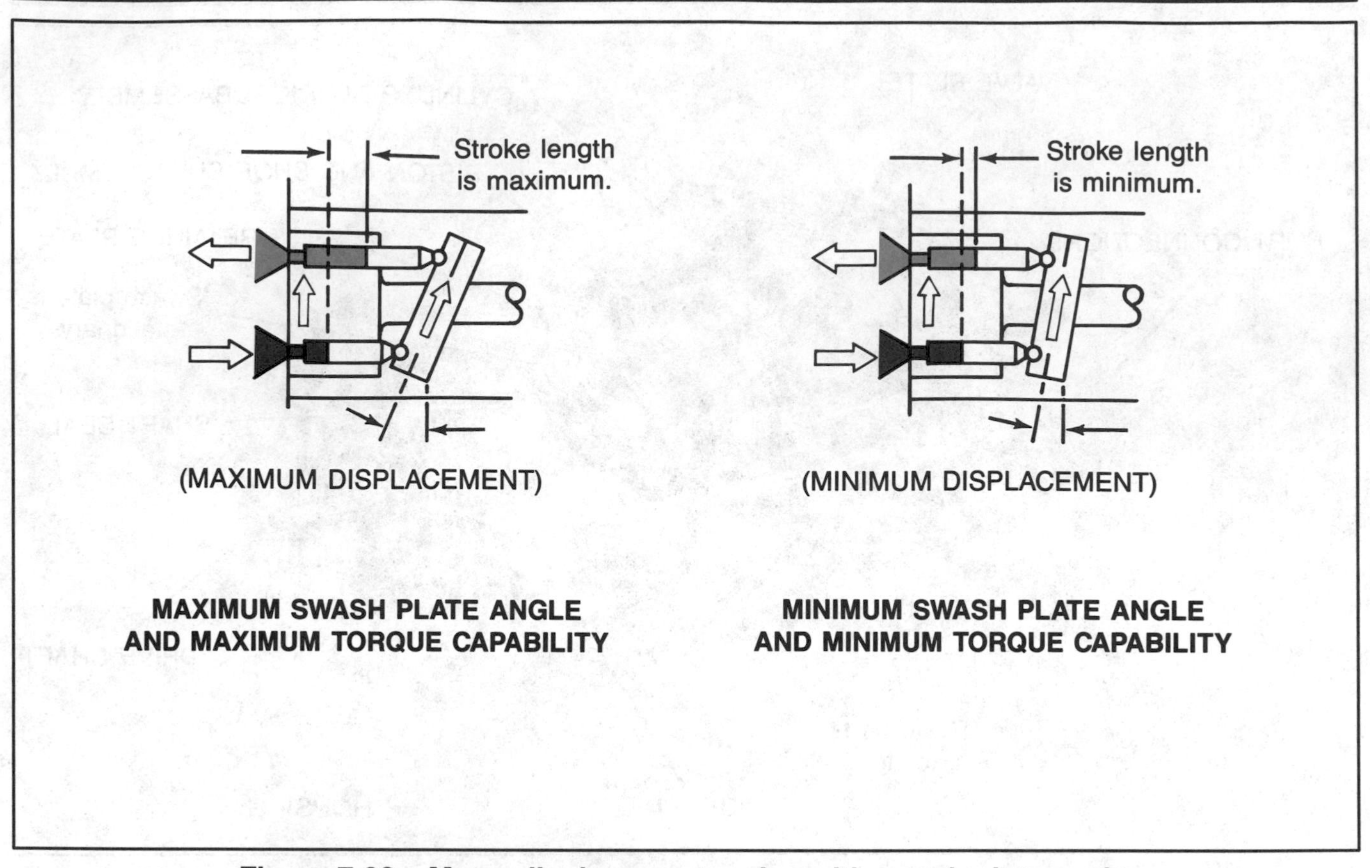

Figure 7-32. Motor displacement varies with swash plate angle.

determine the correct selection of a piston motor type. Information on the in-line piston motor, radial piston motor, and the bent-axis piston motor is covered in this section.

Piston motors are probably the most efficient of the three types of hydraulic motors discussed and are usually capable of the highest speeds and pressures. In aerospace applications in particular, they are used because of their high power to weight ratio. In-line motors, because of their simple construction and resultant lower costs, are finding many applications on machine tools and mobile equipment.

In-line Piston Motors. Piston motors generate torque through pressure on the ends of reciprocating pistons operating in a cylinder block. In the in-line design (Figure 7-29), the motor driveshaft and cylinder block are centered on the same axis. Pressure at the piston ends causes a reaction against a swashplate, driving the cylinder block and motor shaft in rotation. Torque is proportional to the area of the pistons and is a function of the angle at which the swash plate is positioned.

These motors are built in both fixed displacement (Figure 7-30) and variable displacement (Figure 7-31) models. The swash plate angle determines the displacement. In the variable model, the swash plate is mounted in a swinging yoke, and the angle can be changed by various means ranging from a simple lever or handwheel to sophisticated servo controls. As shown in Figure 7-32, increasing the swash plate angle increases the torque capability but reduces the drive shaft speed. Conversely, reducing the angle reduces the torque capability but increases drive shaft speed. Minimum angle stops are usually provided so that torque and speed stay within operating limits.

Radial Piston Motors. In addition to the in-line type, another popular piston motor design is the radial piston motor. There are two derivatives of this type of piston motor.They are the fixed displacement model (Figure 7-33) and the variable displacement model (Figure 7-34). They both operate under the same basic principles and are almost identical in design and construction. Oil is supplied by a pump to the distributor valve spool of the motor. The valve directs the oil to two of the cylinders. The pistons in these cylinders push down on the off center throw of the crankshaft (drum), causing the crankshaft to rotate. The remaining cylinders exhaust their oil back through the valve to the exhaust port. The valve is keyed to the crankshaft so both rotate together and port oil to each of the cylinders in succession. The timing is slightly out of phase between the two so there is no dead

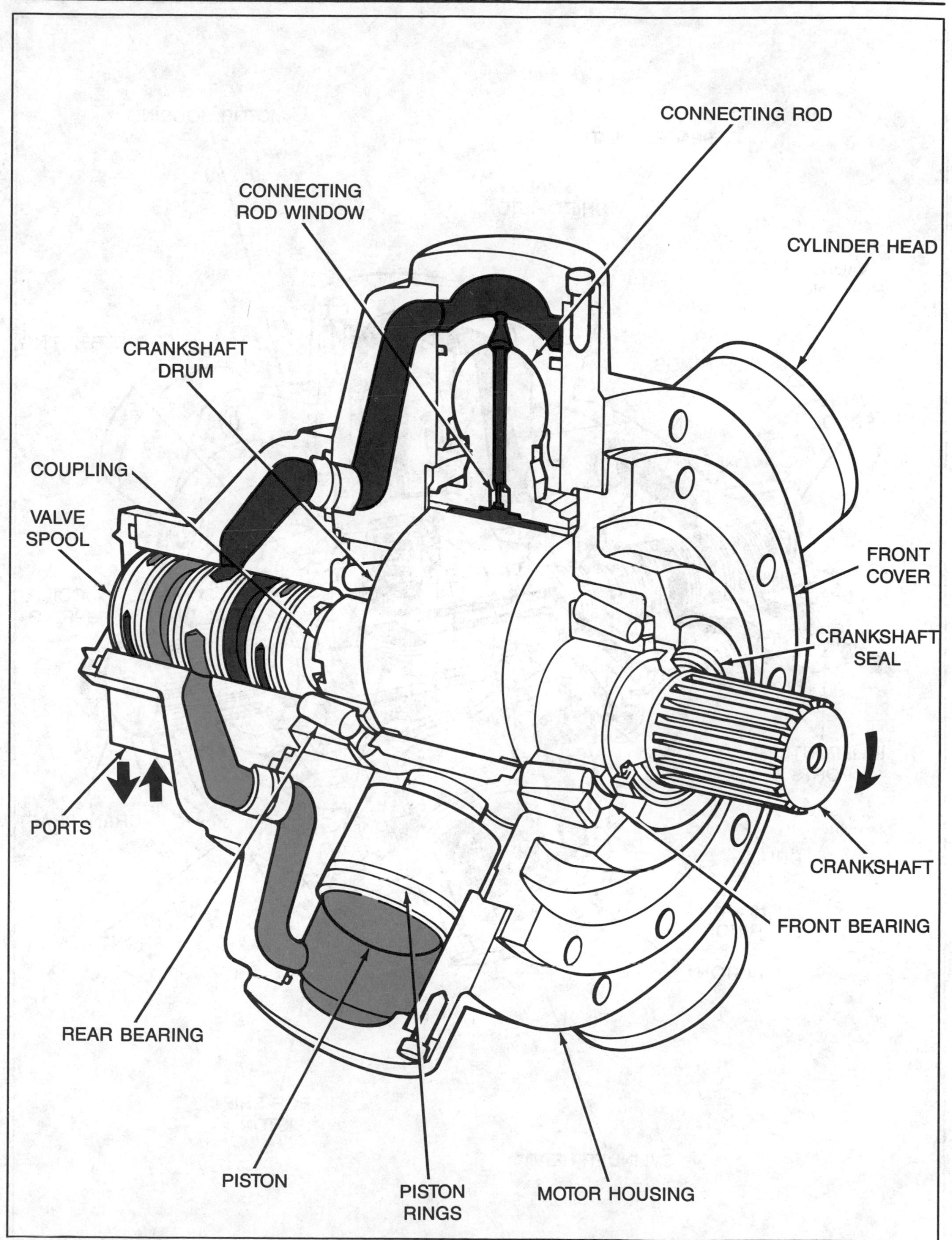

Figure 7-33. Fixed displacement radial piston motor.

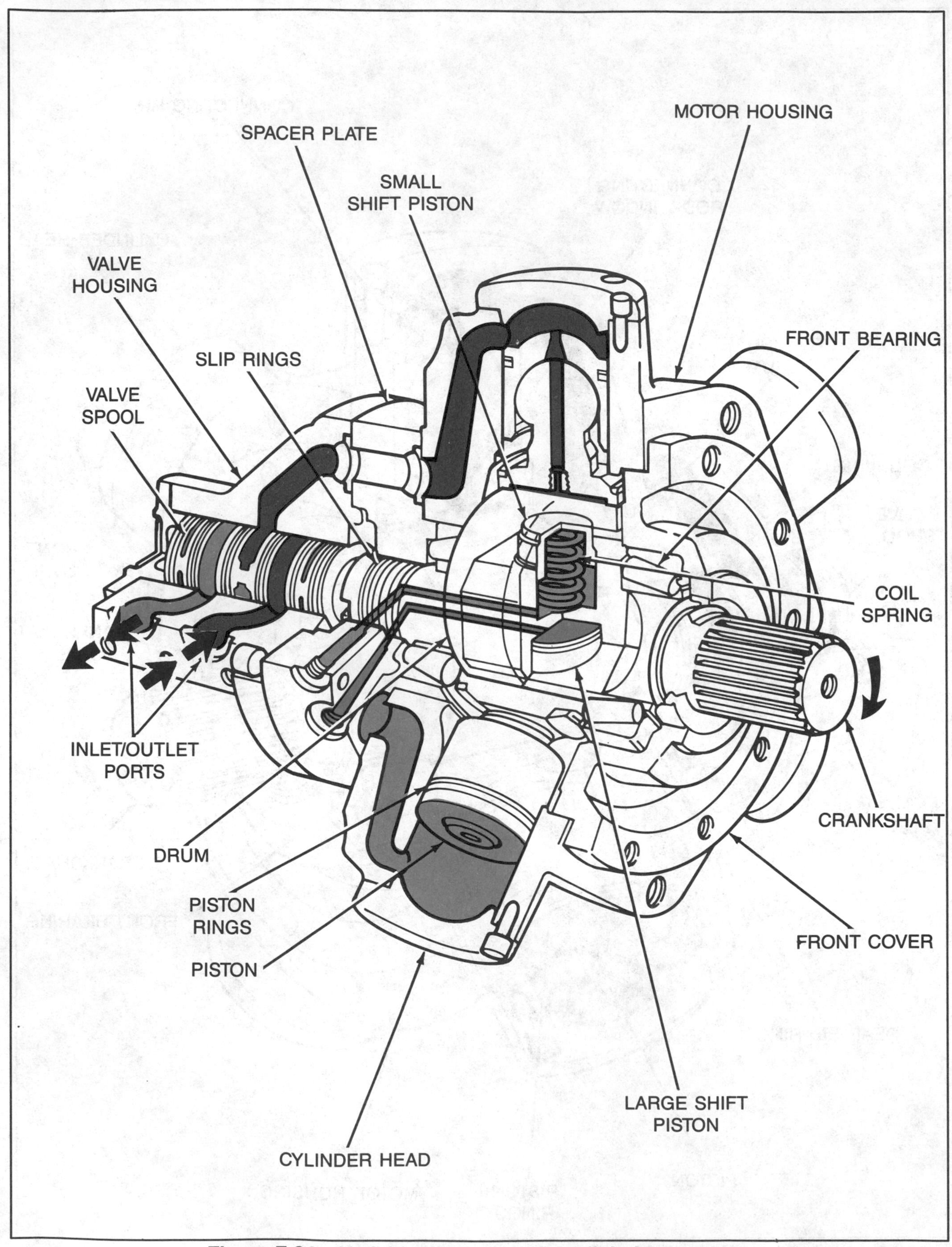

Figure 7-34. Variable displacement radial piston motor.

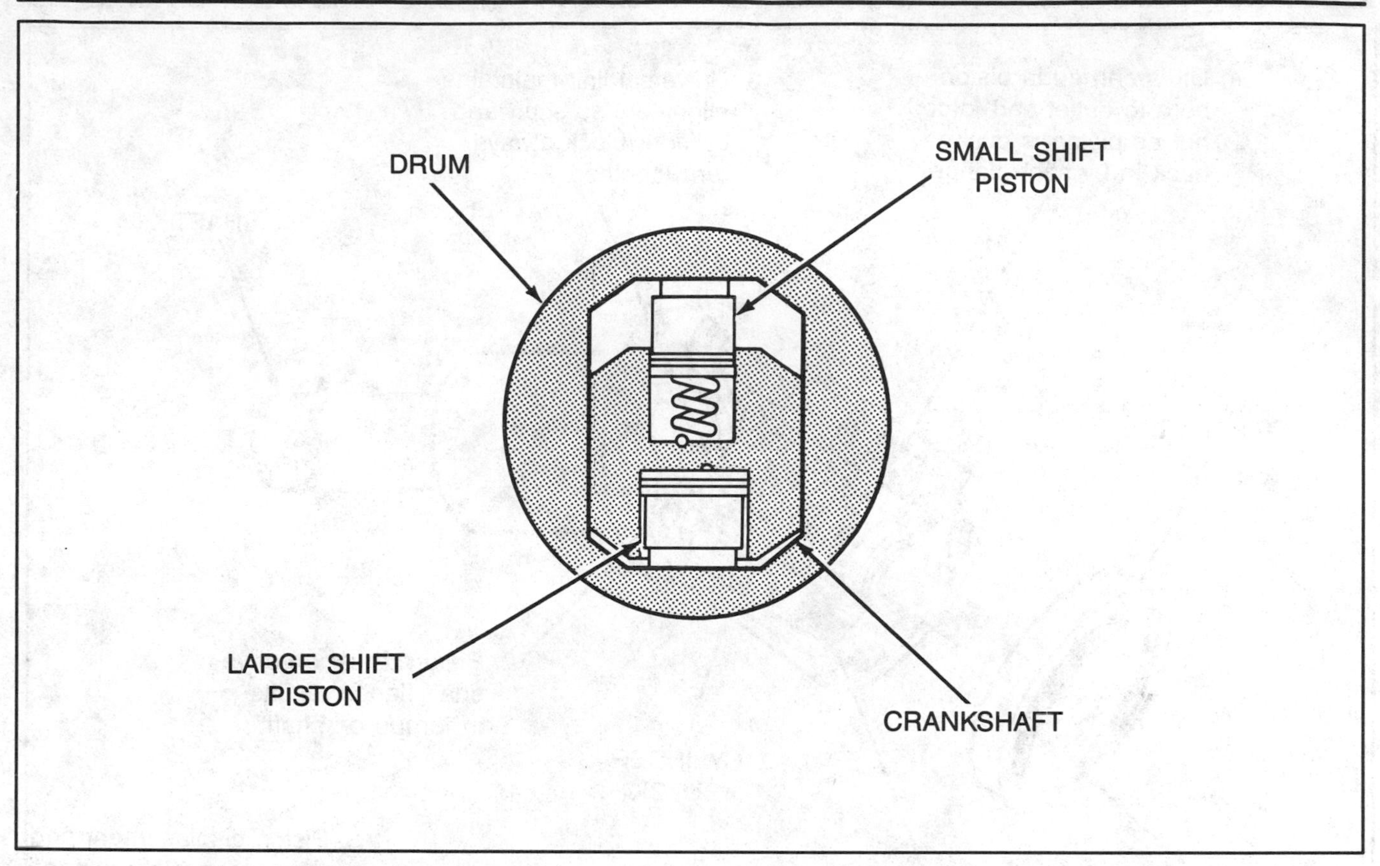

Figure 7-35. Crankshaft for variable displacement radial piston motor.

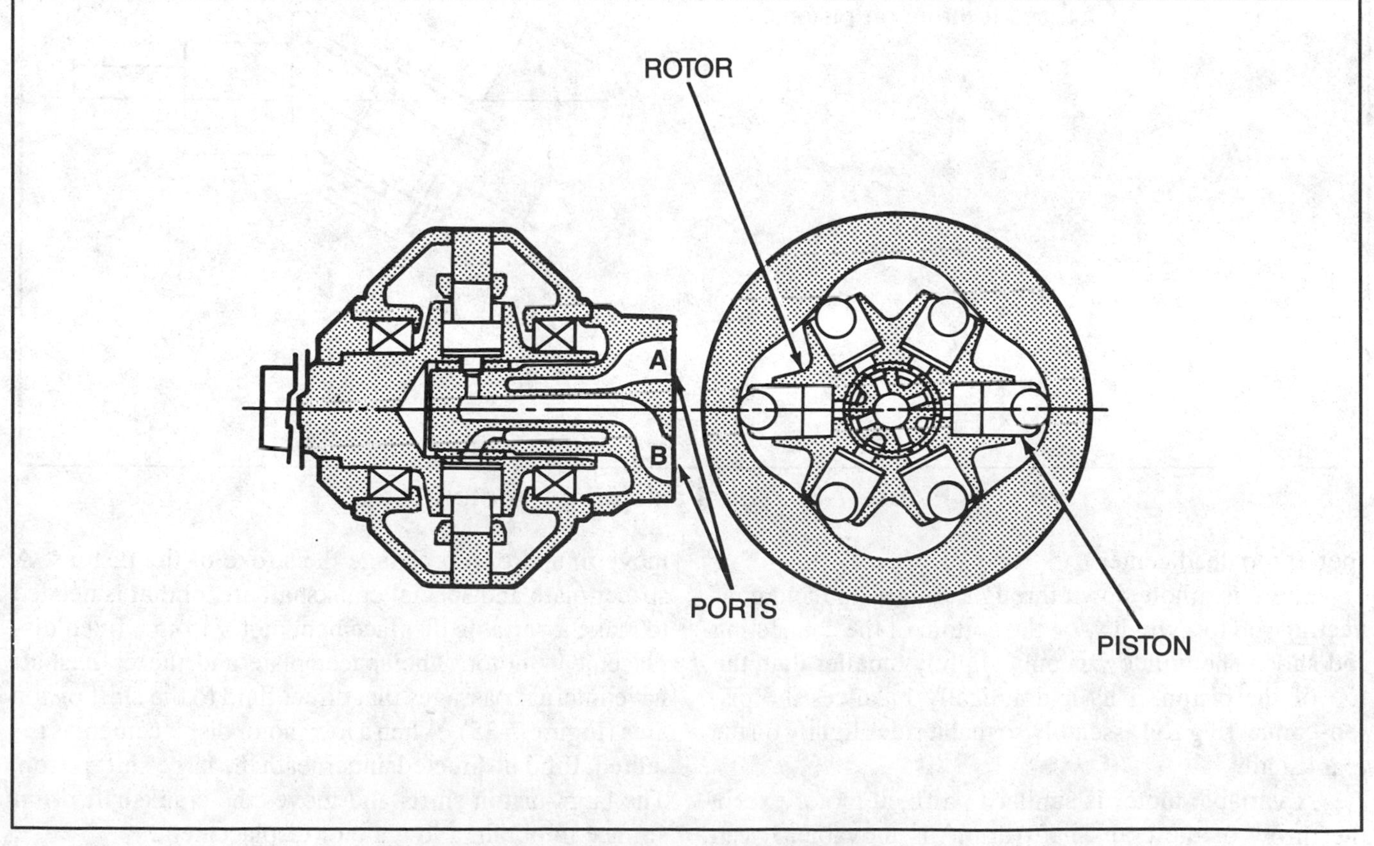

Figure 7-36. Cam-type radial piston motor.

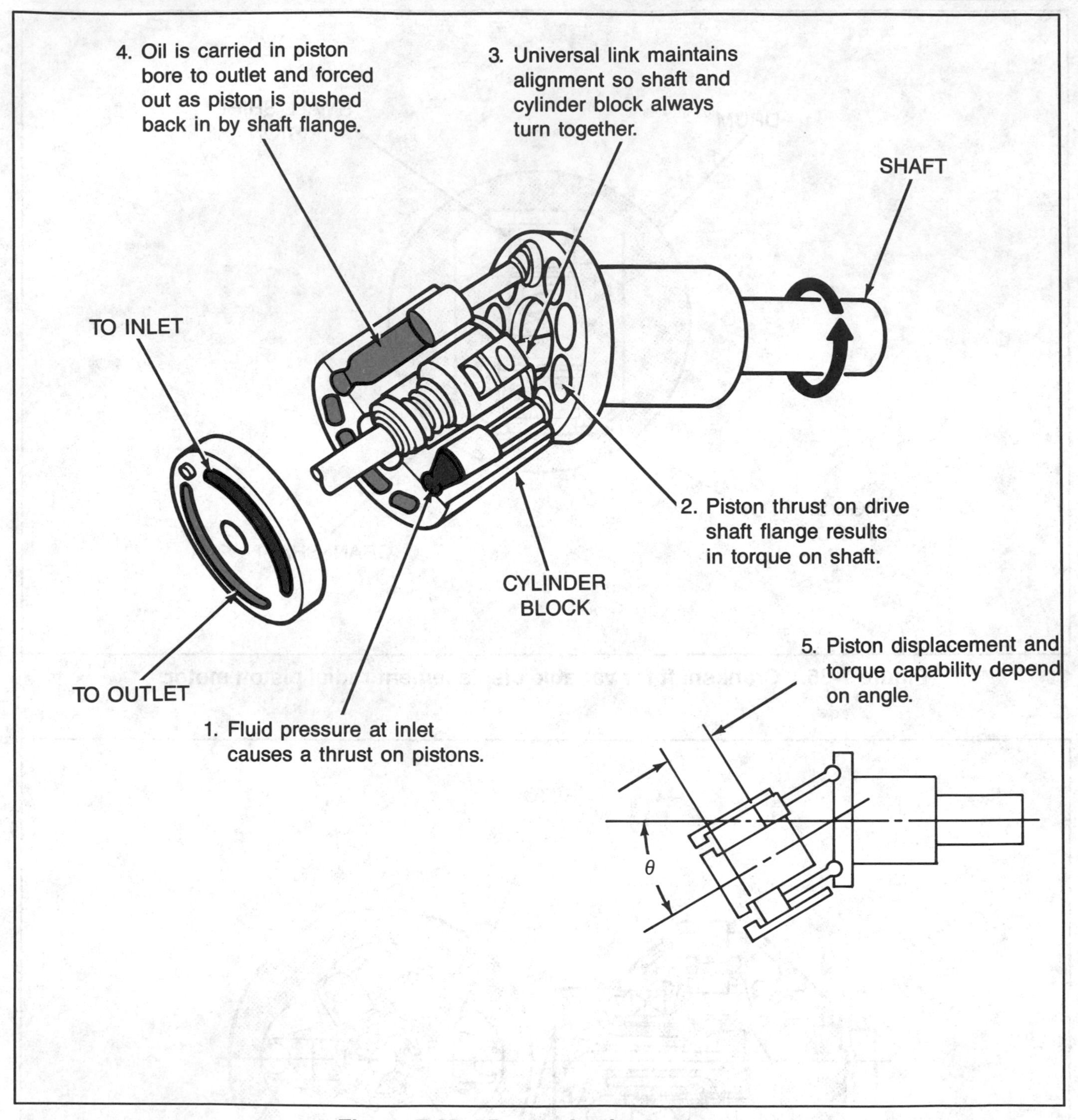

Figure 7-37. Bent-axis piston motor.

spot at top dead center.

There is a hole down through each piston and connecting rod to a window on the bottom of the connecting rod shoe. The window is only slightly smaller than the top of the piston. This hydraulically balances the piston-connecting rod assembly so that it rides lightly on the crankshaft.

A variable motor is similar to a fixed motor except the throw of the crankshaft (drum) is movable. It can move in and out to change the stroke of the pistons. A spacer plate and special crankshaft are all that is needed to make a variable displacement motor from a fixed displacement motor. The spacer plate and the crankshaft have internal passages that direct fluid to the shift piston area (Figure 7-35). When a low motor displacement is required, fluid is directed underneath the large shift piston. The large piston shifts and moves the crankshaft drum surface to obtain a low motor displacement.

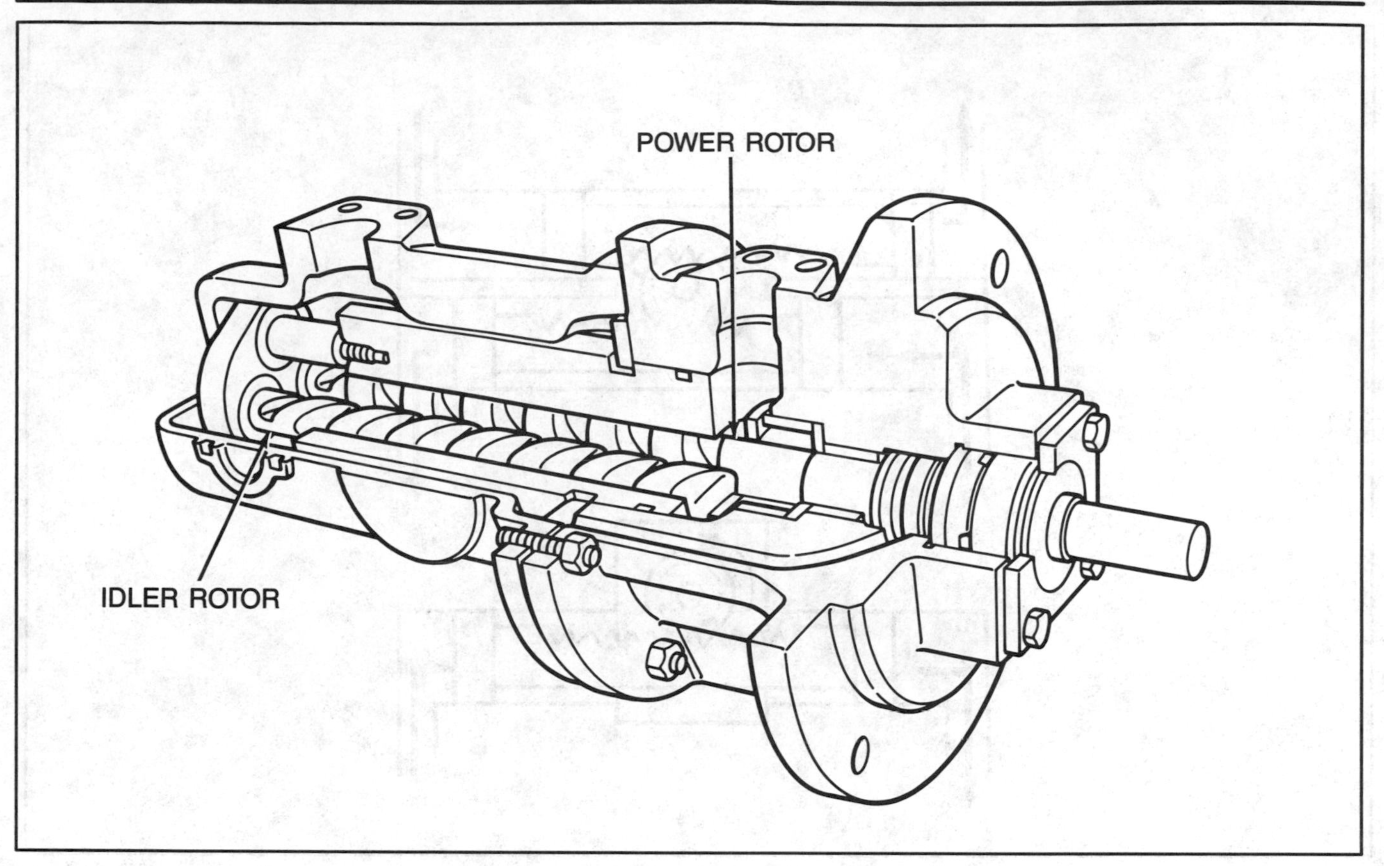

Figure 7-38. Screw motor.

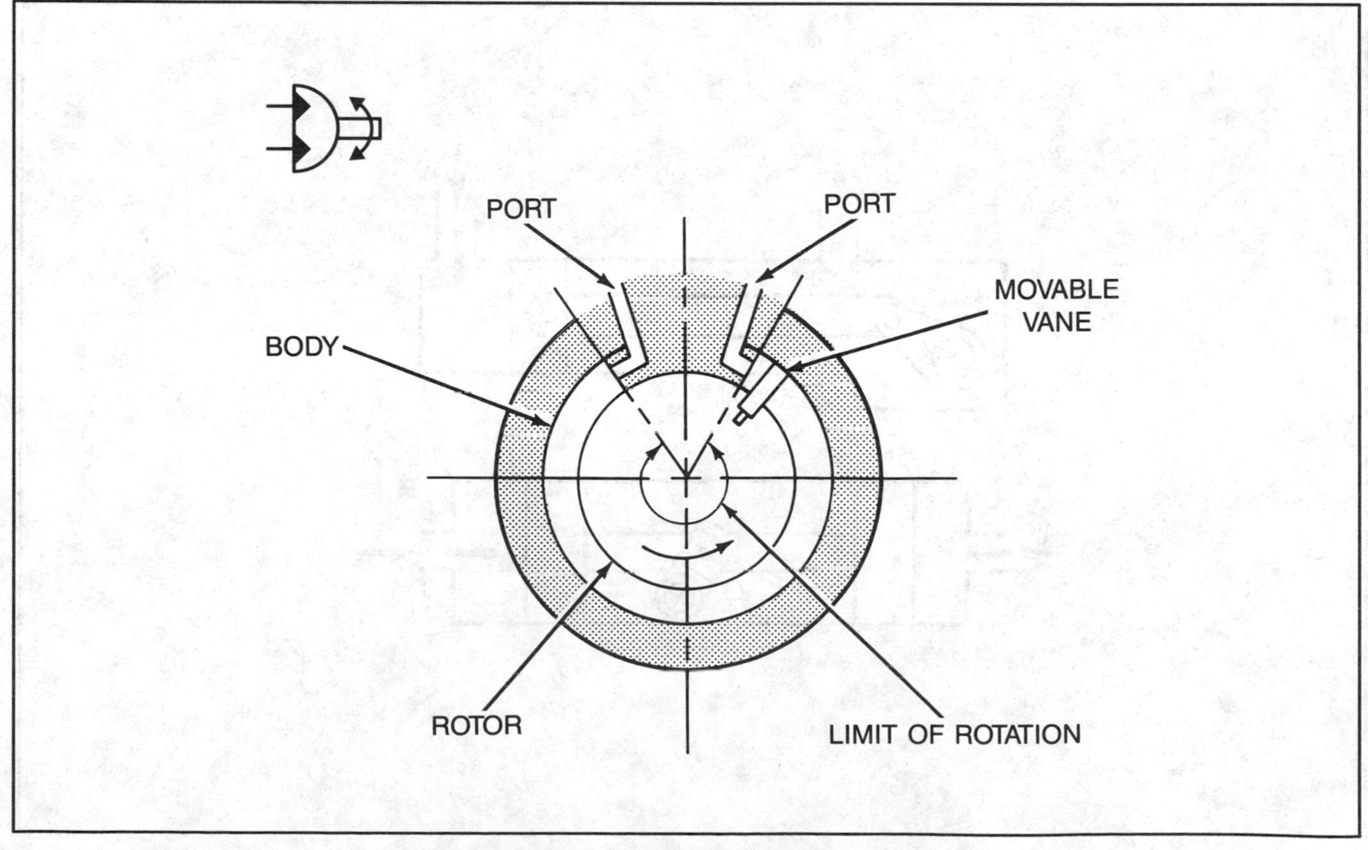

Figure 7-39. Vane-type torque generator.

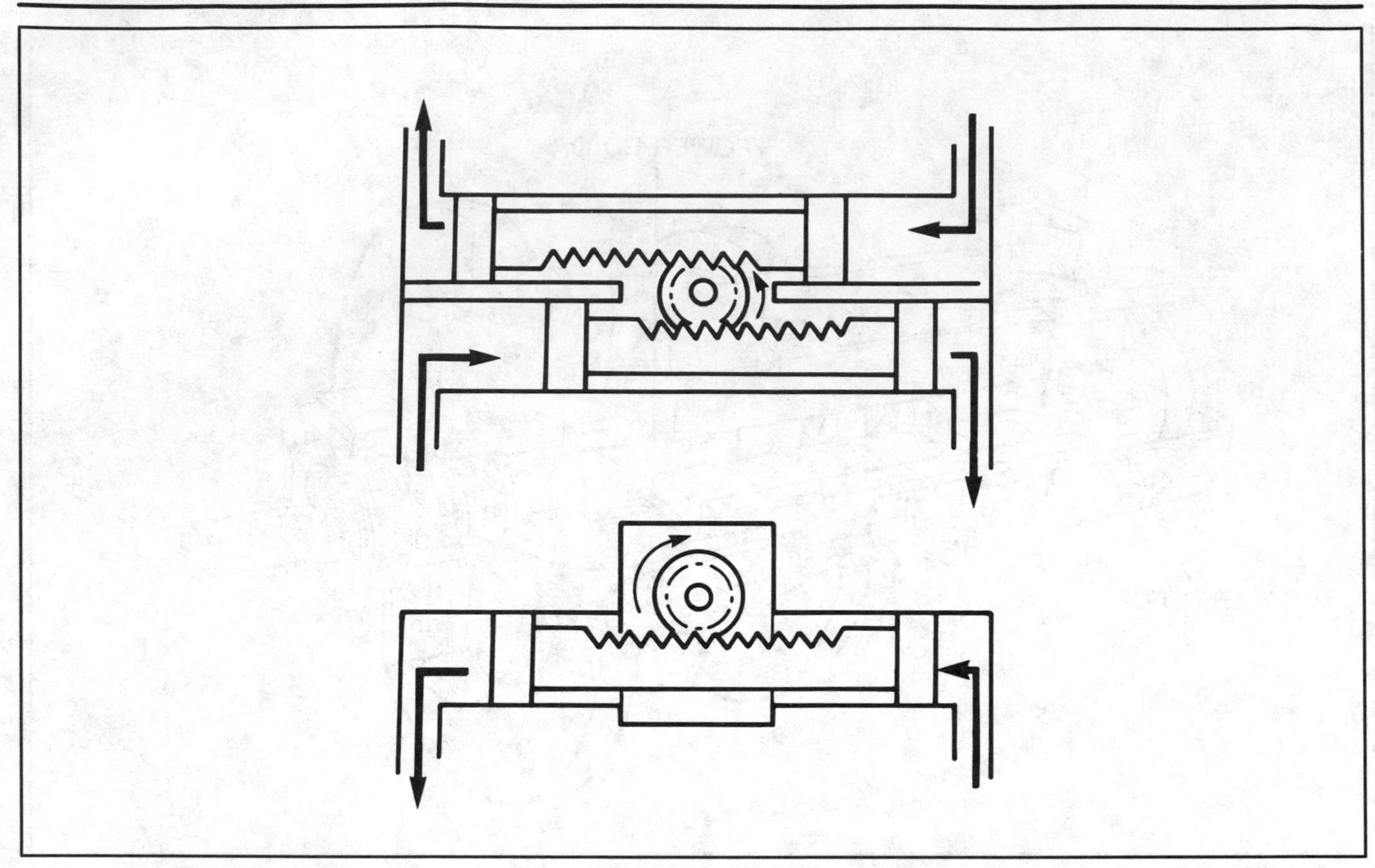

Figure 7-40. Rack actuators.

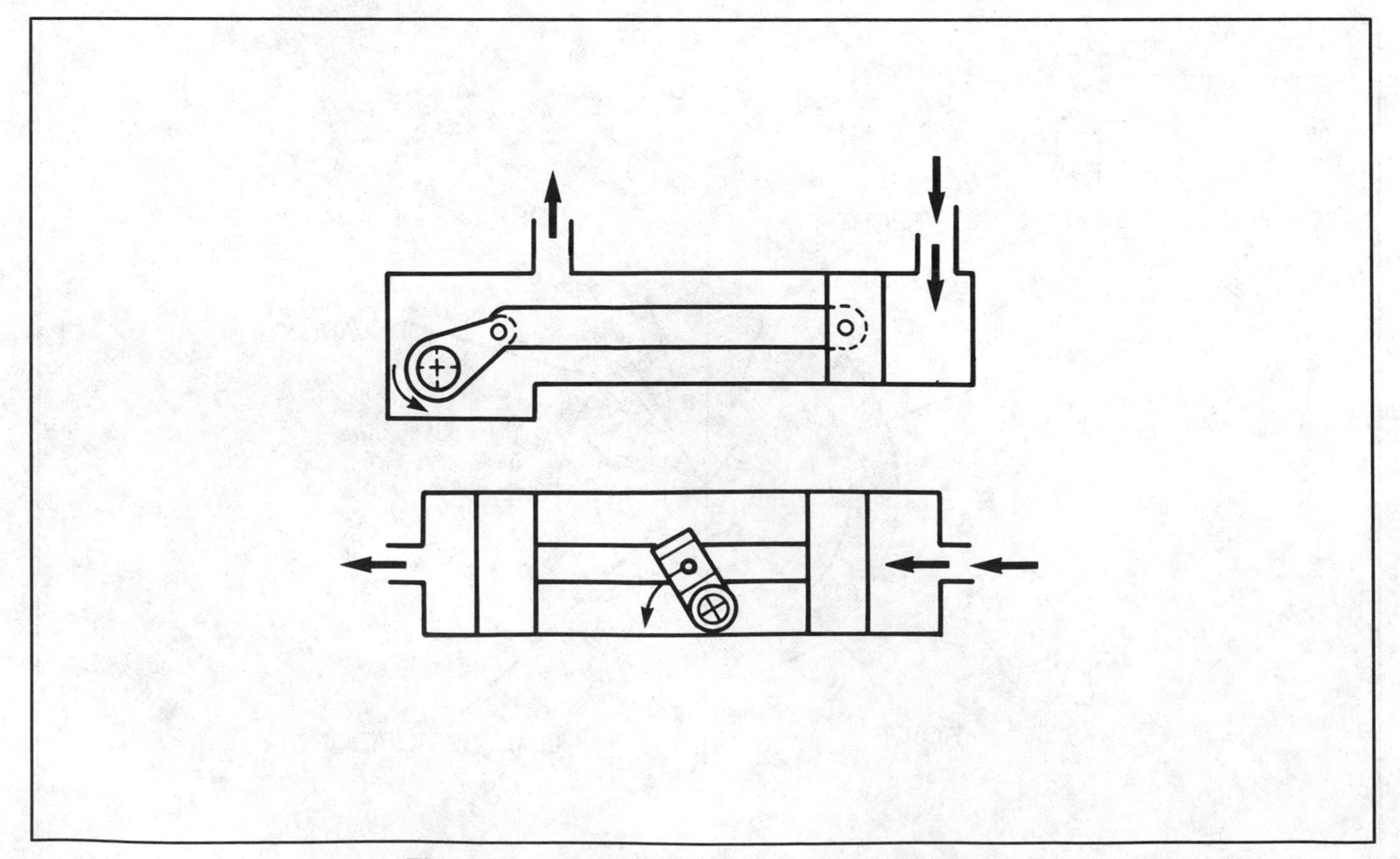

Figure 7-41. Crank action piston actuator.

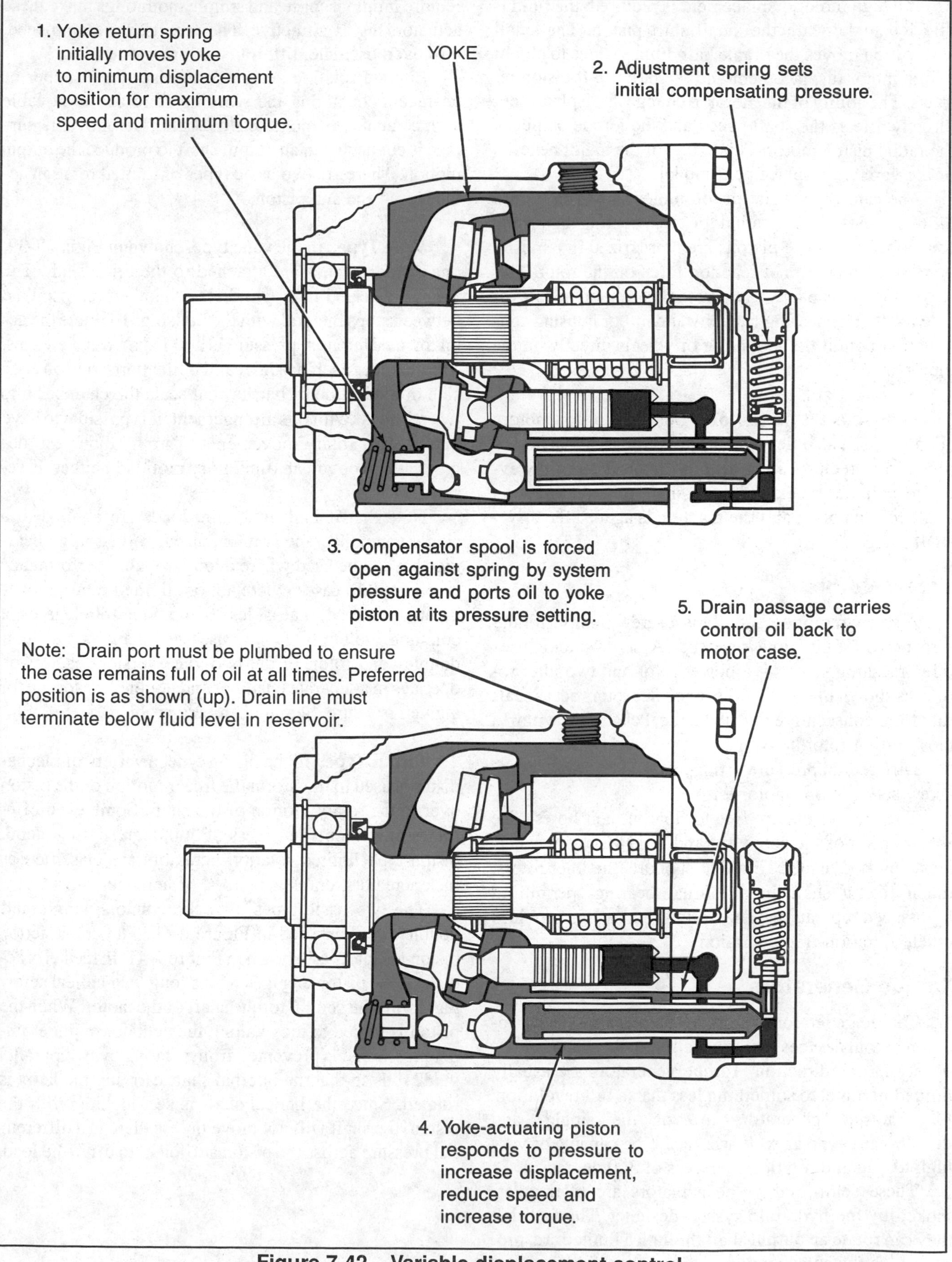
1. Yoke return spring initially moves yoke to minimum displacement position for maximum speed and minimum torque.
YOKE
2. Adjustment spring sets initial compensating pressure.
3. Compensator spool is forced open against spring by system pressure and ports oil to yoke piston at its pressure setting.
5. Drain passage carries control oil back to motor case.
Note: Drain port must be plumbed to ensure the case remains full of oil at all times. Preferred position is as shown (up). Drain line must terminate below fluid level in reservoir.
4. Yoke-actuating piston responds to pressure to increase displacement, reduce speed and increase torque.

Figure 7-42. Variable displacement control.

If a high motor displacement is required, the fluid is directed underneath the small shift piston. The small shift piston moves the crankshaft drum surface to obtain a high motor displacement and the large shift piston retracts. The ability of the motor to change its displacement directly affects the shaft speed and the torque output of the radial piston motor. This is one of the major benefits of the variable displacement model.

The cam-type radial piston motor shown in Figure 7-36 consists of six radial pistons bearing against a four-lobed cam. The pistons are pressurized in opposite pairs so there is no net reaction force on the rotor. This leaves the full capacity of the rotor bearings available for side loads. The bearing capacity and motor housing construction permit the mounting of wheels directly on the motor shaft.

Bent-Axis Piston Motors. Bent-axis piston motors (Figure 7-37) also develop torque through a reaction to pressure on reciprocating pistons. In this design, however, the cylinder block and drive shaft are mounted at an angle to each other and the reaction is against the drive-shaft flange.

Screw Motors

A screw motor essentially is a screw pump with the direction of fluid flow reversed. A screw motor uses three meshing screws—a power rotor and two idler rotors, as shown in Figure 7-38. The idler rotors act as seals that form consecutive isolated helical chambers within a close-fitting rotor housing.

Differential pressure acting on the thread areas of the screw set develops motor torque.

The idler rotors are free to float in their bores. The rotary speed of the screw set and fluid viscosity generates a hydrodynamic film that supports the idler rotors, much like a shaft in a journal bearing, permitting high-speed operation. The rolling screw set provides quiet, vibration-free operation.

Torque Generators

Torque generators or torque actuators are partial rotation output devices which usually cannot rotate continuously in one direction. Torque generators are usually limited in travel to something less than a full revolution. Most torque generators are of the single- and double-vane type as well as a rack type capable of very high torque and rotation in excess of 360 degrees.

These semi-rotary type actuators are often overlooked by the hydraulic system designer. The fact that they can rotate an output shaft through a limited arc, produce high instantaneous torque in either direction, and require minimal space and simple mountings make these actuators highly attractive. They are used to clamp, feed, turn over, oscillate, lift, transfer, or hold tension.

Limited rotation actuators consist of a chamber or chambers containing the working fluid and a movable surface against which the fluid acts. The movable surface is connected to an output shaft to produce the output motion. There are two basic types of limited rotation actuators—vane and piston.

Vane Type. In the vane type, shown in Figure 7-39, one or more vanes are attached to the drive shaft. The vanes are sealed in a cylindrical chamber and oscillate between stops integral with the housing. To rotate the actuator, a differential pressure is applied across the vanes. The chamber itself has inlet and outlet ports, one on each side of a partition or barrier that seals the chamber into two sections. With this arrangement, it is possible to have output shaft rotary movements of up to about 280 degrees and to obtain an oscillatory motion in either direction.

Note that if high hydraulic loads are applied to a single vane, the vane is unbalanced, so that large bearings have to be used where a long, reliable performance is required. In cases where a more limited rotary movement is required, that is, less than about 160 degrees, a common solution is to use an actuator with a double-vane rotor. In the two-vane version, each vane operates in a separate chamber and doubles the torque for a given pressure. Sealing can be a problem.

Piston Type. In the piston type, a variety of mechanisms is used to transform the linear motion of the piston over to the rotary motion of the shaft. Some even offer rotary movements in excess of 360 degrees. In general, piston-type limited rotation actuators are classified as rack actuators, crank actuators, or helix actuators.

The two basic types of rack actuators, single and double, are illustrated in Figure 7-40. The crank action piston actuators are shown in Figure 7-41. In the helix actuators, the piston engages with a long lead helical screw that forms the central torque shaft of the motor. When the piston is forced to move along the actuator cylinder under pressure, it is prevented from rotating by two parallel guide rods so that the internal shaft carrying the helix is rotated. Since the helical shaft is self-locking with the piston, the shaft will not move unless there is a differential pressure across the piston sufficient to drive the load.

Use of Variable Displacement Controls

The compensator control (Figure 7-42) is used to vary the motor displacement in response to the changes in workload. A spring-loaded piston is connected mechanically to the yoke and moves it in response to variations in operating pressure. Any load increase is accompanied by a corresponding pressure increase as a result of the additional torque requirements.

At this point the control automatically adjusts the yoke to increase the displacement to increase the torque without any increase in pressure. Ideally, the compensator regulates the displacement for maximum performance under all load conditions up to the relief valve setting.

QUESTIONS

1. Describe the operating characteristics of single- and double-acting cylinders.
2. With an actual delivery of 3 gpm to the head end of a two-inch diameter cylinder, what is the speed of rod travel?
3. A three-inch (0.76 mm) diameter ram can operate up to 2000 psi (137.88 bar / 13,788.00 kPa). What is the maximum output force? In pounds? In newtons?
4. How much pressure is required for a force output of 14,000 pounds (62,275.00 N) if the effective piston area of the cylinder is 7 square inches (45.16 square centimeters)?
5. What is the primary function of a hydraulic cylinder?
6. Define displacement and torque ratings of a hydraulic motor.
7. A winch requires 50 pound-feet maximum torque to operate. What size hydraulic motor is needed if maximum pressure must be limited to 1500 psi?
8. A 20 pound-inch motor operates with a torque load of 500 pound-inches. What is the operating pressure?
9. Explain how the vanes are held in contact with the cam ring in high performance vane motors.
10. How is torque developed in the in-line-type piston motor?
11. If a hydraulic motor is pressure compensated, what is the effect of an increase in the working load?
12. What type of hydraulic motor is generally most efficient?

CHAPTER 8
DIRECTIONAL VALVES

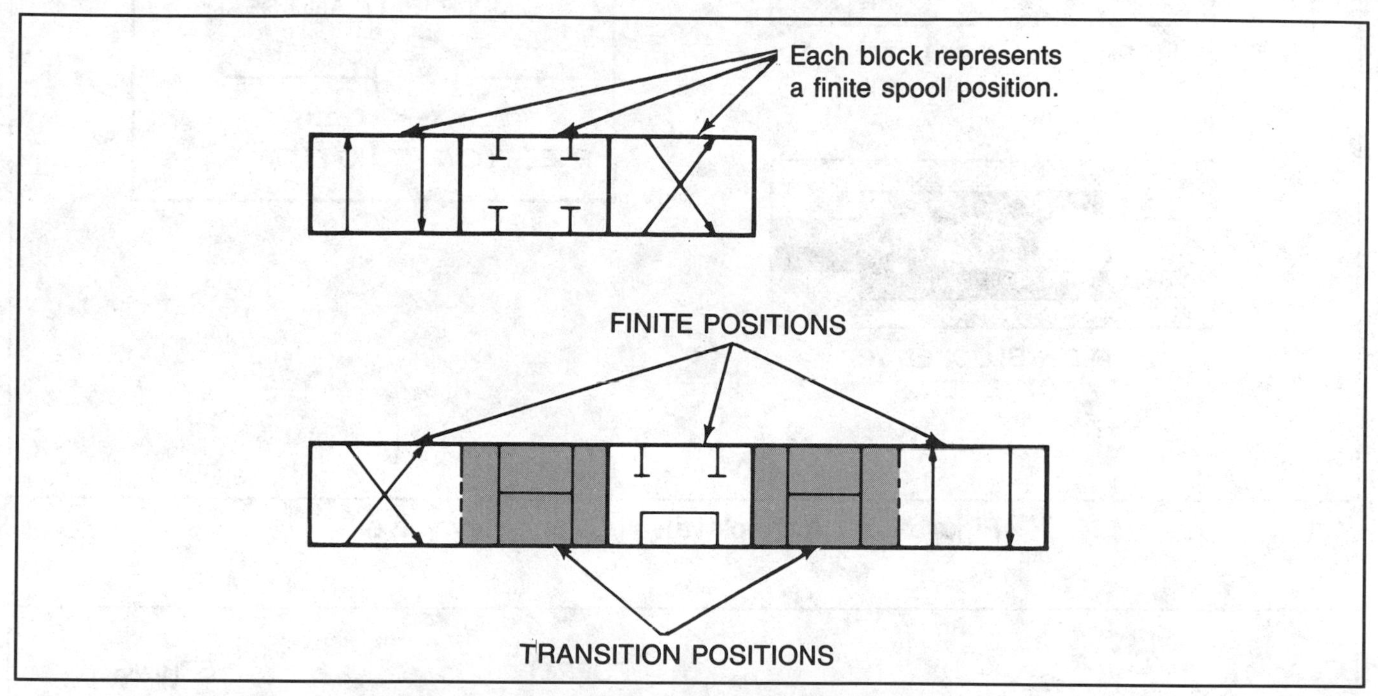

Figure 8-1. Graphic symbol for directional valve.

As the name implies, directional valves start, stop, and control the direction of fluid flow. Although they share this common function, directional valves vary considerably in construction and operation. They are classified according to principal characteristics such as those listed below.

Type of internal valving element — poppet (piston or ball), rotary spool, or sliding spool.

Methods of actuation — manual, mechanical, pneumatic, hydraulic, electrical, or combinations of these.

Number of flow paths — two-way, three-way, and four-way. Vickers references the total number of available flow paths. A four-way, i.e: four flow path valve is illustrated in Figure 8-1. Some manufacturers may refer to the number of active ports rather than the number of active flow paths. In any case, the graphic symbology reflects the function of the component rather than the construction, and will enhance understanding of the components affect in an hydraulic circuit.

Size — nominal size of port or flange connections to the valve or its mounting plate, rated gpm (LPM), or with reference to a standard mounting pattern.

Connections — pipe thread, straight thread, flanged and subplate, or manifold mounted.

This chapter discusses finite positioning directional valves. These valves direct the fluid by opening and closing flow paths in definite valve positions. For each finite position, the graphic symbol for a directional valve contains a separate envelope or square that shows the flow paths in that position. When needed, this symbol also indicates a transition position (Figure 8-1) to clarify a particular spool function.

CHECK VALVES

In its simplest form, a check valve is a one-way directional valve. It allows free flow in one direction, while blocking flow in the other direction. The graphic symbol for a check valve is a ball and seat (Figure 8-2). A light spring, usually equivalent to 5 psi (0.34 bar) (34.47 kPa), holds the poppet in the normally closed position. Other spring pressures are available to suit application requirements. In the free flow direction, the poppet cracks open

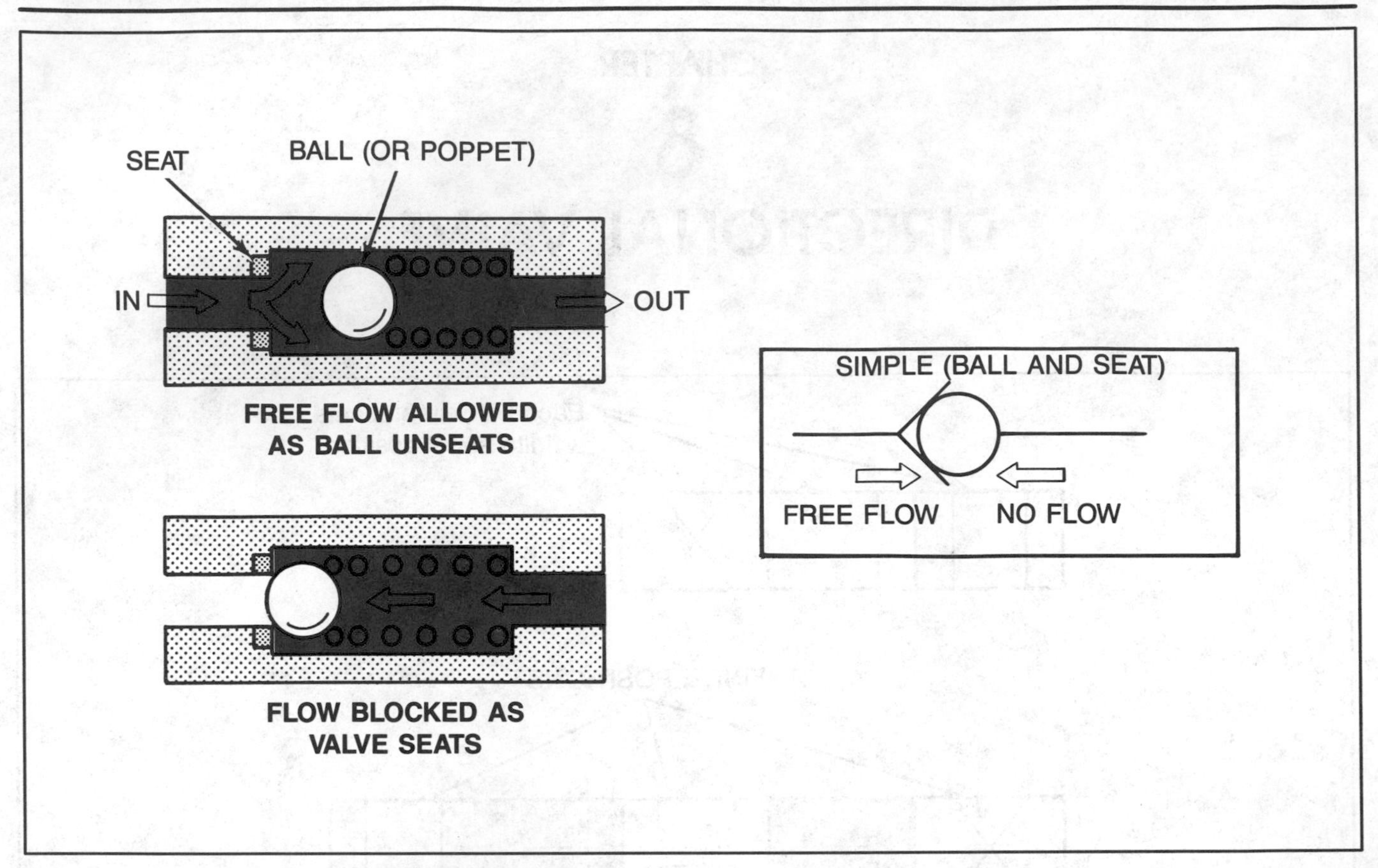

Figure 8-2. A check valve is a one-way valve.

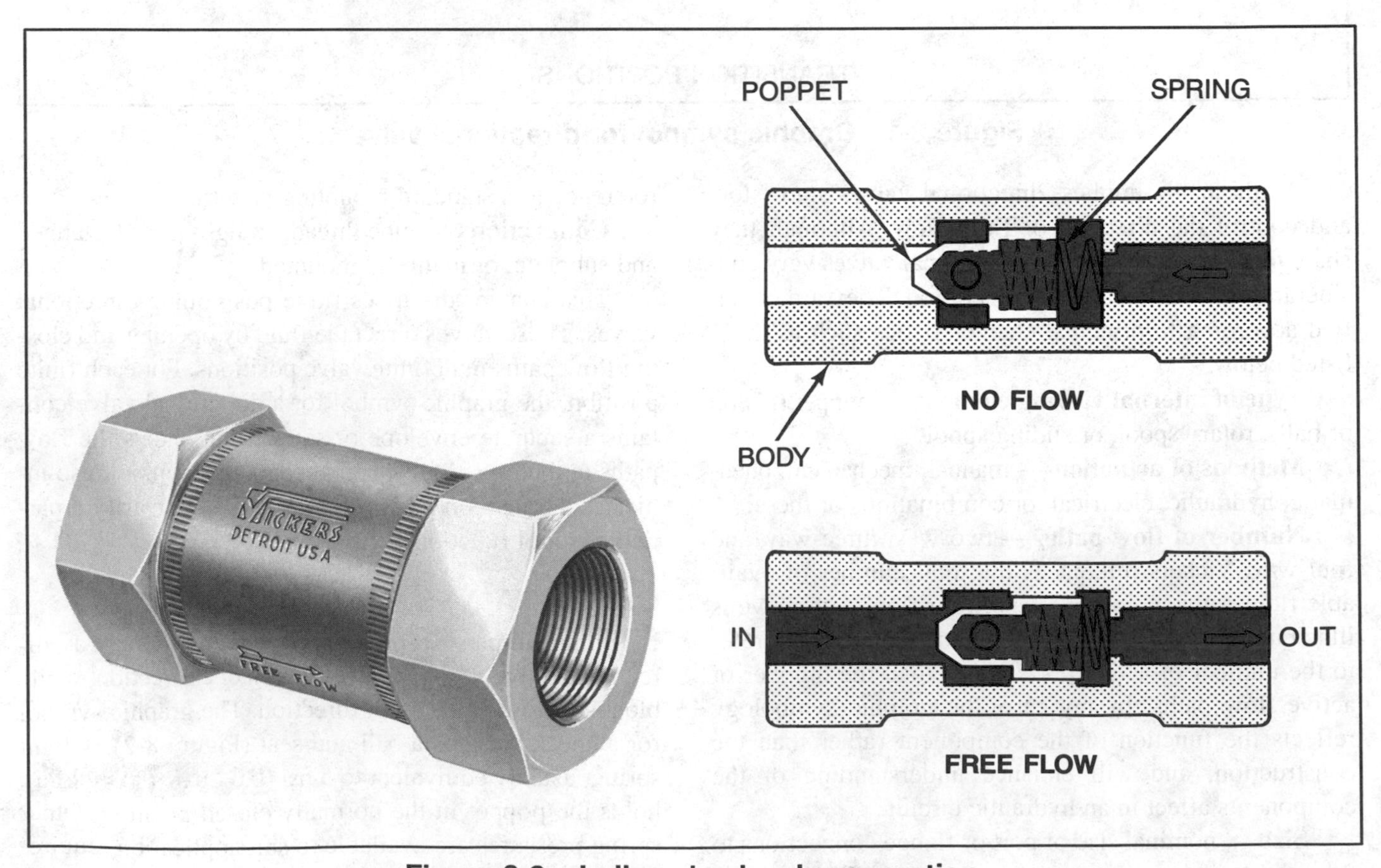

Figure 8-3. In-line check valve operation.

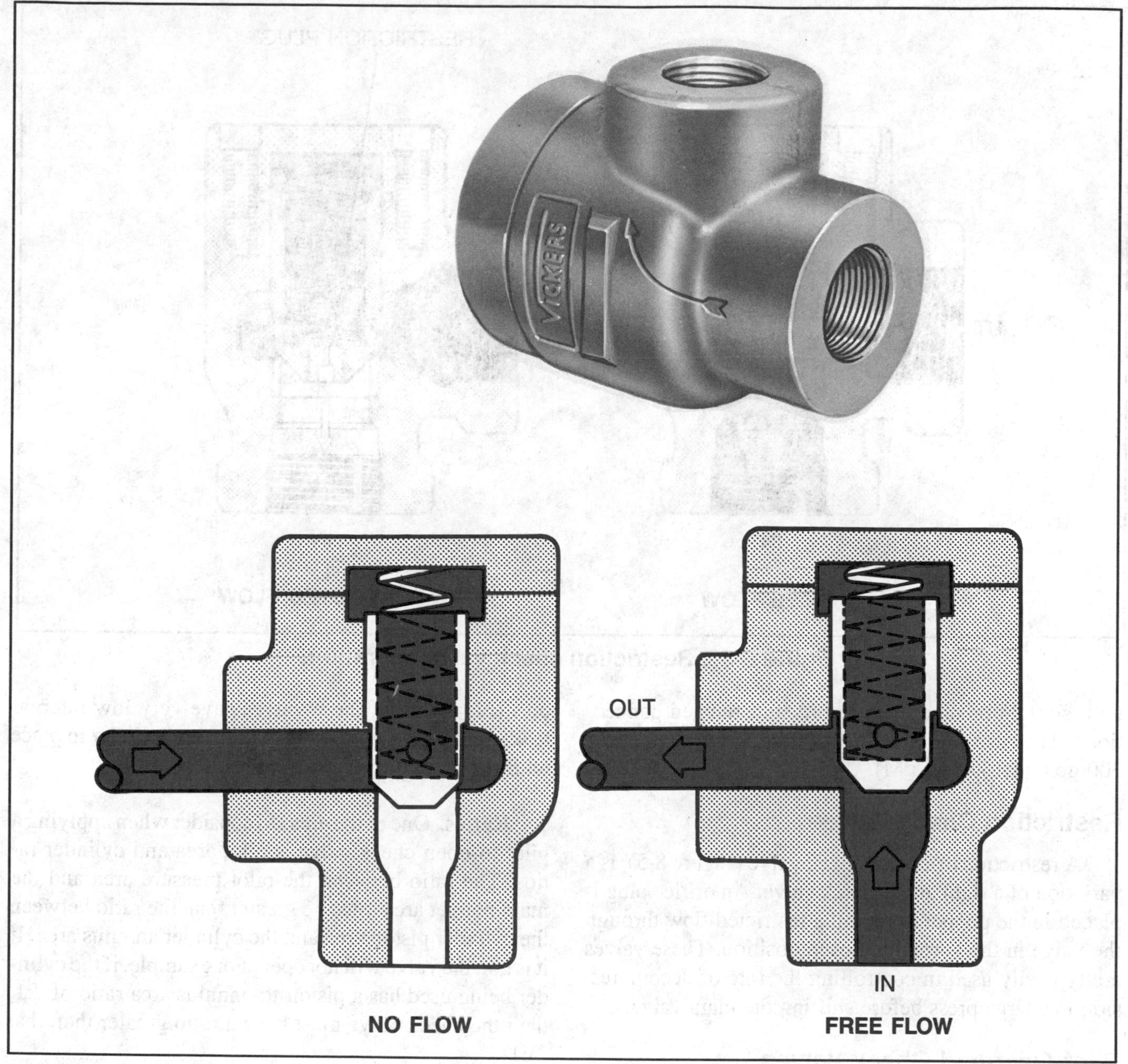

Figure 8-4. Typical right angle check valve.

at a pressure equivalent to the spring rating, allowing fluid to pass through the valve.

In-line Check Valves

To allow fluid to flow straight through the valve, an in-line check valve (Figure 8-3) is installed directly in the hydraulic line with either a straight thread or pipe thread connections. The valves are available in a range of flows, up to approximately 100 gpm (378.53 LPM). One of their design drawbacks is that as the flow through the valve increases, the pressure drop through the valve also increases.

In-line check valves are not only used for blocking flow, but also as a safety bypass for flow surges through filters and heat exchangers. With a higher spring rating, an in-line check valve can be used as a means of generating pilot pressure.

Right Angle Check Valves

Right angle check valves (Figure 8-4) are designed with the inlet and outlet ports at right angles to each other. This design allows for higher flows with less pressure drop.

To improve cycle life, they generally have hardened steel seats and poppets. This type of valve is available

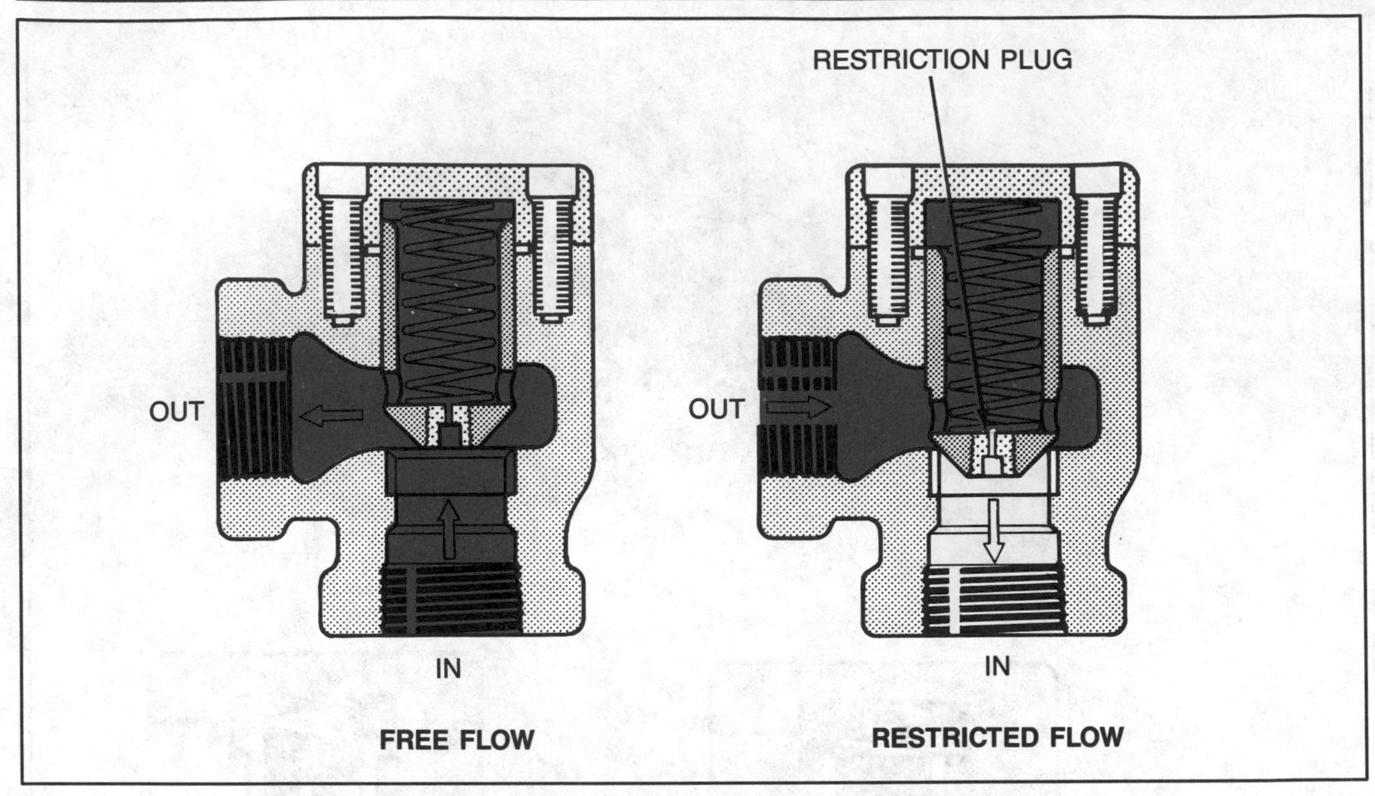

Figure 8-5. Restriction check valve operation.

with straight or pipe thread ports and flanged connections. They are available in flows up to approximately 300 gpm (1,135.59 LPM).

Restriction Check Valves

A restriction or orifice check valve (Figure 8-5), is a variation of a right angle check valve. An orifice plug is placed in the poppet to permit a restricted flow through the valve in the normally closed position. These valves are typically used in controlling the rate of decompression in a large press before shifting the main valve.

Pilot-Operated Check Valves

The check valves discussed to this point have been direct- operated check valves. That is, they are opened by fluid acting directly against the main poppet or element. Pilot-operated check valves are acted upon both directly and by a pilot signal, though not necessarily for the same purpose.

The two basic types of pilot-operated check valves are pilot-to-open and pilot-to-close.

Pilot-to-Open. Similar to the direct-acting check valve, a pilot-to-open check valve is designed to permit free flow in one direction and prevent flow in the reverse direction. The difference is that free reverse flow is permitted when a pilot pressure signal is applied to the pilot port (Figure 8-6). These valves have very low internal leakage and are typically used to lock a cylinder in place until the main directional valve shifts.

Ratios. One of the areas to consider when applying a pilot-to-open check valve is pilot area and cylinder ratios. The ratio between the pilot pressure area and the main poppet area must be greater than the ratio between the cylinder piston area and the cylinder annulus area. If it is not, the valve will not open. For example, if the cylinder being used has a piston-to-annulus area ratio of 2:1, then the check valve must have a ratio greater than that (3:1).

With large pilot-operated check valves or with large differential-area cylinders, this ratio is sometimes difficult to achieve. For those cases, a pressure breaker or decompression-type of pilot-operated check valve (Figure 8-7) can be used. It is designed with a two-stage main decompression poppet that is much smaller than the pilot piston and can have an opening ratio of 12:1 or greater. This poppet opens first, reducing the pressure behind the main poppet that is holding it closed. When the pressure behind the main poppet is low enough, the pilot piston pushes the main poppet into the open position (Figure 8-8).

Pilot-to-Close. Less common is the pilot-to-close check valve. It is designed to permit flow in the free-flow

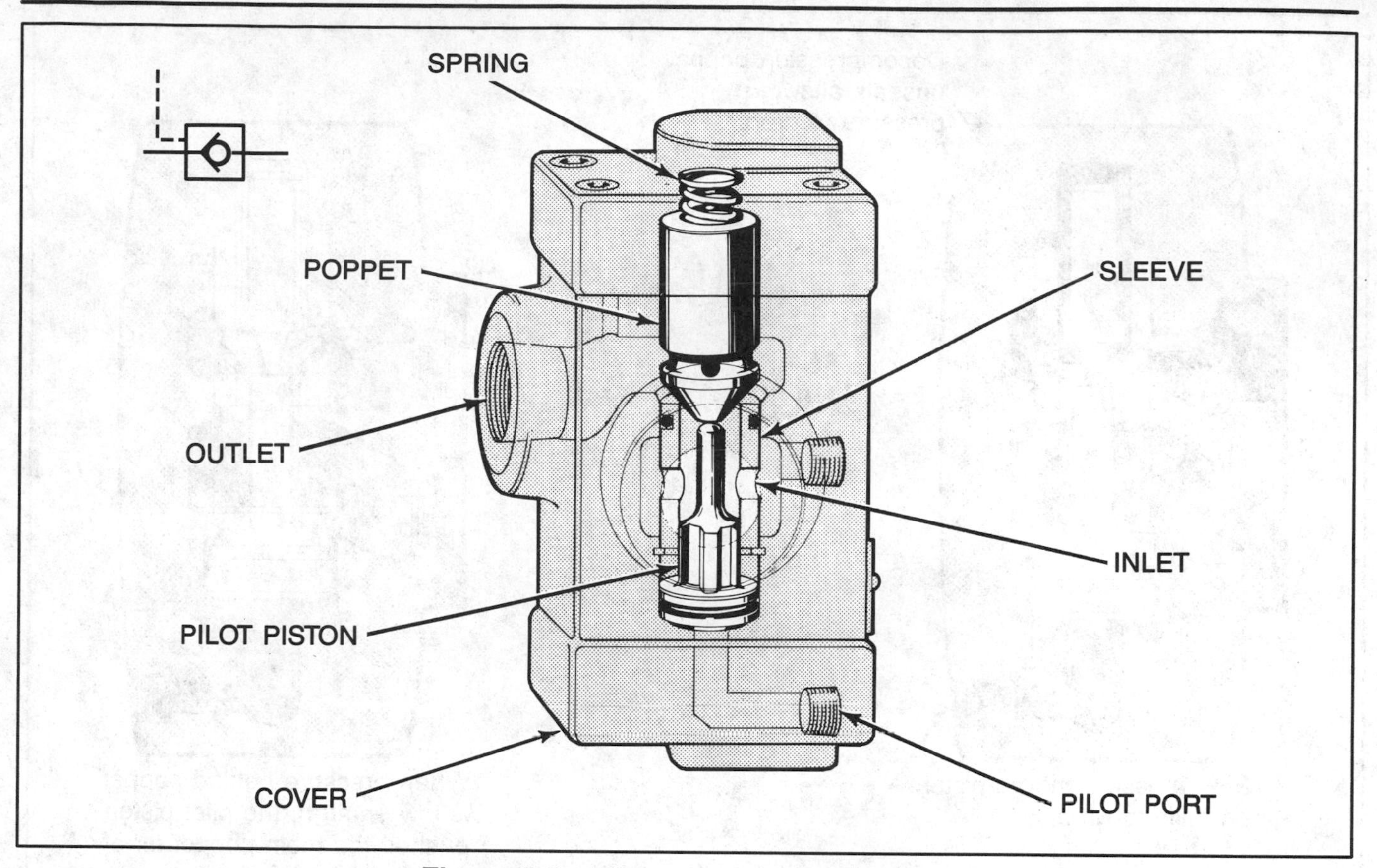

Figure 8-6. Pilot-to-open check valve.

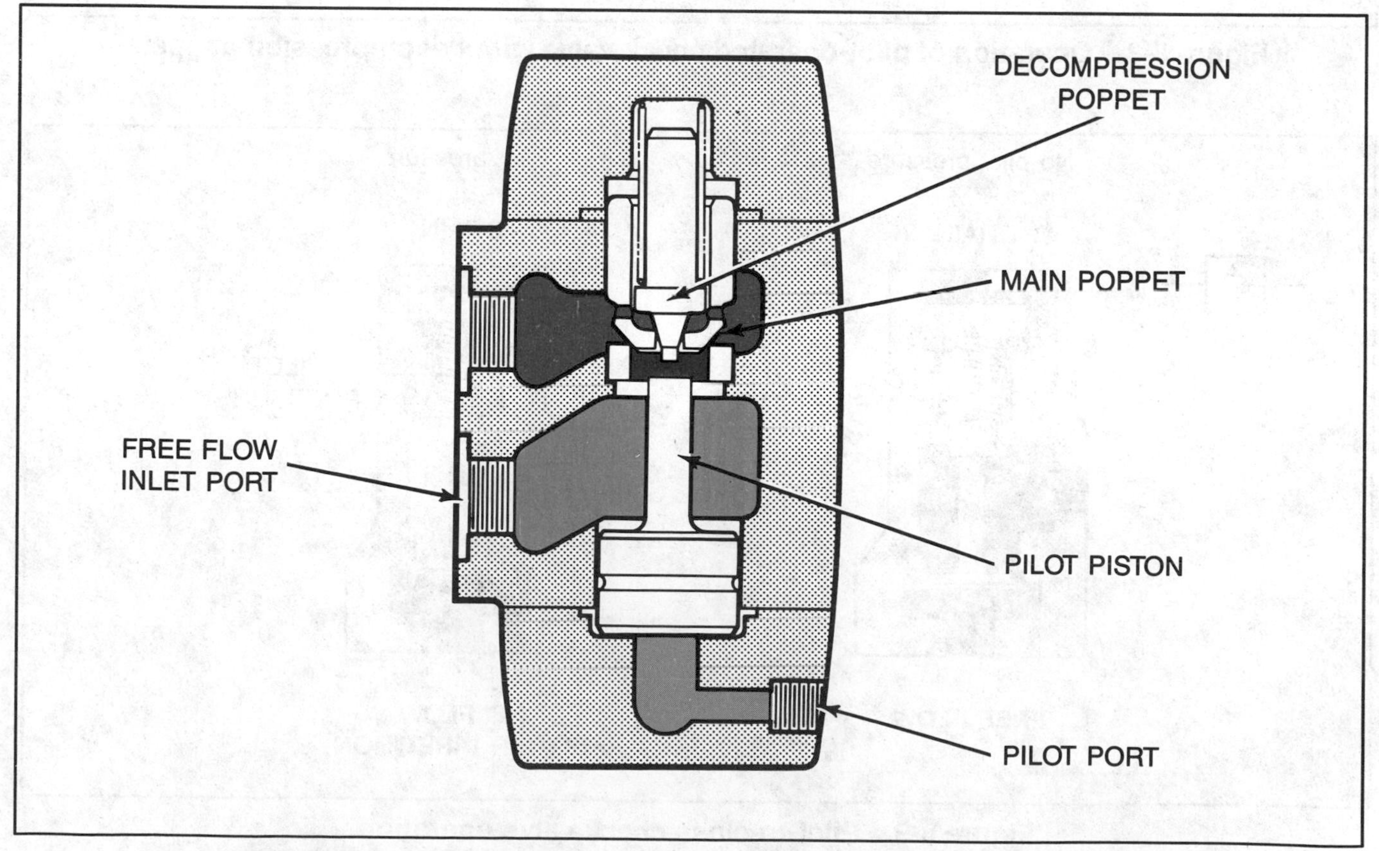

Figure 8-7. Pilot-operated check valve with decompression poppet.

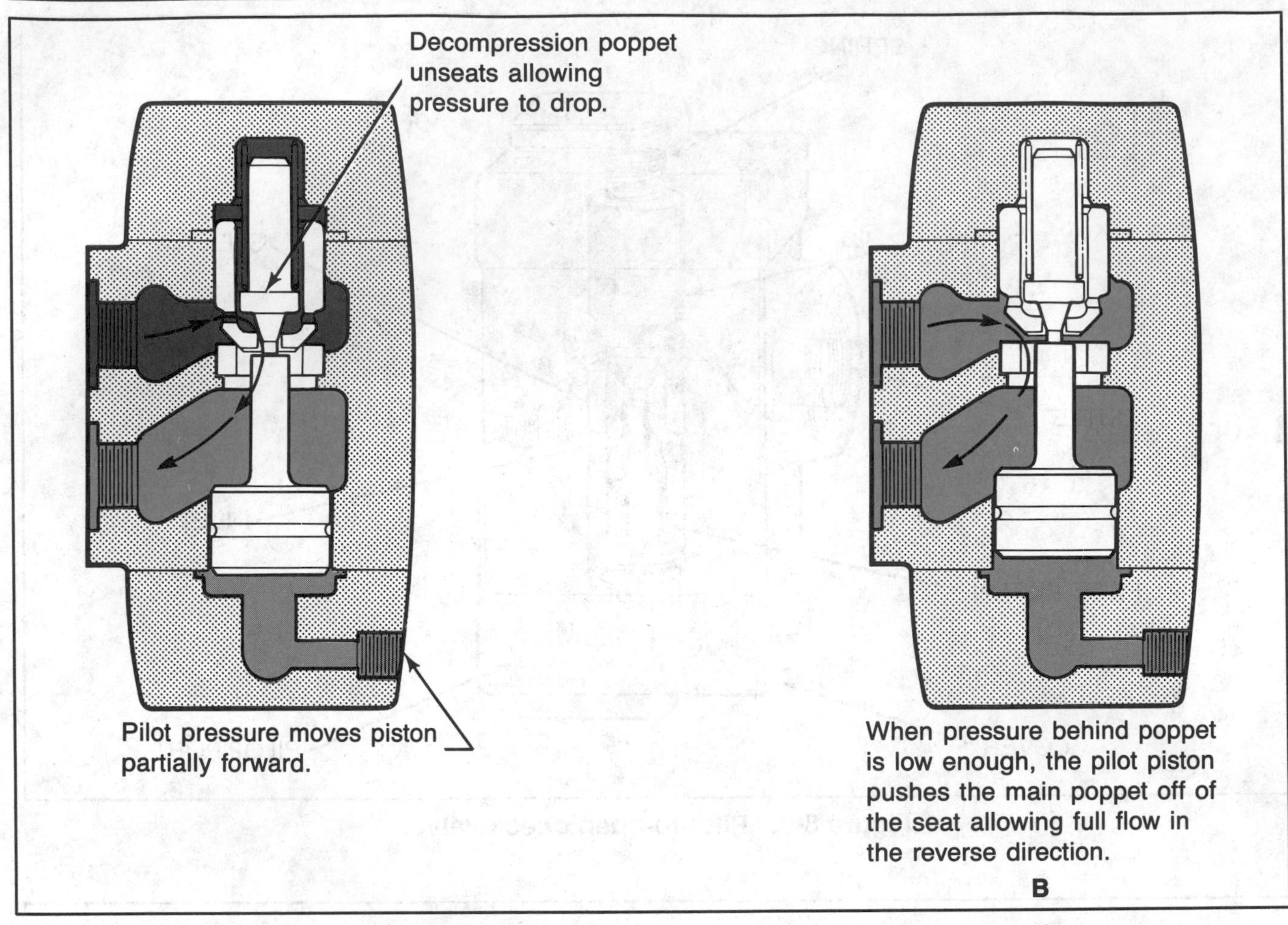

Decompression poppet unseats allowing pressure to drop.
Pilot pressure moves piston partially forward.
When pressure behind poppet is low enough, the pilot piston pushes the main poppet off of the seat allowing full flow in the reverse direction.
B

Figure 8-8. Operation of pilot-operated check valve with decompression poppet.

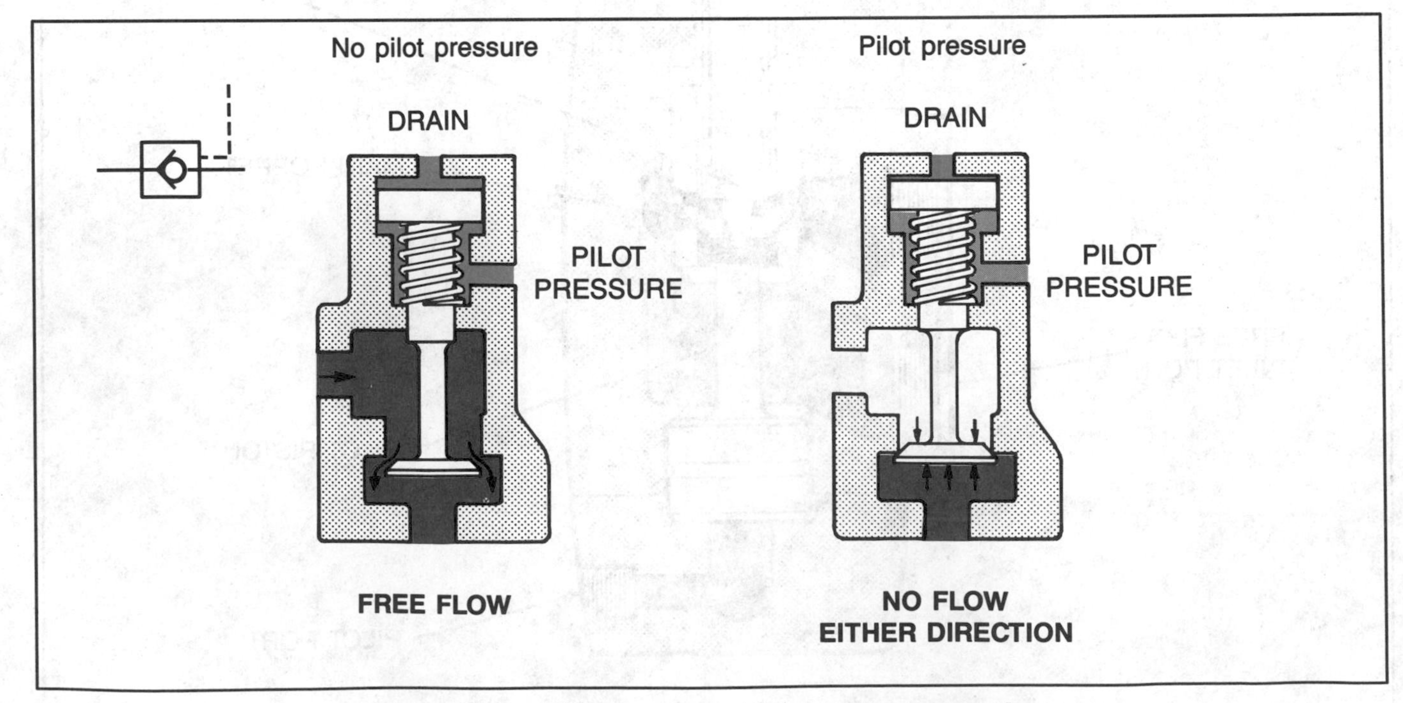

Figure 8-9. Pilot-to-close check valve operation.

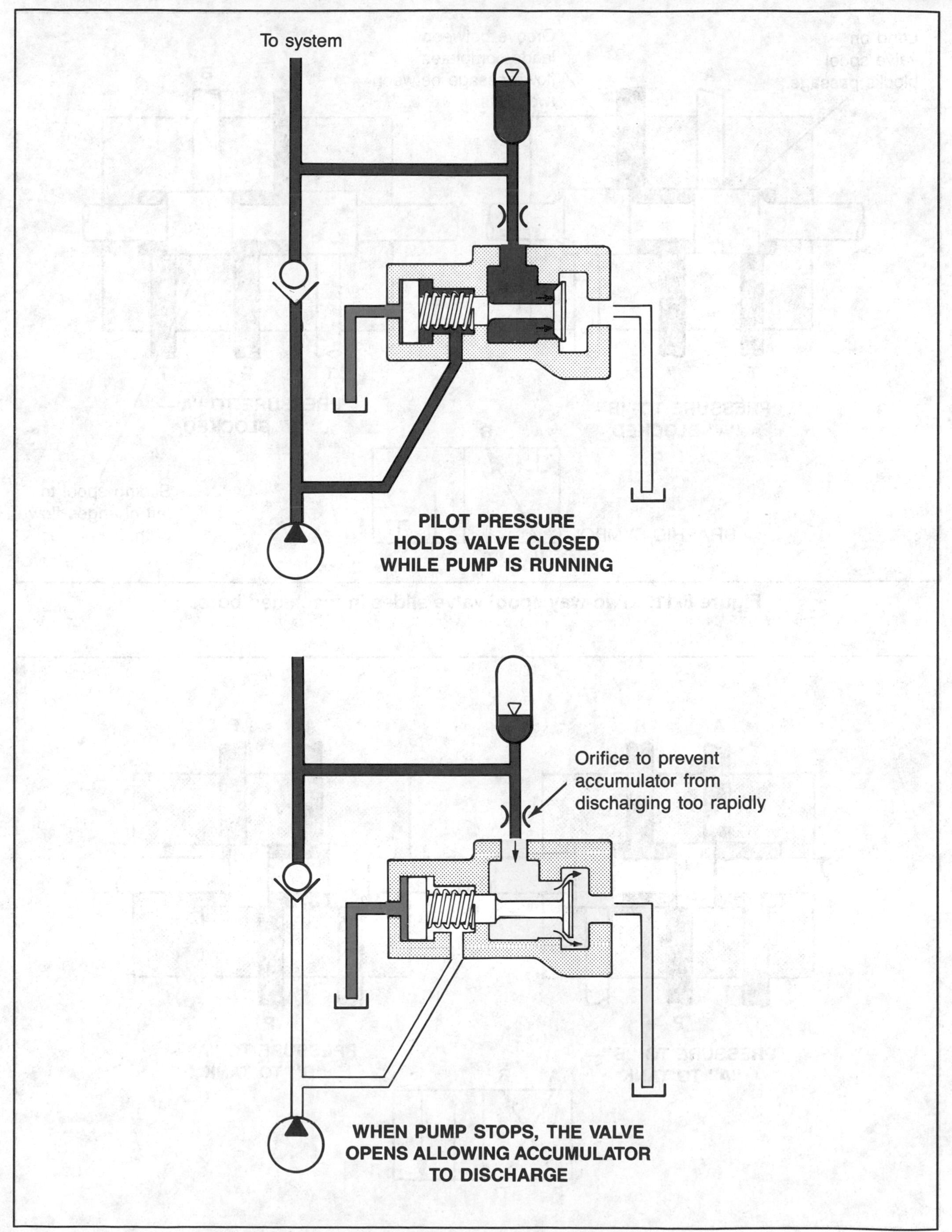
To system
PILOT PRESSURE
HOLDS VALVE CLOSED
WHILE PUMP IS RUNNING
Orifice to prevent
accumulator from
discharging too rapidly
WHEN PUMP STOPS, THE VALVE
OPENS ALLOWING ACCUMULATOR
TO DISCHARGE

Figure 8-10. Automatic accumulator discharge circuit.

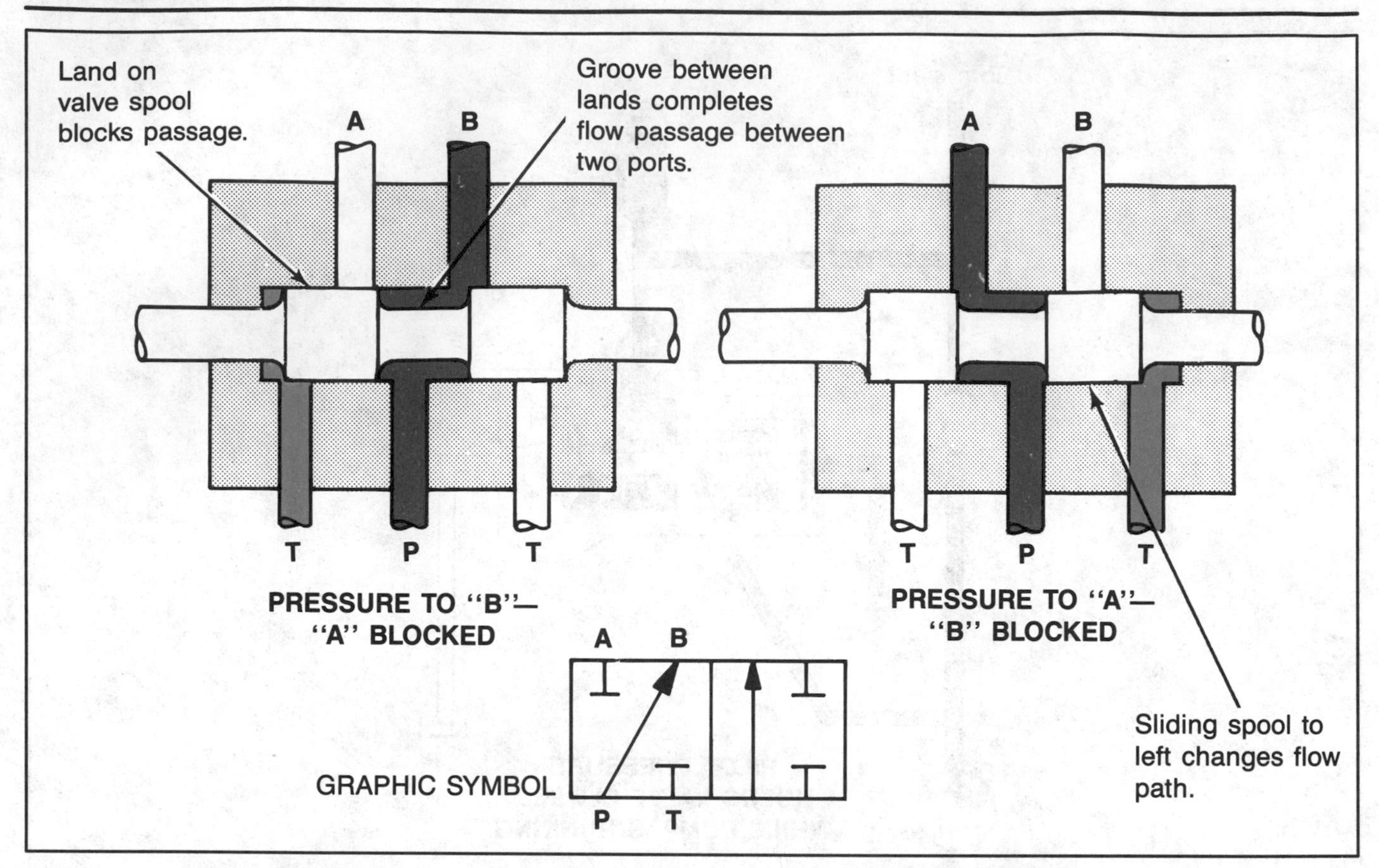

Figure 8-11. Two-way spool valve slides in machined bore.

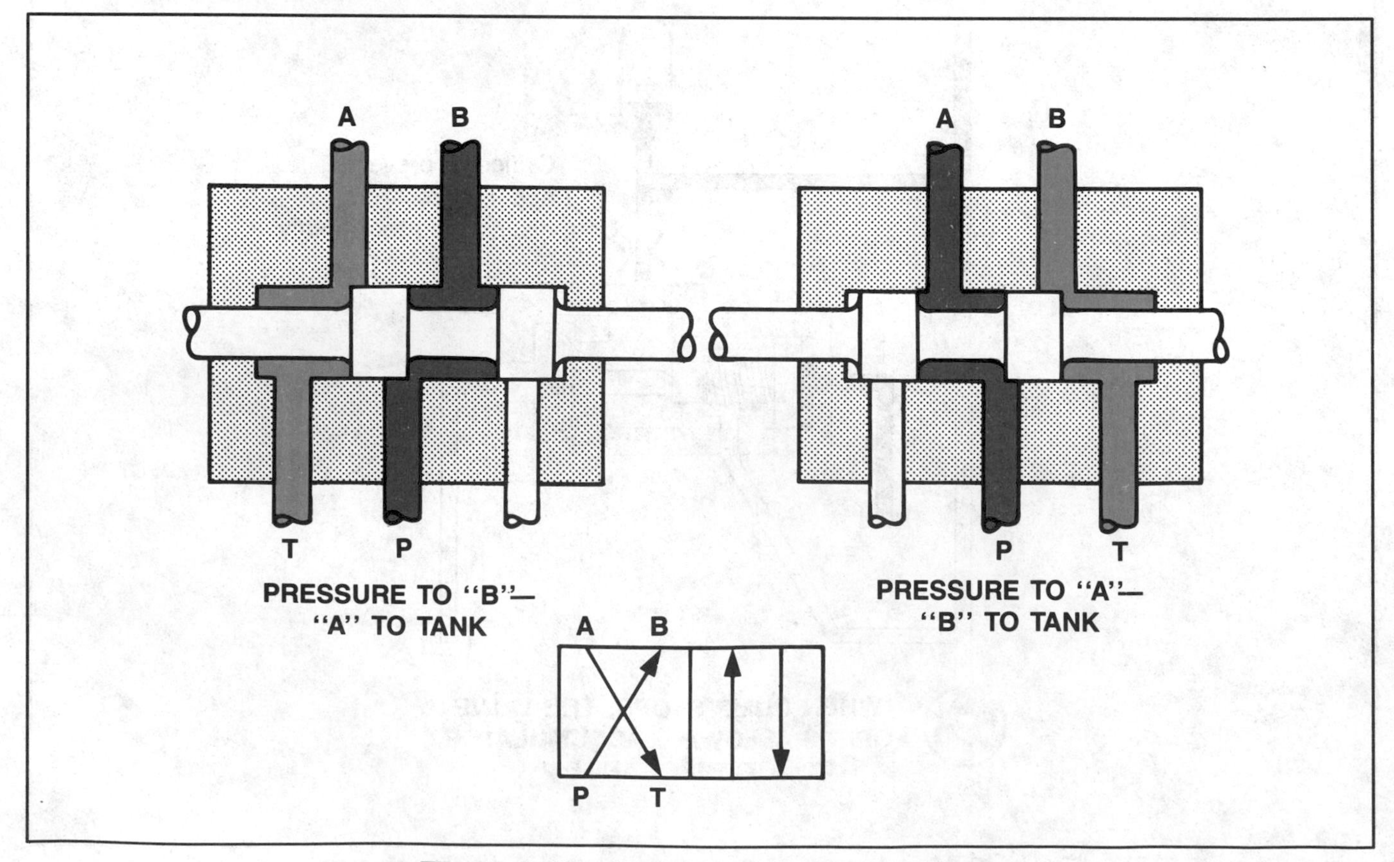

Figure 8-12. Spool type four-way valve.

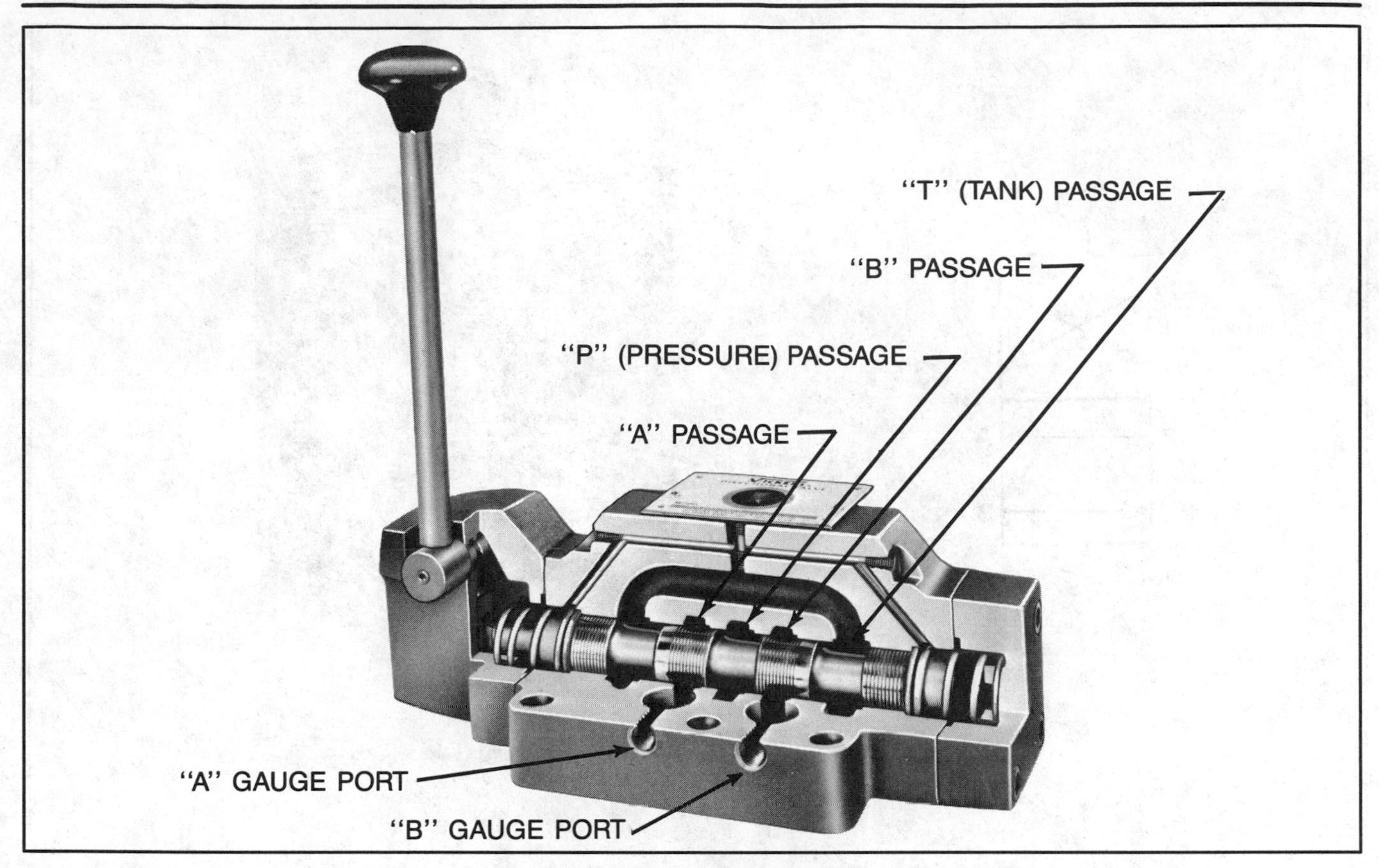

Figure 8-13. Manually actuated four-way valve.

direction until a pilot pressure signal is applied to an auxiliary port (Figure 8-9).

The most common use is in accumulator circuits where, for safety reasons, automatic emptying of the accumulator is desired when a machine is shut down (Figure 8-10).

Mounting Styles

Several different mounting styles are used for check valves. They can be line-mounted with straight thread ports, pipe thread ports, or flanged connections. Subplate- or manifold-mounted versions are available, along with a cartridge design that screws directly into a block or manifold. Finally, sandwich or modular designs can be stacked between a four-way type valve and the subplate.

TWO-WAY, THREE-WAY, AND FOUR-WAY VALVES

The basic function of these valves is to direct flow from the inlet or pressure port to either of two outlet ports. The number of ports to and from which fluid flows determines whether a valve is a two-way, three-way, or four-way.

A two-way valve uses the pressure port and an outlet port in one shifted position and the pressure port and the other outlet port in the other shifted position (Figure 8-11). The tank port is used only as a drain for internal leakage flow.

Three-way valves have special spool configurations that permit flow from the pressure port to two other ports at the same time. These valves are usually found only in special applications or circuits.

A four-way valve selects alternate ports just like the two-way valve, but with this valve, the tank port is used for return flow (Figure 8-12). A four-way valve can be used to move an actuator in either direction.

Regardless of the number of ports, all of these valves are often referred to as four-way valves. They are also what most people commonly think of when using the term directional valve.

DIRECT-ACTING VALVES

An operator or actuator shifts the spool or rotating element of a direct-acting directional valve. The five categories of actuators are: manual, mechanical, pneumatic, hydraulic, and electrical.

Manual Actuator. A manual actuator is usually a simple lever connected to the spool through some kind of linkage (Figure 8-13). There are also some foot-operated valves, but they are generally considered unsafe in today's workplace.

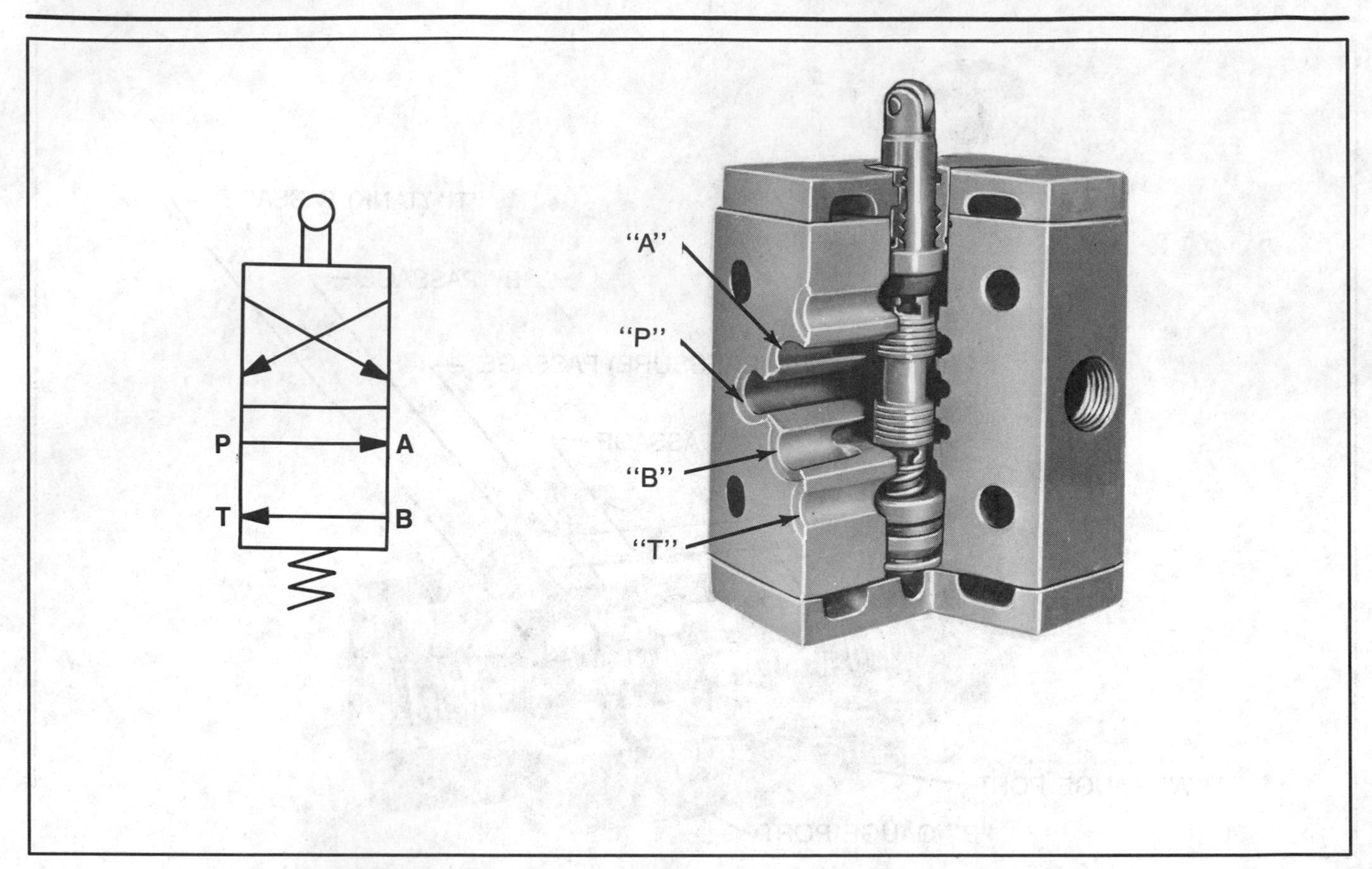

Figure 8-14. Mechanically actuated four-way valve.

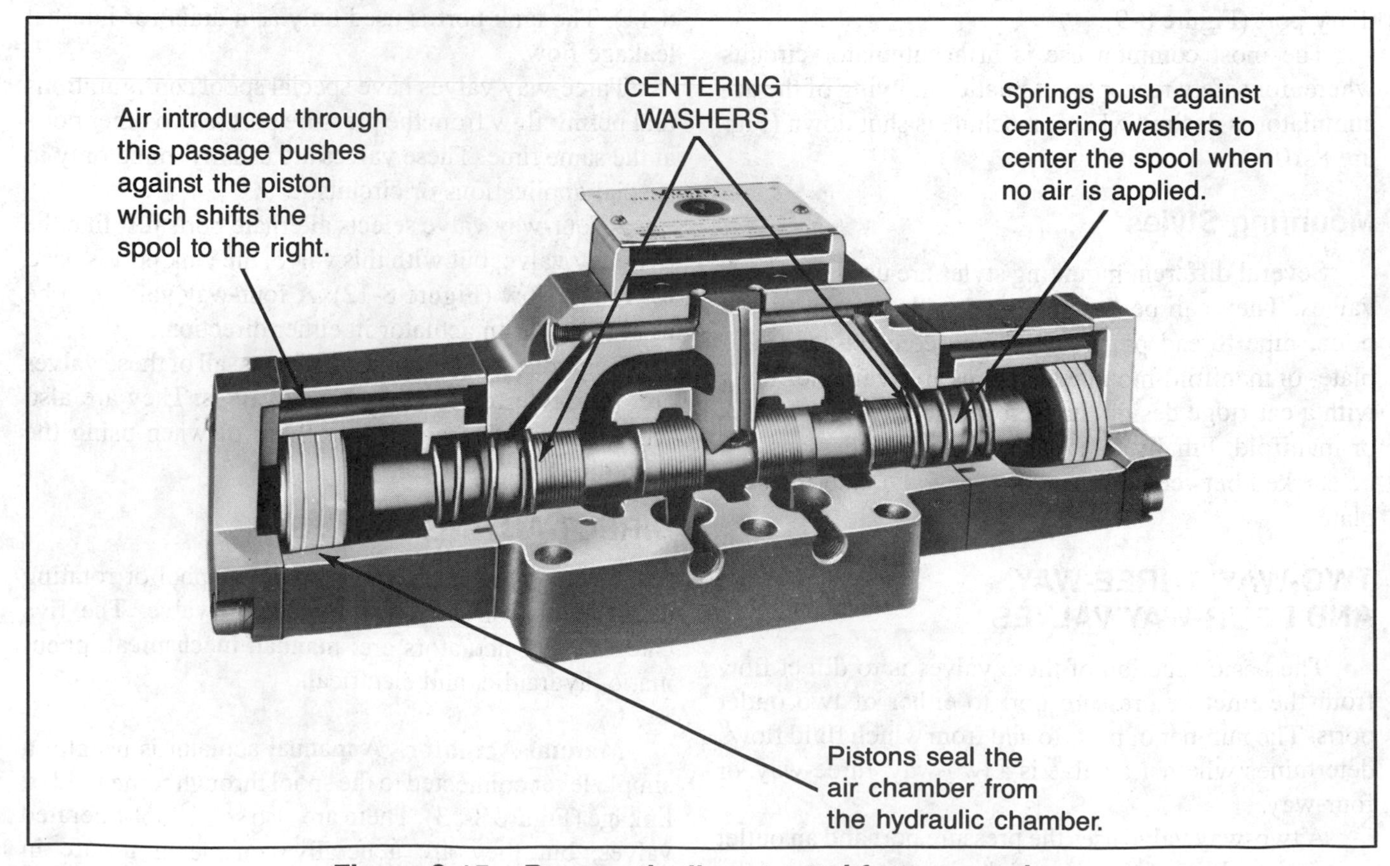

Figure 8-15. Pneumatically actuated four-way valve.

Mechanical Actuator. Mechanical actuators are either wheel or plunger type devices, as shown in Figure 8-14. When moved by some mechanical device such as a cylinder or cam, mechanical actuators cause the spool to move.

Pneumatic Actuator. A pneumatic actuator uses air pressure applied to a piston to shift the valve spool (Figure 8-15). The parts in this type of actuator are usually made of aluminum or other noncorrosive material so that moisture in the air lines will not cause the parts to rust and stick. Usually, a small hole in the actuator housing allows accumulated moisture to drain out.

Because air pressure can be quite low, the actuator piston must be relatively large to overcome spring and flow forces.

Hydraulic Actuator. Hydraulic actuators, like the one shown in Figure 8-16, use pilot oil flow for shifting the valve spool. Because the pilot flow that controls these actuators must be controlled by its own directional valve, hydraulically operated valves cannot be used by themselves.

Electrical Actuator. Electrical actuators are commonly called solenoids. The solenoids discussed in this chapter are known as "on-off" solenoids because, upon receiving an electrical signal, the solenoid is either fully shifted or turned off. Proportional valves, discussed in Chapter 14, have unique solenoid characteristics.

A solenoid is made up of two basic parts: a coil and an armature. Applying electricity to the coil creates a magnetic field that attracts the armature into it. The armature pushes on the spool or solenoid pin as it is pulled into the magnetic field (Figure 8-17).

The two solenoid designs used today are the air gap and wet armature types. In the air gap design (Figure 8-18), air space separates the solenoid from the hydraulic system. To prevent hydraulic fluid leakage, the solenoid pin is sealed by a dynamic seal.

One of the problems of the air gap solenoid is that the dynamic seal eventually wears and starts to leak. In the wet armature design, all of the solenoid's moving parts operate in the system's hydraulic fluid, eliminating the need for a dynamic seal. An encapsulated coil surrounds a core tube that retains the armature and system fluid. Only two seals are used: a static seal where the core tube is screwed into the valve body and a static seal at the manual override pin that only acts dynamically when the manual override is pushed (Figure 8-19).

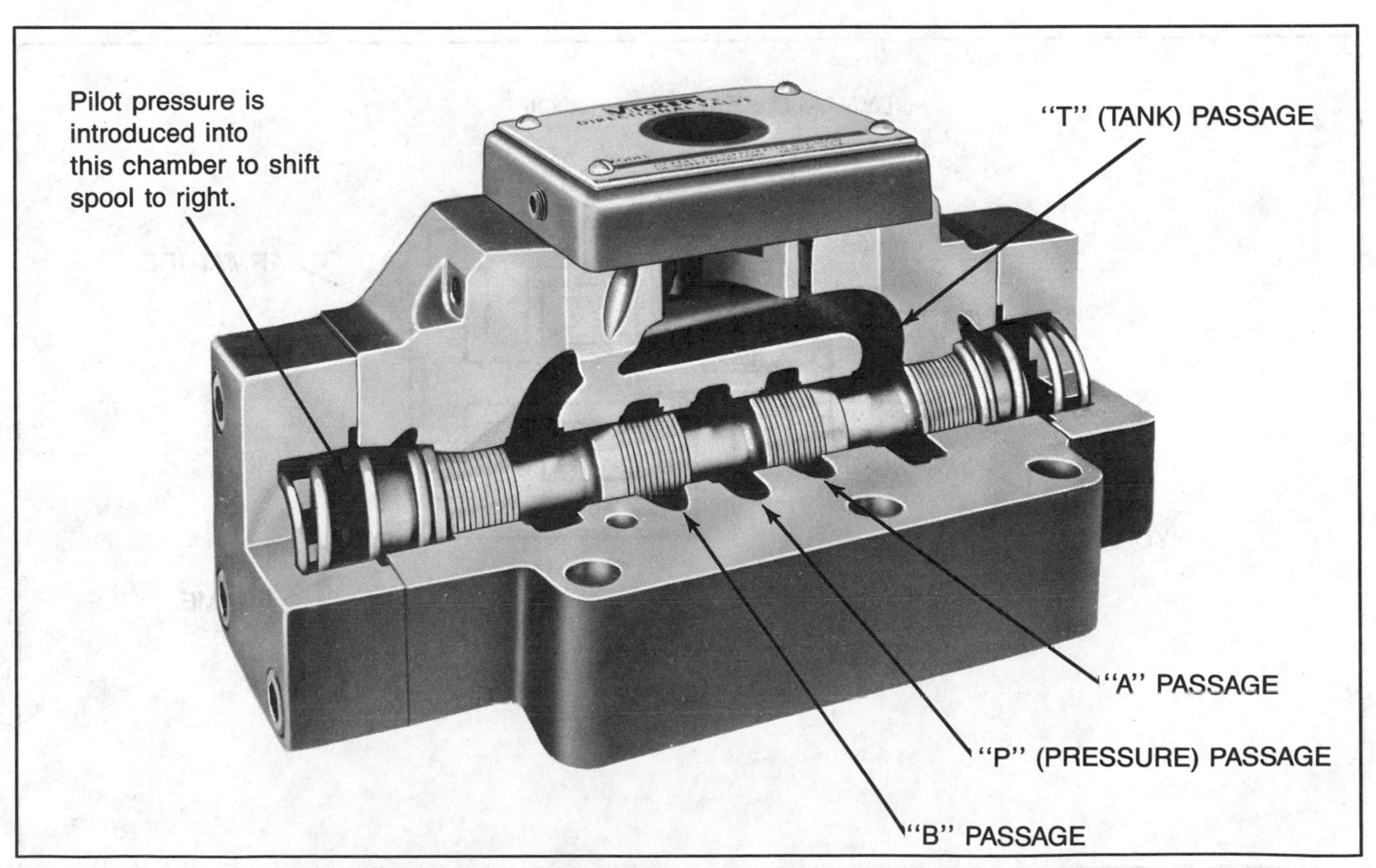
Pilot pressure is introduced into this chamber to shift spool to right.
"T" (TANK) PASSAGE
"A" PASSAGE
"P" (PRESSURE) PASSAGE
"B" PASSAGE

Figure 8-16. Hydraulically actuated pilot-operated valve.

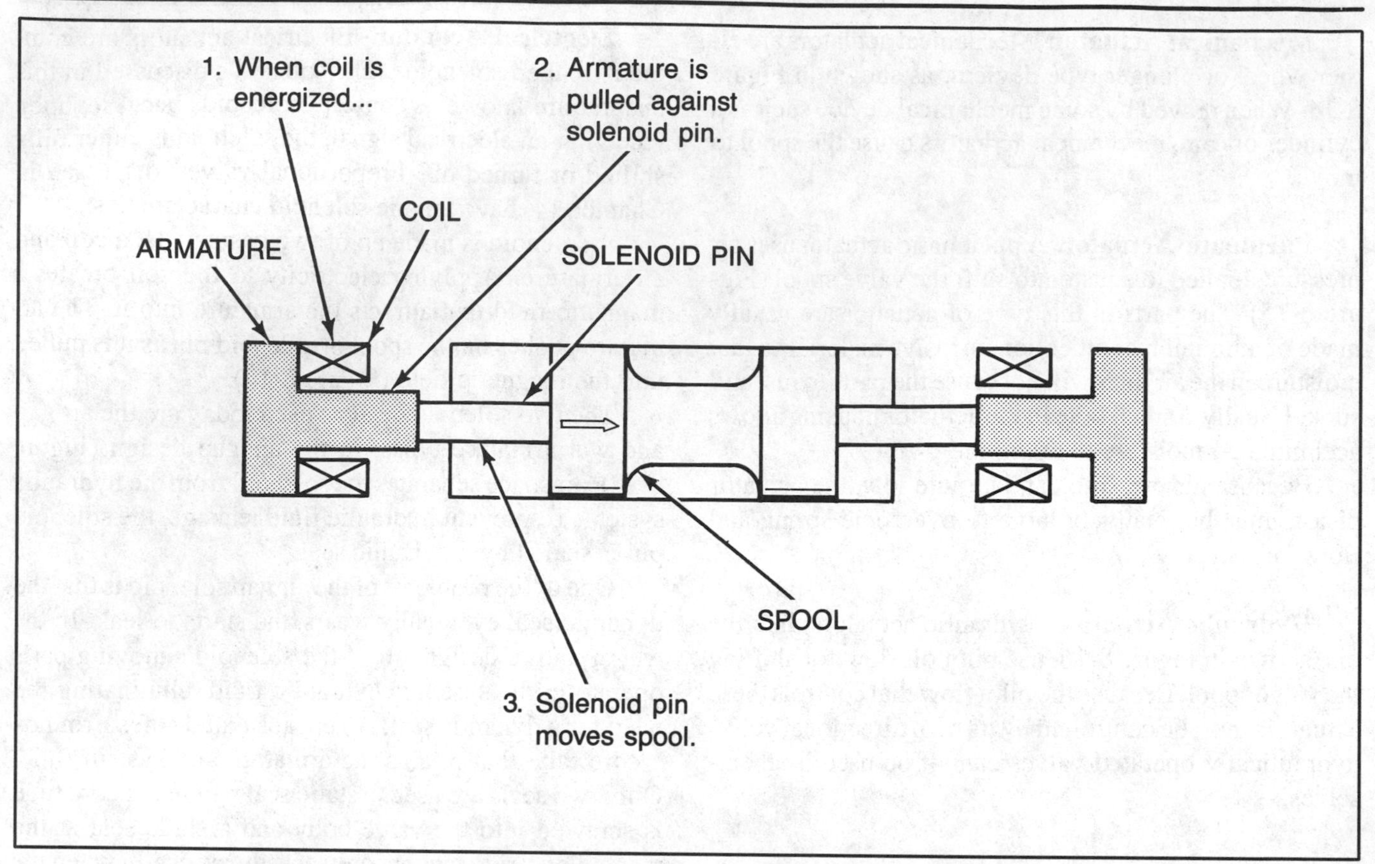

Figure 8-17. Push-type solenoids shift many small valve spools.

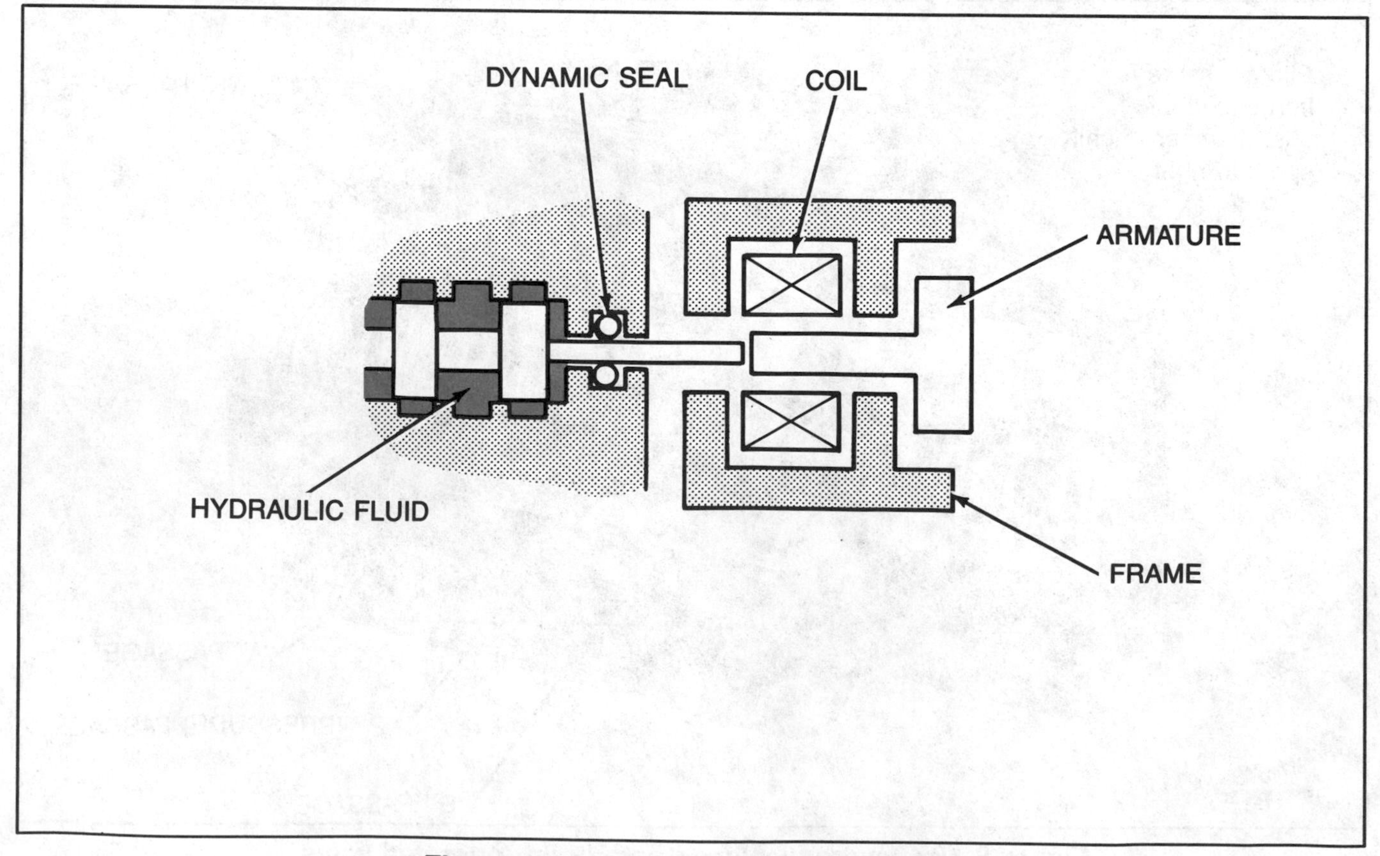

Figure 8-18. Air gap solenoid design.

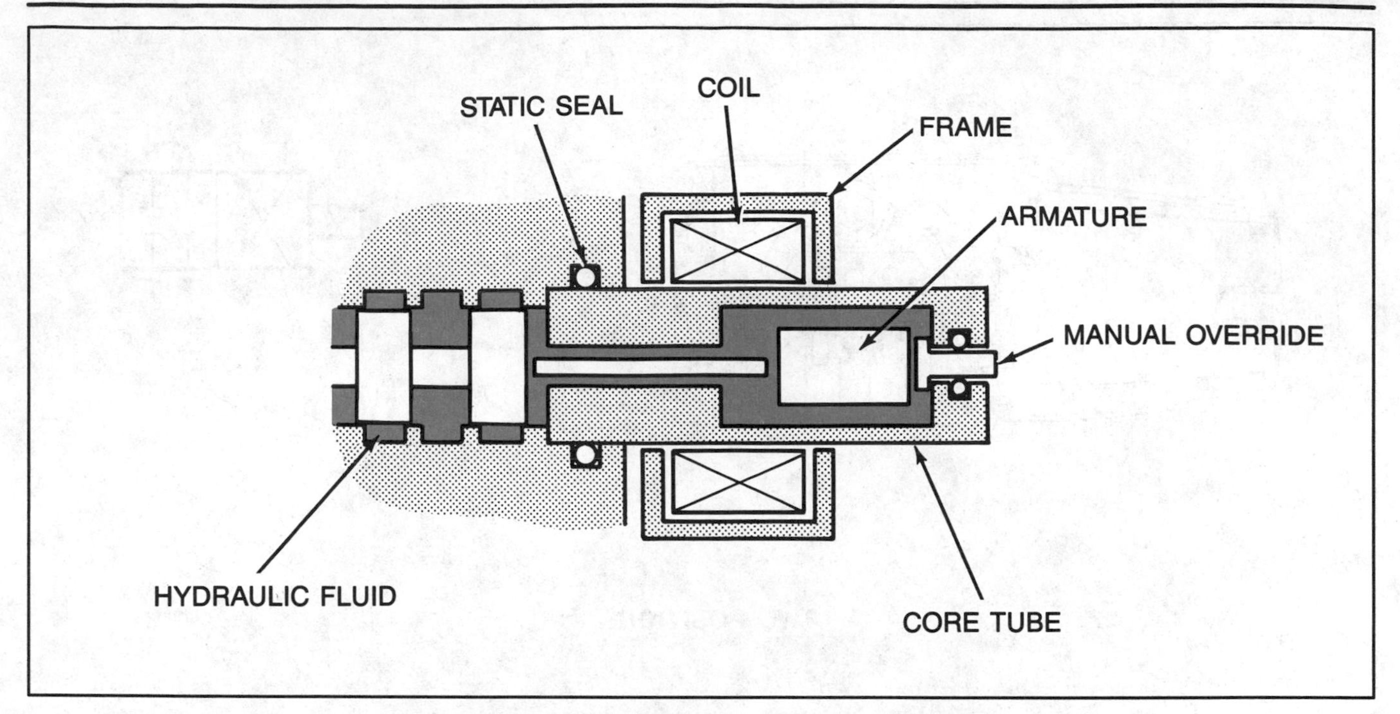

Figure 8-19. Wet armature solenoid design.

Another advantage of the wet armature design is that the fluid acts as a cushion for the spool, push pin, and armature. Although these advantages mean that a wet armature solenoid operates more quietly and usually has a longer life, it requires about 1.6 times as much electrical power as the air gap type.

AC/DC. Solenoids can operate on either alternating current (AC) or direct current (DC). The designs are slightly different, but the operating principle is the same.

When energized, an AC solenoid has a high current draw at the beginning of the stroke (in-rush current) and a lower draw at the end (holding current). If the solenoid cannot complete a shift, it will continue to draw high current and eventually burn out. This can also occur if a solenoid is energized and de-energized at very high cycle rates.

DC solenoids draw a constant current when energized and are designed to handle high, continuous current. This design prevents burnout as a result of incomplete shifting or high cycle rates. They are generally safer than AC solenoids because they operate at much lower voltages. DC solenoids are found more often in mobile applications than in industrial ones.

Regardless of whether a solenoid is AC or DC, it is designed to be held energized constantly without burning out.

Spool Positions

Most four-way valves are available in various spool position conditions. Reference to these conditions relates to the normal or de-energized condition of the valve. The five variations are discussed below.

Spring-Centered. Spring-centered valves (Figure 8-20) are returned to the center position by spring force whenever the actuating force is released. A **three-position, spring-centered** valve has two actuators (except in the case of manually or mechanically actuated valves), while **two-position, spring-centered** valves have only one.

Two-Position, Spring-Offset. This type of valve (Figure 8-21) is normally offset to one extreme position by a spring. It has one actuator that shifts the spool to the other extreme position. The center condition of the spool is referred to, although it is only seen as the spool passes through the center position. It is often shown in the valve symbol.

Two-Position, Actuate-to-Center. Similar to the two-position, spring-offset valve, this valve (Figure 8-22) is normally spring offset to one extreme position. When actuated, however, this valve shifts to a center position rather than the other extreme position.

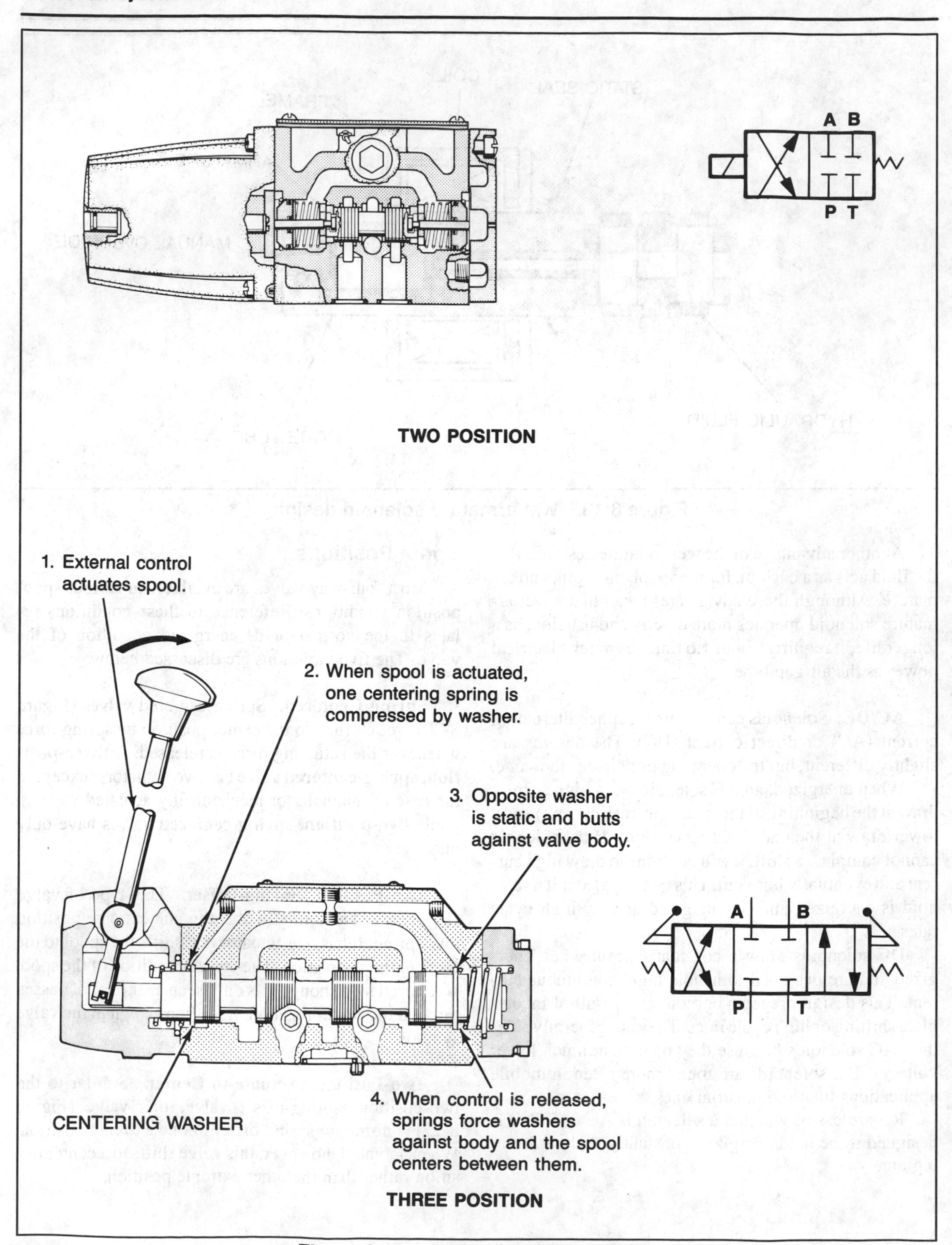

A B
P T
TWO POSITION
1. External control actuates spool.
2. When spool is actuated, one centering spring is compressed by washer.
3. Opposite washer is static and butts against valve body.
A B
P T
CENTERING WASHER
4. When control is released, springs force washers against body and the spool centers between them.
THREE POSITION

Figure 8-20. Spring-centered valves.

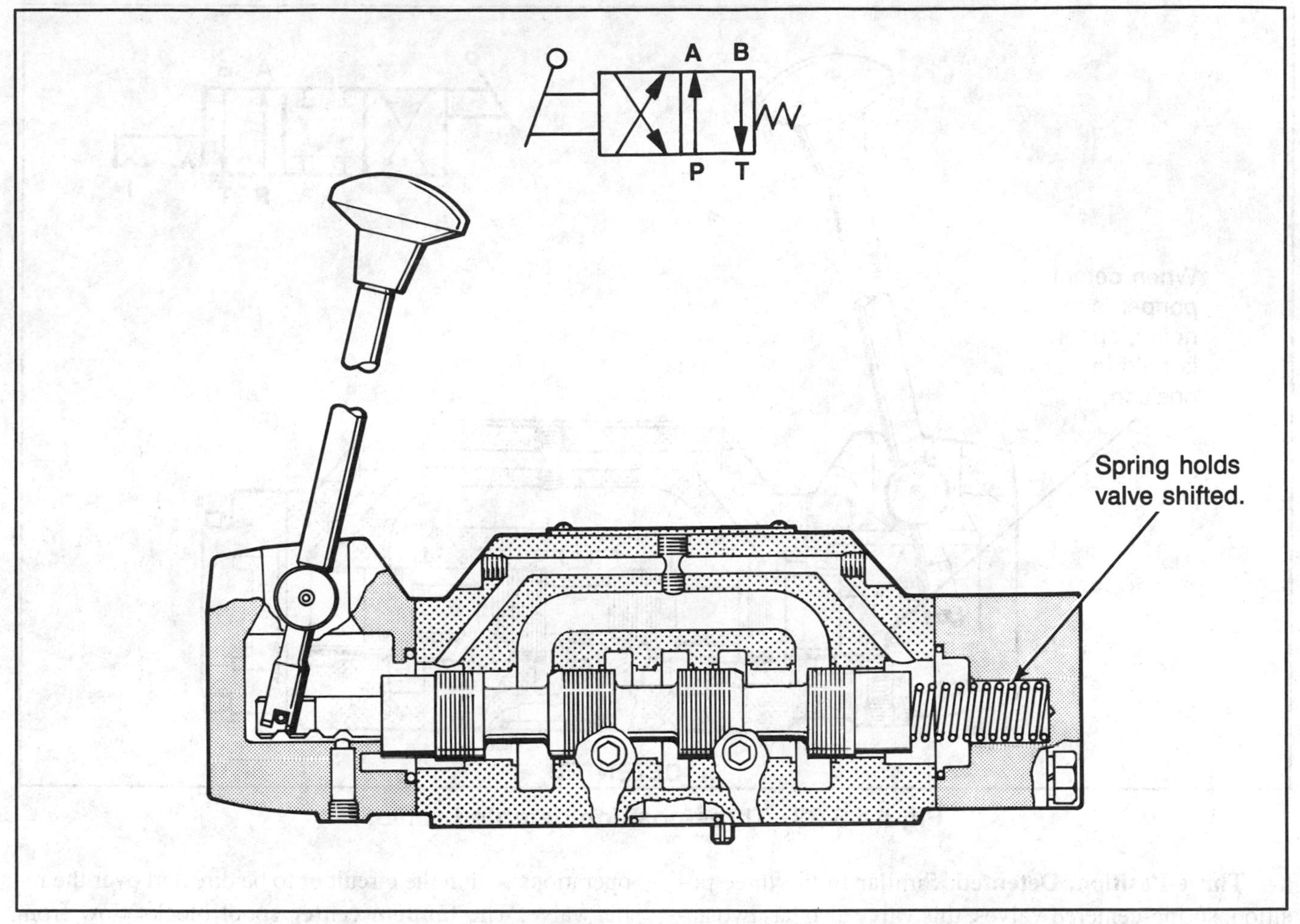

Figure 8-21. Spring-offset valve has two positions.

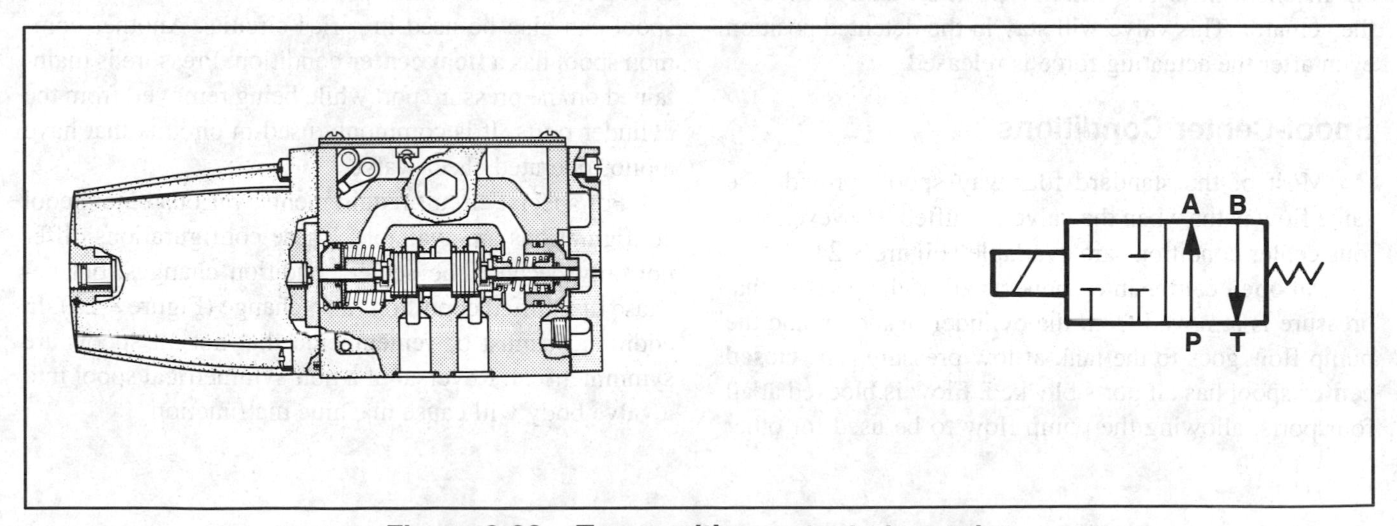

Figure 8-22. Two-position, actuate-to-center.

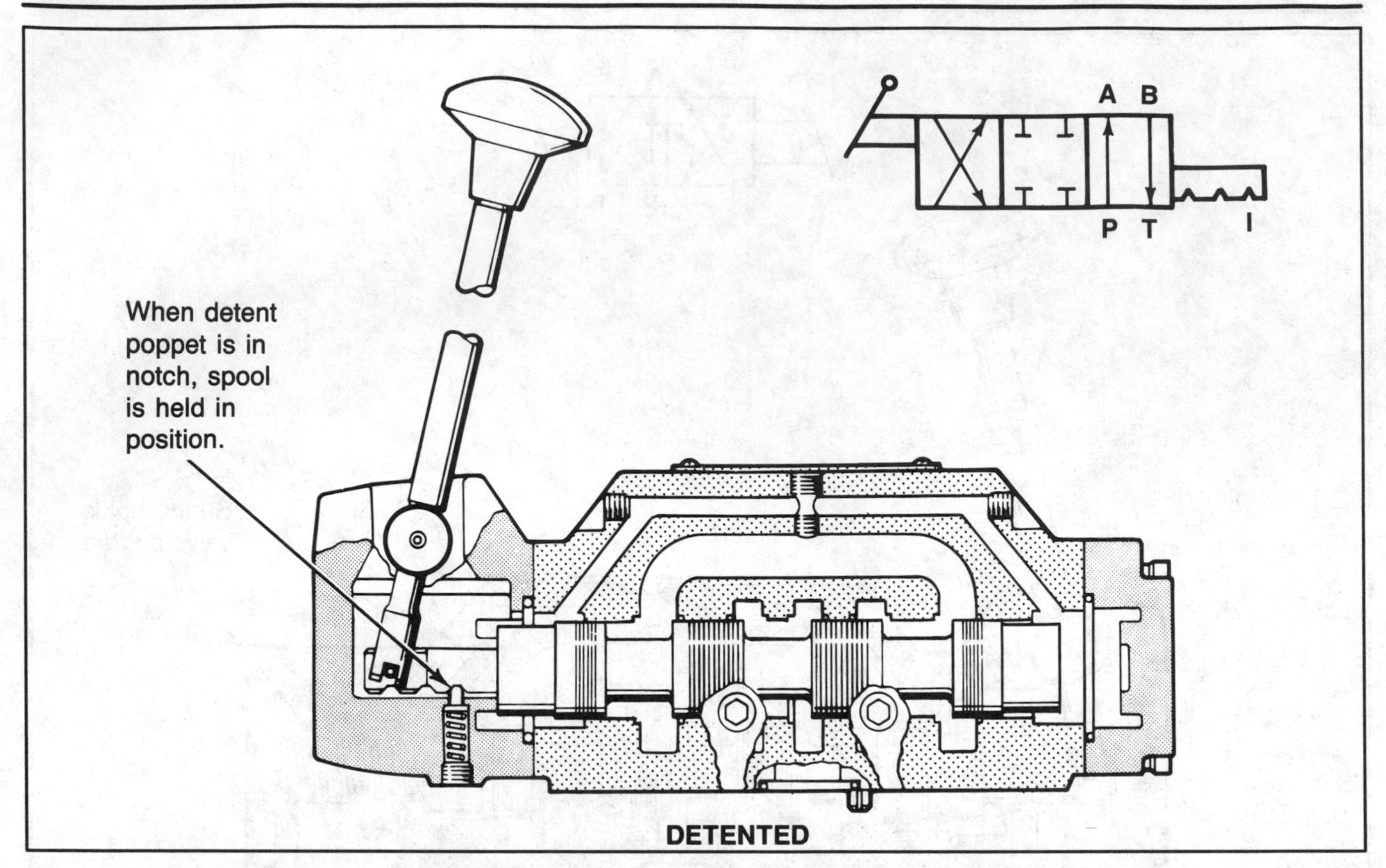

A B
P T
When detent poppet is in notch, spool is held in position.
DETENTED

Figure 8-23. Three-position detented valve.

Three-Position, Detented. Similar to the three-position, spring-centered valves, this valve also has two actuators, shown in Figure 8-23. The spool is held by a detent mechanism in one extreme position until shifted by the actuator. This valve will stay in the detented position even after the actuating force is released.

Spool-Center Conditions

Most of the standard four-way spools provide the same flow paths when the valve is shifted. However, various center conditions are available (Figure 8-24).

An **open center** interconnects all of the ports so that pressure is removed from the cylinder or motor, and the pump flow goes to the tank at low pressure. The **closed center** spool has all ports blocked. Flow is blocked at all four ports, allowing the pump flow to be used for other operations within the circuit or to be directed over the relief valve. The **tandem center** spool blocks flow from the cylinder ports while allowing the pump flow to be directed to the tank at low pressure. The **tandem center** spool can also be used in series circuits. Another common spool has a **float center** condition. Pressure is maintained on the pressure port while being removed from the cylinder ports. It is commonly used in circuits that have a pilot-operated check valve.

For special circuit requirements, various other spool configurations are available. These configurations differ not only because the center condition changes, but because the shifted conditions also change (Figure 8-25). In addition, it must be remembered that not all spools are symmmetrical. Reversal of a non-symmetrical spool into a valve body will cause machine malfunction.

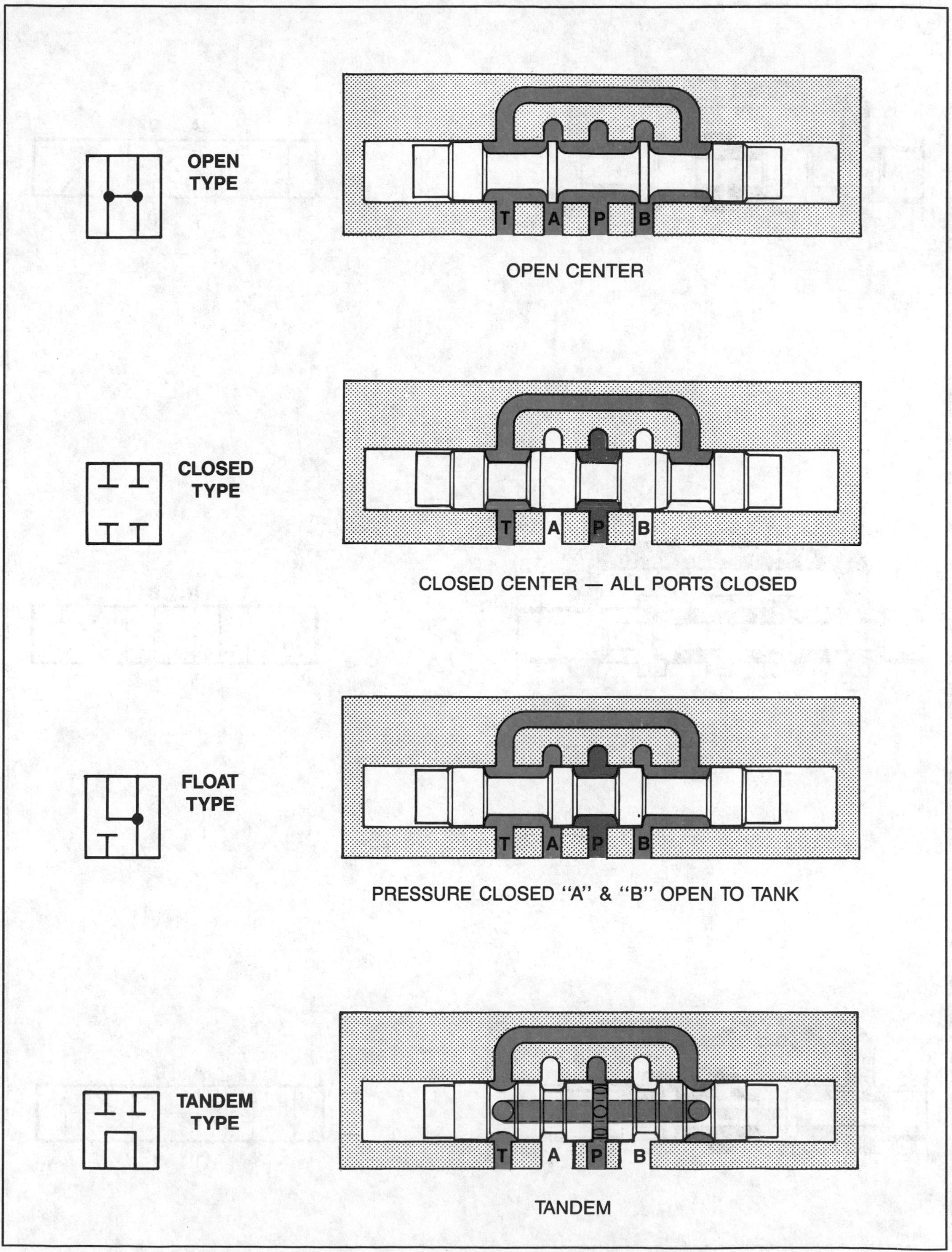

Figure 8-24. Four-way valve center conditions.

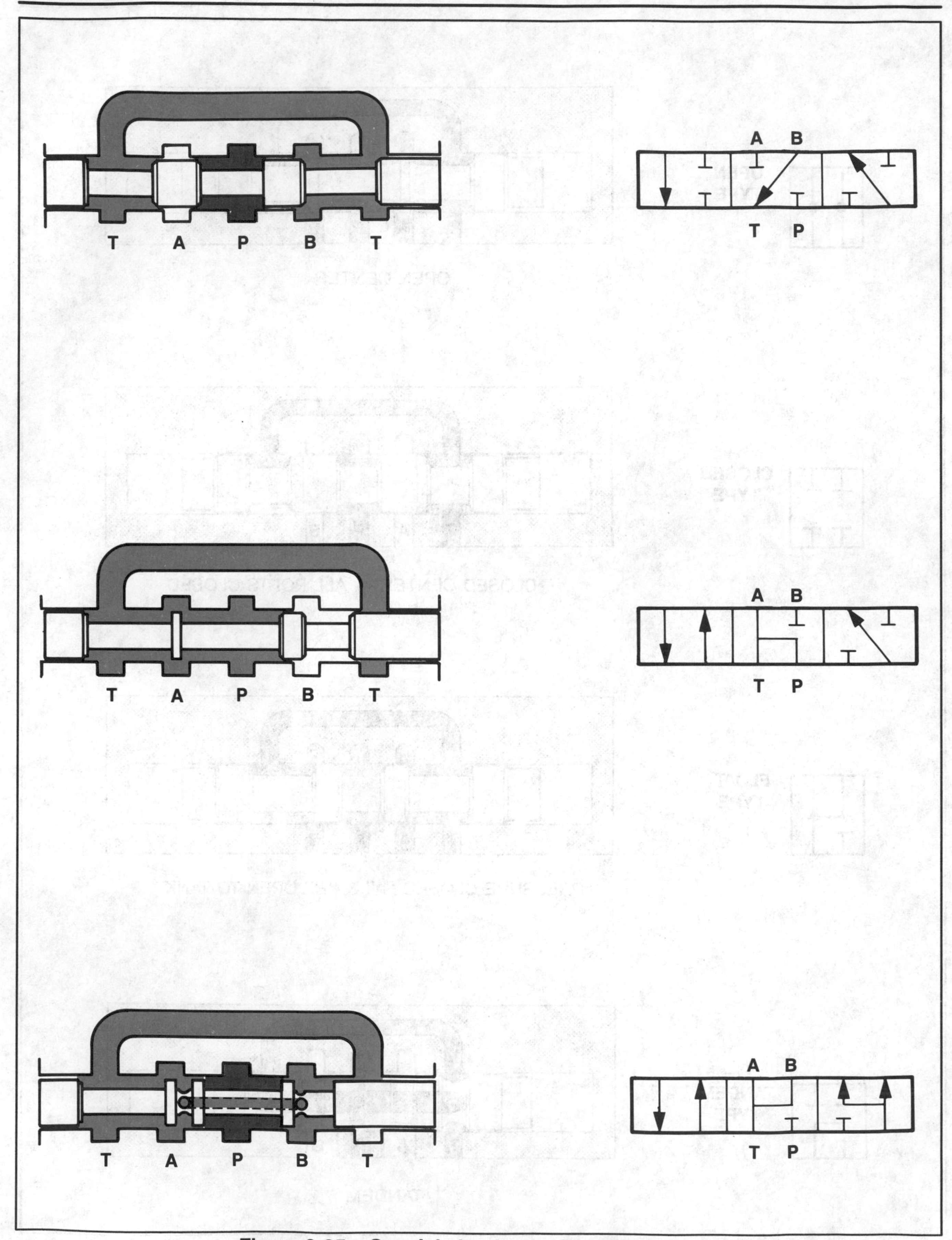

Figure 8-25. Special circuit spool configurations.

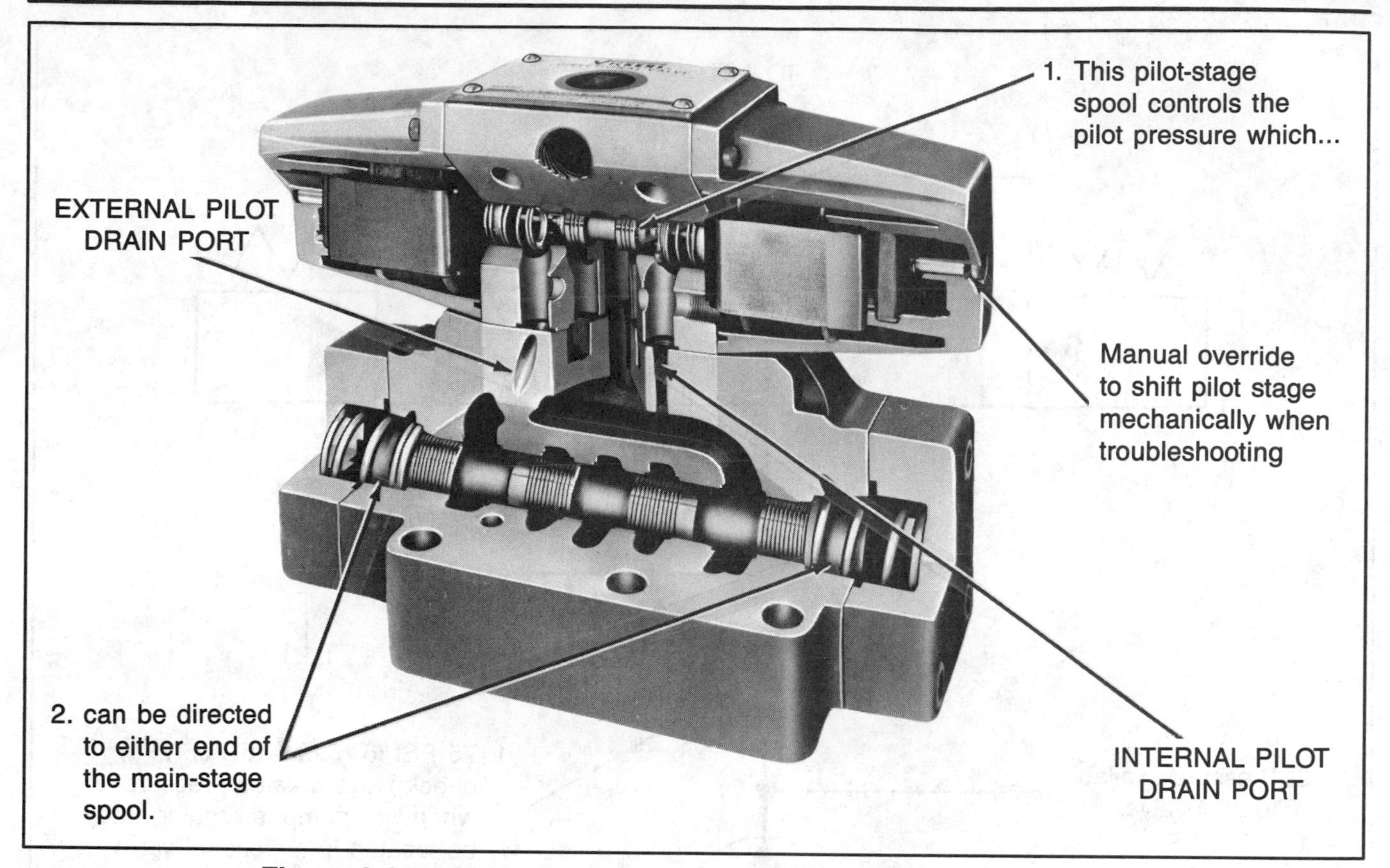
1. This pilot-stage spool controls the pilot pressure which...
EXTERNAL PILOT DRAIN PORT
Manual override to shift pilot stage mechanically when troubleshooting
2. can be directed to either end of the main-stage spool.
INTERNAL PILOT DRAIN PORT

Figure 8-26. Typical solenoid-controlled, pilot-operated valve.

TWO-STAGE VALVES

Using the flexibility and power of hydraulics, two-stage valves can control large volumes of fluid at high pressures. A direct-acting solenoid valve, used in the same capacity, would be very large and require a great amount of electricity. In order to be able to shift the main spool, a pneumatic actuator would also have to be very large in relation to the valve.

Two-stage valves consist of a pilot-operated main stage and a pilot stage that is usually either electrically or pneumatically operated. They are often referred to as either solenoid-controlled, pilot-operated or pneumatically-controlled, pilot-operated. When the pilot valve is shifted, it directs fluid to one end or the other of the main spool while connecting the opposite end to the tank.

The amount of pressure required to shift the main spool is usually 75 to 100 psi (5.17 to 6.89 bar) (517.05 to 689.40 kPa), but this varies depending upon the spool configuration. Normally, the pilot pressure is supplied through an internal passage connected to the pressure port inside the valve (Figure 8-26).

If the pressure port is connected to the tank, as in the case of tandem or open-center spools, a check valve with a heavy spring must be installed in the tank line of the valve to create pilot pressure (Figure 8-27). Because two-stage valves have the option of being internally drained to the main tank port, a check valve on the tank port will not work. An alternative is to obtain pilot pressure from another source and use an externally piloted valve.

Some two-stage valve models create pilot pressure with a check valve built into the main pressure port of the valve (Figure 8-28). Figure 8-29 shows the graphic representations of various two-stage valves.

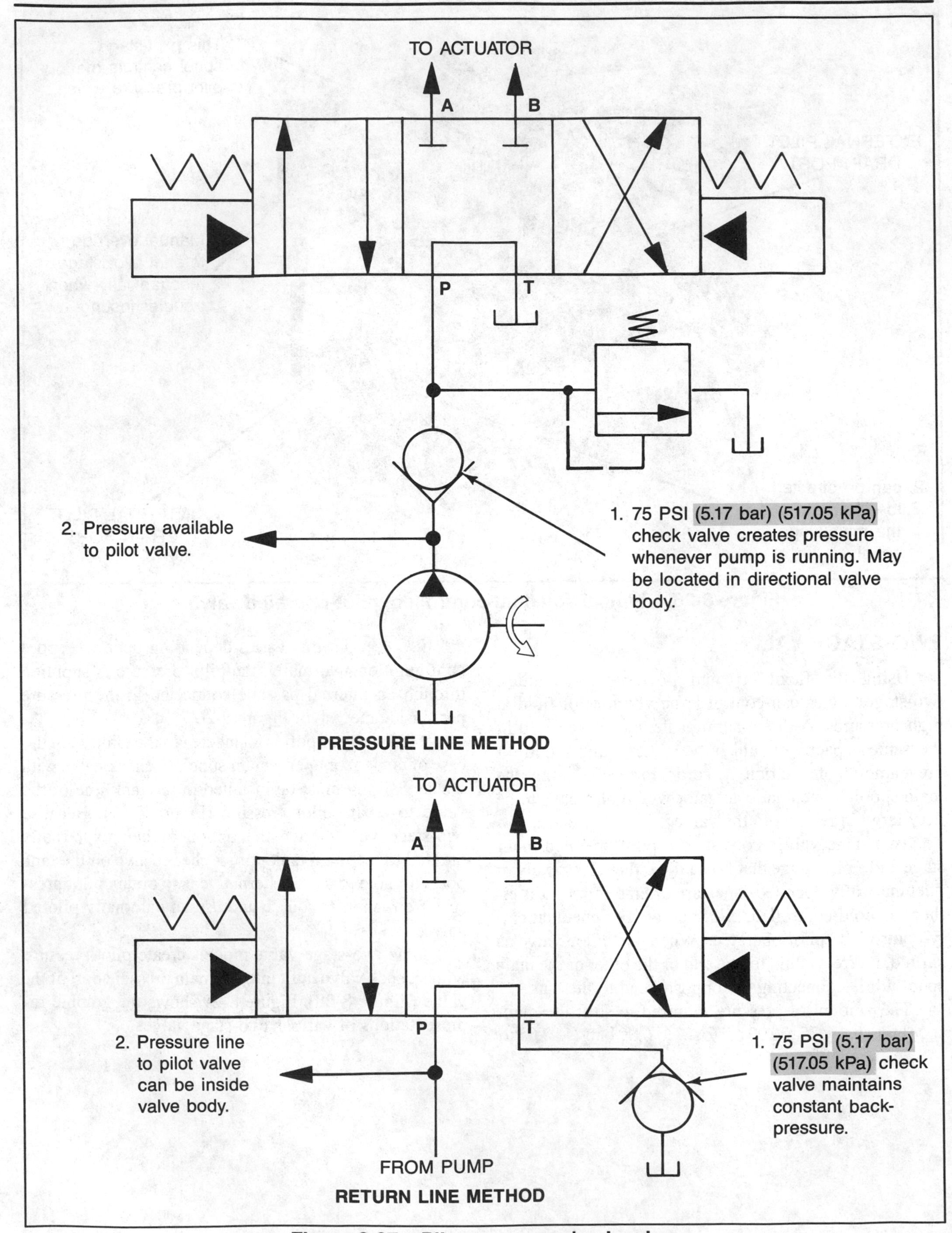
TO ACTUATOR
A
B
P
T
2. Pressure available to pilot valve.
1. 75 PSI (5.17 bar) (517.05 kPa) check valve creates pressure whenever pump is running. May be located in directional valve body.
PRESSURE LINE METHOD
TO ACTUATOR
A
B
P
T
2. Pressure line to pilot valve can be inside valve body.
1. 75 PSI (5.17 bar) (517.05 kPa) check valve maintains constant back-pressure.
FROM PUMP
RETURN LINE METHOD

Figure 8-27. Pilot pressure check valve.

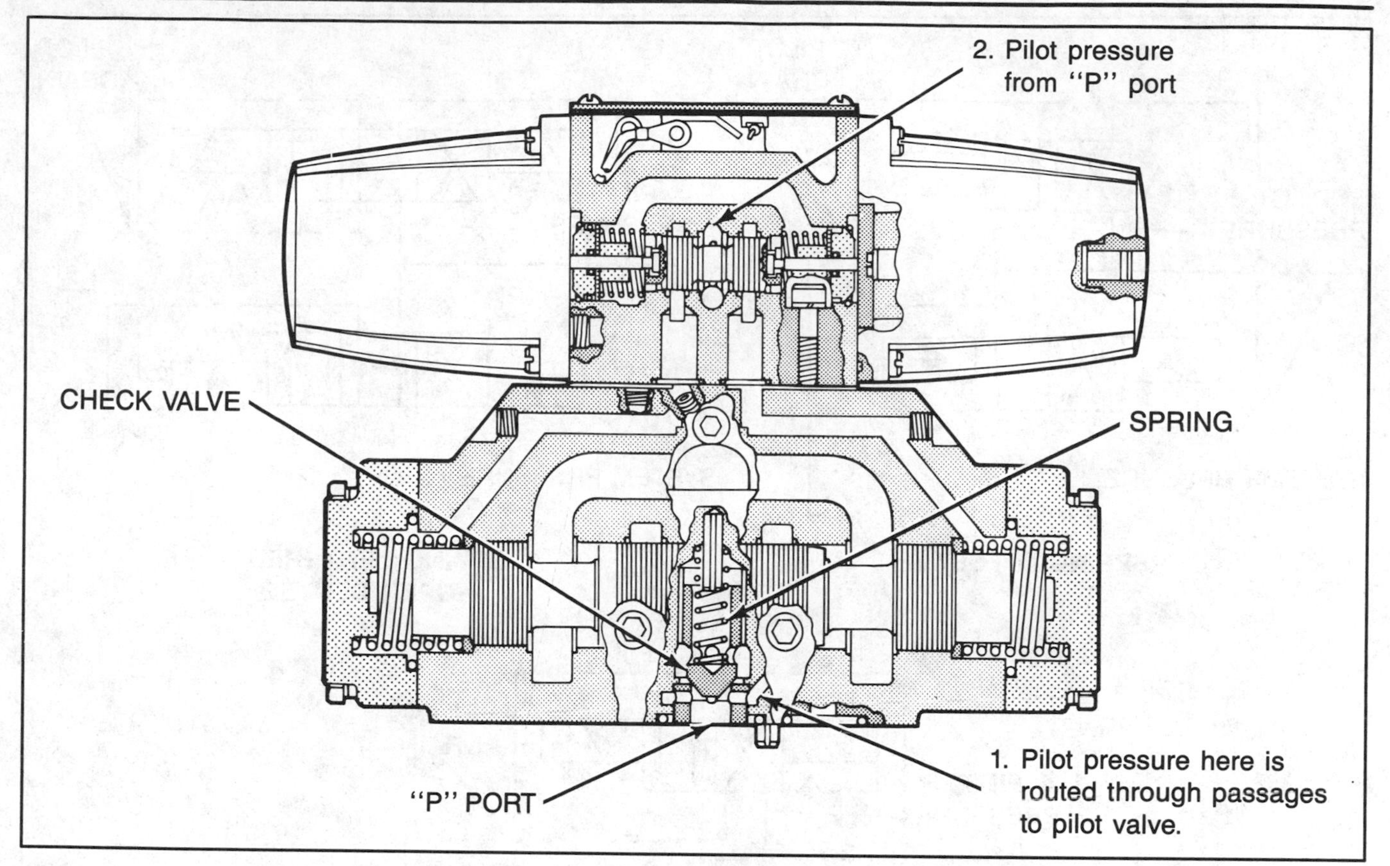

Figure 8-28. Integral pilot pressure check valve.

Pilot Choke Options

To control the shift speed of the main spool, a pilot choke (Figures 8-30 and 8-31) may be incorporated on a two-stage valve. This provides smoother reversals and less hydraulic shock.

Of the various available pilot choke options, one is a fixed orifice fitted between the main stage and the pilot valve. The others are adjustable and control the flow of fluid either to or from the end of the main spool with free flow in the opposite direction.

The meter-out design pilot choke controls the fluid as it is exhausted from the end of the spool. When the valve is energized or de-energized, this type of choke slows down the shifting speed of the main spool. If controlled shifting with fast centering is required, the meter-in design should be used.

Pilot Piston Option

When it is necessary to increase the main spool shifting speed, pilot pistons (Figure 8-32) can be used. The speed increases because the volume of fluid required to shift the small piston is less than the volume required to shift the main spool. This smaller area requires higher pilot pressures.

Spool Stroke Limiter Option

Spool stroke limiters (Figure 8-31) are a simple kind of flow control that restrict the distance the main spool can shift in a given direction.

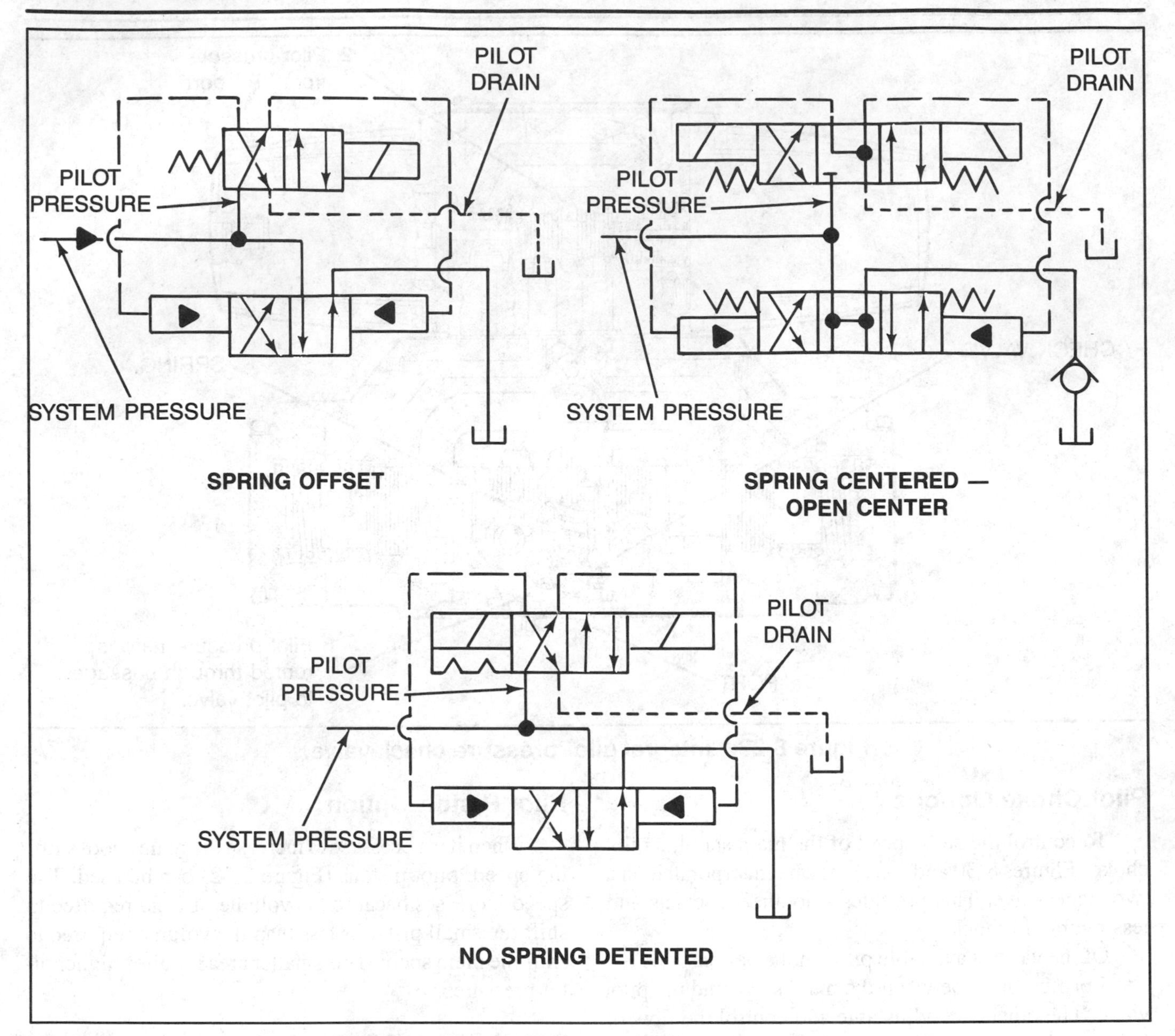

Figure 8-29. Variations of two-stage valves.

MOUNTING STYLES

Sliding spool-type directional valves are available in both line-mounted and subplate- or manifold-mounted designs. Line-mounted versions are more frequently found in mobile applications, while the subplate-mounted design is used for industrial ones.

Mounting Standards

Standard mounting patterns have been developed to make interchangeability between valve manufacturers more convenient. Figure 8-33 shows various mounting pattern interface names for both United States and international standards. There are also nonstandard patterns that are offered by some manufacturers but will not be covered in this book.

Standards Organizations

National Fluid Power Association. The NFPA sets the standards for the United States. Its pattern conforms to ANSI-B93.7, which is controlled by the American National Standards Institute. Valves conforming to this pattern will also fit on the ISO and CETOP patterns discussed below

International Standards Organization. The ISO sets the interface pattern for international standardization.

European Oil Hydraulic and Pneumatic Committee. CETOP is the accepted acronym for this European trade organization that represents the organizations of various European countries, similar to the NFPA.

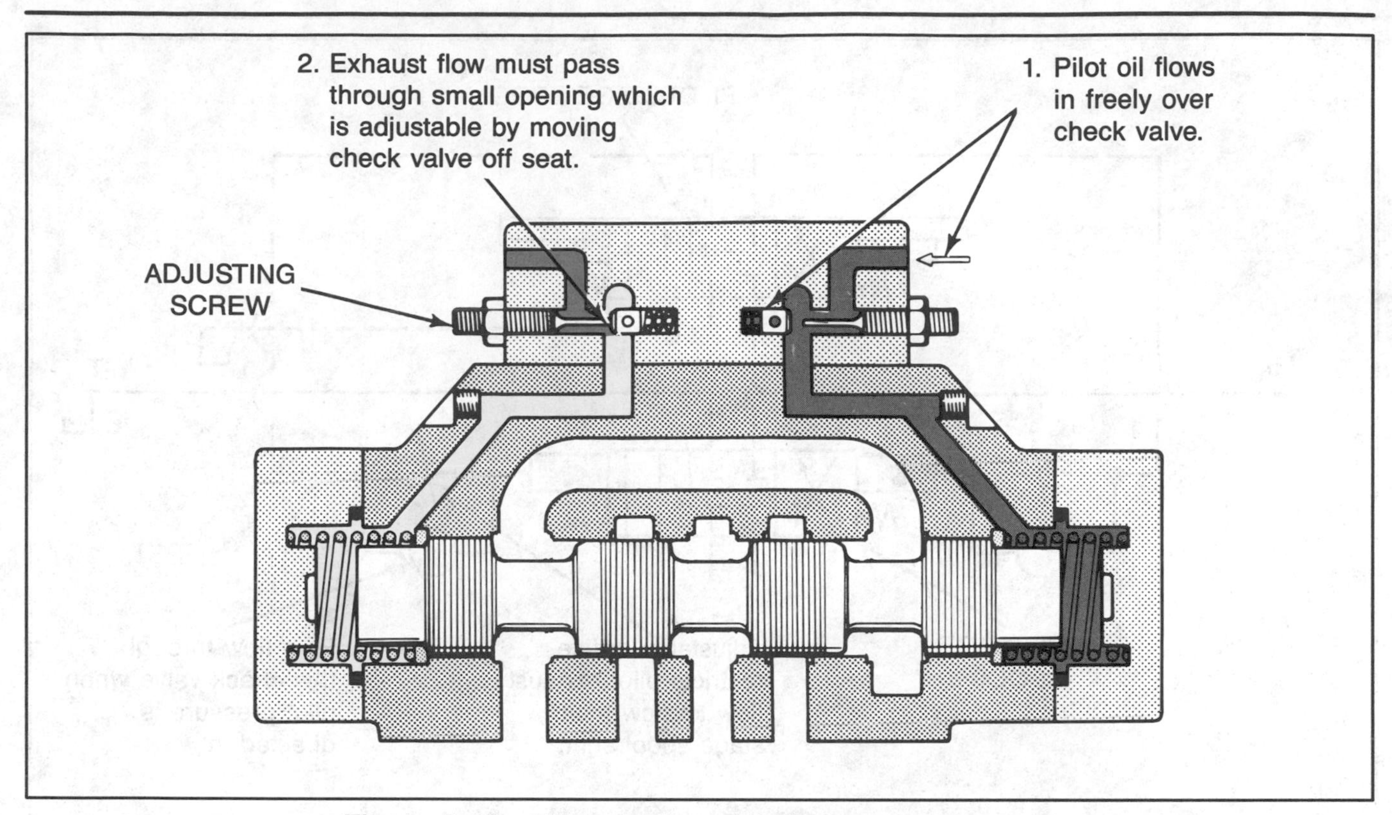
2. Exhaust flow must pass through small opening which is adjustable by moving check valve off seat.
1. Pilot oil flows in freely over check valve.
ADJUSTING SCREW

Figure 8-30. Pilot choke controls spool movement.

ROTARY VALVES

A rotary four-way valve (Figure 8-34) consists of a rotating spool that is fitted into a close-tolerance body. Passages in the spool connect or block the ports in the valve body, and a center position can be incorporated, if required. The graphic symbol for a rotary valve is the same as for a sliding spool valve.

Rotary valves can be line-, panel-, or subplate-mounted. They are actuated manually or mechanically and are generally low-flow valves. Although they can be used for reversing cylinders and motors, rotary valves usually control the pilot flow to the larger, pilot-operated directional valves.

DECELERATION VALVES

Hydraulic cylinders often have built-in cushions that slow down the cylinder pistons at the extreme ends of their travel. An external valve is required when it is necessary to decelerate a cylinder at some intermediate position or to slow down or stop a rotary actuator (motor).

Most deceleration valves are cam-operated with tapered spools. They gradually decrease flow to or from an actuator for smooth stopping or deceleration. A "normally open" valve cuts off flow when its plunger is depressed by a cam. It may be used to slow the speed of a drillhead cylinder at the transition from rapid traverse to feed, or to smoothly stop heavy index tables and large presses.

Some applications require a valve to permit flow during actuation and to cut off flow when the plunger is released. In this case, a "normally closed" valve is used. This type of valve provides an interlocking arrangement that allows flow to be directed to another branch of the circuit when the actuator or load reaches a certain position. Both the "normally open" and the "normally closed" valves are available with integral check valves that permit reverse free flow.

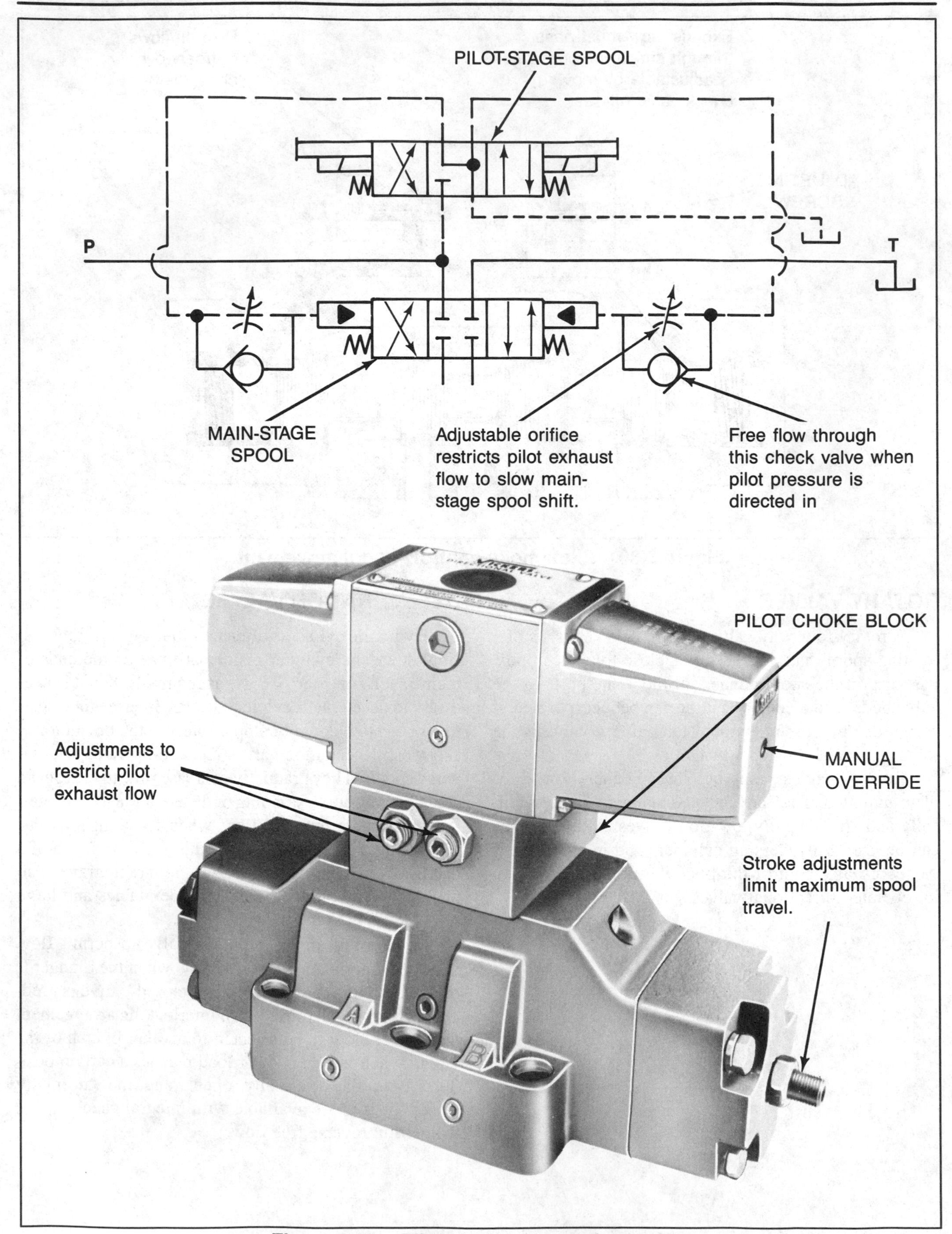
PILOT-STAGE SPOOL
P
T
MAIN-STAGE SPOOL
Adjustable orifice restricts pilot exhaust flow to slow main-stage spool shift.
Free flow through this check valve when pilot pressure is directed in
PILOT CHOKE BLOCK
Adjustments to restrict pilot exhaust flow
MANUAL OVERRIDE
Stroke adjustments limit maximum spool travel.
A
B

Figure 8-31. Pilot choke mounted on valve.

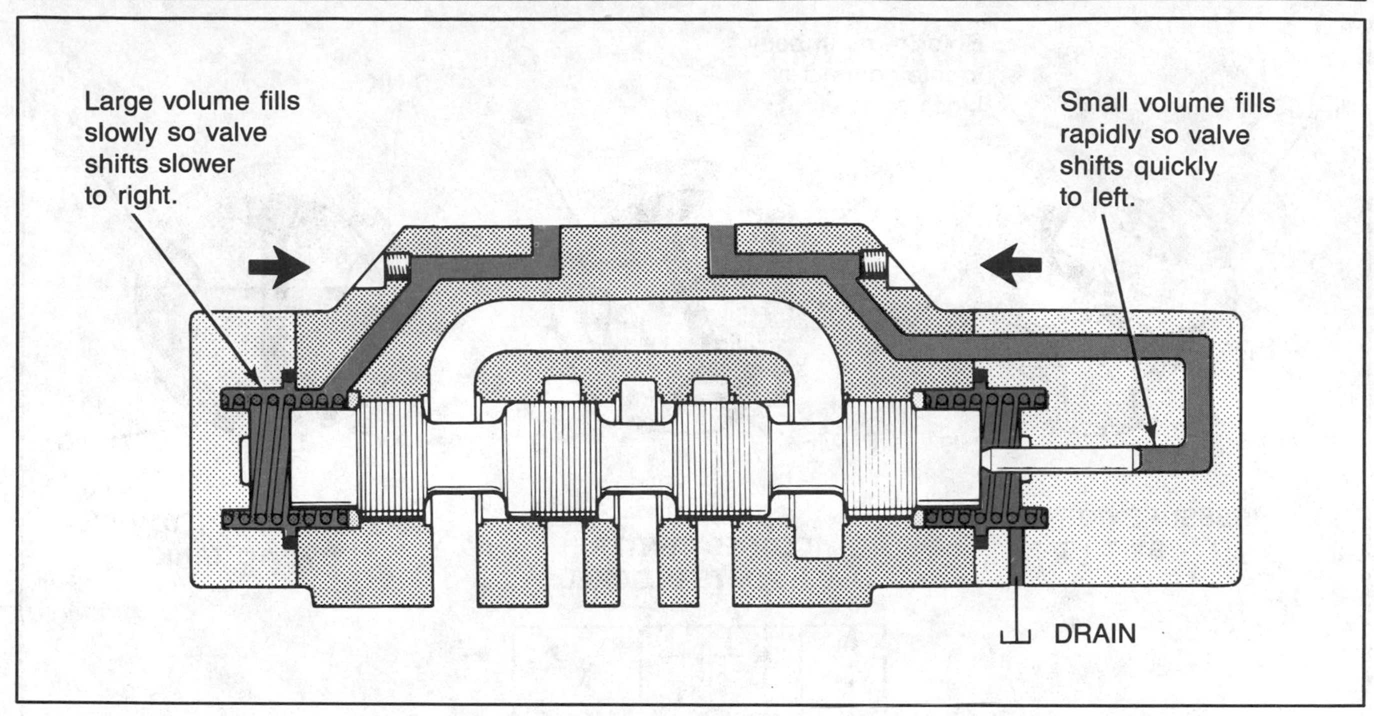

Large volume fills slowly so valve shifts slower to right.
Small volume fills rapidly so valve shifts quickly to left.
DRAIN

Figure 8-32. Pilot piston speeds valve shifting.

NFPA		ISO	CETOP	PORT DIA.
OLD	NEW			
D01	D03	03	3	.260″ (6.6 mm)
D02	D05	05	5	.438″ (11.1 mm)
D04	D07	07	7	.688″ (17.5 mm)
D06	D08	08	8	.906″ (23.0 mm)
D10	D10	10	10	1.312″ (33.3 mm)

Figure 8-33. Pattern interface names for mounting standards.

Tapered Plunger Design

An early design of deceleration valve (Figure 8-35) uses a tapered plunger to reduce flow as it is actuated by the cam. Before the plunger is depressed (View A), free flow is permitted from the inlet to the outlet. Depressing the plunger gradually cuts the flow off (View B). Reverse free flow (View C) is permitted by the integral check valve.

The control range of this valve depends on both the flow volume and cam rise. At nearly maximum volume (with an initial pressure drop through the valve), the plunger stroke is completely controlled. One drawback is that at low flow rates, this control is available only from the point where a pressure drop is created. The adjustable orifice design valve compensates for this by allowing the valve to be tailored to any given flow.

Adjustable Orifice Design

An adjustable orifice design valve (Figure 8-36) features a closely-fitted plunger and sleeve, both of which have rectangular ports or "windows" that control the flow. As the plunger moves inside the sleeve, the ports in each coincide when the open position is reached. Oil entering the inlet flows through the small upper ports in the sleeve and plunger, down through the center of the plunger, and out the large ports to the outlet (View A). When the plunger is depressed, the "window" area is gradually cut off to stop flow (View B). Reverse free flow is allowed by the integral check valve.

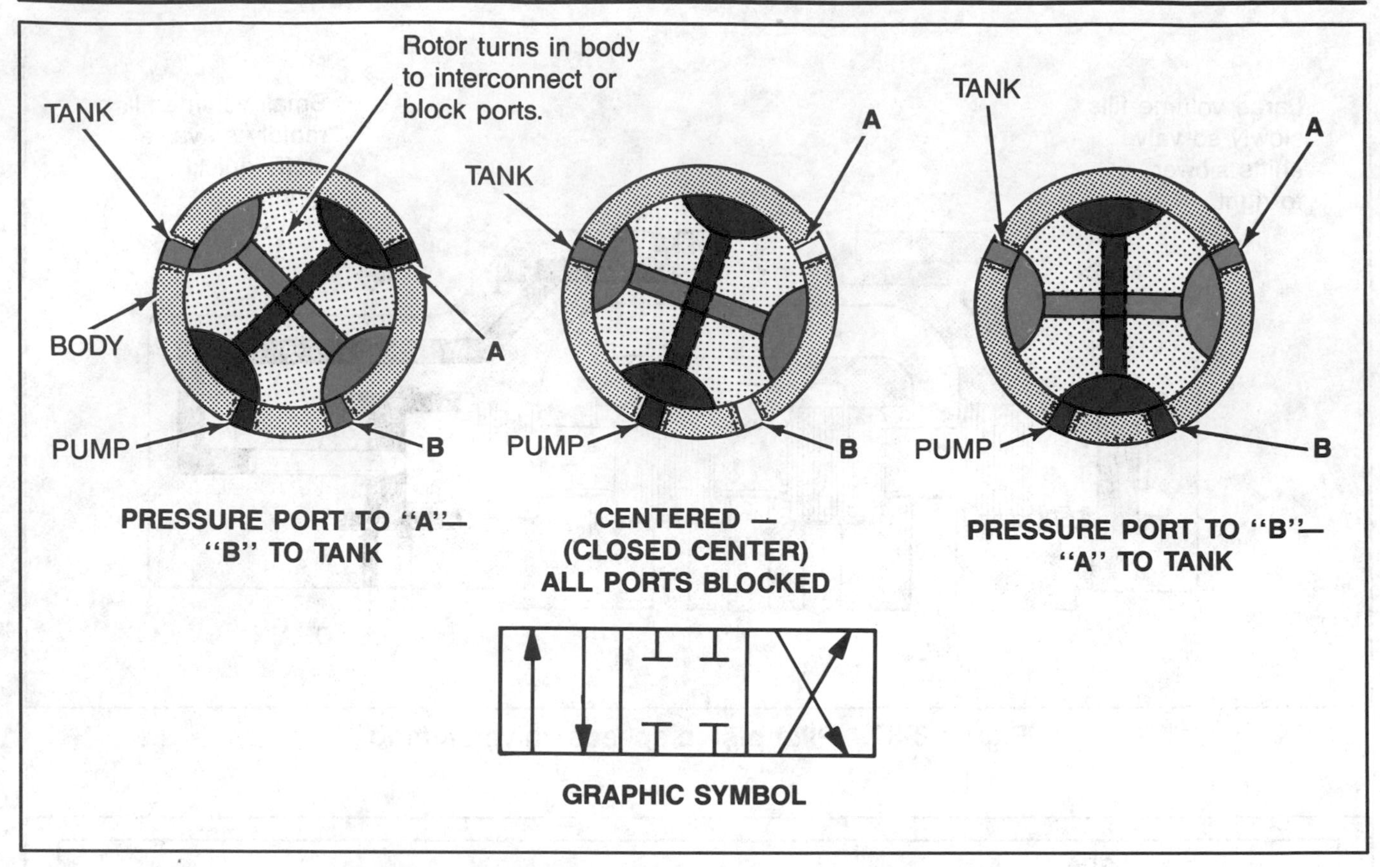

Figure 8-34. Rotary four-way valve operation.

Initial Pressure Drop Adjustment. For precise control throughout the plunger stroke, the width of the port openings is controlled by adjusting screws that turn the sleeve. A low flow rate has a narrow opening, while a higher flow requires a wider one. The adjustment is made by attaching a pressure gauge at the side of the valve and turning the screws to obtain the desired initial pressure drop.

This valve also includes an adjustable orifice that allows some flow with the plunger fully depressed. It consists of a small plunger with a chamfered end and a "vee" notch that can be set to bypass the spool-sleeve closure. During indexing or some similar application, the orifice permits the load to creep to its final position.

The window orifice valve is built in both the pipe-threaded and back-mounted versions. Both valves require a drain so that leakage oil can escape from beneath the plunger.

Typical Applications

Figure 8-37 illustrates a deceleration valve in a typical application. At a preset point, the valve slows a drill-head cylinder from rapid advance speed to feed speed. View A shows rapid advance, with exhaust flow from the cylinder passing unrestricted through the deceleration valve.

In View B, the valve plunger is depressed by a cam. Exhaust flow is cut off at the deceleration valve and passes through the volume control valve, which sets the feed speed. In View C, the directional valve is reversed to return the cylinder.

Whether the plunger is depressed or not, oil from the directional valve passes freely through the deceleration valve's check valve.

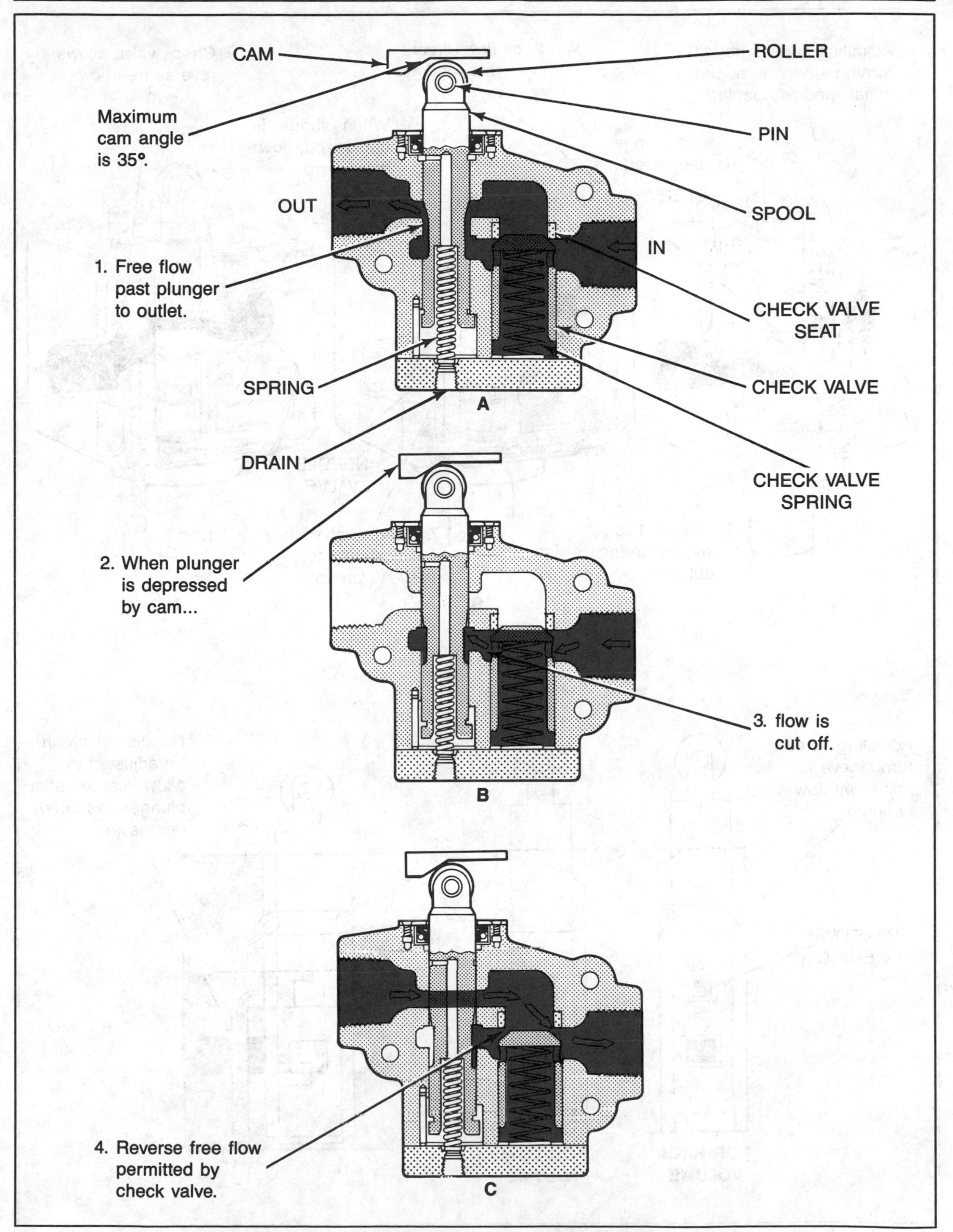

Figure 8-35. Deceleration valve with tapered plunger.

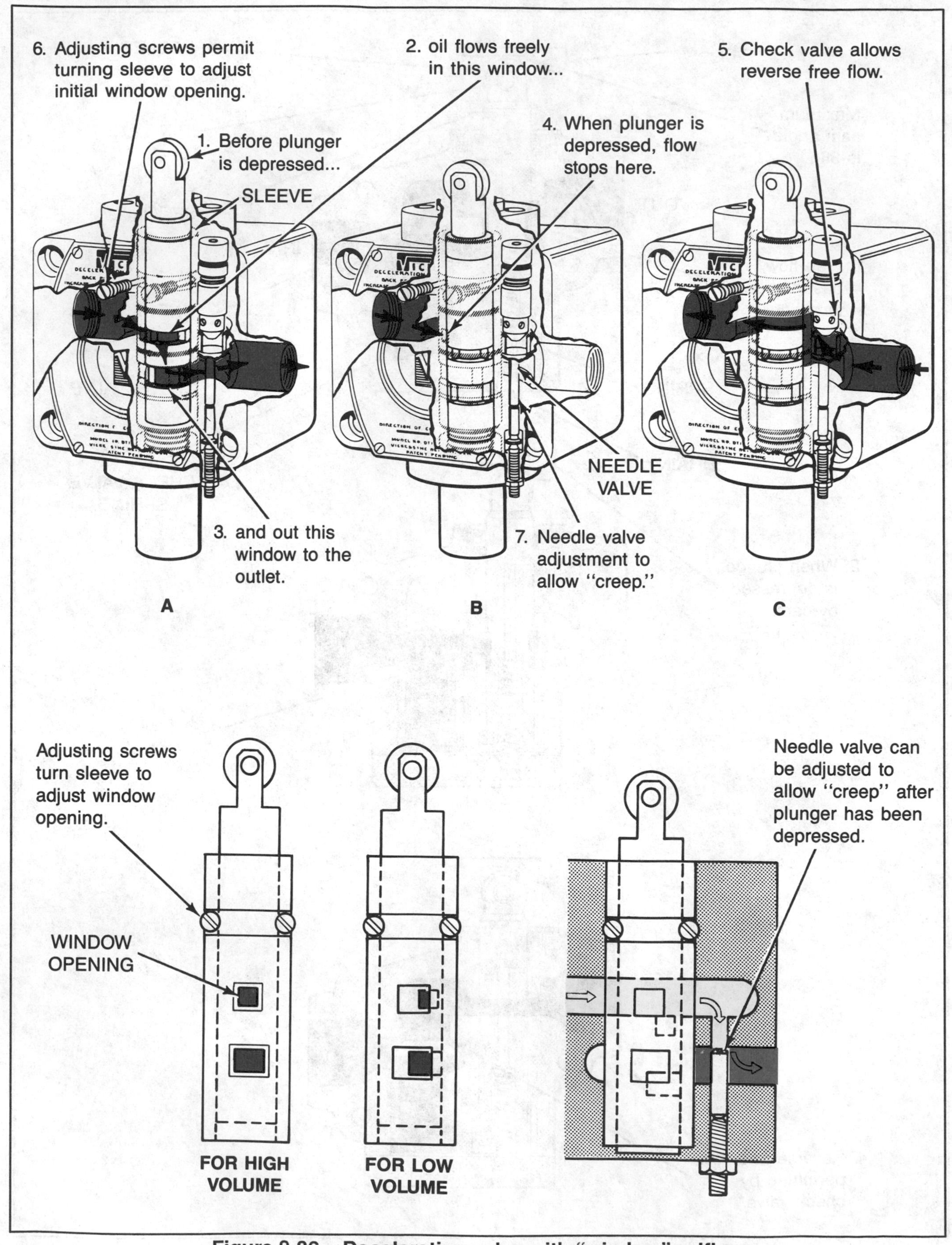

6. Adjusting screws permit turning sleeve to adjust initial window opening.
2. oil flows freely in this window...
5. Check valve allows reverse free flow.
4. When plunger is depressed, flow stops here.
1. Before plunger is depressed...
SLEEVE
NEEDLE VALVE
3. and out this window to the outlet.
7. Needle valve adjustment to allow "creep."
A
B
C
Adjusting screws turn sleeve to adjust window opening.
WINDOW OPENING
Needle valve can be adjusted to allow "creep" after plunger has been depressed.
FOR HIGH VOLUME
FOR LOW VOLUME

Figure 8-36. Deceleration valve with "window" orifice.

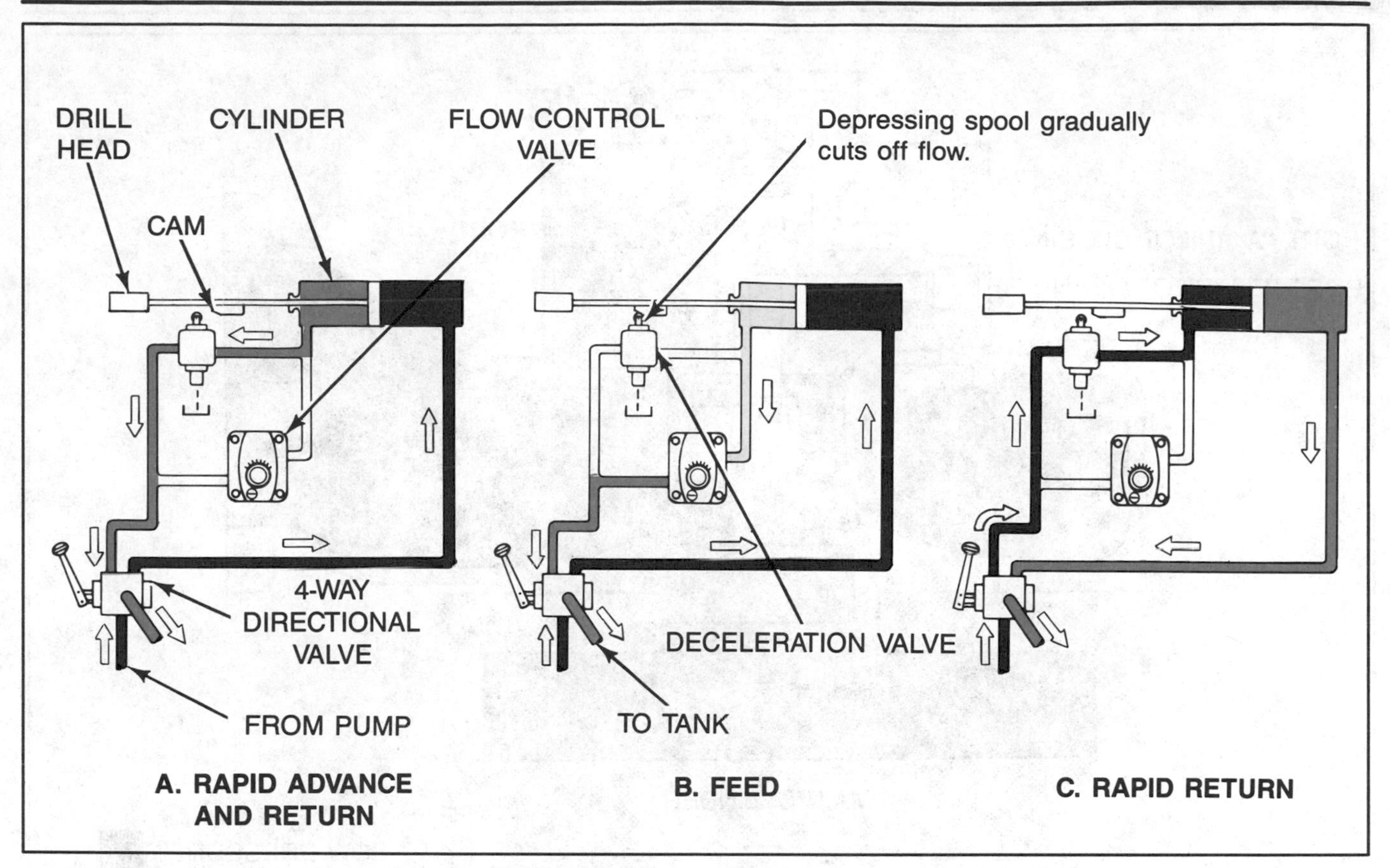

Figure 8-37. Deceleration valve in feed control system.

PREFILL VALVES

Prefill valves are specifically designed to handle large volumes of fluid at low pressure drop. Their flow capacities range from several hundred to several thousand gallons per minute and are typically used on large pressing or die casting machine cylinders. Gravity or atmospheric pressure, pushes the fluid through the valve, eliminating the need for large volume pumps. The five basic operational phases of prefill valves (Figure 8-38) are discussed below. Mounting styles that are available for the prefill valves are shown in Figure 8-39.

Neutral Position

In the neutral position, the valve is held open by a spring. Fluid can flow either into the valve through the prefill and pressure ports before flowing into the press cylinder, or vice-versa. No pressure can be built up in this phase.

Rapid Advance

During rapid advance, the main cylinder advances toward the workpiece. This motion allows fluid to flow from the reservoir and pressure port through the prefill valve and into the cylinder.

Pressing

Pilot pressure is applied to the closing port, shifting the gate assembly into the closed position. Pressure is then applied to the pressure port to perform the pressing function.

Decompression

Decompressing the fluid is necessary before the valve is shifted to return to the cylinder. This avoids a possibly damaging hydraulic shock. The prefill valve remains closed until the pressure drops to approximately 250 psi (17.24 bar) (1723.50 kPa). Decompression valves are available either as an integral part of the prefill valve or as a separate unit.

Return

After decompression, the valve returns to its neutral position, allowing fluid to flow back to the reservoir.

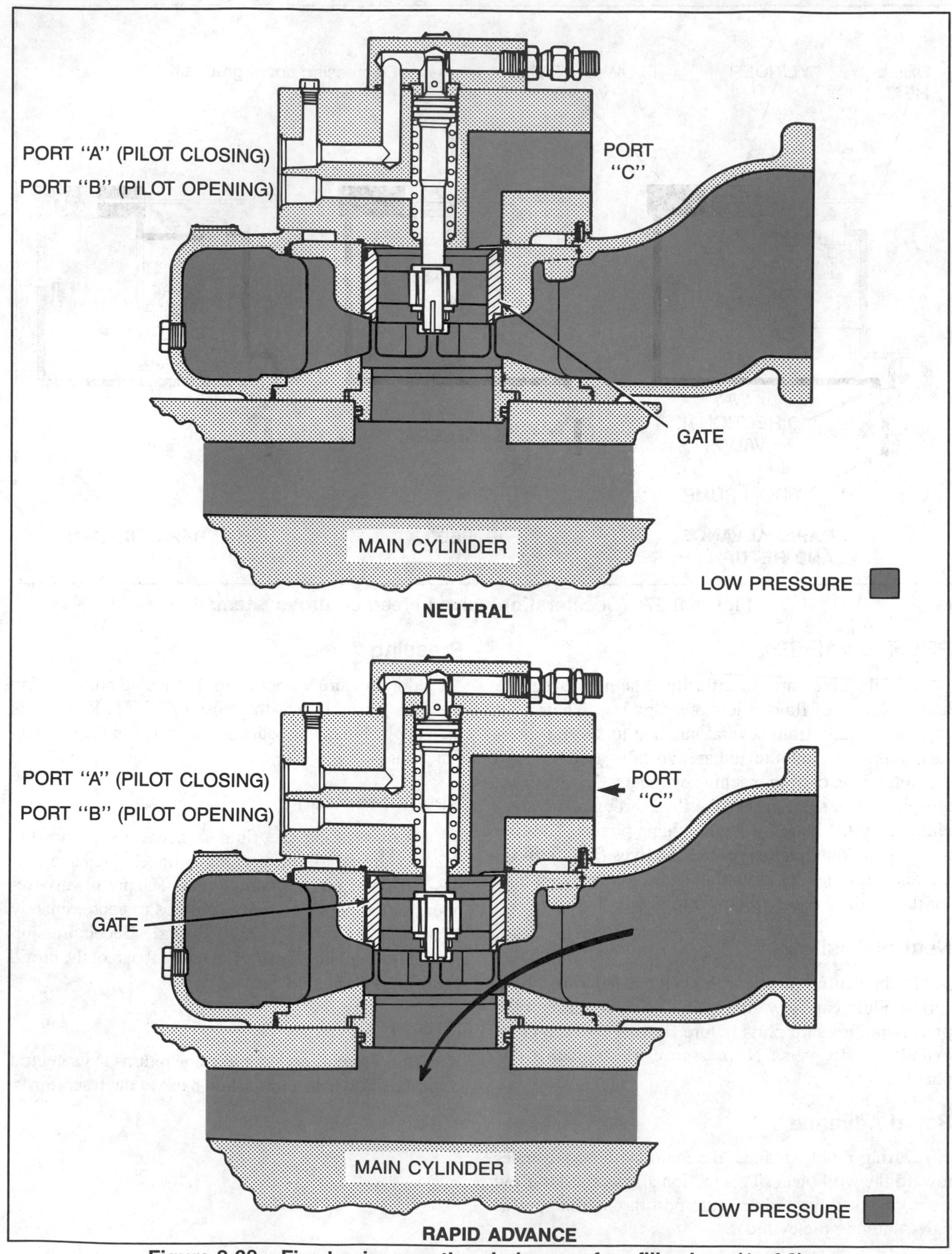

Figure 8-38. Five basic operational phases of prefill valves (1 of 3).

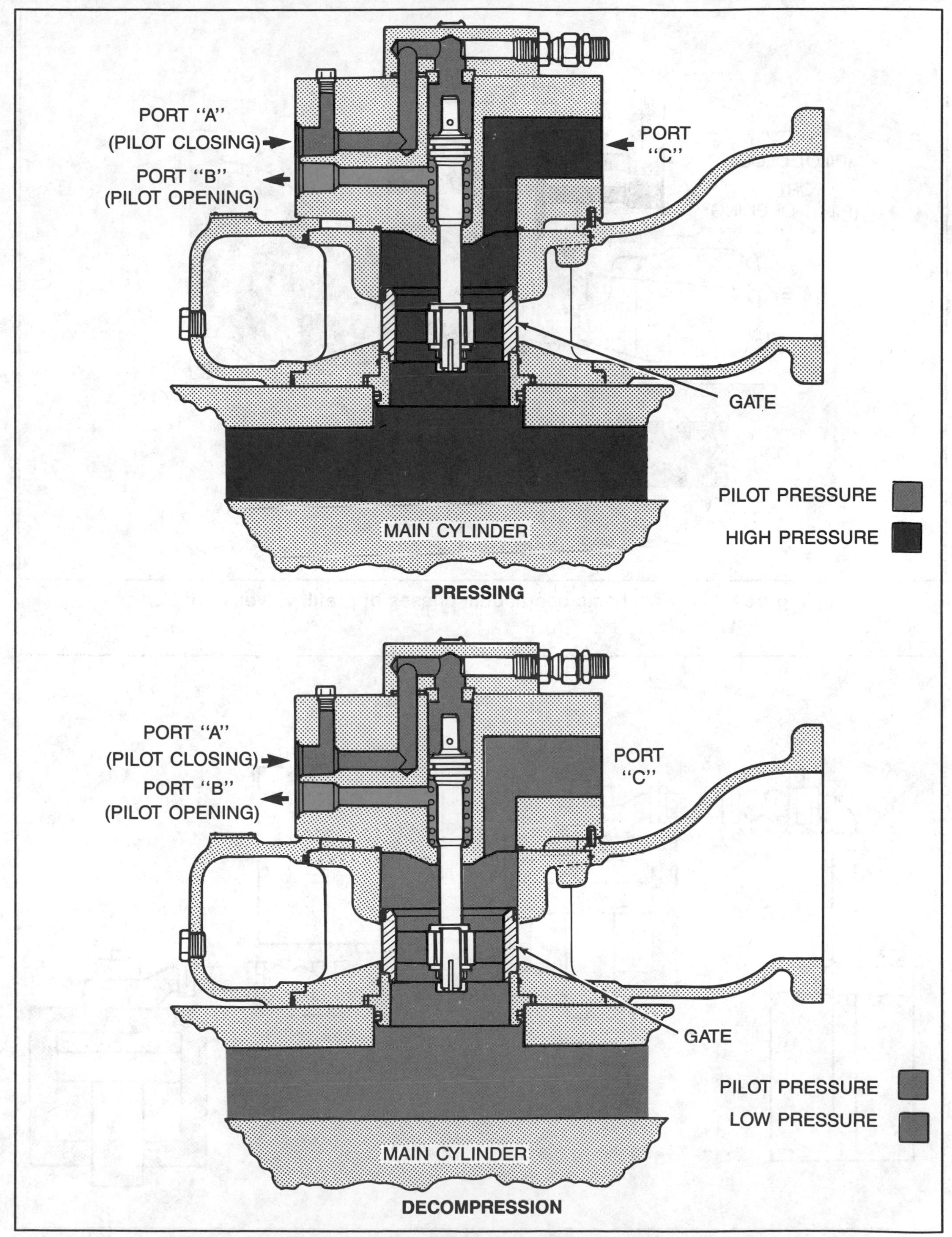

Figure 8-38. Five basic operational phases of prefill valves (2 of 3).

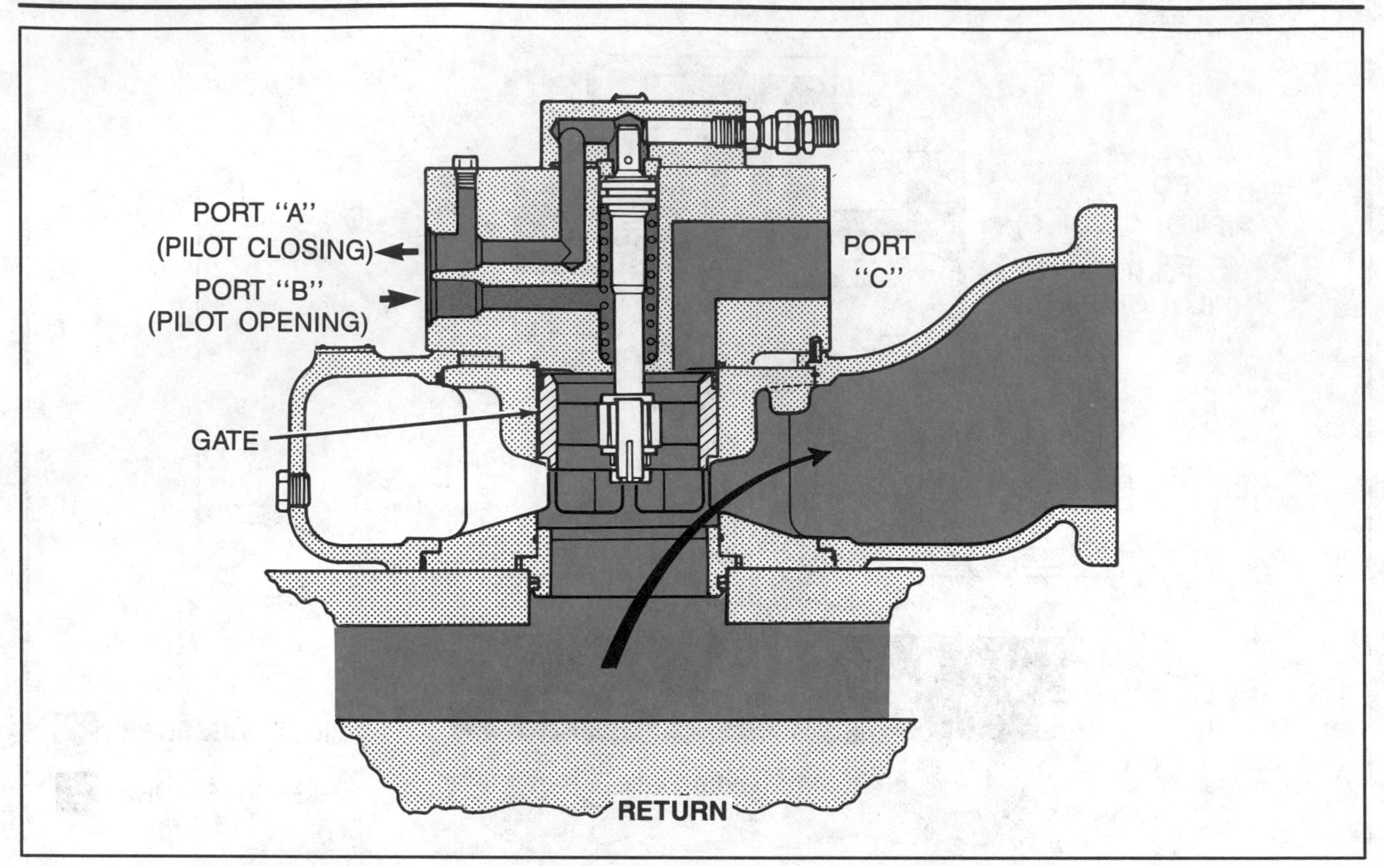

Figure 8-38. Five basic operational phases of prefill valves (3 of 3).

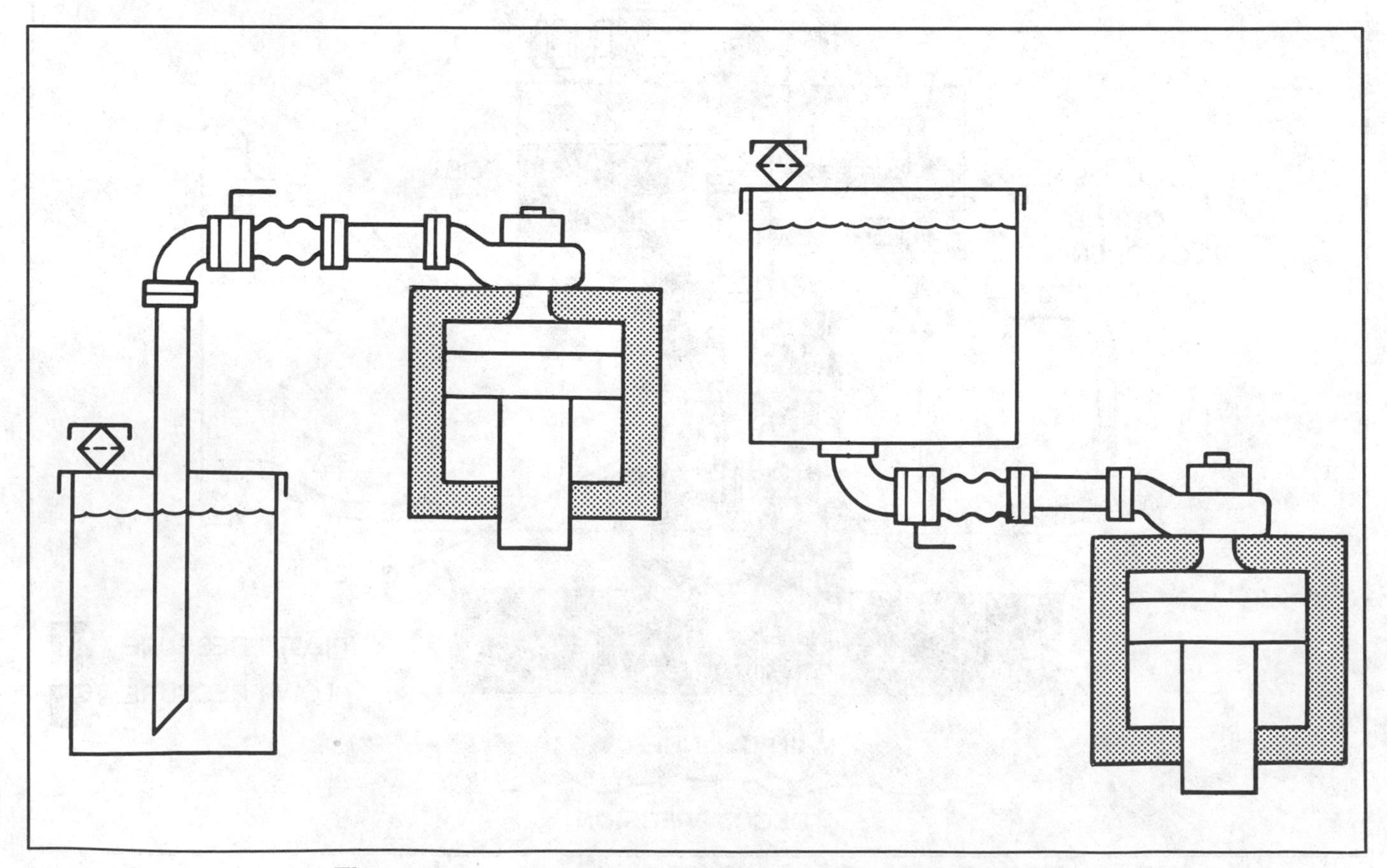

Figure 8-39. Available prefill valve mounting styles.

QUESTIONS

1. What is the function of a directional valve?
2. What is meant by finite positioning?
3. Explain the function of a check valve. Draw the graphic symbol.
4. What is the major concern when applying a pilot-to-open check valve?
5. List five ways to shift a four-way valve.
6. Describe the "centered" condition of the tandem center spool. A float-center spool?
7. How many positions does a spring-offset valve have? A spring-centered valve?
8. Draw the graphic symbol for a two-position spring centered valve, a spring-offset valve, an energize-to-center valve.
9. Describe two ways of obtaining pilot pressure for an open center, two-stage valve.
10. What is the function of pilot chokes? What type would be used to ensure rapid centering in case of an emergency shutdown?
11. What is the main use of rotary valves?
12. What is the purpose of a deceleration valve?
13. What is the maximum recommended cam angle for deceleration valves?
14. How is the adjustable orifice deceleration valve an improvement over the tapered plunger design?
15. List the five basic phases a prefill valve goes through.

CHAPTER

9

BASIC ELECTRICAL PRINCIPLES

This chapter is an introduction to basic electrical concepts and terms. It presents the fundamentals of simple electrical circuits and defines many of the words commonly associated with electricity. Basic circuit elements such as current and voltage are explored as well as the devices and methods used to measure them. The characteristics and operation of several electrical devices are explained and the concept of electrical ground is discussed. Wiring procedures and electrical safety are also covered. Later in this manual, detailed information is provided on more advanced electrical circuitry. In particular, those circuits most often associated with hydraulic systems are examined.

Hydraulics and electricity are often compared because the two types of systems have many similarities. Previous chapters explain that a hydraulic circuit requires a power source (usually a pump), a load device (actuator) and conductors to connect them. Hydraulic circuits differ mainly in the types of devices used to control, direct, and regulate the flow of hydraulic fluid and in the type and capacity of the actuators employed to accomplish the work, which varies depending on the application. An electrical circuit has the same basic requirements. There must be a power source (a battery, generator, etc.), a load device (lights, bells, motors, etc.), and proper connections in between. A wide assortment of devices also are incorporated to control, direct, and regulate the flow of electrical current.

Hydraulic components are usually represented on diagrams by symbols, which have been standardized to ensure that they will have the same meaning to anyone using the diagrams. Electrical components are represented in the same manner with their own standardized symbols. Electrical diagrams are often called schematics. Many of the illustrations in this and other electrical sections of the manual use these electrical schematic symbols. Figure 9-1 illustrates some of the more common symbols.

There are many important differences between hydraulic and electrical systems and circuits. For example, electrical current is invisible while hydraulic fluid is not, and electrical current flows through solid wires while hydraulic fluid flows through hollow lines. Despite these and other differences, hydraulic and electrical circuits are very similar in a theoretical sense. Figure 9-2 shows symbols for electrical components on the left of the illustration with their hydraulic equivalents shown on the right.

Figure 9-3 illustrates a hydraulic circuit and its electrical counterpart. Either type of circuit could be used to perform the same work; only the cost of operation, the physical construction of components, and the medium being controlled are different. Much of the material on hydraulics presented up to this point can be used to understand electrical circuits as well.

BASIC ELECTRICAL CIRCUIT ELEMENTS

Every electrical circuit contains measurable quantities of the following four circuit elements:

- Current
- Voltage
- Resistance
- Power

These circuit elements are interrelated and a change in any one will produce a change in one or more of the others. The interrelationship among current, voltage, resistance, and power is explained later in this chapter.

Current

Electrical current is defined as the directed flow of electrical charges from one point to another around a closed electrical circuit. It is the flow of current that accomplishes the work or purpose of a circuit. Electrical charges can be either negative or positive. Conventional current flow consists of the movement of positive charges from positive to negative around a circuit. The direction of conventional current flow is explained by the Law of Electrical Charges.

The Law of Electrical Charges. The Law of Electrical Charges (Figure 9-4) states that:

1. Like charges repel.

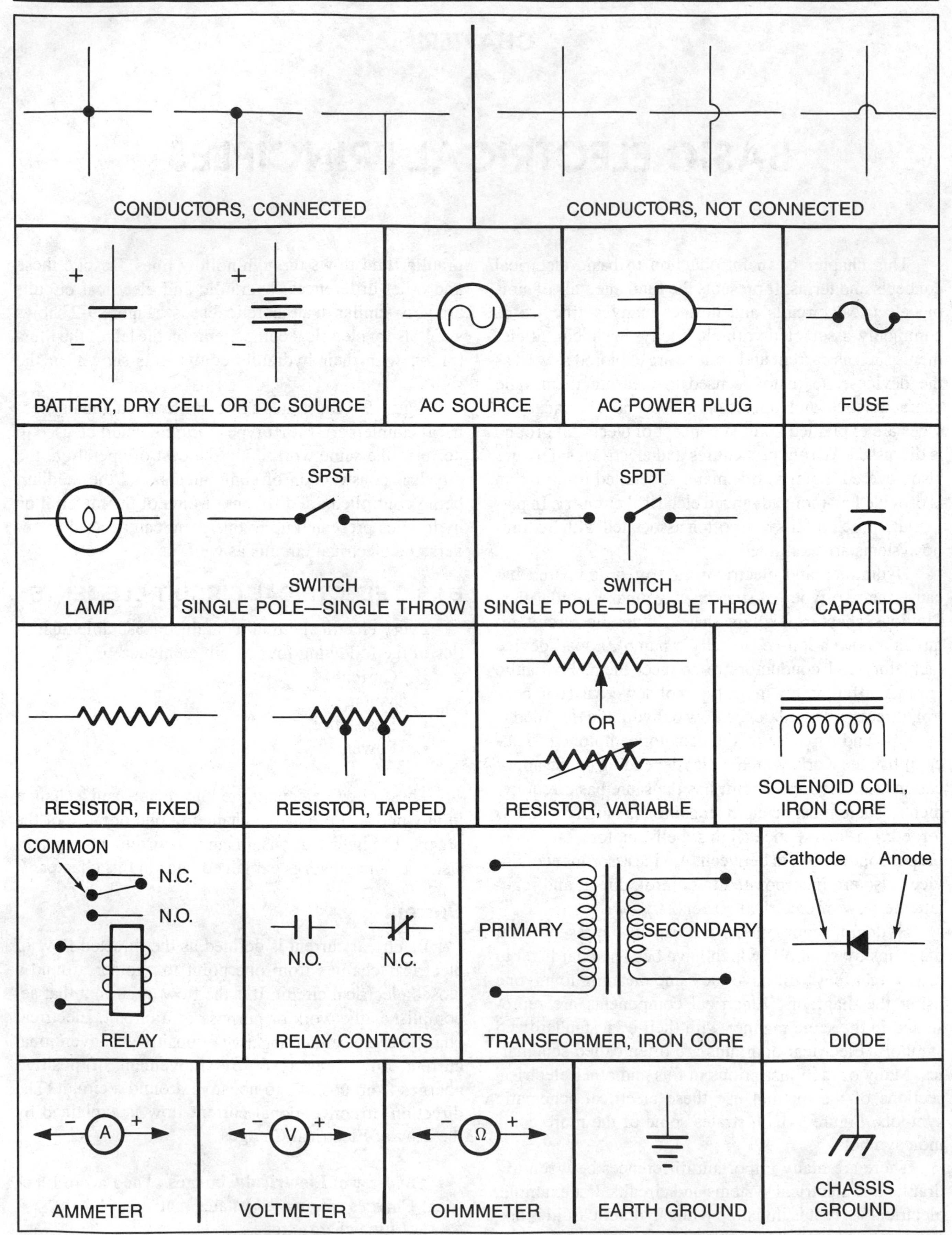

Figure 9-1. Common electrical schematic symbols.

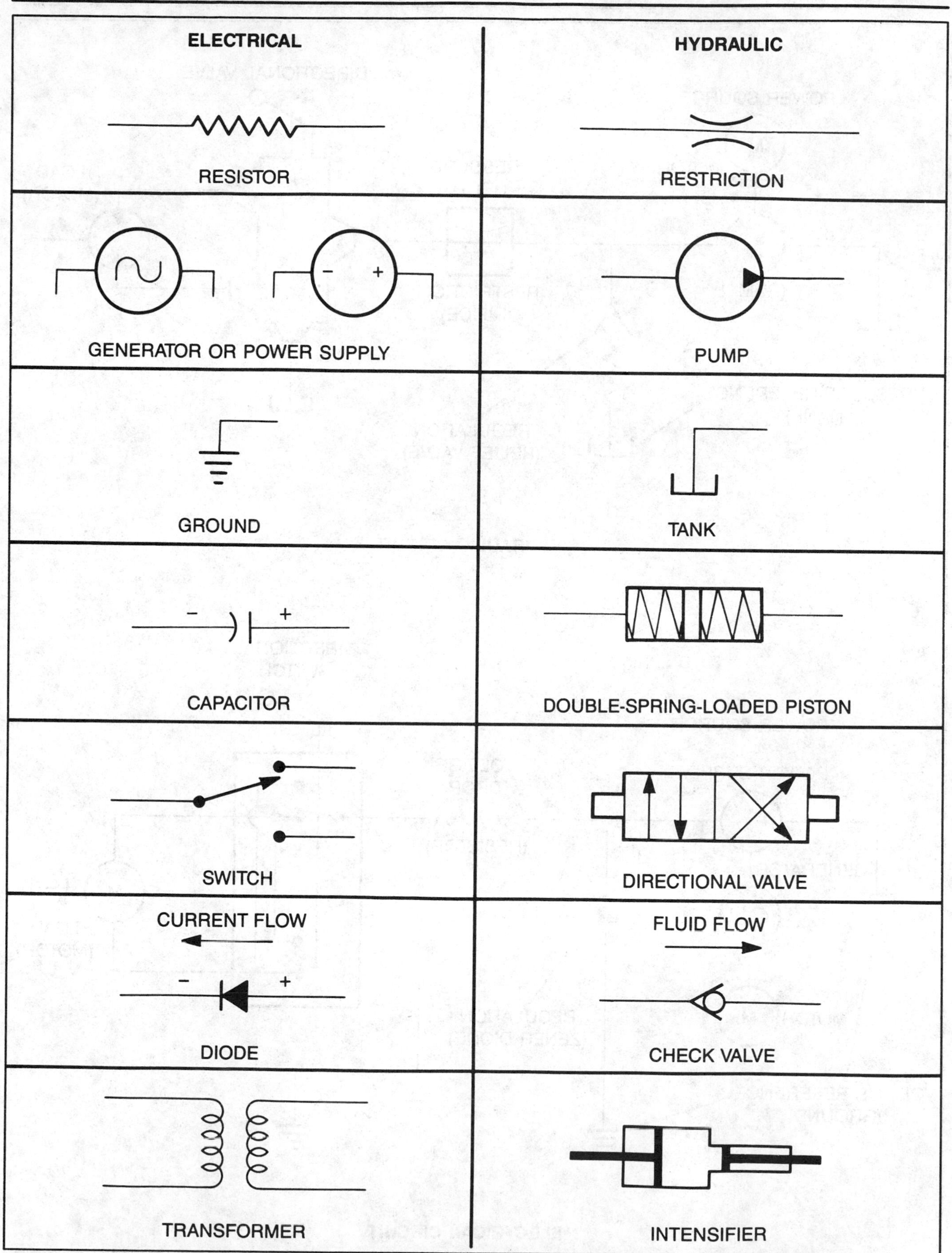

Figure 9-2. Functional equivalence of electrical and hydraulic components.

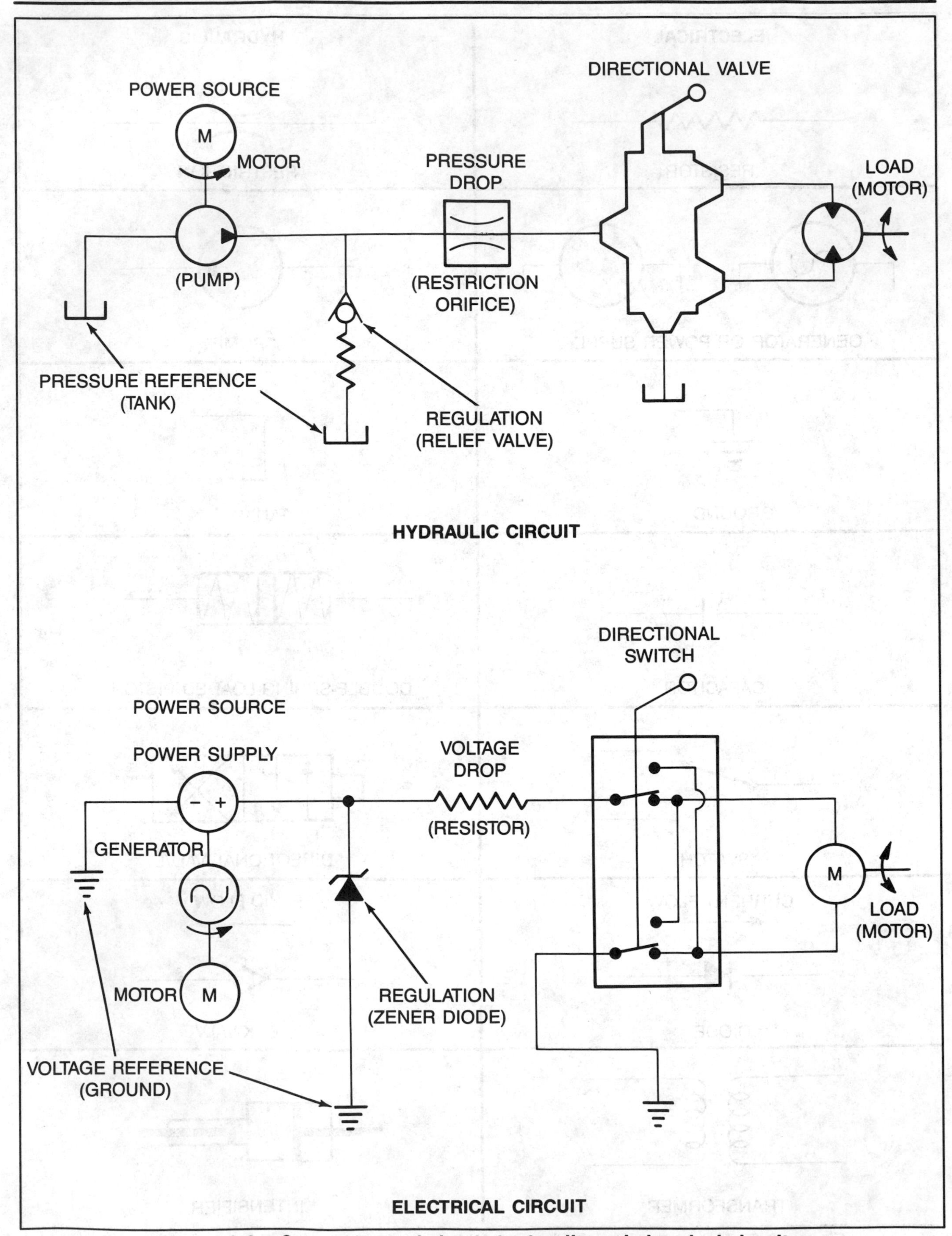

Figure 9-3. Comparison of simple hydraulic and electrical circuits.

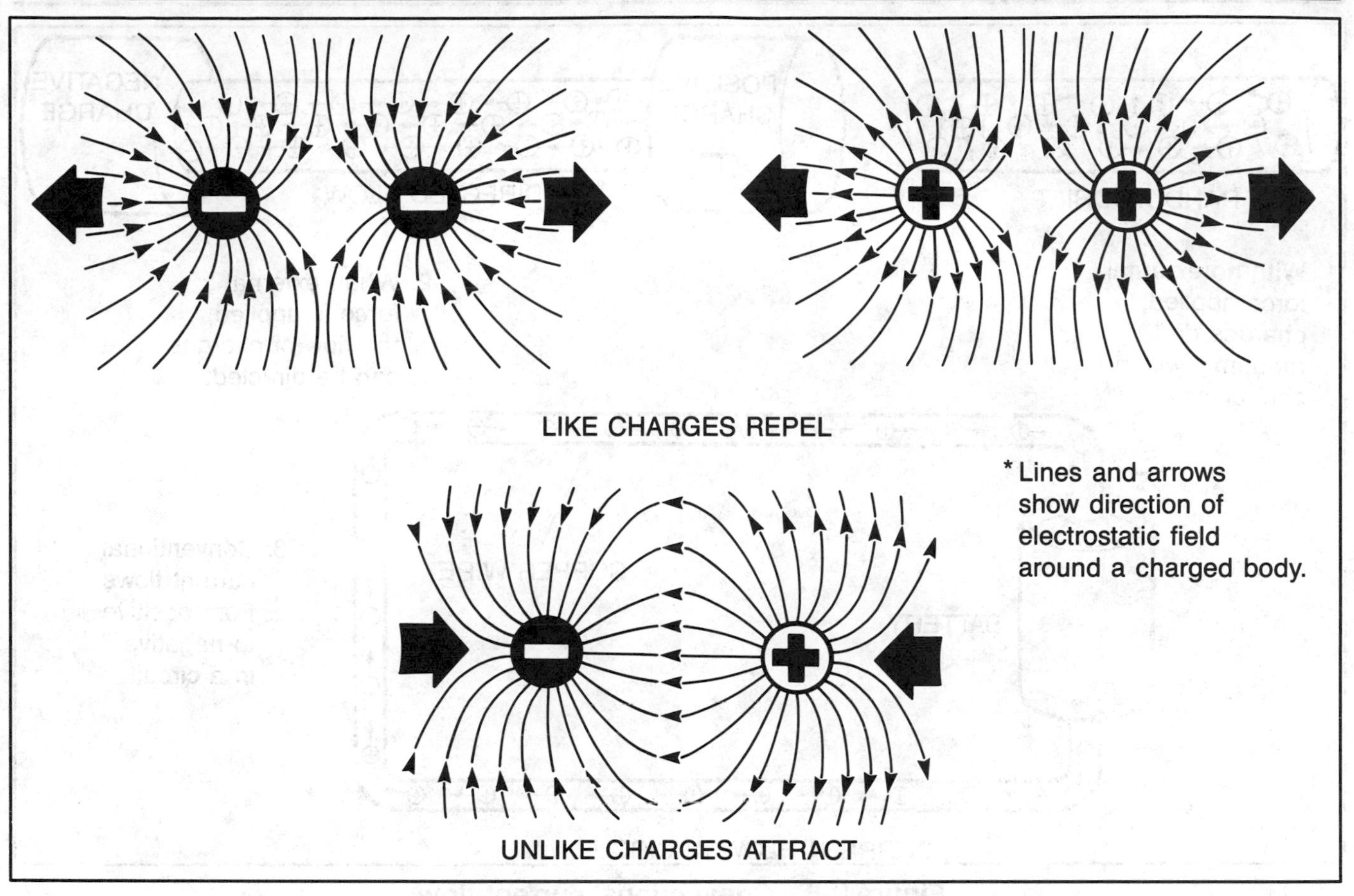
LIKE CHARGES REPEL
* Lines and arrows show direction of electrostatic field around a charged body.
UNLIKE CHARGES ATTRACT

Figure 9-4. The law of electrical charges.

   a. Two positive charges will repel each other.
   b. Two negative charges will repel each other.
2. Unlike charges attract.
   a. A positive charge will be attracted to a negative charge.
   b. A negative charge will be attracted to a positive charge.

The motion of positive charges can be directed and controlled to form conventional current flow. Positive charges will be repelled by the positive power terminal and attracted to the negative power terminal. Therefore, due to the Law of Electrical Charges, conventional current always flows from positive to negative within a circuit. See Figure 9-5.

Unit of Measurement. Current is measured in units called amperes (A). An expression meaning "the current is 2 amperes" would be written I = 2A. Since an ampere is a large quantity of current, current is often measured in smaller units called milliamps (mA), which represent one-thousandth of an amp (0.001A). Therefore, 200mA = 200/1000A = 200 x 0.001A = 0.2A.

Conductors. Conductors form the path by which the current can flow from the power source to the load and back. Conductors are materials (usually metals) that support current flow through them without offering much resistance. Conductors are usually wires, but they are also found in other forms such as the metal chassis or skin of an electrical device, or the foil patterns on a printed circuit board.

Insulators. Insulators are made of high resistance materials that prevent current from flowing through them. The rubber or plastic insulating compound covering a wire, for instance, forms a barrier to current flow that provides protection from electrical shocks and short circuits that could cause serious damage to equipment, and physical hazards to the personnel using it.

Open and Closed Circuits. Current can flow through a circuit only when a complete path exists from the positive terminal to the negative terminal of the power source. Such a circuit is called a **closed circuit** (Figure 9-6A).

If the complete path is accidentally broken or purposely interrupted by switch or relay contacts, current cannot flow in the circuit and work cannot be performed

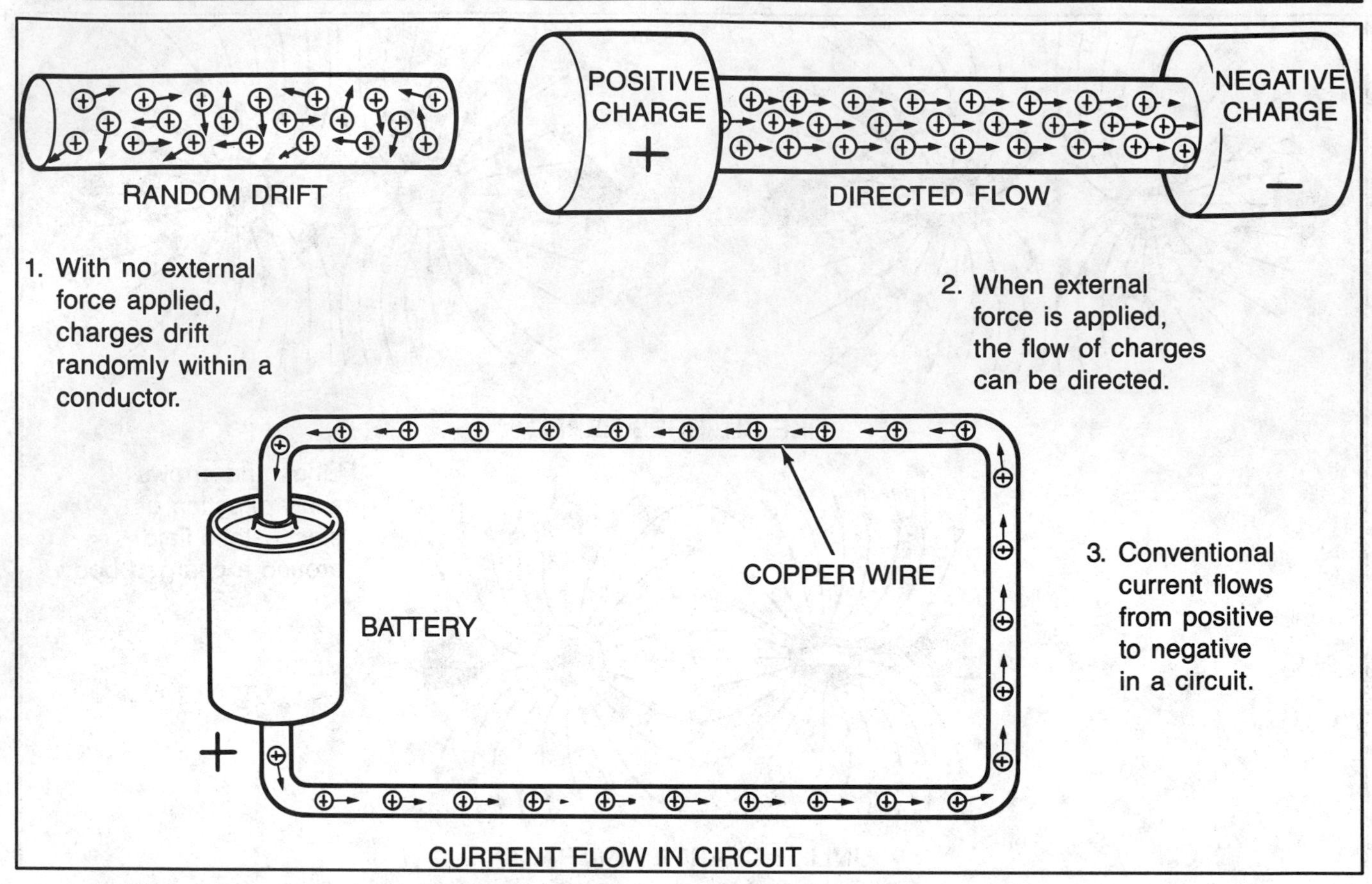
RANDOM DRIFT
POSITIVE CHARGE
+
NEGATIVE CHARGE
–
DIRECTED FLOW
1. With no external force applied, charges drift randomly within a conductor.
2. When external force is applied, the flow of charges can be directed.
–
COPPER WIRE
BATTERY
+
3. Conventional current flows from positive to negative in a circuit.
CURRENT FLOW IN CIRCUIT

Figure 9-5. Conventional current flow.

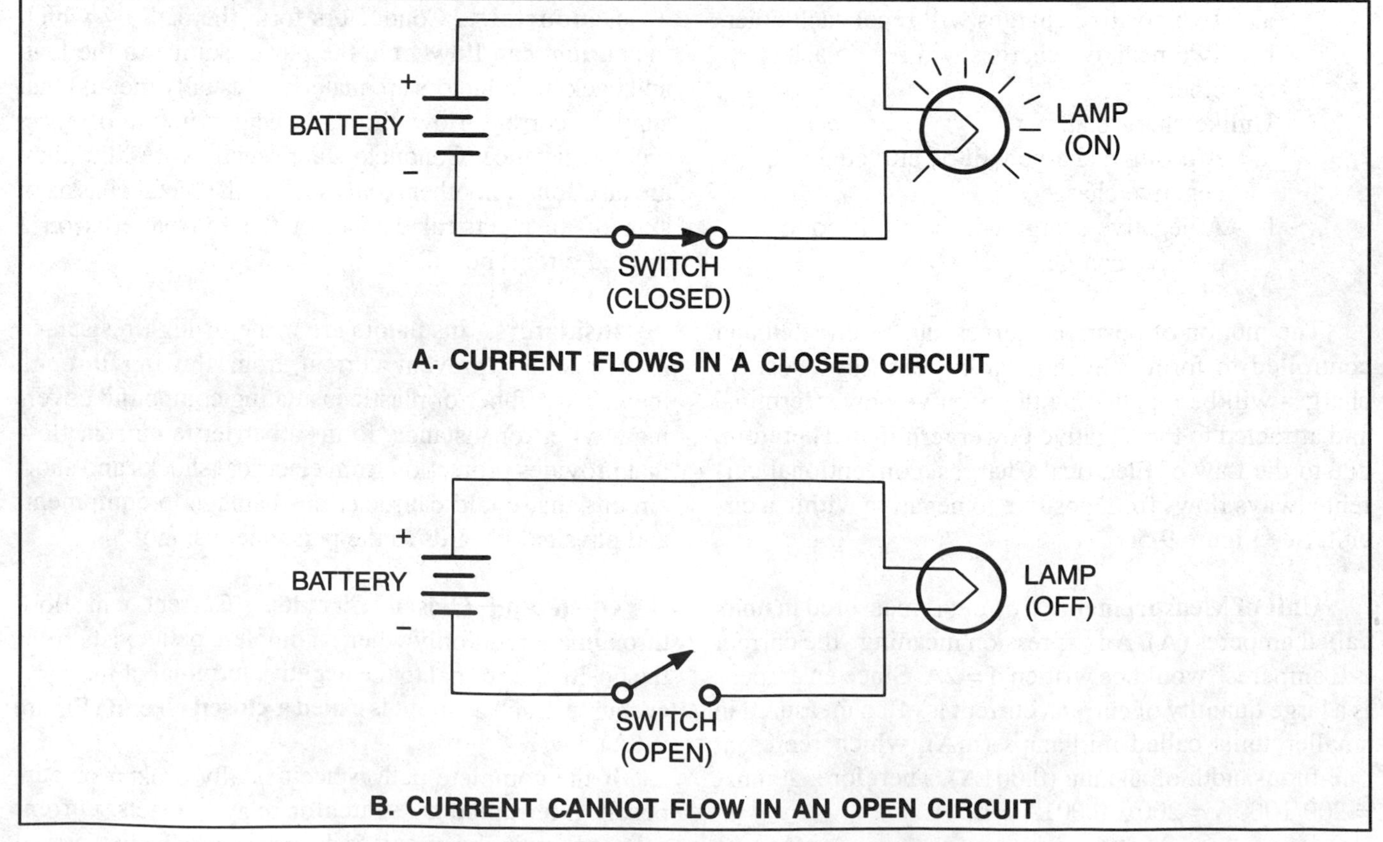
+
BATTERY
–
LAMP (ON)
SWITCH (CLOSED)
A. CURRENT FLOWS IN A CLOSED CIRCUIT
+
BATTERY
–
LAMP (OFF)
SWITCH (OPEN)
B. CURRENT CANNOT FLOW IN AN OPEN CIRCUIT

Figure 9-6. Closed and open circuits.

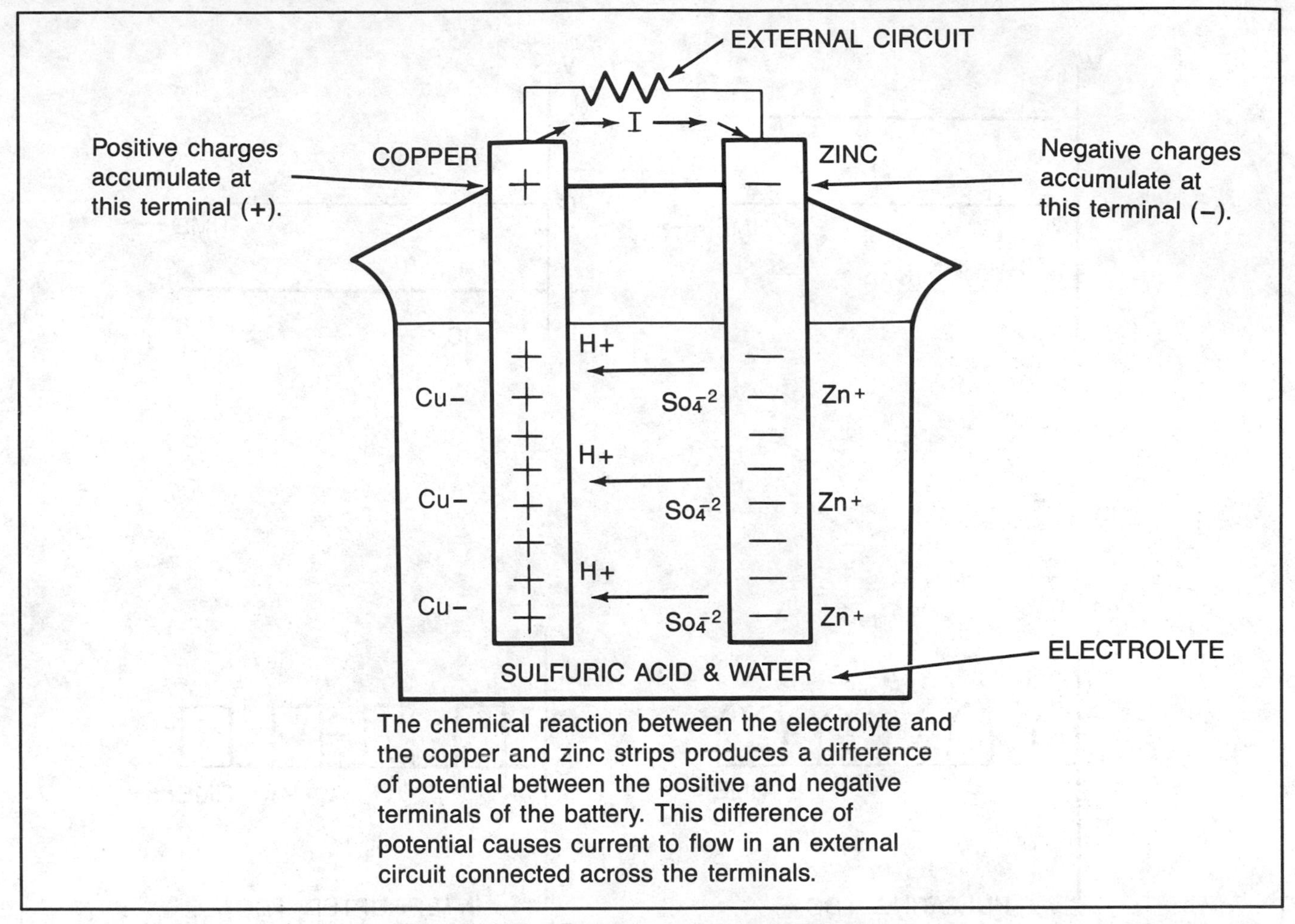
EXTERNAL CIRCUIT
I
Positive charges accumulate at this terminal (+).
COPPER
ZINC
Negative charges accumulate at this terminal (–).
H+
Cu–
So4-2
Zn+
H+
Cu–
So4-2
Zn+
H+
Cu–
So4-2
Zn+
ELECTROLYTE
SULFURIC ACID & WATER
The chemical reaction between the electrolyte and the copper and zinc strips produces a difference of potential between the positive and negative terminals of the battery. This difference of potential causes current to flow in an external circuit connected across the terminals.

Figure 9-7. Battery as a power source.

by the load. A circuit in this state is called an **open circuit** (Figure 9-6B).

Circuit Controls. Circuit controls are devices such as switches, relay contacts, and timers, which can be opened or closed, either automatically or manually. These devices determine whether or not current can flow and which path or paths the current will follow when it does so. There are other devices, such as rheostats, that are included in a circuit to limit the amount of current that can flow. Some of these devices are examined in more detail later in this chapter.

Voltage

Power sources are devices which convert various other forms of energy into electrical energy. For example, a battery changes chemical energy into electricity and a photocell converts light energy into electricity. During these conversions, negative charges are forced to collect at the negative terminal, while the positive charges are moved toward the positive terminal. This movement of charges creates a **difference of potential** between the two terminals, with one having an excess of negative charges (– terminal) and the other having an excess of positive charges (+ terminal). See Figure 9-7.

If a path is connected between the positive and negative terminals of the power source, the electrons will try to reach an equilibrium or balancing of charges. This will cause the movement of positive charges (current flow) from the positive terminal, through the external path, to the negative terminal.

Electromotive Force. The force which causes current to flow is called Electromotive Force (EMF), and the amount of EMF produced by the power source is called **voltage**. The amount of voltage directly depends on the size of the difference of potential which exists between the two terminals of the power source.

Unit of Measurement. Electromotive force is measured in units called volts (V) and is referred to in formulas by the capital letter E. Therefore, the expression, E = 12V, means "the voltage equals twelve volts." Large quantities of voltage are often described in units called

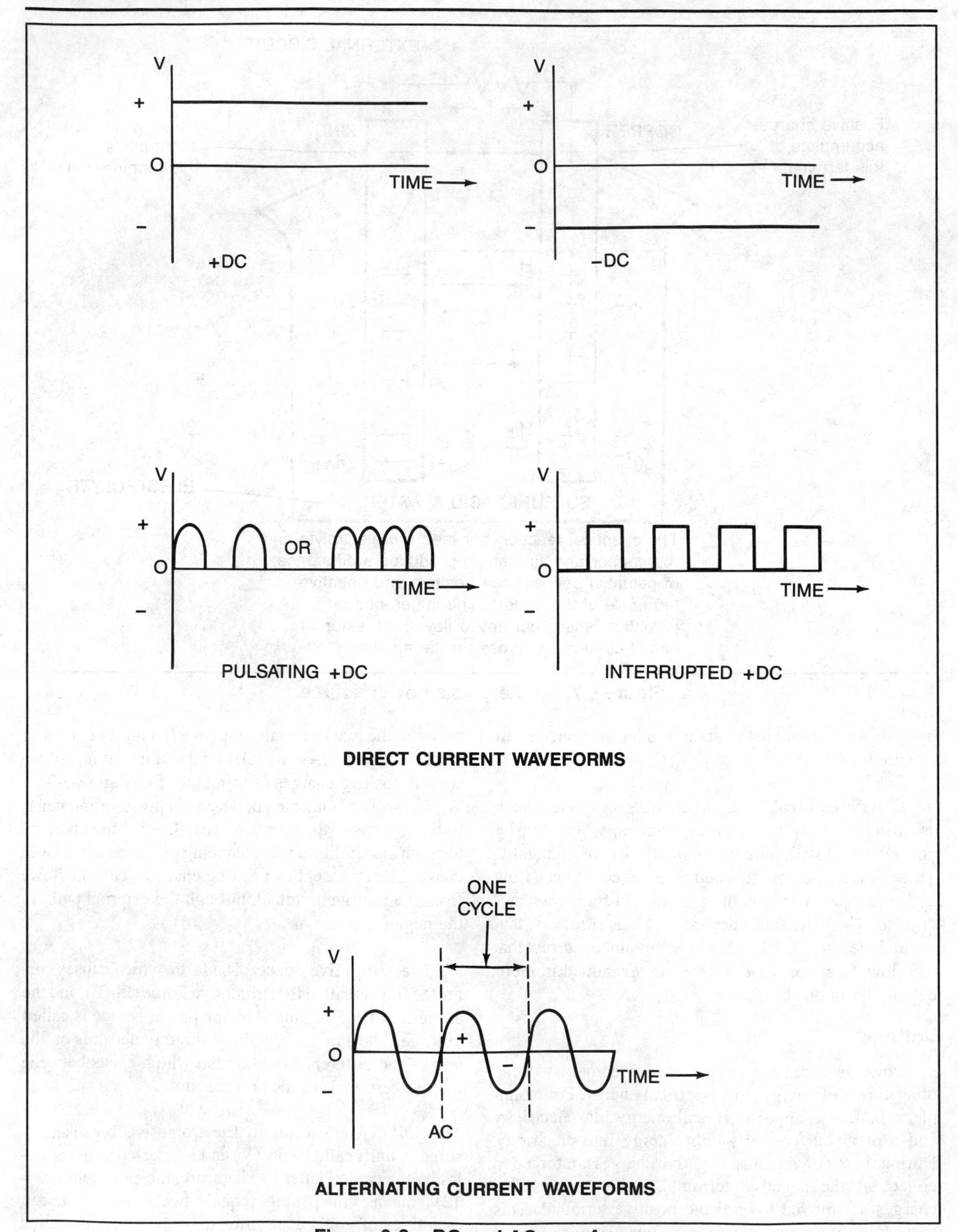

Figure 9-8. DC and AC waveforms.

kilovolts (kV). A kilo represents the number 1000. Therefore, one kilovolt equals 1000 volts, or 1kV = 1000V.

DC and AC Power Sources. Two types of current can be produced by EMF sources: direct current (DC) or alternating current (AC). Most sources produce DC which is a steady level of current, flowing in only one direction through the circuit. An EMF-producing magnetoelectric device called a generator produces AC, which continuously changes from a positive to a negative peak value, and alternately flows into and out of a circuit. See Figure 9-8.

Most of the stationary equipment found in hydraulic applications uses the relatively stable and economical AC voltage supplied by the local power company as its power source. The AC voltage is converted to DC with a rectifier circuit built into the equipment power supply. The easy to use DC voltage is then distributed by internal wiring to power the various electrical circuits within the equipment. These circuits may process signals that are either AC or DC.

Resistance

Resistance is the opposition to current flow offered by components of the circuit. All electrical load components have some resistance value. A light bulb works only because of the resistance of its filament. The resistance produces the heat which causes the filament to glow.

A fuse provides another example of how resistance can be useful in a circuit. A fuse has a fragile length of resistance wire designed to melt when its current rating is exceeded. The melted fuse opens the circuit and prevents current from flowing until the source of the overcurrent condition is found and corrected. **The fuse is then replaced with a new one that has the same current and voltage rating.**

Unit of Measurement. Resistance is measured in units called ohms, which can be expressed in terms of current and voltage. One ohm is equal to the amount of resistance which allows one amp of current to flow when an EMF of one volt is applied. The Greek letter omega (Ω) is used to represent the ohm, and the capital letter R

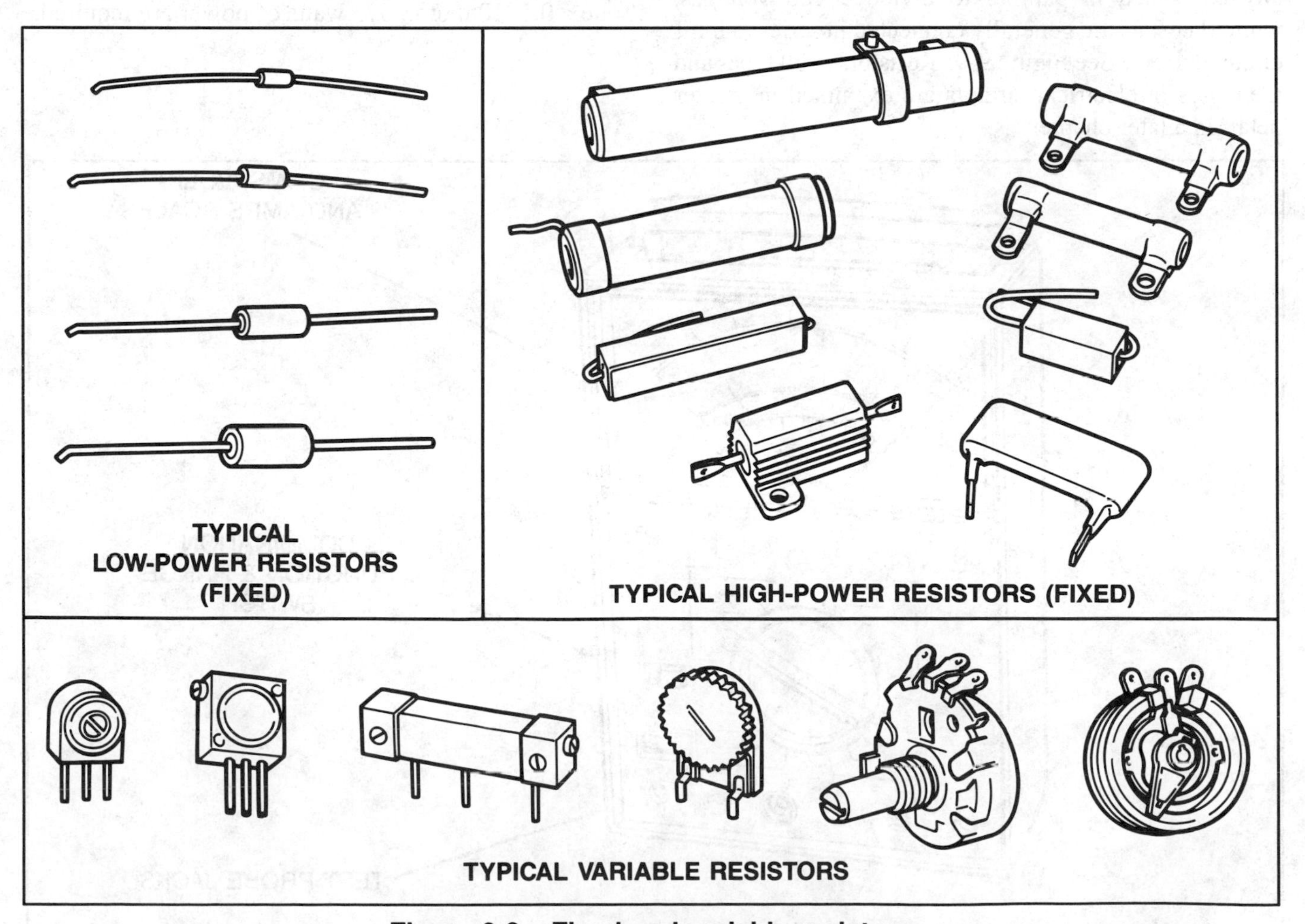

Figure 9-9. Fixed and variable resistors.

represents resistance in circuit formulas. Therefore, "the resistance equals 100 ohms" can be written R = 100Ω.

Resistance values are often quite large. For this reason, the kilohm (kΩ) and the megohm (MΩ) are practical units of measurement for resistance. As mentioned previously, k represents 1000. Therefore, 100,000 ohms can also be expressed as 100kΩ. Mega (M) represents 1,000,000. So, 7,000,000 ohms can also be written as 7MΩ.

Resistors. The resistor is a device designed to present a specific amount of opposition to the current in a circuit. Fixed resistors are manufactured in many sizes and types, but the opposition to current flow which they present to the circuit remains constant. There are three popular types of fixed resistors: carbon composition, wire wound, and deposited film.

Not all resistors have a fixed value; some are variable. A resistor whose value can be changed by rotating a shaft or moving a sliding tap is called a potentiometer if it uses all three available terminals, and a rheostat if it only uses two of the three terminals. Potentiometers are usually connected in parallel to divide circuit voltages, while rheostats are generally connected in series to limit circuit current. See Figure 9-9. Resistors of all types and their uses in electrical circuits are examined in greater detail in a later chapter.

Power

Power is the amount of work that can be done in some standard unit of time (usually one second). Electrical power is consumed by forcing the current through the resistance of the circuit. The opposition to current flow offered by resistive devices in the circuit produces heat that is dissipated or lost into the surrounding air. This lost energy is called a power loss. The fees charged by the local electrical power company are based on the amount of power consumed over a given amount of time and power losses represent a significant portion of these fees.

Unit of Measurement. Electrical power is expressed in units called watts (W). One watt represents the power used when one ampere of current flows due to an EMF of one volt. It is quite common to see the term kilowatts (kW) used when referring to electrical power. Remember, kilo represents 1000, so 1kW is equal to 1000 watts, or 2.5kW = 2,500 watts. In formulas, power is represented by the capital letter P. So, P = 2W is the same as saying "the power equals 2 watts." As mentioned earlier in this manual, 746 watts = 1 horsepower (HP). So, to operate a 0.5 HP motor, 373 watts of power are required.

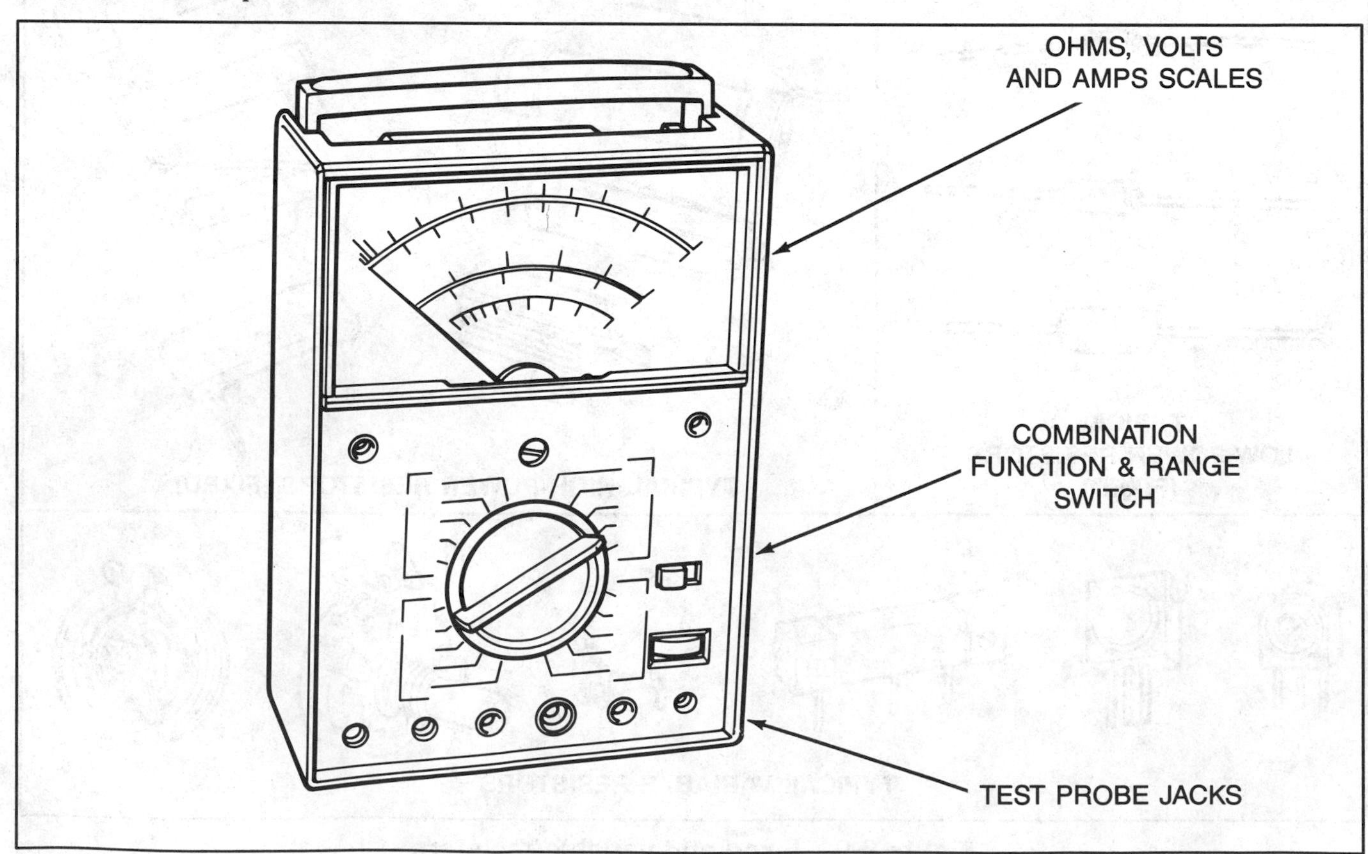

Figure 9-10. Analog meter.

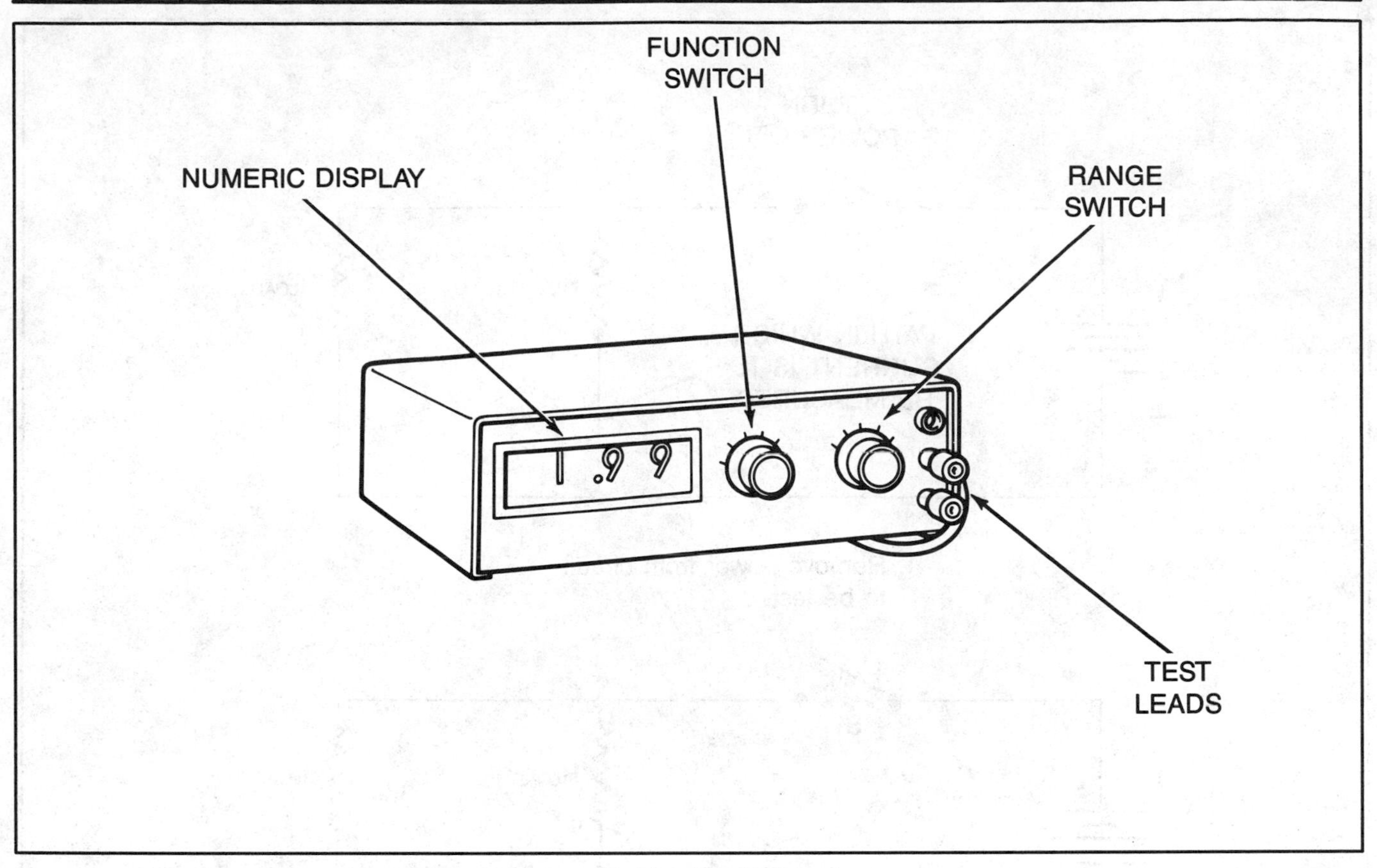

Figure 9-11. Digital meter.

Wattage Rating. The fact that components heat up due to the current passing through them makes it necessary for the circuit designer to consider heat tolerances along with component values for the devices selected for the circuit. **This heat tolerance must also be considered when devices are replaced in a circuit. The replacement part must have the same or better tolerance for heat as the original part.** The heat tolerance is known as the wattage rating of the device and determines how much heat a component can withstand. Generally, the larger the physical size of a component, the higher its wattage rating will be.

MEASURING ELECTRICAL QUANTITIES

When working with electrical circuits, it is often necessary to know exactly how much current is flowing in the circuit, how much voltage is present, or how much resistance a device is offering to the current flow. When adjusting, calibrating, or troubleshooting a circuit, measurements of electrical quantities are made with a device called a meter.

Types of Meters

Some meters are designed to measure only one of the circuit elements. A voltmeter, for example, can only measure the voltage present between two points in the circuit. More commonly, however, a device called a multimeter, or VOM (volt-ohm-milliammeter), is used to make the measurements required. This device is a combination voltmeter, ohmmeter, and ammeter; different sections of the meter are used depending on the measurement needed. A function selector switch is rotated or buttons are pushed to select the type of measurement. A range selector switch is used to adjust the meter for reading different amounts of volts, ohms, or amps.

There are also two kinds of multimeters: analog and digital. The analog meter has a pointer that moves across calibrated scales (Figure 9-10). The measurement is obtained by reading the appropriate scale. It is important to look directly at the meter face to avoid incorrect readings. A digital meter indicates the measurement on a numerical display (Figure 9-11).

Meters are connected to the circuit in different ways depending on the type of measurement being made. There are also several safety precautions that must be observed when using the meter. These safety factors are necessary to protect both the meter and the person using it.

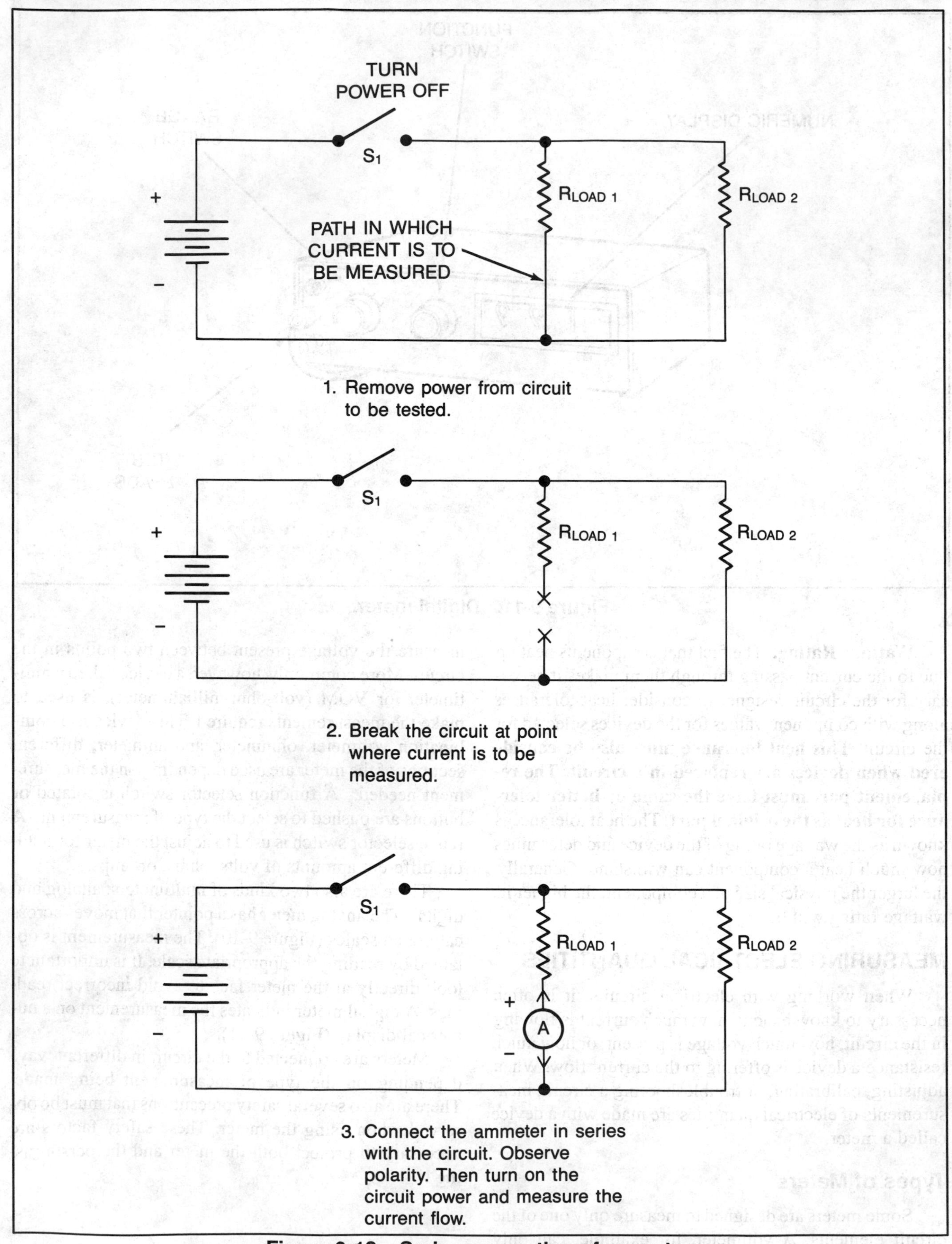
TURN
POWER OFF
S1
+
−
RLOAD 1
RLOAD 2
PATH IN WHICH
CURRENT IS TO
BE MEASURED
1. Remove power from circuit
to be tested.
S1
+
−
RLOAD 1
RLOAD 2
2. Break the circuit at point
where current is to be
measured.
S1
+
−
RLOAD 1
RLOAD 2
+
A
−
3. Connect the ammeter in series
with the circuit. Observe
polarity. Then turn on the
circuit power and measure the
current flow.

Figure 9-12. Series connection of ammeter.

Measuring Current with an Ammeter

Current, or the amount of electrical charge flowing in a circuit, is measured with a device called an ammeter (or milliammeter).

Safety Precautions. Certain precautions must be taken to ensure accurate ammeter readings, to prevent shocks to the technician, and to protect the delicate meter circuits. Never exceed the capabilities of the meter being used. All meters have a current rating that indicates the maximum amount of current that can safely be measured. Some meters have a range selector switch which allows the user to adjust the meter for reading different amounts of current. Set this switch to a range that includes the anticipated current to be measured. The highest range shown on the range switch determines the maximum current that can be safely measured.

Prevent electrical shock by always disconnecting the circuit being tested from the electrical power source before measuring current. **Before disconnecting, be sure, however, that it is hydraulically safe to do so.**

Connecting the Ammeter. Break the circuit at the point where the current flow is to be measured by unsoldering a connection or snipping a wire. Then connect the two meter leads from one side of the break to the other, completing the path for current flow. This is called a **series** connection (Figure 9-12). When a device is connected in series it means that there is only one path for the current flow. Because the ammeter is inserted directly into the circuit, it is designed to offer very little resistance to the current flowing through it. This is done so that normal circuit operation is not disturbed by the connection of the meter to the circuit.

Observing Polarity. The proper placement of the positive and negative meter leads is referred to as **observing polarity**. When measuring direct current (DC), the meter must be inserted in such a way that the current being measured flows into the positive meter lead, through the meter, and out the negative meter lead. If proper polarity is not observed when using an analog meter, the pointer attempts to move backwards on the scale, causing damage to the meter.

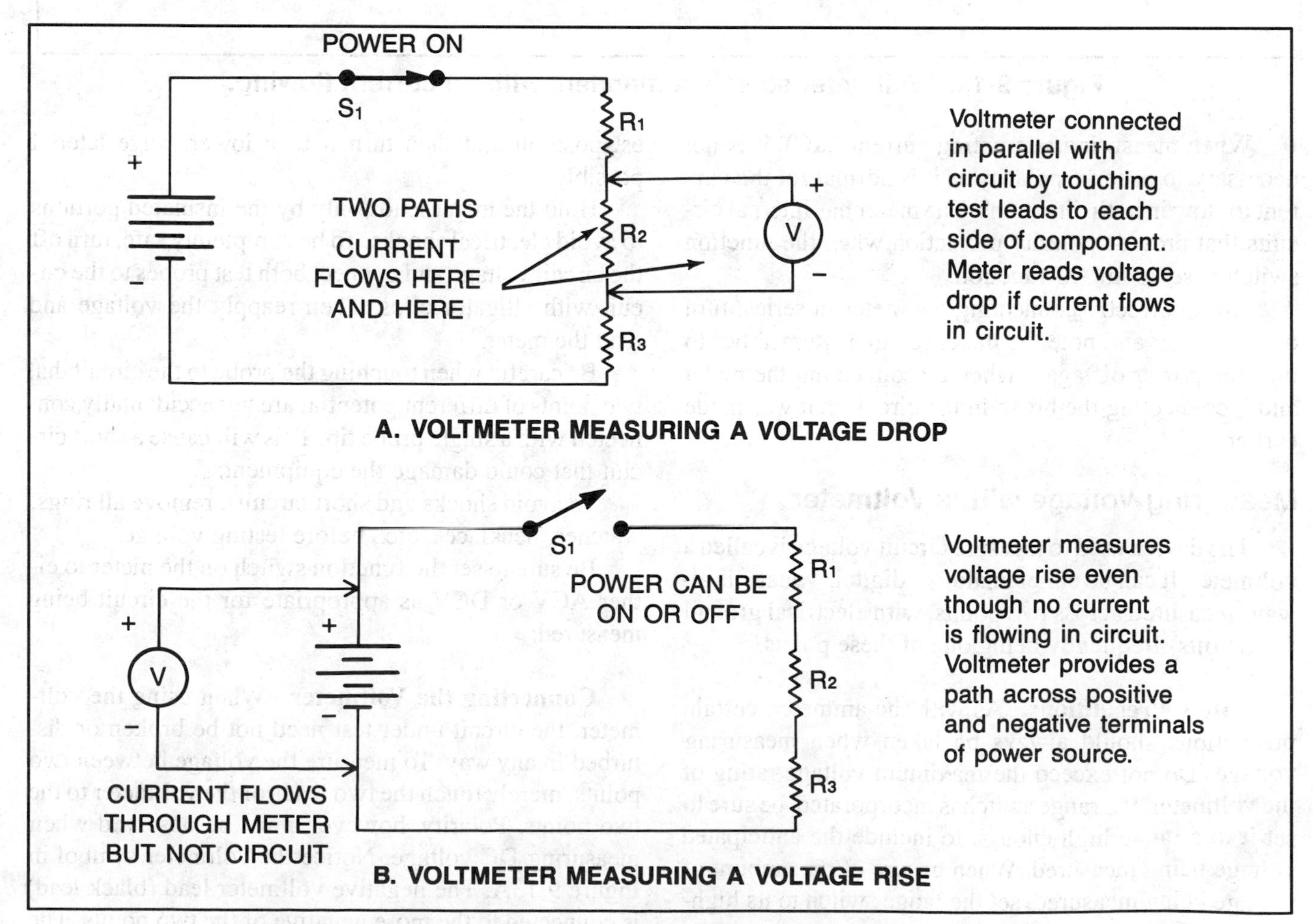

Figure 9-13. Voltmeter measuring voltage drop and voltage rise.

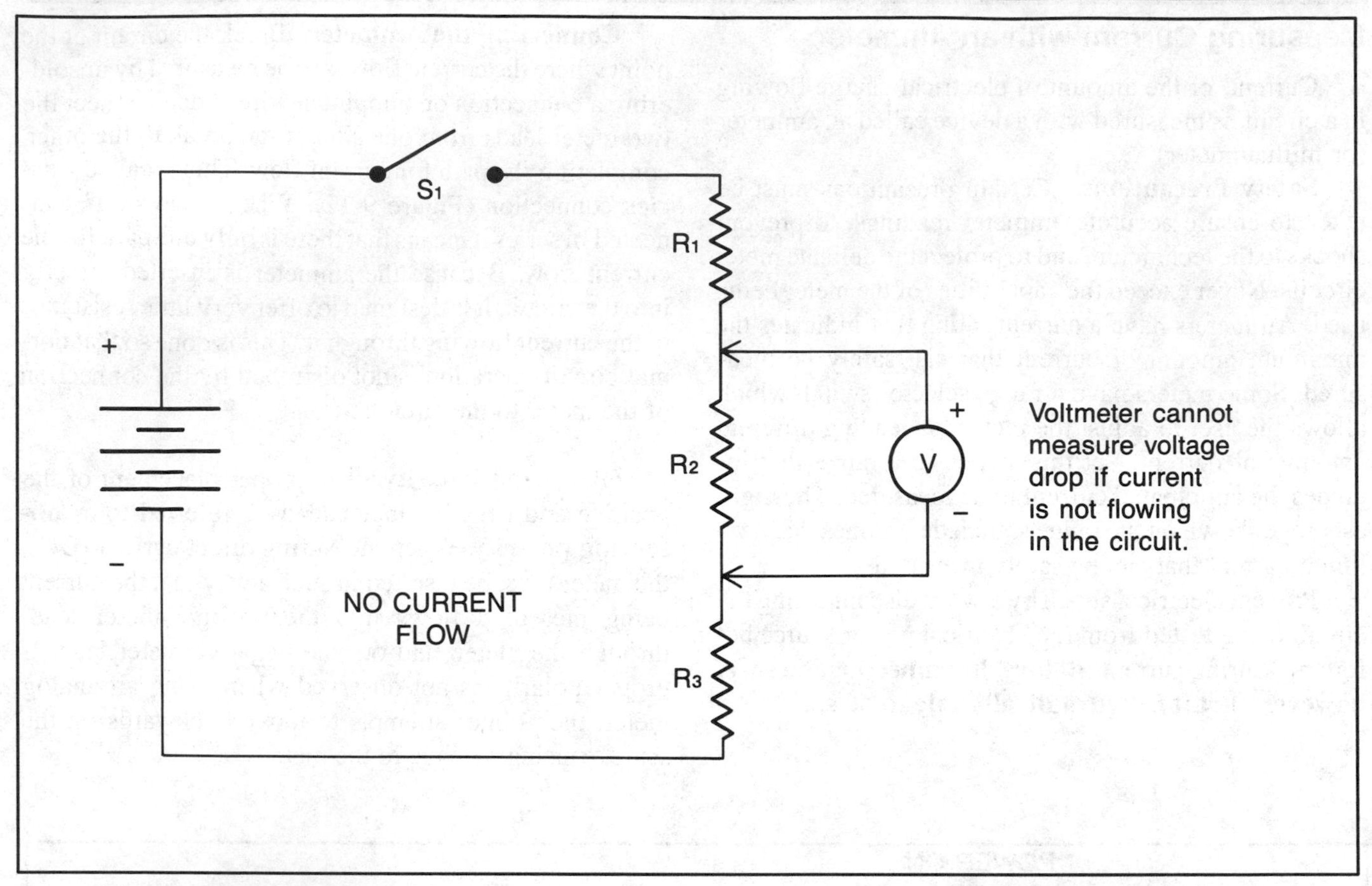

Figure 9-14. Voltmeter across component with no current flowing.

When measuring alternating current (AC), it is not necessary to observe polarity as it is normal for the current to flow in both directions. The meter has internal circuits that provide polarity protection when the function switch is set to an AC function.

After correctly connecting the meter in series, turn on the power and note the meter reading. Remember to turn the power off again when disconnecting the meter and reconnecting the break in the circuit that was made earlier.

Measuring Voltage with a Voltmeter

The device used to measure circuit voltage is called a voltmeter. It can be either analog or digital. Voltage is always measured across two points, with electrical ground (zero volts) frequently being one of these points.

Safety Precautions. As with the ammeter, certain precautions should always be taken when measuring voltage. Do not exceed the maximum voltage rating of the voltmeter. If a range switch is incorporated, be sure to set it to a range high enough to include the anticipated voltage being measured. When unsure of the amount of voltage being measured, set the range switch to its highest position and then turn it to a lower range later if possible.

Hold the meter leads only by the insulated portions to avoid electrical shocks. To be completely safe, turn off the circuit voltage and connect both test probes to the circuit with alligator clips. Then reapply the voltage and read the meter.

Be careful when touching the probe to the circuit that two points of different potential are not accidentally connected with a single probe tip. This will cause a short circuit that could damage the equipment.

To avoid shocks and short circuits, remove all rings, watches, necklaces, etc., before testing voltage.

Be sure to set the function switch on the meter to either ACV or DCV as appropriate for the circuit being measured.

Connecting the Voltmeter. When using the voltmeter, the circuit under test need not be broken or disturbed in any way. To measure the voltage between two points, merely touch the two leads of the voltmeter to the two points. Polarity, however, must be observed when measuring DC voltage. Notice the voltmeter symbol in Figure 9-13A. The negative voltmeter lead (black lead) is connected to the more negative of the two points. The

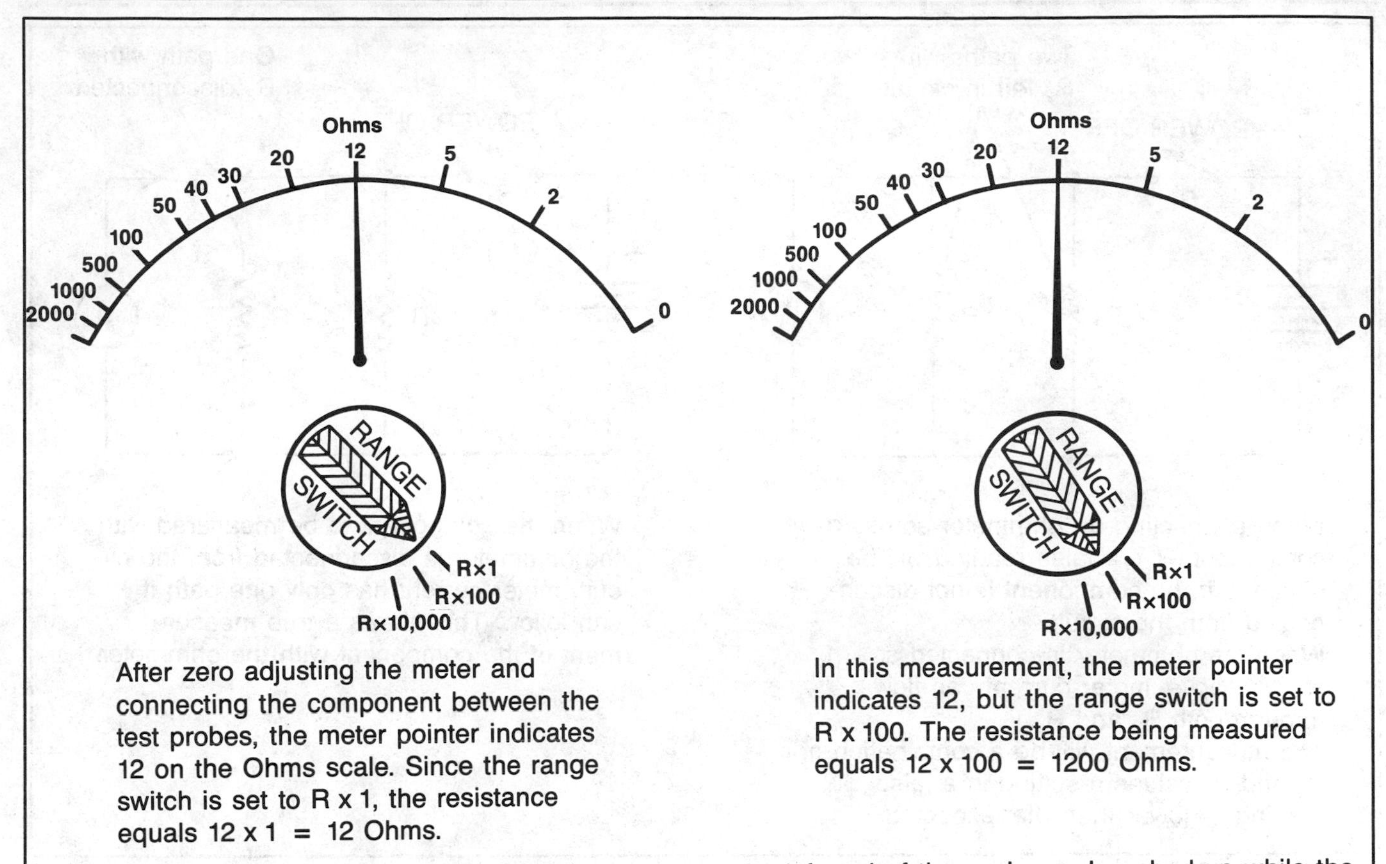

Ohms
12
20
30
40
50
100
500
1000
2000
5
2
0
RANGE SWITCH
R×1
R×100
R×10,000
After zero adjusting the meter and connecting the component between the test probes, the meter pointer indicates 12 on the Ohms scale. Since the range switch is set to R x 1, the resistance equals 12 x 1 = 12 Ohms.
In this measurement, the meter pointer indicates 12, but the range switch is set to R x 100. The resistance being measured equals 12 x 100 = 1200 Ohms.
Note that the Ohms scale is nonlinear. The values on the left end of the scale are bunched up while the values on the right are spread apart. If possible, select a range switch setting that positions the pointer in the middle one-third of the Ohms scale.

Figure 9-15. Ohmmeter measurements.

meter is connected in **parallel** with the points being measured. This means that there are two paths for the current flow, one path through the circuit and another path through the voltmeter. The voltmeter is designed to have a high internal resistance so that very little of the circuit current flows into the extra path created by connecting the meter.

Voltage Rises and Voltage Drops. Both voltage rises and voltage drops can be measured with the voltmeter. A voltage rise indicates a point where electromotive force (EMF) is added to the circuit, such as would be measured across the two power source terminals (Figure 9-13B). A reading will be obtained here even if current is not flowing in the circuit attached to the power source. This happens because a complete (closed) circuit is provided by the voltmeter itself.

A voltage drop is a point in the circuit where electrical energy is being used by the circuit, such as across the load. A voltage reading here is dependent upon current flowing through the load. When no circuit current is flowing, no voltage can be measured and the meter indicates a zero reading (Figure 9-14).

In any circuit in which current is flowing, **the sum of the voltage drops must equal the sum of the voltage rises.**

Measuring Resistance with an Ohmmeter

The device used to measure resistance is the ohmmeter. The ohmmeter has its own internal power source (usually a battery) which is used to force current through the resistance being measured and, as a result, circuit power is not required to operate the meter circuits.

Analog ohmmeters (Figure 9-15) have a nonlinear scale with zero at one end and infinity (∞) at the other. A range switch is provided so that a wide range of values can be accurately measured. The switch is marked with settings such as R x 1 and R x 100. To obtain a resistance measurement, the component is connected between the two test probes. A value is read off the resistance scale and then multiplied by the range switch setting. For ex-

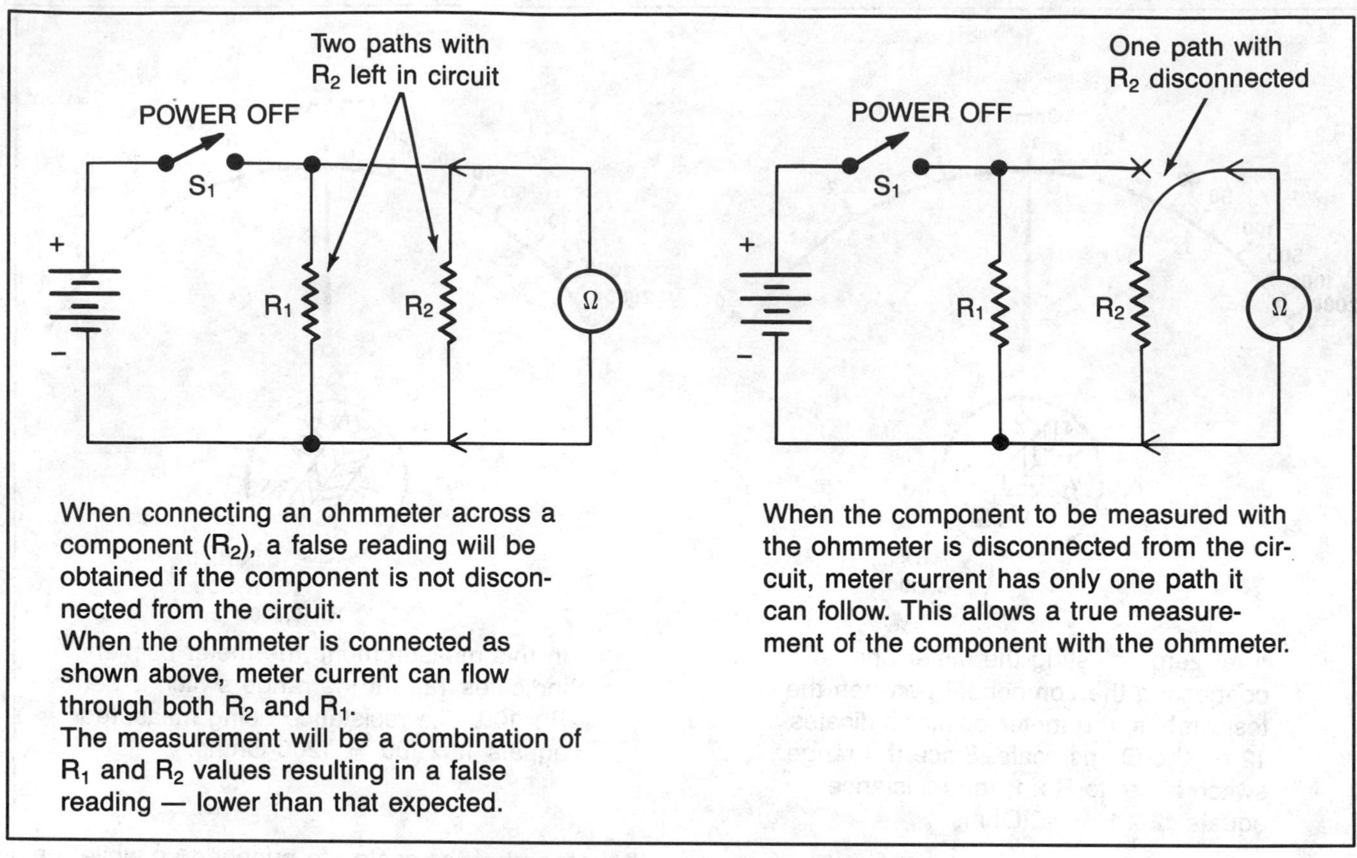

Two paths with
R_2 left in circuit
POWER OFF
S_1
+
-
R_1
R_2
Ω
When connecting an ohmmeter across a component (R_2), a false reading will be obtained if the component is not disconnected from the circuit.
When the ohmmeter is connected as shown above, meter current can flow through both R_2 and R_1.
The measurement will be a combination of R_1 and R_2 values resulting in a false reading — lower than that expected.
One path with
R_2 disconnected
POWER OFF
S_1
+
-
R_1
R_2
Ω
When the component to be measured with the ohmmeter is disconnected from the circuit, meter current has only one path it can follow. This allows a true measurement of the component with the ohmmeter.

Figure 9-16. False ohmmeter readings.

ample, if the reading on the scale was 5 and the range switch setting was R x 1k, the resistance value would be 5 x 1k or 5 kilohms (5000 ohms).

A zero ohms adjust control is included to compensate for the gradually diminishing internal battery voltage. This control must be readjusted each time the range switch is changed to a new position.

Safety Precautions. Since the ohmmeter has its own source of power, all circuit power must be turned OFF before resistance measurements are taken. Attempting to read resistance with circuit power on may damage the circuit, meter, or both! With the circuit voltage removed, there is very little danger involved in measuring resistance. There are, however, several factors that should be considered to ensure accuracy of measurement.

Ohmmeter Accuracy. Always adjust the zero ohms control on the ohmmeter before taking the reading. Touch the two meter probe tips together and then adjust the zero ohms control until the pointer indicates zero ohms. The procedure is the same for a digital meter except that the control is adjusted until the numerical display shows all zeroes. If the meter cannot be adjusted to zero, the internal battery is too weak to operate the ohmmeter and should be replaced with a fresh battery.

If the component to be measured is installed in the circuit, one end of it must be disconnected before making the resistance measurement. The reason for this can be seen in Figure 9-16. Notice that when one end is not disconnected there are two paths for the meter battery current to flow: one path through the component and another path back through the rest of the circuit. The ohmmeter will measure the combination resistance of both paths and the result will be a value different than the one intended. If, however, the component is disconnected at one end, there is only one path for the current to follow-through the component. This method produces the accurate measurement desired.

Touching the probe tips with the fingers while taking a resistance measurement also creates an additional path for current flow—through the body. While this is not dangerous, it does cause false measurements. Hold the probes by the insulated part only.

While probe polarity does not matter when measuring a resistor, other devices, like diodes, must be measured in both directions to determine if the resistance is within tolerance. (Testing a diode with an ohmmeter is covered later in this chapter.)

Finally, always adjust the range switch until the pointer rests within the middle one-third of the ohms

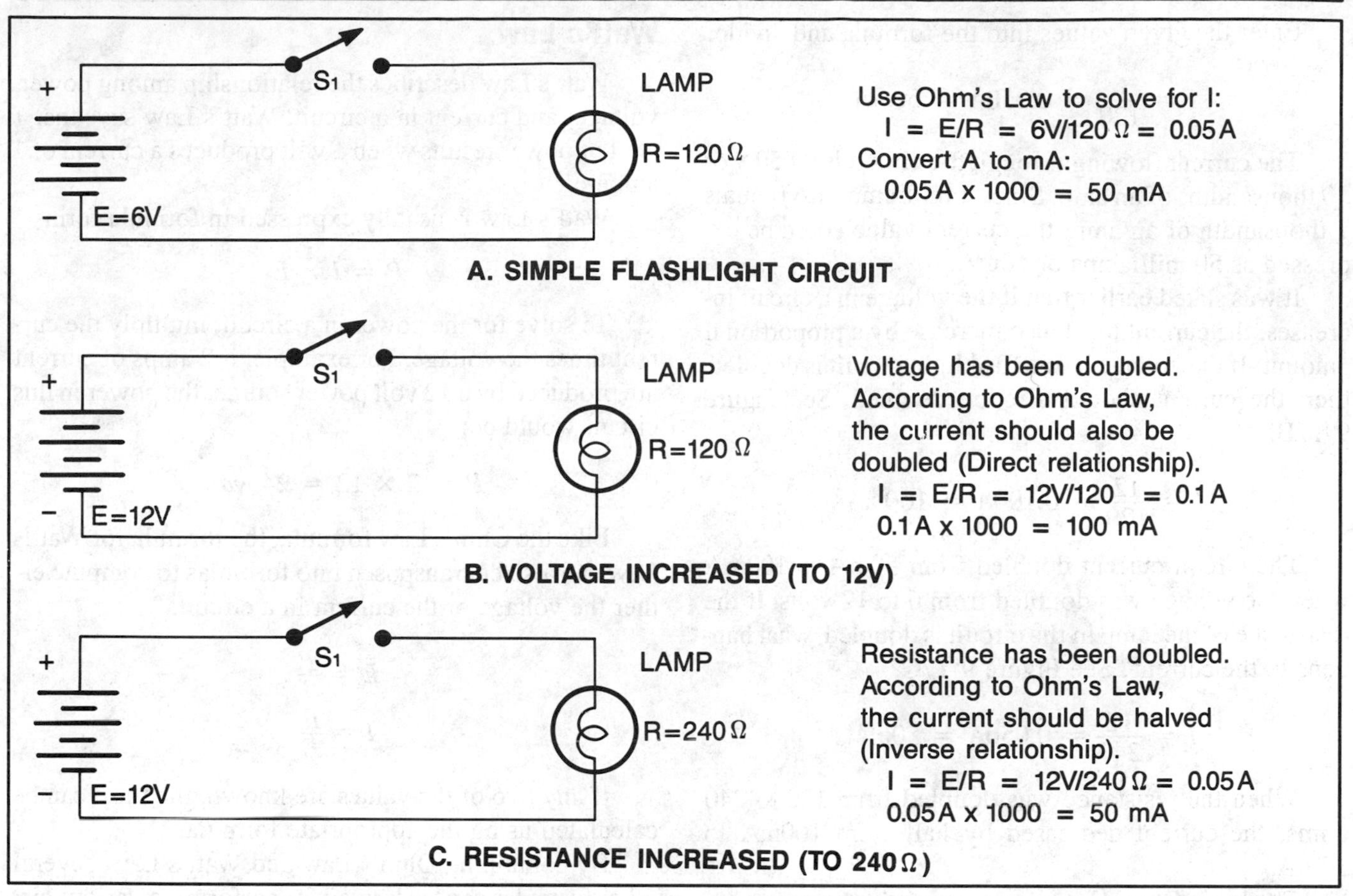
S1
LAMP
+
-
E=6V
R=120 Ω
Use Ohm's Law to solve for I:
I = E/R = 6V/120 Ω = 0.05 A
Convert A to mA:
0.05 A x 1000 = 50 mA
A. SIMPLE FLASHLIGHT CIRCUIT
S1
LAMP
+
-
E=12V
R=120 Ω
Voltage has been doubled.
According to Ohm's Law,
the current should also be
doubled (Direct relationship).
I = E/R = 12V/120 = 0.1 A
0.1 A x 1000 = 100 mA
B. VOLTAGE INCREASED (TO 12V)
S1
LAMP
+
-
E=12V
R=240 Ω
Resistance has been doubled.
According to Ohm's Law,
the current should be halved
(Inverse relationship).
I = E/R = 12V/240 Ω = 0.05 A
0.05 A x 1000 = 50 mA
C. RESISTANCE INCREASED (TO 240 Ω)

Figure 9-17. Ohm's law proof.

scale with the component connected between the two probe tips. This is the most accurate area of the nonlinear ohms scale and should be used if at all possible. Remember to check the zero ohms adjustment after changing range switch settings.

RELATIONSHIPS AMONG CIRCUIT ELEMENTS

As mentioned at the beginning of this chapter, relationships exist among current, voltage, resistance, and power in any electrical circuit. These relationships are expressed as a series of simple formulas known as Ohm's Law and Watt's Law.

Ohm's Law

Ohm's law states that the current in a circuit is directly proportional to the voltage and inversely proportional to the resistance. This means that:

- As the voltagc applied to a circuit increases, the current also increases by a proportional amount. Likewise, if the voltage applied to a circuit decreases, the current also decreases. (If the resistance is held constant.)
- As the resistance in a circuit increases, the current decreases by a proportional amount. Similarly, if the resistance in a circuit decreases, the current will increase. (If the voltage is held constant.)

The relationships should be obvious from the definitions of these electrical circuit elements. Voltage is the force that causes current to flow. If this force is increased, the amount of current flow must also increase. Resistance is the opposition to current flow. If the opposition is increased, the current must decrease.

Ohm's Law is most often expressed as a formula:

$$I = \frac{E}{R}$$

Solving for I. The current in a circuit is equal to the voltage divided by the resistance. For example, consider the simple flashlight circuit of Figure 9-17A. If the battery voltage equals 6 volts and the resistance of the lamp (the resistance of the wires is negligible in this case) equals 120 ohms, how much current will flow when the switch is closed?

Enter the given values into the formula and divide:

$$I = \frac{6}{120} = 0.050A = 50mA$$

The current flowing through the circuit is 0.050A or 50 thousandths of an amp. Since 1 milliamp (mA) equals 1 thousandth of an amp, the current value could be expressed as 50 milliamps or 50mA.

It was stated earlier that if the voltage in a circuit increases, the current must also increase by a proportional amount. If the voltage in the flashlight circuit is doubled, then the current should also be doubled. See Figure 9-17B.

$$I = \frac{12}{120} = 0.100A = 100mA$$

The circuit current doubled from 50mA to 100mA when the voltage was doubled from 6 to 12 volts. If the resistance of the lamp in the circuit is doubled, what happens to the current? See Figure 9-17C.

$$I = \frac{12}{240} = 0.050A = 50mA$$

When the resistance was doubled from 120 to 240 ohms, the current decreased by half from 100mA to 50mA.

The basic Ohm's Law formula can be transposed to also solve for the voltage or the resistance in a circuit. When any two circuit values are known, the third can be calculated using the appropriate formula.

$$E = I \times R$$

$$R = \frac{E}{I}$$

Solving for E. Suppose that the current through a device must be limited to one-half of an amp, and that its resistance is fixed at 60 ohms. How much voltage should be applied to the circuit to meet these requirements?

$$E = 0.5 \times 60 = 30 \textit{ volts}$$

Solving for R. Consider a circuit with a 50 volt power source. How much resistance must be in the circuit to limit the current to one-tenth of an amp?

$$R = \frac{50}{0.1} = 500 \textit{ ohms}$$

Watt's Law

Watt's Law describes the relationship among power, voltage, and current in a circuit. Watt's Law says that 1 watt of power results when 1 volt produces a current of 1 ampere.

Watt's Law is usually expressed in formula form:

$$P = I \times E$$

To solve for the power in a circuit, multiply the current times the voltage. For example, if 2 amps of current are produced by a 12 volt power source, the power in this circuit would be:

$$P = 2 \times 12 = 24 \textit{ watts}$$

Like the Ohm's Law formula, the formula for Watt's Law can also be transposed into formulas to compute either the voltage or the current in a circuit.

$$E = \frac{P}{I}$$

$$I = \frac{P}{E}$$

If any two of the values are known, the third can be calculated using the appropriate formula.

By combining Ohm's Law and Watt's Law, several other formulas can be derived for performing various circuit calculations. For example, two additional formulas for determining power are:

$$P = I^2R$$

$$P = \frac{E^2}{R}$$

The chart in Figure 9-18 is a summary of the information on the four circuit factors or elements and includes definitions, units of measurement, unit symbols, type of meter used to make measurements, and Ohm's and Watt's Law formulas. This information is also shown in Appendix D.

ELECTRICAL DEVICES

This section contains basic information on several electrical devices commonly found in hydraulic systems and is intended as an introduction only. A comprehensive examination of these devices is beyond the scope of this manual. The basic operation and use of the following devices is described:

- Attenuator
- Solenoid

CIRCUIT ELEMENT	DEFINITION	UNIT OF MEASUREMENT	UNIT SYMBOL	MEASURED WITH	FUNCTION SWITCH POSITION	OHM'S OR WATT'S LAW
CURRENT	the flow of electrons from + to – around a circuit	Amperes (amps)	A	Ammeter	ACmA or DCmA	$I=E/R$ $I=P/E$ $I=\sqrt{P/R}$
VOLTAGE (EMF)	the force which causes current to flow	Volts	V	Voltmeter	ACV or DCV	$E=I \times R$ $E=P/I$ $E=\sqrt{P/R}$
RESISTANCE	the opposition to current flow	Ohms	Ω	Ohmmeter	Ohms	$R=E/I$ $R=E^2/P$ $R=P/I^2$
POWER	the rate of doing electrical work	Watts	W	Wattmeter or Calculated	N/A	$P=I \times E$ $P=I^2R$ $P=E^2/R$

Figure 9-18. Electrical circuit element summary chart.

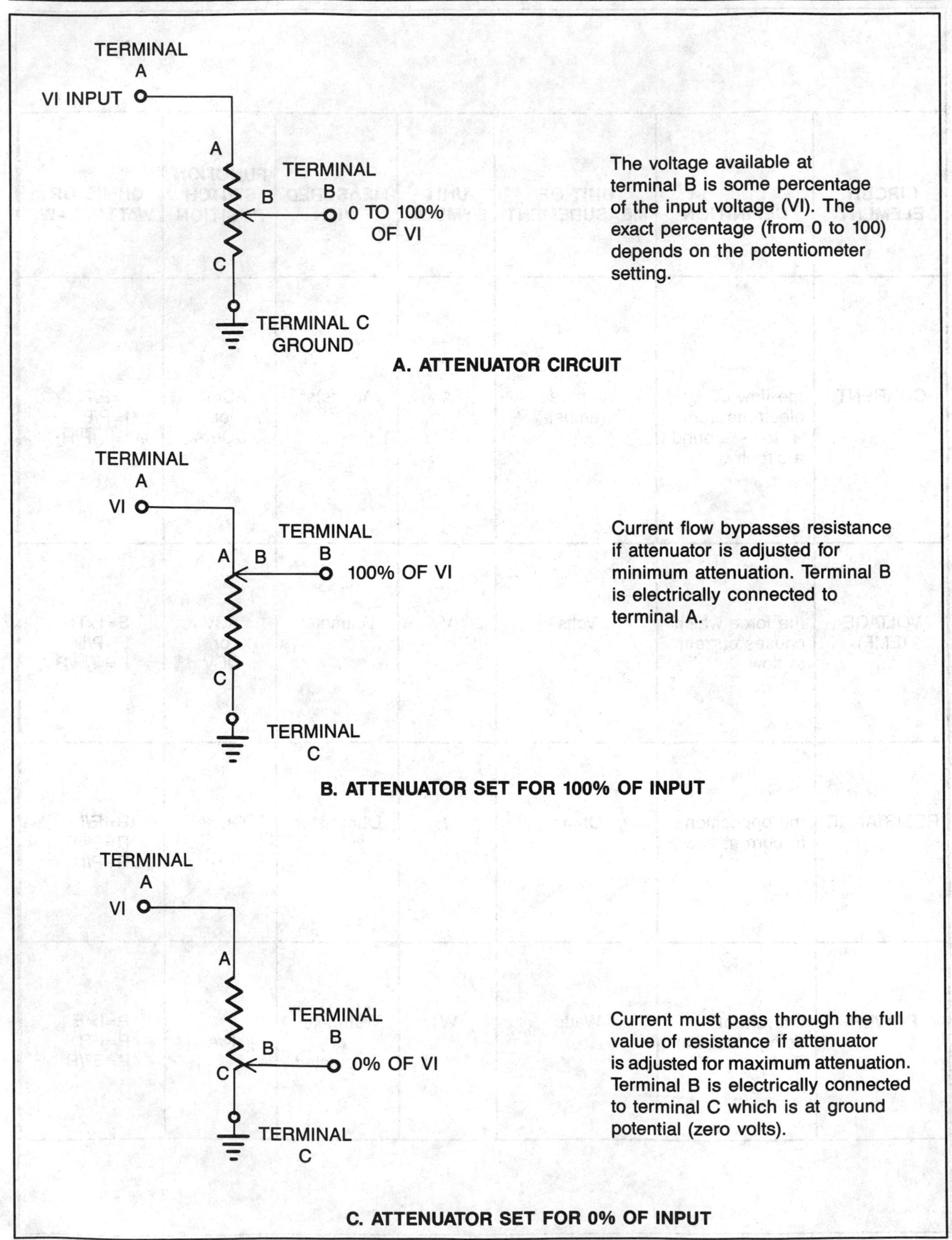
TERMINAL
A
VI INPUT
A
TERMINAL
B
B
0 TO 100%
OF VI
C
TERMINAL C
GROUND
The voltage available at terminal B is some percentage of the input voltage (VI). The exact percentage (from 0 to 100) depends on the potentiometer setting.
A. ATTENUATOR CIRCUIT
TERMINAL
A
VI
TERMINAL
B
A
B
100% OF VI
C
TERMINAL
C
Current flow bypasses resistance if attenuator is adjusted for minimum attenuation. Terminal B is electrically connected to terminal A.
B. ATTENUATOR SET FOR 100% OF INPUT
TERMINAL
A
VI
A
TERMINAL
B
B
C
0% OF VI
TERMINAL
C
Current must pass through the full value of resistance if attenuator is adjusted for maximum attenuation. Terminal B is electrically connected to terminal C which is at ground potential (zero volts).
C. ATTENUATOR SET FOR 0% OF INPUT

Figure 9-19. Attenuator operation.

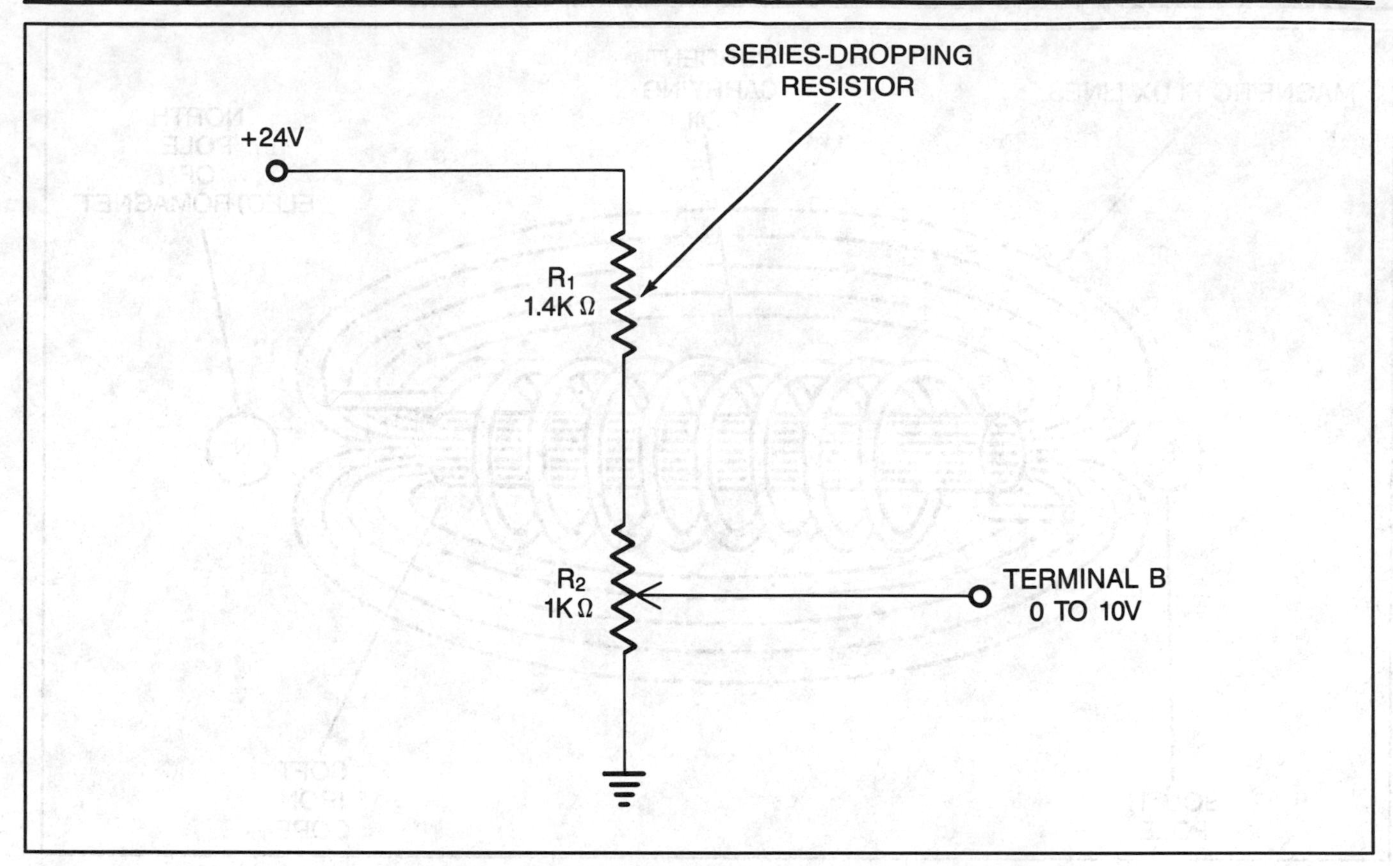

Figure 9-20. Attenuator with series dropping resistor.

- Relay
- Transformer
- Diode rectifier

These devices are available in many different physical sizes, wattage ratings, load capacities, etc., and may serve purposes other than those covered in this manual. The basic principles of operation for a device, however, do not change, regardless of the style.

Attenuator

The attenuator is a variable resistive device used to reduce the value of an electrical signal or voltage. It may also be called a gain control, fader, taper, or pot (potentiometer) depending on its design and application. The volume control on a radio or TV is an attenuator. Changing the setting of this control increases or decreases the resistance in the circuit, and raises or lowers the strength of the signal reaching the speaker to the desired sound level.

Attenuators provide a means of varying the voltage output of one circuit before it is applied as an input to another circuit. In this way, command signals to electrohydraulic valves, feedback signals from transducers, or circuit board supply voltages can be adjusted and fine tuned to allow optimal performance of the equipment.

Potentiometer Operation. Figure 9-19A illustrates a typical attenuator circuit and also shows the most common schematic symbol for a potentiometer. Notice the three connections or terminals on this device. The voltage or signal to be attenuated (VI) is shown at terminal A. Terminal B is the connection where the output from the circuit is reduced to a percentage of VI (between 0 and 100). The arrow symbol indicates that point B can be moved anywhere between point A and point C. Terminal C is connected to a ground (zero volts) point.

If the potentiometer is set so that point B is at point A (Figure 9-19B), the current flow does not pass through the resistance of the potentiometer and VI is not attenuated. One hundred percent of VI is present at terminal B.

If the potentiometer is set so that point B is at point C (Figure 9-19C), the current must pass through all of the resistance of the potentiometer and VI is attenuated to zero percent. Point B is electrically connected to ground (at Point C) and no voltage is present at terminal B.

Any portion of VI from 0 to 100 percent may be present at terminal B depending on how the potentiometer is adjusted.

For example, if point B is positioned at the midpoint between A and C, 50 percent of the voltage is available at terminal B.

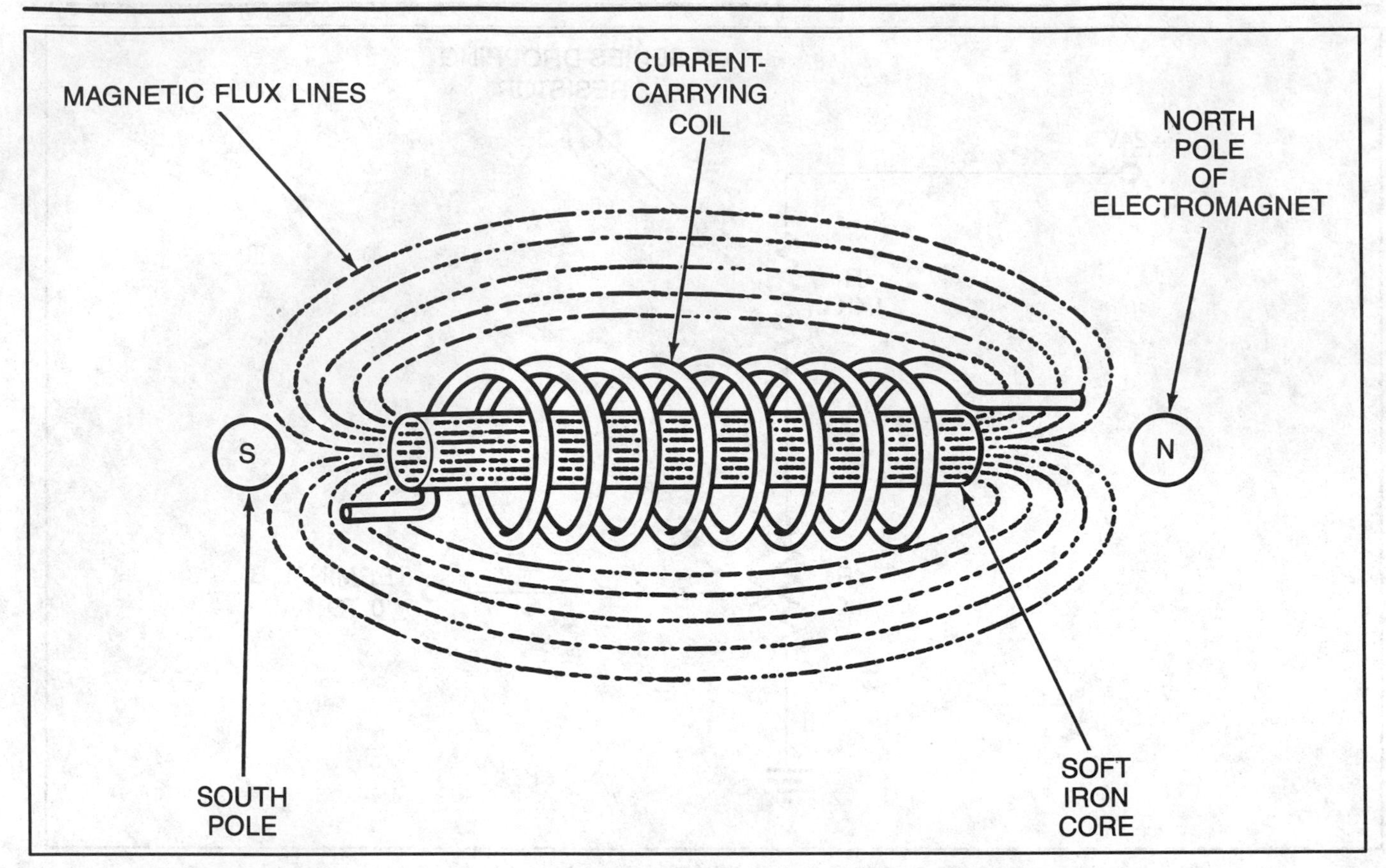

Figure 9-21. Magnetic field around solenoid coil.

Series Dropping Resistor. The circuit shown in Figure 9-20 illustrates how a fixed resistor can be connected in series with the potentiometer to drop the voltage down to a predetermined level before allowing the potentiometer to provide adjustment over the remaining voltage. In this case, the desired voltage range at terminal B is from 0 to 10 volts and this output voltage must never exceed 10 volts. The voltage input to the circuit is 24 volts. This means that R_1 must reduce the input voltage by 14 volts (24-10 = 14). The total resistance of R_1 and R_2 in series is 1.4 kilohms + 1.0 kilohms = 2.4 kilohms. The current flowing through both of these resistors can be found using Ohm's Law:

$$I = \frac{E}{R} = \frac{24}{2400} = 0.01A = 10mA$$

The voltage dropped by R_1 can then be calculated using another Ohm's Law formula:

$$E = I \times R = 0.01 \times 1400 = 14.0V$$

In this circuit, the fixed resistance value of R_1 was selected to drop the supply voltage by 14 volts from 24 to 10 volts. R_2 allows adjustment of the remaining voltage between zero and 10 volts.

Solenoid

A solenoid is an electromechanical device that converts electrical energy into linear mechanical motion. In hydraulics,solenoids are most often used to actuate directional valves and are typically on/off devices. Solenoids are available for both AC and DC operation and in all standard voltages. They can be divided into air gap and wet armature types. In order to understand the operation of solenoids it is necessary to briefly examine the principle of electromagnetism.

Electromagnetism. Electric current flowing through a conductor produces a circular magnetic field around the conductor. The direction of the magnetic field depends on the direction of current flow through the wire. Normally, this magnetic field is weak and does not affect the surrounding components. However, if the conductor is formed into a coil consisting of many loops of the conducting wire (Figure 9-21), the strength of the magnetic field will increase. The field, made up of many magnetic flux lines, forms around the outside and through the center of the coil.

As long as current is flowing through the wire, the coil acts as an electromagnet and will have north and south poles just like any other magnet. The strength of the magnetic action can be increased by adding more

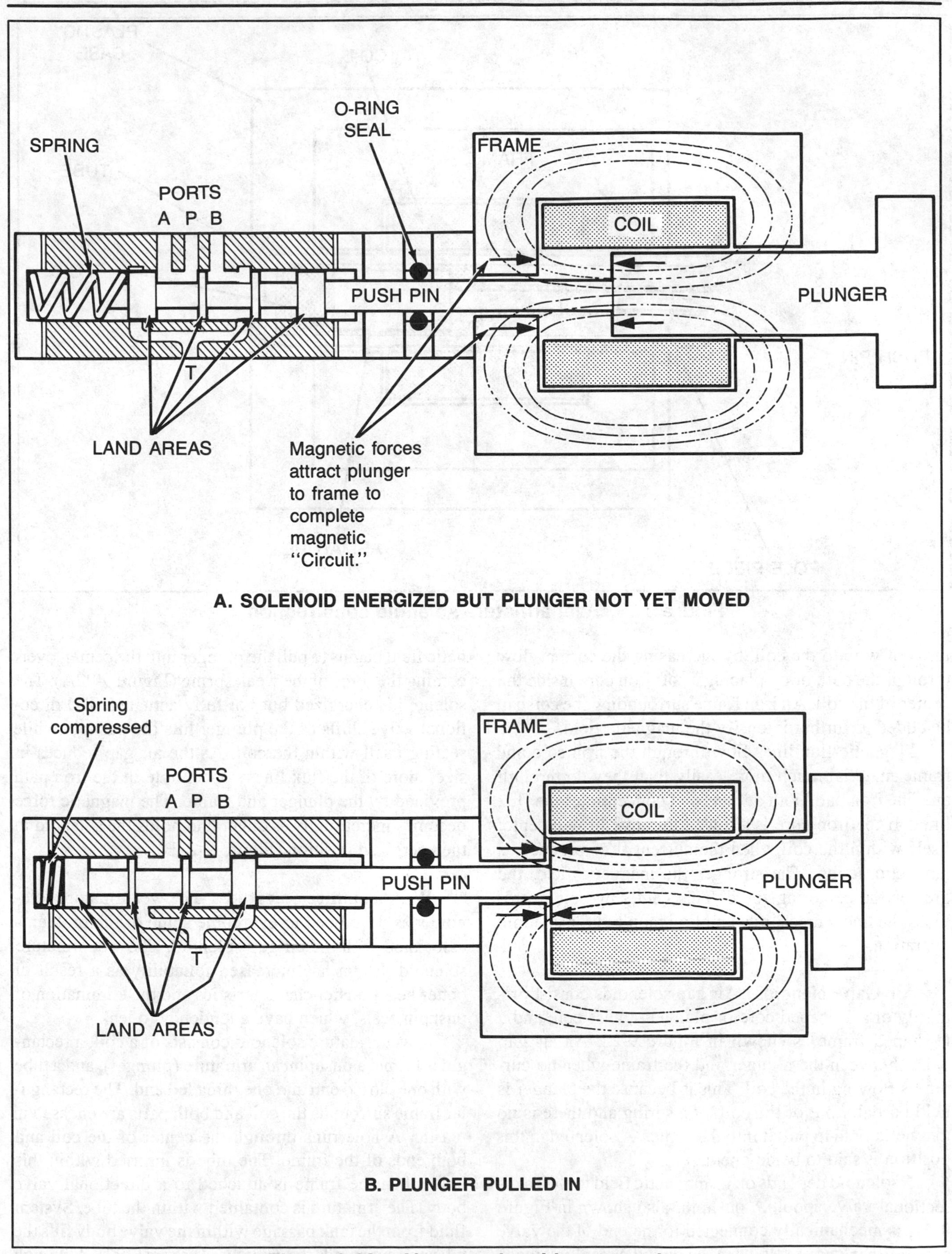

Figure 9-22. Air gap solenoid construction.

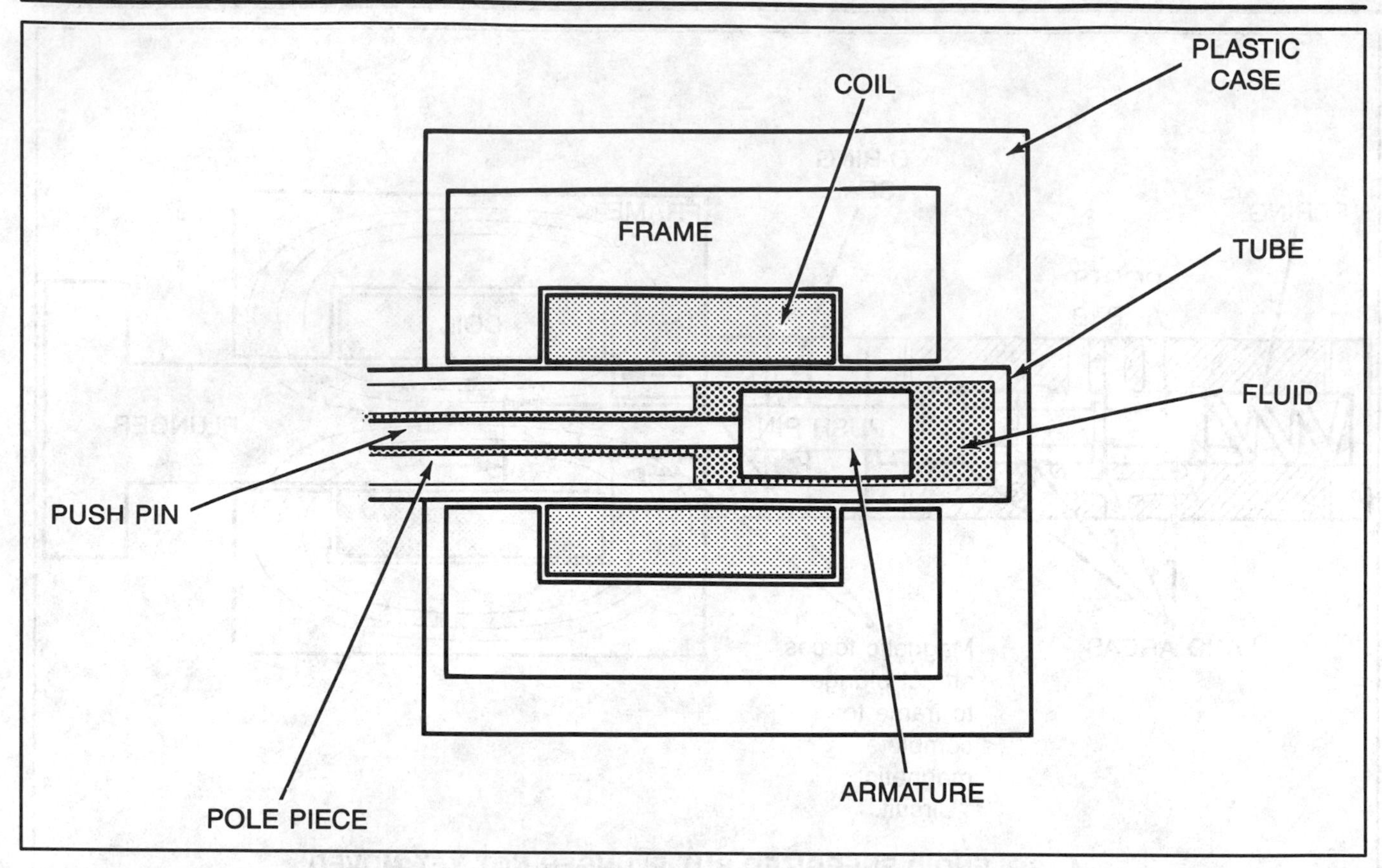

Figure 9-23. Wet armature solenoid construction.

turns of wire to the coil, by increasing the current flow through the coil, or by placing a soft iron core inside the center of the coil. An iron frame surrounding the coil can be added to further intensify the magnetic field.

Magnetic flux lines flow through the iron core and frame material much more easily than they do through air. The iron path both attracts and concentrates the flux lines. If the iron core is allowed to move, it will center itself within the coil when the current flow produces a magnetic action. The stronger the magnetic field, the greater the centering force. This centering action provides the linear mechanical motion required for solenoid operation.

Air Gap Solenoids. Air gap solenoids consist primarily of a T-shaped core or plunger, a wire coil, and a C-shaped frame as shown in Figure 9-22. An air gap exists between the plunger and the frame when no current is flowing in the coil. This is because the plunger is held partially out of the coil by a spring and there is no magnetic field to pull it into the center. A solenoid in this position is said to be de-energized.

A solenoid depends on its magnetic field to shift a directional valve spool. A pushpin, also shown in Figure 9-22, is mechanically connected to the end of the valve spool. When current flows in the coil, the resulting magnetic field begins to pull the plunger into the center, overcoming the force of the weak spring (Figure 9-22A). The solenoid is energized but not fully actuated. The directional valve shifts as the plunger hits the pushpin while seating itself within the coil. As the air gap reduces in size, more of the flux lines concentrate in the iron path provided by the plunger and frame. The magnetic force becomes increasingly stronger and holds the solenoid in the energized position (Figure 9-22B).

Wet Armature Solenoids. The wet armature solenoid was developed to overcome some of the problems associated with the air gap solenoid. The wet armature solenoid design has increased reliability as a result of better heat transfer characteristics and the elimination of pushpin seals, which have a tendency to leak.

A wet armature solenoid consists of a coil, a rectangular frame, a pushpin, an armature (plunger), and a tube with one closed end and one threaded end. The rectangular frame surrounds the coil and both parts are encased in plastic. A hole runs through the center of the coil and both ends of the frame. The tube is inserted within this hole when the frame is attached to a directional valve body. The armature is contained within the tube. System fluid from the tank passage within the valve body fills the tube and surrounds the armature (Figure 9-23). Although the wet armature solenoid differs in construction, it operates according to the same electromagnetic principles as the air gap solenoid.

DC Operation. DC operation of a solenoid is simpler than AC operation because the current rises to a constant level after coil energization and flows in only one direction. The magnetic field expands and then reaches a steady state condition as the plunger is pulled into the coil. This condition exists until the coil is de-energized. The only resistance to the flow of current is the fixed resistance of the conducting material used to make the coil.

AC Operation. When AC is used as the control power to energize a solenoid, the current is opposed by a factor known as impedance. Impedance is produced by the fixed resistance of the wire used to make the coil winding and, also, by the inductive reactance of the coil, which exists whenever the magnetic field is expanding outward or collapsing inward. Inductive reactance is the opposition developed in a coil to any change of current passing through the coil. The amount of inductive reactance depends on the coil size and the frequency of the AC applied to the coil. It is measured in ohms. Earlier in this chapter it is explained that AC flows first in one direction and then the other, completing this cycle at a rate of 60 times per second. To understand how the AC cycle and impedance affect solenoid operation, it is necessary to examine one complete cycle of the AC applied to the coil.

When the cycle is at zero, the magnetic field does not yet exist and there is only the fixed resistance of the wire in the coil to oppose current flow. The plunger is partially out of the coil. Impedance is at a minimum and, therefore, a high inrush current occurs in the coil windings.

The magnetic field expands outward as the AC cycle rises from zero to a positive peak. Solenoid force is greatest at the peak or maximum value and pulls the plunger into the coil, actuating the directional valve to which it is mechanically connected by the pushpin. Impedance is at a maximum when the plunger is fully centered within the coil. At this time the current is undergoing the most change and that causes the inductive reactance and the impedance to increase to their maximum value. When the impedance is at its maximum value, the AC current is at its minimum value. This minimum amount of current is known as the holding current for the solenoid and is several times smaller than the peak inrush current experienced when the coil is first energized.

As the AC cycle progresses through the positive peak value and returns to zero, the magnetic field collapses, the solenoid force decreases and the plunger is pushed out of the center of the coil by the load (usually a spring-biased spool).

The magnetic field expands and the solenoid force builds as the AC cycle increases in the negative direction until the plunger is again pulled into the center of the coil at the negative peak value.

As the AC cycle returns to zero, the magnetic field and the solenoid force decrease and the plunger is once again forced out of the center of the coil by the load.

The continuous alternation of the current produces a constant motion of the plunger in and out of the coil and a chattering noise known as AC hum. This motion is highly undesirable and must be minimized by the use of shading coils.

Shading coils are built into all solenoids that are designed to be powered by AC. In the air gap solenoid, the shading coils consist of copper wire loops attached to the frame. In wet armature solenoids, the shading coil is a copper wire ring installed into the pushpin end of the tube. As the magnetic field from the main coil begins to collapse inward, the moving magnetic flux lines produce a current flow in the shading coil. This current lags behind the main coil current (it is increasing as the main coil current is decreasing) but supports a magnetic field in the shading coil that is strong enough to hold the solenoid in the energized position during the periods when the AC cycle is passing through zero. As a result, the AC hum is effectively reduced to a minimum.

AC Solenoid Failure. Heat is the main problem associated with solenoid failures. Excessive heat in the coil windings melts the thin coat of insulation on the conducting wire and causes the loops of the winding to short circuit. A short circuit creates an unintended path for current flow and causes the coil to malfunction. Solenoids used on industrial directional valves are of the continuous-duty type, which can be held energized indefinitely without overheating. The heat dissipating ability of these solenoids is good enough to dissipate most of the heat produced by the coil's low holding current.

Heat problems with solenoids usually occur under one of the following three conditions:

- Blocked plunger or armature
- Low line voltage
- Excessive ambient temperature

Most solenoid electrical failures are a result of a mechanical problem. Either the valve spool is blocked and cannot move when the solenoid is energized, or the plunger (or armature) of the solenoid binds up during energization and does not fully center itself within the coil. When either of these conditions occurs, a continuous high inrush current is present in the coil windings. The coil cannot dissipate the tremendous amount of heat generated by the high current and burns itself out within a short period of time.

Valve spools may stick as a result of contamination of the hydraulic fluid or because of burrs which may form between the spool and the valve body. Valves may also become blocked due to a valve base that is not flat. When the mounting bolts are tightened, the base may warp enough to bind up the spool and cause coil failure when the solenoid is energized.

In an air gap solenoid, plunger movement may become restricted if the iron laminations spread apart due to the wear of repeated contact with the pushpin. Also, if a solenoid is disassembled for cleaning, and then reassembled with the plunger not in its original position, different wear patterns may cause the plunger to bind during actuation.

Failures also occur if the voltage supplied to the coil is too low to produce sufficient solenoid force. In this situation, the plunger is unable to fully center itself within the coil in the designed time frame. Inrush current is present for a longer period of time and, as a result, the coil overheats every time it is energized and eventually fails. Low voltage conditions may occur during brownouts or periods of heavy demand on the local power source.

If the air temperature surrounding a solenoid is excessively high, due to poor ventilation or installation too near another heat source, the solenoid's ability to release heat into the air is greatly reduced. If a solenoid is operated for prolonged periods under these conditions, it will probably fail.

Occasionally, an electrical circuit component failure or an incorrect wiring hookup may cause both solenoids of a double-solenoid valve to energize at the same time. These valves have a solenoid installed at each end of the valve spool. The flow paths through the valve depend on which solenoid is energized. If both solenoids are accidentally selected, one solenoid may seat completely and block spool movement in the other direction. Even though the solenoid on the other end of the valve spool has been energized, it cannot move and soon fails due to overheating.

Solenoid Manual Override. Both air gap and wet armature solenoids come equipped with a manual override for checking movement of the directional valve spool without energizing the solenoid. A small metal pin protrudes from the air gap solenoid cover or the wet armature tube. Pressing the override pin causes the plunger or armature to contact the pushpin which moves the spool. If the plunger or spool is blocked, the override

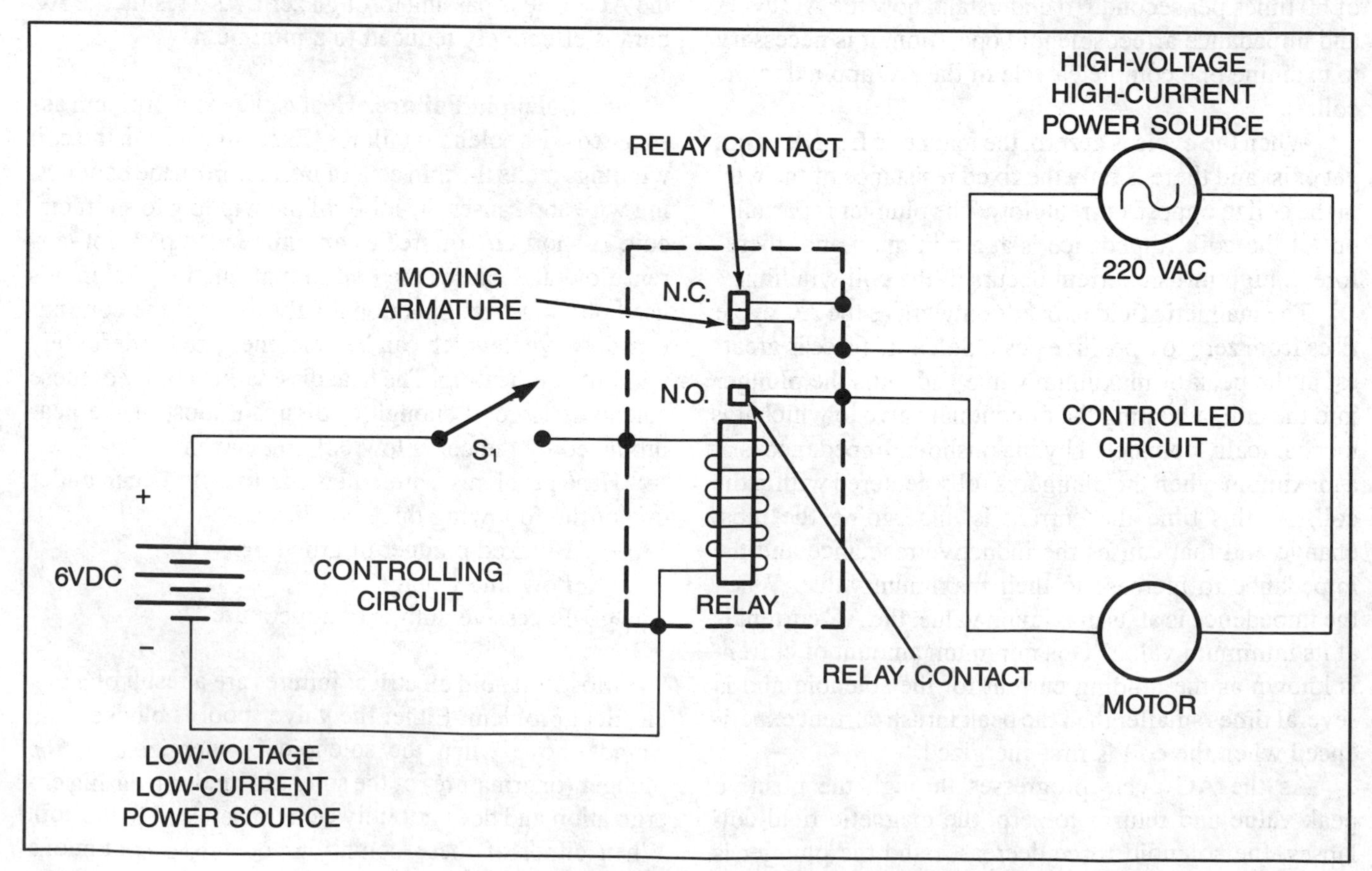

Figure 9-24. Simple relay circuit.

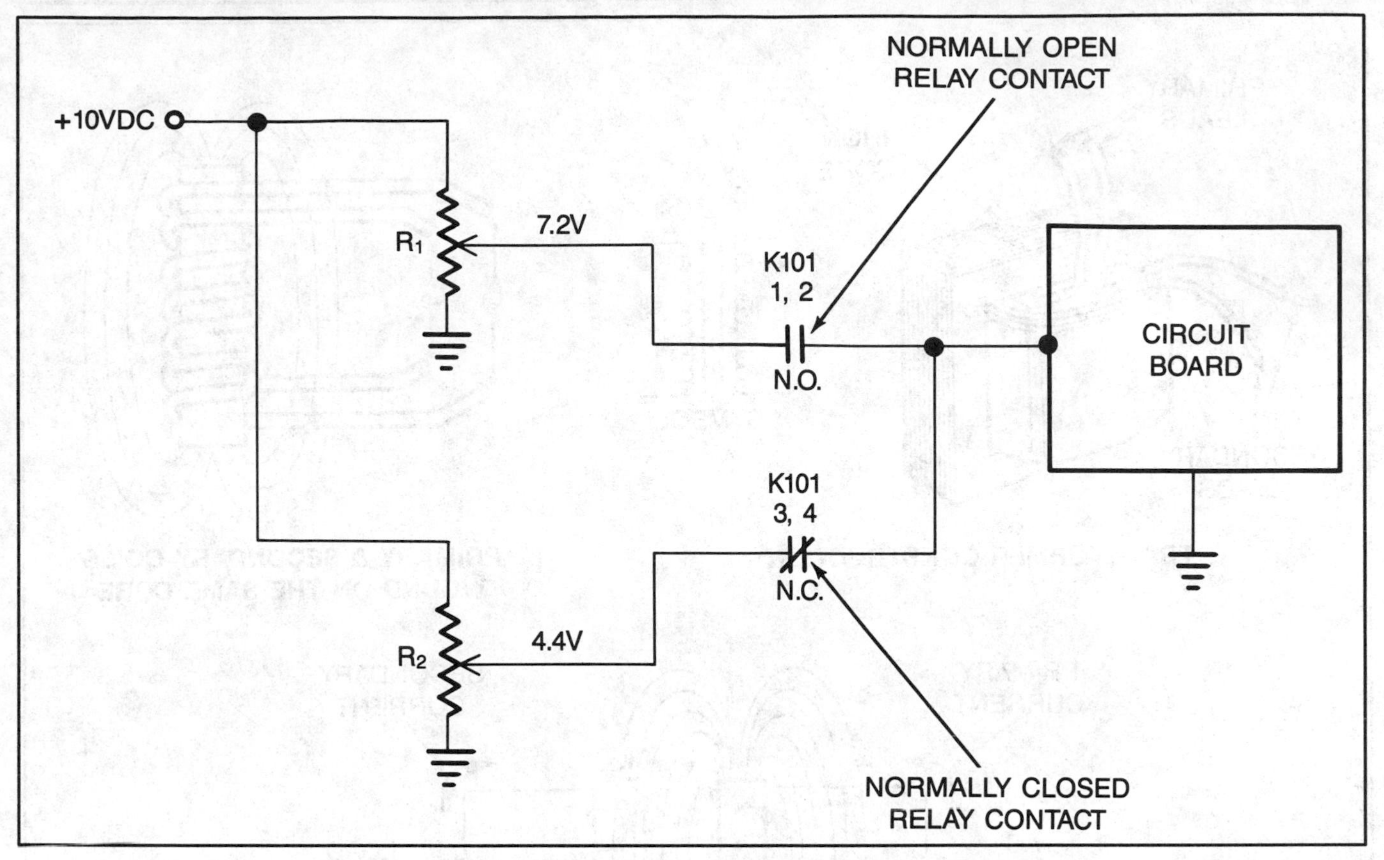

Figure 9-25. Relay controlling two input voltages.

cannot be pushed completely into the cover. The override may also be used to check or adjust hydraulic system operation without energizing the complete electrical control system.

Relay

The relay, which operates much like a solenoid, is an electromagnetic device that consists of a coil with a soft iron core and electrical contacts mounted on, but insulated from, the moving armature or plunger. The relay allows one circuit to control another.

As shown in Figure 9-24, the controlling circuit usually contains the relay solenoid, switch contacts, and a low voltage power source. The switch contacts in the coil circuit act as a pilot device and may be the contacts of a pushbutton switch, a pressure switch, a timer, a limit switch, or the contacts of another relay. The controlled circuit contains the relay contacts, a load device, and a separate power source.

Relay Operation. When current is supplied to the coil, the resulting magnetic field pulls the armature and the contacts toward the coil. The relay is in the energized position when the armature is fully seated. Contacts that were open are now closed. These contacts are referred to as the normally open (N.O.) contacts. Contacts that were closed are now open. These contacts are called the normally closed (N.C.) contacts. Therefore, the terms "normally open" and "normally closed" refer to the condition of the contacts when the relay coil is **de-energized**. A relay may have one or both types of contacts and may also have multiple sets of contacts for controlling several circuits at once.

When the coil circuit is opened, current cannot flow in the coil, the magnetic field collapses, and the relay de-energizes. A spring attached to the armature returns the contacts to their normal positions.

Relay Applications. The relay can be used to open and close high-voltage or high-current circuits with relatively little voltage and current in the coil circuit. The controlling circuit is electrically isolated from the controlled circuit. In other words, current from the controlling circuit does not flow in the controlled circuit. The link between the two circuits is strictly magnetic. The relay is also useful for remote control of circuits. The switch contacts in the coil circuit may be located at one point and the other circuit components located some distance away. Because the controlling circuit is usually operated by a low-current power source, relatively small gauge wires can be used to connect the remote switch device.

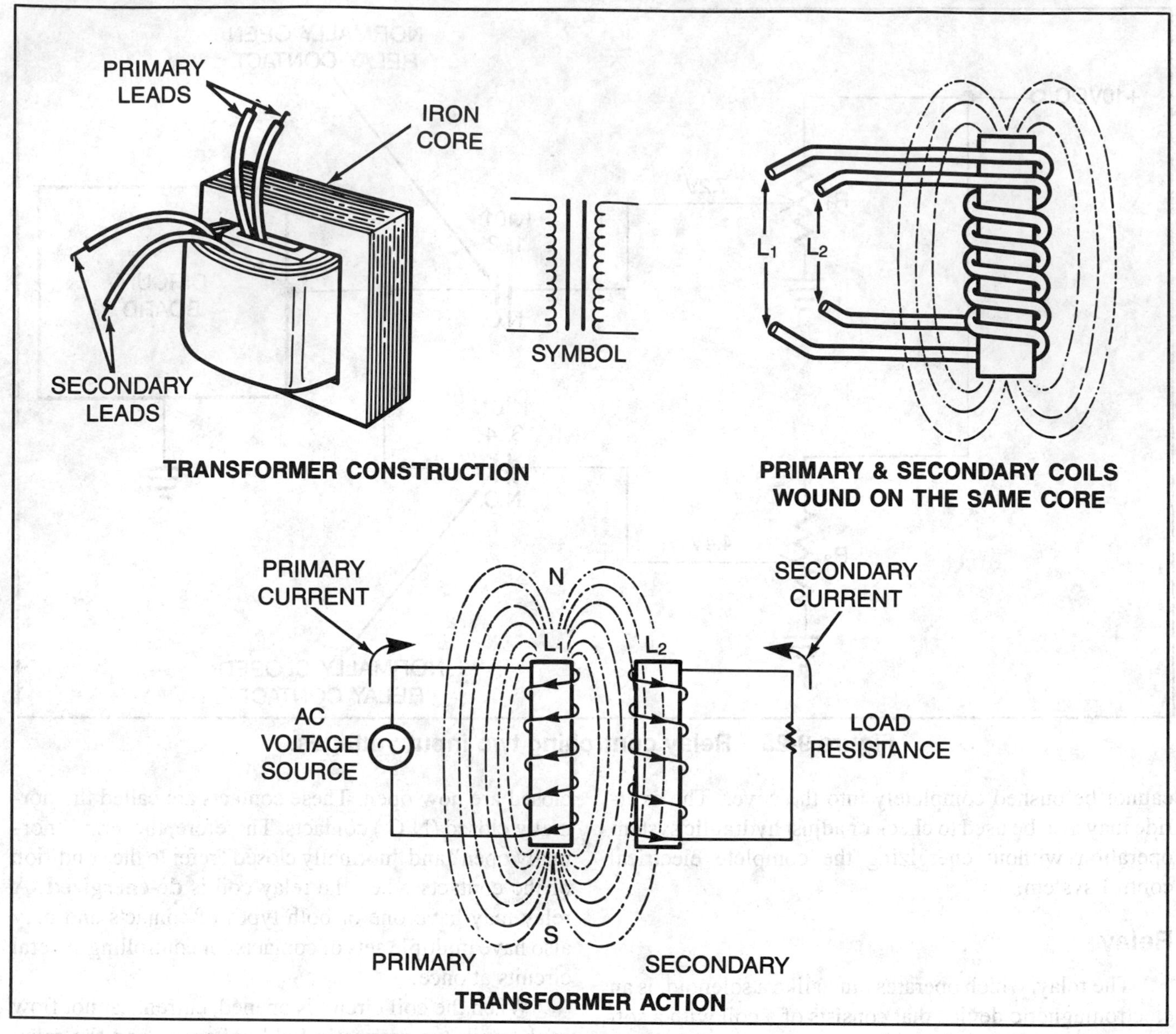

Figure 9-26. Iron core transformer.

Figure 9-25 shows another application of a relay. In this case, relay contacts are used to switch between two different voltage levels that are applied to a circuit. The two voltage levels may produce two different speeds for a device driven by the circuit board. Attenuators are used to develop the 7.2 volts and the 4.4 volts from the 10 volt source. The upper relay symbol represents a set of normally open relay contacts and the lower relay symbol shows a set of normally closed relay contacts on the same relay. When the relay is de-energized, 4.4 volts are supplied to the board. When the relay is energized, 7.2 volts are applied.

A relay which activates or deactivates heavy electrical equipment such as motors and heaters is known as a contactor. The contacts of this device are physically strong enough to switch heavy current flow. The energizing circuit is generally controlled by a pushbutton switch located on an operator's console.

Relays can also be constructed to introduce a time delay when energized (ON DELAY) or de-energized (OFF DELAY). The time delay can be either fixed or variable and may range from a fraction of a second to as much as several minutes. Time delay relays are used primarily for sequence control of a system.

Relay Ratings. Two separate ratings are needed to describe a relay for a given application. The electrical contacts are rated for the voltage and current that they can safely switch. The relay is also rated for the voltage and current required to properly energize the coil. If a time delay is desired, this value must also be specified.

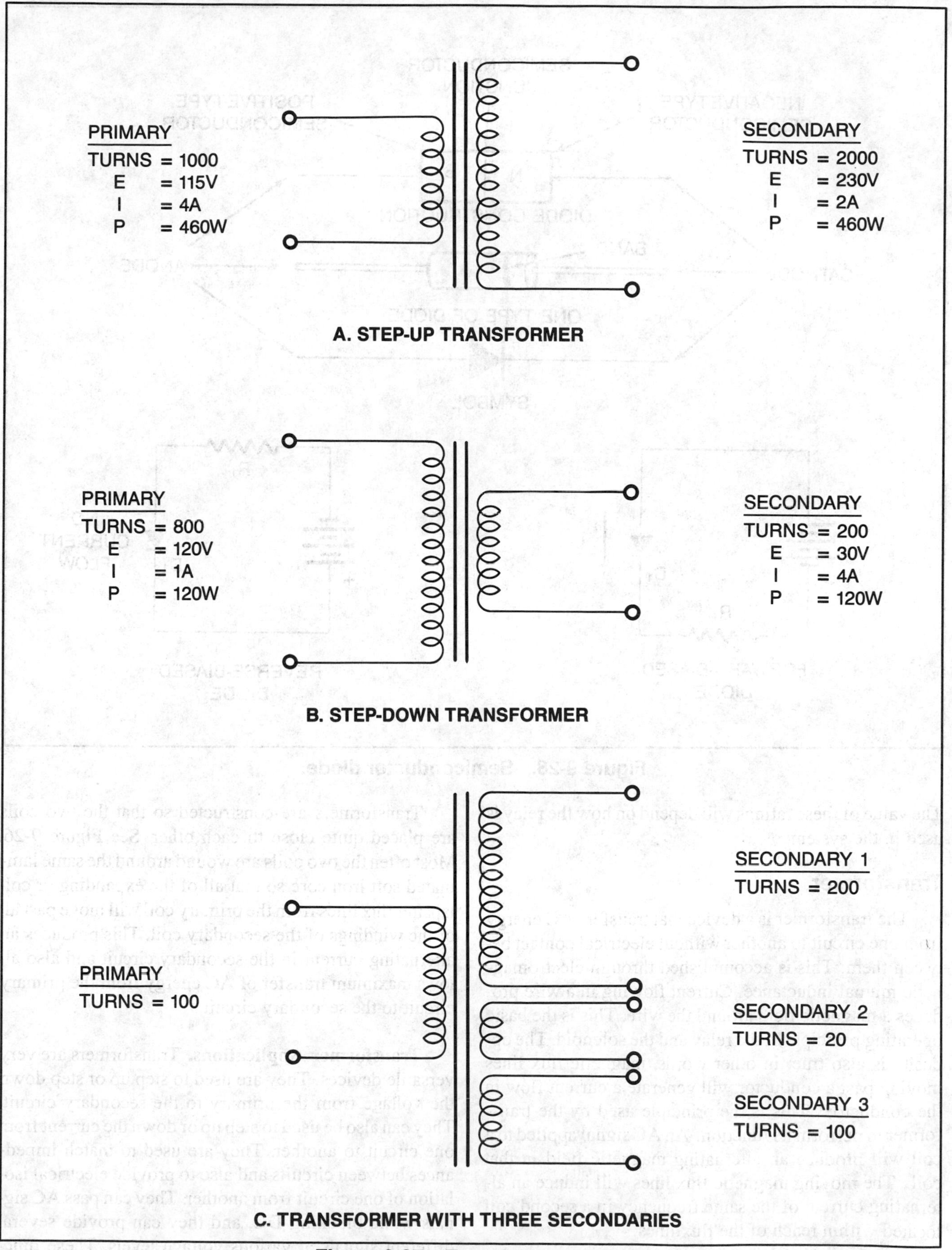
PRIMARY
TURNS = 1000
E = 115V
I = 4A
P = 460W
SECONDARY
TURNS = 2000
E = 230V
I = 2A
P = 460W
A. STEP-UP TRANSFORMER
PRIMARY
TURNS = 800
E = 120V
I = 1A
P = 120W
SECONDARY
TURNS = 200
E = 30V
I = 4A
P = 120W
B. STEP-DOWN TRANSFORMER
PRIMARY
TURNS = 100
SECONDARY 1
TURNS = 200
SECONDARY 2
TURNS = 20
SECONDARY 3
TURNS = 100
C. TRANSFORMER WITH THREE SECONDARIES

Figure 9-27. Transformer types.

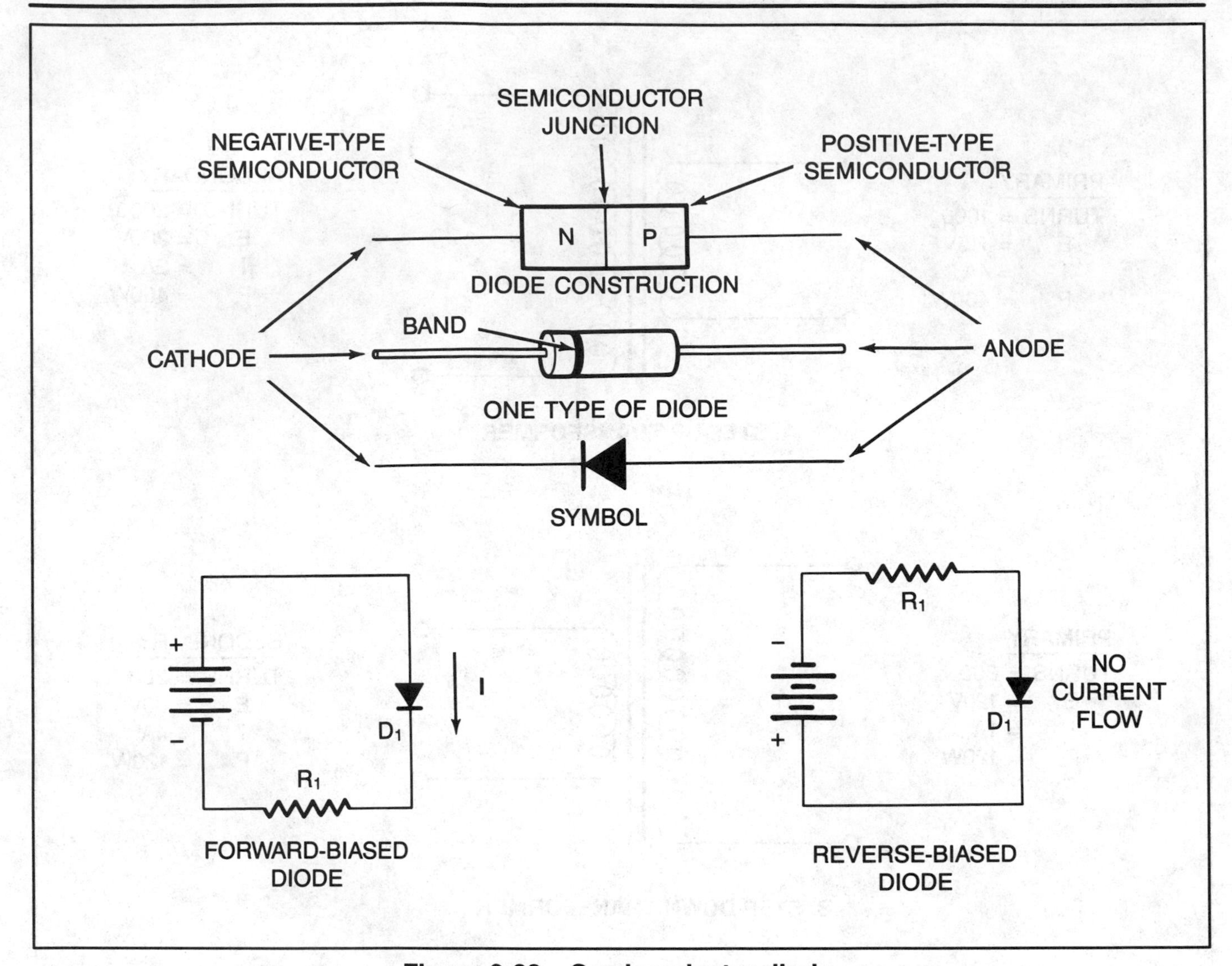

Figure 9-28. Semiconductor diode.

The value of these ratings will depend on how the relay is used in the system.

Transformer

The transformer is a device that transfers AC energy from one circuit to another without electrical contact between them. This is accomplished through electromagnetic mutual inductance. Current flowing in a wire produces a magnetic field around the wire. This is the basic operating principle of the relay and the solenoid. The opposite is also true; in other words, magnetic flux lines moving past a conductor will generate a current flow in the conductor. This is the principle used by the transformer to perform its function. An AC signal applied to a coil will produce an alternating magnetic field in that coil. The moving magnetic flux lines will induce an alternating current of the same frequency in a second coil located within reach of the flux lines.

Transformers are constructed so that the two coils are placed quite close to each other. See Figure 9-26. Most often the two coils are wound around the same laminated soft iron core so that all of the expanding or collapsing flux lines from the primary coil will move past all of the windings of the secondary coil. This produces an alternating current in the secondary circuit and also allows maximum transfer of AC energy from the primary circuit to the secondary circuit.

Transformer Applications. Transformers are very versatile devices. They are used to step up or step down the voltage from the primary to the secondary circuit. They can also be used to step up or down the current from one circuit to another. They are used to match impedances between circuits and also to provide electrical isolation of one circuit from another. They can pass AC signals while blocking DC, and they can provide several different signals at various voltage levels. These func-

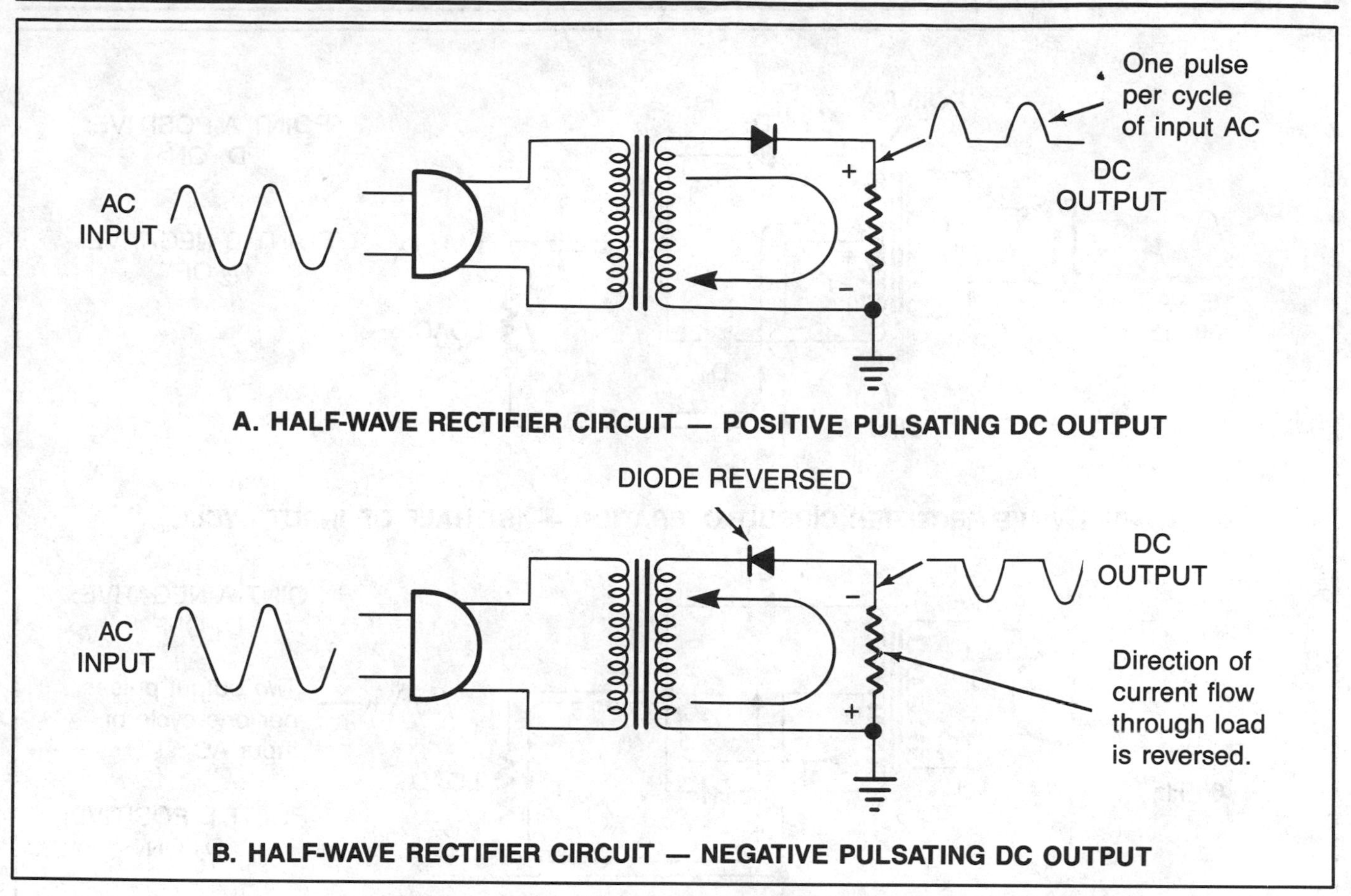

Figure 9-29. Half wave rectifier circuits.

tions are determined by the "turns ratio" of the two coils used in the transformer. Turns ratio is defined as the ratio between the number of turns in the secondary winding and the number of turns in the primary winding.

A step-up transformer is constructed so that there are more turns of wire in the secondary coil than in the primary coil. A transformer made this way will provide more voltage in the secondary circuit than that supplied to the primary circuit. See Figure 9-27A. This increase in the secondary voltage comes at the expense of a decrease in the secondary current. The total circuit power does not change. (Remember that P = I x E. If E increases, but I decreases by the same proportion, P remains the same.)

In Figure 9-27A, the number of turns in the secondary winding is 2000, while the number of turns in the primary winding is only 1000. This is a ratio of 2000:1000 or 2:1. This means that the applied AC voltage will be stepped up by a ratio of 2 to 1. In this example, the 115 volts applied to the primary circuit is increased to 230 volts in the secondary circuit. Since the voltage is doubled, the current must be halved from 4 amps in the primary coil to 2 amps in the secondary coil. Note that the power in the primary is equal to 4A x 115V = 460 watts. The power in the secondary must equal the power in the primary, 2A x 230V = 460W. A step-up transformer steps up the voltage but steps down the current.

In a step-down transformer, the voltage is stepped down by some ratio while the current is stepped up by the same ratio. The transformer is constructed so that the number of turns in the secondary winding is less than the number of turns in the primary winding. In Figure 9-27B the number of secondary turns is 200 while the number of turns in the primary winding is 800. The turns ratio is 200:800 or 1:4. This means that the applied AC voltage will be stepped down by a ratio of 1 to 4. For every 4 volts applied to the primary, 1 volt will be produced in the secondary. If the applied voltage is 120 volts, then the secondary voltage will be 30 volts. Since the secondary voltage is stepped down to one-fourth of its primary value, the current in the secondary must be quadrupled, from 1A to 4A. The power in the primary equals 1A x 120V = 120W. The power in the secondary equals 4A x 30V = 120W.

An isolation transformer is constructed so that the number of turns in the primary and secondary windings is equal. The turns ratio is 1:1. The voltage and current is neither stepped up nor stepped down. This type of transformer is used to isolate the voltage and current in one circuit from the voltage and current in another circuit

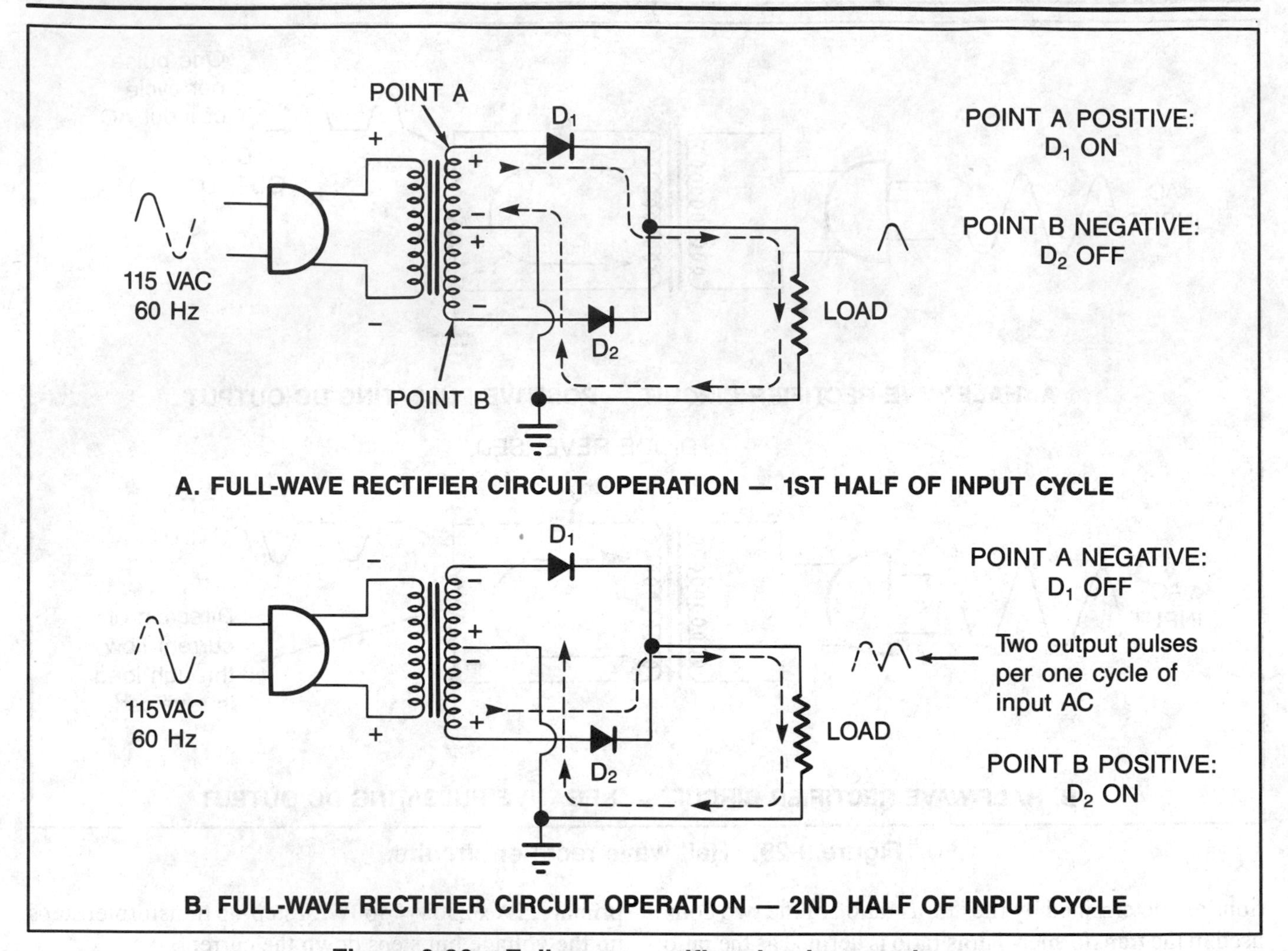

Figure 9-30. Full wave rectifier circuit operation.

while passing an AC signal from one to the other. The signal is coupled from primary to secondary by the magnetic field through mutual inductance.

Transformers can also be made with more than one secondary winding. See Figure 9-27C. One secondary is a step-up, another is a step-down, and the third is for isolation. Transformers can produce any desired change in the AC signal applied to the primary circuit simply by winding the primary and secondary coils in the proper ratios.

Diode Rectifier

Although most electrical equipment is connected to an AC source, the circuits within the equipment often require DC voltages for operation. The process of converting the incoming AC voltage into DC pulses is called rectification. The diode is the device used to produce this change from AC to DC.

Diodes are constructed of semiconductor material. This means that the material is normally neither a good conductor nor a good insulator. This material can be given good conducting ability by a process called doping in which certain impurities are added to the base material. There are two types of semiconductor material used in making diodes negative-type and positive-type. Negative or N-type material contains extra negative charges within the structure of the material. Positive or P-type material is made so that the structure of the material contains extra positive charges. The negative side of a diode is known as the cathode and the positive side is known as the anode. For ease of identification, the end of the diode marked with a band of color is the cathode end. The two types of semiconductor material are placed back to back within the diode case and the area where they meet is called the semiconductor junction (Figure 9-28).

Because of the way a diode is constructed, current can flow across the junction and through the diode in only one direction and only under certain circumstances. As shown in Figure 9-28, current flows through the diode from the anode to the cathode only when the anode has a positive electrical potential with respect to the cathode. Current cannot flow from cathode to anode and cannot

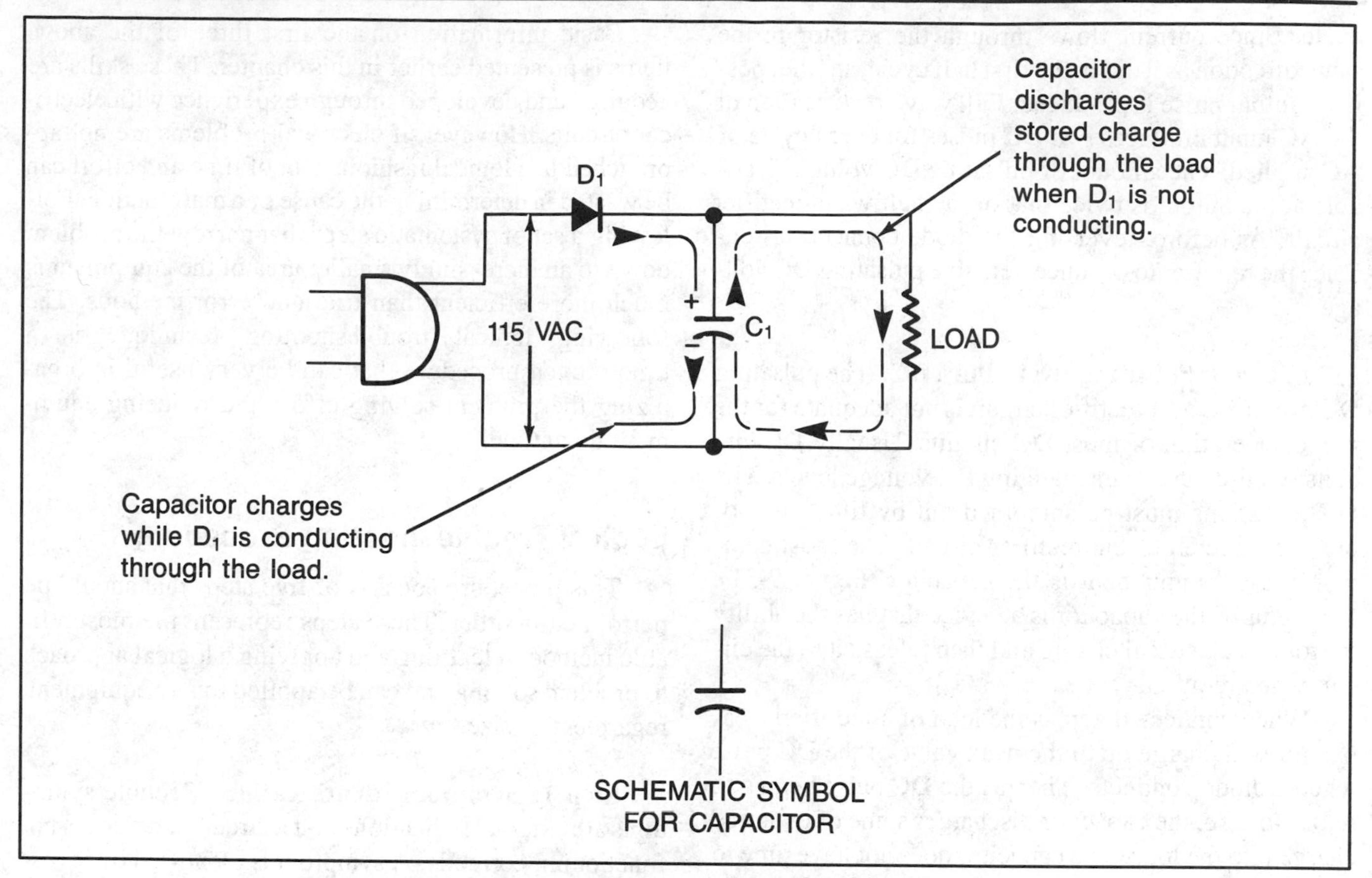

Figure 9-31. Capacitor used as filter for rectifier circuit.

flow when the cathode is more positive than the anode. Note that when viewing the schematic symbol for a diode, current flows out of the arrowhead if the proper biasing conditions are met. Biasing refers to the application of positive and negative voltage to the ends of the diode. It is the ability to allow current flow in one direction while blocking it in the other that enables a diode to rectify AC into DC pulses.

Half Wave Rectification. Figure 9-29A shows a transformer, diode, and load resistor in a simple rectifier circuit. When the AC voltage is in the positive half cycle, the anode of the diode is more positive than the cathode. Current flows down through the load resistor and from anode to cathode through the diode. The current flowing through the resistor produces a positive pulse across the load.

When the AC input voltage crosses zero and progresses through the negative half cycle, the anode of the diode becomes negative with respect to the cathode and current cannot flow through the diode or in the circuit. A positive pulse is produced for only one-half of each AC cycle and no pulses are produced during the other half cycle. This is called half wave rectification and the resulting output of the rectifier is called pulsating DC because it only flows in one direction and does not alternate between positive and negative.

If the diode connections were reversed, as shown in Figure 9-29B, the diode would only conduct current during the negative half cycle of the AC wave and the pulses developed by the load would be negative DC voltage appearing for one-half cycle. (If the cathode is more negative than the anode, it's the same as saying the anode is positive with respect to the cathode; the diode will conduct in either case.) The pulses are negative because the current flows through the load resistor in the opposite direction.

Full Wave Rectification. By using two diodes and a center tapped transformer, full wave rectification of the input AC voltage can be accomplished. As shown in Figure 9-30A, during the first half cycle of the AC input, conditions are correct for diode D1 to conduct and current flows through the load resistor producing a positive output pulse. Diode D2 is improperly biased during this period and cuts off current flow through that part of the circuit.

When the AC input is in the negative half cycle, as shown in Figure 9-30B, diode D1 cuts off, but the conditions are right for diode D2 to conduct. This allows current to flow through the load resistor during this half cycle. Since current flows through the resistor in the same direction as it did in the first half cycle, another positive output pulse is produced. Full wave rectification of the AC input produces two DC pulses for every cycle of AC applied. The amount of pulsating DC voltage available at the output is twice that of the half wave rectifier circuit. As before, reversing the diode connections enables the rectifier to produce negative pulsating DC voltage.

Filtering Pulsating DC to Pure DC. The pulsating DC produced by a rectifier circuit is not adequate for the proper operation of most DC circuits. Usually DC circuits require pure or unchanging DC voltage levels. The DC pulsations must be smoothed out by filter components connected to the rectifier circuit. The most commonly used component is the capacitor (Figure 9-31). Very simply, the capacitor is a device that has the ability to store an electrical charge and then release it to the circuit when required.

When connected across the load of a rectifier, a capacitor will charge up to the peak value of the DC pulse when a diode conducts. Then as the DC pulse begins to return to zero, the capacitor discharges some of its stored charge into the load. The capacitor does not have time to completely discharge before the next peak is felt, so the output is held at a value very close to the peak value. This action is repeated for each pulse and helps to maintain the output at a constant average level. If the capacitor is large enough, the DC output voltage remains essentially free of variations.

TROUBLESHOOTING ELECTRICAL DEVICES

Electrical troubleshooting is the process of locating the cause of malfunctions in electrical circuits. This section contains some general troubleshooting information as well as specific tests for determining the status of the previously described electrical devices.

Skill in troubleshooting electrical equipment and circuits requires:

- Enough knowledge of electrical principles to understand how a circuit or device should function.
- Skill in reading and interpreting electrical schematics, diagrams, product data, etc.
- Skill in operating test equipment and interpreting test measurements.
- An ability to analyze problems in a logical manner.

Basic information on the first three of the above items is presented earlier in this chapter. These skills are acquired and developed through experience with electrical circuits. However, if electrical problems are not approached in a logical fashion, a lot of time and effort can be wasted in determining the cause of a malfunction. Following a set of systematic steps that narrow the problem down to an increasingly smaller area of the equipment is much more efficient than trial and error methods. The following logical troubleshooting technique is a time-proven procedure that can be very useful in organizing the problem-solving effort and reducing equipment downtime.

Logical Troubleshooting Procedure

This procedure consists of five steps that should be performed in order. These steps represent the most reliable method of learning and applying a logical approach to problem solving and can be applied to any equipment, regardless of size.

Step 1: Symptom Identification. Trouble symptoms are external indications that a circuit or device is not functioning correctly. A symptom is identified by investigating the problem with the senses—sight, sound, smell, and touch. For example, a visual inspection of the equipment may reveal that a circuit component has overheated and changed color, or that an indicator lamp which should be on is not. Perhaps there is a peculiar odor that leads to the discovery of melted insulation or a chattering noise that indicates a solenoid is about to fail. If there are controls or adjusting knobs, moving their position may change the problem or have no effect on the equipment at all, either of which may be a significant factor in finding the problem. Sometimes, even the fact that the equipment is completely inoperable is a good symptom.

If someone else was operating the equipment when it failed, ask if anything unusual was noticed prior to the failure. Funny noises, things that don't look quite right, and improper sequence of operation are all symptoms that may lead to the cause of the problem.

If there are no immediately identifiable symptoms, try operating the equipment after determining that it is safe to do so. Notice what works and what doesn't. Identify the point in a sequence where the problem occurs. Look, listen, smell, and touch during this investigation and note anything unusual, no matter how small. It just may be the key to finding the cause of the malfunction.

The time spent on Step 1 is generally more important than that spent on any other step in the process. Frequently, a thorough search for trouble symptoms leads directly to the cause of the malfunction.

Step 2: Symptom Analysis. The purpose of this step is to identify the functions where symptoms indicate a malfunction. Use the symptom information obtained in Step 1 along with schematic and functional block diagrams and knowledge of how the equipment is supposed to operate to make logical technical deductions.

For example, if the equipment with the malfunction is a plastic injection molding machine, the possible functions might include injection forward, injection return, screw run, back pressure control, clamp close, clamp pressurization, prefill shift, mold protect, and clamp open. The symptoms observed in Step 1 indicate the clamp will not pressurize. In analyzing the symptoms obtained in Step 1, the problem might be narrowed down to clamp close, clamp pressurization, or prefill shift, any one of which may contain the faulty circuit. At this point in the procedure, without using any test equipment, several functions have been eliminated and the problem is isolated to just a few possible circuits.

Step 3: Isolate the Single Faulty Function. In this step, test equipment is used to decide which of the possible faulty functions is actually the cause of the malfunction. In making these tests, follow these guidelines:

- Make only those tests that are safe to make.
- Make the least difficult tests first.
- Test those functions first that will eliminate one or more of the other possible faulty functions.

For example, if an ohmmeter reading can determine the fault, do not take a voltmeter reading as that requires power on the equipment. If half the machine has to be disassembled to reach a test point, perform a simpler test first. Test at a midway point in the circuitry if possible. A good reading at the midway point eliminates the preceding functions and indicates that the problem is in the remaining circuits. A faulty signal at the midway point means the problem is located in the functions that process the signal before the midway point.

In the injection molding example, testing in the clamp pressurization circuits at the point where the clamp-fully-closed signal is input either eliminates that function or confirms that the cause of the problem is a clamp that is not fully closed and, therefore, cannot be pressurized.

Continue testing inputs and outputs of the suspect functions until the single faulty function is identified and confirmed.

Step 4: Isolate the Faulty Circuit. In this step, the single malfunctioning circuit within a functional group of circuits is located. Use the accumulated symptom and test data to close in on the single faulty circuit. Follow the guidelines from Step 3, but apply them this time to the circuits related to the faulty function. Use schematic and block diagrams to locate test points.

Using the injection molding machine example again, assume that the clamp-fully-closed signal is not present at the input to the clamp pressurization circuits. Test within the clamp close circuits until a single faulty circuit is identified. Perhaps the first test reveals that the output of the clamp-fully-closed circuit is bad. A check of the inputs to this circuit indicates that the input from a clamp-closed limit switch is bad, but all other inputs are good. The problem is now identified as being associated with one of the relatively few parts contained in a single circuit.

Step 5: Locate/Verify Cause of Malfunction. The tests made in this step identify the failing part within the faulty circuit. Test the circuit until the cause of the malfunction is found. Examine and test the faulty part to verify that it has caused the problem and produced the observed symptoms.

In checking out the clamp-fully-closed circuit, for example, remove the suspect limit switch from the circuit and test with an ohmmeter to determine that the switch contacts are closing correctly to complete the circuit. Connect the ohmmeter across the contacts of the switch and actuate the switch arm several times while the meter reading is checked. If the contacts close properly, the meter should read zero ohms when the arm is in one position and infinity when the arm is in the other position. If the meter pointer does not move when the switch arm is actuated, disassemble and examine the switch. Suppose this last examination reveals that the mechanical linkage connecting the switch arm to the contacts is broken. The cause of the malfunction has been found and a final analysis shows that this cause explains the observed symptoms. However, the procedure is not complete until verification is conducted. In this case, install a new limit switch in the circuit and operate the equipment to confirm that the problem has indeed been fixed.

Electrical circuits are designed, built, and operated in a logical manner. A logical troubleshooting technique takes advantage of this fact and should be used whenever possible to reduce machine downtime.

The next part of this section outlines some basic electrical tests that can be conducted on the devices presented earlier in this chapter. Although they are not described below, mechanical inspections of the devices should also be performed as part of any troubleshooting test. Also, if spare parts are available, substitution of a known good part for a suspect part is often the quickest method of returning equipment to operation. The suspect part can then be tested and either repaired or discarded.

Testing a Potentiometer

Since a potentiometer is a variable resistance device, it should be disconnected from its circuit and tested with an ohmmeter if it is suspect. Only two of the three leads need to be disconnected for this test.

To test a potentiometer, use the following procedure:

1. Determine the expected resistance value from a schematic diagram for the circuit. The value may also be printed on the case of the device.
2. Connect the ohmmeter across the ends of the potentiometer and confirm that the reading matches the expected value.
3. Remove a test lead from one end and move it to the middle terminal. The middle teminal <u>may not</u> be the wiper. (It is often not.)
4. Rotate the shaft or turn the screw that varies the resistance of the device. The ohmmeter reading should indicate zero ohms at one end of the shaft rotation and the full expected resistance value of the potentiometer at the other end. It should also show a smooth change in resistance as the shaft is turned.
5. Move the lead that is still connected to an end terminal over to the other end.
6. Rotate the shaft again while looking for the same smooth transition from zero to maximum resistance.

Be very careful when adjusting small potentiometers installed on printed circuit boards as they are quite fragile and can easily be broken if rotated beyond the end stops.

Testing a Solenoid Coil

When a solenoid is thought to be faulty:

1. Remove it from the machine (plug opened ports on valves if necessary).
2. Disassemble and visually examine the solenoid for signs of overheating or mechanical problems.
3. Test the solenoid coil by attaching an ohmmeter (set to a low resistance range) across the coil terminals. A relatively low reading (a few thousand ohms or less) should be observed on the meter if the coil is good. It should not read zero ohms as this indicates that the coil windings are shorted to each other, probably as a result of melted insulation. If the ohmmeter reads infinity, it means that the coil has opened up and is defective.

Testing a Relay

To test a suspect relay:

1. Remove it from the equipment.
2. Carefully examine it for signs of a mechanical problem.
3. If none is noted, check the relay coil in the same way as a solenoid coil. The electrical contacts can be tested with an ohmmeter in the same way that any switch contacts are tested. The meter should read zero when the contacts are closed and infinity when they are open.

Manually actuate the relay armature to conduct these tests and remember to test both the normally open and normally closed contacts.

Testing a Transformer

When it is determined by voltage readings or symptom information that a transformer may be the cause of a malfunction, check the primary and secondary coil resistance with an ohmmeter. Disconnect one end of the primary winding (coil) and one end of the secondary winding from the rest of the circuit prior to testing. If the failure is the result of an open winding, the ohmmeter will read infinity when connected across the defective winding.

If the failure is caused by shorted turns within a winding, the problem is more difficult to diagnose because the ohmmeter will indicate a very low resistance. Since a winding consists of a length of conductor wound into a coil, the resistance readings are normally quite low anyway. If shorted turns are suspected:

1. Use the expected primary and secondary operating voltages to determine the approximate turns ratio. Divide the secondary voltage into the primary voltage. For example, 120 volts divided by 24 volts equals a ratio of 5:1.
2. Use this ratio to compare the measured primary resistance to the measured secondary resistance. In the example, if the primary resistance is 20 ohms, then the secondary resistance should be approximately 20/5 = 4 ohms. However, this is only if the same gauge wire is measured in both instances.

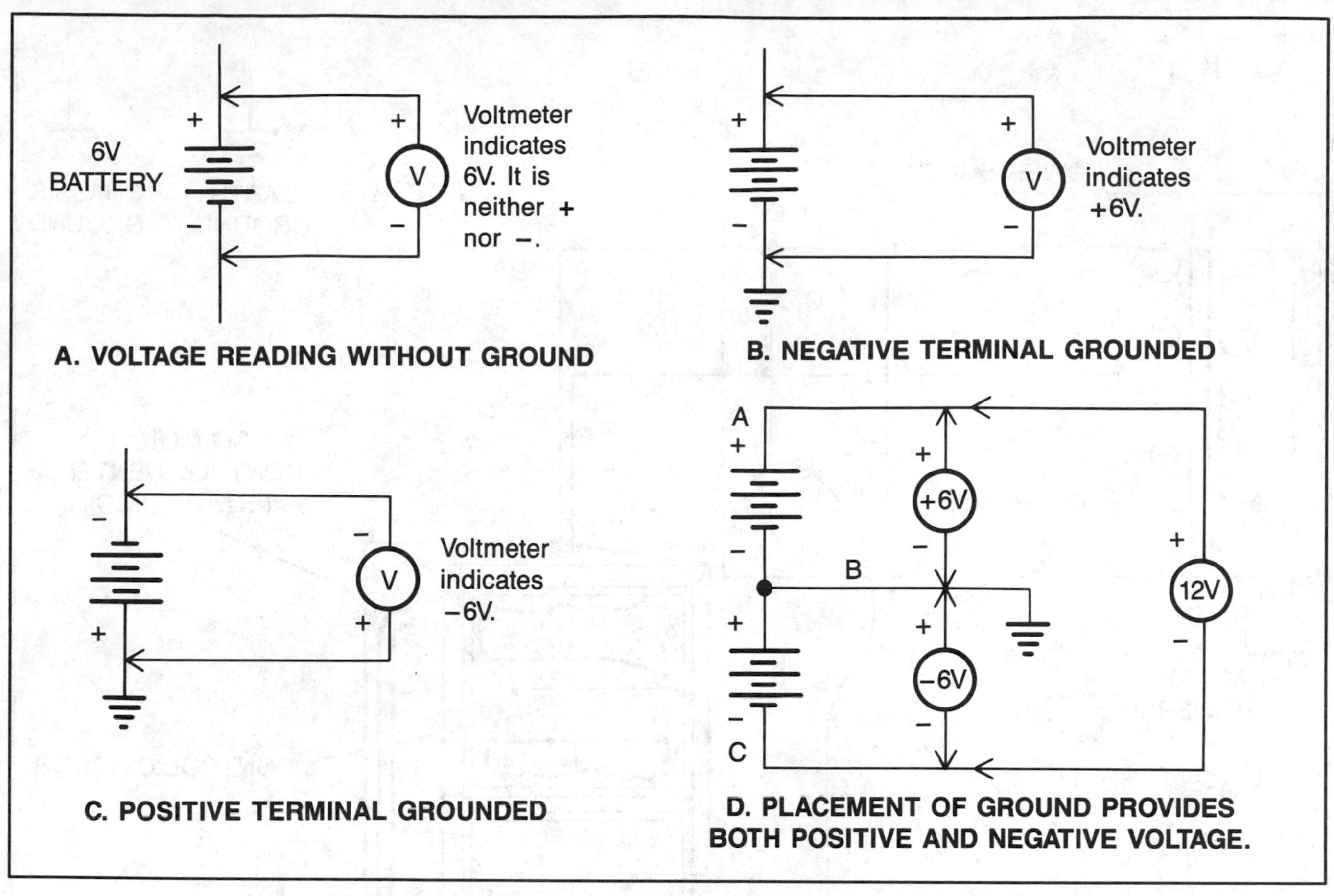

Figure 9-32. Significance of zero reference point.

Be sure to adjust the zero ohms control prior to making the measurement and hold the test probes by the insulated portion only.

It may be quite difficult to determine whether the reading is accurate as the measurement is so near the low end of the ohms scale. Compare the readings to those obtained from a replacement transformer, if one is available. To positively verify that the transformer is faulty, it may be necessary to substitute a known good transformer for the suspect one.

Testing a Diode

A simple resistance check with an ohmmeter can be used to test a diode's ability to pass current in one direction only.

To test a suspect diode:

1. Remove one end of the diode from the circuit.
2. Connect the positive ohmmeter lead to the anode and the negative lead to the cathode. When the ohmmeter is connected this way, the diode is forward biased and the measured reading should be very low. The ohmmeter should be set for the appropriate diode test range.
3. Reverse the ohmmeter connections. When the negative ohmmeter lead is attached to the anode and the positive lead is attached to the cathode, the diode is reverse biased and the meter should read a high resistance.

A good diode should read low resistance when forward biased and high resistance when reverse biased. If the diode reads a high resistance in both directions, it is probably open. If the readings are low in both directions, the diode is shorted.

It is also quite possible for a defective diode to show a difference in forward and backward resistance. In this case, the ratio of forward to backward resistance, known as the diode's front to back ratio, is the important factor. The actual ratio depends on the type of diode. As a rule of thumb, a small signal diode should have a ratio of several hundred to one, and a power rectifier can operate with a ratio as low as ten to one.

ELECTRICAL GROUND

Every electrical circuit has a point of reference to which all circuit voltages are compared. This reference point is called ground and circuit voltages are either posi-

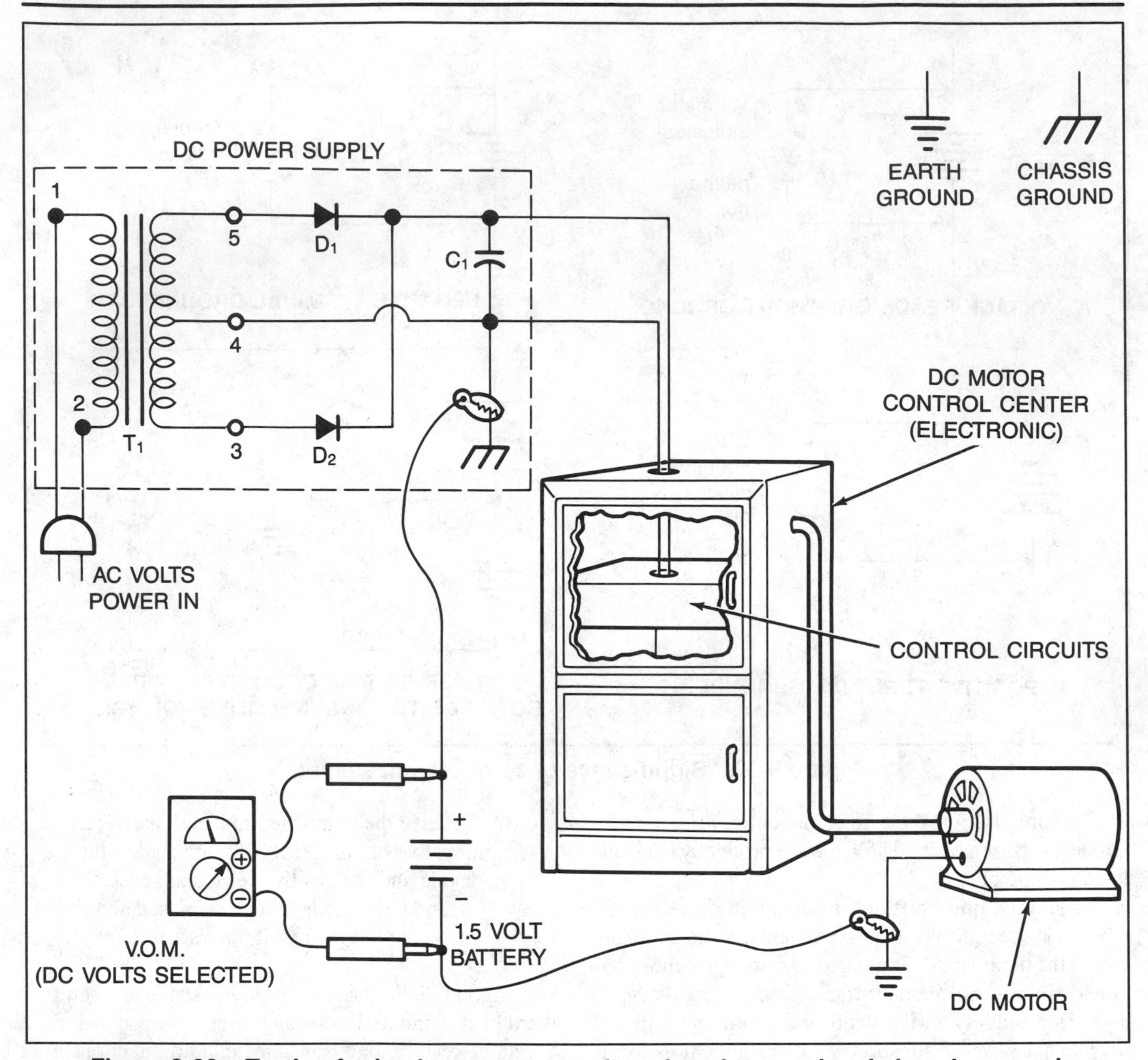

Figure 9-33. Testing for inadvertent connection of earth ground and chassis ground.

tive or negative with respect to ground. When voltage measurements are taken, the difference of potential between a point in the circuit and a ground point is measured by the voltmeter. This type of ground is referred to as chassis or common ground. Connections to ground are also made for safety reasons and, in this case, ground refers to earth ground.

Earth Ground

In the early days of electricity, ground referred to the earth itself and since then it has represented a point of zero potential or zero volts. Connections to earth ground are made primarily for safety and protection purposes. A short circuit within a device that connects live voltage to the frame could cause a serious shock to anyone touching it. However, if the frame is connected to earth ground, it is held at the safe potential of zero volts as the earth itself absorbs the voltage.

A third prong on grounded power plugs connects most stationary equipment in use today to earth ground through the electrical wiring system. Some equipment is connected to earth ground by a conductor that goes from the metal frame of the equipment to a long copper rod driven into the earth. In the home, appliances are often grounded by connecting the conductor to a water pipe running into the ground. In any case, the frames of all equipment connected to earth ground are at the same zero volt potential. This prevents shocks that might oc-

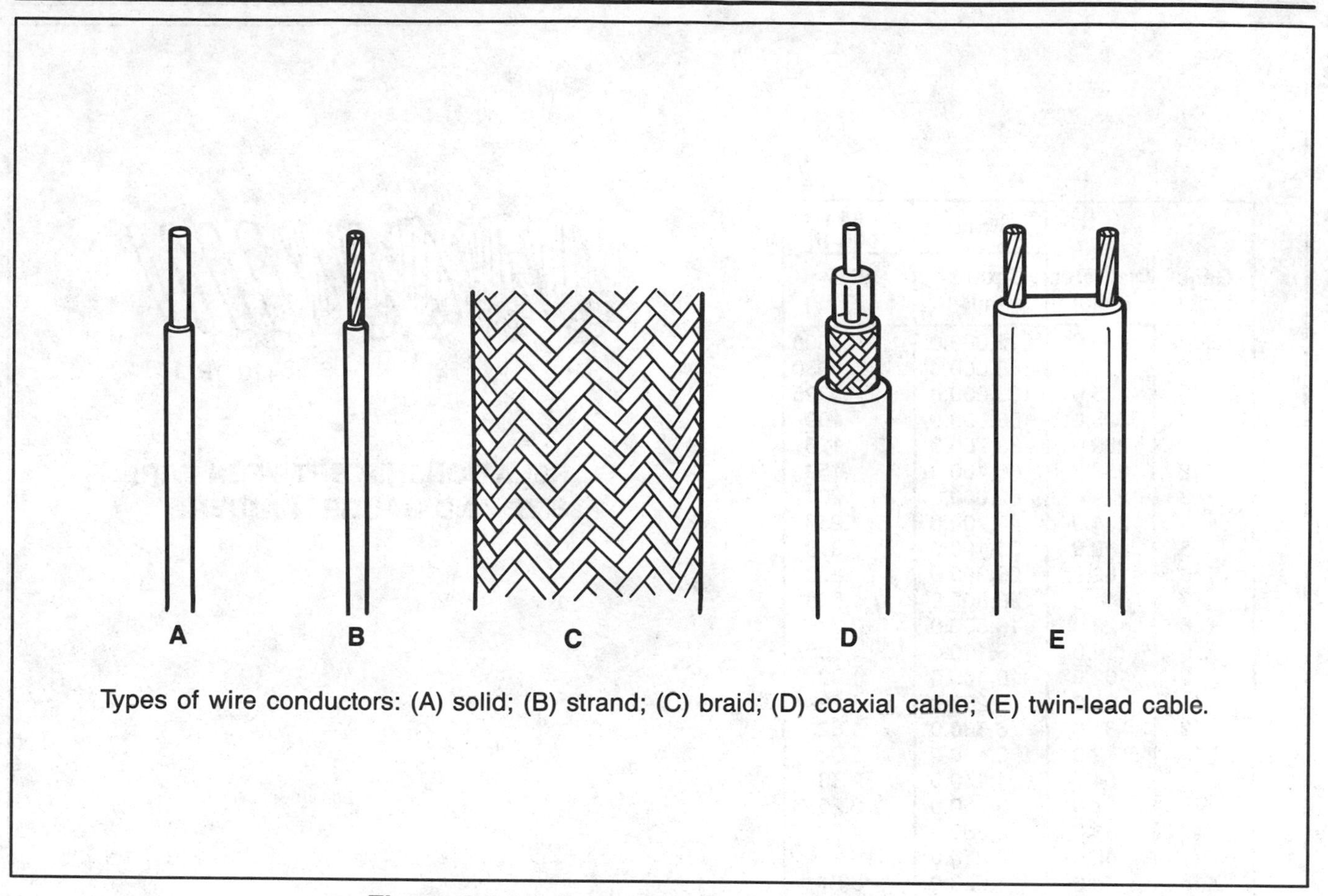

Figure 9-34. Common types of conductors.

cur should a person touch two pieces of ungrounded equipment at the same time.

Chassis or Common Ground

In some cases, electrical circuits used today are not connected directly to earth ground, but still require a point of reference or a common point to which elements of each circuit are connected. For example, a portable battery-operated transistor radio does not have a ground conductor connecting it with the earth. A strip of conducting foil on the internal circuit board is used as the common point. In an automobile battery, the negative terminal is generally connected to the engine block or chassis frame by a heavy cable. The connecting point, as well as every other point on the metal frame, is considered to be a ground for the electrical circuits of the vehicle. The rubber tires insulate the vehicle from the earth ground. Ground in these devices is simply a zero reference point within an electrical circuit and is referred to as chassis ground. All voltages in the circuit are measured with respect to this common point.

Importance of a Zero Reference Point

Without a zero reference point, which may be either earth or chassis ground, voltage could not be expressed as a positive or negative value. The schematic diagrams in Figure 9-32 illustrate this point. Figure 9-32A shows a voltmeter connected to the two terminals of a 6V dry cell battery. Without a ground in the circuit, the voltage being measured is simply 6V between the two terminals. It is neither positive nor negative.

In Figure 9-32B, the negative battery terminal is connected to ground. The voltmeter measures the difference of potential between the positive terminal and the ground point. The voltage measured in this case is +6V because the ungrounded terminal is 6V more positive than the ground or zero reference point.

Figure 9-32C shows that -6V is measured by the voltmeter when the positive terminal of the battery is connected to the zero reference point (ground). The ungrounded battery terminal is now -6V more negative than the reference point.

In Figure 9-32D, the schematic shows two 6V batteries connected in series. The voltage between point A and point C is 12V. When a ground is placed at point B, between the two batteries, +6V is available between point

Gage number	Diameter (mils)	Area: Circular mils	Ohms per 1000 ft. 25°C (=77°F)
0000	460.0	212,000.0	0.0500
000	410.0	168,000.0	.0630
00	365.0	133,000.0	.0795
0	325.0	106,000.0	.100
1	289.0	83,700.0	.126
2	258.0	66,400.0	.159
3	229.0	52,600.0	.201
4	204.0	41,700.0	.253
5	182.0	33,100.0	.319
6	162.0	26,300.0	.403
7	144.0	20,800.0	.508
8	128.0	16,500.0	.641
9	114.0	13,100.0	.808
10	102.0	10,400.0	1.02
11	91.0	8,230.0	1.28
12	81.0	6,530.0	1.62
13	72.0	5,180.0	2.04
14	64.0	4,110.0	2.58
15	57.0	3,260.0	3.25
16	51.0	2,580.0	4.09
17	45.0	2,050.0	5.16
18	40.0	1,620.0	6.51
19	36.0	1,290.0	8.21
20	32.0	1,020.0	10.4
21	28.5	810.0	13.1
22	25.3	642.0	16.5
23	22.6	509.0	20.8
24	20.1	404.0	26.2
25	17.9	320.0	33.0
26	15.9	254.0	41.6
27	14.2	202.0	52.5
28	12.6	160.0	66.2
29	11.3	127.0	83.4
30	10.0	101.0	105.0
31	8.9	79.7	133.0
32	8.0	63.2	167.0
33	7.1	50.1	211.0
34	6.3	39.8	266.0
35	5.6	31.5	335.0
36	5.0	25.0	423.0
37	4.5	19.8	533.0
38	4.0	15.7	673.0
39	3.5	12.5	848.0
40	3.1	9.9	1,070.0

A. WIRE GAUGE NUMBERS AND CHARACTERISTICS

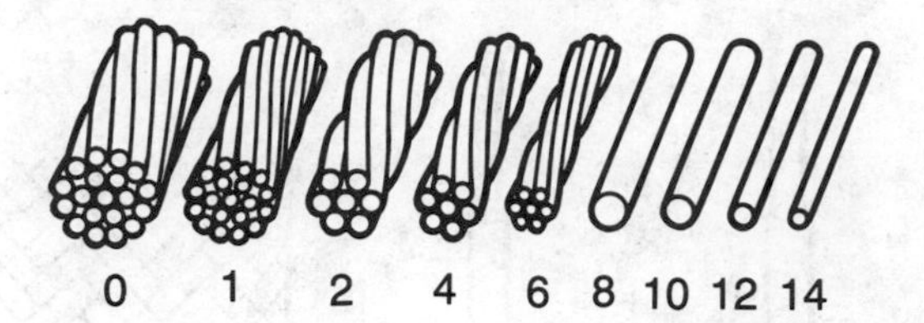

B. RELATIONSHIP BETWEEN WIRE SIZE AND GAUGE NUMBER

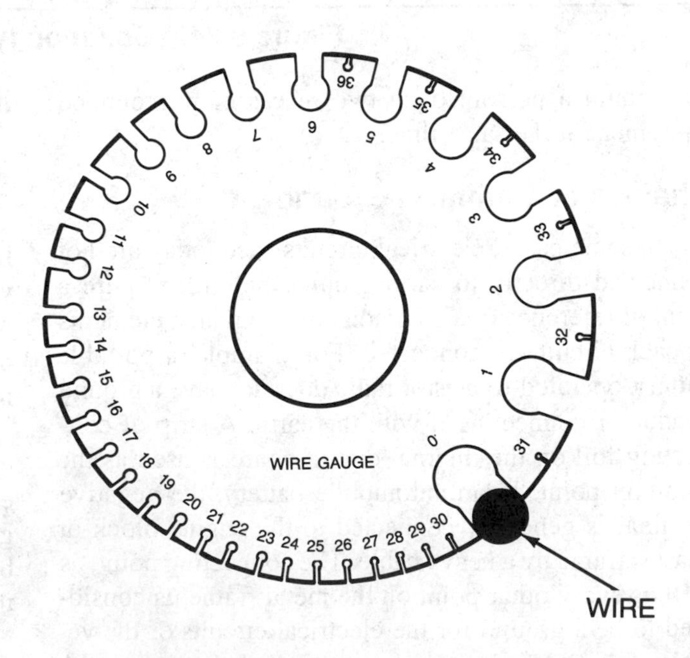

C. AMERICAN WIRE GAUGE (AWG)

Figure 9-35. Wire gauge numbers—characteristics and measurement.

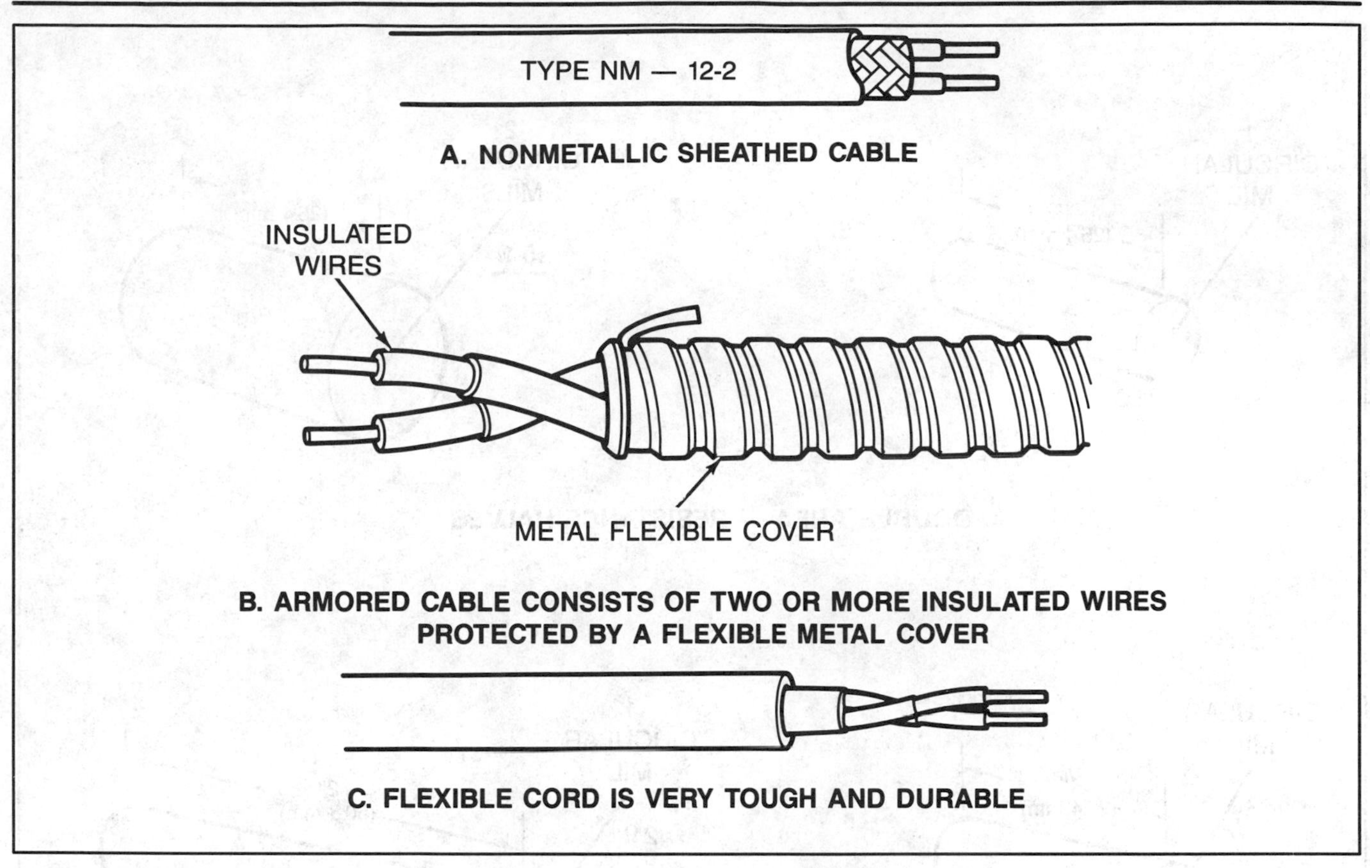
TYPE NM — 12-2
A. NONMETALLIC SHEATHED CABLE
INSULATED WIRES
METAL FLEXIBLE COVER
B. ARMORED CABLE CONSISTS OF TWO OR MORE INSULATED WIRES PROTECTED BY A FLEXIBLE METAL COVER
C. FLEXIBLE CORD IS VERY TOUGH AND DURABLE

Figure 9-36. Types of cable.

A and point B, and 6V is available between point C and point B. Many modern electronic circuits require both positive and negative voltage for proper operation. This would be impossible without a zero reference point in the circuit.

Isolation Between Earth and Chassis Ground

Industrial equipment often requires both an earth and a separate chassis ground for proper operation. The earth ground represents an actual potential of zero volts, while the chassis ground is used only as a reference point and may be at some potential above or below the earth ground. In these cases, it is important that the earth ground and the chassis ground are not connected together at any point in the equipment. However, during installation or repairs, the chassis ground may be inadvertently connected to the earth ground. A simple way to check for this condition uses a 1.5V "D" cell battery and holder, connecting wires, and a voltmeter. **The equipment must be turned OFF to make this test.**

As shown in Figure 9-33, the battery is installed between the chassis ground and the earth ground. The voltmeter, set to measure 1.5VDC, is connected across the battery. If a connection exists between the chassis and the earth ground, it will place a short circuit across the battery and the voltmeter will indicate zero volts. If this is the case, temporarily disconnect one end of the battery to keep it from discharging while looking for the improper connection between the grounds. When the connection is found, remove it and reconnect the battery and the meter. The voltmeter should read the battery potential of 1.5V. If the voltmeter reading is still zero volts, an improper connection still exists in the equipment. Repeat the test until the voltmeter reads the battery voltage. Remember to disconnect the battery when the test is complete.

ELECTRICAL WIRING

Electrical wires or conductors form the path through which electrical charges are transferred from one point in a circuit to another. Current flows quite easily through a conductor. Insulators, such as the rubber, plastic or nylon coverings on the surface of conductors, confine the flow of electrical charges within the conductor to protect against electrical shocks and short circuits. The current flow in an insulator is so small that for practical purposes it is considered to be zero.

The use of conductors and their insulation is governed primarily by the NATIONAL ELECTRICAL CODE (NEC), sponsored by the National Fire Protection Association. The NEC lists the minimum safety precautions needed to safeguard persons, buildings and their

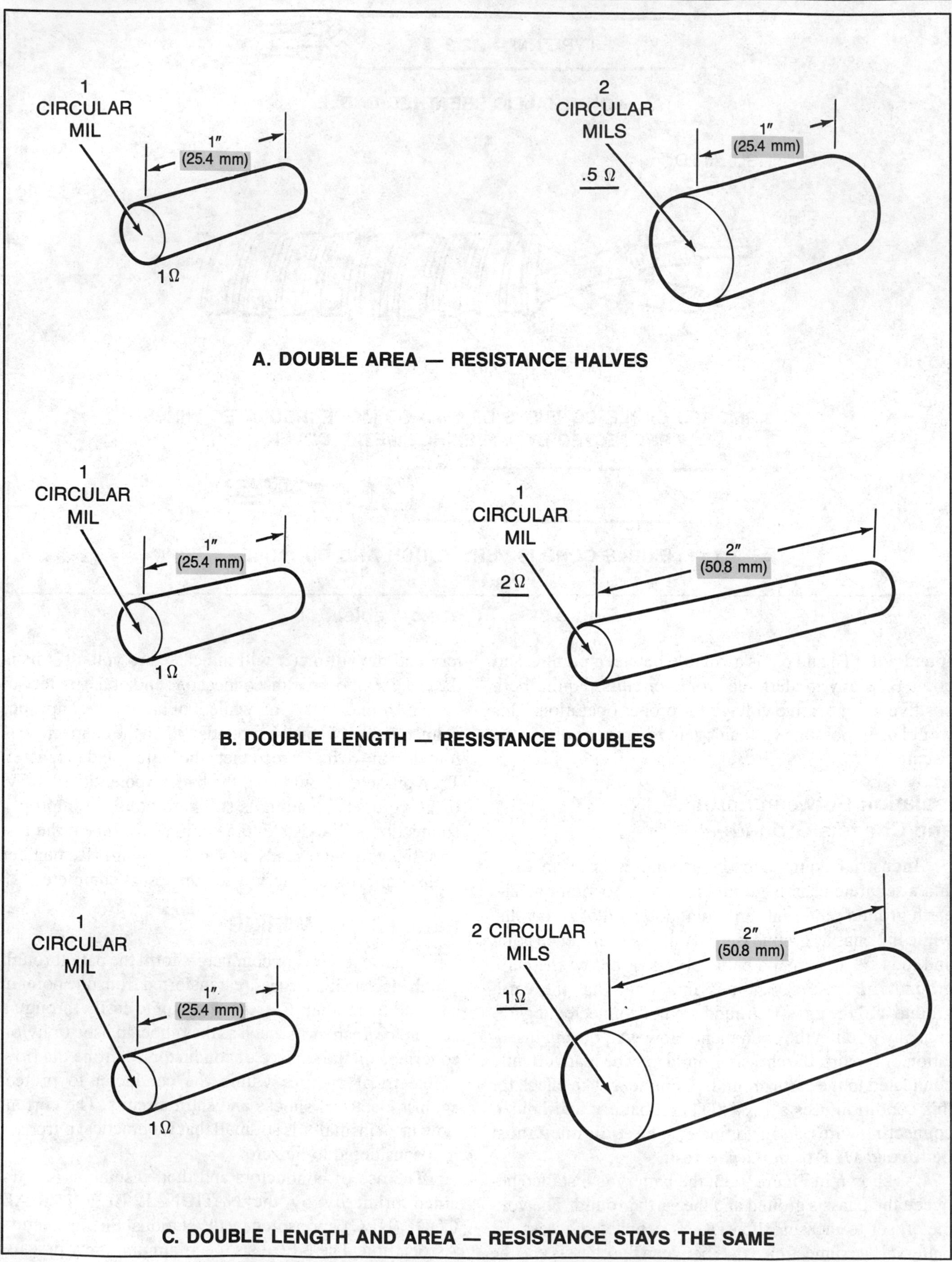

Figure 9-37. The effect of length and area on the resistance of conductors.

Wire size	In conduit, cable, or buried directly in the earth		Single conductors in free air		
	Types T, TW	Types RH, RHW, THW	Types T, TW	Types RH, RHW, THW	Weather-proof
	A	B	C	D	E
14	15	15	20	20	30
12	20	20	25	25	40
10	30	30	40	40	55
8	40	45	55	65	70
6	55	65	80	95	100
4	70	85	105	125	130
2	95	115	140	170	175
1/0	125	150	195	230	235
2/0	165	175	225	265	275
3/0	195	200	260	310	320

Figure 9-38. Ampacity of copper wires.

contents from hazards arising from the use of electricity. Various sections of the code cover electrical conductors and equipment installed in industrial plants. While the code is advisory in nature, it should be considered as the authority in all questions regarding the safe and proper use of electricity. Compliance with the NEC, coupled with proper electrical maintenance procedures, results in an installation that is relatively free from electrical hazards.

This section describes the most common conductors and insulating materials, as well as some typical devices used to make wiring connections. It also covers some important wiring considerations like conductor ampacity, voltage drop on long wire runs, and electromagnetic interference.

Conductors and Insulating Materials

Wires and cables are the most common conductors used to carry electrical current through all kinds of circuits and systems. Wires and cables are made in a wide variety of types and construction that are suited to many different applications (see Figure 9-34). Differences include the size of the wire used, the kind of insulation covering the wire, the type of outer covering, and the number and makeup of the conductors contained within a wire or cable.

Wire Size. Because the diameters of round conductors or wires are usually only a small fraction of an inch , wire diameter is expressed in mils, which are thousandths (0.001) of an inch, to avoid the use of decimals. A diameter of 0.025 inch is equivalent to 25 mils. The cross-sectional area of a wire is usually expressed in units called circular mils. The area in circular mils of a wire is equal to the square of the diameter. A wire having 25 mils of diameter has a cross-sectional area of 25 x 25 = 625 circular mils of area.

Wires are manufactured in specific sizes, called gauge numbers. Figure 9-35A shows characteristics of wire sizes ranging from 0000 to 40. Gauge numbers are listed in the first column. In successive columns are listed the diameter in mils, circular-mil area, and the resistance of 1000 feet (304.80 m) of that size copper wire at 77°F (25°C). Note that the gauge number and the diameter of the wire vary in an inverse relationship (see Figure 9-35B). The higher the gauge number, the thinner the wire. Also, note that the thicker wire sizes have the lowest resistances. Thicker conductors have more electrical charges available to support the flow of current. Figure 9-35C shows an American Wire Gauge that can measure wire sizes from No. 36 to No. 0. The wire is measured by the slot into which it will fit, not the hole behind the slot.

Stranded Wire. When considerable flexibility is needed in a conductor, stranded wire is used. Instead of being one solid wire, the conductor consists of many strands of fine wire twisted together (see Figure 9-34B). The gauge number assigned to such a conductor is determined by the total cross-sectional area of all the individual strands added together.

Cable. For many purposes, it is desirable to group two or more insulated wires together in the form of a cable. A cable that contains two No. 14 wires is known as 14-2 cable. If a cable has two insulated No. 14 wires and a bare uninsulated grounding wire, it is called 14-2 with ground. Cables are available in many types including nonmetallic-sheathed cable and armored cable. Nonmetallic-sheathed cable, shown in Figure 9-36A, contains insulated wires covered with an outer jacket of moisture- and fire-resistant material which may be fibrous or made of plastic or nylon. As shown in Figure 9-36B, armored cable contains insulated conductors protected by a spiral armor made of galvanized steel.

Flexible Cord. If conductors need to be moved about, as on equipment where the conductors are constantly being flexed by the motion of actuators, etc., flexible cord is used for convenience and to prevent the conductors from breaking (see Figure 9-36C). There are many kinds of flexible cord but the type most often used for industrial applications is Type SO. Constant service Type SO flexible cable uses conductors with many finely stranded wires to provide long cable life. This type of cord is made with a special oil-resistant outer jacket to prevent deterioration of the cable from exposure to harsh fluids.

Types of Insulation. The most common insulation materials are Type T (for thermoplastic), Type R (for rubber), and Type X (for cross-linked synthetic polymer). Additional letter designations are added to these base types to represent other properties of these insulators. These include: W for insulators that can be used in wet or dry locations, H for insulators that can withstand more heat than the basic types, and N for insulators that have a final extruded jacket of nylon. For example, Type THWN insulation is nylon jacketed thermoplastic insulation that can withstand more heat than Type T and can be used in either wet or dry locations. Many different kinds of insulators and outer jacketings are available and new types are currently under development. All existing types are listed in the NEC along with various considerations regarding their use in different kinds of installations.

Wire Resistance. The resistance of a wire depends on the material from which it is made, its cross-sectional area, and its length. The most widely used conductor material is copper. It has very good conducting ability and is much more economical to use than slightly better conductors such as gold and silver. Conductors are sometimes made of aluminum when copper is in short supply, but its use is not recommended for most industrial applications.

Wire resistance varies inversely with the circular-mil area of a wire. As shown in Figure 9-37A, increasing the diameter of a wire decreases its resistance (as long as other factors, like temperature and length, are held constant). It should be noted that temperature has only a slight effect on conductor resistance and is usually not an important consideration in most applications.

Wire resistance varies directly with the length of a wire. As shown in Figure 9-37B, increasing the length of a wire increases its resistance (as long as temperature and area are held constant). It takes energy to produce current flow along the length of wire. Longer wires require more energy to support the flow of current, and therefore have more resistance.

Figure 9-37C shows the effect on the resistance of a conductor when both the length and the area are doubled. If the length of a conductor must be increased, but the overall wire resistance must not change, increase the the cross-sectional area of the wire in the same proportion as the change in wire length. In other words, use a smaller gauge number conductor for the wire run.

Conductor Ampacity

Current flowing through a conductor produces heat as it overcomes the resistance of the wire. Heat does not harm the copper conductor, but does harm its insulation. As the temperature of a wire increases, the insulation may become damaged by the heat in various ways dependent on the degree of overheating and the kind of insulation. Some kinds melt, some harden, some burn. In any case, insulation loses its usefulness if overheated, leading to eventual breakdown and fires. With sufficient amperage the conductor itself may get hot enough to start a fire. It is therefore most necessary to carefully limit the amperage to a maximum value, one that is safe for any given size and type of wire.

The maximum number of amperes that a wire can safely carry continuously is called the ampacity of the wire. The ampacity of a given wire is related to its ability to radiate heat into the surrounding atmosphere. The NEC carefully specifies the ampacity of each kind and size of wire for various conditions and methods of installation.

Amperes	Volt-amperes at 115 volts	No. 14	No. 12	No. 10	No. 8	No. 6	No. 4	No. 2	No. 1/0	No. 2/0
1	115	450	700	1,100	1,800	2,800	4,500	7,000		
2	230	225	350	550	900	1,400	2,200	3,500		
3	345	150	240	350	600	900	1,500	2,300	3,750	
4	460	110	175	275	450	700	1,100	1,750	2,750	3,500
5	575	90	140	220	360	560	880	1,400	2,250	2,800
7½	860	60	95	150	240	375	600	950	1,500	1,900
10	1,150	45	70	110	180	280	450	700	1,100	1,400
15	1,725	30	45	70	120	180	300	475	750	950
20	2,300	22*	35	55	90	140	225	350	550	700
25	2,875	18*	28*	45	70	110	180	280	450	560
30	3,450	15*	25*	35	60	90	150	235	340	470
35	4,025	...	20*	30*	50	80	125	200	320	400
40	4,600	...	17*	27*	45	70	110	175	280	350
45	5,175	...	...	25*	40*	60	100	155	250	310
50	5,750	...	...	22*	35*	55	90	140	225	280
60	6,900	...	...	...	30*	45*	75	120	185	240
70	8,050	...	...	...	25*	40*	65	100	160	200
80	9,200	...	...	...	...	35*	55*	85	140	180
90	10,350	...	...	...	...	30*	50*	75	125	160
100	11,500	...	...	...	...	28*	45*	70*	115	140

In this table, the figures below each size wire represent the maximum distance in feet which that size wire will carry the amperage in the left-hand column, with 2% voltage drop. All distances are one-way; in a circuit 100 ft long, of course 200 ft of wire is used, but the one-way distance is listed.

*High temperature insulation required.

Figure 9-39. Voltage drop/wire length table (copper wires—2% voltage drop).

Figure 9-38 shows ampacities for different sizes of copper wire covered with various kinds of insulation and installed in different ways. Note that the larger wires (smaller wire gauge numbers) have a higher ampacity than smaller wires. Larger wires have a greater surface area from which to radiate heat. Also, note that wires with insulation types containing an H in their designations have higher ampacity ratings. These insulations are made to withstand greater amounts of heat than the other types without breaking down. Finally, note that conductors located within enclosed spaces have lower ampacity ratings than conductors suspended in free air. Conductors confined in conduit, cable, or buried in the earth cannot dissipate as much heat as single conductors in free air.

Conductor Voltage Drop

Heat is generated in a conductor as energy is used in overcoming conductor resistance to produce a flow of current. The voltage that is lost in forcing current through a conductor is known as voltage drop. It is wasted power; it merely heats the wires. Wire sizes must be selected for given combinations of voltage and current so that they are big enough to prevent the development of dangerous temperatures, and also big enough to avoid wasted power in the form of excessive voltage drop.

Electrical devices are most efficient when they operate on the voltage for which they were designed. If an electric motor is operated on a voltage 5 percent below its rated voltage, its power output drops by 10 percent. The output decreases at a rate greater than the reduction in operating voltage. For the sake of efficiency, it should be clear that voltage drop must be kept to a minimum. A commonly accepted standard is a voltage drop no greater than 2 percent from power source to load. For a 115 volt circuit, the voltage drop should not exceed 2.3 volts. In some circuits, even a 2 percent voltage drop may drastically affect circuit performance.

The problem of voltage drop increases with long wire runs. For example, assume that a device requiring 460 watts of power is operated at a distance of 50 feet (15.24 m) from a 115 volt power source. At this distance,

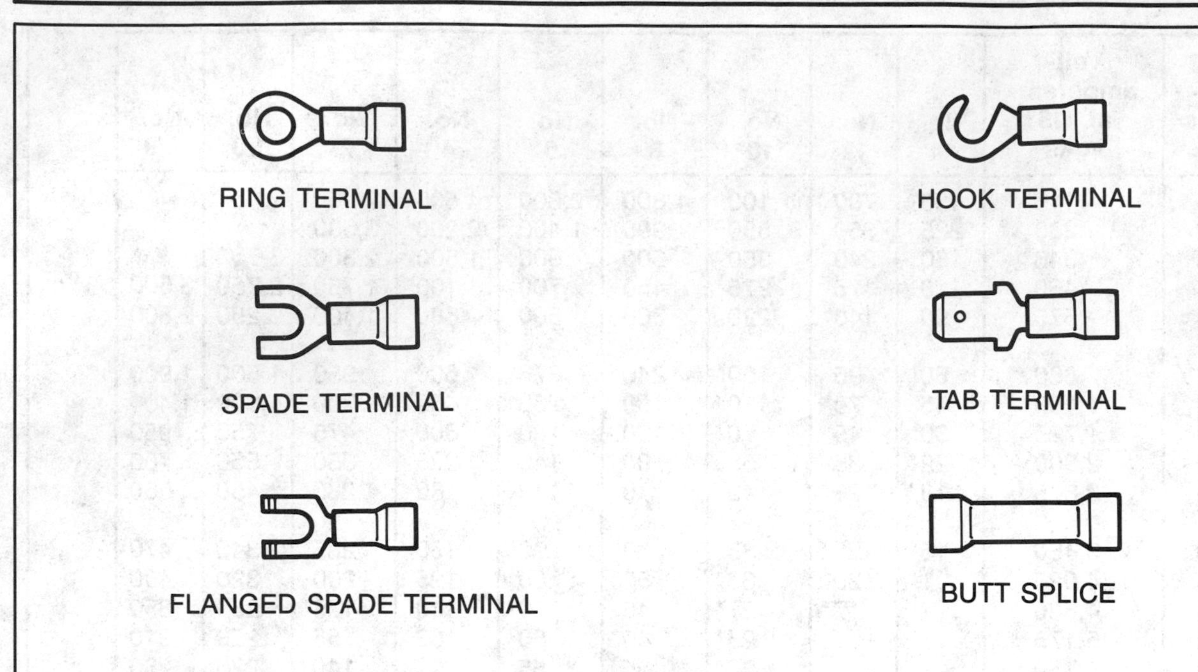

A. TYPES OF SOLDERLESS CONNECTORS

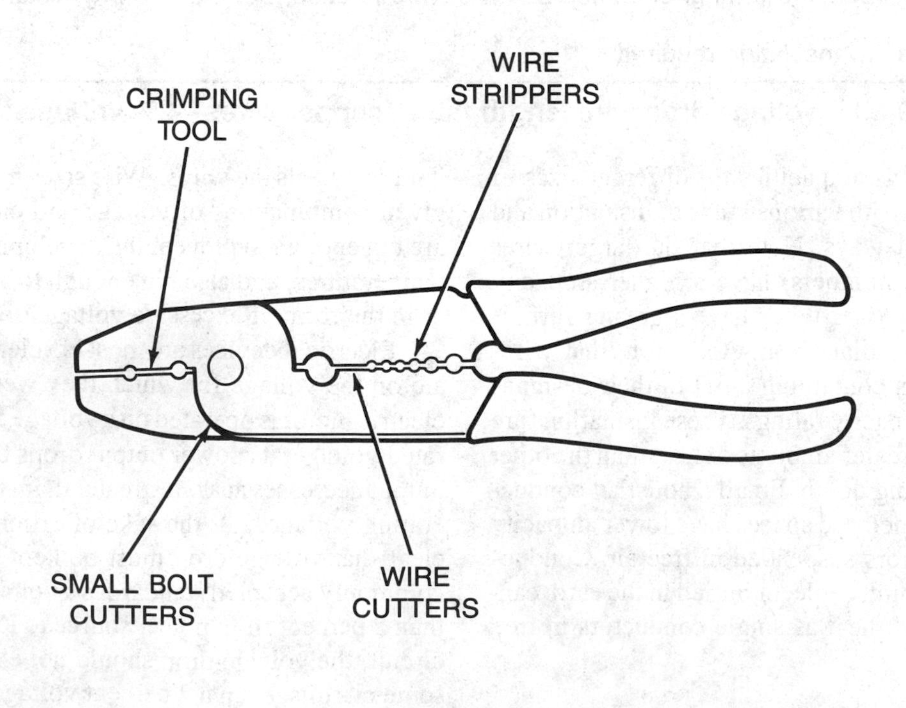

B. CRIMPING TOOL

Figure 9-40. Crimp-type solderless connectors.

100 feet (30.48 m) of wire is needed. The voltage drop can be calculated using Ohm's Law, E = I x R, if the current and resistance are known. The current can be found by dividing the power by the voltage, I = P/E or 460/115 = 4 amps.

Taking No. 14 wire as a random size, the table shown in Figure 9-35A shows that 1000 feet (304.80 m) of No. 14 wire has a resistance of 2.58 ohms per 1000 feet (304.80 m). To find the resistance of 100 feet (30.48 m), divide 2.58 by 10 (1000/100 = 10) to get a result of 0.258. The voltage drop equals 4 x 0.258 = 1.03 volts. This is well below the recommended voltage drop of 2.3 volts.

Now, assume that the distance between the same device and its power source is increased to 500 feet (152.40 m). At this distance, 1000 feet (304.80 m) of No. 14 wire is required. The calculation for voltage drop in this case becomes 4 x 2.58 = 10.3 volts. This is considerably above the recommended voltage drop of 2.3 volts for a 115 volt source.

Voltage Drop Formulas. Suppose that the device in the example above had to be located at the longer distance of 500 feet (152.40 m) and that the voltage drop had to be kept below 2.3 volts. What could be done to solve the problem of the voltage drop being too large? Earlier in this section, it was stated that as the cross-sectional area of a conductor increases, its resistance decreases. If a larger diameter wire is used in the example circuit, the voltage drop can be brought down to the recommended level. The formula shown below is frequently used to solve problems of this type:

$$\textit{Circular mils} = \frac{\textit{Distance}\,^* \times \textit{Amperage} \times 22}{\textit{Voltage Drop}}$$

* Distance in this formula means the length of the circuit in feet, not the amount of wire required in the circuit.

Substituting the values from the example problem in the formula gives the following result:

$$\textit{Circular mils} = \frac{500 \times 4 \times 22}{2.3} = 19,130.4$$

Referring to the table shown in Figure 9-35 note that none of the gauge numbers have exactly the calculated circular-mil area of 19,130. The first even wire gauge number having a circular-mil area greater than 19,130 is No. 6 with a circular-mil area of 26,300. (Odd gauge numbers are not usually used for wiring; they are primarily used for motor and transformer windings.) Since the circular-mil area of No. 6 wire is well above the calculated value, the voltage drop for this circuit would be within limits if No. 6 wire is used.

If the actual voltage drop of a given size wire is to be determined, the formula can be transposed to read:

$$\textit{Voltage Drop} = \frac{\textit{Distance} \times \textit{Amperage} \times 22}{\textit{Circular Mils}}$$

To determine the number of feet any given size of wire will carry any given amperage, the formula can be transposed another way to read:

$$\textit{Distance} = \frac{\textit{Voltage Drop} \times \textit{Circular Mils}}{\textit{Amperage} \times 22}$$

All of the above formulas apply to circuits using direct current or single-phase alternating current. Substitute 19 for 22 in these formulas as a correction factor for 3-phase alternating current circuits.

Voltage Drop/Wire Length Tables. For most purposes, voltage drop/wire length tables are available for determining the one-way distance that a given size wire will carry a given amperage with a given voltage drop (see Figure 9-39). To use the table, find the amperage in the left-hand column, or the volt-amperes (watts in DC circuits) in the second column. Next, find the wire gauge number across the top. Read the distance in feet (meters) where the two values meet. Using the values from the previous example, note that the maximum distance No. 6 wire will carry 4 amps with a voltage drop of only 2 percent is 700 feet (213.36 m).

Electrical Connections

Connections between conductors and other components, such as limit switches, sensors, terminal blocks, etc., are a necessity in most electrical circuits. These connections must be mechanically secure and must also provide a good conducting path. For small diameter conductors, such connections are often made by crimp-type solderless electrical connectors. The connector is crimped onto the end of a conductor which has been stripped of its insulation. The connector is then attached to the component by a screw terminal. In the case of larger diameter conductors, compression-type solderless connectors must be used. Solderless means that the connector is secured to the conductor by a quick, mechanical means. The time-consuming and laborious process of soldering is eliminated. For smaller conductors, the mechanical force required to secure the connector to the conductor is provided by special hand-held tools. For larger wires, special electric, hydraulic, or pneumatic tools are used.

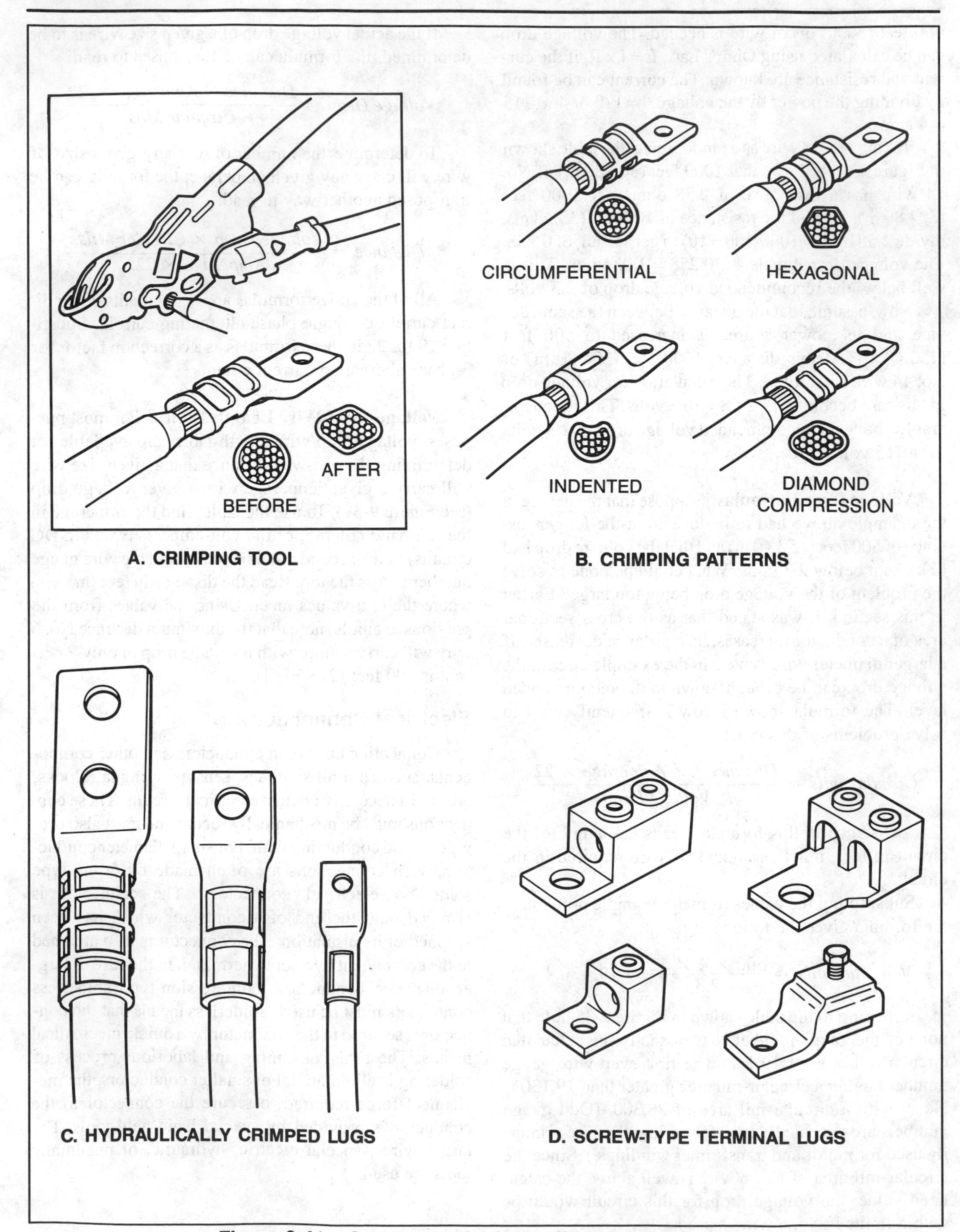

Figure 9-41. Connectors for large diameter conductors.

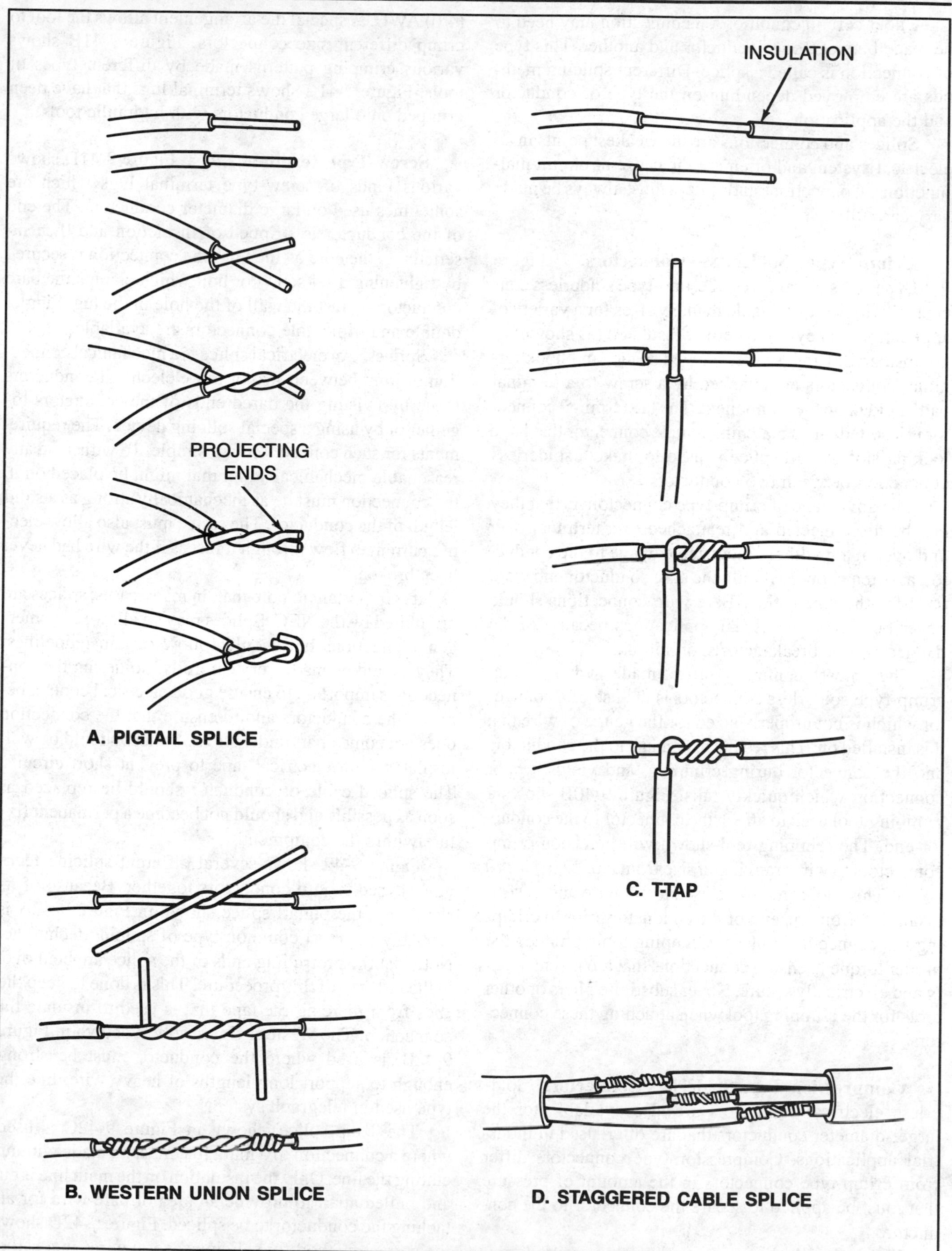

Figure 9-42. Types of splicing techniques.

Under certain conditions, a connection may need to be made between one conductor and another. This type of connection is called a splice. Different splicing methods are employed depending on the type of conductor and the application.

Splices and connections are the weakest points in an electrical system and often are the cause of circuit malfunctions. Connections and splices must always be made very carefully.

Crimp-Type Solderless Connectors. Figure 9-40A shows several types of crimp-type solderless connectors which are available in many sizes for a variety of applications. They can be either insulated (as shown) or noninsulated. Ring, spade, flanged spade, and hook terminal connectors are attached by a screw to a terminal pad or block on the component. The tab terminal connector is inserted into a matching female connector that has a locking slot. A butt splice is used to make a solderless connection between two conductors.

The advantage of crimp-type connectors is that they can be disconnected and reattached to a terminal over and over again without incurring damage to the conductor, as often happens if only the bare conductor end were used for the connection. Bare wire connections should never be used in industrial applications because of the danger of wire breakage or short circuits.

The most common error made when using crimp-type solderless connectors is the use of a connector which is not properly sized for the gauge of wire that it is installed on. This results in damage to the conductors and the connector during crimping, and yields a poor connection which quickly fails. Figure 9-40B shows a crimping tool used to attach the connector to the conductor end. The crimping tool shown will precision crimp connectors to wires ranging in size from No. 22 to No. 10 AWG. This tool can also be used to cut wire and to strip insulation from the end of the conductor prior to crimping the connector in place. Crimping tools produce the proper torque to ensure connections that are mechanically and electrically secure. Never substitute pliers or other tools for the crimping tool when attaching these connectors.

Compression-Type Solderless Connectors. Solderless electrical connectors are also available for the larger diameter conductors that are often used in industrial applications. Compression-type connectors differ from crimp-type connectors in the amount of pressure that must be applied to secure the connector to the conductor.

Figure 9-41A shows one tool used to secure terminals onto wires ranging in size from No. 8 to No. 0000 (4/0) AWG. A special die arrangement allows the tool to crimp different size connectors. Figure 9-41B shows various crimping patterns made by different types of tools. Figure 9-41C shows terminal lugs that have been crimped onto large conductors with hydraulic tools.

Screw-Type Terminal Lugs. Figure 9-41D shows various kinds of screw-type terminal lugs which are sometimes used on large diameter conductors. The end of the conductor is stripped of insulation and then inserted into the hole on the lug. The connection is secured by tightening a set screw or bolt which clamps the bare conductor against the wall of the hole in the lug. Single, double and triple hole connectors are available.

Splices. An electrical splice is a mechanical connection or joint between two or more electrical conductors made by twisting the bared ends of the conductors together or by using a special splicing device. The requirements for such connections are simple. To withstand any reasonable mechanical strain that might be placed on it, the connection must be as mechanically strong as a solid length of the conductor. The splice must also allow electric current to flow through it just as if the wire had never been broken.

It is important to note that, in many cases, splices are prohibited by the NEC. Splices may be allowed in emergency situations but should follow certain guidelines. They should be made very carefully. Soldering the connection is important to ensure good electrical contact between the conductors and to ensure that the connection does not come apart under stress. Splices should be well insulated with electrical tape to prevent short circuits. The spliced cable or conductor should be replaced as soon as possible and should not become a permanent fixture within the equipment.

Figure 9-42 shows several different splicing techniques used to join conductors together. Because of its simplicity, the pigtail splice shown in Figure 9-42A is probably the most common type of splicing technique. Note that the protruding ends of the splice are bent over in the last step of this procedure. This is done to keep the ends from piercing the tape that is used to insulate the connection. The Western Union splice, shown in Figure 9-42B, is used where the conductor must be strong enough to support long lengths of heavy wire, like the type used in telegraph systems.

The T-tap splice, shown in Figure 9-42C is used where a connection to a main line must be made without cutting the line. Only the insulation on the main line is cut and pulled apart to provide an area of bare wire for attaching the conductor to be spliced. Figure 9-42D shows the correct method for splicing several wires in a cable.

The splices are staggered so that a large bump in the cable is not created when the connections are taped.

Electromagnetic Interference and Noise

Electromagnetic Interference (EMI) and noise are interference-type problems within electrical circuits and associated wiring that can cause erratic or improper operation of equipment. These problems are often hard to locate and correct. EMI and noise can usually be minimized by good circuit design and by proper wiring installation methods. Careful maintenance, repair and replacement procedures must also be followed to prevent EMI and noise problems from occurring after installation of the equipment.

Electromagnetic interference is defined as a modification of signals contained in circuit wiring caused by a strong magnetic field radiated from some other electrical or electronic device or conductor. Previously, it was explained that current flowing through a conductor produced a circular magnetic field around the conductor. It was also explained that devices like transformers, relays and solenoids have coils that intentionally produce a magnetic field as part of normal operation. Devices or conductors carrying a high current produce strong magnetic fields that extend outward. As the flux lines from these magnetic fields pass through the insulation of nearby wires and components, a current flow is induced that can modify or change the signals already flowing in the conductor. This induced current (EMI) can interfere with command signals going to a device or with feedback signals coming from a device.

Noise is defined as an unwanted disturbance superimposed upon a useful signal which tends to obscure its information content. Random noise is noise that is unpredictable. Noise can be generated in a circuit whenever high currents are switched on or off, such as when relay contacts open and close or when motor armatures make and break contact with brushes. Dirty contacts can cause intermittent current flow and are often the source of many noise problems. The rapid increase or decrease in circuit current produced under such circumstances generates magnetic fields that can cause high amplitude pulses of current (noise spikes) to appear on conductors and cables running to other circuits. These noise spikes, even though very short in duration, add unwanted interference to signals normally present in circuits connected to these lines.

Shielded Conductors

The harmful effects of EMI and noise can be reduced by using shielded conductors and cables and by following proper wiring practices. A shield is a metallic sheath (usually copper or aluminum) applied over the insulation of a conductor during the manufacturing process. It may be a very thin sheet of solid conducting material or a braided wire mesh. It is usually covered by an outer jacket of insulation. The conducting path provided by the shield is generally connected to ground at some point in the circuit. The shield reduces the effects of noise or electromagnetic interference by absorbing random magnetic flux lines before they reach the inner conductor.

When installing a shielded cable or conductor, **never ignore the shield connection.** Always follow the manufacturer's installation instructions on connecting the end(s) of the shield. Improper connection of the shield can lead to faulty operation of the equipment.

ELECTRICAL SAFETY

Effective safety measures are a blend of common sense, knowledge of basic electrical and hydraulic principles, and knowledge of how a system or circuit operates, including any dangers associated with that operation.

General safety information and safety practices are listed below for personnel who work on electrical circuits and components. The list is by no means all inclusive, is not intended to alter or replace currently established safety practices, and does not include safety practices for hydraulic equipment.

When working with electrical equipment consider the following:

- Injuries associated with electrical work may include electrical shocks, burns, and puncture, laceration, or abrasion wounds.
- Current flow through the body can be fatal. As little as 0.01A produces muscle paralysis and extreme breathing difficulty in the average person. Permanent physical damage and death can result from 0.1A flowing through the heart.
- The amount of current received from an electrical shock depends on the voltage applied and the resistance of that part of the body through which the current flows. For most individuals, 0.1A can be produced by about 30V. Use extreme caution when working in circuits that include voltages higher than 30V.
- Most electrical shocks are unexpected. These shocks, even if they are not particularly dangerous, may cause you to jerk your hand into heavier currents or hit some sharp object. Always check to see that the power is turned off before placing your hand in a circuit.

- Never put both hands in a live circuit as this provides a path for current flow through the heart. Keep one hand behind you or in your pocket when taking measurements with a meter.
- Never work on live circuits when wet as this lowers the body resistance and increases the chance for a fatal shock.
- Never work alone on electrical equipment. Shocks above 0.01A can paralyze your muscles and leave you unable to remove yourself from the source of current flow. Always be sure someone else is around to help in an emergency.
- Use the proper equipment for circuit testing, remembering to check for correct settings of function and range switches, proper insulation on test probes, etc.
- Before starting work on an electrical circuit, remove all watches, rings, chains, and any other metal jewelry that may come in contact with an electrical potential or get caught in moving mechanical parts.
- Understand the circuit you are working on, and think out beforehand what you will be doing in the circuit. If your knowledge is not adequate for the task, ask for assistance.

Remember that electrical current can be compared to a rattlesnake without the warning rattle. It is invisible, deadly, and ready to strike when least expected. However, the danger associated with electrical work can be reduced if proper safety precautions and practices are observed and applied.

QUESTIONS

1. Identify the three basic components found in any electrical circuit.
2. Name the four measurable circuit elements contained in an electrical circuit.
3. Define current. What is the unit of measurement for current? In what direction does current flow through a circuit?
4. What is the Law of Electrical Charges?
5. What is the purpose of a conductor? What is the purpose of an insulator?
6. What is a closed circuit? What is an open circuit?
7. Define voltage or EMF. What is the unit of measurement for voltage? What is a voltage drop? What is a voltage rise?
8. Describe the difference between a DC voltage and an AC voltage.
9. Define resistance. What is the unit of measurement for resistance? What is the difference between a fixed resistor and a variable resistor?
10. Convert 2,500,000 ohms to megohms. Convert 6.3 kilohms to ohms.
11. Define power. What is the unit of measurement for power? One horsepower is the equivalent of how many watts? Explain what is meant by the wattage rating of a device.
12. Explain the difference between an analog and a digital meter. What is the name for a combination voltmeter, ohmmeter, and ammeter?
13. How is an ammeter connected to a circuit? How is a voltmeter connected to a circuit? What is meant by observing polarity?
14. What is the most important thing to remember about using an ohmmeter? Why?
15. An ohmmeter is set to the R x 10K range. The pointer indicates 5.5 on the ohms scale. How much resistance is being measured by the ohmmeter?
16. Describe how a meter is adjusted for zero ohms.
17. State Ohm's Law and give the formula. If the voltage applied to a circuit is increased, how is the current in the circuit affected? If the resistance in a circuit is increased, how does this affect the current in the circuit?
18. The voltage applied to a circuit is 120V. The resistance in the circuit equals 1200 ohms. How much current will flow in this circuit? Give answer in milliamps.
19. If a circuit contains 1000 ohms of resistance and the current in the circuit cannot exceed 0.1A, what is the maximum voltage that can be applied to the circuit?
20. The power source for a circuit supplies 30V. The current in this circuit must not exceed 50mA. What is the smallest value of resistance that the circuit can contain and still limit the current to 50mA?
21. State Watt's Law and give the formula. If the current in a circuit is decreased, how is the power affected? If the voltage in a circuit is increased, how is the power affected?
22. What is the purpose of an attenuator? Why would a series dropping resistor be used in a circuit?
23. What is a solenoid? How are solenoids most often used in hydraulic applications?
24. Describe the difference between an air gap solenoid and a wet armature solenoid.
25. Describe the main cause of solenoid failure.
26. What is the purpose of a relay? What is meant by N.O. and N.C. contacts? Explain one way in which a relay might be used.
27. What is the function of a transformer? What is meant by the turns ratio of a transformer? Name two kinds of transformers and explain how they are constructed with respect to turns ratio.
28. Why is a diode called a unidirectional device? How is this characteristic of a diode used to change AC into pulsating DC?
29. What is the difference between a half wave and a full wave rectifier?
30. Name the component most often used to filter pulsating DC to pure DC and explain how it works.
31. Name the four requirements for developing skill in troubleshooting electrical circuits.
32. Name the five steps of the logical troubleshooting procedure.
33. Describe how to test a potentiometer.
34. Describe how to test a solenoid coil.
35. Describe how to test a relay.
36. Describe how to test a transformer.
37. How is a diode tested? What is meant by the front to back ratio of a diode?
38. Name the two types of electrical ground and explain the difference between them.

39. Explain why a zero reference point is so important.
40. Describe a method of testing for isolation between earth and chassis ground in electrical equipment.
41. What is the cross-sectional area (in circular-mils) of a wire with a diameter of 30 mils?
42. Which conductor has the larger cross-sectional area No. 14 AWG or No. 8 AWG?
43. Describe the characteristics of THW insulation.
44. Increasing the diameter of a wire (increases/decreases) its resistance?
45. As the length of a wire increases, its resistance (increases/decreases)?
46. What is meant by the ampacity rating of a conductor and why is it important?
47. A 575 watt device must be operated at a distance of 340 feet (103.63 m) from a 115 volt power supply. The voltage drop must not exceed 2 percent. What size wire should be used for this circuit?
48. What is the voltage drop in a 115 volt circuit 45 feet (13.72 m) long if No. 12 wire is used to carry 15 amps? Is the calculated voltage drop within 2 percent?
49. When using shielded wire or cable, how should the shield be connected to the circuit?
50. Why is it so dangerous to put both hands in an electrical circuit? What is the reason for not working alone?
51. How much current produces muscle paralysis? How much current will cause death? What two things determine the amount of current received in an electrical shock?

CHAPTER 10
PRESSURE CONTROLS

Pressure control valves perform functions such as limiting maximum system pressure or regulating reduced pressure in certain portions of a circuit, and other functions wherein their actuation is a result of a change in operating pressure. Their operation is based on a balance between pressure and spring force. Most are **infinite positioning** that is, the valves can assume various positions between fully closed and fully open, depending on flow rate and pressure differential.

Pressure controls are usually named for their primary function, such as relief valve, sequence valve, brake valve, etc. They are classified by size, pressure operating range, and type of connections. The valves covered in this chapter are typical of the pressure controls used in most industrial systems.

RELIEF VALVES

The relief valve is found in virtually every hydraulic system. It is a normally-closed valve connected between the pressure line (pump outlet) and the reservoir. Its purpose is to limit pressure in the system to a preset maximum by diverting some or all of the pump's output to tank when the pressure setting is reached.

Direct Acting Relief Valve

A direct acting relief valve (Figure 10-1) may consist of nothing but a ball or poppet held seated in the valve body by a heavy spring. When pressure at the inlet is insufficient to overcome the force of the spring, the valve remains closed. When the preset pressure is reached, the ball or poppet is forced off its seat and allows flow through the outlet to tank for as long as pressure is maintained.

In most of these valves, an adjusting screw is provided to vary the spring force. Thus, the valve can be set to open at any pressure within its specified range.

Cracking Pressure. The pressure at which the valve first begins to divert flow is called the cracking pressure. As flow through the valve increases, the poppet is forced farther off its seat causing increased compression of the spring. When the valve is bypassing its full rated flow, the pressure can be considerably higher than the cracking pressure.

Full-Flow Pressure. Pressure at the inlet when the valve is passing its maximum volume is called full-flow pressure.

Pressure Override. The difference between full-flow pressure and cracking pressure is sometimes called pressure override. In some cases, pressure override may not be objectionable. In others, it can result in considerable wasted power due to the fluid lost through the valve before its maximum setting is reached. It can permit maximum system pressure to exceed the ratings of other components. Where it is desirable to minimize override, a pilot operated relief valve should be used.

Pilot Operated Relief Valve

A pilot operated relief valve (Figure 10-2) operates in two stages. The pilot stage in the upper valve body contains the pressure limiting valve, a poppet held against a seat by an adjustable spring. The port connections are made to the lower body, and diversion of the full-flow volume is accomplished by the **balanced piston** in the lower body.

Balanced Piston. The balanced piston is so named because in normal operation (Figure 10-3A) it is in hydraulic balance. Pressure at the inlet port acting under the piston is also sensed on its top by means of an orifice drilled through the large land. At any pressure less than the valve setting, the piston is held on its seat by a light spring.

When pressure reaches the setting of the adjustable spring, the poppet is forced off its seat limiting pressure in the upper chamber (Figure 10-3B).

The restricted flow through the orifice into the upper chamber results in an increase in pressure in the lower chamber. This unbalances the hydraulic forces and tends to raise the piston off its seat. When the difference in pressure between the upper and lower chambers is sufficient to overcome the force of the light spring (approxi-

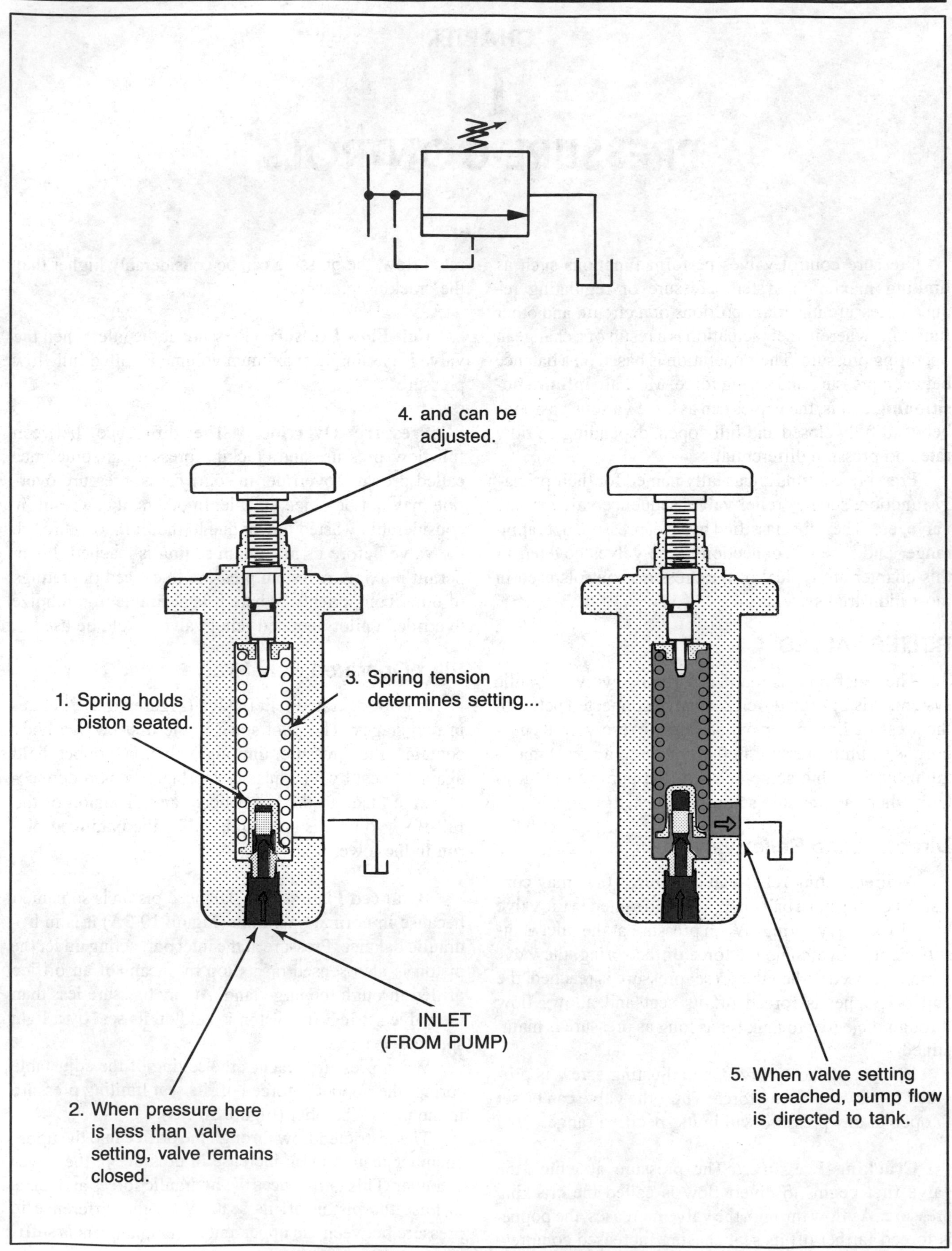
4. and can be adjusted.
1. Spring holds piston seated.
3. Spring tension determines setting...
INLET (FROM PUMP)
2. When pressure here is less than valve setting, valve remains closed.
5. When valve setting is reached, pump flow is directed to tank.

Figure 10-1. Direct acting relief valve.

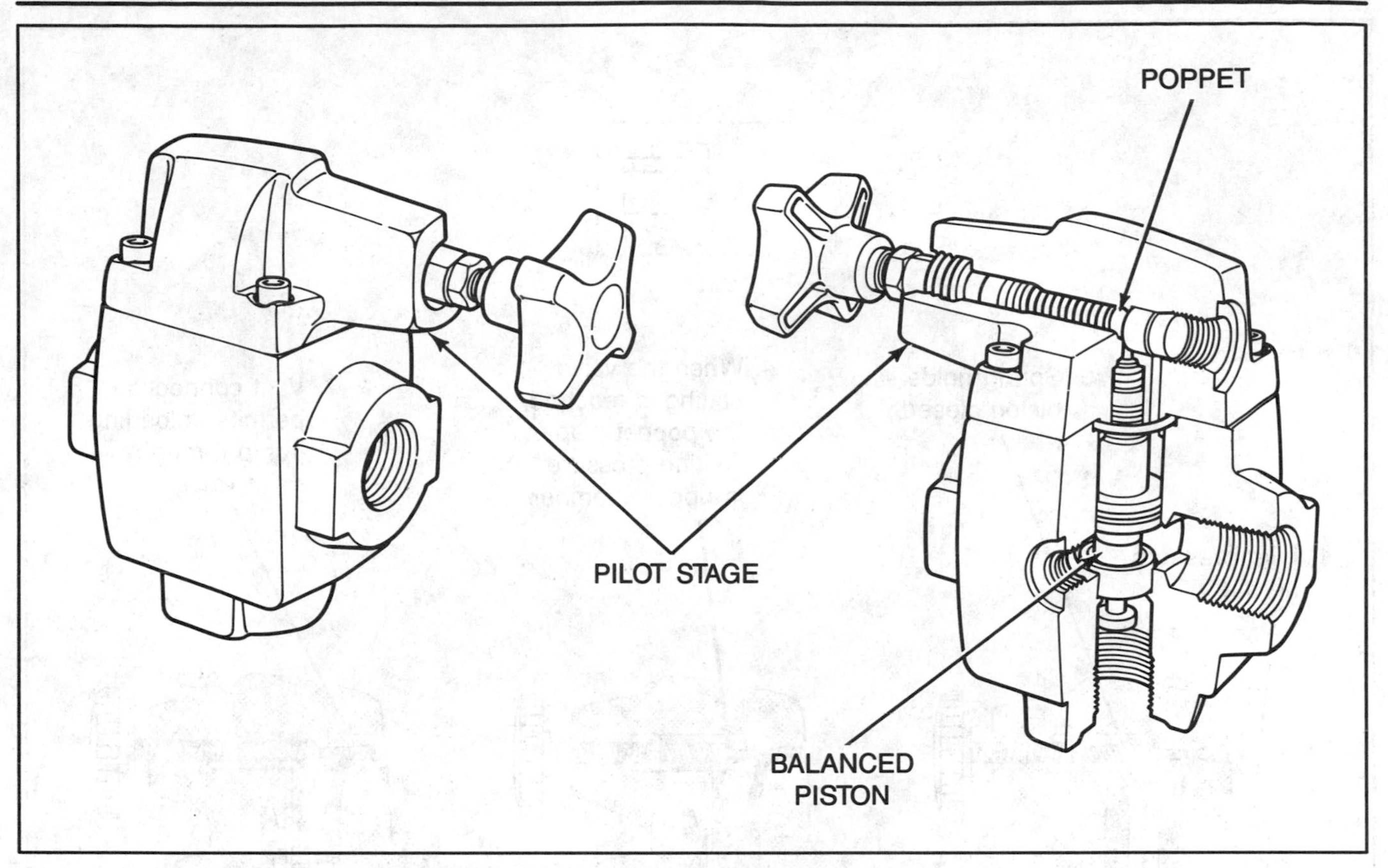

Figure 10-2. Pilot operated relief valve.

mately 20 psi (1.38 bar) (137.88 kPa)), the large piston unseats permitting flow directly to tank (Figure 10-3C). Increased flow through the valve causes the piston to lift further off its seat but since this compresses only the light spring very little override is encountered.

Vent Connection. Pilot operated relief valves may be remotely controlled by means of an outlet port from the chamber above the piston. When this chamber is vented to tank, the only force holding the piston on its seat is that of the light spring, and the valve will open fully at approximately 20 psi (1.38 bar) (137.88 kPa). See Figure 10-4.

Occasionally, this standard spring is replaced by a heavier one permitting vent pressures of approximately 80 psi (5.52 bar) (551.52 kPa) when required for pilot pressure. A second benefit of the heavier vent spring is that it causes faster and more positive seating of the piston.

Remote Control. It is also possible to connect a direct acting relief valve to the vent connection to control pressure from a remote location (Figure 10-5). To exercise control, the remote valve must be set for a lower pressure than the integral pilot stage. An application of remote pressure control is illustrated in Chapter 19.

Multiple Preset Pressures. Multi-pressure, solenoid operated relief valves are also available that provide an ability to electrically select one pressure from a set of given preset pressures. Bi-pressure relief valves can be used to select either of two preset pressures or one pressure and a vent. Tri-pressure relief valves (Figure 10-6A) can be used to electrically select any one of three preset pressures, or two pressures and a vent. The main stage in either valve is a balanced piston type relief valve, and the intermediate stages are of the spring-loaded poppet type. Heads in the intermediate stages contain springs with different pressure ratings.

As shown in Figure 10-6B, pilot head No. 3 will control the pressure when solenoid A is energized. Pilot head No. 1 controls the pressure when solenoid B is energized. With both solenoids de-energized, pilot head No. 2 will control the pressure.

Electrically Modulated. An electrically modulated relief valve provides the capability to modulate system pressure using a remote electrical controller. The pressure setting of the valve is approximately proportional to the input current; increasing current provides increasing pressure.

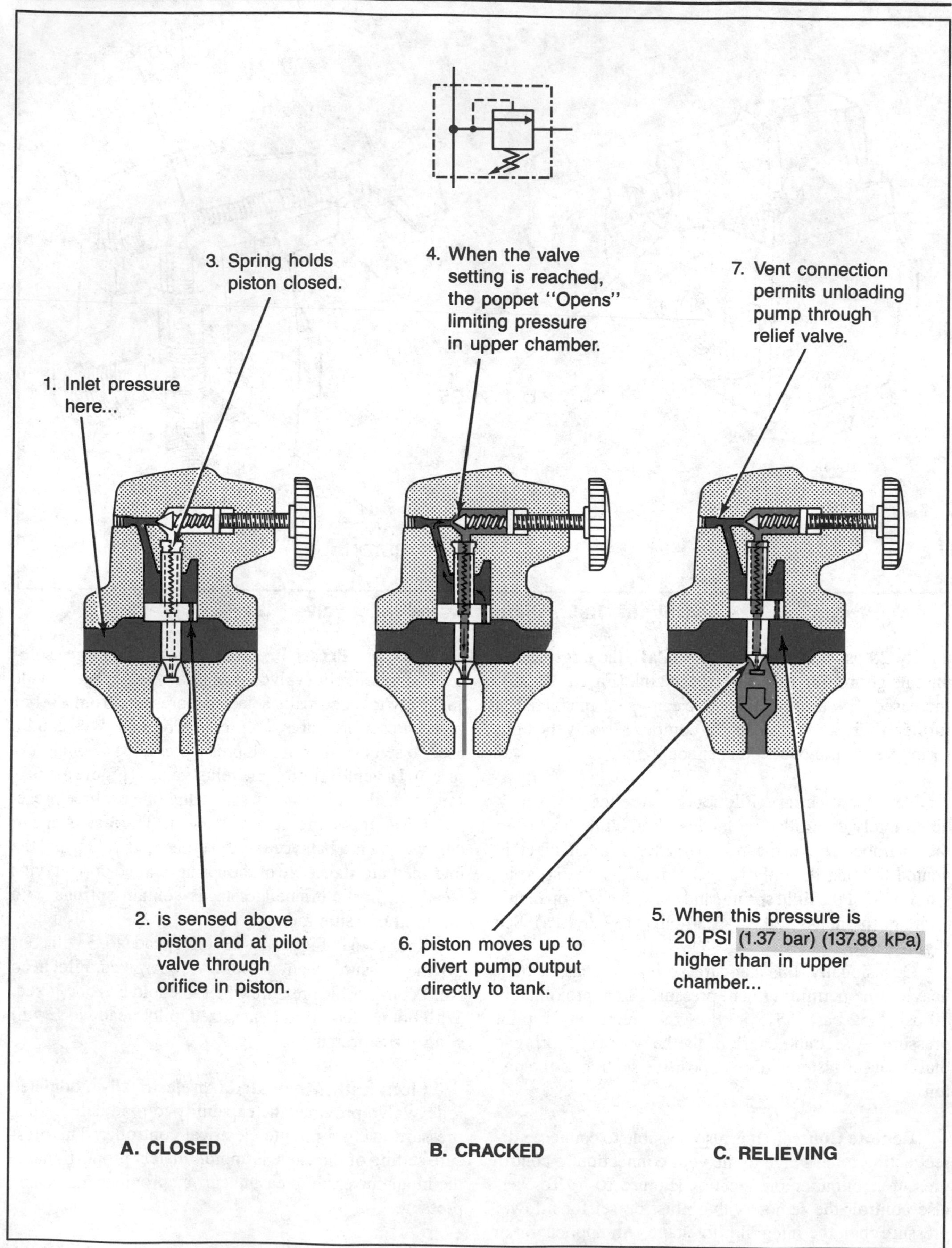
3. Spring holds piston closed.
4. When the valve setting is reached, the poppet "Opens" limiting pressure in upper chamber.
7. Vent connection permits unloading pump through relief valve.
1. Inlet pressure here...
2. is sensed above piston and at pilot valve through orifice in piston.
6. piston moves up to divert pump output directly to tank.
5. When this pressure is 20 PSI (1.37 bar) (137.88 kPa) higher than in upper chamber...
A. CLOSED
B. CRACKED
C. RELIEVING

Figure 10-3. Operation of balanced piston relief valve.

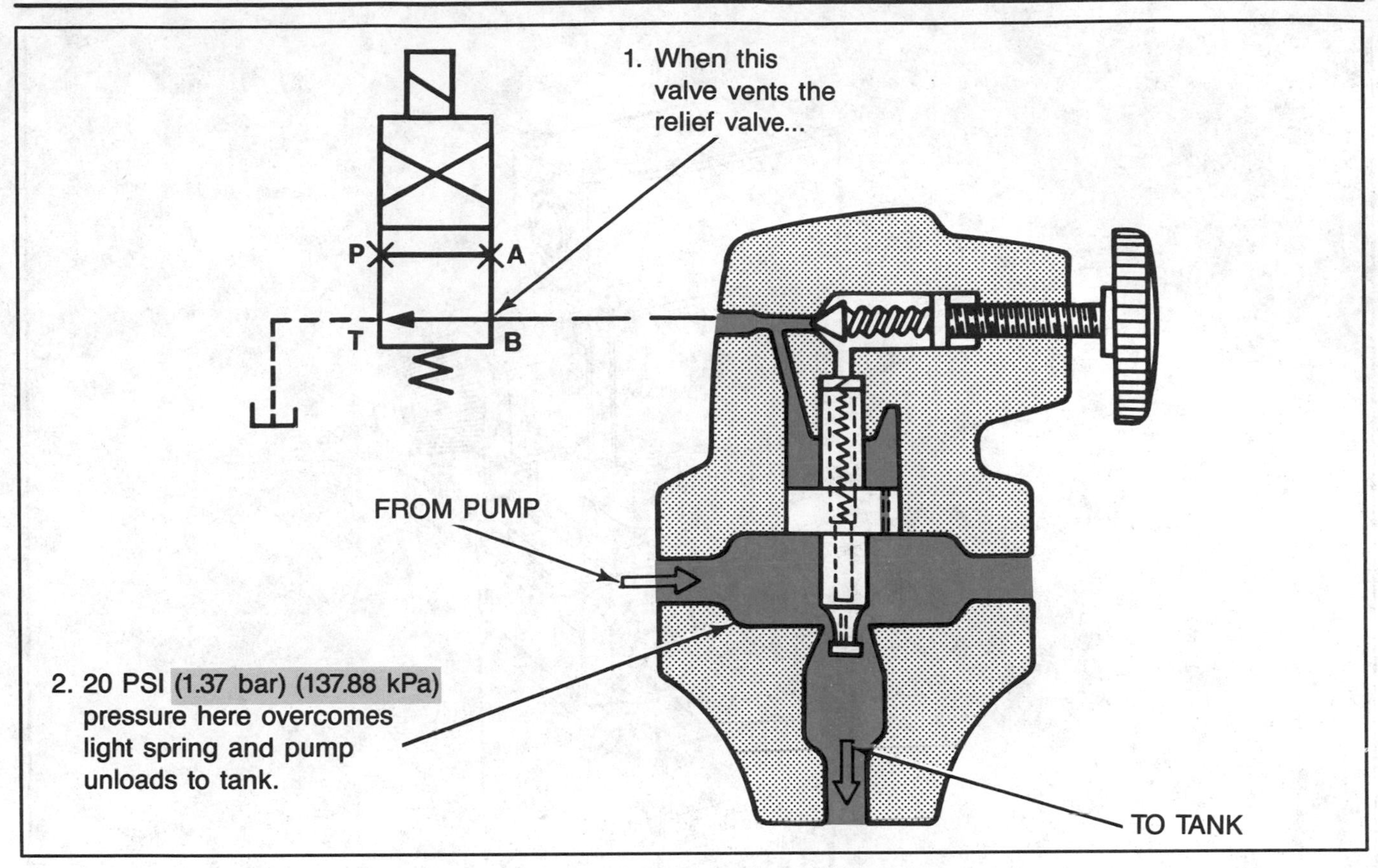

1. When this valve vents the relief valve...
P
A
T
B
FROM PUMP
2. 20 PSI (1.37 bar) (137.88 kPa) pressure here overcomes light spring and pump unloads to tank.
TO TANK

Figure 10-4. Venting the relief valve.

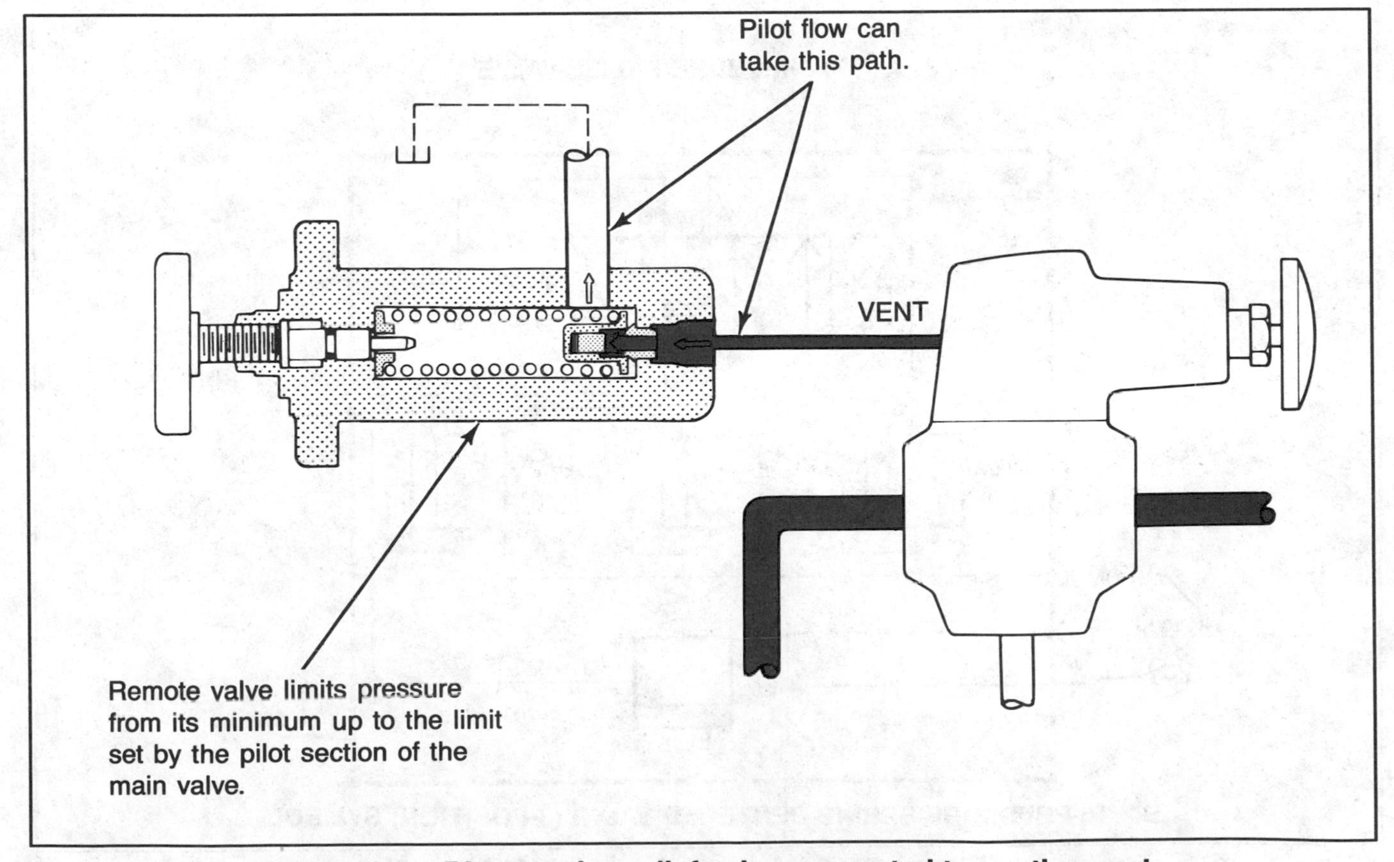

Pilot flow can take this path.
VENT
Remote valve limits pressure from its minimum up to the limit set by the pilot section of the main valve.

Figure 10-5. Direct acting relief valve connected to venting port.

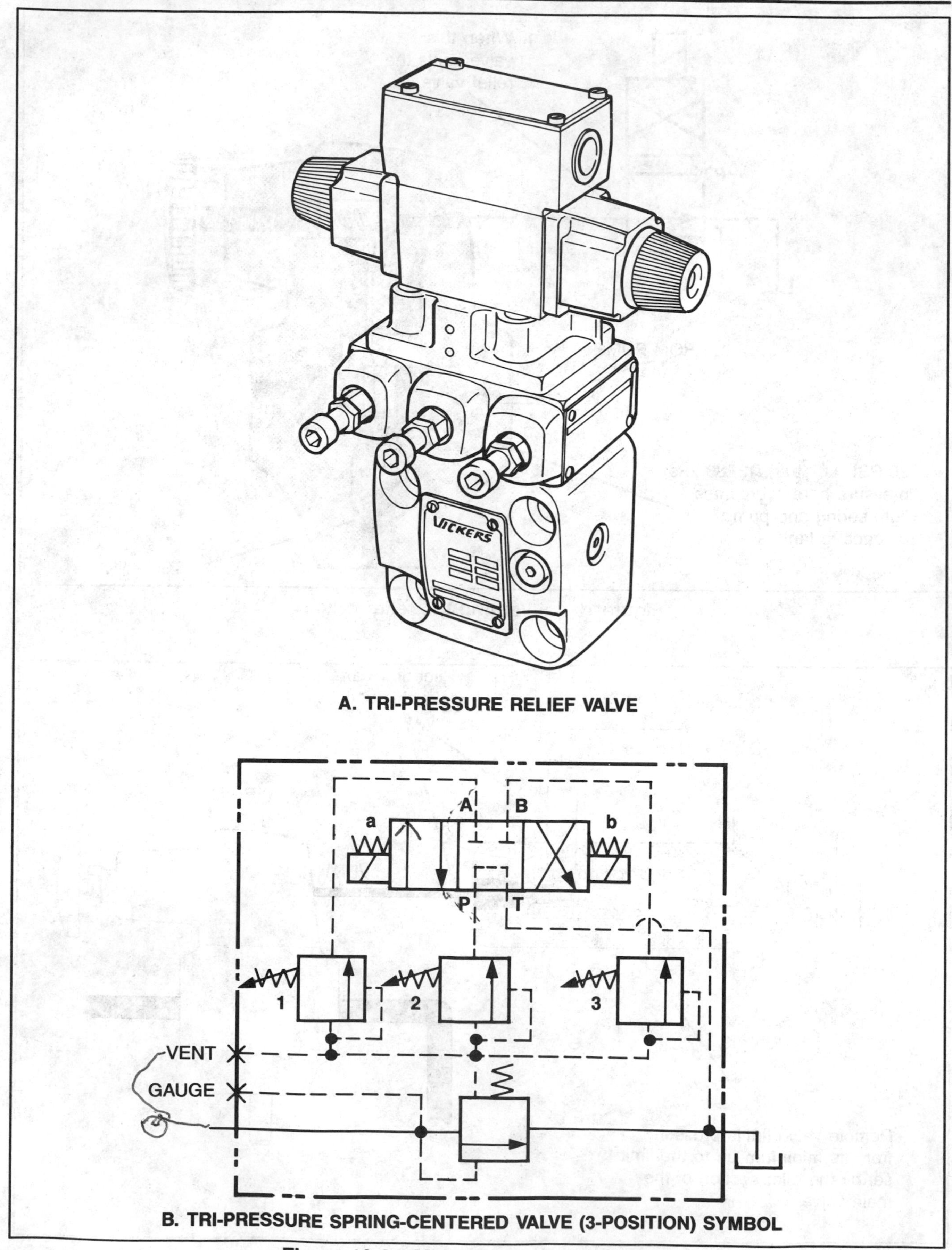

Figure 10-6. Multi-pressure relief valve.

KACG-6 valve

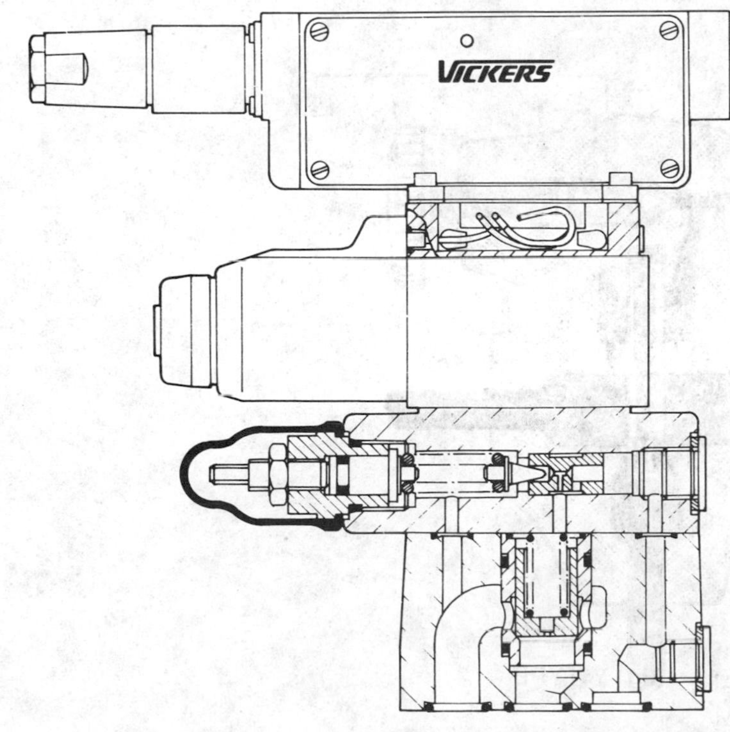

Manual Pilot Internally Drained to Port T; Electrical Pilot Drained to Side Drain Port

Symbol for KACG models.
For KCG models omit 7-pin plug (▲) and amplifier.

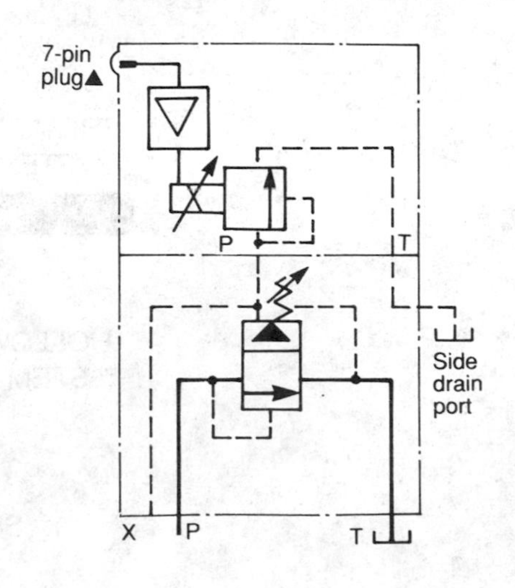

Figure 10-7. Proportional pressure relief valve.

An electrically modulated relief valve (Figure 10-7A) consists of three basic parts: a main stage, an intermediate body, and the electrically modulated pilot. The main stage is similar to the standard relief valve. The intermediate body contains standard relief valve pilot parts to provide manual adjustment and has a mounting pad which accepts the electrically modulated pilot. The pilot contains a flapper valve assembly, a blocking valve, and a ball-type check valve.

This valve provides a self-contained unit that modulates system pressure electrically over a wide range without external feedback devices. The valve has the ability to control system pressure from any distance or location. Modulating signals may be derived from potentiometers, analog computers, power supplies or any other source that will provide the necessary drive current for the pilot stage.

Figure 10-7B is a functional diagram of the electrically modulated relief valve. With zero current to the valve, flow and pressure conditions are as follows: oil flows into the pressure inlet of the valve and the light spring holds the main piston closed against the seat until the pressure below the piston builds up to about 80 psi (5.52 bar) (551.52 kPa). The 80 psi (5.52 bar) (551.52 kPa) pressure differential lifts the piston, porting oil to the reservoir. Also, the 80 psi (5.52 bar) (551.52 kPa) causes oil to flow through the self-cleaning filter screen and the control orifice, into the area above the piston and the valve body passages. This flow continues through the orifice formed by the flapper and nozzle which develops a back pressure on the top of the main piston and increases the original 80 psi (5.52 bar) (551.52 kPa) to approximately 150 psi (10.34 bar) (1034.10 kPa). The flow continues into the pilot cover, through the 5 psi (0.34 bar) (34.47 kPa) ball check and out the pilot drain.

The following action occurs when current to the coil is increased:

The increased current strengthens the magnetic field and attracts the flapper with a greater force toward the nozzle.

The modulating orifice decreases in size, increasing the pressure at the upstream side of the orifice. This increase in pressure is reflected back to the top of the main piston and adds to the light spring force. A higher pressure is then required at the inlet port to lift the piston.

PILOT OPERATED SEQUENCE VALVE

The construction of a pilot operated sequence valve differs from the pilot operated relief valve in that the drain passage from the pilot stage is external rather than

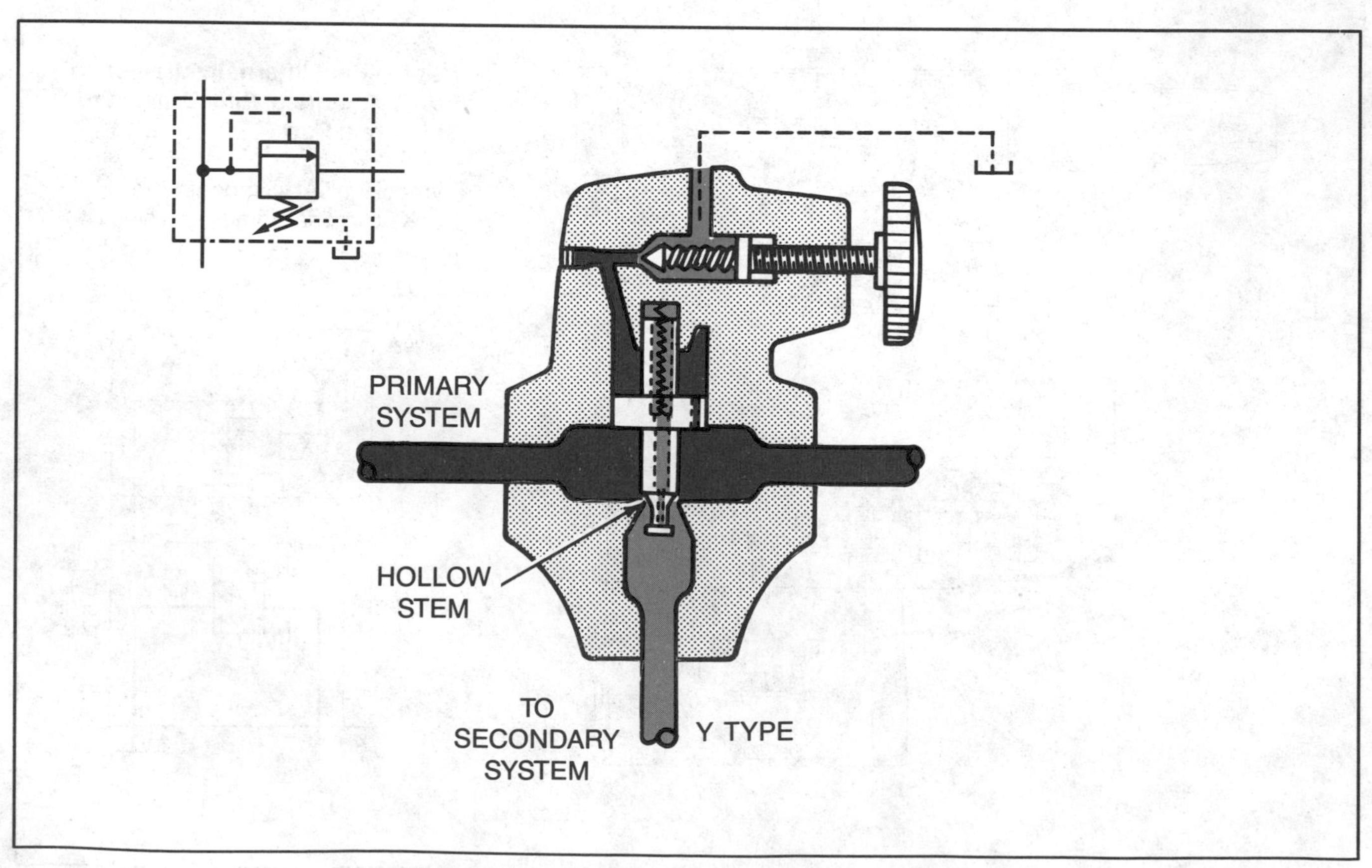

Figure 10-8. Two-stage sequence valve.

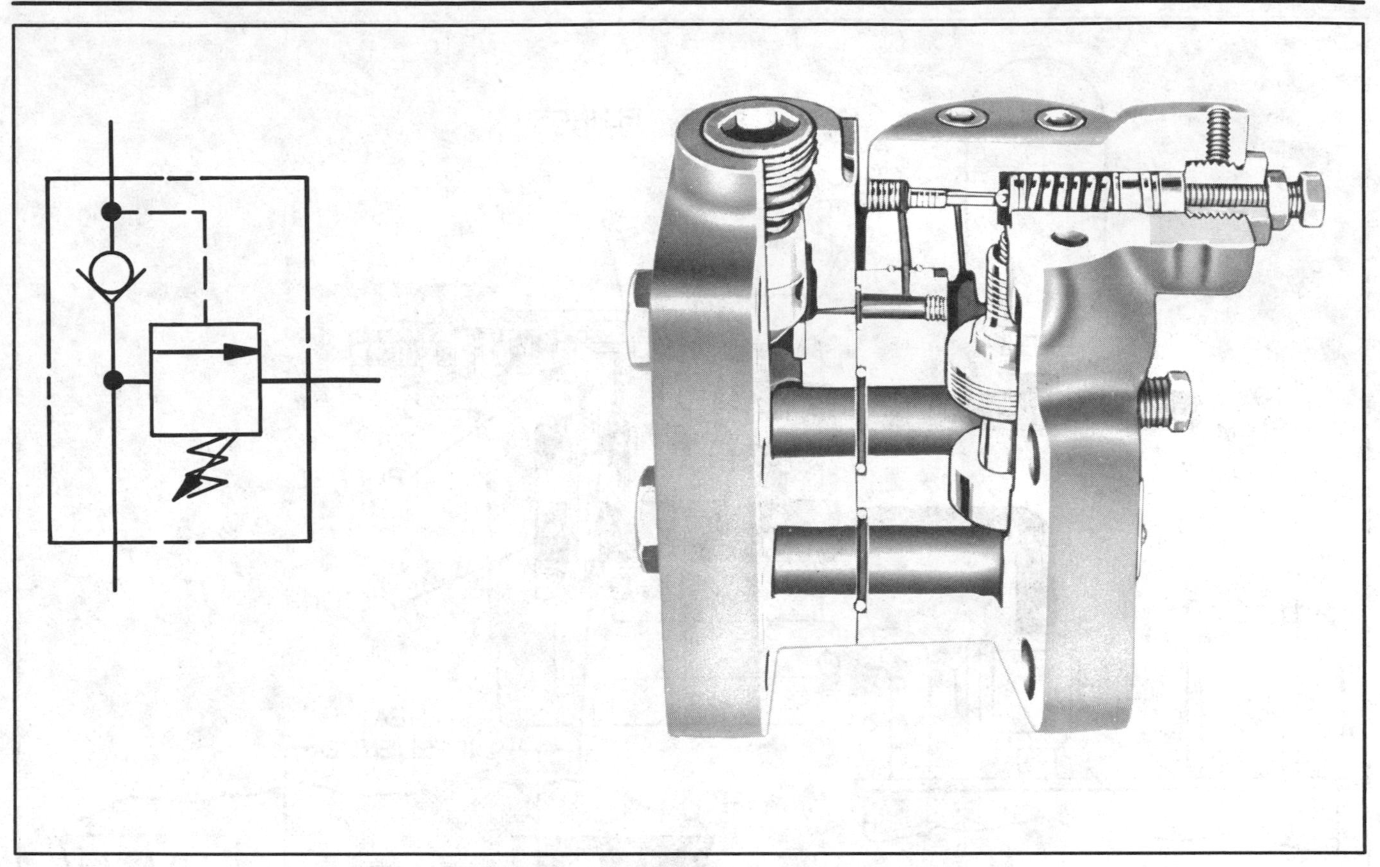

Figure 10-9. Unloading relief valve.

through the stem of the balanced piston. In operation, the primary system passage is connected below the piston land, with the secondary system connected to the bottom port. Sequencing occurs when the primary system pressure is about 20 psi (1.38 bar) (137.88 kPa) higher than the pilot valve cracking pressure.

Operation

Referring to Figure 10-8, the stem of the balanced piston is hollow. Secondary system pressure below the piston is sensed at the top of the stem, and balances the pressure under the stem. Secondary system pressure then has no effect on the piston movement.The piston thus remains in infinite positioning and maintains the preset pressure in the primary system. As the piston opens at the preset pressure, flow is routed to the secondary system. Reverse flow is not possible. If required, a check valve is used to permit flow from secondary to primary.

UNLOADING RELIEF VALVE

An unloading relief valve (Figure 10-9) is used in accumulator charging circuits to (1) limit maximum pressure and (2) unload the pump when the desired accumulator pressure is reached.

In construction, it contains a compound, balanced piston relief valve, a check valve to prevent reverse flow from the accumulator and a pressure-operated plunger which vents the relief valve at the selected pressure.

Operation

Figure 10-10A illustrates the flow condition when the accumulator is charging. The relief valve piston is in balance and is held seated by its light spring. Flow is through the check valve to the accumulator.

In Figure 10-10B, the preset pressure has been reached. The relief valve poppet has unseated limiting pressure above the piston and on the poppet side of the plunger. Further increase in system pressure acting on the opposite end of the plunger has caused it to force the poppet completely off its seat, in effect, venting the relief valve and unloading the pump. The check valve has closed permitting the accumulator to maintain pressure in the system.

Because of the difference in area between the plunger and poppet seat (approximately 15 percent), when pressure drops to about 85 percent of the valve setting, the poppet and piston reseat and the cycle is repeated.

PRESSURE REDUCING VALVES

Pressure reducing valves are normally-open pressure controls used to maintain reduced pressures in certain portions of the system. They are actuated by pressure

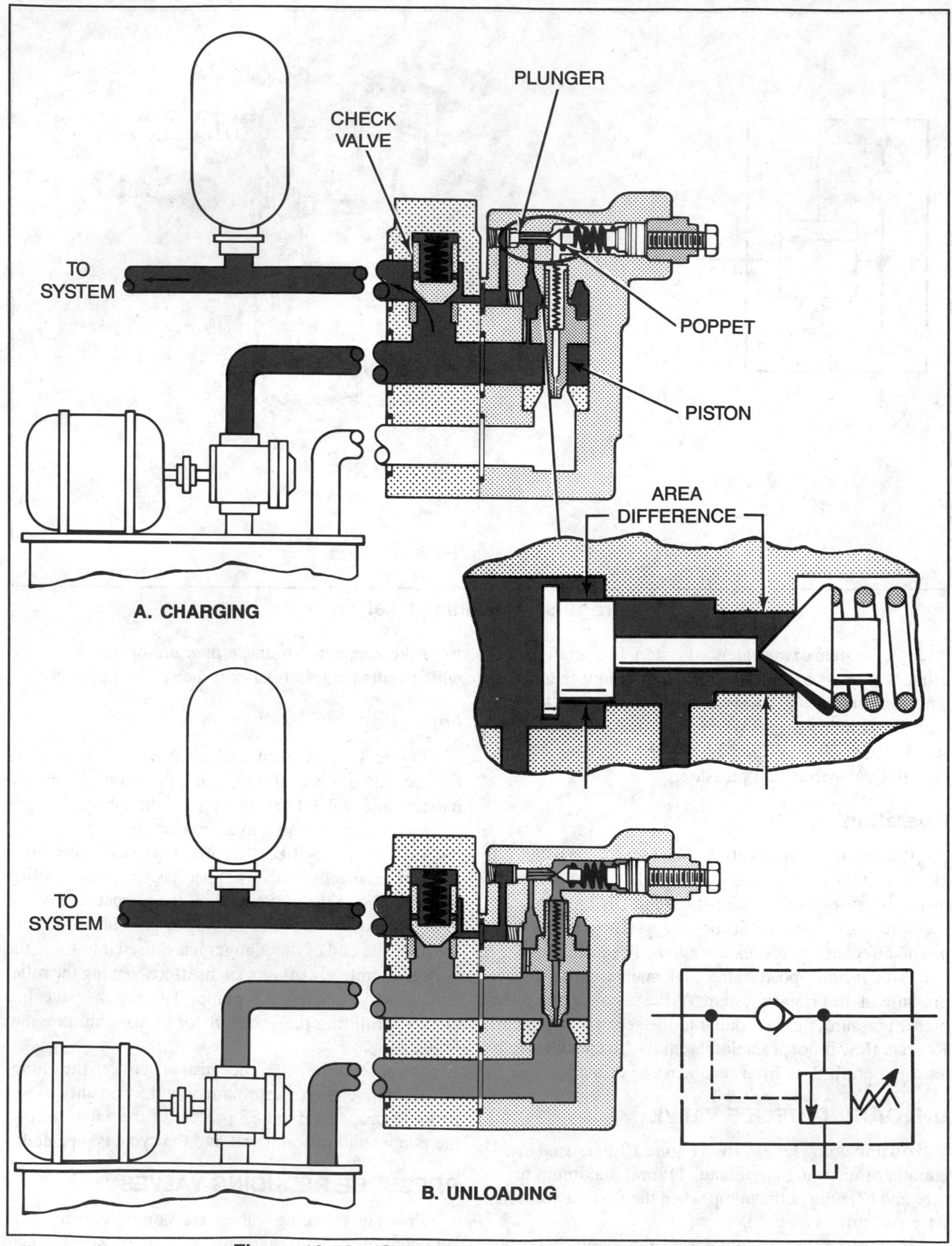

Figure 10-10. Operation of unloading relief valve.

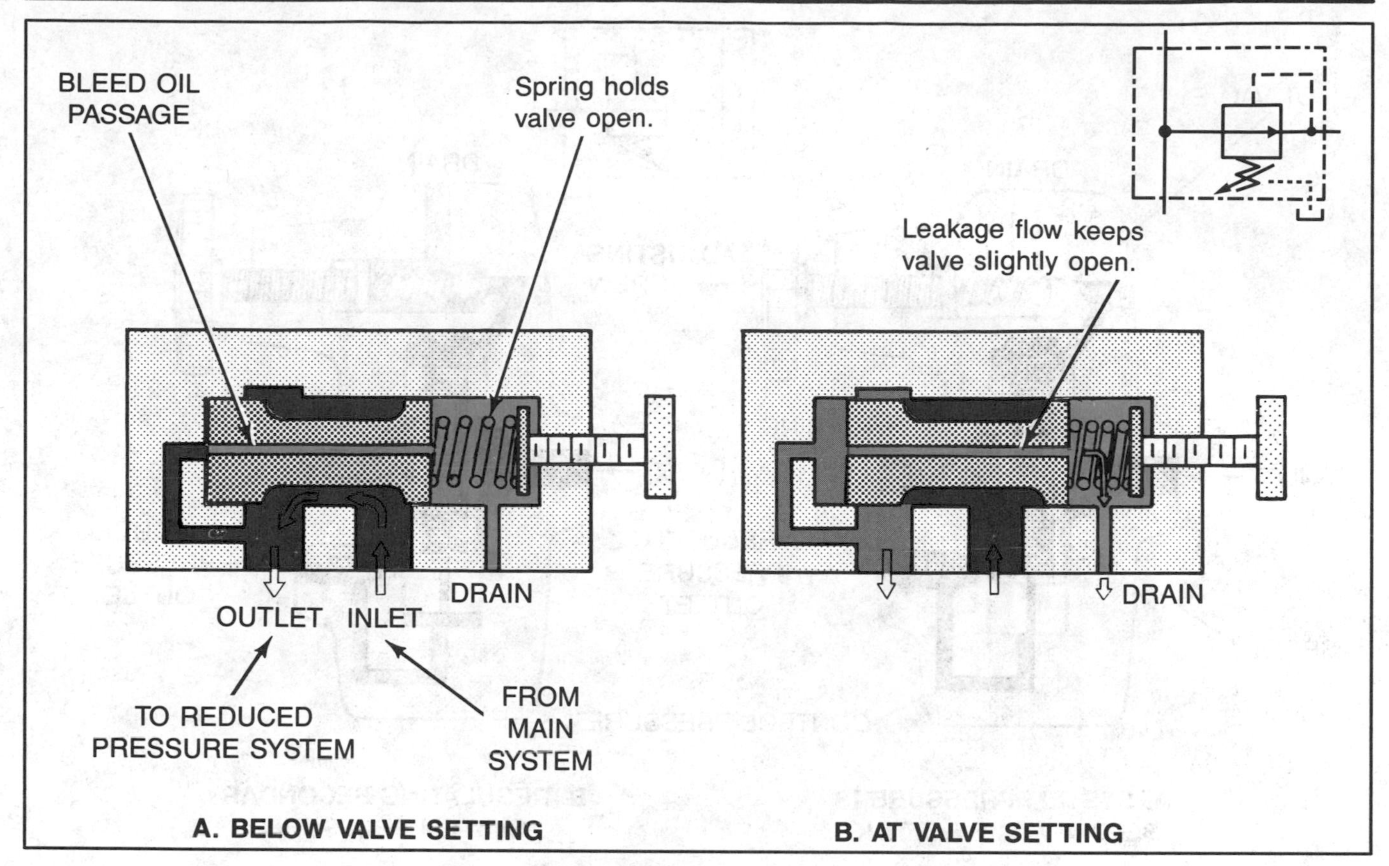

Figure 10-11. Direct acting pressure reducing valve.

sensed in the branch circuit and tend to close as it reaches the valve setting, thus preventing further buildup. Both direct acting and pilot operated versions are in use.

Direct Acting Pressure Reducing Valves

A typical direct acting valve is shown in Figure 10-11. It uses a spring-loaded spool to control the downstream pressure.

If the main supply pressure is below the valve setting, fluid will flow freely from the inlet to the outlet. An internal connection from the outlet passage transmits the outlet pressure to the spool end opposite the spring (Figure 10-11A).

When the outlet pressure rises to the valve setting (Figure 10-11B), the spool moves to partly block the outlet port. Only enough flow is passed to the outlet to maintain the preset pressure. If the valve closes completely, leakage past the spool could cause pressure to build up in the branch circuit. Instead, a continuous bleed to tank is permitted to keep it slightly open and prevent downstream pressure from rising above the valve setting. A separate drain passage is provided to return this leakage flow to tank.

Pilot Operated Pressure Reducing Valves

The pilot operated pressure reducing valve (Figure 10-12) has a wider range of adjustment and generally provides more accurate control. The operating pressure is set by an adjustable spring in the pilot stage in the upper body. The valve spool in the lower body functions in essentially the same manner as the direct acting valve discussed previously.

Figure 10-12A shows the condition when supply pressure is less than the valve setting. The spool is hydraulically balanced through an orifice in its center, and the light spring holds it in the wide-open position.

In Figure 10-12B, pressure has reached the valve setting and the pilot valve is diverting flow to the drain passage limiting pressure above the spool. Flow through the orifice in the spool creates a pressure difference that moves the spool up against the spring force. The spool partially closes the outlet port to create a pressure drop from the supply to the branch system.

Again, the outlet port is never entirely closed. When no flow is called for in the branch system, there is still a continuous flow of some 60-90 cubic inches (983.23-1474.84 cubic centimeters) per minute through the spool orifice and the pilot valve to drain.

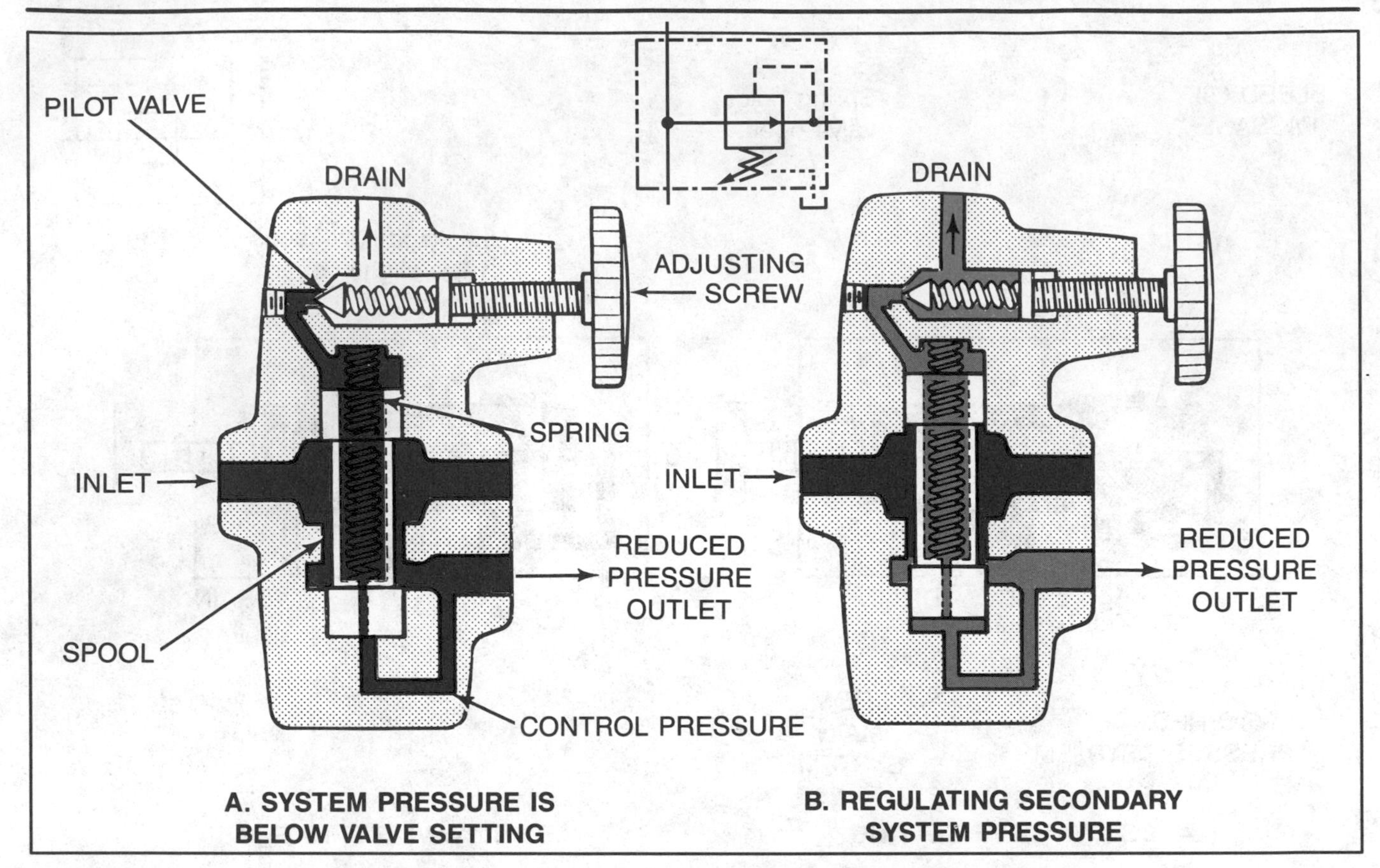

Figure 10-12. Pilot operated pressure reducing valve.

Optional Reverse Free Flow Check

The valve illustrated in Figure 10-12 will handle reverse flow only if the system pressure is less than the valve setting. If reverse flow pressure is higher, a bypass check valve is required. This is an integral part of the valve shown in Figure 10-13.

DIRECT ACTING, SPOOL-TYPE, PRESSURE CONTROL VALVES

A direct acting sliding spool-type pressure control valve is shown in Figure 10-14. The spool operates within a valve body and is held in the closed position by an adjustable spring. Operating pressure sensed through a passage in the bottom cover opposes the spring load. The spool area is such that with the heaviest spring normally used, the valve would open at approximately 125 psi (8.62 bar) (861.75 kPa). To extend their pressure range, most models include a small piston or plunger in the bottom cover to reduce the pressure reaction area to 1/8 (1/16 in the 2000 psi (137.88 bar) (13,788.00 kPa) range) of the area of the spool end. When operating pressure exceeds the valve setting, the spool is raised and oil can flow from the primary to the secondary port.

A drain passage is provided in the top cover to drain the spring chamber. This drain also removes leakage oil from the space between the spool and piston by means of a passage drilled lengthwise through the spool.

Depending on the assembly of the top and bottom covers, this valve can be used as a back pressure valve, unloading valve, sequence valve, counterbalance valve, or brake valve. It is built with an optional integral check when reverse flow is required from the secondary to the primary port.

Back Pressure Valve

Figure 10-15 illustrates the valve assembled for back pressure operation. The line that requires back pressure is connected to the primary port and the secondary port is connected to tank. This application permits the valve to be internally drained and the top cover is assembled with the drain passage aligned with the secondary port. The bottom cover is assembled so that operating pressure is sampled internally from the primary port making it necessary to maintain adequate back pressure to keep the valve open.

In Figure 10-15A, the system pressure against the piston is too low to overcome the spring and the valve remains closed. In Figure 10-15B, pressure has shifted the spool to allow flow to the secondary port and to tank at the pressure determined by the spring setting.

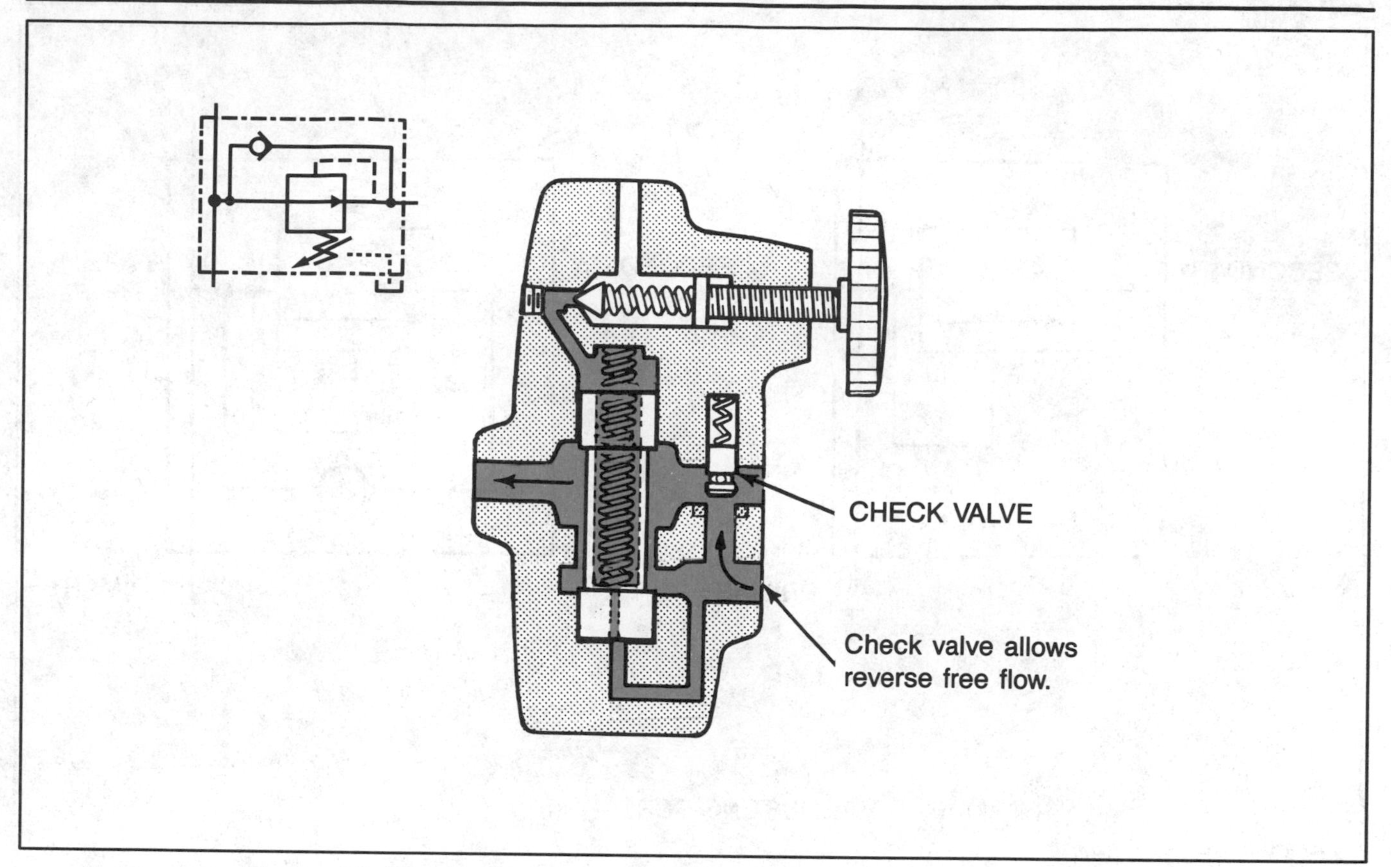
CHECK VALVE
Check valve allows
reverse free flow.

Figure 10-13. Pressure reducing valve with internal check valve.

Unloading Valve

To use the same valve as an unloading valve (Figure 10-16), the lower cover is assembled to block the internal operating pressure passages. An external pressure source is used to move the spool and divert pump delivery to the secondary port. The drain connection remains internal, since the secondary port is still connected to the tank.

Note the operating difference between the unloading and back pressure valves. The back pressure valve operates in balance, being held open at one of an infinite number of positions by the flow of oil through it. Back pressure maintained at the primary port is determined by the spring adjustment. With the unloading valve (Figure 10-16), however, the primary port pressure is independent of the spring force because the remote pressure source operates the spool. As long as the control pressure is at least 150 psi (10.34 bar) (1034.10 kPa) above the spring setting, free flow is permitted from the primary to the secondary port.

Sequence Valve

A sequence valve is used to cause actions to take place in a system in a definite order, and to maintain a predetermined minimum pressure in the primary line while the secondary operation occurs. Figure 10-17 shows the valve assembled for sequencing. Fluid flows freely through the primary passage to operate the first phase until the pressure setting of the valve is reached. As the spool lifts (Figure 10-17B), flow is diverted to the secondary port to operate a second phase. A typical application is clamping from the primary port and feeding a drill head from the secondary after the work piece is firmly clamped.

To maintain pressure in the primary system, the valve is internally operated. However, the drain connection must be external, since the secondary port is under pressure when the valve sequences. If this pressure were allowed in the drain passage, it would add to the spring force and raise the pressure required to open the valve.

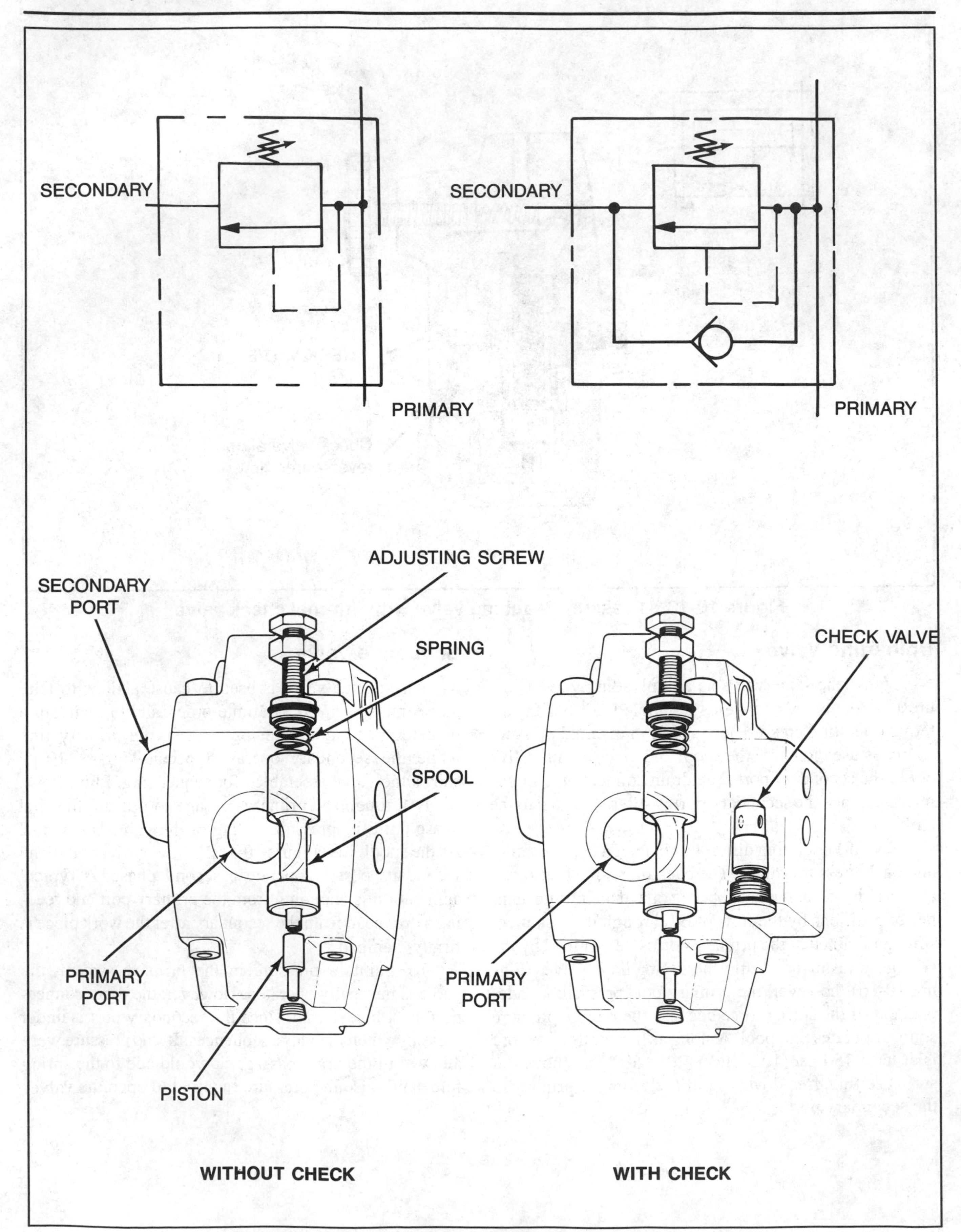

Figure 10-14. Direct acting spool-type valve.

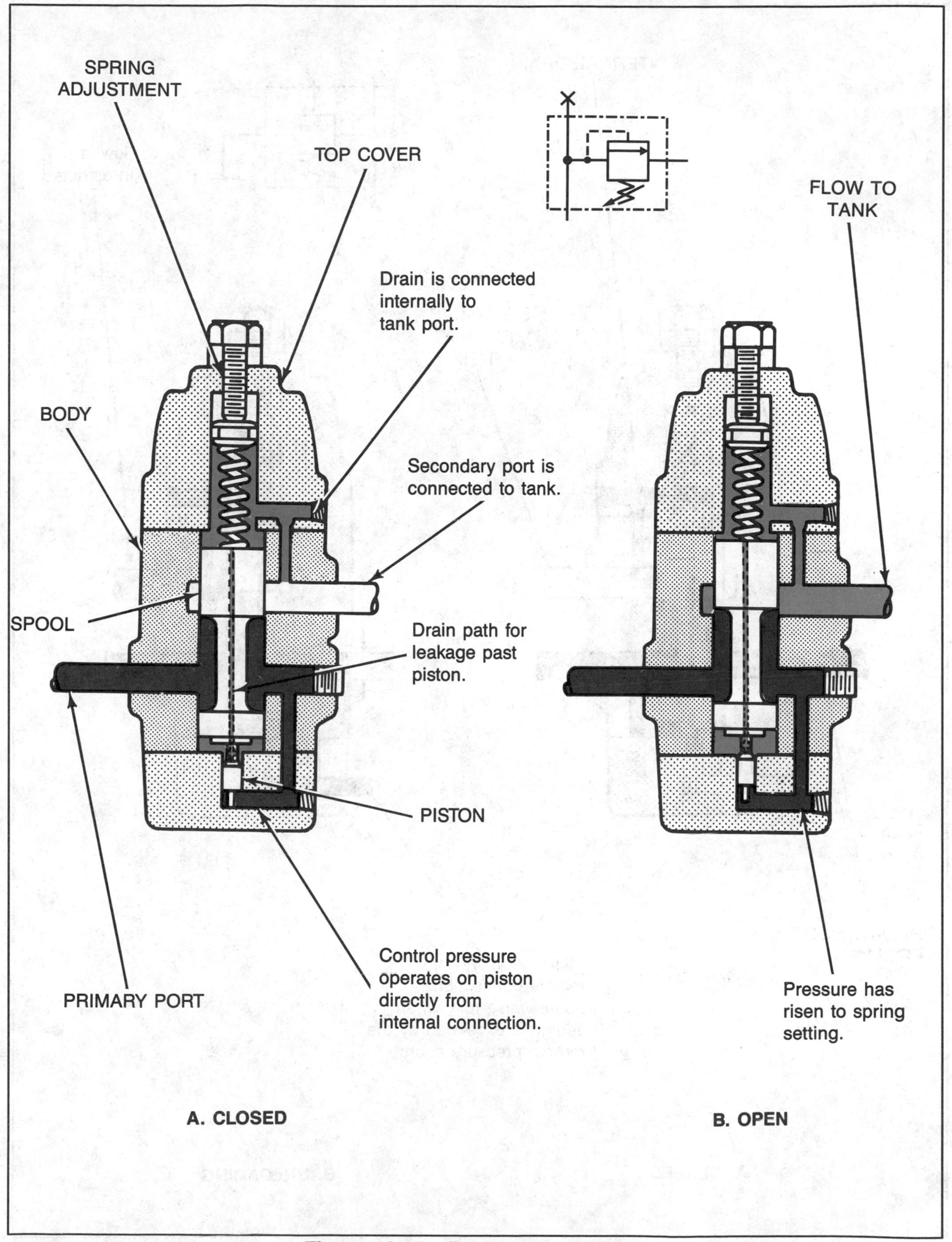

Figure 10-15. Back pressure valve.

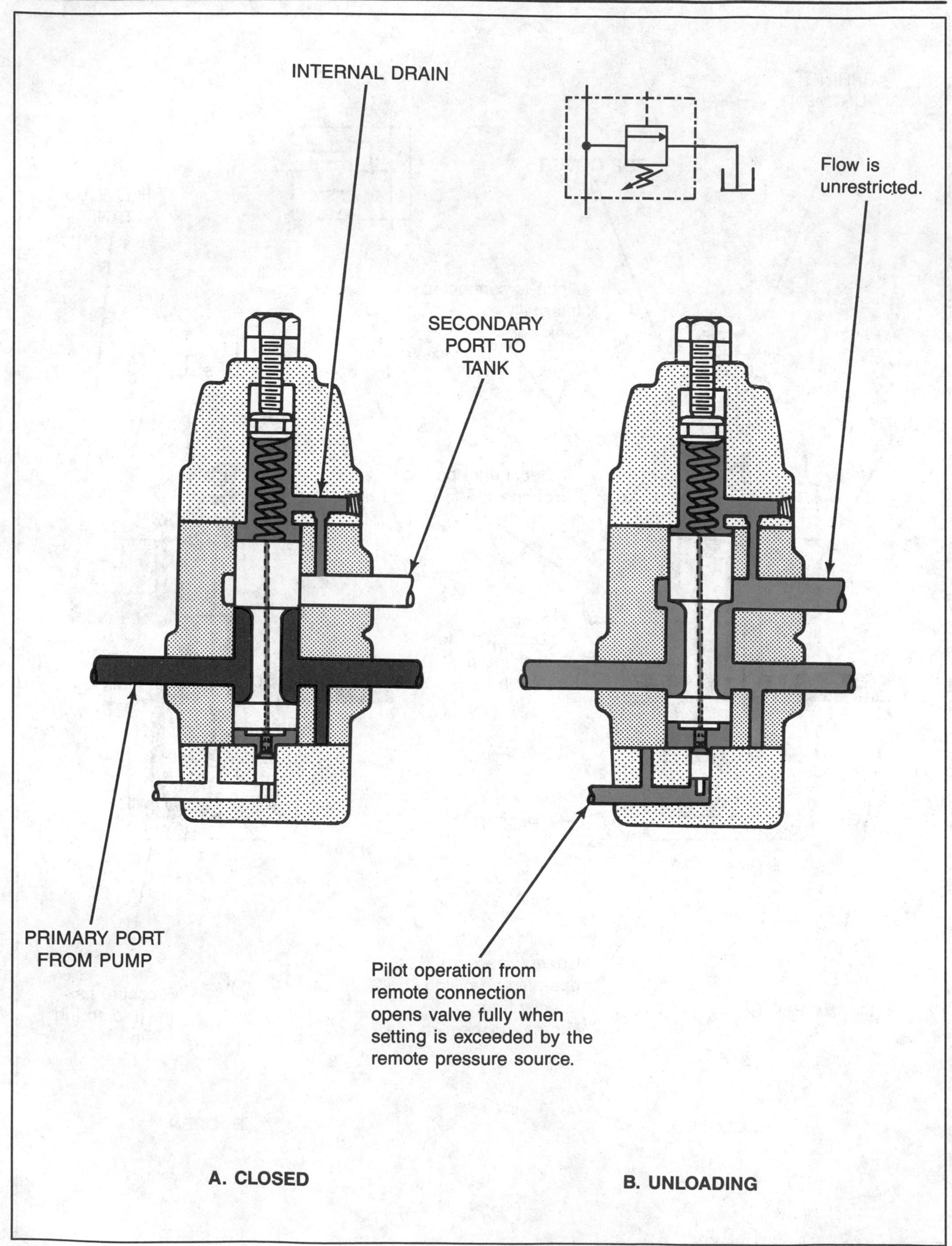
INTERNAL DRAIN
Flow is unrestricted.
SECONDARY PORT TO TANK
PRIMARY PORT FROM PUMP
Pilot operation from remote connection opens valve fully when setting is exceeded by the remote pressure source.
A. CLOSED
B. UNLOADING

Figure 10-16. Unloading valve.

Sequence Valve With Integral Check. The sequence valve is suitable for systems where it can be installed upstream from the directional valve. If it is installed downstream (in a cylinder line), some provision must be made for return free flow when the cylinder is reversed. A bypass check valve can be used, or the valve can be replaced with a valve (Figure 10-18) which has an integral check valve for return flow. The operation otherwise is identical.

Remote Control. In some systems, it is desirable to provide an interlock so that sequencing does not occur until the primary actuator reaches a definite position. In these applications, the bottom cover on the sequence valve is assembled for remote operation. A cam-operated directional valve blocks the control pressure from the piston in the bottom cover until the clamp cylinder reaches the prescribed position. Only then is the sequence valve permitted to shift and direct flow to the second operation.

Counterbalance Valve

A counterbalance valve is used to maintain control over a vertical cylinder so that it will not fall freely because of gravity. The primary port of the valve is connected to the lower cylinder port and the secondary port to the directional valve (Figure 10-19). The pressure setting is slightly higher than is required to hold the load from falling.

When the pump delivery is directed to the top of the cylinder, the cylinder piston is forced down causing pressure at the primary port to increase and raise the spool, opening a flow path for discharge through the secondary port to the directional valve and subsequently to tank. In cases where it is desired to remove back pressure at the cylinder and increase the force potential at the bottom of the stroke, this valve too can be operated remotely.

When the cylinder is being raised (Figure 10-19B), the integral check valve opens to permit free flow for returning the cylinder.

The counterbalance valve can be internally drained. In the lowering position (Figure 10-19A), when the valve must be open, its secondary port is connected to tank. In the reverse condition, it does not matter that load pressure is effective in the drain passage, because the check valve bypasses the spool.

Brake Valve

A brake valve is used in the exhaust line of a hydraulic motor to (1) prevent overspeeding when an overrunning load is applied to the motor shaft and (2) prevent excessive pressure buildup when decelerating or stopping a load.

When the valve is used as a brake valve, it has a solid spool (no drain hole through the center), and there is a remote operating pressure connection in the bottom cover directly under the spool (Figure 10-20). This connection is teed into the supply line to the motor. The internal control connection also is used under the small piston and senses pressure from the primary port of the valve which is connected to the motor exhaust port.

Accelerating. When the load is being accelerated, pressure is maximum at the motor inlet and under the large area of the brake valve spool holding it in the full open position permitting free flow from the exhaust port of the motor (Figure 10-20A).

Running. When the motor gets up to speed, load pressure still holds the brake valve open **unless the load tries to run away**. If this happens, the pressure falls off at the motor inlet and in the remote control pressure passage (Figure 10-20B). The spring force tends to close the valve thus increasing the back pressure. This in turn raises the driveline pressure to the motor and under the small piston holding the valve at the proper metering position to maintain constant motor speed.

Braking. When the directional valve is shifted to neutral, inertia causes the motor to continue rotating. Until the motor stops turning, it will operate as a pump, drawing fluid from the reservoir through the directional valve and circulating it back through the brake valve.

At this time, pressure at the motor outlet tending to bring it to a stop will be whatever is required under the small piston to overcome the brake valve setting.

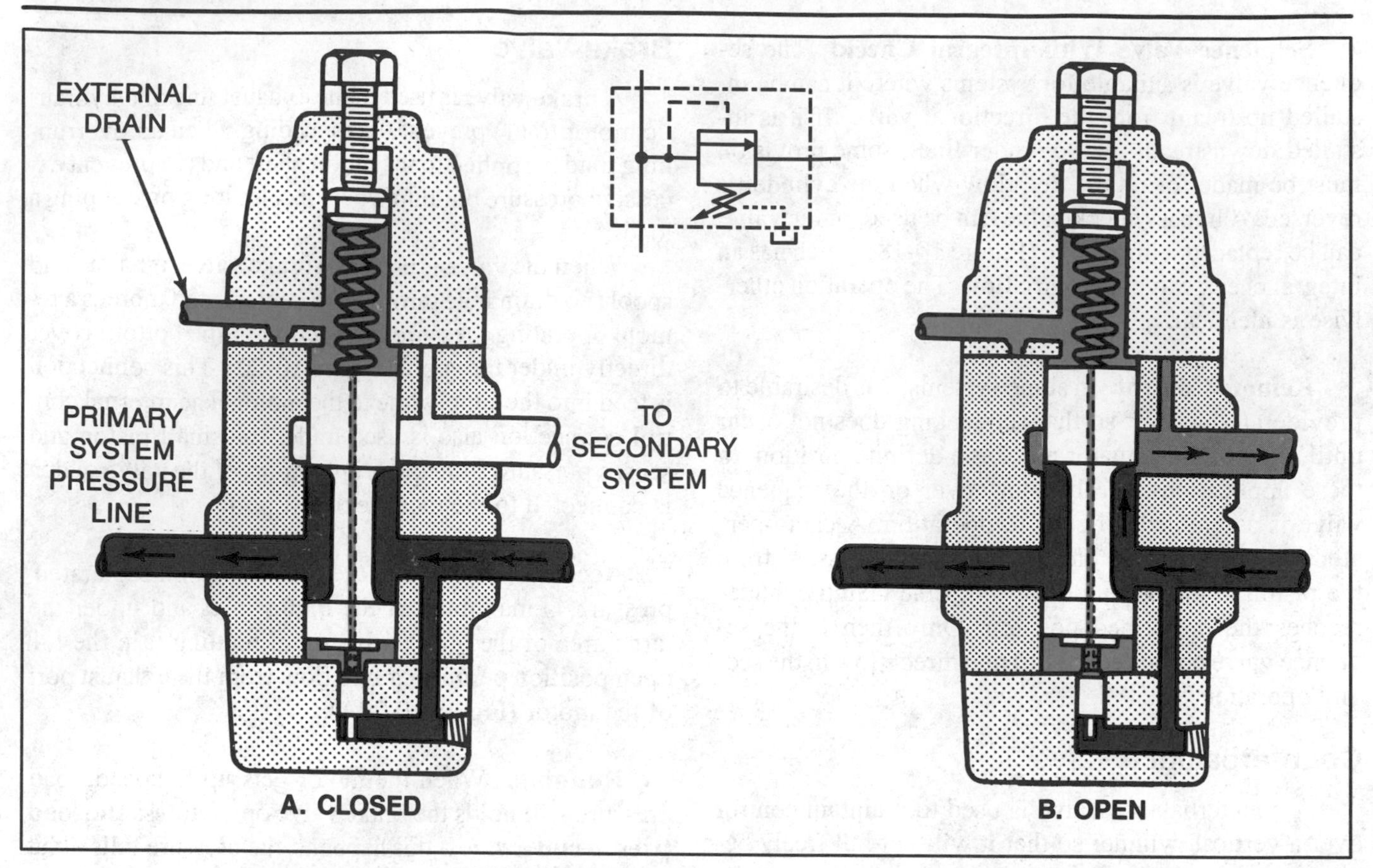

Figure 10-17. Sequence valve.

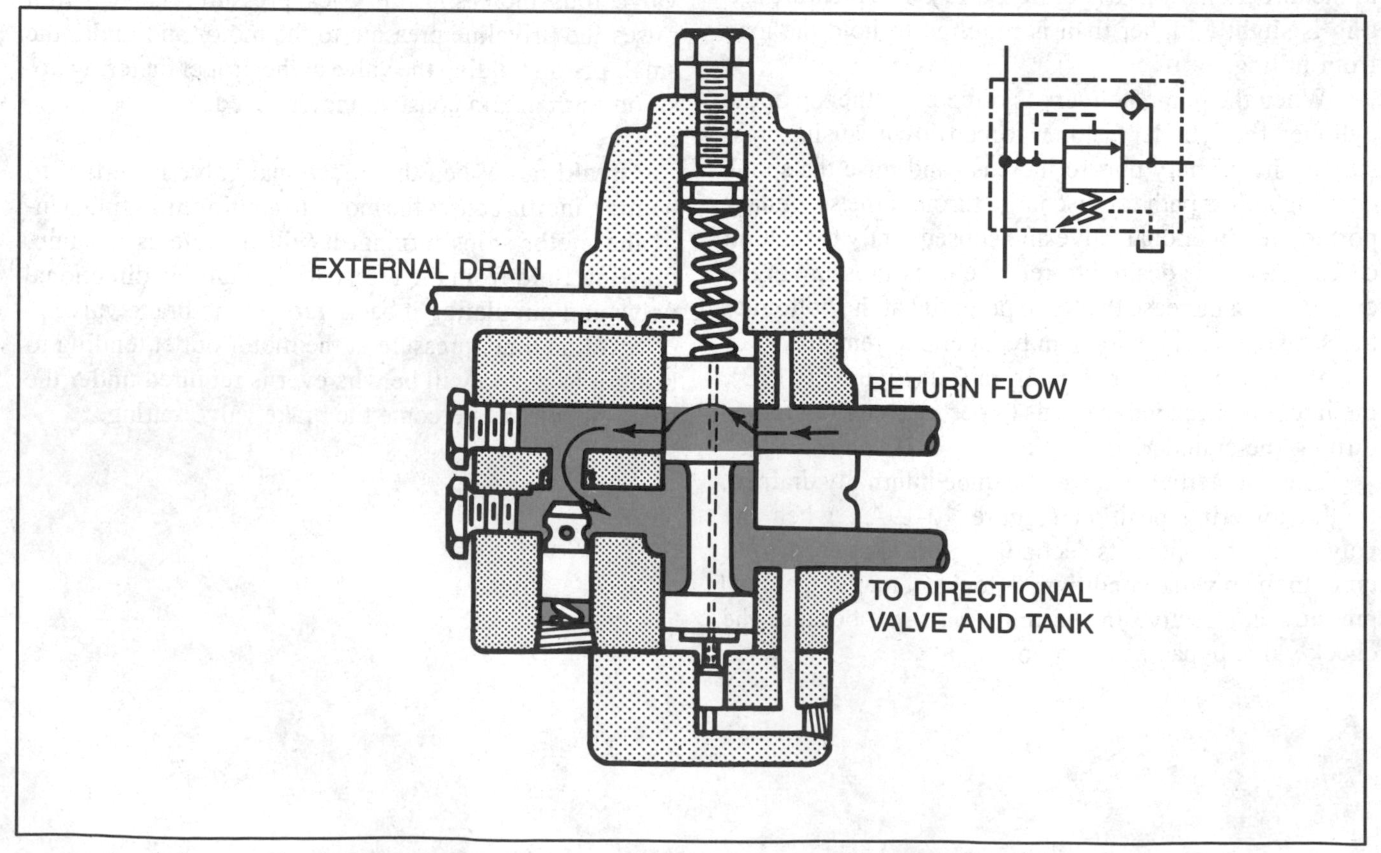

Figure 10-18. Sequence valve with integral check permits reverse free flow.

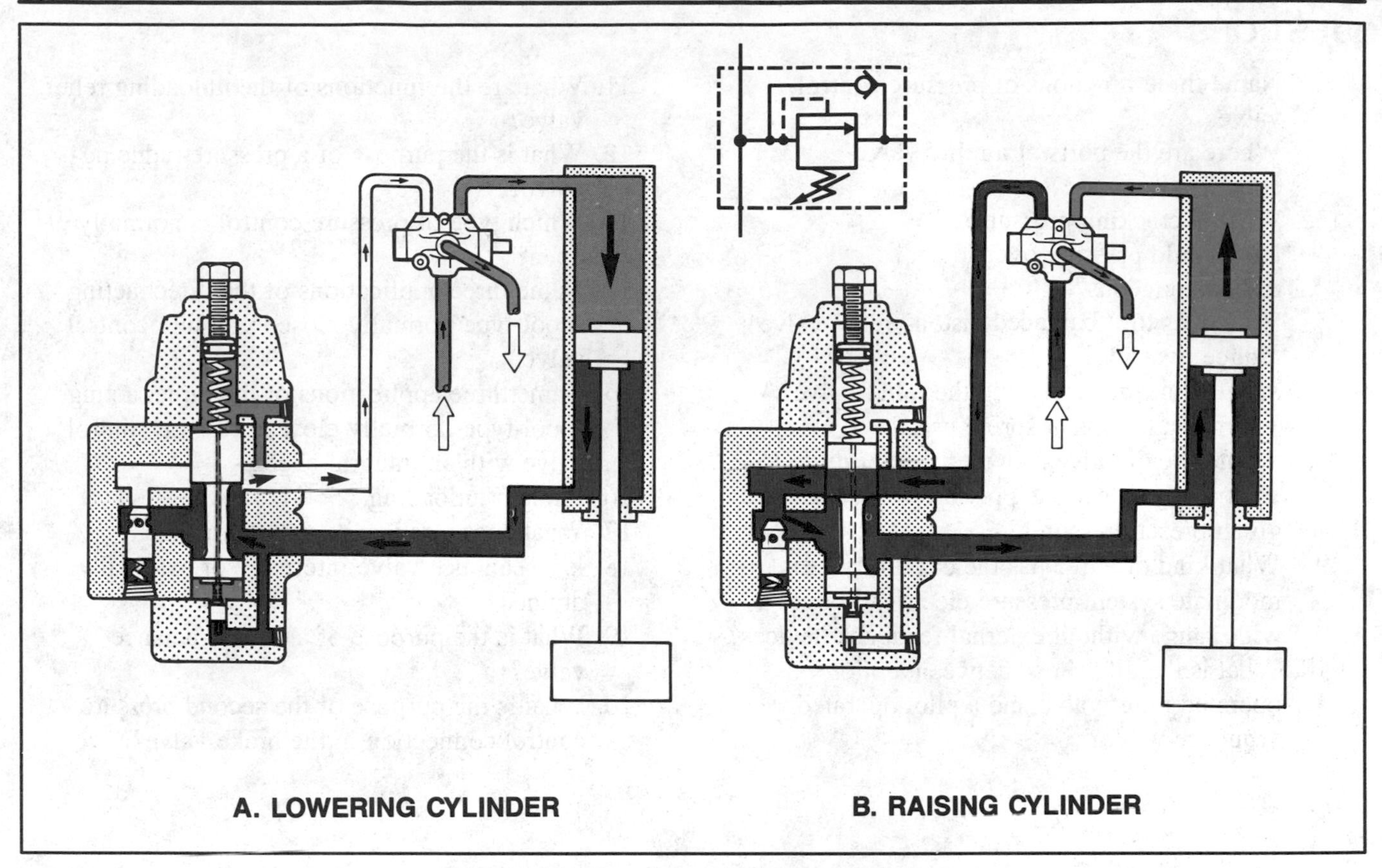
A. LOWERING CYLINDER
B. RAISING CYLINDER

Figure 10-19. Counterbalance valve.

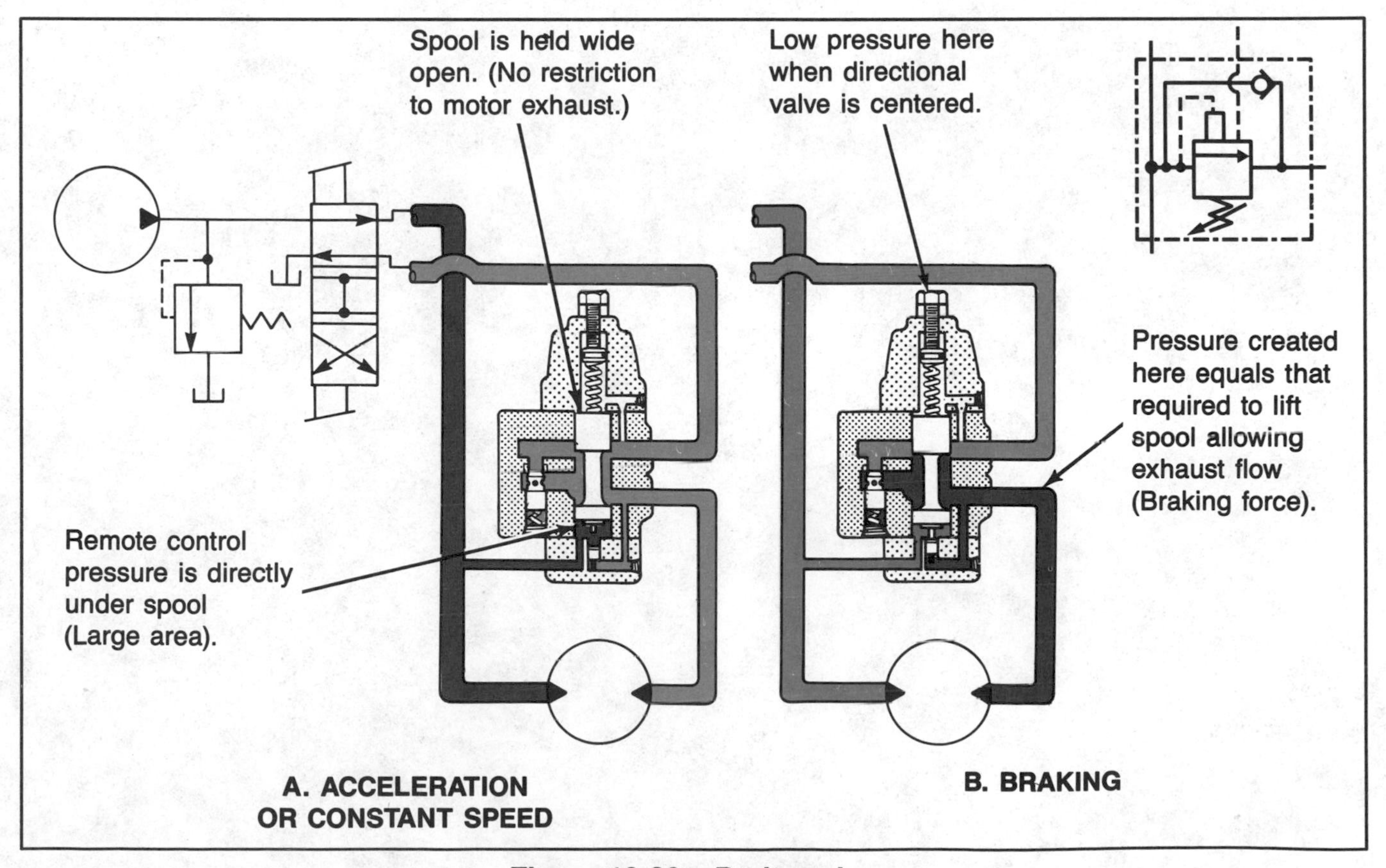
Spool is held wide open. (No restriction to motor exhaust.)
Low pressure here when directional valve is centered.
Pressure created here equals that required to lift spool allowing exhaust flow (Braking force).
Remote control pressure is directly under spool (Large area).
A. ACCELERATION OR CONSTANT SPEED
B. BRAKING

Figure 10-20. Brake valve.

QUESTIONS

1. Name three functions of pressure control valves.
2. Where are the ports of a relief valve connected?
3. What is cracking pressure?
4. How could pressure override be disadvantageous?
5. How does the "balanced piston" relief valve reduce override?
6. What is meant by venting the relief valve?
7. When is a high vent spring used?
8. What type of valve provides the ability to electrically select one pressure from a set of given preset pressures?
9. What kind of valve has the capability to modulate system pressure electrically over a wide range without external feedback devices?
10. What is the difference between a pilot operated relief valve and a pilot operated sequence valve?
11. What are the functions of the unloading relief valve?
12. What is the purpose of a pressure reducing valve?
13. Which type of pressure control is normally open?
14. Name three applications of the direct acting spool-type normally closed pressure control valve.
15. Name three applications of the direct acting spool-type normally closed pressure control valve with an integral check.
16. Explain unloading.
17. What does a sequence valve do?
18. Is a sequence valve internally or externally drained?
19. What is the purpose of a counterbalance valve?
20. What is the purpose of the second pressure control connection in the brake valve?

CHAPTER 11

FLOW CONTROLS

By controlling the rate of flow in a hydraulic circuit, it's possible to control the speed of hydraulic cylinders or motors. A cylinder's speed is determined by its size and the flow rate of the oil going into or out of it. A large diameter cylinder would hold more oil and take longer to complete its stroke; a smaller one would move faster. Changing the flow rate from the pump would also change the extension time of the cylinder. Changing either the cylinder or pump size to regulate speed would be impractical, especially if a speed change is desired in mid-stroke.

A more typical method is to use a flow control valve. This valve, in its simplest form, is nothing more than a variable orifice, and could be as basic as a needle valve. By varying the size of the opening, one can vary the amount of oil entering the cylinder and thus control its speed. A kitchen faucet is an example of a simple flow control valve.

Three factors affect flow rate: pressure, fluid temperature, and orifice size. If any one of these factors is increased, the flow rate increases.

FLOW CONTROL METHODS

Control of flow-in hydraulic circuits can be accomplished with a meter-in circuit, with a meter-out circuit, or with a bleed-off circuit. Each type of circuit has advantages and disadvantages depending on the type of application.

Meter-In Circuit

In meter-in operation, the flow control valve is placed between the pump and actuator (Figure 11-1). In this way, it controls the amount of fluid going into the actuator. Pump delivery in excess of the metered amount is diverted to tank over the relief valve. If the pump delivered 10 gpm (37.85 LPM) and the cylinder speed needed to be reduced by half, the flow control could be adjusted to pass only 5 gpm (18.93 LPM). However, since it's a fixed displacement pump, allowing only 5 gpm (18.93 LPM) to pass through the flow control into the cylinder means that the other five gallons (18.93 liters) have no choice but to go over the relief valve at the relief valve setting. If the relief valve happened to be set at 1000 psi (68.94 bar) (6894.00 kPa), a pressure gauge at point A would read 1000 psi (68.94 bar) (6894.00 kPa). The pressure found at point B would be determined by the work load on the cylinder. The difference between the two readings is called the pressure drop across the orifice; that is, the pressure required to push the 5 gpm (18.93 LPM)through the orifice.

Meter-in circuits can only be used with opposing loads. If the load in Figure 11-2 should tend to run away, it would pull the cylinder piston ahead of the oil supply; and since the exhaust flow has a free path back to tank, the meter-in circuit could not prevent the load from running away.

Figure 11-3 shows three locations where a flow control could be placed in a meter-in circuit. Note that for two of the locations, check valves are included to allow the cylinder to freely extend or retract.

Because the excess flow goes over the relief valve to tank, the meter-in circuit is somewhat inefficient.

Meter-Out Circuit

In Figure 11-4, the flow control is on the outlet side of the cylinder to control the flow coming out. This is known as a meter-out circuit. If the flow control were closed completely, the oil could not exhaust from the cylinder and it could not move. Regulating the size of the opening controls the flow rate and thus the speed of the cylinder.

If this happened to be a 2 to 1 ratio cylinder and the outlet flow rate was controlled to 3 gpm (11.36 LPM), the flow rate into the other side would be 6 gpm (22.71 LPM). But, with a 10 gpm (37.85 LPM) pump, the other four gallons (15.14 liters) would be forced to return to tank over the relief valve. The pump would be operating at the relief valve setting regardless of how easy or difficult it was to move the load.

As mentioned above, metering into the cylinder is fine with an opposing load, but if the load tends to run away, a better way is to meter out. In fact, a meter-out circuit works if the load pushes or pulls. In any case, both

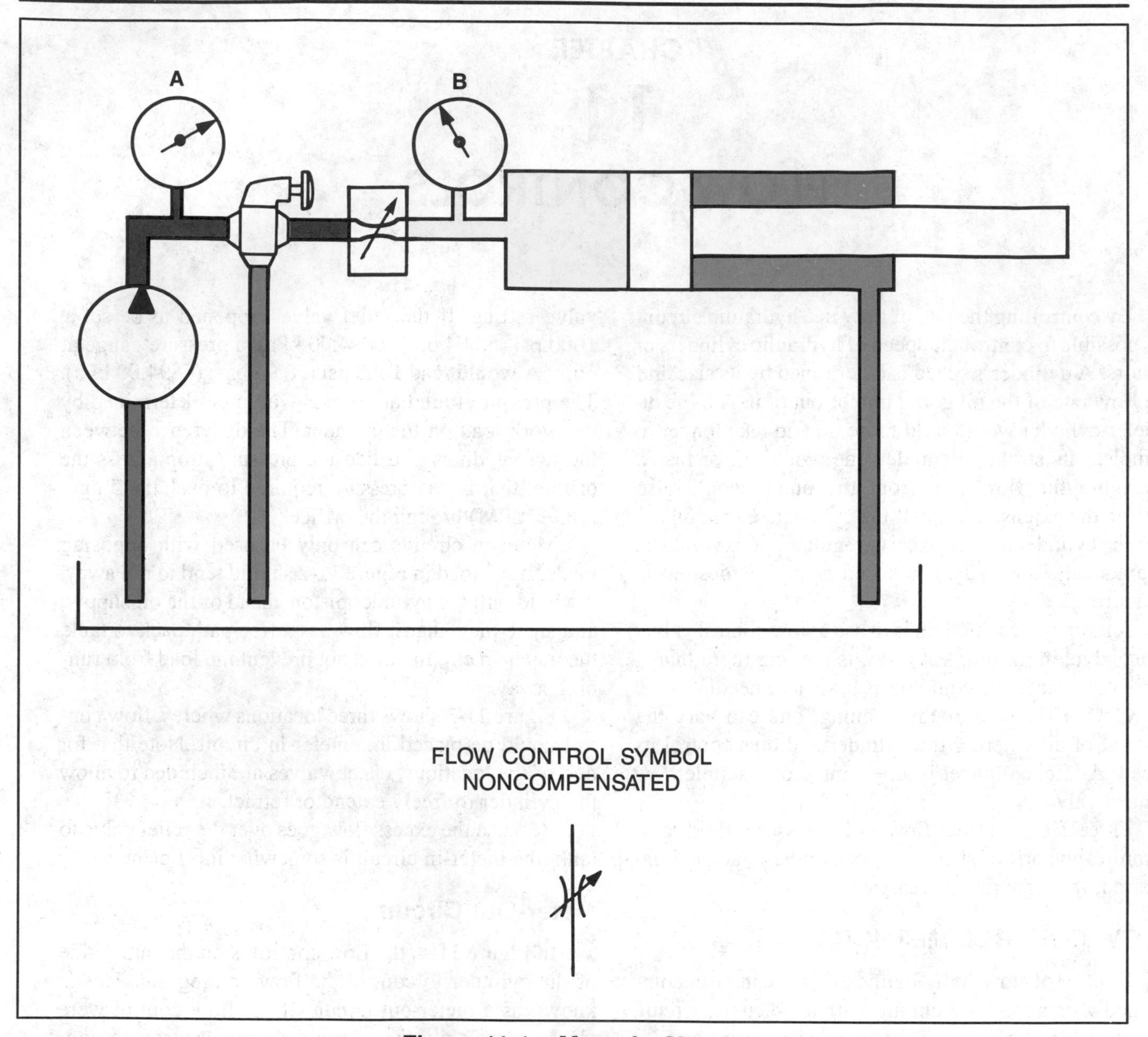

Figure 11-1. Meter-In Circuit.

circuits require the pump to operate at the relief valve setting.

As shown in Figure 11-5, there are also three possible locations for the flow control valve in a meter-out circuit. This circuit, like the meter-in circuit, is considered inefficient due to the fact that the pump must operate at the relief valve setting.

Pressure Intensification in Meter-Out Circuits. In meter-out circuits with differential cylinders, it is possible to develop rod end pressures much higher than the system relief valve setting. In Figure 11-6, with no load on the cylinder, the rod end pressure is double the relief valve setting. To understand the reason for this the situation must be analyzed using basic physical principles.

A basic law of physics states that when an object is in motion at a steady speed, all the forces acting on the object are equal. In the example illustrated in the figure, all forces to the right must equal all forces to the left, or for every force there must be an equal and opposite force (Newton's Third Law). The cylinder is moving at a steady speed due to the flow control metering out 4 gpm (15.14 LPM); therefore, the laws just described apply. If 4 gpm (15.14 LPM) is metered out, 8 gpm (30.28 LPM) must be going into the cap end which has twice the area. The remaining 2 gpm (7.57 LPM) has no choice but to go through the relief valve. Therefore, the cap end pressure is equal to the relief valve setting of 1000 psi (68.94 bar) (6894.00 kPa). This develops a force to the right equal to 2000 lb (8896.44 N) (F = PA). Since there is no load on

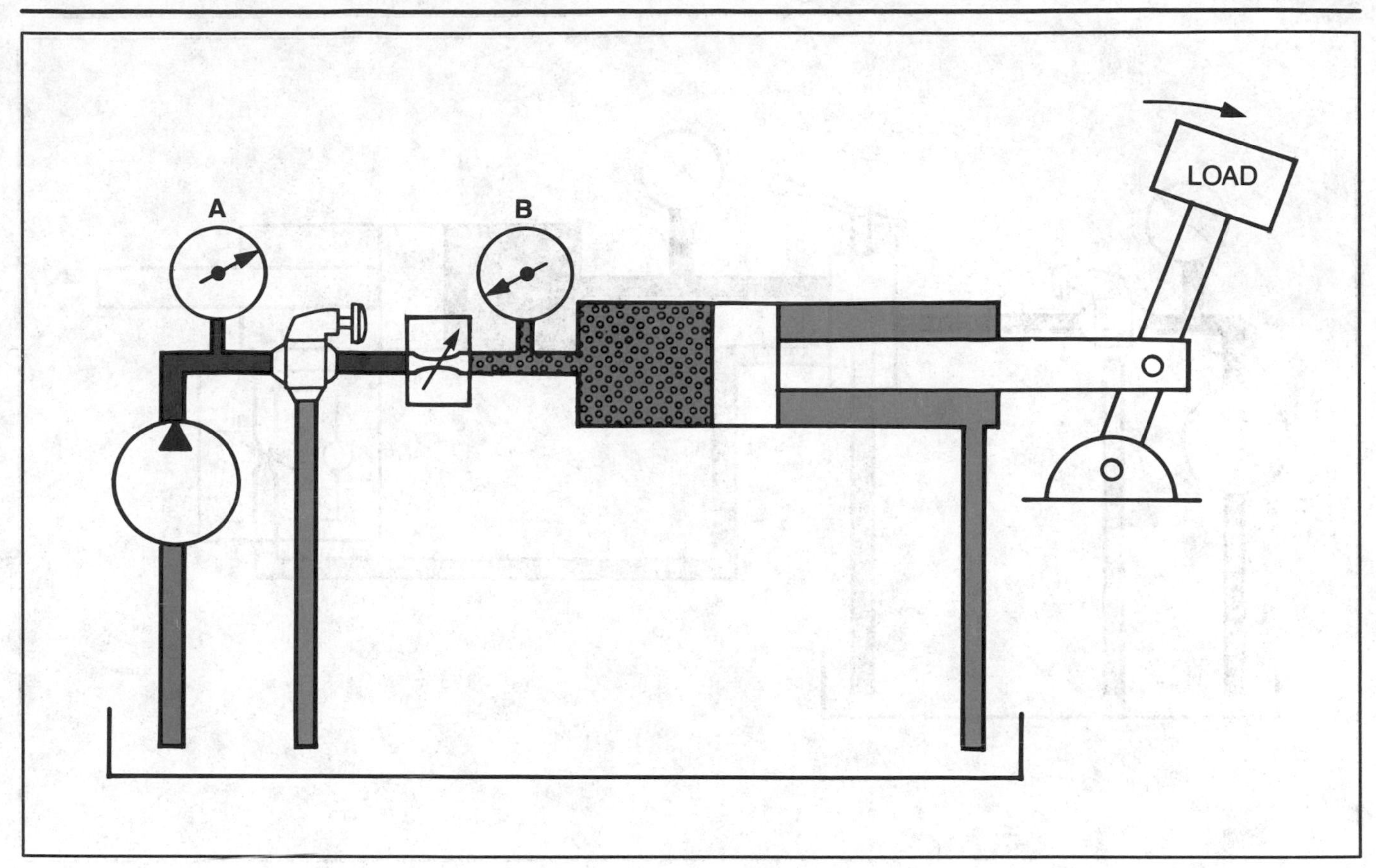

Figure 11-2. Meter-In Circuits Do Not Control Runaway Loads.

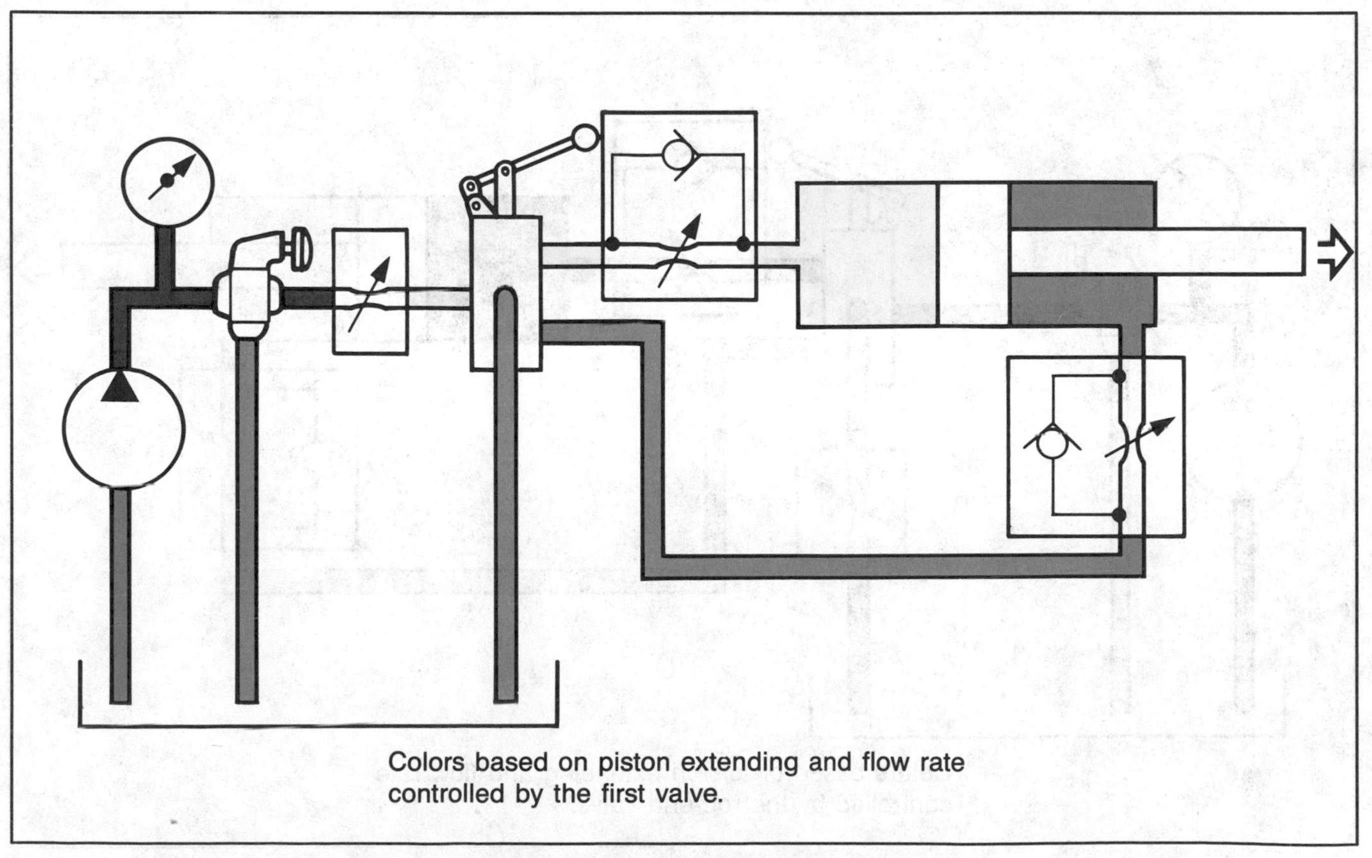

Colors based on piston extending and flow rate
controlled by the first valve.

Figure 11-3. Locations for Meter-In Applications.

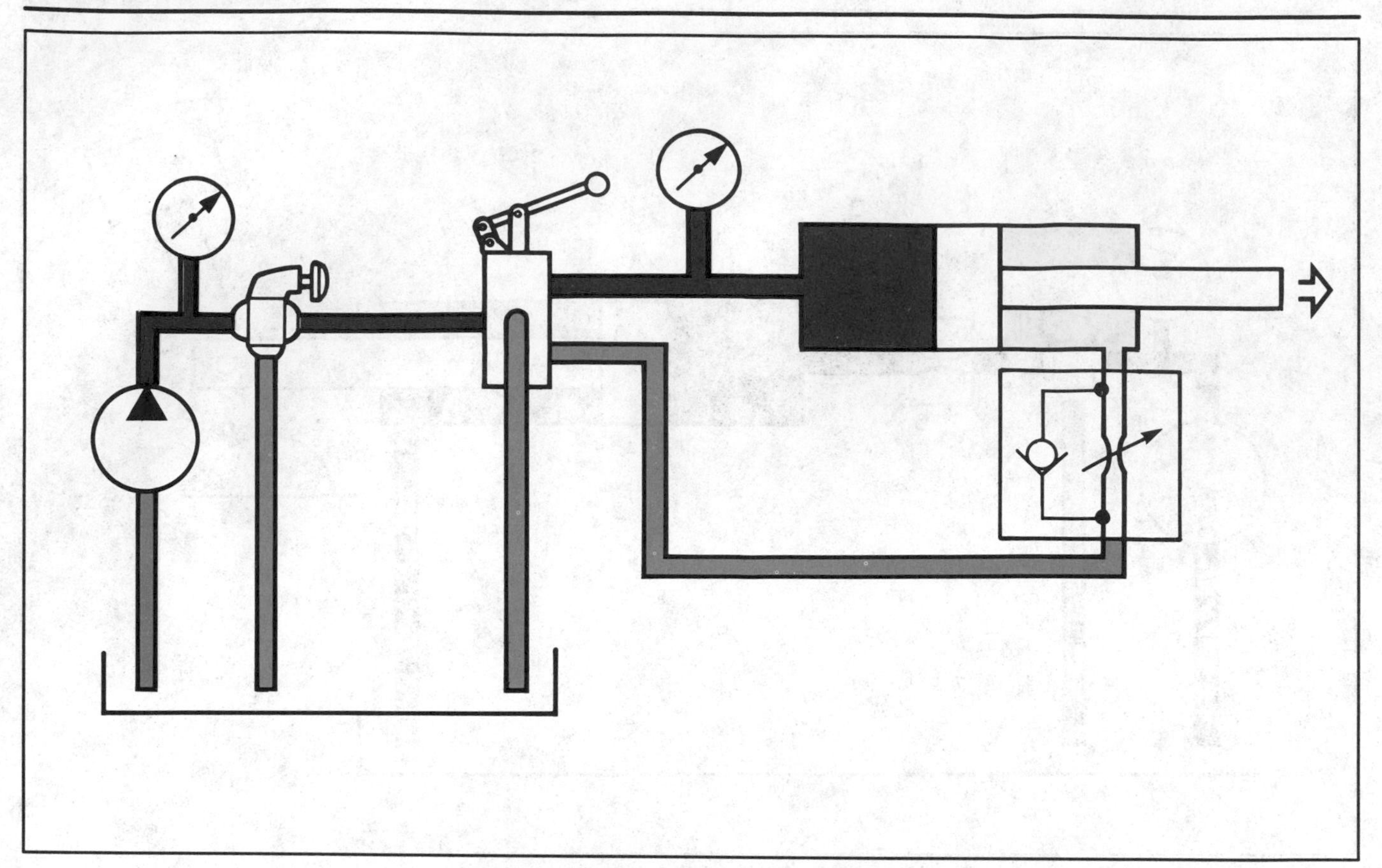

Figure 11-4. Meter-Out Circuit.

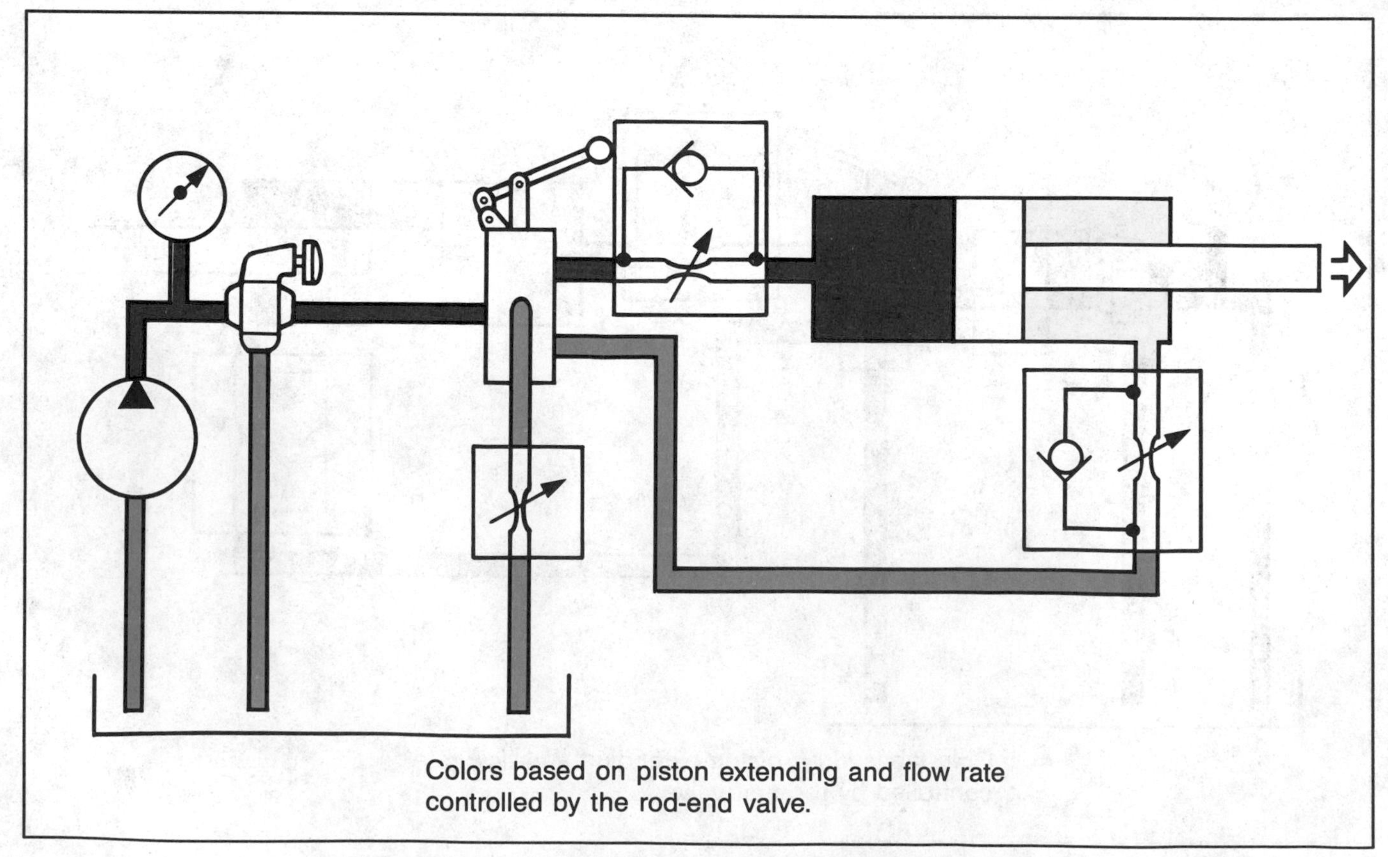

Colors based on piston extending and flow rate
controlled by the rod-end valve.

Figure 11-5. Locations for Meter-Out Applications.

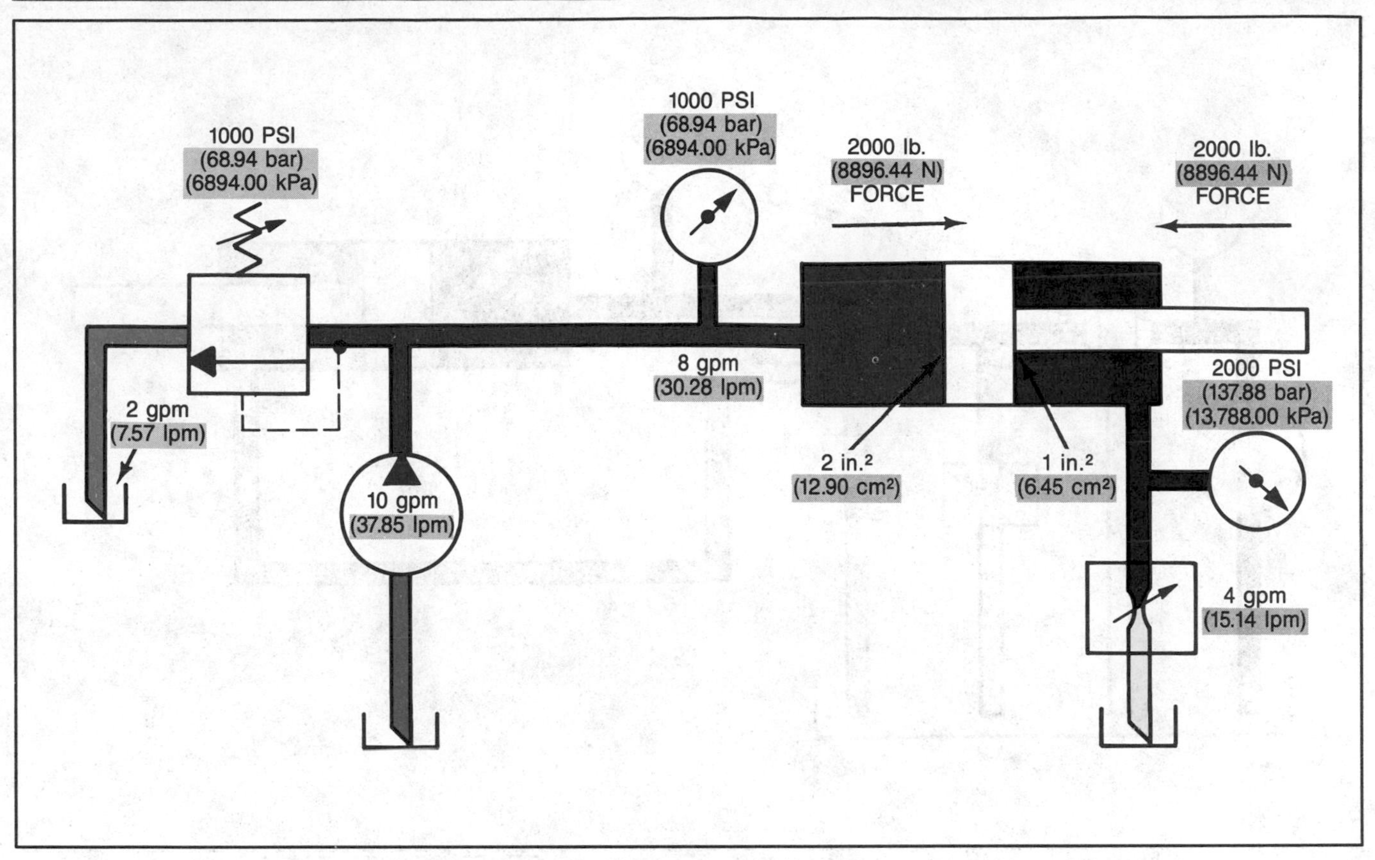

Figure 11-6. Pressure Intensification in Meter-Out Circuits.

the cylinder, the force to the left is determined solely by the oil pressure around the rod due to the resistance (restriction) offered by the flow control. Therefore, the restricted oil flow is holding back with 2000 lb (8896.44 N) of force. With an area of one square inch, the pressure developed is 2000 psi (137.88 bar) (13,788.00 kPa) (P = F/A).

If an external load is applied to the cylinder so as to pull it to the right (overrunning load) with a force of 500 lb (2224.11 N), this would add to the 2000 lb (8896.44 N) force on the cap end. Therefore, to balance this force, the rod end pressure would increase to 2500 psi (172.35 bar) (17,235.00 kPa).

Bleed-Off Circuit

The third method of applying a flow control is shown in Figure 11-7. The flow control is simply teed off the main line to control cylinder speed. For instance, with the flow control completely closed, the full flow from the pump would go into the cylinder. However, the moment the flow control is opened up, some bleed off of that pump delivery occurs and the cylinder starts to slow down. Adjusting the size of the opening will bleed off any amount necessary to control how fast the cylinder moves. In this case, unlike the meter-in or meter-out circuits, there is no excess flow going over the relief valve and the pump operates at only the pressure that is needed to move the work load on the cylinder, which saves energy. The bleed-off circuit will not prevent a load from running away. As with a meter-in circuit, it can be used with opposing loads only.

The bleed-off circuit is not as accurate as the meter-in and meter-out circuits because the measured flow goes to the tank rather than into the cylinder. This makes the cylinder speed subject to changes in the pump delivery and hydraulic system leakages which occur as the work load pressure changes. To minimize these effects, it is recommended to bleed off no more than half the pump delivery and to avoid using a bleed-off circuit completely where there's a wide fluctuation in the work load pressure. Figure 11-8 shows three possible locations for flow control placement in a bleed-off circuit.

Summary

The meter-in and meter-out circuits are accurate but not very efficient. The bleed-off circuit is efficient but not very accurate. The meter-out circuit can control overrunning as well as opposing loads while the other two methods must be used with opposing loads only. The choice of flow control method and the location of the flow control in the circuit are dependent on the type of application being controlled.

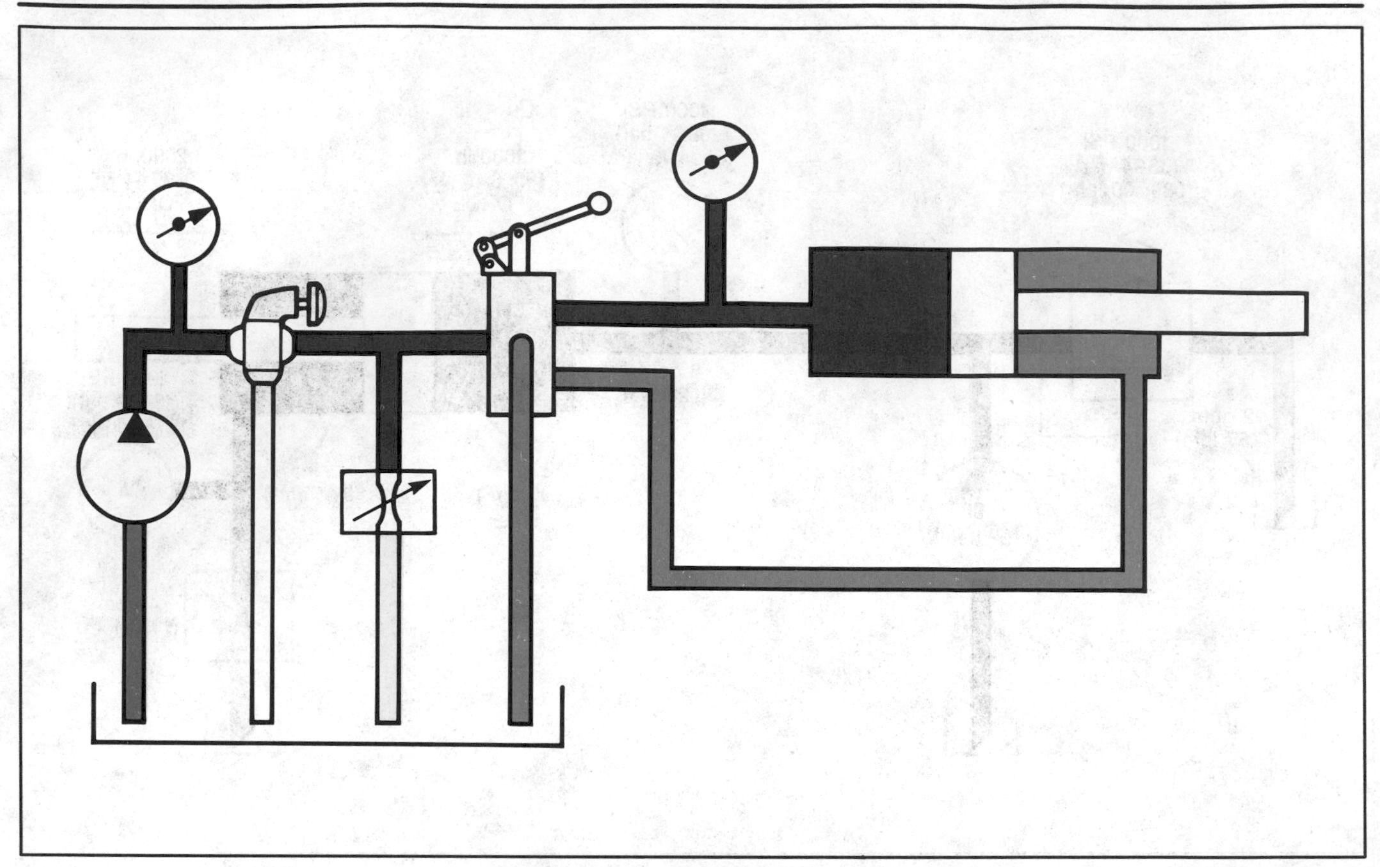

Figure 11-7. Bleed-Off Circuit.

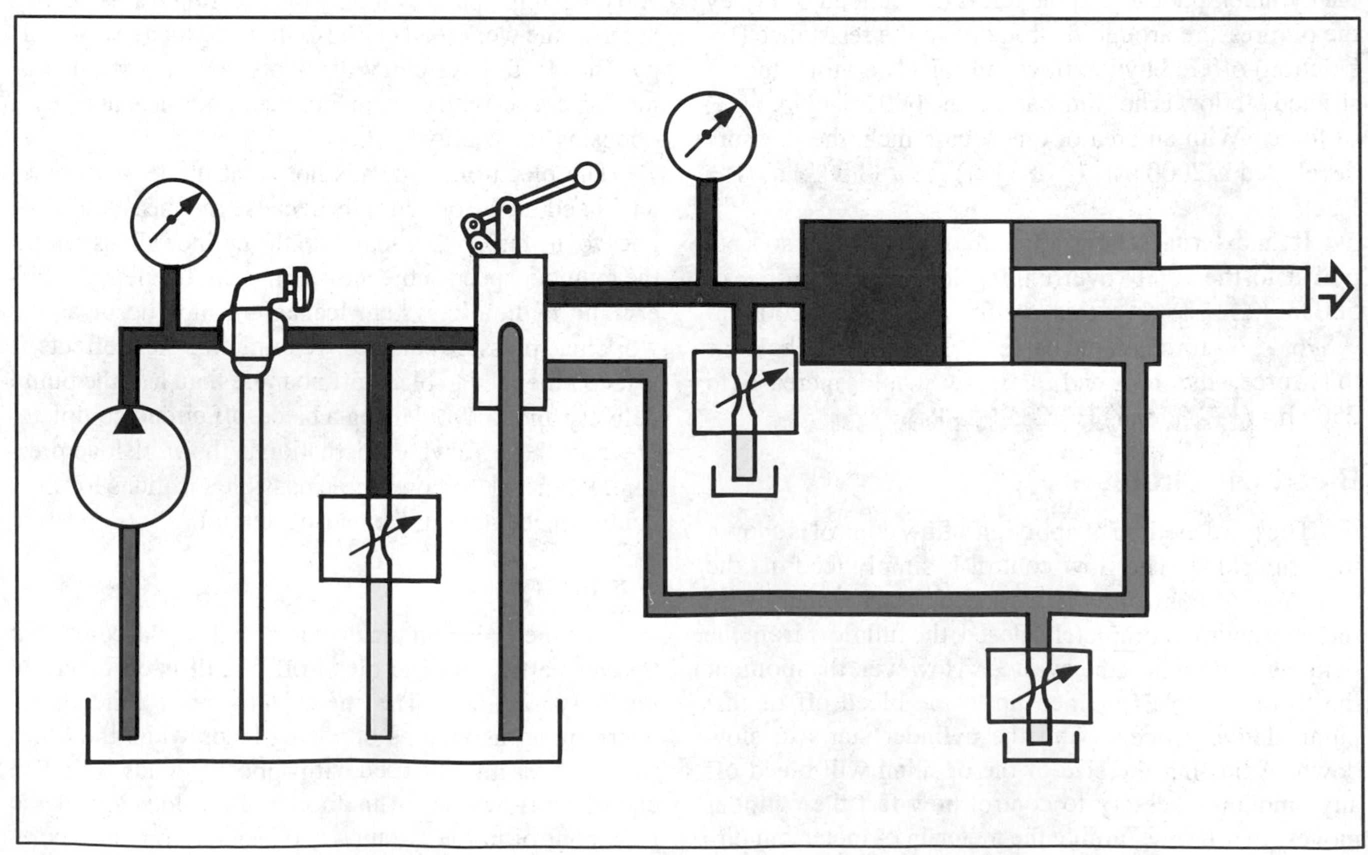

Figure 11-8. Locations for Bleed-Off Applications.

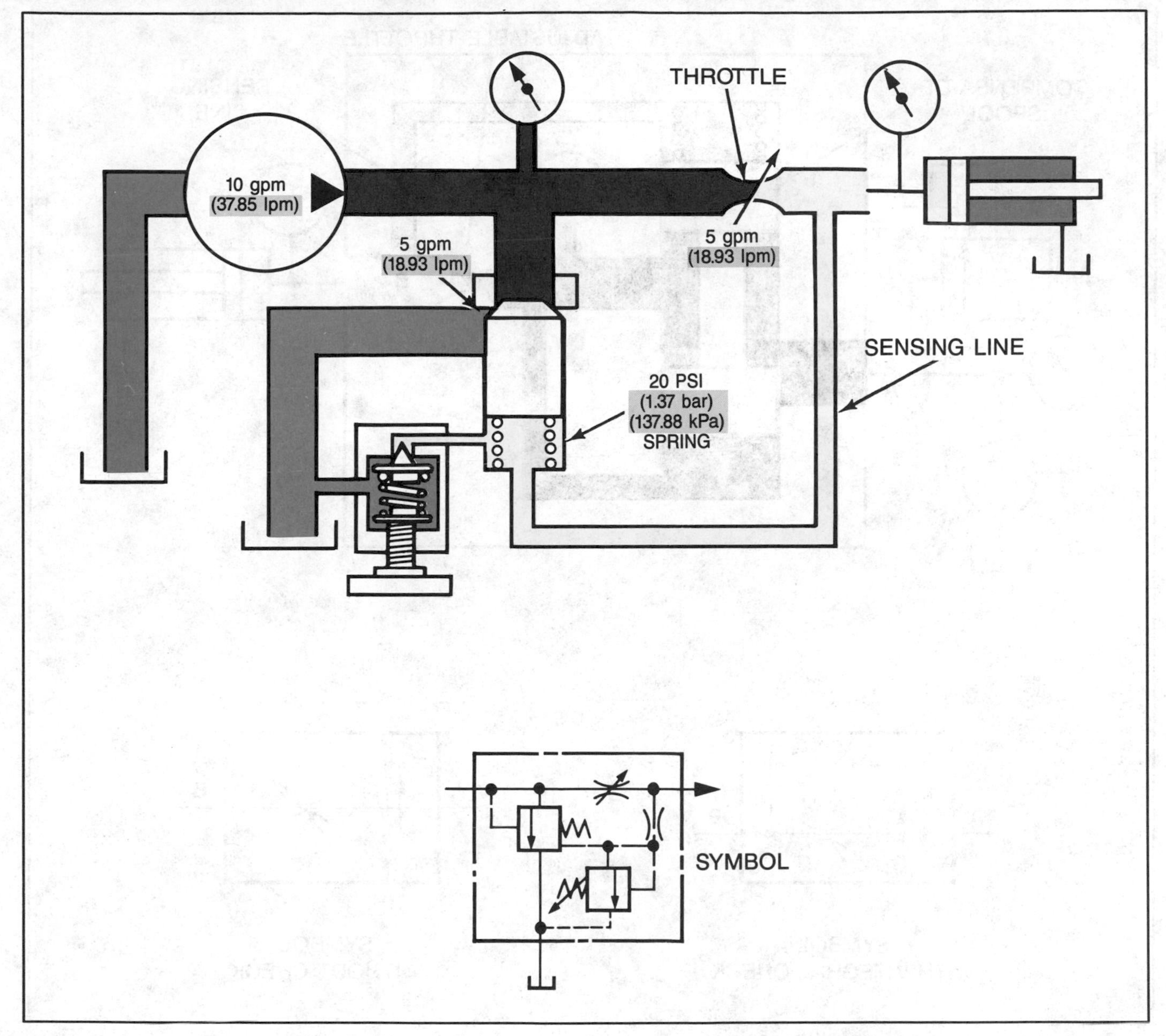

Figure 11-9. Pressure-Compensated Flow Control.

TYPES OF FLOW CONTROLS

Flow controls are of two types: nonpressure compensated and pressure compensated. Pressure-compensated flow controls are either the bypass or restrictor type.

Nonpressure-Compensated Flow Control

The flow control valve used in the preceding examples was a simple variable orifice, such as a needle valve. That makes a good flow control valve as long as the work load does not change very much. However, if the load on the cylinder changes, the amount of fluid flowing through the needle valve will change. It will increase as the load goes down and decrease as the load increases. That's because flow through an orifice, or any other type of opening, increases as the pressure drop across the orifice increases.

In the meter-in example, the relief valve was set at 1000 psi (68.94 bar) (6894.00 kPa). If the needle valve had been adjusted to pass 5 gpm (18.93 LPM) to the cylinder when the work load was 200 psi (13.79 bar) (1378.80 kPa), and then the work load increased to 400 psi (27.58 bar) (2757.60 kPa), the flow would then be less than 5 gpm (18.93 LPM). That's because the pressure drop across the needle valve would have dropped from 800 psi (55.15 bar) (5515.20 kPa) to only 600 psi (41.36 bar) (4136.40 kPa). Less pressure drop, less flow (and vice-versa).

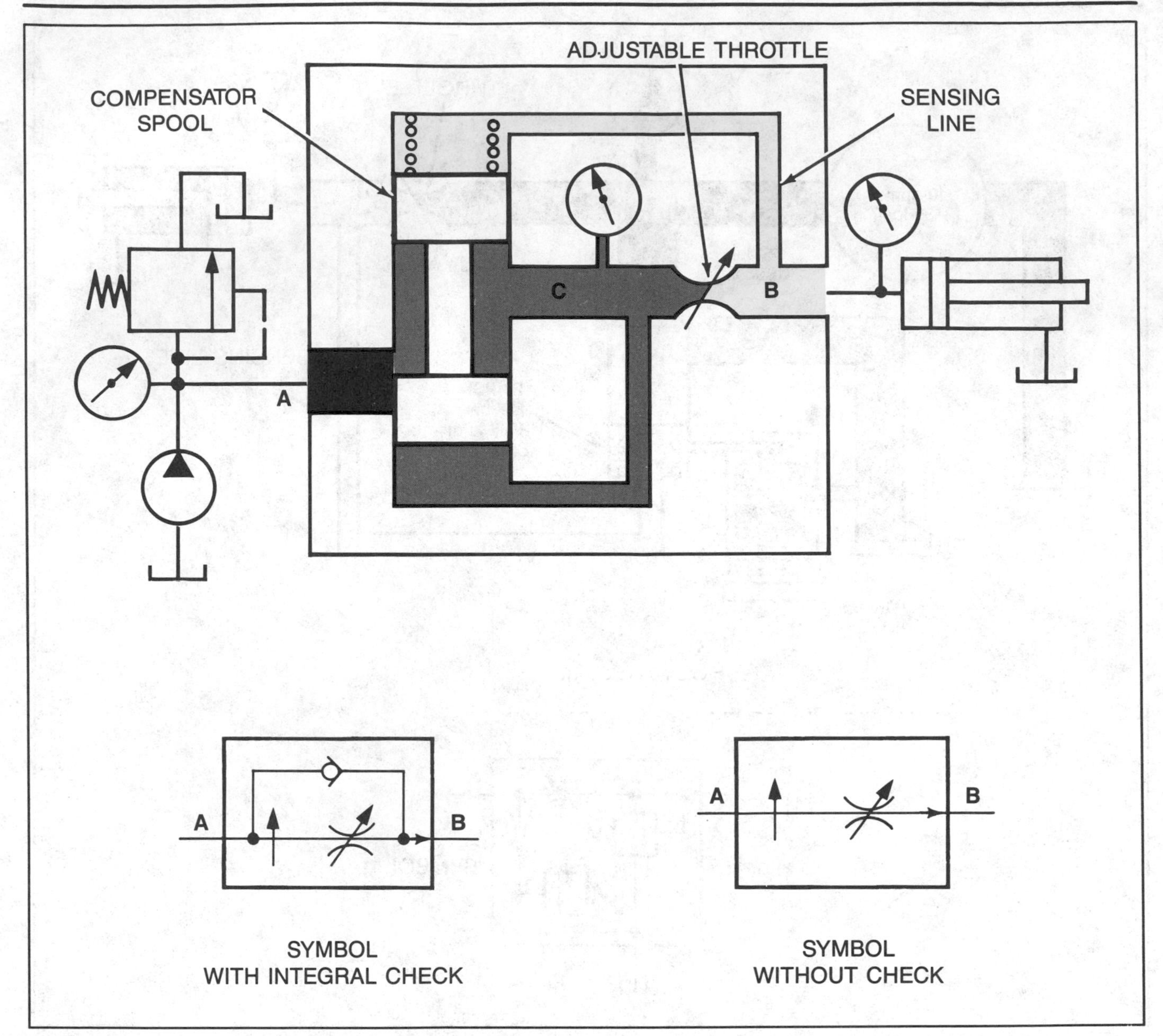

Figure 11-10. Restrictor-Type, Pressure-Compensated Flow Control.

Pressure-Compensated Flow Control

A pressure-compensated flow control valve automatically compensates for pressure changes and maintains its setting even as the work load changes.

Bypass Type. Consider a hydraulic circuit with only a wide open adjustable throttle and a spring-loaded bypass valve. Because the spring is very light, this valve will start to open (crack) when the pressure at its inlet reaches 20 psi (1.38 bar) (137.88 kPa). This system won't move a very heavy load. However, if a pilot line is connected between the outlet of the adjustable throttle and the spring chamber of the bypass valve (Figure 11-9), then the cylinder load pressure will add to the 20 psi (1.38 bar) (137.88 kPa) spring pressure to hold the bypass valve closed. As long as the pressure difference across the throttle is less than 20 psi (1.38 bar) (137.88 kPa), the bypass valve will remain closed no matter how high the load pressure goes. However, as soon as the throttle starts to close to limit the amount of fluid flowing to the cylinder, there will be an increase in the pressure ahead of the throttle.

It takes more pressure to squeeze fluid through a small opening than through a large opening. When the pressure difference between the inlet and outlet of the throttle reaches 20 psi (1.38 bar) (137.88 kPa), the bypass valve will start to open and it will pass just enough fluid to the tank to keep the pressure drop across the throttle at 20 psi (1.38 bar) (137.88 kPa).

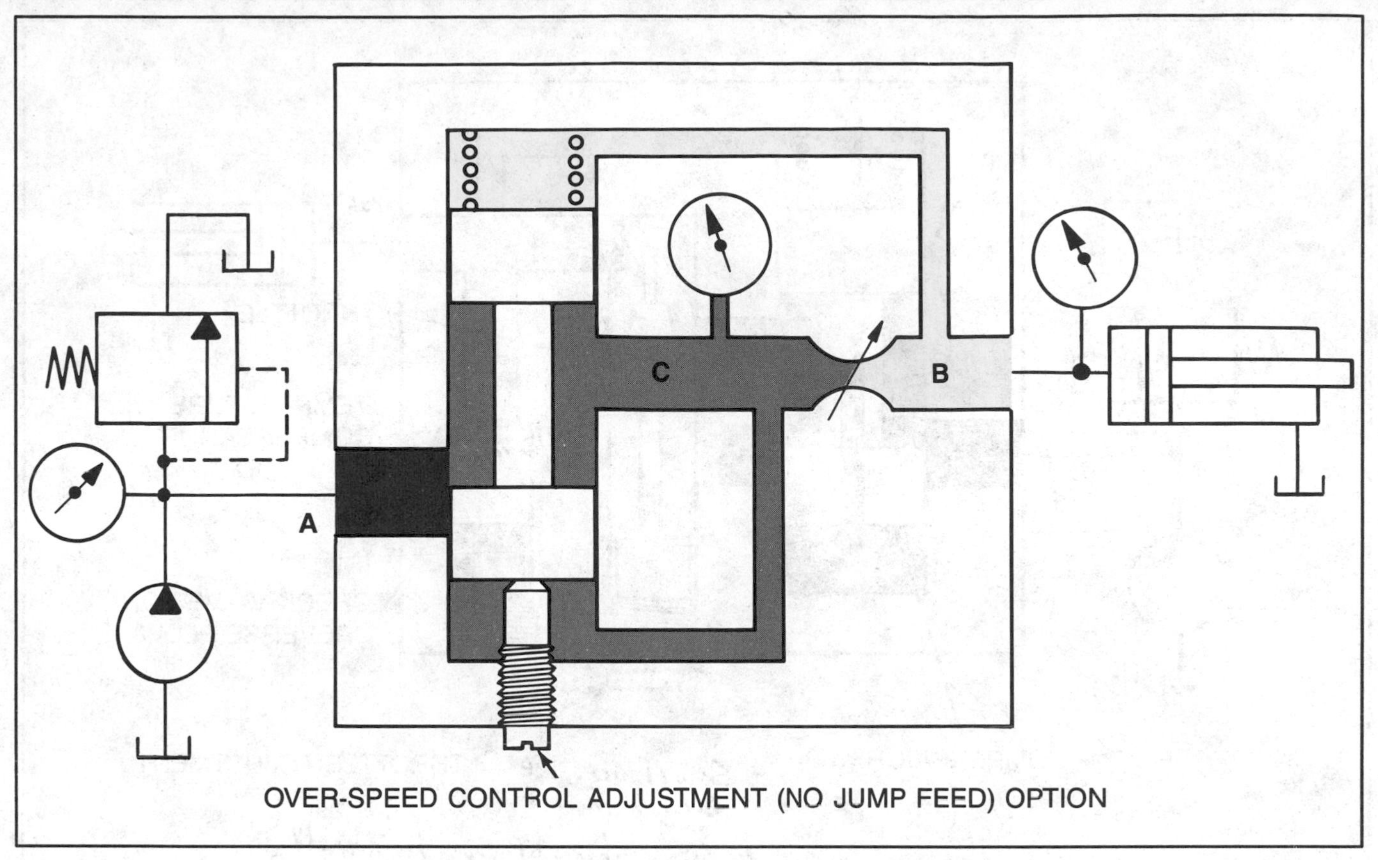

Figure 11-11. No Jump Feed Adjustment.

Increasing or decreasing the orifice size permits an increase or decrease in the cylinder speed as needed. Keeping the pressure difference across the orifice the same, regardless of changes in the work load pressure, results in a pressure-compensated flow control. This control can be modified to include a built-in relief valve for overload protection (see Figure 11-9). To do that, the pressure in the spring chamber is limited by means of a small poppet and an adjustable spring. If the relief valve is set for 980 psi (67.56 bar) (6756.12 kPa), it limits the pressure in the spring chamber to 980 (67.56 bar) (6756.12 kPa). Add to that the 20 psi (1.38 bar) (137.88 kPa) for the spring, and the most pressure available is 1000 psi (68.94 bar) (6894.00 kPa).

Packaging these elements in one unit results in a combination overload relief valve and pressure-compensated flow control. Because of the relief valve, it would have to be mounted between the pump and a directional valve. It has the advantage of saving energy in that it doesn't operate at the relief valve setting when controlling speed. The disadvantage is that it can only be used in meter-in applications, because the excess fluid is bypassed directly to the reservoir.

Restrictor Type. There is a pressure-compensated flow control that can work in a meter-in, meter-out or a bleed-off function. In this type, like the bypass type, a constant pressure drop across an orifice is maintained, but in a somewhat different manner. This valve is a restrictor-type control, and its compensator spool moves up and down in response to work load changes to keep the pressure difference constant (Figure 11-10). The adjustable orifice controls the flow rate into the cylinder; it's called a throttle.

A 20 psi (1.38 bar) (137.88 kPa) spring holds the compensator spool in the normally open position. If the pressure underneath the spool reaches 20 psi (1.38 bar) (137.88 kPa) higher than the pressure on the top and tries to exceed it, the spool will move up, tending to close off flow into the valve. Again, assume a 10 gpm (37.85 LPM) pump but only 5 gpm (18.92 LPM) are wanted into the cylinder. When the pump is turned on, flow through the throttle is restricted and pressure builds up at the throttle and under the compensator spool, moving it up against the spring. As it moves up, the opening gets smaller and restricts the flow entering the valve to what 20 psi (1.38 bar) (137.88 kPa) pressure difference can push through the throttle (in this case, 5 gpm (18.92 LPM)). The excess flow again must return to tank over the relief valve.

If the relief valve were set for 1000 psi (68.94 bar) (6894.00 kPa), the pressure at point A (when controlling

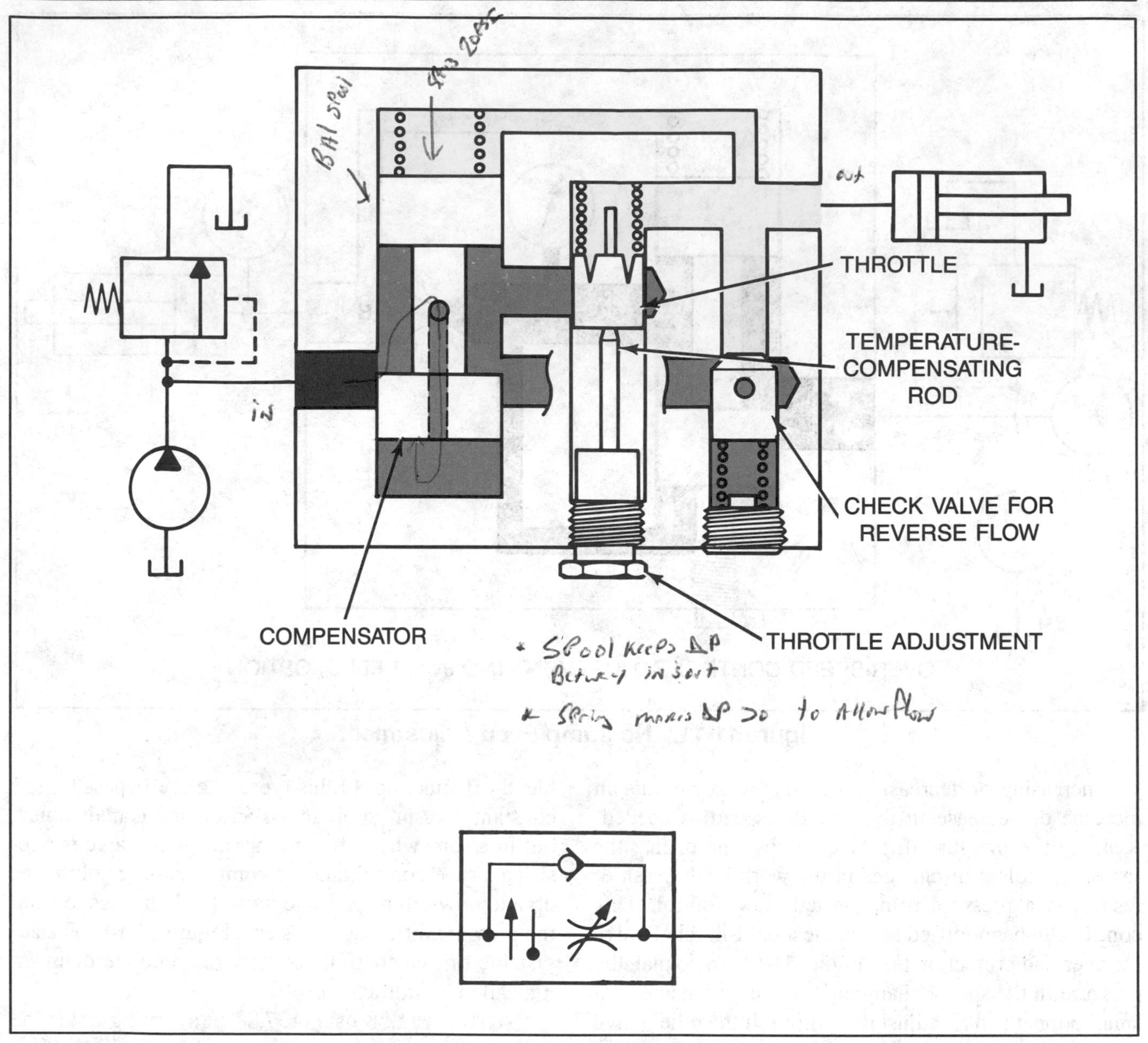

Figure 11-12. Pressure- and Temperature-Compensated Flow Control.

flow rate and speed) would be 1000 psi (68.94 bar) (6894.00 kPa). The pressure at point B would be determined by whatever the work load is. If the work load pressure were 100 psi (6.89 bar) (689.40 kPa), there would be 100 psi (6.89 bar) (689.40 kPa) at B and 1000 psi (68.94 bar) (6894.00 kPa) at A.

With a sensing line from the work load side of the cylinder, whatever pressure is needed to move the cylinder adds to the 20 psi (1.38 bar) (137.88 kPa) spring. Therefore, the pressure at C will always balance out the spring force plus the work load pressure, or in this example, 120 psi (8.27 bar) (827.28 kPa). The pressure difference across the throttle is 20 psi (1.38 bar) (137.88 kPa), and the flow will be 5 gpm (18.93 LPM). If the work load went to 400 psi (27.58 bar) (2757.60 kPa), the pressure at C would go to 420 (28.95 bar) (2895.48 kPa). The pressure difference is still 20 psi (1.38 bar) (137.88 kPa). The flow rate is still 5 gpm (18.92 LPM), and the speed remains the same. This restrictor-type pressure-compensated flow control can be used in meter-in, meter-out or bleed-off applications to control speeds accurately, even with varying work loads. If unrestricted reverse flow is desired, a flow control with an integral check valve is used or an external check valve must be installed across the flow control.

No Jump Feed Adjustment. In a circuit with a centered directional control and a stationary load, there is no flow through the flow control when it is mounted in the actuator line. In this situation, the compensator is held

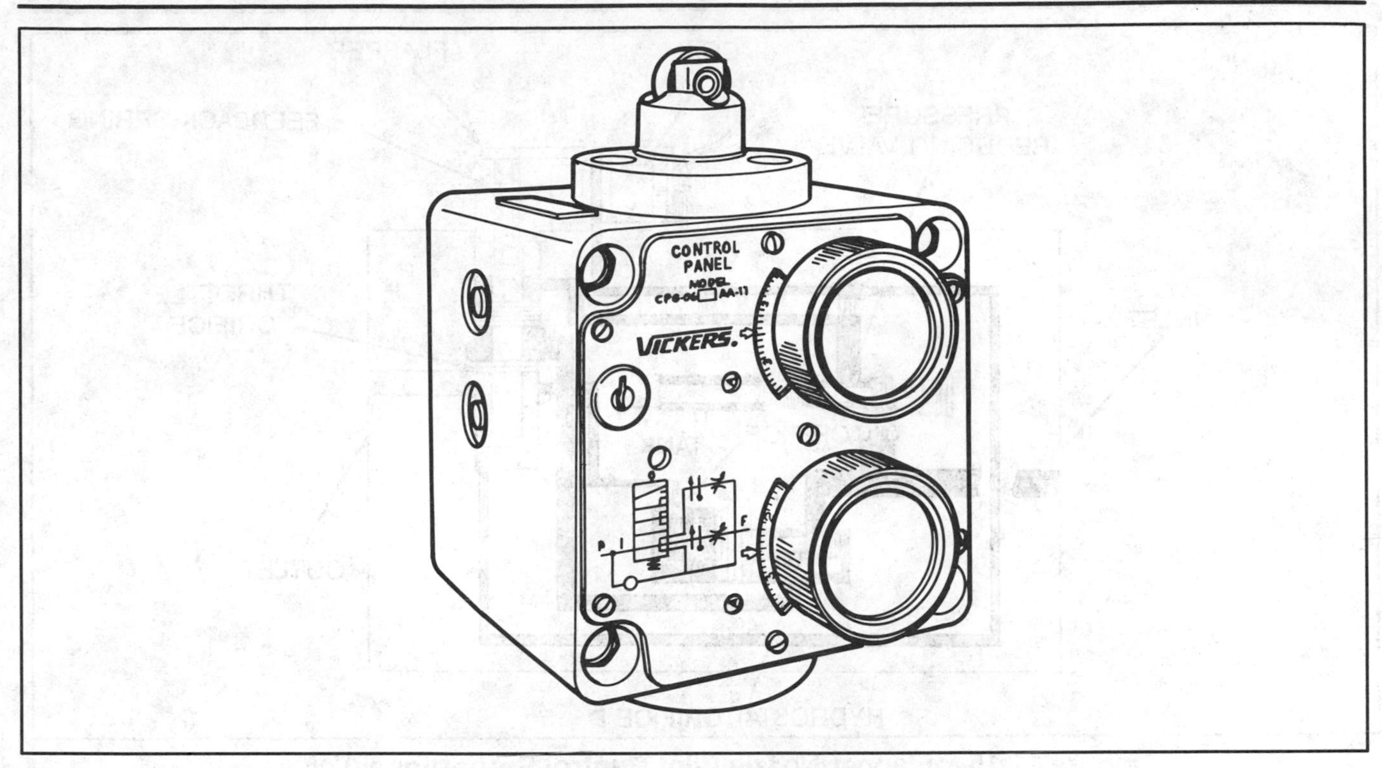

Figure 11-13. Feed Control Panel.

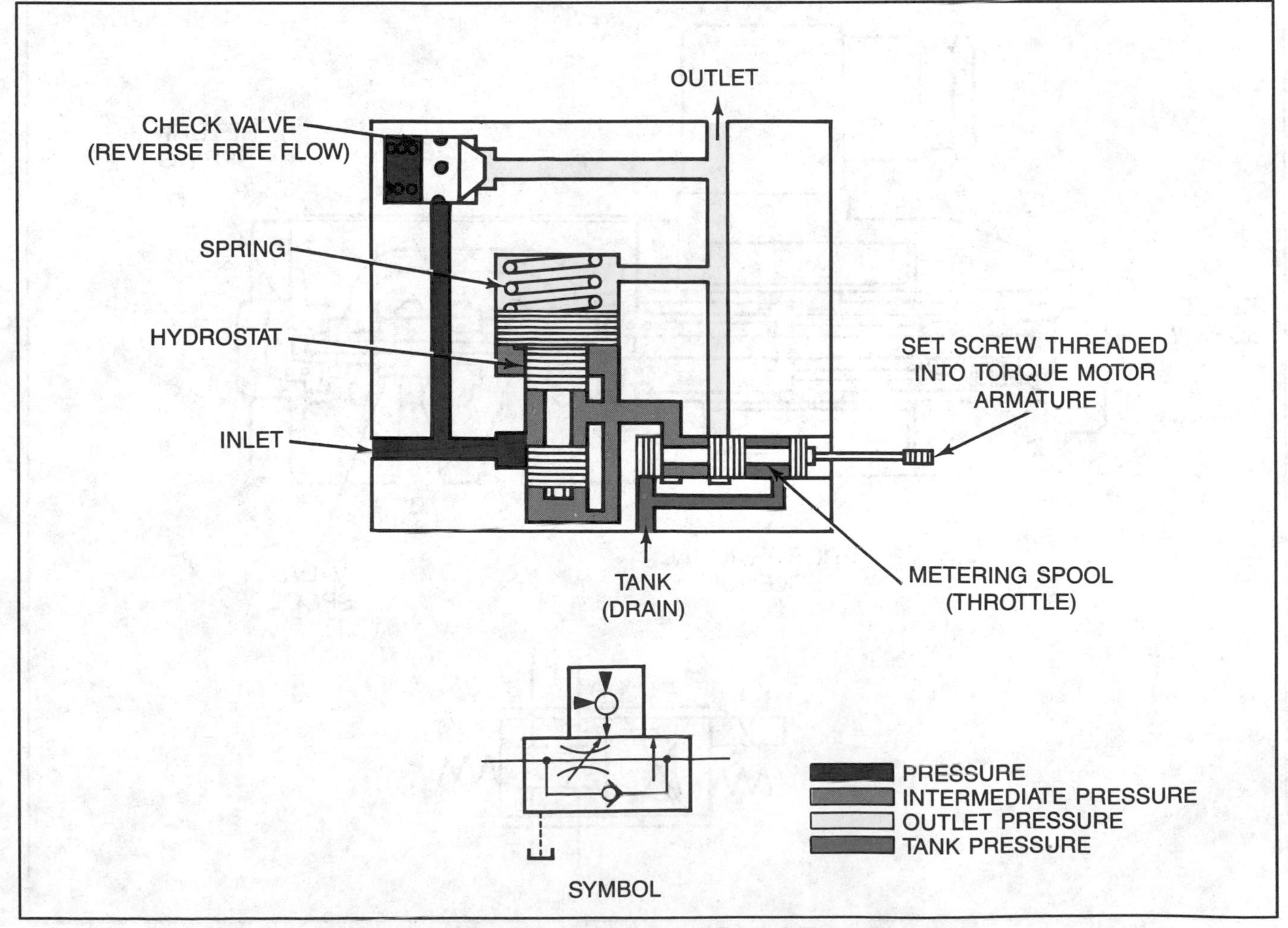

Figure 11-14. Servo Actuated Spool-Type Throttle.

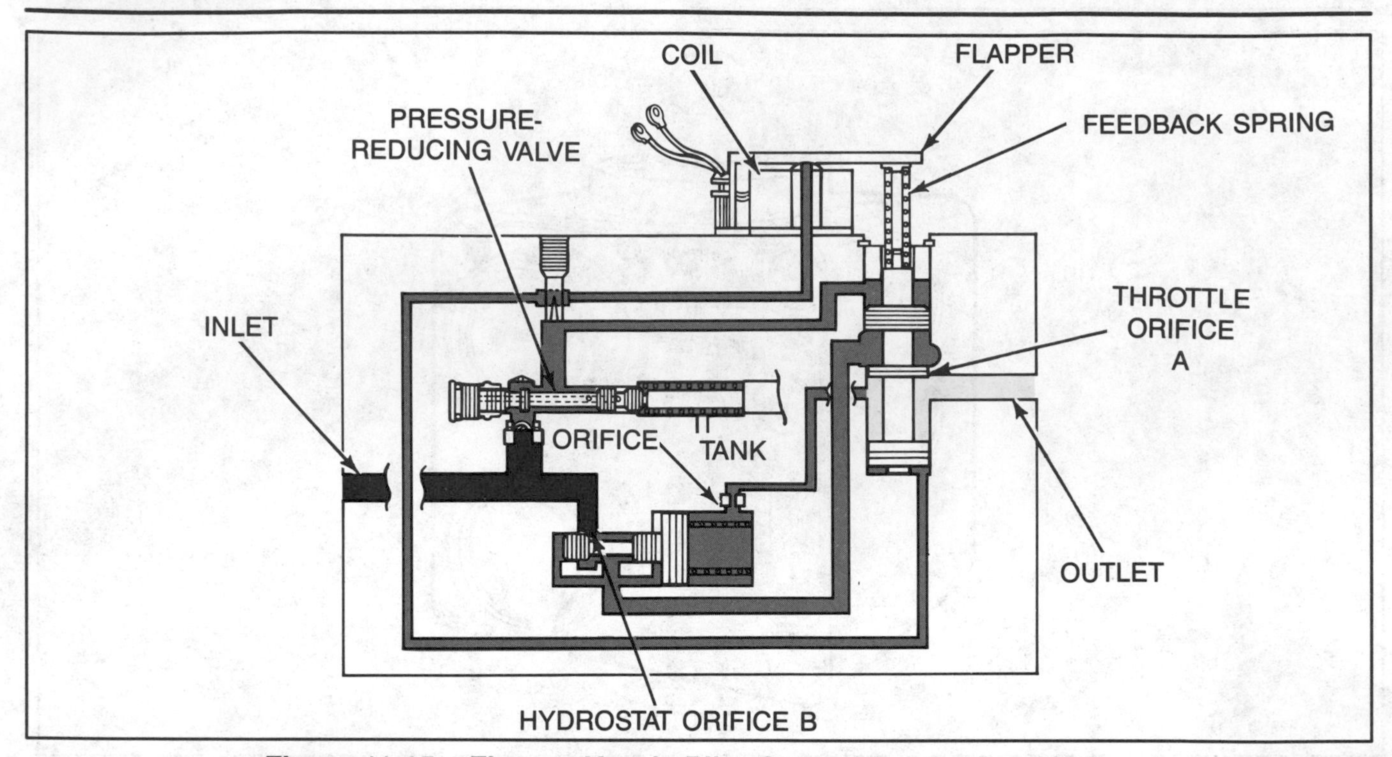

Figure 11-15. Flapper Nozzle Pilot Control Proportional Valve.

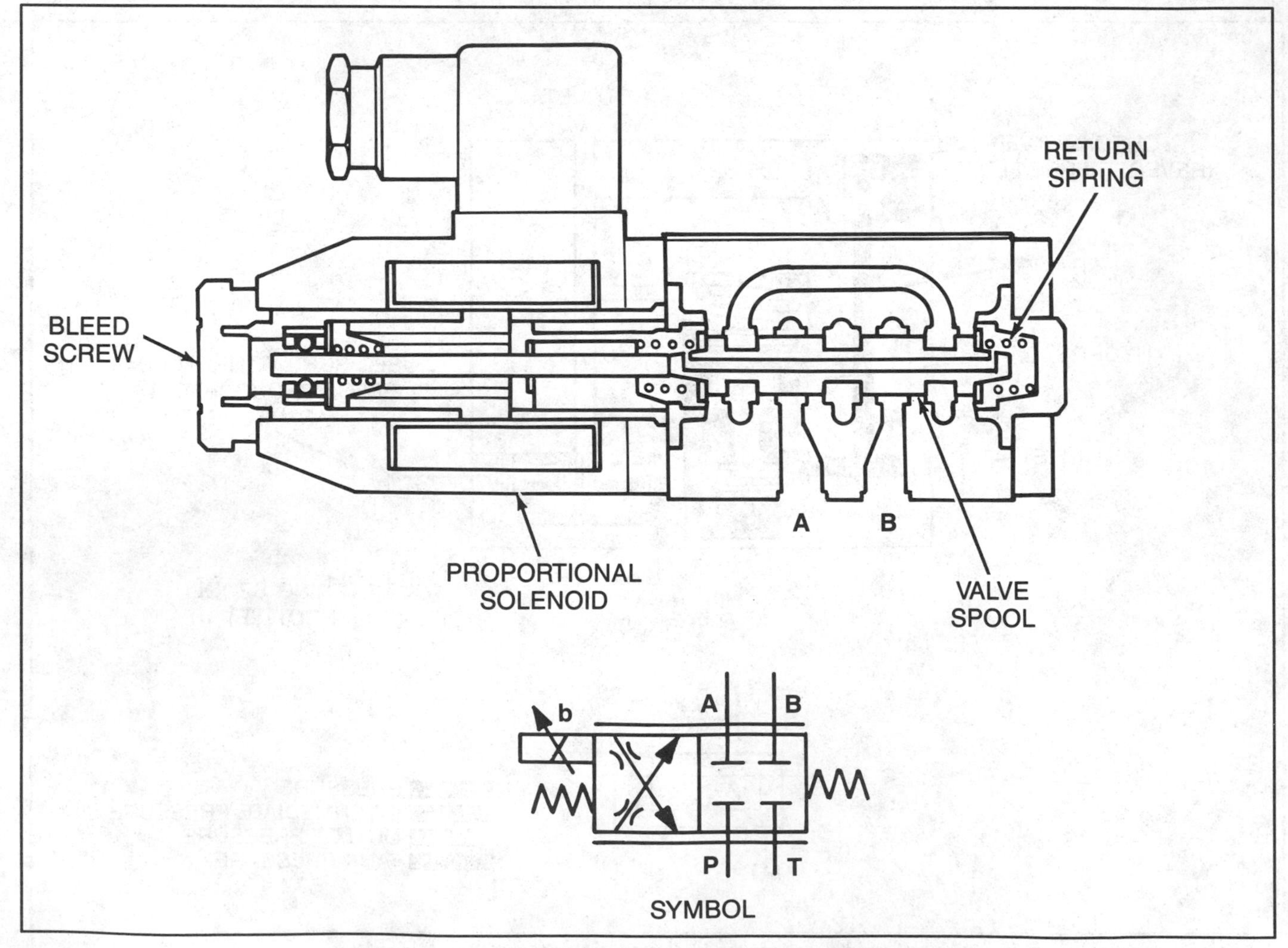

Figure 11-16. Proportional Solenoid Throttle

wide open by the spring. When the directional control is shifted to move the load, the load may have a tendency to jump ahead (momentarily overspeed). This is because with the compensator wide open there is a momentary increase in flow beyond the adjusted flow rate, until the compensator moves to its controlled position. This requires some amount of time—however small. To reduce the distance the compensator has to move and therefore the tendency to jump, a stroke adjustment is used, as shown in Figure 11-11. This is a mechanical means of limiting how far the compensator will open, thereby reducing the distance and time it requires for the compensator to regain control.

TEMPERATURE COMPENSATION

It's well known that the temperature of oil affects its viscosity. More hot, thin oil will go through an orifice at a given pressure than cold, thick oil. Any change in the temperature of the oil changes its viscosity, which in turn changes the rate of flow through the flow control and the speed of the cylinder.

Figure 11-12 shows a pressure-compensated flow control with a temperature compensating throttle. Its operation is essentially the same as the valve shown in Figure 11-10. Note the compensator spool; instead of having the sensing passage in the valve body, pressure is sensed to the bottom of the spool through a passage drilled in the spool. Instead of using the usual throttle symbol, a more realistic one is the cup-shaped device with "V" notches for better metering. It is held by a small spring against the shoulder of an aluminum alloy rod which extends through the cup into the oil flow. If it was set for 6 gpm (22.71 LPM) and the temperature went up, the oil would become a little thinner and tend to flow faster through the throttle. However, the increased temperature also causes the aluminum rod to expand and close the throttle opening to compensate for the change in oil viscosity. Thus, even with the thinner oil, the speed stays essentially the same.

There are three possible locations for the flow control. Where it is actually located is determined by the requirements of the particular application.

FEED CONTROL PANELS

A feed control panel (Figure 11-13) is a mechanically operated directional control valve which also includes two pressure-compensated flow control valves. It allows a cylinder to have rapid advance with coarse and fine feed cycles. A cam, actuated by the cylinder, depresses the spool and directs oil through a coarse feed throttle. As the cylinder progresses further, a longer cam depresses the spool even more to shift the flow to a fine feed throttle. On completion of its forward travel, the cylinder returns rapidly to its original position. Throttle openings are adjusted with knobs, or the panel can be remotely controlled via electrical signals. The panel is installed in the circuit on a subplate; no oil lines are connected directly to the valve body.

PROPORTIONAL FLOW CONTROL VALVES

Proportional flow control valves provide the ability to control hydraulic flow rates from any distance or location. They can be activated by a remotely located device and some can be programmed for complete automatic cycles. They are called proportional flow control valves because the amount of output hydraulic flow is proportional to the magnitude of the electrical input command signal. Proportional valves are covered in detail in Chapter 14.

Types of Proportional Flow Control Valves

There are three basic types of proportional flow control valves:

- Torque Motor Actuated Spool-Type Throttle
- Flapper Nozzle Pilot Control
- Proportional Solenoid Throttle Valve

The use of any type of proportional valve requires that the equipment contain an electrical power supply and electronic controls.

The torque motor actuated spool-type throttle (Figure 11-14) is an electrically controlled valve that varies fluid flow in approximate proportion to its input current. The throttle spool is linked to the armature of a torque motor and moves in response to signals to the torque motor. Operation is otherwise the same as a pressure-compensated flow control valve.

Either cylinders or small hydraulic motors can be controlled in applications where precise feed speeds, controlled acceleration, or remote electrical programming are required. It is a pressure-compensated restrictor-type valve that can be used in meter-in, meter-out or bleed-off applications.

The flapper nozzle pilot control proportional valve (Figure 11-15) is a pressure and temperature compensated control that can be used in meter-in, meter-out or bleed-off applications.

The proportional solenoid throttle (Figure 11-16) is a restrictor-type nonpressure-compensated valve. A separate hydrostat is required for pressure compensation. It can also be used in meter-in, meter-out or bleed-off applications.

QUESTIONS

1. Name two ways of regulating flow to an actuator.
2. What are the three methods of applying flow control valves?
3. Under what conditions would you use each of the three methods named in Question #2 above?
4. What is the difference between a bypass- and restrictor- type flow control?
5. What is pressure compensation?
6. When might temperature compensation be needed?
7. What is the advantage of the flow control and relief valve over a conventional flow control?
8. Why are remote flow control valves called proportional flow control valves?
9. Name three types of proportional flow control valves.
10. What two things are required in all equipment using proportional control valves?

CHAPTER 12 CARTRIDGE VALVES

Refinements in hydraulic system development have led to greater use of manifold blocks. A manifold block greatly reduces the number of fittings required for the interconnecting lines between components in a system. This eliminates many potential leakage points and reduces fluid waste from leakage. A cartridge valve is inserted into a standardized cavity in a manifold block and held in place with either self-contained screw threads or a cover secured with bolts to complete the cartridge valve design concept.

This chapter contains information on the construction, operation, and application of two types of cartridge valves: slip-in cartridge valves and screw-in cartridge valves. Most slip-in cartridges are poppet-type elements that are normally controlled by another valve to provide a complete hydraulic function (such as a directional, pressure, or flow control valve). Screw-in cartridges can be either spool or poppet-type elements. With a few exceptions, a single screw-in cartridge element provides a complete hydraulic valve function.

ADVANTAGES OF CARTRIDGE VALVES

Cartridge valves provide several advantages over conventional line- or subplate-mounted spool-type directional, pressure, and flow control valves. In many applications, the advantages include:

- Greater system design flexibility
- Lower installed cost
- Smaller package size
- Better performance and control
- Improved reliability
- Higher pressure capability
- More efficient operation
- Elimination of external leakage and reduction of internal leakage
- Greater contamination tolerance
- Faster cycle times
- Lower noise levels

Cartridge valves offer a design alternative rather than a replacement for conventional sliding spool valves. Often, the most economical system employs combinations of screw-in and slip-in cartridge valves with conventional sliding spool valves, all mounted on a common manifold.

BASIC OPERATION OF SLIP-IN CARTRIDGE VALVES

Slip-in cartridge valves are similar to poppet check valves and consist of an insert assembly that slips into a cavity machined into a manifold block. A control cover bolted to the manifold secures the insert within the cavity. As you can see in Figure 12-1, the insert includes a sleeve, a poppet, a spring, and seals.

The cartridge valve insert can be viewed as the main stage of a two-stage valve. It has two main flow ports, "A" and "B." Drilled passages in the manifold connect the "A" and "B" ports to other cartridges or to the operating hydraulic system. Similarly, a drilled pilot passage in the manifold connects the control port "X" as desired.

Notice the orifice in the drilled passage between the "X" port and the spring chamber "AP." The purpose of this orifice is to reduce the speed at which the valve poppet opens and closes. Various orifice sizes are available to optimize or tune cartridge response in relation to that of the entire hydraulic system. The hydraulic system designer can select the orifice size that provides maximum operating speeds with minimum hydraulic shock.

Area Ratios

As Figure 12-2 indicates, the cartridge valve insert has three areas ("A_A," "A_B," and "A_{AP}") that affect the opening or closing of the valve poppet in the sleeve. "A_A" is the effective area of the poppet exposed to the "A" port. "A_B" is the effective area of the poppet exposed to the "B" port. "A_{AP}" is the effective area of the poppet exposed to the spring chamber "AP." The "A_{AP}" area is always equal to the sum of the "A_A" and "A_B" areas.

The area ratio of an insert is the ratio of the "A_A" area to the "A_{AP}" area. Slip-in cartridge valves are available in three common area ratios:

- 1:1 area ratio, where "A_{AP}" equals the "A_A" area

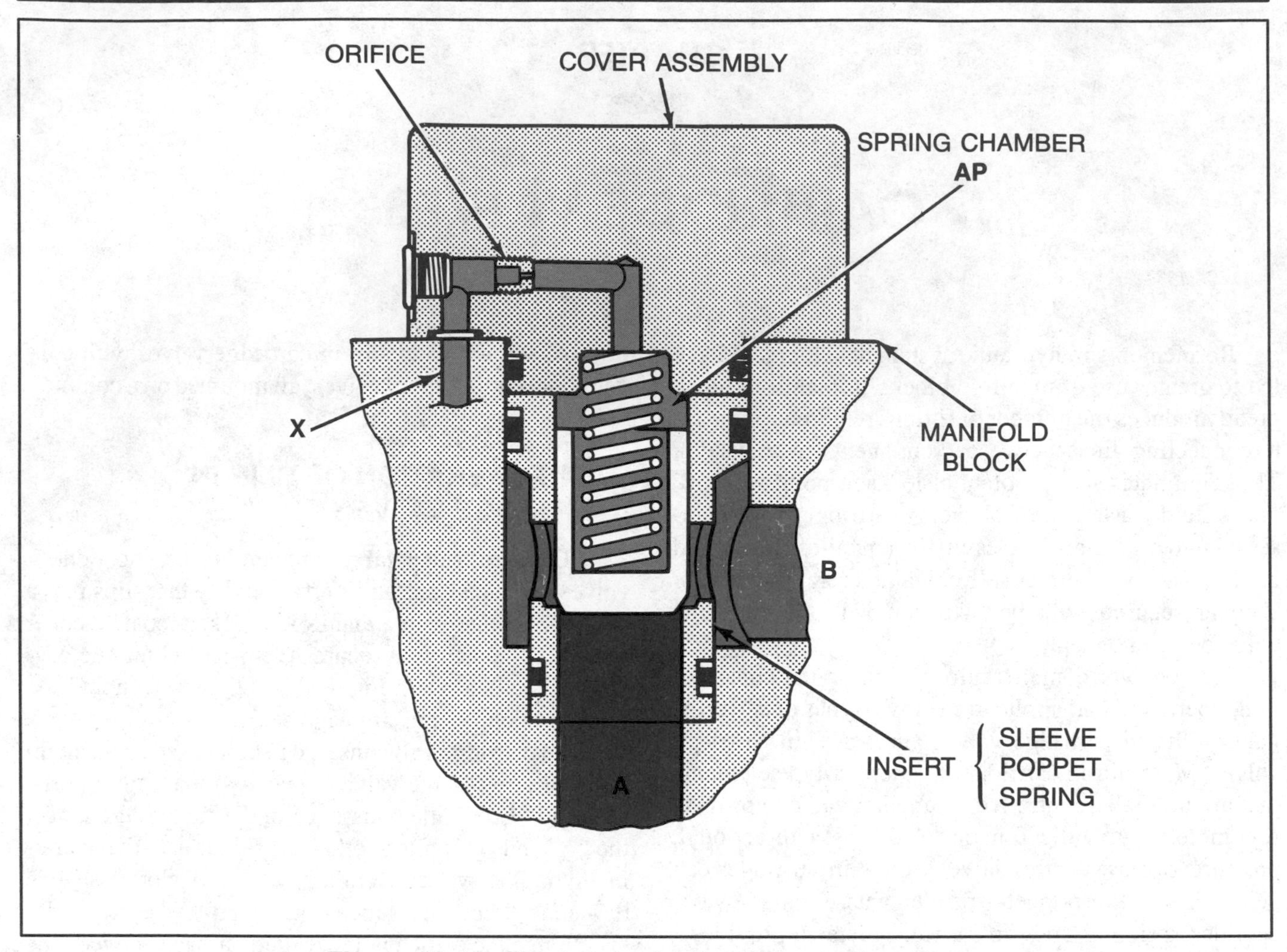

Figure 12-1. Slip-in cartridge valve construction.

- 1:1.1 area ratio, where "A_{AP}" is 1.1 times the "A_A" area
- 1:2 area ratio, where "A_{AP}" is twice the "A_A" area.

Poppet Differences

Figure 12-2A–D shows the construction of the various inserts, the three area ratio relationships, and the associated graphic symbols. As you can see, both the poppet nose design and the sleeve seat change configuration with each area ratio. Figure 12-2A (area ratio = 1:1) shows a square-cornered poppet on the bottom and a sleeve that is chamfered at the point where it touches the poppet.

The insert shown in Figure 12-2B (area ratio = 1:1.1) has a chamfered poppet and a large-diameter, sharp-cornered seat in the sleeve. This means there is a relatively small area "A_B" of the poppet on which the pressure in the "B" port can act. The area amounts to only ten percent of the "A" port area "A_A."

The poppet shown in Figure 12-2C (area ratio =1:2) also has a chamfered nose, but the sleeve has a sharp-edged seat with a smaller diameter than that of Figure 12-2B. Figure 12-2D shows a 1:2 area ratio insert with a metering notch in the poppet. The poppet seat area is the same as the standard 1:2 poppet. However, the nose of the poppet is extended and has a "V" notch cut in it to provide damping or flow metering.

These differences prevent the exchange of sleeves and poppets among the various area ratio inserts.

Closing Versus Opening Forces. To design or troubleshoot a circuit that uses slip-in cartridge valves, you need to know how to determine whether the valve should be open or closed under different circuit conditions. You also need to know how to calculate the pressure required to open or to close the valve. These calculations must take into account the fact that the pressure acts on three areas: "A_A," "A_B," and "A_{AP}." In addition, three different area ratios (1:1, 1:1.1, and 1:2) and different springs are available.

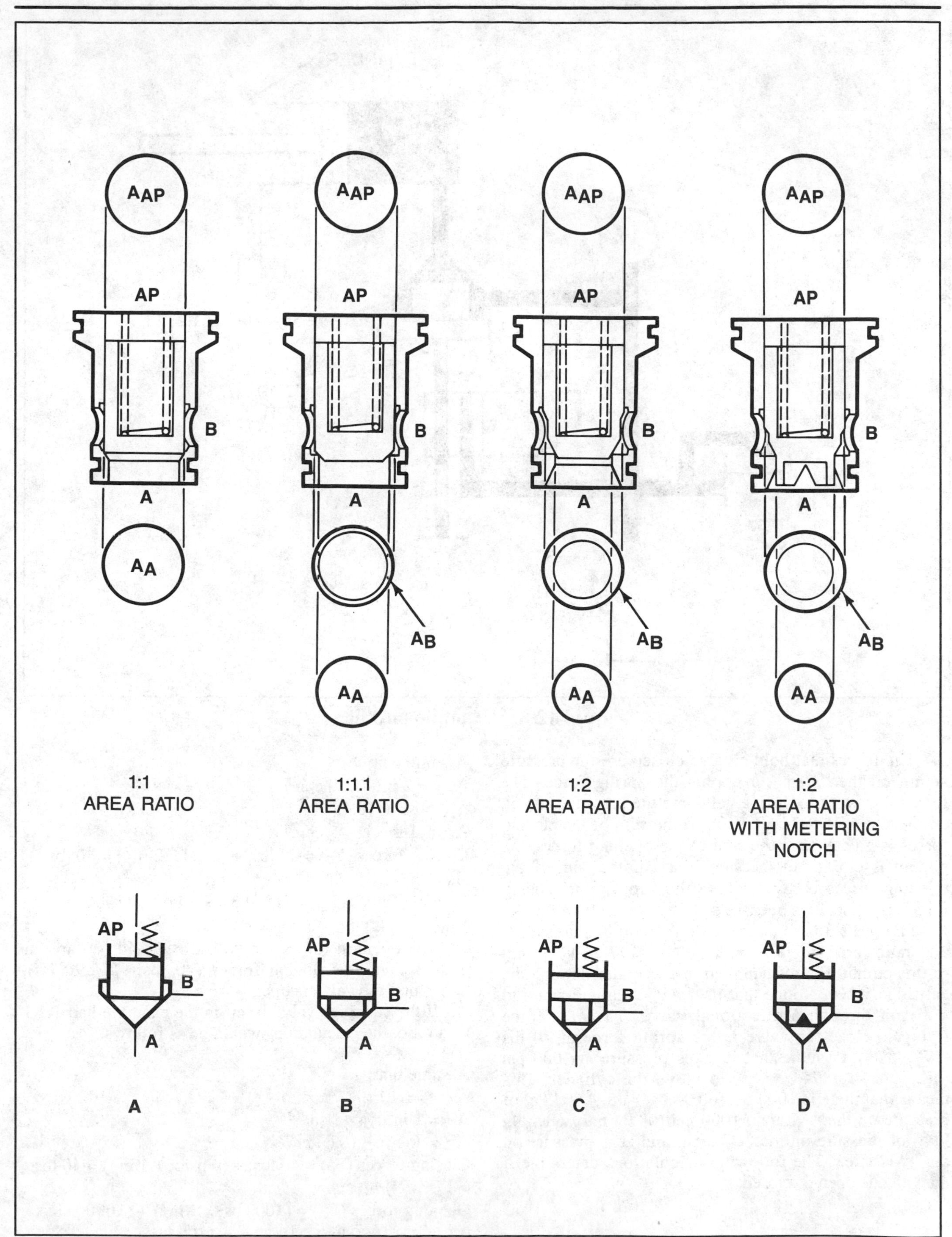

Figure 12-2. Spool area ratio differences.

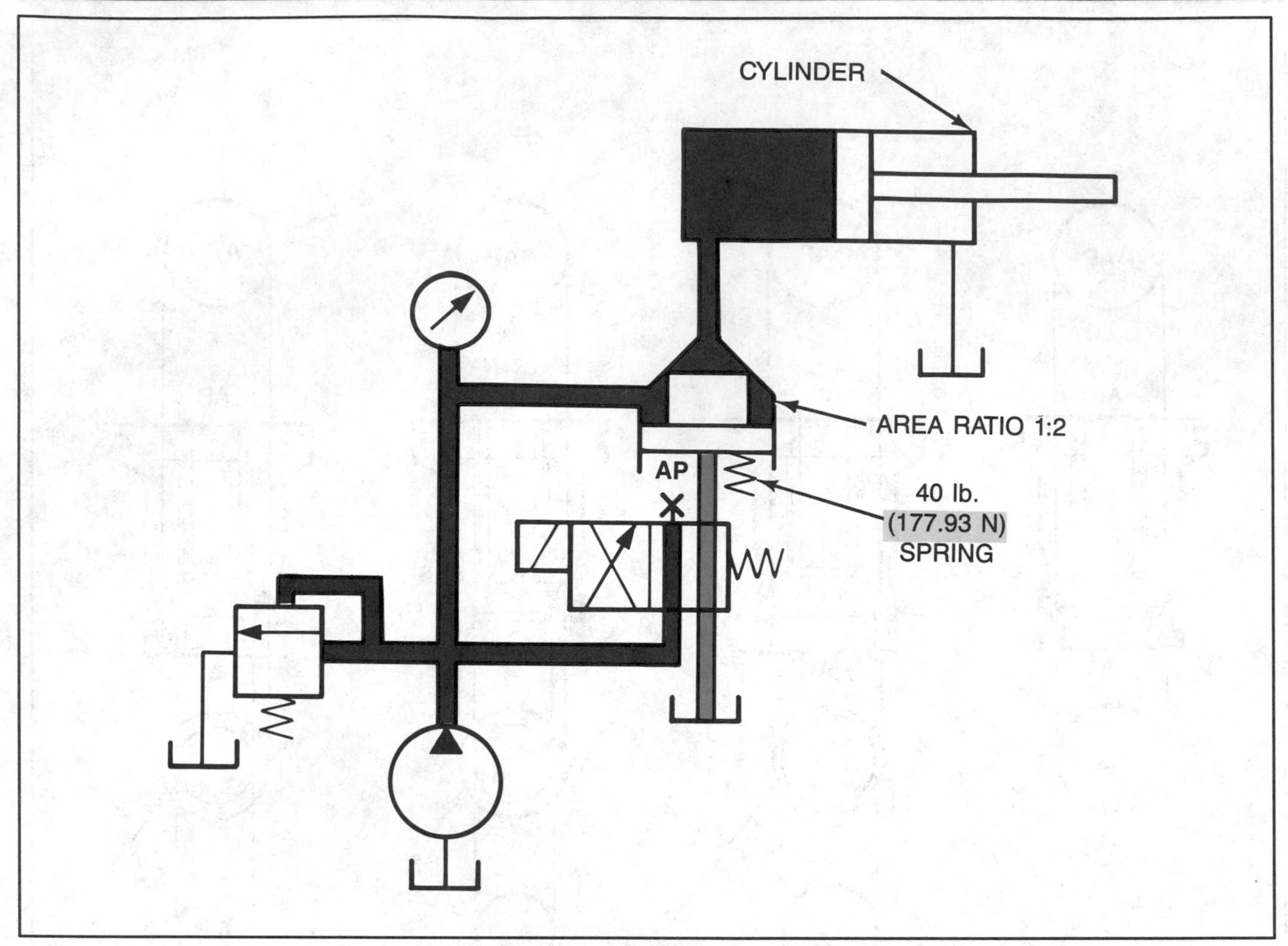

Figure 12-3. Example circuit.

The forces that hold a valve closed are the pressure acting on the "A_{AP}" area plus the spring force. The forces that work to open a valve are the pressures acting on the "A" and "B" port areas. If the sum of the closing forces is greater than the sum of the opening forces, the poppet is closed. In the same way, if the sum of the opening forces is greater than the sum of the closing forces, the poppet is open.

Figure 12-3 is an example of a circuit that uses a 1:2 area ratio cartridge valve with a 40 lb. (177.93 N) spring. In this circuit, flow is supplied to a cylinder through the cartridge valve. Notice that the "AP" area is drained to the tank through a directional valve. The only force trying to close the valve is the spring force of 40 lbs (177.93 N). Assume that it takes pressure at 1000 psi (68.94 bar / 6894.00 kPa) to move the cylinder. This means that there is 1000 psi (68.94 bar / 6894.00 kPa) of pressure on the "A" area, 1000 psi (68.94 bar / 6894.00 kPa) of pressure on the "B" area, and zero pressure on the "AP" area. The following calculations determine if the valve is open or closed.

Assume that:
A_{AP} = 2 in^2 (12.9 cm^2)
A_A = 1 in^2 (6.45 cm^2)
A_B = 1 in^2 (6.45 cm^2)
Closing forces (F_C) = 40 lbs. + (0 psi × 2 in^2) = 40 lbs. (177.93 N)
Opening forces (F_O) = (1000 psi × 1 in^2) + (1000 psi 1 in^2) = 2000 lbs. (8896.44 N)

The opening forces (2000 lbs. (8896.44 N)) are far greater than the closing forces (40 lbs. (177.93 N)). Therefore, the valve is open.

If the cartridge valve insert in the example had 1:1.1 area ratio, the calculations would be as follows:

Assume that:
A_{AP} = 1.1 in^2 (7.1 cm^2)
A_A = 1 in^2 (6.45 cm^2)
A_B = 0.1 in^2 (0.65 cm^2)
Closing forces (F_C) = 40 lbs. + (0 psi × 1.1 in^2) = 40 lbs. (177.93 N)
Opening forces (F_O) = (1000 psi × 1 in^2) + (1000 psi 0.1 in^2) = 1000 lbs. + 100 lbs. = 1100 lbs. (4893.04 N)

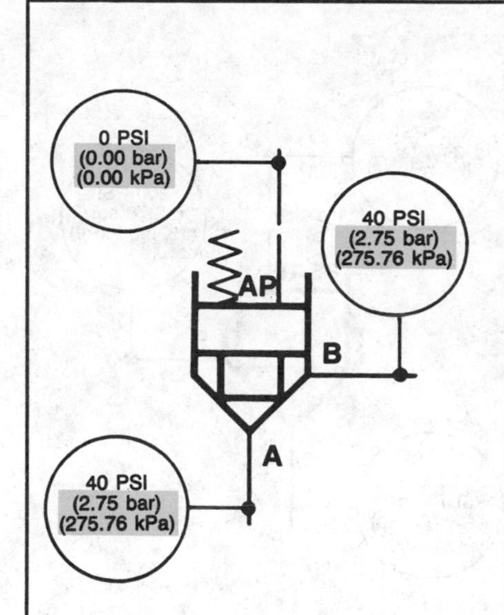

1:2 Area Ratio

Let area ratio equal areas:
1 in^2
2 in^2

Therefore:

$A_A = 1\ in^2$

$A_{AP} = 2\ in^2$

$A_B = A_{AP} - A_A$
$= 2\ in^2 - 1\ in^2$
$= 1\ in^2$

Let Spring Force = 70 lbs.

Closing Forces:

F_C = Spring Force + (P_{AP} x A_{AP})

F_C = 70 lbs. + (0 PSI x 2 in^2)
= **70 lbs.**

Opening Forces:

F_O = (P_A x A_A) + (P_B x A_B)

F_O = (40 PSI x 1 in^2) +
(40 PSI x 1 in^2)

F_O = 40 lbs. + 40 lbs.
= **80 lbs.**

Therefore, the poppet is wide open.

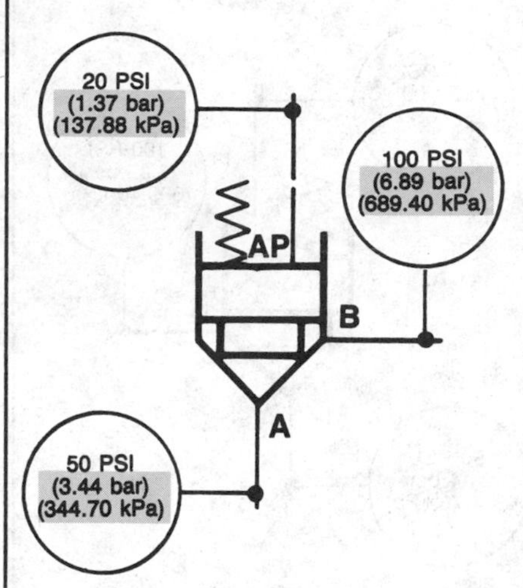

1:1.1 Area Ratio

Let area ratio equal areas:
1 in^2
1.1 in^2

Therefore:

$A_A = 1\ in^2$

$A_{AP} = 1.1\ in^2$

$A_B = A_{AP} - A_A$
$= 1.1\ in^2 - 1\ in^2$
$= 0.1\ in^2$

Let Spring Force = 40 lbs.

Closing Forces:

F_C = Spring Force + (P_{AP} x A_{AP})

F_C = 40 lbs. + (20 PSI x 1.1 in^2)

F_C = 40 lbs. + 22 lbs.
= **62 lbs.**

Opening Forces:

F_O = (P_A x A_A) + (P_B x A_B)

F_O = (50 PSI x 1 in^2) +
(100 PSI x 0.1 in^2)

F_O = 50 lbs. + 10 lbs.
= **60 lbs.**

Therefore, the poppet is closed.

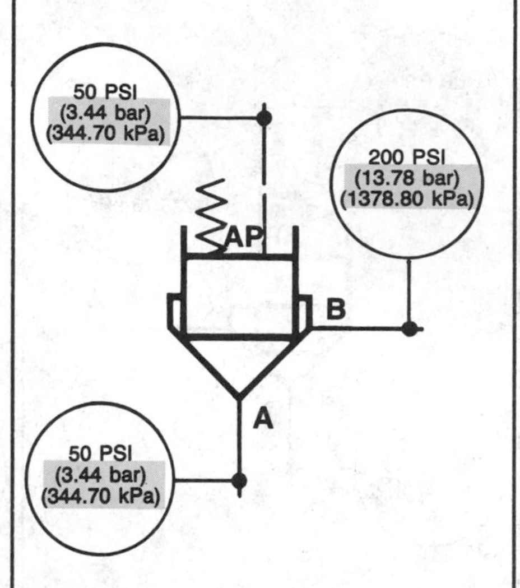

1:1 Area Ratio

Let area ratio equal areas:
1 in^2
1 in^2

Therefore:

$A_A = 1\ in^2$

$A_{AP} = 1.1\ in^2$

$A_B = A_{AP} - A_A$
$= 1\ in^2 - 1\ in^2$
$= 0\ in^2$

Let Spring Force = 40 lbs.

Closing Forces:

F_C = Spring Force + (P_{AP} x A_{AP})

F_C = 40 lbs. + (50 PSI x 1 in^2)

F_C = 40 lbs. + 50 lbs.
= **90 lbs.**

Opening Forces:

F_O = (P_A x A_A) + (P_B x A_B)

F_O = (50 PSI x 1 in^2) +
(200 PSI x 0 in^2)

F_O = 50 lbs. + 0 lbs.
= **50 lbs.**

Therefore, the poppet is held closed.

Figure 12-4. Valve closing and opening forces. (English)

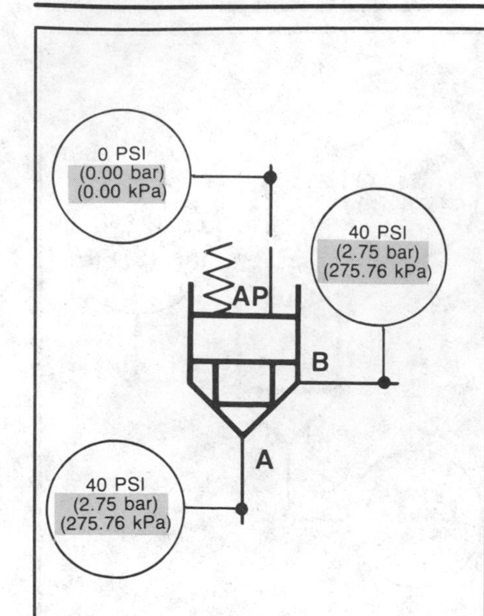

1:2 Area Ratio

Let area ratio equal areas:
645.16 mm^2
1290.32 mm^2

Therefore:

A_A = 645.16 mm^2

A_{AP} = 1290.32 mm^2

$A_B = A_{AP} - A_A$
= 1290.32 mm^2 − 645.16 mm^2
= 645.16 mm^2

Let Spring Force = 311.37 N

Closing Forces:

F_C = Spring Force + $(P_{AP} \times A_{AP})/10$

F_C = 311.37 N + (0 bar x 1290.32 mm^2)/10
= **311.37 N**

Opening Forces:

$F_O = (P_A \times A_A)/10 + (P_B \times A_B)/10$

F_O = (2.76 bar x 645.16 mm^2)/10 + (2.76 bar x 645.16 mm^2)/10

F_O = 178.06 N + 178.06 N
= **356.12 N**

Therefore, the poppet is wide open.

20 PSI
(1.37 bar)
(137.88 kPa)
100 PSI
(6.89 bar)
(689.40 kPa)
AP
B
A
50 PSI
(3.44 bar)
(344.70 kPa)

1:1.1 Area Ratio

Let area ratio equal areas:
645.16 mm^2
709.68 mm^2

Therefore:

A_A = 645.16 mm^2

A_{AP} = 709.68 mm^2

$A_B = A_{AP} - A_A$
= 709.68 mm^2 − 645.16 mm^2
= 64.52 mm^2

Let Spring Force = 177.93 N

Closing Forces:

F_C = Spring Force + $(P_{AP} \times A_{AP})/10$

F_C = 177.93 N + (1.38 bar x 709.68 mm^2)/10

F_C = 177.93 N + 97.94 N
= **275.87 N**

Opening Forces:

$F_O = (P_A \times A_A)/10 + (P_B \times A_B)/10$

F_O = (3.45 bar x 645.16 mm^2)/10 + (6.89 bar x 64.52 mm^2)/10

F_O = 222.58 N + 44.45 N
= **267.03 N**

Therefore, the poppet is closed.

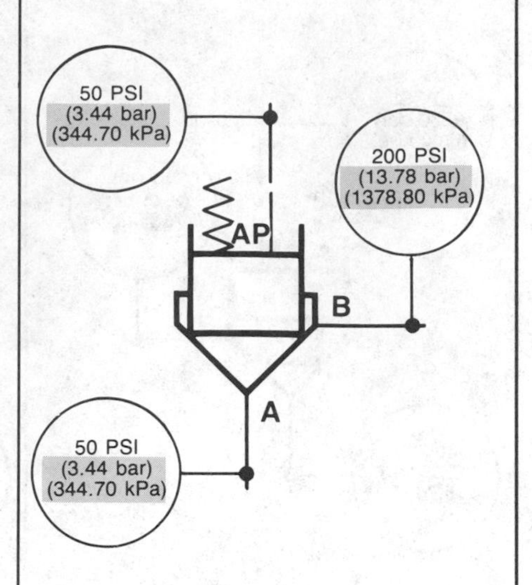

1:1 Area Ratio

Let area ratio equal areas:
645.16 mm^2
645.16 mm^2

Therefore:

A_A = 645.16 mm^2

A_{AP} = 645.16 mm^2

$A_B = A_{AP} - A_A$
= 645.16 mm^2 − 645.16 mm^2
= 0 mm^2

Let Spring Force = 177.93 N

Closing Forces:

F_C = Spring Force + $(P_{AP} \times A_{AP})/10$

F_C = 177.93 N + (3.44 bar PSI x 645.16 mm^2)/10

F_C = 177.93 N + 222 N
= **400 N**

Opening Forces:

$F_O = (P_A \times A_A)/10 + (P_B \times A_B)/10$

F_O = (3.45 bar x 645.16 mm^2)/10 + (13.79 bar x 0 mm^2)/10

F_O = 222.41 N + 0 N
= **222.41 N**

Therefore, the poppet is held closed.

Figure 12-5. Valve closing and opening forces. (Metric)

The opening forces (1100 lbs. (4893.04 N)) are much greater than the closing forces (40 lbs. (177.93 N)). Therefore, the valve is open.

You will find other examples of closing and opening forces in Figures 12-4 and 12-5.

DIRECTIONAL CONTROL SLIP-IN CARTRIDGE VALVES

The following section examines the operation of the slip-in cartridge valves that control the direction of the flow of hydraulic fluid.

Check Valve Operation

When a 1:2 area ratio cartridge valve is used as a check valve, as shown in Figure 12-6A, it is necessary to connect the "AP" chamber to the "B" port. If this is not done, any pressure at the "B" port great enough to compress the spring will open the valve. To operate as a check valve, the cartridge should open only when pressure at the "A" port is greater than pressure at the "B" port plus the spring force.

In Figure 12-6B, you can see that a check valve function can also be created by connecting the "AP" chamber to the "A" port. This provides free flow from port "B" to port"A." However, there can be leakage from port "A" through the pilot line to the "AP" chamber and then through the clearance between the poppet and sleeve to the "B" port. To avoid this, the arrangement in Figure 12-6A always should be used for check valve functions.

Four-Way Directional Control Valve Operation

By definition, four-way directional control valves provide four flow paths, two at a time. One application of a four-way valve is to extend, retract, and stop the piston of a double-acting cylinder. Figures 12-7, 12-8, and 12-9 show how cartridge valves can be installed to operate together as a single four-way valve.

Figure 12-7 shows a circuit in which the piston extends when the "AP" areas of two inserts are connected to the tank.

To retract the piston (Figure 12-8), the pump output must be directed to the rod end while the cap end is connected to the tank. Another pair of cartridge valves provides these flow paths. The "AP" areas are open to the tank at this time.

Figure 12-9 illustrates a circuit in which the four cartridge valves from the previous two figures are piloted by one double-solenoid, spring-centered spool valve. Pilot pressure is obtained by teeing off the main system pres-

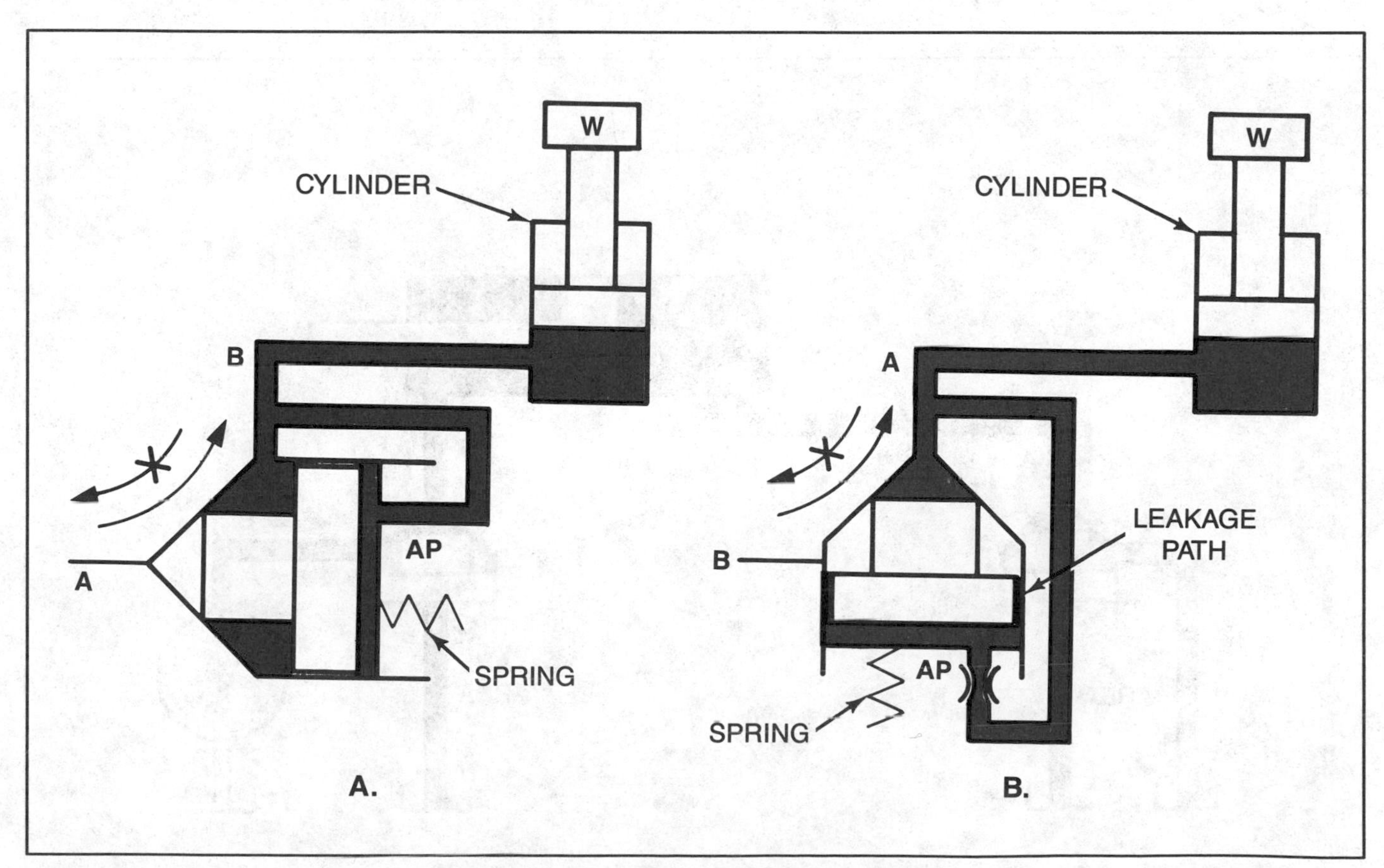

Figure 12-6. Check valve operation.

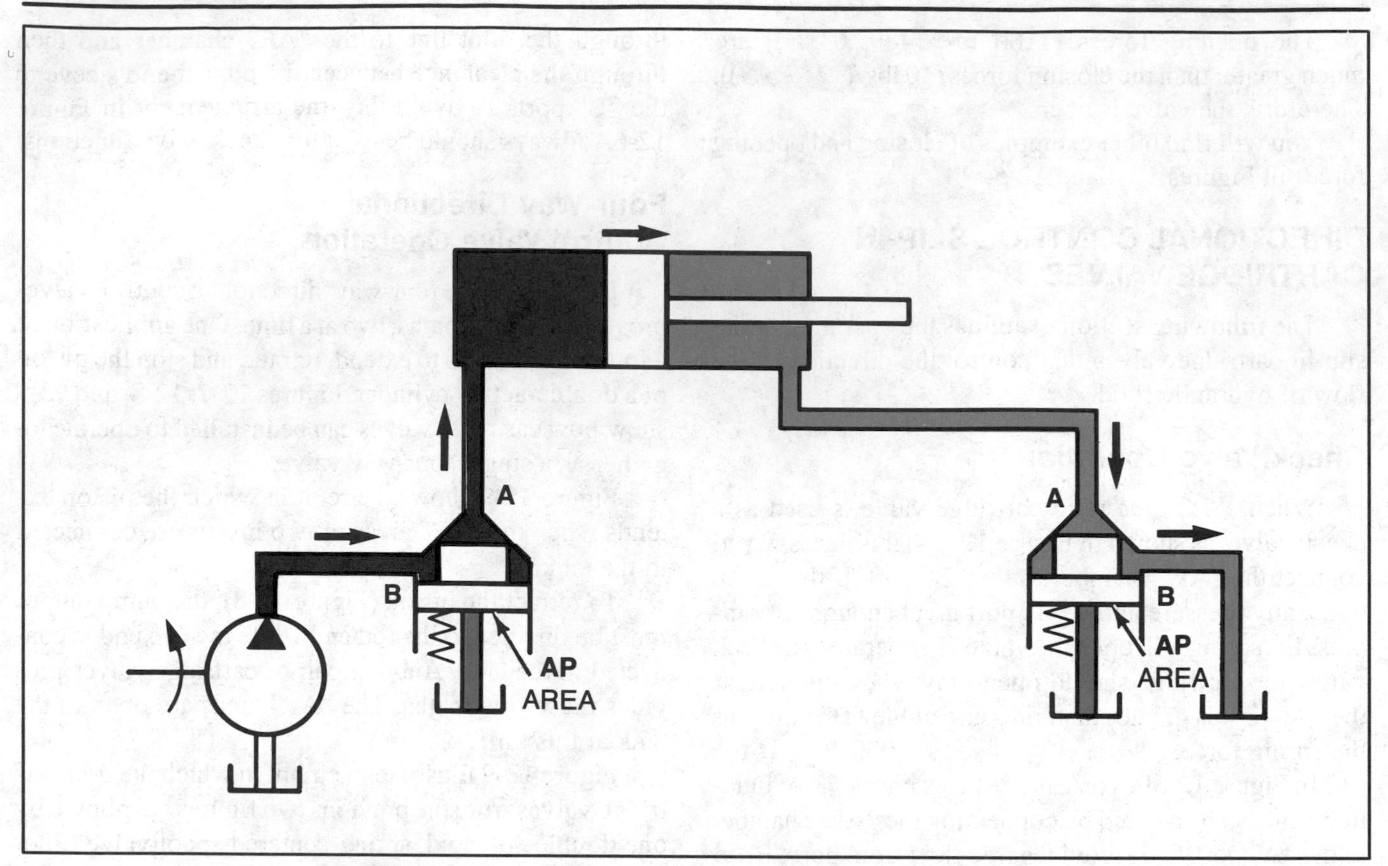

Figure 12-7. Directional control valve operation.

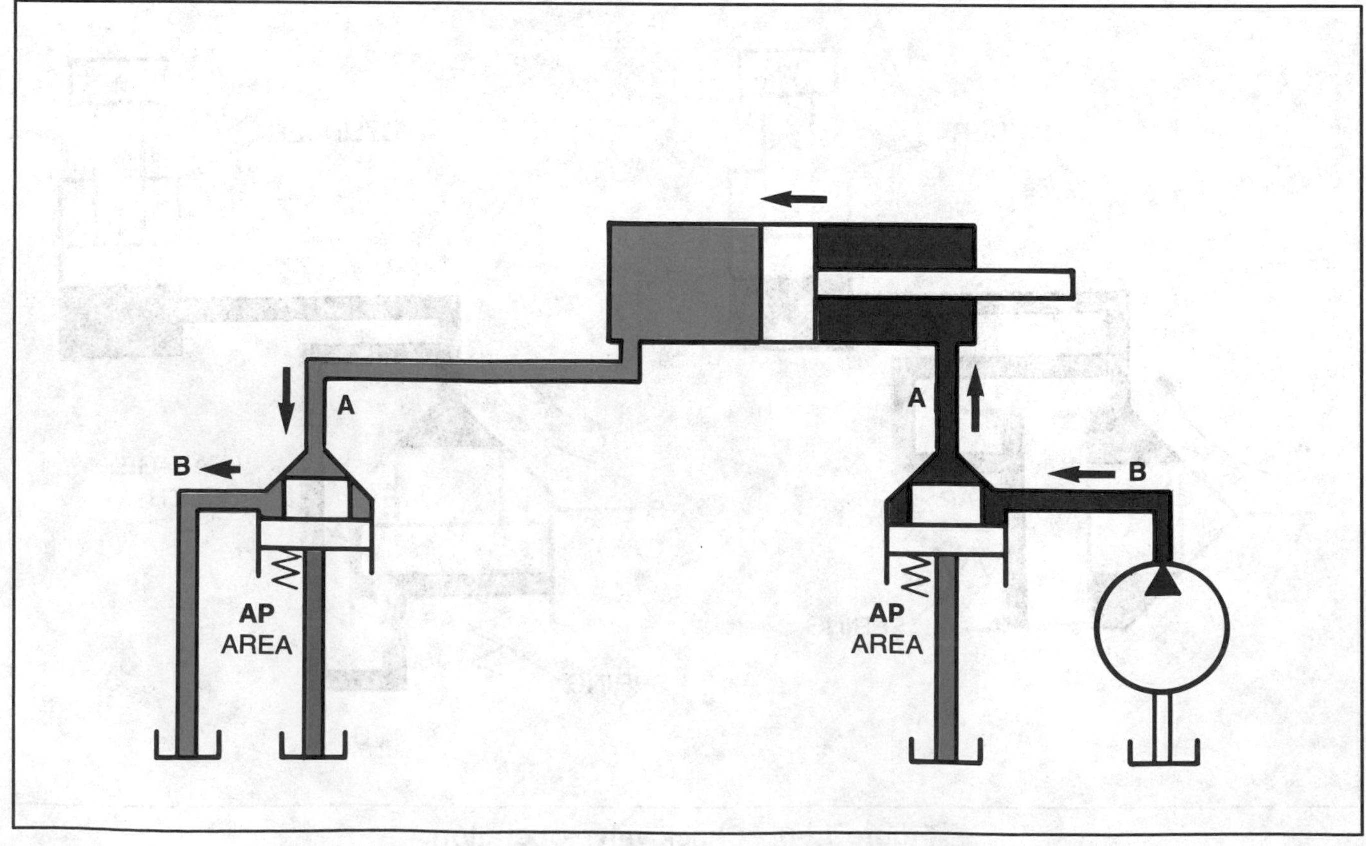

Figure 12-8. Cartridge directional control valve retract piston.

sure line and is applied to each cartridge pilot chamber when the pilot valve solenoids are de-energized. The pilot chambers of cartridge valves 1 and 3 are interconnected, as are the pilot chambers of cartridges 2 and 4, so that the paired valves open or close simultaneously. This essentially forms a closed center circuit, because all four cartridges are closed when both solenoids of the pilot valve are de-energized.

Energizing solenoid S-1 drains the "AP" areas of cartridges 2 and 4 while maintaining pilot pressure on the "AP" areas of cartridges 1 and 3. System pressure acting over the "B" area of cartridge 2 forces the valve open, directing pump flow to the cap end of the cylinder.

Return flow from the cylinder rod end passes to the tank over cartridge 4 when rod end pressure is high enough to overcome the spring acting against the cartridge valve poppet. The piston of the cylinder moves to the right. Energizing solenoid S-2 drains the "AP" areas of cartridges 1 and 3 and applies pressure to the "AP" areas of cartridges 2 and 4. This causes the piston to move to the left.

Conventional spool valves must be sized for the highest flow at any one port. Cartridge valves, on the other hand, are sized to handle only the flow required through their individual ports. Economies are realized and system size is optimized.

Large system flows can be controlled with a small pilot directional valve and four cartridge valves. This arrangement is similar to a spool-type directional valve with three distinct positions, which is shown on the right in Figure 12-9. The parallel paths, the crossed arrow paths, and a center condition are shown. Notice that in the center condition, a check valve appears in each cylinder port line. This is done to indicate that when neither solenoid is energized, an external force pushing on the piston rod could cause valve 2 to open. This could happen when the cylinder cap end pressure exceeds the system pressure. In such a case, the system pressure acting on the "B" area, plus slightly more than system pressure acting on the "A" area, combine to overcome system pressure and the spring acting on the "AP" area. (If the rod is pulled instead of pushed, valve 3 rather than valve

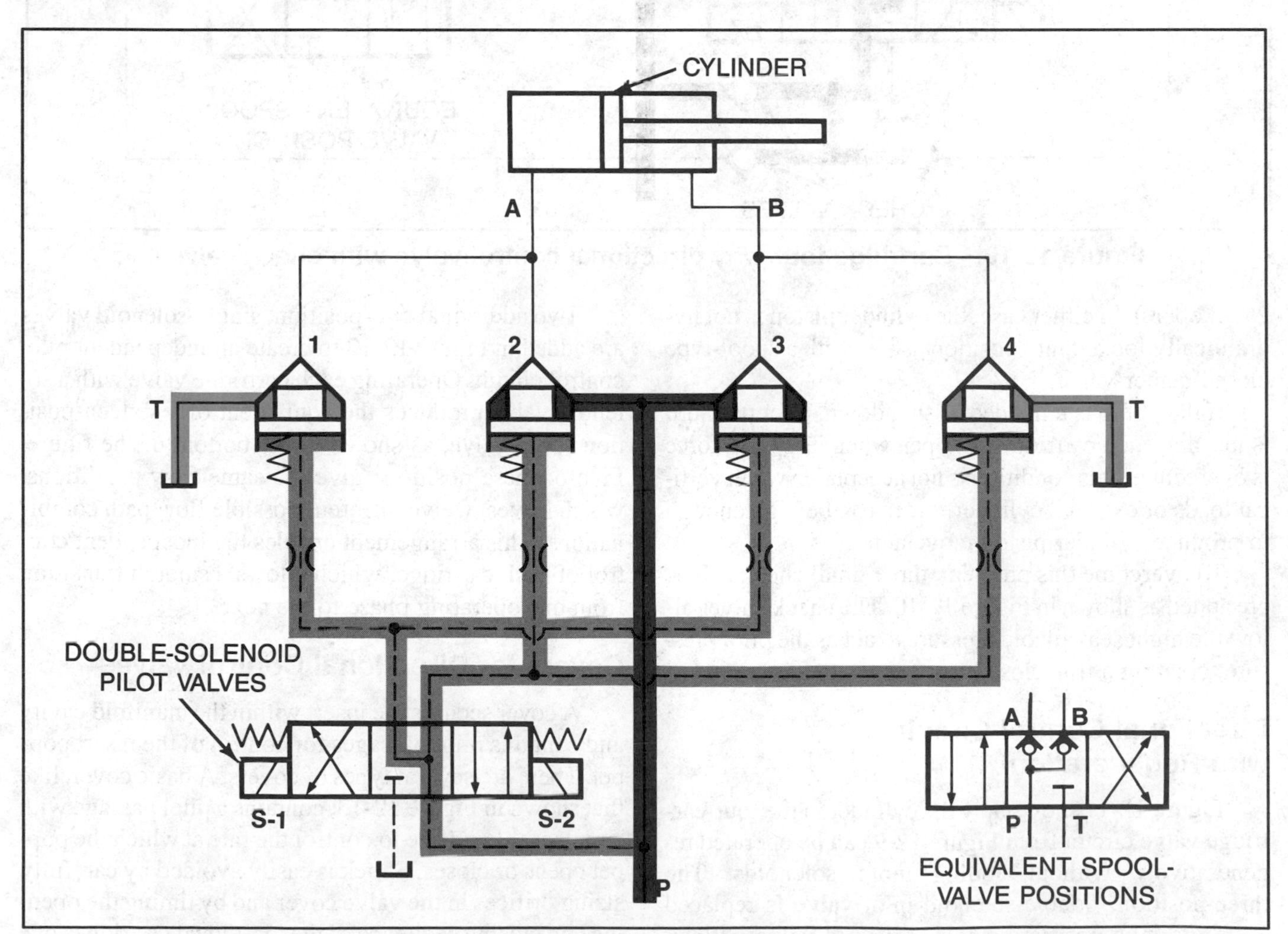

Figure 12-9. Complete cartridge four-way directional control valve.

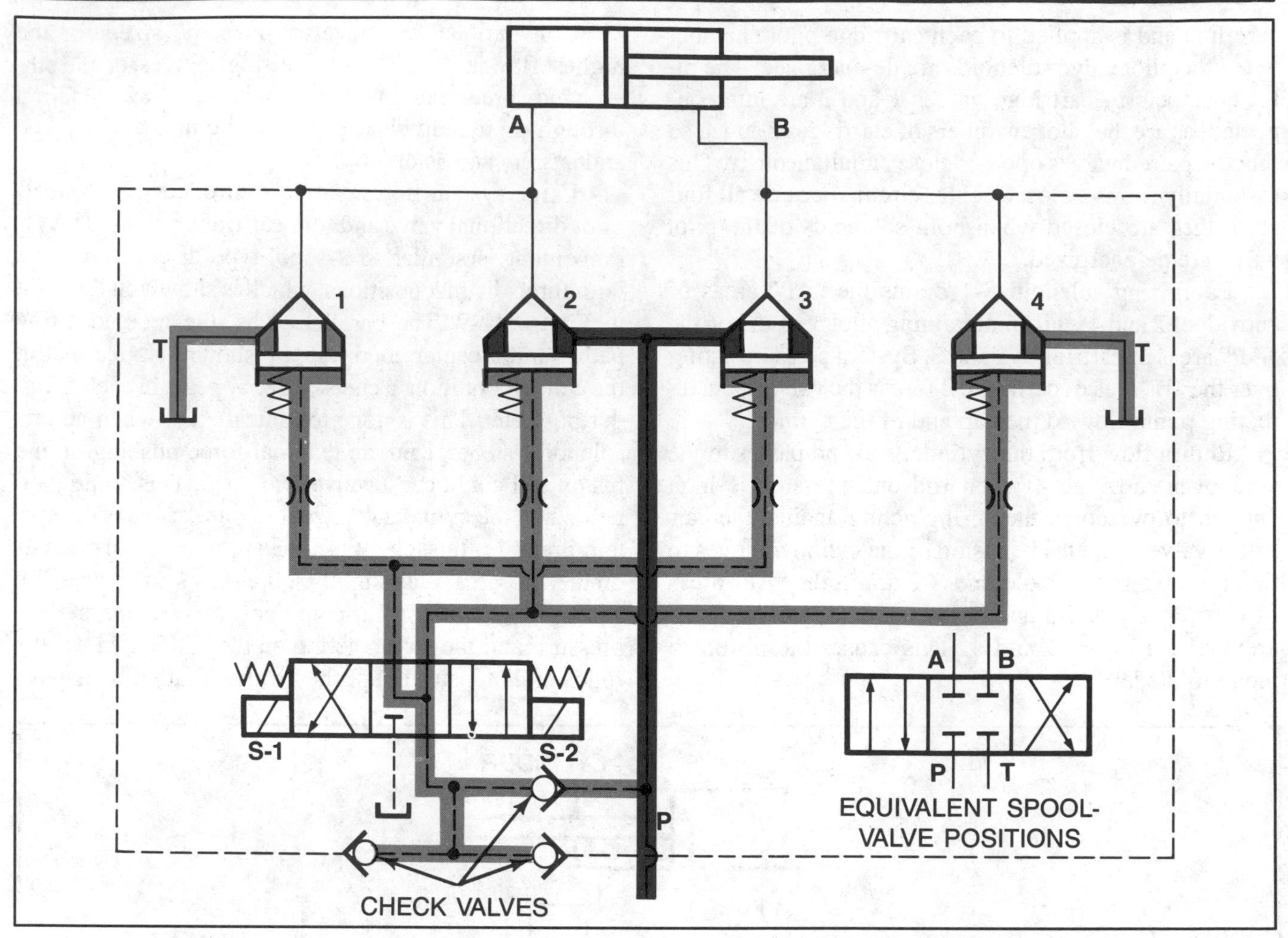

Figure 12-10. Cartridge four-way directional control valve with check valves.

2 will open.) In either case, the cylinder piston is not hydraulically locked into position as it is with a spool-type closed-center valve.

If the pump is unloaded or shut down, pilot pressure is lost and each cartridge can open when its spring force is overcome. This condition is not acceptable when vertical loads or external cylinder forces can be high enough to produce cylinder piston movement.

To overcome this problem, three small check valves are added as shown in Figure 12-10. The check valves allow the highest available pressure to act as the pilot pressure, creating a true closed-center system.

Directional Control Circuit with Regeneration

Figure 12-11 shows how the cylinder in the four-cartridge valve circuit from Figure 12-9 can be operated regeneratively without adding more solenoids. The three-position, double-solenoid pilot valve is replaced with two, two-position, single- solenoid valves. When the two solenoids are energized simultaneously, the cylinder operates in a regenerative manner.

Two additional two-position, single-solenoid valves are added in Figure 12-12 to create an independent pilot control circuit. Operating each cartridge valve with a solenoid valve produces the equivalent of a sixteen-position spool valve, as shown at the bottom of the figure. Five of these positions give the same flow conditions, which leaves twelve different possible flow path combinations. This arrangement enables the independent control of each cartridge, which allows a smooth transition from one operating phase to the next.

Covers for Directional Control Valves

A cover secures the insert within the manifold cavity and contains pilot passages for control of the insert poppet. There are several types of covers. A basic cover, like that shown in Figure 12-13, contains a pilot passage with a replaceable orifice to control the rate at which the poppet opens or closes. Shock is easily avoided by carefully sizing orifices in the valve cover and by timing the opening and closing sequence of the individual cartridges of a directional control valve. In conventional spool-type valves, the timing of the opening or closing of all four

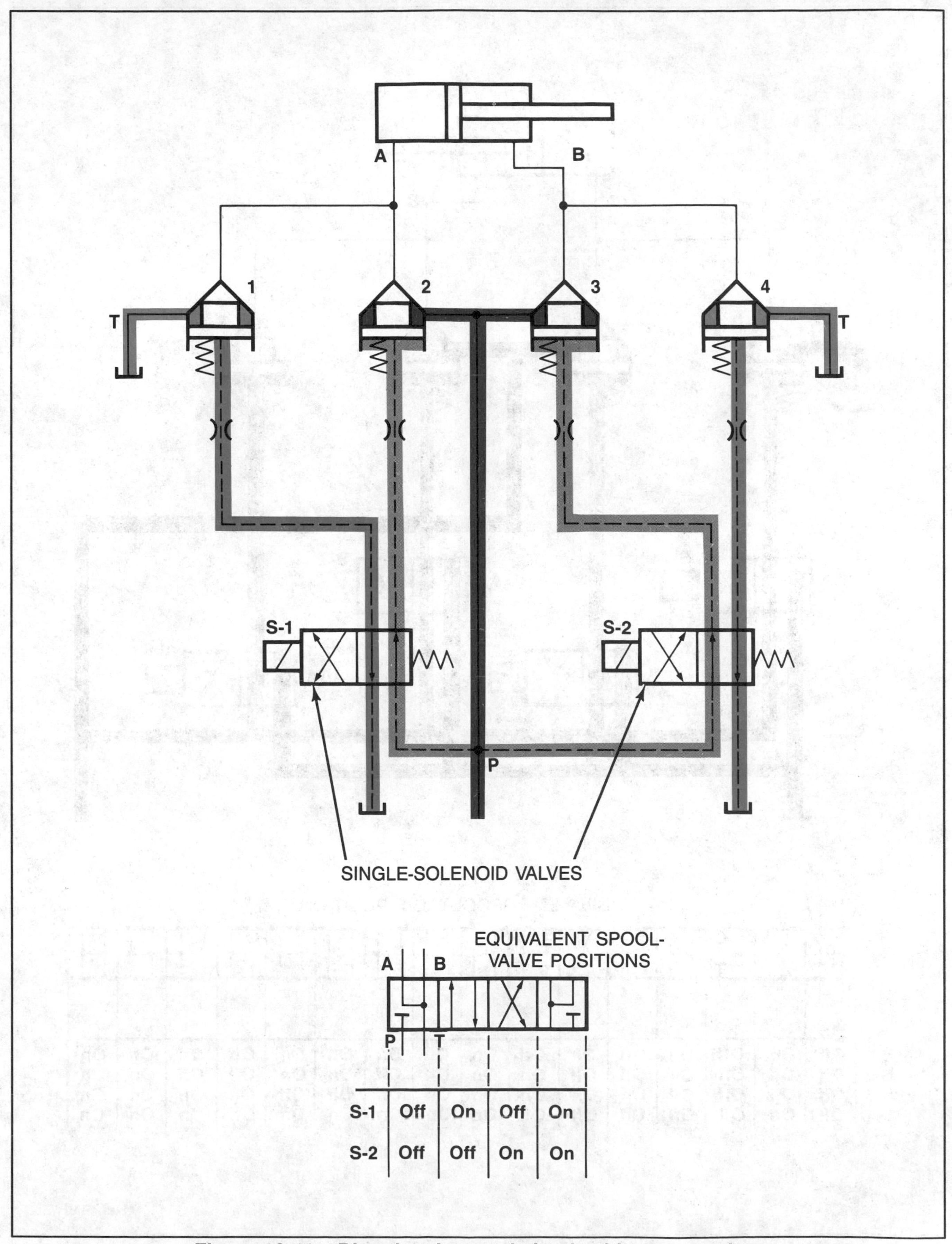

Figure 12-11. Directional control circuit with regeneration.

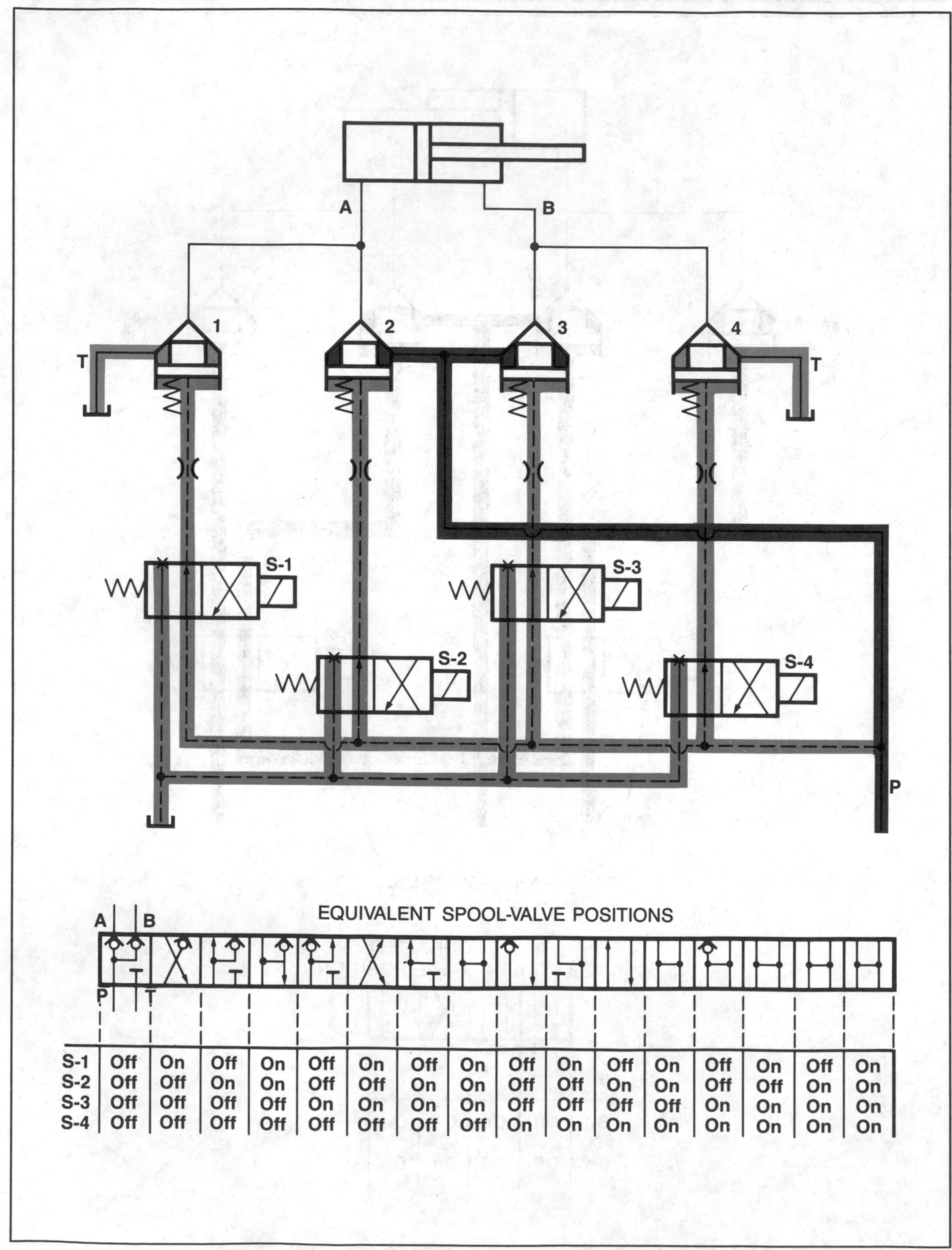

S-1	Off	On	Off	On	Off	On	Off	On	Off	On	Off	On	Off	On	Off	On
S-2	Off	Off	On	On	Off	Off	On	On	Off	Off	On	On	Off	Off	On	On
S-3	Off	Off	Off	Off	On	On	On	On	Off	Off	Off	Off	On	On	On	On
S-4	Off	Off	Off	Off	Off	Off	Off	Off	On	On	On	On	On	On	On	On

Figure 12-12. Independent directional control circuit.

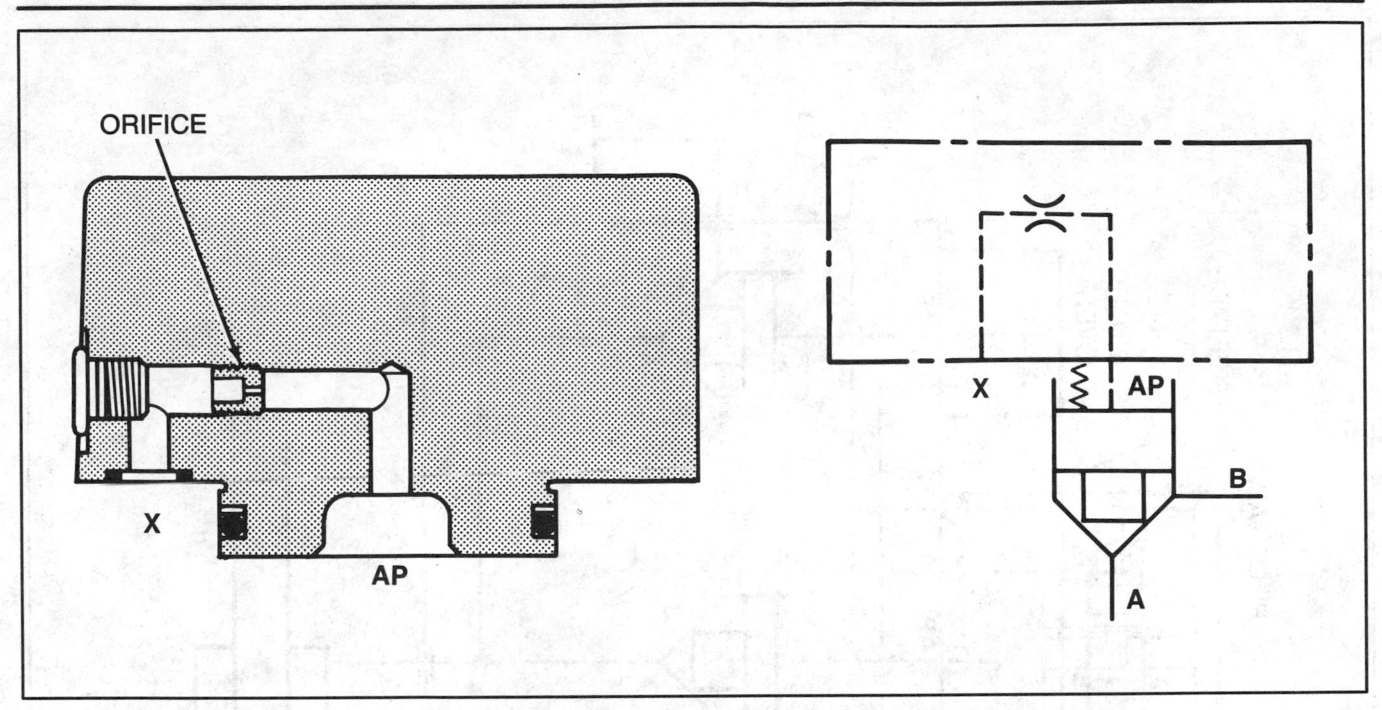

Figure 12-13. Standard cover.

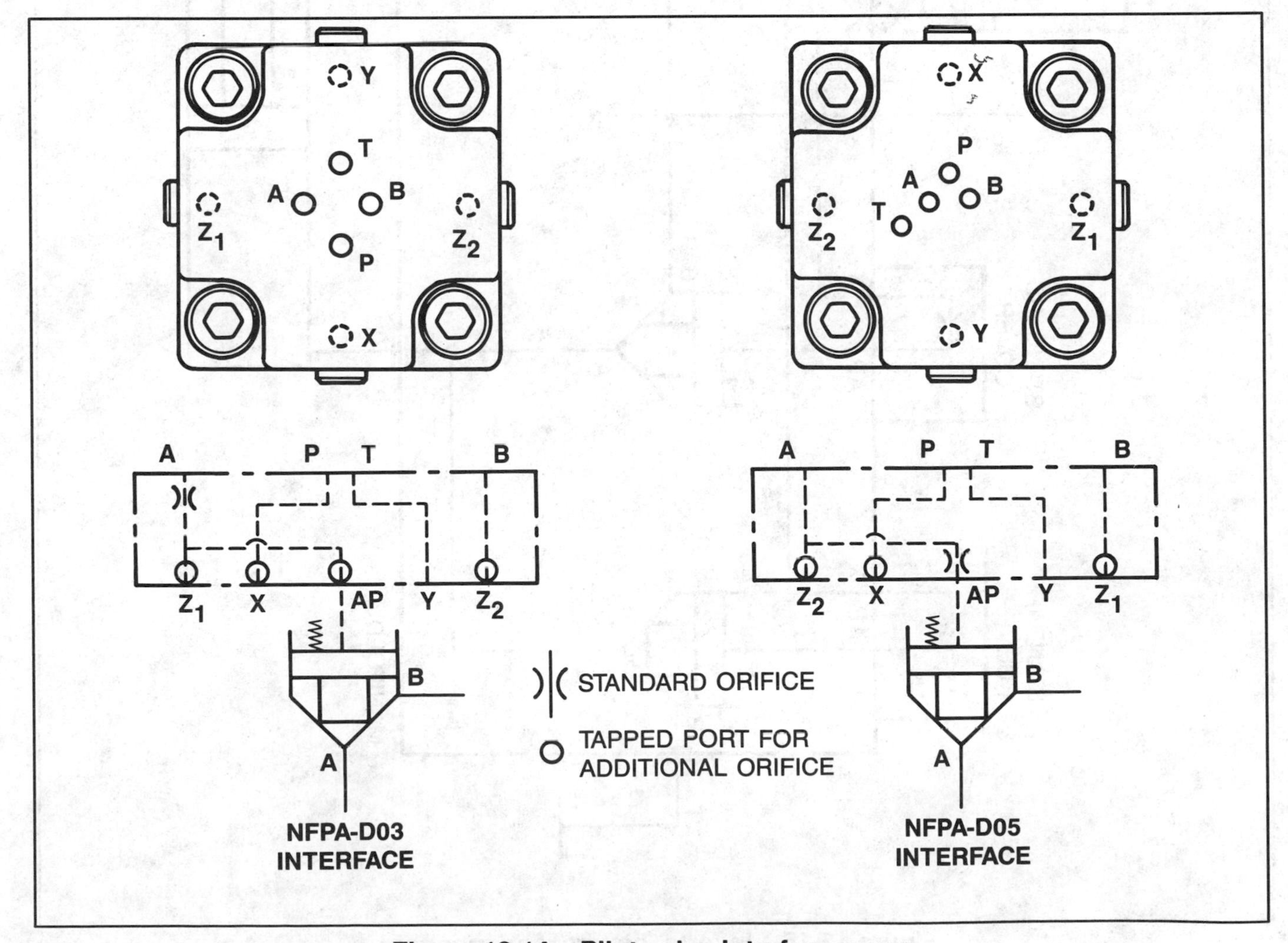

Figure 12-14. Pilot valve interface cover.

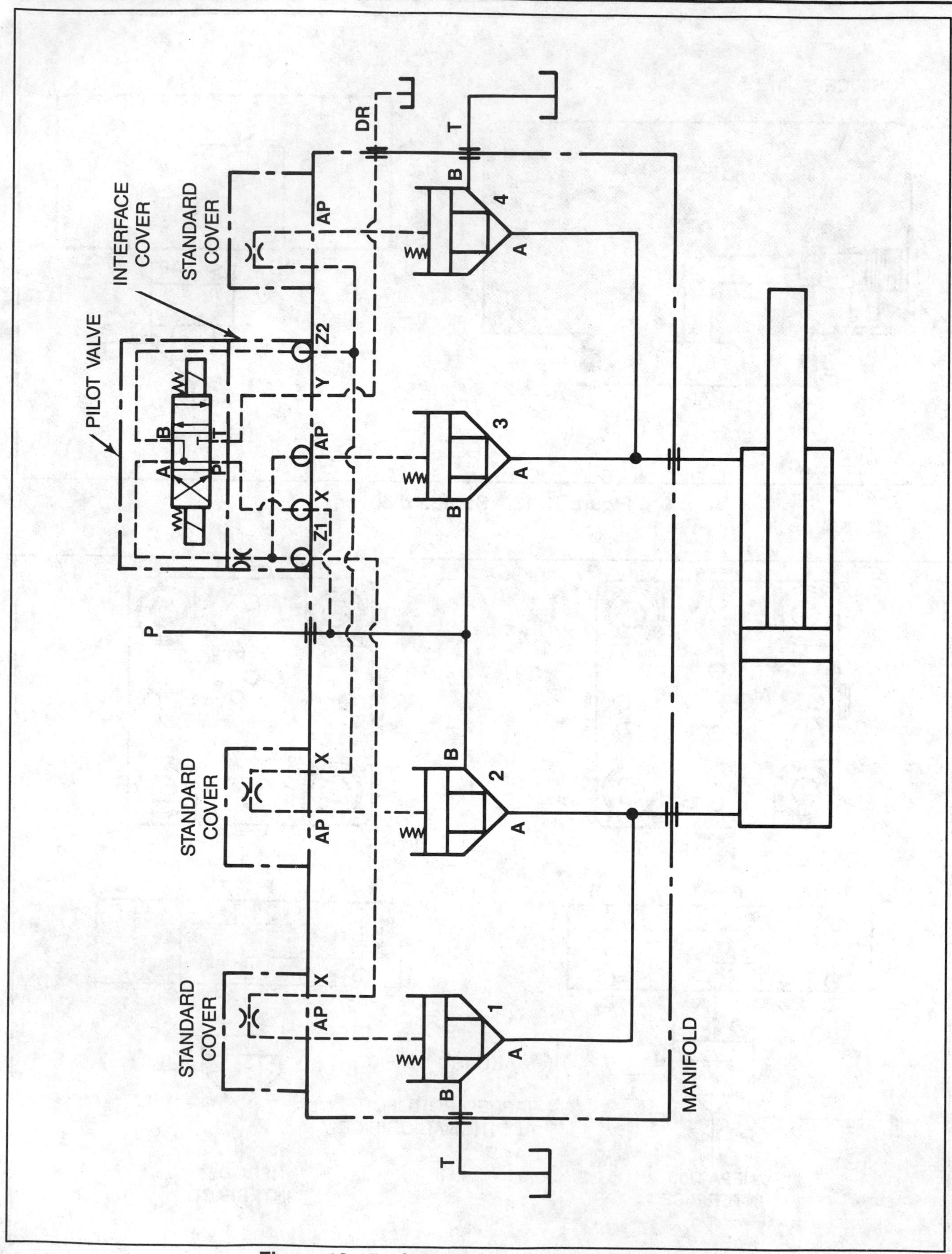

Figure 12-15. Inserts, covers, and manifold.

ports occurs simultaneously, which makes it more difficult to avoid shock.

Pilot Valve Interface Cover

Figure 12-14 contains the two covers that have a pilot valve interface. One has a NFPA-D03 interface, and the other has a NFPA-D05 interface. Auxiliary ports "Z_1" and/or "Z_2" may be connected in the manifold to other valve pilots to provide simultaneous operation of more than one cartridge valve with one pilot valve.

Of course, it is possible to fit any type of solenoid valve arrangement (such as a spring offset, detented, or spring centered valve) on the cover plate, depending upon the particular circuitry requirements.

Figure 12-15 illustrates a complete manifold, including the covers and inserts. This is the same cartridge-type four-way that appears in Figure 12-9.

Shuttle Valve Cover

A shuttle valve cover (Figure 12-16) takes the higher of two pilot pressures and directs it to pilot spring area "AP." The "X" and "Y" ports feed fluid to each end of the shuttle. The spring chamber area "AP" and the "Z_2" port are connected to the center section of the shuttle.

Shuttle valve covers are also available with a NFPA-D03 pilot valve directional interface (Figure 12-17). When the solenoid is de-energized as shown, the highest pressure in either port "X" or "Y" is transmitted through the shuttle valve to the poppet spring area to close the poppet. When the solenoid is energized, the poppet spring chamber is drained and fluid can travel in either direction by simply overcoming the spring holding the poppet closed.

Pilot-Operated Check Valve Cover

A pilot-operated cartridge check valve cover (Figure 12-18) can be applied much like other pilot-operated check valves. Three pilot ports are used.

- The "X" port is drilled in the manifold to the pilot signal.
- The "Y" port is drilled in the manifold to connect with the "B" port of the insert kit.
- The "Z_1" port is connected to a drain line.

When there is no pilot signal to the "X" port in the cover, a spring, acting against a fluted piston, holds the ball against the right-hand seat, blocking the "Z_1" drain port and opening pilot port "Y" to the "AP" port. The insert poppet functions as a standard check valve with free flow from port "A" to port "B." The "AP" area, pressurized from the "B" port through the "Y" port passage, blocks reverse flow from "B" to "A."

When a pilot pressure signal that is at least 30 percent of the pressure at pilot port "Y" is applied to the pilot piston through the "X" port, the pilot piston moves the ball against the left-hand seat and blocks pilot port "Y" from the "AP" area of the insert. At the same time, the "AP" area opens to the "Z_1" port and to the drain. Reverse flow from the "B" port to the "A" port is now possible, if the area ratio of the insert is anything other than 1:1.

PRESSURE CONTROL SLIP-IN CARTRIDGE VALVES

The following section concerns the operation of slip-in cartridge valves used for pressure-relieving and -reducing valve functions.

Relief Valve

A slip-in cartridge relief valve works like a traditional relief valve. As Figure 12-19 indicates, it is a two-stage or pilot-operated relief valve. A traditional relief valve pilot sensing passage is a 0.040-inch (1 mm) hole through the spool. A slip-in cartridge relief valve pilot sensing passage is a drilled hole in the manifold that connects the "A" port to the "X" port and the control orifice in the cover.

Notice the three pilot connections at the cover interface. "X" is the pilot sensing connection, "Y" is the drain connection, and "Z" is the vent connection. Flow through the main stage is from "A" to "B." The "Y" port may be connected to "B" or to a separate drain. Like a conventional relief valve, any back pressure in the drain port adds directly to the setting of the valve. The "Z" port may be drilled in the manifold to connect to a vent control or to a remote pressure pilot. If it is not needed, it is simply not drilled in the manifold.

Three pressure ranges are available by changing the spring in the pilot section. The main-stage spring remains the same in all cases.

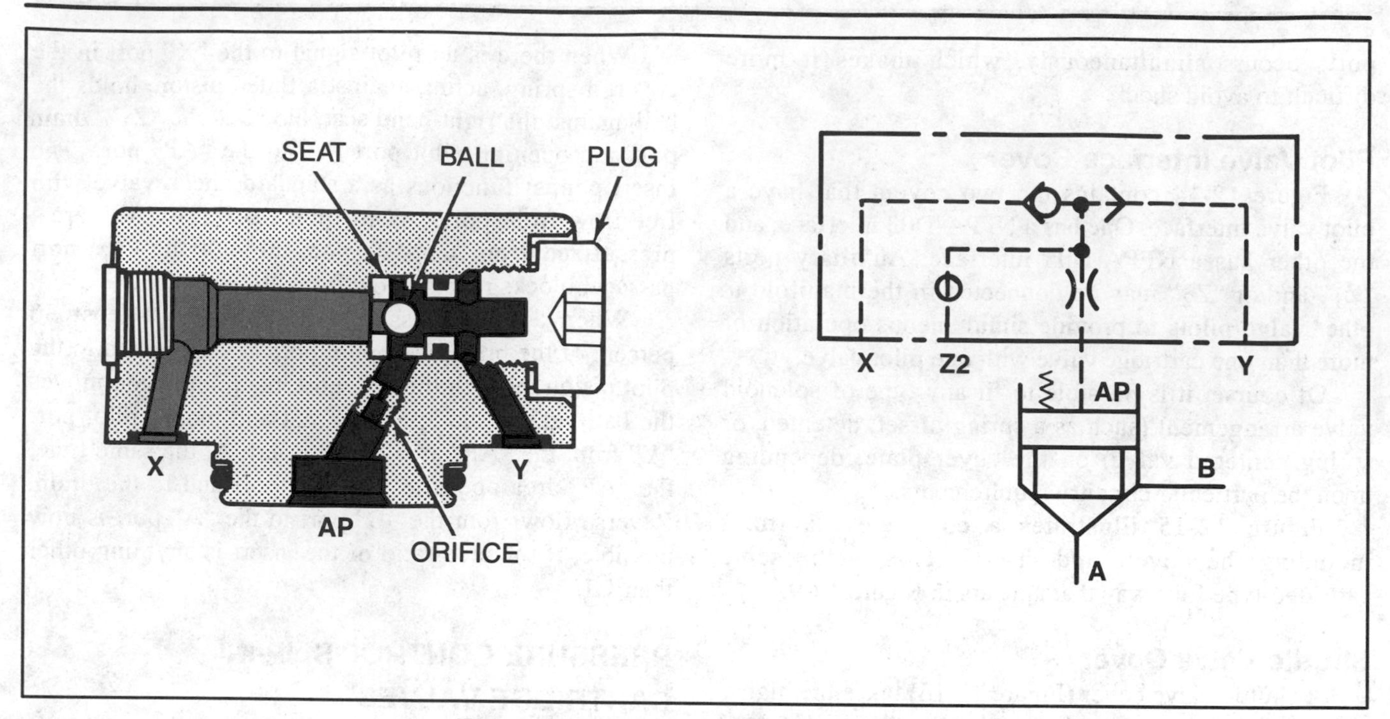

Figure 12-16. Shuttle valve cover.

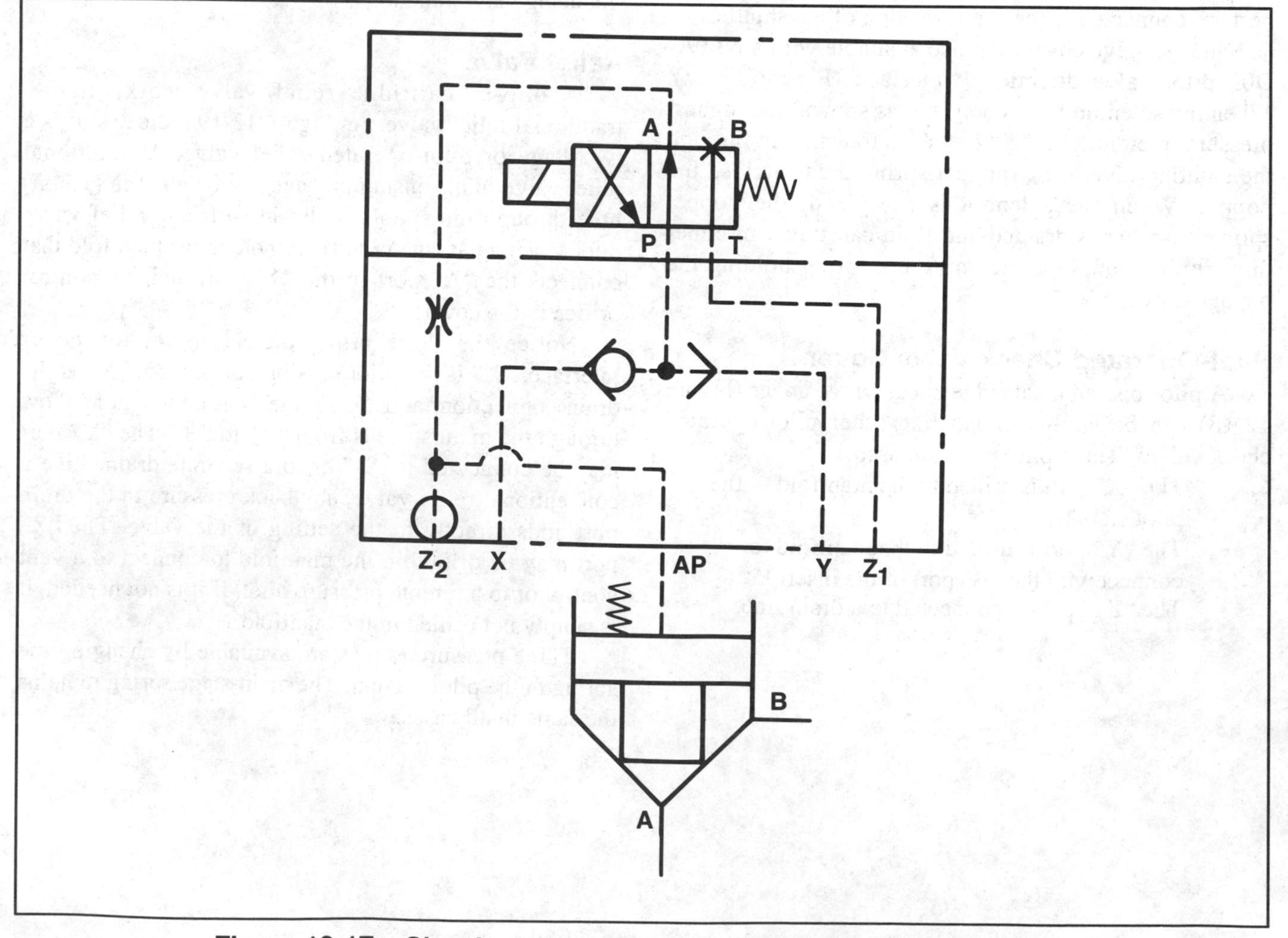

Figure 12-17. Shuttle valve cover with directional valve interface.

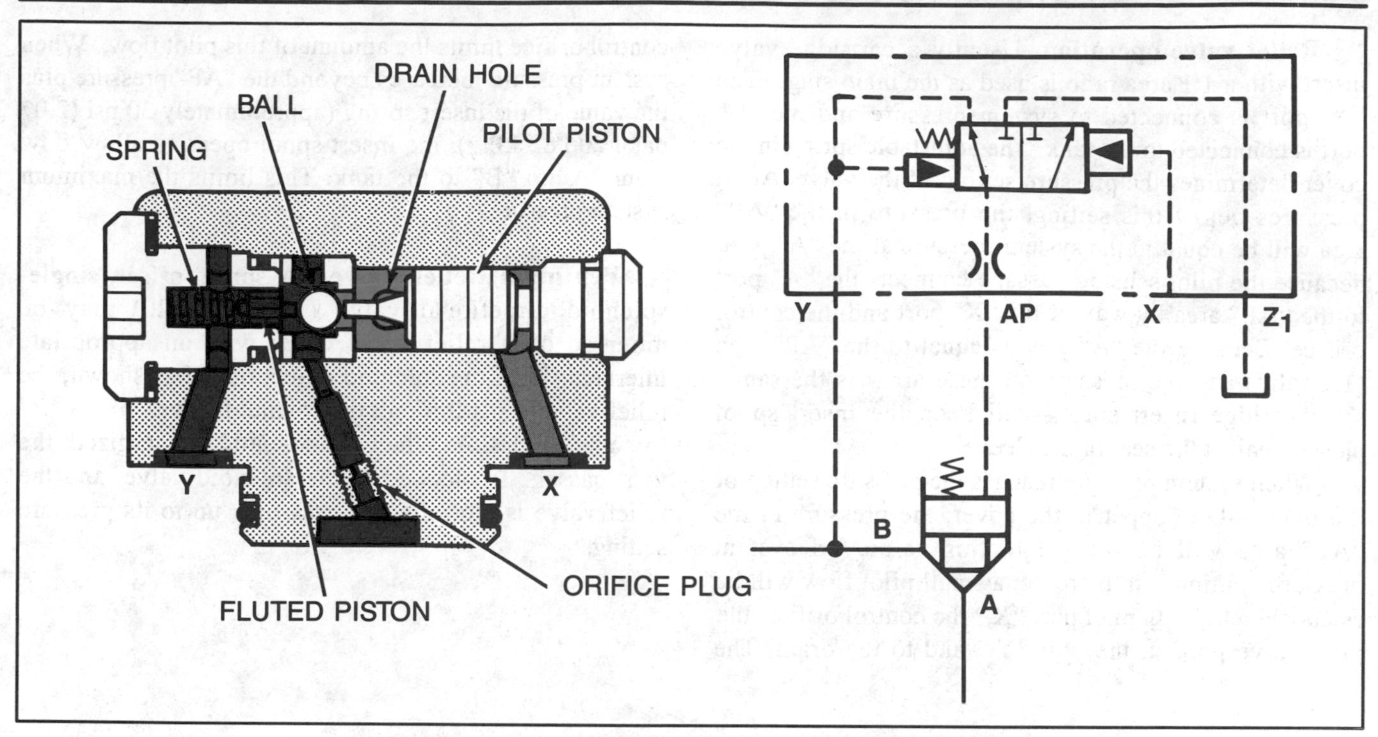

Figure 12-18. Pilot-operated check cover.

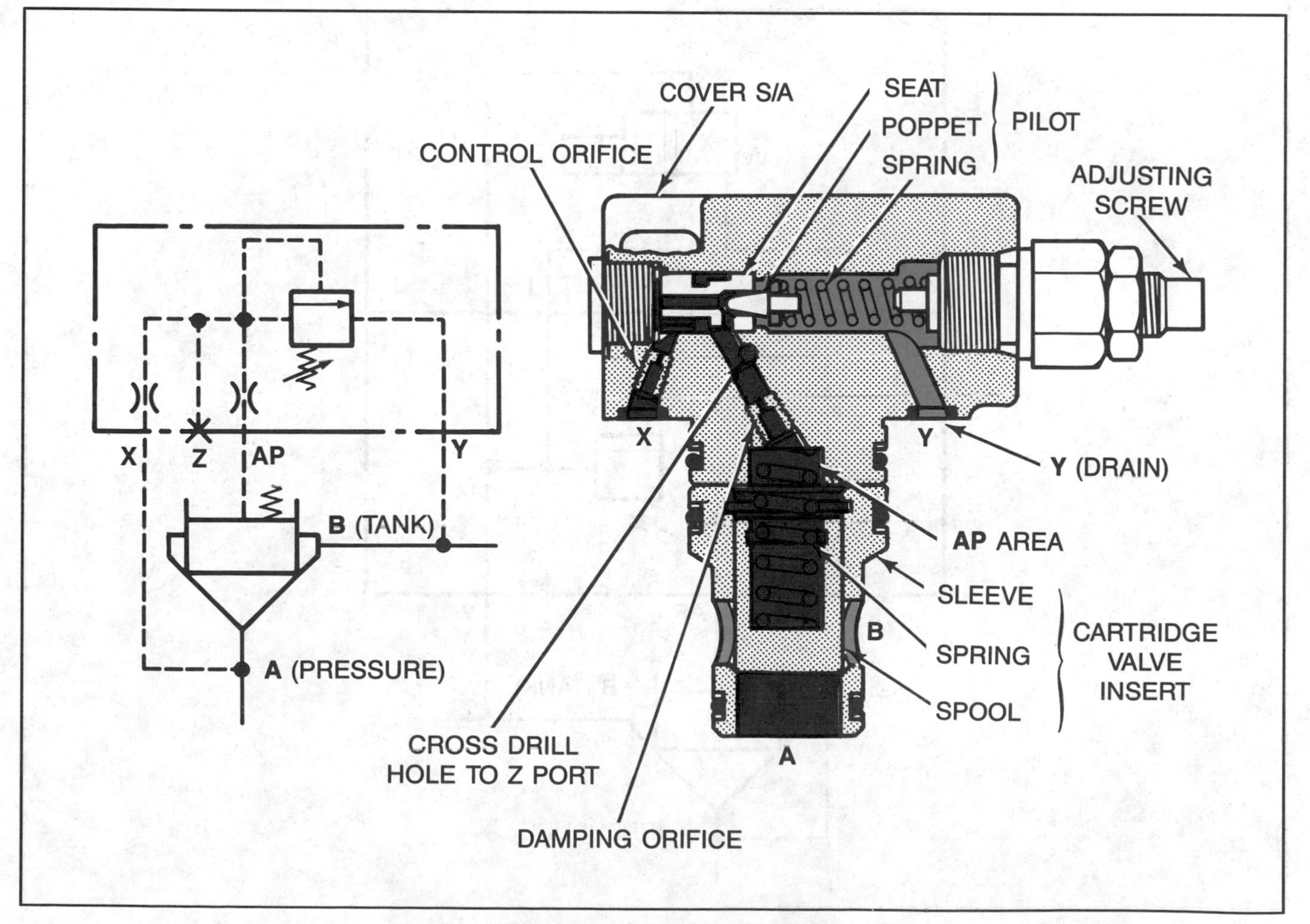

Figure 12-19. Pilot-operated relief valve.

Relief valve operation. Usually a cartridge valve insert with a 1:1 area ratio is used as the main stage. The "A" port is connected to system pressure and the "B" port is connected to the tank. The adjustable spring in the cover determines the pressure setting of the valve. At all pressures below this setting, the pressure in the "AP" area will be equal to the system pressure at the "A" port, because the pilot sensing passage connects the "A" port to the "AP" area by way of the "X" port and the control orifice. Because the "A" area is equal to the "AP" area (1:1 ratio) and the pressure on these areas is the same, the cartridge insert spring will keep the insert spool closed against the seat in the sleeve.

When system pressure reaches the pressure setting of the pilot valve poppet in the cover, the pressure in the "AP" area will be limited to this value. If system pressure continues to increase, a small pilot flow will be established through pilot port "X," the control orifice, the pilot valve poppet, the port "Y," and to the drain. The control orifice limits the amount of this pilot flow. When system pressure increases beyond the "AP" pressure plus the value of the insert spring (approximately 30 psi (2.07 bar / 206.82 kPa)), the insert spool opens to allow flow from "A" to "B" to the tank. This limits the maximum system pressure.

Venting a Relief Valve. A spring-offset, single-solenoid directional valve (Figure 12-20) may be mounted on a relief valve cover with an appropriate interface. With the solenoid de-energized as shown, the relief valve vents through the solenoid valve, "P" to "B" to "Y" (the drain). When the solenoid is energized, the vent passage is blocked at the solenoid valve, and the relief valve is closed at all pressures up to its pressure setting.

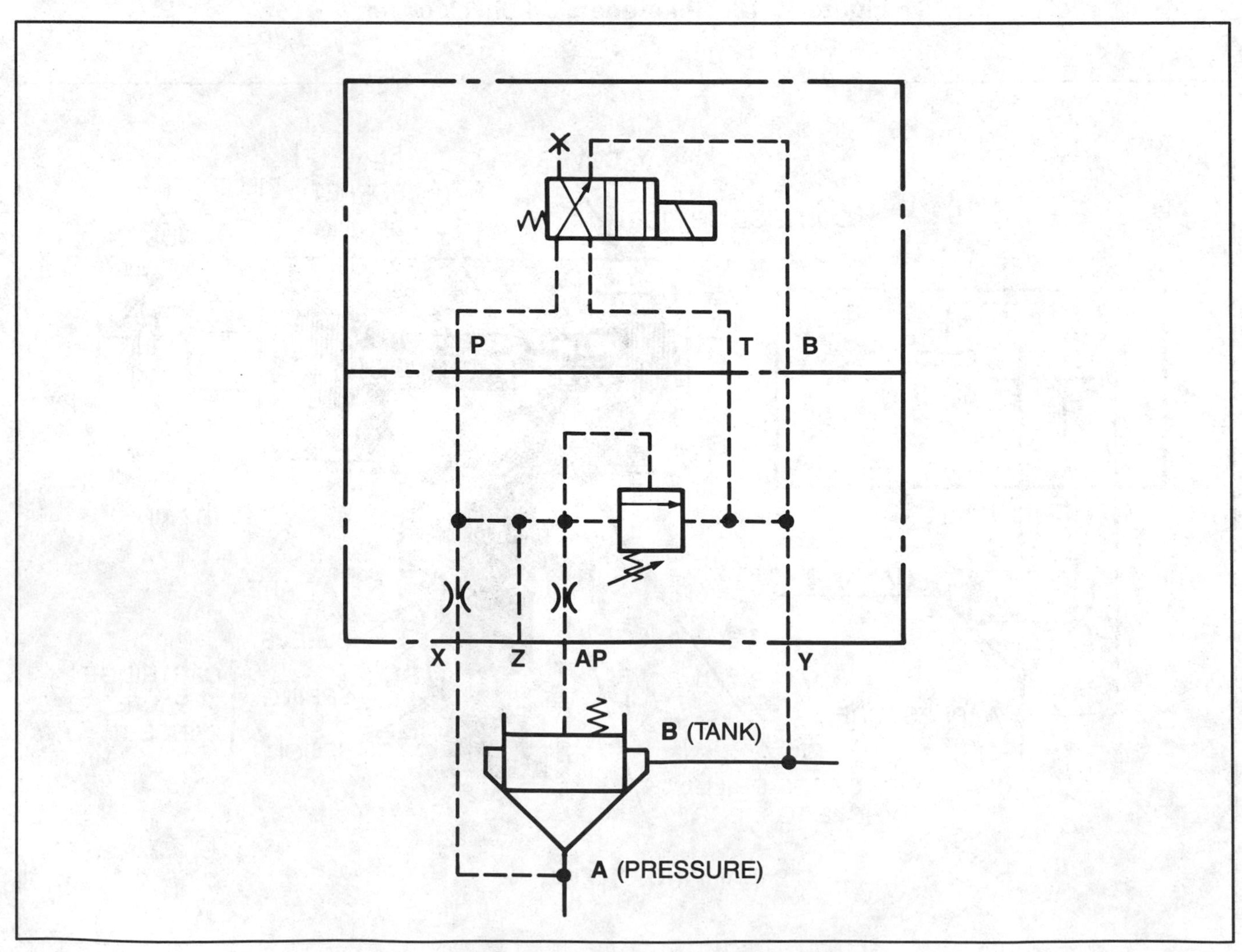

Figure 12-20. Venting the relief valve.

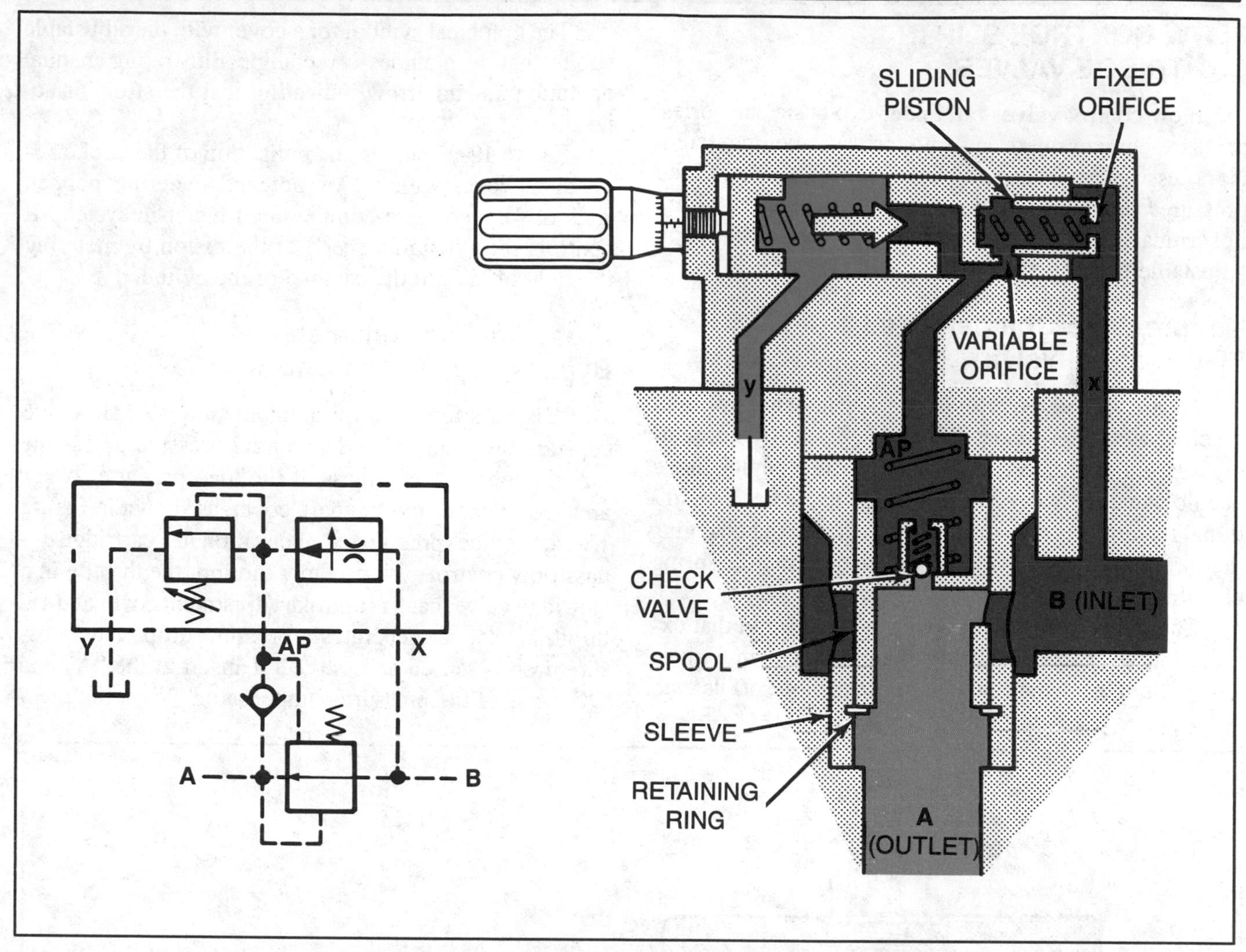

Figure 12-21. Pressure-reducing valve.

Reducing Valve

A slip-in cartridge pressure-reducing valve (Figure 12-21) is similar to a traditional pressure-reducing valve. It has a pilot valve in the cover that consists of an adjustable screw, a spring, and a poppet. The cover also contains a flow control to limit the pilot flow through the pilot valve to drain "Y." The cartridge insert is the main stage and consists of a sliding spool held by the insert spring in a normally open position within a sleeve.

Reducing valve operation. The "B" port is the high- pressure inlet, and the "A" port is the reduced-pressure outlet. At all pressures below the setting of the pilot valve poppet in the cover, the pressure in the "AP" area, the "B" port, and the "A" port will be equal, and the insert spring will hold the spool against a retaining ring in the sleeve. This is the normal open position of the pressure-reducing valve. When pressure reaches the pressure setting of the pilot valve poppet, the pressure in the "AP" area will be limited to this value. If pressure at the "B" and "A" ports continue to increase, a small pilot flow will be established from the "B" port through pilot port "X," the pressure-compensated flow control, and the pilot valve poppet to drain "Y."

When the increasing pressure at the "A" port equals the "AP" pressure plus the value of the insert spring, the insert spool will close sufficiently to limit the "A" port outlet pressure to this value, while the pressure at inlet port "B" can continue to increase. Any pressure spike at the "A" port that momentarily closes the spool would be relieved through a small check valve incorporated in the spool. The pressure-compensated flow control in the cover maintains a constant pilot flow across the pilot valve poppet, which reduces the pressure override. If a fixed orifice were used, this pilot flow would increase as the pressure at port "B" increased.

FLOW CONTROL SLIP-IN CARTRIDGE VALVES

Flow control valves fall under two basic categories: pressure-compensated and nonpressure-compensated. Nonpressure-compensated valves are used where load pressures remain relatively constant and feed rates are not critical. They may be as simple as a fixed orifice or an adjustable needle valve.

Nonpressure-Compensated Flow Control Valve

A cartridge valve with stroke adjustment cover and notched metering poppet (Figure 12-22) provides nonpressure-compensated control of flow rates much like a needle valve. However, the cartridge valve has an additional feature: it can be turned on or off without changing the throttle setting by venting or pressurizing the spring chamber.

The stroke adjustment cover contains a pin that extends into the cartridge spring chamber. The pin forms a stop to limit the distance the poppet can move off its seat.

The graphical symbol for a cover with the adjustable stroke feature includes a rectangle illustrating manual operation and an arrow indicating that the stroke is adjustable.

Figure 12-23 shows an application of the stroke adjustment cover with a "V" notched metering poppet. This is a nonpressure-compensated meter-in system. It controls the extending speed of the piston by metering the flow of oil into the cap end of the cylinder.

Pressure-Compensated, Bypass-Type Flow Control

A bypass flow control holds the adjusted flow rate constant by keeping the differential pressure across the throttle constant regardless of the load pressure. A 1:1 area ratio poppet and standard cover, as shown in Figure 12-24, provides pressure compensation in a cartridge bypass flow control system. Flow through the throttle in a cartridge valve that has a stroke adjustment cover and 1:2 throttling poppet generates a pressure drop. This pressure drop is sensed across the 1:1 insert at the "A" and "X" ports of the pressure compensator. When the pres-

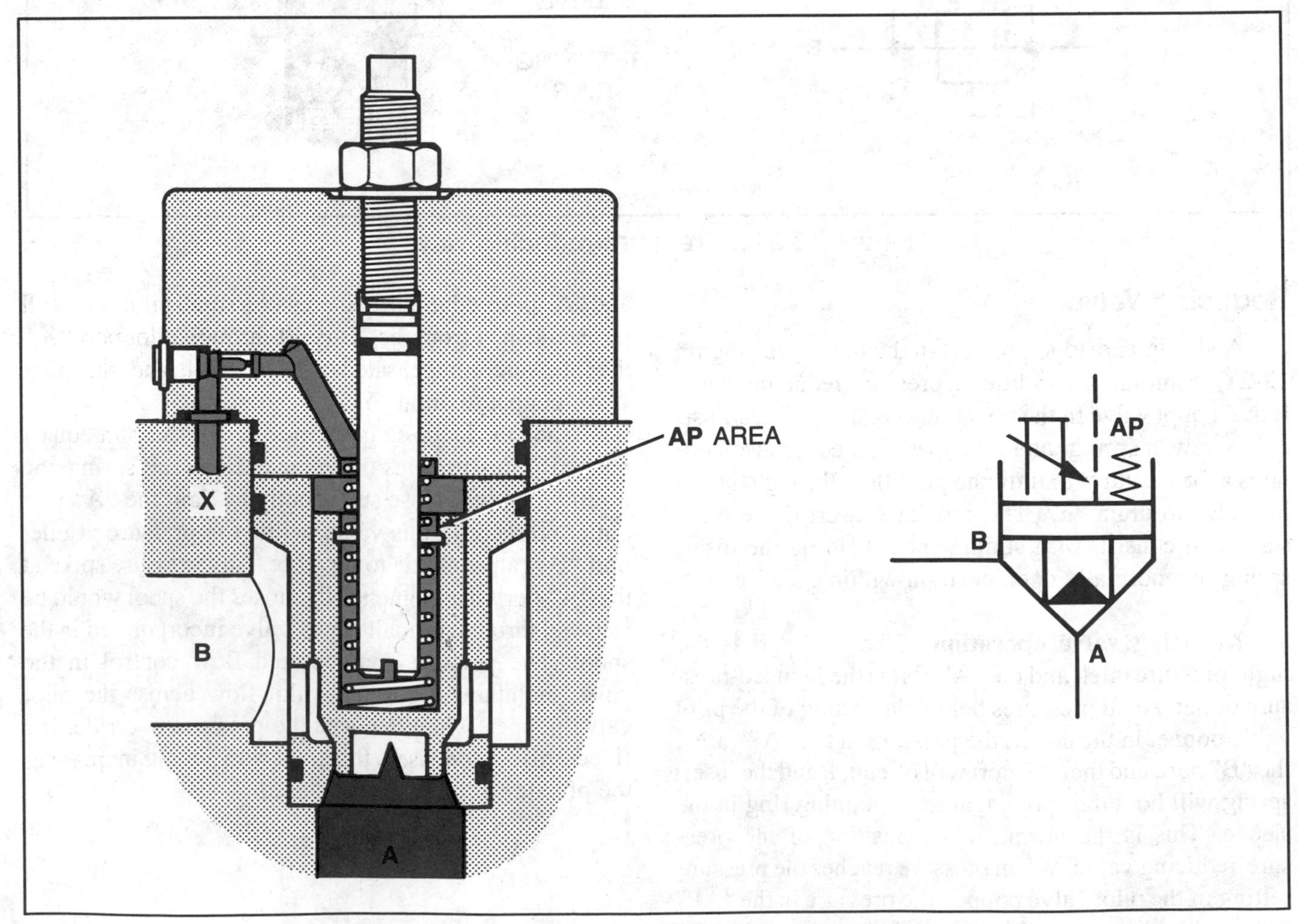

Figure 12-22. Nonpressure-compensated flow control valve.

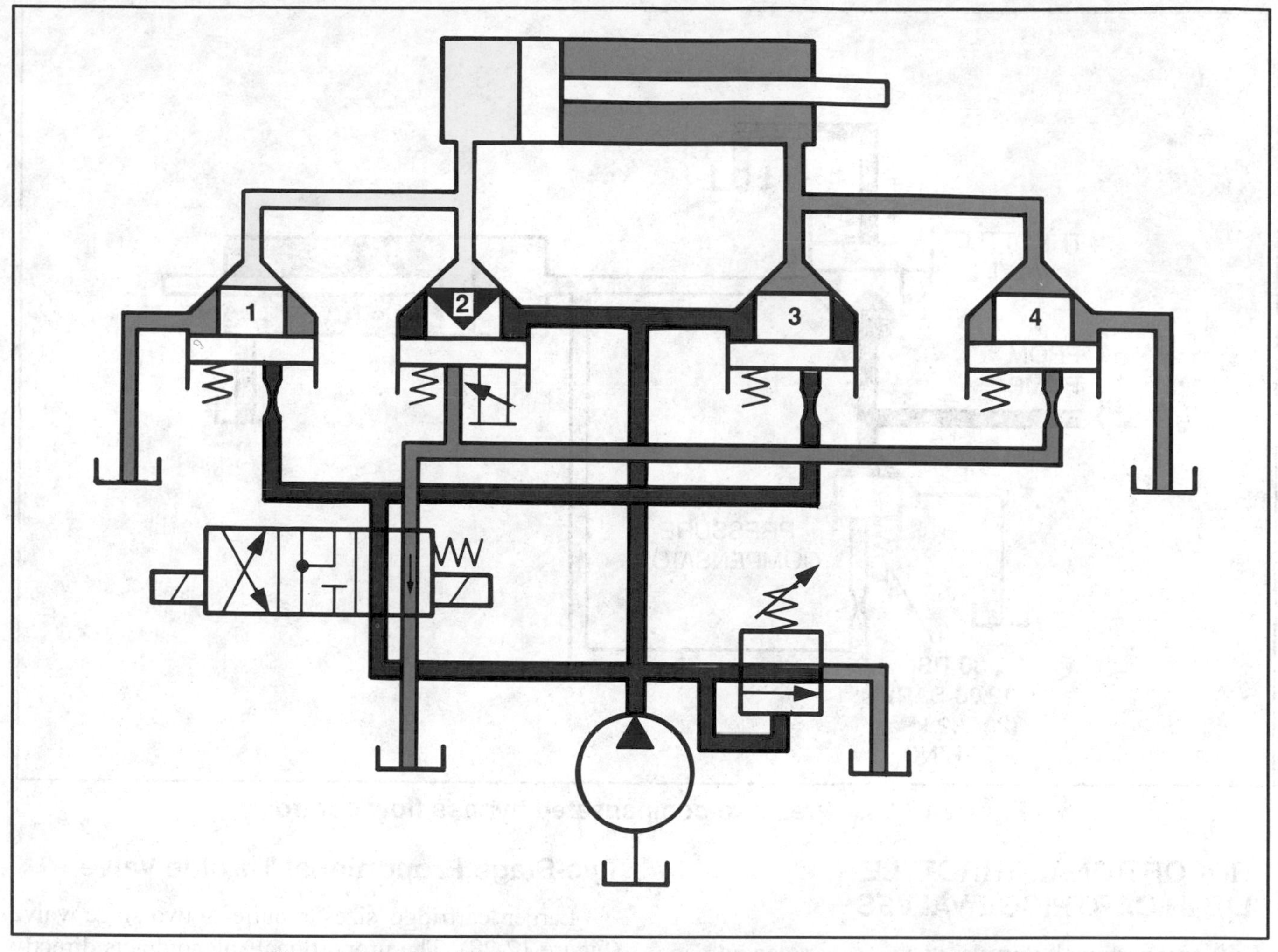

Figure 12-23. Nonpressure-compensated meter-in circuit.

sure drop equals the pressure value of the insert spring (approximately 30 psi (2.07 bar / 206.82 kPa)), the poppet opens and bypasses excess oil to the tank. A load increase on the cylinder would increase the pressure at the "X" port. This raises the pressure level at port A by an equal amount, which holds the pressure drop across the throttle constant. This, in turn, keeps the flow through the throttle valve at a constant rate.

Pressure at the pump is approximately 30 psi (2.07 bar / 206.82 kPa) higher than load pressure. **No maximum pressure limit** is provided by this arrangement.

Figure 12-25 shows that a maximum pressure limitation can be provided in a bypass pressure-compensated flow control valve by replacing the basic cover of the 1:1 insert pressure compensator valve with a relief valve cover. The pilot relief valve limits pressure in the "AP" area of the poppet. Maximum system pressure is limited to this pressure plus the pressure value of the compensator spring.

Pressure-Compensated Restrictor-Type Flow Control

A pressure-reducing valve insert with a standard cover acts as a pressure compensator when connected in series with a 1:2 metering insert (Figure 12-26). The restrictor flow control may be applied in meter-in, meter-out, or bleed-off circuits.

The outlet pressure at the "B" port of the throttle is applied to the "AP" area of the pressure compensator by way of the load-sensing line. The throttle restricts the flow, so the pressure at the "A" ports of the throttle and compensator will increase. When this pressure equals the pressure in the "AP" area of the compensator spool plus the pressure value of the compensator insert spring, the compensator spool will close sufficiently to limit the pressure to this value. Therefore, the pressure difference across ports "A" and "B" of the throttle valve will always be equal to the pressure value of the compensator insert spring. With a constant pressure difference across the throttle assured, the flow rate will be constant until the throttle is adjusted for a larger or smaller flow.

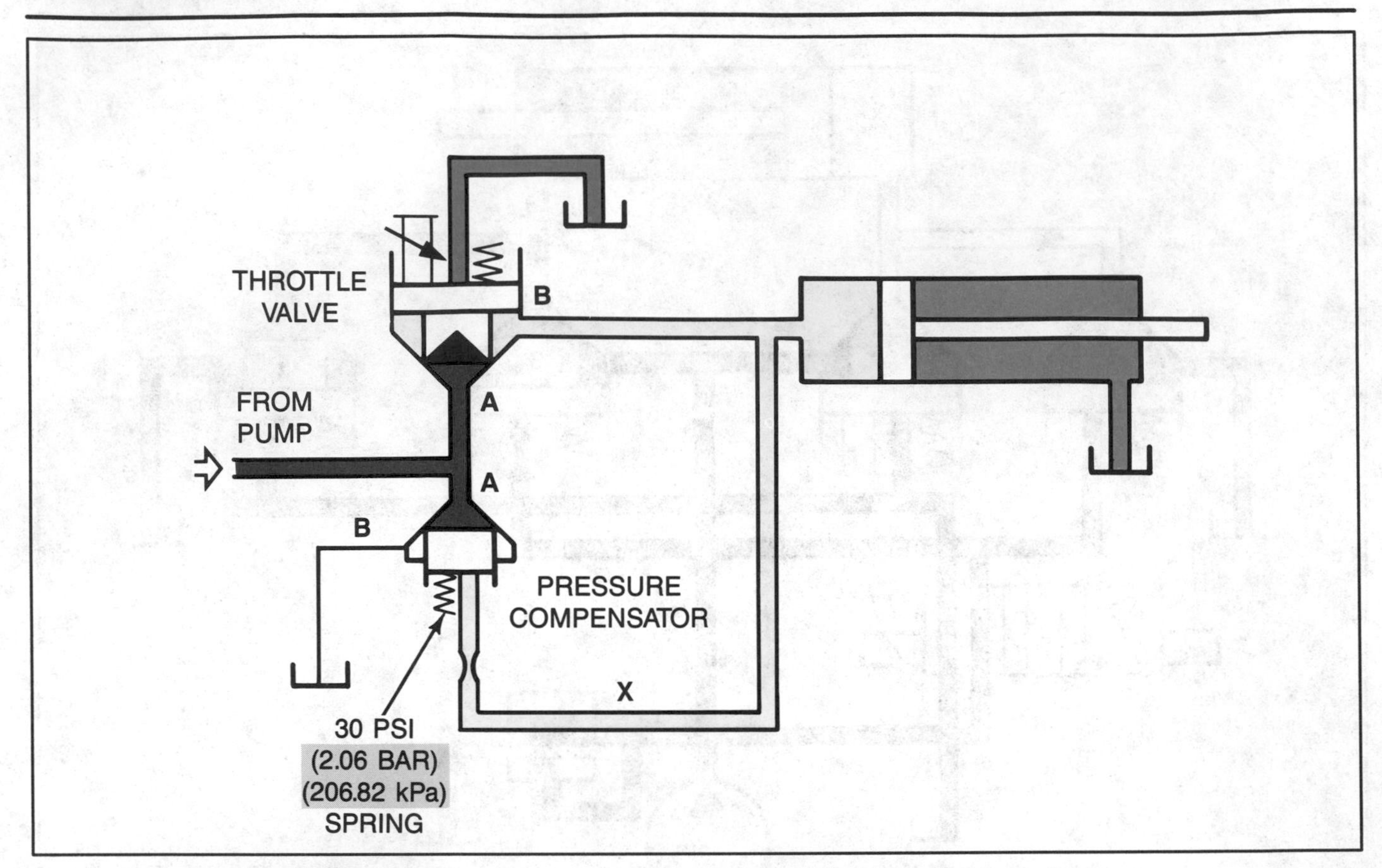

Figure 12-24. Pressure-compensated bypass flow control.

PROPORTIONAL THROTTLE SLIP-IN CARTRIDGE VALVES

A proportional throttle valve is a solenoid-controlled, normally closed valve that opens in proportion to the electric input signal to the solenoid. The following section discusses direct-acting and two-stage proportional throttle slip-in cartridge valves.

Direct-Acting Proportional Throttle Valve

As shown in Figure 12-27, flow through a direct-acting proportional throttle valve is controlled from port "A" to port "B." The spool is hydraulically balanced by a drilled passage that connects the "A" area to an annulus area that is equal to the "A" area. A bias spring keeps the spool normally closed when there is no electrical signal to the solenoid. A position sensor, called a linear variable differential transformer (LVDT), is placed on the armature of the solenoid to monitor its movement. (Refer to Chapter 14 for more information on the LVDT.) An input signal to the solenoid opens the valve until the LVDT signal balances with the input signal. Therefore, the valve opening is proportional to the solenoid input signal.

Two-Stage Proportional Throttle Valve

Larger cartridge sizes require a two-stage valve (Figure 12-28). The proportional solenoid acts directly on a spring-offset pilot spool. When the solenoid is de-energized, pilot oil taken from the "A" port is directed to the top of the main spool. Therefore, the main spool is hydraulically balanced and is held closed by the main spool spring. As a signal is applied to the solenoid, the pilot spool moves against its spring to partially open the pilot oil to drain. This causes the pilot pressure to drop in the main spool spring chamber. Pressure at the "A" port then opens the main spool. As the main spool opens, a position sensor (LVDT) generates a signal that is compared to the solenoid input signal. When these two signals are equal, the main spool provides an opening that is proportional to the solenoid input signal.

SCREW-IN CARTRIDGE VALVES

Screw-in cartridge valves are a second group of hydraulic cartridge valves used to control pressure, direction, and flow. As you learned earlier in this chapter, these valves perform similar functions, but operate differently from slip-in cartridge valves. The following section examines the ways in which screw-in valves differ from their slip-in valve counterparts.

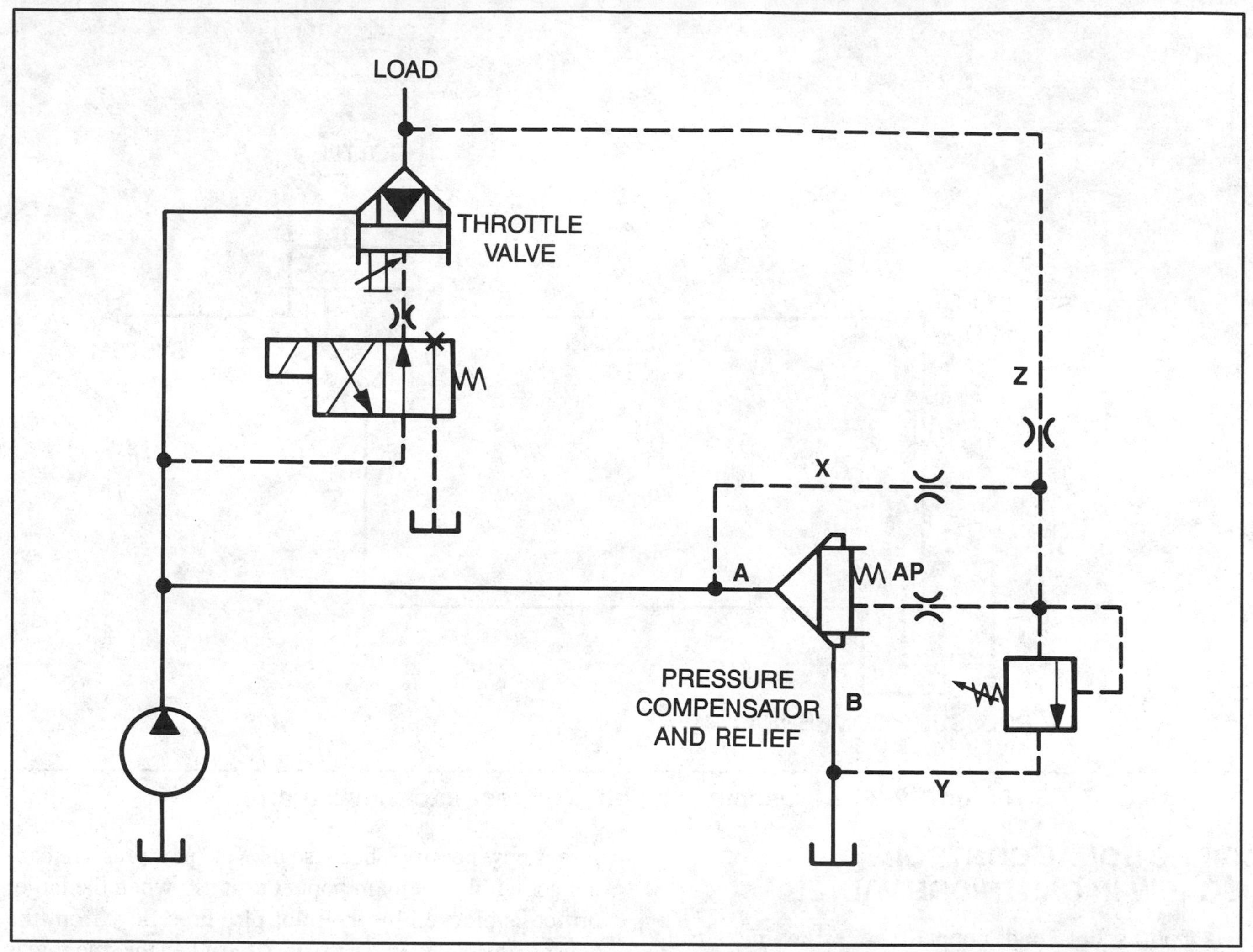

Figure 12-25. Pressure-compensated bypass flow control with maximum pressure limitation.

Unique Features of Screw-In Cartridge Valves

As you recall, a primary difference between screw-in and slip-in cartridge valves concerns the degree to which they perform their assigned hydraulic control functions. Typically, slip-in valves depend on a pilot valve to perform a complete control function. Most screw-in cartridge valves perform a complete control function themselves.

Slip-in and screw-in cartridge valves also differ in design aspects. While the majority of the slip-in cartridge valves are poppet-type, screw-in cartridge valves feature both poppet and spool elements. Like all cartridge valves, screw-in cartridge valves can be fitted into a manifold block or used as an individual valve assembly. However, unlike slip-in valves, the exterior of a screw-in valve insert has threads that screw into a machined manifold or individual cavity. The thread design differs from that of a slip-in valve, which is not threaded and has a cover that holds the valve element in place.

Screw-in cartridge valves share the flexibility characteristic of slip-in valves. They have standard, common parts that make them easily interchangeable and simpler to maintain than other types of valves. As Figure 12-29 indicates, screw-in cartridge valves and their corresponding cavities have two-way, three-way, three-way short, or four-way functions. These functions refer to valves and valve cavities with two ports, three ports, three ports with one acting as a pilot port (three-way short), and four ports. Many different valve functions can fit into a common-sized cavity. To illustrate their flexibility, all the different types of valves shown in Figure 12-30 can fit into the same two-way valve cavity.

In the next section, we will examine some of the basic design characteristics and operation of a variety of screw-in cartridge valves that provide directional, pressure, and flow control functions.

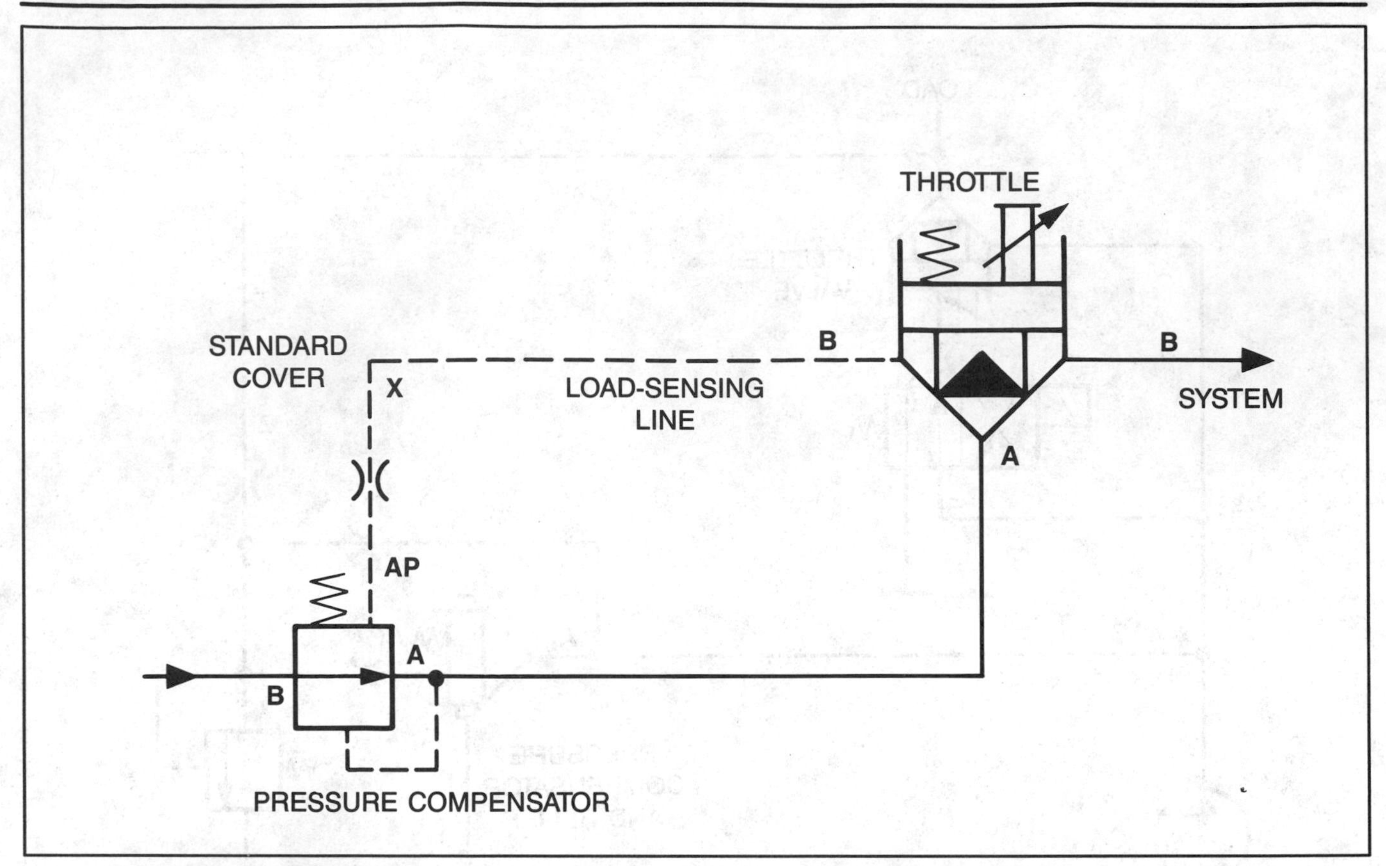

Figure 12-26. Pressure-compensated restrictor flow control.

DIRECTIONAL CONTROL SCREW-IN CARTRIDGE VALVES

Both spool- and poppet-type screw-in cartridge valves are used for directional control functions. The types of screw-in valves used to control the direction of hydraulic fluids include a variety of solenoid-controlled, pilot-operated valves, manual rotary valves, pilot-operated valves, check valves, and shuttle valves.

Two-Way Directional Control Valves

Figure 12-31 is a poppet-type, solenoid-controlled, pilot- operated, normally closed two-way valve. The valve has two ports (in and out), with the "in" port on the side and the "out" port at the bottom. When the solenoid is not energized, the flow from the "in" to the "out" port is blocked, because the small orifice in the side of the main poppet allows the "in" port pressure to act within the poppet. Free reverse flow from the "out " to the "in" port occurs whenever the "out" port pressure exceeds the "in" port pressure combined with the armature spring pressure.

When the solenoid is energized, the armature moves up and lifts the pilot pin off its seat within the main poppet. This exposes an orifice larger than the one at the side of the main poppet. Free flow from the "in" to the "out" port is now possible, because the "in" port pressure cannot act within the main poppet as it can when the larger orifice is blocked by the pilot pin. Free flow from the "out" to the "in" port is also possible when the solenoid is energized. This flow occurs because a small check disc in the bottom of the main poppet closes the larger orifice when the "out" port pressure exceeds the "in" port pressure.

The solenoid-operated spool-type valve shown in Figure 12-32 is a normally closed valve with two ports. Depending on its spool position, this valve can block or permit flow in both directions, unlike a poppet, which can block flow in one direction only. As the figure indicates, when the solenoid is not energized, the valve is closed and flow is blocked in both directions. When the solenoid is energized, the hydraulically-balanced spool moves to allow flow in both directions.

A poppet-type, solenoid-controlled, pilot-operated, normally open valve with two ports is shown in Figure 12-33. This valve allows unrestricted reverse flow. When the solenoid is not energized, there is free flow from the "in" port at the side to the "out" port at the bottom. As seen on other valves we've discussed, a check disc in the bottom of the poppet allows flow out of the large orifice, which causes pressure to decrease inside the poppet. The check disc closes when flow begins to

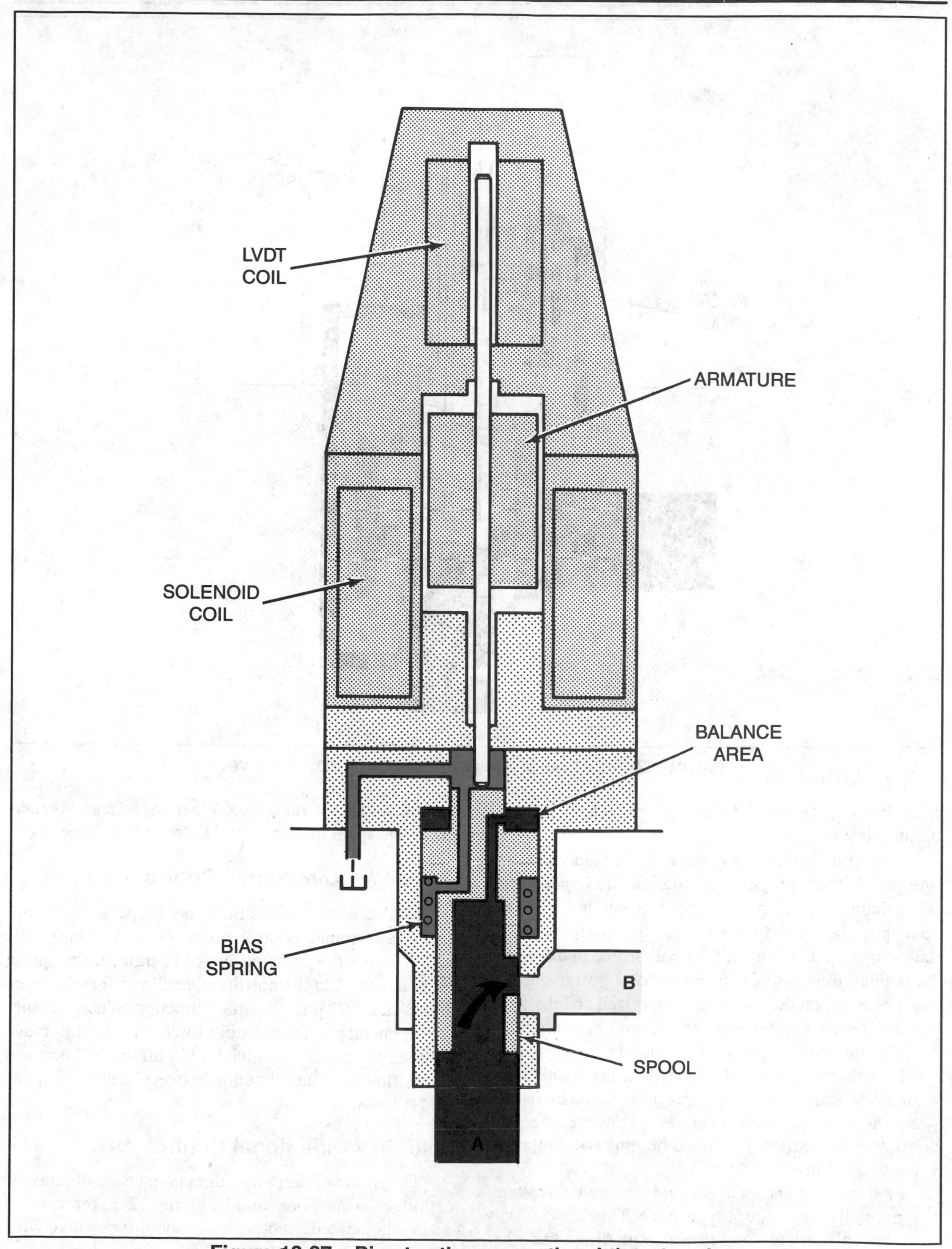

Figure 12-27. Direct-acting proportional throttle valve.

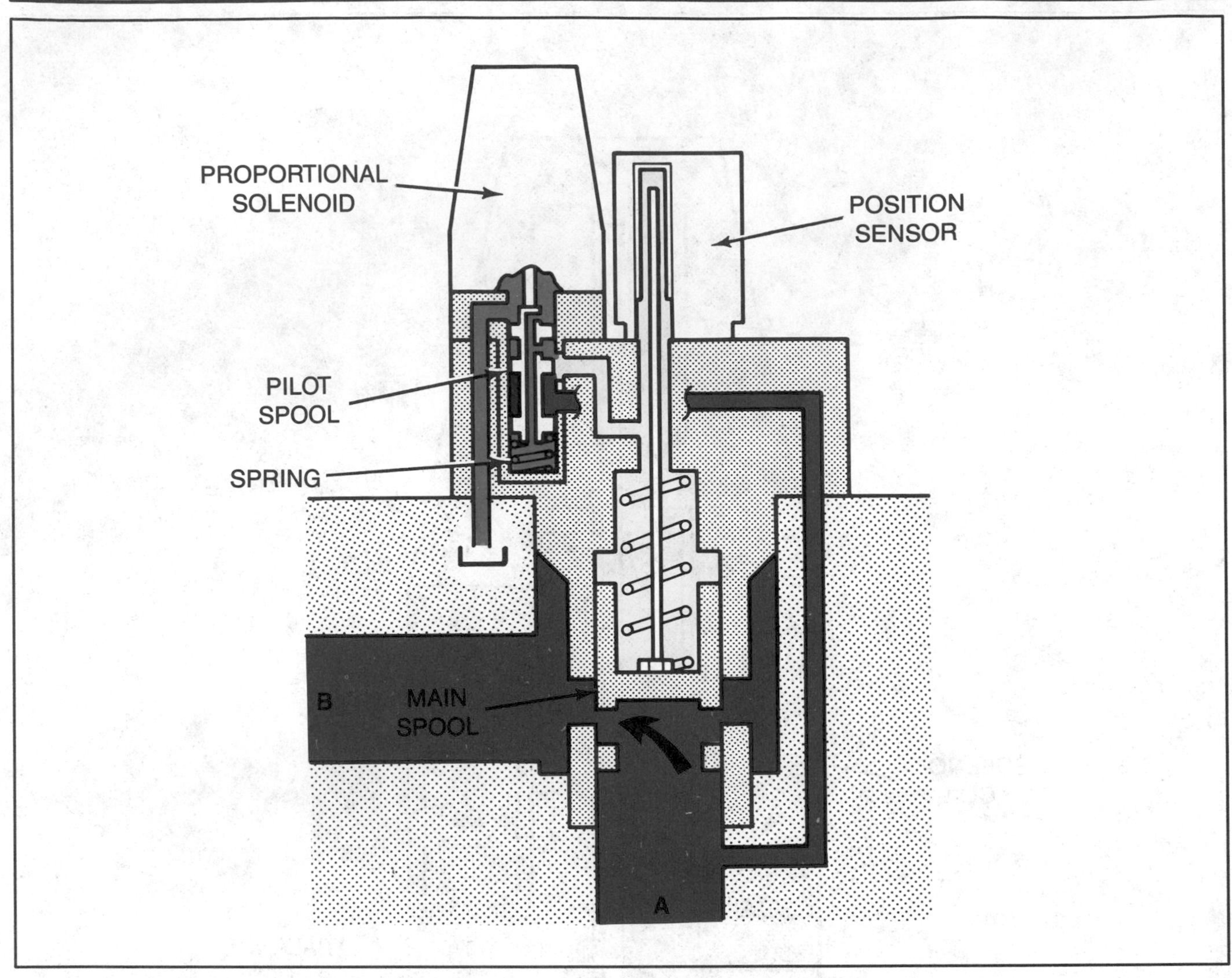

Figure 12-28. Two-stage proportional throttle valve.

come in from the "out" port, which prevents a pressure buildup inside the poppet.

When the solenoid is energized, the solenoid armature pushes the pilot pin down to close the larger orifice in the main poppet. This blocks flow from the "in" to the "out" port, because pressure at the "in" port acts within the main poppet by way of the small orifice in the side of the poppet. Reverse flow from the "out" port to the "in" port occurs when the solenoid is energized and the "out" port pressure is greater than the "in" port pressure by an amount equal to the spring and solenoid pressure values.

Note that if this valve is mounted in any position other than vertical, the valve poppet may not be on its seat when the solenoid is not energized. However, because the flow is unrestricted at this time, this will not affect valve operation.

Figure 12-34 shows a solenoid-actuated, direct-acting, normally open, spool-type valve. This is a two-way cartridge valve that allows flow in both directions when the solenoid is not energized. When the solenoid is energized, the spool moves to block flow through the valve.

Three-Way Directional Control Valve

A three-way, two-position, spool-type, solenoid-operated directional control valve is shown in Figure 12-35. When the solenoid is not energized, the spring shifts the spool to the position that allows flow between the "B" and "C" ports in either direction. When the solenoid is energized, the solenoid forces the spool to move to its second position, which blocks off the "C" port and allows flow in either direction between the "B" port and the "A" port.

Four-Way Directional Control Valve

There are also four-way directional control screw-in cartridge valves. For example, Figure 12-36 represents a solenoid-operated, two-position, spool-type valve with

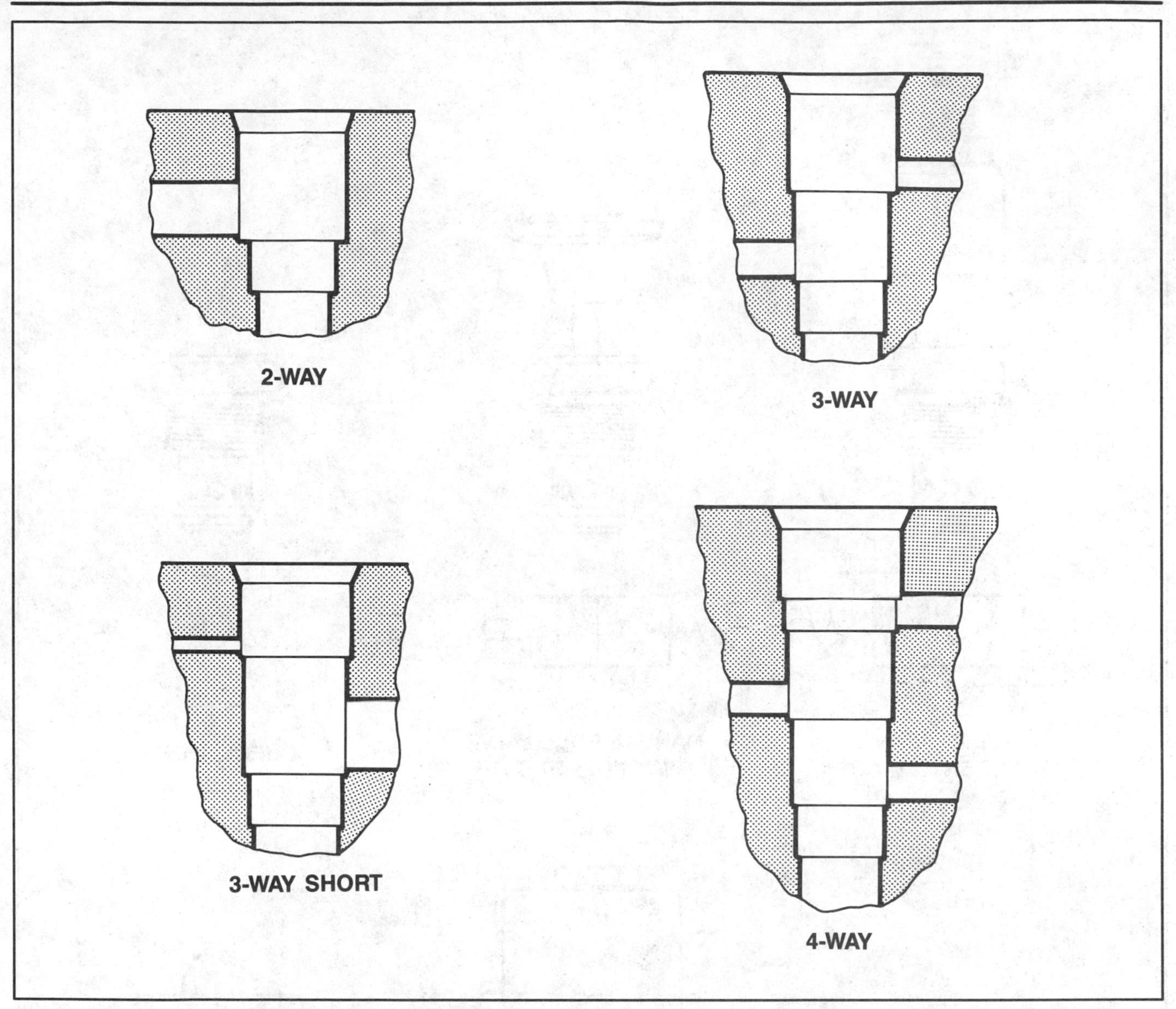

Figure 12-29. Two-way, three-way, and four-way screw-in cartridge valve functions.

four ports. When the solenoid is not energized, the spring shifts the spool to the position that connects the "P" port to the "C1" port, and the "C2" port to the "T" port. When the solenoid is energized, the spool moves to connect the "C1" port to the "T" port, and the "P" port to the "C2" port.

Manual Directional Control Valve

In addition to the solenoid-controlled, pilot-operated valves, there are manually operated screw-in cartridge valves that perform directional control functions. Figure 12-37 shows a two-position, manual rotary spool-type valve with three ports. Spool position and, consequently, flow direction are changed by turning the manual knob 90°. One spool position allows flow between the "C" and "B" ports and blocks the "A" port; the second position allows flow between the "A" and "B" ports and blocks the "C" port. When the knob is replaced by a lever with a detent, the valve becomes a three-position valve with all ports blocked in the neutral position.

Pilot-Operated Directional Control Valve

Pilot-operated directional control valves are another group of directional control valves. Figure 12-38 shows a pilot- operated valve with three ports: "A," "B," and "C." The valve's sliding spool has two positions and is spring-offset. When the spring pressure is greater than the pilot pressure, there is flow between the "A" and "B" ports, and the "C" port is blocked.

The spring chamber is internally drained to the "A" port. Therefore, the pilot pressure has to be greater than the spring pressure plus any pressure in the "A" port to

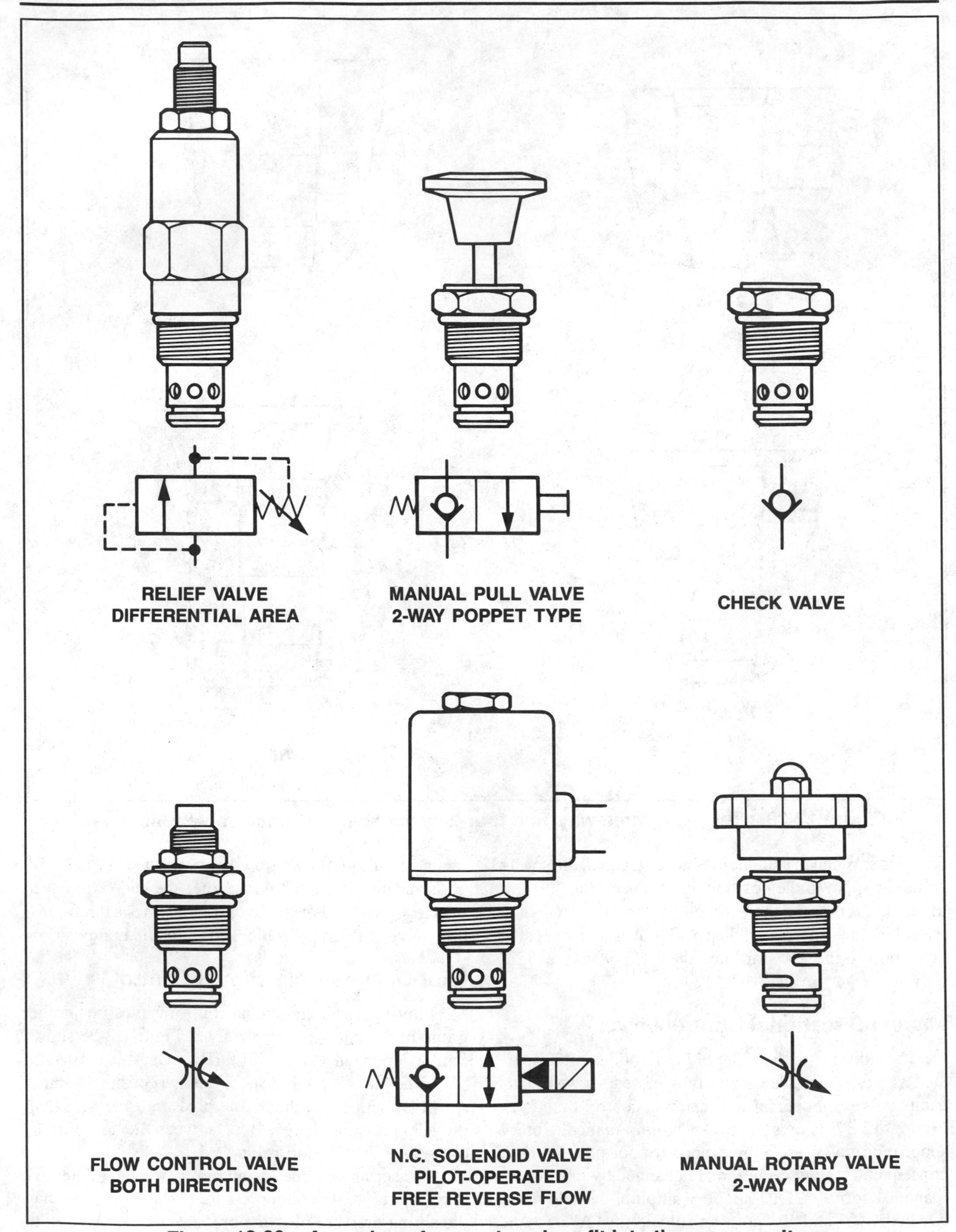

Figure 12-30. A number of screw-in valves fit into the same cavity.

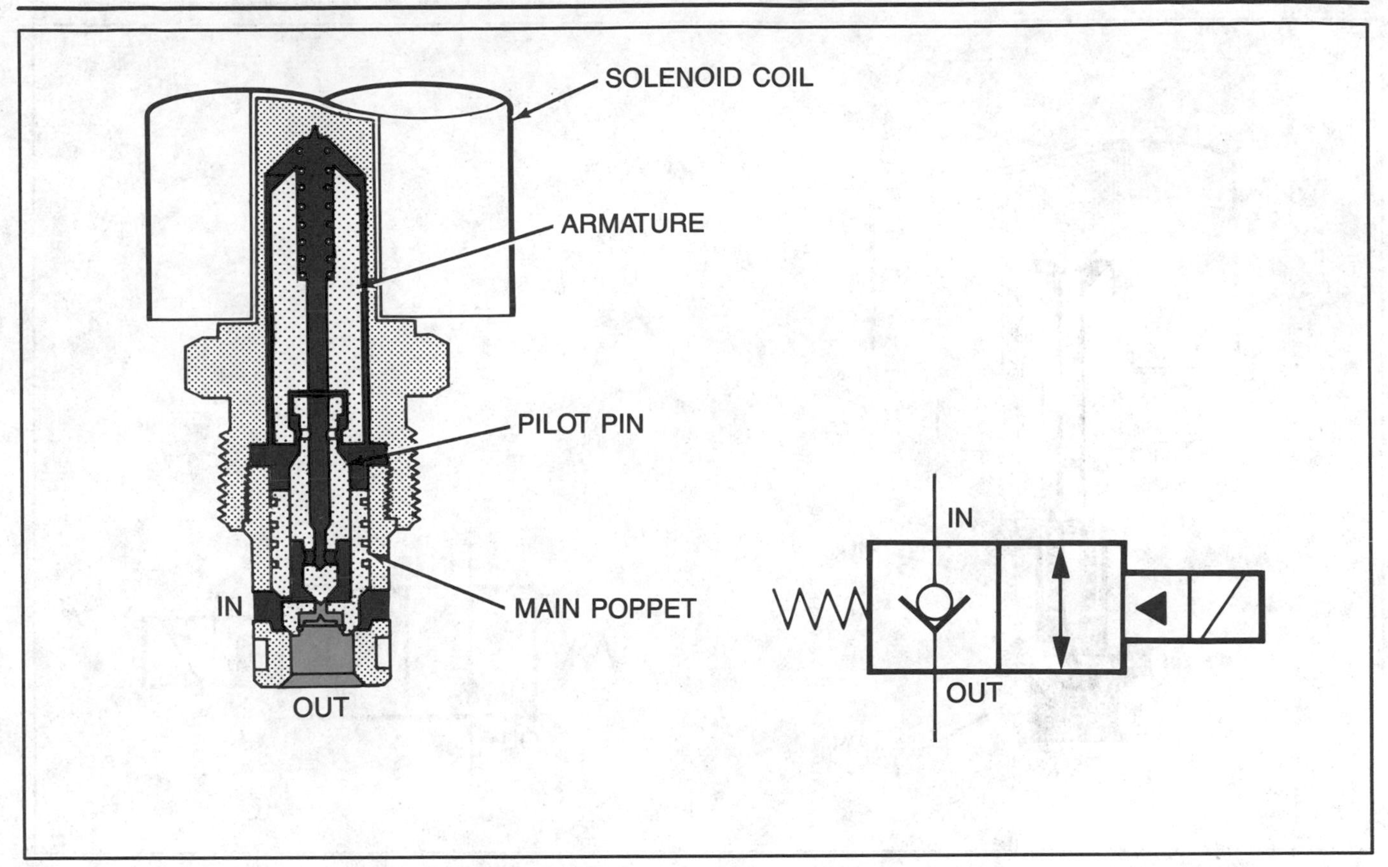

Figure 12-31. Solenoid-controlled, pilot-operated, normally closed, two-way poppet-type valve.

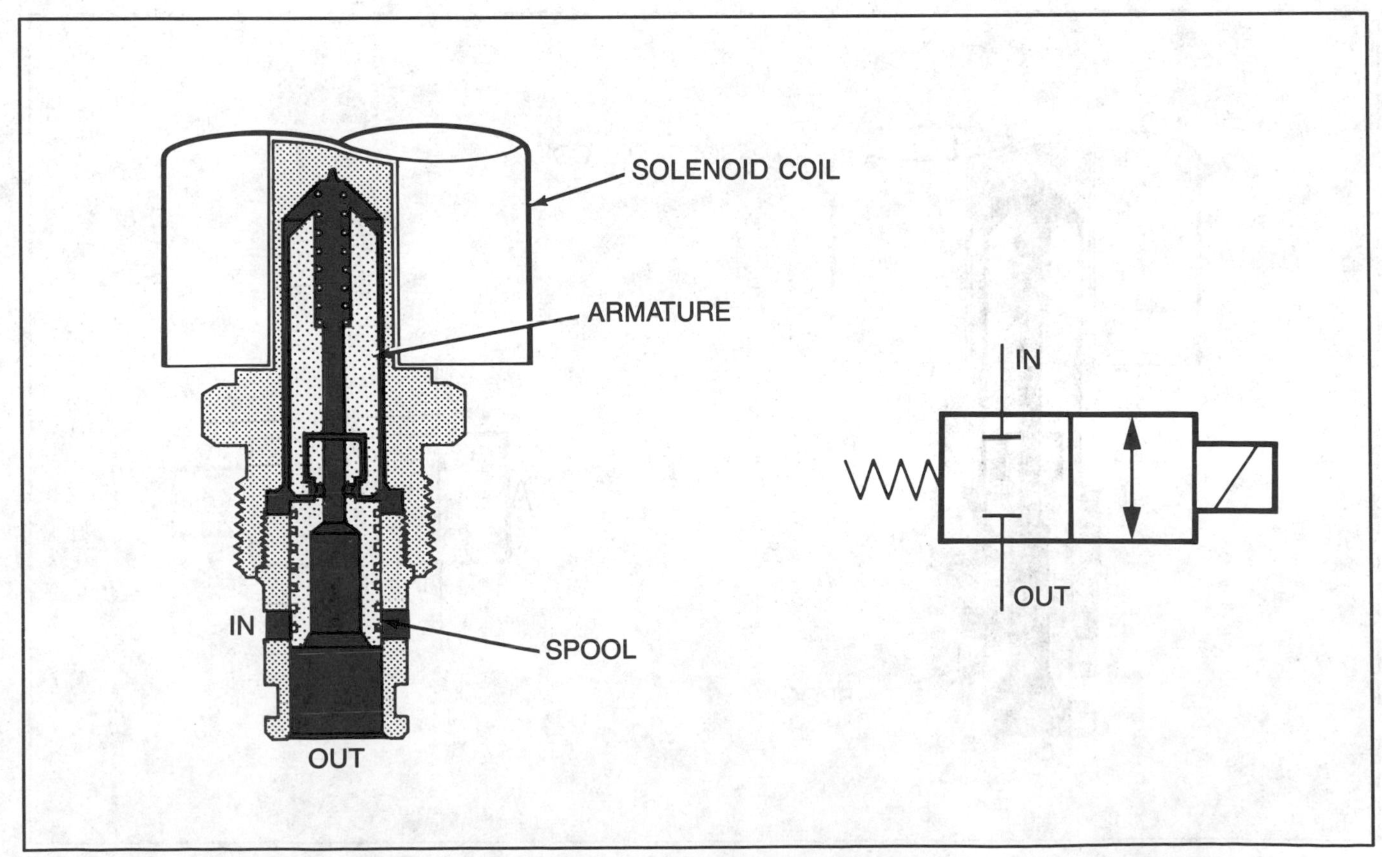

Figure 12-32. Solenoid-operated, normally closed, two-way spool-type valve.

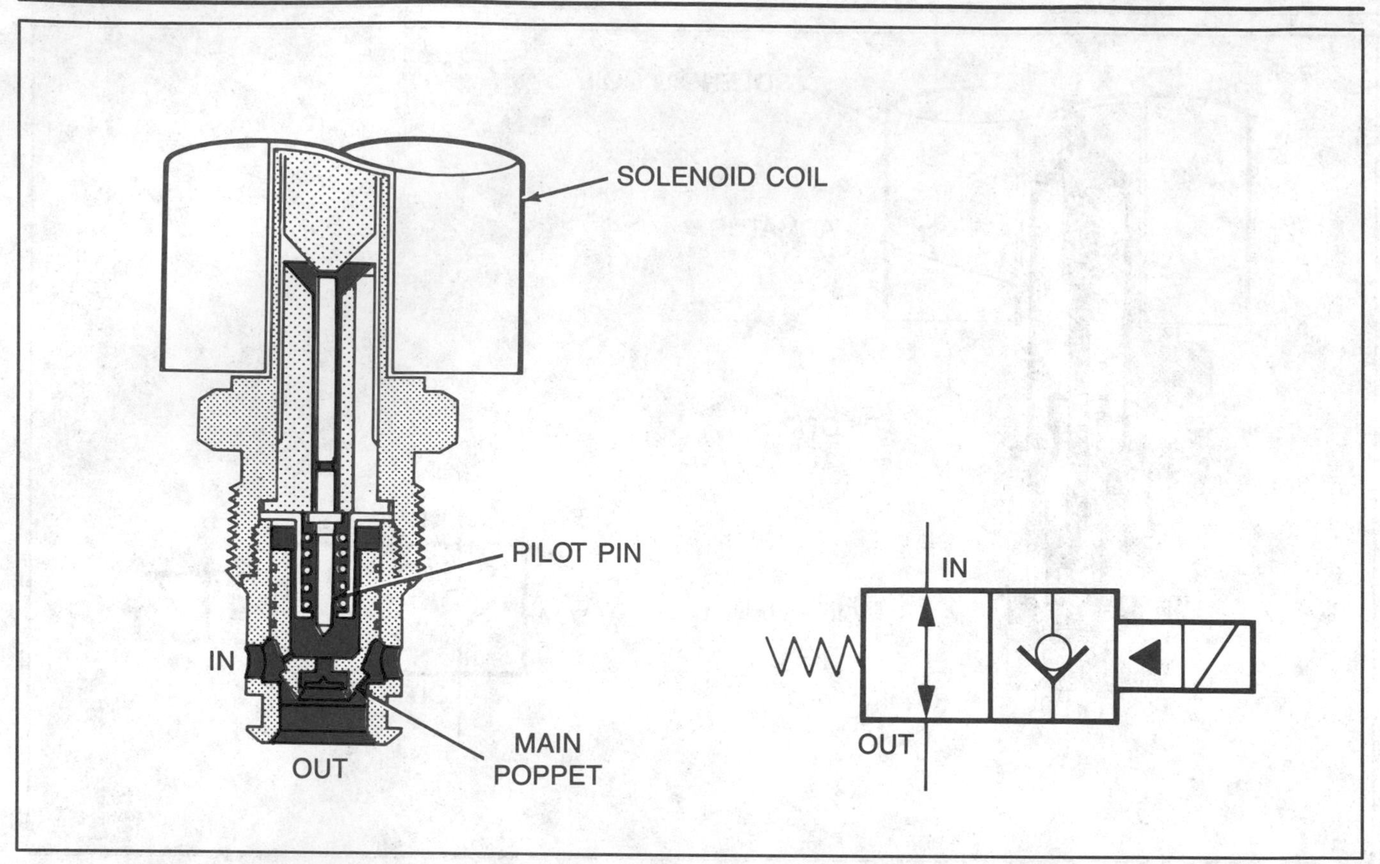

Figure 12-33. Solenoid-controlled, pilot-operated, normally open, two-way poppet-type valve.

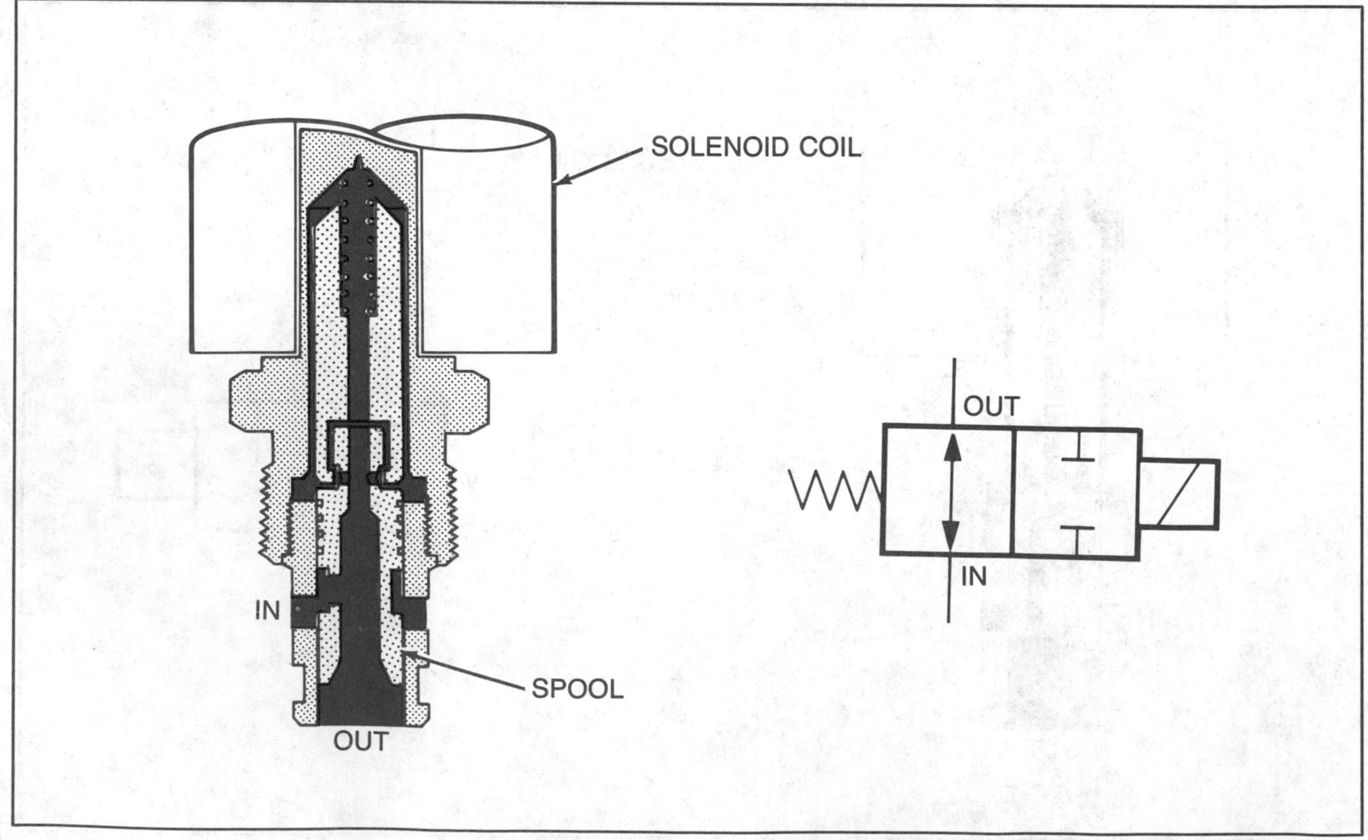

Figure 12-34. Solenoid-operated, direct-acting, normally open, two-way spool-type valve.

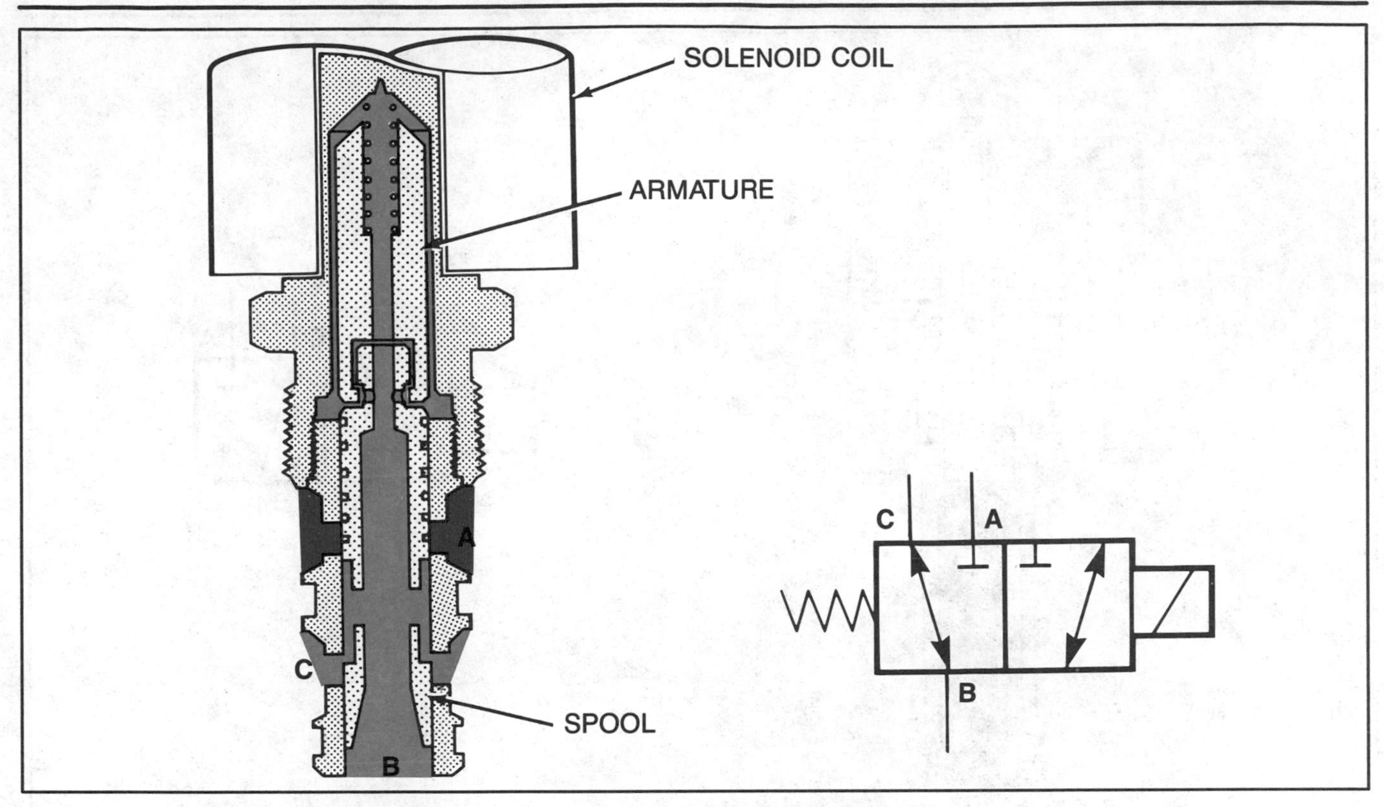

Figure 12-35. Solenoid-operated, three-way, two-position spool-type valve.

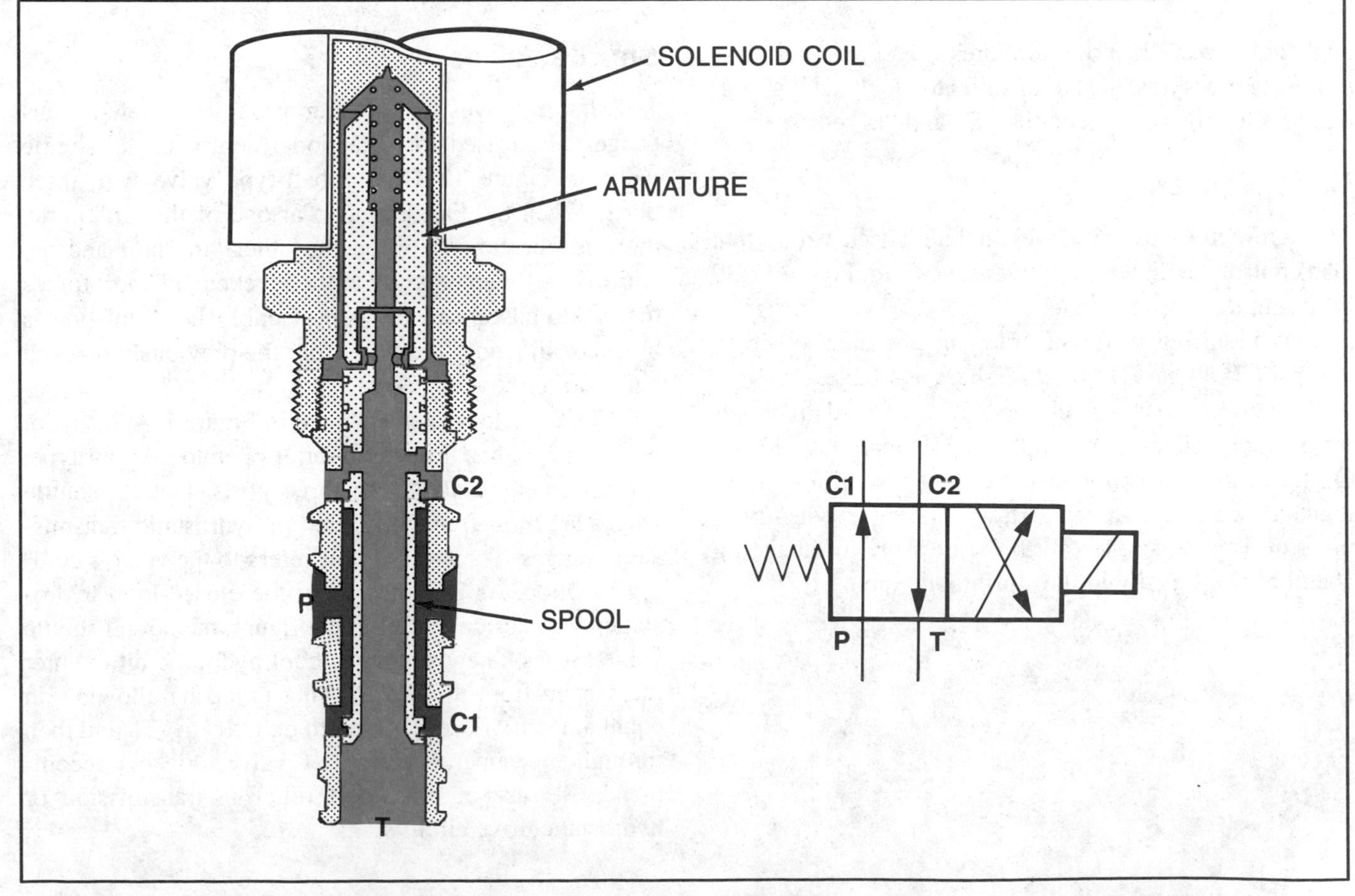

Figure 12-36. Solenoid-operated, direct-acting, four-way, two-position spool-type valve.

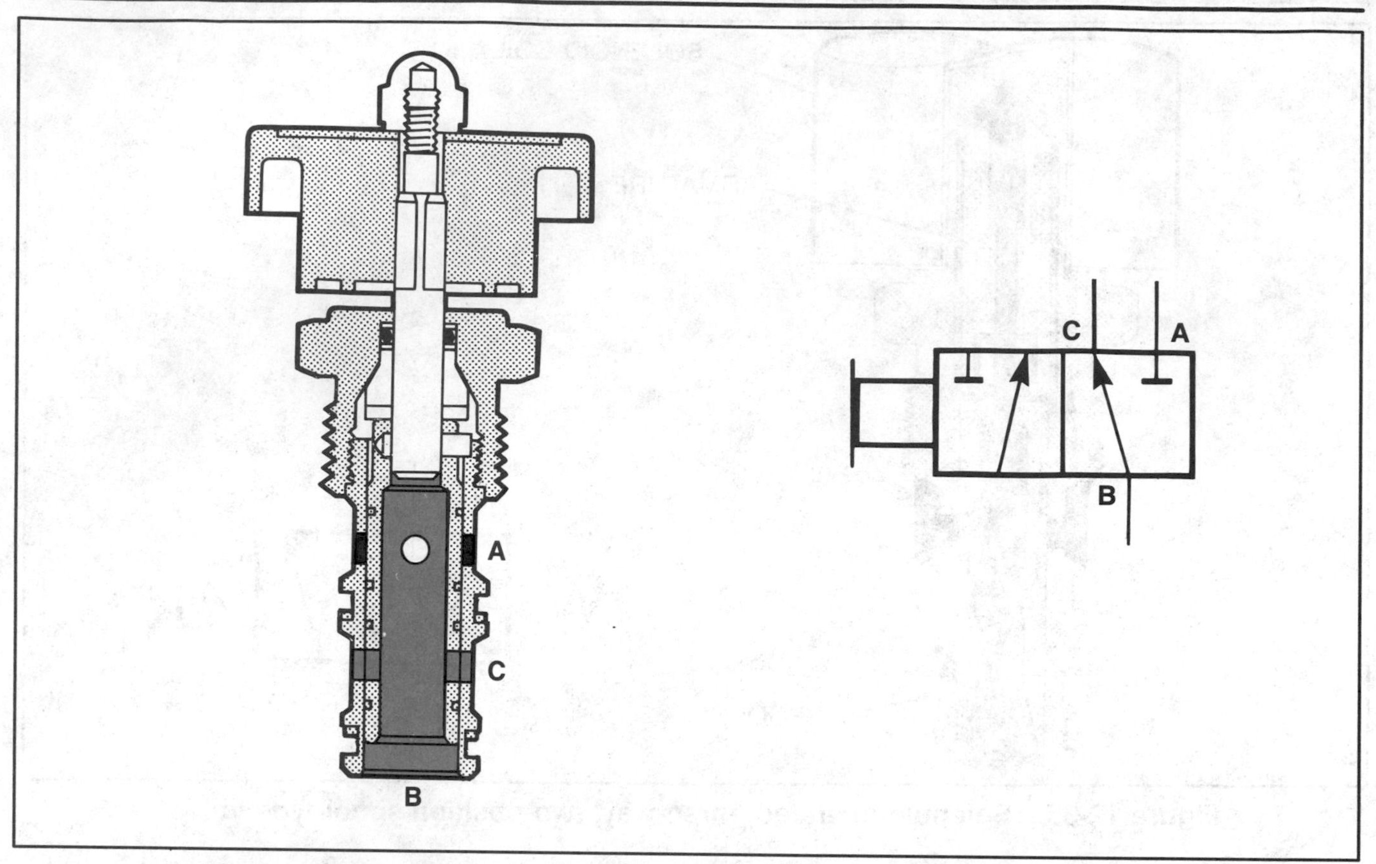

Figure 12-37. Manually-operated rotary spool-type directional valve.

shift the valve. When the pilot pressure overcomes these opposing pressures, the spool shifts to block the "A" port and to allow flow between the "C" and "B" ports.

Check Valves

Screw-in cartridge valves include check valves that also control the direction of hydraulic fluid. Figure 12-39 is a simple poppet-type check valve. Several different springs requiring different cracking pressures of up to 300 psi (20.68 bar /2068.20 kPa) are available.

Screw-in cartridge check valves can also be pilot-operated, like the example shown in Figure 12-40. On this valve, the pilot piston area is four times larger than the seat area. Reverse flow is possible when pilot pressure is at least one-fourth the pressure in the spring chamber combined with the spring pressure.

Shuttle Valves

Shuttle valves are another group of screw-in cartridge valves used for directional controls. The shuttle valve in Figure 12-41 is a ball-type valve with three ports. When the ball is seated at one of the "in" ports, there is free flow between the other "in" port and the "out" port. When pressure at the blocked "in" port forces the ball to unseat, the previously unblocked " in" port is blocked, allowing free flow from the previously blocked "in" port to the "out" port.

The directional valve shown in Figure 12-42 is a hot oil shuttle valve. This is a pilot-operated, spool-type, spring-centered valve with three ports. Hot oil shuttle valves are most frequently used in hydrostatic transmission systems. The term hot oil refers to the valve's cooling function. As is indicated in the closed-loop hydrostatic transmission circuit in the figure, the hot oil shuttle valve causes a new supply of cool hydraulic oil to enter the system from the replenishing pump by allowing an equal amount of used, hot oil to exit "P" to "T" and then through the supercharge relief valve. Vickers recommends the use of cross-port relief in transmission or hydrostatic drive circuits.

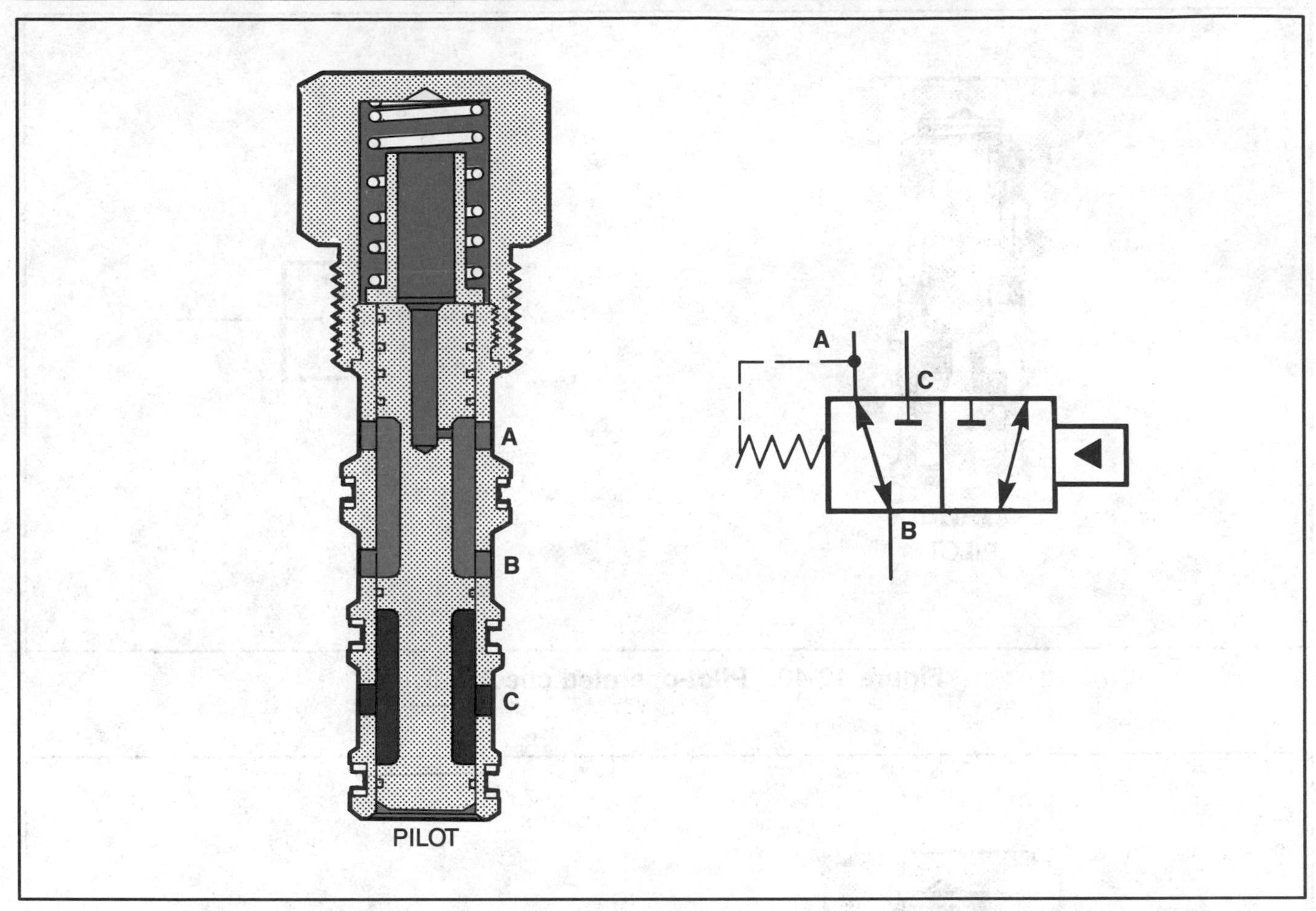

Figure 12-38. Pilot-operated, three-way, two-position, spring return spool-type valve.

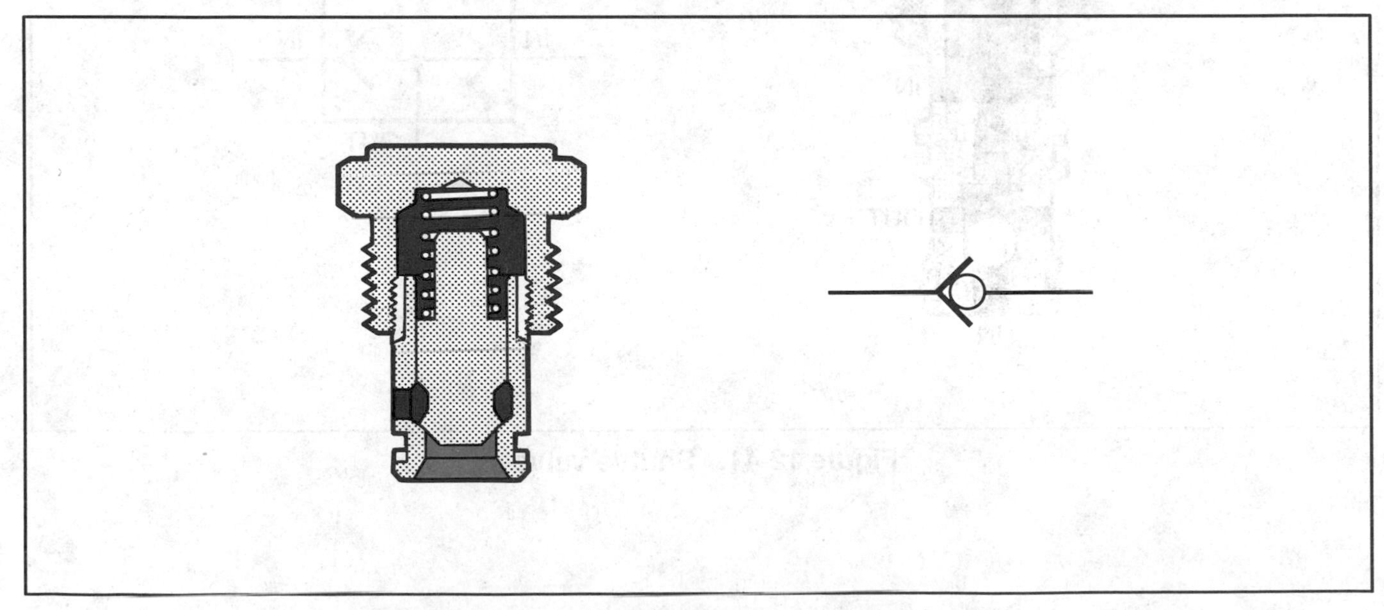

Figure 12-39. Simple poppet-type check valve.

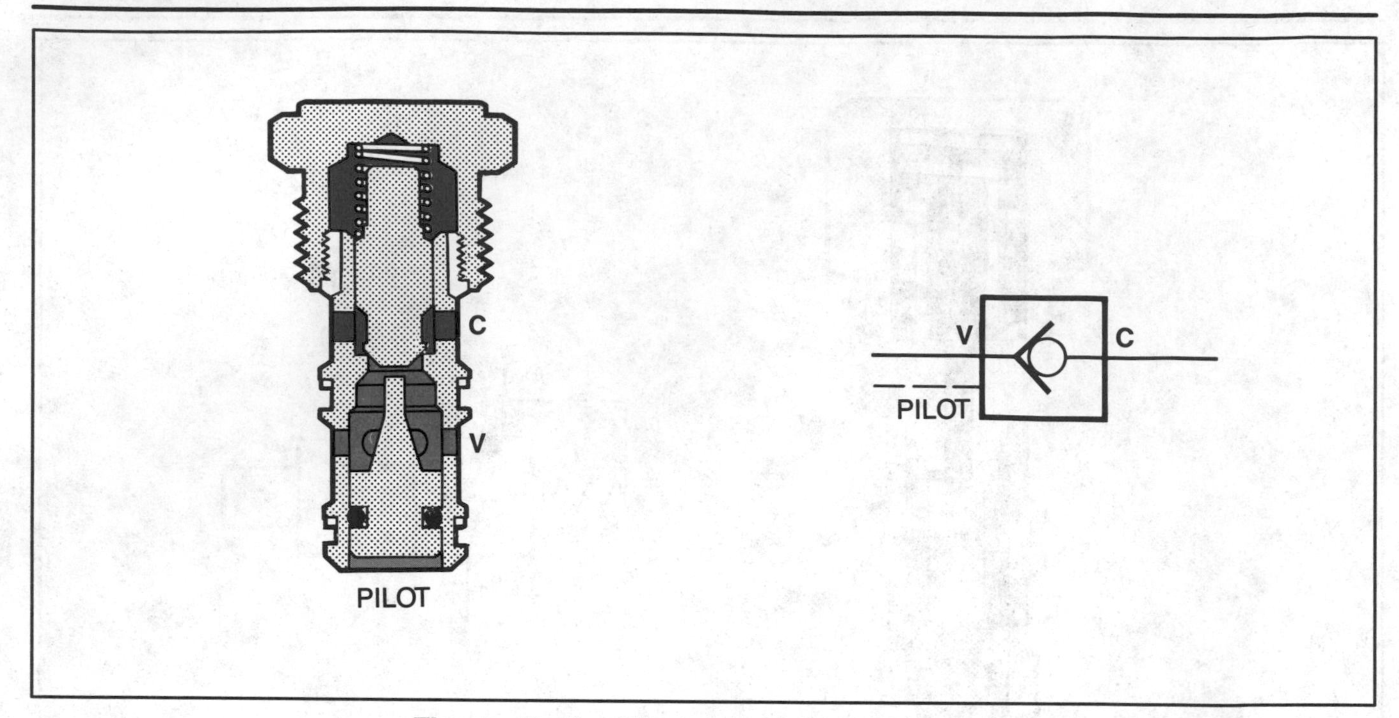

Figure 12-40. Pilot-operated check valve.

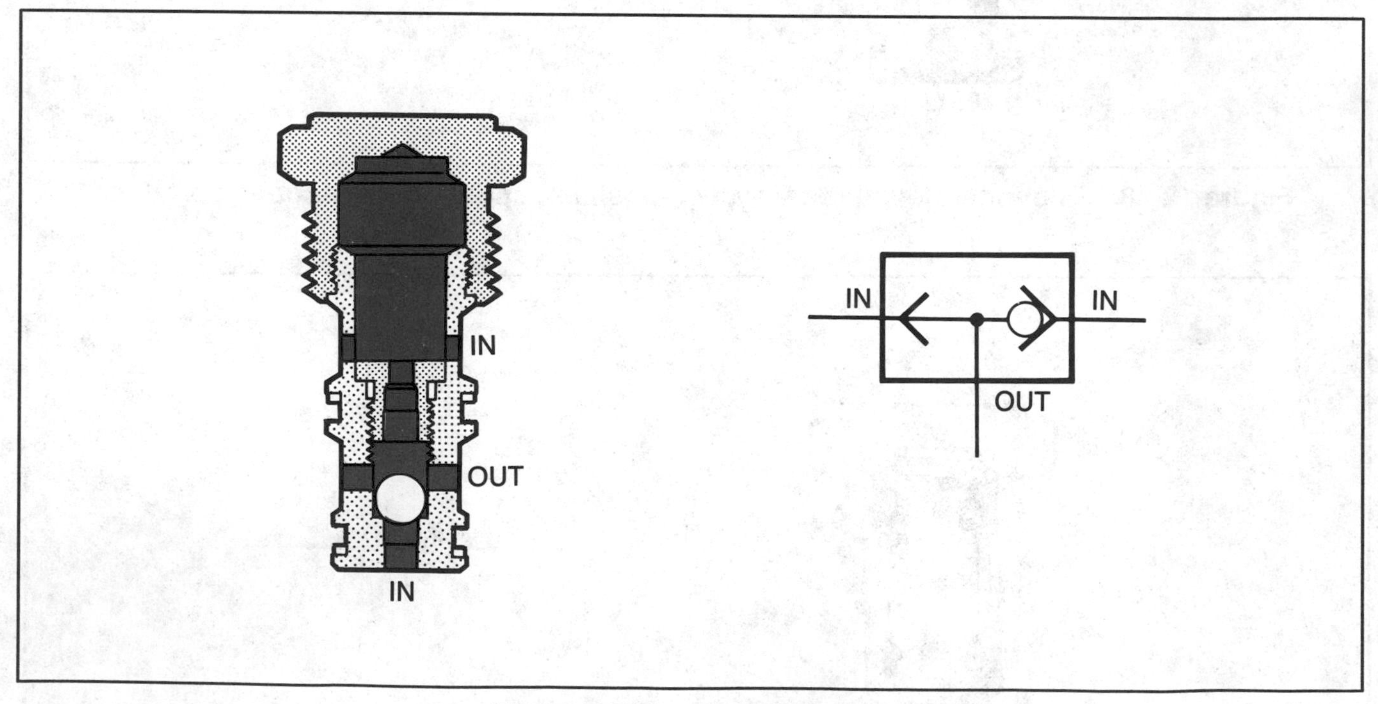

Figure 12-41. Shuttle valve.

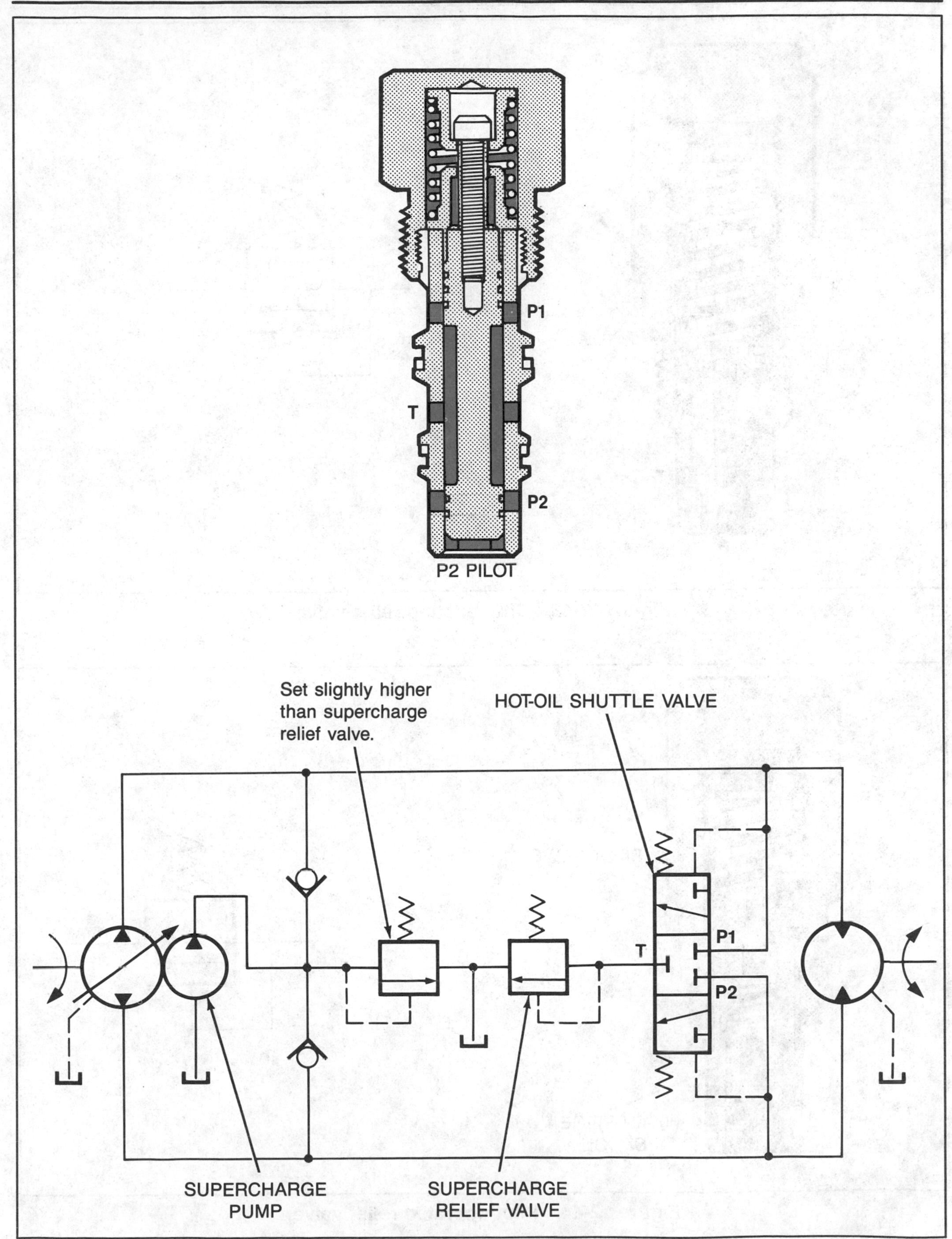

Figure 12-42. Hot oil shuttle valve used in a closed-loop hydrostatic transmission.

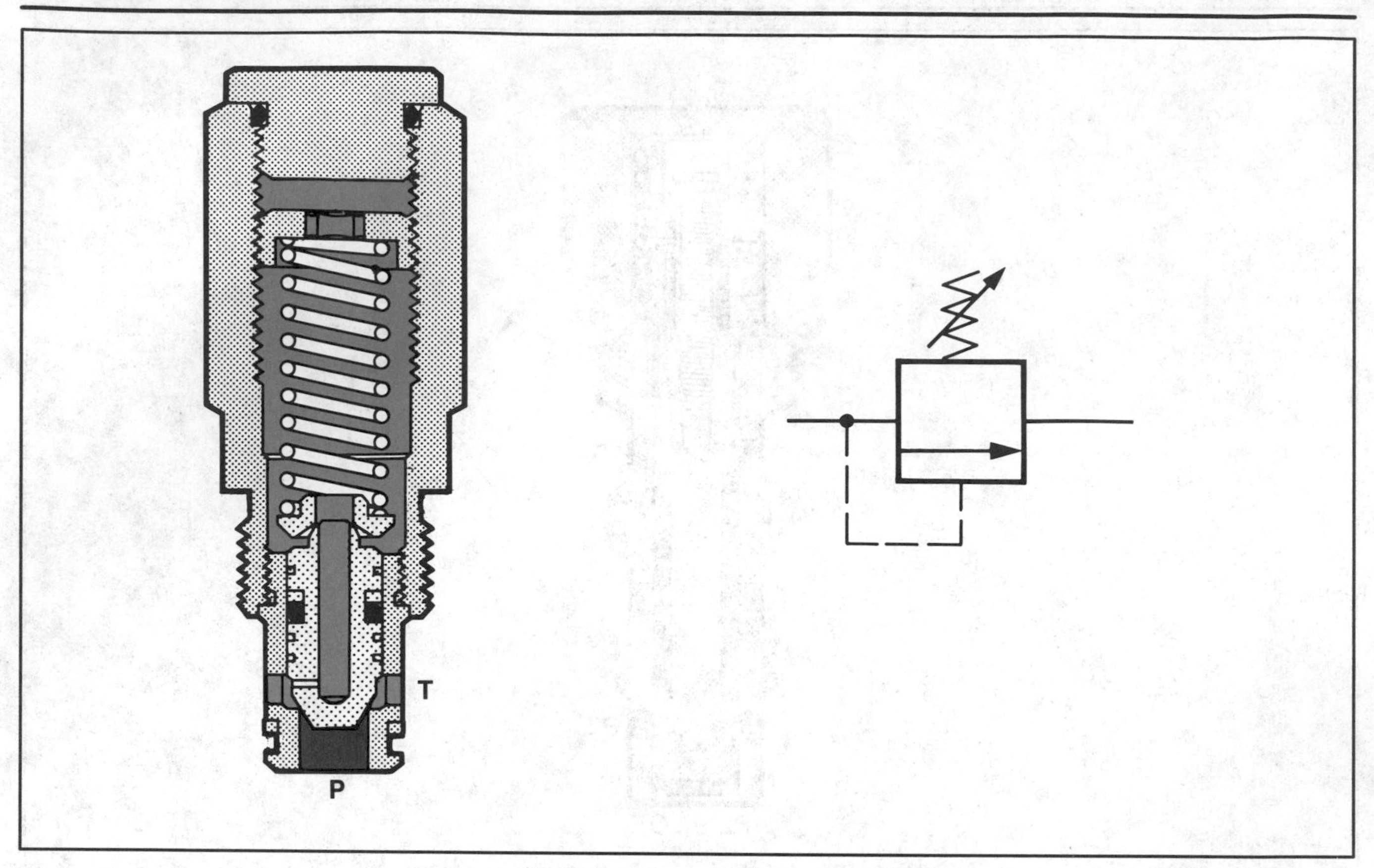

Figure 12-43. Direct-acting relief valve.

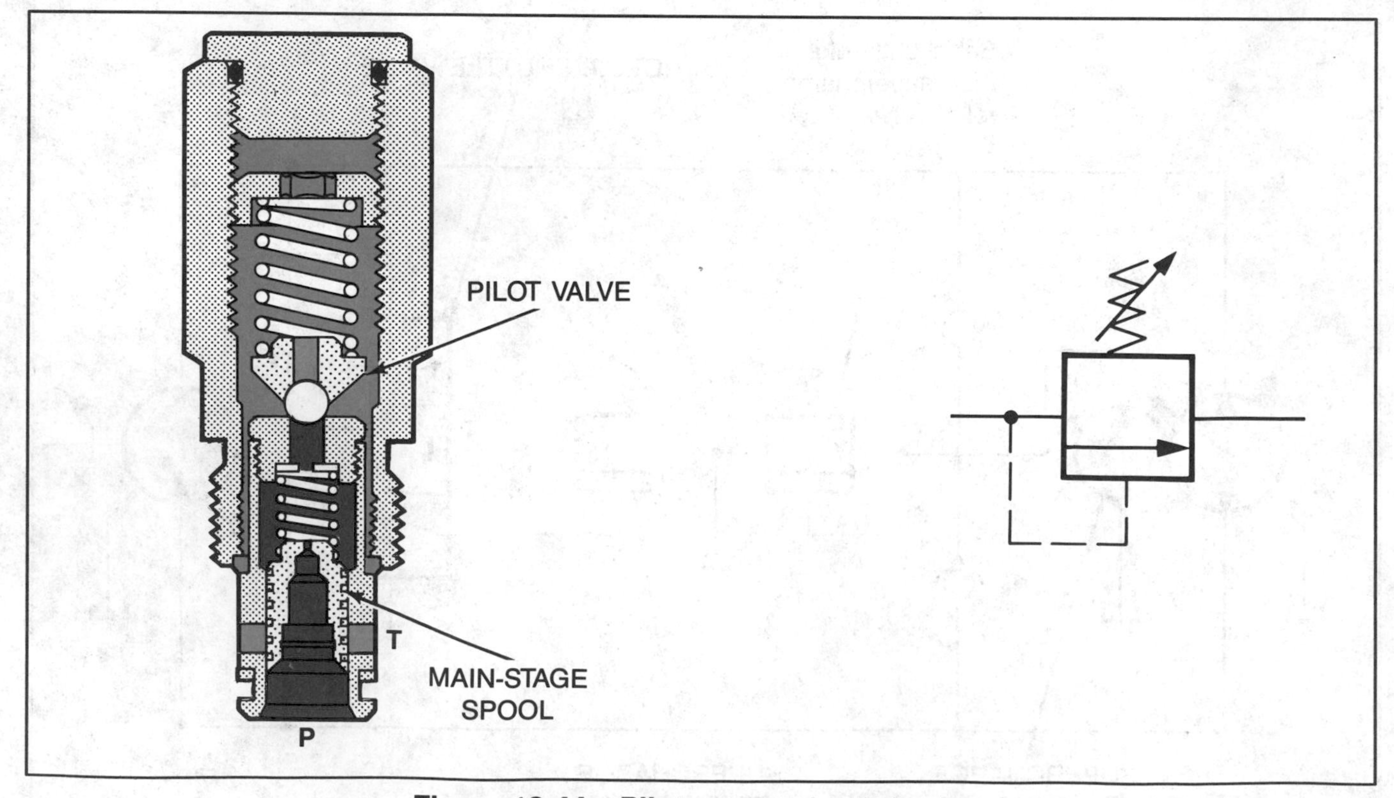

Figure 12-44. Pilot-operated relief valve.

PRESSURE CONTROL SCREW-IN CARTRIDGE VALVES

The types of screw-in cartridge valves used to control pressure include pressure relief valves, pressure-reducing and -relieving valves, pressure-sequence valves, and pressure-unloading valves.

Relief Valves

Figure 12-43 represents a simple, normally closed, direct-acting relief valve. When system pressure acting at the "P" port overcomes the valve spring's pressure setting, the valve opens to direct oil from "P" to "T."

Figure 12-44 is a normally closed, pilot-operated relief valve. Again, system pressure is located at the "P" port, and the "T" port is connected to the tank. The main stage spool remains hydraulically balanced until the system pressure reaches the pilot valve pressure setting. Once the system pressure overcomes the pilot valve pressure and the light spring pressure, the main stage spool moves up to direct oil to the tank.

Reducing and Relieving Valves

A cartridge valve can also serve as both a pressure-reducing and pressure-relieving valve. In Figure 12-45, the valve is a direct-acting, normally open pressure-reducing and -relieving valve. The primary system port is the lower side port, the regulated pressure is the bottom port, and the tank port is the top side port.

On this valve, when pressure at the regulated port is lower than the valve's pressure setting, the valve spool moves down to allow unrestricted flow from the primary pressure port to the regulated pressure port. This blocks flow to the tank. When pressure at the regulated pressure port reaches the pressure setting, the valve spool moves away from the regulated port to partially restrict the flow, thereby limiting the pressure at the regulated pressure port. When pressure at the regulated pressure port exceeds the pressure setting, the valve spool moves further up to open the flow from the regulated flow to the tank. The spool's movement also blocks the primary pressure port.

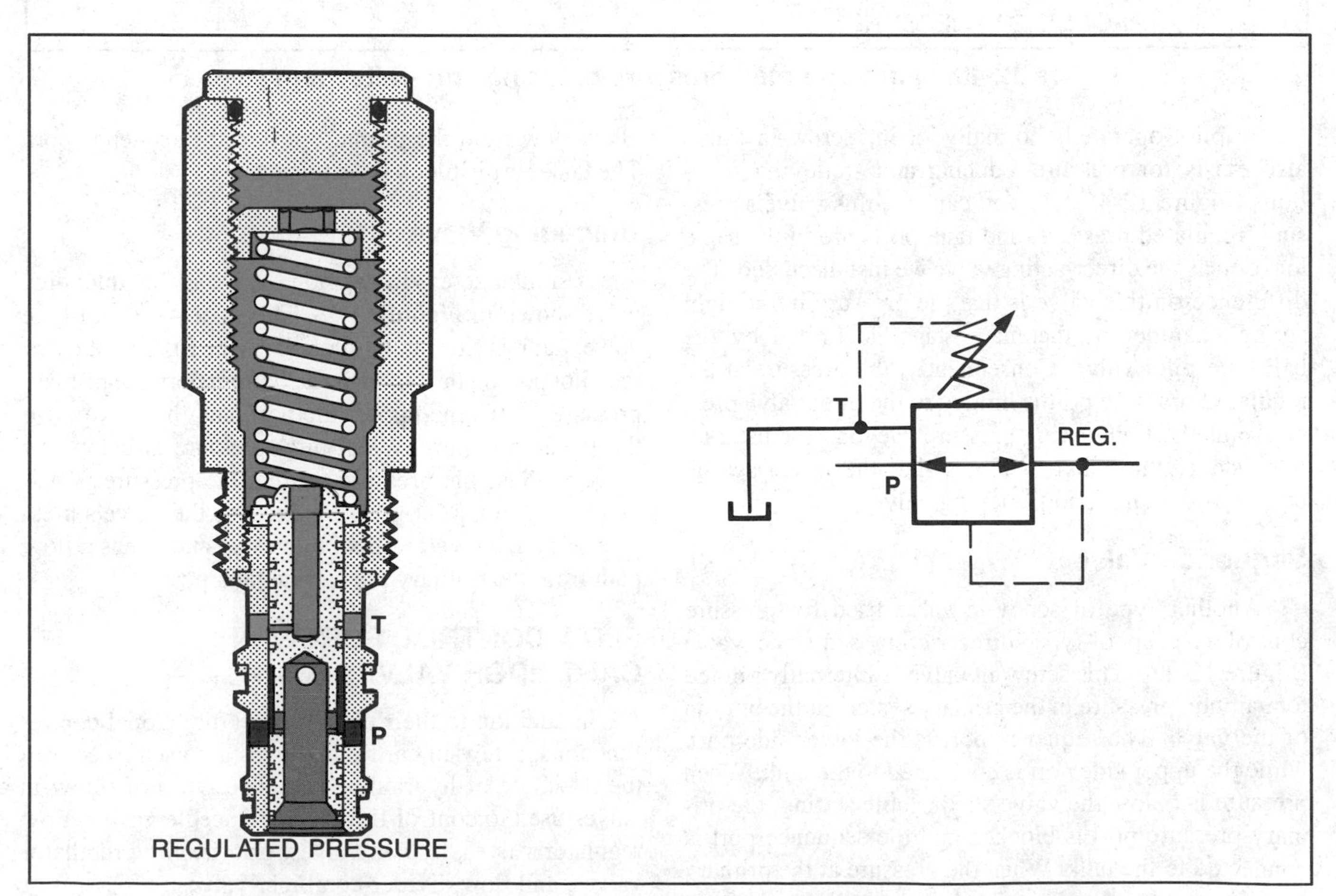

Figure 12-45. Direct-acting pressure-reducing and -relieving valve.

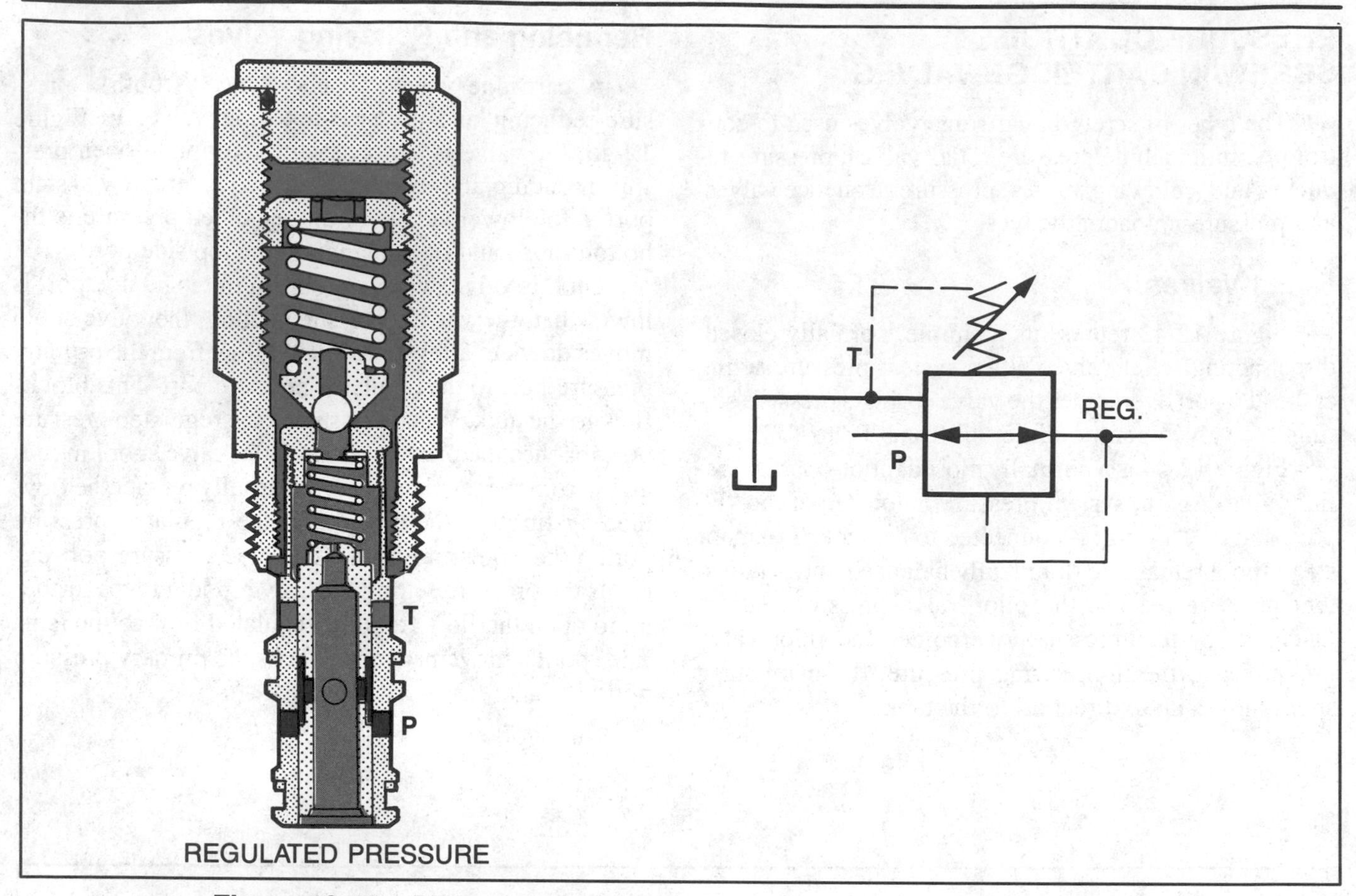

Figure 12-46. Pilot-operated pressure-reducing and -relieving valve.

A pilot-operated, normally open, screw-in valve also exists for pressure-reducing and -relieving functions (Figure 12-46). As you can see, this valve's pressure, regulated pressure, and tank ports are in the same location as the direct-acting valve we just discussed. The difference on this valve is that the pressure in the light spring chamber of the main stage is limited by the ball-type pilot valve. Consequently, the pressure at the regulated pressure port is limited to the pilot valve pressure plus the light spring pressure. Beyond the use of a pilot stage, this valve operates like the direct-acting, pressure-reducing and -relieving valve.

Sequence Valve

Another type of screw-in valve used for pressure control is a spool-type, direct-acting sequence valve (Figure 12-47). This screw-in valve is internally piloted to maintain pressure in the primary system at the bottom of the valve. The sequence port is the lower side port, while the upper side port is connected to the tank. When pressure is below the valve's adjustable setting, the primary pressure port is blocked and the sequence port is connected to the tank. When the pressure at the primary port reaches the valve's pressure setting, the spool lifts to allow flow from the primary port to the sequence port. The tank port is blocked as this occurs.

Unloading Valve

A similar screw-in cartridge valve is the unloading valve shown in Figure 12-48. This is also a spool-type valve, but it is externally piloted. As the figure indicates, the pilot port on this valve is the bottom port, the primary pressure port is the upper side port, and the tank port is the lower side port. When pilot pressure is below the valve's adjustable pressure setting, the pressure port is blocked. When pilot pressure exceeds the valve's pressure setting, the valve spool moves up and opens a flow path from the primary port to the tank port.

FLOW CONTROL SCREW-IN CARTRIDGE VALVES

In addition to their pressure and directional control functions, screw-in cartridge valves are used to control the flow rate of hydraulic fluid. The types of screw-in valves used to control flow include needle valves, flow regulator valves, bypass valves, priority flow regulator valves, and flow divider-combiner valves.

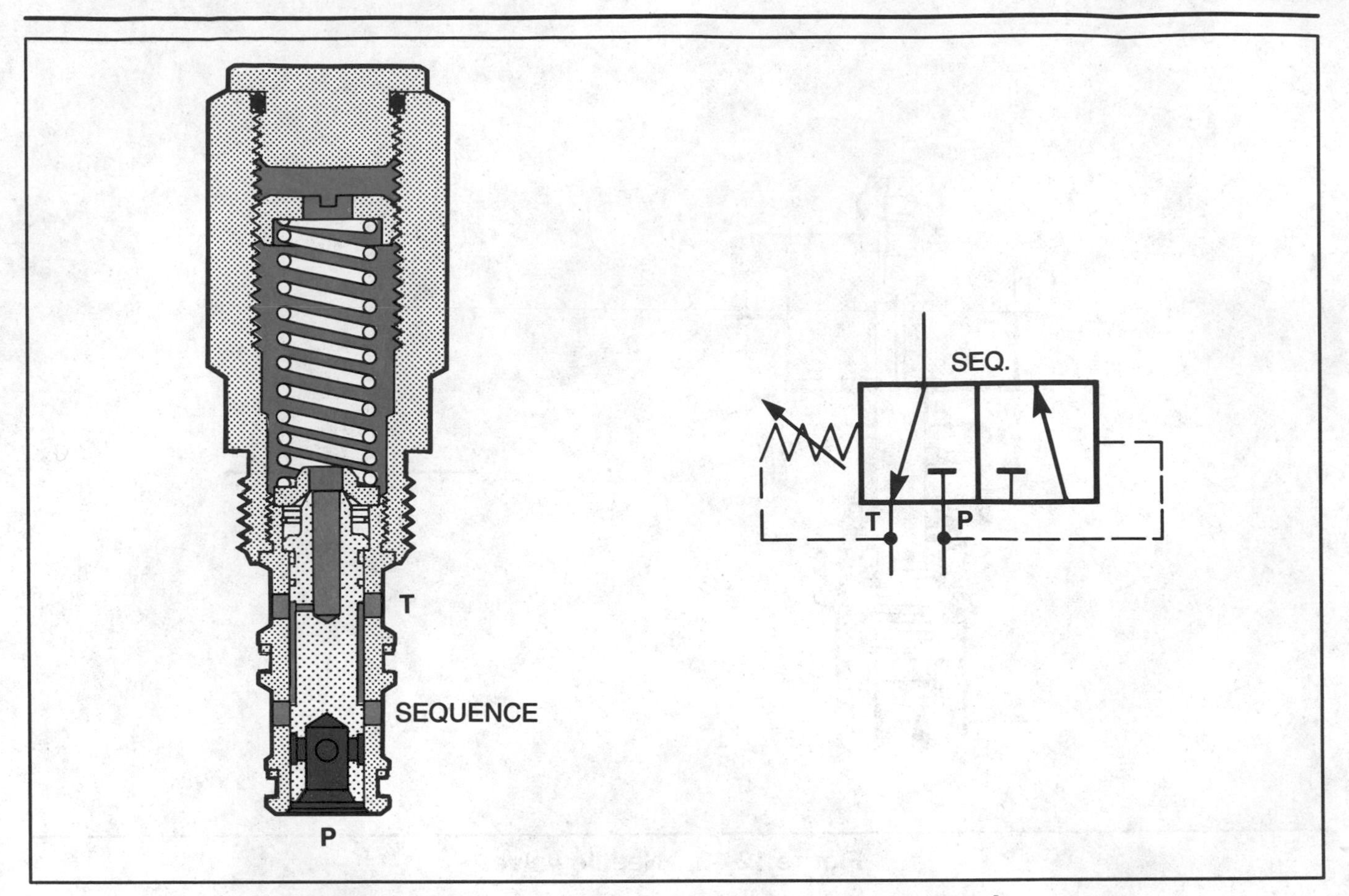

Figure 12-47. Direct-acting pressure sequence valve.

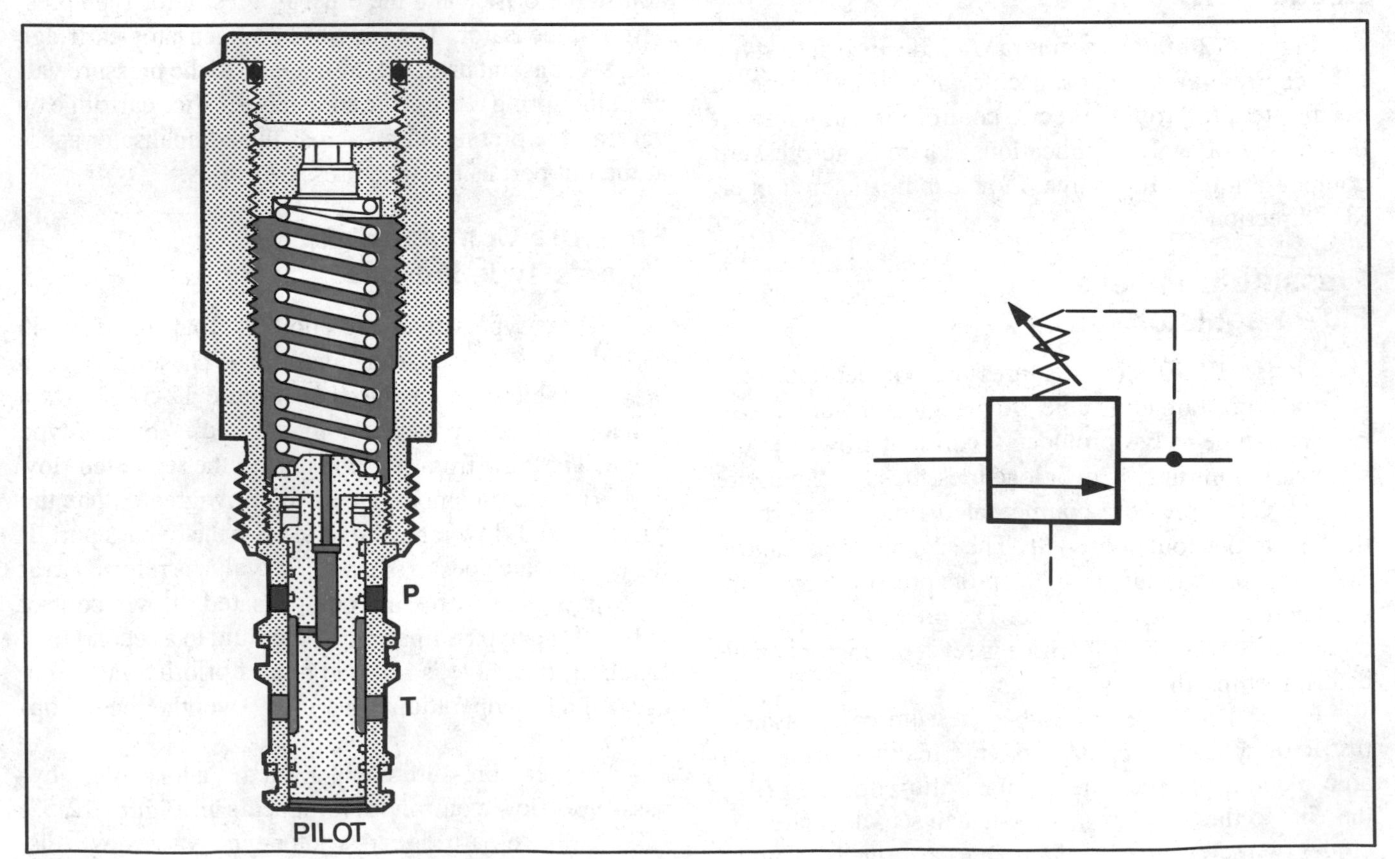

Figure 12-48. Externally-piloted unloading valve.

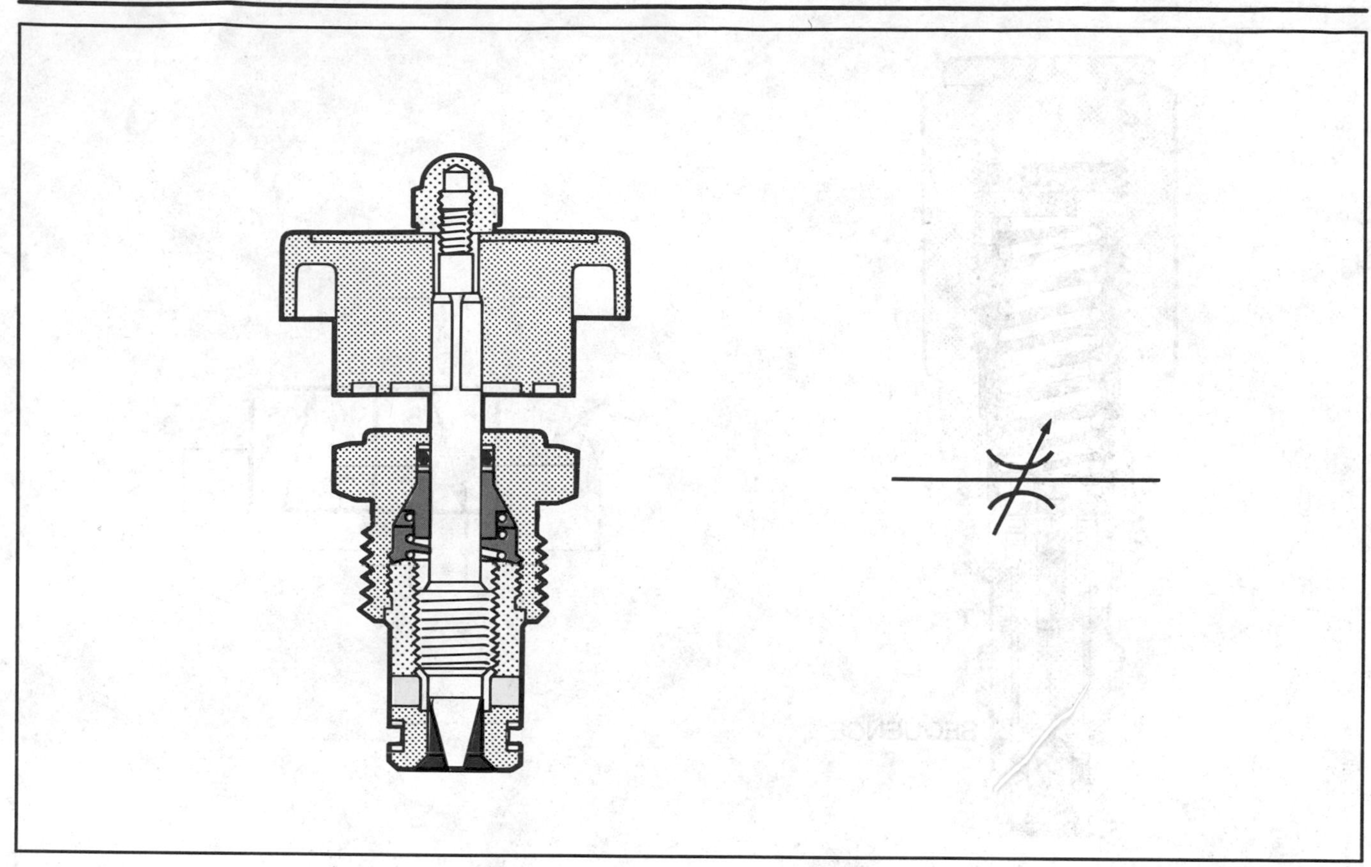

Figure 12-49. Needle valve.

Needle Valve

Figure 12-49 represents a variable restrictor-type flow control valve called a needle valve. It has a tapered needle stem to provide precise control for many restriction or shutoff valve applications. There is no pressure compensation on this valve. Flow can be metered in either direction.

Pressure-Compensated Flow Regulator Valves

Figure 12-50 shows a pressure-compensated, restrictor-type, nonadjustable flow regulator valve. This restrictor-type valve provides a constant flow rate despite variations in system or load pressures. As the figure indicates, the pressure-compensated flow travels from the "in" to the "out" port only. The size of the passage at the "out" port modulates to keep the pressure drop constant across the control orifice. The pressure drop maintained across the control orifice is set at the factory and is determined by the valve spring.

Figure 12-51 shows another pressure-compensated, restrictor-type flow control valve. This valve is adjustable. Again, the pressure compensation operates from the "in" to the "out" port. As you can see, this valve includes two screw-in cartridges. One cartridge is a simple, adjustable orifice and the other is a restrictor-type pressure compensator. The pressure compensator cartridge keeps a constant pressure drop (equal to the pressure value of the spring) across the adjustable orifice cartridge by varying the pressure drop across the compensator spool at the out port as the load pressure varies.

Pressure-Compensated, Bypass-Type Valves

Bypass-type valves are another group of screw-in cartridge valves used to regulate flow. These valves can be adjustable or nonadjustable. Figurc 12-52 shows a nonadjustable, pressure-compensated, bypass-type valve. The flow from the "in" port to the regulated flow port is pressure-compensated. Any flow greater than the valve's fixed flow rate is diverted out the bypass port. If the bypass line goes to the tank, the valve is referred to as a bypass-type, pressure-compensated flow control valve. If the bypass line carries the fluid to a second load function, the valve is referred to as a priority valve, because the first operation has priority over the second operation.

Another pressure-compensated, adjustable, bypass-type flow control valve appears in Figure 12-53. Like the two-cartridge restrictor-type valves we discussed earlier, this valve includes two cartridges: one

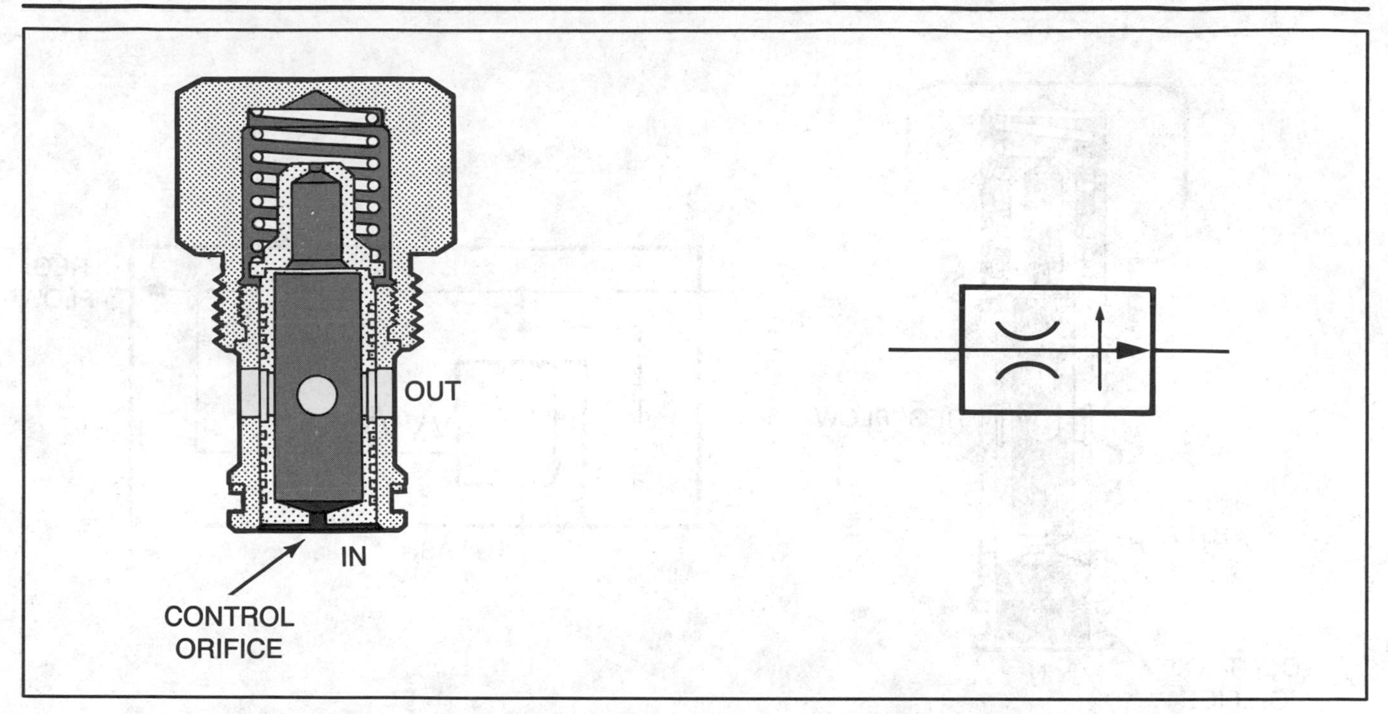

Figure 12-50. Pressure-compensated, nonadjustable restrictor-type control valve.

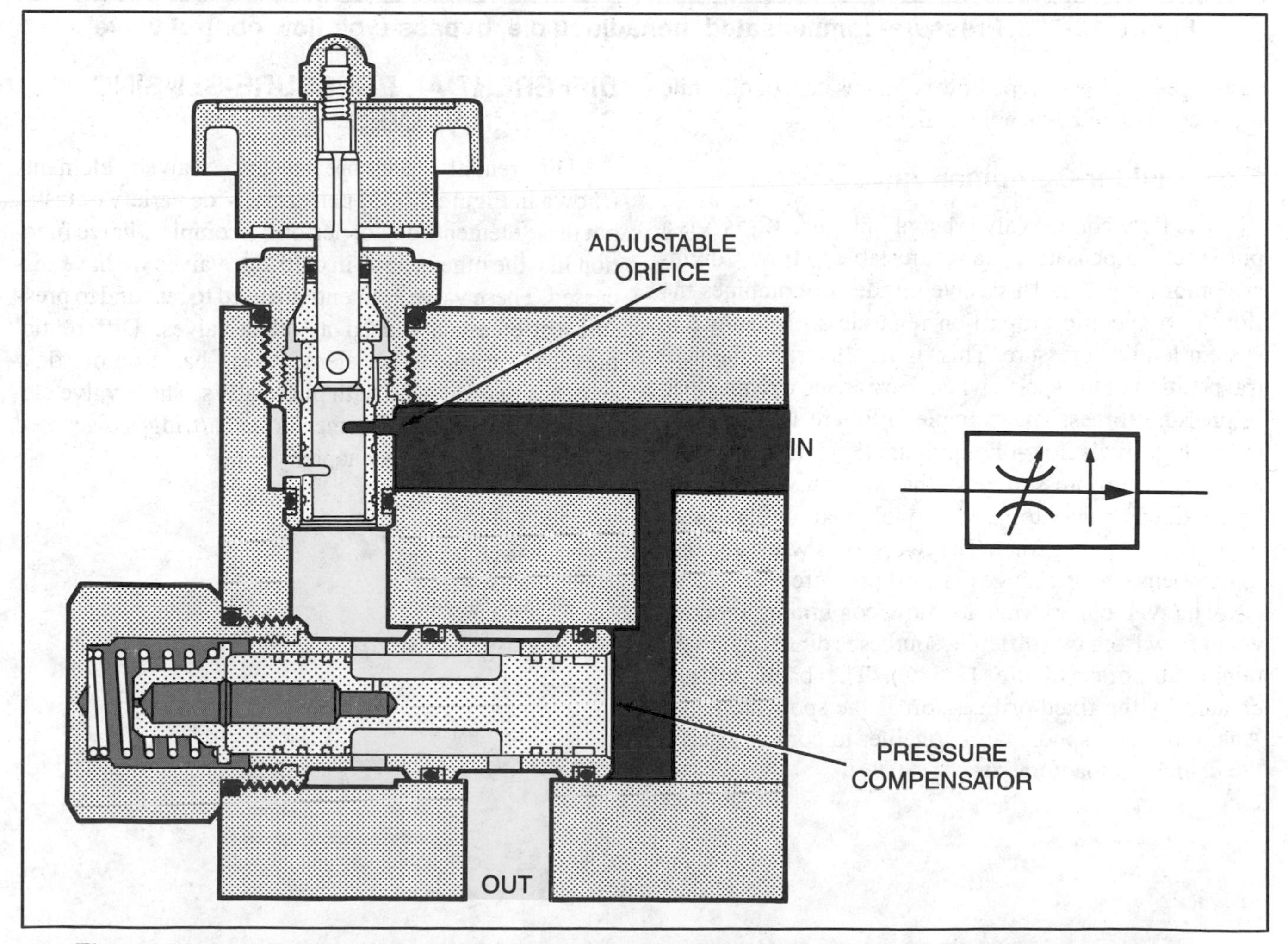

Figure 12-51. Pressure-compensated, flow restrictor valve with two screw-in cartridges.

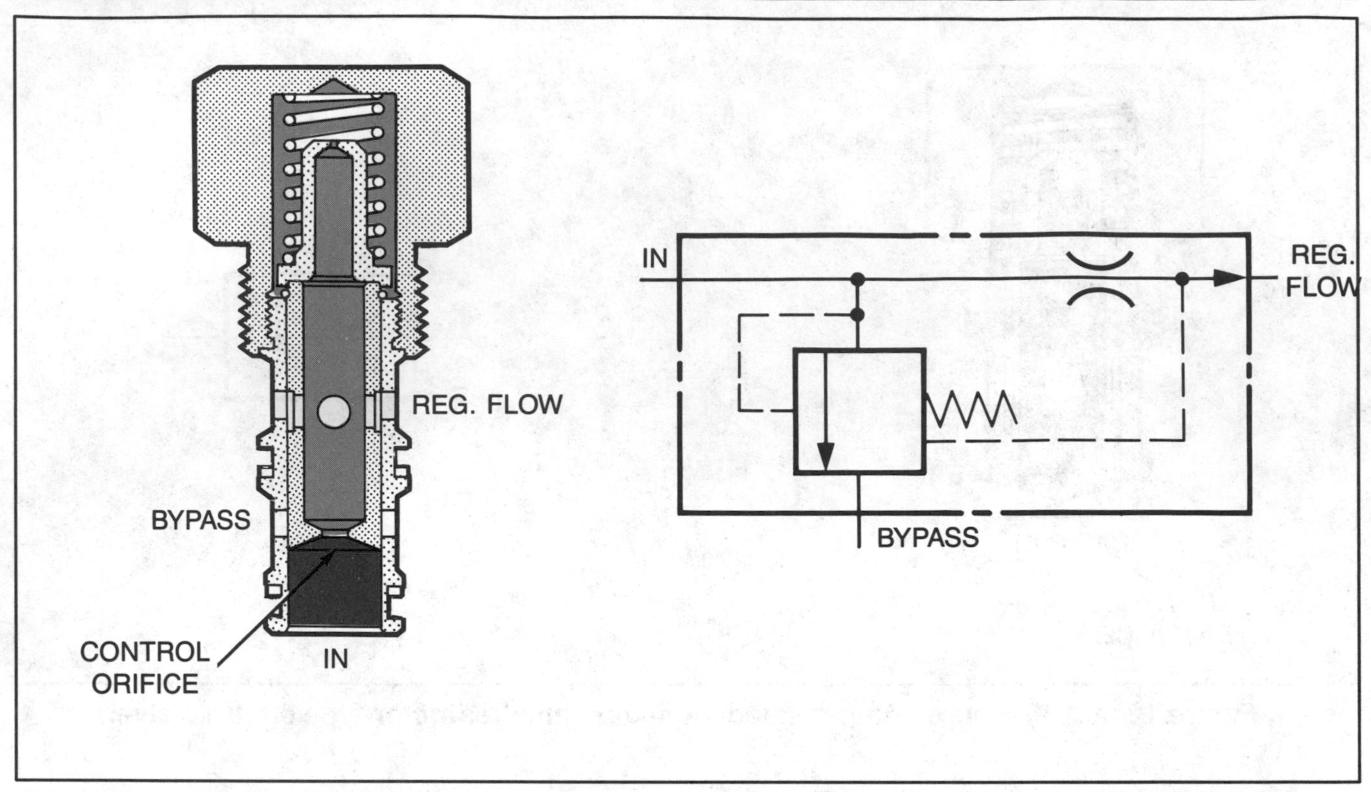

Figure 12-52. Pressure-compensated, nonadjustable, bypass-type flow control valve.

cartridge is a bypass-type flow compensator spool, while the second is an adjustable orifice.

Flow Divider-Combiner Valve

The flow control valve shown in Figure 12-54A is a pressure-compensated, nonadjustable, flow divider-combiner valve. This valve divides or combines the flow in a specific proportion regardless of changes in system load or pressure. This figure illustrates the neutral position of the spools when there are no existing load or pressure forces. The example in Figure 12-54B illustrates the valve's flow-dividing function. When flow enters the system pressure port, the back pressure created by the fixed orifices forces the spools apart until the ends become hooked together. The two spools work together to compensate for changes in load pressure.

The valve performs its flow/combining function when flow from two different sources is directed into the regulated ports (Figure 12-54C). The back pressure created by the fixed orifices forces the spools together. Again, the two spools work together to compensate for the change in load pressure.

DIFFERENTIAL PRESSURE-SENSING SCREW-IN VALVES

Differential pressure-sensing valve elements (shown in Figure 12-55) perform a wide variety of tasks, but these elements do not perform a complete valve function like the other screw-in cartridge valves we have discussed. These valve elements are used to respond to pressure differentials sensed at other valves. Differential pressure-sensing valve elements are basic on/off flow switches that are used with pilot valves. These valve elements operate much like the slip-in cartridges discussed in the first part of this chapter.

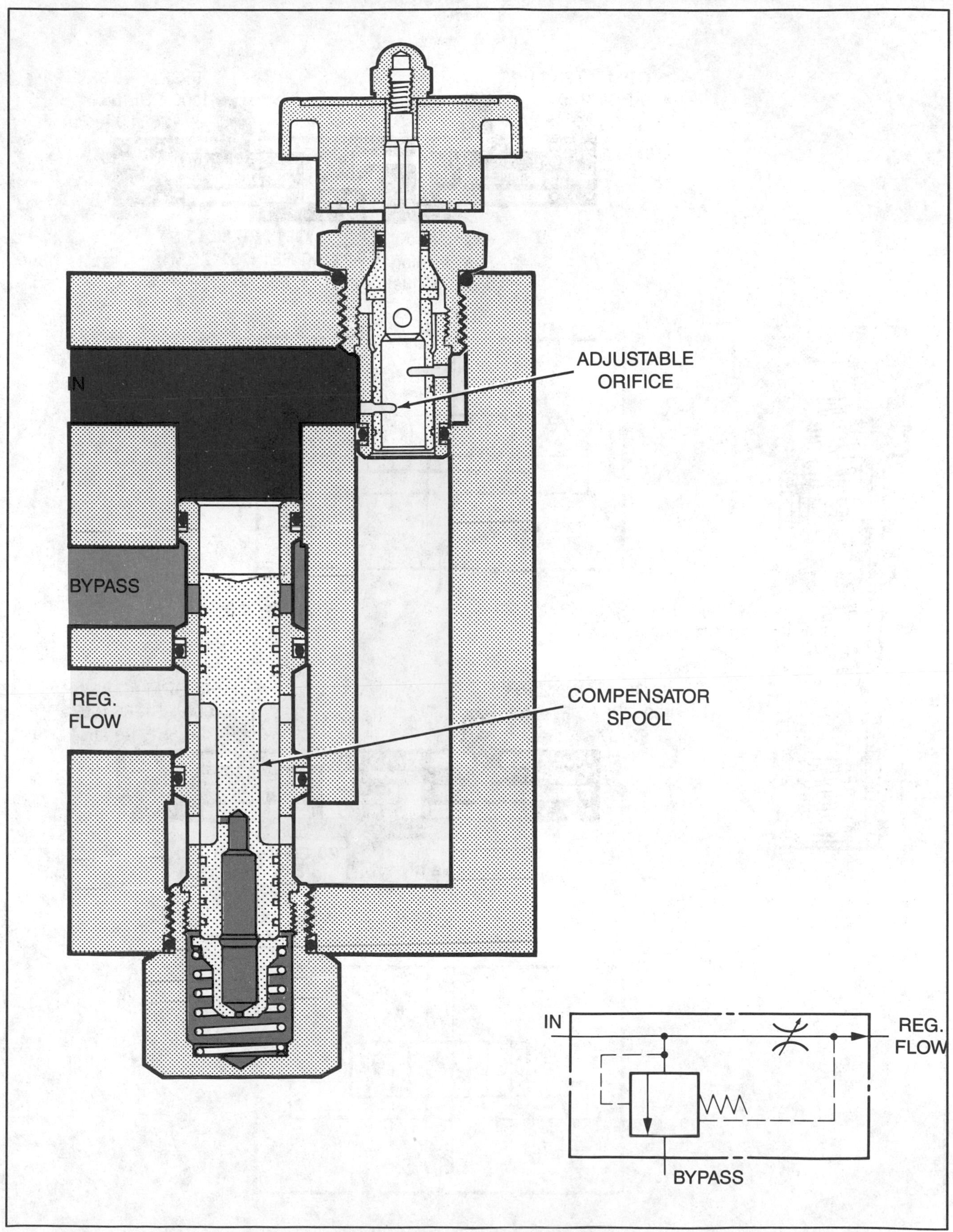

Figure 12-53. Pressure-compensated, nonadjustable, bypass-type flow control valve with two screw-in cartridges.

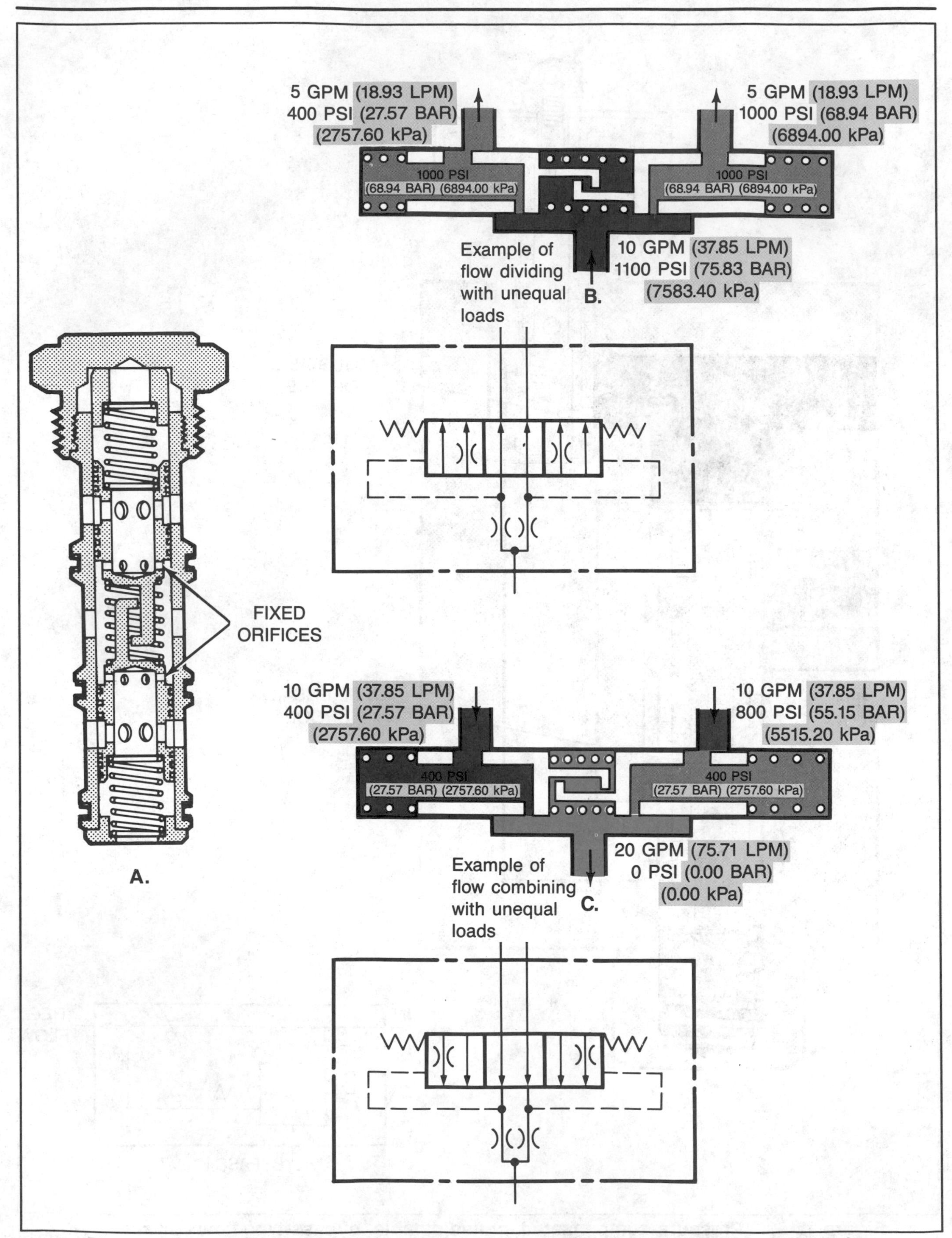

Figure 12-54. Pressure-compensated, nonadjustable, flow divider-combiner valve.

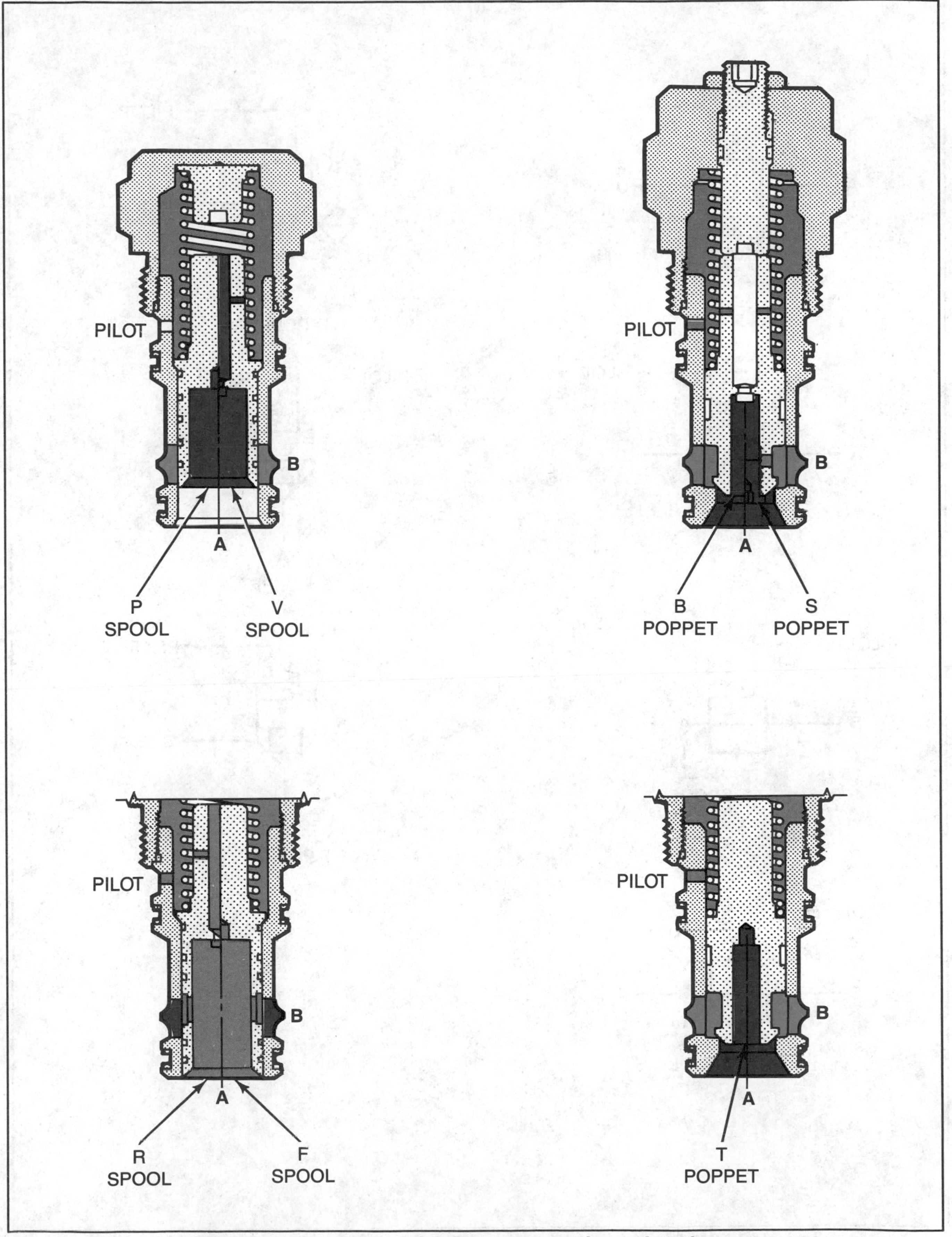

Figure 12-55. Differential pressure sensing valve element.

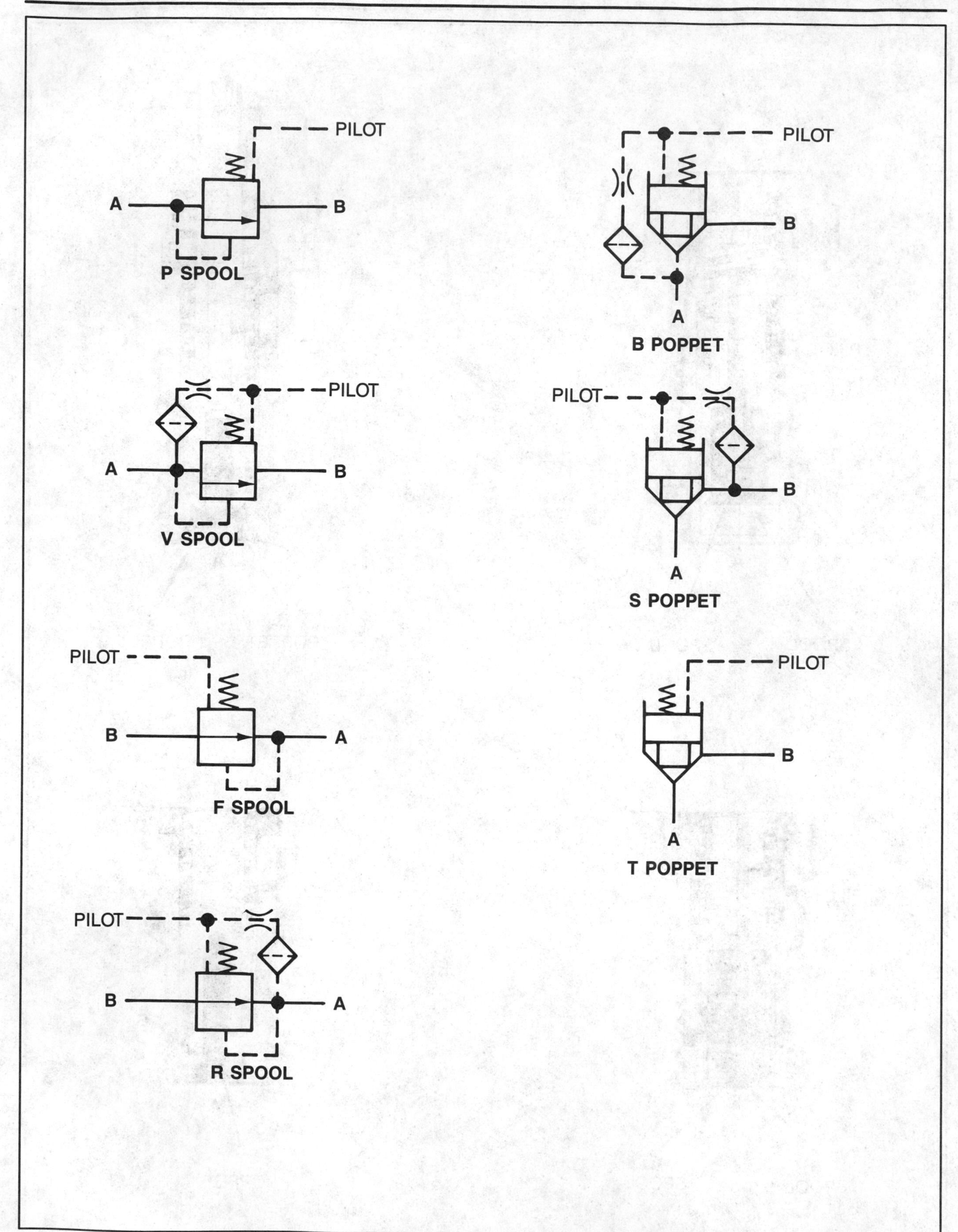

Figure 12-56. Graphic symbols for differential pressure sensing valves.

QUESTIONS

1. Describe three major differences between slip-in and screw-in cartridge valves.
2. List five benefits in using cartridge valves.
3. Slip-in cartridge valves are most similar to what other type of conventional valve?
4. What is the significance of the "A_A," "A_B," and the "A_{AP}" areas of a slip-in cartridge valve?
5. How do you determine the area ratio of a valve insert?
6. Are the poppets and sleeves of slip-in-type valves with different area ratios interchangeable?
7. What three variables must be taken into account when determining opening and closing forces?
8. Identify the forces that work to open a valve and those that work to close the valve.
9. What determines whether a valve is open or closed?
10. When using a 1:2 area ratio insert as a check valve, why is it better to connect the "AP" chamber to the "B" port rather than to the "A" port?
11. Explain how sizing a cartridge valve differs from sizing a conventional spool-type valve.
12. What is the benefit of an independent pilot control circuit in which each slip-in cartridge valve is operated by a solenoid valve?
13. Briefly describe the two main functions of a basic slip-in valve cover.
14. Explain the additional feature of a pilot valve interface cover.
15. Describe the difference between a conventional relief valve and a slip-in cartridge relief valve.
16. Explain the difference between a nonpressure-compensated flow control slip-in cartridge valve and a typical needle valve that has a similar function.
17. Describe how to obtain a maximum pressure limitation on a slip-in, pressure-compensated, bypass-type flow control valve.
18. Describe a proportional throttle valve and identify the two types of slip-in cartridge valves in this valve group.
19. What hydraulic system functions can screw-in cartridge valves perform?
20. What do the two-way, three-way, three-way short, and four-way screw-in cartridge valve functions signify?
21. What pressure is required at the pilot port to permit reverse flow through a screw-in pilot-operated check valve?
22. Describe the operation of a screw-in shuttle valve.
23. Briefly describe the operation of a screw-in cartridge that provides a direct-acting, normally open, pressure-reducing and -relieving valve function.
24. How does a screw-in cartridge that is used as a pressure-compensated, nonadjustable flow regulator valve maintain a constant pressure drop across the control orifice?

CHAPTER 13
ELECTRONICS—OP AMPS

Many modern electrohydraulic devices, such as proportional and servo valves, servo pumps, pressure and flow control valves, and speed control circuits, require relatively high-power electrical signals for control and/or positioning. Typically, however, the source of the command or control signal is a device capable of delivering only low-power signals. For example, the input signal to control a proportional valve may come from such sources as a:

- Potentiometer
- Temperature sensor
- Pressure transducer
- Tachogenerator
- Mircoprocessor controlled device, etc.

Since all of these sources produce a low power signal, it is necessary to raise or amplify the electrical level of the signal (in terms of voltage or current, or both) before it is capable of operating the proportional valve.

Because of heat handling and physical size limitations, it is not practical to increase the capacity of the originating device in order to deliver higher power signals. Fortunately, when a low power signal is applied to the input of the electronic circuit known as an amplifier, it can produce the required command signal for electrohydraulic devices at its output. See Figure 13-1.

This chapter provides a brief introduction to amplifier operation and terminology and a more detailed explanation of the Op Amp (Operational Amplifier), which is a special type of amplifier that has several characteristics especially suitable for electrohydraulic applications.

DEFINITION OF AN AMPLIFIER

An amplifier is a relatively simple circuit used to raise the level (or increase the amplitude) of an electronic signal. The amplified output signal is proportional to the input signal. Various types of amplifiers are used in electrical equipment, but all of them perform this same basic function.

TRANSISTORS AS AMPLIFIERS

A semiconductor device known as a transistor forms the basic element of most amplifier circuits. Transistors perform many important functions in electronics in addition to amplification and are constructed and configured in many different ways. This chapter does not elaborate on transistor theory or application other than to provide some basic understanding of amplifier operation and terminology.

The schematic symbol for one type of transistor, called an NPN type, is shown in Figure 13-2. The transistor has three terminals labeled emitter (e), base (b), and collector (c) in the figure. The collector is connected to a positive voltage supply, the emitter is connected to a negative voltage, and the base is connected to a voltage slightly higher than the emitter. The main current flow path through the transistor is from the collector to the

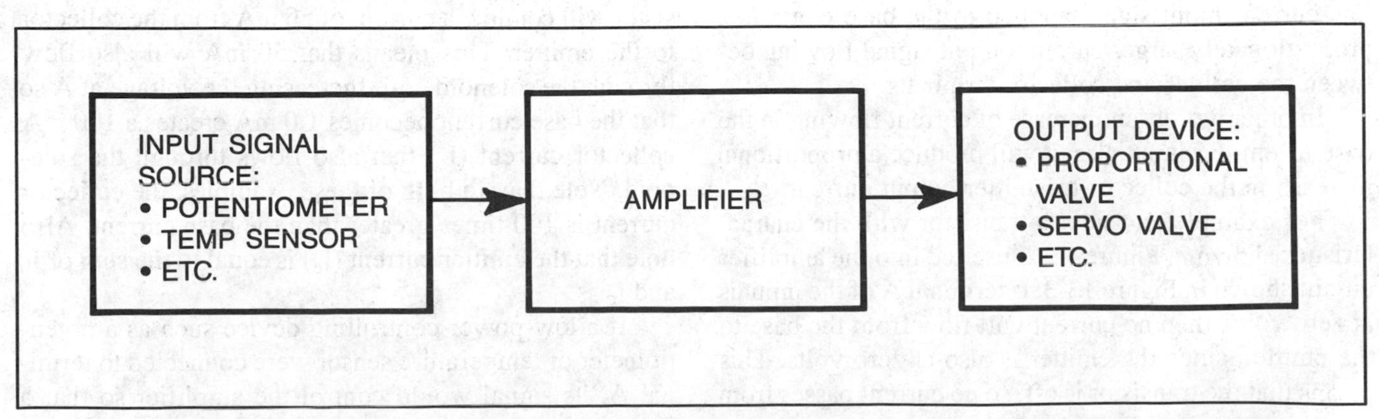

Figure 13-1. Functional block diagram of input device, amplifier and output device.

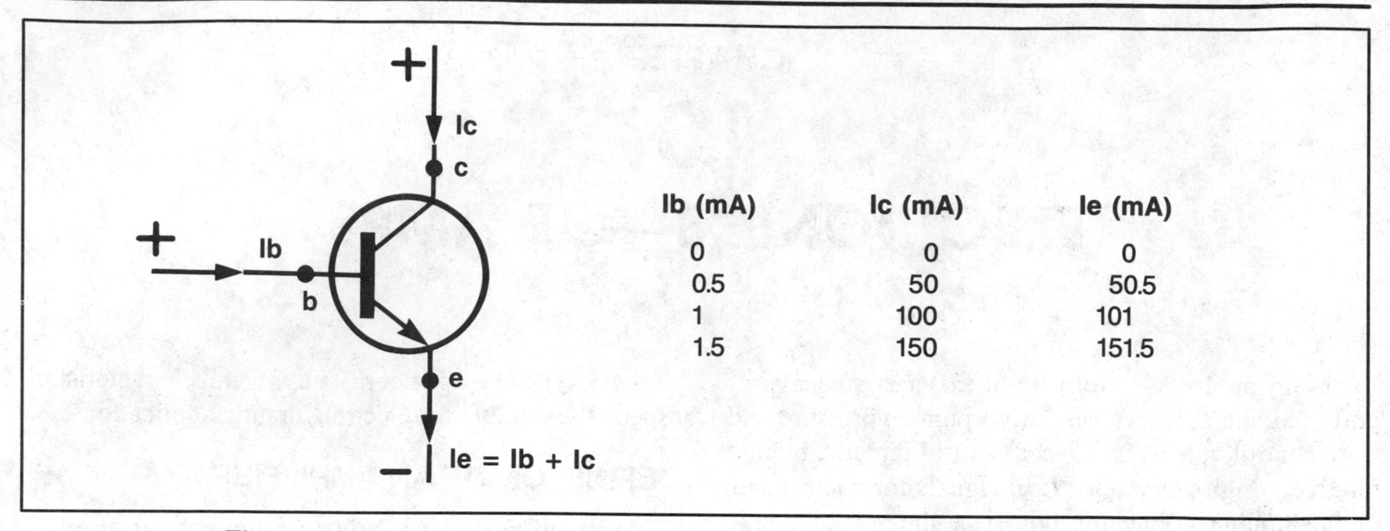

Ib (mA)	Ic (mA)	Ie (mA)
0	0	0
0.5	50	50.5
1	100	101
1.5	150	151.5

Figure 13-2. Schematic symbol and characteristics for a transistor.

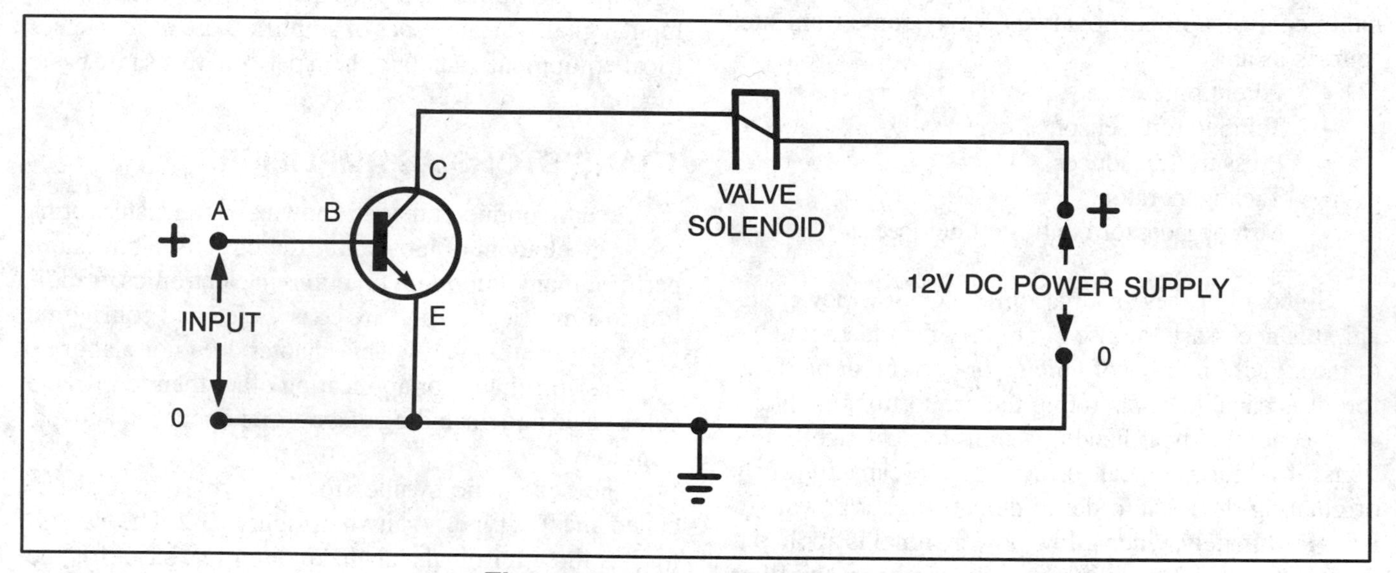

Figure 13-3. Amplifier circuit.

emitter, with the base acting as a controller in much the same way that a small pilot flow can control a much greater flow of fluid through a hydraulic valve.

In the most common amplifier configuration, a low-current input signal applied to the base controls a proportionately larger current output signal flowing between the emitter and collector terminals.

In other words, an increase of current flowing in the base-to-emitter input circuit will produce a proportional increase in the collector-to-emitter output current.

For example, consider a transistor with the characteristics shown in Figure 13-2 inserted into the amplifier circuit shown in Figure 13-3. If terminal A of the input is at zero volts, then no current will flow from the base to the emitter, since the emitter is also at zero volts. This means that the transistor is off, so no current passes from the collector to the emitter. In this case, no current flows through the valve solenoid.

If a small positive voltage sufficient to create a base current (I_b) of 0.5 mA is applied to terminal A, the transistor will conduct a current of 50 mA from the collector to the emitter. This means that 50 mA will also flow through the solenoid coil. Increasing the voltage at A so that the base current becomes 1.0 mA creates a 100 mA collector current (I_c) that also flows through the solenoid. Note that, in both of these examples, the collector current is 100 times greater than the base current. Also note that the emitter current (I_e) is equal to the sum of I_b and I_c.

If a low-power controlling device such as a potentiometer or temperature sensor were connected to terminal A, its signal would control the amplifier so that a high-current output would flow through the transistor

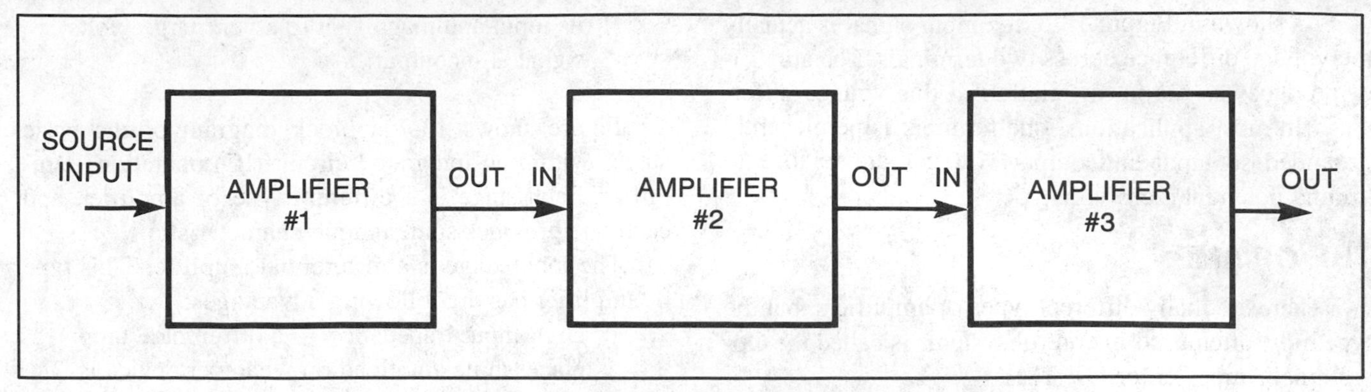

Figure 13-4. Stages of amplification.

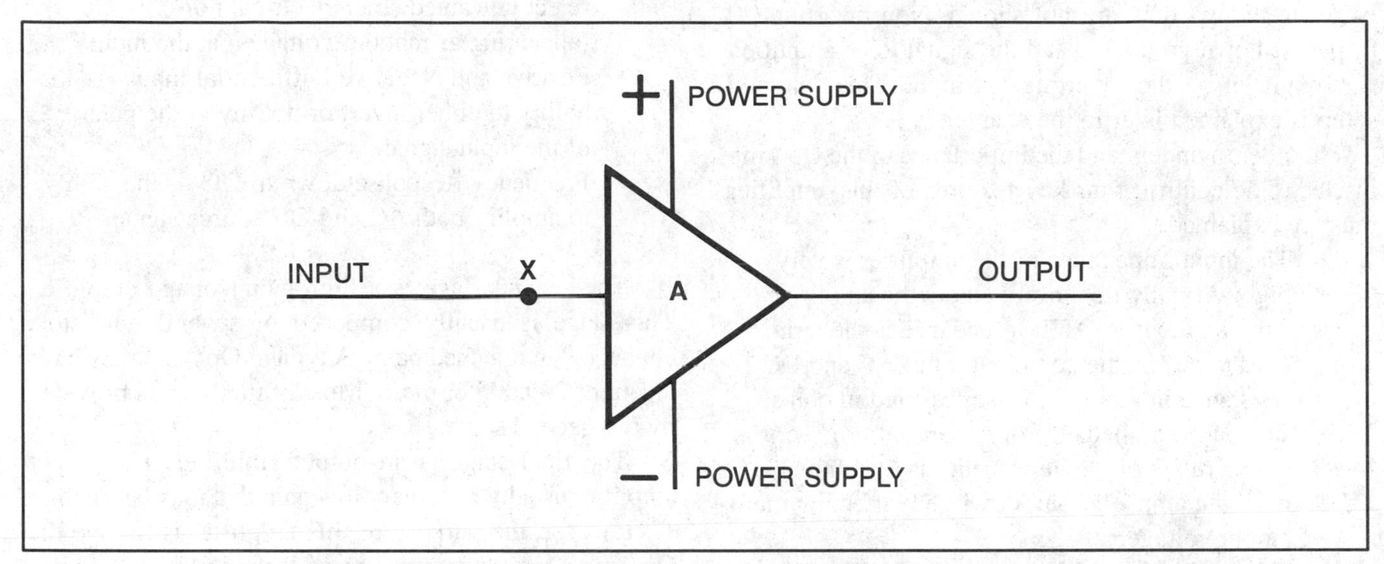

Figure 13-5. Schematic symbol for an amplifier circuit.

and the valve solenoid. In actual practice, the amplifier would consist of many more components than those shown in Figure 13-3 and would usually consist of two or more stages of amplification (see Figure 13-4). The output of the first amplifier is used as the input to the second amplifier. For the purpose of this chapter, it is not necessary to understand the detailed circuitry of the amplifier, and it should be regarded simply as a device that produces an amplified output proportional to its input.

The schematic symbol for an amplifier circuit is shown in Figure 13-5. The input is connected to the base of the triangle, and the output is taken from the apex. The power supply connections are always there in practice, but are not always drawn on the symbol. Point X is known as the Summing Junction (SJ), and as will be explained later, all inputs connected to this point are added together.

GAIN

The gain of the amplifier is defined as the ratio of output to input:

$$GAIN = A = \frac{OUTPUT\ VOLTAGE}{INPUT\ VOLTAGE}$$

The gain, A, can then be used as a multiplication factor with the input voltage. The gain may be as high as 10^6 or 1,000,000. An amplifier may also invert the input signal; that is, a small, positive-voltage input produces a large, negative-voltage output. Therefore, to find the output voltage of an inverting amplifier circuit.

$$A = \frac{-\ OUTPUT\ VOLTAGE}{INPUT\ VOLTAGE}$$

The output of a noninverting amplifier can be found by:

$$A = \frac{OUTPUT\ VOLTAGE}{INPUT\ VOLTAGE}$$

The value of A is normally found in the characteristics supplied with a particular amplifier type, but it varies depending on factors like circuit configuration, temperature, or power supply voltage.

As shown in Figure 13-3, the input signal is actually the voltage difference across two terminals. The amplifier produces an output in relation to this voltage difference. In most applications, one terminal is permanently grounded so input and output voltages are relative to ground potential (zero volts).

THE OP AMP

There are many different types of amplifiers, but the type most often used in control systems is called the Operational Amplifier, or **OP AMP**.

Figure 13-6 shows the schematic symbol for an Op Amp. The power supply connections are assumed to exist, even though they are not shown. Non-inverting (+) in-put is shown grounded, and the signal to be amplified is brought in on the inverting (–) input. This common setup is explained later in the chapter.

In order to understand the importance of the Op Amp in electronic control, four key features of this amplifier must be explained:

- The input impedance of the amplifier is very high; virtually no current enters the amplifier from its input. NOTE: Impedance is the term used to define the combination of AC and DC resistance in a circuit. It is measured in ohms.
- The output impedance of the amplifier is very low; a relatively large current may be drawn from the amplifier output without affecting the output voltage.
- The Op Amp is a **Differential Amplifier**. It amplifies any voltage **difference** applied between its two inputs.
- The **gain** of the Op Amp is very high. The Op Amp can take a tiny (i.e. microvolt) signal at its input and magnify it to a very large (volts) signal at its output.

Figure shows 13-7 a block diagram of the basic stages within an integrated circuit (IC) operational amplifier. Each stage is a different type of amplifier, and each one provides some unique characteristic.

The input stage is a differential amplifier. This type of amplifier has the following advantages:

- High Input Impedance — a differential amp places almost no load on whatever device is connected to it.
- High Common-Mode Rejection — the ability to reject unwanted stray electrical noise (appearing as random voltages) at the input.
- Positive and Negative Differential Inputs — the ability to either invert or not invert the polarity of the input signal.
- Frequency Response down to DC — the ability to amplify both AC and DC source signals.

The second stage is a high-gain voltage amplifier. This stage is usually composed of several transistors connected in matched pairs. A typical Op Amp may have a gain of 200,000 or more. Most of that gain is provided by this second stage.

The final stage is an output amplifier. The output amplifier usually has a very low gain that may be small as 1. However, the purpose of this amplifier is to give the Op Amp a very low output impedance. A low output impedance means that the Op Amp will itself consume very little of the power it outputs, sending almost all of it to the load.

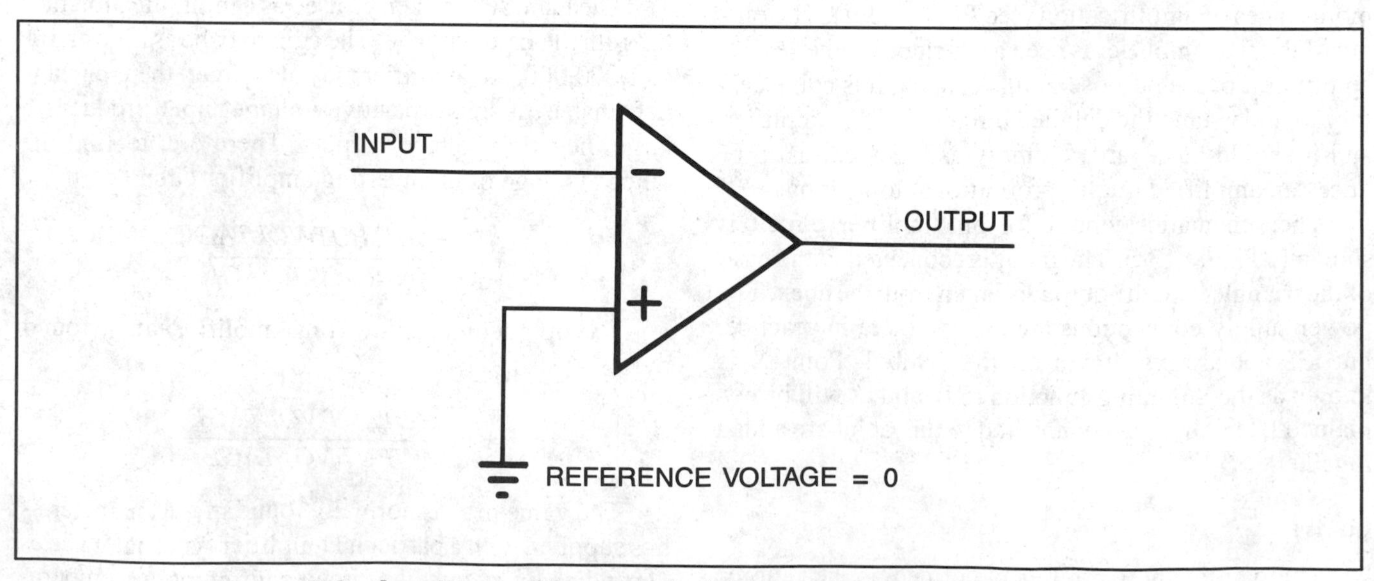

Figure 13-6. Amplifier with grounded input terminal for zero reference.

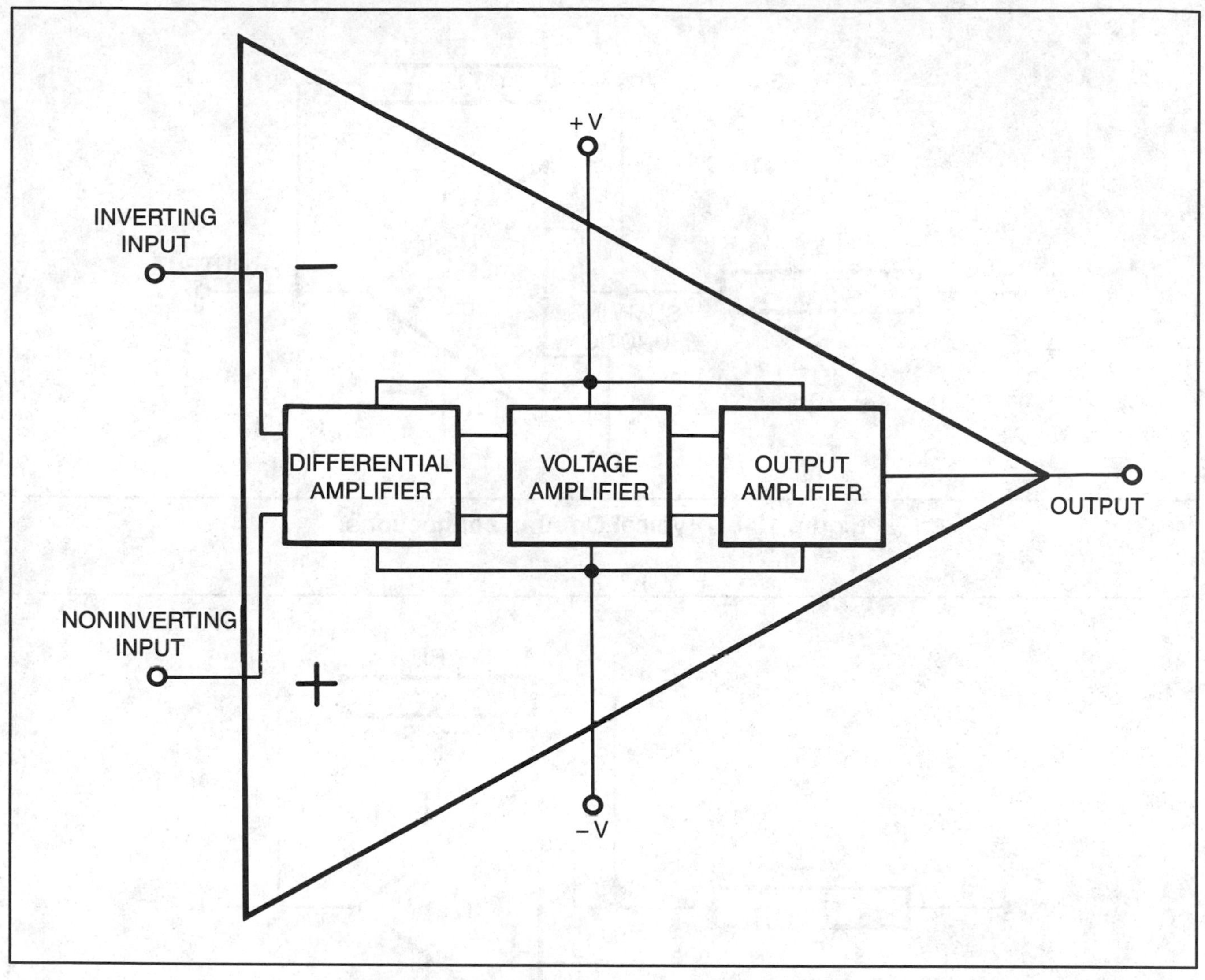

Figure 13-7. Basic operational amplifier.

Due to its small physical size, the IC Op Amp can output about 1 watt of power. However, most valve solenoids require from 5 to 40 watts of power for operation. Therefore, the output of the Op Amp is usually sent on to a Driver circuit, which is composed of large power transistors that can physically handle the current requirements of the solenoid. Op Amps can also be built from discrete components (not in an integrated circuit chip) and can, therefore, use power transistors as components of the output amplifier stage.

Open-Loop Gain

When an Op Amp is actually connected as shown in Figure 13-6, the signal applied to the input is amplified at the full gain of the Op Amp. For example, suppose that the Op Amp has a gain of 1,000,000 and is powered by a ±10 volt power supply. The maximum voltage output of the Op Amp is limited by its power supply to ±10 volts.

Since,

$$OUTPUT\ VOLTAGE = A \times INPUT\ VOLTAGE$$

Then,

$$INPUT\ VOLTAGE = \frac{OUTPUT\ VOLTAGE}{A}$$

Therefore,

$$INPUT\ VOLTAGE = \frac{10\ VOLTS}{1{,}000{,}000} = .000010\ VOLTS$$

This means that if we input as little as 10 microvolts, the Op Amp will give maximum output! Many applications call for amplification of very tiny voltages, but they are usually not encountered in electrohydraulics. When used in this open loop manner, the Op Amp is of little use, since the gain it gives is far too large. Fortunately, the gain of an Op Amp can be very closely controlled through the use of external feedback components.

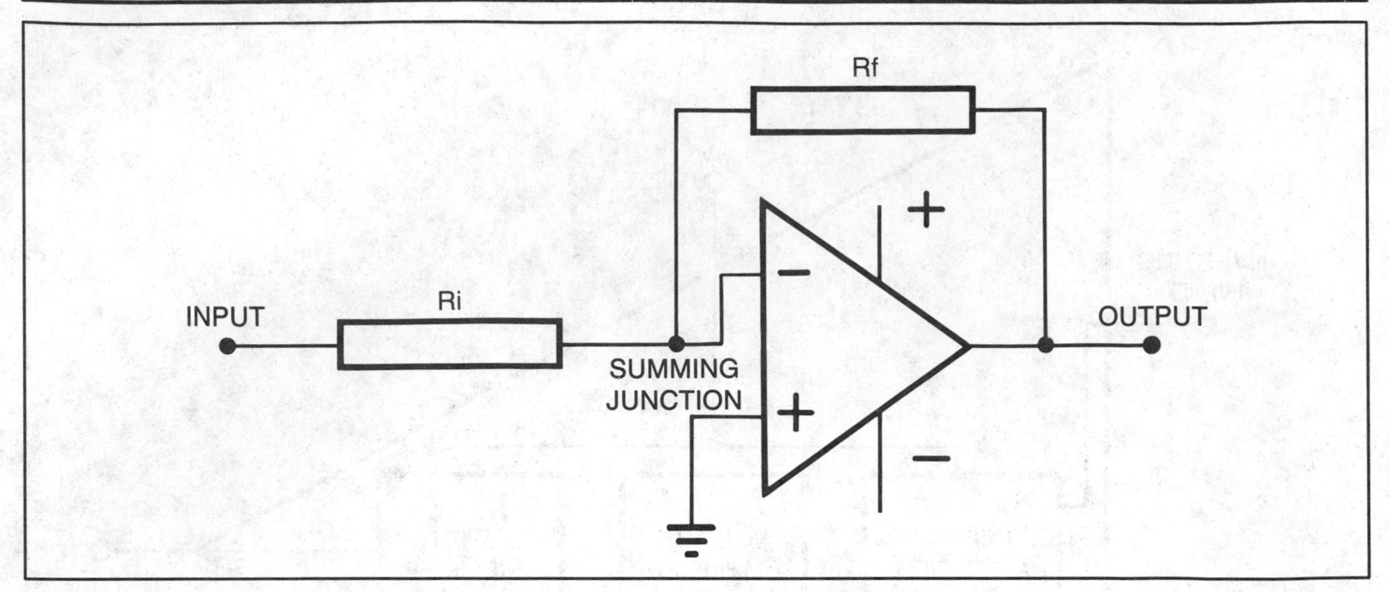

Figure 13-8. Typical Op Amp connections.

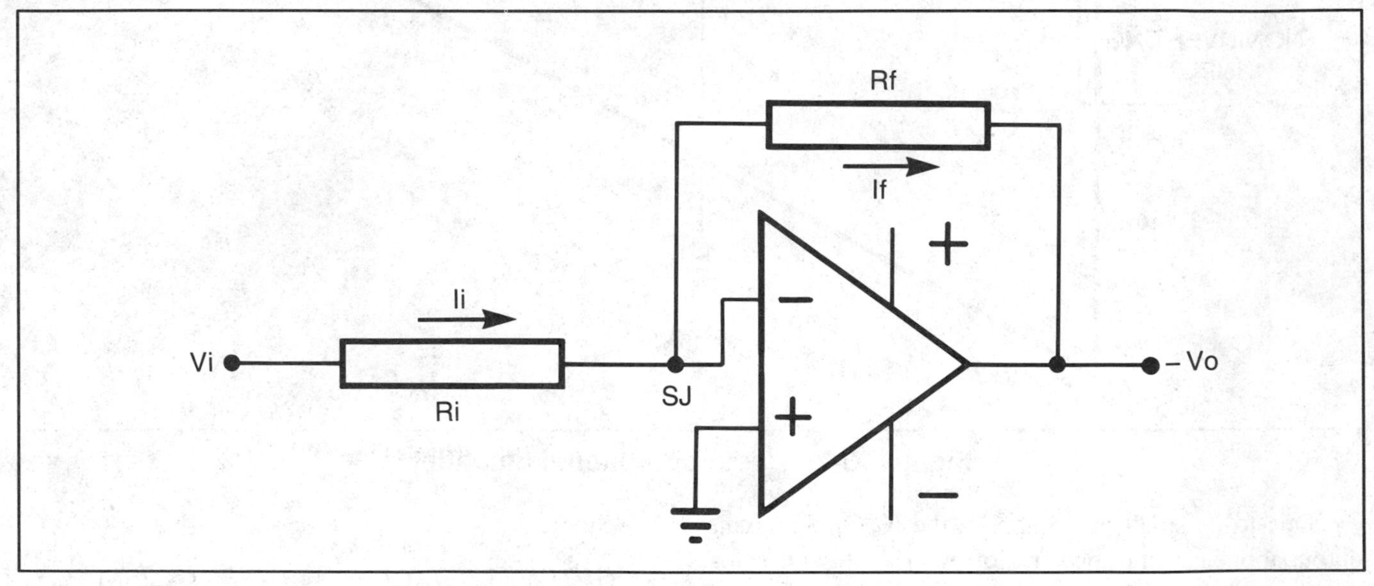

Figure 13-9. Typical Op Amp with voltage applied to input.

Closed-Loop Gain

A more tpyical Op Amp circuit is shown in Figure 13-8. Note that one resistor (R_i) is connected in series with the amplifier input, and a second resistor (R_f) connects the output to the summing junction. R_i is known as the input resistor, and R_f is the feedback resistor. This is known as the **Closed-Loop** configuration, since the feedback resistor forms a loop from output to input.

When connected in this manner, the Op Amp does a very useful thing. It will now attempt to maintain identical voltages at its inputs by changing its output. For example, since the positive (+) input is connected to ground (0 volts), the Op Amp will change its output so that the negative (–) input and the summing junction connected to it are also at 0 volts.

Op Amps are usually set up to invert the signal being amplified by connecting it to the negative input. This is done to simplify amplifier design. While it is possible to connect an input signal to the positive input and get a noninverted output, our discussion concentrates on the simpler and more common inverting arrangement.

Suppose that a voltage (V_i) is applied to the input as shown in Figure 13-9. Assuming that the summing junction is at zero volts, the current flow through the input resistor (I_i) can be calculated using Ohm's Law:

$$I_i = \frac{V_i}{R_i}$$

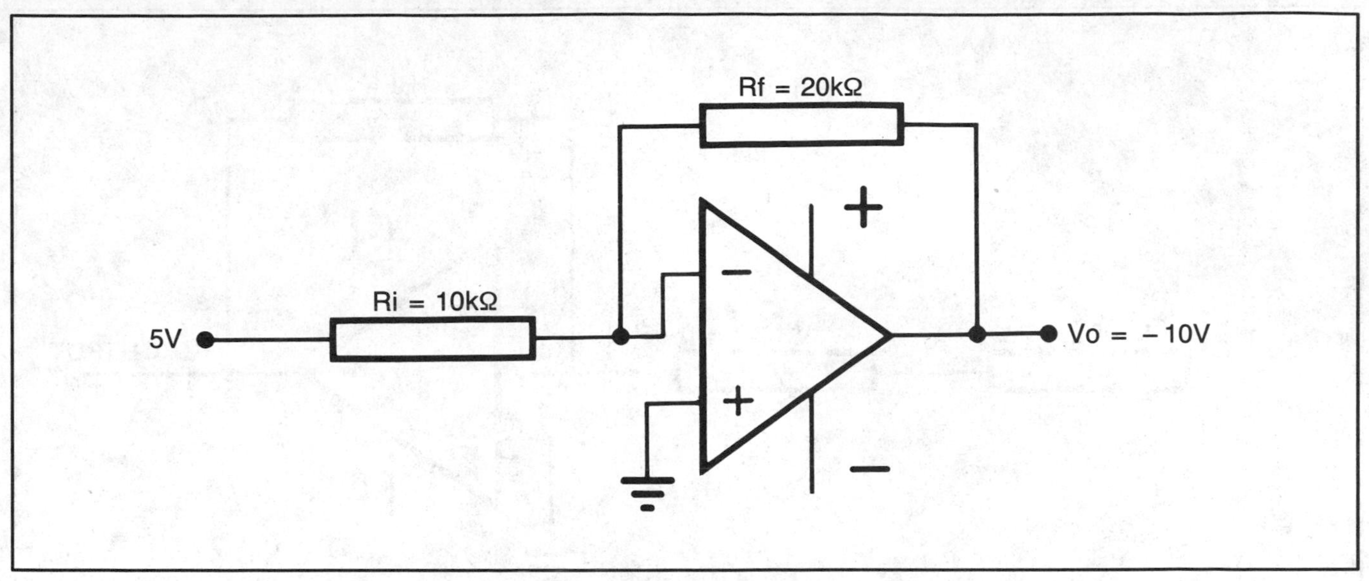

Figure 13-10. Example circuit.

Since no current can enter the input to the amplifier, and by using Kirchoff's Law (the algebraic sum of the currents at a junction is zero) at the summing junction, then:

$$I_i + I_f = 0$$
$$\text{so, } I_i = -I_f$$

Again using Ohm's Law:

$$I_i + I_f = 0$$
$$\text{so, if } I_i = -I_f$$
$$\text{then } \frac{V_i}{R_i} = \frac{-V_o}{R_f}$$
$$\text{or } V_0 = \frac{-R_f}{R_i} \times V_i$$

The multiplication factor between input and output is therefore R_f/R_i, which is known as the **Closed-Loop Gain** of the amplifier. Note that this gain value can be varied by changing the value of either R_f or, more commonly, R_i.

For example, refer to Figure 13-10. Suppose that in this circuit:

V_i = +5V
R_i = 10 kilohms (10,000)
R_f = 20 kilohms (20,000)

$$\text{then } V_o = -\frac{R_f}{R_i} \times V_i$$
$$V_o = -\frac{20,000}{10,000} \times 5$$
$$V_o = -10 \textit{ volts}$$

The gain of the amplifier is therefore -2. If the value of R_i is changed to 5 kilohms (5000) then:

$$V_o = -\frac{R_f}{R_i} \times V_i$$
$$V_o = -\frac{20,000}{5000} \times 5$$
$$V_o = -20 \textit{ volts}$$

so the gain is now -4.

Adjusting the value of R_i therefore adjusts the amplifier gain. If R_i is a variable resistor, the amplifier can be adjusted to give the required maximum output voltage when the maximum input signal voltage is applied. In practice, a fixed and a variable resistor would normally be used in series on the input to limit the maximum permissible gain (Figure 13-11).

For example, if R_i equals 5000 ohms, R_f equals 20,000 ohms, and VR can be adjusted from zero ohms to 5000 ohms, then the effective input resistance (R_i + VR) would be adjustable between 5000 and 10,000 ohms. This permits a gain adjustment of -2 to -4.

While the voltage gain of an amplifier may not be particularly high and may even be less than 1 in some cases, the significant point is that the current drawn from the input source is very small:

If V_i = 5 volts and
R_i = 10,000 ohms

$$I_i = \frac{V_i}{R_i} = 0.5 \textit{ mA}$$

However, the current drawn from the amplifier output can be relatively large, since this current is provided by the power supply.

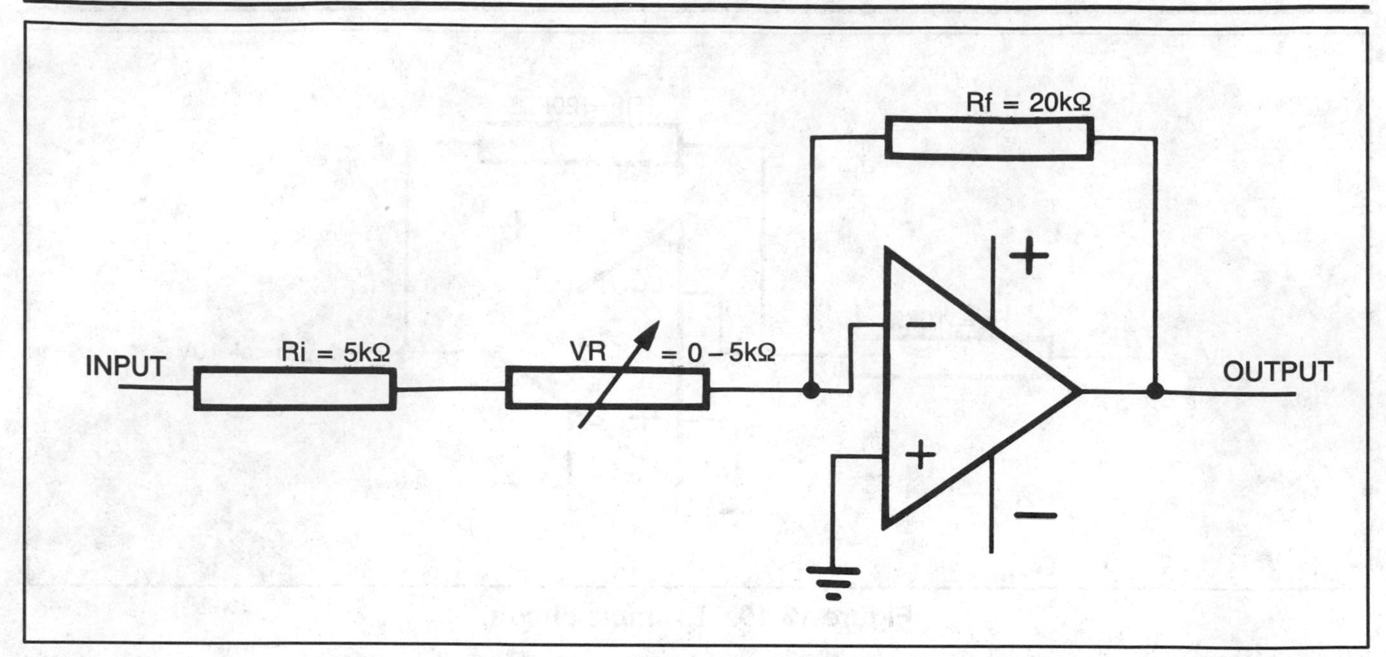

Figure 13-11. Op Amp with variable resistor at the input.

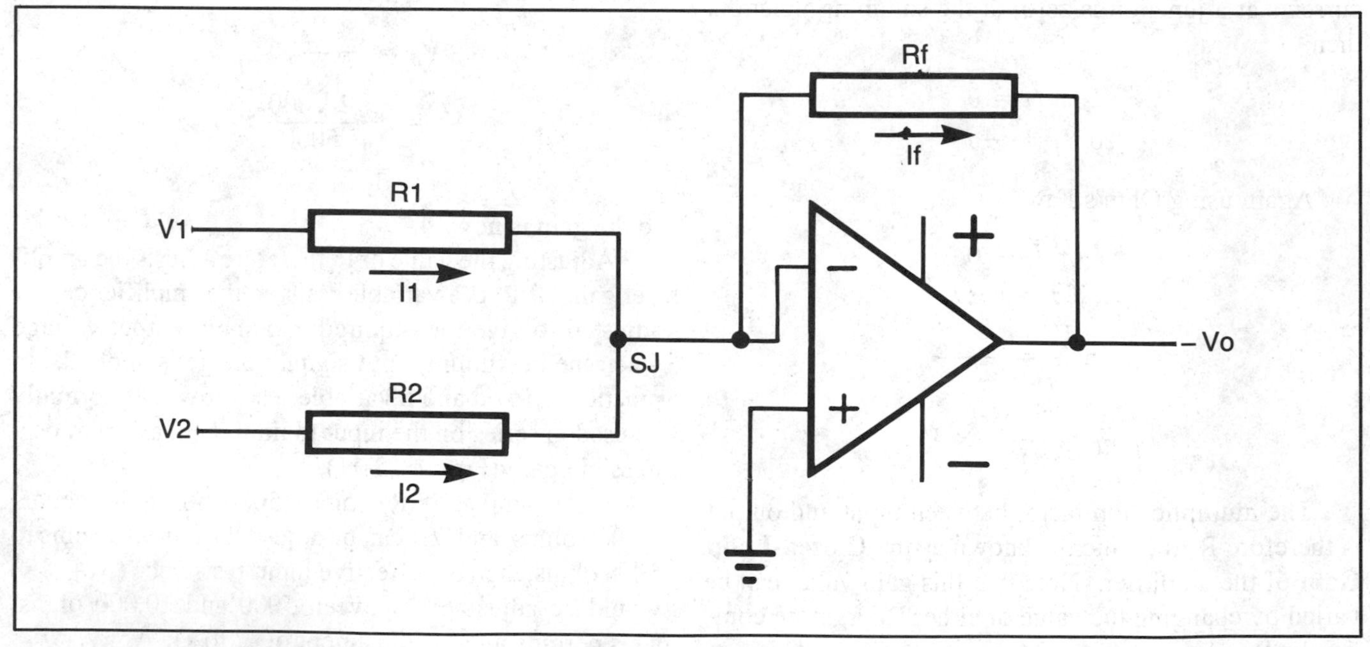

Figure 13-12. Op Amp with two inputs at the summing junction.

The important features of this arrangement are:

- The output voltage will normally be the opposite polarity to the input (a positive at the input produces a negative at the output and vice-versa).
- The output voltage will be proportional to the input voltage
(output voltage = input voltage x gain)
- The input current to the amplifier arrangement will be very small.
- The output current may be relatively large, since it is provided by the power supply.

If the amplifier has two inputs connected to the summing junction as shown in Figure 13-12, Kirchoff's Law can be applied at the summing junction. For example:

$$I_1 + I_2 + I_f = 0$$

$$\text{So, } \frac{V_1}{R_1} + \frac{V_2}{R_2} = -\frac{V_o}{R_f} \text{ (from Ohm's Law)}$$

$$\text{and } -V_o = \frac{R_f}{R_1} \times V_1 + \frac{R_f}{R_2} \times V_2$$

The amplifier output is the sum of the two input signals (V_1 and V_2) multiplied by their individual gains (R_f/R_1 and R_f/R_2). If R_1 equals R_2, the gains will be equal. In other cases, the gains may be different to make one signal more dominant than the other. The amplifier may have more than two inputs, but the same principle applies whatever the number. In other words:

$$-V_o = (A_1 \times V_1) + (A_2 \times V_2) + (A_3 \times V_3) + ...$$

where A = individual input gain.

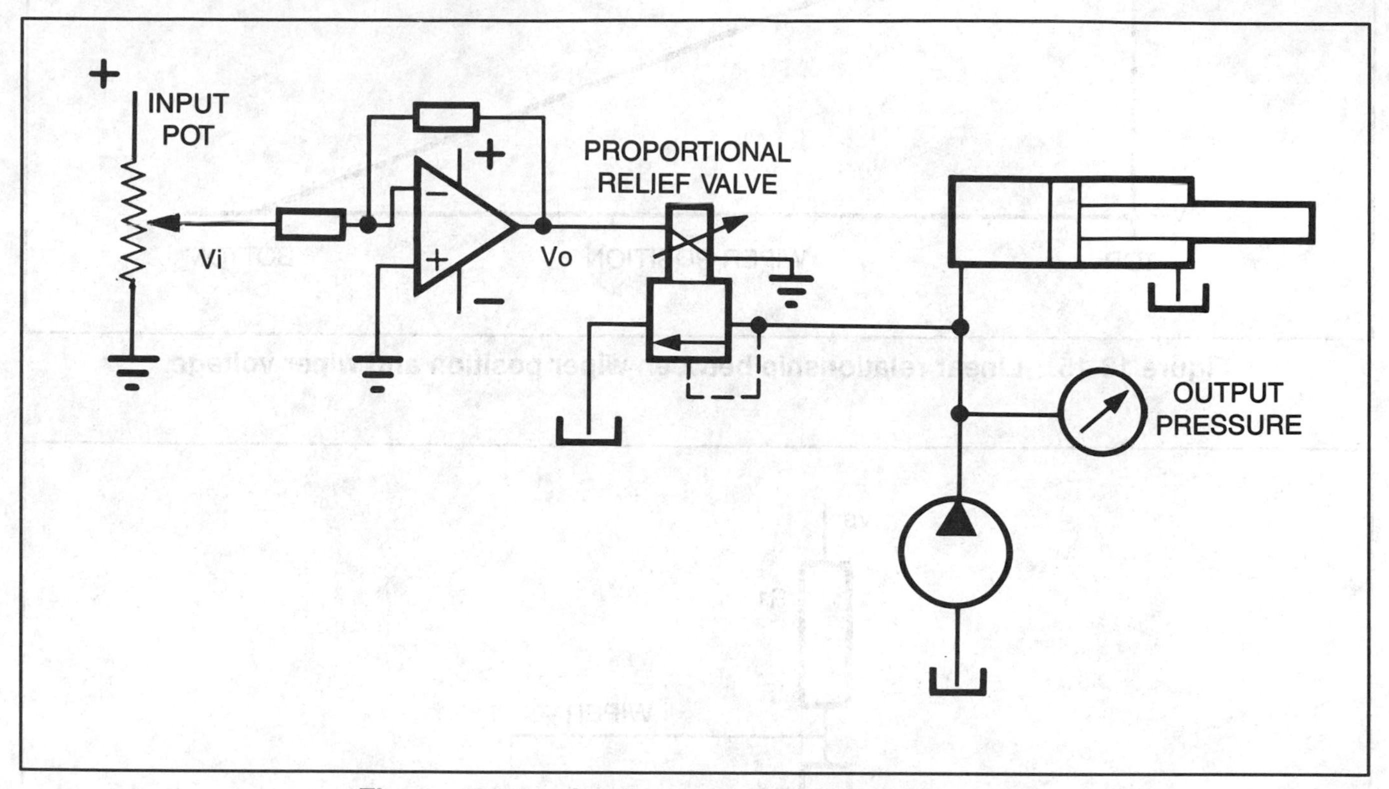

Figure 13-13. Op Amp with input potentiometer.

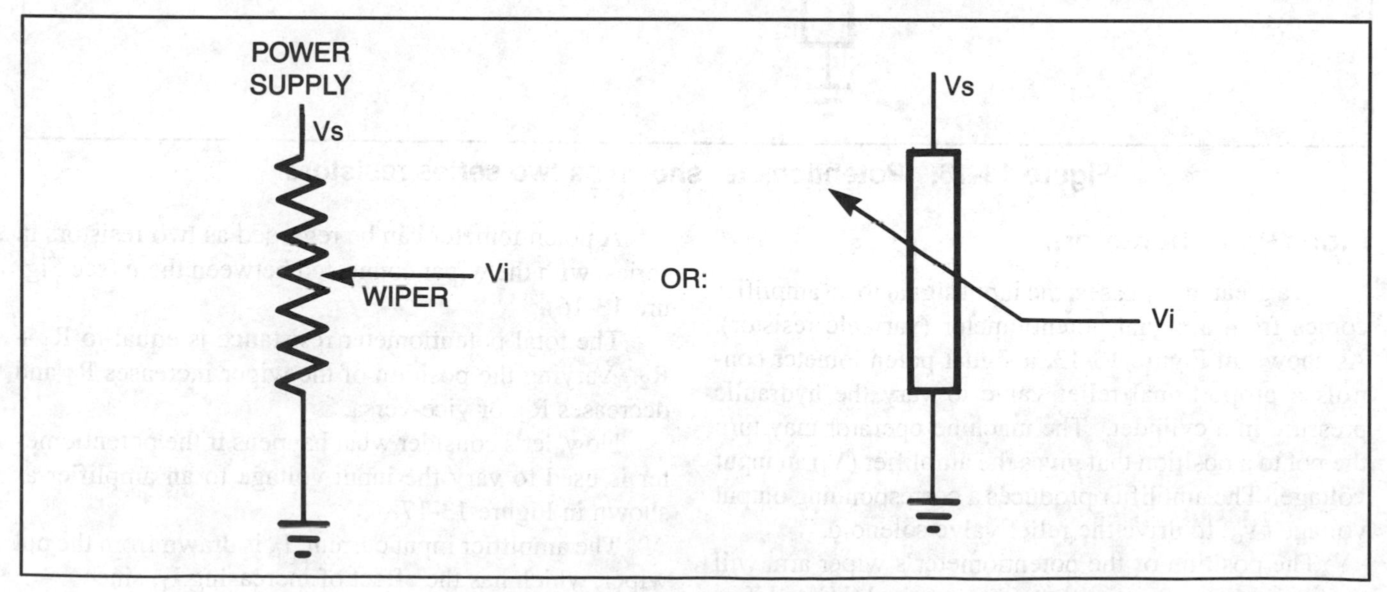

Figure 13-14. Voltage divider.

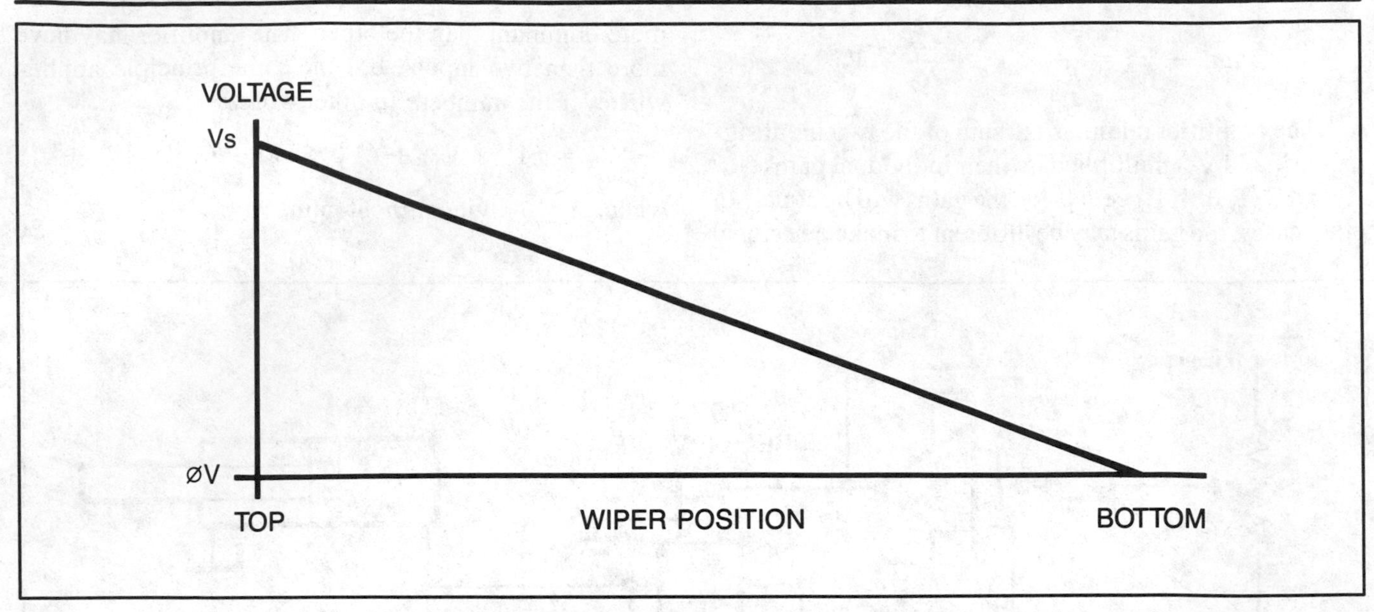

Figure 13-15. Linear relationship between wiper position and wiper voltage.

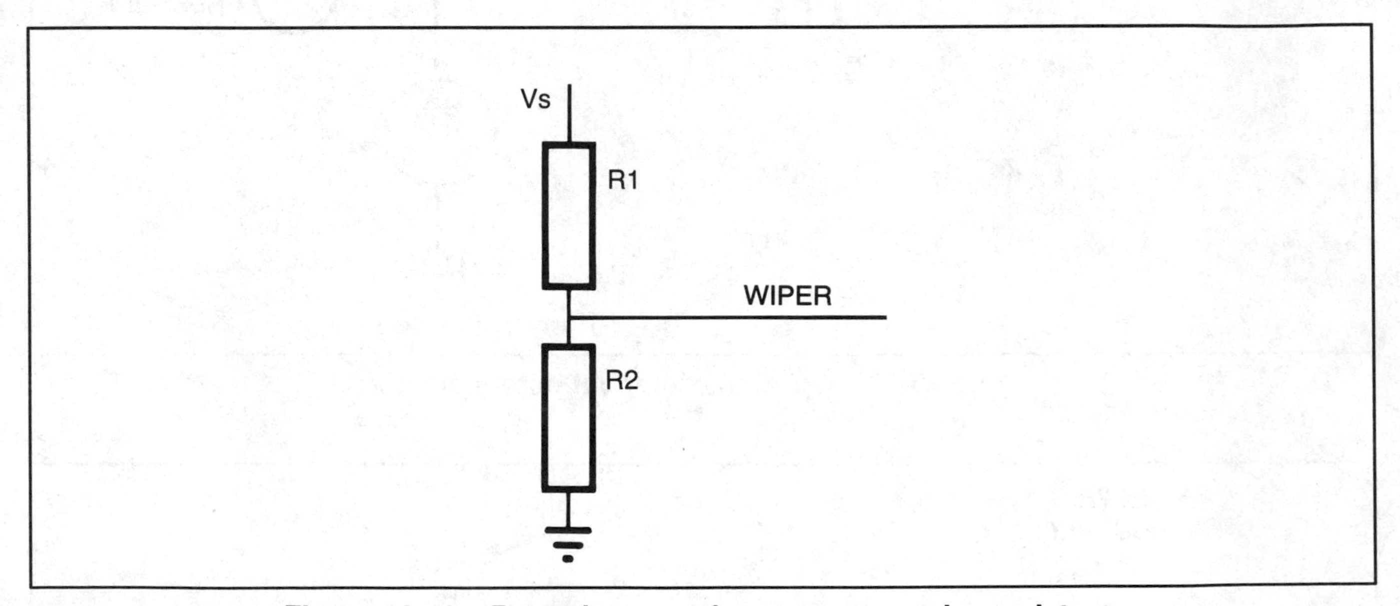

Figure 13-16. Potentiometer shown as two series resistors.

Input Potentiometers

In a great many cases, the input signal to an amplifier comes from a signal potentiometer (variable resistor). As shown in Figure 13-13, a signal potentiometer controls a proportional relief valve to vary the hydraulic pressure in a cylinder. The machine operator may turn the pot to a position that gives the amplifier (V_i) an input voltage. The amplifier produces a corresponding output voltage (V_o) to drive the relief valve solenoid.

The position of the potentiometer's wiper arm will determine the voltage at the wiper terminal (V_i) and if no current is drawn from the wiper, a linear relationship exists between the wiper position and the wiper voltage, as shown in Figure 13-15.

A potentiometer can be regarded as two resistors in series, with the wiper connected between them (see Figure 13-16).

The total potentiometer resistance is equal to R_1 + R_2. Varying the position of the wiper increases R_1 and decreases R_2, or vice-versa.

Now, let's consider what happens if the potentiometer is used to vary the input voltage to an amplifier as shown in Figure 13-17.

The amplifier input current, I_i, is drawn from the pot wiper, which has the effect of increasing I_1, since:

$$I_1 = I_i + I_2$$

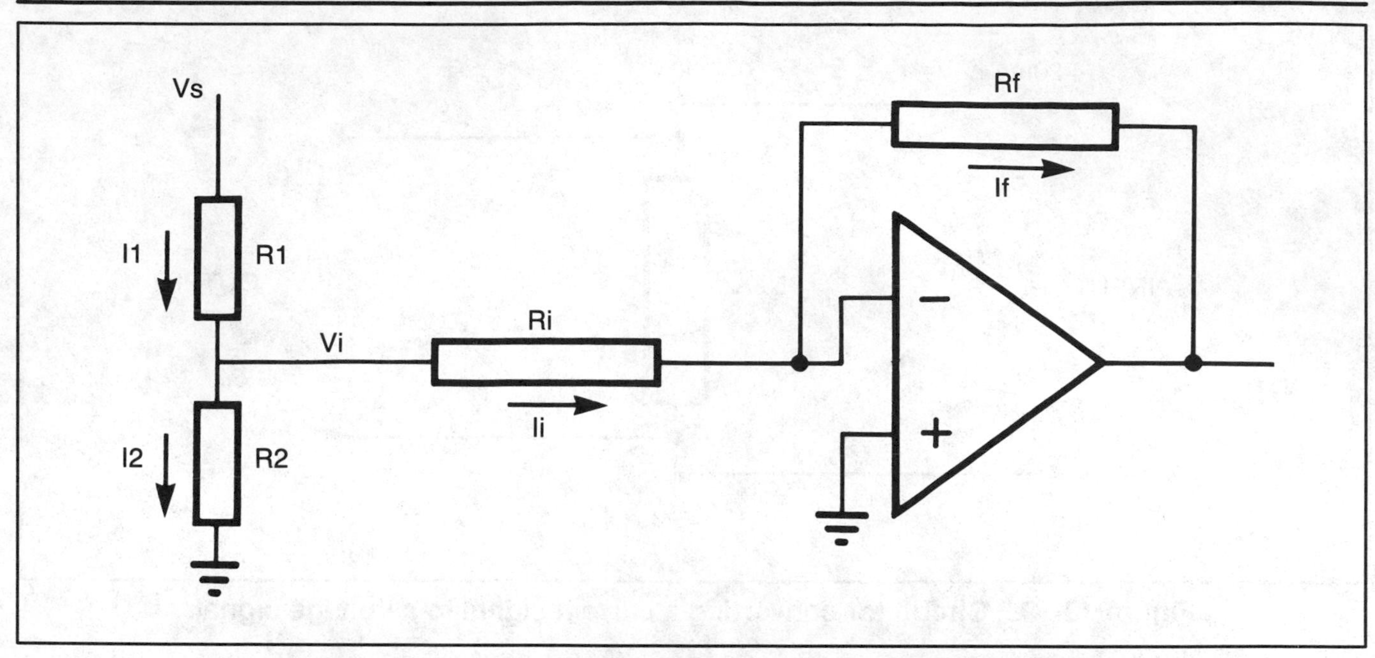

Figure 13-17. Potentiometer used to vary the input voltage.

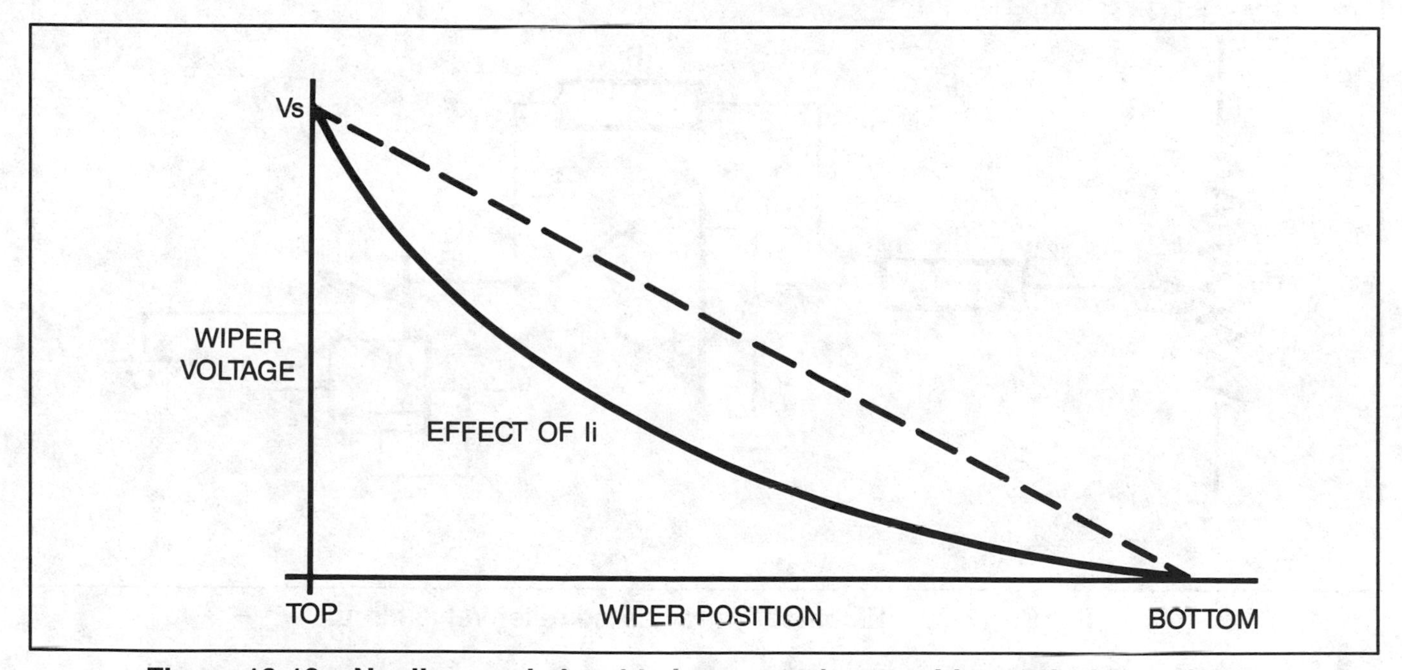

Figure 13-18. Nonlinear relationship between wiper position and wiper voltage.

An increased current through R1 creates a large voltage difference across R_1, and so V_i drops. This produces a nonlinear relationship between the wiper position and the wiper voltage as shown in Figure 13-18.

To reduce this nonlinearity, the wiper current (Ii) must be kept small in relation to I_1 and I_2. For practical reasons, however, I_1 and I_2 must be kept small to reduce power loss and to avoid heat generation; this means that I_i must be smaller still. I_i can be reduced by increasing the value of the amplifier input resistor R_i. A corresponding increase of R_f maintains the amplifier gain.

Although one might be tempted to simply make R_i and R_f very high resistances, there are practical limits to the values of these resistors. It was stated earlier that the inputs of an Op Amp draw virtually no current from the source. In fact, they do draw a tiny amount (nanoamps) of current, called input leakage current. Normally, it is an amount so small that it can be ignored. However, if R_i and R_f are too high in resistance, I_i will be too small.

If I_i is too small, then the assumption that no current enters the amplifier no longer holds true. In other words, the input leakage current starts to become significant rel-

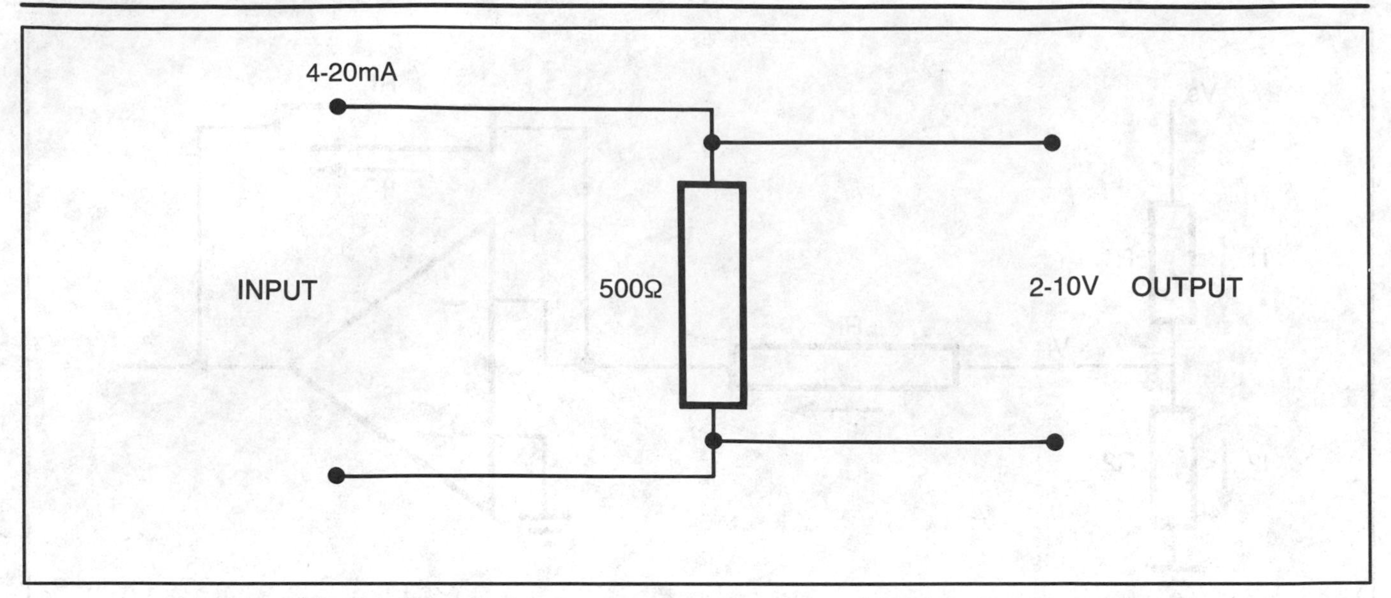

Figure 13-19. Circuit for converting a current signal to a voltage signal.

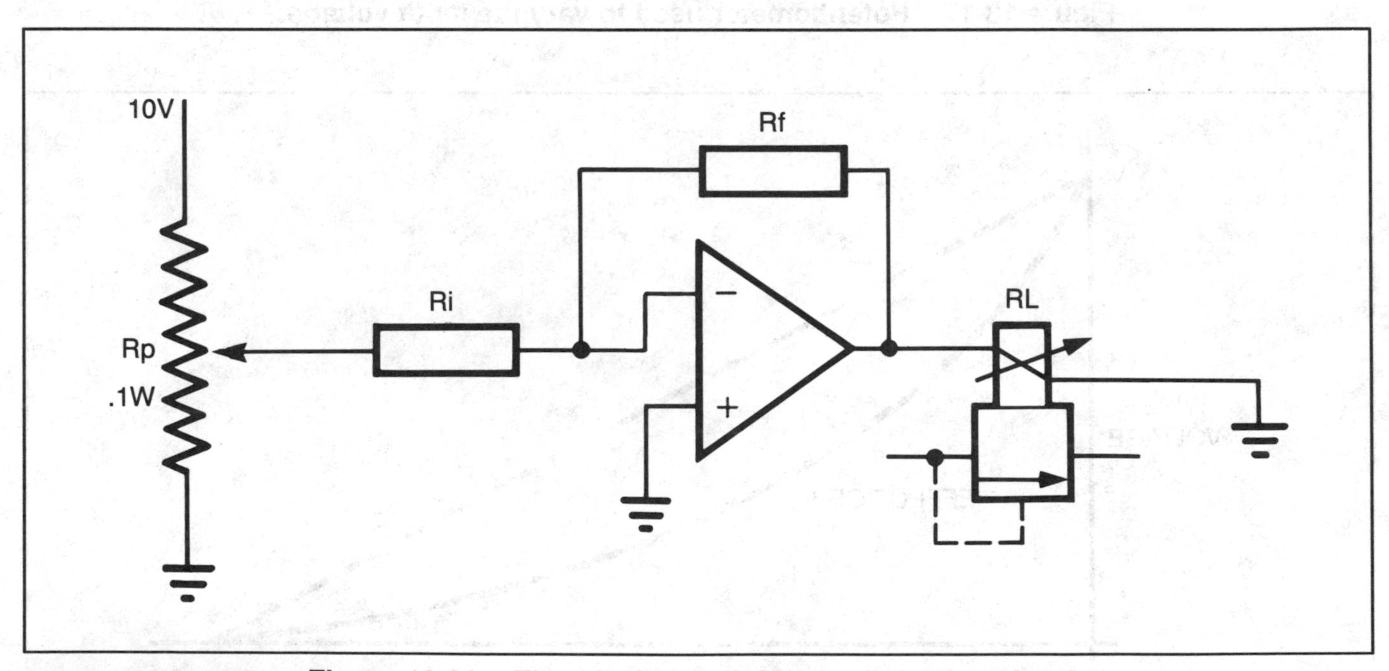

Figure 13-20. Electrically modulated relief valve circuit.

ative to I_i causing nonlinearities in the amplifier performance.

In practice, choosing a total potentiometer resistance ($R_1 + R_2$) that is no greater than one-tenth of the input resistance (R_i) maintains the nonlinearity within 2 to 3 percent, which is good enough for most applications.

The rule of thumb is:

$$POT\ RESISTANCE = 0.1 \times INPUT\ RESISTANCE$$

In many process control applications, a current input signal is preferred to a voltage signal to avoid voltage drops down long wires. Commonly, a 4 to 20mA signal is used. The 4mA represents a zero signal and 20mA represents a maximum signal. The 4mA signal is chosen to represent zero for two reasons:

1. It makes the system less sensitive to electrical noise, because any current less than 4mA is ignored.
2. It provides an indication of malfunction. If the input drops below 4mA it indicates a fault in the system (such as a broken lead), and the system is usually designed to react accordingly.

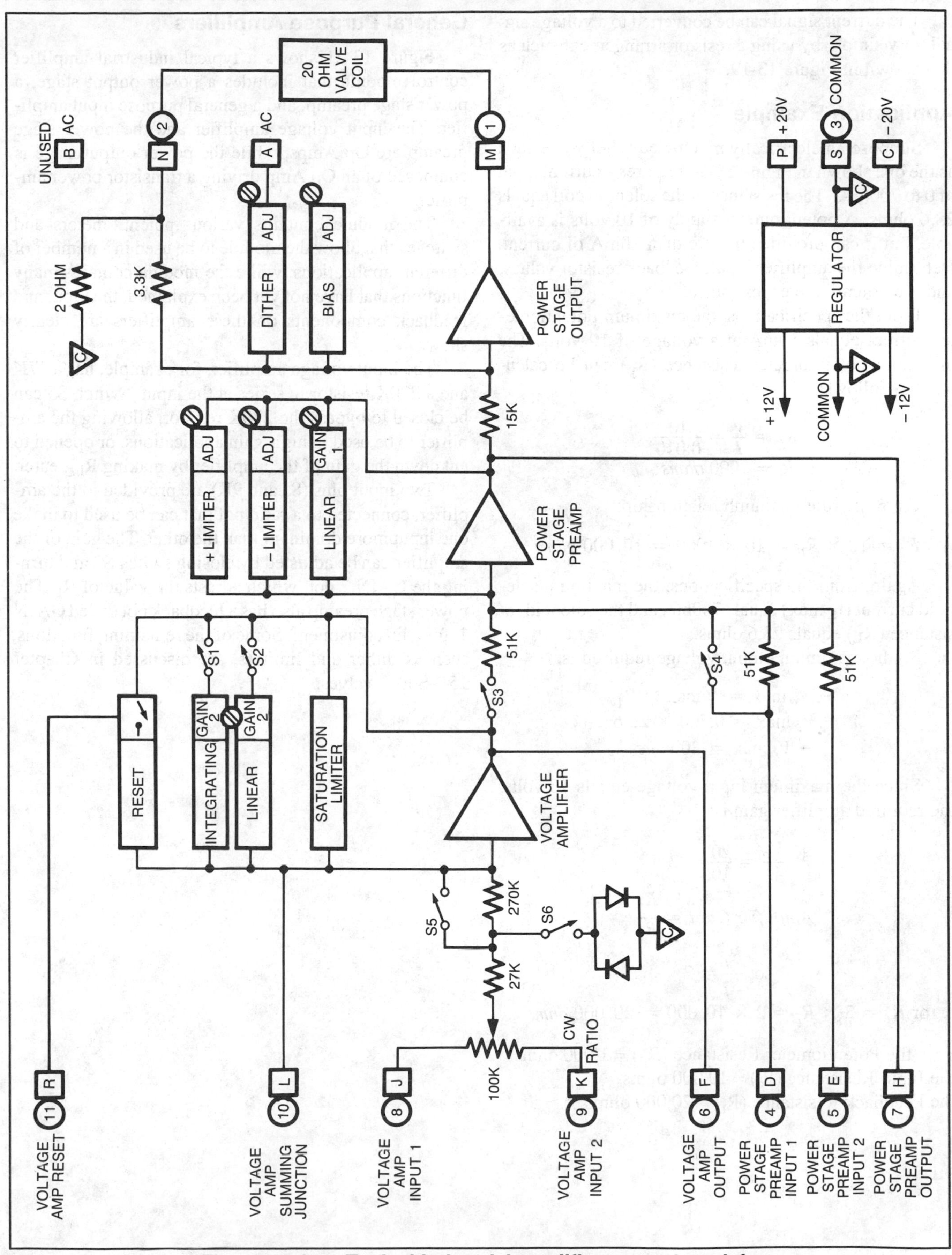

Figure 13-21. Typical industrial amplifier control module.

The current signal can be converted to a voltage signal very simply by using a resistor arrangement such as that shown in Figure 13-19.

Application Example

Suppose an electrically modulated relief valve such as the one shown in Figure 13-20 requires a current input of 0 to 700mA. The resistance of the solenoid coil equals 28.6 ohms. A potentiometer supply of 10 volts is available, but it can provide no more than 10mA of current. Determine the amplifier input, feedback resistor values, and the potentiometer resistance.

From the specifications, the maximum potentiometer current equals 10mA at a voltage of 10 volts. The minimum potentiometer resistance (R_p) can be calculated as follows:

$$R_p = \frac{V}{I} = \frac{10}{0.010}$$
$$R_p = 1000 \; ohms$$

From the rule-of-thumb relationship:

$$R_i = 10 \times R_p = 10 \times 1000 = 10,000 \; ohms$$

Again, from the specifications, the maximum solenoid current (I_L max) equals 700mA and the solenoid resistance (R_L) equals 28.6 ohms.

So the maximum output voltage required is:

$$-V_o \max = I_L \max \times R_1$$
$$-V_o \max = 0.700 \times 28.6$$
$$-V_o \max = 20 \; volts$$

Since the maximum input voltage equals 10 volts, the required amplifier gain is:

$$\frac{V_o}{V_i} = \frac{20}{10} = 2$$

$$Amplifier\ Gain = \frac{R_f}{R_i}$$

$$\text{so, } \frac{R_f}{R_i} = 2$$

$$\text{or } R_f = 2 \times R_i = 2 \times 10,000 = 20,000 \; ohms$$

So, the Potentiometer Resistance (R_p) = 1,000 ohms
the Input Resistance (R_i) = 10,000 ohms
the Feedback Resistance (R_f) = 20,000 ohms

General Purpose Amplifiers

Figure 13-21 shows a typical industrial amplifier control module that includes a power output stage, a power stage preamp, and a general purpose input amplifier. The input voltage amplifier and the power stage preamp are Op Amps, while the power output stage is composed of an Op Amp driving a transistor power amplifier.

The module contains various potentiometers and switches that allow the module to be used in a number of different applications. while the module contains many functions that have not yet been explained, the input and feedback components on these amplifiers are clearly shown.

The input voltage amplifier, for example, has a 27K and a 270K resistor in series at the input. Switch S5 can be closed to bypass the 270K resistor, allowing the amplifier to be used in high-gain applications, or opened to cut down the gain of the amplifier by making R_i greater.

Two input pins (8J and 9K) are provided to the amplifier, connected to a ratio pot that can be used to make one input more dominant than the other. The gain of the amplifier can be adjusted by closing switch S2 and turning the GAIN 2 pot, which adjusts the value of R_f. The power stage preamp also has a feedback pot called GAIN 1, for gain adjustment. Some of the remaining functions, such as dither and limiters, are discussed in Chapter 15—Servo Valves.

QUESTIONS

1. Name five input signal sources for controlling a proportional valve.
2. Why is an amplifier needed between the input signal source and the electrohydraulic device being controlled?
3. What device forms the basic element of most amplifier circuits?
4. Name the three terminals found on a transistor.
5. Where is the input signal applied in the most common amplifier configuration?
6. Define gain.
7. If the input voltage to an amplifier is 0.05 volts and the output voltage is 5.0 volts, what is the gain?
8. If an amplifier has a gain of 50 and the input voltage is 2 volts, what is the output voltage?
9. What are the four key characteristics of operational amplifiers?
10. Name the three basic stages within an integrated circuit operational amplifier.
11. What elements determine the operational characteristics of an operational amplifier?
12. What is the formula for the multiplication factor when determining closed-loop gain in an Op Amp?
13. If the input voltage (V_i) to an Op Amp is 4 volts, the value of the input resistor (Ri) is 5000 ohms, and the value of the feedback resistor (R_f) is 20,000 ohms, what is the output voltage (V_o)?
14. What is the closed-loop gain value in Question 13?
15. What device is commonly used to vary the value of R_i or R_f and, therefore, the value of closed-loop gain in an operational amplifier circuit?
16. What is the rule of thumb for choosing a potentiometer resistance at the input to an operational amplifier to reduce nonlinearity of the output signal?
17. Why is a current input signal preferred to a voltage signal in many process control applications?
18. With a current input signal, why is 4mA chosen to represent a zero signal?

CHAPTER 14
PROPORTIONAL VALVES

Proportional valves fill a gap between conventional solenoid valves and servo valves. Like conventional solenoid valves, proportional valves are simple in design and relatively easy to service. However, unlike conventional on/off DC solenoid valves, they can assume an infinite number of positions within their working range. Proportional valves have many of the control features, without the design complexity and high cost, of the more sophisticated servo valves. They are used in applications that require moderately accurate control of hydraulic fluid.

PROPORTIONAL SOLENOID VALVES

Proportional valves control and vary pressure, flow, direction, acceleration, and deceleration from a remote position. They are adjusted electrically and are actuated by proportional solenoids rather than by a force or torque motor.

This chapter concentrates on three types of proportional valves:

- Proportional pressure control valves, including relief and reducing valves with electronically adjusted pressure settings.
- Proportional flow control valves, which are proportional valves with electronically varied flow rate through the valve.
- Proportional direction control valves, which have electronic controls for flow rate as well as for flow direction.

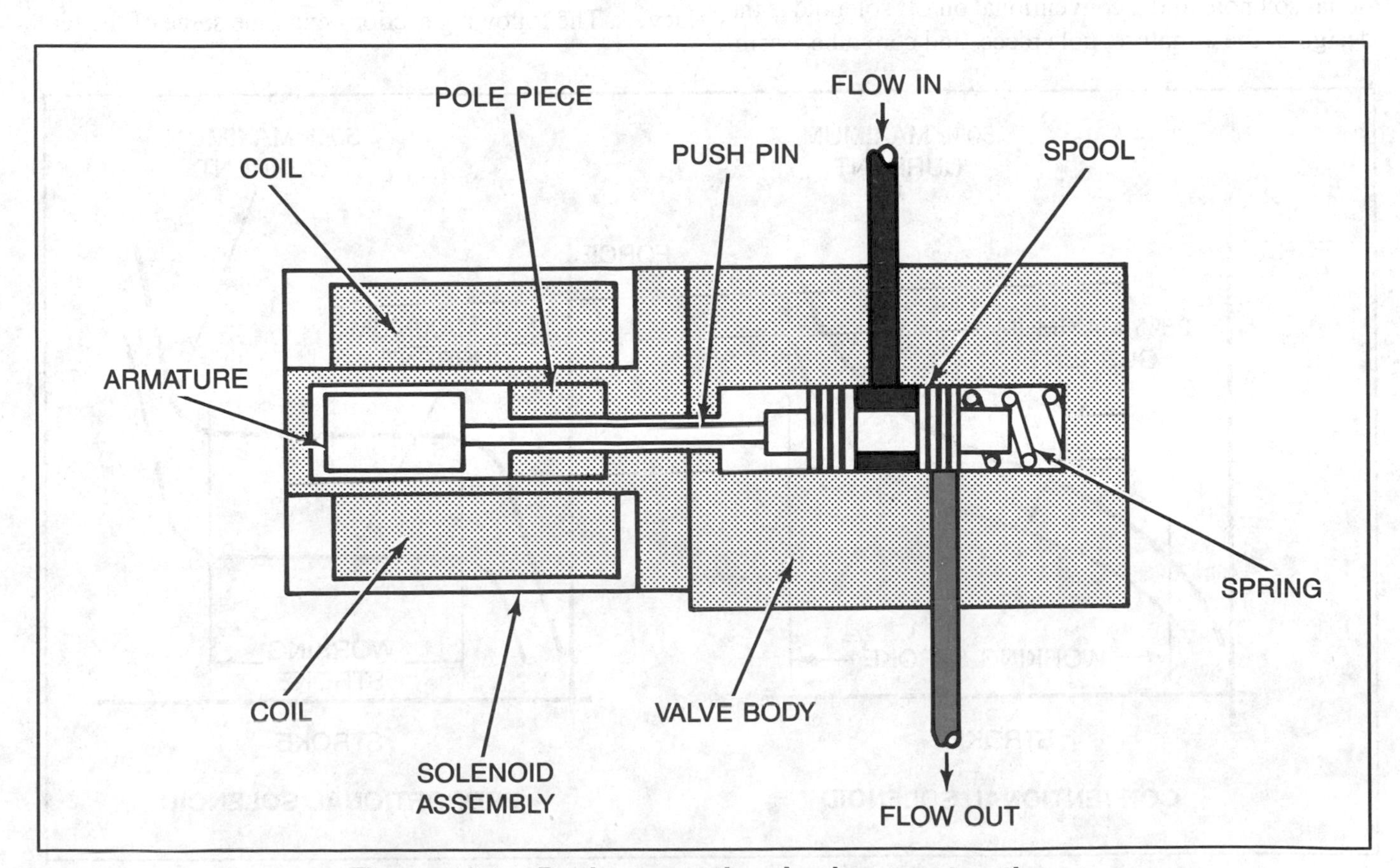

Figure 14-1. Basic proportional valve construction.

Remote Control of Hydraulic Flow

Figure 14-1 is a basic proportional solenoid valve that provides remote control of hydraulic flow. As you can see, proportional valves are spool-type valves. Typically, the spool is spring-centered, or spring-offset as shown in Figure 14-1.

Output Flow Proportional to Input Signal

Unlike conventional on/off solenoid valves, the proportional solenoid current can be varied to make the spool move variable distances. The term proportional describes the valve's operation. The valve spool moves in proportion to an electrical signal applied to the solenoid; thus, the electrical signal is converted to a mechanical spool motion. In other words, the output flow is proportional to the input signal. By varying the input signal, the solenoid adjusts the spool movement to vary the flow through the valve.

Constant Force Solenoid

On a solenoid, a magnetic force is created when a current is passed through the solenoid coil. This force pulls the solenoid armature toward the pole piece. A push pin attached to the armature then transmits the force to the valve spool. The major difference between a proportional solenoid and a conventional on/off solenoid is the design of the armature, pole piece, and core tube assembly. The proportional solenoid is shaped in a manner that delivers a more constant force over the entire working stroke.

In Figure 14-2, you can see that the proportional solenoid delivers a constant force, regardless of armature position. The coil current alone determines the amount of force transmitted to the valve spool. The solenoid force moves the spool until a balance is achieved between the solenoid force and the valve's spring force.

By varying the current, the solenoid can force the spool to assume a position anywhere within its working range. This operation is called spring feedback, because the valve spring is the only feedback for the solenoid force. Because this operation does not take into account other forces that might affect spool position (such as frictional and flow forces) it is not appropriate for applications that require a high degree of valve performance.

ASSOCIATED ELECTRONIC DEVICES AND CONTROLS

Proportional solenoid valves are used with electronic control amplifiers, which provide the power necessary to operate the valve and perform additional functions. Proportional valves can be used in both open- and closed-loop control systems. These two control methods perform a similar function at two different performance levels. The following sections examine some of the fea-

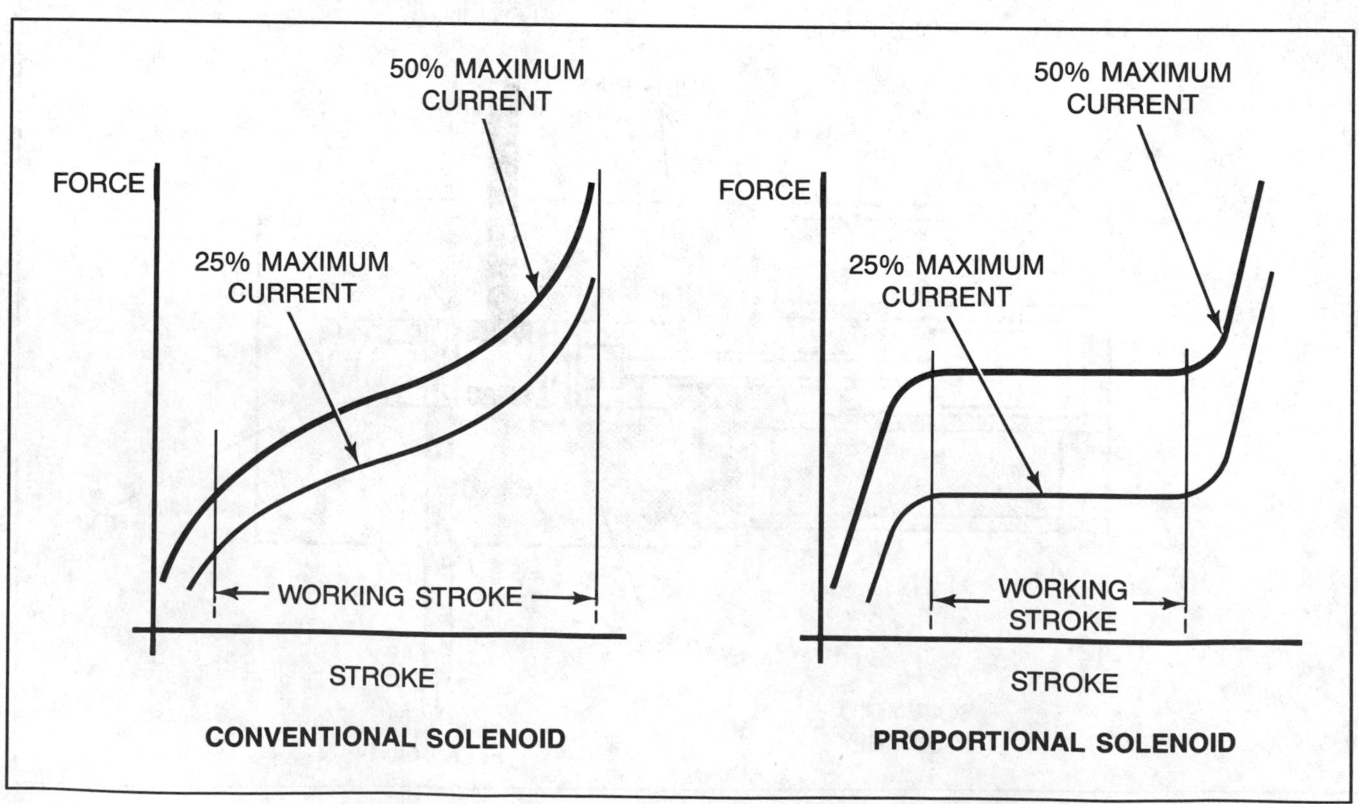

Figure 14-2. Conventional versus proportional solenoid force.

tures of electronic control amplifiers and how proportional valves operate within feedback and nonfeedback control systems.

Electronic Control Amplifier

There are many different electronic control amplifiers for proportional valves, but each one provides a portion of one group of functions. The initial input signals that control proportional solenoid valves come from a variety of sources, including:

- Potentiometers
- Temperature sensors
- Pressure transducers
- Tachogenerators
- Microprocessors

These devices are all low power sources in terms of voltage and/or current. Their power must be amplified before they can operate a proportional valve. To increase power to the required level, a small voltage from one of the sources listed above is input to an amplifier. The amplifier produces a correspondingly larger current flow that is transmitted to the solenoid. This current flow enables the valve spool to move.

Control Amplifier Functions

Several important control amplifier functions, such as deadband compensation, gain, and dither, are described in this section.

Deadband Compensation. Deadband is primarily caused by the spool overlapping the valve ports. Spool overlap is shown in Figure 14-3A. As the figure indicates, when the spool is in the centered position, the spool must be moved a certain amount before oil can flow either from "P" to "A" or from "P" to "B." In the example shown, an input signal of 0.3V must be applied to the solenoid to move the spool past the deadband region and to allow flow through the ports. (In Chapter 15, you will learn that close tolerance manufacturing practices reduce servo valve overlap to almost zero.)

Many proportional valve amplifiers have deadband compensation adjustments to electronically boost the amplifier's output when it approaches the deadband region so that the amplifier "skips over" it. Deadband compensation produces the spool motion shown in Figure 14-3C. This adjustment dramatically improves valve performance in the deadband region, often producing almost ideal results.

Gain. An amplifier's gain, expressed as a multiplication factor, is a measure of the ratio between the amplifier's small input signal and its large output current flow to the valve. The output voltage is proportional to the input voltage. Gain is determined in the following way:

$$Gain = A = \frac{output\ voltage}{input\ voltage}$$

Amplifier gain can be in the area of 10^6. In other words:

$$output\ voltage = input\ voltage \times 1,000,000$$

However, smaller gains are more typical in actual use.

Dither. Dither is a high-frequency (50-100Hz), low-amplitude AC signal used to offset the effects of a condition known as hysteresis. Hysteresis is caused by friction between a proportional valve spool and bore, as well as by the inertia of the spool itself. Friction can be affected by manufacturing tolerance, thermal expansion, wear, fluid viscosity, contamination, and other various factors.

When there is friction in a proportional valve, the solenoid force has to overcome the spool spring force and the additional frictional force. Because frictional forces are not evenly distributed within the valve, the same input signal can produce a different amount of spool movement when the signal is increasing compared to when the signal is decreasing (Figure 14-4). To offset the effects of hysteresis, a dither AC signal is superimposed onto the valve's DC signal (Figure 14-5).

+ and – Limiter. Some amplifiers have a + and – limiter, which allows adjustment of the maximum power output to the proportional valve. This limitation prevents valve solenoid damage caused by the accidental application of too much power.

Ramp Functions. As you have learned, one advantage of proportional valves is that spool speed, in addition to position, can be controlled electronically. This means that the valve can also control the speed, acceleration, and deceleration of the final control element, such as a cylinder.

To control the speed of the valve spool movement, a gradually increasing or decreasing signal (a ramp function) is fed to the control amplifier. The ramp function is illustrated in Figure 14-6. If a switch applies a stepped input to the input of the ramp function (View A), it will output a voltage that is shaped like that shown in View B. Amplifiers like the one shown in the figure contain ramp functions with separately adjustable acceleration and de-

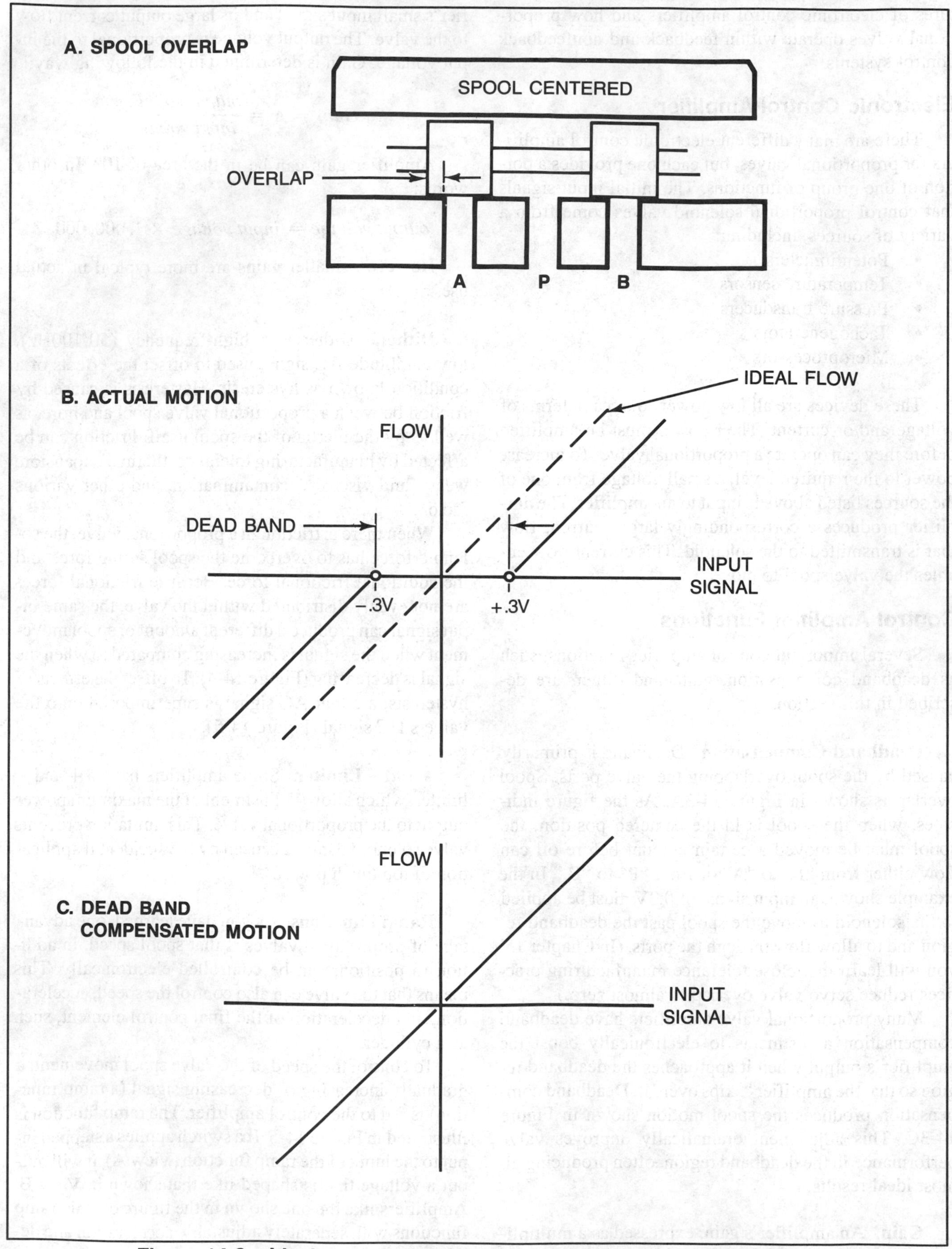

Figure 14-3. Ideal, actual, and deadband compensated spool motion.

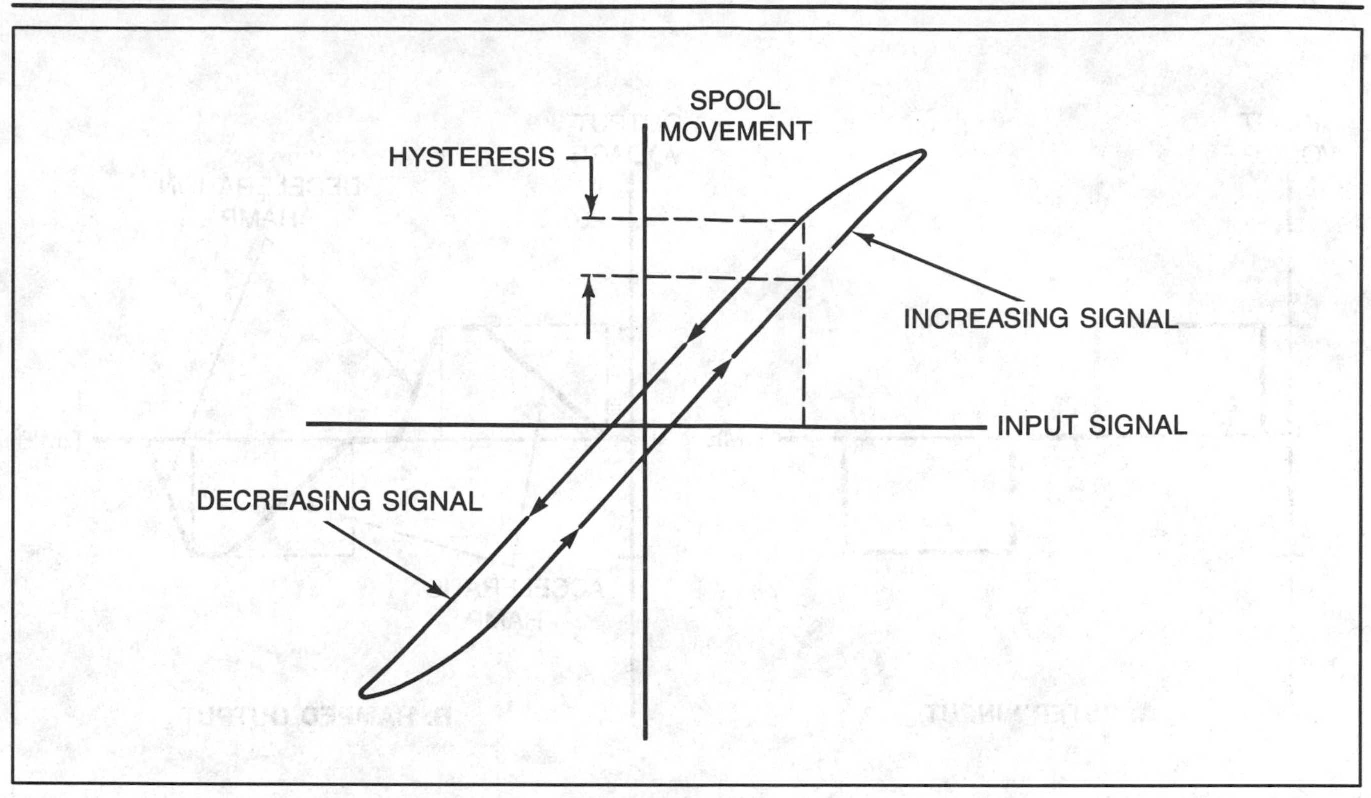

Figure 14-4. Hysteresis causes the same input signal to produce a different amount of spool movement when the signal is increasing compared to when it is decreasing.

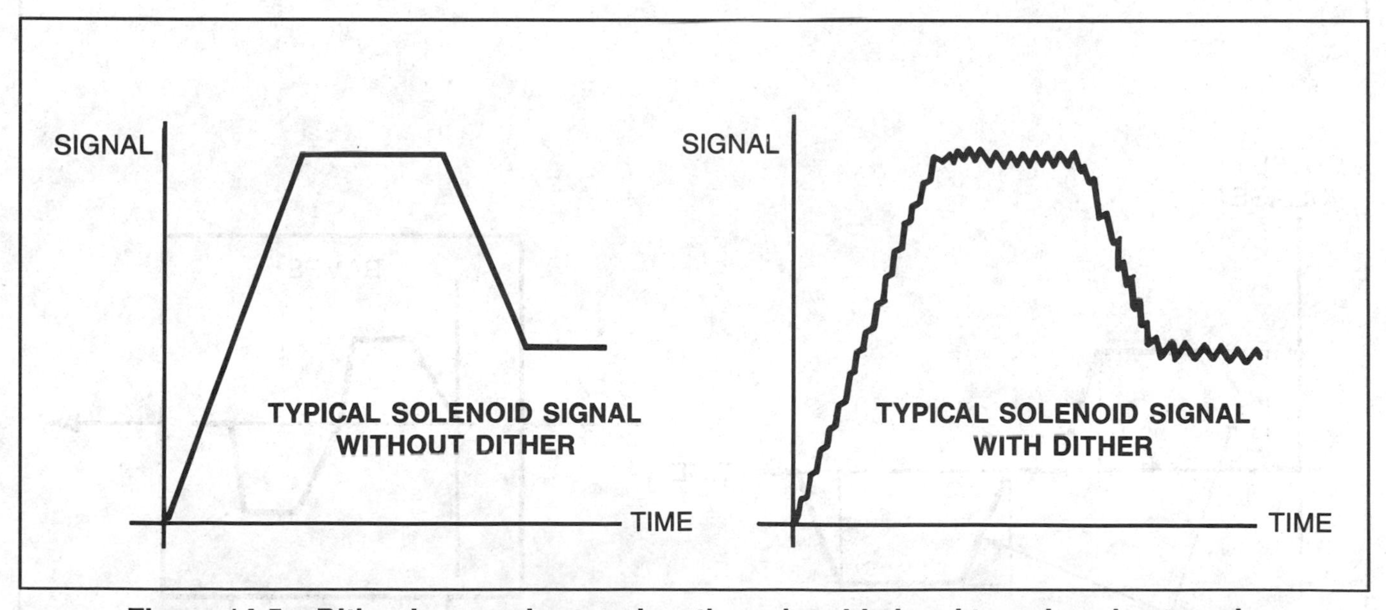

Figure 14-5. Dither is superimposed on the solenoid signal to reduce hysteresis.

celeration ramps; others have only one adjustment tha-thas the same effect on both the acceleration and deceleration side. View C of Figure 14-6 illustrates how adjusting the acceleration ramp results in faster acceleration.

View D of Figure 14-6 shows a typical symbol for a ramp function in an electronic block diagram. The symbol shows the waveform produced and indicates the presence of potentiometers for acceleration and deceleration adjustments.

Drive Enables. Some amplifier cards have an enable function, which requires a specified voltage to be present at the enable connection before the output stage of the card will operate. This can be connected to an emergency stop switch so that if the enable signal is lost, the amplifier immediately produces a zero output and the

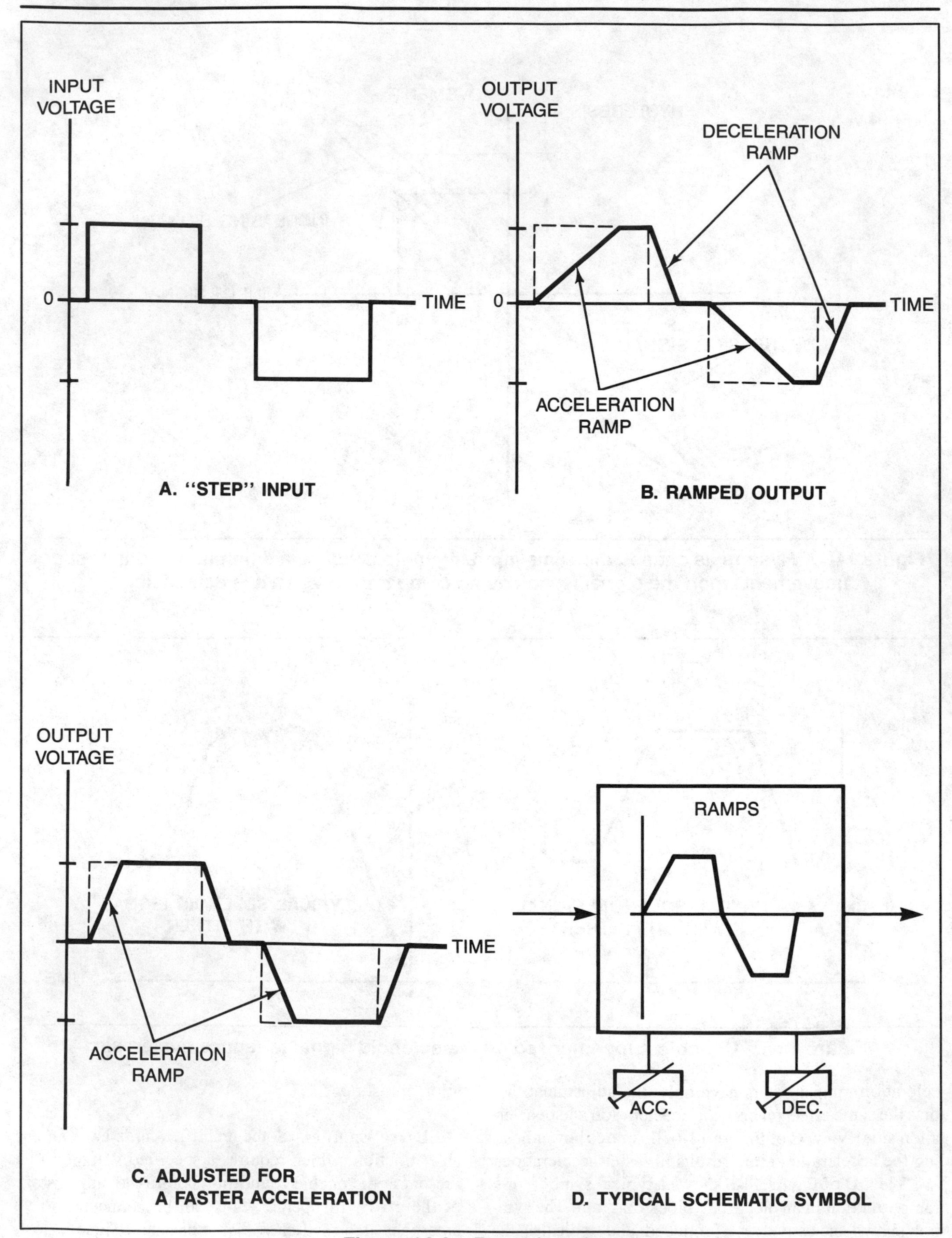

Figure 14-6. Ramp function.

valve responds accordingly. Normally, it is not recommended that an emergency stop switch be in the power supply line, because the stored charge in capacitors may maintain the valve signal for some time after the switch is opened.

Current Feedback. In a solenoid, passing a current through the coil generates heat. This heat increases the coil's resistance. For example, a solenoid coil may have a resistance of 26 ohms at 68°F (20°C) and a resistance of 38 ohms at 185°F (85°C). When the solenoid resistance changes, the solenoid current and the valve setting also change. A 50-percent increase in coil resistance as the coil heats up results in a 33-percent reduction in the valve setting.

To compensate for the negative effects of heated coils, some amplifiers have current feedback, in which a low value (1-2 ohms) current feedback resistor is added in series with the solenoid coil. The amplifier feedback is taken from a point between the solenoid coil and the current feedback resistor and fed back to the summing junction. This allows the solenoid current to be proportional to the input voltage and independent of solenoid resistance.

Feedback Connection. Figure 14-7 is a signal arrangement for a typical power amplifier that is used with some proportional valves. In the upper left of the figure (pins z22 and b14) are the feedback connections for a linear variable differential transformer (LVDT), which we will discuss later in this chapter.

Pulse Width Modulation. When an infinitely variable DC signal is used to operate a proportional valve, the control amplifier has to reduce the voltage from the power supply down to the voltage required by the valve solenoid at any given time. As shown in Figure 14-8, the control amplifier's output transistor acts like a variable resistor. The full solenoid coil current, which may be several amps, also passes through the amplifier's output transistor. High current combined with a relatively large voltage drop produces heat, which wastes energy. Also, a relatively large heatsink is required to dissipate the heat created, which requires considerable space on the electronic card.

A technique called pulse width modulation is used in some amplifiers to prevent the creation of heat caused by large voltage drops. With pulse width modulation, the output transistor is used as an on/off switch to feed the valve solenoid with a series of on/off pulses at a constant power supply voltage. The transistor is either fully on or fully off and, therefore, generates much less heat than that created with a DC output signal. (When the transistor is fully on, the voltage drop across it is very low, so it consumes very little power. When the transistor is off, no current flows through it, so it consumes no power at all.)

The pulses are kept at a constant frequency (typically 1 kHz), which is so fast that the solenoid cannot respond to the individual pulses. However, it does respond to the average voltage level of the pulses. The length of the on time in relation to the off time determines the voltage level (Figure 14-9).

Response time can be further improved by setting the pulsed voltage to twice the solenoid's rated voltage, so that the maximum signal level is achieved when the on and off pulses are equal in duration (Figure 14-10).

Nonfeedback Control

Proportional solenoid valves and electronic control amplifiers can be used in nonfeedback systems for applications that require smooth control of actuator speed, but only moderately accurate flow control. As you can see in Figure 14-11, a nonfeedback control arrangement has no position sensor to measure the actual movement of the spool. In addition, no feedback signal is sent back to the amplifier for comparison with the command signal. As you learned in our earlier discussion of spring feedback, the valve's spring force, which indicates the spool movement, is the only feedback provided.

Feedback Control

When more accurate control is needed, a position sensor can be attached to the spool. The sensor sends a signal back to a summing junction, where it is compared to the original input signal. This greatly improves valve performance.

As the block diagram in Figure 14-11 shows, the input signal is sent first to the amplifier, where the power is amplified. Next, the amplifier transmits an output signal to the solenoid, which then transmits a force to the proportional valve spool, causing the spool to move. There, a position sensor, typically an LVDT, measures the actual valve spool movement. The feedback signal (voltage) is sent back to the summing junction at the amplifier.

At the summing junction, the feedback signal is compared to the input signal. The difference between the two signals produces an error signal. The error signal leaves the summing junction and goes to the voltage amplifier. The amplifier output to the solenoid then changes to reflect the new error signal. This loop continues until the feedback signal balances with the input signal, and the spool reaches the specified position.

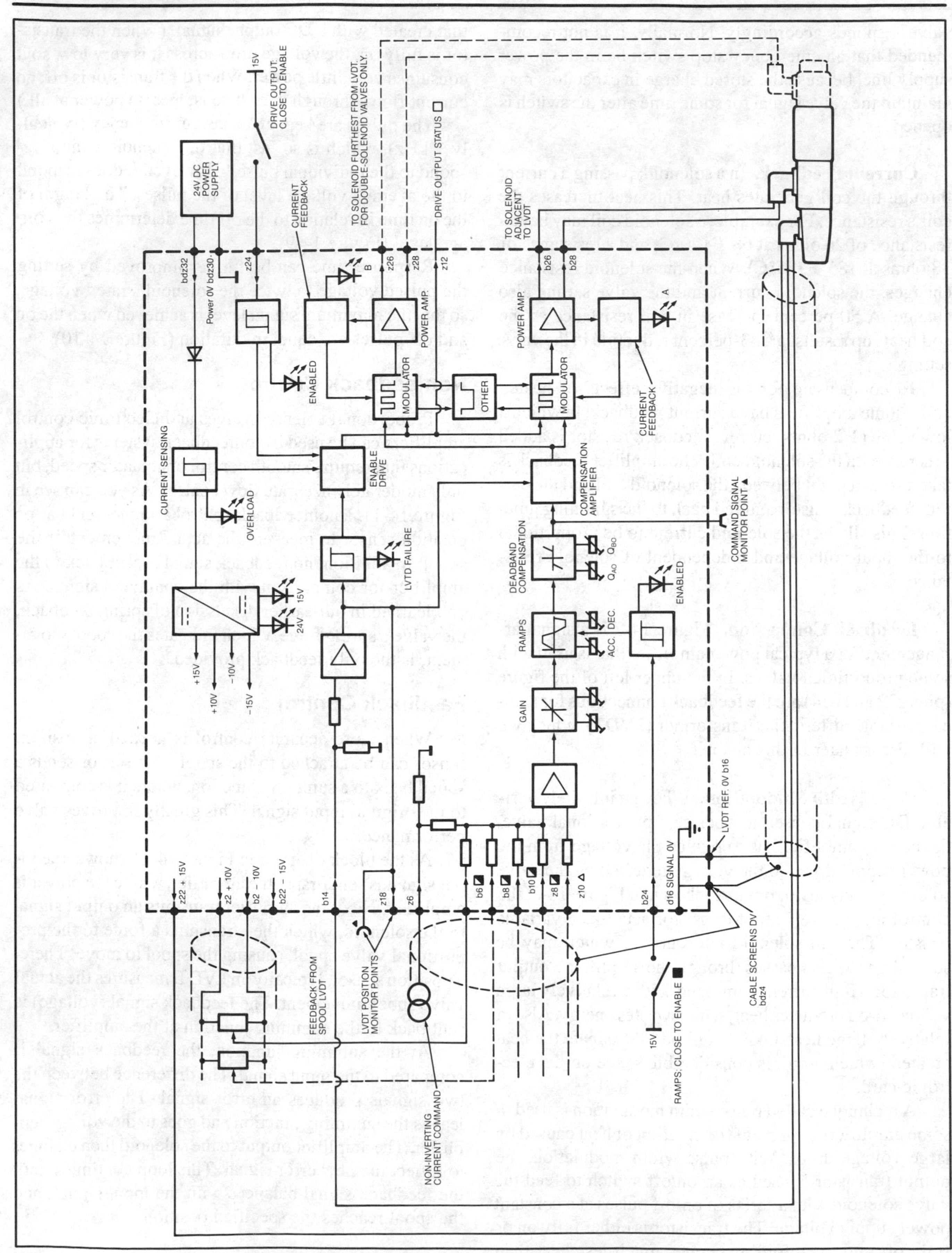

Figure 14-7. Typical signal arrangement for a power amplifier.

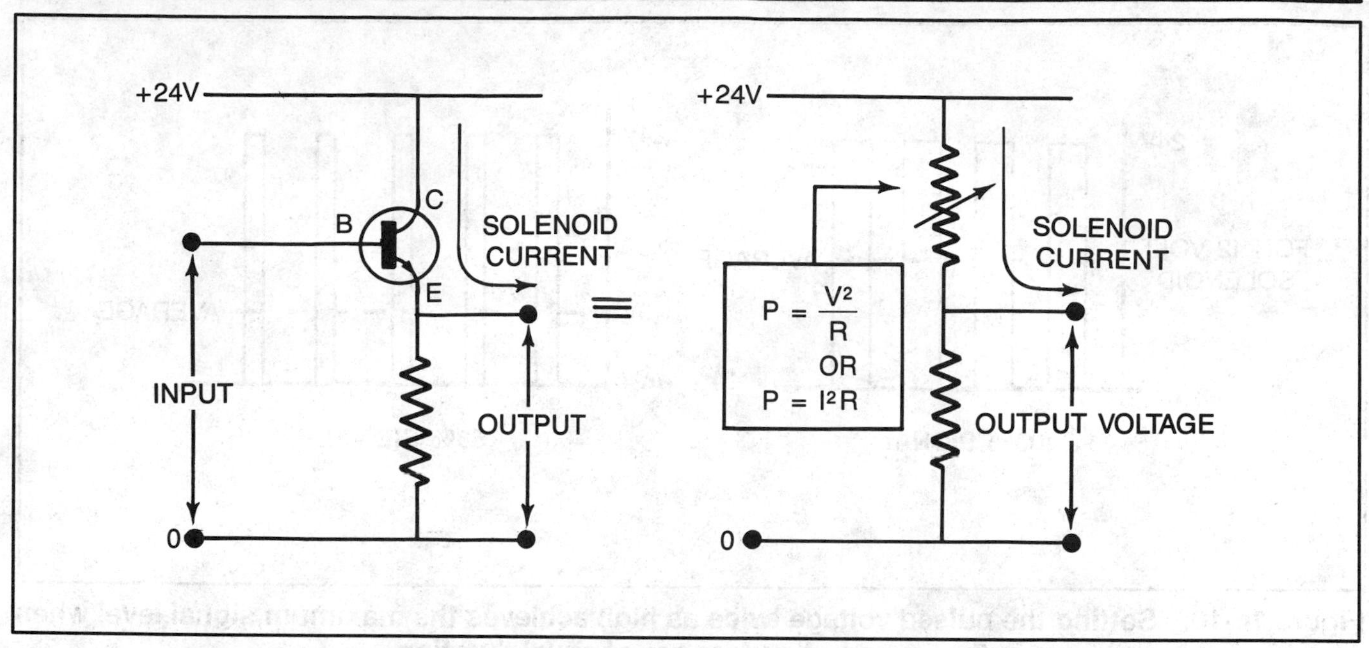

Figure 14-8. The amplifier's output transistor acts like a variable resistor.

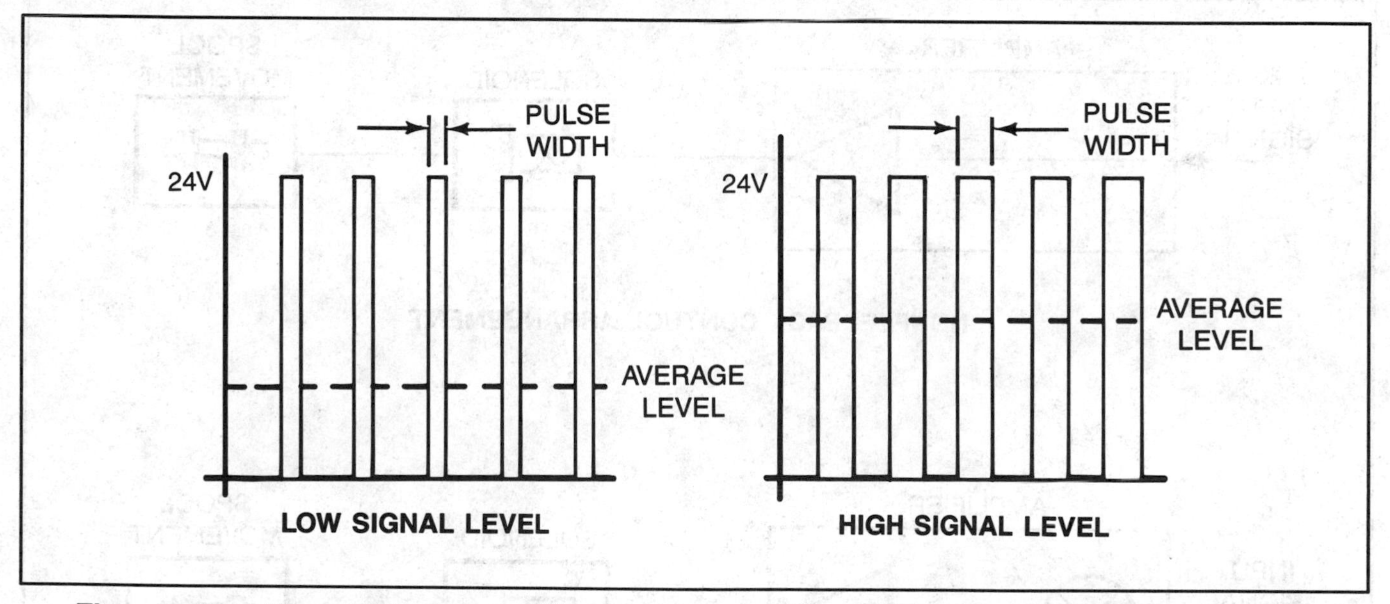

Figure 14-9. The width of the on time in relation to the off time determines voltage level.

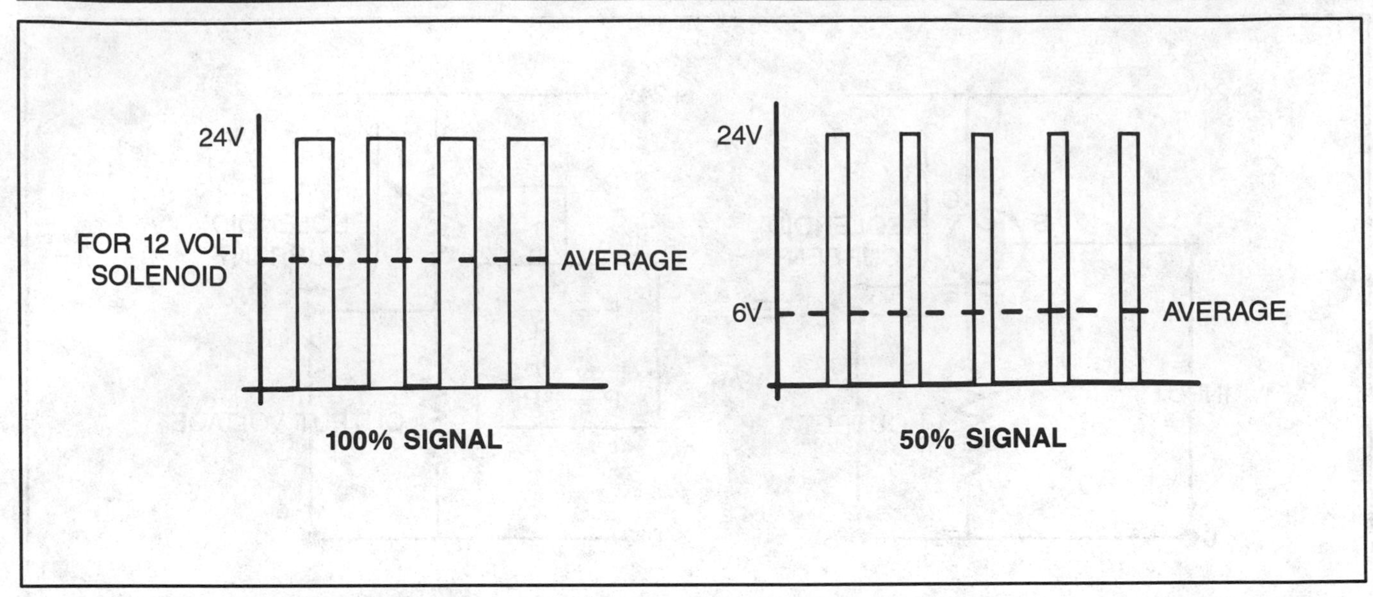

Figure 14-10. Setting the pulsed voltage twice as high achieves the maximum signal level when the on and off pulses are of equal duration.

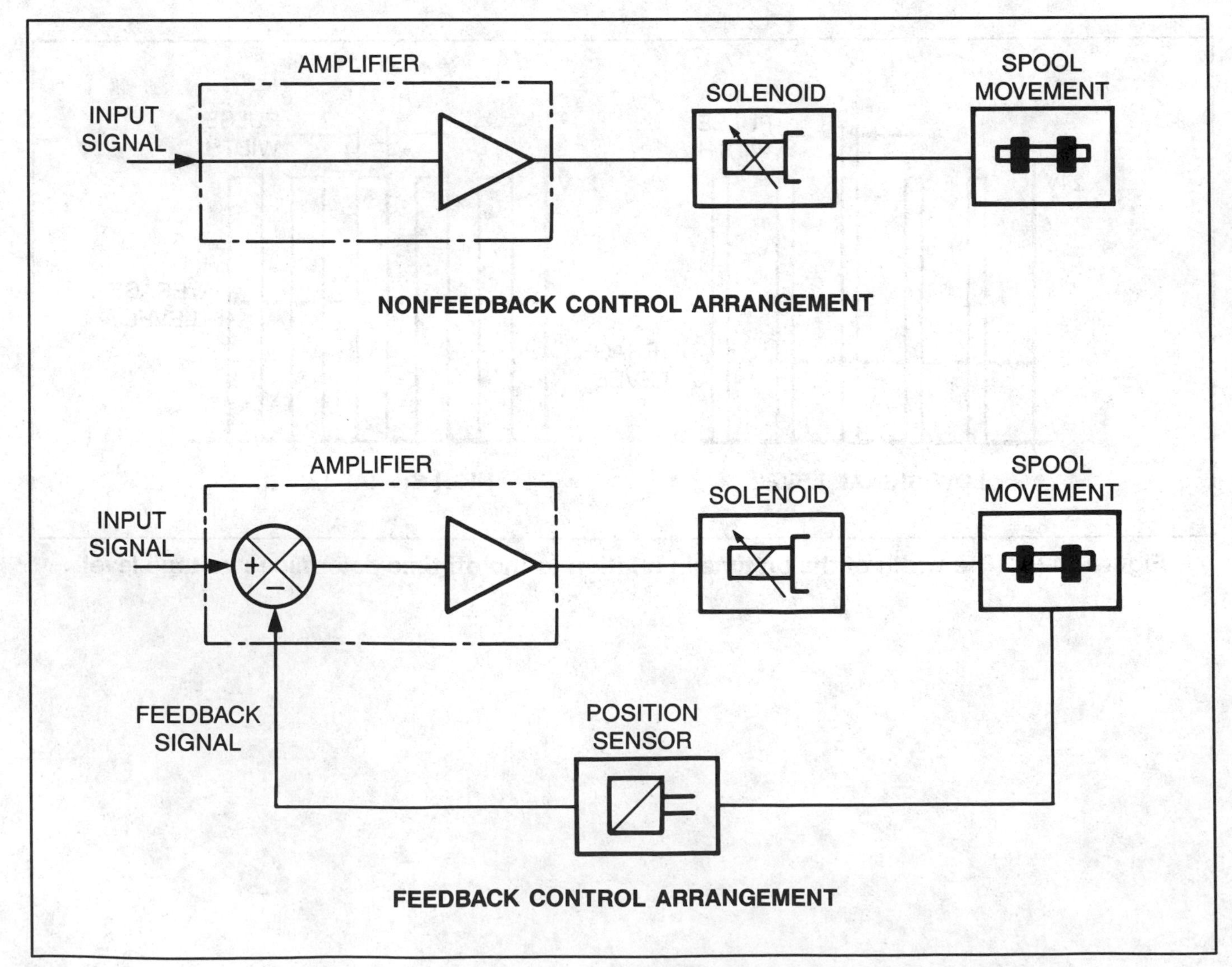

Figure 14-11. Nonfeedback and feedback control arrangements.

Required Power Supply

Most proportional solenoid valves require a properly installed DC power supply to provide power to the amplifier. Typically, an amplifier card has two pins for each power connection to increase the reliability of the contact.

The power can be supplied by a battery or by a rectified AC supply. This varies according to the specific application. With either supply, a smoothing capacitor is usually recommended for installation on the power supply connections at the amplifier card. This helps to reduce any rippling effects of a rectified AC supply and filters out any voltage spikes or noise that may be induced in an unshielded power supply line.

Areas of Application. Proportional solenoid valves can be used to vary pressure, flow, or both flow and direction in many different applications. They can be used to reduce shock caused by rapid pressure changes and the quick starts and stops of a heavy mass. Application examples include controlling hydraulic motors, single- and double-acting cylinders that move loads, and variable pumps.

BASIC HYDRAULIC PRINCIPLES OF PROPORTIONAL VALVES

A proportional solenoid valve must be properly engineered into a system to perform up to its capabilities. This requires attention to factors that are not present for conventional on/off solenoid valves, such as how flow and pressure act on the valve's variable orifices.

Two Flow Paths Through the Valve

In normal operation, a proportional valve has two flow paths. In Figure 14-12, you can see that there will be flow either from the P port to the A port and the B port to the T port, or from P to B and A to T.

Symmetrical and Nonsymmetrical Spools

A proportional valve can have a symmetrical or a nonsymmetrical spool. A symmetrical spool restricts the two flow paths equally, which enables the valve to both meter in and meter out fluid. On a nonsymmetrical spool, the main restriction is in the B to T or the A to T flow path, so the valve will meter out fluid only.

Symmetrical and nonsymmetrical spools are used for different types of applications. In Figure 14-13, a proportional valve is used to control a hydraulic motor, which is an equal-area actuator. A symmetrical spool, which provides two restrictions in a series with the actuator, is the practical choice to use, because the actuator area is equal on both sides.

When the proportional valve spool is in such a position that fluid flows from P to A and B to T and the motor is turning at a constant speed (no acceleration or deceleration), the four pressure gauges shown in Figure 14-13 indicate the following pressures:

P_1—indicates main system pressure. Usually determined by the pump relief valve or compensator setting.

P_2—indicates pressure needed to turn the load, plus the backpressure on motor outlet port.

P_3—indicates backpressure on the motor outlet port created by restricting exhaust flow B to T across the valve, plus the backpressure in tank line.

P_4—indicates backpressure in the tank return line caused by pipework, filters, and other restrictions.

Equation for Determining Flow Rate

Flow through each restricted opening in a proportional valve is determined by an equation called the sharp edge orifice equation. This equation is:

$$Q = Cd \times A \times \sqrt{\frac{2 \times \Delta P}{\varrho}}$$

Where:
Q = flow rate
Cd = orifice discharge coefficient
A = orifice area
P = pressure drop across orifice
ϱ = fluid density

While you won't need to calculate flow rates on a regular basis, you should understand that flow rate through the valve is proportional to the square root of the pressure drop across the valve:

Q is proportional to $\sqrt{\Delta P}$ and
ΔP is proportional to Q^2.

In other words, if you <u>double</u> through the valve, the pressure drop across the valve <u>quadruples</u>.

Using the example of Figure 14-13, the following can be stated based on the flow rate equation:

$Q_{(P-A)}$ is proportional to $\sqrt{\Delta P_{(P-A)}}$

Where:
$Q_{(P-A)}$ = flow through the valve P to A
And:
$\Delta P_{(P-A)}$ = pressure drop across the valve from P to A.

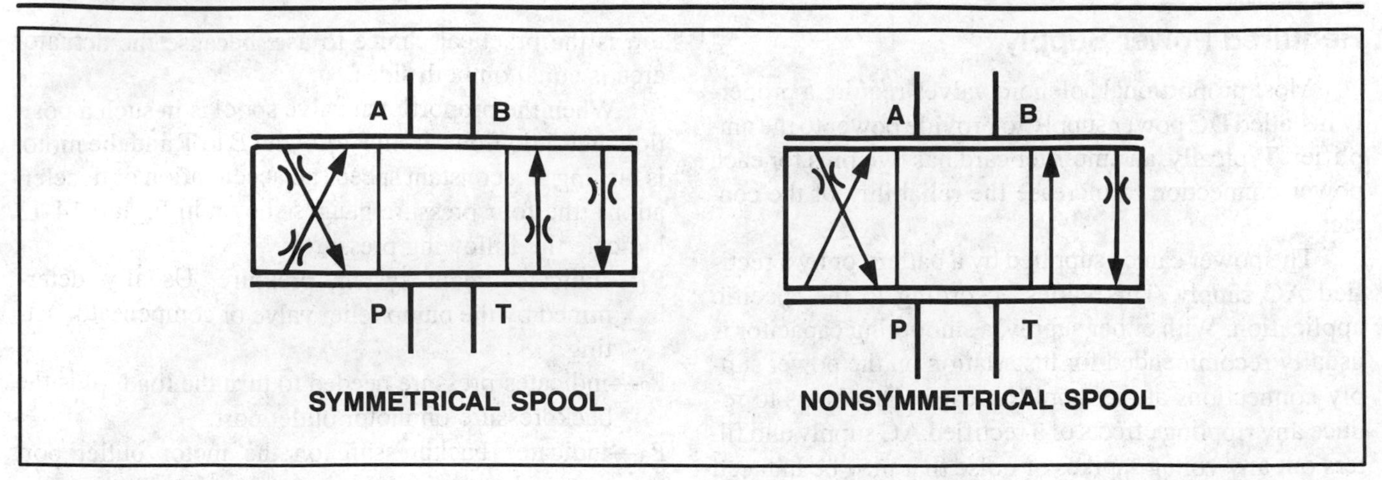

Figure 14-12. Symbols for the flow paths through symmetrical and nonsymmetrical spools.

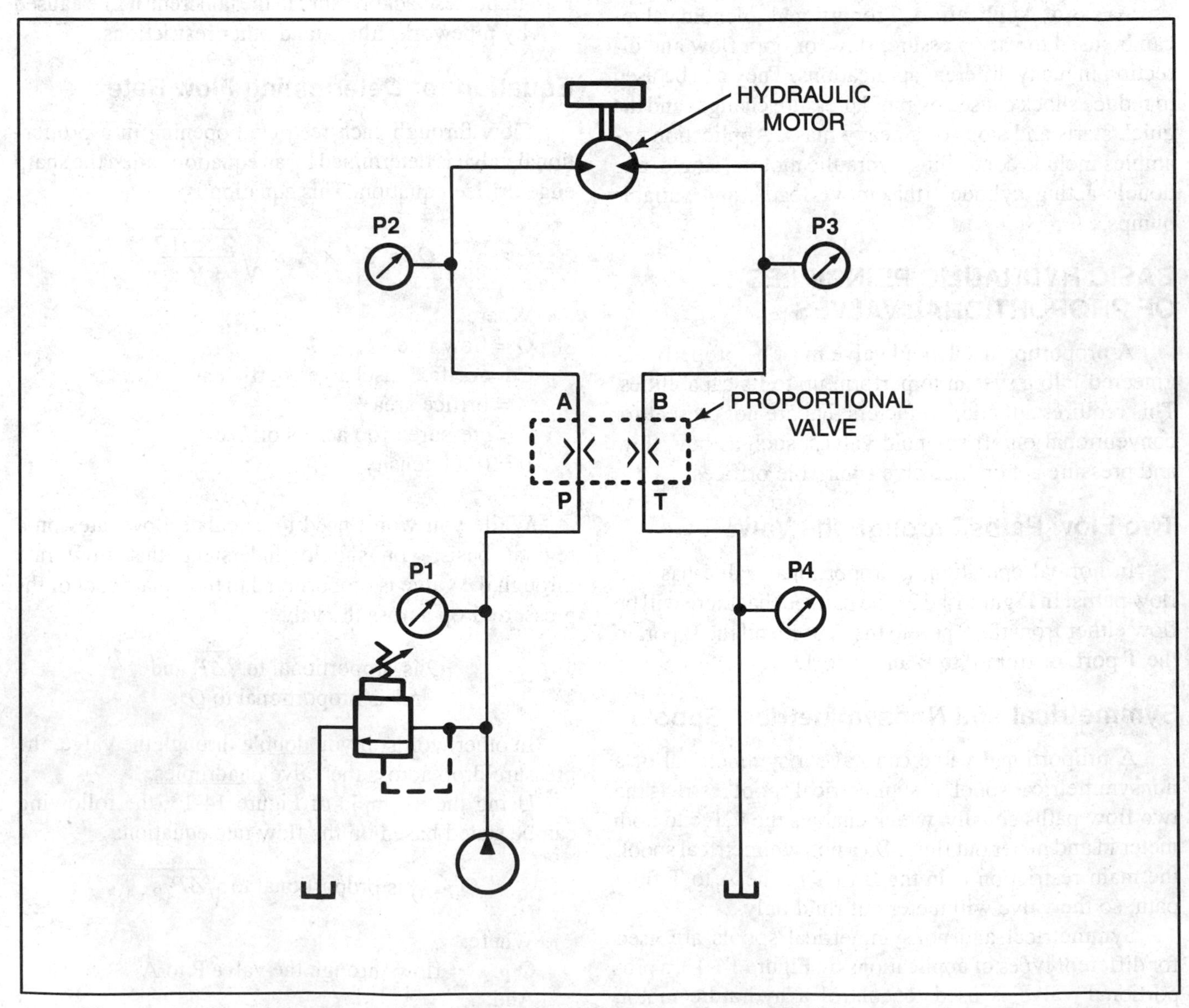

Figure 14-13. A symmetrical spool proportional valve controls an equal area actuator.

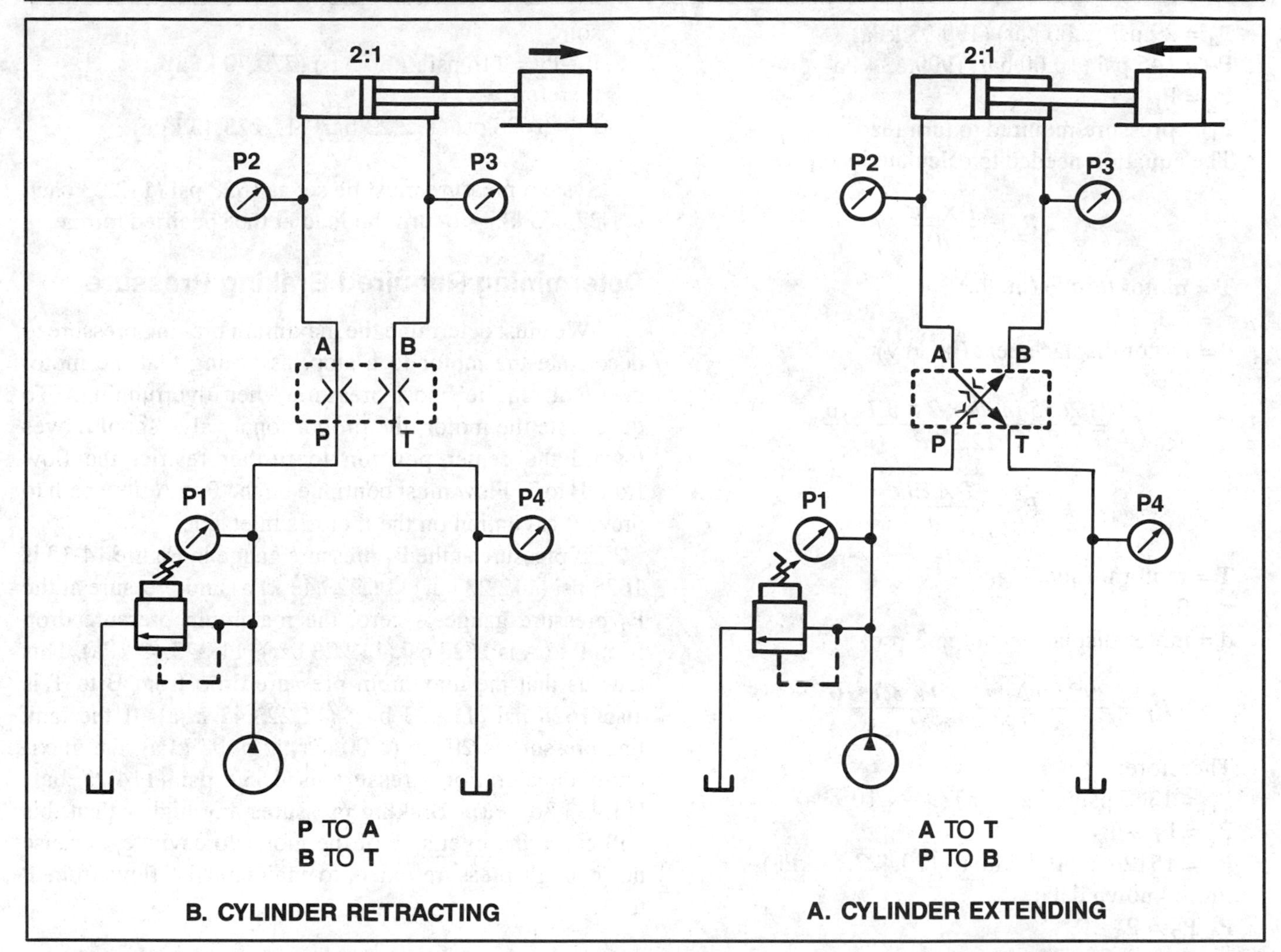

Figure 14-14. A symmetrical spool proportional valve controls a differential area cylinder.

In the same way:

$$Q_{(B-T)} \text{ is proportional to } \sqrt{\Delta P_{(B-T)}}$$

Where:
$Q_{(B-T)}$ = flow through the valve B to T
And:
$\Delta P_{(B-T)}$ = pressure drop across the valve from B to T.

Since the motor in Figure 14-13 has equal inlet and outlet flows (discounting effects of leakage), then:

$$Q_{(P-A)} = Q_{(B-T)}$$

and with equal restrictions on both flow paths,

$$\Delta P_{(P-A)} = \Delta P_{(B-T)}$$

or:

$$P_1 - P_2 = P_3 - P_4$$

Therefore, when a symmetrical spool is used to control an equal-area actuator, the pressure drop across each flow path through the valve is the same.

Determining Required System Pressure

Now, let's assume that the hydraulic motor shown in Figure 14-13 has a displacement of 12.2 in^3 (199.92 cm^3) per revolution and that it needs to turn a load at a torque of 2655 in. lb. (299.96 N m). What system pressure is required in this application?

From available catalog data, we know that the proportional valve has a 116 psi (8.00 bar / 799.70 kPa) pressure drop across each flow path when the motor is running at maximum speed. With a backpressure of, for example, 29 psi (2.00 bar / 199.93 kPa) in the tank return line, we can determine the system pressure that must be maintained. In these calculations the two restrictions, $P_{(P-A)}$ and $P_{(B-T)}$ must be taken into account.

Known data:
P_3–P_4 = 116 psi (8.00 bar / 799.70 kPa)

P_4 = 29 psi (2.00 bar) (199.93 kPa)
P_3 = 145 psi (10.00 bar) (999.63 kPa)
$P_2 = P_L + P_3$
P_L = pressure required to turn load.
The equation needed to calculate load pressure is:

$$P_L = \frac{T \times 2\pi}{d}$$

T = motor torque (in. lb.)
π = 3.1416
d = motor displacement (in^3/rev)

$$P_L = \frac{2655 \text{ in. lb} \times 2 \times 3.1416}{12.2 \text{ in}^3}$$

$$P_L = \frac{T \times 20\pi}{d}$$

T = motor torque (N m)
π = 3.1416
d = motor displacement (cm3/rev)

$$P_L = \frac{299.96 \text{ N m} \times 20 \times 3.1416}{199.92 \text{ cm}^3}$$

Therefore:
P_L = 1367 psi (94.24 bar) (9424.10 kPa)
$P_2 = P_L + P_3$
P_2 = 1512 psi (104.24 bar) (10,423.73 kPa)
From known data:
$P_1 - P_2 = P_3 - P_4$
so:
$P_1 - P_2$ = 116 psi (7.80 bar) (779.70 kPa)
therefore:
P_1 = 1628 psi (112.23 bar) (11,223.43 kPa)

System pressure must be set at 1628 psi (112.23 bar) (11,223.43 kPa) to turn the load at the specified torque.

Determining Required Braking Pressure

We must determine the maximum braking pressure to decelerate the motor to a stop, assuming that the motor does not require boost pressure when overrunning. To decelerate the motor, the proportional valve spool moves toward the center position to further restrict the flow from B to T. Flow must continue on the P to A flow path to prevent cavitation on the motor's inlet side.

If pressure at the P_1 pressure gauge in Figure 14-13 is 1628 psi (112.23 bar) (11,223.43 kPa) and pressure at the P_2 pressure gauge is zero, the maximum pressure drop from P to A is 1628 psi (112.23 bar) (11,223.43 kPa). This tells us that the maximum pressure drop from B to T is also 1628 psi (112.23 bar) (11,223.43 kPa). If the tank line pressure is 29 psi (2.00 bar) (199.93 kPa), the maximum deceleration pressure is 1657 psi (114.23 bar) (11,423.36 kPa). Braking pressures any higher than this will cause the inlet side of the motor to cavitate, because not enough pressure exists to maintain the flow from P to A.

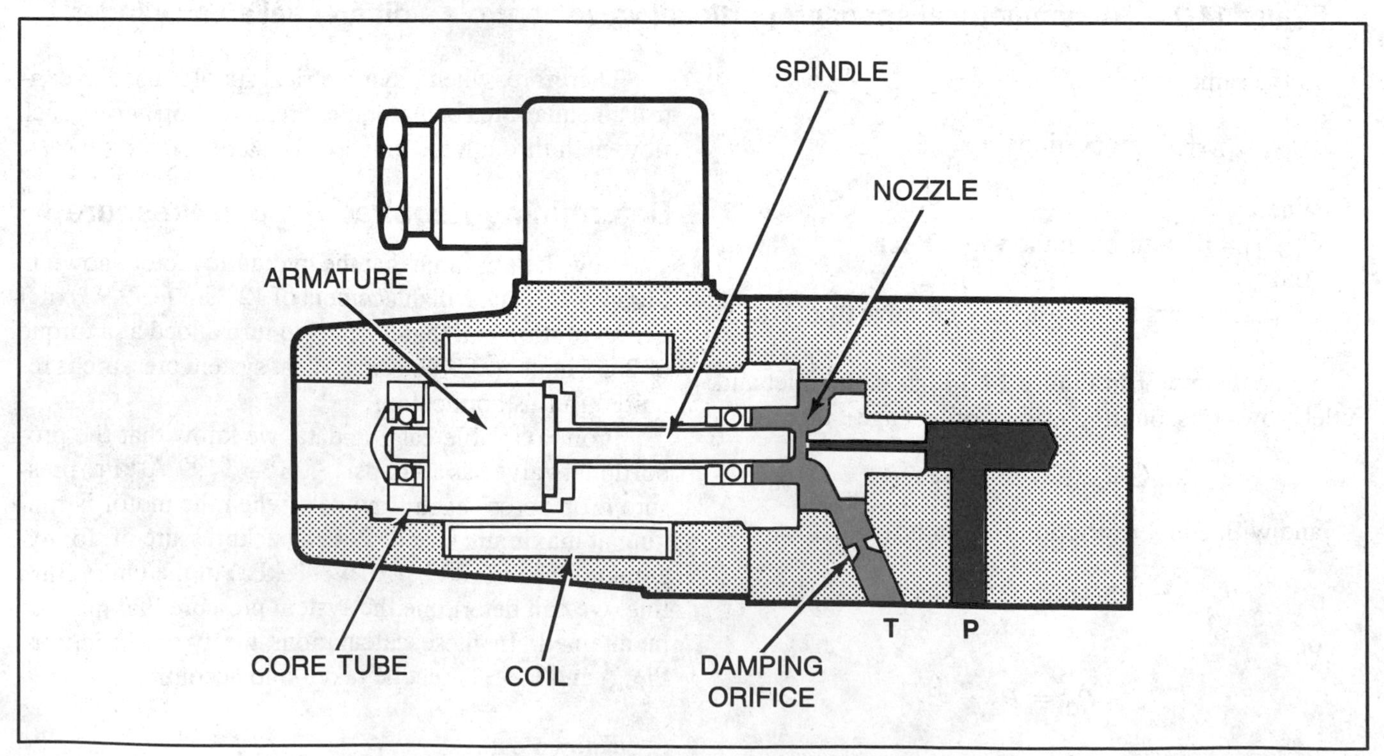

Figure 14-15. Nozzle-type pressure relief valve.

It might seem appropriate to use a meter-out spool for this application. However, a meter-out spool used in the same system only restricts flow on the outlet side of the motor. This prevents cavitation, but if system pressure at the P port is maintained during deceleration, the flow continues to drive the motor. Having pressure on both sides of the motor at the same time is unacceptable for some motor designs. Therefore, in general, the symmetrical spool, sometimes used with additional braking and anticavitation arrangements, is the best choice for motor drive applications.

Now that we have discussed flow and pressure as they affect an equal-area motor, we can address the same issues for a differential-area cylinder. In Figure 14-14A, a proportional valve is used to control a differential-area cylinder. To extend the cylinder, the valve spool moves into a position that allows flow from P to A and from B to T.

If a symmetrical spool is used and the cylinder's area ratio is 2:1, the flow on the full bore side is twice the flow on the rod side of the cylinder. Therefore, the inlet pressure drop from P to A is four times the outlet pressure drop from B to T:

$$Q_{(P-A)} = 2 \times Q_{(B-T)}$$

and: ΔP is proportional to Q^2
then: $\Delta P_{(P-A)} = 4 \times \Delta P_{(B-T)}$.

In the same way, as the cylinder retracts (Figure 14-14B), the outlet pressure drop from A to T is four times the inlet pressure drop from P to B:

$$Q_{(A-T)} = 2 \times Q_{(P-B)}$$

and: ΔP is proportional to Q^2
then: $\Delta P_{(A-T)} = 4\ \Delta P_{(P-B)}$.

Now that we have discussed some of the basic principles of proportional valve operation, we can examine the three different types of valves: proportional pressure control valves, proportional flow control valves, and proportional directional control valves.

PROPORTIONAL PRESSURE CONTROL VALVES

Proportional pressure control valves include relief and reducing valves with the same electrically controlled parts. In most pressure control applications, the pilot stage is a small electronically controlled valve, while the main stage is a regular relief or reducing valve sized for the required flow rate. Types of pressure control valves include nozzle-type relief valves, poppet-type relief valves with linear variable differential transformer (LVDT) feedback, and plate-type relief valves.

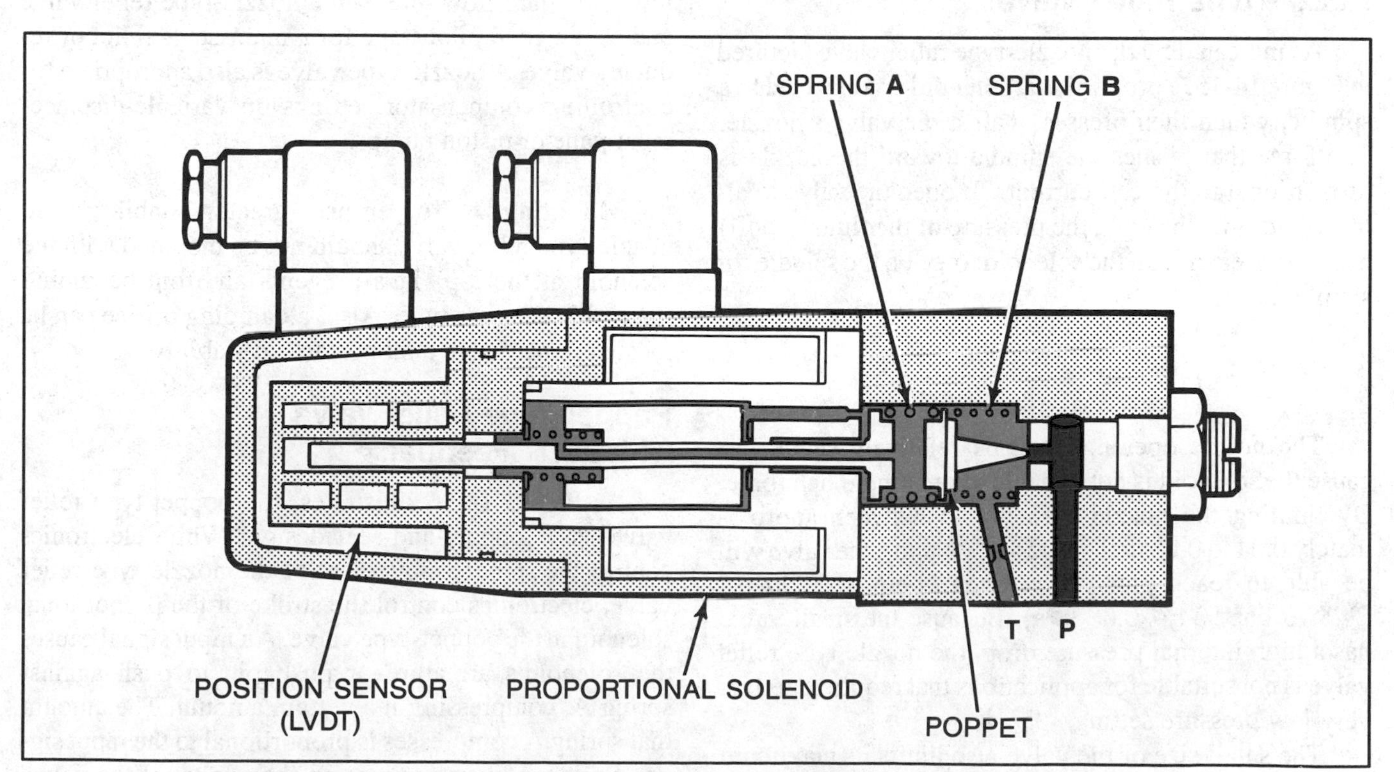

Figure 14-16. Poppet-type pressure relief valve with LVDT feedback.

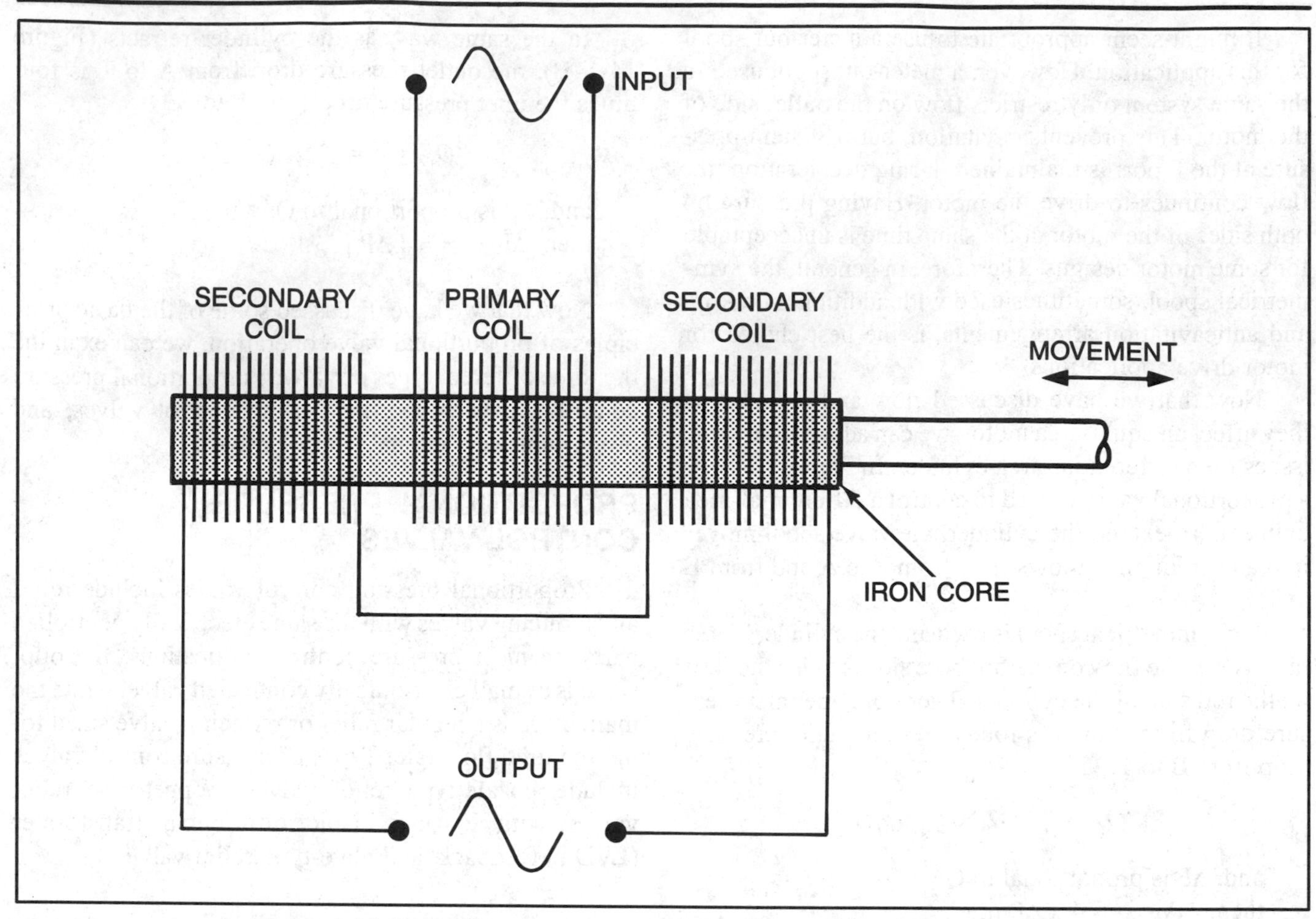

Figure 14-17. Linear variable differential transformer (LVDT).

Nozzle-Type Relief Valve

As indicated by the nozzle-type relief valve pictured in Figure 14-15, a proportional solenoid is connected to a spindle, which then presses against the valve's nozzle. The force that pushes the spindle toward the nozzle is proportional to the coil current. To open the valve to allow oil to flow through, the pressure of the fluid (P port) has to be greater than the solenoid force on the spindle. In simple terms:

$$\textit{pressure} = \frac{\textit{solenoid force}}{\textit{nozzle area}}$$

The nozzle opening must be relatively small, because the solenoid is not capable of creating much force. By limiting the nozzle diameter to between approximately 0.04 to 0.08 inch (1.02 to 2.04 mm), the valve will be able to reach pressure settings of up to 5000 psi (344.70 bar / 34,470.00 kPa). Because the small valve has a high internal pressure drop, the nozzle-type relief valve is not suitable for applications that require a zero or very low pressure setting.

The small size of the valve also limits its maximum flow rate to about 0.8 gpm (3.03 LPM). The relatively low maximum flow rate of the nozzle-type relief valve makes it a good pilot stage for a main-stage relief or reducing valve. A nozzle-type valve is also appropriate for controlling compensator settings on variable-displacement vane or piston pumps.

Mounting. To ensure greater stability, the nozzle-type relief valve should not be mounted with the solenoid at the top. This prevents air from becoming trapped in the core tube. Also, a damping orifice can be located in the valve tank port to aid stability.

Poppet-Type Relief Valve with LVDT Feedback

As Figure 14-16 illustrates, the poppet-type relief valve has a poppet-and-seat design. While electronics control the spindle movement of the nozzle-type relief valve, electronics control the stroke of the proportional solenoid in the poppet-type valve. An input signal causes the solenoid's armature or push pin to push against spring A, compressing it a certain amount. The amount that spring A compresses is proportional to the input signal. Spring A exerts a force on the poppet of the valve.

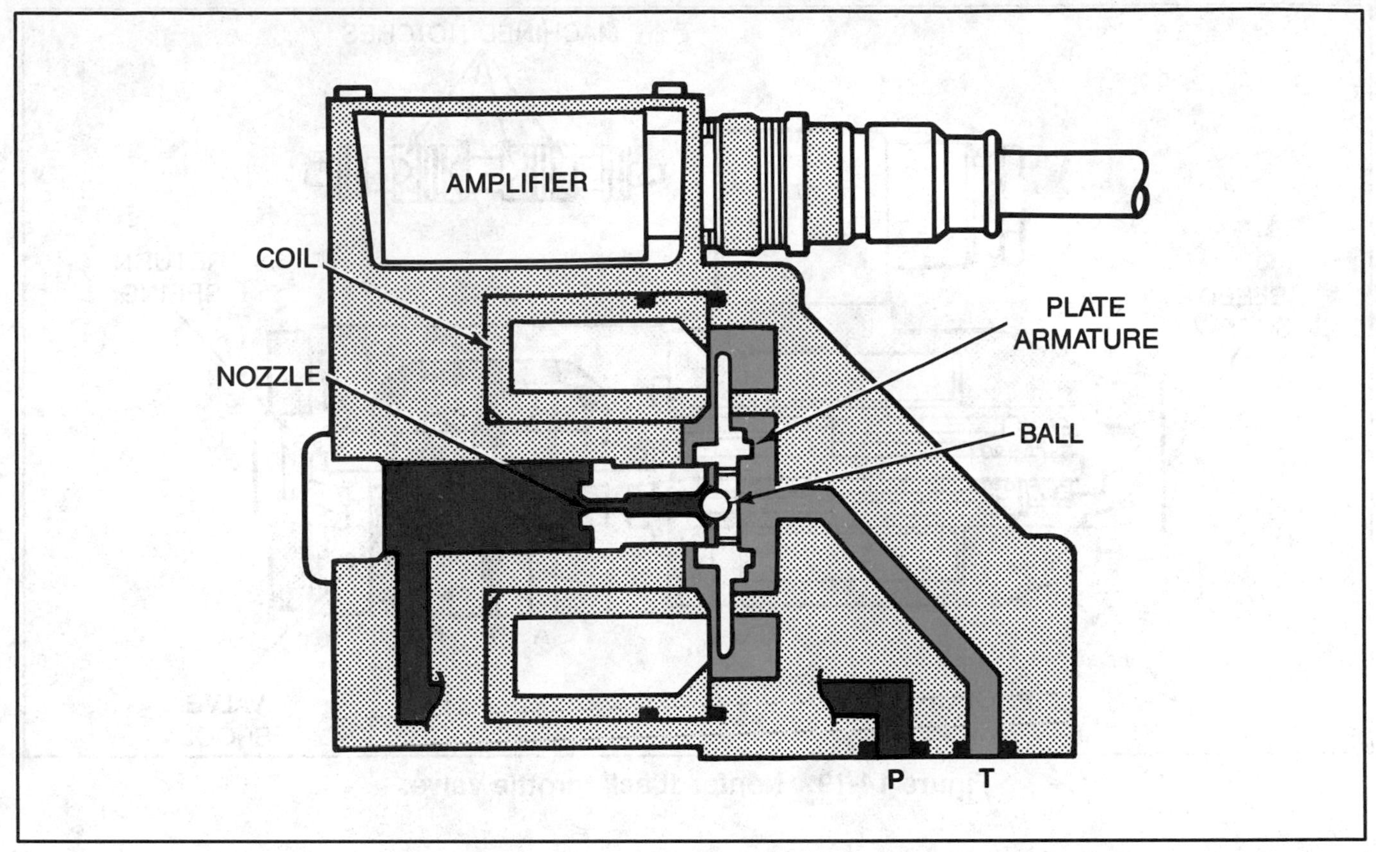

Figure 14-18. Plate-type pressure relief valve.

This force determines the valve's pressure setting. Light spring B serves to keep the poppet square with the seat. The maximum flow rate through this valve is about 1.3 gpm (4.92 LPM).

Linear Variable Differential Transformer. To control the solenoid's armature stroke, a position sensor called a linear variable differential transformer (LVDT) is used. The LVDT is placed on the solenoid's armature to monitor its movement. The LVDT sends a feedback signal to the control amplifier, where it is compared to the input signal. This method ensures more accurate positioning, provides for automatic correction of any disturbances, and reduces hysteresis.

The LVDT is the most common position sensor used with spool-type valves such as proportional solenoid valves. In Figure 14-17, you can see that an LVDT has one primary coil wrapped around a soft iron core, with two secondary coils also wrapped around the core. The iron core is connected to the solenoid's push pin. A high-frequency AC signal is fed to the primary coil. This power supply creates a varying magnetic field in the iron core, which also induces voltage in the two secondary coils through transformer action.

When the core is centrally positioned and the two secondary coils are connected in opposition, the voltages in the secondary coils are balanced. Therefore, the two voltages cancel out each other to produce a zero output. As the iron core moves away from the center, the voltage induced in one secondary coil increases while the voltage in the second coil decreases. This imbalance produces an output. The magnitude of the output is proportional to the distance the iron core moved from center, and phase shifts indicate the direction of the core's movement. The LVDT's output signal is fed to a phase-sensitive rectifier, which produces a DC voltage that accurately reflects the distance and direction of the core's movement.

Plate-Type Relief Valve

The plate-type pressure relief valve has a plate-type armature (Figure 14-18). Like the other valves we have discussed, a current is passed through the valve's solenoid coil to create a magnetic field. On this valve, the magnetic field pulls the plate armature toward the coil. A small steel ball is held in the center of the plate, near the valve's nozzle opening.

As the armature is pulled toward the coil, the steel ball at the center of the armature is pushed against the nozzle. This reduces or blocks flow through the nozzle. To create a flow through the nozzle, the fluid pressure must be greater than the force created by the coil current.

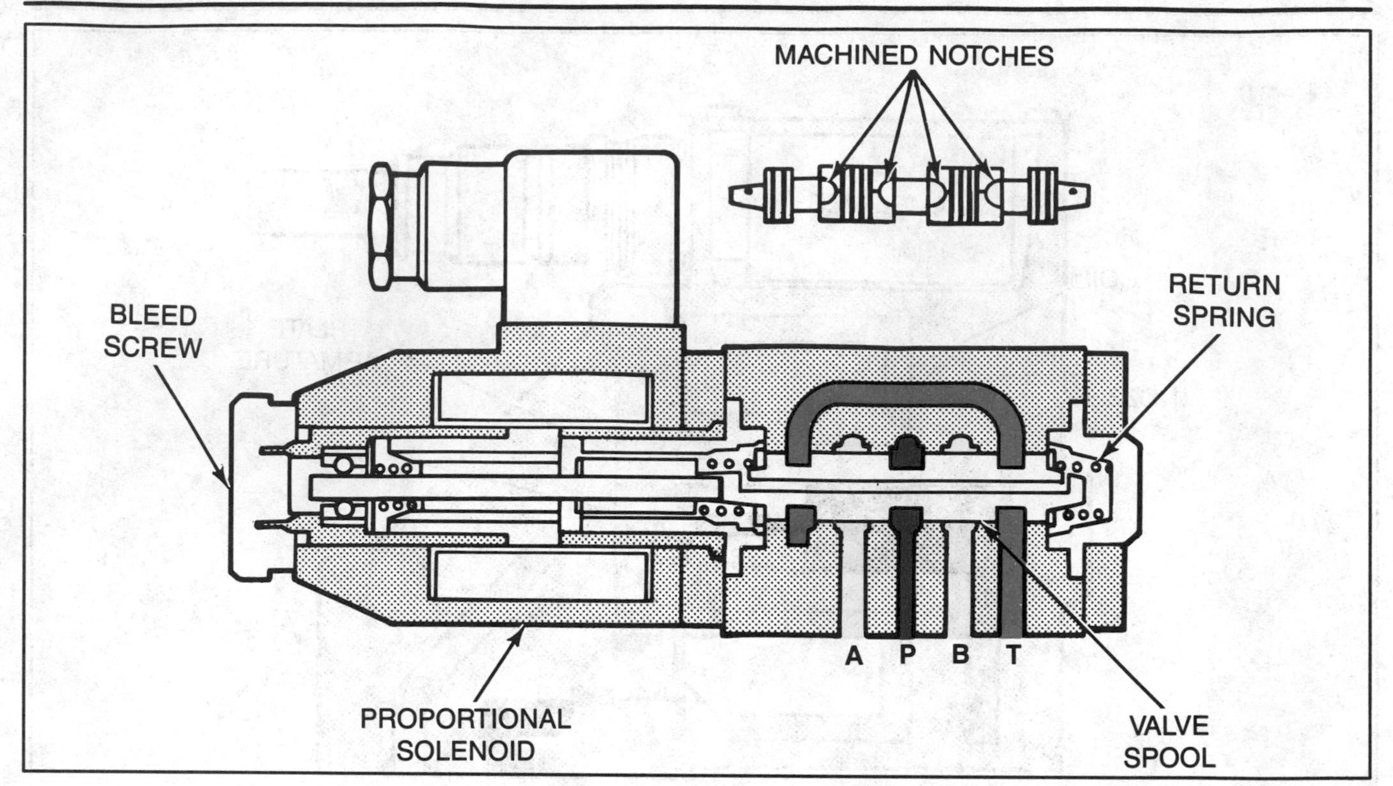

Figure 14-19. Nonfeedback throttle valve.

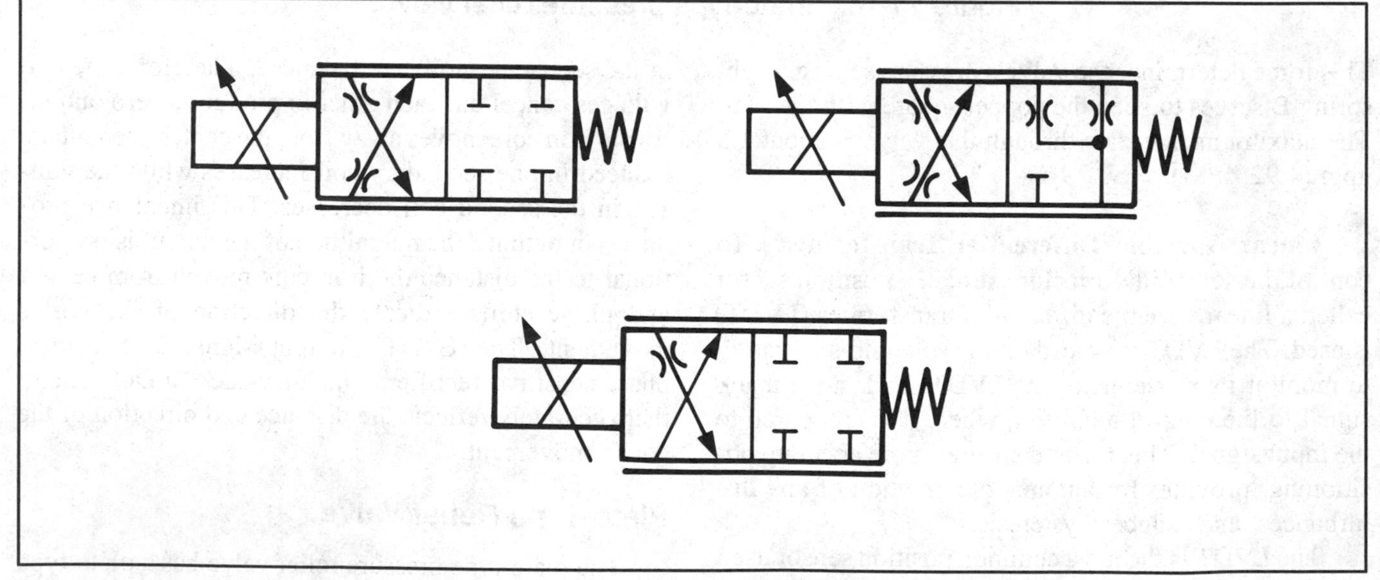

Figure 14-20. Three spool options for a nonfeedback throttle valve.

Therefore, the coil current determines how much pressure is required to allow flow through the nozzle. A variety of ball and nozzle sizes can be fitted for different pressure requirements.

A major benefit of the plate-type relief valve is the absence of springs, which results in less hysteresis, faster response times, and better resolution. Also, when the valve plate is energized, there is no mechanical contact between the plate armature and the valve body or nozzle.

Applications Summary

We have described three types of proportional pressure relief and pressure reducing proportional valves. Now, we will discuss the applications in which each pressure control valve works best.

Nozzle-Type Proportional Solenoid Valve. This valve is the simplest and least expensive of the three

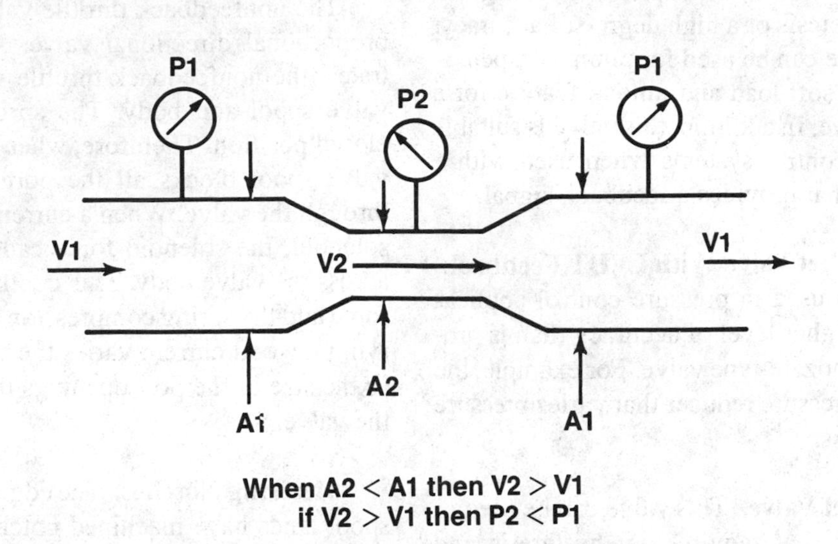

A. BERNOULLI'S THEOREM

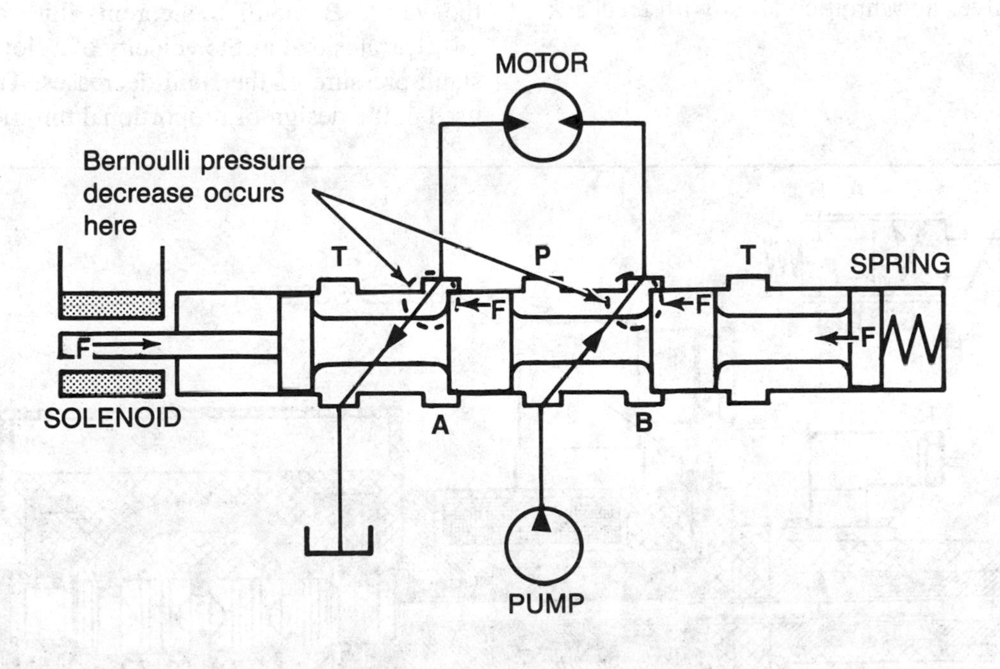

B. BERNOULLI FORCES OPPOSING SOLENOID FORCE

Figure 14-21. Bernoulli forces.

pressure valves discussed. It is designed for applications that require control of the electronic pressure setting, but do not need low hysteresis or a high degree of accuracy. The nozzle-type valve can be used for pump compensator control and as a soft load and unload feature for a main-stage relief valve. In addition, this valve is suitable for use in pressure control systems when used with a pressure transducer that provides a feedback signal.

Poppet-Type Relief Valve With LVDT Feedback. This valve usually is used in pressure control applications that require a higher level of accuracy than is provided by the simpler nozzle-type valve. For example, the valve can control a pressure reducer that varies pressure on an upstroking press.

Plate-Type Relief Valve. This valve can be used in pressure applications that require low hysteresis and smooth linear control. Its onboard electronics make it compatible with microprocessor controls.

PROPORTIONAL FLOW CONTROL VALVES

Proportional flow control valves control the flow of hydraulic fluid. A proportional solenoid varies the valve settings for actuator speed, acceleration, and deceleration. Proportional flow control valves include nonfeedback throttle valves and throttle valves with feedback.

Nonfeedback Throttle Valve

The nonfeedback throttle valve is very similar to a proportional directional valve. As Figure 14-19 illustrates, the nonfeedback throttle valve has a directional valve spool and body. The spring is extended in the closed position. Therefore, when it is not energized, the valve spool blocks all the ports to prevent any flow through the valve. When a current is passed through the solenoid, the solenoid force causes the spool to move across the valve body. This continues until the solenoid force and the spring compression force are balanced. Varying the coil current varies the spool's position as well as the size of the port openings that allow flow through the valve.

Metering Notches. The edges of the throttle valve's spool lands have machined notches. These notches improve the valve's sensitivity, without reducing the maximum flow rate, when metering flow through very small openings. As you can see in Figure 14-20, three spools have different notches on them, allowing different combinations of blocked and restricted flows. In addition, the spools can be sized to meet different flow rate requirements.

As is true for all throttling valves, the flow rate through the proportional throttle valve is determined by the size of the valve opening and the pressure drop across the valve. Bernoulli's theorem, illustrated in Figure 14-21, states that as the velocity of a flow increases, the static pressure of the fluid decreases. This principle is used in the design of proportional throttle valves.

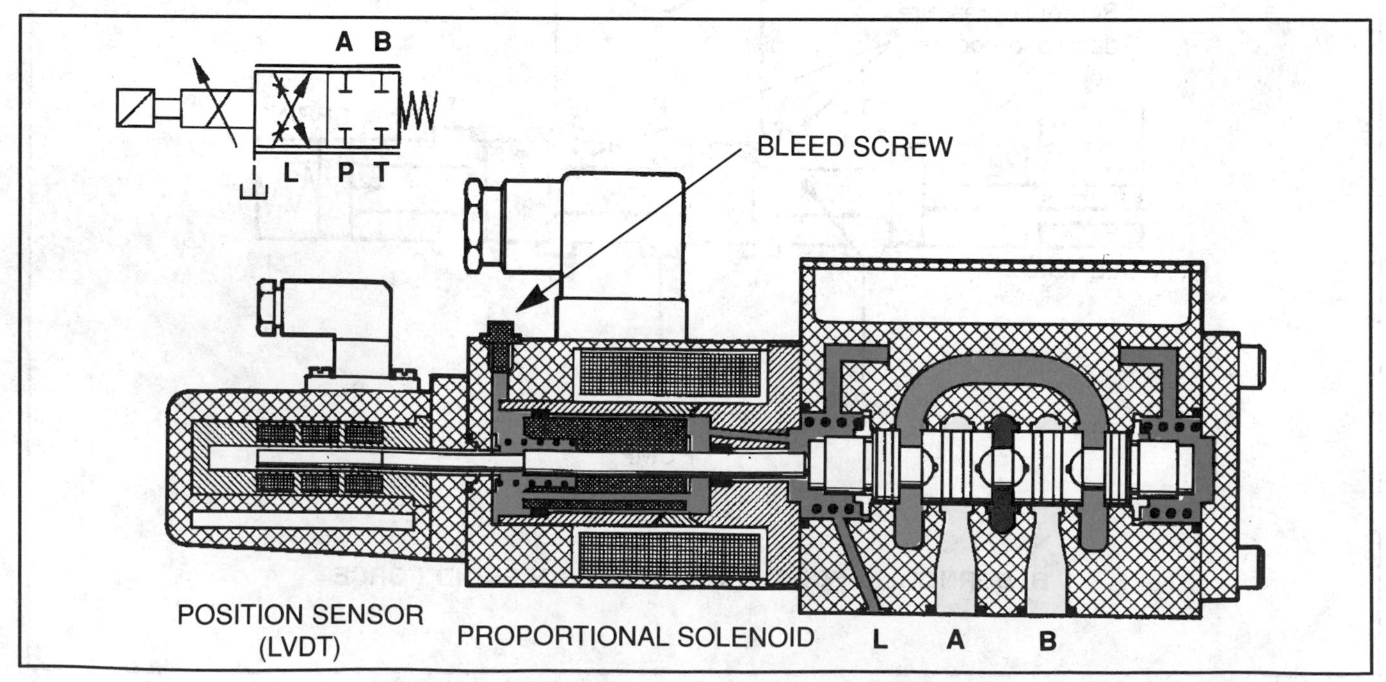

Figure 14-22. Throttle valve with feedback.

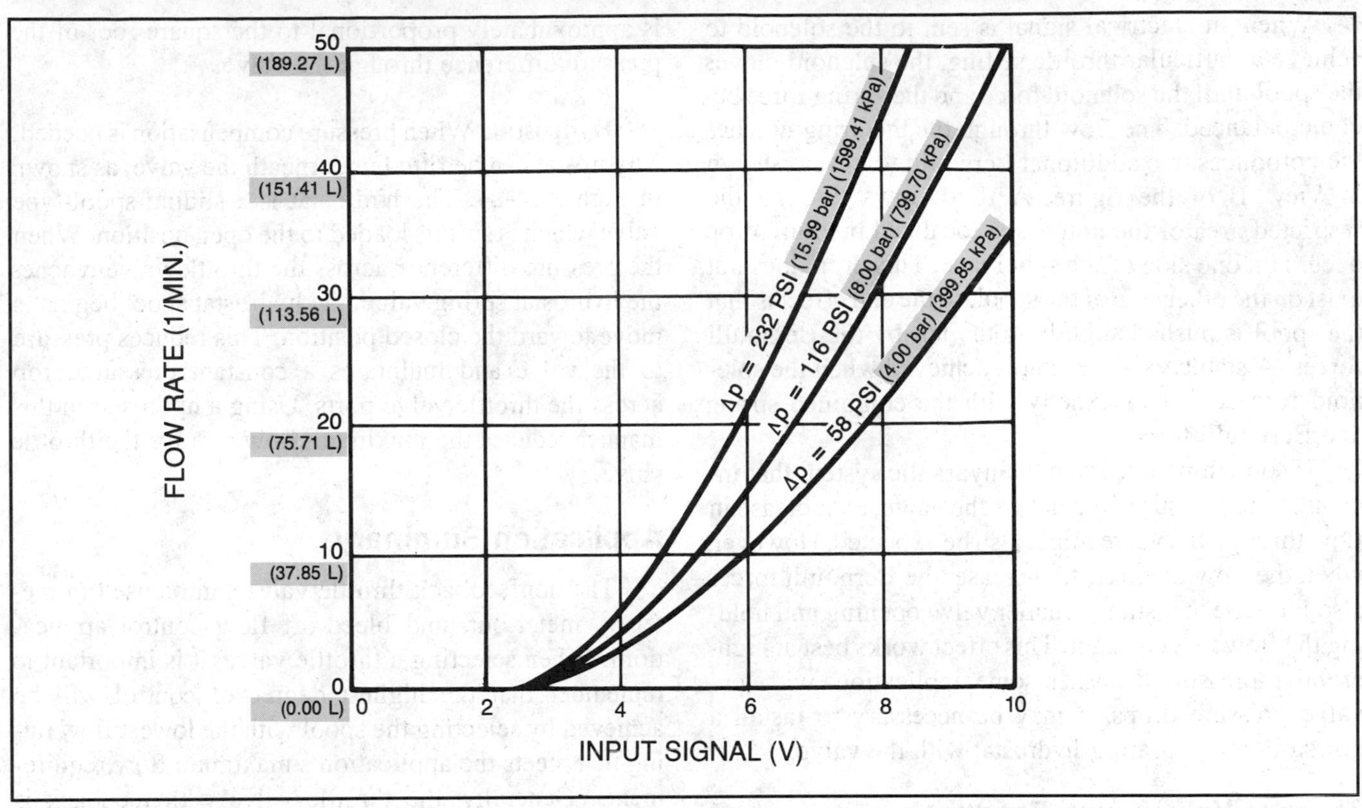

Figure 14-23. Flow rate is proportional to the square root of the pressure difference.

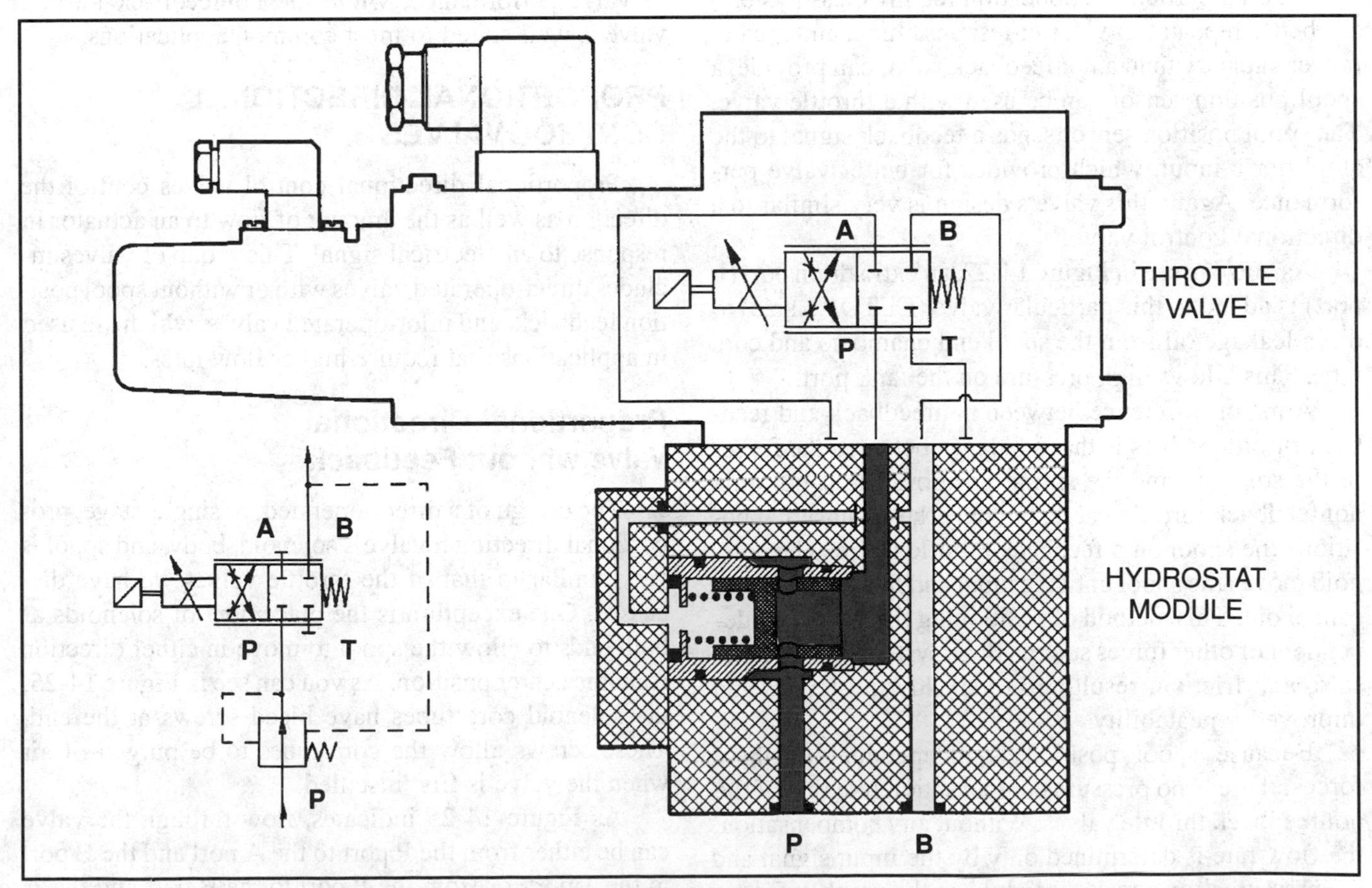

Figure 14-24. Hydrostat module fitted beneath a proportional throttle valve for pressure compensation.

When an electrical signal is sent to the solenoid to achieve a particular throttle setting, the solenoid moves the spool until the solenoid force and the spring force become balanced. The flow through the metering notches then produces two additional Bernoulli forces, as shown in View B of the figure. A fluid flows through the restricted area of the notches, a localized pressure drop occurs on one side of the spool face. This drop does not exist on the other side of the spool, so the end effect is that the spool is pushed slightly to the left by the Bernoulli forces. A stable valve setting is achieved when the solenoid force balances exactly with the combined spring and Bernoulli forces.

If something occurs in the hydraulic system that increases the pressure drop across the valve, an increase in flow through the valve might also be expected. However, when the flow attempts to increase, the Bernoulli forces also increase, causing a smaller valve opening and holding the flow rate constant. This effect works best at higher valve pressure drops. In some applications with low valve pressure drops, it may be necessary to install a pressure-compensating hydrostat with the valve.

Throttle Valves with Feedback

When a particular application requires less hysteresis, better repeatability, faster response time, and greater power capacity than a nonfeedback valve can provide, a spool position sensor can be used with a throttle valve. The spool position sensor sends a feedback signal to the amplifier's input, which provides for better valve performance. Again, this valve's design is very similar to a directional control valve.

As you can see in Figure 14-22, an extra drain port (L port) is added to this particular valve (CETOP 5 size) to drain leakage oil from the spool end chambers and core tube. This allows high pressure on the tank port.

A major difference between nonfeedback and feedback throttle valves is that a balance between the forces of the solenoid and the spring positions the spool on a nonfeedback throttle valve, but only the input signal positions the spool on a feedback throttle valve. The solenoid moves the spool until the input and feedback signals cancel out. This method of positioning the spool is independent of other forces such as flow, system or load pressure, and friction, resulting in very low hysteresis and improved repeatability.

Because spool position is independent of these forces, there is no pressure compensation like that of the nonfeedback throttle valves. Without any compensation, the flow rate is determined only by the input signal and the pressure drop across the valve. As the graph in Figure 14-23 illustrates, an input signal produces a flow rate that is approximately proportional to the square root of the pressure difference through the valve.

Hydrostat. When pressure compensation is needed, a hydrostat can be fitted underneath the valve, as shown in Figure 14-24. The hydrostat is a sliding spool-type valve which is spring-loaded to the open position. When the pressure difference across the throttle valve reaches the hydrostat spring value, the hydrostat spool begins to move toward the closed position. This reduces pressure to the valve and maintains a constant pressure drop across the throttle valve ports. Using a hydrostat in this manner reduces the maximum flow rate for the throttle valve.

Application Summary

The nonfeedback throttle valve can be used in meter-in, meter-out, and bleed-off flow control applications. When selecting a throttle valve, it is important to remember that the highest degree of control will be achieved by selecting the spool with the lowest flow rating that meets the application's maximum flow requirements. Generally, the throttle valve with feedback is more suitable for applications that require a high degree of valve performance, while the nonfeedback throttle valve is well-suited to most common applications.

PROPORTIONAL DIRECTIONAL CONTROL VALVES

Proportional directional control valves control the direction as well as the amount of flow to an actuator in response to an electrical signal. This group of valves includes direct-operated valves with or without spool position feedback, and pilot-operated valves, which are used in applications that require higher flow rates.

Proportional Directional Valve without Feedback

The design of a direct-operated, or single-stage, proportional directional valve's solenoid, body, and spool is very similar to that of the throttle valves we have discussed. One exception is the placement of solenoids at both ends to allow the spool to move in either direction from the center position. As you can see in Figure 14-25, the solenoid core tubes have bleed screws at the end. These screws allow the core tubes to be purged of air when the valve is first installed.

As Figure 14-25 indicates, flow through the valve can be either from the P port to the A port and the B port to the T port, or from the P port to the B port and the A port to the T port. The actual flow paths used depend on

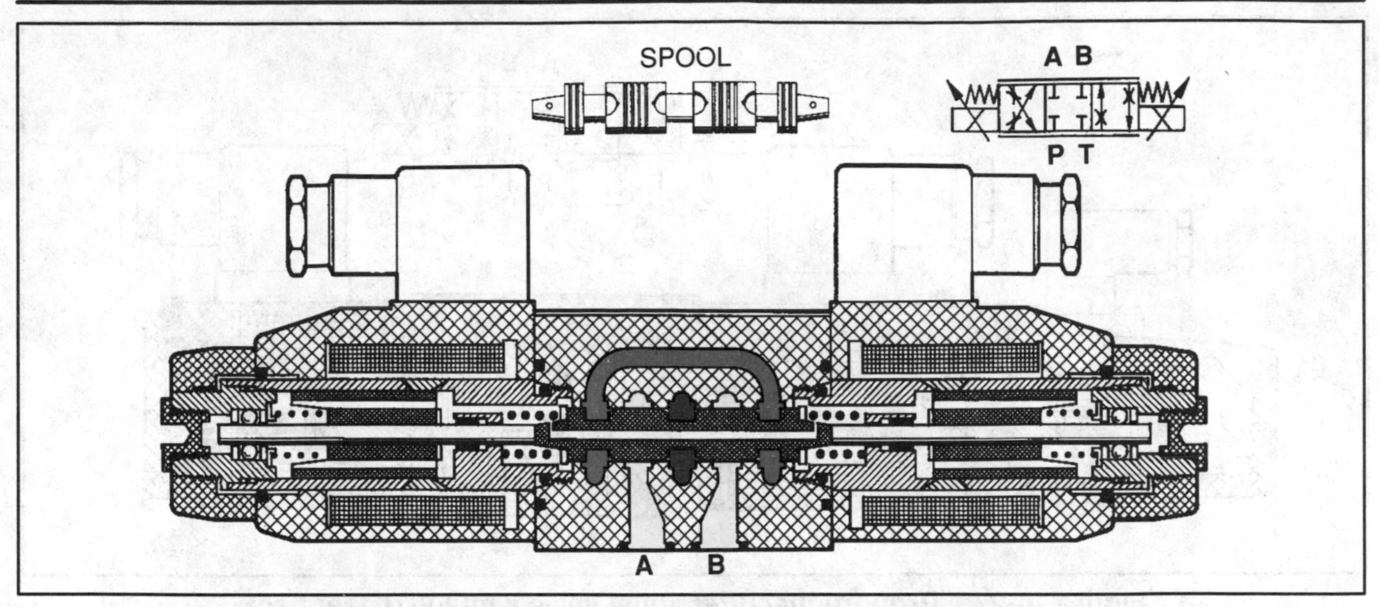

Figure 14-25. Direct-operated proportional directional valve without feedback.

CETOP 3	CETOP 5	NOMINAL FLOW (1/MIN)	METERING	SYMBOL
03F		3 (11.36 L)	IN & OUT	A B P T
07N		7 (26.50 L)	IN & OUT	
13N		13 (49.20 L)	IN & OUT	
20N		20 (75.70 L)	IN & OUT	
28S		28 (106.00 L)	OUT ONLY	
	30N	30 (113.56 L)	IN & OUT	
	50N	50 (189.26 L)	IN & OUT	
	65S	65 (246.05 L)	OUT ONLY	

Figure 14-26. Nonfeedback proportional directional valves have several spool options.

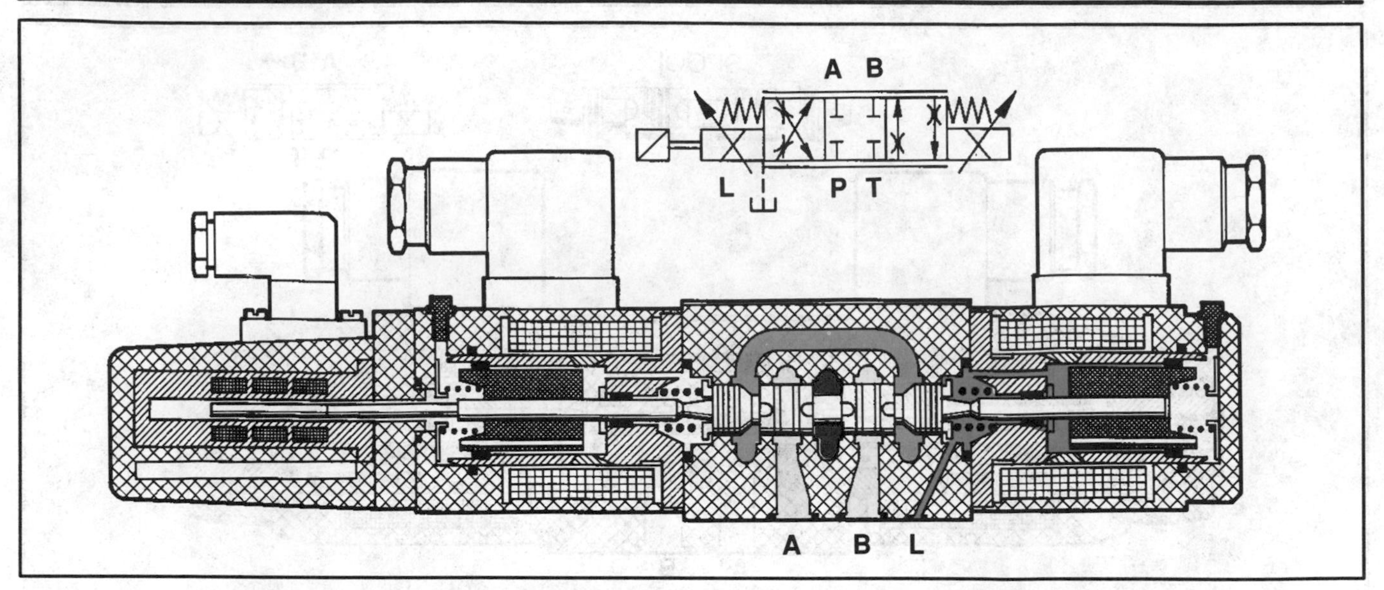

Figure 14-27. Proportional directional valve with LVDT feedback.

which solenoid is energized. The level of the input signal to the energized solenoid specifies the amount of flow through the ports.

Once again, different spool constructions and sizes can be used to match the requirements of different applications. Two different spools have different flow conditions in the center position; one blocks flow through all ports, while the other allows the A and B ports to be bled to the tank. Also, spools with different flow ratings, which we will discuss next, can be selected for different maximum flow rates.

In Figure 14-26, the spool names each contain a letter. These letters indicate the type of notches on the spool: F indicates fine metering notches, N indicates standard notches, and S indicates standard notches that meter out only. The F- and N-type spools have notches on all four spool lands, which permits flow metering on both the inlet and outlet sides of the actuator. When the flows of the two paths are equal, there is no bias toward metering in or metering out. However (for example), when the valve is used to control an unequal-area cylinder, the flow rates are not equal; one of the metering in or metering out flows dominate.

The S-type spool meters only on the two lands that control the flow to the tank port. There is very little metering in the P to A or P to B flow paths. The lands on S-type spools are similar to the lands on a normal switching valve. The pressure drop is considerably less on this spool, because there is less restriction than on the metering lands of the other types of notched spools.

Like throttle valves, the nonfeedback, direct-operated directional valve provides some degree of pressure compensation. As the pressure drops across the valve and the flow rates both increase, the Bernoulli forces increase to present a greater opposition to the solenoid force. This results in smaller port openings, which prevents an increase in flow rate.

Proportional Directional Valve with Feedback

For applications that require more accurate spool positioning, an LVDT position sensor can be used with a direct-operated, proportional directional valve, as shown in Figure 14-27. The spool movement is controlled by the input signal and, therefore, is independent of forces that work on the spool, such as flow, pressure, and friction. For this reason, the valve does not make compensations for pressure like the nonfeedback directional control valve. Like throttle valves with feedback, this valve's flow rate is determined by both the input signal and the pressure drop across the valve. Again, the flow rate is approximately proportional to the square root of the pressure difference.

Up to a certain point, the valve's maximum flow rate is determined by the maximum pressure drop across it. However, the combined forces of flow, pressure, and friction eventually become so large that they overcome the solenoid force, and the valve stops working properly. Large pressure drops are normally avoided when systems are designed; but when a proportional valve is used, for example, to decelerate an actuator or to control an overrunning load, large pressure drops are unavoidable.

Hydrostat. To keep the valve's pressure drop constant, a hydrostat can be used with a proportional directional valve with feedback. The hydrostat can be installed in the valve's pressure inlet line. The hydrostat spool senses pressure at port P and, with a shuttle valve,

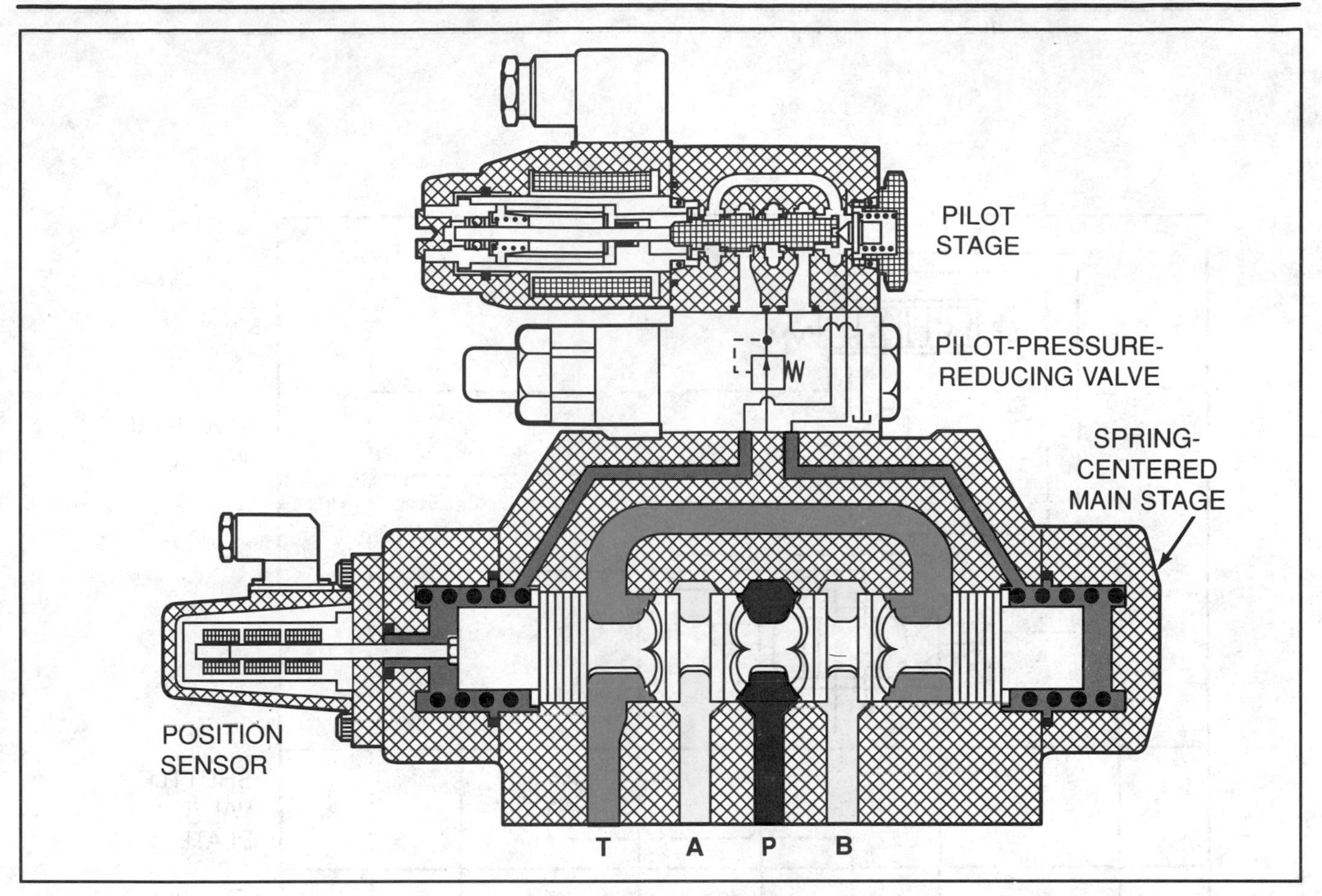

Figure 14-28. Pilot-operated proportional directional valve with feedback.

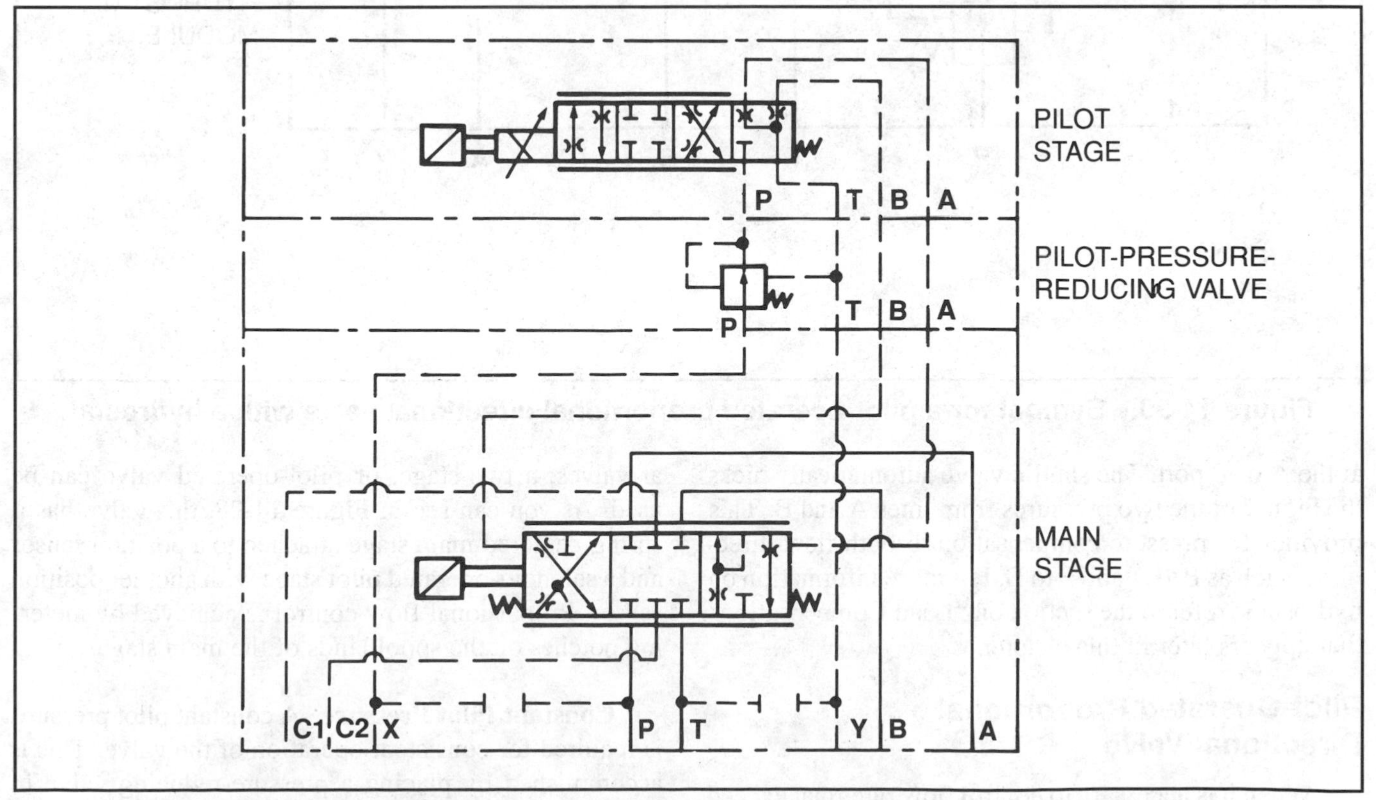

Figure 14-29. Symbol for a pilot-operated proportional directional valve.

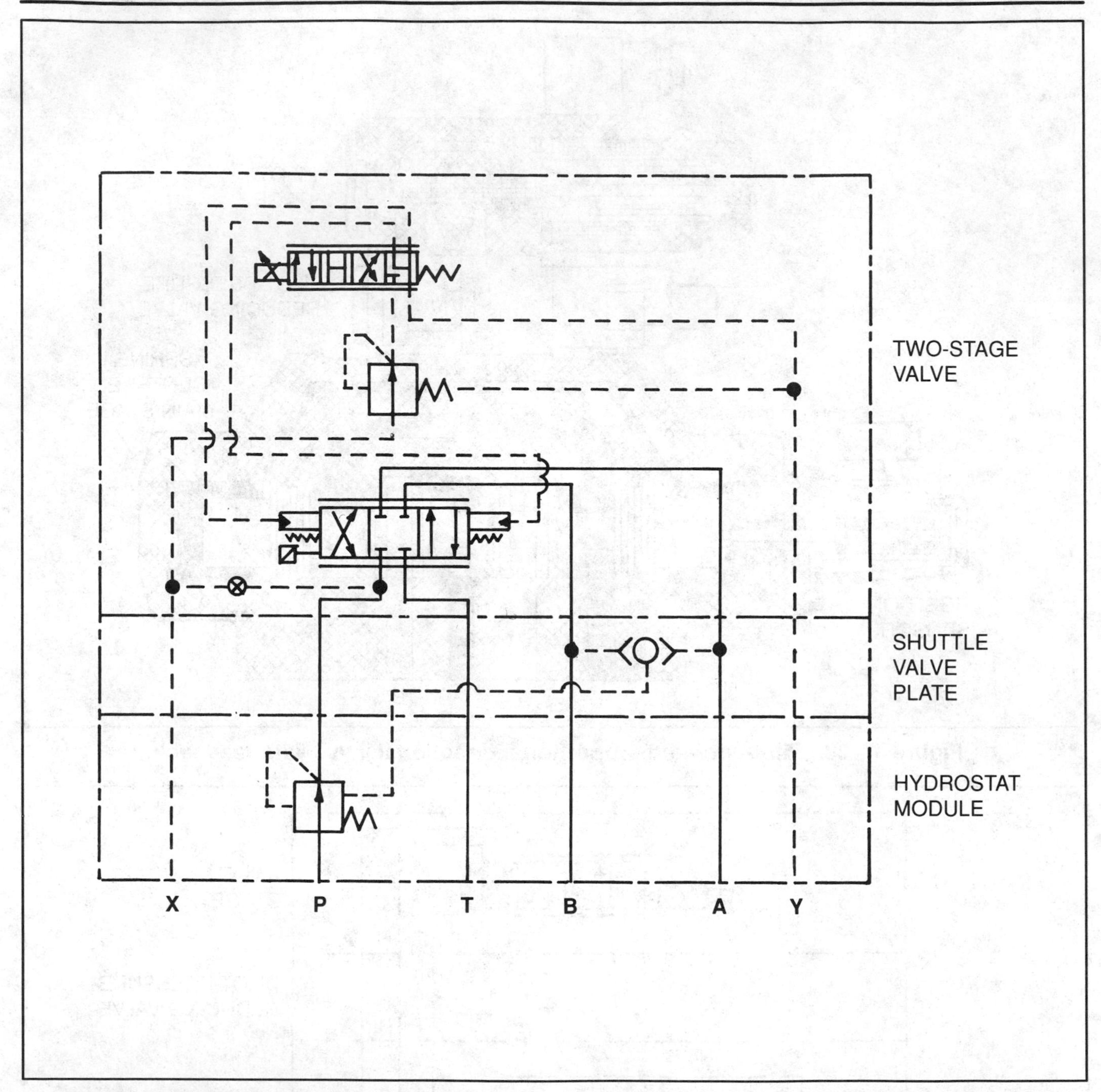

Figure 14-30. Symbol for a pilot-operated proportional directional valve with a hydrostat.

at the A or B port. The shuttle valve automatically picks the higher of the two pressures from lines A and B. This provides for pressure compensation in both flow directions, such as P to A and P to B. For more information on hydrostats, refer to the section on "Load Compensation" that appears later in this chapter.

Pilot-Operated Proportional Directional Valve

When it is necessary to control flow rates that exceed the maximum for direct-operated proportional directional valves, a two-stage, or pilot-operated valve can be used. As you can see in Figure 14-28, this valve has a spring-centered main stage attached to a position sensor and a solenoid-operated pilot stage with another position sensor. Proportional flow control is achieved by metering notches on the spool lands of the main stage.

Constant Pilot Pressure. A constant pilot pressure is required for consistent operation of the valve. This is accomplished by placing a pressure-reducing valve in the pilot pressure feed line to the pilot stage. The pressure

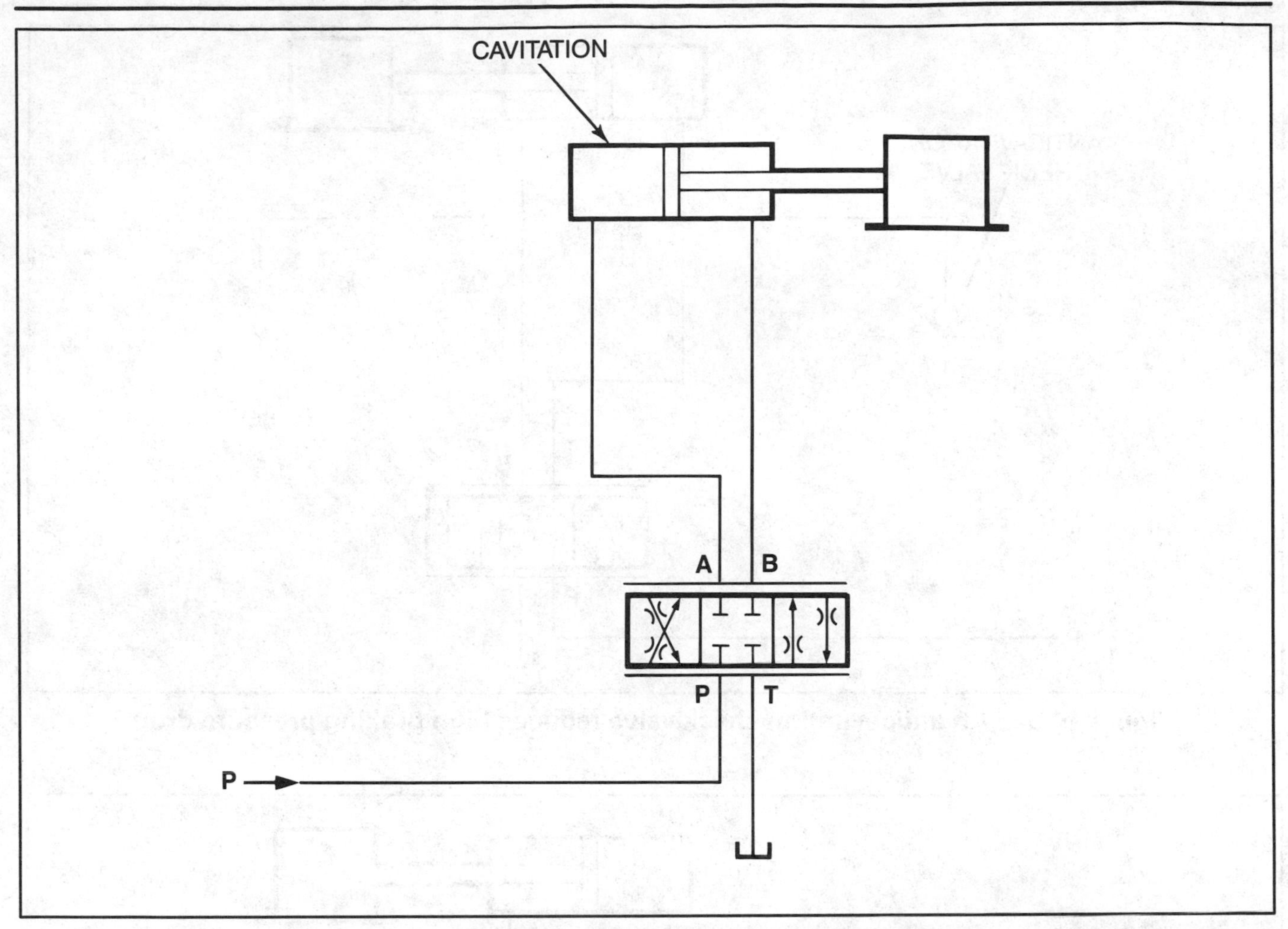

Figure 14-31. Full bore side cavitates with the forward deceleration of a positively loaded cylinder.

reducer is fitted between the valve's pilot and main stages.

As the symbol in Figure 14-29 represents, the pilot spool has four possible controlling positions. When positioned to the extreme right, the pilot spool is in the fail-safe position, which vents both end chambers of the main spool. This allows the springs on the main stage to center the spool, which stops the actuator movement. Under normal operating conditions, the pilot spool operates within the other three positions. The all-ports-blocked position is the null position.

Like other two-stage valves, options are available for internal pilot, external pilot, and drain by fitting and removing the appropriate plugs within the valve body.

Hydrostat. The valve's main stage has two extra ports for sensing pressure when the valve is used with a hydrostat. With an internal drilling in the main spool, the pressure at the A port is sensed at C1, and port B pressure is sensed at C2 (Figure 14-30). The hydrostat is located in the pressure inlet connection to the valve. It senses pressure in the P port and either the A or B port, depending on the main spool movement. Pressure compensation is achieved either P to A or P to B, with a constant pressure drop maintained across the valve spool. As with other valves we have discussed in this chapter, there are two optional spool center conditions. One spool type blocks all ports in the center position, while another spool allows A and B to be bled to the tank when centered.

Application Summary

A direct-operated directional valve without spool position feedback can be used in applications with a moderate accuracy requirement. A direct-operated directional valve with spool position feedback should be used in applications that require more accurate spool positioning. Pilot-operated directional valves with feedback can be used in applications that require accurate control of higher flow rates.

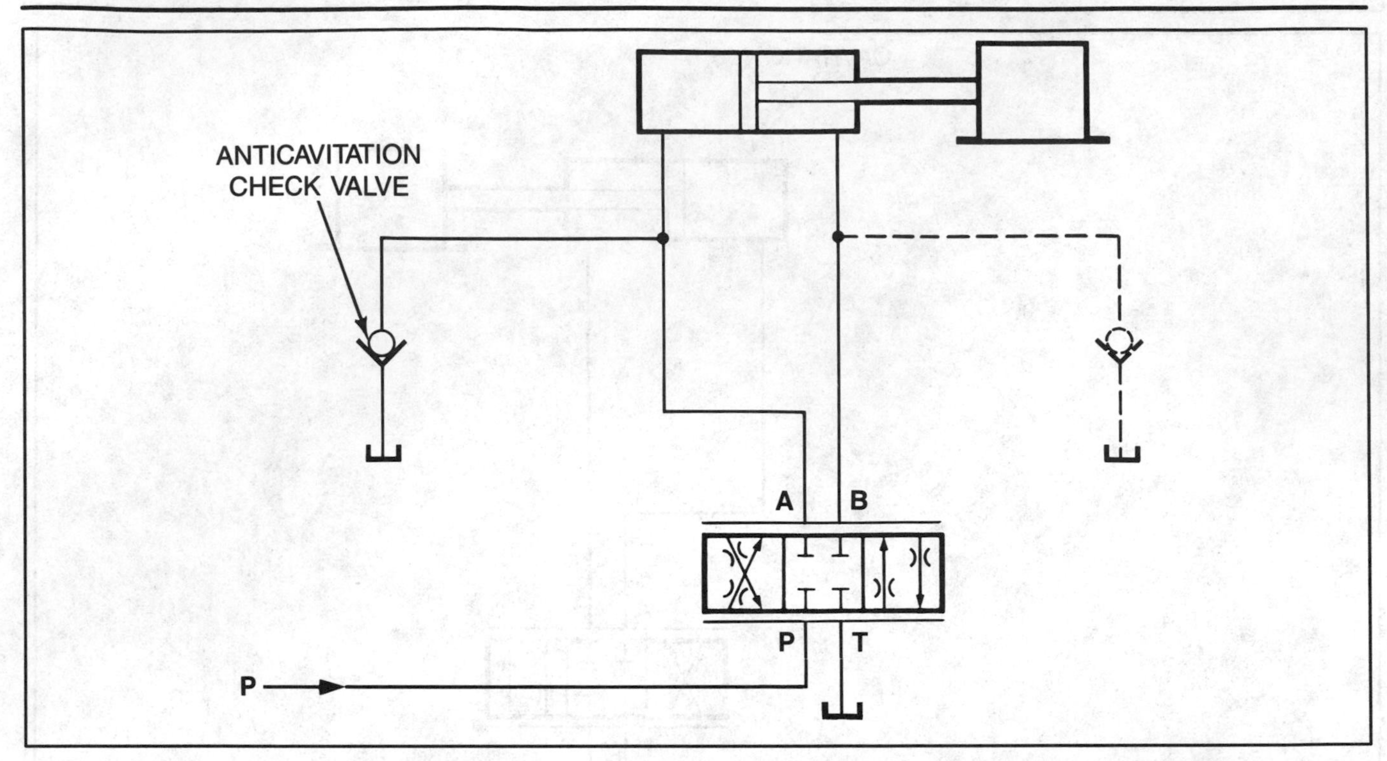

Figure 14-32. An anticavitation check valve reduces high braking pressure drop.

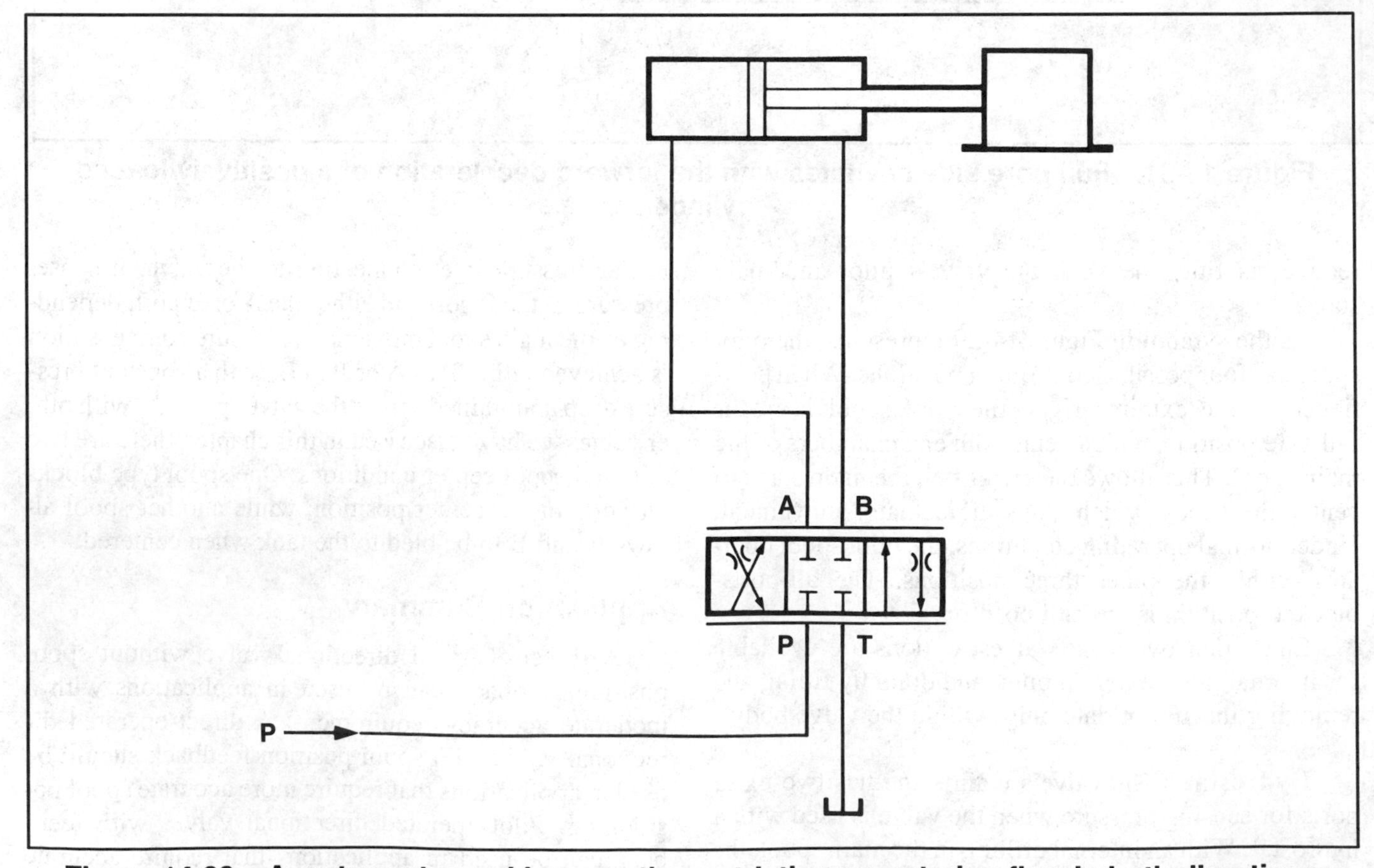

Figure 14-33. A meter-out spool lessens the restriction on entering flow in both directions.

APPLICATION GUIDELINES

The following section contains several sample applications of proportional valves. These examples illustrate some of the major considerations involved in the implementation of proportional valves.

Example Number One

Figure 14-31 represents an application of a symmetrical spool-type valve that controls a positively loaded cylinder with all ports blocked in the center position When all flow paths are equally restricted, the main control is meter-in on the advancing stroke, and meter-out as the stroke retracts.

Like all valves that block all ports in the center position, the piston tends to creep out when the valve is centered. This happens when the spool leaks to one side of the piston, causing it to move slightly.

To decelerate the forward stroke, the spool moves back toward its center position to restrict the flow on the rod side of the cylinder piston. This also restricts the flow on the full-bore side of the cylinder piston. The flow rate on the cylinder's full-bore side is larger than the flow on the rod side. As you recall, the pressure drop is proportional to the square of the flow. Therefore, this high braking pressure on the rod side of the cylinder causes an extremely high pressure drop in the full-bore flow path. This large pressure drop can cause cavitation.

One solution is to install an anticavitation check valve, as shown in Figure 14-32. This reduces the pressure drop on the full-bore side and prevents cavitation.

A second solution to this problem is to use a meter-out spool instead of a meter-in spool, as shown in Figure 14-33. This allows the inward flow to the cylinder to remain almost totally unrestricted in both the forward and retracting directions. However, this solution does not fully consider the problem of intensified pressure on the rod side of the cylinder piston when decelerating. Also, the meter-out spool cannot be used in a system with some types of hydrostats.

Another factor to consider is the sudden loss of power during a power failure or emergency stop conditions. When ramp signals are fed to a proportional valve to decelerate a high inertia load, a sudden loss of power causes the spool to move to the center position very quickly, which will cause peak pressures. The pressure will be greatest on the rod side of the cylinder piston when the cylinder is extended. To avoid peak pressures, relief and anticavitation valves can be installed, as illustrated in Figure 14-34.

Differential creep is unwanted piston movement caused by leakage into lines A and B. To reduce its effect, a spool that bleeds A and B to the tank can be used (Figure 14-35). However, under light load conditions, a high backpressure or peak pressure in the tank can still cause the piston to creep out of position.

Example Number Two

Figure 14-36 is a schematic of a negatively loaded cylinder, which means that gravity acts on the load. For this type of cylinder, it is important to consider the effects of the load acting on the full-bore area or the load acting on the rod side of the cylinder piston. In the arrangement shown in Figure 14-36, the main restriction is meter-in on the forward stroke and meter-out when the cylinder retracts. Because the load naturally decelerates as it moves upward, anticavitation check valves are normally not required.

One problem might occur when the cylinder is held in a partially or fully extended position. The spool would move to its center position, in which all ports are blocked. Line A could leak to the tank, causing the piston to move downward. The amount that leaks depends on variations in the load, fluid temperature and viscosity, valve wear, and other influences.

Therefore, an electronic adjustment of the spool's null position to compensate for the leak would not sufficiently solve the problem. A better solution to this leakage problem is to install a pilot-operated check valve (Figure 14-37).

The pilot-operated check valve must be piloted independently. This ensures that the valve will stay open at all times while the cylinder is moving, especially during deceleration. Another important consideration for a negatively loaded cylinder is that, if it requires smooth deceleration when lowered, a spool with all ports blocked in the center position is better than a spool that has A and B bled to tank in the center position.

If there is leakage across the proportional valve spool, however, the cylinder might move as soon as the pilot-operated check valve is opened. In addition, if the pilot-operated check valve is installed a good distance from the proportional valve, fluid compression in the piping might cause the cylinder to drop momentarily as the pilot-operated check valve is opened.

Therefore, in some applications, it may be necessary to give a small lifting signal to the proportional valve before a lowering signal, the check valve opening speed could be made slower, or a counterbalance valve could be used.

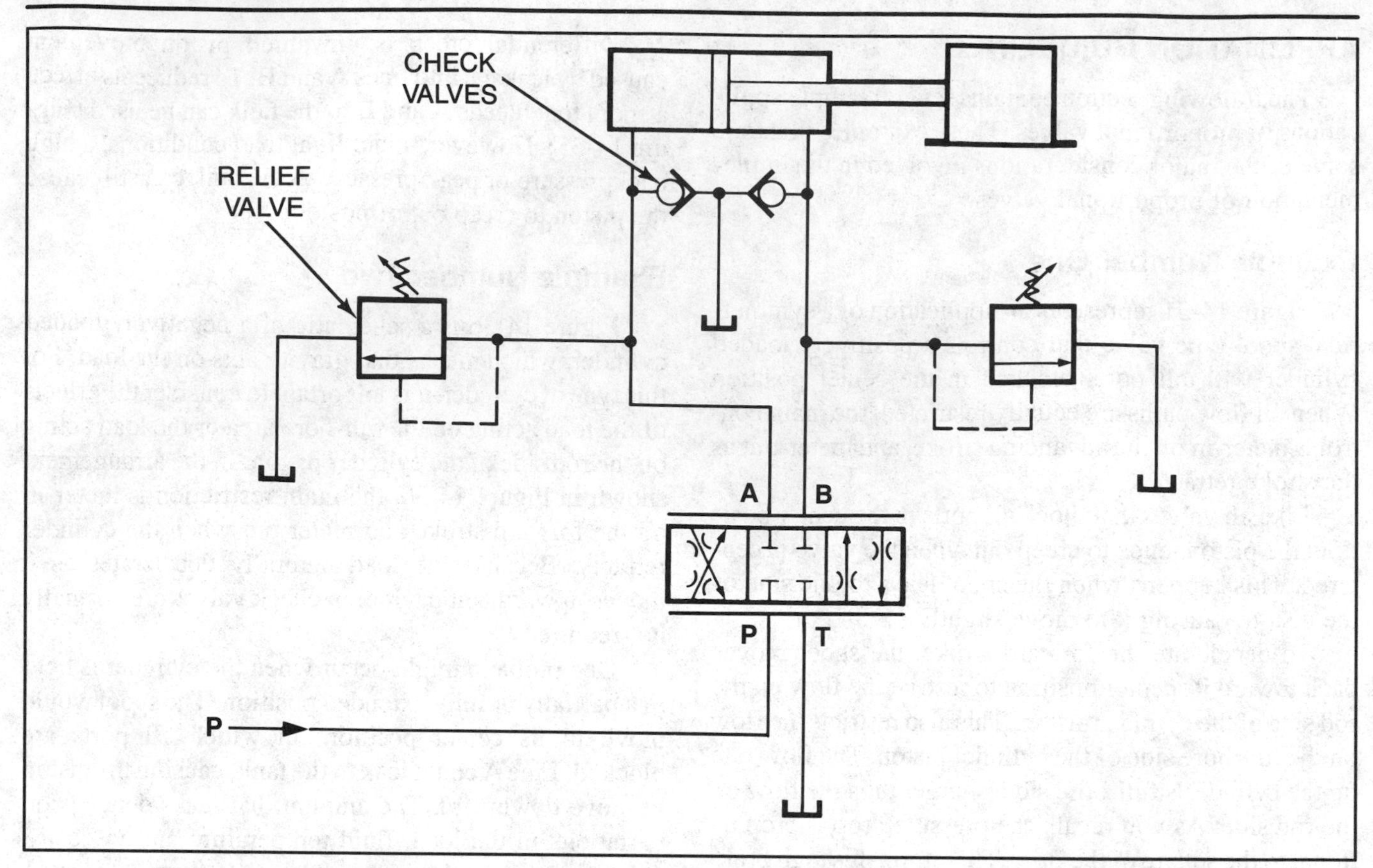

Figure 14-34. Relief and anticavitation check valves prevent peak pressures.

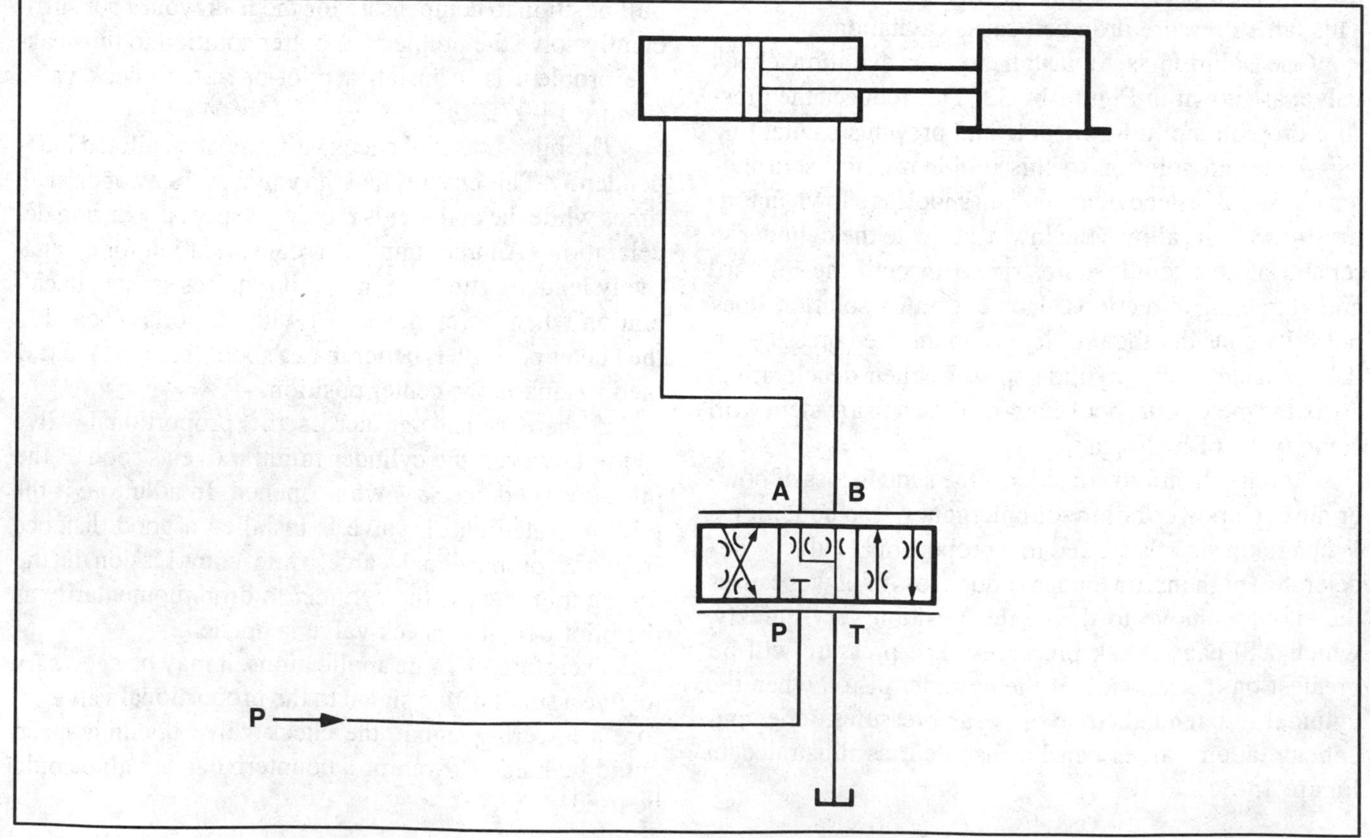

Figure 14-35. A type 33 spool bleeds A and B to the tank in the center position to offset differential creep.

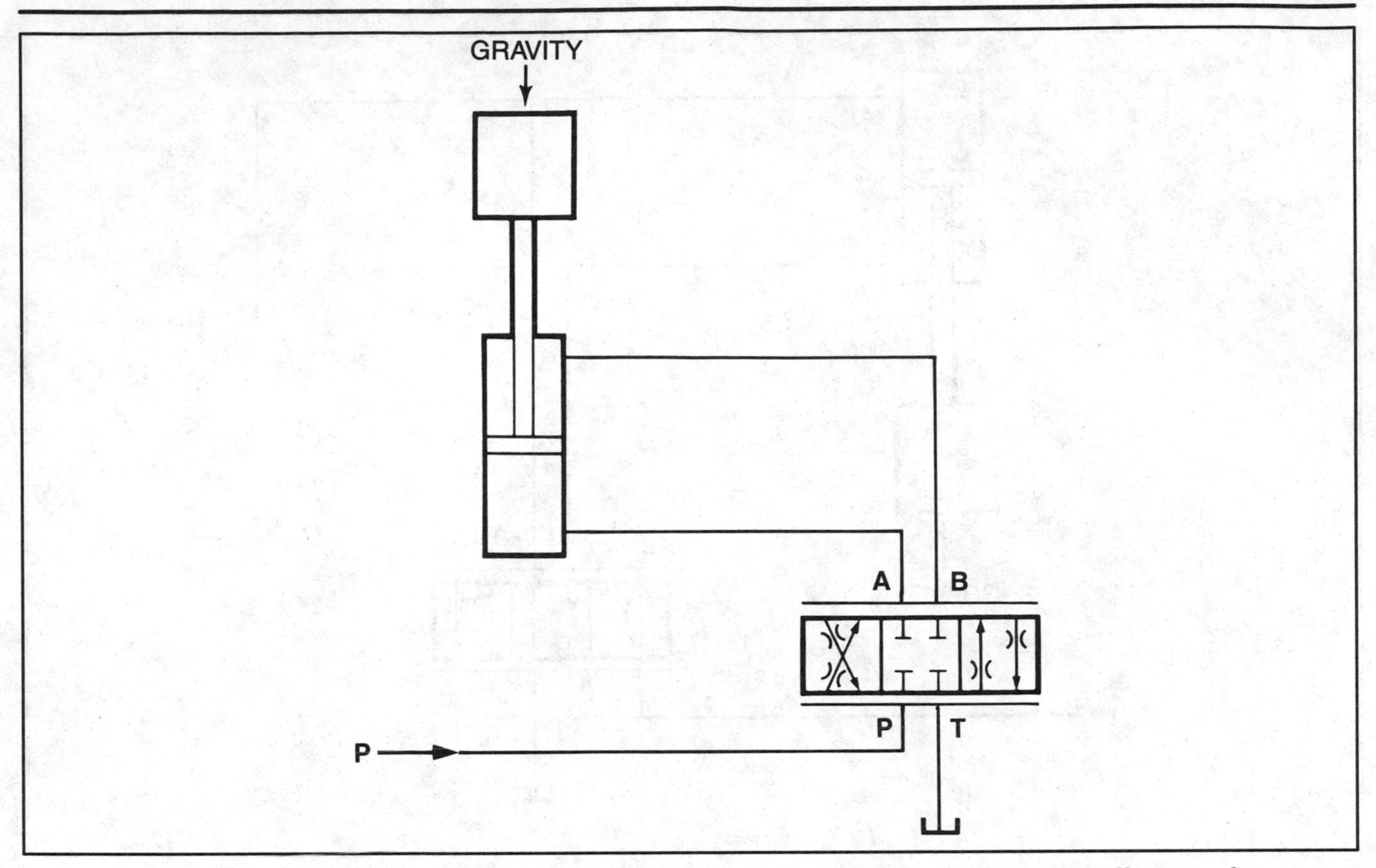

Figure 14-36. The load acts on the rod side of the cylinder in the extending stroke.

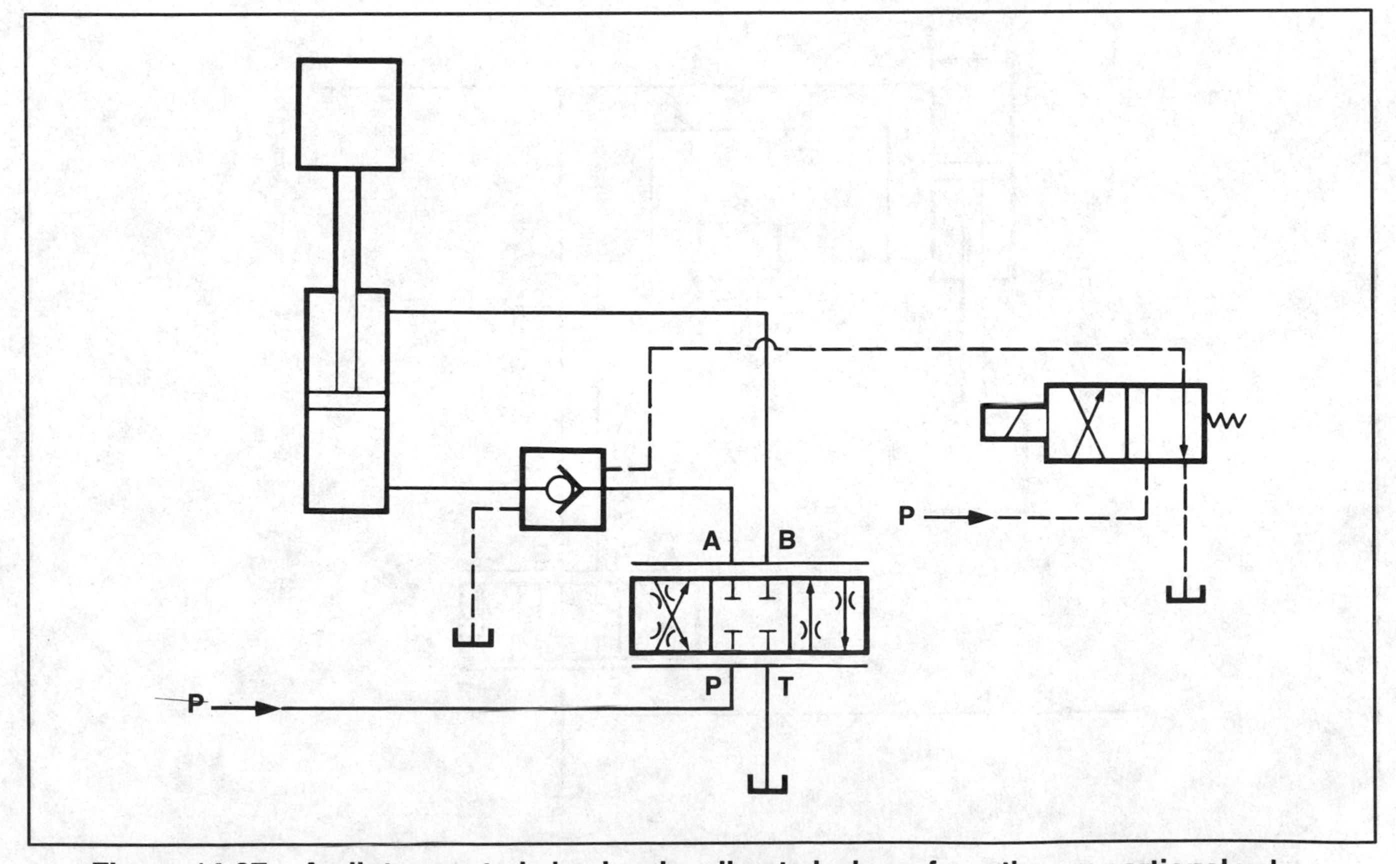

Figure 14-37. A pilot-operated check valve diverts leakage from the proportional valve.

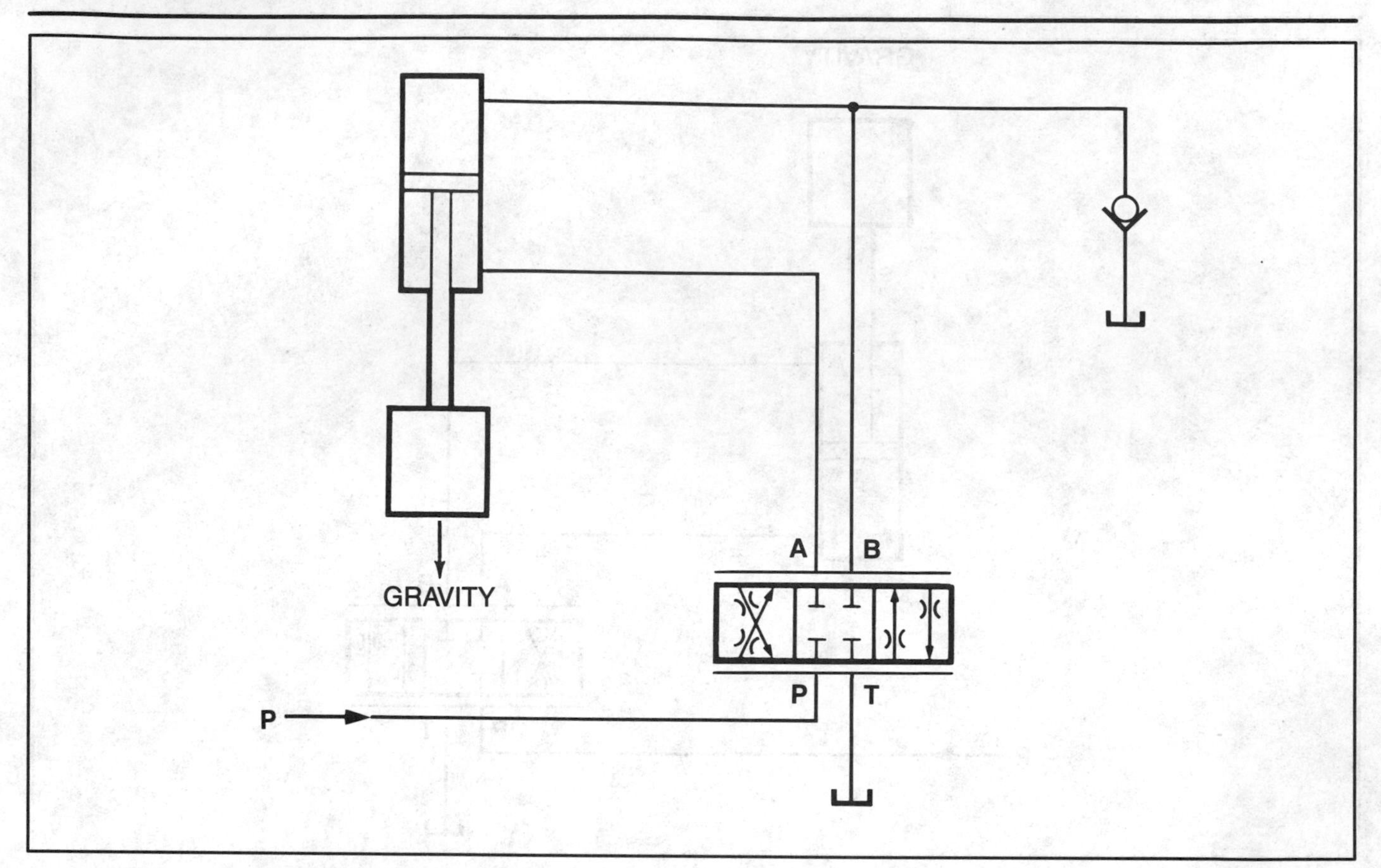

Figure 14-38. The load acts on the rod side of the cylinder in the extending stroke.

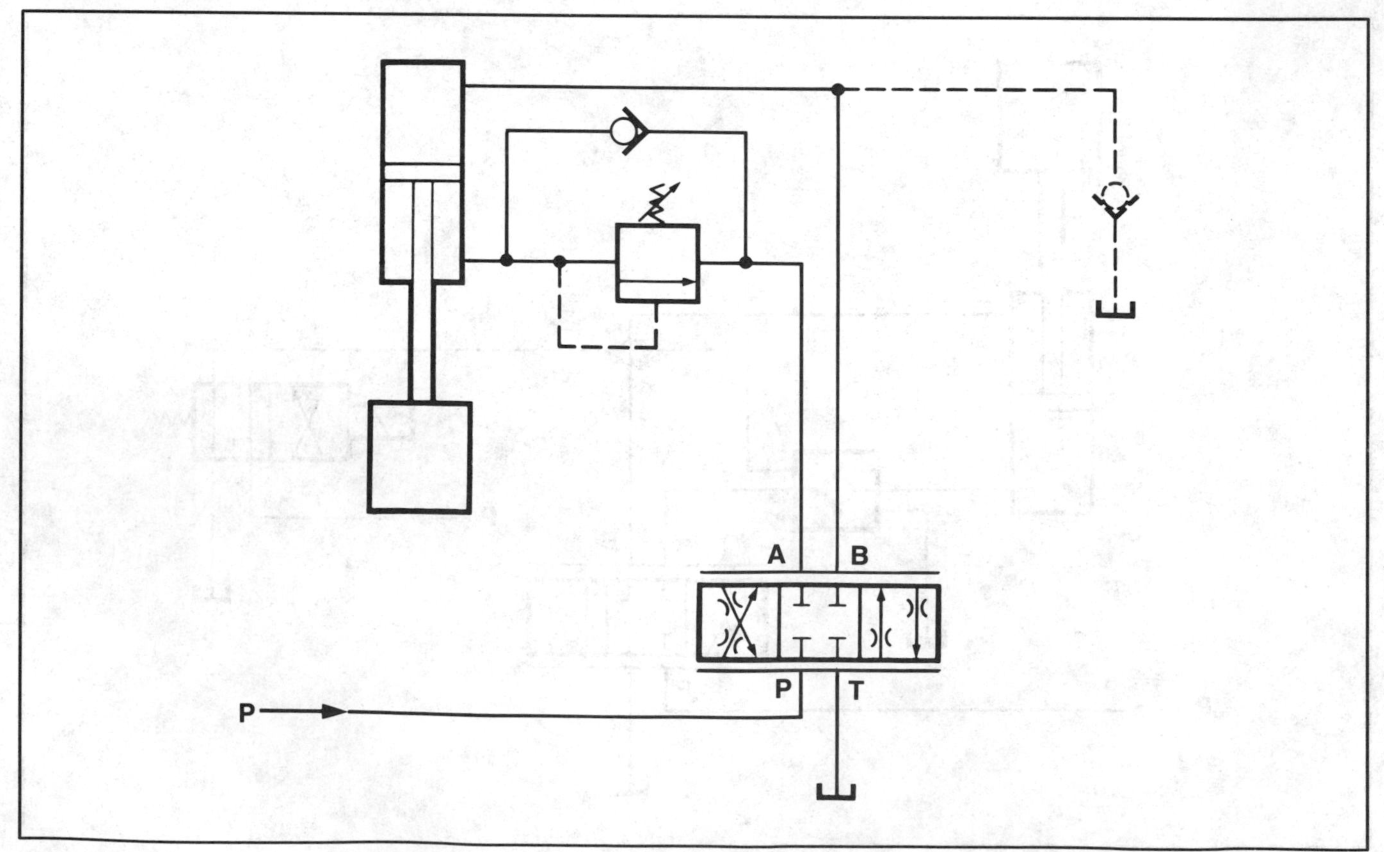

Figure 14-39. A counterbalance valve prevents uncontrolled lowering of the cylinder.

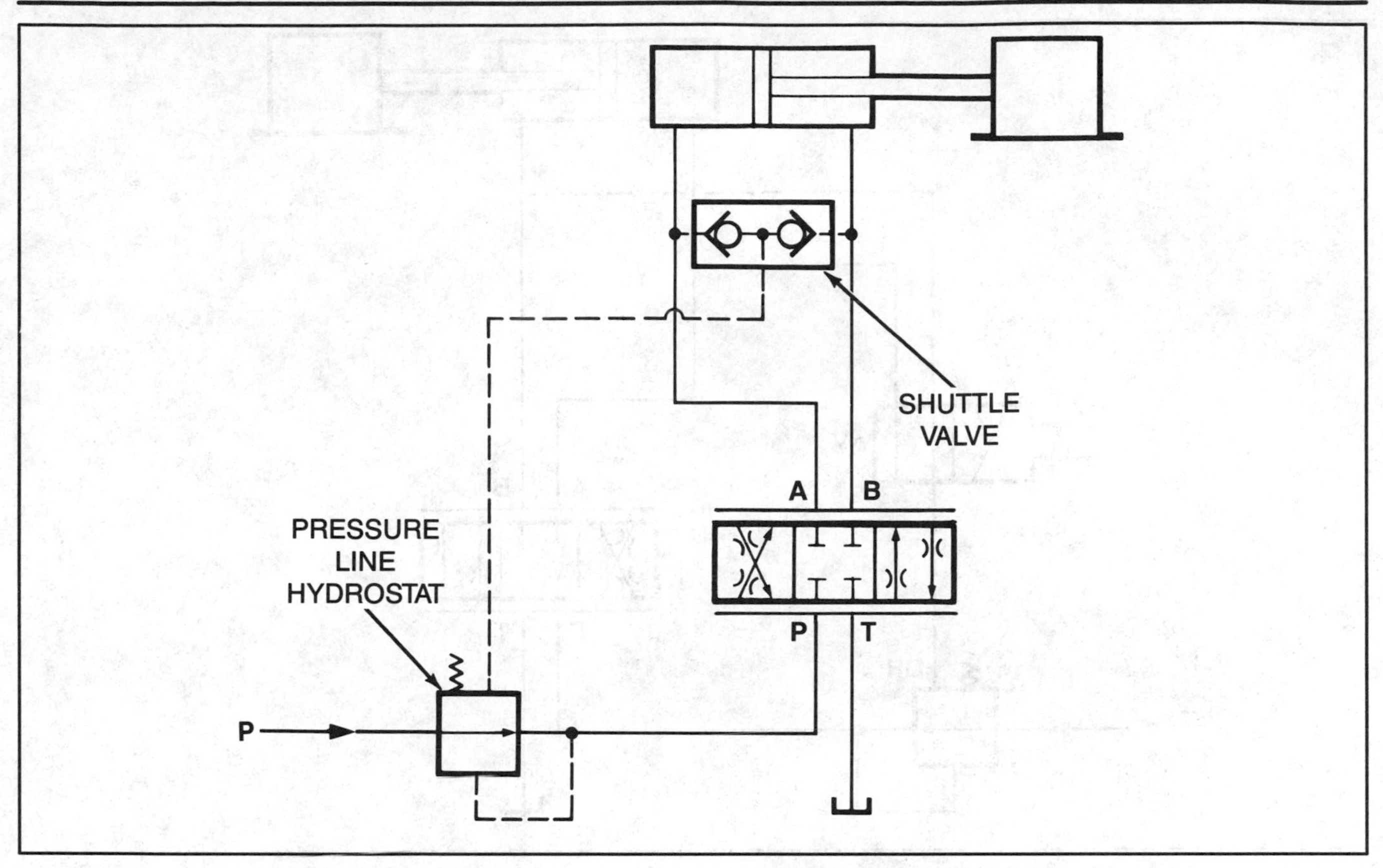

Figure 14-40. Pressure line hydrostat and shuttle valve.

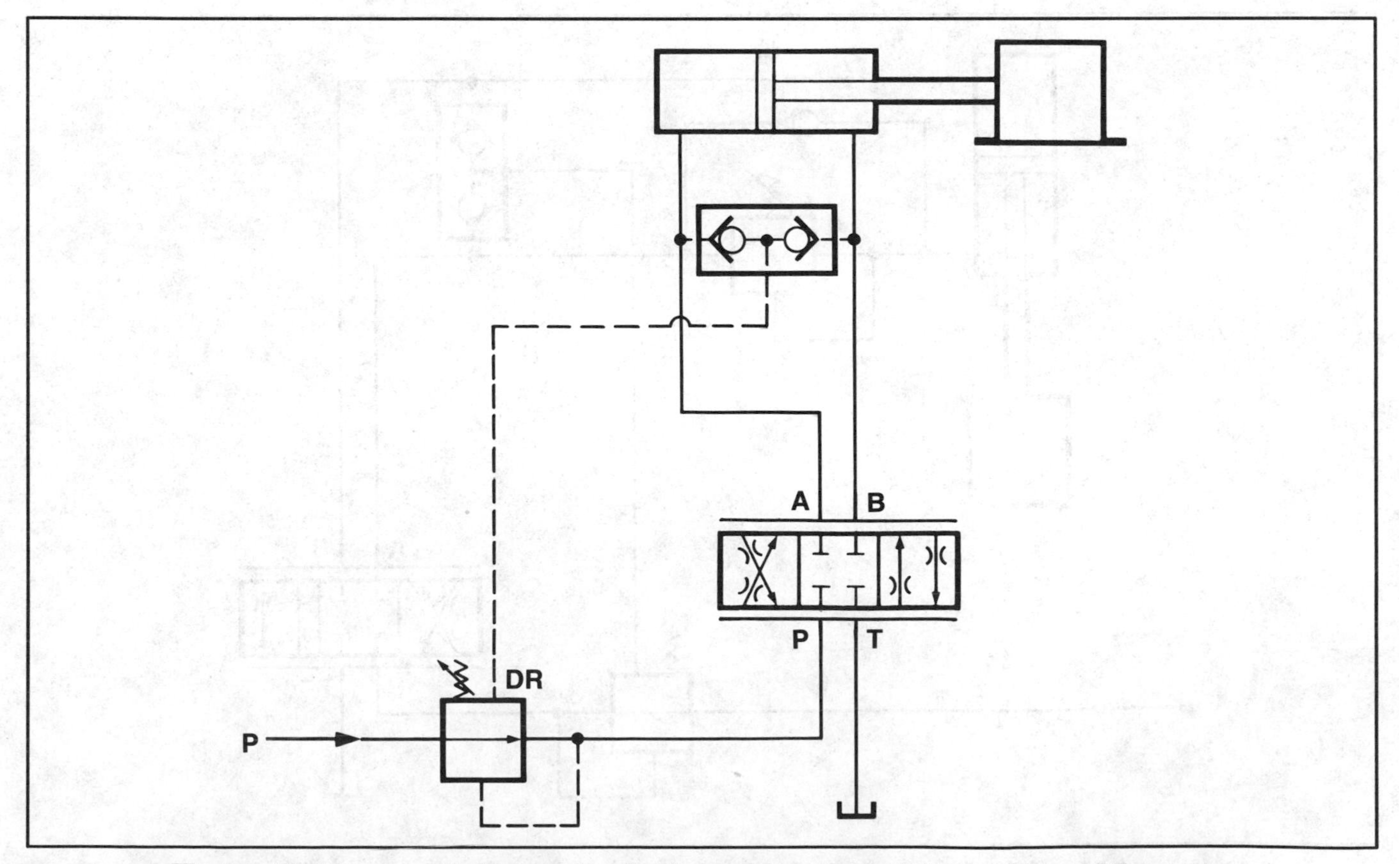

Figure 14-41. A variable spring hydrostat increases the maximum flow rate.

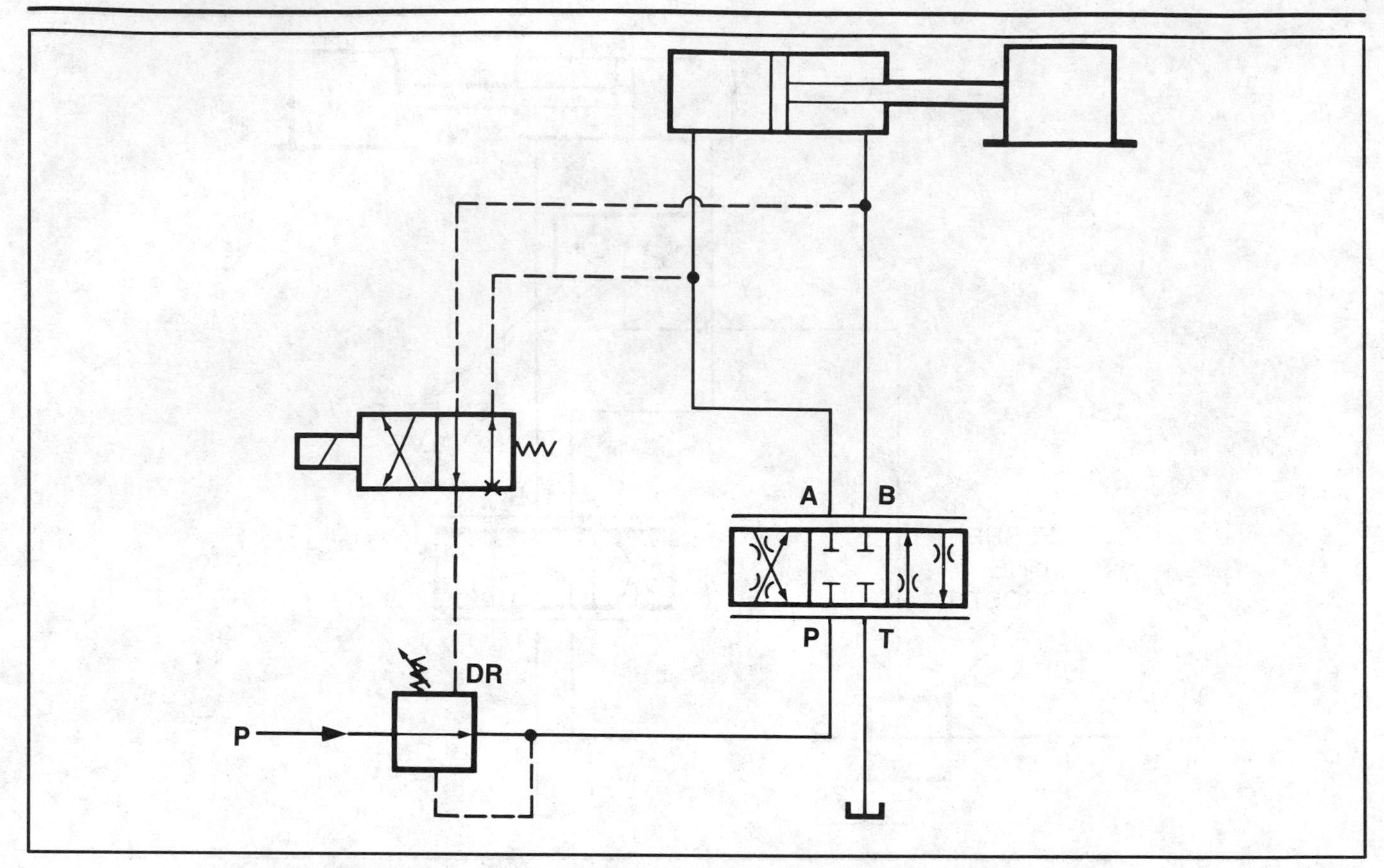

Figure 14-42. A solenoid valve provides load compensation.

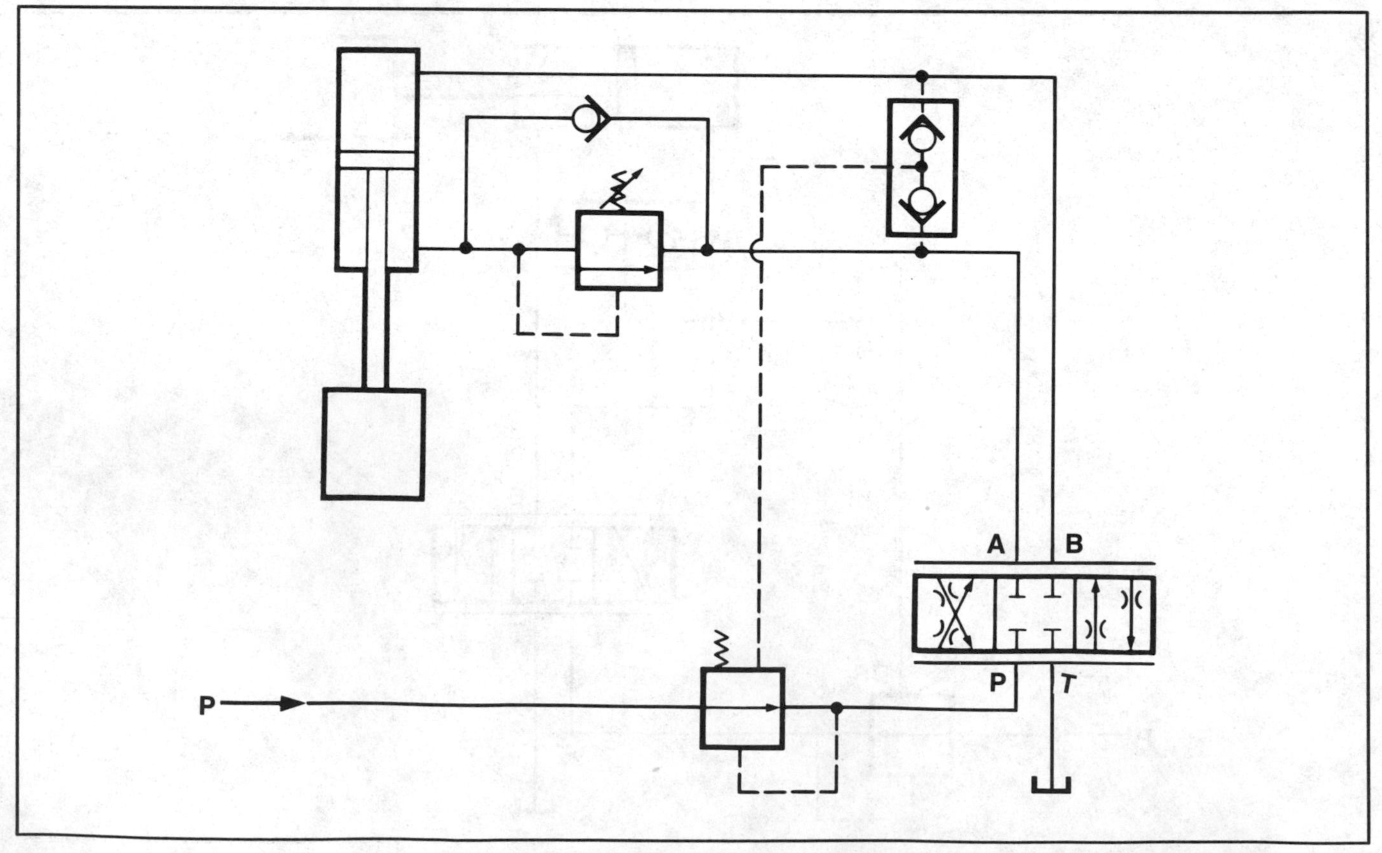

Figure 14-43. A counterbalance valve is used for a negatively loaded cylinder.

Example Number Three

When the load acts on the rod side of the cylinder piston, like in Figure 14-38, several different factors must be considered. In particular, the forward extension of the stroke requires special attention.

In the schematic shown in Figure 14-38, the main restriction is the meter-in when the cylinder is extended and the meter-out when the cylinder retracts. However, as we discussed earlier, metering in while the cylinder is lowered can cause cavitation. Therefore, this application requires an anticavitation check valve.

While simply using a meter-out spool might prevent cavitation in some cases, each application of this kind should be carefully checked out for intensified pressure in the rod side of the cylinder piston.

In some cases, safety considerations require the use of a counterbalance valve (Figure 14-39). This valve ensures that the cylinder is not lowered uncontrollably when the proportional valve is fully opened. Also, the use of a counterbalance valve makes the cylinder positively loaded, so the anticavitation check valve is not required, unless the application demands very rapid deceleration as the cylinder is lowered.

A counterbalance valve would still restrict the A to T flow path when the cylinder is lowered. Therefore, when analyzing circuit operation, the backpressure created across the valve would need to be added to the counterbalance valve setting.

LOAD COMPENSATION

As you have learned, many proportional solenoid valve applications require some degree of pressure compensation. This is especially true for applications that require a constant actuator speed during fluctuations in load or system pressure.

As you recall, nonfeedback-type valves compensate for pressure differences to a certain extent. When the built-in pressure compensation does not sufficiently reduce pressure, a hydrostat can be used with the proportional solenoid valve.

In Figure 14-40, a pressure line hydrostat senses the pressure at the P port and either the A or B port of the valve. A built-in shuttle valve picks the higher of the two pressures from A and B and feeds it back to the hydrostat spool. The hydrostat spool then makes any adjustments necessary to maintain a constant pressure drop across the proportional valve. The pressure difference will be equal to the hydrostat spring. This provides for meter-in pressure compensation when the cylinder moves in either direction.

When the cylinder moves forward, the A port pressure must be higher than the B port pressure. This is generally true as long as the load does not overrun, except during deceleration. As the cylinder decelerates in the forward direction, pressure in the B port very likely will be higher than in the A port, unless frictional forces acting on the load are high. When this happens, the hydrostat spool moves toward the fully open position to reduce the effect of cavitation on the full-bore side of the cylinder.

A hydrostat used for load compensation affects a system's pressure requirements. If the hydrostat in Figure 14-40 has a 58 psi (4.00 bar / 399.85 kPa) spring and the cylinder has a 2:1 area ratio, when the cylinder is retracting with a zero load, the pressure drop from P to B will be 58 psi (4.00 bar / 399.85 kPa). The pressure drop from A to T will be 232 psi (15.99 bar / 1599.41 kPa) (remember, P is proportional to Q^2), and the pressure in the full-bore side of the cylinder will be at least 232 psi (15.99 bar / 1599.41 kPa). This will require a pressure of 464 psi (31.99 bar / 3198.82 kPa) on the rod-side of the cylinder to overcome the pressure on the other side. Therefore, a system pressure of 522 psi (35.99 bar / 3598.67 kPa) is required (464 psi (31.99 bar / 3198.82 kPa) plus 58 psi (4.00 bar / 399.85 kPa) drop from P to B). If the cylinder is loaded in the retract direction, the load pressure must be added to the total system pressure requirement.

Using a hydrostat also affects a proportional valve's maximum flow rate. A hydrostat has a lowering effect on flow rate, because the maximum flow is the amount that gets through the valve at a pressure difference equal to that of the hydrostat spring. A hydrostat with a variable spring can be used to increase the maximum flow rate. This creates a conventional pressure-reducing valve.

Instead of connecting the DR port of the reducing valve to the tank, it can be connected to lines A and B using a shuttle valve, as shown in Figure 14-41. Therefore, by varying the setting on the reducing valve, the spring setting of the hydrostat and the pressure drop of the proportional valve can be adjusted. However, the amount of wasted power goes up with the hydrostat setting. In addition, the continual pilot flow from the reducing valve's DR port establishes a specific minimum flow capability.

If a solenoid valve replaces the shuttle valve, the sensing line can be positively selected at any time (Figure 14-42).

To achieve proper compensation, negatively loaded cylinders must have a counterbalance valve in the negative load direction when used with a meter-in compensator (Figure 14-43).

It is possible to build virtually any combination of proportional valve/hydrostat arrangements using independently mounted pressure-reducing valves. For example, a full reducing or relief valve assembly can be used if variable hydrostat pressure is required. If the load changes direction while the cylinder is moving, such as during deceleration, the hydrostat sensing line can be independently switched using a solenoid valve.

The same general principles apply when using load sensing with a variable pump. Flow must be metered into the actuator when the load pressure is sensed. Therefore, the meter-out spool should not be used for load sensing.

SUMMARY

Proportional solenoid valves provide solutions to design problems that cannot be resolved using only conventional on/off DC solenoid valves. The replacement of hydraulic valves with electrical control devices simplifies the variable control of pressure, flow, and direction. In many applications, proportional solenoid valves are implemented in systems that require various combinations of conventional, cartridge, and proportional valves.

QUESTIONS

1. List three major functions of proportional solenoid valves.
2. Explain the expression "output flow is proportional to the input signal" and how a proportional solenoid actuates a proportional valve spool.
3. Explain the significance of a constant-force solenoid.
4. Describe deadband compensation.
5. Describe hysteresis and how a control amplifier compensates for it.
6. Explain ramp functions and list the variables they are designed to control.
7. Explain the purpose of current feedback.
8. Describe pulse width modulation.
9. Explain the difference between feedback and nonfeedback proportional valve operation.
10. Explain the difference between symmetrical and nonsymmetrical proportional valve spools.
11. Describe the relationship between flow rate and pressure drop through a proportional valve.
12. Describe three types of proportional pressure valves.
13. Explain the operation of an LVDT.
14. Describe metering notches and their purpose.
15. How can cavitation be prevented when controlling an unequal-area actuator, such as a cylinder?
16. Describe some of the special considerations for controlling a negative load with a proportional valve.
17. Describe load compensation and why it is needed.
18. How can a hydrostat affect pressure drop across a proportional valve?

CHAPTER 15
SERVO VALVES

A servo valve is a directional valve that may be infinitely positioned to provide control of both the amount and the direction of fluid flow. A servo valve coupled with the proper feedback sensing devices provides very accurate control of the position, velocity, or acceleration of an actuator.

The **mechanical** servo valve or **follow valve** has been in use for several decades. The **electrohydraulic** servo valve is a more recent arrival on the industrial scene.

MECHANICAL SERVO VALVE

The mechanical servo valve is essentially a force amplifier used for positioning control. It is illustrated schematically in Figure 15-1.

The control handle or other mechanical linkage is connected to the valve spool. The valve body is connected to and moves with the load. When the spool is actuated, it allows flow to a cylinder or piston to move the load in the same direction as the actuated spool. The valve body follows the spool. Flow continues until the body is centered, or neutral, with the spool. The effect is that the load always moves a distance proportional to the spool movement. Any tendency to move farther would reverse oil flow, moving it back into position.

The mechanical servo valve is often referred to as a booster. The hydraulic boost is capable of considerably greater force than the mechanical input, with precise control of the distance moved.

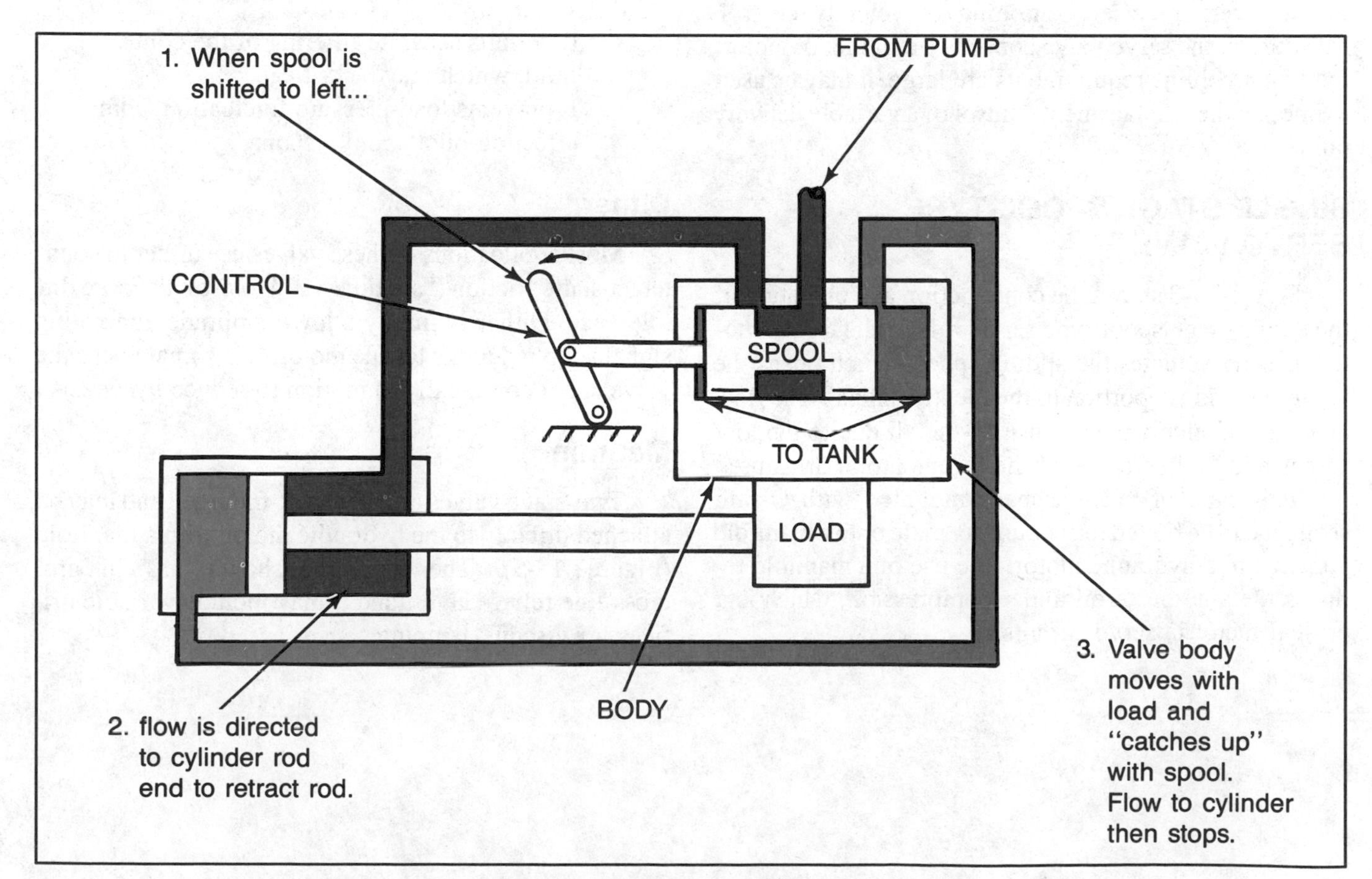

1. When spool is shifted to left...
FROM PUMP
CONTROL
SPOOL
TO TANK
LOAD
BODY
2. flow is directed to cylinder rod end to retract rod.
3. Valve body moves with load and "catches up" with spool. Flow to cylinder then stops.

Figure 15-1. Mechanical servo valve uses "follow valve."

Perhaps the most significant application of the mechanical servo valve is in power steering. Some of the first hydraulic steering units were developed by Harry Vickers, founder of Vickers, Inc. Power steering today is almost universal on full-size passenger cars and widely used on trucks, buses, and other large vehicles. At the present time, the many design variations of power steering systems all operate on this same principle.

ELECTROHYDRAULIC SERVO VALVES

An electric signal to a torque motor or similar device, which directly or indirectly positions a valve spool, essentially operates electrohydraulic servo valves. The signal to the torque motor (Figure 15-2) may come from a simple potentiometer, an electronic controller, or other source. This signal, fed to the servo valve through a servo amplifier, commands the load to move to a specific position or to assume a specific velocity. The amplifier also receives an electrical signal fed back by a tachometer generator, potentiometer, or other transducer connected to the load. This feedback is compared with the original command input; any resultant deviation is relayed to the torque motor as an error signal, causing a correction to be made.

The various types of electrohydraulic servos can provide very precise positioning or velocity control. Most often, the servo valve controls a cylinder or motor; but when volume requirements are large, it may be used to operate the displacement control of a variable delivery pump.

SINGLE-STAGE SPOOL-TYPE SERVO VALVE

Figure 15-3 shows the construction and operation of the single-stage spool-type servo valve. The torque motor directly actuates the sliding spool, which opens the valve ports in proportion to the electric signal. The flow capacity of such valves is usually small due to the low forces and limited travel of the torque motor armature.

This type of valve is back-mounted with O-ring seals. It can be bolted to a mounting plate or to a manifold attached to a hydraulic motor. The use of a manifold reduces the amount of oil under compression, which is a critical factor in servo circuits.

TWO-STAGE SPOOL-TYPE SERVO VALVES

Two-stage spool-type servo valves (Figure 15-4) are used where larger flow rates are desired. In this design, the torque motor actuates a pilot valve inside a ported sleeve. The pilot valve, when shifted, directs fluid to shift the main valve spool. The main valve spool allows flow to the actuator.

Mechanical Feedback

The mechanical feedback linkage in this valve lets the pilot valve act as a follow valve. Movement of the main spool is transmitted back to the pilot valve sleeve to effectively center the pilot valve when the main spool has moved the desired increment. The feedback linkage fulcrum is variable so that the ratio of main spool movement to that of the pilot spool can be as much as 5 1/2 to 1.

Control Pressure

Control pressure for this valve is usually taken from a separate source, such as supply pressure, by incorporating a pressure-reducing valve and accumulator. The separate source is preferred because:

- It provides more flexibility for trimming the system.
- It permits separate filtering of the control fluid, which may be critical.
- It prevents load-pressure fluctuation from affecting pilot-spool response.

Dither

Most applications of these valves use dither to counteract static friction ("stiction") and to provide more dirt tolerance. Dither is simply a low-amplitude alternating signal supplied to the torque motor; this signal keeps the valve spool continually in motion to reduce hysteresis.

Mounting

Two-stage valves also are back-mounted and may be attached directly to the hydraulic motor with a manifold (Figure 15-5). The manifold shown has integral cross-line relief valves, and it may include variable orifices for viscous damping.

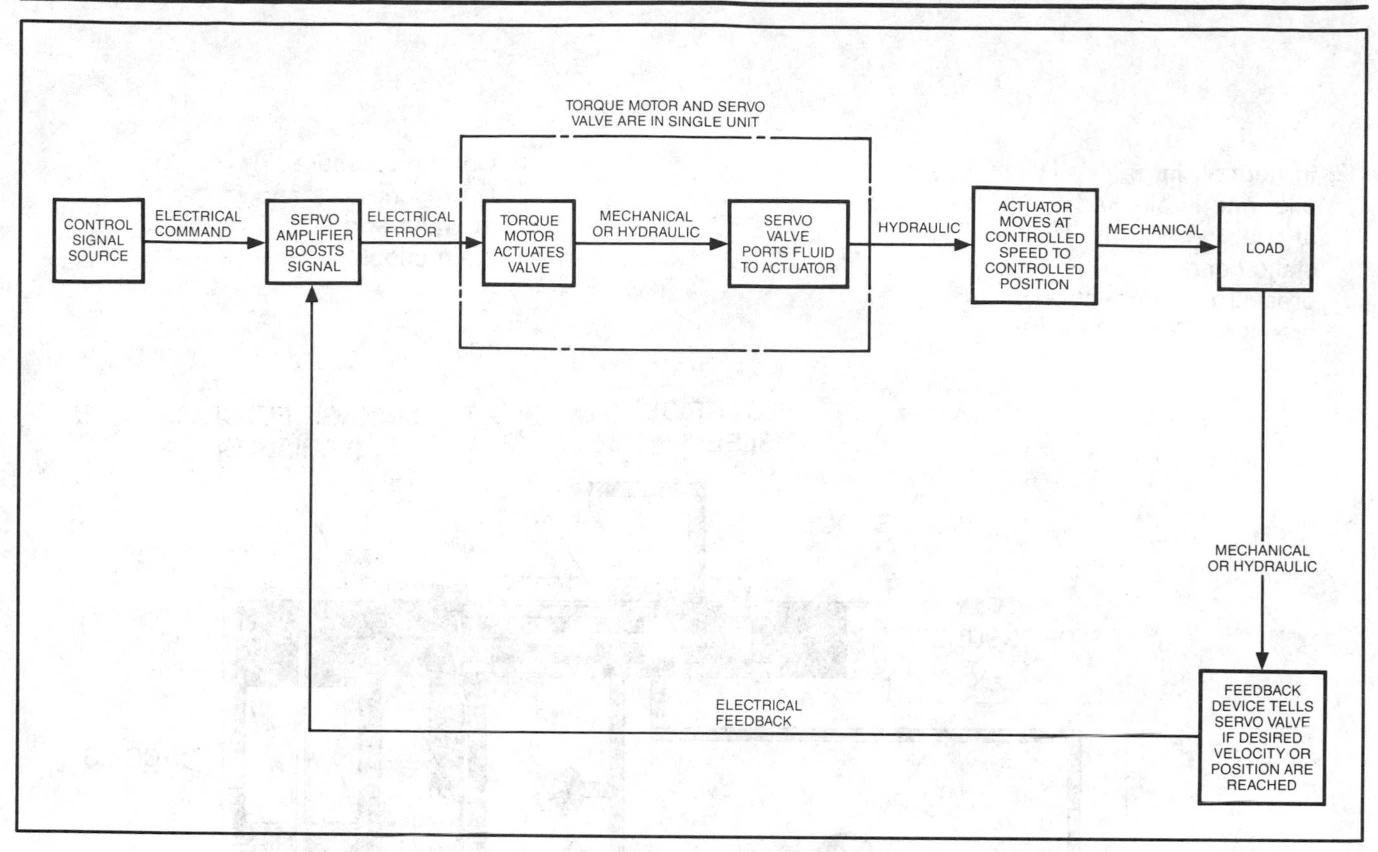

Figure 15-2. Block diagram of servo valve system.

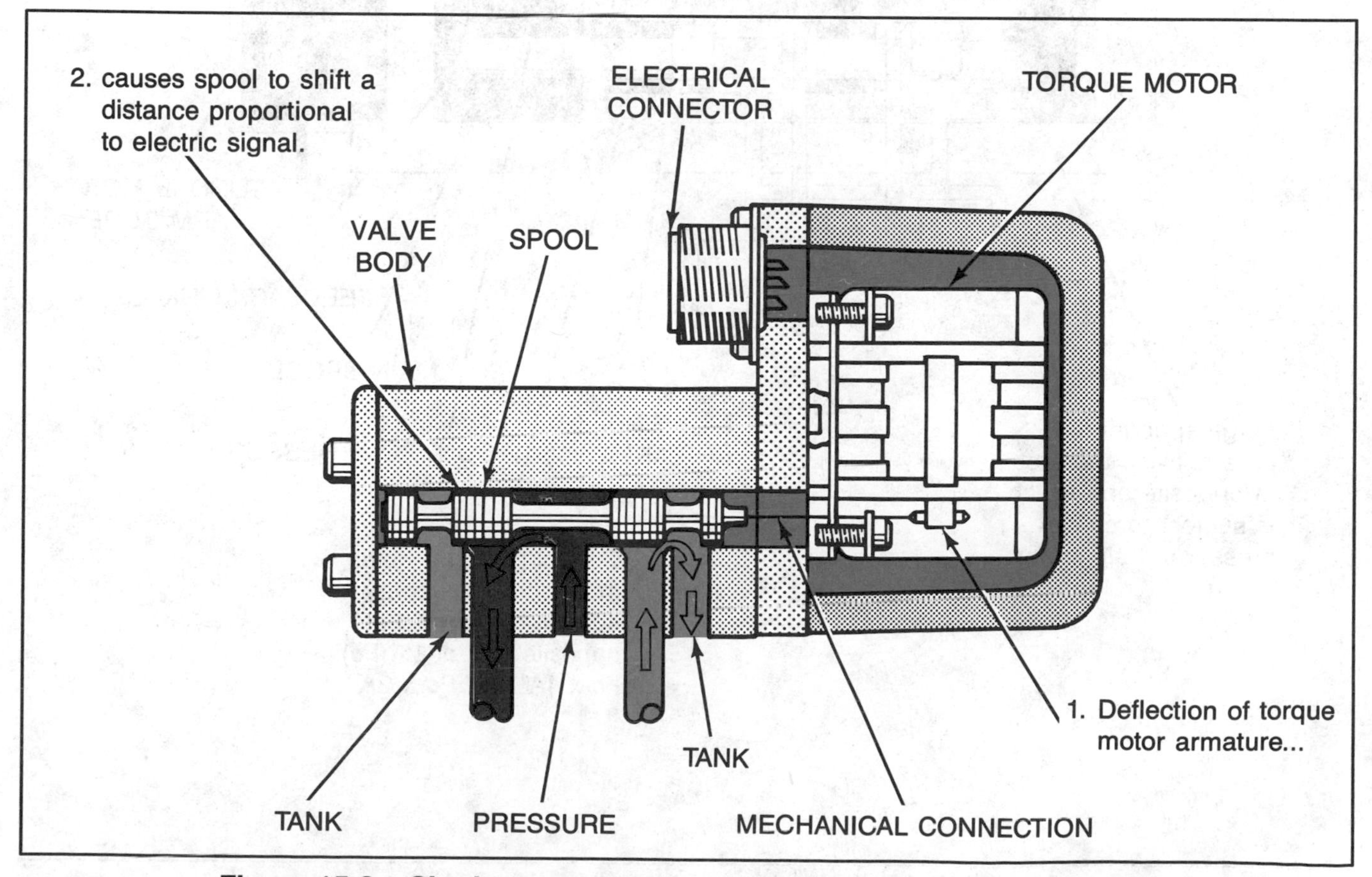

Figure 15-3. Single-stage spool-type servo valve is shifted directly.

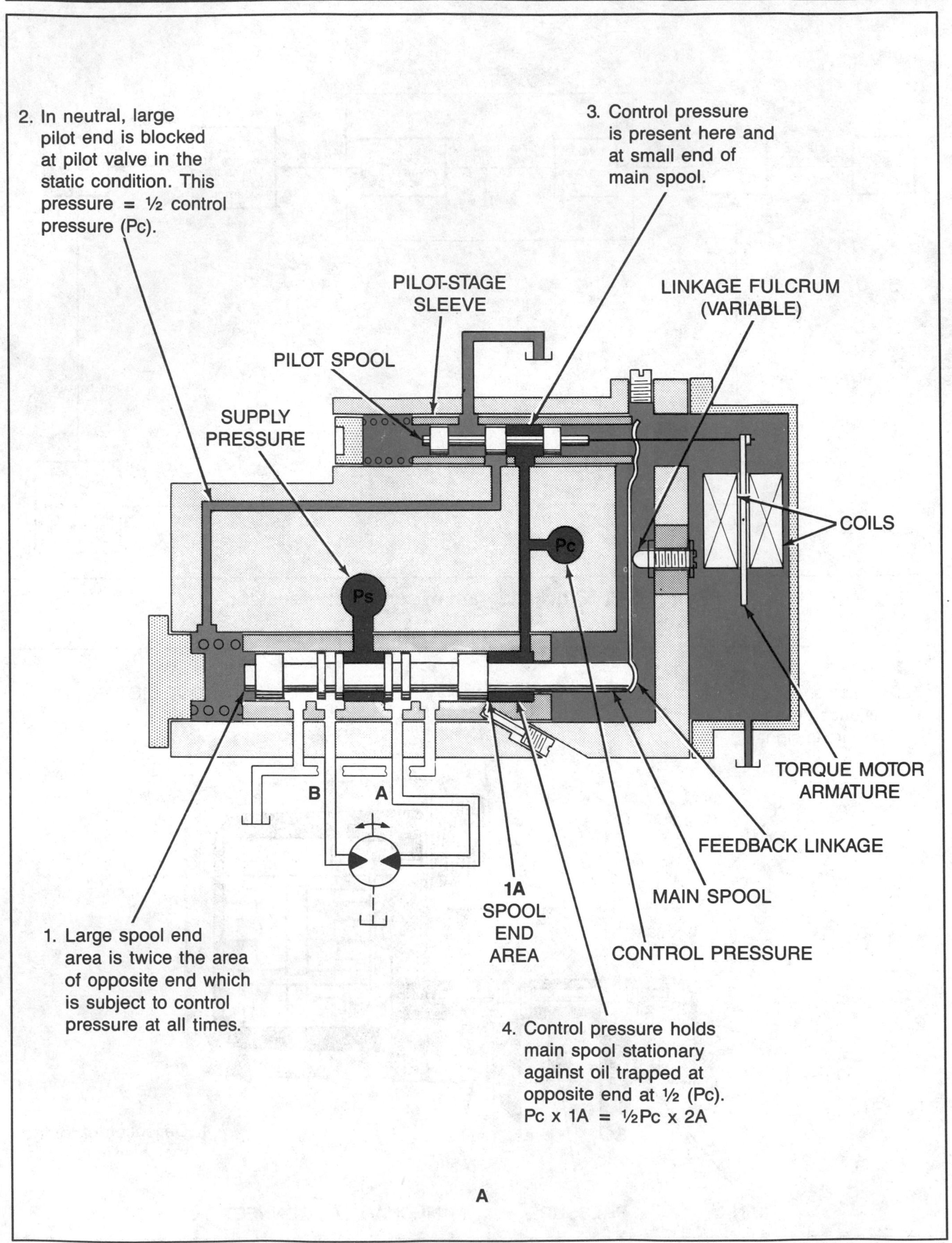
2. In neutral, large pilot end is blocked at pilot valve in the static condition. This pressure = ½ control pressure (Pc).
3. Control pressure is present here and at small end of main spool.
PILOT-STAGE SLEEVE
LINKAGE FULCRUM (VARIABLE)
PILOT SPOOL
SUPPLY PRESSURE
Pc
Ps
COILS
B
A
TORQUE MOTOR ARMATURE
FEEDBACK LINKAGE
1A SPOOL END AREA
MAIN SPOOL
CONTROL PRESSURE
1. Large spool end area is twice the area of opposite end which is subject to control pressure at all times.
4. Control pressure holds main spool stationary against oil trapped at opposite end at ½ (Pc).
Pc x 1A = ½Pc x 2A
A

Figure 15-4A. Two-stage servo valve is pilot operated.

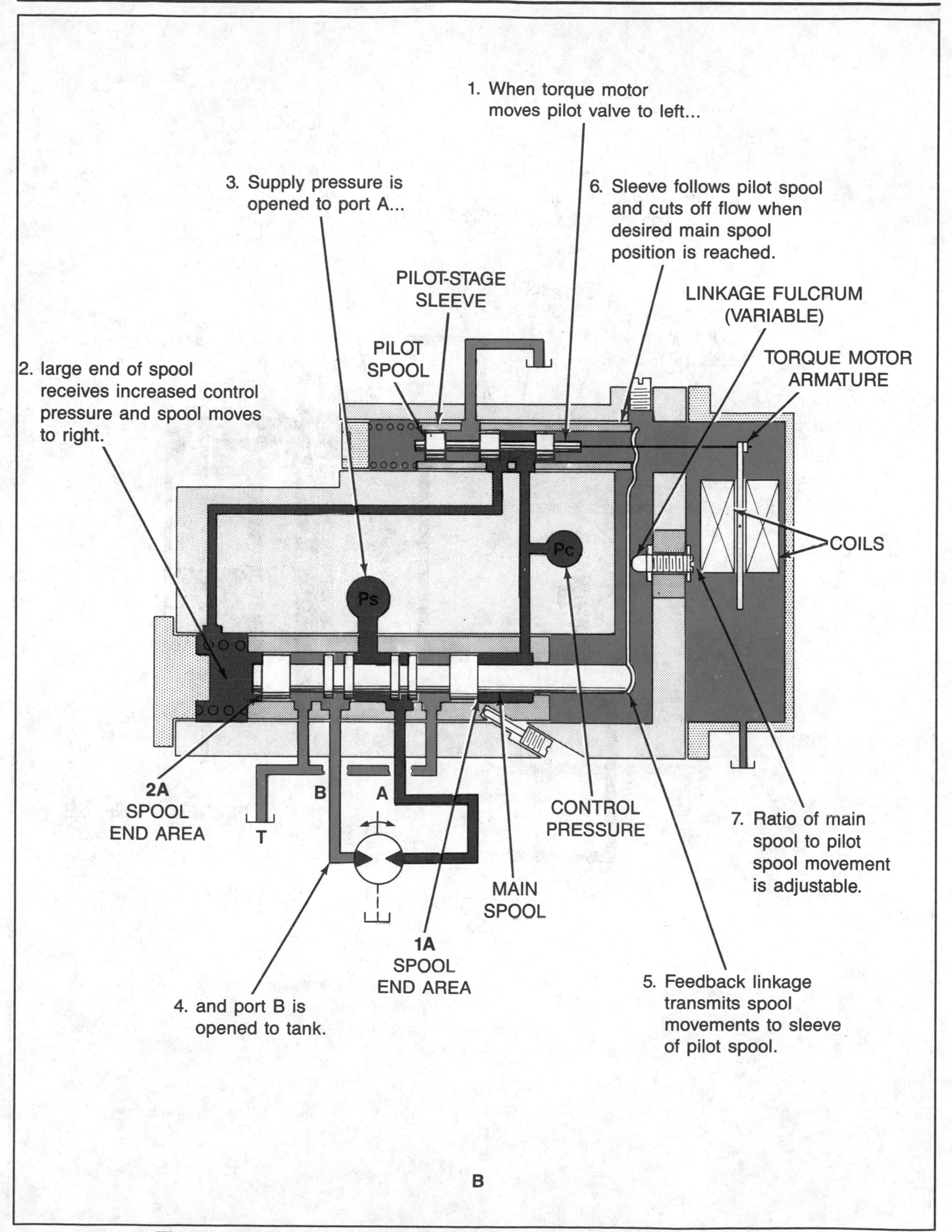

1. When torque motor moves pilot valve to left...
3. Supply pressure is opened to port A...
6. Sleeve follows pilot spool and cuts off flow when desired main spool position is reached.
PILOT-STAGE SLEEVE
LINKAGE FULCRUM (VARIABLE)
PILOT SPOOL
TORQUE MOTOR ARMATURE
2. large end of spool receives increased control pressure and spool moves to right.
Pc
COILS
Ps
2A SPOOL END AREA
T
B
A
CONTROL PRESSURE
7. Ratio of main spool to pilot spool movement is adjustable.
MAIN SPOOL
1A SPOOL END AREA
4. and port B is opened to tank.
5. Feedback linkage transmits spool movements to sleeve of pilot spool.
B

Figure 15-4B. Two-stage servo valve is pilot operated (continued).

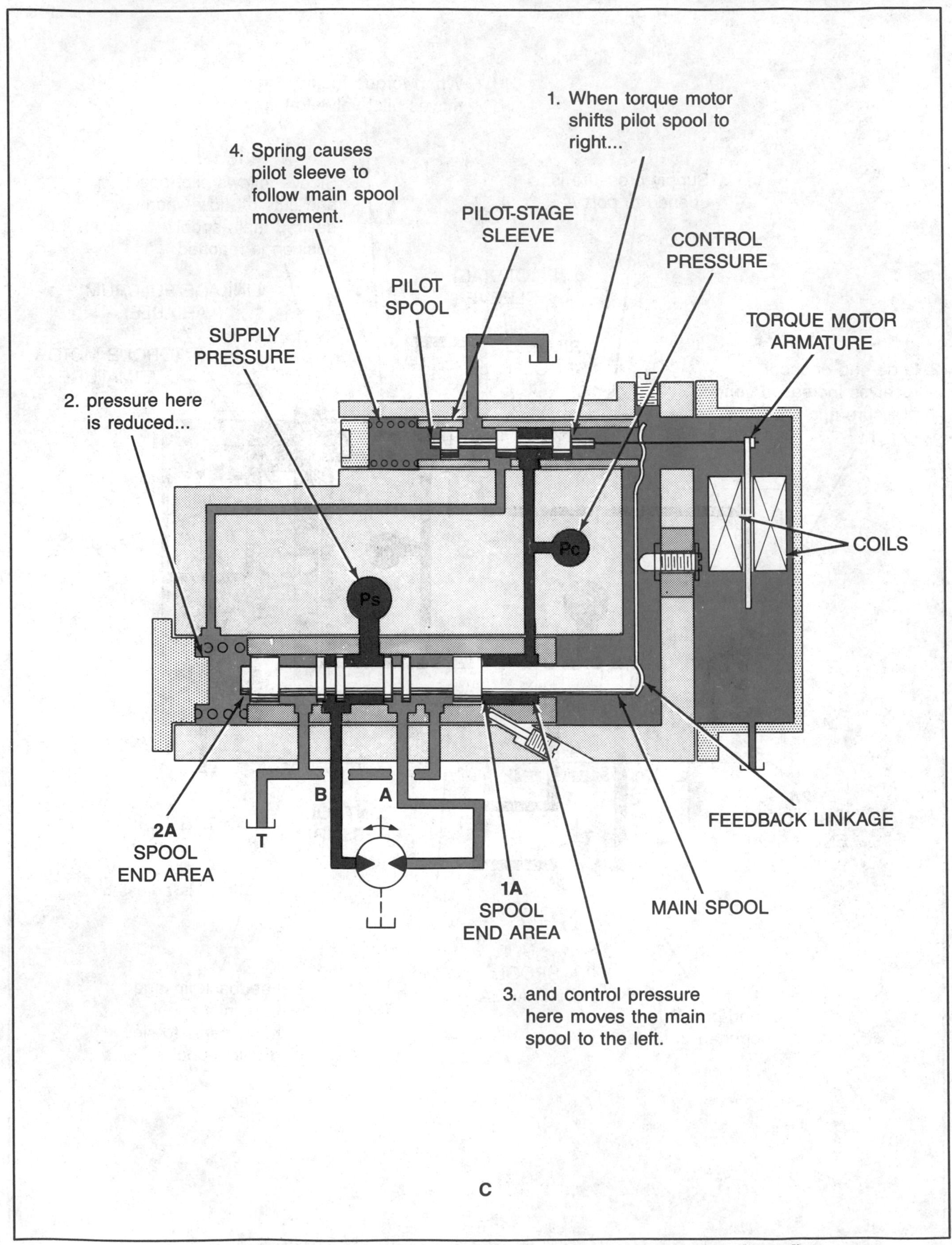

Figure 15-4C. Two-stage servo valve is pilot operated (continued).

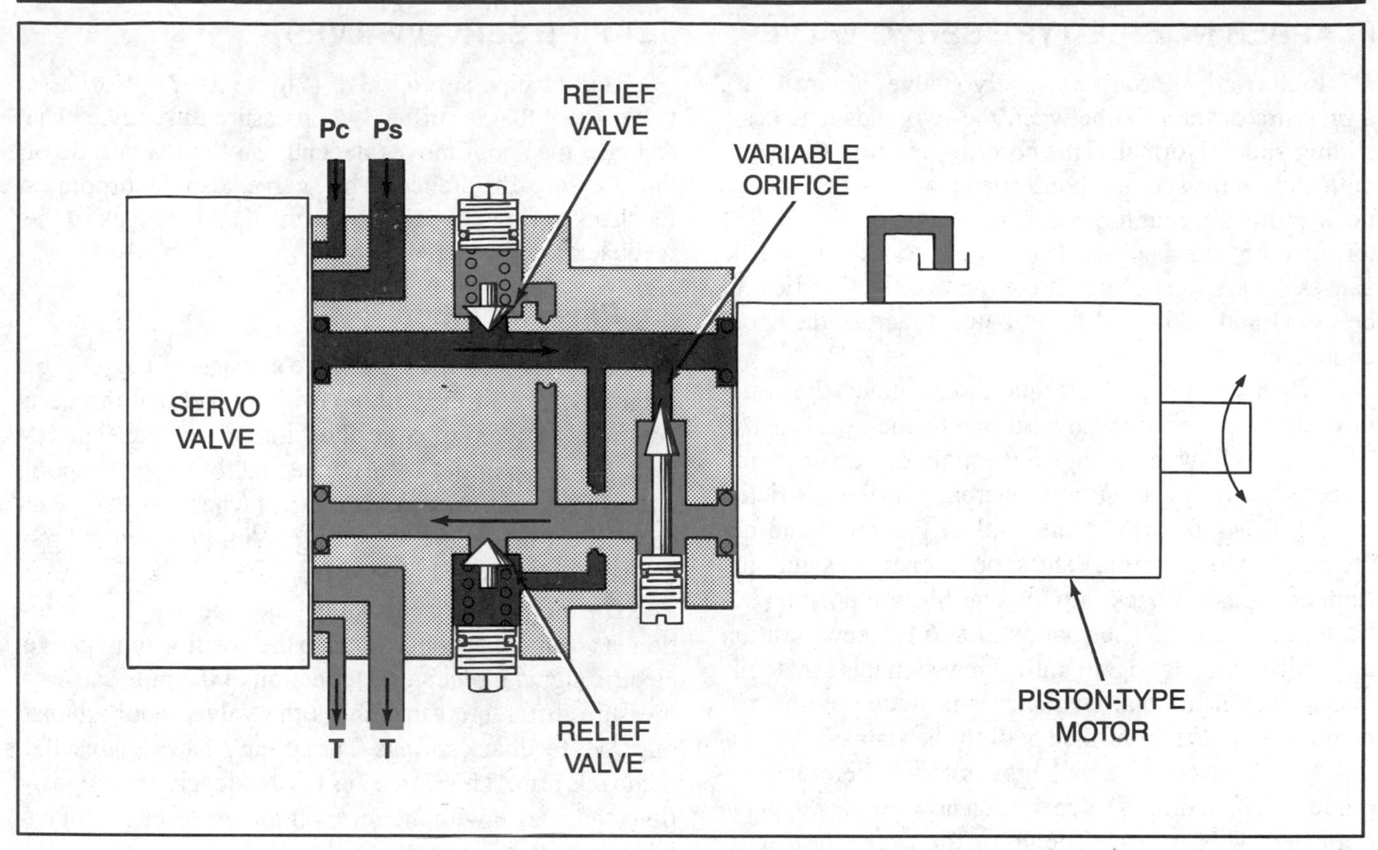

Figure 15-5. Servo valve manifold to piston motor.

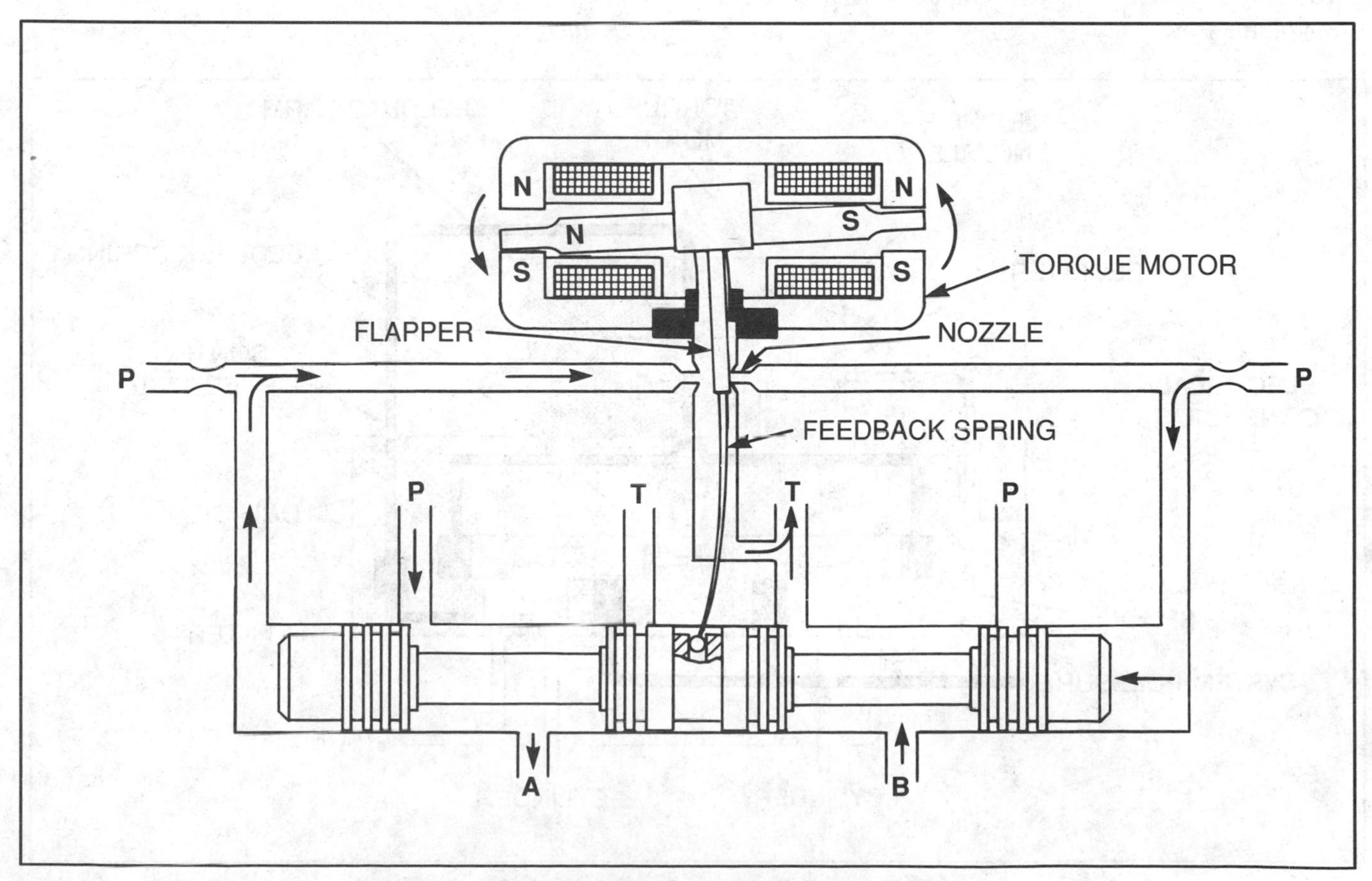

Figure 15-6. Flapper nozzle-type servo valve.

FLAPPER NOZZLE-TYPE SERVO VALVE

In the flapper nozzle-type servo valve (Figure 15-6), a pressure differential between the two ends actuates a sliding spool. Normally, the control pressure is equal at both ends of the spool. A controlled amount of oil continuously flows through orificed passages to nozzles that terminate at the flapper; it then goes to exhaust. This causes the pressure drops across the two fixed orifices to be equal and maintains force balance across the spool ends.

When a signal to the torque motor moves the armature, the flapper moves toward one of the nozzles. The balance of flow is changed through the orifices and nozzles, causing pressure to increase in the restricted nozzle. Pressure rises at this end of the spool and decreases on the other end. The spool then moves until the flapper is pushed back into the equilibrium position between the nozzles. The spool shifts to the new position and will not change again until a new signal is sent to the torque motor. Internal feedback is provided by the mechanical connection of the spool to the flapper.

The distance the spool moves (and, therefore, the amount of oil it meters), depends on how far the flapper is deflected. This, in turn, depends on the size of the electrical signal to the torque motor. A high-input signal produces a high-volume flow; a low-input signal produces a low-volume flow.

JET PIPE SERVO VALVES

The jet pipe servo valve (Figure 15-7) also has a valve spool that is shifted by a pressure difference. The distance the spool moves depends on the magnitude of the pressure difference. This valve also incorporates feedback springs, providing mechanical feedback to the feedback arm.

Jet Pipe Operation

The pilot section of the valve consists of the jet pipe (Figure 15-7), a tube with an orificed end that directs a continuing stream of control oil into a receiver. The receiver has two outlet ports connected to the valve spool ends. Pressure in these ports is equal when the jet pipe is centered in the receiver opening. With pressure at both ends equal, the spool is centered.

The torque motor can deflect the pipe in either direction in an amount proportional to the positive or negative electric signal it receives. Deflection of the pipe causes a pressure differential that shifts the valve spool against one of its feedback springs. The spring pushes against the feedback arm. This force on the feedback arm repositions the nozzle with the receiver and restores a balance of force across the spool.

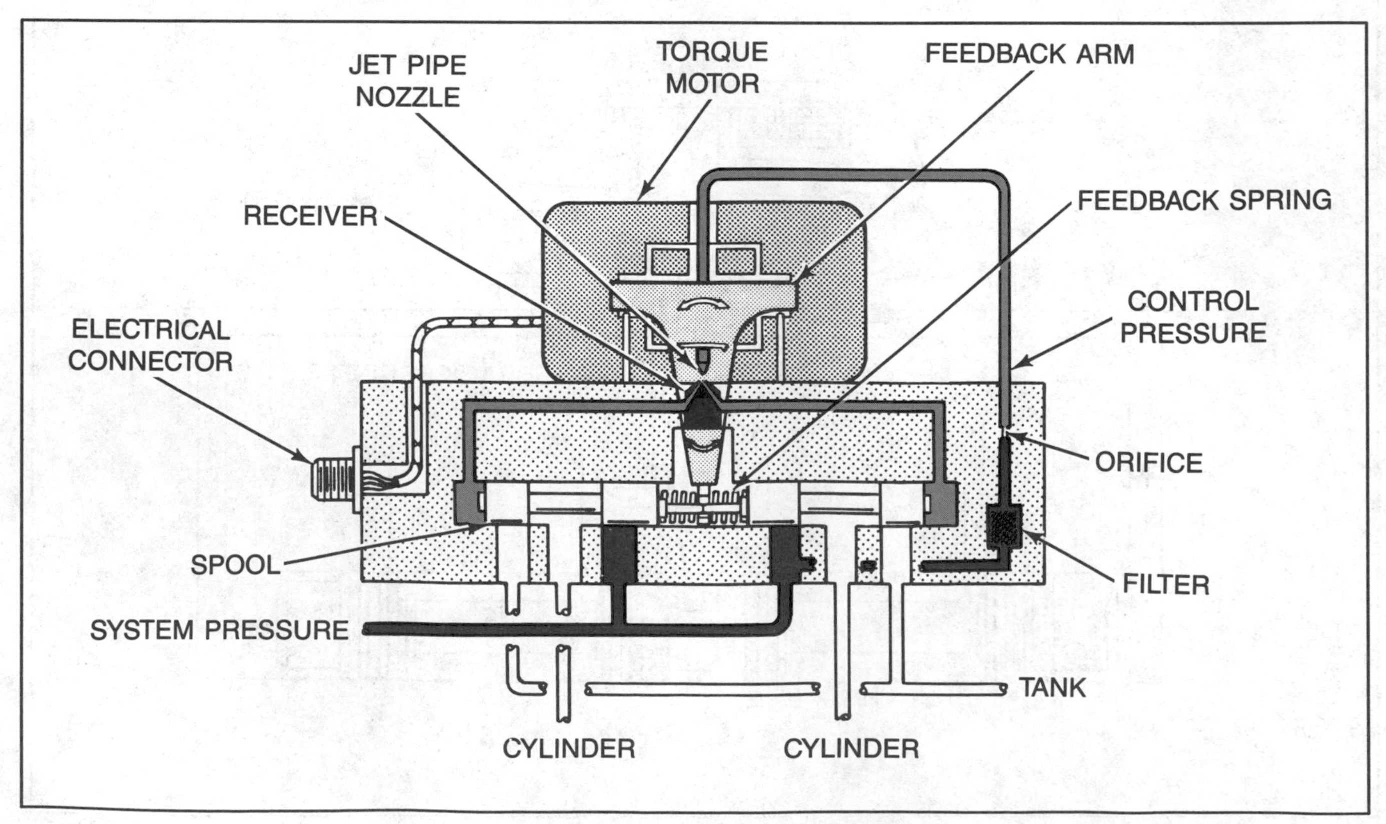

Figure 15-7. Jet pipe servo valve.

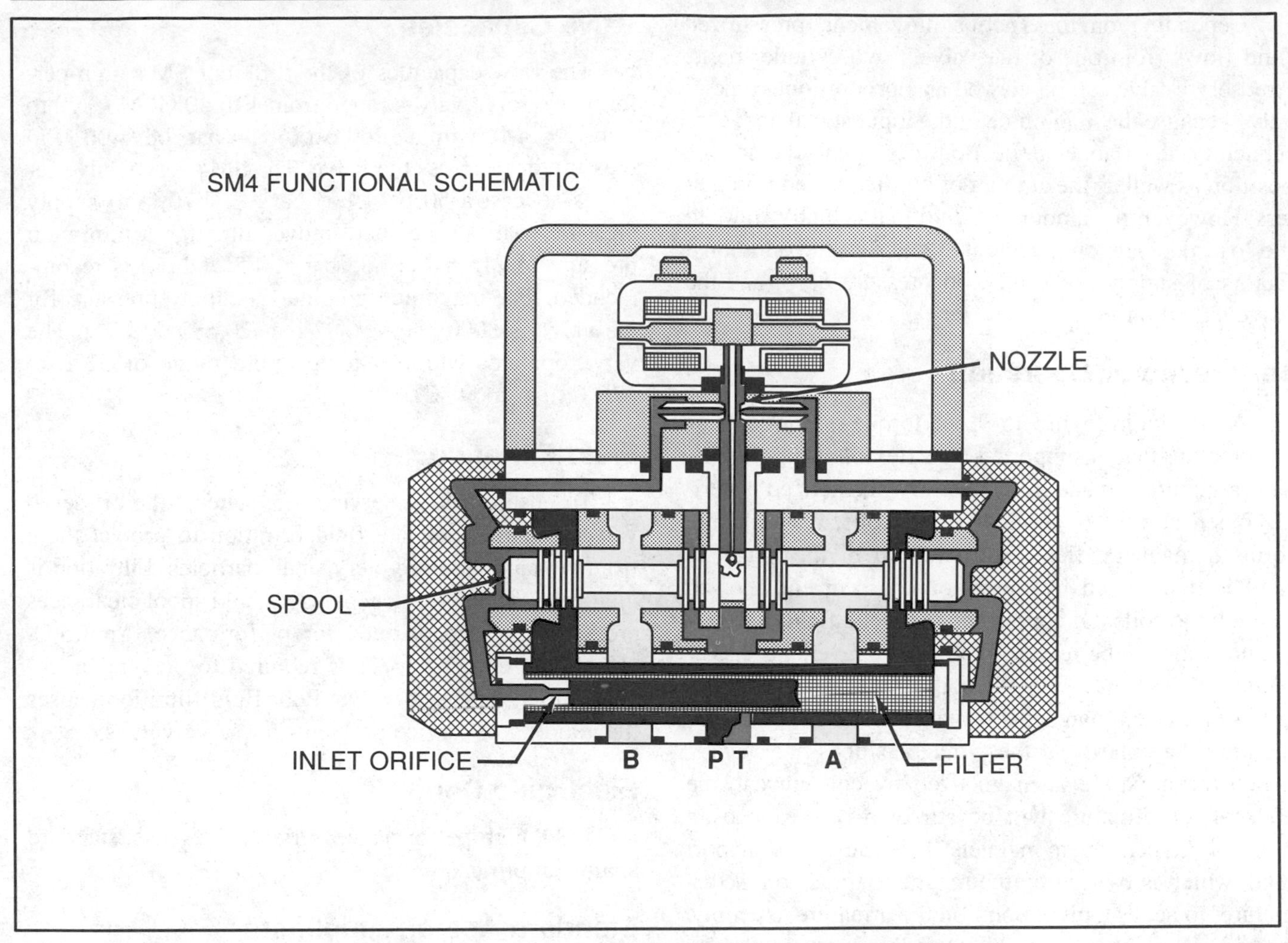

Figure 15-8. SM4 high performance servo valve.

HIGH-PERFORMANCE SERVO VALVES WITH TORQUE MOTORS

The symmetrical design and mechanical operation of the Vickers SM4 high-performance servo valve (Figure 15-8) are similar to the flapper nozzle servo valve. As you can see in the figure, the SM4 valve has a pilot stage with a flapper and two nozzles, and a four-way sliding spool valve on the main stage. As with the flapper nozzle servo valve, the deflection of the flapper in the high-performance servo valve creates a pressure differential between the two nozzles on the pilot stage, which causes the main-stage spool to move.

The SM4 high-performance servo valve, with a typical positional accuracy of less than 0.0001 inch (0.00254 mm), is designed for operations, such as robotics applications, that require extremely precise motion control. Unlike the other servo valves presented in this chapter, the SM4 servo valves can use digital control electronics. This more efficient and reliable control method enables the valve to provide exceedingly accurate motion control as well as rapid control of position, velocity, and acceleration in closed-loop systems.

The SM4 servo valve itself can be equipped with the electronic circuitry that controls and drives the valve, develops diagnostic information, and receives commands from and sends information to a host device, such as a programmable logic controller or a personal computer.

SM4 Operation

Like the flapper nozzle servo valve, the SM4 servo valve is actuated by a torque motor. Coils in the torque motor transmit a magnetic field to the armature and establish the direction and distance that the armature moves. The armature movement forces the flapper to move toward one of the two nozzles on the pilot stage. This restricts fluid flow through one of the nozzles, which causes pressure to increase behind the nozzle on that side of the valve. The two ends of the spool have equal pressures when the flapper is centered between the nozzles. An increase in control pressure on one side of the spool creates an imbalance, which causes the spool to move away from the end with the higher pressure.

Depending on the spool's movement, pressurized fluid flows from one of the valve's two cylinder ports. Any servo valve can be viewed as a proportional type of valve because the magnitude of the input signal and, consequently, the flapper deflection, determine the spool's position as well as the amount of oil the opened port meters. However, the amount of fluid that actually flows to the load depends on the input signal and the additional factors of supply pressure, load pressure drop, and the valve's flow rating.

Torque Motor Operation

As shown in Figure 15-9, the torque motor contains an armature that is supported by a flexible support tube. This armature is made from an easily magnetized material. When current is applied to the coils, the armature forms a magnetic field. The strength of the magnetic field is determined by the amount of current allowed through the coils. Of course, the magnetic field in the armature can also be reversed by simply reversing the direction of current flow in the coils.

A permanent magnet is located at each end of the armature. The polarity of these magnets does not change. When the armature is magnetized by coil current, the ends of the armature will be attracted to the opposite poles of the permanent magnets. This causes the support tube, which is twisted by the magnetic forces on the armature, to act as a pivot point on the armature assembly, The whole armature assembly "tips," causing the flapper to block off one of the nozzles. If current is removed from the coil, the support tube flexes back to the center position, centering the flapper between the nozzles once again.

Mechanical Feedback

As Figure 15-8 illustrates, the SM4's pilot-stage flapper is attached to the sliding spool on the valve's main stage. When the deflected flapper restricts one of the two nozzles, the main-stage spool moves away from the blocked nozzle side, where pressure is higher. The spool movement causes the feedback spring to be deflected. This creates a force on the pilot-stage flapper that opposes the force created by the armature movement.

When the feedback spring force balances with the armature force, the flapper returns to a position that maintains force balance between the spool ends. At that time, the main-stage spool stops and remains in the same position until the input signal changes again. This highly dependable method of mechanical spool feedback eliminates the need for complex spring arrangements, levers, pilots and other complicated feedback mechanisms.

Valve Capacities

The flow capacities of the different SM4 high-performance servo valves range from 1 to 40 GPM (3.79 to 151.41 LPM), with a 1000 psi (68.94 bar /6894.00 kPa) pressure drop across the valve. The SM4 servo valve requires at least a 50 psi (3.45 bar / 344.70 kPa) supply pressure to make the spool move, although a minimum pressure of 215 psi (14.82 bar /1482.21 kPa) is recommended. The maximum continuous supply pressure for the valve is 3000 psi (206.82 bar /20,682.00 kPa). The valve operates within a temperature range of 32°F to 176°F (0°C to 80°C).

Fluid Filtration

To ensure a long service life, state-of-the-art servo valves require adequate fluid filtration to protect them from contamination by very small particles. Filtration is vital because the nozzles, passages, and spool clearances are critically dimensioned for performance. An I.S.O. cleanliness code of 15/11 is required to prevent blockages within the servo valve. Poor fluid filtration causes the majority of service problems for servo valves.

Electronic Controls

SM4 high-performance servo valves use standard analog amplifiers.

Standard Analog Amplifiers

One method of electronic control for the SM4 servo valve uses standard analog command signals and amplifiers. This control method includes an amplifier that sends an analog error signal to the servo valve. A position transducer, or similar sensing device, measures the load position and feeds back that information to the amplifier's summing junction. There, the command and feedback signals are compared to produce a new error signal. This process is repeated until the command and feedback signals cancel out.

SERVO VALVE PERFORMANCE

You have learned that the servo valve is a high-performance valve. This final section compares three performance aspects of the servo valve with those of a proportional valve. This comparison includes a look at zero-lap and overlap conditions, hysteresis, and linearity.

Zero-lap versus Overlap

Figure 15-12A illustrates the performance of a typical proportional valve. As you will recall from Chapter 14, there is a deadband region within which an input

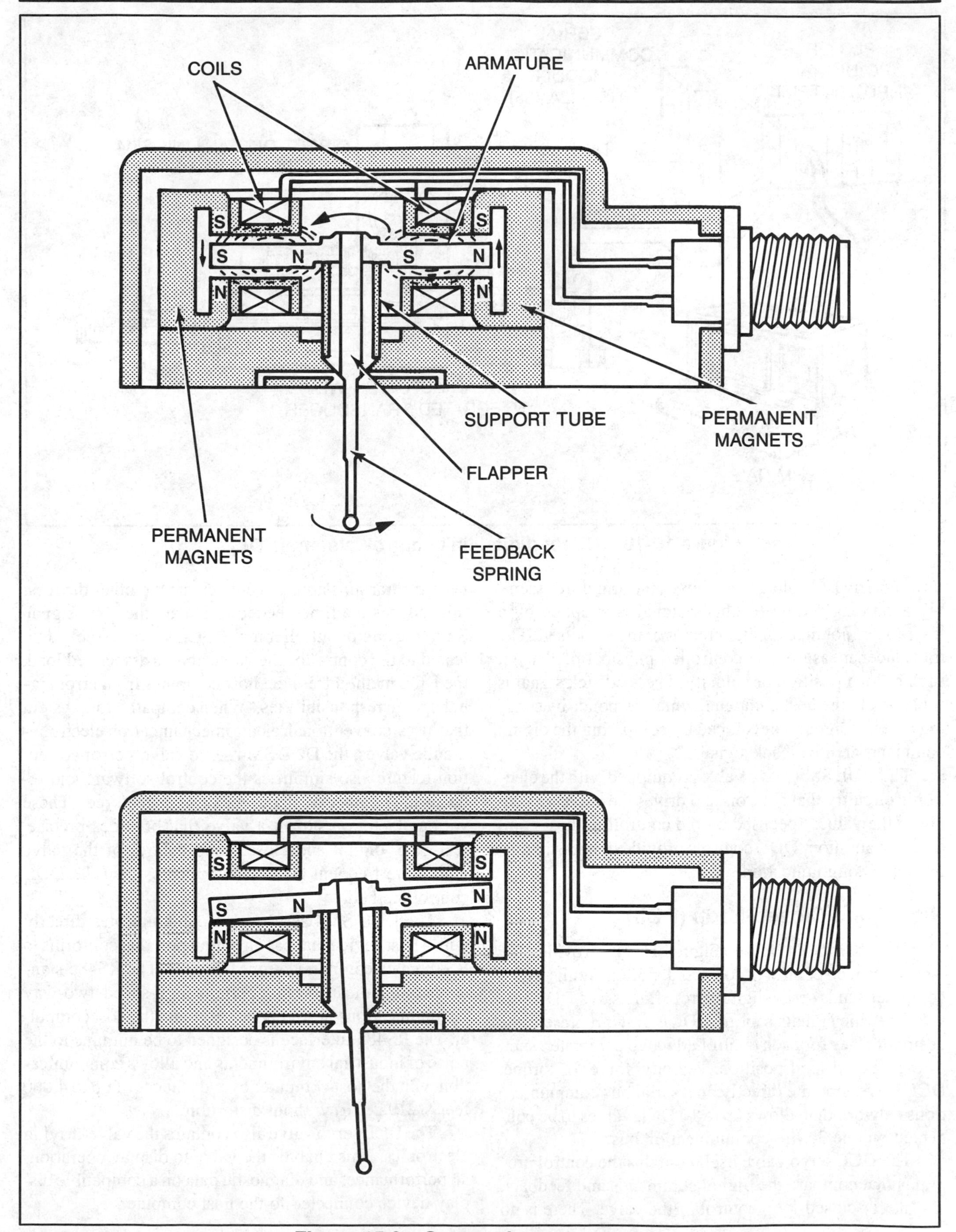

Figure 15-9. Servo valve torque motor.

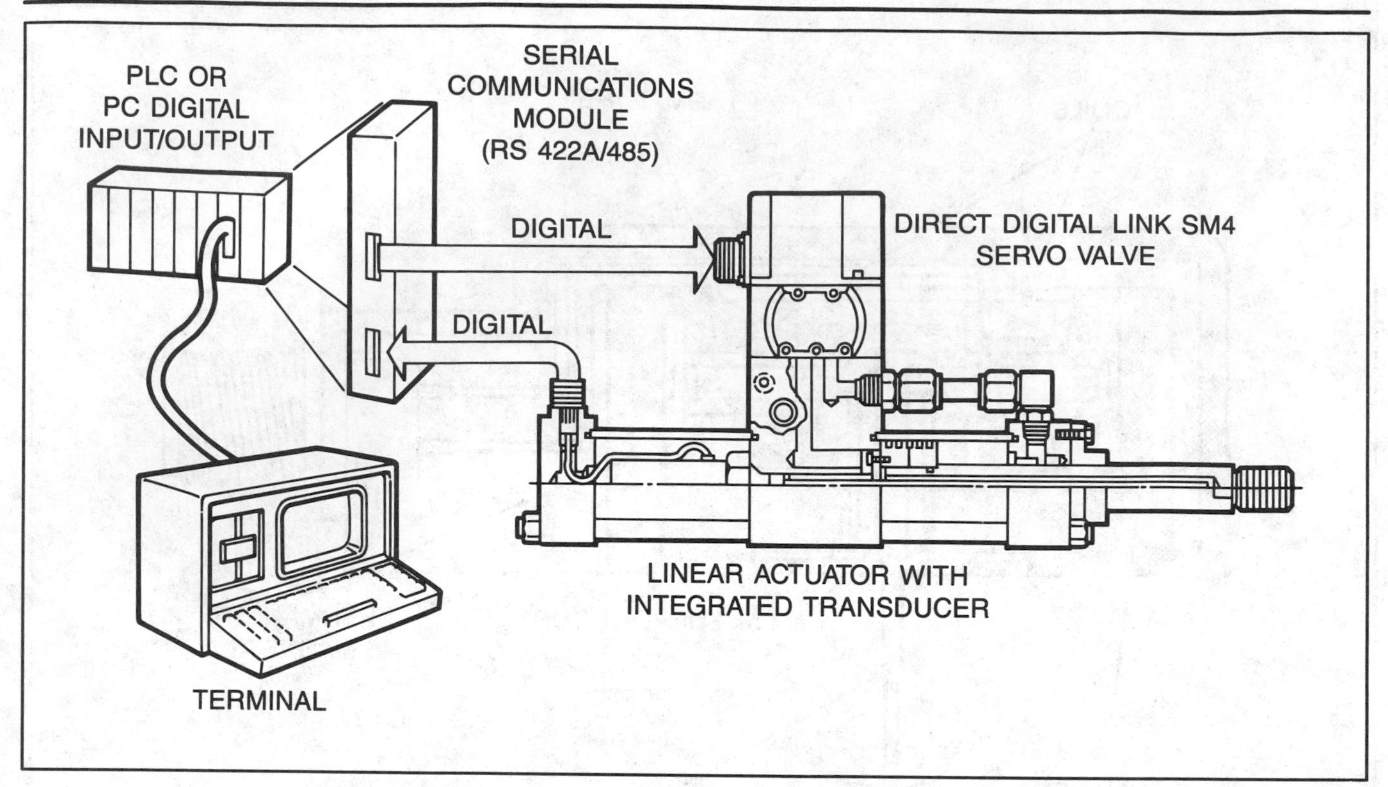

Figure 15-10. Direct digital link control system (DDL).

as a specific DC voltage and moves the armature assembly accordingly. The feedback signal is supplied by a transducer mounted on the actuator or the load itself. The transducer measures the controlled physical parameter, such as load position or velocity. The feedback signal is sent back to the host computer, which responds by sending an error signal it developed by comparing the digital command and feedback signals.

The DDL SM4 servo valve is equipped with the electronic circuitry that controls and drives the valve according to the values specified by the controlling host computer. In addition, DDL controls provide diagnostic routines and communication.

Digital and Closed-Loop (DCL)

The DCL SM4 valve (Figure 15-11) provides the most advanced control of force and motion. While a host computer sends serial digital error signals to a DDL servo valve, the signals sent to a DCL valve represent the desired values for each controlled valve parameter, such as the desired load position, velocity, force, or torque. DCL valves operate directly on a serial bus communications network that allows up to 16 DCL valves to be connected on one 3-wire communication bus.

The DCL servo valve itself contains the control programs that compare the digital commands and feedback signals for closed-loop control of the valve. There is no need for transmission of feedback data to other devices; this reduces the time needed to update the error signal. Since the microcontroller in the DCL valve is solely dedicated to the control of the valve and its associated load, the DCL method fress the host computer from error signal update responsibilities. When compared to a system that uses conventional electromechanical or electrohydraulic valves, the DCL SM4 servo valve's error correction feature also simplifies the control software and reduces the number of electronic devices required. These various factors result in a valve that provides precise, real-time, digital and closed-loop control of the valve with more frequent error signal updates than the DDL control method.

The DCL SM4 servo valve communicates directly with a host device that has an output port that conforms to RS485 data communication standards. The RS485 standard communications link permits high-speed, two-way communications between the valve and the host computer. The RS485 interface is designed to be immune to the noise of industrial environments and allows communication with the host computer over distances of up to 4,000 feet (1219.20 m) without distortion.

The DCL servo valve also contains the valve-driving electronics. This enables the valve to display operational, performance, and diagnostic data on a compatible display device connected to the host computer.

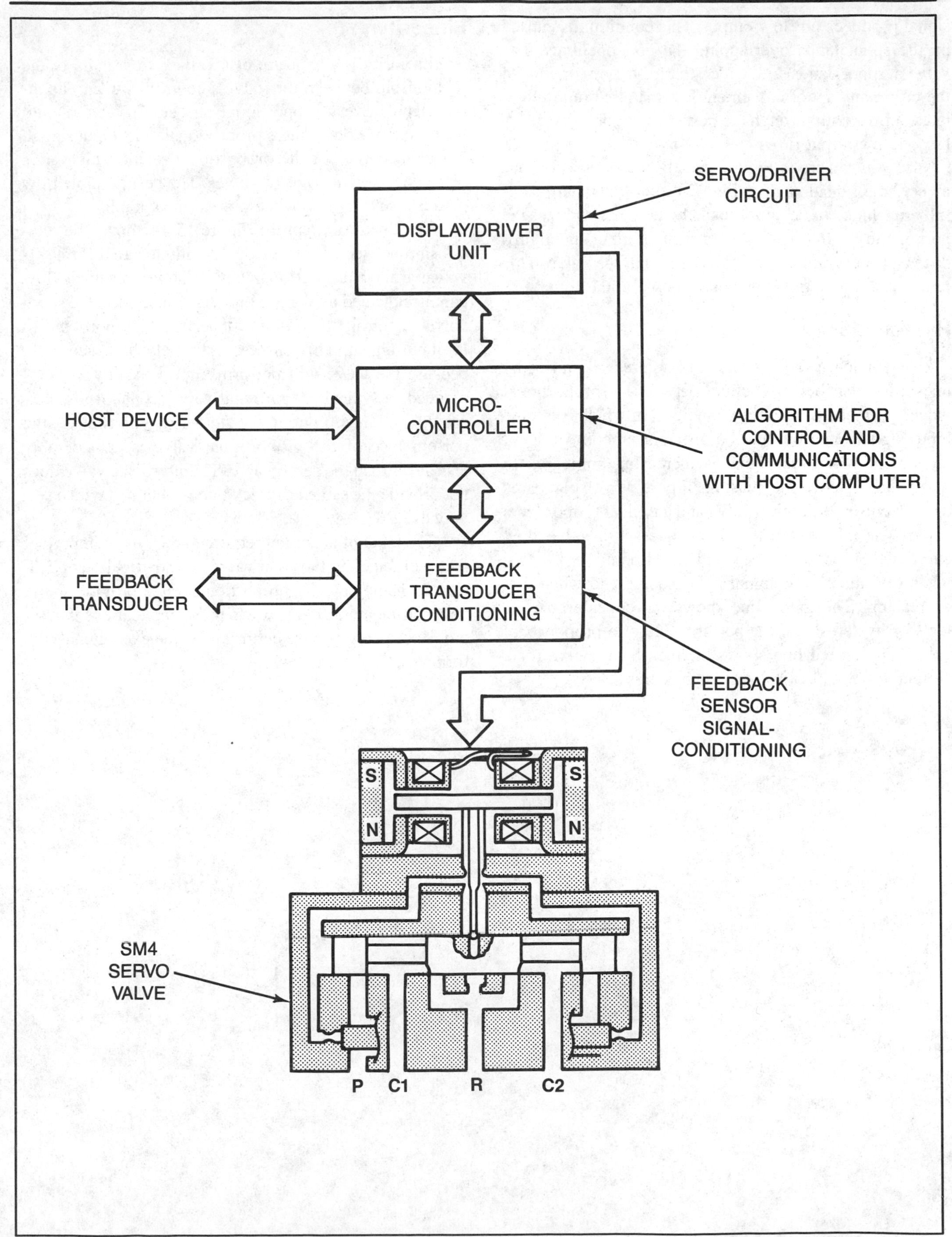

Figure 15-11. Block diagram of a digital and closed loop (DCL) SM4 servo valve.

signal produces no flow output. This condition is caused by the spool lands overlapping the port openings. The valve's spool lands are wider than the port openings; therefore, any spool movement less than that required to make a flow path through the port does not produce a corresponding output flow.

As you can see in Figure 15-12B, the servo valve has a very small deadband region. The spool lands are carefully machined to be almost exactly the same width as the port openings. In this way, even a small input signal produces a proportional output flow through the valve. This produces the straight flow curve shown in the figure.

Hysteresis

As you learned in Chapter 14, hysteresis is caused by unevenly distributed friction between a proportional valve spool and bore. It is a condition in which the same input signal level produces a different amount of spool movement when the signal is increasing compared to when the signal is decreasing. Figure 15-13 compares the hysteresis for a servo valve and a typical proportional valve. The red line in the figure is a servo valve flow curve, which indicates the small variation in output flow when the same input signal is increasing and when it is decreasing. The green line shows the considerable effects of hysteresis on the performance of a proportional valve. The dotted line shows the ideal flow curve for a perfect valve with absolutely no hysteresis.

Linearity

Linearity is a measure of a valve's flow gain, or the relationship between the valve's control flow and its input signal. Every proportional and servo valve should have a control flow that is proportional to its input signal, but the accuracy of the proportional relationship varies between different types of valves. The best possible flow gain is a straight line when plotted on a graph.

The linearity graph in Figure 15-14 shows the relationship between control flow and input signal for a servo valve and a proportional valve. The servo valve's flow gain is indicated by the red line. As you can see, the servo valve's control flow is almost perfectly proportional to its input signal. This ensures extremely high accuracy, because the valve will perform almost exactly as commanded. A typical proportional valve's linearity is indicated by the green line in the figure. As you can see, the performance of the proportional valve is considerably less linear than the servo valve. Therefore, the valve does not provide the same accuracy that is obtained with a servo valve.

The level of accuracy required is the most important factor in selecting the right valve for a particular application. In many cases, a proportional valve provides more than sufficient accuracy, while those applications that require extremely precise control need the exceptional performance of a servo valve.

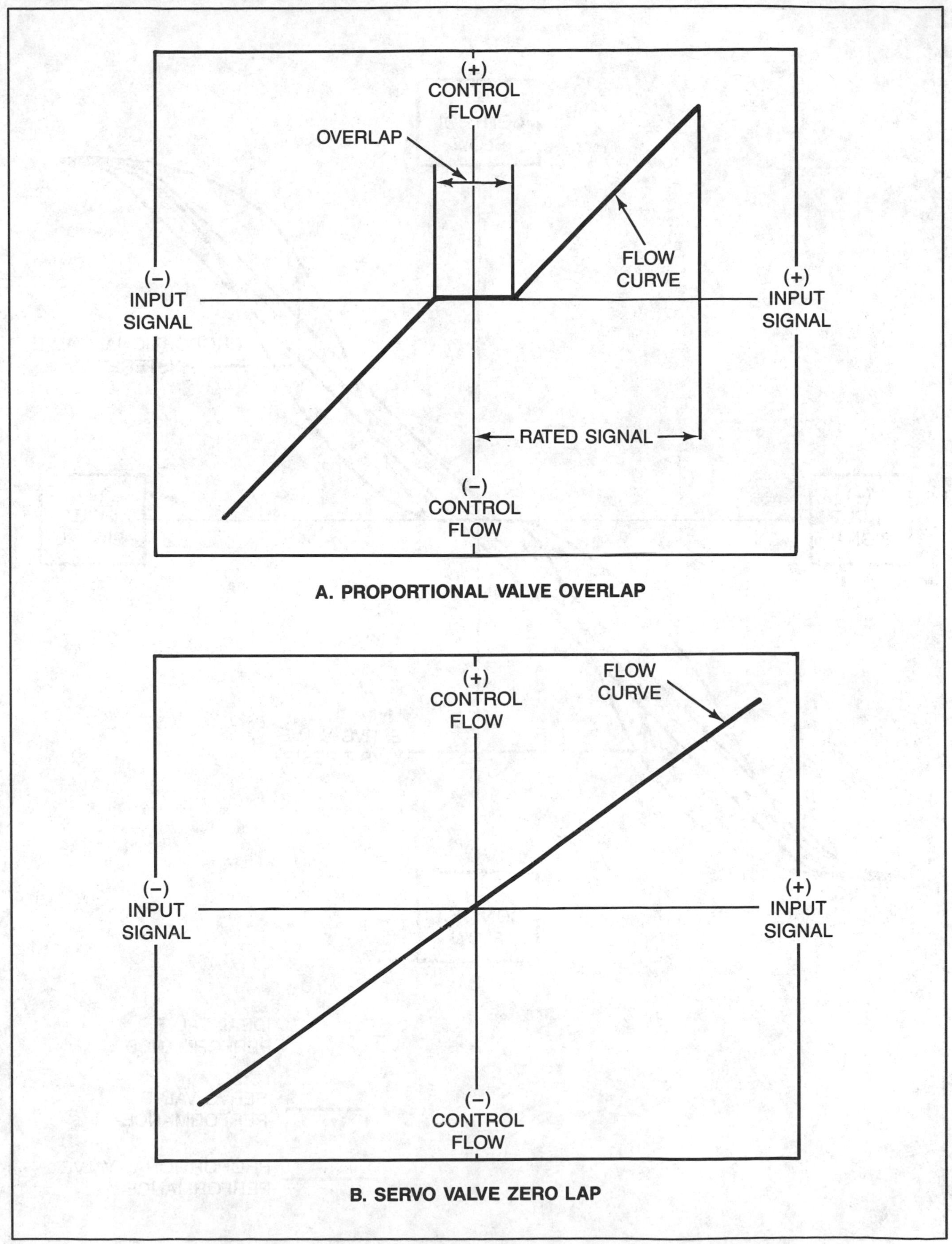

Figure 15-12. Overlap versus zero lap valve performance.

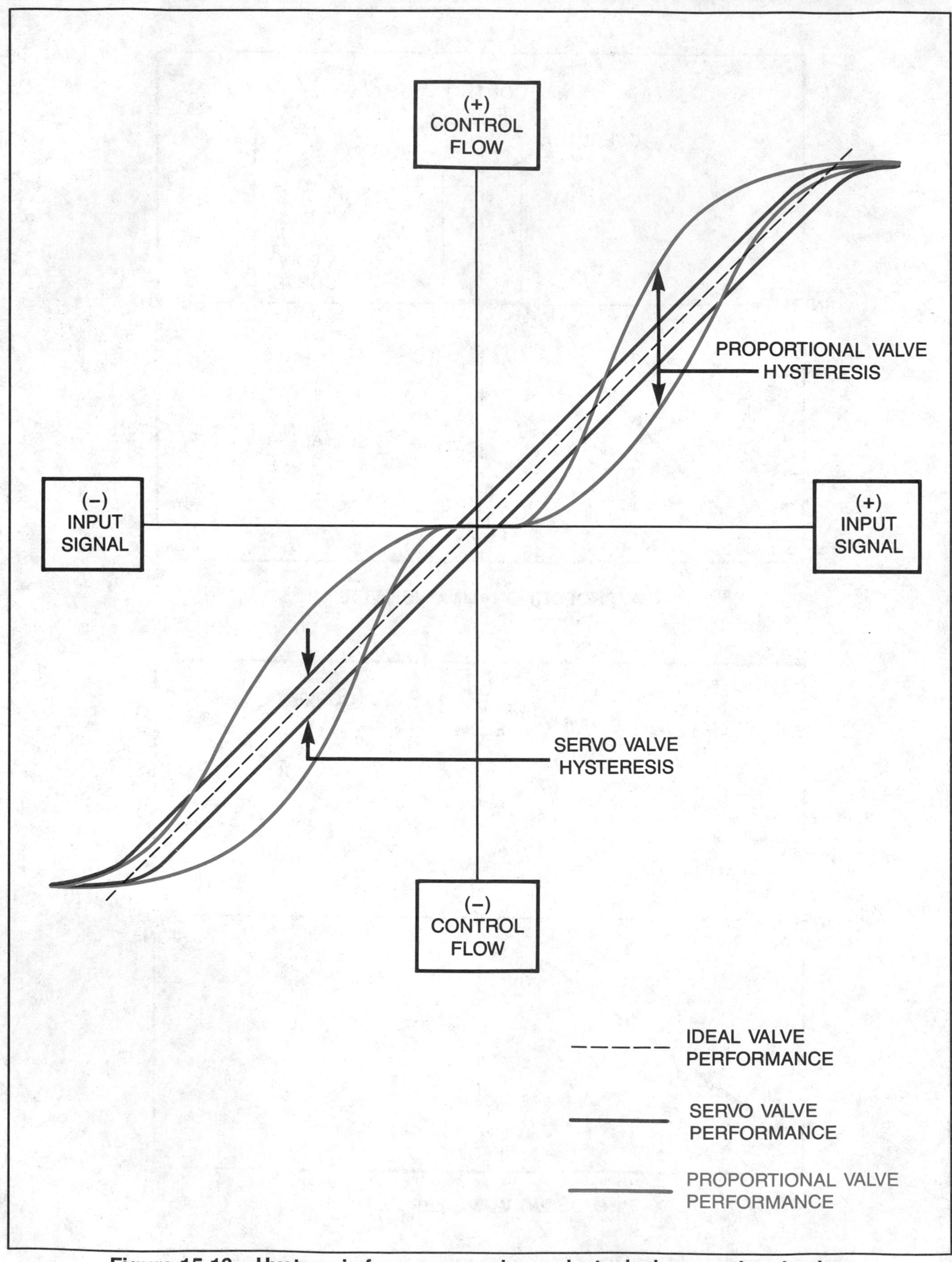

Figure 15-13. Hysteresis for a servo valve and a typical proportional valve.

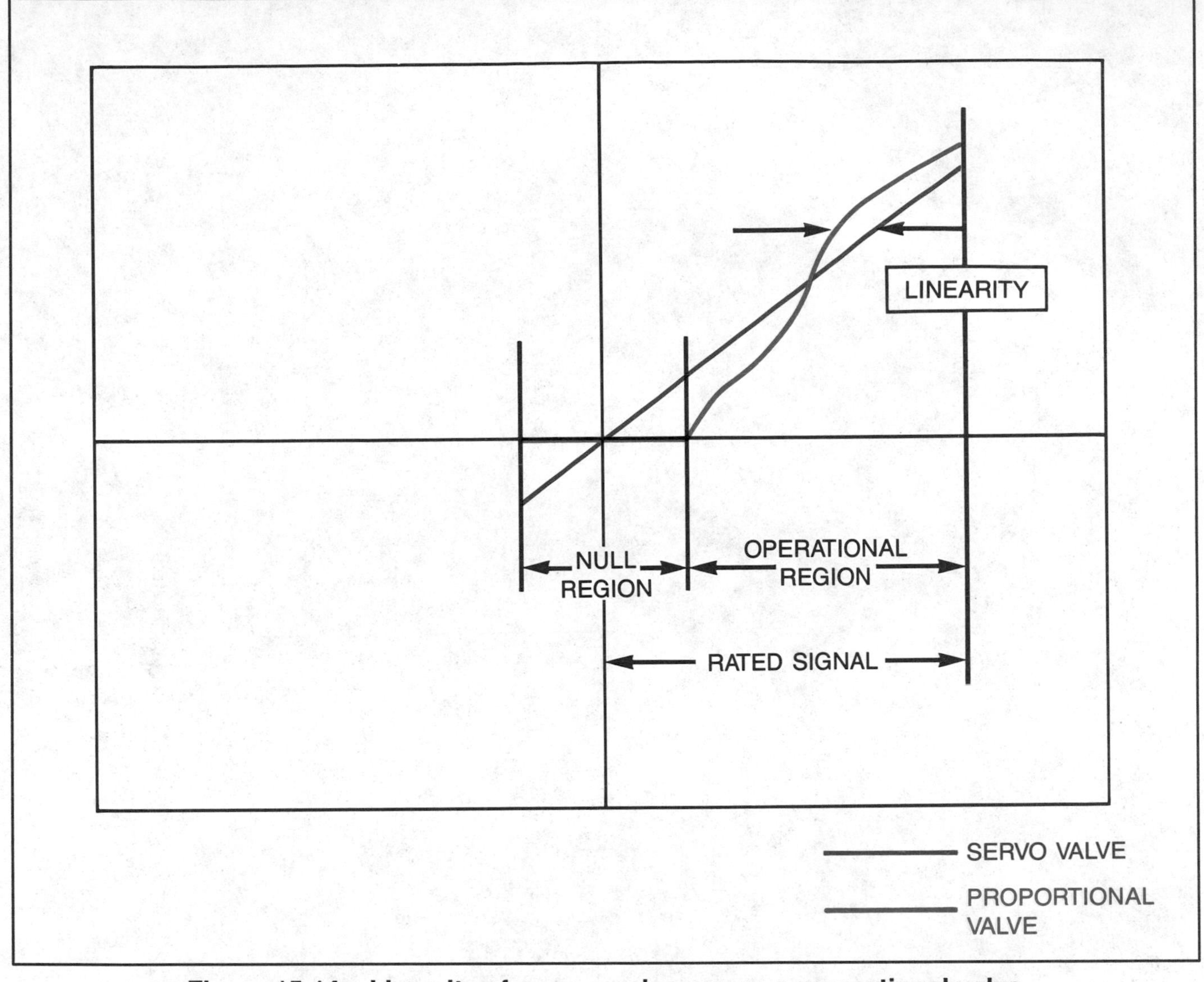

Figure 15-14. Linearity of a servo valve versus a proportional valve.

QUESTIONS

1. In a mechanical servo valve, what part of the servo valve moves with the load? What part moves with the control?
2. In a single-stage electrohydraulic servo, how is the valve spool actuated?
3. What primary feature makes a servo valve different from an ordinary directional valve? What is the purpose of this feature?
4. Explain dither—what it is, how it is applied, and why it is needed.
5. How is the spool actuated in jet pipe and flapper nozzle-type servo valves?
6. The SM4 high-performance valve is similar in design and operation to what other type of servo valve?
7. List and briefly describe the three major types of electronic controls available for SM4 servo valves.
8. Describe zero overlap and how it affects servo valve performance.
9. Explain lincarity, comparing that of a servo valve with that of a typical proportional valve.

CHAPTER 16

BASIC DIGITAL ELECTROHYDRAULIC DEVICES

Digital electrohydraulic devices are being used in more and more hydraulic systems every day. These devices provide electrical control of hydraulic components such as valves and servos. Digital refers to the type of electrical signal used to actuate these electrohydraulic devices.

The choice of a controlling device used in a given hydraulic circuit is influenced by such factors as safety, flexibility, dependability, initial cost and ease of maintenance. Electrical controls provide a flexibility and convenience that cannot be attained with any other kind of control. In certain applications, digital electrical controls may be the only way to obtain the desired level of performance.

This chapter will provide a basic overview and introduction to many of the concepts encountered in digital electronics as applied to hydraulic components. It covers topics such as the difference between digital and analog devices, basic logic circuits, the binary number system, and forms of data representation. This chapter also briefly covers general digital applications in hydraulic systems including microprocessors, industrial controls and relay logic. The chapter concludes with some basic information on specific digital electrohydraulic devices currently used in the industry.

DIGITAL TECHNIQUES

Basically, there are two types of electronic signals, analog and digital. Analog signals are the most familiar of the two simply because digital electronics is such a relatively new field and analog devices have been in use for a long time. For this reason, both kinds of signals are briefly examined in this chapter. The emphasis, however, is on digital techniques and applications because the increased use of digital devices in hydraulic circuits is clearly a trend for the future.

Definition of an Analog Signal

An analog signal can be defined as an AC or DC voltage or current that varies smoothly or continuously. Analog signals do not change abruptly or in increments.

The most common type of analog signal is the sine wave shown in Figure 16-1A. The voltage or current changes at a regular rate from a positive to a negative value. The positive and negative alternations are symmetrical. Radio signals, audio tones, and the AC power supplied by the local power company are all examples of electronic signals or voltages in the form of sine waves.

Signals do not have to be sinusoidal in nature for them to be considered analog signals. The randomly changing AC signal shown in Figure 16-1B and the varying negative DC voltage illustrated in Figure 16-1C are also analog signals because they change at a fairly smooth rate. Electronic circuits that process analog signals are called linear circuits.

Definition of a Digital Signal

Digital signals, like those shown in Figure 16-2, almost always vary between two distinct and fixed voltage levels. The digital signal is normally in the form of a series of pulses that rapidly change from one voltage level to the other in discrete steps or increments. Figure 16-2A shows a signal changing between a positive 5 volt DC level and zero (ground). In Figure 16-2B, the levels are zero (ground) and negative 5 volts DC. Digital signals can also switch between a positive and a negative value as shown in Figure 16-2C.

Because of the speed at which the signals switch from one level to the other, digital signals are described as two-state signals. At any given time, they are either high or low, on or off, up or down. This fast switching characteristic is standard for all digital signals. The circuits that process digital signals are called digital, logic or pulse circuits.

Contrasting Analog and Digital Signals

In the following simple examples, familiar devices and ideas are used to further illustrate the difference between analog and digital methods and techniques.

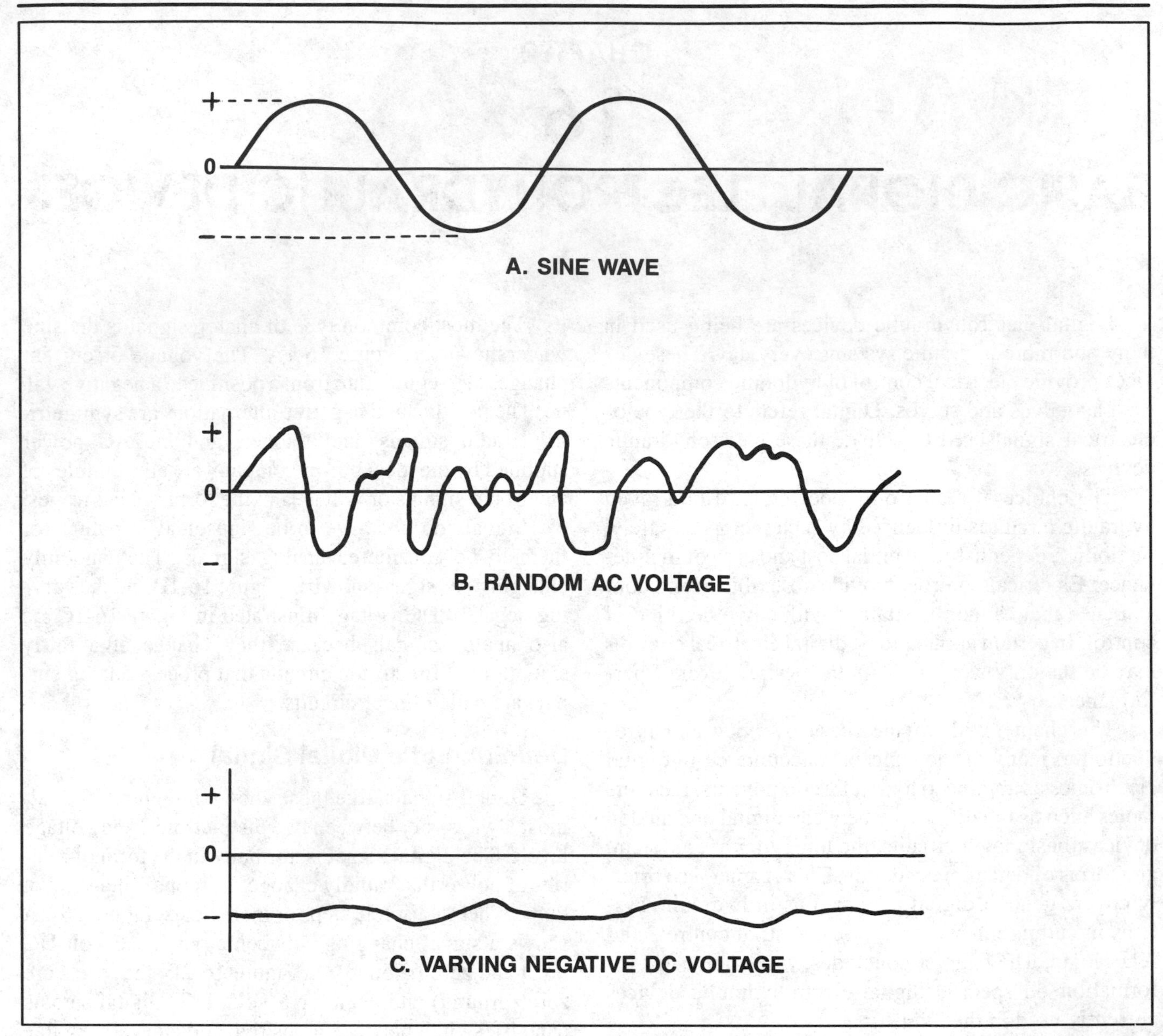

Figure 16-1. Analog signals.

A light bulb can be either an analog or a digital device depending on how it is used. If the current through the filament of the bulb is varied between maximum and minimum, the brightness level will vary at a proportional rate between full brightness and off. There are virtually an infinite number of brightness levels that can be attained. When used in this way, the light bulb is an analog device.

The light bulb can also be used as a digital device if it's brightness level is only allowed to change in discrete steps. The most common way to use a bulb as a digital device is to permit only two brightness levels, usually full brightness and off. When used this way, the bulb has two states and because of this ON/OFF characteristic the bulb is called binary in nature. The term binary refers to any two-state device or signal.

A watch with hour and minute sweep hands is an analog device. It continuously indicates the time by the positions of the hands on a calibrated dial. To tell the exact time, the hand positions must be estimated. The ability to read the time accurately is limited by the precision of the dial calibration. On a digital watch the time is read directly from decimal number display readouts that change in discrete increments of hours, minutes and seconds. The accuracy is improved because there is no need to estimate the exact time.

A speedometer with a moving needle is an analog device. A speedometer with an LED readout is a digital device. The mechanical channel selector on a television set is a digital device because it changes in discrete positions, i.e. 2, 4, 7, 9, etc. The volume control on the same television set is analog because its position can be

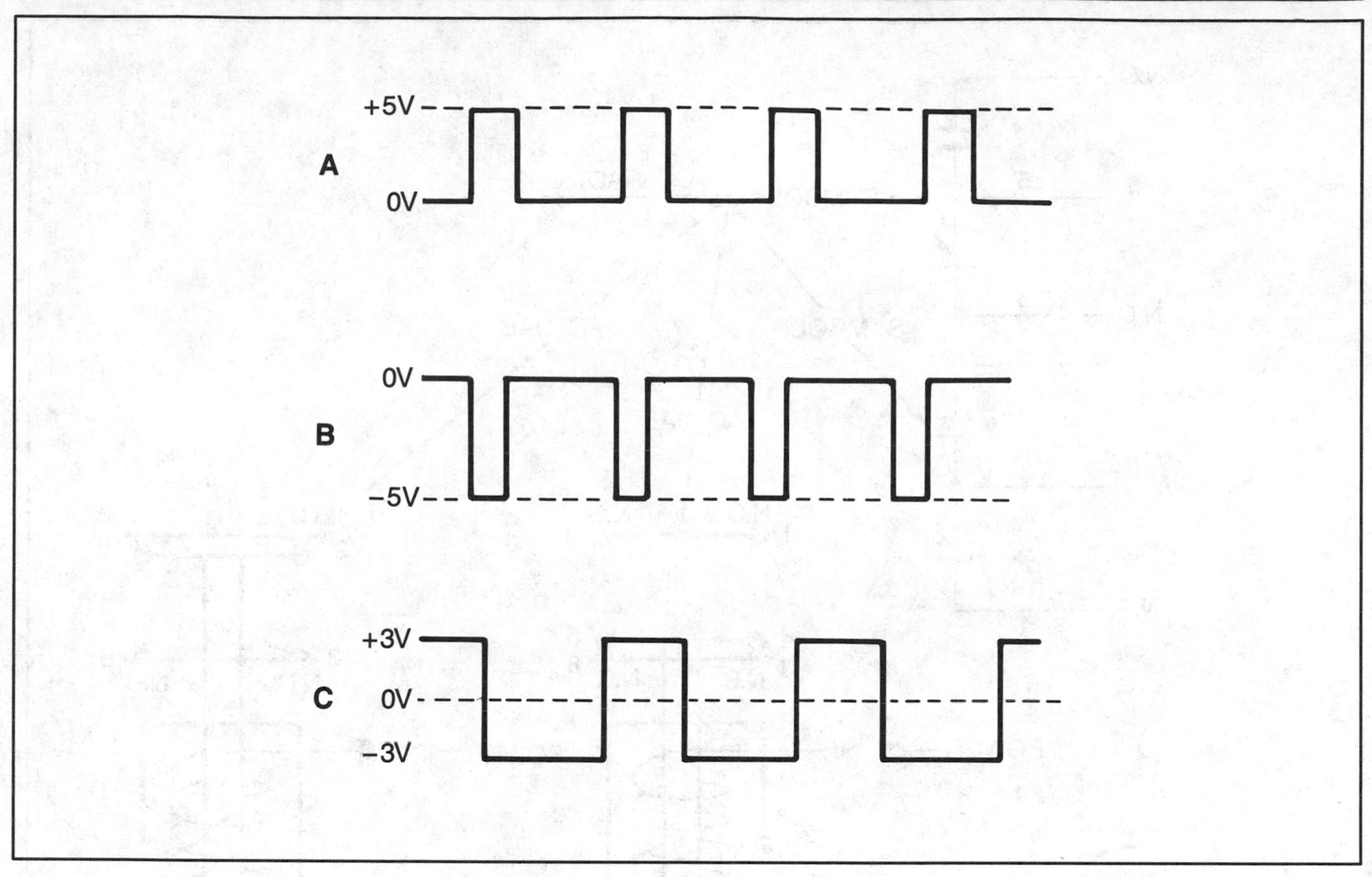

Figure 16-2. Digital signals.

changed continuously over a wide range to achieve sound levels from completely off to extremely loud. Any type of switch or relay is a digital device because it has two or more distinct positions. The glass thermometer is an analog device. So is a compass. Money is digital. A pointer-type pressure gauge is analog.

General Digital Applications

Digital devices and techniques are being employed in almost every area of electronics from communications to test instruments. Relay logic circuits, microprocessors, and industrial controls are devices that are of particular interest to the hydraulics industry.

Relay Logic. A relay can be considered digital in nature because it is basically an ON/OFF device. Figure 16-3 shows a relay and its associated contacts in both the energized and de-energized state. When current is allowed to flow in the coil circuit of the relay, a magnetic field is created that moves the plunger to the energized position. A set of common contacts are connected to the plunger. When the relay is de-energized, the common contacts are connected to the N.C. (normally closed) contacts. In the illustration, contacts 2 and 3 are connected and contacts 5 and 6 are connected. When the relay is energized, the common contacts are connected to the N.O. (normally open) contacts. As shown in the drawing, contacts 1 and 2 and contacts 4 and 5 are connected. The relay acts like an electrically operated switch.

Relays have been in use in industrial applications for many years and have been quite reliable. The life span of a relay is measured in millions of actuations. Relays are still widely used today for switching or controlling circuits using high voltages and currents, and for controlling high voltage devices with low voltage control circuits. Figure 16-4 shows a typical basic relay logic circuit. The relays in this circuit operate on 115 volts AC (single phase) and control a 480 volt AC, three-phase motor. The operation of this circuit is described below.

Starting with Line A in the figure, Fuse 1 protects the circuit from overcurrent conditions. Circuit power is provided by turning on SW1. This energizes the Master Control Relay (MCR). On Line B, when the MCR is energized, contacts MCR-1 and MCR-2 close, providing power to the rest of the circuit. On Line E, the motor will be started by Motor Starter 1M if conditions on Line C are correct. There are three overload contacts after 1M (one for each phase) that are normally closed. These contacts open if an overcurrent condition exists in the motor itself.

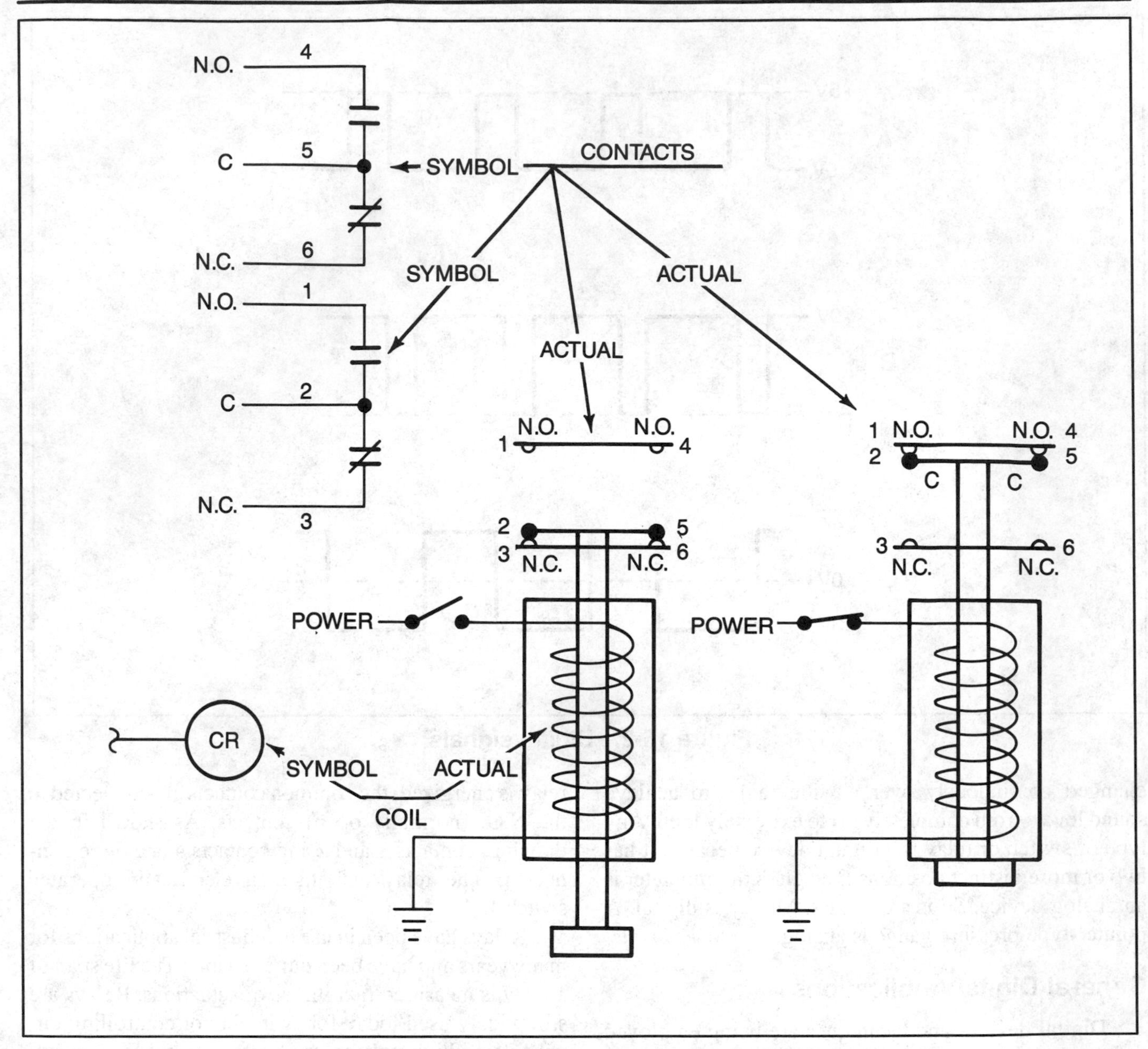

Figure 16-3. Basic relay construction.

In order to energize 1M, the 1LS Safety Gate Proximity Switch on Line C must be closed and SW2, the Manual Stop Switch, must not be depressed. When SW3, the Manual Start Switch, is depressed, 1M is energized. Immediately, seal-in (or holding) contact 1M-1 (on Line D) closes to seal-in the 1M relay so that 1M will stay energized when the Start Switch (SW3) is no longer depressed.

The circuit also has red and green lamps to indicate whether the motor is ON or OFF. The red light (Motor Off) is in series with the normally closed (N.C.) contact 1M-2 on Line G. When relay 1M is de-energized, this contact is closed and the red light is ON. When 1M energizes, the 1M-2 contact opens and turns off the red light. The normally open contact 1M-3 closes when 1M is energized and this allows the green light to illuminate indicating that the motor is ON.

Therefore, the motor is energized by the following logical conditions:

Manual Stop not Depressed
AND
Manual Start Depressed **OR** Seal-in Contact Closed
AND
Safety Gate Closed
AND
All 3 Overload Contacts Closed

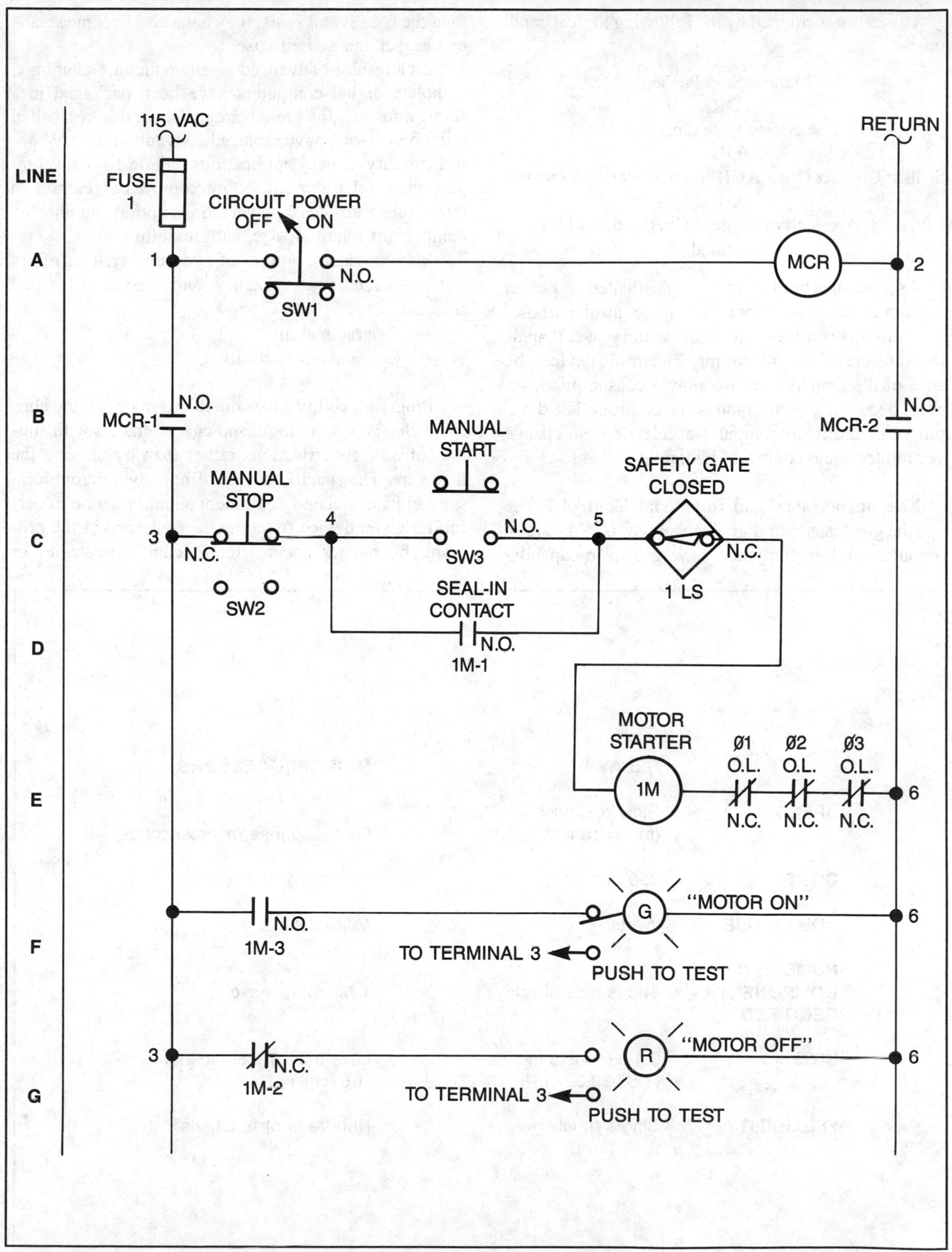

Figure 16-4. 115 volt, single phase circuit is used to control a 480 volt, three-phase motor.

The motor is stopped by the following logical conditions:

Manual Stop Pushed
OR
Safety Gate Open
OR
Seal-in Contact Open **AND** Manual Start not Pushed
OR
Phase 1 Overload or Phase 2 Overload or Phase 3 Overload.

As machines became more sophisticated and faster, relays proved to be impractical for control purposes. Relay circuits consumed more space and power than the machines they were controlling. Electronic devices, because of their small size, low power consumption, and speed, began to appear in industrial controls. The development of the microcomputer accelerated the change-over to electronic control of hydraulic systems.

Microprocessors and Industrial Controls. Perhaps the greatest use of digital techniques is in the area of computers. While computers have grown in capability over the last several years, they have also become smaller, cheaper and easier to use.

As a result of advanced semiconductor technology, complete digital computers have been packaged in a single miniature integrated circuit. These devices, called microprocessors, have replaced conventional relay control circuitry in many applications. Figure 16-5 is a comparison of relay control vs. microprocessor control in terms of speed, cost, power consumption, number of components required, size, and flexibility.

Microprocessor control of hydraulic systems offers improved accuracy, efficiency, and versatility for two reasons:

- Programmability
- Computational ability

Programmability gives the microprocessor the ability to change system logic and characteristics with simple software modifications rather than by altering the hardware. The specific details of how the microprocessor and its associated equipment should operate to perform a given device function are contained in the program. If changes occur later in control strategies or

	RELAYS		MICROPROCESSORS
SPEED	Slow response (milliseconds)		Fast response (microseconds)
COST	\$50 and up		\$7.50 to \$20
POWER USE	Watts		Milliwatts
NUMBER OF COMPONENTS REQUIRED	Thousands of relays	=	1 Microprocessor
SIZE	0.1-6.0 cubic inches (1.63-98.32 cm^3)		Less than 0.1 cubic inches (1.63cm^3)
FLEXIBILITY	Single function		Hundreds of functions

Figure 16-5. Comparison between relays and microprocessors.

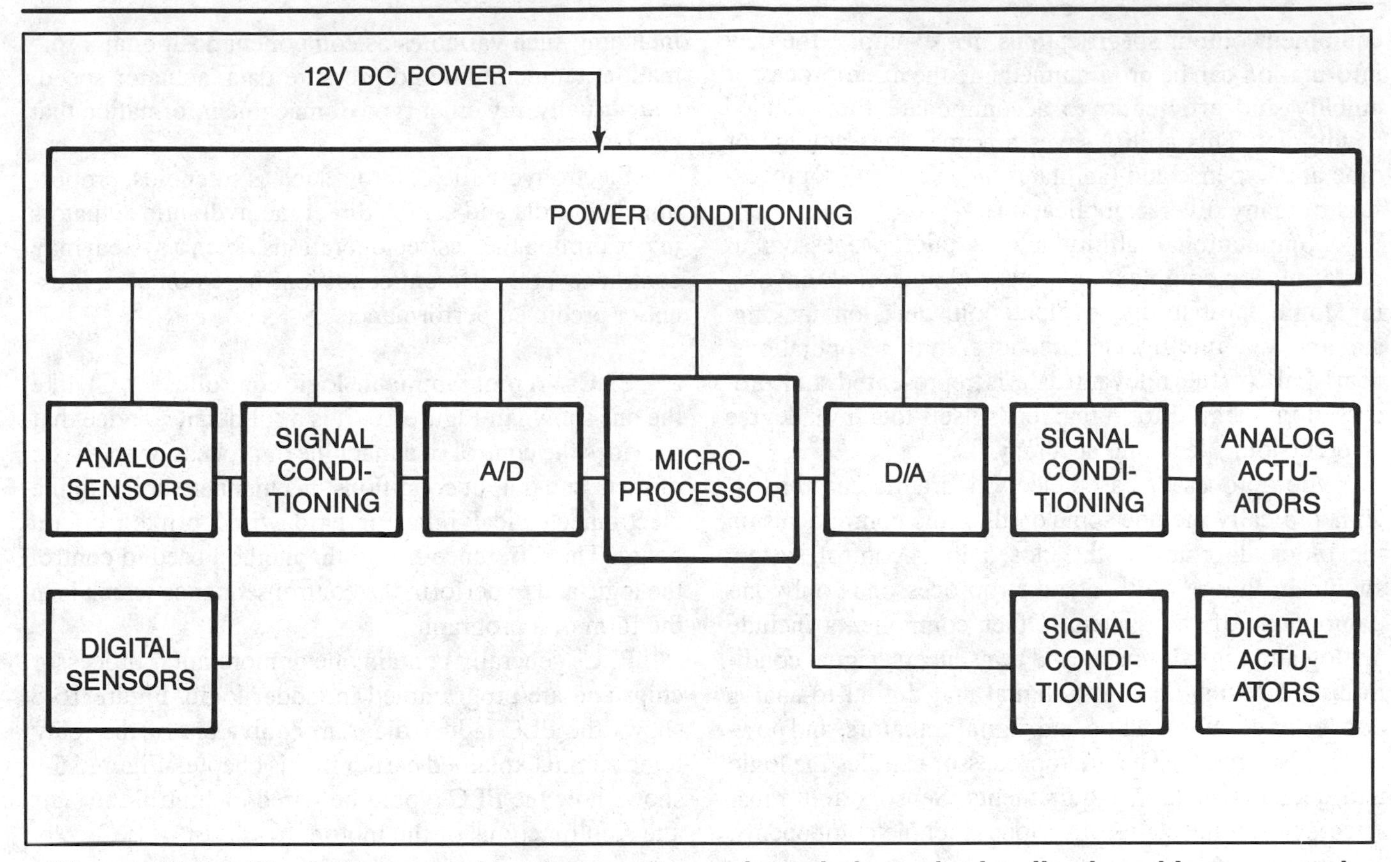

Figure 16-6. Block diagram of a microprocessor based electrohydraulic closed loop control system.

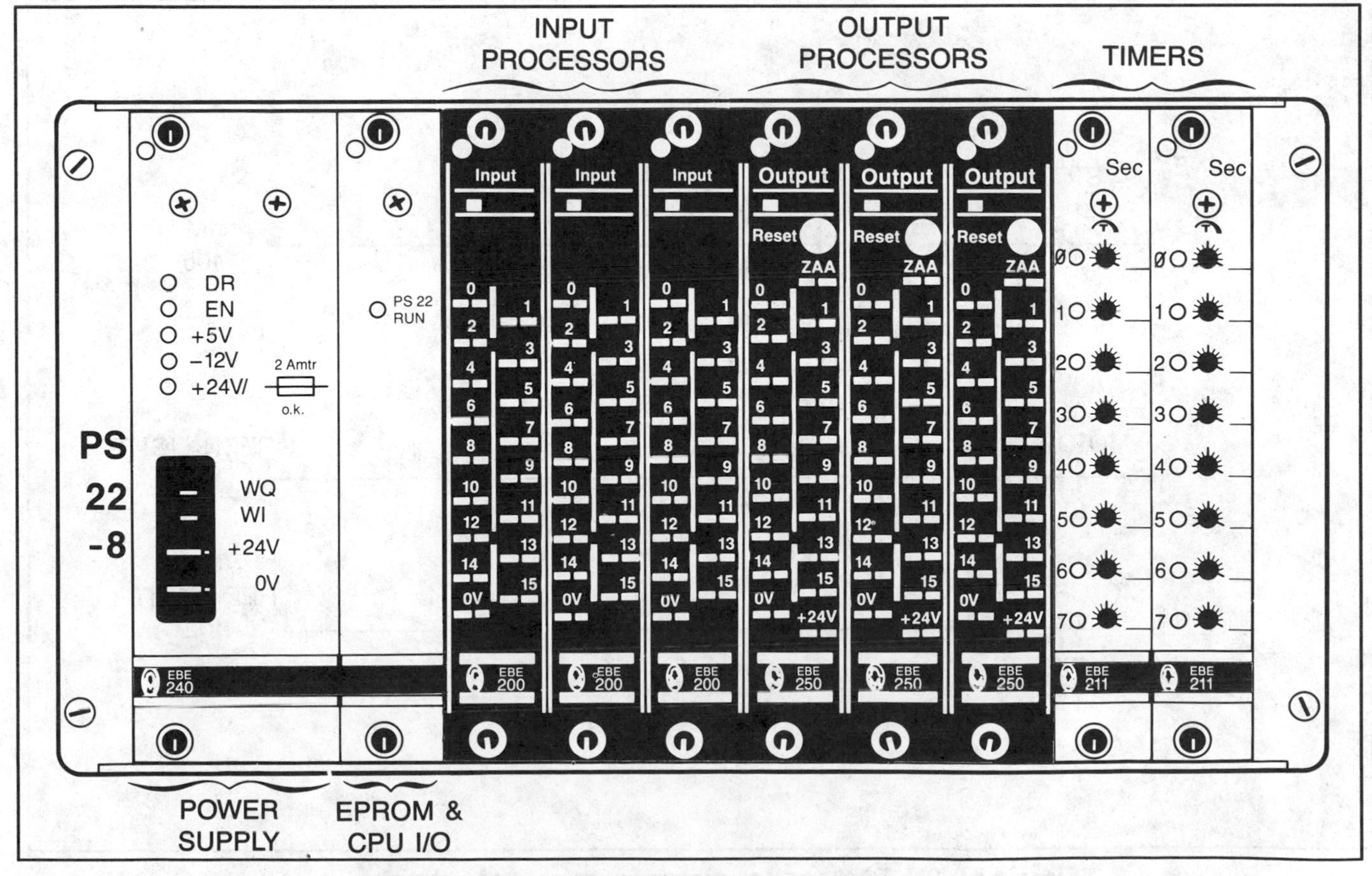

Figure 16-7. Programmable logic controller.

equipment output specifications, for example, the new information can be programmed into the microprocessor quickly and efficiently to accommodate the changed conditions. This ability saves a tremendous amount of time and expense and facilitates the use of microprocessors in many diverse applications.

Computational ability allows microprocessors to perform very complex and powerful operations involving logic, probability, computation, decision making, and artificial intelligence functions. In these operations, nearly all of the relevant data is represented and processed in a digital form that lends itself to a high degree of precision, speed and accuracy.

Microprocessor based electrohydraulic control systems typically include some or all of the components on the block diagram of the closed loop control system shown in Figure 16-6. The microprocessor is only one component of the system. Other components include analog and digital sensors and transducers, signal conditioning circuits, analog-to-digital and digital-to-analog converter circuits, analog and digital actuators, and power supply circuits. The microprocessor handles the logic and decision making requirements. Sensors and transducers report the status of various machine components, including such variables as component positional information, temperature and pressure data, actuator speed, and virtually any other type of machine information that can be sensed.

Electrohydraulic controls such as solenoids, proportional controls and servos direct the hydraulic actuators in performing the desired operations. Such a system may control several different conditions based on past, present or predicted performance.

PLCs. A programmable logic controller (PLC), like the one shown in Figure 16-7, is a solid-state device that provides the control of a machine or process on the basis of input and output conditions in a manner similar to the electromechanical relay or hard wired printed circuit board. The difference lies in the method used to control the logic and to perform the control sequence which is in the form of a program.

PLCs generally contain one or more microprocessor chips and are programmed in ladder logic. Figure 16-8 shows the PLC ladder diagram equivalent of the relay logic circuit explained earlier in this chapter. Figure 16-9 shows how the PLC would be wired to control the start and stop functions of the motor.

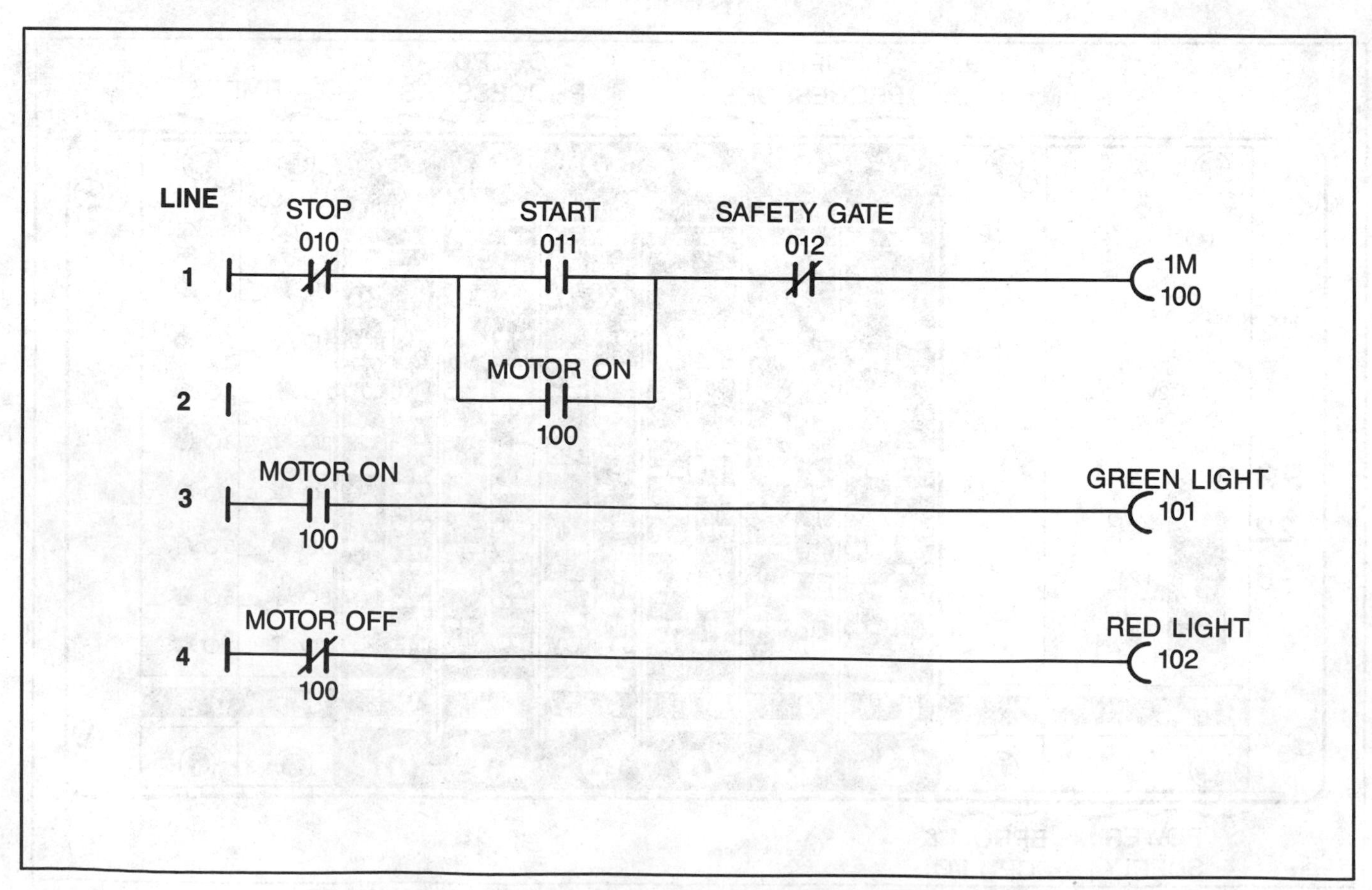

Figure 16-8. PLC ladder diagram for relay logic circuit.

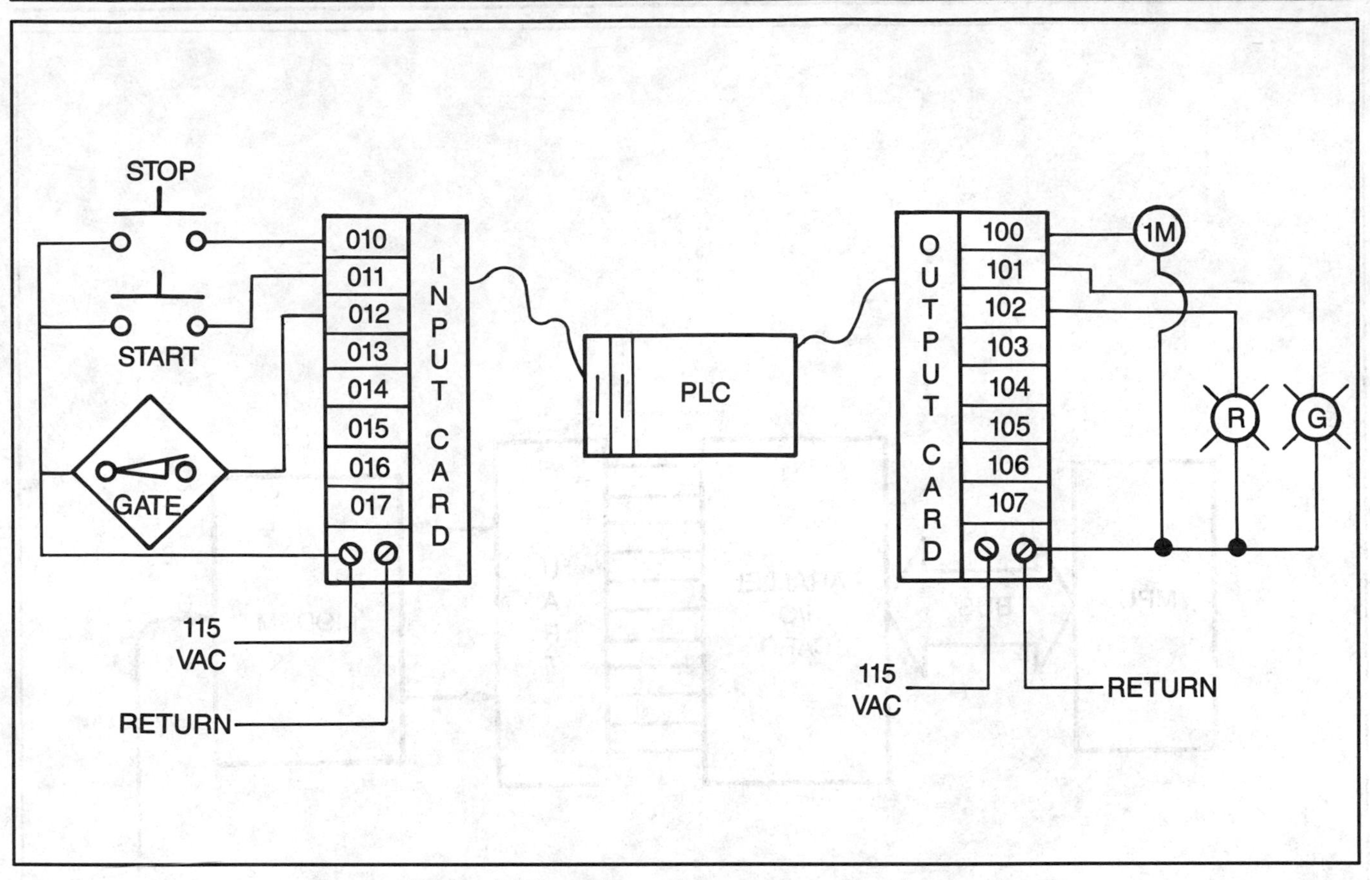

Figure 16-9. PLC wiring diagram for relay logic circuit.

With a programmable controller, the sequence of the logic is converted from the symbology of a circuit diagram to data stored on memory chips through an electronic device referred to as a memory loader or programmer. It is comprised of a simplified keyboard with identification characters representing relay ladder diagram symbols and operational instructions. When a chip has been programmed, all of the logic circuitry has essentially been transferred onto solid-state devices. Because these devices can be changed or reprogrammed quite easily, the PLC has the versatility to quickly modify a process or a set of operational parameters that control the machine's function.

A microprocessor is one of many components of a programmable logic controller. It serves as the heart of the microcomputer allowing it to perform arithmetic calculations, make comparisons, and remember what it has done. The use of a microprocessor lends an intelligent capability to an ordinary ON/OFF controller. Other components include input and output conditioning circuits, power supplies, analog to digital and digital to analog converters, and safety and alarm circuits.

PLC inputs can be digital ON/OFF signals from switches, contacts, etc.; analog signals from thermocouples, instruments and transducers, or digital words from other computers or PLCs. Outputs can be digital ON/OFF voltages or currents to relays, lights, starters, and other control devices, or analog voltages and currents to such devices as small motors. PLC outputs may also be digital words to LED displays or other PLCs and computers.

PLCs are available in various sizes starting with the smallest versions that can handle three inputs and outputs. The small PLC is basically used to replace or substitute for a few relays and costs around $200. The largest PLCs can handle 3000 to 6000 inputs and outputs, have full computer capability, and run entire process facilities. These versions cost anywhere from $10,000 to $100,000.

Digital Data Transmitters. Digital data transmitters are devices that move data between digital devices. They can be built into the digital device or located in separate enclosures. For example, most personal computers have a built-in RS-232 interface which moves data serially (one bit at a time) from the computer to its printer. The term, RS-232, refers to an industry standard which dictates the voltages, number of wires, and definitions of the electrical signals which must be used if the interface is called an RS-232. If data transmission is to be from one computer to another, a modem is used.

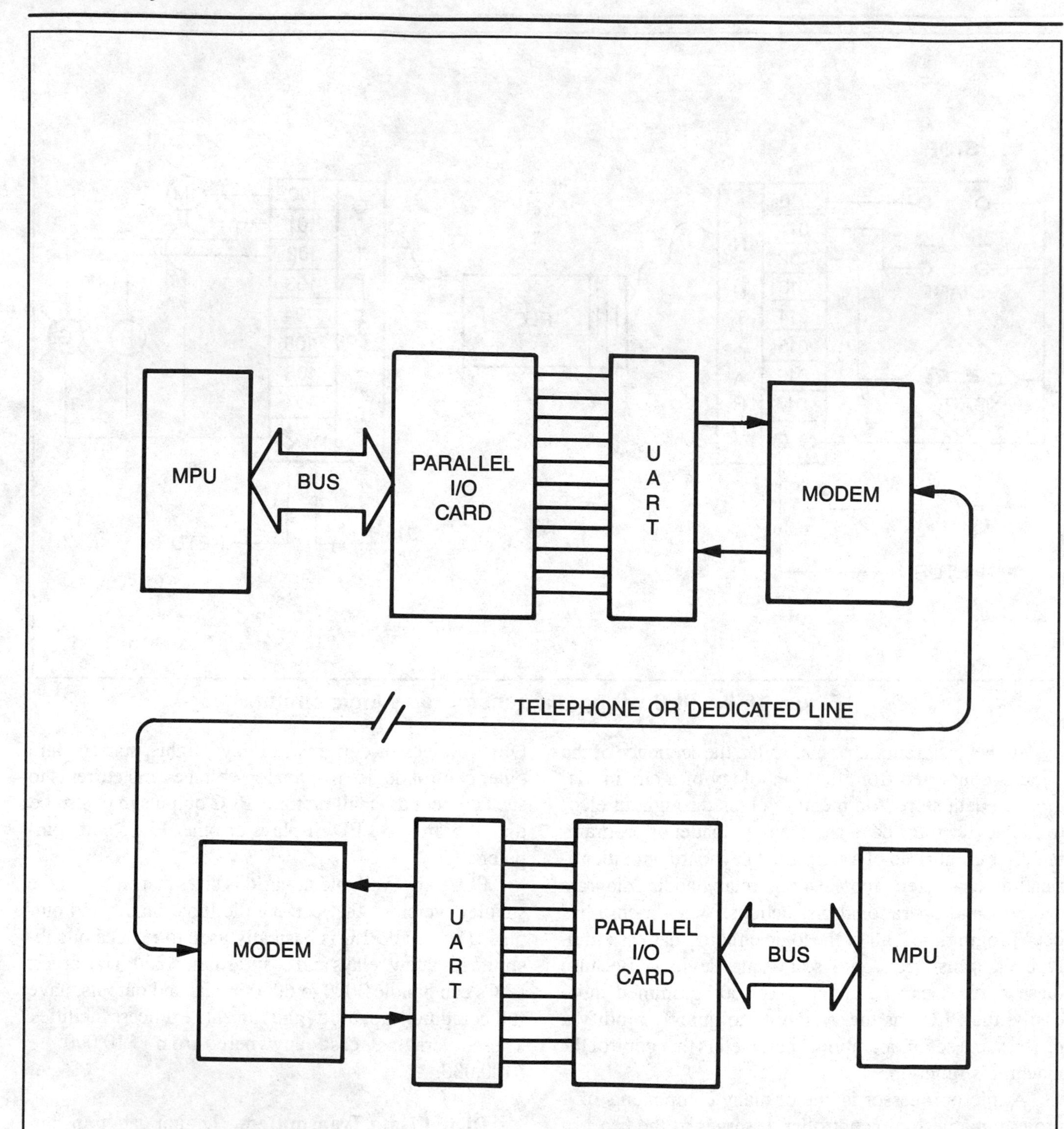

Using modems to communicate over long distances. The modems send and receive digital signals by using different tones for logic ''1''s and logic ''0''s. NOTE: UART is an acronym for Universal Asynchronous Receiver Transmitter. MPU is an acronym for Microprocessing Unit.

Figure 16-10. Modem used for data transmission.

A modem, or Modulator-Demodulator, is a device used to provide data communications between two separate microprocessing units (Figure 16-10). The modulator portion of the modem converts digital signals into tones that can be transmitted over a telephone line or over any long distance serial line. For example, the modulator might convert the binary output of a circuit into a high tone (1270 Hz) for logic 1 and a low tone (1070 Hz) for a logic 0. The demodulator portion of the modem converts the received tones from the telephone line into signals that are once again compatible with the digital circuits of the receiving microprocessor.

There are many other types of digital data transmission and the method used depends on cost, environment, and speed requirements. The RS-232 interface is low cost, good for up to 50 feet (15.24 m) of transmission line, low speed, voltage driven, and typically used in an office environment. Parallel interfaces cost more but operate at a very high speed. They are also good for short distances, allow computers to communicate with each other, and are found in office environments. The RS-485 interface also costs more, but is current driven, is good for thousands of feet (meters) of distance, has excellent immunity to electrical noise, and is typically used in industrial environments. Fiber optics are high in cost, but are very high speed, are totally immune to electrical noise and interference, and are able to handle large amounts of traffic in a very small diameter cable. They are used where absolute data transmission reliability is required.

Other, more sophisticated digital data transmission methods include radio signals, microwave, and other industry transmission standards such as RS-422.

Serial and Parallel Transmission of Data. Digital numbers are transmitted, processed or otherwise manipulated in two basic ways. In the serial methods of data handling, each bit of a binary word or number is transmitted serially one at a time. In a parallel system, all bits of a word or a number are transmitted simultaneously.

Figure 16-11 shows a binary number represented in a serial data format that is being sent over a transmission line to a receiving circuit. The binary number exists as a series of voltage levels representing the binary 1s and 0s. Preceding the binary number is a start bit that indicates to the receiver that data is being sent. Following the binary number is a stop bit to indicate to the receiver that the binary number has ended. All of these voltage level changes occur at a single point in a circuit or on a single data line. Each bit of the word exists for a specific inter-

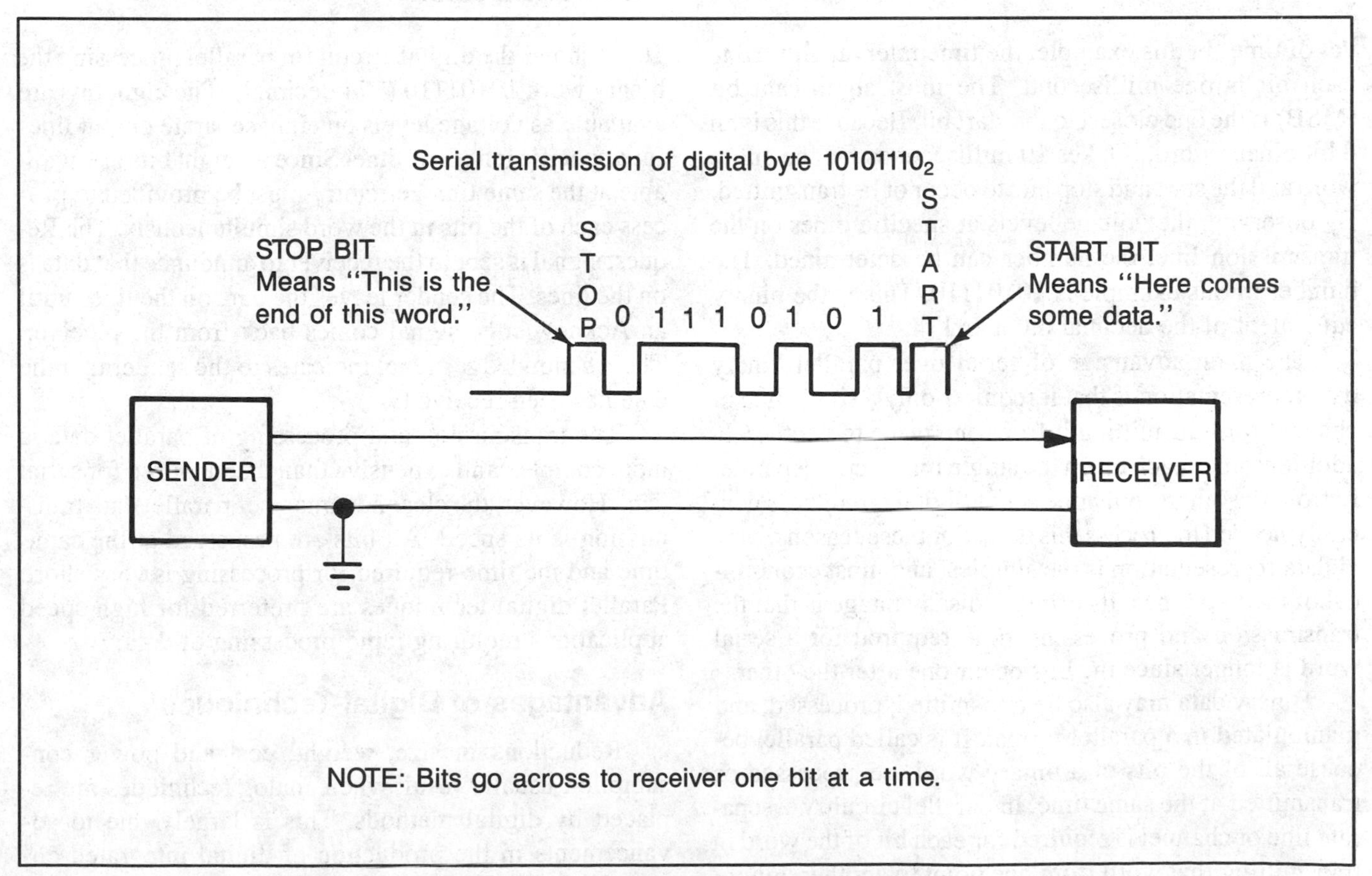

Figure 16-11. Serial data transmission.

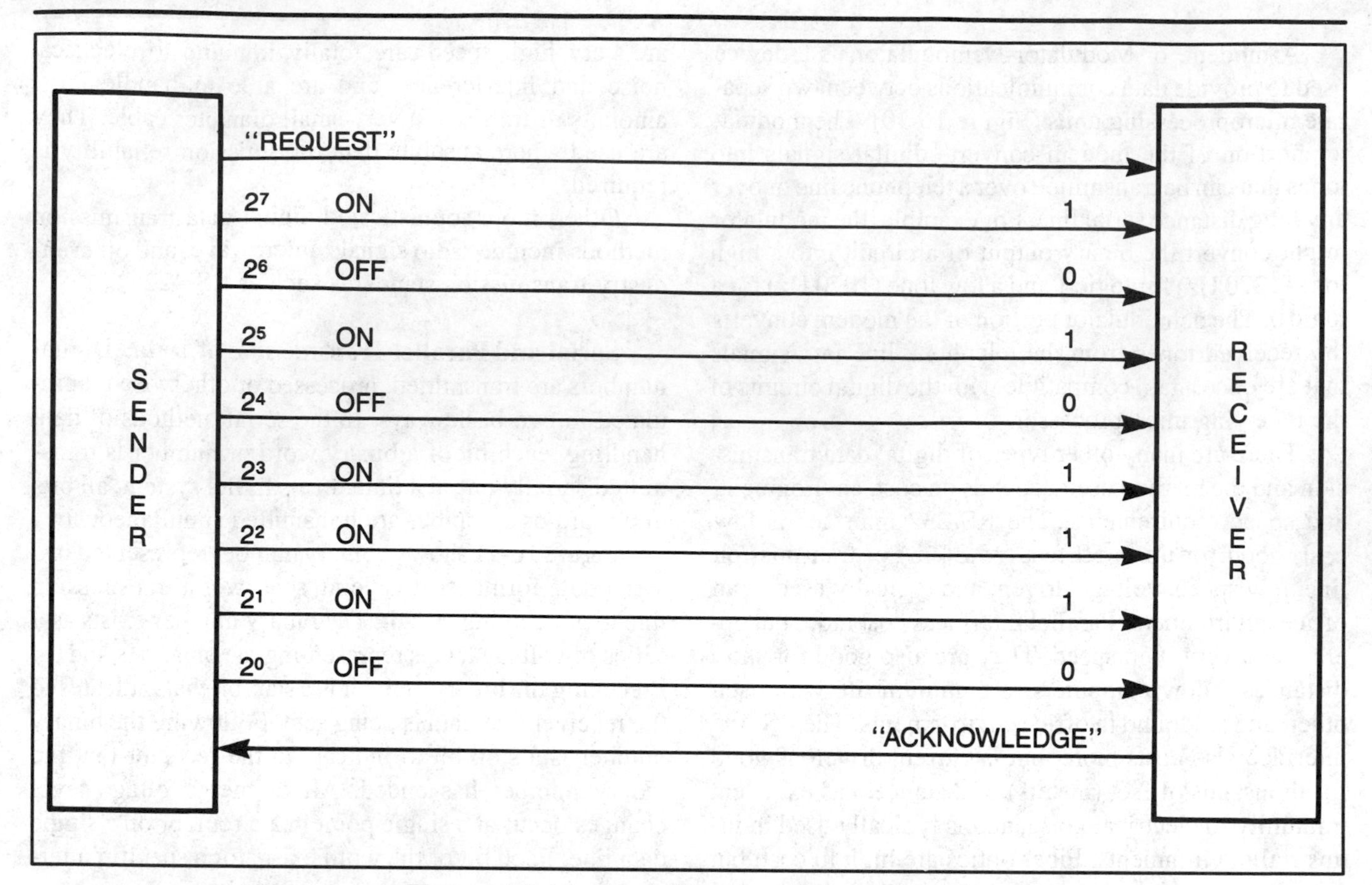

Figure 16-12. Parallel data transmission.

val of time. In this example, the time interval allotted to each bit is one millisecond. The most significant bit (MSB) is the one closest to the start bit. Because this is an 8 bit binary word, it takes 10 milliseconds for the entire word and the start and stop bits to occur or be transmitted. By observing the voltage levels at specific times on the transmission line, the number can be determined. The number in this example is 10101110. This is the binary equivalent of the decimal number 174.

The main advantage of serial over parallel binary data representation is that it requires only a single line or channel for transmitting it from one place to another. In addition, since each bit on the single line occurs separately from the others, only one set of digital circuitry is generally needed to process this data. For these reasons, serial data representation is the simplest and most economical of the two types. Its primary disadvantage is that the transmission and processing time required for a serial word is longer since the bits occur one after the other.

Binary data may also be transmitted, processed, and manipulated in a parallel format. It is called parallel because all of the bits of a binary word are processed or transmitted at the same time. In parallel circuitry, a separate line or channel is required for each bit of the word in transmitting that word from one point to another. Figure 16-12 shows the digital circuit for parallel processing the binary word 10101110 (174 decimal). The eight bits are available as voltage levels on eight separate output lines for a specific length of time. Since all eight bits are available at the same time, circuitry must be provided to process each of the bits in the word simultaneously. The Request signal is sent to the receiver to announce that data is on the lines. The sender leaves the data on the lines until an Acknowledge signal comes back from the receiver. The Acknowledge signal indicates to the sender that the data has been received.

The transmission and processing of parallel data is more complex and expensive than that required for serial data. However, the clear advantage of parallel data transmission is its speed. All bits are processed at the same time and the time required for processing is very short. Parallel digital techniques are preferred for high speed applications requiring rapid processing of data.

Advantages of Digital Techniques

Reductions in size, weight, cost and power consumption usually result when analog techniques are replaced by digital methods. This is largely due to advancements in the production of digital integrated circuits or "chips." Other advantages include:

- Greater accuracy—digital techniques allow greater precision and resolution in representing quantities or in making measurements than do analog methods. Accuracy can be verified using parity checking methods.
- Greater dynamic range—dynamic range is the difference between the upper and lower data values that a system or instrument can handle. Analog systems are limited because of component capabilities and noise to a range of something less than 100,000 to 1. With digital techniques practically any desired dynamic range can be obtained.
- Greater stability—analog circuits are subject to the effects of drift and component tolerance problems. Temperature and other environmental factors affect resistor, capacitor, and inductance values. Transistor control voltages may vary causing nonlinear operation and distortion. Component imperfections and aging cause drift and resultant problems. Digital methods greatly minimize or eliminate these problems.
- Convenience—digital techniques make instruments and equipment more convenient to use. The direct decimal display of data is not only more convenient, but the error of reading or interpolating analog meters or in setting analog dials is eliminated.
- Automation—many electronic processes can be fully automated if digital techniques are used. Special control circuits or a digital computer can automatically set up, control, and monitor many operations. Data is readily recorded, stored, and displayed.
- New approaches—digital circuits make it possible to do some things that have no analog equivalent.

Basic Logic Circuits

The two basic types of digital logic circuits are decision- making circuits and memory circuits. In all digital circuits, decisions are made based on given information

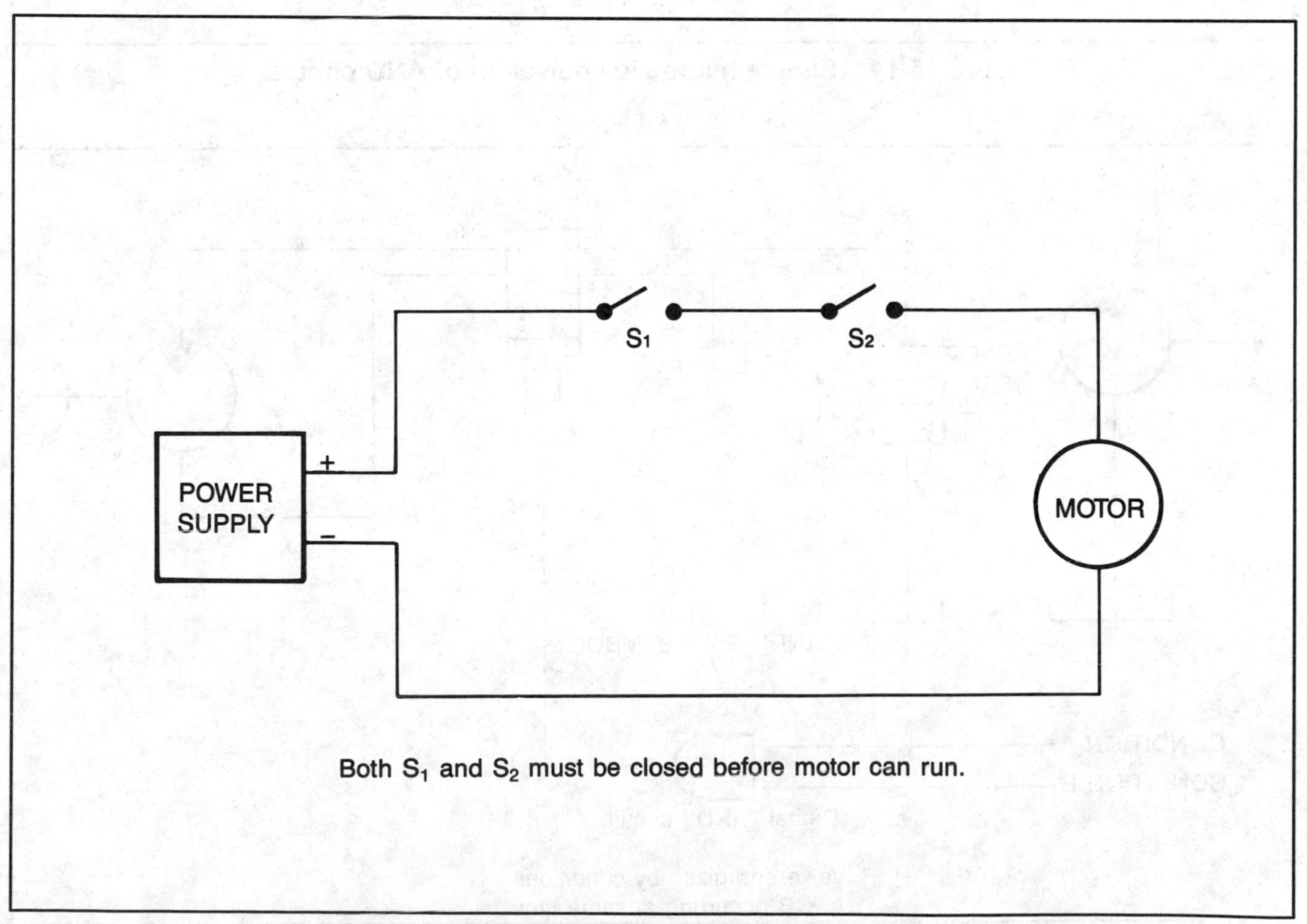

Figure 16-13. Simple electrical AND circuit.

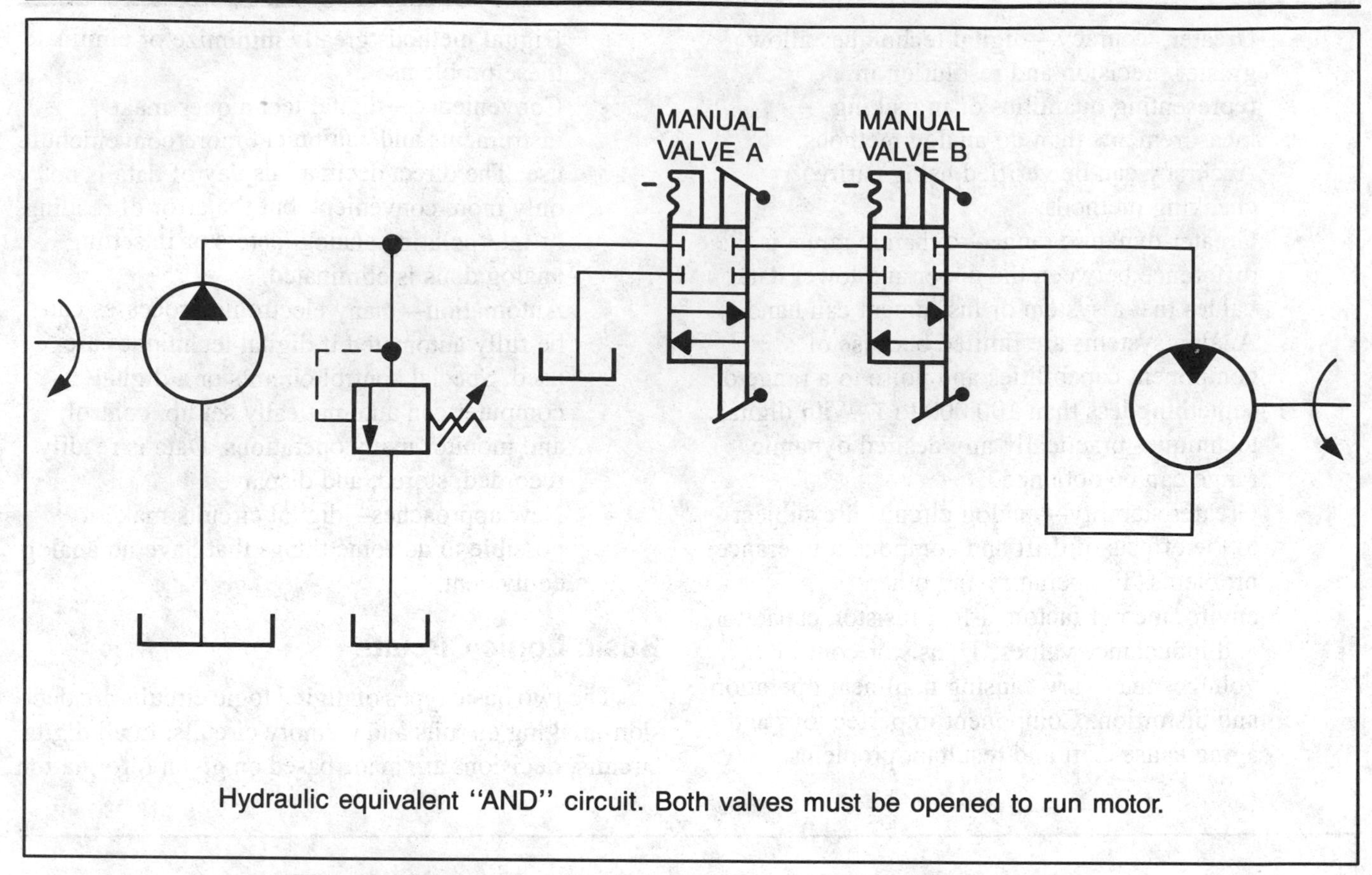

MANUAL
VALVE A
MANUAL
VALVE B
Hydraulic equivalent "AND" circuit. Both valves must be opened to run motor.

Figure 16-14. Simple hydraulic equivalent of AND circuit.

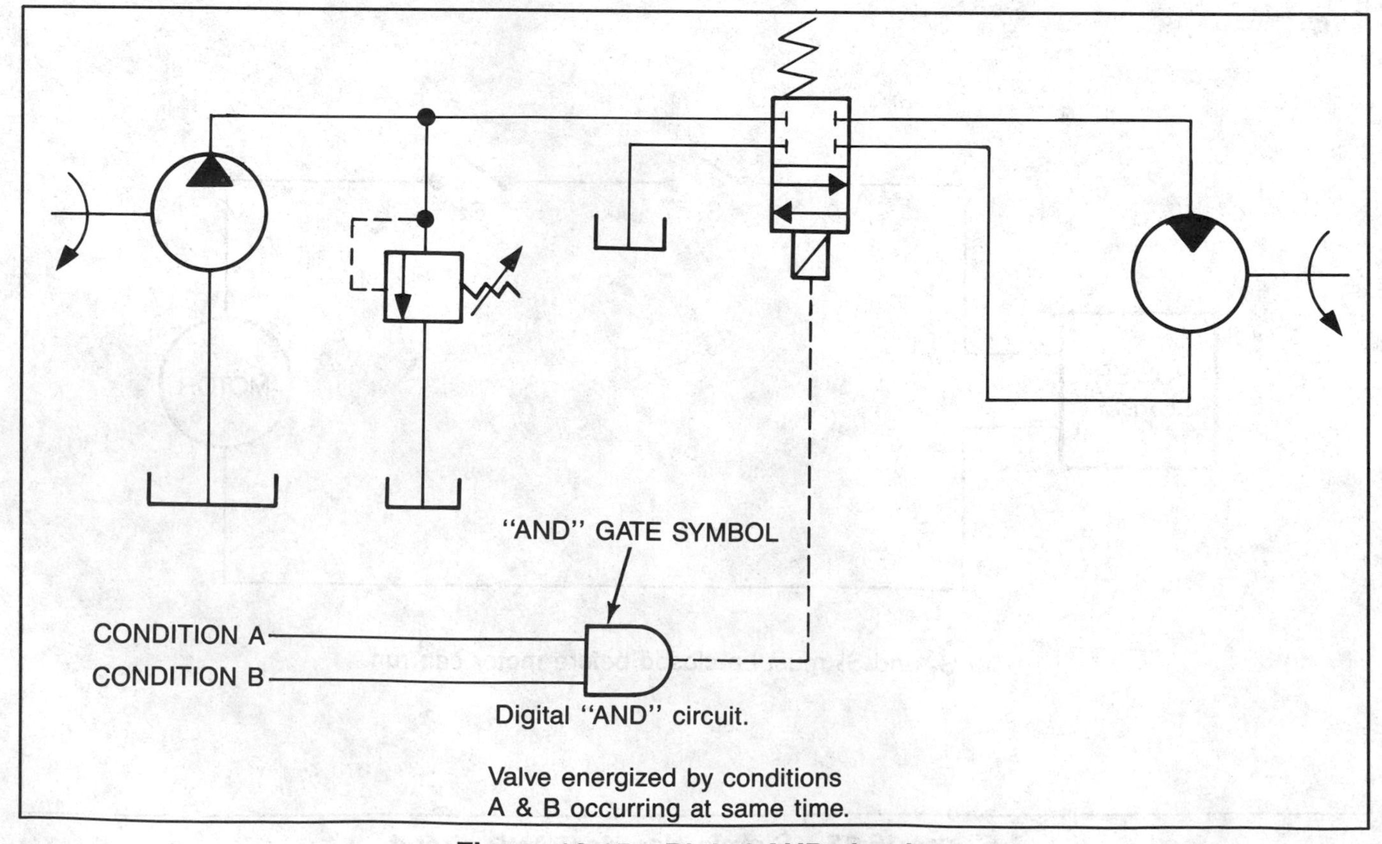

Figure 16-15. Digital AND circuit.

or information is stored for later use. Decision-making logic circuits are called gates and usually have two or more inputs and a single output. Both inputs and outputs are in binary form. The gates are combined in a variety of ways to form combinational logic circuits that can perform many complicated decision-making functions.

Memory circuits store data in binary form and are commonly called flip-flops. Flip-flops store a single bit of data and are usually combined to form sequential circuits which can store, count, and shift larger amounts of binary data. The emphasis in this section is on the two basic decision-making circuits or gates, the AND gate and the OR gate.

AND Circuit. A logic gate has two or more inputs and a single output. The gate makes its decision based upon the input states and its particular function, then generates the appropriate binary output. The AND gate operates in such a manner that its output state is a binary 1 if and only if all its inputs are a binary 1 at the same time. If any or all inputs are binary 0, the output of the AND gate is a binary 0. This function can be described by a table of input and output conditions known as a truth table. A truth table for a two input AND gate is:

Inputs		Output
A	B	C
0	0	0
0	1	0
1	0	0
1	1	1

Figure 16-13 shows a simple electrical equivalent of the AND circuit. Current from the power supply flows through the motor only when both of the electrical switches are closed. If either switch is opened, the motor cannot run. Figure 16-14 shows the hydraulic equivalent of the AND circuit. Manually actuated valves take the place of the switches in the hydraulic circuit, but both valves must allow the flow of fluid through them before the motor can turn.

Figure 16-15 shows a digital circuit for controlling a solenoid valve with an AND gate. Both inputs to the AND gate must be a binary 1 before the solenoid can be actuated. The inputs are logic level pulses of voltage that represent the binary states. They are transmitted to the AND gate by digital circuits in the system controller that determine when the solenoid should be actuated. In this case, two conditions must be met before the solenoid is actuated.

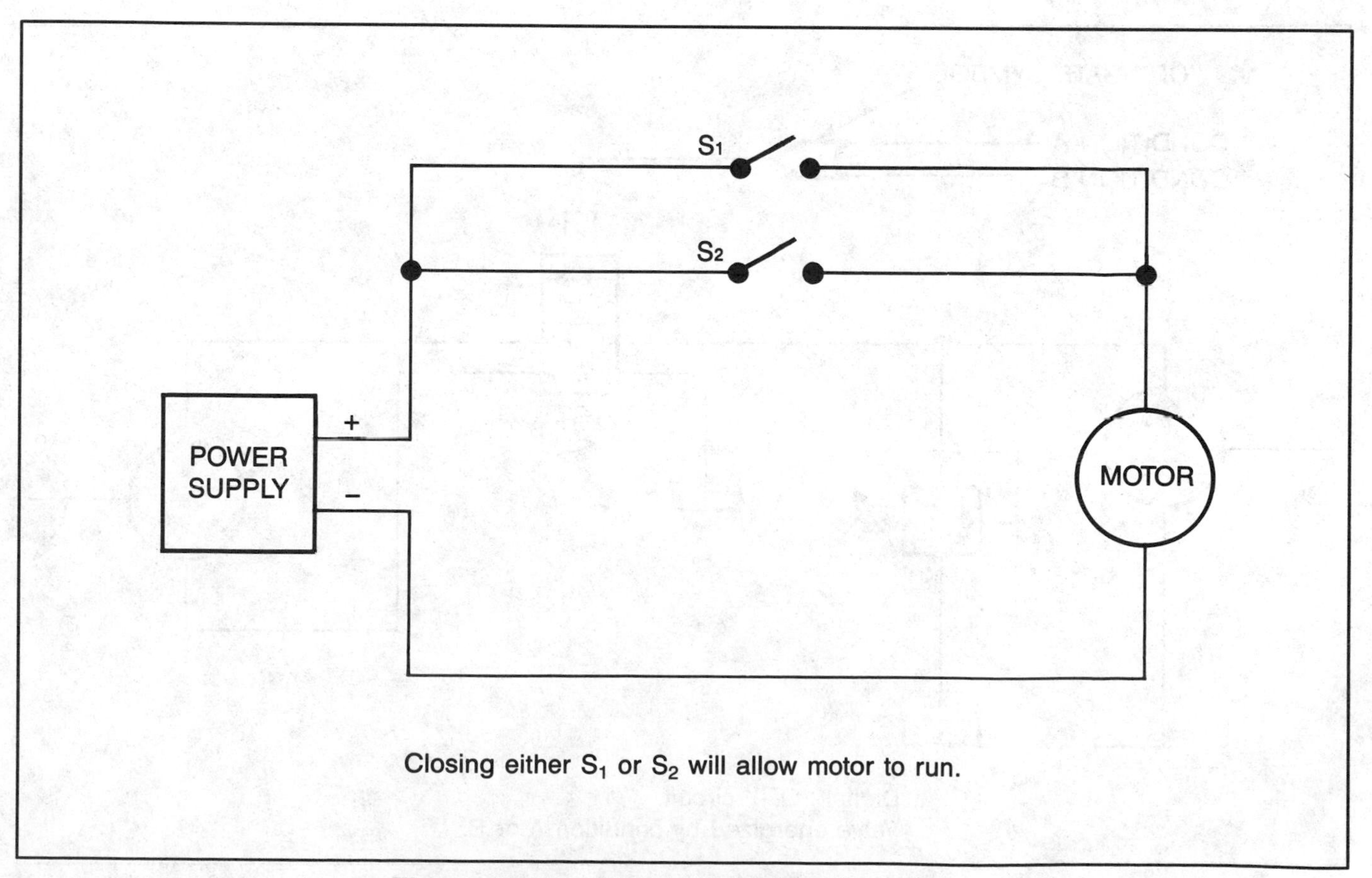

Figure 16-16. Simple electrical OR circuit.

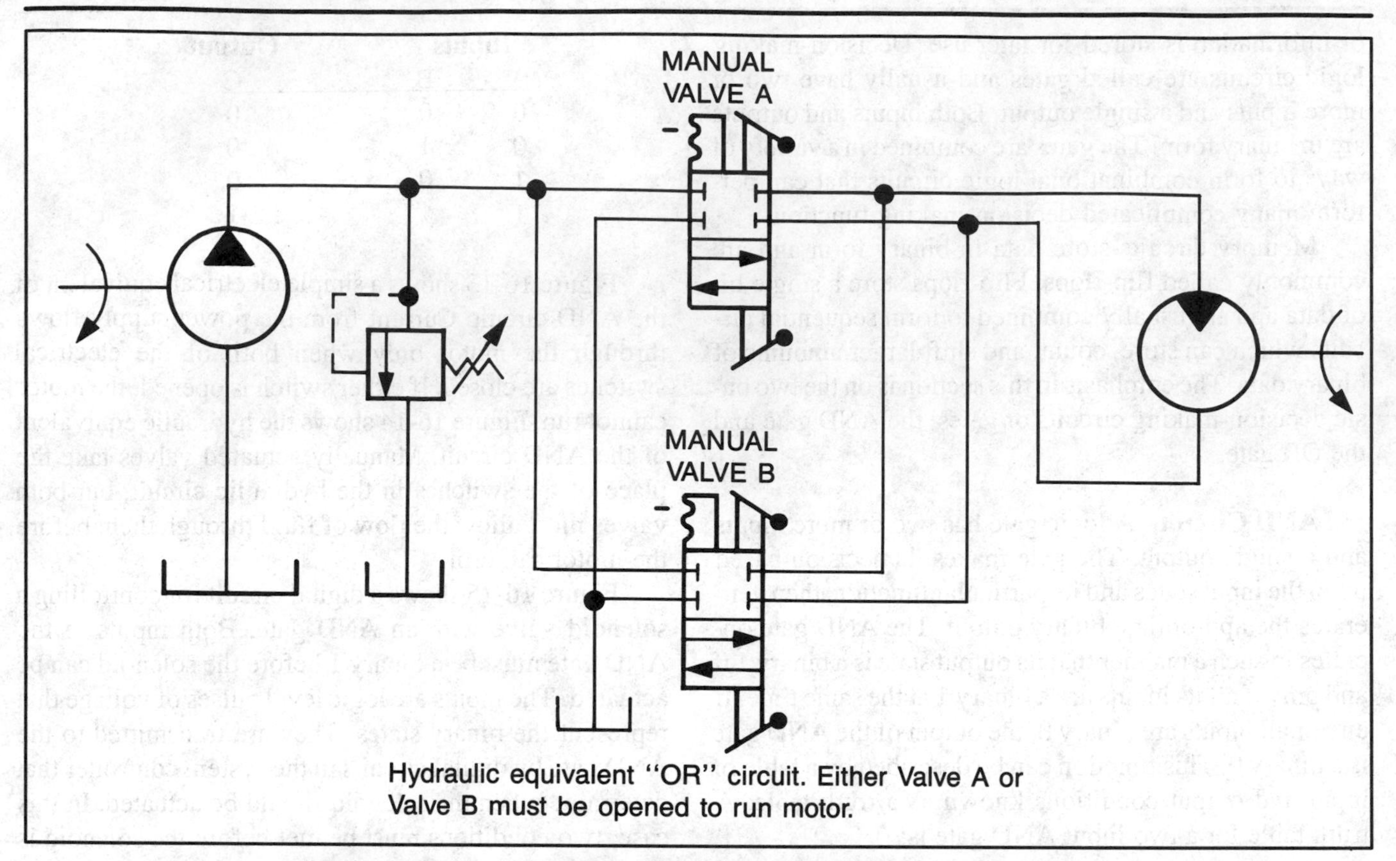

Figure 16-17. Simple hydraulic equivalent of OR circuit.

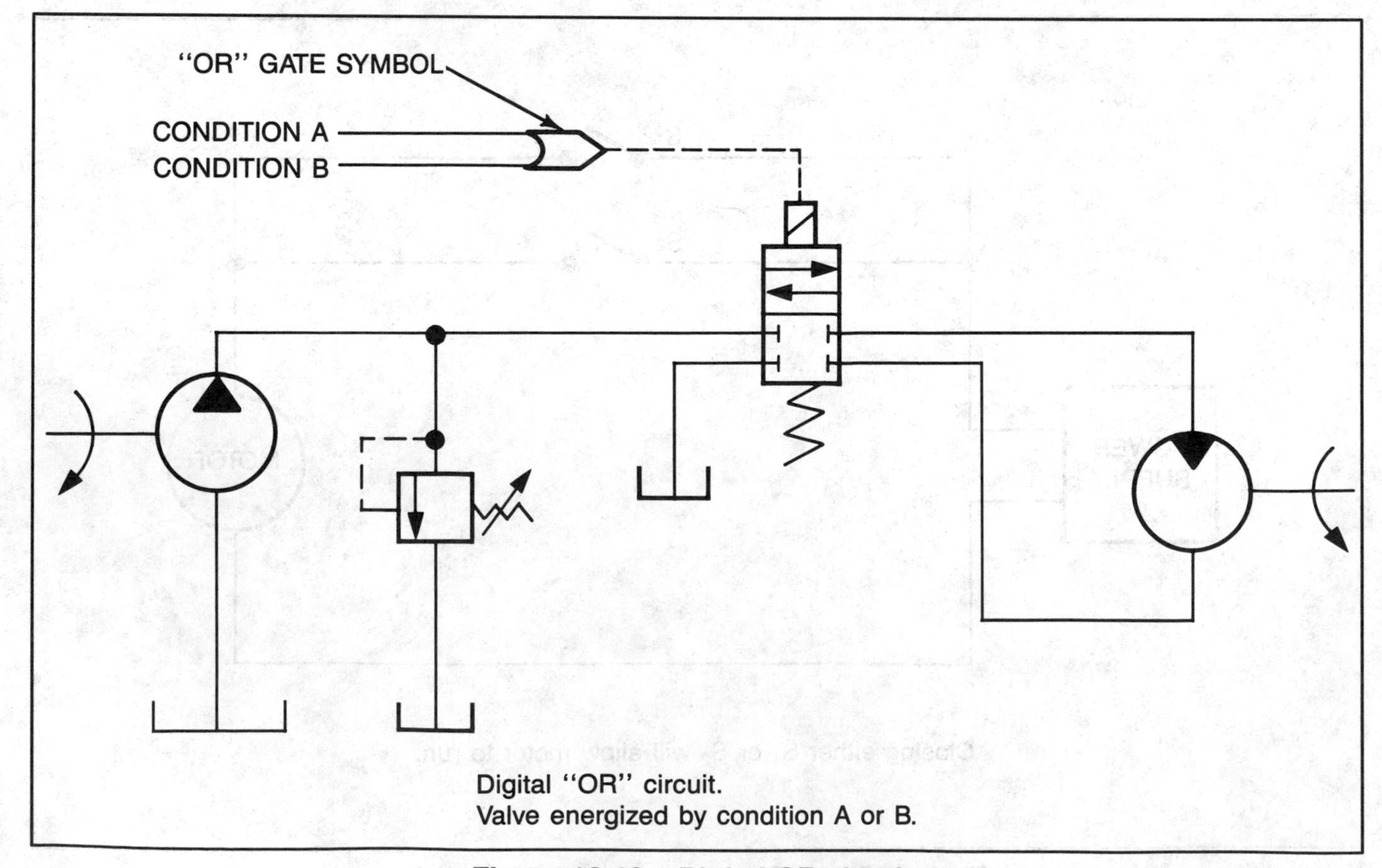

Figure 16-18. Digital OR circuit.

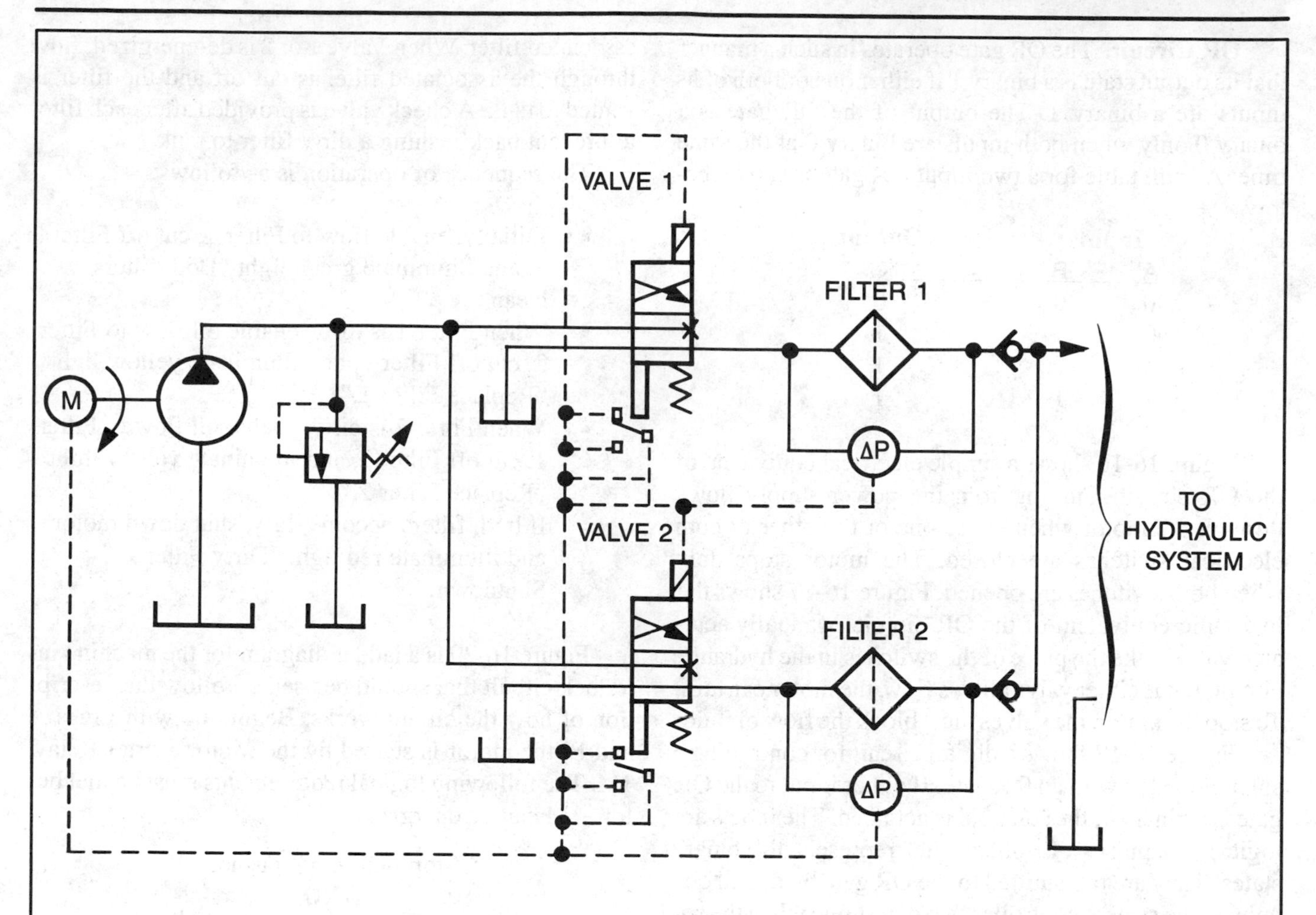

ΔP = Differential Pressure Switch. Normal Condition = Clean Filter = Low Differential Pressure. Contacts time delayed to avoid triggering due to pressure surges.

Valves 1 and 2 — Solenoid operated, spring offset with monitor switch. Switch off = Valve off = Oil flow to Filter off.

The above circuit will be interlocked to automatically switch to a clean filter and to shut down the motor if both filters are dirty.

Figure 16-19. Machine interlock circuit.

OR Circuit. The OR gate operates in such a manner that its output state is a binary 1 if either one or both of its inputs are a binary 1. The output of the OR gate is a binary 0 only when both inputs are binary 0 at the same time. A truth table for a two input OR gate is shown below.

Inputs		Output
A	B	C
0	0	0
0	1	1
1	0	1
1	1	1

Figure 16-16 shows a simple electrical equivalent of the OR circuit. Current from the power supply flows through the motor when either one or the other or both electrical switches are closed. The motor stops only when both switches are opened. Figure 16-17 shows the hydraulic equivalent of the OR circuit. Manually actuated valves take the place of the switches in the hydraulic circuit, but if either valve allows flow, the motor can turn. To stop the motor, the valves must block the flow of fluid.

Figure 16-18 shows a digital circuit for controlling a solenoid valve with an OR gate. If either input to the OR gate is a binary 1, the solenoid is actuated. The inputs are logic level pulses of voltage that represent the binary states. They are transmitted to the OR gate by digital circuits in the system controller that determine when the solenoid should be actuated. In this case, either of two separate signals could actuate the solenoid.

Machine Interlock Circuit. The following description of a machine interlock circuit is provided as an example of how AND and OR functions can be combined to accomplish a specific task. Following a discussion of how the circuit operates, the logical AND and OR functions are summarized.

The interlock circuit is shown in Figure 16-19 automatically switches to a clean filter when the operational filter becomes dirty. The interlock also shuts down the motor if both filters are dirty.

The circuit has an electric motor driving a hydraulic pump that provides flow to a hydraulic system. The relief valve protects the system from overpressure conditions. Valves 1 and 2 are solenoid operated, spring offset valves with a monitor switch. If the monitor switch is in the off position, the valve is off and the oil flow to its associated filter is cut off. There is also a differential pressure switch across each filter that actuates if the pressure difference reaches a level that indicates the filter needs changing.

Valves 1 and 2 select Filter 1 and 2 respectively. When Valve 1 or 2 is energized, it directs flow through its associated filter. When Valve 1 or 2 is de-energized, flow through the associated filter is cut off and the filter is vented to tank. A check valve is provided after each filter to prevent backflushing a dirty filter to tank.

The sequence of operation is as follows:

- Initially, enable flow to Filter 1, cut off Filter 2, and illuminate green light "Both filters clean."
- When Filter 1 is dirty, enable oil flow to Filter 2, cut off Filter 1, and illuminate yellow light "Replace Filter 1."
- When Filter 2 is dirty, enable oil flow to Filter 1, cut off Filter 2, and illuminate yellow light "Replace Filter 2."
- If both filters become dirty, shut down motor and illuminate red light "Dirty Filter Shutdown."

Figure 16-20 is a ladder diagram for the machine interlock circuit that should be used to follow the description of how the circuit works. Beginning with Lines A and B, the motor is started by the Motor Starter Relay, 1M. The following logical requirements must be met before 1M can be energized:

Stop button not pushed
AND
Start button pushed OR 1M-1 "Motor Started" contacts are closed
AND
Valve 1 OR Valve 2 ON

On Line C, another set of Pump Start contacts selects Valve 1 by energizing Relay 1TDR. Maintenance Switch prevents valve 1 from being selected if someone is working on it. On Line D, when 1TDR energizes, 1TDR-1 contacts close to hold 1TDR energized, as long as P1 stays closed (differential pressure not high). When Filter 1 gets dirty, the differential pressure goes up and P1 contacts open, deselecting Valve 1. At the same time, the P1 switch contacts on Line G close. Since Filter 2 hasn't been used yet, the P2 switch contacts are closed. Since Valve 1 has been deselected, 1TDR-2 contacts are closed and 2TDR is energized, selecting Valve 2.

Lines K through N show the 3CR and 4CR latches, which "remember" when a filter has experienced high differential pressure. For example, if P1 closes, indicating Filter 1 is dirty, the 3CR relay is energized. The 3CR-1 contacts close to hold the relay energized until the Maintenance Switch is opened (turned off) for filter replacement. The 4CR circuit operates in an identical manner for Filter 2. If Filter 1 is replaced while Filter 2 is be-

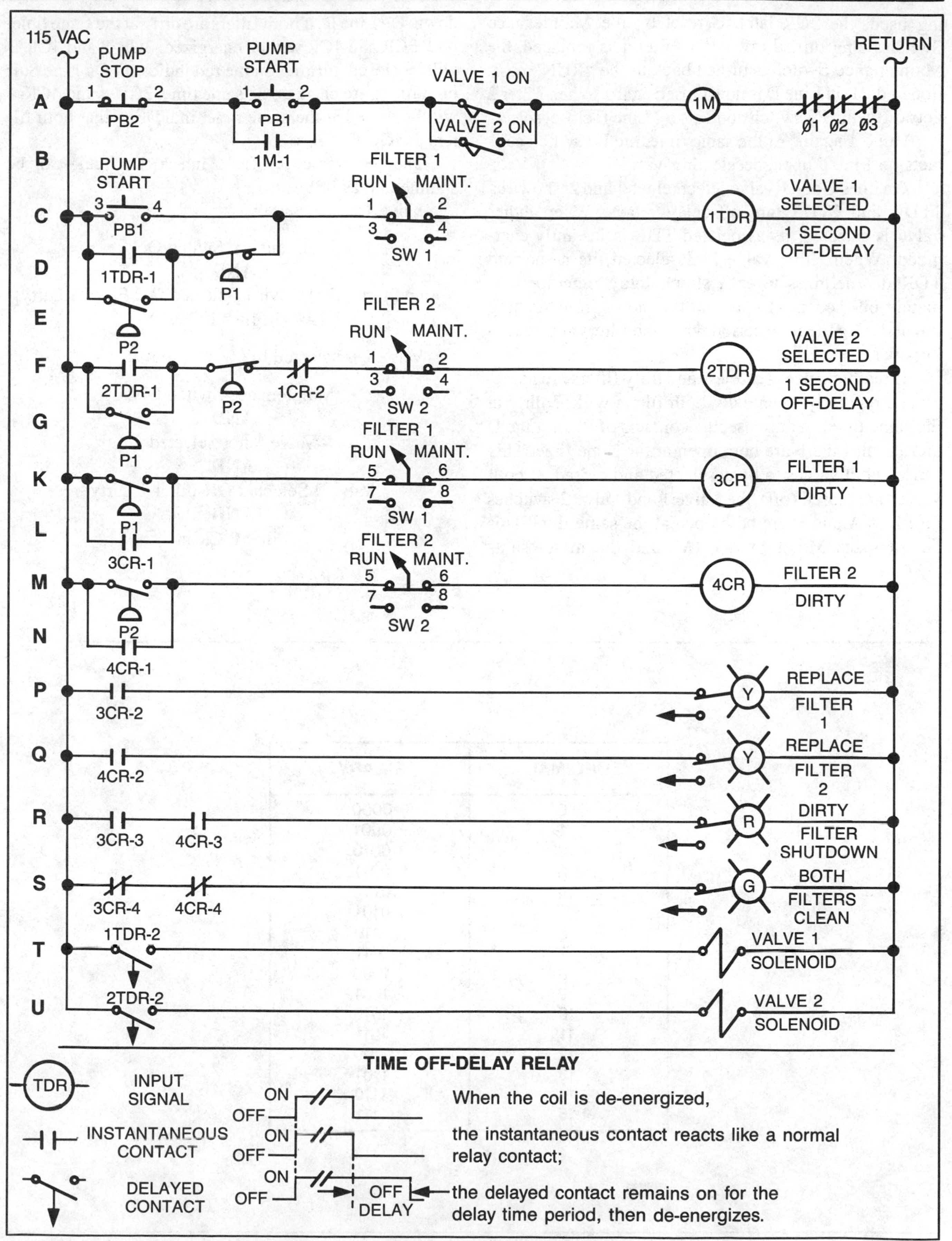

Figure 16-20. Ladder diagram of machine interlock circuit.

ing used, the 3CR latch is reset by the Maintenance Switch being turned off. After Filter 1 is replaced, the Maintenance Switch is turned back to the "RUN" position and P1 on Line D is now closed again. When Filter 2 gets dirty, the P2 switch contacts on Line E close, selecting Valve 1 again. At the same time, the P2 switch contacts on Line F open, deselecting Valve 2.

On lines C and F valve select relays 1 and 2, shown as 1TDR and 2TDR, are off-delay relays. When either valve is selected, its associated TDR is instantly energized. When either valve is deselected, its associated TDR de-energizes **after a short delay** time(approximately one second) to prevent the motor from shutting down during the time period when the filters are switching over.

If maintenance is careless and dirty filters are not replaced regularly, eventually both filters will be dirty at the same time. In this case, the contacts of P1 on Line D and P2 on Line F are both open at the same time. This causes both valves to be deselected and closed. If both valves are closed (off), the Valve 1 and Valve 2 switches on Lines A and B are both open at the same time. This de-energizes Motor Starter 1M, and the motor shuts down. On Line R, if both filters are dirty at the same time, both 3CR and 4CR will be energized. 3CR-3 and 4CR-3 will be closed, turning on the red indicator. On Line S, if both filters are clean at the same time, 3CR-4 and 4CR-4 will be closed at the same time, indicating that both filters are clean.

The requirements of the interlock circuit can be summed up as follows:

Valve 1 is selected by:

Maintenance Switch ON
AND
Pump Start **OR** ([Valve 1 Selected **OR** Filter 2 Dirty]
AND Filter 1 Clean)

Valve 2 is selected by:

Maintenance Switch ON
AND
Valve 1 not Selected
AND
Valve 2 Selected **OR** Filter 1 Dirty
AND
Filter 2 Clean

DECIMAL	BINARY
0	0000
1	0001
2	0010
3	0011
4	0100
5	0101
6	0110
7	0111
8	1000
9	1001
10	1010
11	1011
12	1100
13	1101
14	1110
15	1111

Figure 16-21. Binary and decimal equivalents.

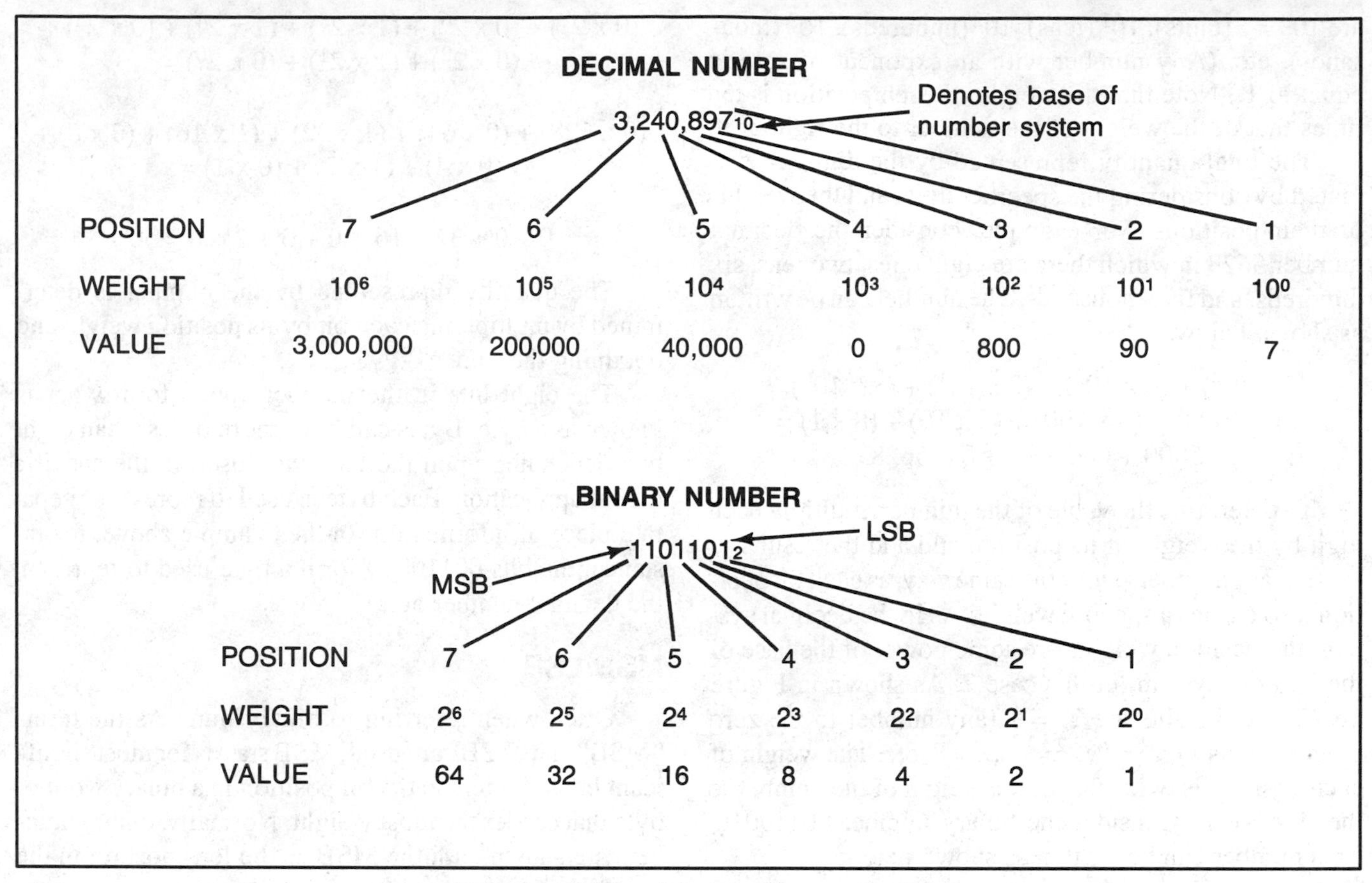

Figure 16-22. Bit position values.

"Dirty Filter Shutdown" indicator is illuminated by:

Filter 1 Dirty
AND
Filter 2 Dirty

"Both Filters Clean" indicator is illuminated by:

Filter 1 not Dirty
AND
Filter 2 not Dirty

BINARY NUMBERS

All digital circuits, instruments, and systems work with numbers that represent specific quantities. The most familiar type of numbers are decimal numbers. In the decimal number system, the ten digits 0 through 9 are combined in a certain way so that they indicate a specific quantity. In the binary number system, only two digits, 0 and 1, are used. The binary digits, or bits, when properly arranged can also represent any decimal number. For example, the binary number 110010 represents the decimal quantity 50.

The basic distinguishing feature of a number system is its base. The base indicates the number of digits used to represent quantities in that number system. The decimal number system has a base of 10 because ten digits, 0 through 9, are used to represent quantities. The binary number system has a base of 2 since only the digits 0 and 1 are used in forming numbers. Figure 16-21 shows binary and decimal equivalents for the decimal numbers 0 through 15.

All modern digital techniques are based on the binary number system. Since there are only two numbers in this system, data can be represented by hardware devices that only have two states—on and off. Two-state devices include switches, relay contacts, and transistors. These devices are significantly simpler, less costly, faster, and more reliable than devices which would need 10 states to represent the ten different possible digits of a decimal number.

Bit Position Values

The decimal and binary number systems are positional or weighted number systems. This means that each digit or bit position in a number carries a particular weight in determining the magnitude of that number. A decimal number has positional weights of units, tens, hundreds, thousands, etc. Each position has a weight that is some power of the number system base (10 in this case). As shown in Figure 16-22, the positional weights are $10^0 = 1$ (units), 10^1 (tens), 10^2 (hundreds), 10^3 (thousands), etc. (Any number with an exponent of zero is equal to 1.) Note that the weight of each position is ten times that of the weight of the number to the right of it.

The total quantity represented by the digits is evaluated by considering the specific digits and the weights of their positions. For example, consider the decimal number 5628 in which there are eight ones, two tens, six hundreds, and five thousands. The number can be written as shown below.

$$(5 \times 10^3) + (6 \times 10^2) + (2 \times 10^1) + (8 \times 10^0) =$$
$$(5 \times 1000) + (6 \times 100) + (2 \times 10) + (8 \times 1) =$$
$$5000 + 600 + 20 + 8 = 5628$$

To determine the value of the number, multiply each digit by the weight of its position and add the results.

Binary numbers work the same way, as each bit position also carries a specific weight. As in the decimal system, the position weights are some power of the base of the number system, in this case 2. As shown in Figure 16-22, the weights are $2^0 = 1$ (any number to the zero power equals 1), $2^1 = 2$, $2^2 = 4$, $2^3 = 8$, etc. The weight of each position is twice that of the weight of the number to the right of it. Consider the binary number 00110010. This number can be written as shown below.

$$(0 \times 2^7) + (0 \times 2^6) + (1 \times 2^5) + (1 \times 2^4) + (0 \times 2^3) + (0 \times 2^2) + (1 \times 2^1) + (0 \times 2^0) =$$

$$(0 \times 128) + (0 \times 64) + (1 \times 32) + (1 \times 16) + (0 \times 8) + (0 \times 4) + (1 \times 2) + (0 \times 1) =$$

$$0 + 0 + 32 + 16 + 0 + 0 + 2 + 0 = 50$$

The quantity represented by the number is determined by multiplying each bit by its position weight and obtaining the sum.

The eight bits in the number above form what is known as a byte. Bytes can have more or less than eight bits depending upon the hardware used in the specific digital application. Each byte is used to represent a separate piece of information. In the example above, for instance, the bits 00110010 form a byte used to represent the decimal number 50.

MSB/LSB

Often when referring to binary numbers the terms "MSB" and "LSB" are used. MSB stands for most significant bit and refers to the bit position in a binary word or byte that carries the most weight. Normally, binary numbers are shown with the MSB as the left-most bit in the

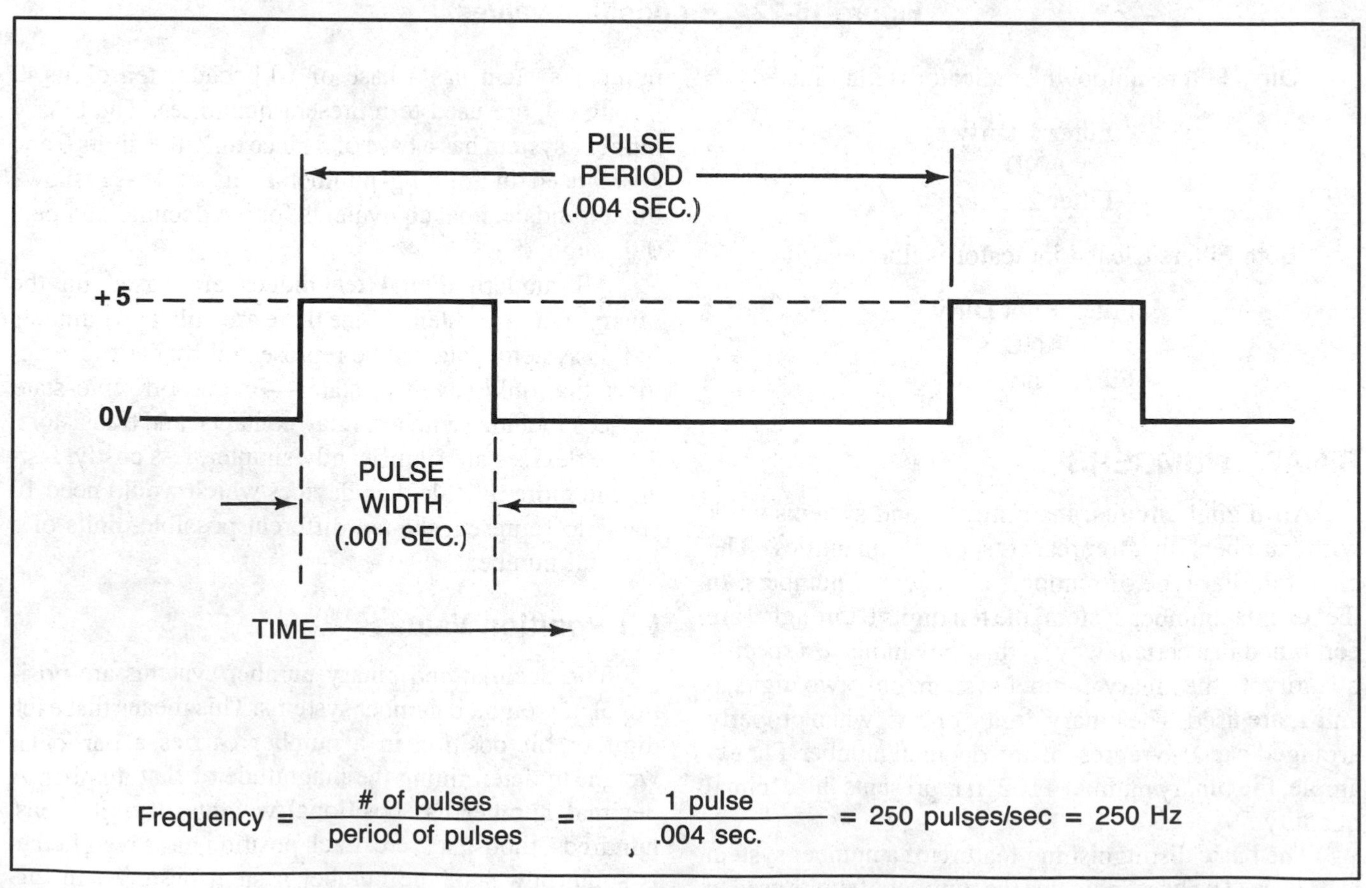

Figure 16-23. Relationship between pulse width and frequency.

byte (Figure 16-22). LSB stands for least significant bit and refers to the bit position that carries the least weight.

DATA REPRESENTATION

Data representation refers to the way in which binary numbers are represented and manipulated by electronic-components and circuits. To represent a bit in a binary word, an electronic component must be capable of assuming two distinct states. One of the states will represent a binary 0 and the other a binary 1.

Electromechanical devices like switches and relays are ideal for representing binary data. A closed switch or relay contact can represent a binary 1 while the open switch or contact can represent a binary 0. These logic representations could just as easily be reversed, as long as the way in which the bit value is represented is consistent within the system. Electromechanical devices are still used in places where static binary conditions are required or very low speed operation can be tolerated.

In early computers, vacuum tubes replaced electromechanical devices for representing binary data because they could attain much higher switching speeds. Vacuum tubes are large in size, consume a lot of electrical power and generate a lot of heat, and because of these limitations, they were replaced by transistors in most equipment.

A transistor can readily assume two distinct states, conducting and cut-off. When a transistor is cut-off it acts as an open circuit, and when it is conducting heavily, it acts as a very low resistance much like a closed contact. Transistors are found in digital equipment both as discrete components or grouped together in integrated circuit chips.

Magnetic cores which can be magnetized in either of two directions are often used to store binary bits in memory. The two directions represent the two binary states, 0 and 1. The direction in which the core has been magnetized can be sensed at a later time when the bit value is needed by the computer.

Logic Levels

Two-state devices, whether electromechanical, electronic, or magnetic, are the basic element for representing bits of digital information. The exact relationship between the state of the device and the bit condition represented by this state is arbitrary. The digital hardware is not concerned with whether a transistor is conducting or cut-off as much as it is with the voltage levels associated

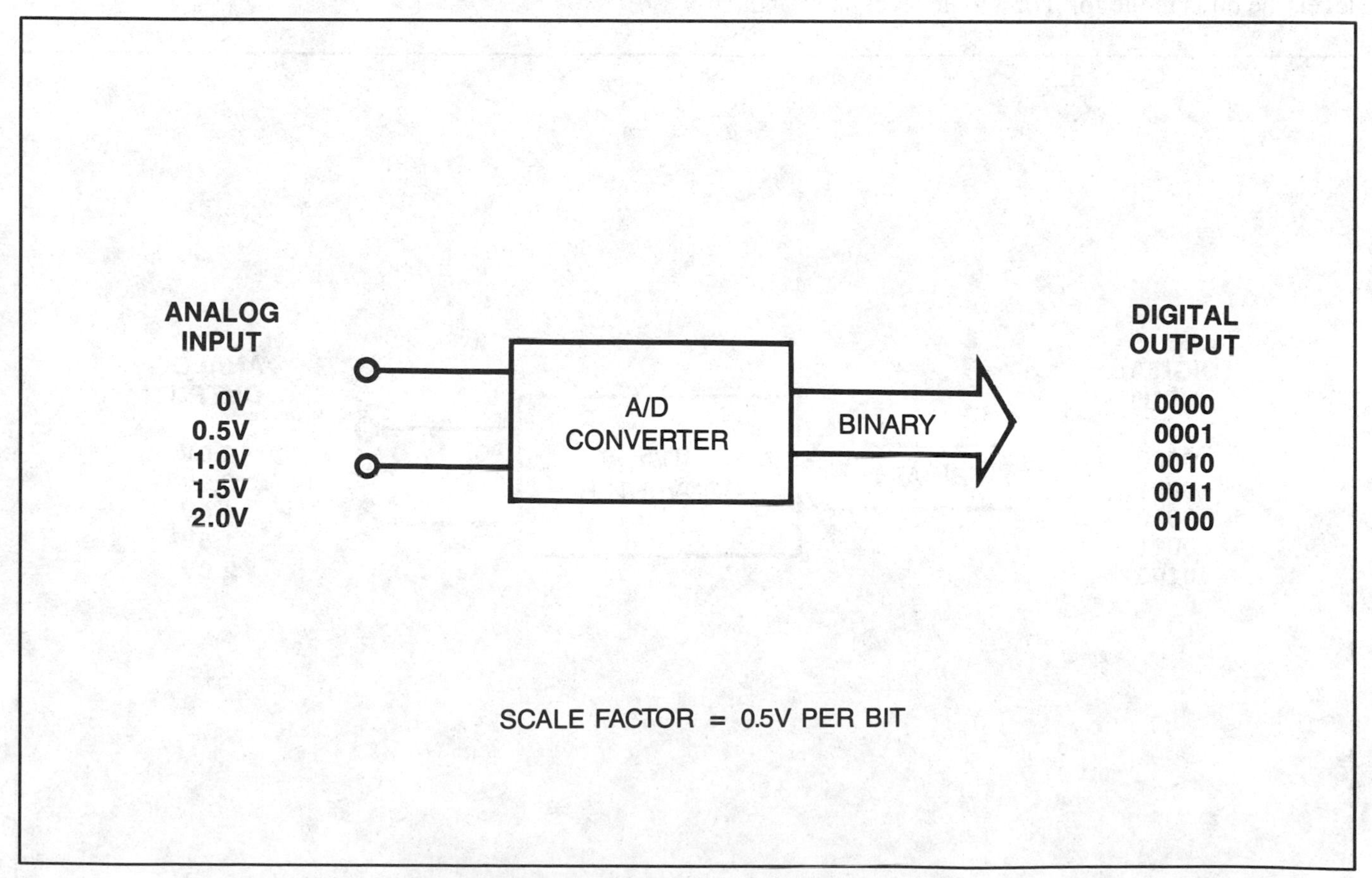

Figure 16-24. Analog-to-digital converter.

with the transistor states. The transistor simply controls these voltage levels which are determined when the circuitry is designed. For example, a binary 0 may be represented by 0 volts or ground. A binary 1 might be represented by a +5 volts. Depending on the equipment power supplies, the design of the circuitry, and the application, almost any voltage level assignments may be used.

Positive Logic

When the most positive of two possible voltage levels is assigned to represent a binary 1, positive logic is being used by the circuitry. More often than not, positive logic is employed in digital equipment, however, negative logic level assignments are also sometimes used but are not discussed here in an effort to keep the explanation of logic levels as simple as possible. Several examples of positive logic level assignments are shown below.

Positive Logic Examples

binary 0 = +0.2V binary 0 = -6.0V binary 0 = 0V
binary 1 = +3.4V binary 1 = 0Vbinary 1 = +15V

Pulse Width and Frequency

When a bit of binary data is represented by a voltage level, the bit is in the form of a logic level pulse and the length of time from the beginning to the end of the pulse is known as the pulse width.

Frequency is defined as the number of pulses per second of time. For example, if each pulse requires 4 ms of time, then 250 pulses could occur each second. The frequency would be 250. If these pulses were transmitted serially, 250 bits per second could be transmitted. If the same information were transmitted in a parallel format, 250 binary bytes could be transmitted per second. Figure 16-23 shows the relationship between pulse width, period, and frequency.

Analog to Digital (A/D) Conversion

Microprocessor based electrohydraulically controlled systems often contain devices that provide feedback or inputs to the system in the form of analog signals. Because these analog signals cannot be processed by digital microprocessor circuits, the analog to digital converter circuit receives the inputs from the external devices and converts them into digital signals in the form of binary numbers before sending them to the control circuits for processing.

The function of an A/D converter is illustrated in Figure 16-24. Each value of input voltage generates a specific binary output of 1s and 0s. However, the binary

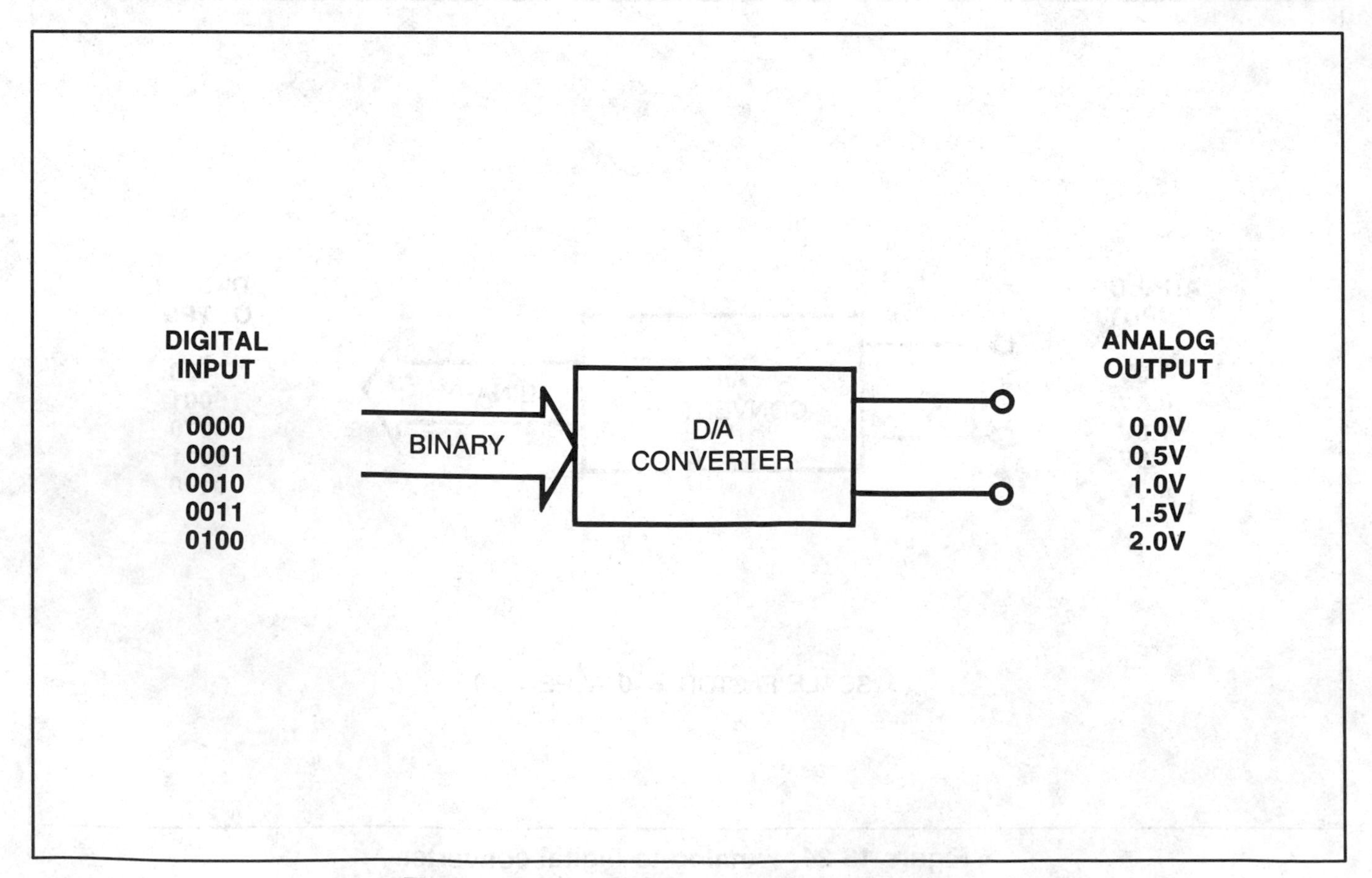

Figure 16-25. Digital-to-analog converter.

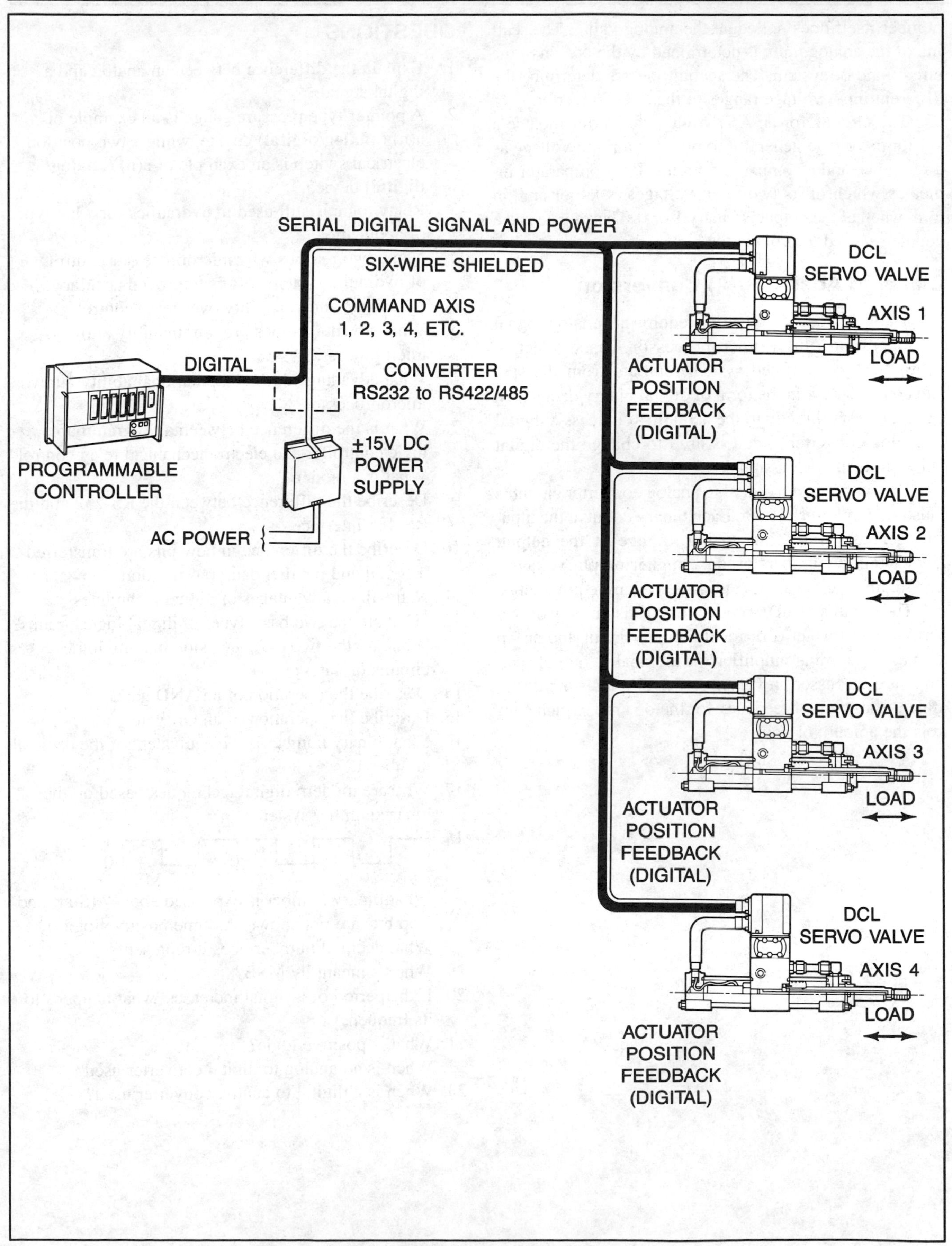

Figure 16-26. Digital and closed loop system.

number itself does not equal the analog value. The scaling of the analog input is determined by the design of the circuit and/or system. The scaling factors determine the different input voltage ranges to the A/D converter.

The circuits for an A/D voltage converter include a sawtooth voltage generator to produce a ramp voltage as a reference and a comparator circuit. The comparator indicates which of its two input voltages is larger and, in turn, whether the output is binary 1 or 0. Then a gate and a counter is used to provide the binary data.

Digital to Analog (D/A) Conversion

Many hydraulic actuators require an analog signal for operation. In a microprocessor based electrohydraulically controlled system, outputs from the system controller are in the form of digital binary data which cannot be sent directly to the actuating devices. A digital to analog converter circuit is used to change the digital pulses into an analog signal.

The function of a digital to analog converter circuit is illustrated in Figure 16-25. Each binary count at the input will generate a specific analog voltage at the output. Higher binary numbers produce higher output voltages. The scaling factor can be set for the system requirements.

The circuits of a D/A converter consist mainly of two parts: a resistor network for the binary input and an op amp as a summing amplifier for the analog output. (Op amps are discussed in Chapter 13.) The resistors determine the amount of feedback for the op amp, which controls the amount of gain.

QUESTIONS

1. Explain the difference between an analog and a digital signal.
2. A pointer-type pressure gauge is an example of a(n) **(analog/digital)** device, while a two position electrical switch is an example of a(n) **(analog/ digital)** device.
3. Relay logic is still used in hydraulics for what type of applications?
4. Name two reasons why microprocessor control of hydraulic systems offers improved accuracy, efficiency, and versatility over relay control.
5. What advantage does programmability allow a microprocessor?
6. What advantage does computational ability allow a microprocessor?
7. What is the difference between a programmable logic controller and electromechanical relay control?
8. What is a modem?
9. Describe the difference between the RS-232 and the RS-485 interface.
10. Describe the difference in how bits are transferred in serial and parallel data transmission.
11. Name three advantages of digital techniques.
12. What are the two basic types of digital logic circuits?
13. What are the two basic decision making logic circuits or gates?
14. Describe the operation of an AND gate.
15. Describe the operation of an OR gate.
16. What binary number is the equivalent of the decimal number 12?
17. Why are modern digital techniques based on the binary number system?
18. 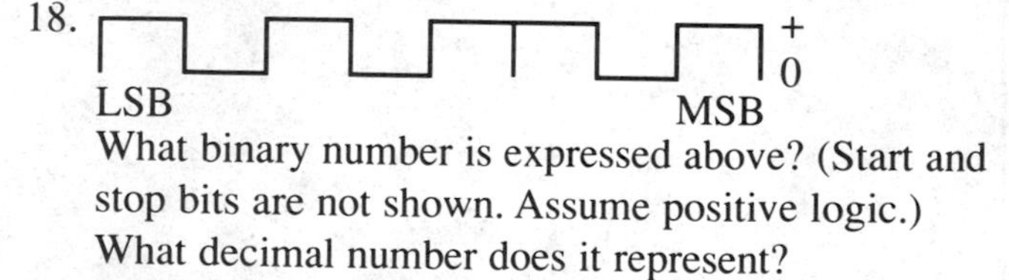

What binary number is expressed above? (Start and stop bits are not shown. Assume positive logic.) What decimal number does it represent?
19. What is meant by MSB?
20. If the period of a signal increases, what happens to its frequency?
21. What is positive logic?
22. When is an analog to digital converter used?
23. When is a digital to analog converter used?

NOTES

NOTES

NOTES

NOTES

NOTES

NOTES

CHAPTER 17

HYDRAULIC PUMPS

In a hydraulic system, the pump converts mechanical energy into hydraulic energy (hydraulic horsepower) by pushing fluid into the system.

All pumps work on the same principle, generating an increasing volume on the intake side and a decreasing volume at the discharge side; but the different types of pumps vary greatly in methods and sophistication. When determining hydraulic horsepower, a common formula is used.

Hydraulic Horsepower = GPM × PSI × 0.000583

Hydraulic Kilowatts = LPM × BAR × 0.001667

DISPLACEMENT

The flow capacity of a pump can be expressed as displacement per revolution or output in gpm.

Displacement is the volume of liquid transferred in one revolution. It is equal to the volume of one pumping chamber multiplied by the number of chambers that pass the outlet per revolution. Displacement is expressed in cubic inches per revolution.

DELIVERY IN GPM (LPM)

Although a pump can be rated as a 10 gpm (37.85 LPM) unit, it may actually pump more than that, under no-load conditions and less than that at its rated operating pressure. The pump's delivery is also proportional to drive shaft speed.

Most manufacturers provide a table or graph (Figure 17-1) showing pump deliveries, horsepower requirements, drive speeds, and pressures under specific test conditions.

VOLUMETRIC EFFICIENCY

Theoretically, a pump delivers an amount of fluid equal to its displacement during each cycle or revolution. In reality, the actual output is reduced because of internal leakage. As pressure increases, the leakage from the outlet back to the inlet (or to the drain) also increases, causing a decrease in volumetric efficiency. Volumetric efficiency is equal to the actual output divided by the theoretical output. It is expressed as a percentage.

$$\text{Efficiency \%} = \frac{\text{Actual output}}{\text{Theoretical output}} \times 100$$

Pump delivery in gpm (lpm) of various displacements at increasing pressures, constant speed (1800 rpm)			Horsepower (kw) Input at 1800 rpm		
0 PSI (0.00 bar) (0.00 kPa)	500 PSI (34.47 bar) (3447.00 kPa)	1000 PSI (68.94 bar) (6894.00 kPa)	0 PSI (0.00 bar) (0.00 kPa)	500 PSI (34.47 bar) (3447.00 kPa)	1000 PSI (68.94 bar) (6894.00 kPa)
1.8 (6.81)	1.5 (5.68)	1.1 (4.16)	.20 (0.14)	0.9 (0.67)	1.5 (1.11)
2.7 (10.22)	2.4 (9.08)	2.0 (7.57)	.25 (0.18)	1.2 (0.89)	2.2 (1.64)
3.7 (14.01)	3.4 (12.87)	3.0 (11.36)	.25 (0.18)	1.4 (1.04)	2.6 (1.93)
5.3 (20.06)	5.0 (18.93)	4.7 (17.79)	.30 (0.22)	1.9 (1.41)	3.6 (2.68)
8.2 (31.04)	7.9 (29.90)	7.5 (28.39)	.35 (0.26)	2.8 (2.08)	5.2 (3.87)
11.5 (43.53)	11.0 (41.64)	10.6 (40.12)	.40 (0.29)	3.7 (2.75)	7.0 (5.22)

Figure 17-1. Typical performance table.

As an example, if a pump presumably delivers 10 gpm (37.85 LPM) but actually delivers only 9 gpm (34.07 LPM) at 1000 psi (68.94 bar / 6894.00 kPa), the volumetric efficiency of that pump at that speed and pressure is 90 percent.

$$Efficiency = \frac{9}{10} = 0.9 \text{ or } 90 \text{ percent}$$

PUMP RATINGS

A pump is generally rated by its maximum operating pressure capability and output flow at a given drive speed.

The pressure rating of a pump is determined by the manufacturer and is based upon reasonable service-life expectancy under specified operating conditions. It is important to note that there is no standard industry-wide safety factor included in this rating. Operating at higher pressure may result in reduced pump life or more serious damage.

TYPES OF PUMPS

There are two basic types of pumps. The first is the **nonpositive displacement** pump. This pump design is used mainly for fluid transfer in systems where the only resistance found is created by the weight of the fluid itself and friction.

Most nonpositive displacement pumps (Figure 17-2) operate by centrifugal force. Fluids entering the center of the pump housing are thrown to the outside by means of a rapidly driven impeller. There is no positive seal between the inlet and outlet ports, and pressure capabilities are a function of drive speed.

Although it provides a smooth, continuous flow, the output from this type of pump is reduced as resistance is increased. In fact, it is possible to completely block off the outlet while the pump is running. For this and other reasons, nonpositive displacement pumps are seldom used in power hydraulic systems today. These properties make it a more likely choice for a water pump in a car engine, dishwasher, or washing machine.

It could also be used as a supercharge pump for a positive displacement pump.

The **positive displacement** pump is most commonly used in industrial hydraulic systems. A positive displacement pump delivers to the system, a specific amount of fluid per stroke, revolution, or cycle. This type of pump is classified as fixed or variable displacement.

Fixed displacement pumps have a displacement which cannot be changed without replacing certain components. With some, however, it is possible to vary the size of the pumping chamber (and the displacement) by

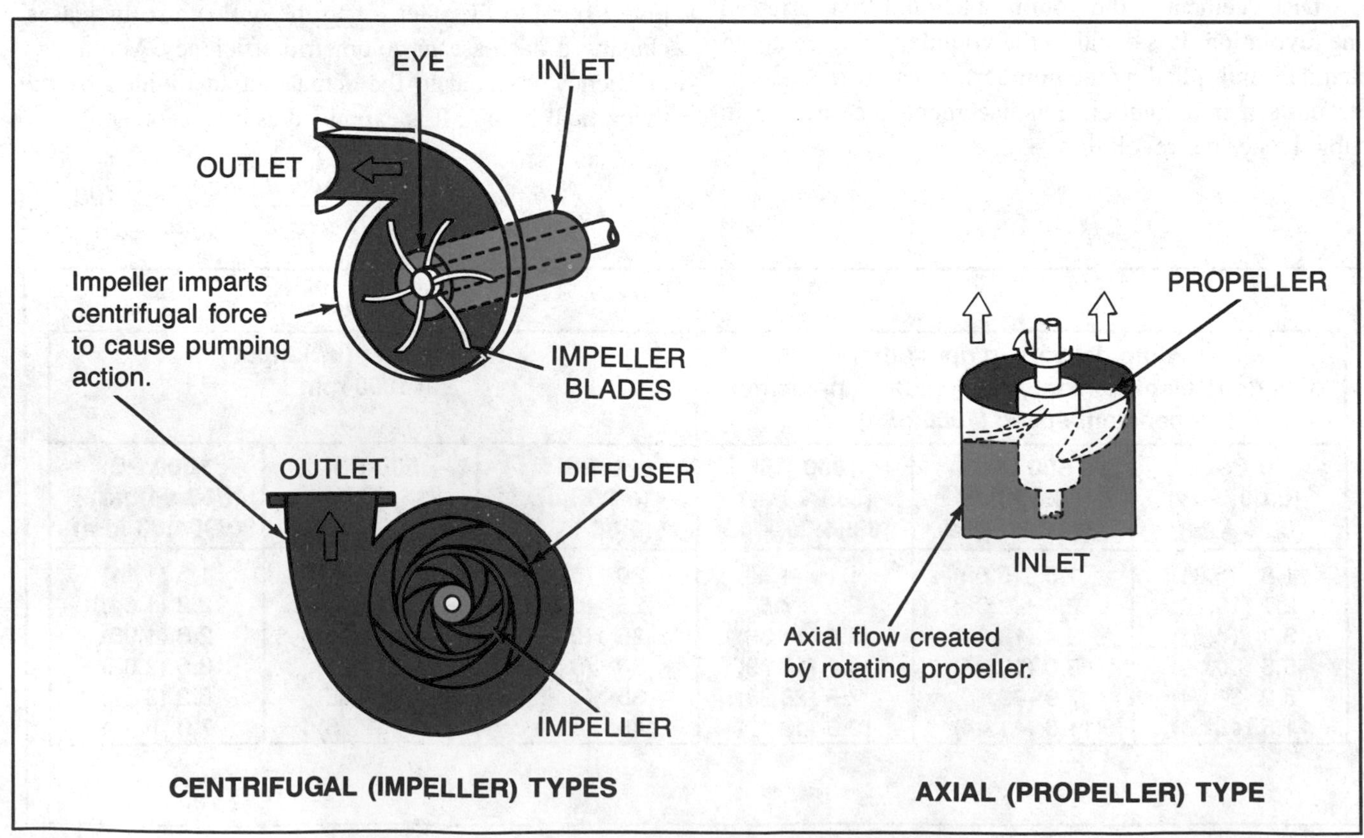

Figure 17-2. Nonpositive displacement pumps.

using external controls. These pumps are known as **variable displacement** pumps.

Certain vane pumps and piston units can be varied from maximum to zero delivery. Some are capable of reversing their flow as the control crosses a center or neutral position.

Figure 17-3 shows that fluid enters the pump through a check valve on the intake stroke. On the forward stroke, the check valve closes, sealing off the inlet. As the piston moves forward, the displaced fluid must flow through the pump outlet.

The pressure is determined by the work load, and except for leakage losses, the output is independent of outlet pressure. This makes the positive displacement pump more appropriate for use in the transmission of power.

This chapter covers the three best-known positive displacement pumps: gear pumps, vane pumps, and piston pumps.

GEAR PUMPS

A gear pump develops flow by carrying fluid between the teeth of two meshed gears. Powered by the drive shaft, the gear known as the drive gear turns the second gear, which is called the driven or idler gear. The pumping chambers formed between the gear teeth are enclosed by the pump housing, or center section, and side plates (often called wear or pressure plates).

Gear pumps are referred to as unbalanced because high pressure at the pump outlet imposes an unbalanced load on the gears and the bearings (Figure 17-4). Large bearings incorporated into the designs counteract these loads. The gears can handle a hydraulic pressure of up to 3600 psi (248.18 bar / 24,818.00 kPa) with adequate compensation for axial loading.

As fixed displacement pumps, gear pumps provide two ways of altering volume levels. The first is to replace the existing gears with gears of different dimensions. The second is to vary the volume by changing the speed at which the drive gear turns.

External Gear Pumps

As shown in Figure 17-4, the gears are placed side by side in an external gear pump. A partial vacuum is created at the inlet as the gear teeth unmesh, drawing fluid into the chambers formed between the teeth. The chambers carry the fluid around the outside of the gears, where it is forced out as the teeth mesh again at the outlet.

As the gears mesh together, any fluid left in the chamber develops a high level of pressure. Decompression notches machined into the side plates relieve this fluid and the corresponding pressure. Any unrelieved fluid is channeled into a groove used to lubricate the bearings.

External gear pumps are available as single, multiple, or through-drive versions (Figure 17-5). The through-drive is a single pump with an auxiliary mounting pad and coupling spline in the rear cover. With these accessories, other pumps can be mounted and driven in tandem. This arrangement provides separate inlet and outlet ports for each pump and permits isolation of each circuit. It also allows the use of different fluids for each system.

The double version has two single pumps, each with its own outlet port, but sharing a common inlet port and input shaft. A double pump can serve two separate hydraulic circuits or supply a single circuit with greater volume.

Multiple units consist of two or more pumping sections driven by a common input shaft. The double pumps are furnished with a single common inlet, while the triple and quadruple pumps normally have one less inlet than the total number of sections. All inlet ports are internally connected and each pump section has a separate outlet. Multiple pumps save installation costs and space and also offer less chance of leakage.

Certain rating limitations should be observed when multiple-pump configurations are used with pumps that have different displacements. The high speed is limited by the lowest high-speed rating of the combination. The low speed is limited to the highest low speed rating of the combination.

When outlet flows are combined, the pressure limit represents the lowest pressure level of the combination.

Multiple-pump configurations with the same displacement use ratings that are common to single pumps. These pressure ratings are also limited by drive shaft torsional loading. The shaft loadings for each pump should be reviewed to ensure that the combined torsional loads do not exceed the drive shaft torsional capability.

The **lobe pump** (Figure 17-6) operates on the same principle as the external gear pump, but has a higher displacement because each of the two gears has only three "teeth" that are much wider and more rounded than those found on a regular external gear pump. This makes the lobe pump better suited for use with shear-sensitive fluids, as well as fluids with entrained gases or particles. This type of pump has a relatively low pressure capability and tends to deliver a more pulsating flow.

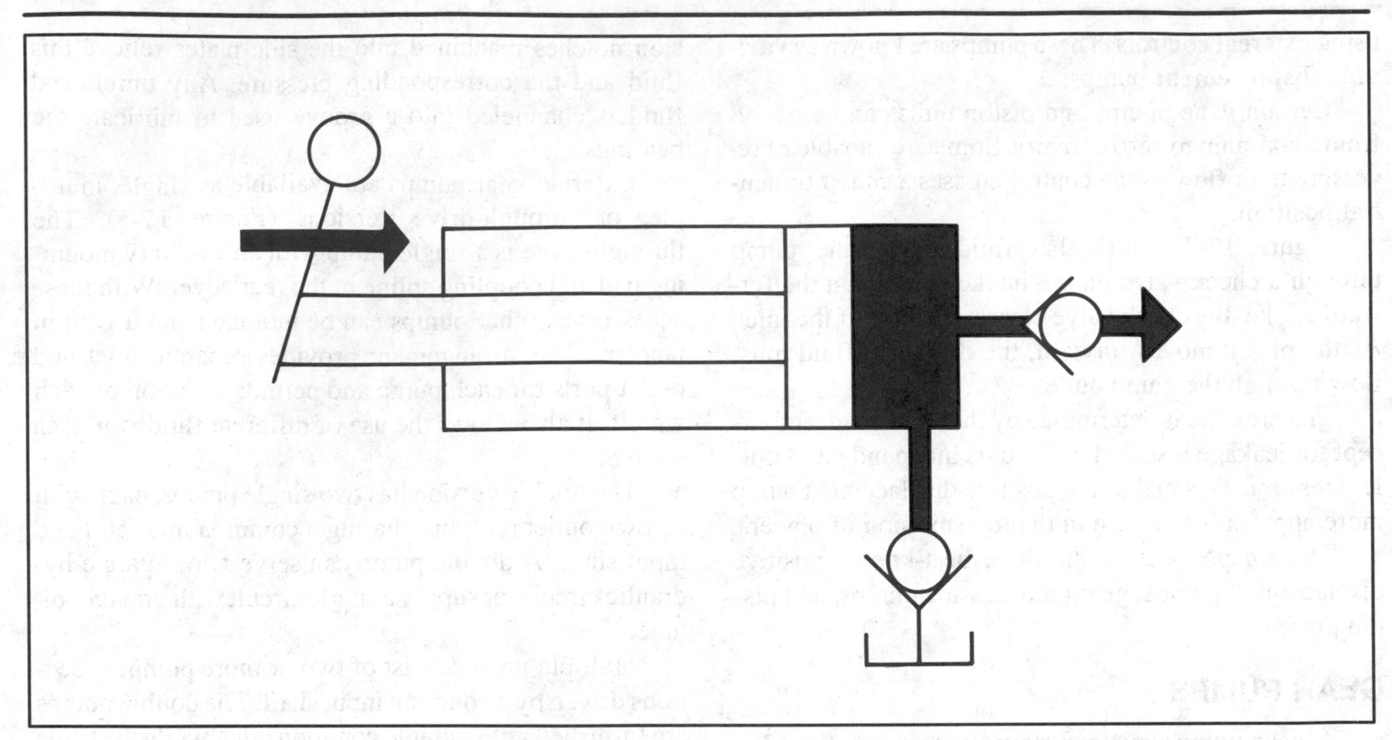

Figure 17-3. Demonstrating the principle of positive displacement pumps.

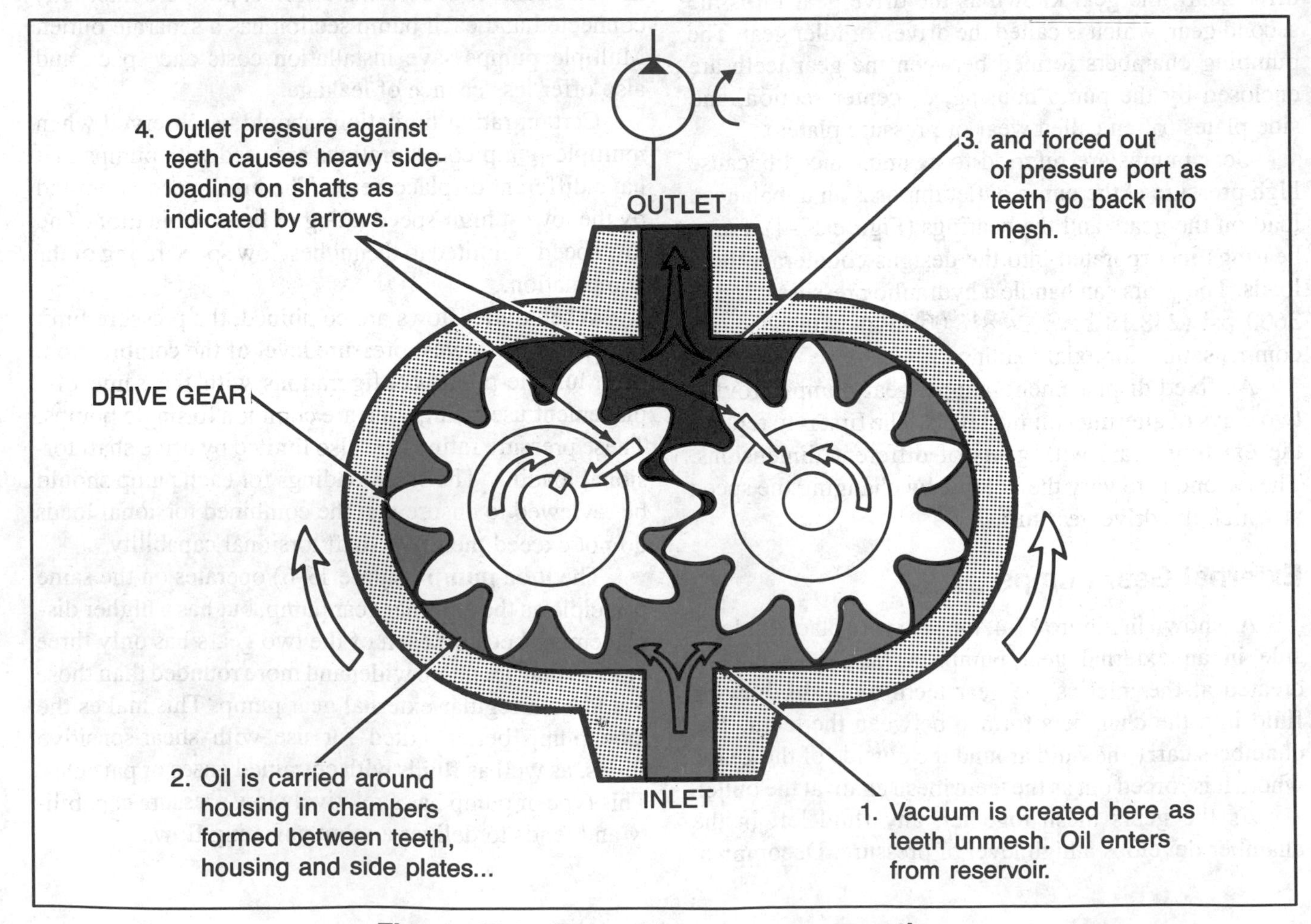
4. Outlet pressure against teeth causes heavy side-loading on shafts as indicated by arrows.
3. and forced out of pressure port as teeth go back into mesh.
OUTLET
DRIVE GEAR
2. Oil is carried around housing in chambers formed between teeth, housing and side plates...
INLET
1. Vacuum is created here as teeth unmesh. Oil enters from reservoir.

Figure 17-4. External gear pump operation.

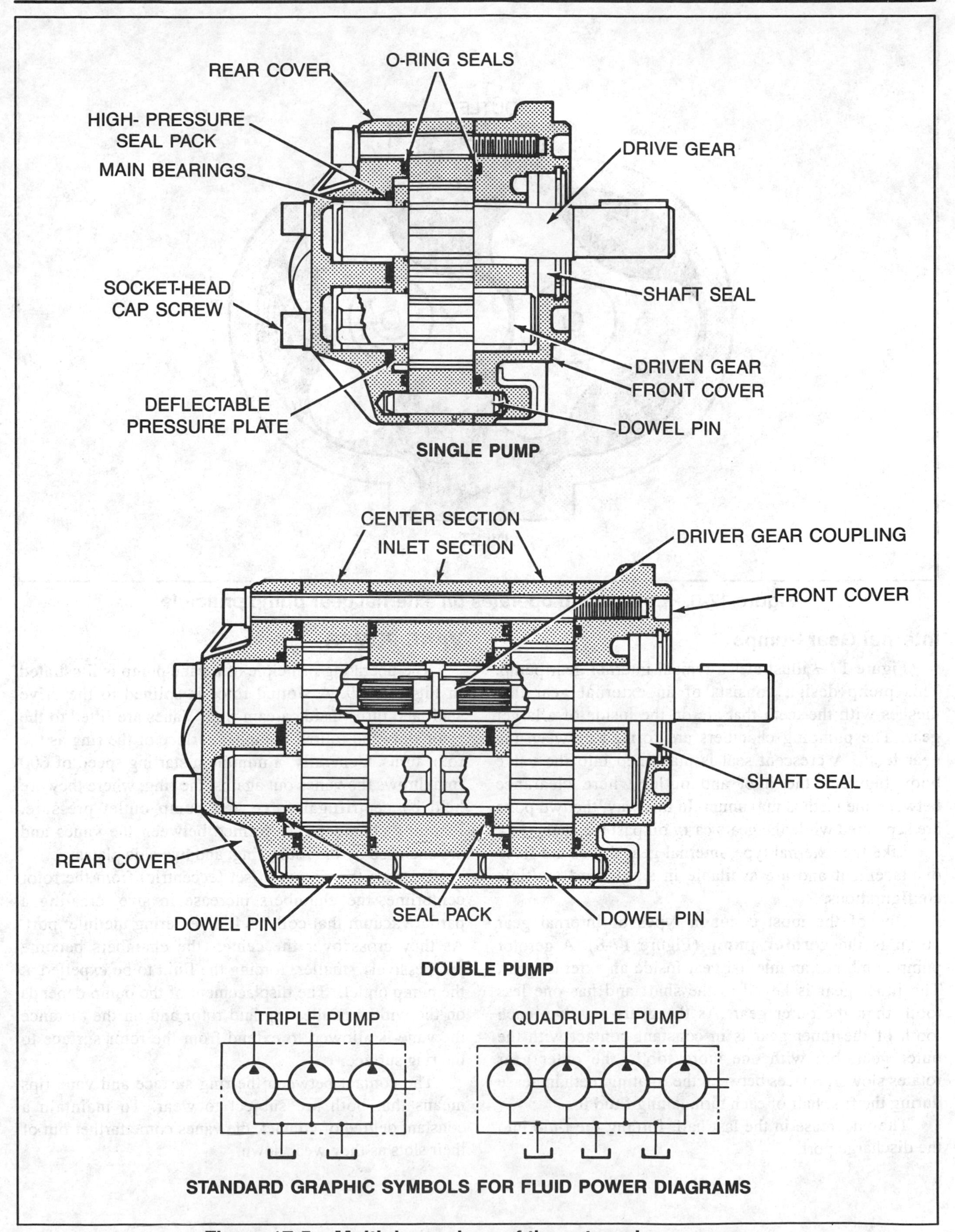

Figure 17-5. Multiple versions of the external gear pump.

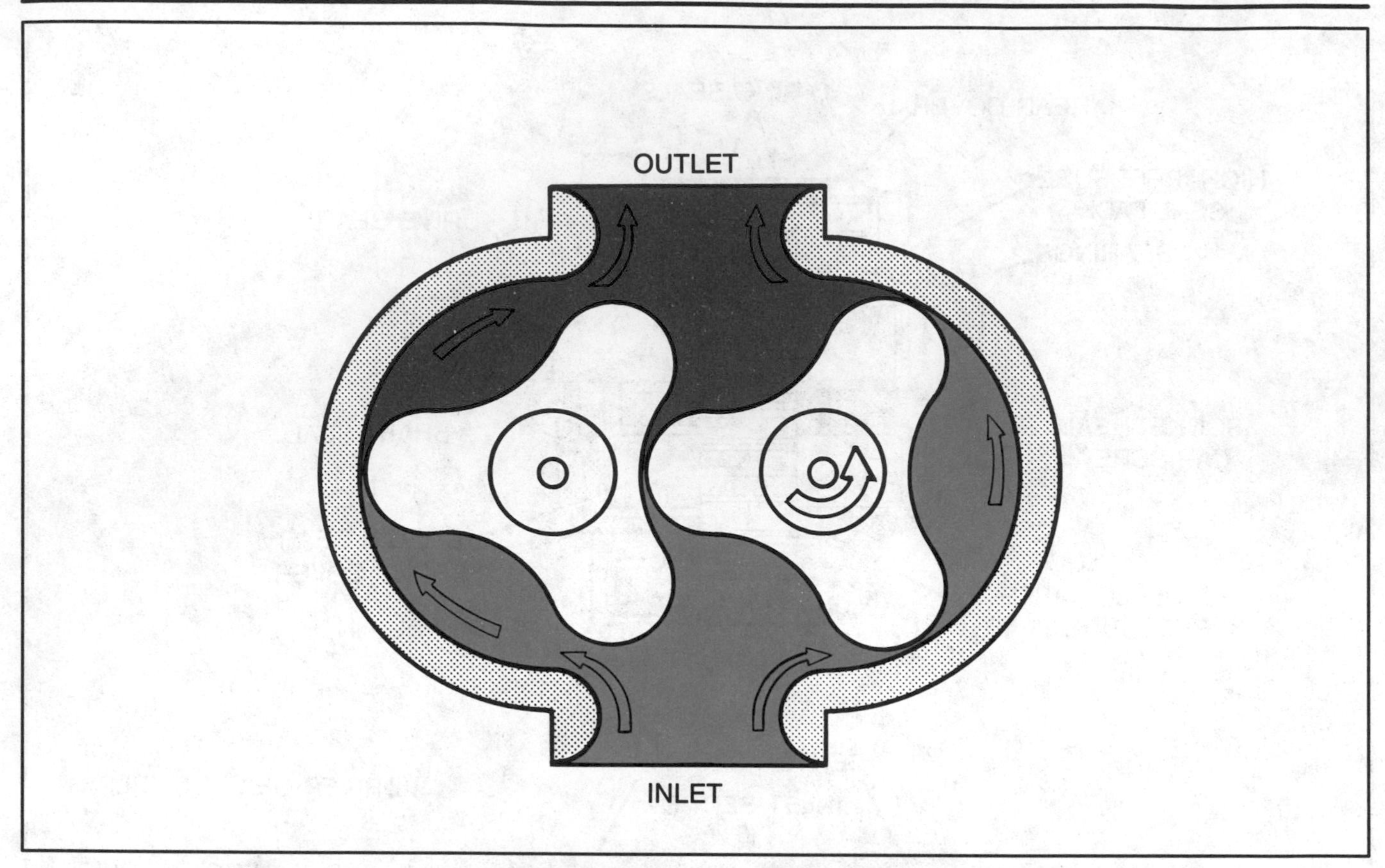

Figure 17-6. Lobe pump operates on external gear pump principle.

Internal Gear Pumps

Figure 17-7 illustrates a typical internal gear pump. This pump design consists of an external gear that meshes with the teeth that are on the inside of a larger gear. The pumping chambers are formed between the gear teeth. A crescent seal is machined into the valve body between the inlet and outlet where clearance between the teeth is maximum. In this way, the two ports are separated while the gears carry oil past the seal.

Like the external type, internal gear drives are fixed displacement and are available in single and multiple configurations.

One of the most common types of internal gear pump is the **gerotor pump** (Figure 17-8). A gerotor pump combines an internal gear inside an external gear. The inner gear is keyed to the shaft and has one less tooth than the outer gear. As the gears revolve, each tooth of the inner gear is in constant contact with the outer gear, but with one more tooth, the outer gear rotates slower. Spaces between the rotating teeth increase during the first half of each turn, taking fluid in.

They decrease in the last half, forcing the fluid into the discharge port.

VANE PUMPS

The operating principle of a vane pump is illustrated in Figure 17-9. A slotted rotor is splined to the drive shaft and turns inside a cam ring. Vanes are fitted to the rotor slots and follow the inner surface of the ring as the rotor turns. Generally, a minimum starting speed of 600 rpm throws the vanes out against the ring, where they are held by centrifugal force and pump outlet pressure. Pumping chambers are formed between the vanes and are enclosed by the rotor, ring, and two side plates.

Because the ring is offset (eccentric) from the rotor centerline, the chambers increase in size, creating a partial vacuum that collects fluid entering the inlet port. As they cross over the center, the chambers become progressively smaller, forcing the fluid to be expelled at the pump outlet. The displacement of the pump depends on the widths of the ring and rotor and on the distance the vane is allowed to extend from the rotor surface to the ring surface.

The contact between the ring surface and vane tips means that both are subject to wear. To maintain a constant degree of contact, the vanes come farther out of their slots as they wear down.

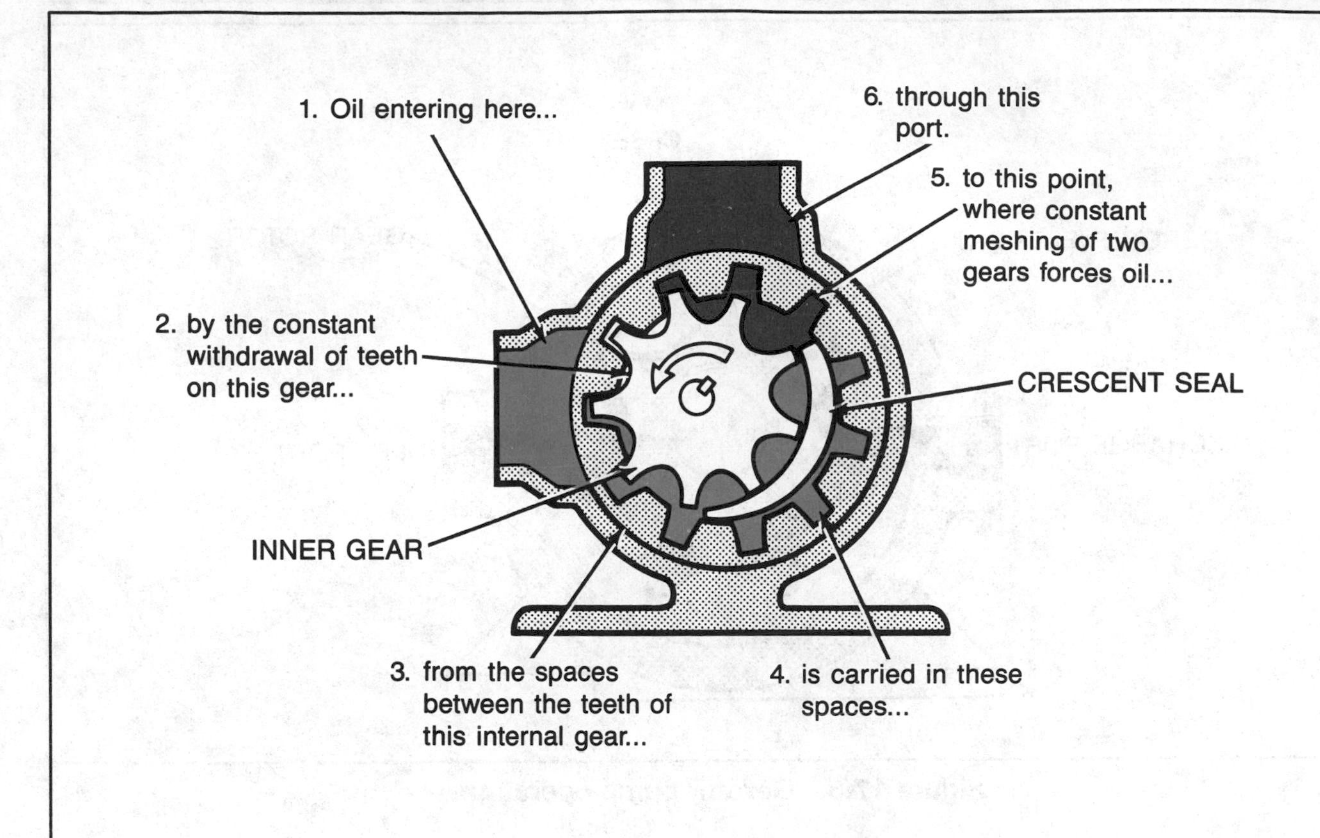
1. Oil entering here...
2. by the constant withdrawal of teeth on this gear...
3. from the spaces between the teeth of this internal gear...
4. is carried in these spaces...
5. to this point, where constant meshing of two gears forces oil...
6. through this port.
CRESCENT SEAL
INNER GEAR

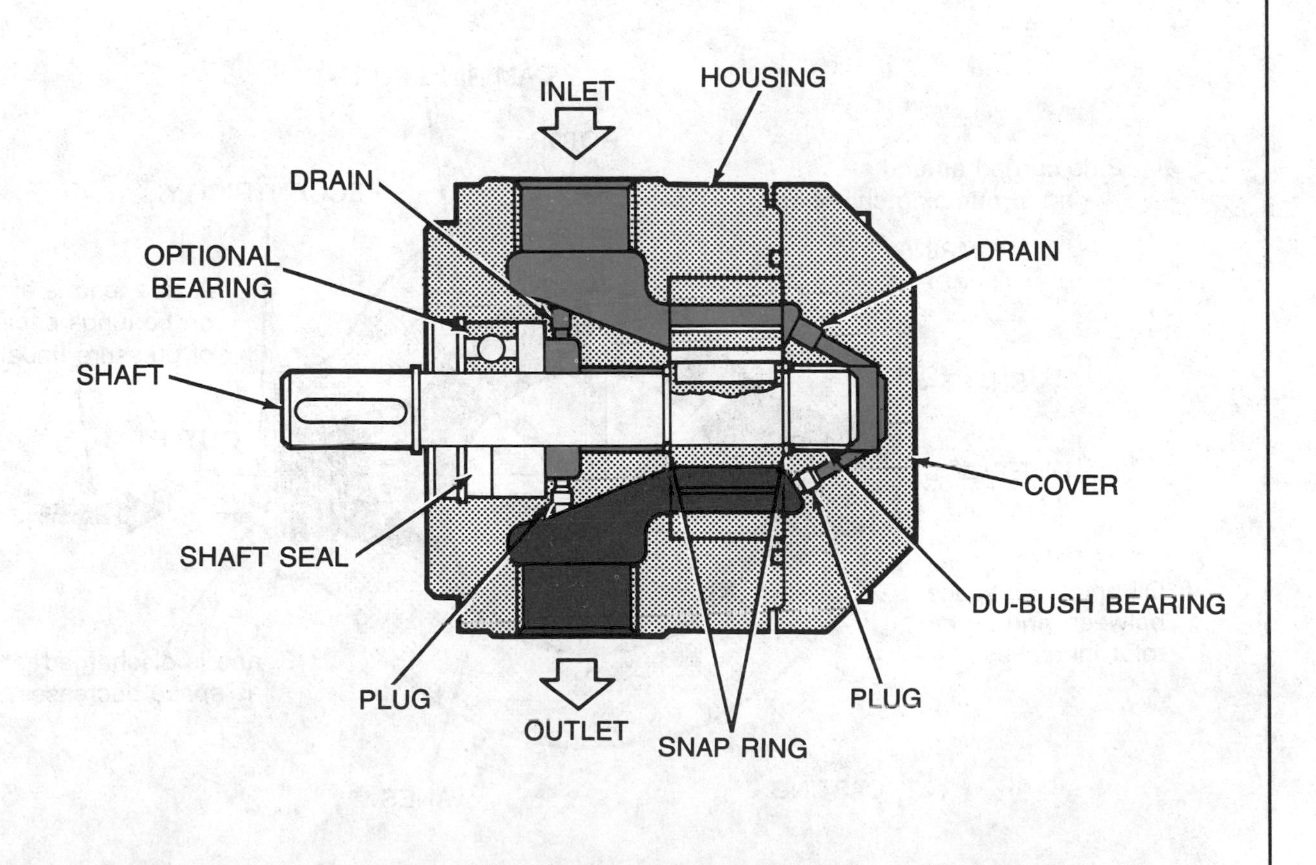
INLET
HOUSING
DRAIN
OPTIONAL BEARING
DRAIN
SHAFT
COVER
SHAFT SEAL
DU-BUSH BEARING
PLUG
PLUG
OUTLET
SNAP RING

Figure 17-7. Internal gear pump operation.

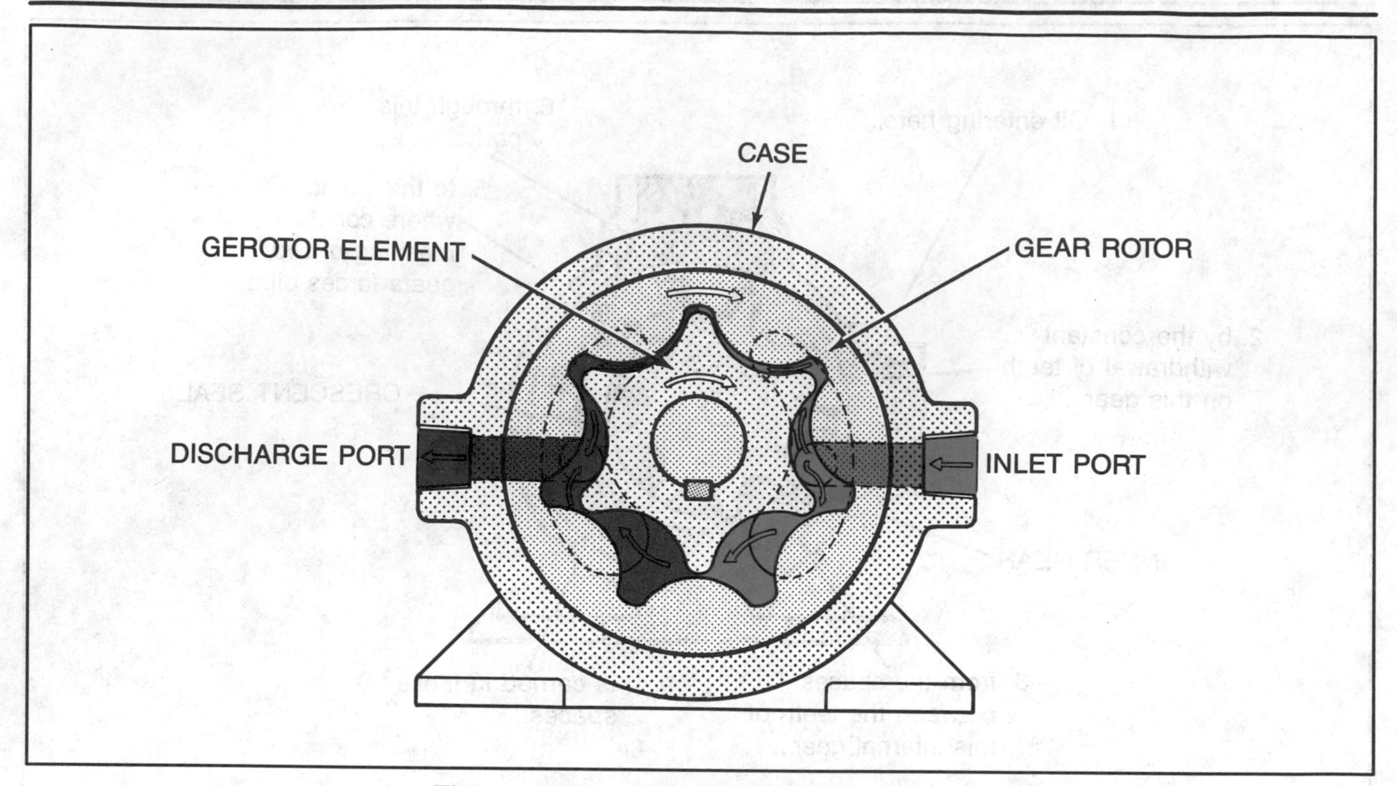

Figure 17-8. Gerotor pump operation.

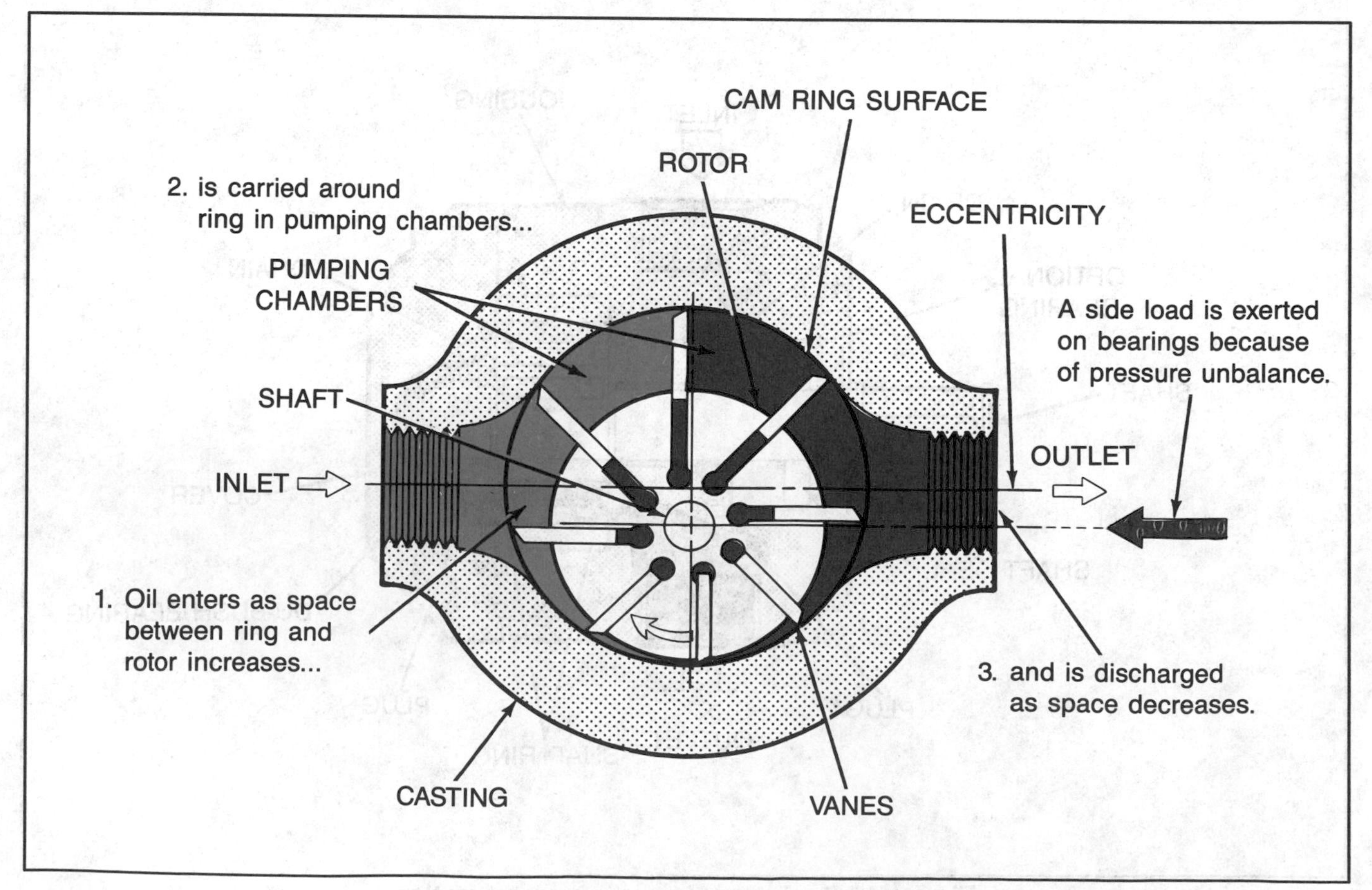

Figure 17-9. Vane pump operation (unbalanced).

Vane pumps cover the low to medium-high volume ranges with operating pressures up to 3000 psi (206.82 bar / 20,682.00 kPa). They are reliable, efficient, and easy to maintain. Along with this high efficiency, vane pumps have a low noise level and a long life.

Unbalanced Design

An unbalanced pump construction with a round cam ring design (Figure 17-9) has a somewhat limited pressure capability because of its unbalanced hydraulic loading (side-loading).

This type of pump is generally found in the variable volume design (Figure 17-10), which is a rotary vane type pump with adjustable maximum operating pressure and maximum displacement stop.

Displacement of an unbalanced vane pump can be changed or reduced to zero through external controls, such as a handwheel or pressure compensation. The control will move the cam ring relative to the rotor centerline, reducing or increasing the size of the pumping chamber. This is known as changing the "throw." The cam ring is held in the eccentric position by a compensator spring. As system pressure reaches the compensator setting, outlet pressure pushes against the inside surface of the cam ring, causing it to pivot left (Figure 17-10). This action moves the cam ring to an almost concentric or neutral position. The pump delivers just enough flow to make up for leakage and maintain the compensator pressure setting.

The maximum displacement stop is diametrically opposed to the compensator control. It provides manual adjustment of the maximum eccentricity of the ring and, therefore, maximum gpm (LPM).

The principle advantage of the pressure compensated variable vane pump over the fixed displacement type is energy savings (no wasted power). As an example, when no flow is required in the system, the pump automatically adjusts to that requirement.

Other available pump control options include:

- Load sensing
- Load sensing with pressure limiter
- Pressure compensator with unloading valve.
- Pressure compensator with electrical solenoid dual or triple pressure settings
- Pressure compensator with remote hydraulic control

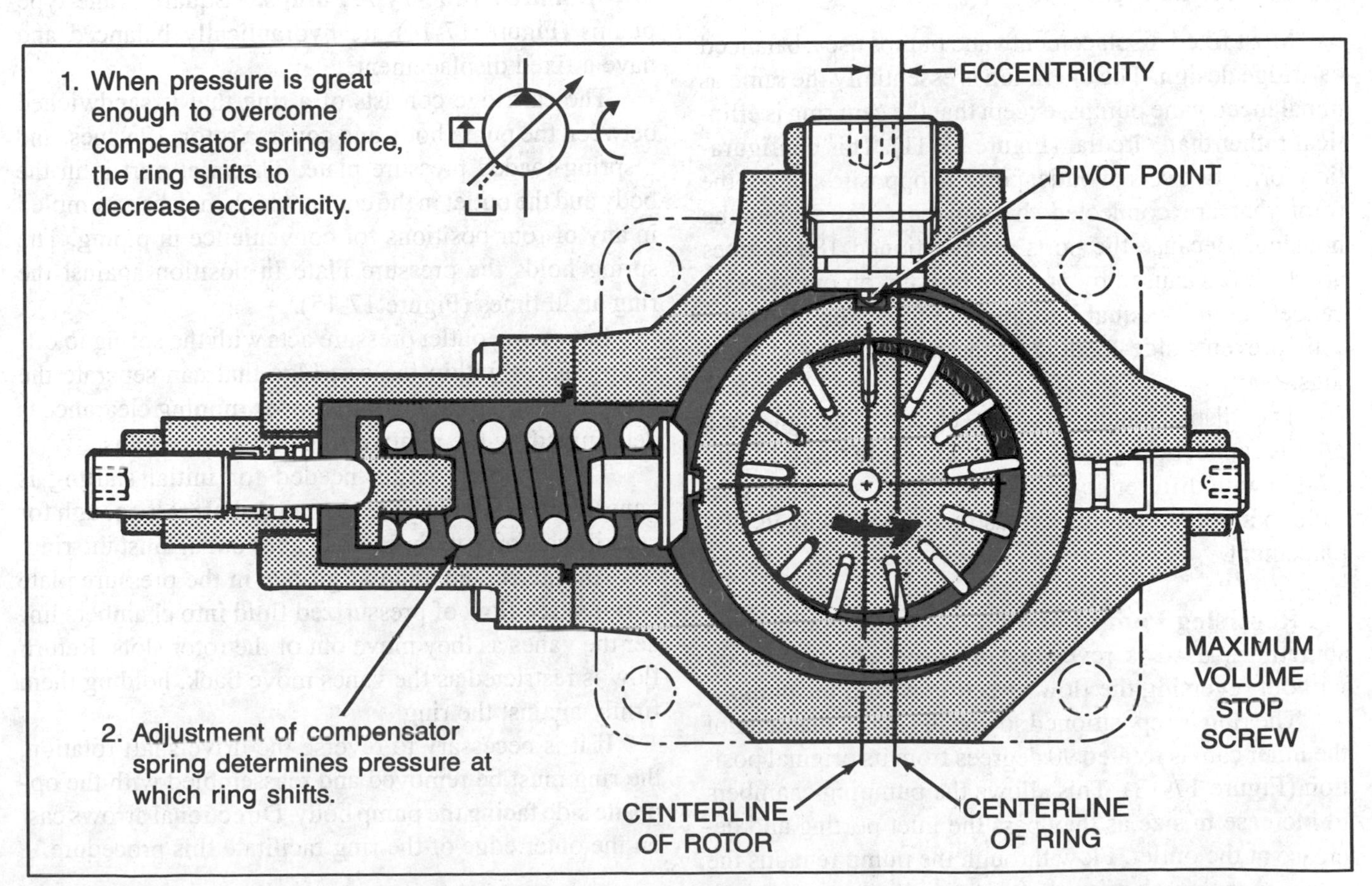
1. When pressure is great enough to overcome compensator spring force, the ring shifts to decrease eccentricity.
ECCENTRICITY
PIVOT POINT
MAXIMUM VOLUME STOP SCREW
2. Adjustment of compensator spring determines pressure at which ring shifts.
CENTERLINE OF ROTOR
CENTERLINE OF RING

Figure 17-10. Unbalanced variable displacement vane pump (pressure compensated).

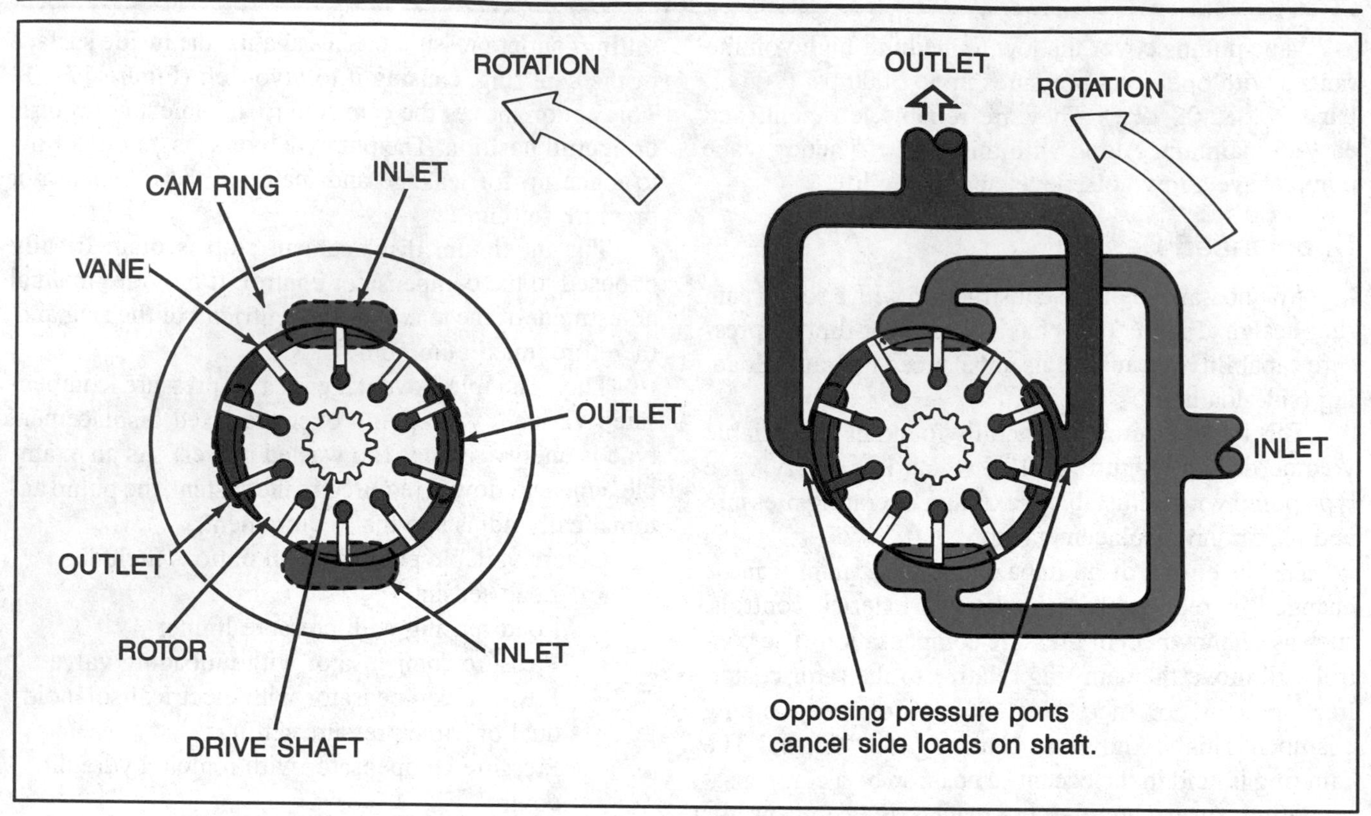

Figure 17-11. Balanced vane pump principle with an elliptical cam ring.

Balanced Design

Most fixed displacement vane pumps use a balanced cartridge design. The operation is essentially the same as unbalanced vane pumps, except that the cam ring is elliptical rather than circular (Figure 17-11). This configuration forms two sets of outlet ports on opposite sides of the rotor that are connected through passages within the housing. Because the ports are positioned 180 degrees apart, forces caused by pressure buildup on one side are canceled out by equal but opposite forces on the other. This prevents side-loading of the drive shaft and bearings.

The displacement of most balanced-design vane pumps cannot be adjusted. Interchangeable rings (Figure 17-12) with different cams are available, making it possible to modify a pump, increasing or decreasing its displacement.

Reversing Pump Rotation. Another modification sometimes used is reversing the drive shaft direction without reversing the flow direction within the pump.

The ring is repositioned so that the major diameter of the inner cam is rotated 90 degrees from its original position (Figure 17-13). This allows the pumping chambers to increase in size as they pass the inlet porting and decrease at the outlet. Flow through the pump remains the same, even though drive shaft rotation has been reversed.

"Square" Vane-Type Pumps. "Square" vane-type pumps (Figure 17-14) are hydraulically balanced and have a fixed displacement.

The cartridge consists of a ring that is sandwiched between the pump body and cover, a rotor, 12 vanes, and a spring-loaded pressure plate. The inlet port is in the body and the outlet in the cover, which may be assembled in any of four positions for convenience in piping. The spring holds the pressure plate in position against the ring at all times (Figure 17-15).

Increasing outlet pressure acts with the spring to offset pressures within the cartridge that can separate the pressure plate from the ring. Proper running clearance is determined by the relative ring and rotor widths.

The pumping action needed for initial starting is generated by spinning the rotor and shaft fast enough for centrifugal force to throw the vanes out against the ring.

An interrupted annular groove in the pressure plate permits free flow of pressurized fluid into chambers under the vanes as they move out of the rotor slots. Return flow is restricted as the vanes move back, holding them firmly against the ring.

If it is necessary to reverse the drive shaft rotation, the ring must be removed and reassembled with the opposite side facing the pump body. Directional arrows cast on the outer edge of the ring facilitate this procedure.

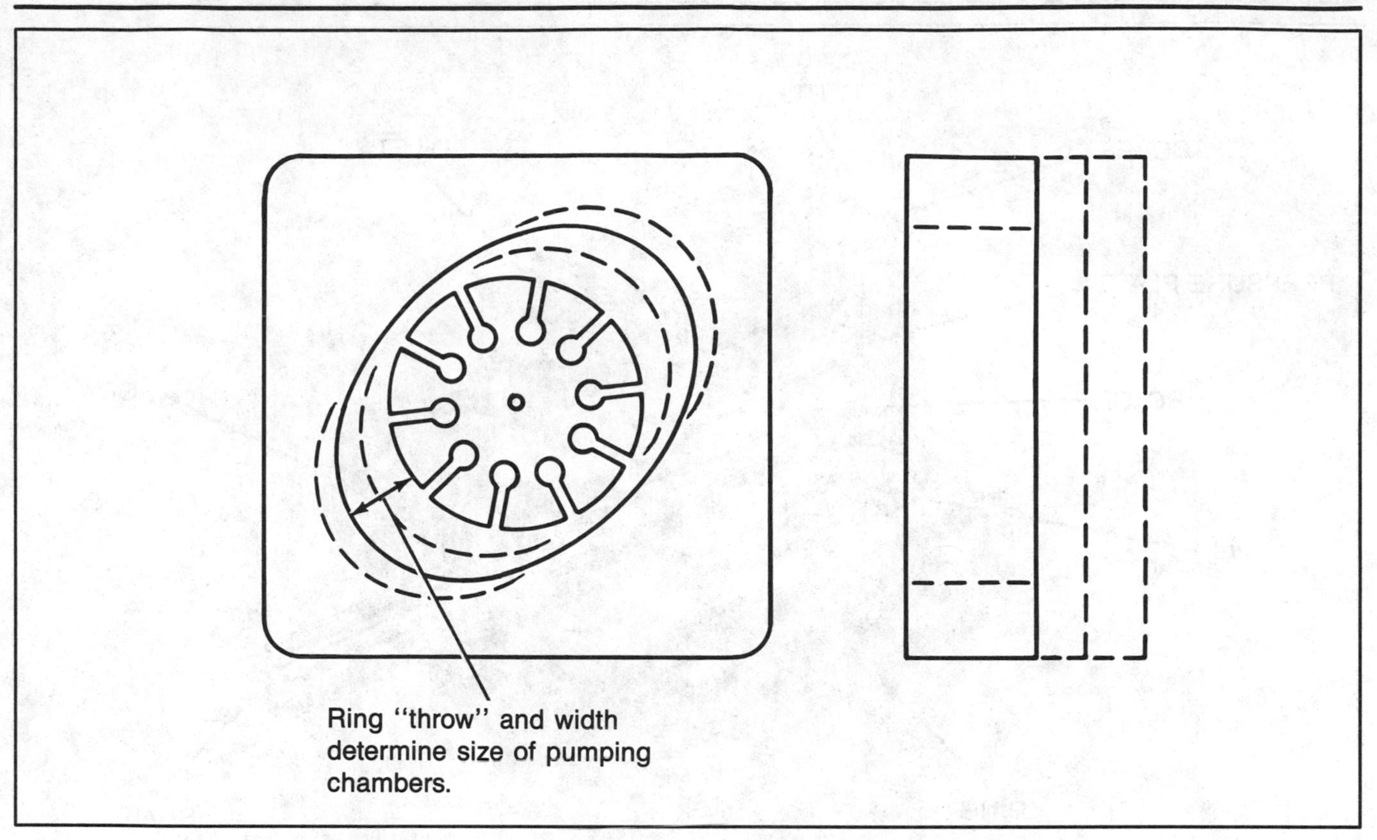
Ring "throw" and width
determine size of pumping
chambers.

Figure 17-12. Variations in vane pump displacement.

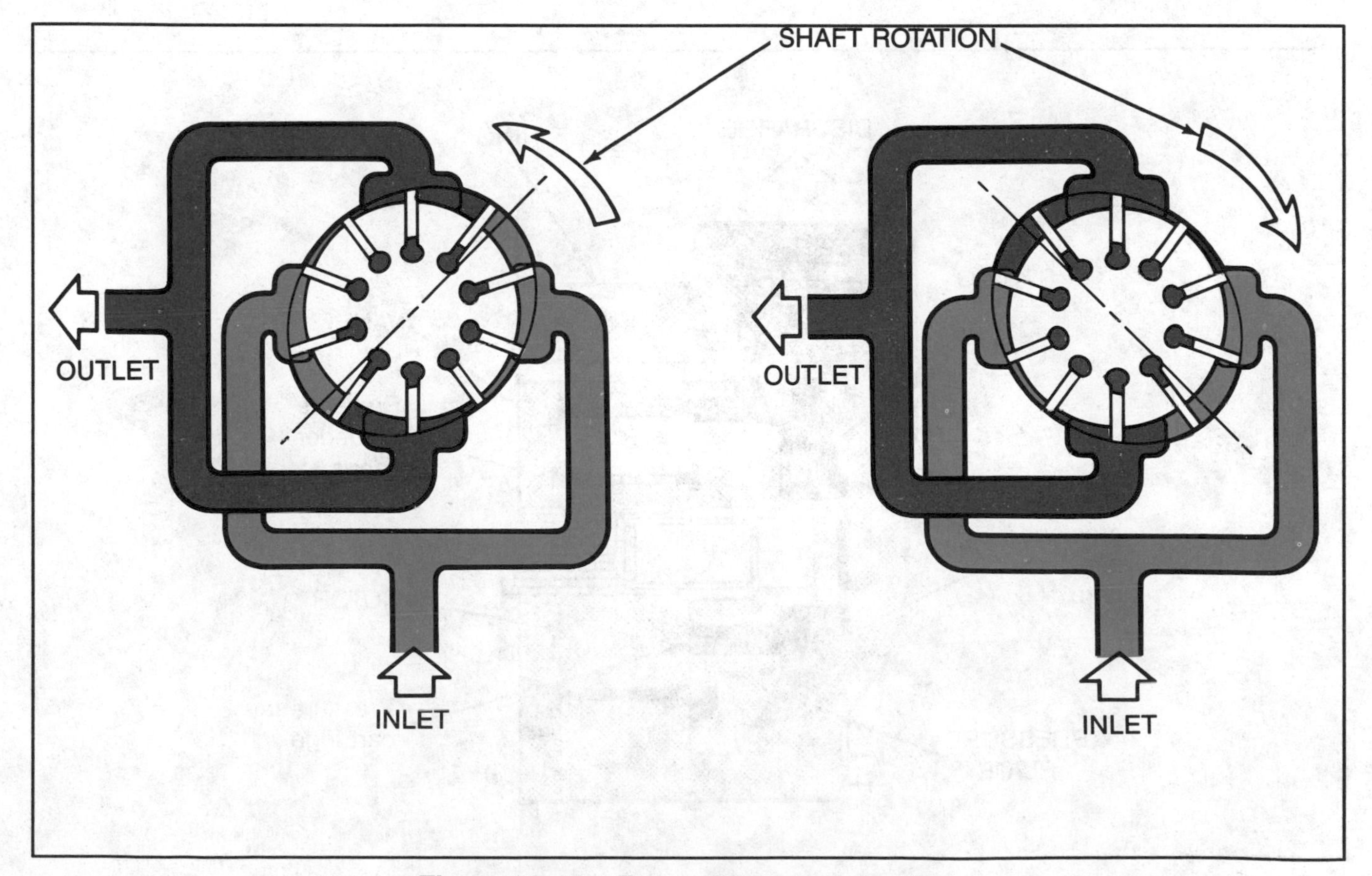
SHAFT ROTATION
OUTLET
OUTLET
INLET
INLET

Figure 17-13. Reversing pump rotation.

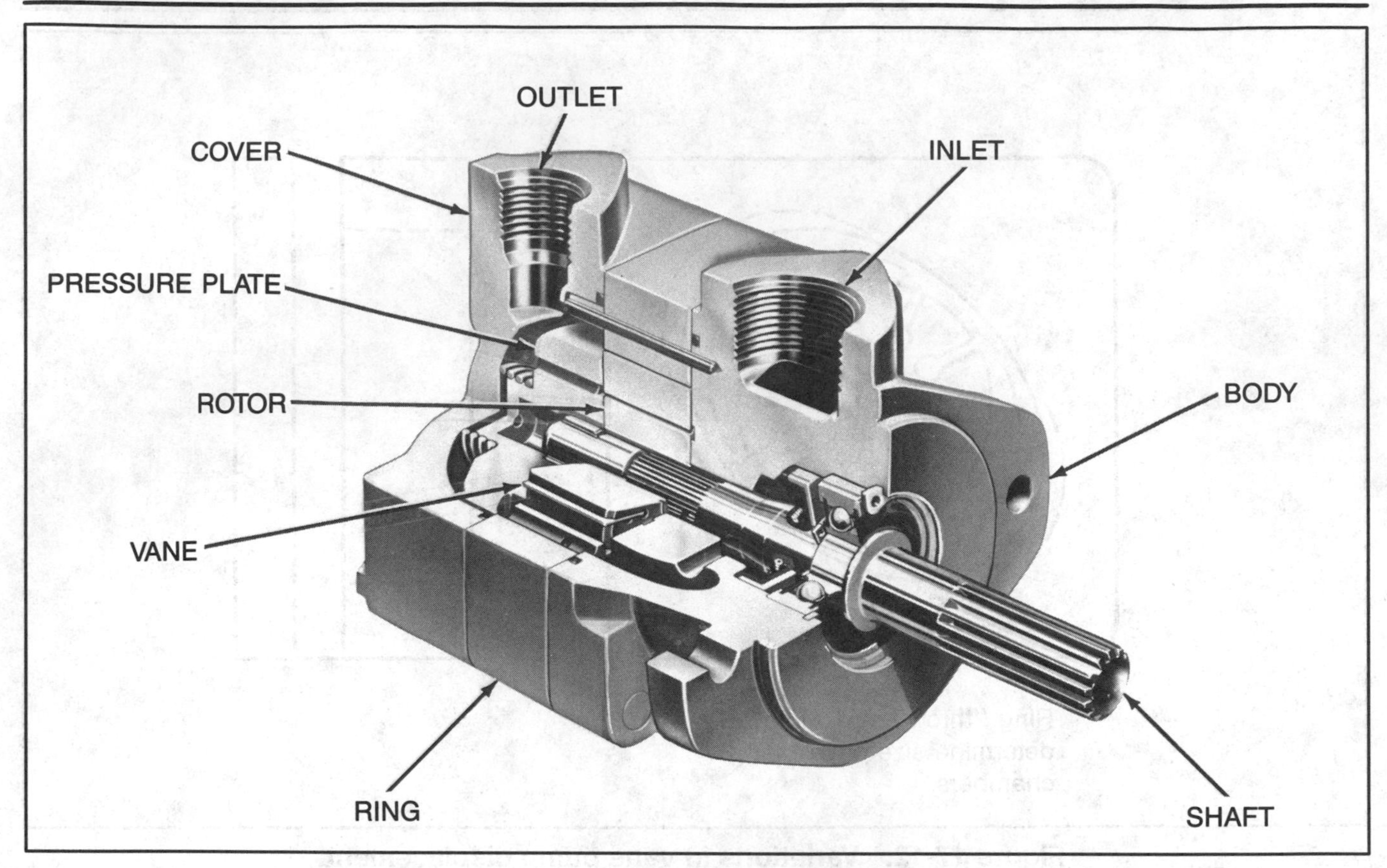

Figure 17-14. "Square" design vane pump.

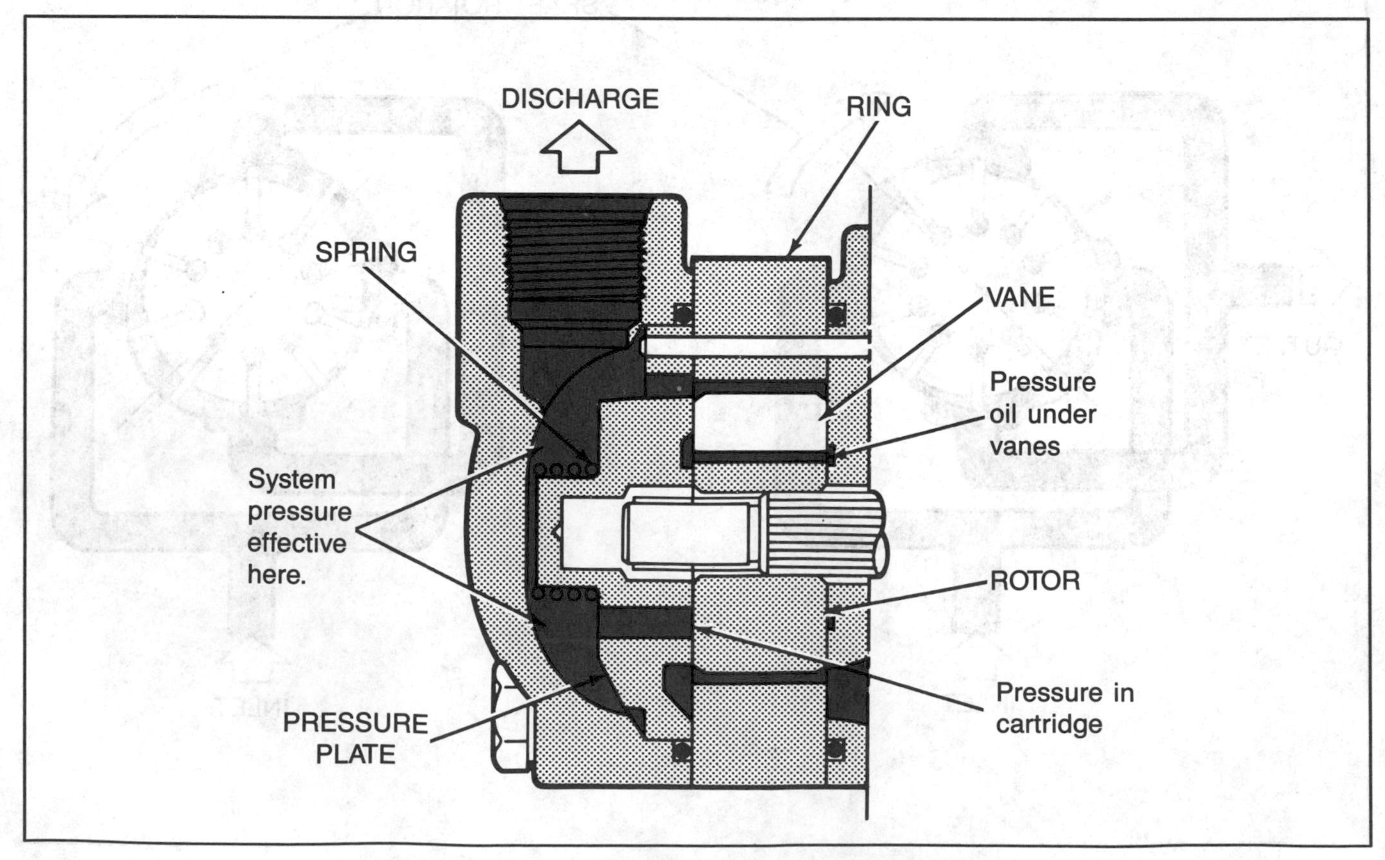

Figure 17-15. Spring-loaded pressure plate.

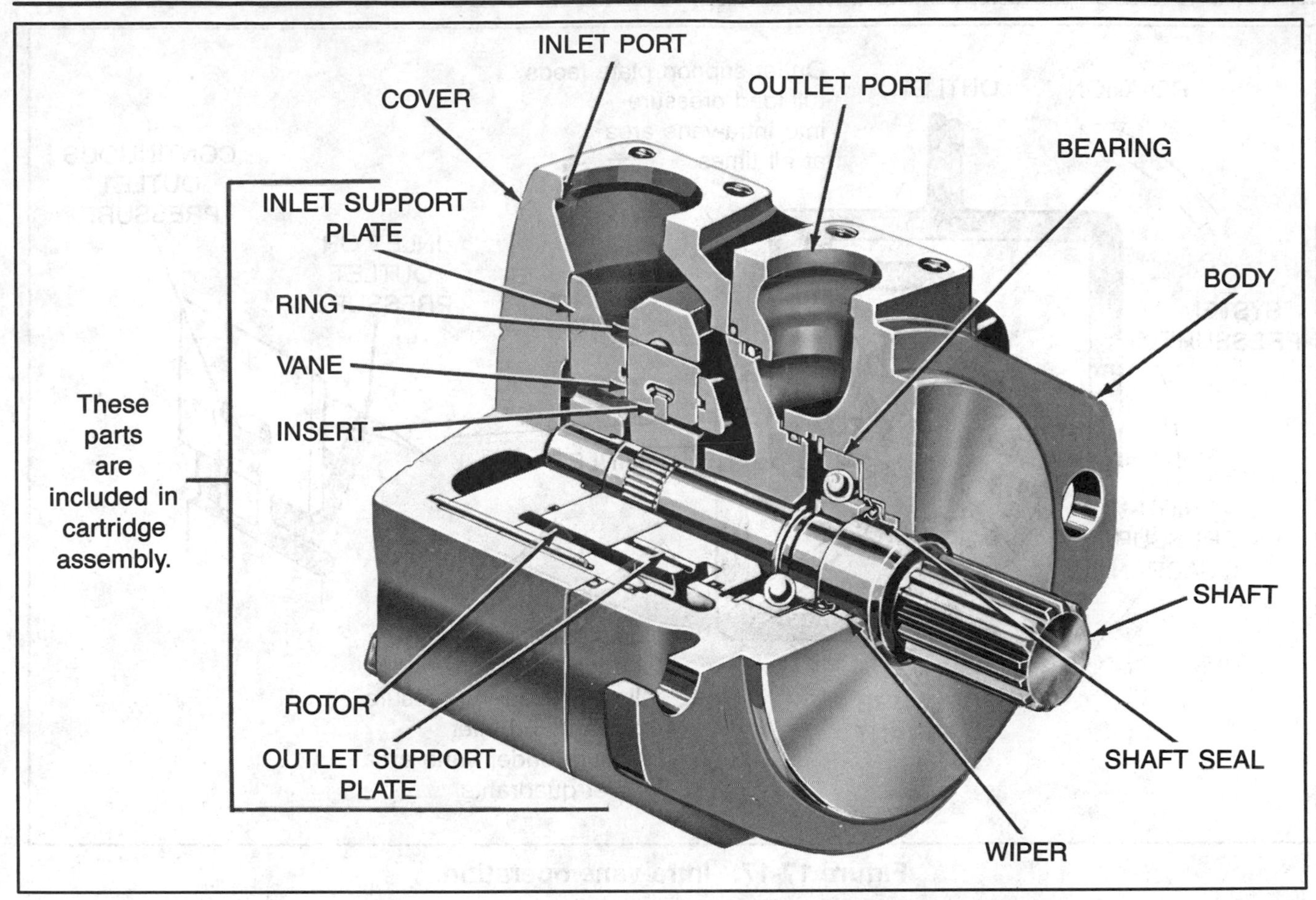

Figure 17-16. High-performance vane pump construction.

Square vane pumps are available in a variety of sizes and cartridges with different displacements for each.

"High Performance" Vane Pumps. The high performance series of balanced vane pumps is capable of higher pressure (2500 psi (172.35 bar / 17,235.00 kPa)) and greater speeds (1800 rpm). A typical single pump of this design is shown in Figure 17-16. Operation is essentially the same as the corresponding square vane pumps, except for some important design differences that are discussed below.

Intra-Vane Design. High-performance pumps incorporate intra-vanes, or small inserts, in the vanes to vary the outward force from pressure in the high- and low-pressure quadrants (Figure 17-17).

Square vane pumps use outlet pressure on the underside of the vanes at all times. In the sizes and pressure ranges available in the high-performance units, this feature could result in high loading and wear between the vane tip and the inlet portion of the cam ring. To avoid this, holes drilled through the rotor segments equalize pressure above and below each vane at all times.

Outlet pressure is constantly applied to the small area between the vane and intra-vane. This pressure, along with centrifugal force, holds the vanes in contact with the ring in the inlet quadrants to assure proper "tracking."

Preassembled Cartridge. For fast replacement, the cartridge used in high-performance pumps (Figure 17-18) is preassembled.

The cartridge consists of a ring, rotor, vanes, vane inserts, outlet support plate, inlet support plate, locating pins, and attaching screws.

They are assembled as right- or left-hand rotating units, but can be reassembled for opposite rotation, if required. Arrows and locating pins serve as guides. When properly assembled, flow direction remains the same in both the right- and left-hand rotating units.

Port Positions. Like the square pumps, the high-performance pumps are built so that the relative positions of the ports can be easily changed to any one of four combinations. This is done by removing the tie bolts and rotating the cover.

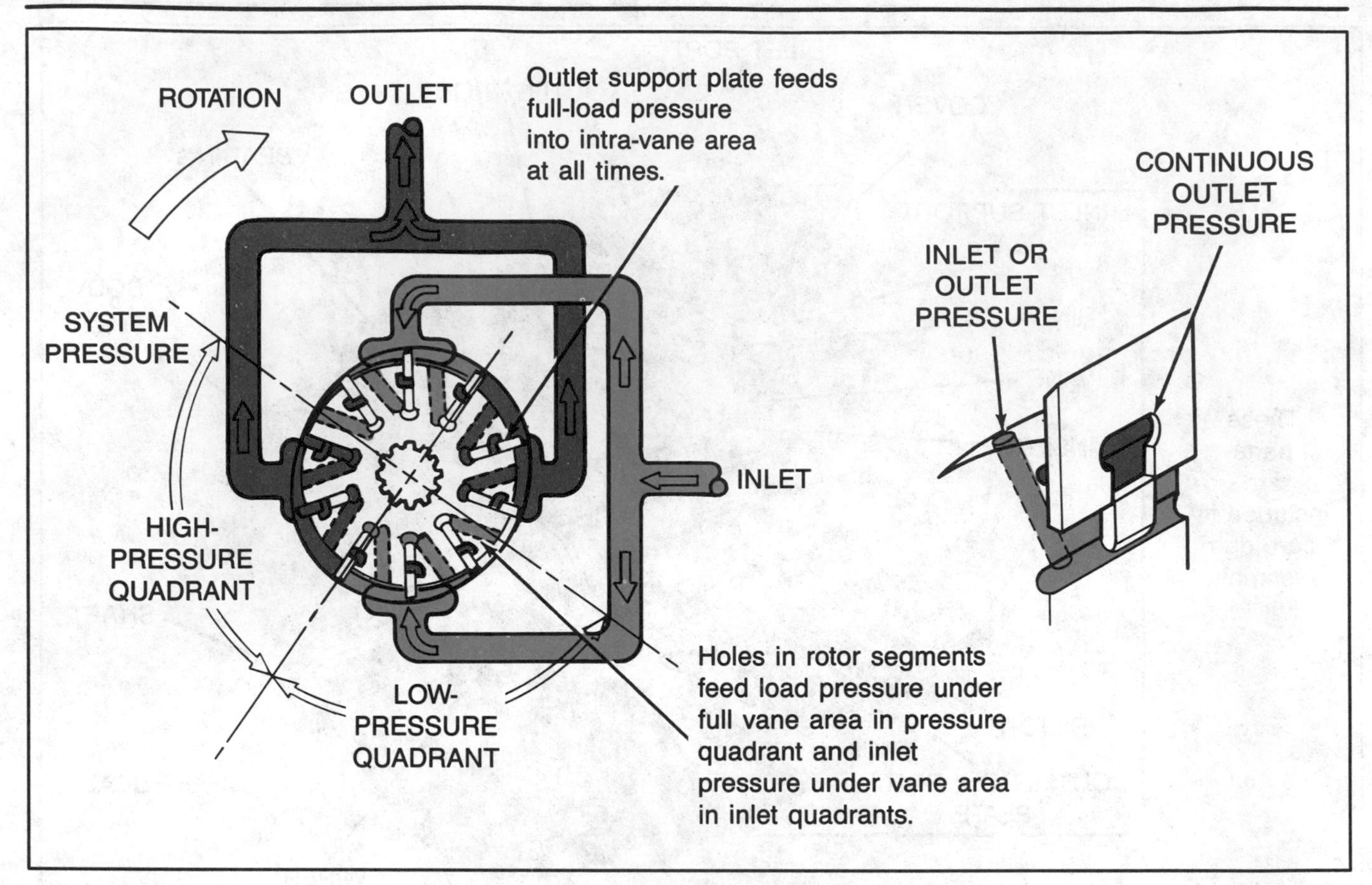

Figure 17-17. Intra-vane operation.

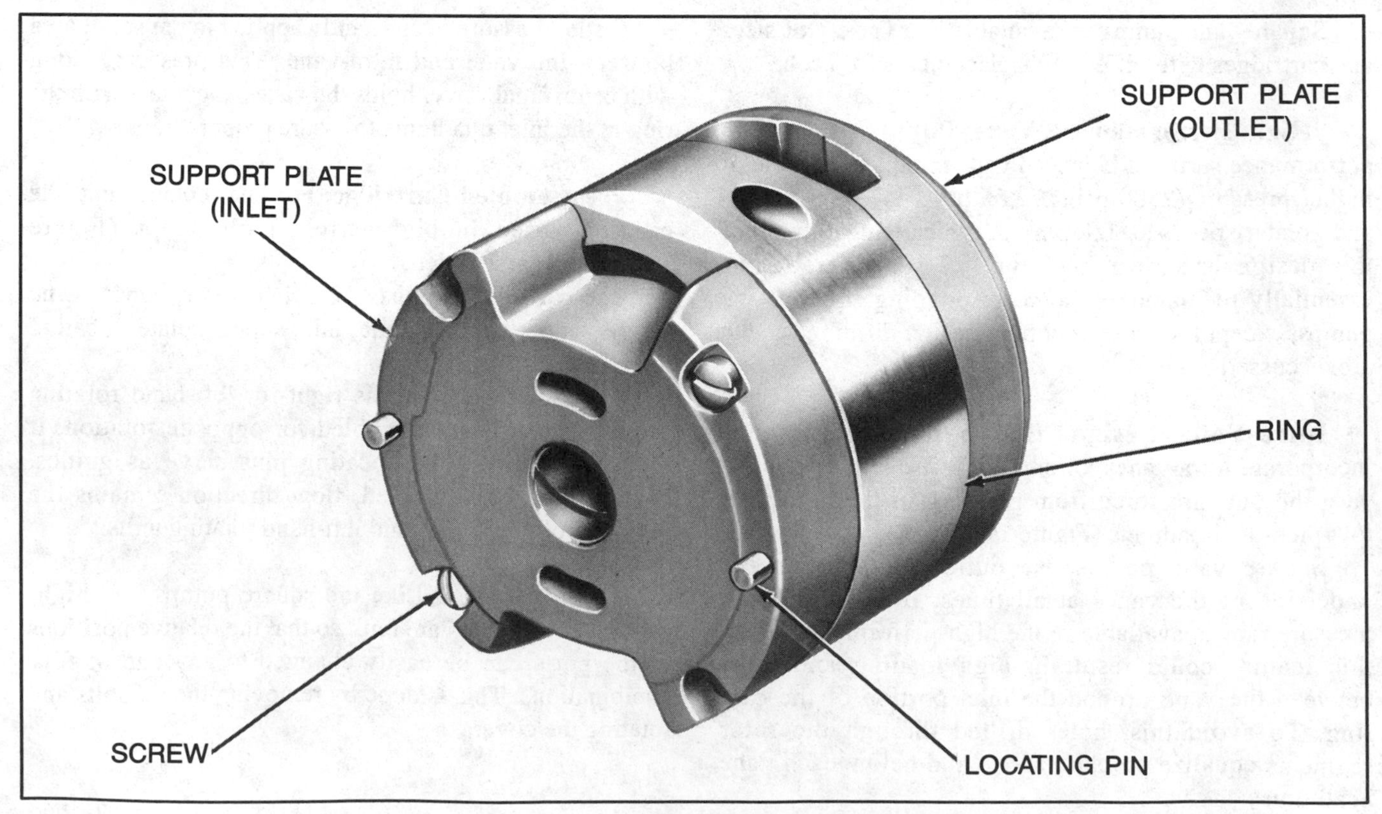

Figure 17-18. Preassembled cartridge.

Balanced Variable Vane Pumps

This type of pump, shown in Figure 17-19, has all the features of a balanced vane pump, including the two pumping actions per revolution which make it a more compact unit.

Although there is only one rotor, there are two vanes, mounted side-by-side, in each rotor slot each running on its own eccentric cam ring. In effect, there are two pumps in one housing. When the two rings, which can be rotated up to 90 degrees relative to each other are aligned, the two pumping elements act together to produce the maximum flow rate.

Because of its construction, this pump is easily adapted to through-drive operation with the possibility of adding multiple units and combinations (variable/variable or variable/fixed).

Double Pump Assemblies

Double pumps (Figure 17-20) provide a single power source capable of serving two separate hydraulic circuits or providing greater volume through combined delivery. Most double pumps have a common inlet in a center housing. The outlet for one unit, usually the larger of the two is in the shaft-end body. The second outlet is in the cover. Some types of double pumps have separate inlets, although they can be mounted in multiples. Both kinds need only one drive motor, however, double pumps that have separate inlets require separate piping.

Cartridge construction for double pumps is essentially the same as in single units, making numerous combinations of sizes and displacements possible.

PISTON PUMPS

All piston pumps operate on the principle that a piston reciprocating in a bore will draw fluid in as it is retracted and expel it as it moves forward. The two basic designs are radial and axial, both available as fixed- or variable-displacement models. A radial pump has the pistons arranged radially in a cylinder block (Figure 17-21), while the pistons in the axial units are parallel to each other and to the axis of the cylinder block (Figure 17-22). Axial piston pumps may be further divided into in-line (swash plate) and bent-axis types.

Operating Characteristics

Piston pumps are highly efficient units, available in a wide range of capacities. They are capable of operating in the medium- to high-pressure range (1500-3000 psi (103.41-206.82 bar / 10,341.00-20,682.00 kPa)), with some going much higher.

Because of their closely fitted parts and finely machined surfaces, cleanliness and good quality fluids are vital to long service life.

Radial Piston Pumps

In a radial pump, the cylinder block rotates on a stationary pintle inside a circular reaction ring or rotor. As the block rotates, centrifugal force, charging pressure, or some form of mechanical action causes the pistons to follow the inner surface of the ring, which is offset from the centerline of the cylinder block. Porting in the pintle permits the pistons to take in fluid as they move outward and discharge it as they move in.

Pump displacement is determined by the size and number of pistons (there may be more than one bank in a single cylinder block) and the length of their stroke. In some models, the displacement can be varied by moving the reaction ring to increase or decrease piston travel.

Several types of external controls are available for this purpose.

Axial Piston Pumps

In axial piston pumps, the pistons reciprocate parallel to the axis of rotation of the cylinder block. The simplest type of axial piston pump is the swash plate in-line design (Figure 17-23).

The cylinder block in this pump is turned by the drive shaft. Pistons fitted to bores in the cylinder block are connected through piston shoes and a shoe plate, so that the shoes bear against an angled swash plate.

As the block turns (Figure 17-24), the piston shoes follow the swash plate, causing the pistons to reciprocate. The ports are arranged in the valve plate so that the pistons pass the inlet as they are pulled out and pass the outlet as they are forced back in.

Like radial piston pumps, the displacement of axial piston pumps is determined by the size and number of pistons, as well as the stroke length. Stroke length is determined by the angle of the swash plate.

In variable displacement models of the in-line pump, the swash plate is installed in a movable yoke (Figure 17-25).

"Pivoting" the yoke on pintles changes the swash plate angle to increase or decrease the piston stroke (Figure 17-22). The yoke can be positioned by any of several means, including manual control, servo control, pressure compensator control, and load sensing and pressure limiter control. Figure 17-25 shows a pump with a compensator control. Maximum angle on this unit is limited by construction to 17.5 degrees.

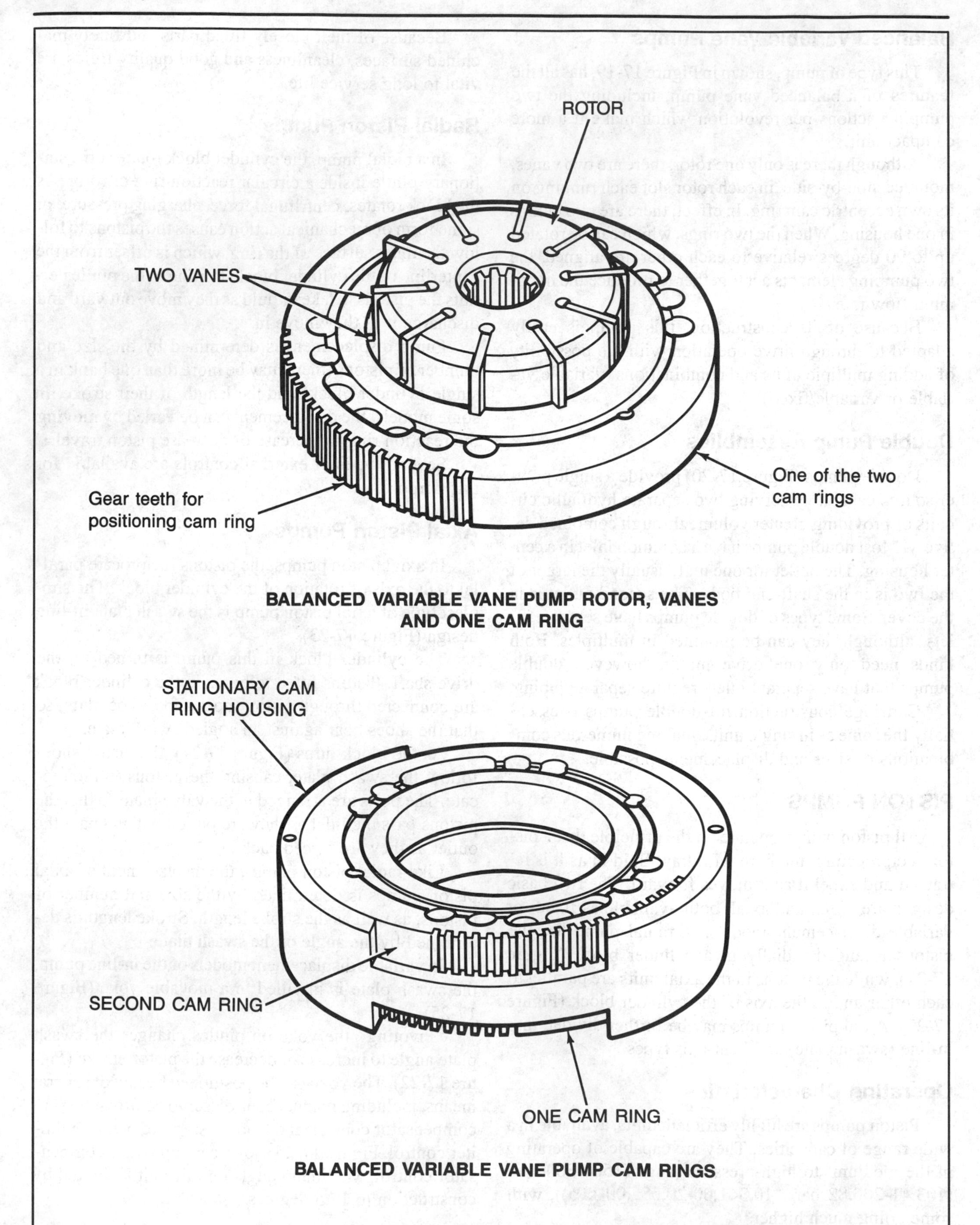

Figure 17-19. Balanced variable vane pump cam rings, rotor, and vanes.

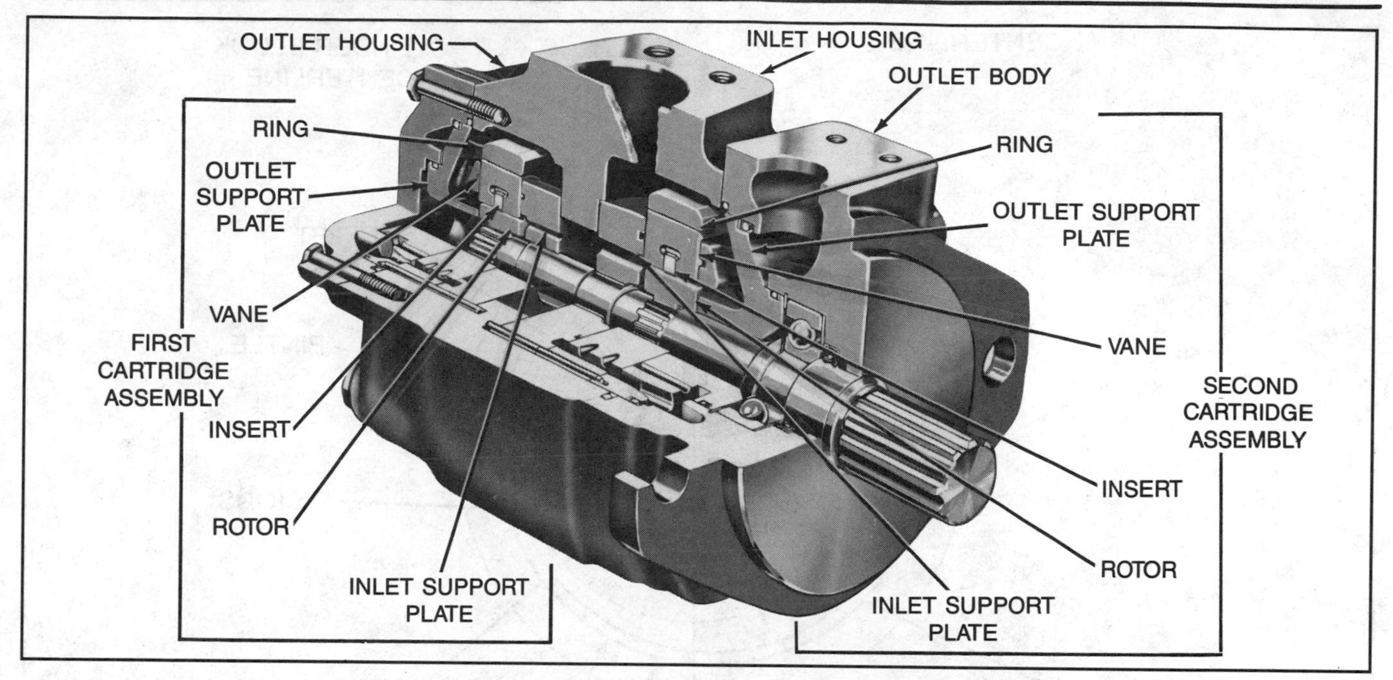

Figure 17-20. Double pump construction.

Pressure Limiting Compensator Operation: A pressure limiting compensator is shown schematically in Figure 17-26 with representation of an in-line piston pump rotating group with a stroking piston (large diameter) and "bias" piston (small diameter). The bias piston uses pump outlet pressure to keep the pump"biased" toward full displacement. A directional control valve symbol is also shown, with pump outlet flow connected to the valve inlet, or "P", port.

The compensator consists of a housing containing a control spool, a load spring, end caps and a load spring adjustment mechanism. The adjustment sets a preload in the load spring, which acts against the control spool, to determine the pressure setting for the pump.

System pressure (pump outlet pressure,) is fed to the control spool area and, through small passages, to the right end of the spool. As long as system pressure is below the load spring sitting, the control spool will remain to the right, and flow will be prevented from passing through the valve. Any small leakage that may pass the spool will be drained to the pump housing (case drain), which is connected to the reservoir.

As system pressure rises and approaches the setting of the load spring, the control spool begins to shift left, allowing fluid to pass the spool land into the control piston area. Because the control piston diameter (and area) is larger than the bias piston diameter, the pump yoke will begin to reduce its stroke. It will continue to reduce its stroke as the system pressure rises, until the output flow of the pump is low enough to maintain the system pressure at the load spring setting, or until the pump flow is reduced to zero.

When system pressure decreases, the spool begins to move to the right, closing flow to the control piston and allowing fluid in the control piston to vent to the pump housing. The yoke begins to stroke out, providing more pump output flow. The compensator adjusts the yoke in this manner, regulating the correct amount of pump flow to maintain system pressure at or below load spring setting.

Load Sensing and Pressure Limiting Compensator Operation: A compensator containing a pressure limiting and a load sensing operation are shown in Figure 17-27. The pressure limiting spool, the lower spool in the figure, is identical to the pressure limiting spool of Figure 17-26.

The upper spool in the figure is a load sensing spool. It operates basically the same as the pressure limiting spool, with the exception that the spring is light, providing for a load setting of 200-400 psi system pressure. In addition to the spring, however, pressure is sensed downstream of the control valve, and is fed to the spring chamber. The combined load against the load sensing spool is the system pressure plus 200-400 psi.

When system pressure is low, the load sensing spool regulates the flow of the pump so pump outlet pressure is 200-400 psi above the system requirement. Therefore, the pressure drop across the directional control valve is maintained at high pressure when the system only requires a low pressure.

As system pressure rises, pump outlet flow is adjusted to maintain this pressure difference across the directional valve. If system pressure rises to the level of the pressure limiting setting, the pressure limiting spool becomes operational, and the pump reduces its flow to protect the system.

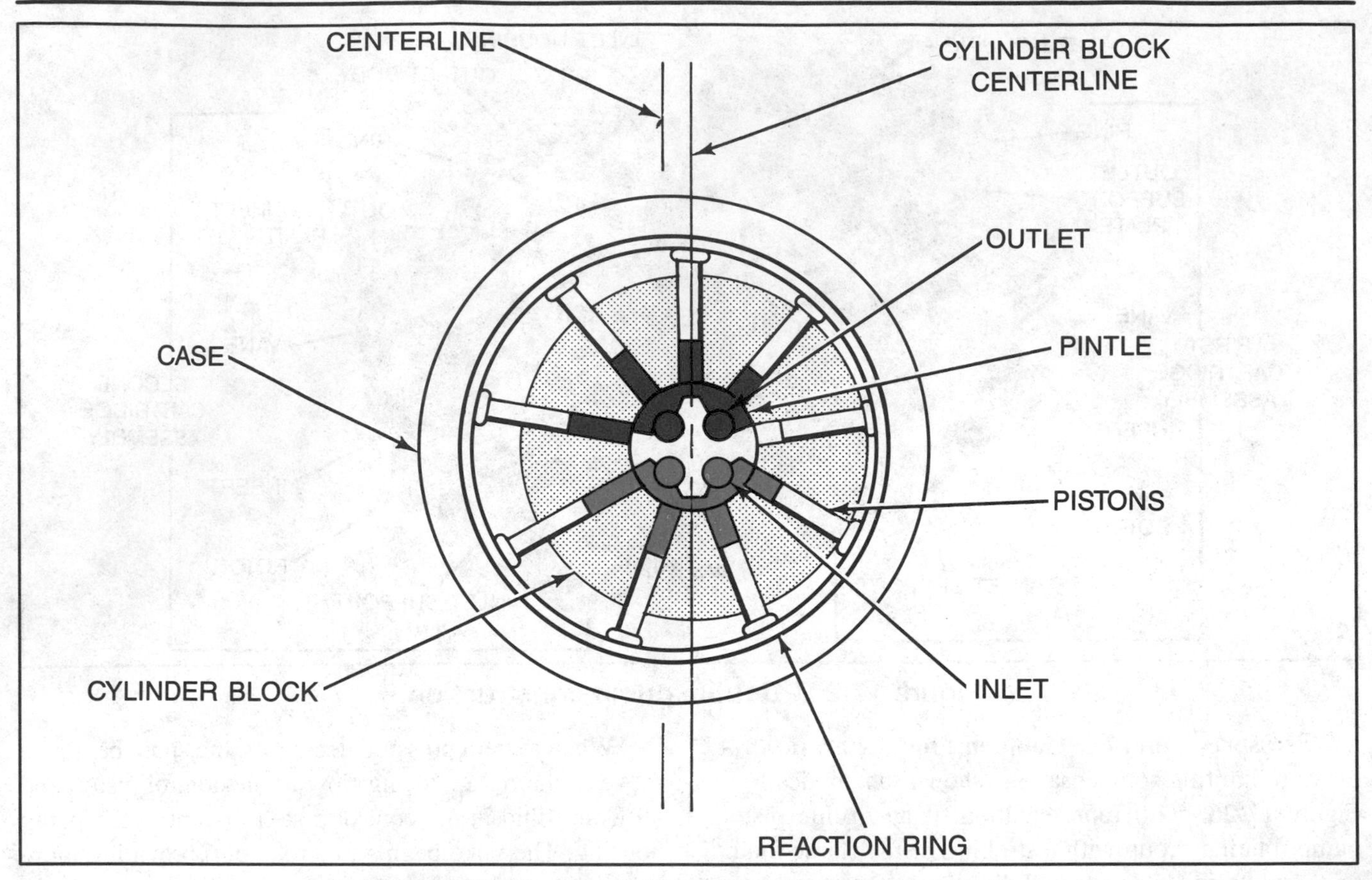

Figure 17-21. Operation of radial piston pump.

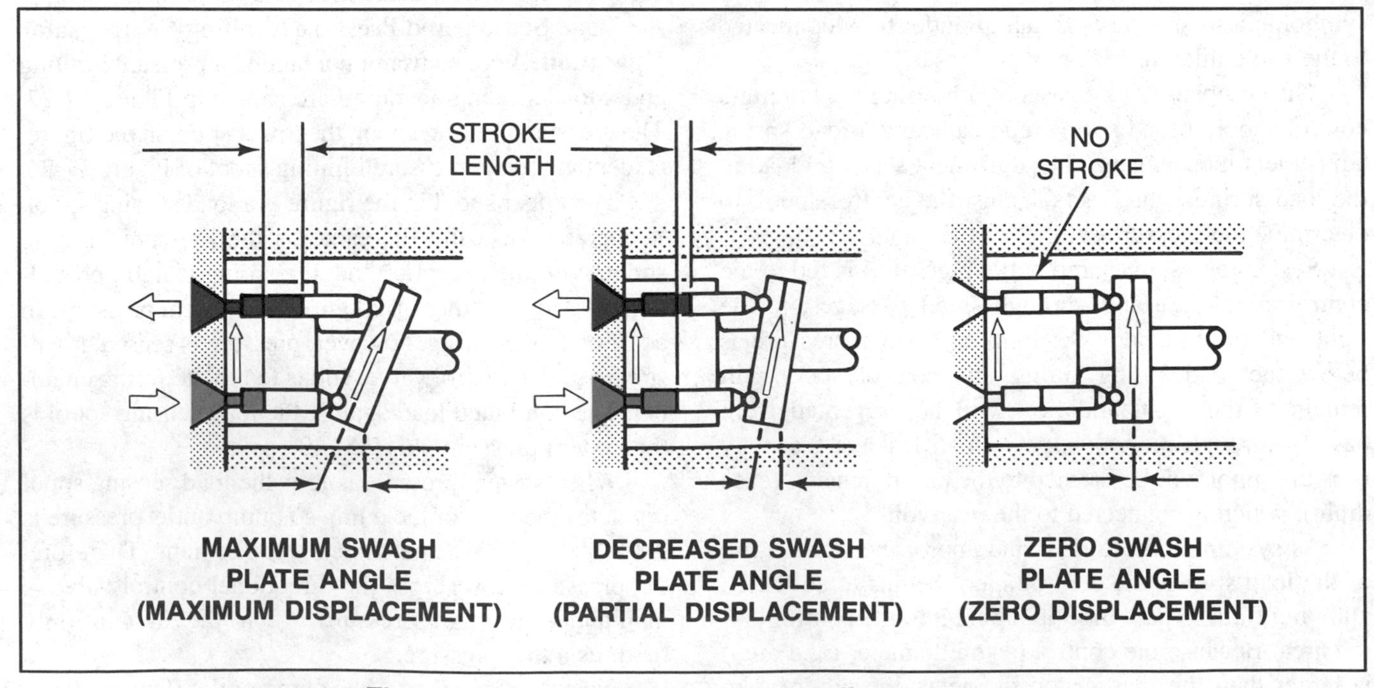

Figure 17-22. Variation in pump displacement.

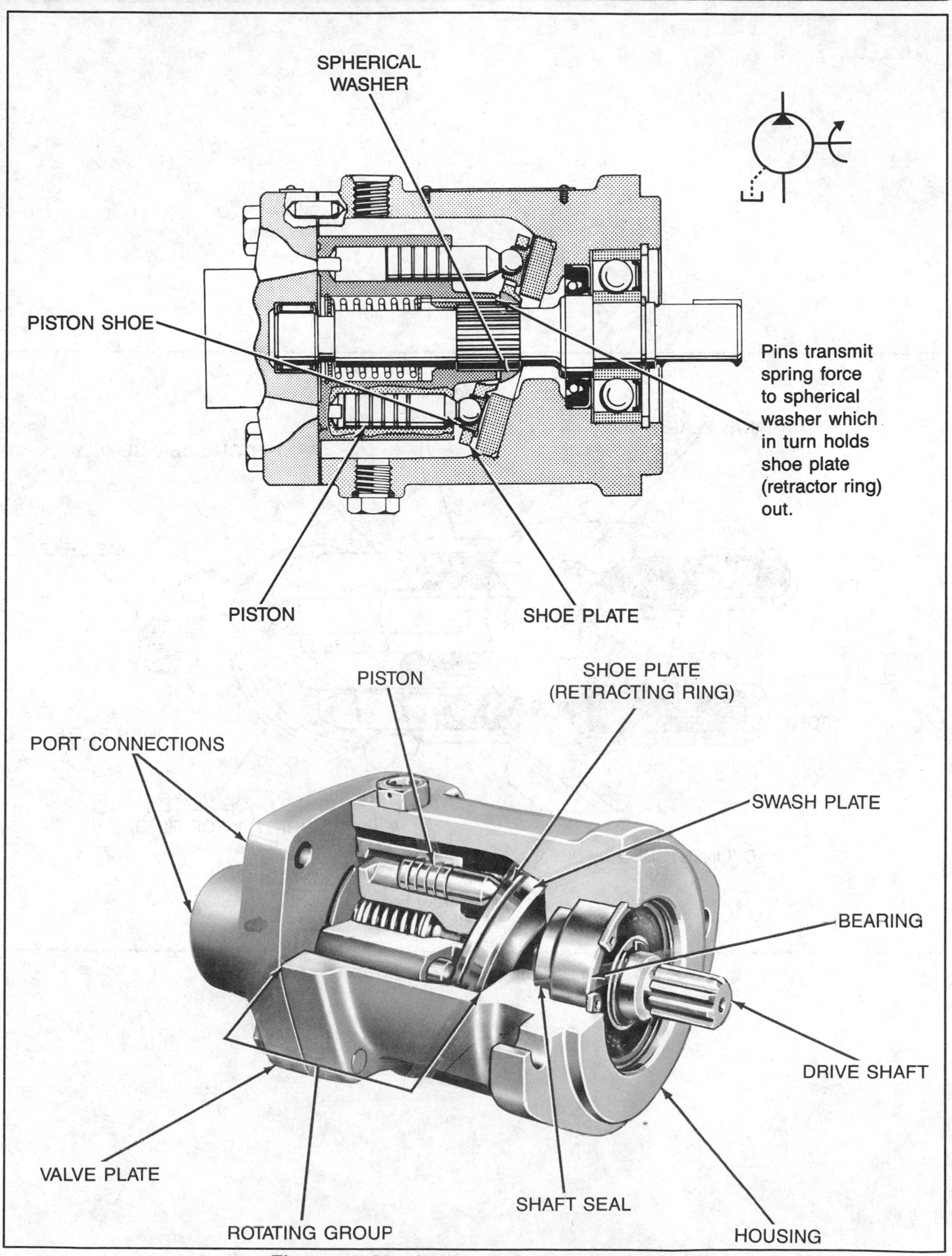

Figure 17-23. In-line design piston pump.

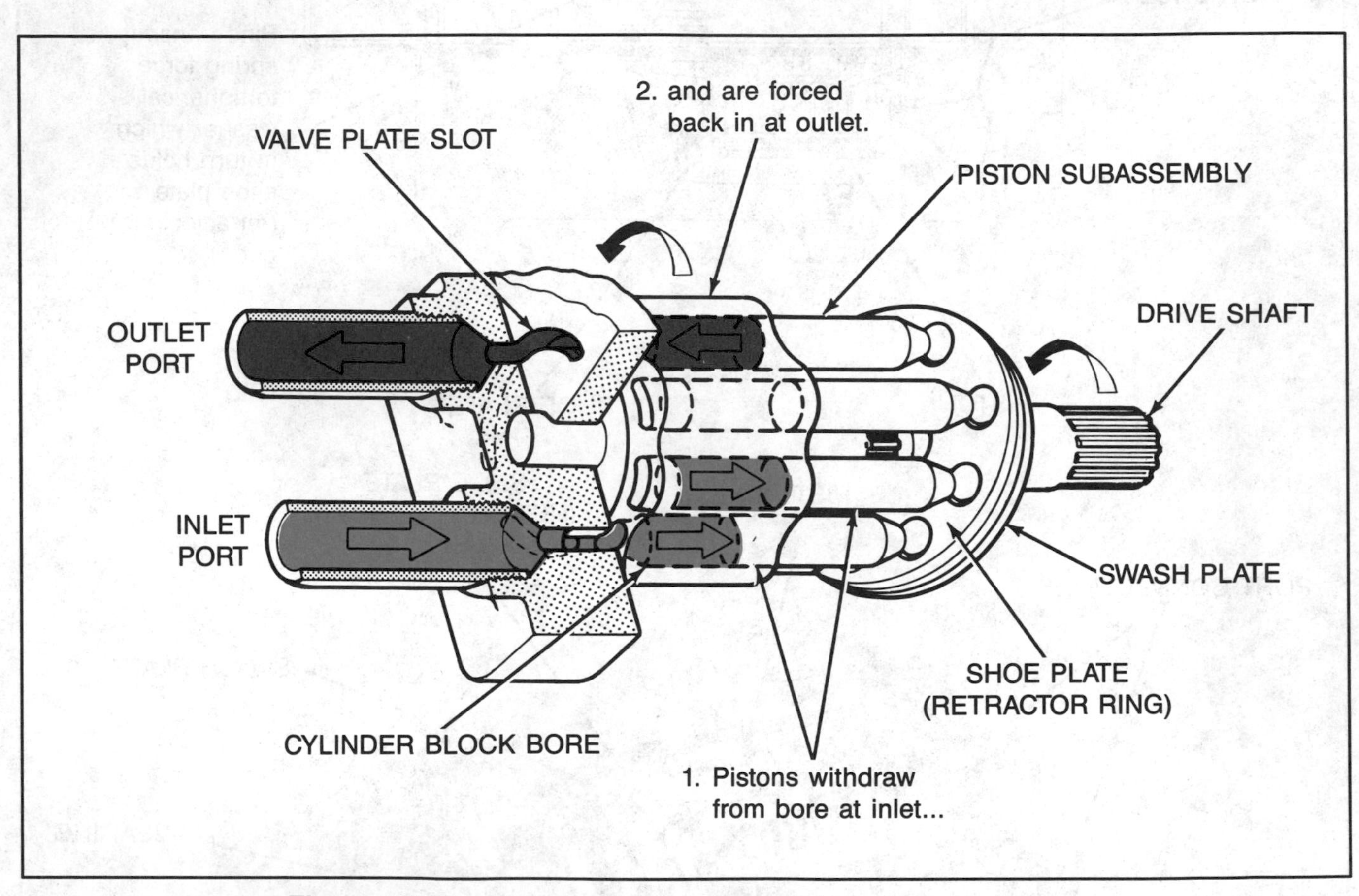

Figure 17-24. Swash plate causes pistons to reciprocate.

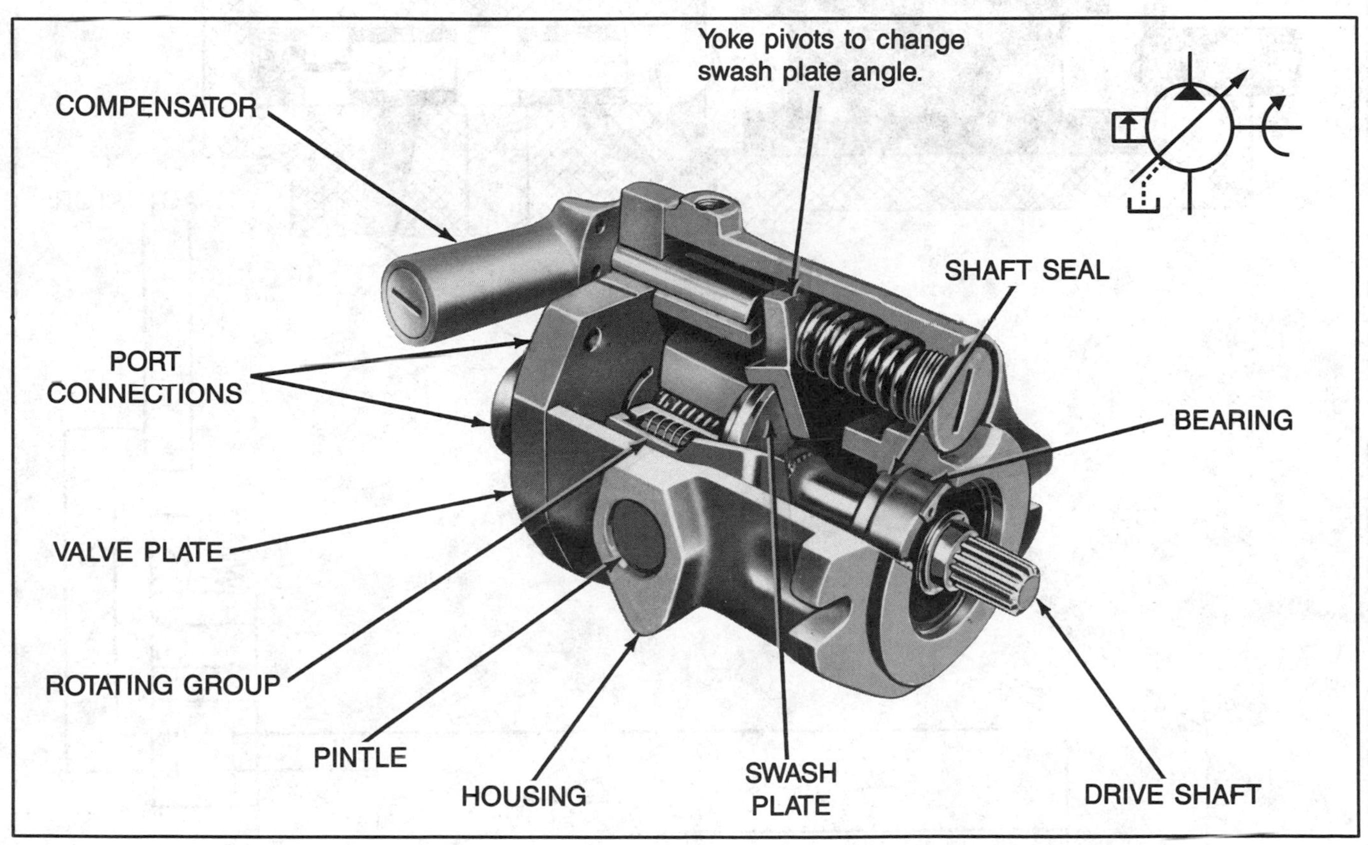

Figure 17-25. Variable displacement version of in-line piston pump.

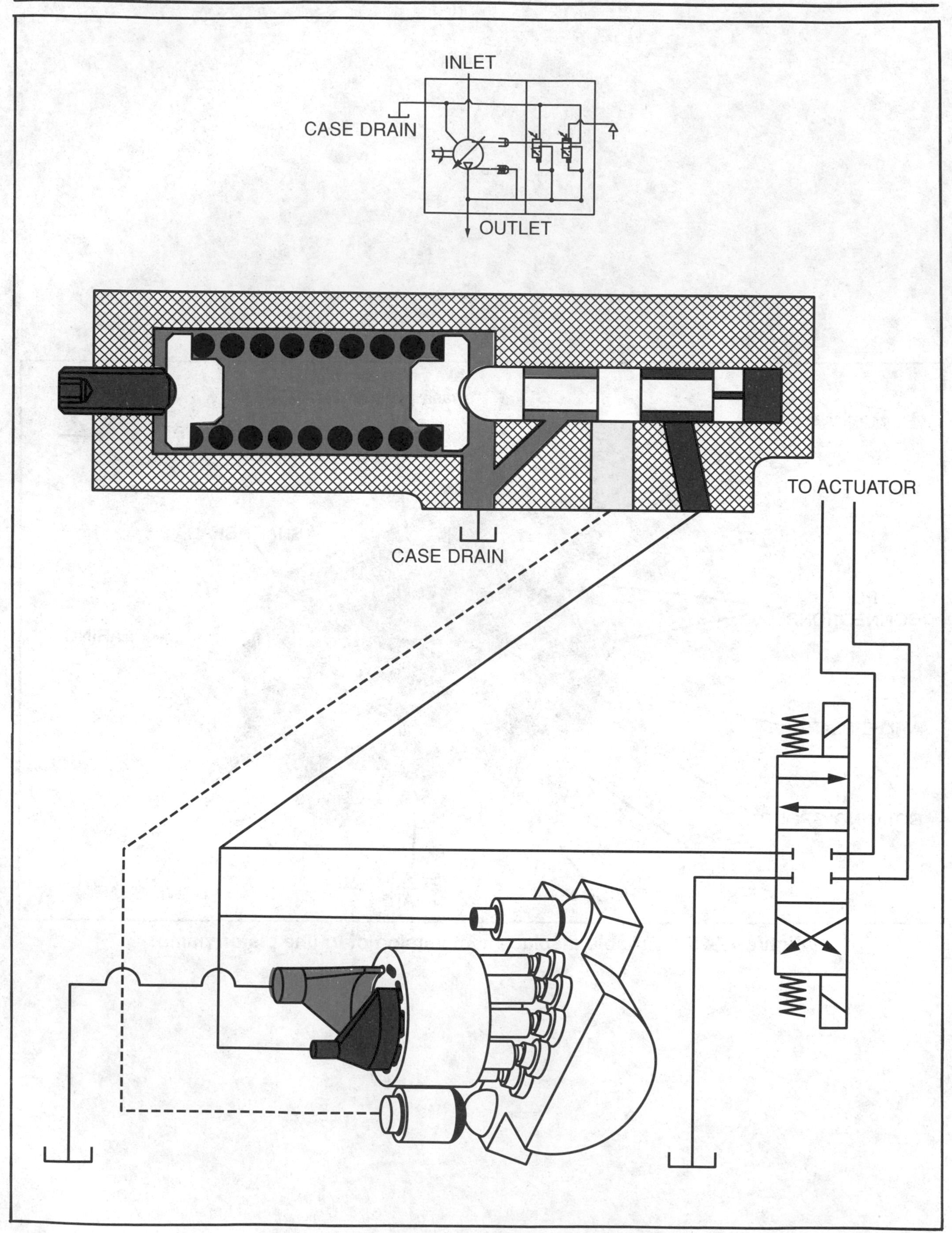

Figure 17-26. Pressure limiting compensator.

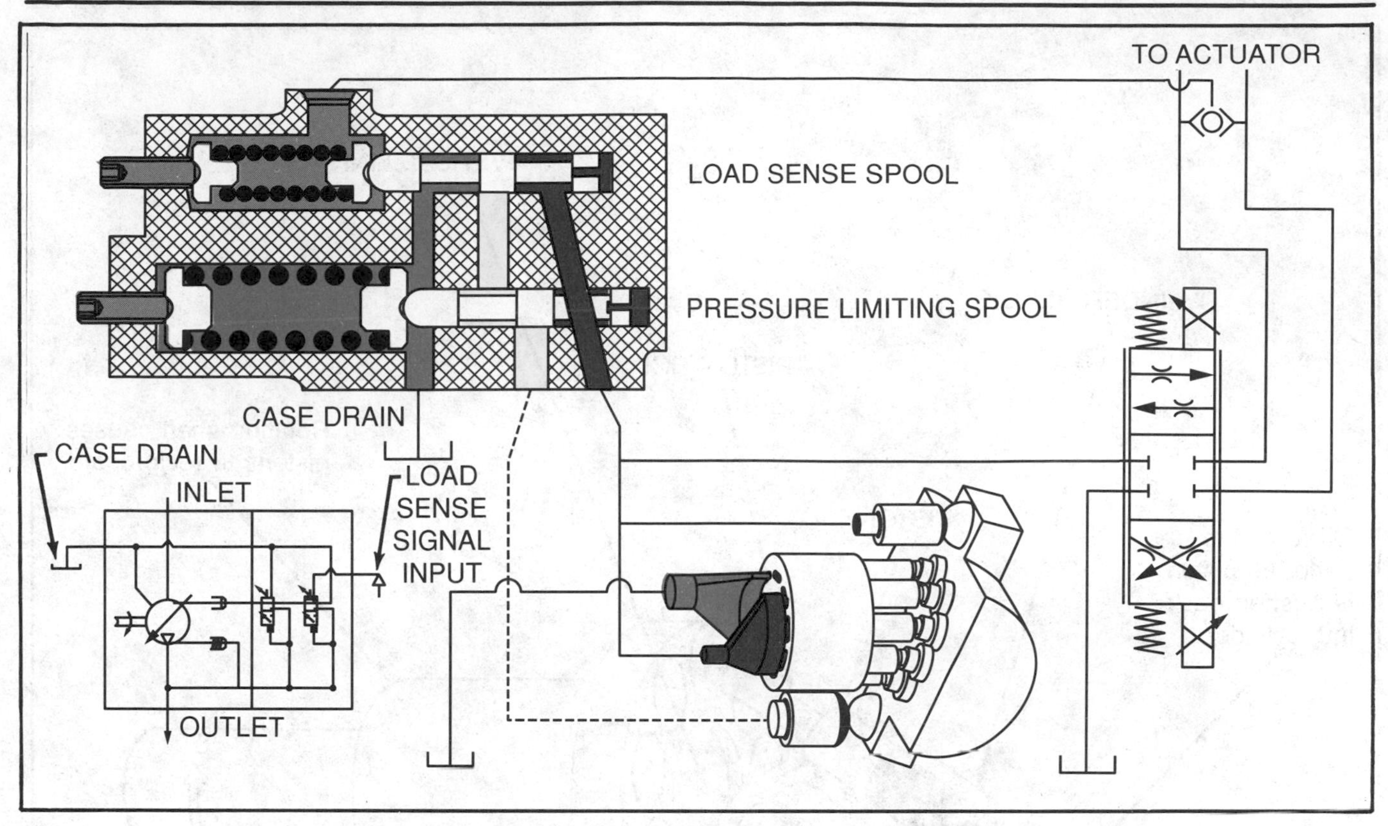

Figure 17-27. Load sensing and pressure limiting compensator.

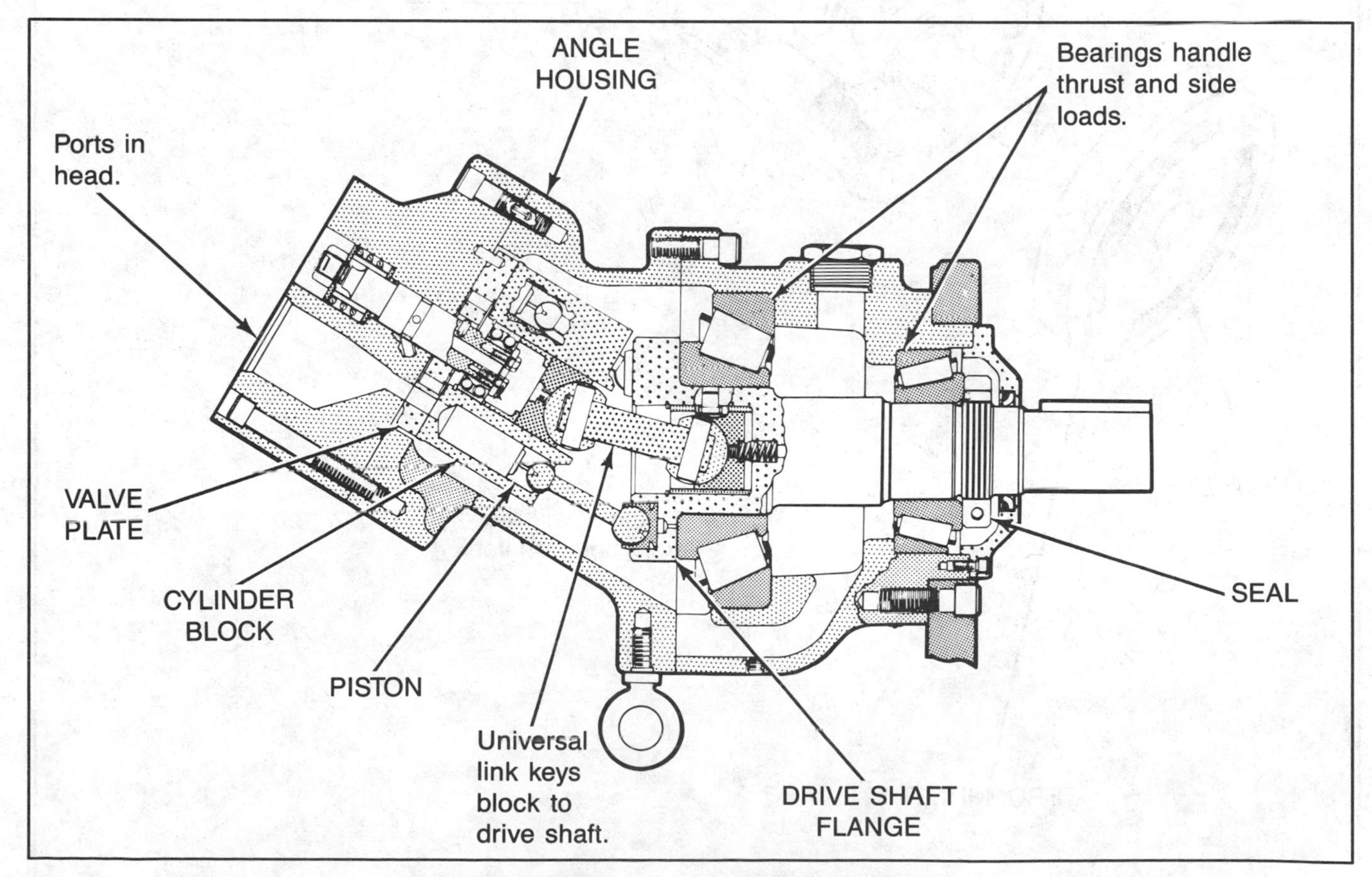

Figure 17-28. Bent-axis piston pump.

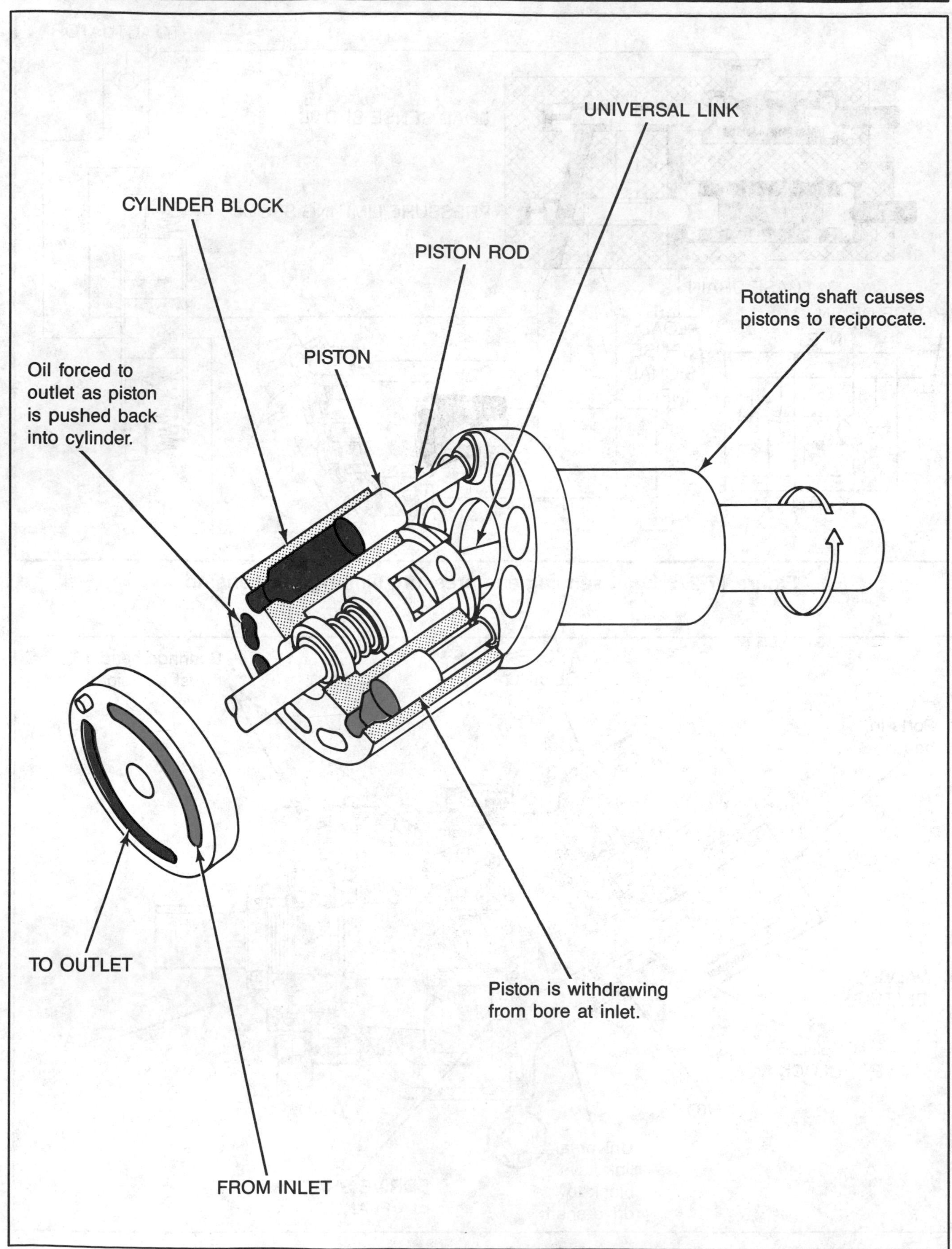
UNIVERSAL LINK
CYLINDER BLOCK
PISTON ROD
Rotating shaft causes pistons to reciprocate.
PISTON
Oil forced to outlet as piston is pushed back into cylinder.
TO OUTLET
Piston is withdrawing from bore at inlet.
FROM INLET

Figure 17-29. Pumping action in bent-axis pump.

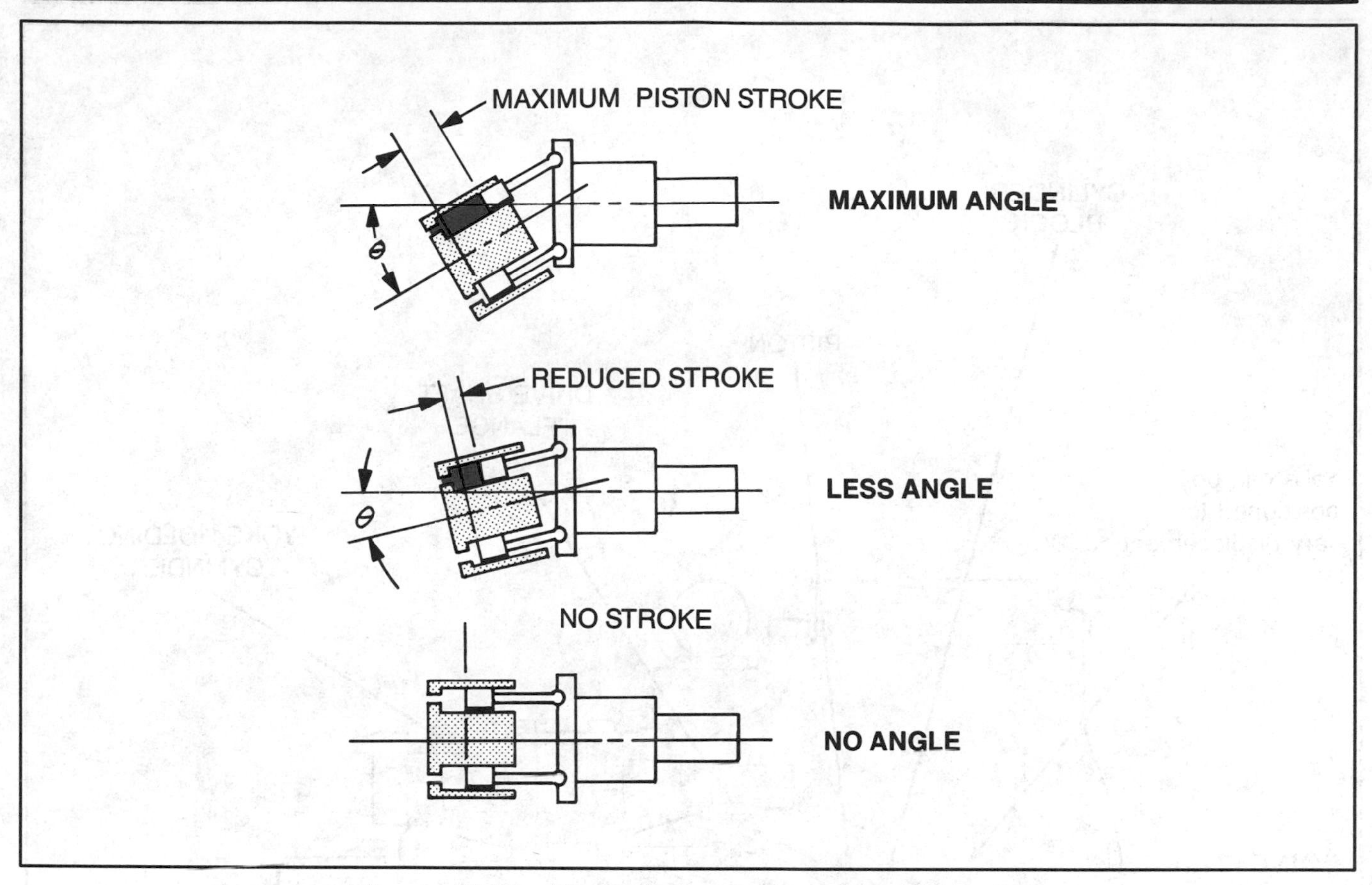

Figure 17-30. Displacement changes with angle.

When the system directional valve is only partially shifted, it restricts the pump outlet flow and causes the pressure to increase on the left-hand side of the load-sensing spool. When this pressure exceeds the load pressure and the pressure setting of the load-sensing spring, the load-sensing spool moves to the right, sending a small amount of oil into the yoke-actuating piston to reduce the pump outlet flow. This continues until the pump outlet pressure is again equal to the load-pressure and load-sensing spring. The yoke position is maintained for reduced flow until either the orifice formed by the system directional valve or the load pressure changes.

It is important to note that the pump operates at a reduced displacement, determined by a variable orifice in the system and at a pressure equal to the load pressure plus load sensing spring. By contrast, a pressure compensator control also reduces pump displacement to equal the flow requirements of the load. This only occurs, however, at the pressure setting of the compensator (maximum pressure), regardless of the load pressure.

When the pressure at the outlet of the pump reaches the pressure limiter setting, the pump outlet flow is reduced in the same manner as with a pressure compensator control, limiting maximum system pressure.

Bent-Axis Piston Pumps

In a bent-axis piston pump (Figure 17-28), the cylinder block turns with the drive shaft, but at an offset angle. The piston rods are attached to the drive shaft flange by ball joints and are forced in and out of their bores as the distance between the drive shaft flange and cylinder block changes (Figure 17-29). A universal link keys the cylinder block to the drive shaft to maintain alignment and assure that they turn together. The link does not transmit force, except to accelerate and decelerate the cylinder block and to overcome resistance of the block revolving in the oil-filled housing.

The displacement of this pump varies between 0 and 30 degrees, depending on the offset angle (Figure 17-30).

Fixed displacement models (Figure 17-28) are usually available with 23 or 30 degree angles. In the variable displacement construction (Figure 17-31), a yoke with an external control is used to change the angle. With some controls, the yoke can be moved over center to reverse the direction of flow from the pump.

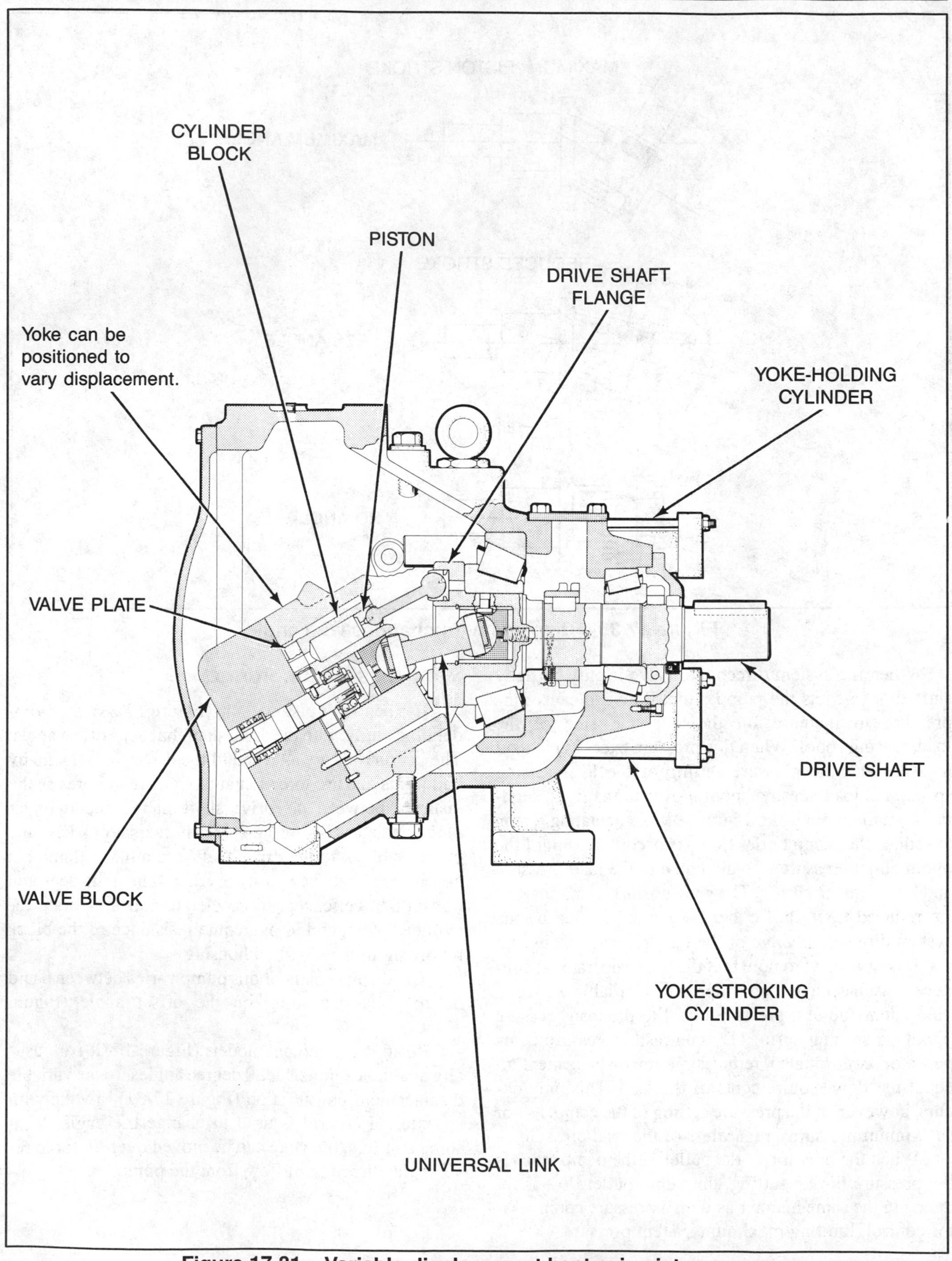

Figure 17-31. Variable displacement bent-axis piston pump.

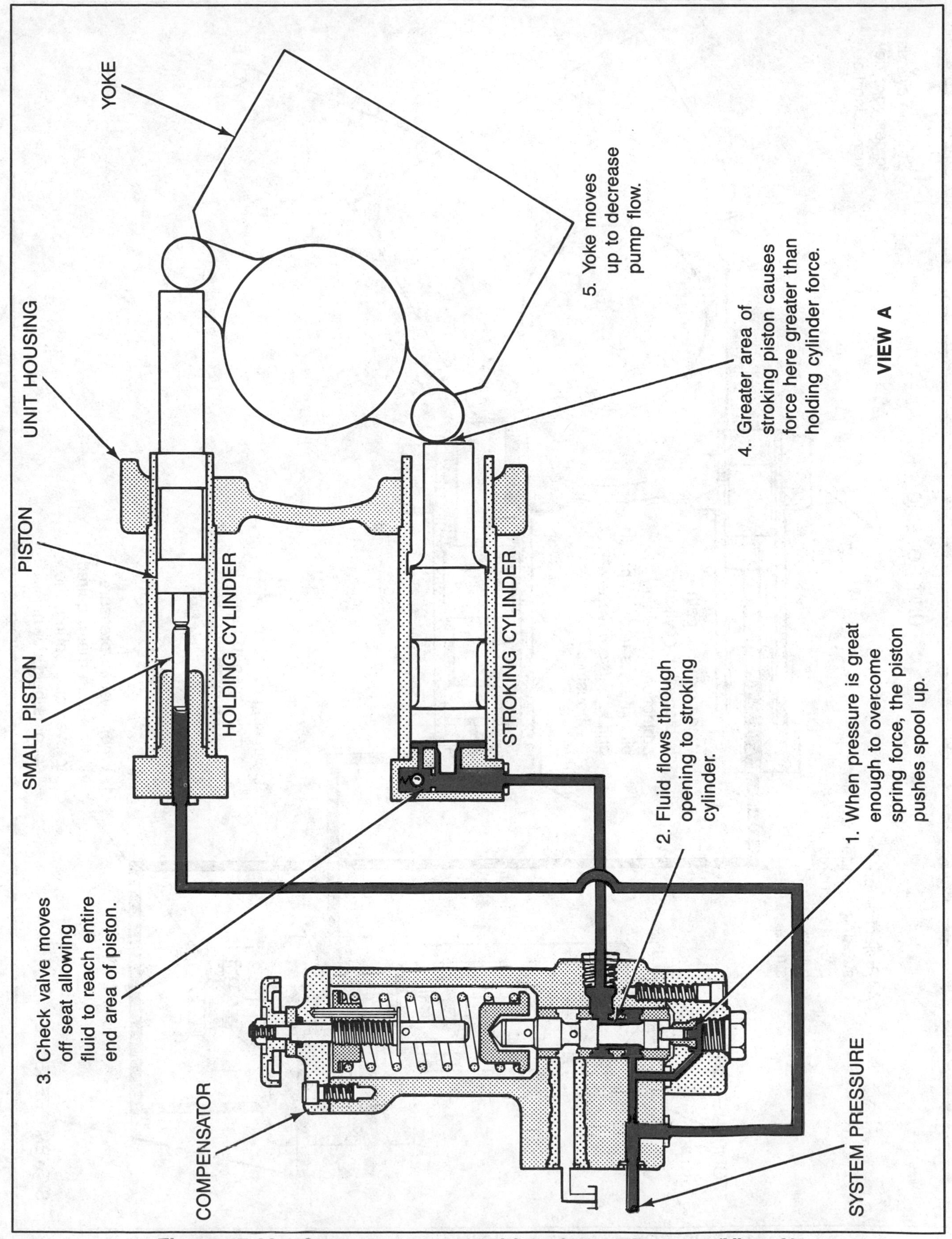
YOKE
UNIT HOUSING
PISTON
SMALL PISTON
HOLDING CYLINDER
STROKING CYLINDER
3. Check valve moves off seat allowing fluid to reach entire end area of piston.
COMPENSATOR
2. Fluid flows through opening to stroking cylinder.
1. When pressure is great enough to overcome spring force, the piston pushes spool up.
SYSTEM PRESSURE
5. Yoke moves up to decrease pump flow.
4. Greater area of stroking piston causes force here greater than holding cylinder force.
VIEW A

Figure 17-32. Compensator control for a bent-axis pump (View A).

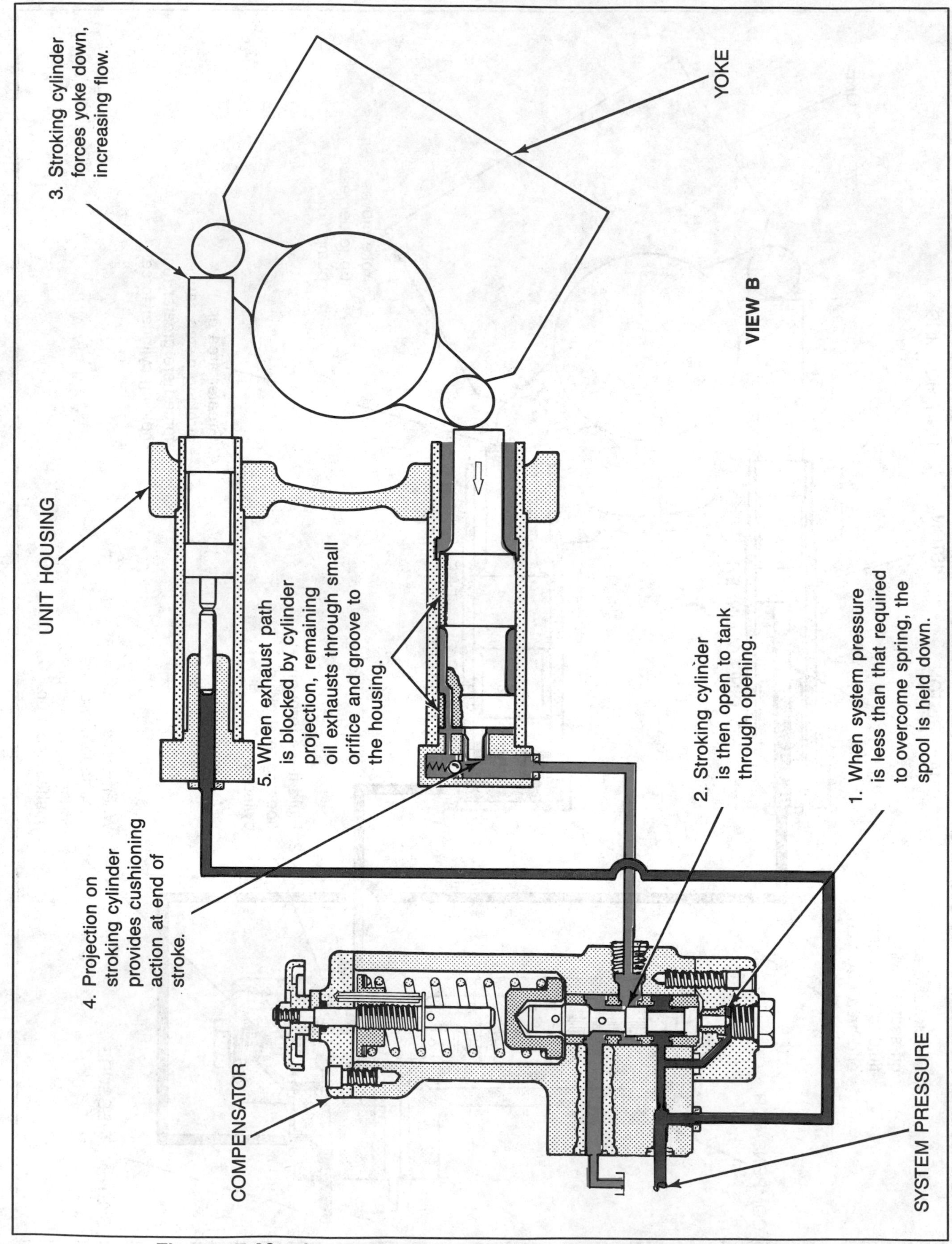

3. Stroking cylinder forces yoke down, increasing flow.
YOKE
VIEW B
UNIT HOUSING
5. When exhaust path is blocked by cylinder projection, remaining oil exhausts through small orifice and groove to the housing.
2. Stroking cylinder is then open to tank through opening.
1. When system pressure is less than that required to overcome spring, the spool is held down.
4. Projection on stroking cylinder provides cushioning action at end of stroke.
COMPENSATOR
SYSTEM PRESSURE

Figure 17-32. Compensator control for a bent-axis pump (View B).

Compensator Operation

Various methods are used to control variable displacement bent-axis pumps. Typical of these are the handwheel, pressure compensator, and servo.

Figure 17-32 shows a pressure compensator control for a bent-axis pump. In view A, the system pressure is sufficient to overcome the spring force of the compensator. As a result, the spool lifts, allowing fluid to flow into the stroking cylinder. Although the holding cylinder also has system pressure applied, the area of the stroking cylinder piston is much greater. So, to decrease flow, the force developed moves the yoke up.

View B shows the yoke moving down as system pressure drops below the level required to overcome the compensator spring force.

QUESTIONS

1. What are the basic characteristics of positive displacement pumps?
2. What are two ways of expressing pump size?
3. What is the purpose of decompression notches in a gear pump?
4. Which type of pump has automatic compensation for some of its wear?
5. What are two ways of altering volume levels in a gear pump?
6. What type of pumps are available in variable displacement models?
7. Explain the principle of a balanced design vane pump?
8. What holds the vanes extended in a vane pump.
9. What is the purpose of the intra-vane design?
10. What are two ways of positioning a yoke in a variable displacement, in-line pump?
11. How can displacement be varied in an axial piston pump?
12. What causes the pistons to reciprocate in an axial piston pump? In a bent-axis pump?
13. Why wouldn't a centrifugal pump be used to transmit pressure?
14. A 5 gpm (18.93 LPM) pump delivers 3.5 gpm (13.25 LPM) at 3000 psi (206.82 bar / 20,682.00 kPa). What is its volumetric efficiency?
15. What tends to limit the pressure capability of a gear pump?

CHAPTER 18
ACCESSORIES

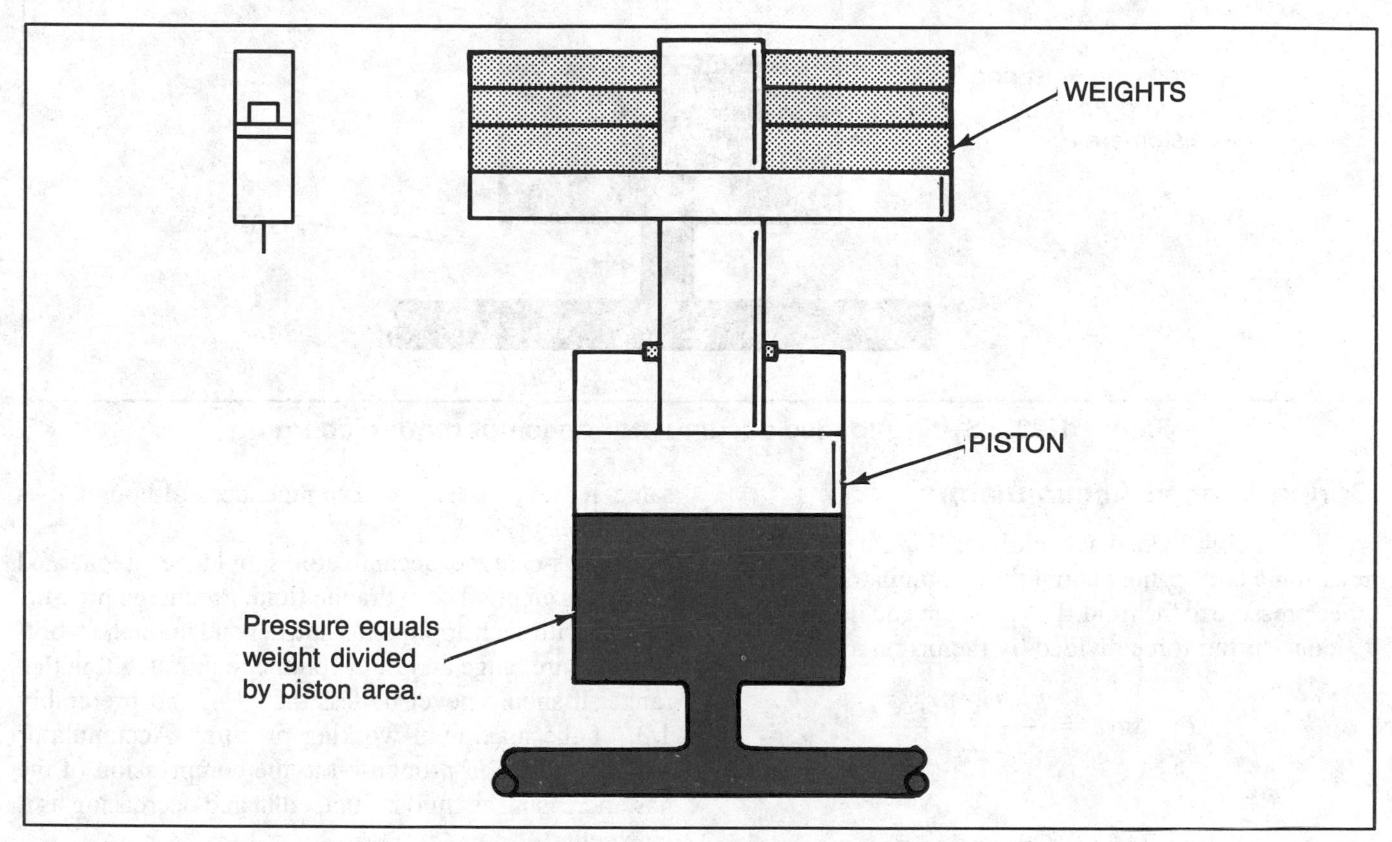
WEIGHTS
PISTON
Pressure equals
weight divided
by piston area.

Figure 18-1. Weighted accumulator produces constant pressure.

This chapter deals with the various accessories used to perform special functions in hydraulic systems: accumulators, pressure intensifiers, pressure switches, instruments for measuring pressures and flows, and sound damping devices.

ACCUMULATORS

Unlike gases, which are compressible and can be stored for a period of time, hydraulic fluids are usually incompressible. Accumulators provide a way to store these fluids under pressure.

Hydraulic fluid enters the accumulator chamber and acts on a piston or bladder area to either raise a weight or compress a spring or a gas. Any tendency for pressure to drop at the accumulator inlet forces fluid back out into the system.

Weight-Loaded Accumulator

The weight-loaded design (Figure 18-1) was the first type of accumulator built. Adding or removing weights on a vertical ram or piston varies the pressure, which is always equal to the weight imposed divided by the piston or ram area exposed to the hydraulic fluid. This is the only type of accumulator where pressure is constant, whether the chamber is full or nearly empty. Weight-loaded accumulators are heavy and bulky, making their use limited. They are generally found on heavy presses where constant pressure is required, or in applications where unusually large volumes are necessary.

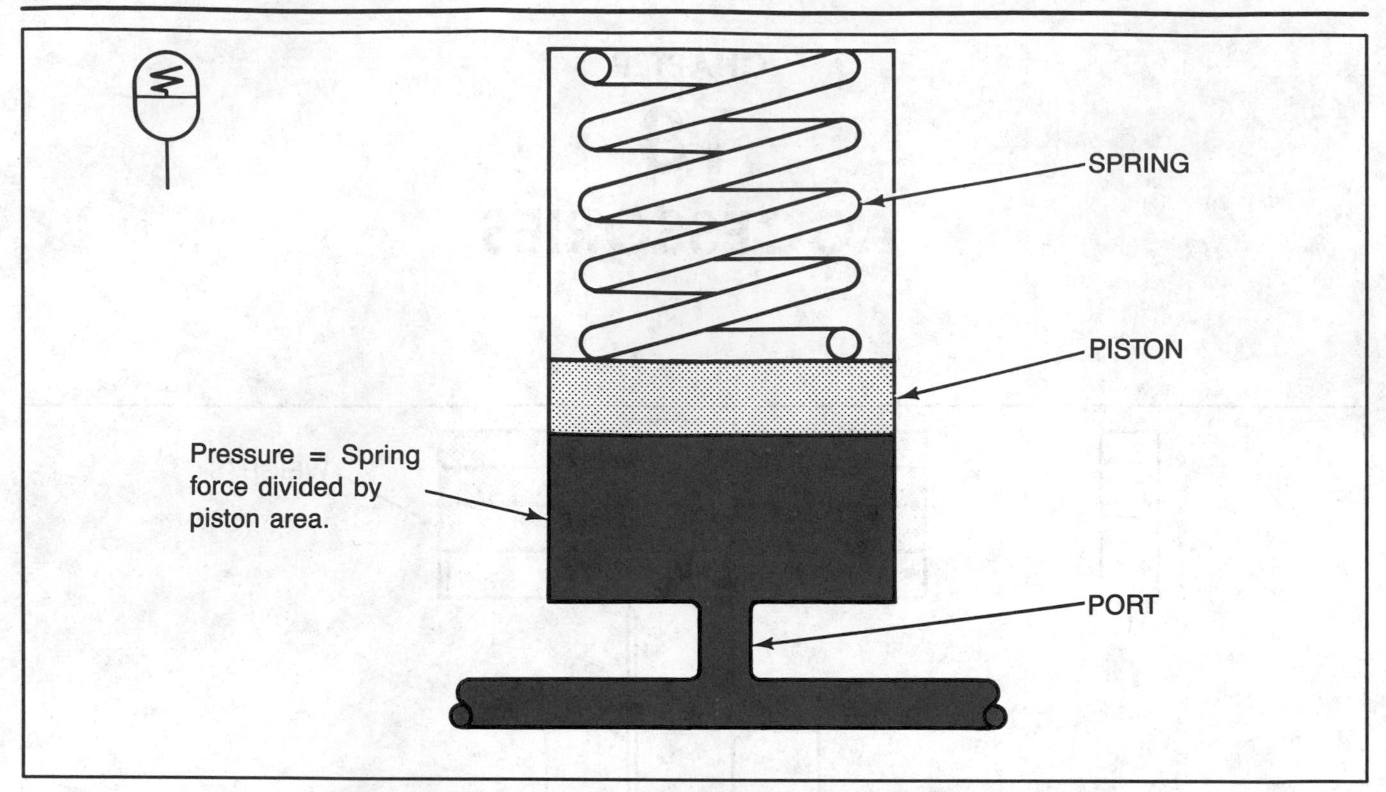
SPRING
PISTON
Pressure = Spring force divided by piston area.
PORT

Figure 18-2. Spring-loaded accumulator does not require charging.

Spring-Loaded Accumulators

In a spring-loaded accumulator (Figure 18-2), compressing a coil spring behind the accumulator piston applies pressure to the fluid. The pressure equals the instantaneous spring force divided by the piston area.

$$Pressure = \frac{Spring\ Force}{Area}$$

Where:

$$Spring\ Force = Spring\ Constant\ x\ Compression\ Distance$$

Since the spring force increases as fluid enters the changer and decreases as it is discharged, the pressure is not constant.

Spring-loaded accumulators can be mounted in any position. However, the spring force (the pressure range) is not easily adjusted. Where large quantities of fluid are required, the forces involved makes spring sizes impractical.

Gas-Charged Accumulator

The most commonly used accumulator has a chamber that is precharged with an inert gas, usually dry nitrogen. Oxygen should never be used because of its tendency to burn or explode when compressed with oil. For the same reason, air is not recommended, although it is sometimes used.

A gas-charged accumulator should be precharged when it is emptied of hydraulic fluid. Precharge pressures vary with each application and depend upon the working pressure range and fluid volume required within that range. It should never be less than 1/4, and preferably 1/3, of the maximum working pressure. Accumulator pressure varies in proportion to the compression of the gas, increasing as fluid is pumped in and decreasing as it is expelled.

Piston Type Accumulator. Using a free piston (Figure 18-3) is another method of separating the gas charge from the hydraulic fluid. Similar in construction to a hydraulic cylinder, the piston is under gas pressure on one side and constantly tries to force the oil out of the opposite side of the chamber. Here too, pressure is a function of the compression and varies with the volume of oil in the chamber.

Diaphragm or Bladder Type. Many accumulators incorporate a synthetic rubber diaphragm or bladder (Figure 18-4) to contain the gas precharge and separate it from the hydraulic fluid. Since certain fire-resistant fluids may not be compatible with conventional diaphragm or bladder materials, making the proper selection is important.

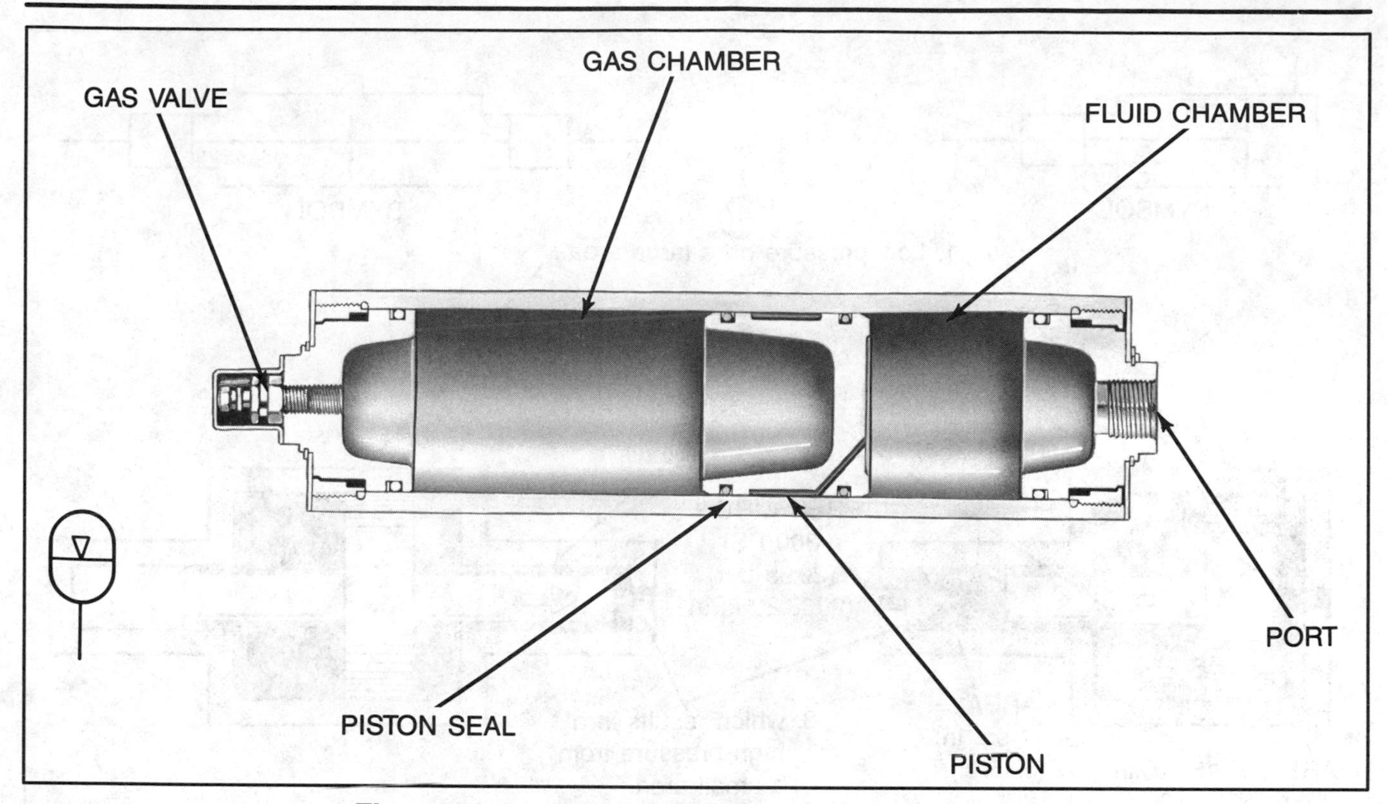

Figure 18-3. Piston accumulator is gas-charged.

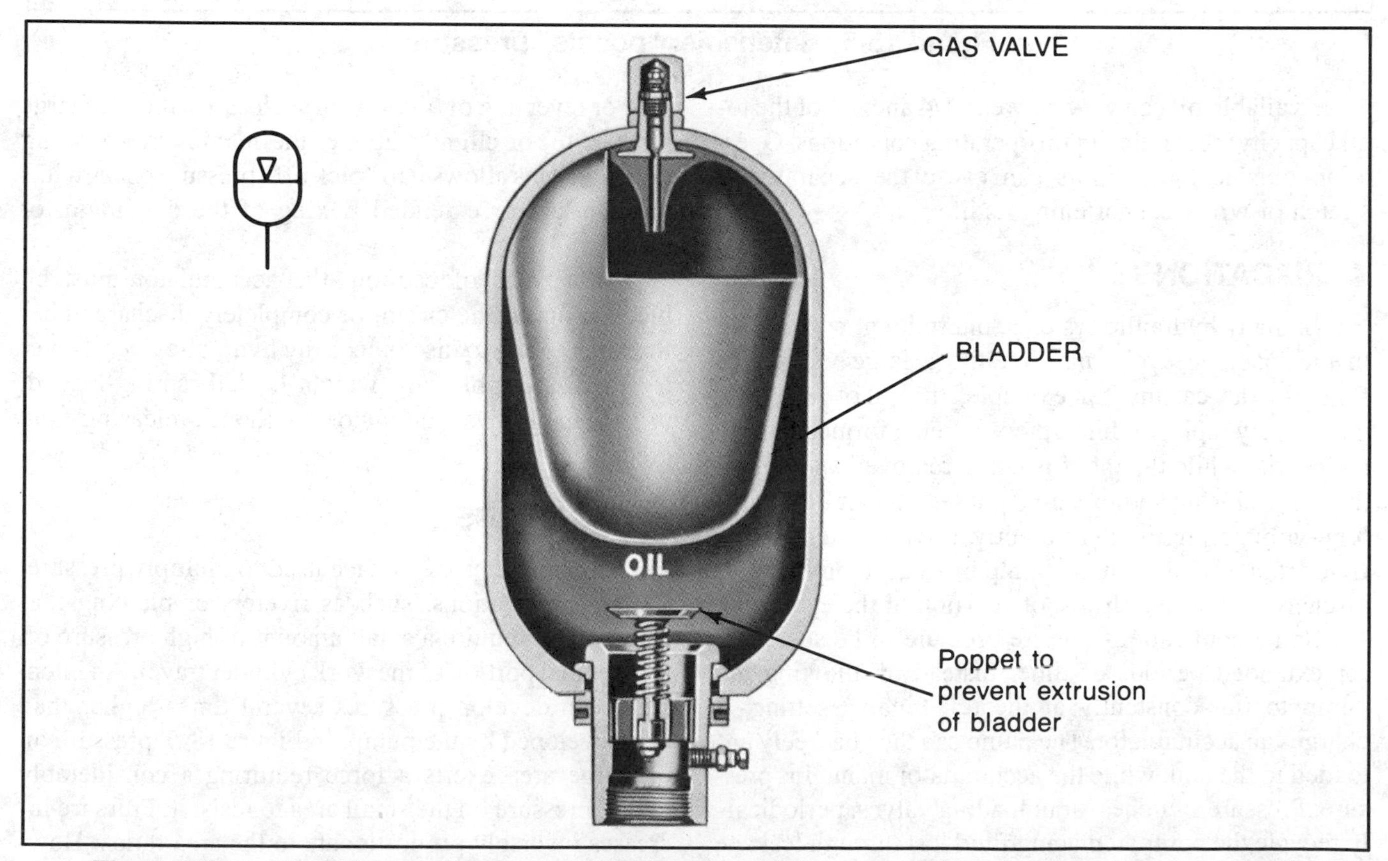

Figure 18-4. Bladder-type accumulator uses rubber separator between gas and liquid.

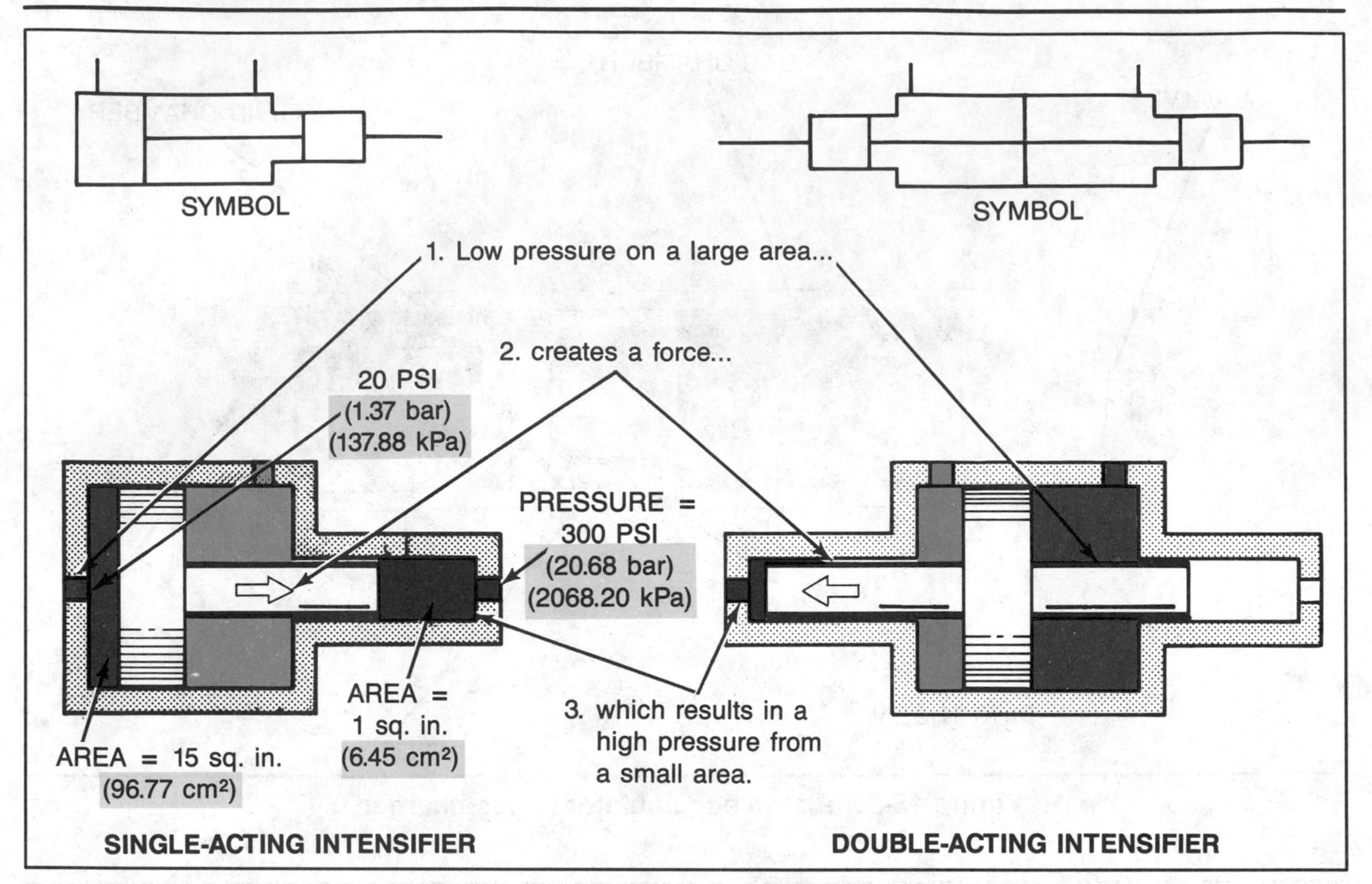

Figure 18-5. Intensifier "boosts" pressure.

Available oil can vary between 1/4 and 3/4 of the total capacity, depending upon operating conditions. Operation outside these limits can cause the separator to stretch or wrinkle, shortening its life.

APPLICATIONS

In many hydraulic systems, intermittent work, done in the machine cycle, may require a large volume of fluid. In die casting for example, the "shot" cylinder moves very rapidly while a piece is being formed, but remains idle while the piece is being removed and during the mold-closing and -opening phases. Rather than use a high-volume pump intermittently, such a system stores fluid from a small-volume pump in an accumulator and discharges it during the "shot" portion of the cycle.

Some applications require pressure to be sustained for extended periods of time. Instead of allowing the pump to run constantly at the relief valve setting, it charges an accumulator. The pump can then be freely unloaded to the tank while the accumulator maintains pressure. Pressure switches, or unloading valves, periodically recycle the pump, replacing fluid lost through leakage or valve actuation.

Accumulators may also be installed in a system to absorb shock or pressure surges due to the sudden stopping or reversing of oil flow. In such cases, the precharge is close to, or slightly above, the maximum operating pressure. This allows it to "pick off" pressure peaks without constant or extended flexing of the diaphragm or bladder.

As a word of caution, the accumulator must be blocked out of the circuit or completely discharged before attempting to disconnect any hydraulic lines. Never try to disassemble any weight-loaded, spring-loaded, or gas-charged accumulator without releasing the precharge.

INTENSIFIERS

An intensifier is a device used to multiply pressure. Certain applications, such as riveters or piercing machines, may require a small amount of high pressure oil for the final portion of the work cylinder travel. An intensifier can develop pressures several times higher than that developed by the pump. In Figure 18-5, pressure on the large area exerts a force requiring a considerably higher pressure on the small area to resist it. Pressure increases inversely proportionate to the area ratios. However, the volume of fluid discharged at high pressure will be proportionately less than that required at the large end.

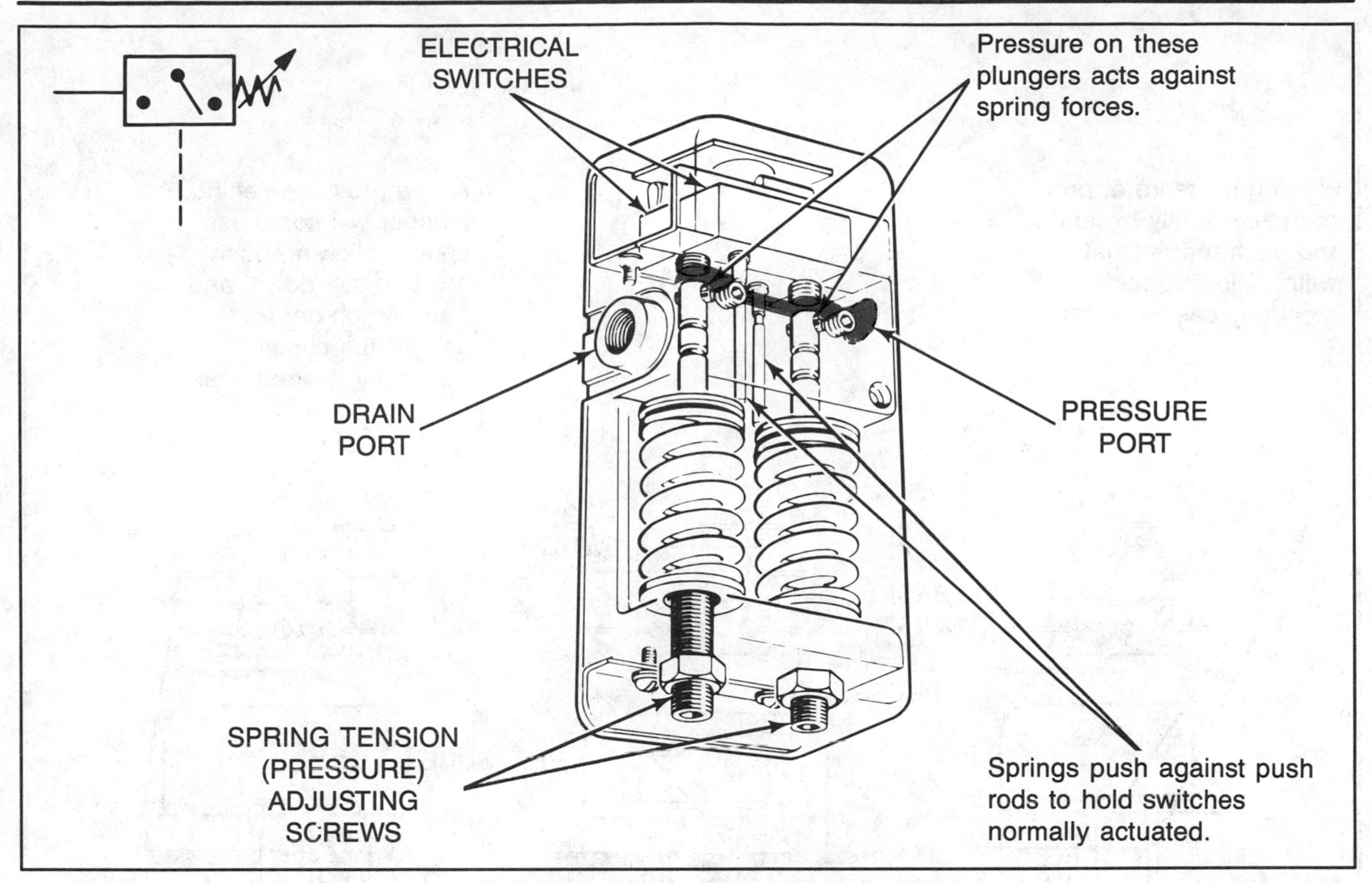

Figure 18-6. Typical pressure switch.

PRESSURE SWITCHES

Pressure switches (Figure 18-6) open or close electrical circuits at selected pressures, actuating solenoid-operated valves or other devices used in the system.

Figure 18-7 shows the operating principle of a pressure switch. A push rod operates each of the two electrical switches by bearing against a plunger that is controlled by hydraulic and spring forces. Turning the adjusting screw to increase or decrease the spring force, alters the operating pressure of each switch.

In this design, the springs actuate the switches when the unit is assembled. This leaves thc normally open contacts closed and vice versa.

When the preset pressure is reached, the plunger compresses the spring, allowing the push rods to move down. This causes the snap-action switches to revert to their normal condition. Using both switches in conjunction with an electrical relay maintains system pressures within widely variable high and low ranges.

INSTRUMENTS

Flow rate, pressure, and temperature measurements are required to evaluate the performance of hydraulic components. These factors are also helpful in setting up or troubleshooting a hydraulic system. Due to the difficulty of installing a flow meter in the circuit, flow measurements are often determined by timing the travel or rotation of an actuator. Gauges and thermometers determine pressure and temperature.

Pressure Gauges

Pressure gauges adjust pressure control valves to required values. They also determine the forces being exerted by a cylinder or the torque of a hydraulic motor.

Two principal types of pressure gauges arc the Bourdon Tube and Schrader types. The Bourdon Tube gauge (Figure 18-8) is a sealed tube formed in an arc. When pressure is applied at the port opening, the tube tends to straighten, actuating linkage to the pointer gear and moving the pointer to indicate the pressure on a dial.

With the Schrader Gauge (Figure 18-9), pressure is applied to a spring-loaded sleeve and piston. When pressure moves the sleeve, it actuates the gauge needle through the linkage.

Most pressure gauges read zero at atmospheric pressure. They are calibrated in pounds per square inch (bar / kilopascals), ignoring atmospheric pressure throughout their range.

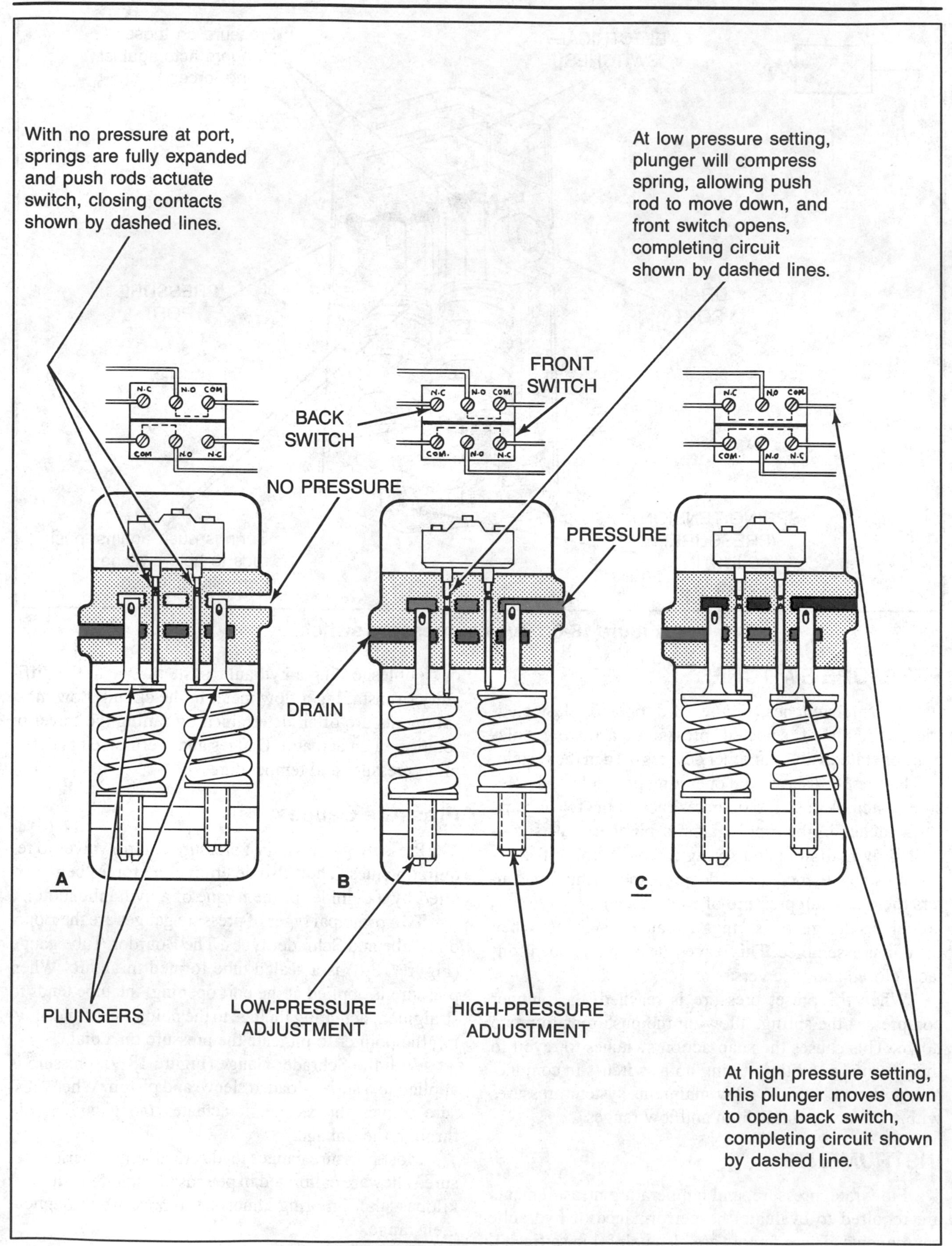

With no pressure at port, springs are fully expanded and push rods actuate switch, closing contacts shown by dashed lines.
At low pressure setting, plunger will compress spring, allowing push rod to move down, and front switch opens, completing circuit shown by dashed lines.
N.C
N.O
COM
FRONT SWITCH
BACK SWITCH
NO PRESSURE
PRESSURE
DRAIN
A
B
C
PLUNGERS
LOW-PRESSURE ADJUSTMENT
HIGH-PRESSURE ADJUSTMENT
At high pressure setting, this plunger moves down to open back switch, completing circuit shown by dashed line.

Figure 18-7. Pressure switch operation.

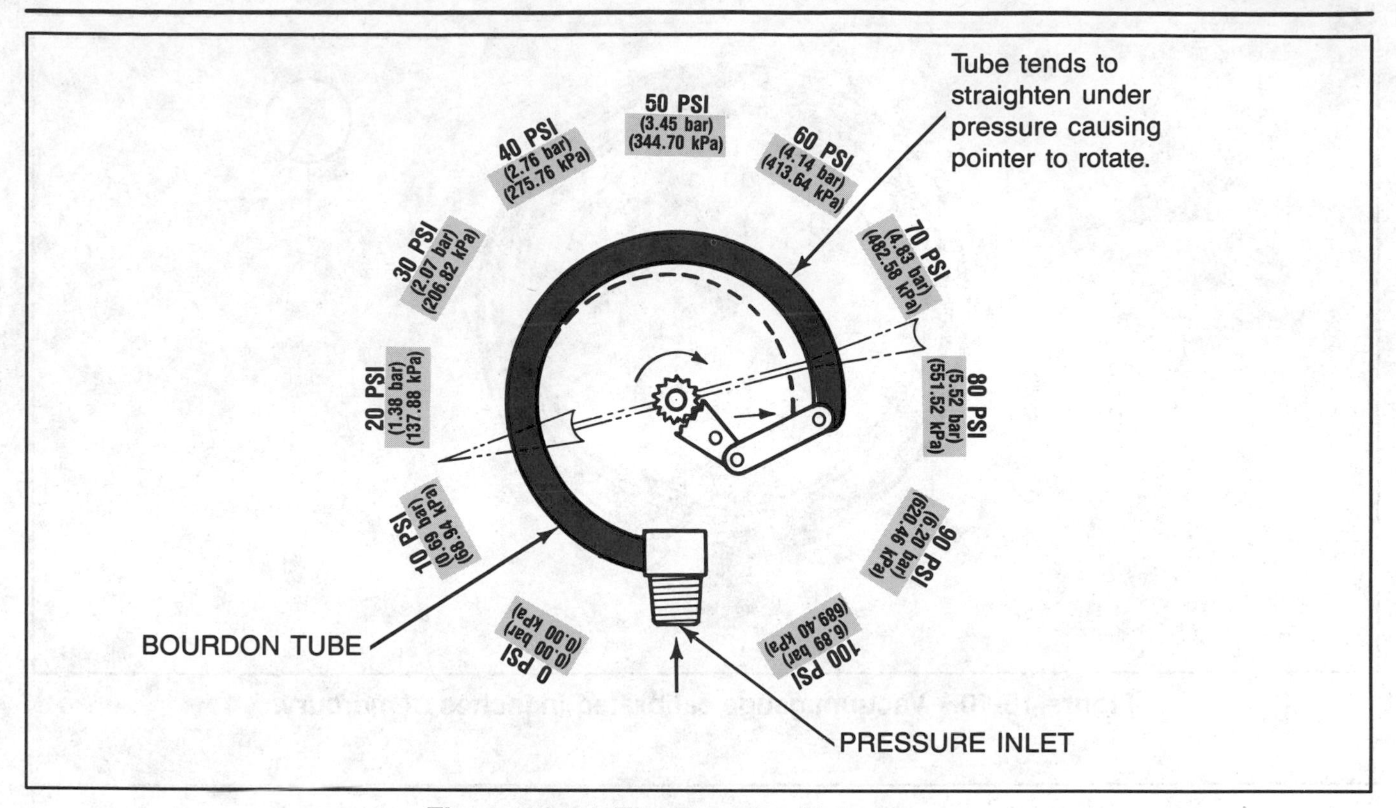

Figure 18-8. The Bourdon Tube gauge.

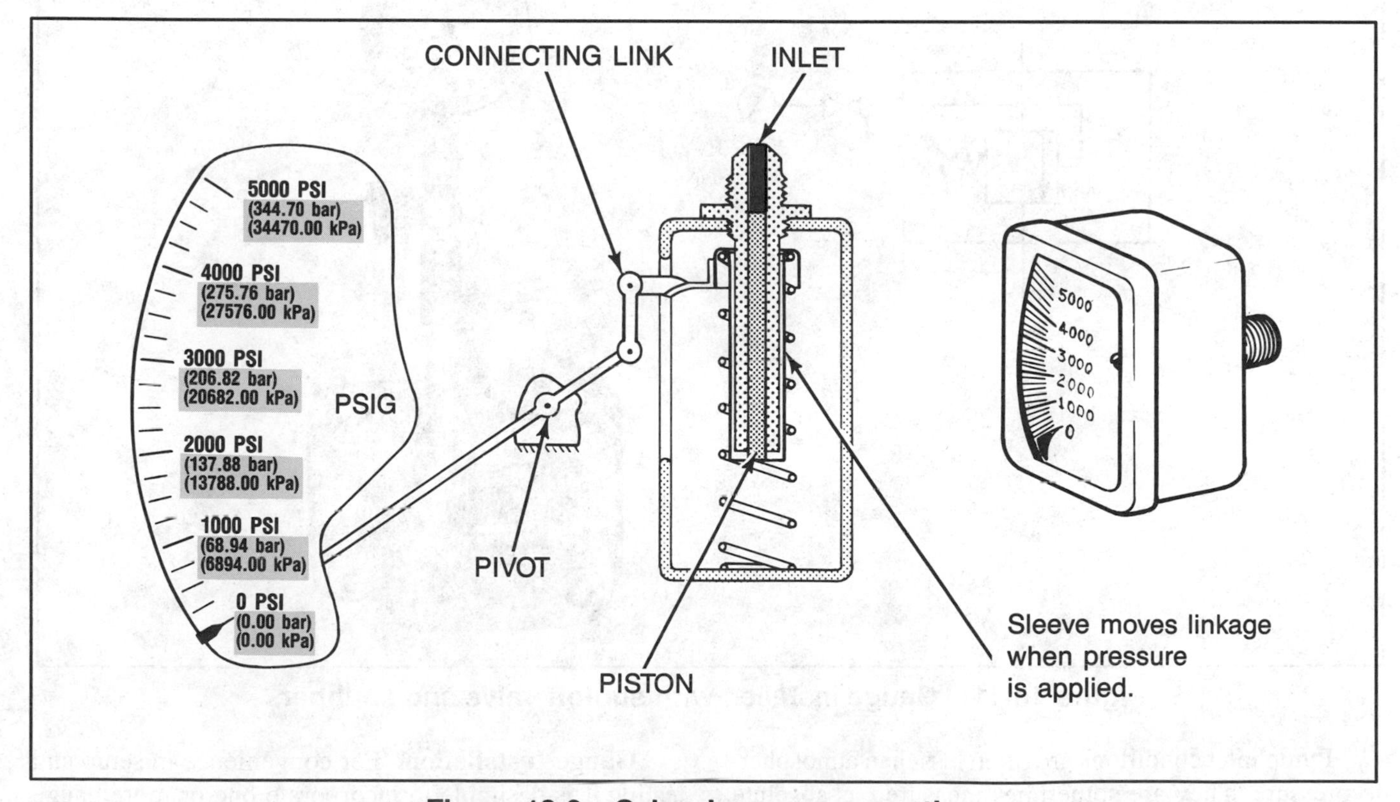

Figure 18-9. Schrader gauge operation.

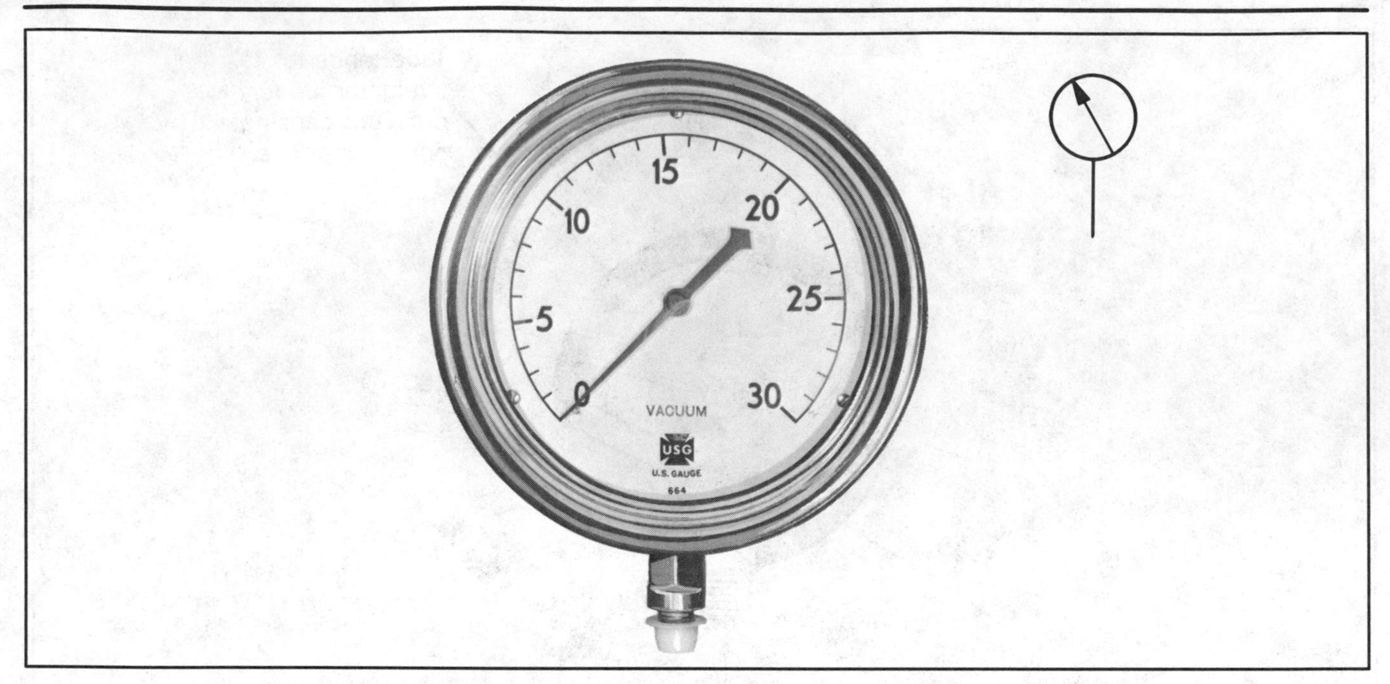

Figure 18-10. Vacuum gauge calibrated in inches of mercury.

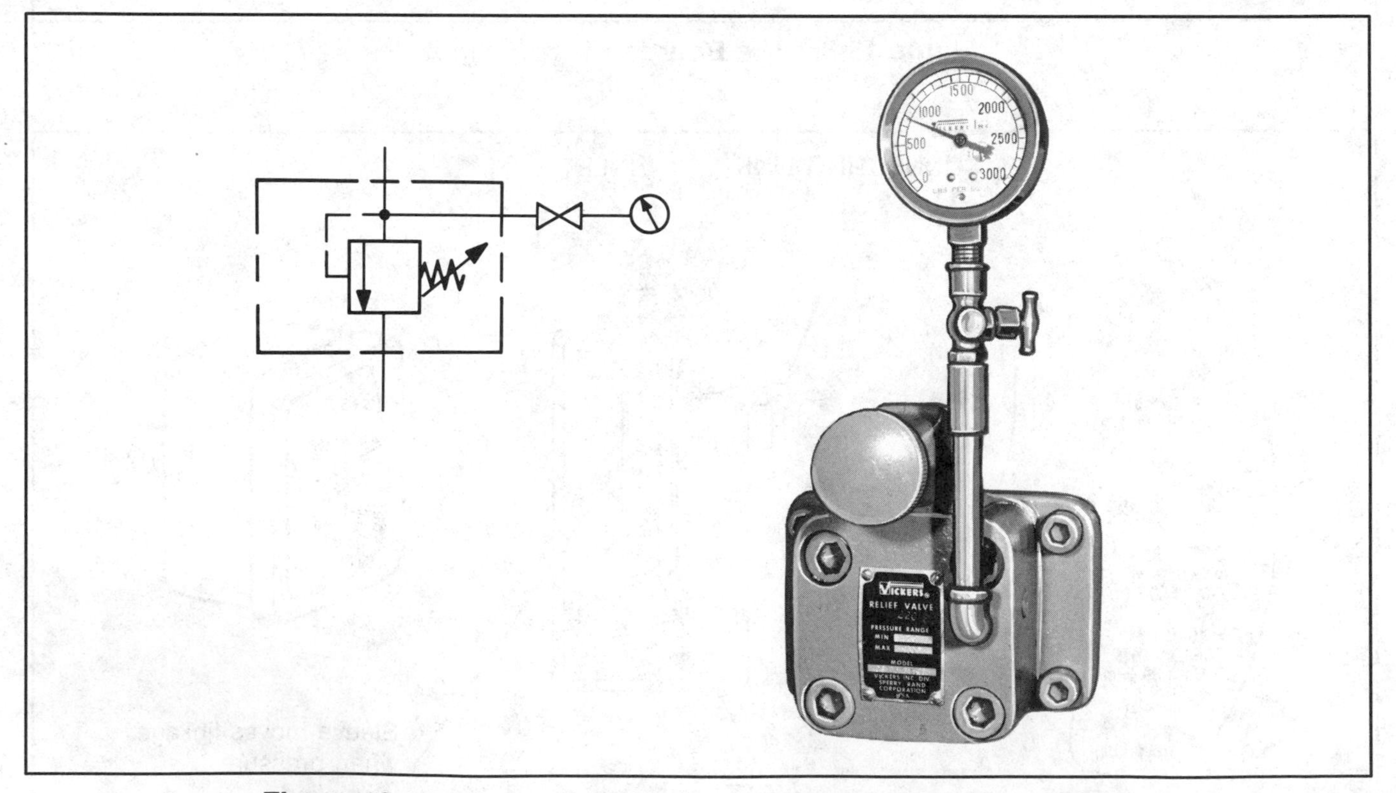

Figure 18-11. Gauge installed with shutoff valve and snubber.

Pump inlet conditions are often less than atmospheric pressure. They are sometimes measured as absolute pressure (psia (bar / kPa)), but are more often calibrated in inches of mercury (millimeters of mercury). Thirty inches of mercury (762 mm of mercury) is considered a perfect vacuum. Figure 18-10 shows a vacuum gauge calibrated in inches of mercury.

Gauge Installation. For convenience in setup and testing,it is desirable to incorporate one or more gauge connections in a hydraulic system. This is true even though gauge ports are included in most relief valves and in some other hydraulic components.

When a gauge is permanently installed on a machine, a shutoff valve and snubber (Figure 18-11) are

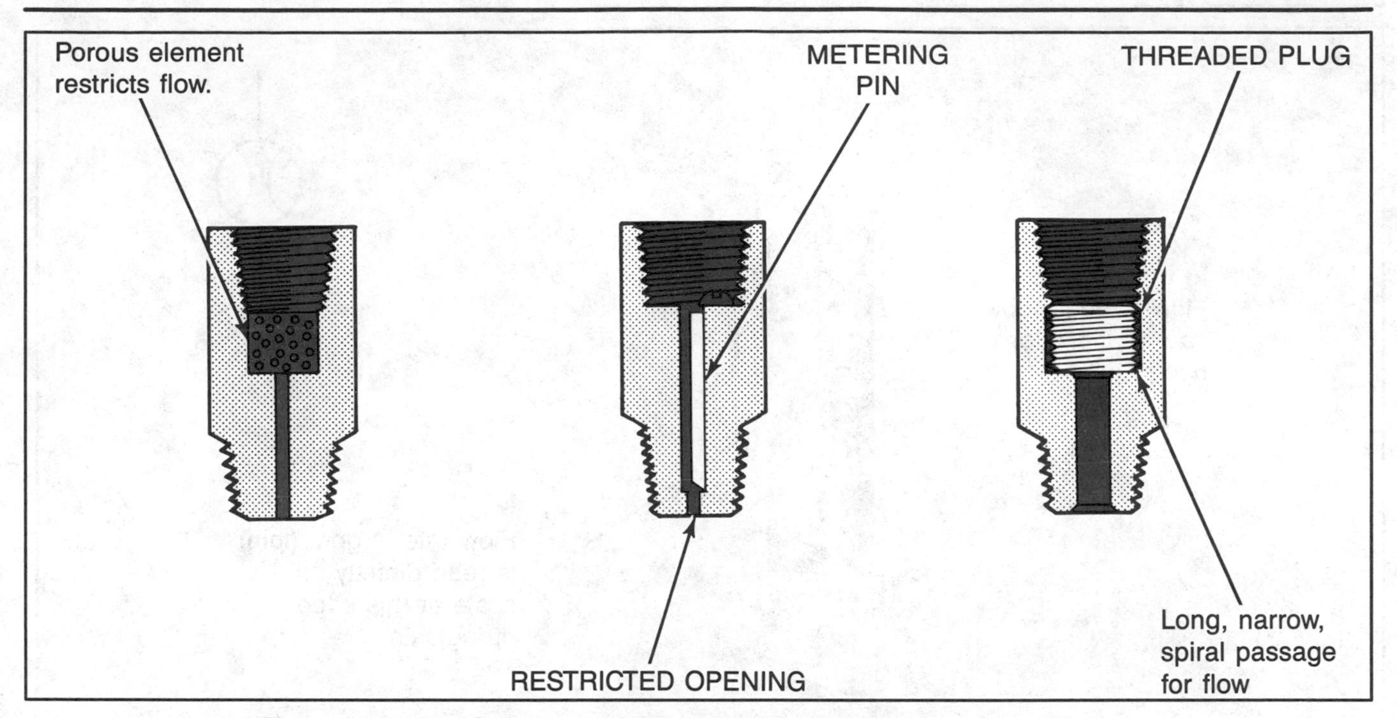

Figure 18-12. Gauge snubbers guard against pressure surges.

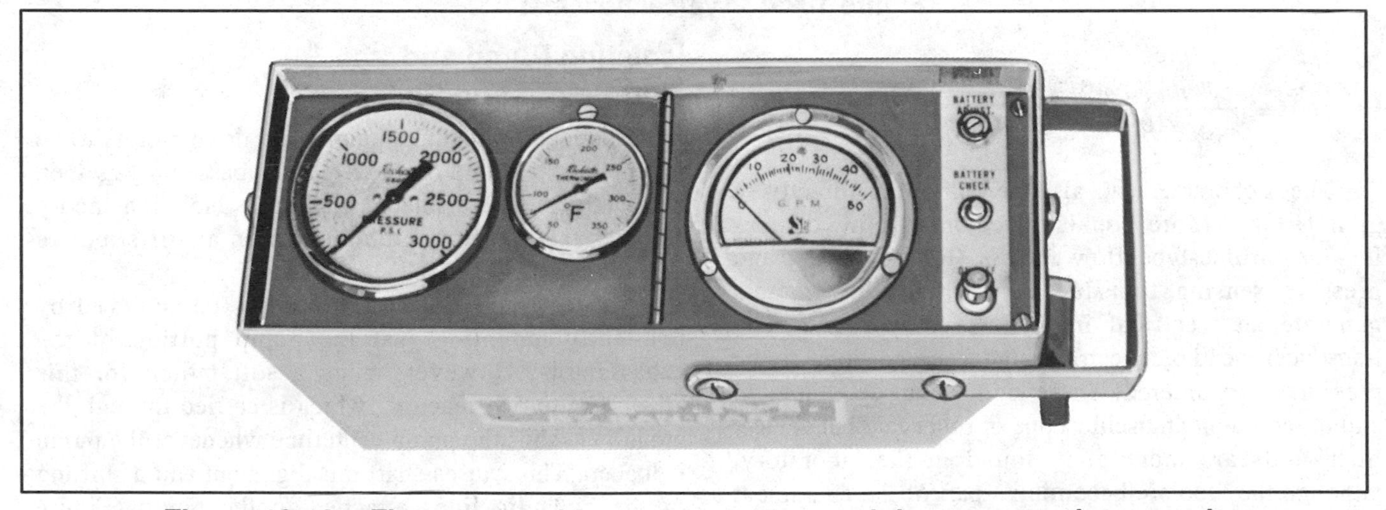

Figure 18-13. Flow meter with pressure gauge and thermometer in one unit.

usually installed with it. The shutoff valve prolongs gauge life by isolating it from the system, except when making a reading.

The snubber (Figure 18-12) prevents the gauge from oscillating and protects it from pressure surges. A small coil (approximately two inches (51 mm) in diameter) of 1/8 inch (3 mm) tubing makes an excellent gauge-damping device when commercially made units are not available.

Flow Meters

Flow meters are usually found on test stands, but portable units are available. Some include the flow meter, pressure gauge, and thermometer in a single unit (Figure 18-13). They are seldom connected permanently on a machine. Coupled into the hydraulic piping, flow meters help to check the volumetric efficiency of a pump and determine leakage paths within the circuit.

A typical flow meter (Figure 18-14) consists of a weight in a calibrated vertical tapered tube. Oil pumps into the bottom and out the top, raising the weight to a height proportional to the flow. For more accurate measurement, a fluid motor of known displacement can be used to drive a tachometer. The gpm (LPM) flow is:

$$gpm = \frac{rpm \times displacement\ (in^3/rev)}{231}$$

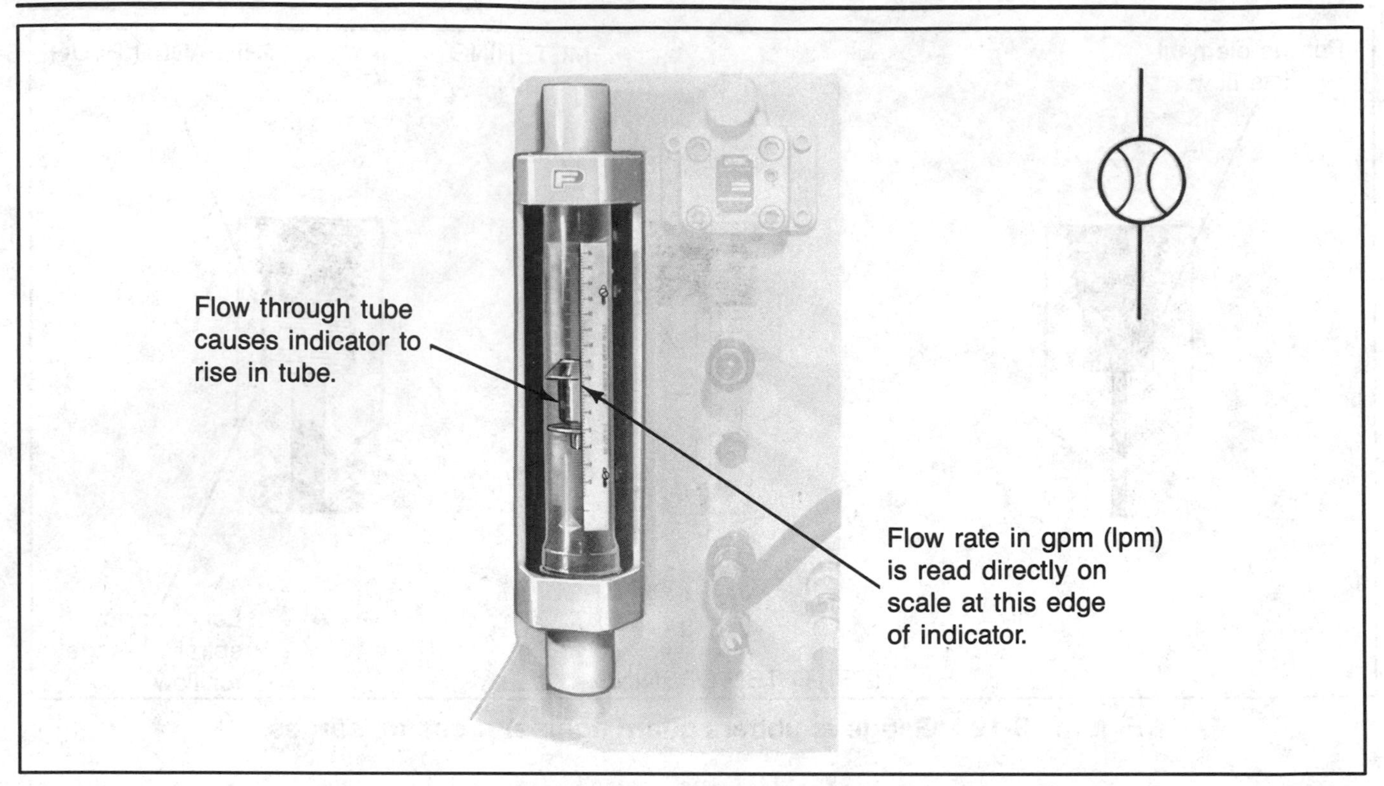
Flow through tube
causes indicator to
rise in tube.
Flow rate in gpm (lpm)
is read directly on
scale at this edge
of indicator.

Figure 18-14. Typical flow meter.

$$LPM = \frac{rev/min \times cm^3/rev}{1000}$$

The tachometer can also be calibrated directly in gpm (LPM). More sophisticated measuring devices include turbine-type flow meters (Figure 18-15) and pressure sensing transducers. Turbine flow meters generate an electrical impulse as they rotate. The transducers send out electrical signals proportional to the pressures encountered. These signals can be calibrated and observed on an oscilloscope or other readout device. Such units are more often found in the laboratory, although they too are becoming a part of the equipment used by fluid power technicians in setting up and maintaining equipment.

SOUND DAMPING DEVICES

Today's machines operate at higher pressures than those of past years. These higher pressures create elevated noise levels, which can be minimized by using sound-damping devices.

These include isolating pumps and electric motors from their mounting plates, using proper drive couplings, and using hoses for noise isolation.

Isolating Pump and Motor Mounting Plates

This method requires pumps and drive motors to be mounted on a common base. This subassembly is then resiliently mounted on the machine. Isolation theory assumes that isolators are mounted on a stiff structure (Figure 18-16).

Additional control can sometimes be achieved by resiliently mounting just the pump portion of the subassembly. However, using a soft mount for this allows the torque reaction, which is carried through the mounts, to shift the pump centerline whenever the pump is loaded. This causes shaft misalignment and a shifting of the hydaulic lines attached to the pump (Figure 18-17).

Drive Couplings

Drive couplings (Figure 18-18) provide vibration isolation between the pump and its drive and reduce the effect of pump misalignment.

For isolation purposes, couplings with rubber-like material in the drive train are favored. However, many units are all metal and provide adequate isolation. Pump misalignment can cause noise by producing high loads that must be carried by the pump and motor bearings. When good alignment practices are followed, almost any commercial flexible couplings will accommodate the misalignments that occur without such loading. When good alignment cannot be provided, or when the torque

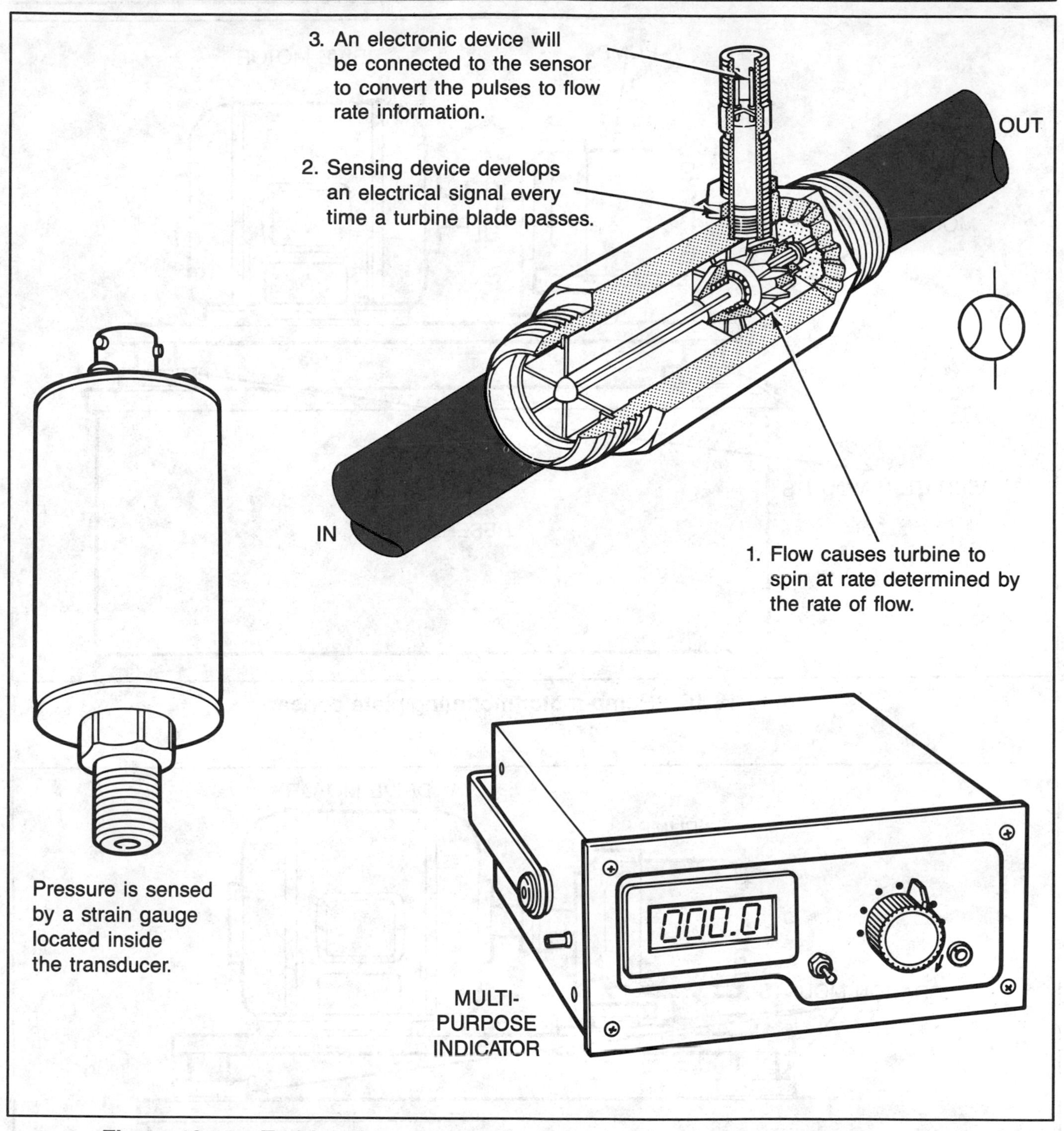

3. An electronic device will be connected to the sensor to convert the pulses to flow rate information.
2. Sensing device develops an electrical signal every time a turbine blade passes.
OUT
IN
1. Flow causes turbine to spin at rate determined by the rate of flow.
Pressure is sensed by a strain gauge located inside the transducer.
000.0
MULTI-PURPOSE INDICATOR

Figure 18-15. Turbine flow meter and pressure transducer with digital readout.

reaction causes misalignment because only the pump is resiliently mounted, two couplings separated by a short shaft should be used.

Isolation With a Hose

When the noise source is resiliently mounted, a flexible hose must be used to accomplish isolation (Figure 18-19). Stiff lines, attached to the noisemaker, will interfere with the isolating action of the mounts.

If used incorrectly, however, flexible lines will actually increase noise rather than decrease it.

Since the hose is so responsive to fluid pulsations, it can be a strong sound radiator if long lengths are used. When bent, the hose acts like the Bourdon Tube, straightening as pressure generates force. These pressure pulsations convert into cyclic forces that can cause vibration in the lines and other machine elements. Similarly, pressure changes the length of the hose. If such changes are

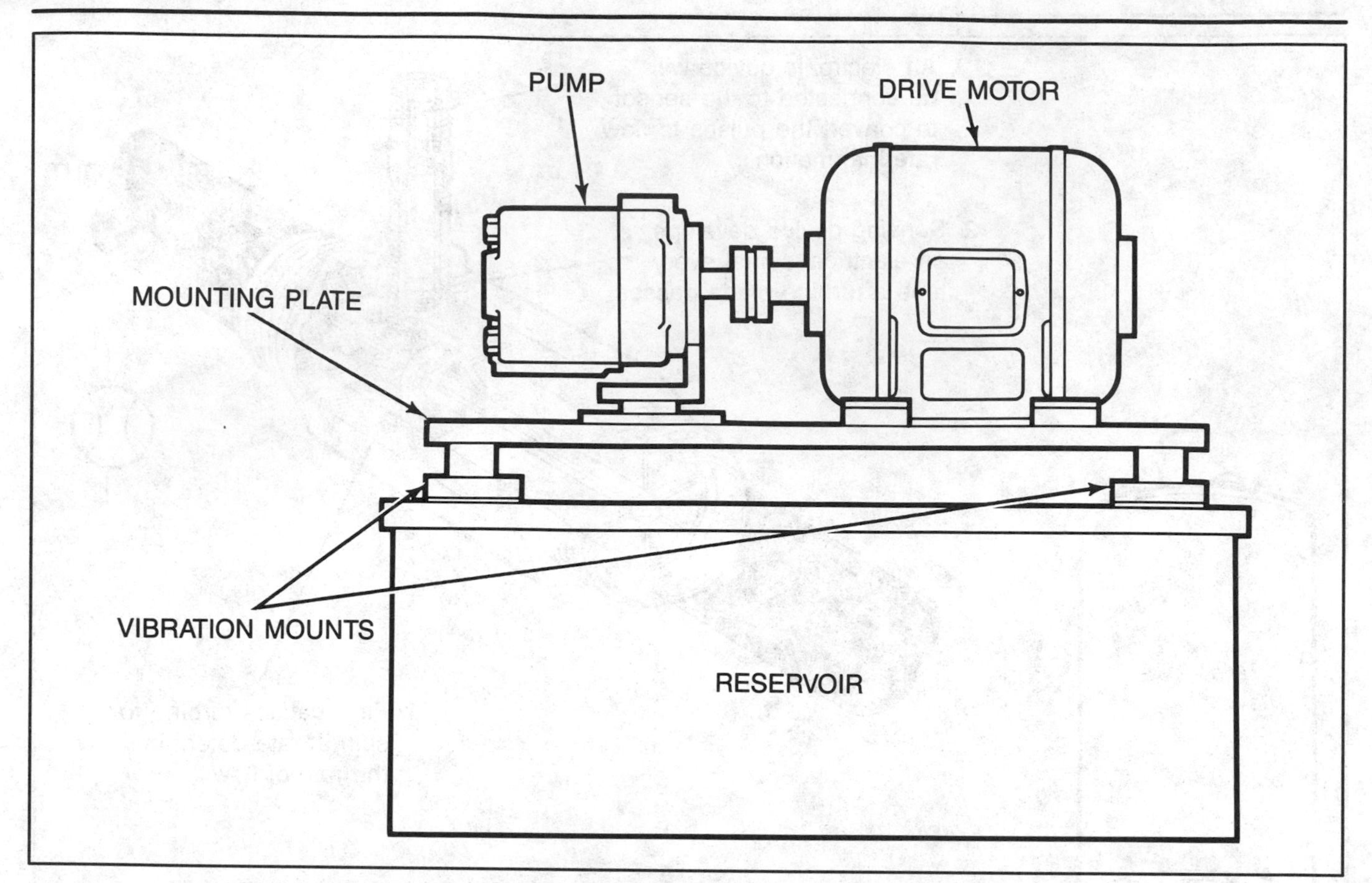

Figure 18-16. Pump-motor mounting plate concept.

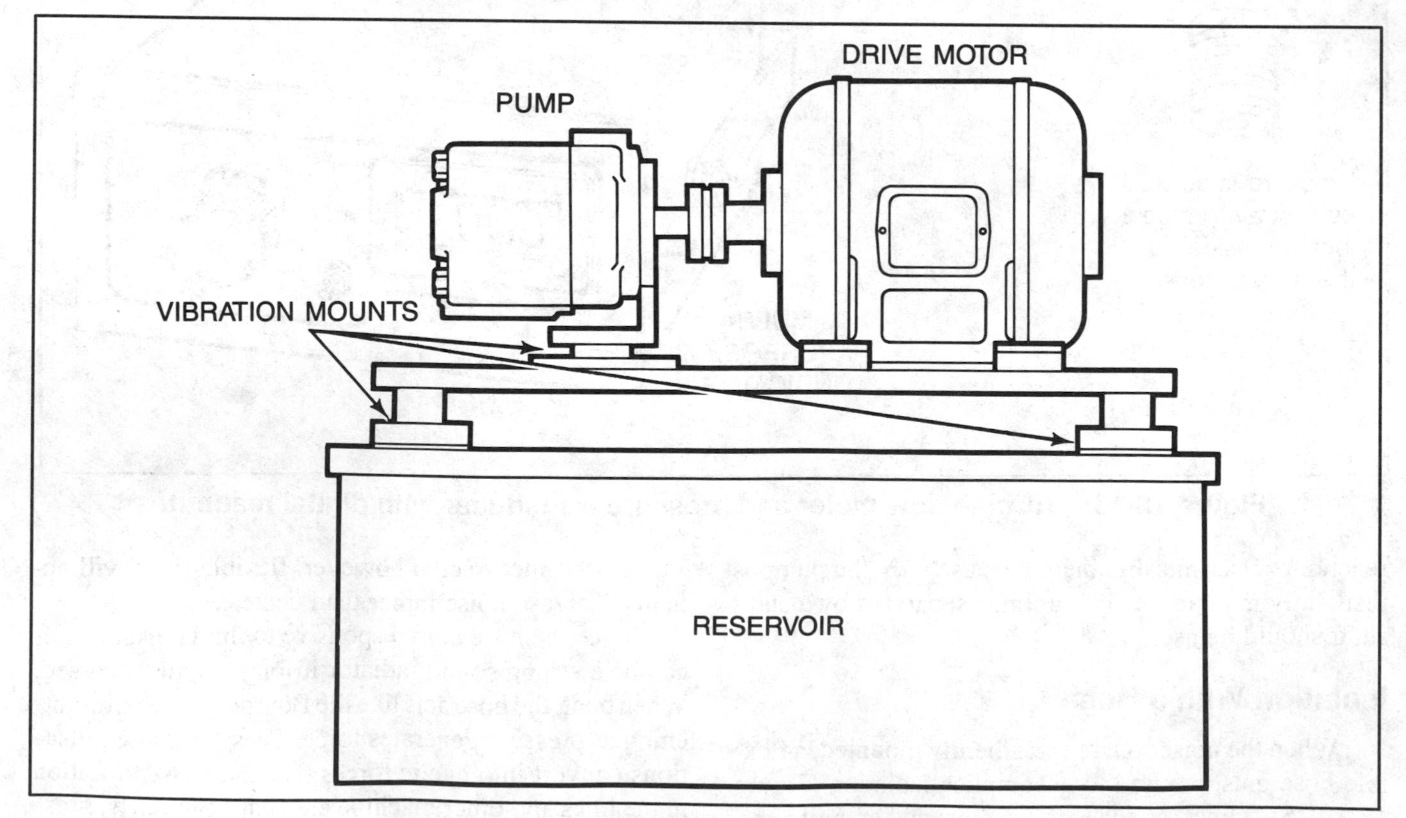

Figure 18-17. Additional isolation of pump.

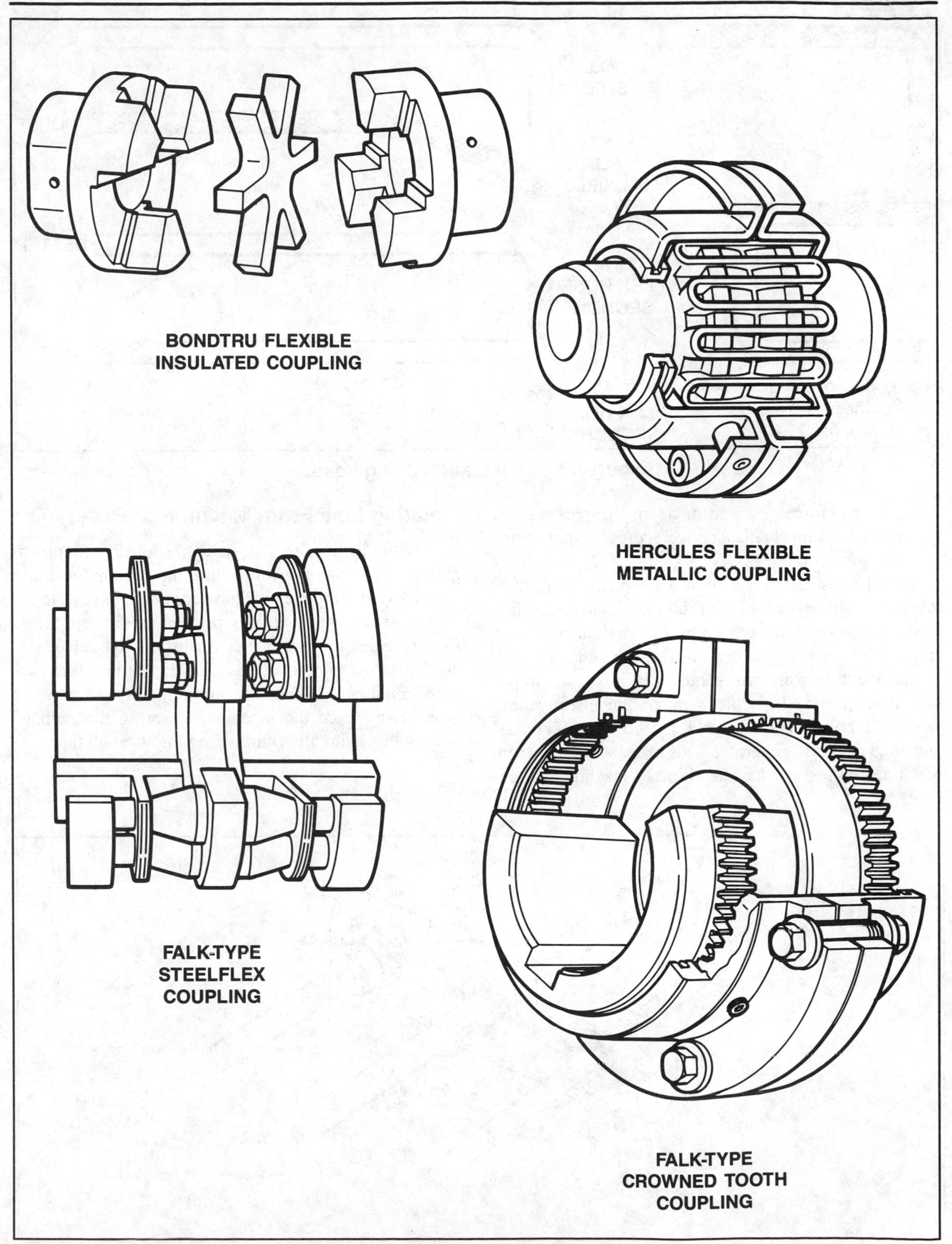

Figure 18-18. Acceptable drive couplings.

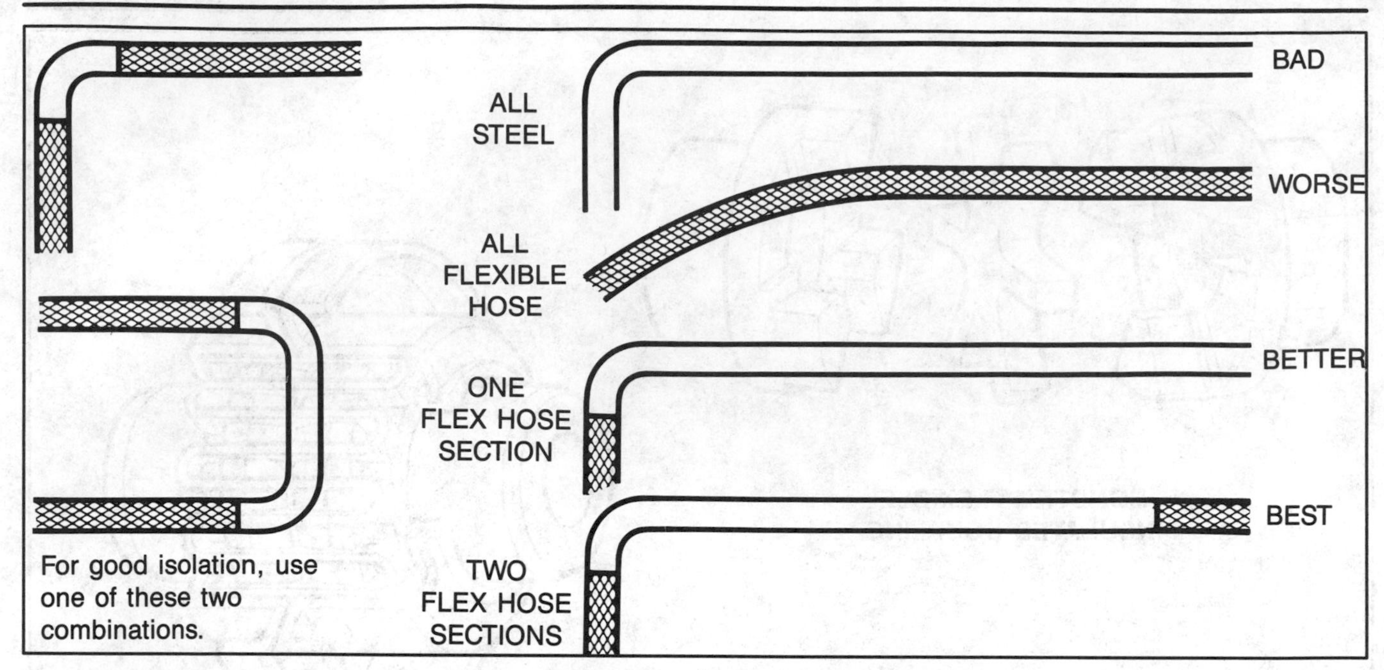

Figure 18-19. Isolation using hose.

unrestrained, forces proportionate to the pressure are generated. This mechanism converts pressure pulsations into vibration forces.

The best isolation is achieved when two short, flexible hoses, either parallel or at right angles to each other, are joined by rigid line or fittings. In a straight-run line, a single, flexible hose located at the noise generator may be sufficient. In some cases, however, it is better to put one at each end. To achieve maximum effectiveness when one hose section is used, the hose should be attached to the noise generator. This keeps vibrations from reaching the rest of the line or other machine components.

Isolating Line From Machine to Reservoir

Hydraulic lines are frequently responsible for propagating noise energy (line vibrations) from the noise source to the components. These components react to the energy, and radiate sound. To prevent the line vibrations from reaching the machine elements, the line can be isolated using commercially available, resilient line supports (Figure 18-20). These supports suspend tubes or hoses away from the machine, lowering the radiated noise when hydraulic fluid is flowing through them.

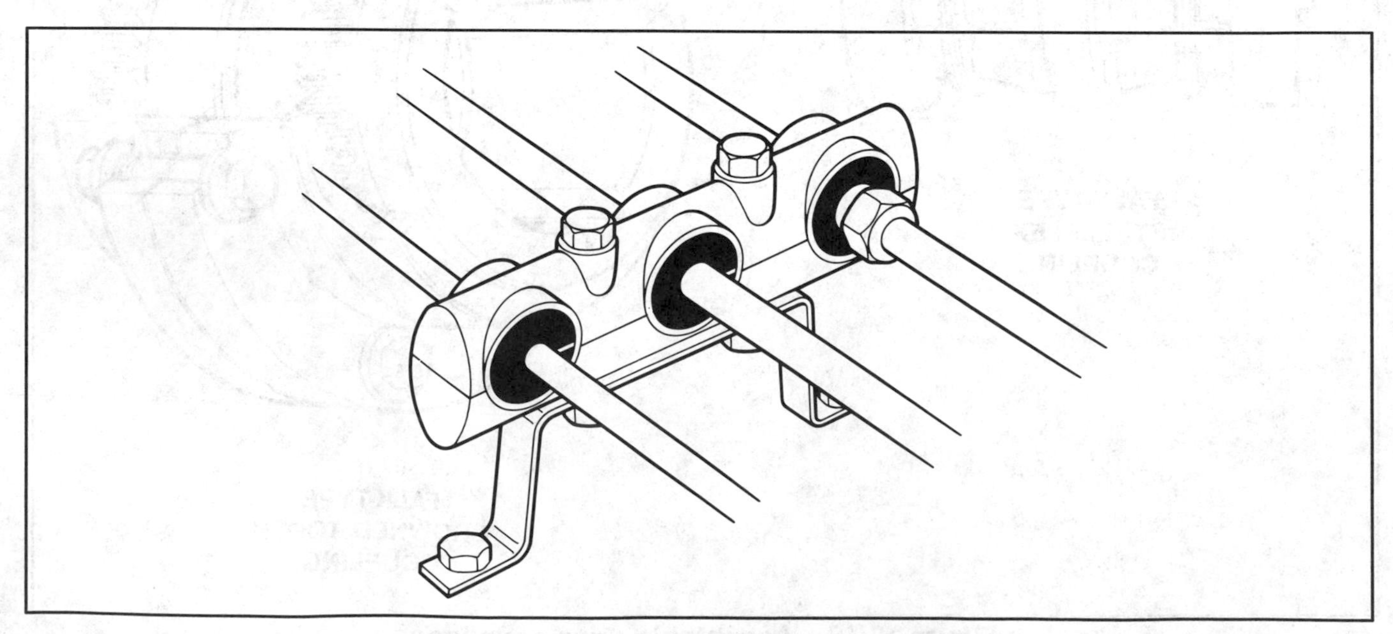

Figure 18-20. Resilient line supports.

QUESTIONS

1. Which type of accumulator operates at a constant pressure? How can the pressure be changed?
2. What type of gas is preferred for gas-charged accumulators?
3. How is pressure created in the free-piston accumulator?
4. Name two functions of an accumulator.
5. What is the purpose of an intensifier?
6. How is a pressure switch operated?
7. Give three situations where a pressure gauge might be required.
8. How are vacuum gauges calibrated?
9. Knowing the displacement and speed of a hydraulic motor, how do you calculate the gpm (LPM) flowing through it?
10. Name three types of sound damping devices.

CHAPTER 19 SYSTEMS

This chapter begins with a section explaining the operation of several generic hydraulic circuits that perform functions such as unloading, venting and clamping within hydraulic systems. A description of a complete plastic injection molding system follows these circuit descriptions to illustrate how circuits are combined to form systems.

INDUSTRIAL HYDRAULIC CIRCUITS

The circuits described in this chapter are typical of systems used in industrial machinery and illustrate the basic principles of applying hydraulics to various kinds of work.

Because there are so many different applications of the principles and components described in this manual, only a sample will be illustrated here. Many of the circuits are presented in cutaway or pictorial diagrams for ease in following oil flow. Graphical diagrams are shown for all circuits to aid in understanding the use of symbols.

Unloading Circuits

An unloading circuit exists when a pump outlet is diverted to the tank at low pressure during part of the cycle. This may occur because load conditions exceed the available input power or simply to avoid wasting power and generating heat during idle periods.

Two-Pump Unloading System. While a cylinder is advancing at low pressure, it is often desirable to combine the delivery of two pumps for more speed. When the high speed is no longer required or the pressure rises to a point where the combined volume would exceed the input horsepower (wattage), the larger of the two pumps is unloaded.

Low Pressure Operation. Figure 19-1A, shows the arrangement of components in an unloading system with the flow condition at low pressure. Oil from the large volume pump passes through the unloading valve and over the check valve to combine with the low volume pump output. This condition continues as long as system pressure is lower than the setting of the unloading valve.

High Pressure Operation. In Figure 19-1B, system pressure exceeds the setting of the unloading valve which opens, permitting the large volume pump to discharge to the tank at little or no pressure. The check valve closes, preventing flow from the pressure line through the unloading valve.

In this condition, much less power is used than if both pumps had to be driven at high pressure. However, the final advance is slower because of the smaller volume output to the system. When motion stops, the small volume pump discharges over the relief valve at its setting.

Automatic Venting

In systems where it is not necessary to maintain pressure at the end of a cycle, it is possible to unload the pump by automatically venting the relief valve. Figure 19-2 shows a system of this type, using a cam-operated pilot valve.

Mid-Stroke Extending (Figure 19-2A). The machine cycle begins when the solenoid of the spring offset directional valve is energized. The pump outputs to the cap-end of the cylinder, while the vent line from the directional valve is blocked at the cam-operated pilot valve. (Note that the pilot valve has only two flow paths instead of the usual four.)

Mid-Stroke Retracting (Figure 19-2B). At the end of the extension stroke, the limit switch is contacted by the cam on the cylinder, de-energizing the solenoid. Pump output is then applied to the rod-end of the cylinder by the directional valve. The relief valve vent connection is still blocked.

Automatic Stop (Figure 19-2C). At the end of the retraction stroke, the cam on the cylinder opens the vent in the pilot valve. The relief valve vent port is connected to the line from the cap-end of the cylinder and the valve is vented through a check valve, the directional valve, and a tank line check valve. Pilot pressure for the directional valve is maintained at a value determined by the

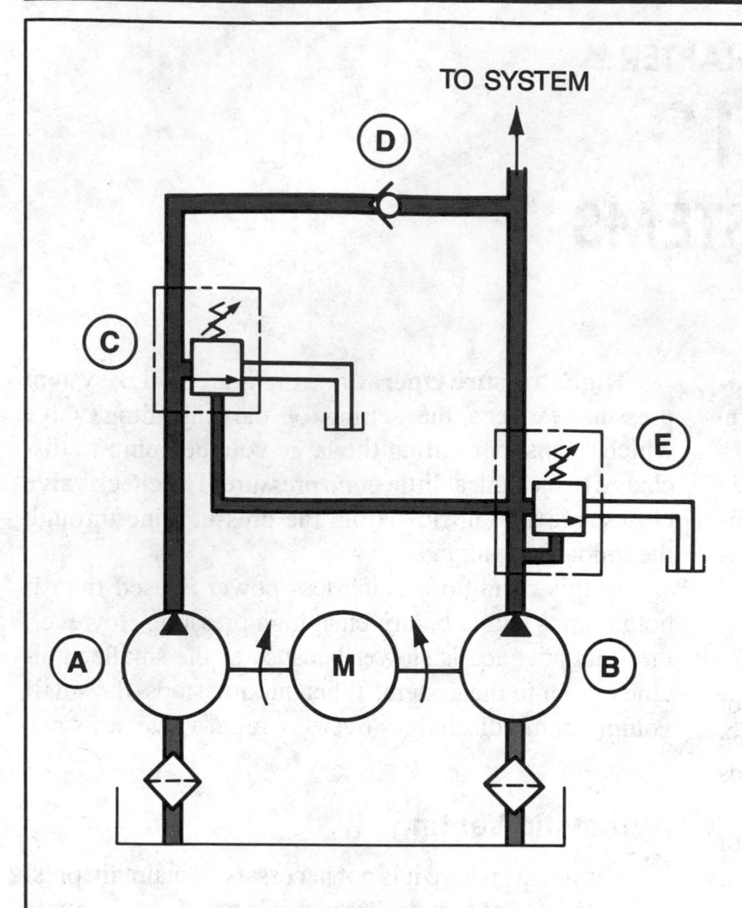

A. LOW-PRESSURE OPERATION

System pressure is less than the adjusted settings of pressure control valves (C) and (E). Therefore, both (C) and (E) are in their normally closed positions. Delivery of pump (B) is directed into the system through (E). Delivery of pump (A) is directed through (C) and check valve (D) and combines with delivery of (B) to also be directed into the system.

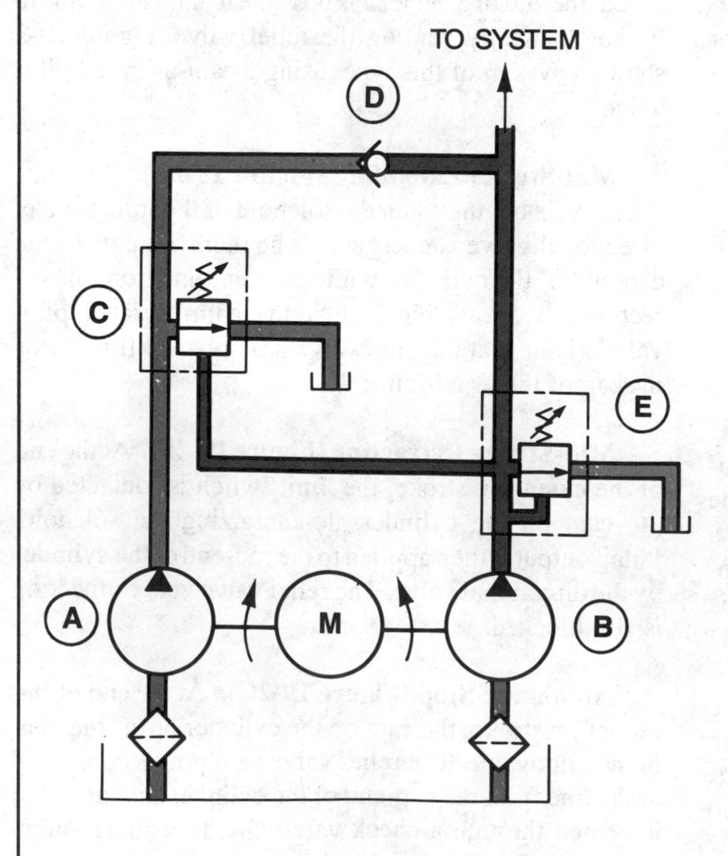

B. HIGH-PRESSURE OPERATION

System pressure is equal to the setting of relief valve (E) but higher than the setting of unloading valve (C). Delivery of pump (B) returns to the tank through valve (E) at its pressure setting. Delivery of pump (A) returns to the tank through valve (C) at a very low pressure (unloaded) as (C) is held wide open by system pressure.

Figure 19-1. Unloading circuit.

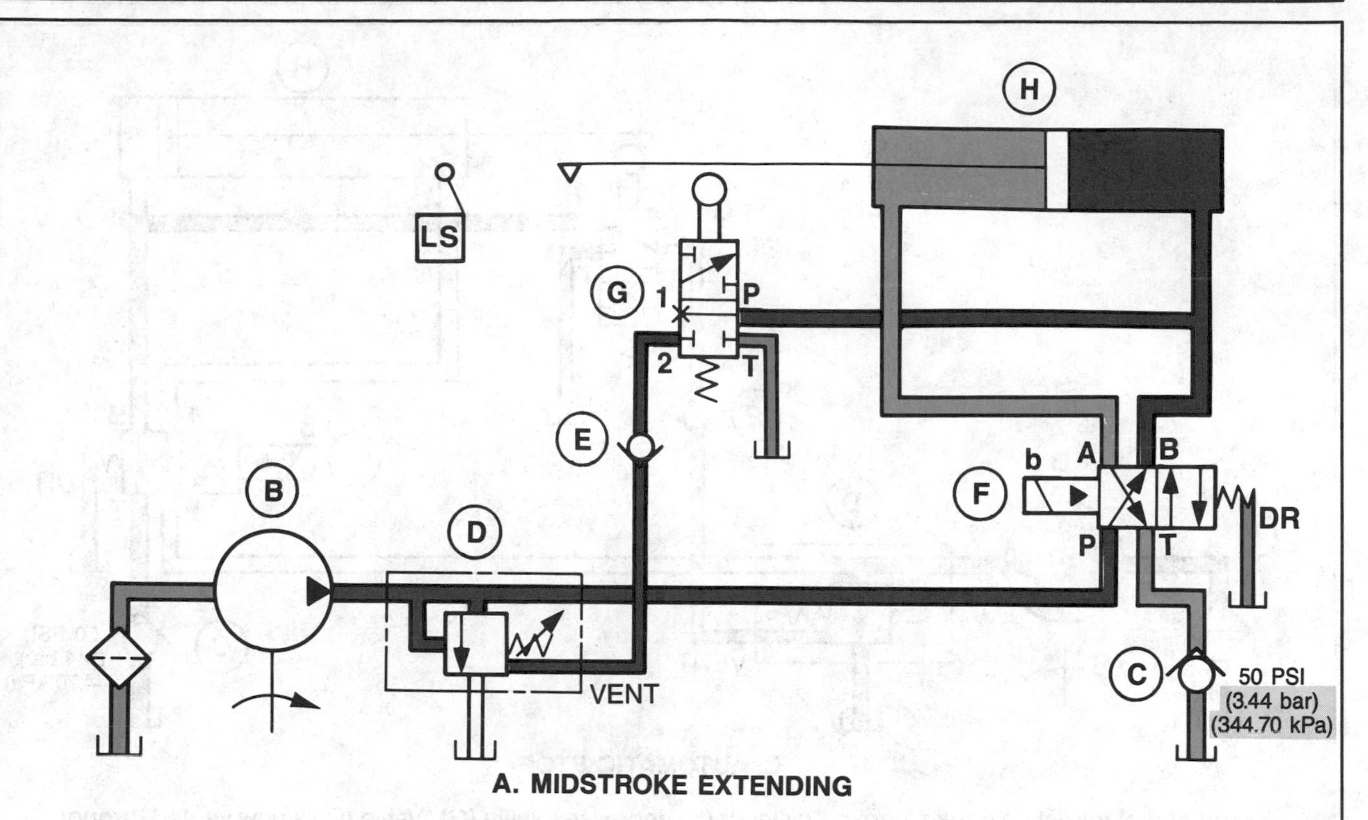

A. MIDSTROKE EXTENDING

Solenoid "b" of valve (F) is held energized during the extending stroke. Vent line from valve (D) is blocked at valve (G). Delivery of pump (B) is directed through (F) into cap end of cylinder (H). Discharge from rod end of (H) flows to tank through valves (F) and (C).

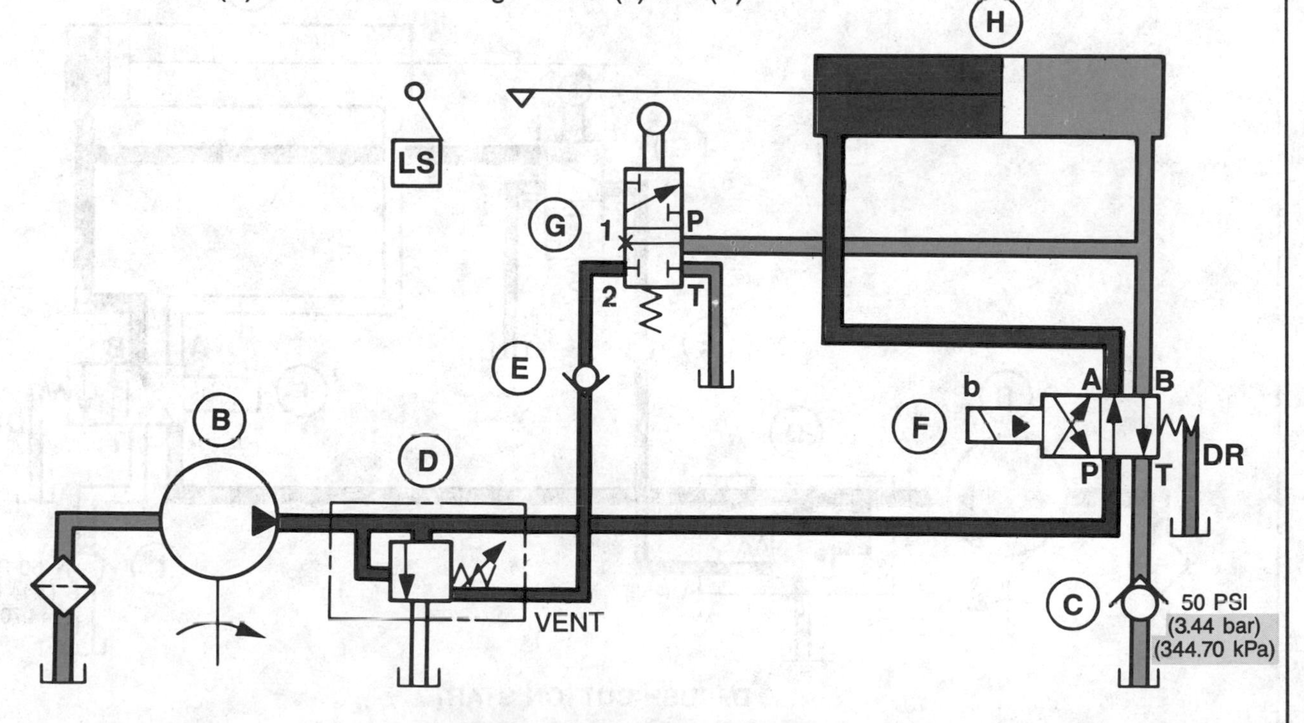

B. MIDSTROKE RETRACTING

At end of extension stroke, cam on cylinder (H) contacts limit switch LS. This causes solenoid "b" of valve (F) to be de-energized. (F) shifts to the spring offset position and directs delivery of pump (B) into rod end of (H). Discharge from cap end of (H) flows to tank through valves (F) and (C).

Figure 19-2. Automatic venting at cycle end. (1 of 2)

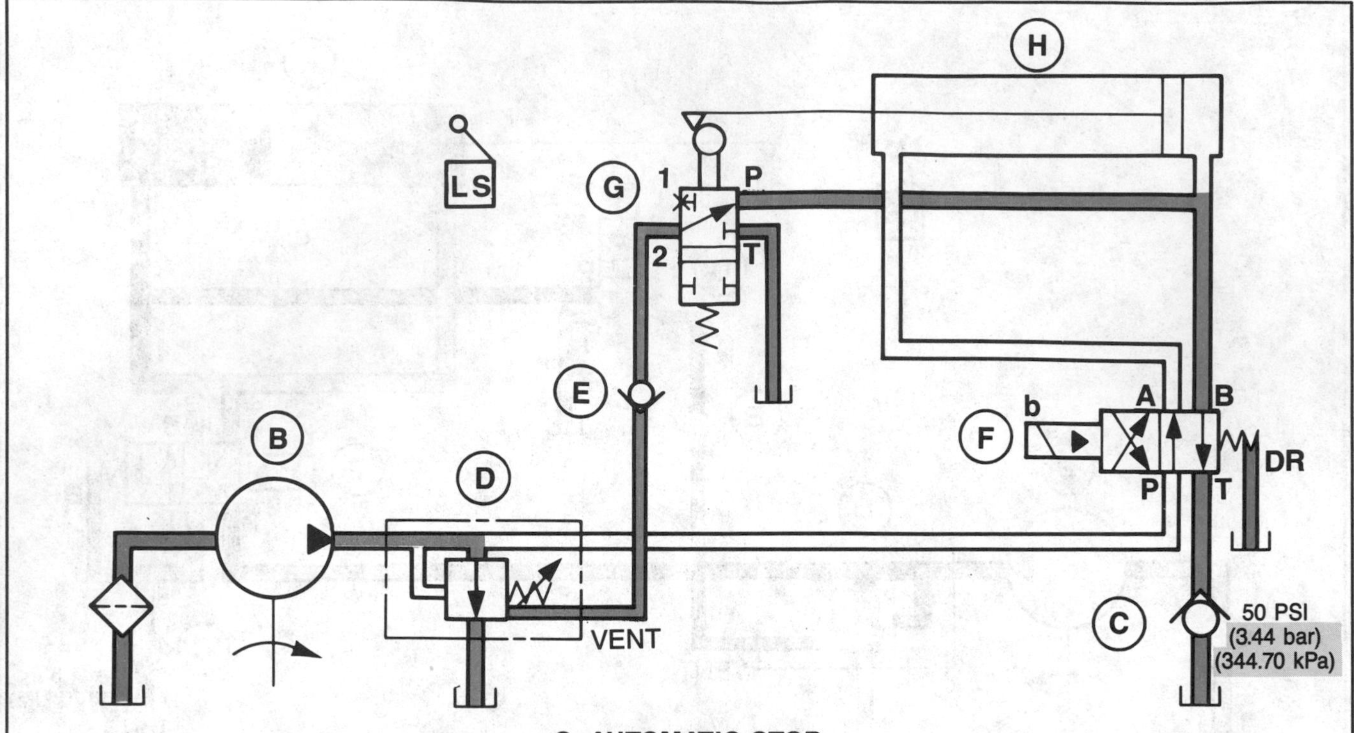

C. AUTOMATIC STOP

At end of retraction stroke, cam on cylinder (H) depresses valve (G). Valve (D) is now vented through valves (E), (G), (F), and (C). Delivery of pump (B) returns to tank over valve (D) at low pressure. Pressure drop through (C) assures pilot pressure for operation of (F).

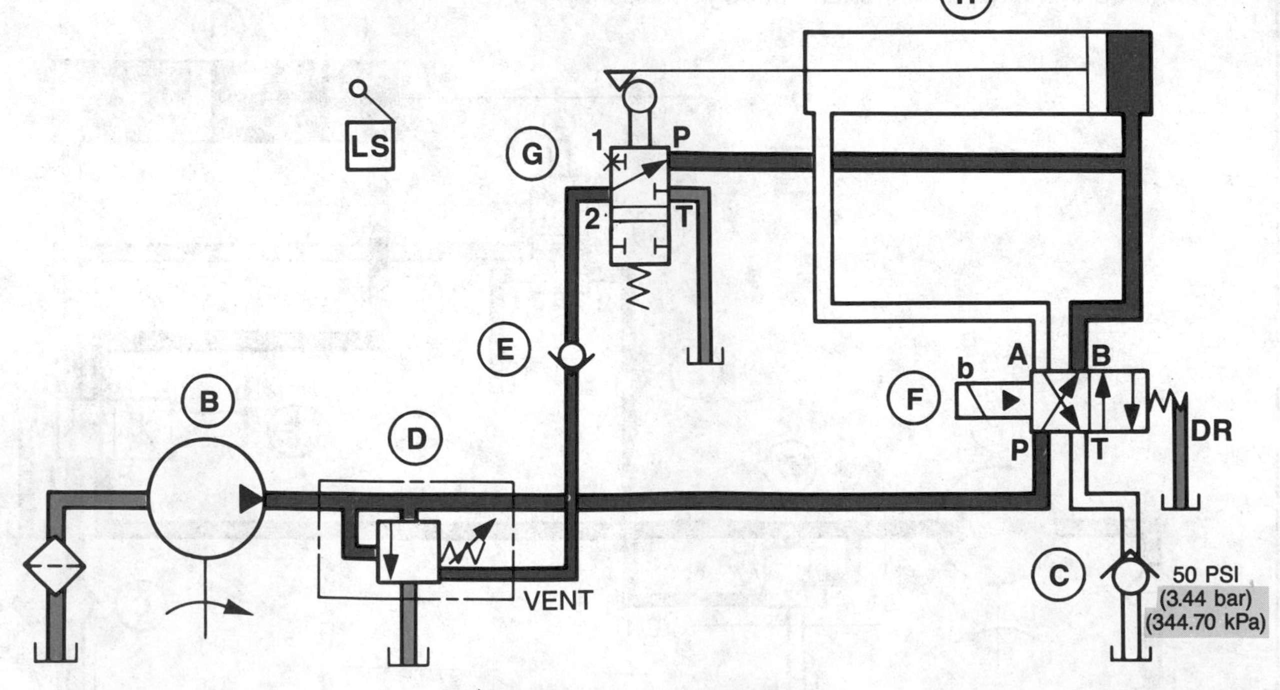

D. PUSH-BUTTON START

Depressing a push button causes solenoid "b" of valve (F) to be held energized. (F) shifts to connect cap end of cylinder (H) to pump (B), and rod end of (H) to tank. Pilot flow from vent of (D) stops when check valve (E) closes. Pressures equalize through balance hole in hydrostat of (D) causing it to start to close. Acceleration of (H) takes place during the closing of the hydrostat of (D).

Figure 19-2. Automatic venting at cycle end. (2 of 2)

spring loads in the balanced piston of the relief valve, vent line check valve, and the tank line check valve. (In this circuit, a high vent spring in the relief valve would have eliminated the need for the tank line check valve.)

Push Button Start (Figure 19-2D). When the start button is depressed, it energizes the solenoid. The directional valve shifts to direct the pump output into the cap-end of the cylinder. This causes the vent line check valve to close, de-venting the relief valve. When pressure builds up again, the cycle is repeated.

Accumulator Pump Unloading—Electric Control

When a preset pressure is reached in an accumulator charging circuit, the pump is unloaded. As pressure drops to a predetermined minimum, the pump cuts back in to recharge the accumulator.

A spring offset, solenoid-operated, directional valve (Figure 19-3), is actuated by a pressure switch. It is used to vent and de-vent the relief valve, as required.

Charging (Figure 19-3A). The two micro-switches in the pressure switch are interconnected to an electric relay. This is done so that, at the low pressure setting, the solenoid is energized and the relief valve vent connection is blocked. The pump output flows into the system through the relief and check valve, where it charges the accumulator.

Unloading (Figure 19-3B). When pressure reaches the maximum setting of the pressure switch, the solenoid is de-energized and the relief valve is vented to unload the pump into the tank. The check valve closes, preventing back flow from the accumulator and maintaining pressure in the system.

Accumulator Safety Circuits—Bleed-Off

When the pump is shut down, the circuit in Figure 19-4 is used to automatically bleed off a charged accumulator. This prevents accidental operation of an actuator and makes it safe to open the system for service. The bleed-off is accomplished through a spring offset directional valve and a fixed restriction.

The solenoid for the directional valve is actuated by the prime mover switch so that it energizes whenever the pump is started (Figure 19-4A). During normal operation, this blocks the bleed passage.

When the pump is shut down (Figure 19-4B), the spool spring shifts the directional valve and opens the accumulator to the tank through the restriction.

The manual valve shown in Figure 19-4 is used to control the accumulator discharge rate into the system. The auxiliary relief valve is set slightly higher than the system relief valve, limiting any pressure rise from heat expansion of the gas charge. The accumulator must have a separator, i.e., diaphragm, bladder or piston to prevent loss of gas preload each time the machine is shut down.

Reciprocating Circuits

Conventional reciprocating circuits provide reversal through the use of a four-way directional valve, piped directly to a cylinder or motor. When a differential cylinder is used, retracting speed is faster than extending speed because of the rod volume.

A regenerative circuit is a nonconventional, reciprocating hookup that occurs when oil from the rod-end of the cylinder is directed into the cap-end to increase speed.

Regenerative Advance

The principle of the regenerative circuit is shown in Figure 19-5. Note that the "B" port on the directional valve, which would normally connect to the cylinder, is plugged and the rod-end of the cylinder is connected directly to the pressure line (Figure 19-5A). With the valve shifted to connect the "P" port to the cap-end (Figure 19-5B), flow out of the rod end joins pump delivery, increasing the cylinder speed. In the reverse condition (Figure 19-5C), flow from the pump is directed to the rod end. Exhaust flow from the cap-end returns to the tank through the directional valve.

If the ratio of cap-end area to rod-end annular area in the cylinder is 2:1, the cylinder will advance and retract at the same speed. However, because the same pressure in the rod-end (effective over half the cap-end area), opposes the cylinder's advance, the pressure during advance will be double the pressure required for a conventional hookup. With a higher ratio of areas, extending speed will increase proportionally.

Regenerative Advance with Pressure Changeover to Conventional Advance. The regenerative principle also can be used to increase advance speed with a changeover to conventional advance to double the final force (Figure 19-6). In this system, a pressure control valve that is normally closed, plugs the "B" port of the directional valve during regenerative advance. When the pressure setting of the valve is reached, it opens, routing oil from the rod-end to the tank through the directional valve. The 5 psi (0.34 bar / 34.47 kPa) check valve permits the oil to join pump delivery during regenerative advance, but prevents pump delivery from taking this route

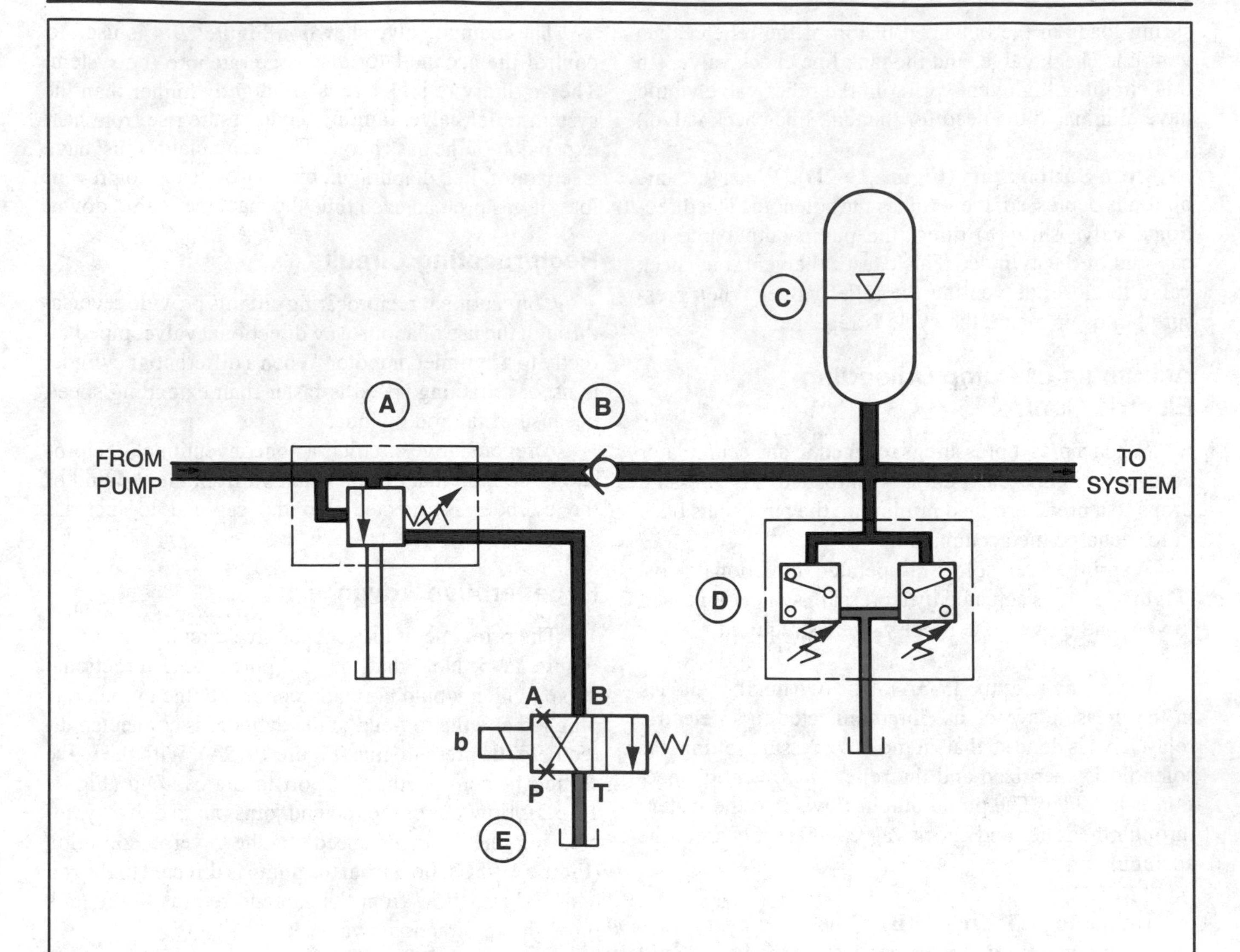

A. CHARGING

An accumulator can be used to:

- Maintain the pressure during a holding operation.
- Augment pump delivery during short periods of large volume demand.
- Absorb hydraulic shock.

This circuit shows one method of unloading the pump when the accumulator is fully charged. It consists of relief valve (A), check valve (B), accumulator (C), dual pressure switch (D) and directional valve (E). Pressure setting of (A) is higher than the high setting of (D).

The electric control circuit performs the following operations: 1) energizes solenoid (Eb) when pump motor is started; 2) de-energizes (Eb) when system pressure reaches the high setting of switch (D); 3) energizes (Eb) when system pressure reduces to the low setting of switch (D); 4) de-energzies (Eb) when pump motor is stopped.

View A shows circuit condition when system pressure is below the low setting of switch (D). Solenoid (Eb) is energized to shift valve (E) and blocks the vent connection of valve (A). Valve (A) is devented and pump delivery is directed through valve (B) into the system. Accumulator (C) is charged with fluid if system volumetric demand is less than delivery rate of the pump.

Figure 19-3. Accumulator pump unloading (electric control). (1 of 2)

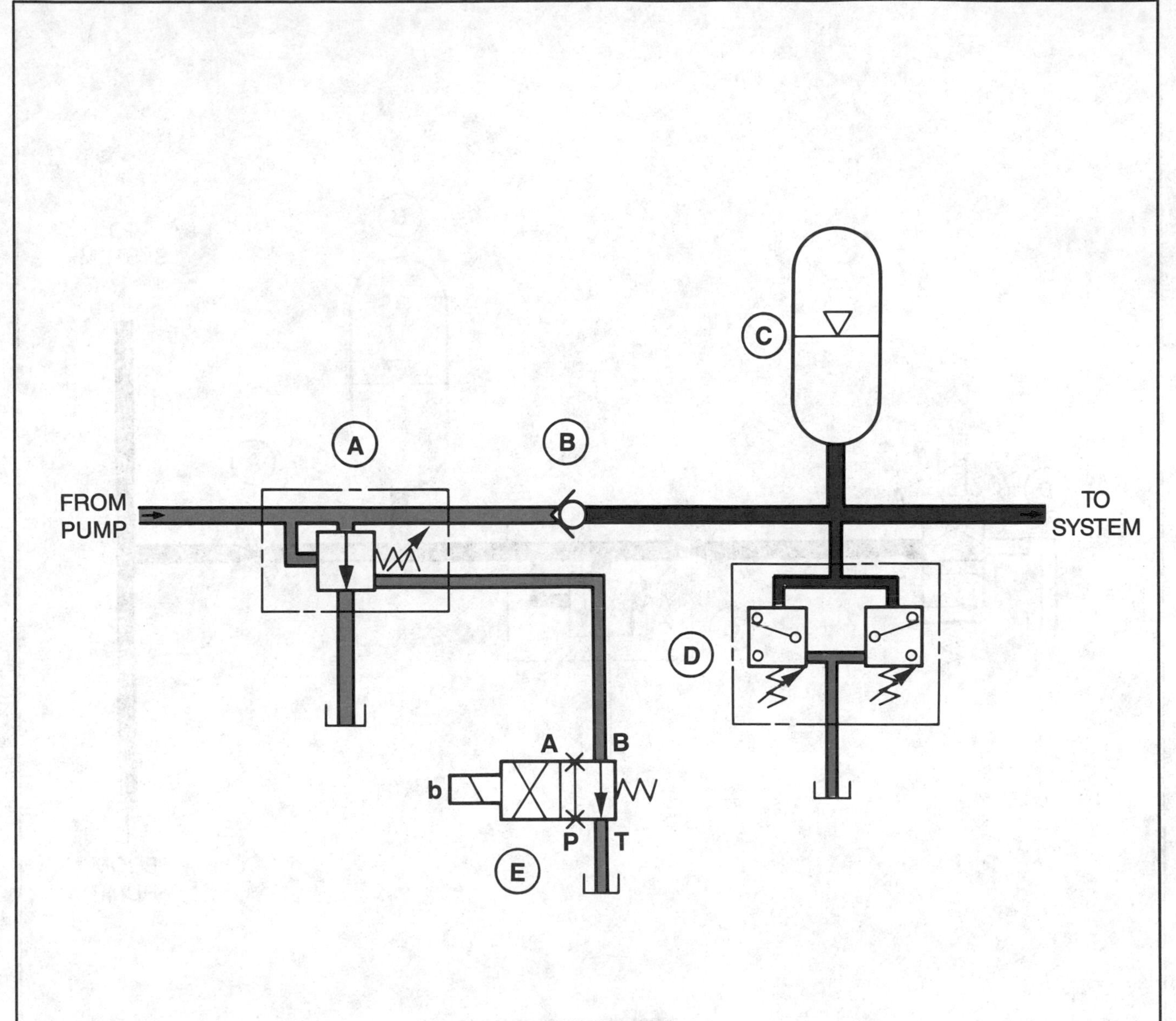

B. UNLOADING

View B shows circuit condition when accumulator (C) is charged and system pressure has reached the high setting of switch (D). Solenoid (Eb) is de-energized to vent valve (A). The pump is unloaded, its delivery being returned freely to tank through valve (A). Check valve (B) closes to permit accumulator (C) to hold pressure and maintain a volume supply in the system.

Charging and unloading continue automatically until pump motor is stopped. The dual pressure switch provides means to adjust the pressure difference between pump "cut-in" and pump "cut-out." The high setting of switch (D) is the maximum pressure control for the system with overload protection provided by valve (A).

Figure 19-3. Accumulator pump unloading (electric control). (2 of 2)

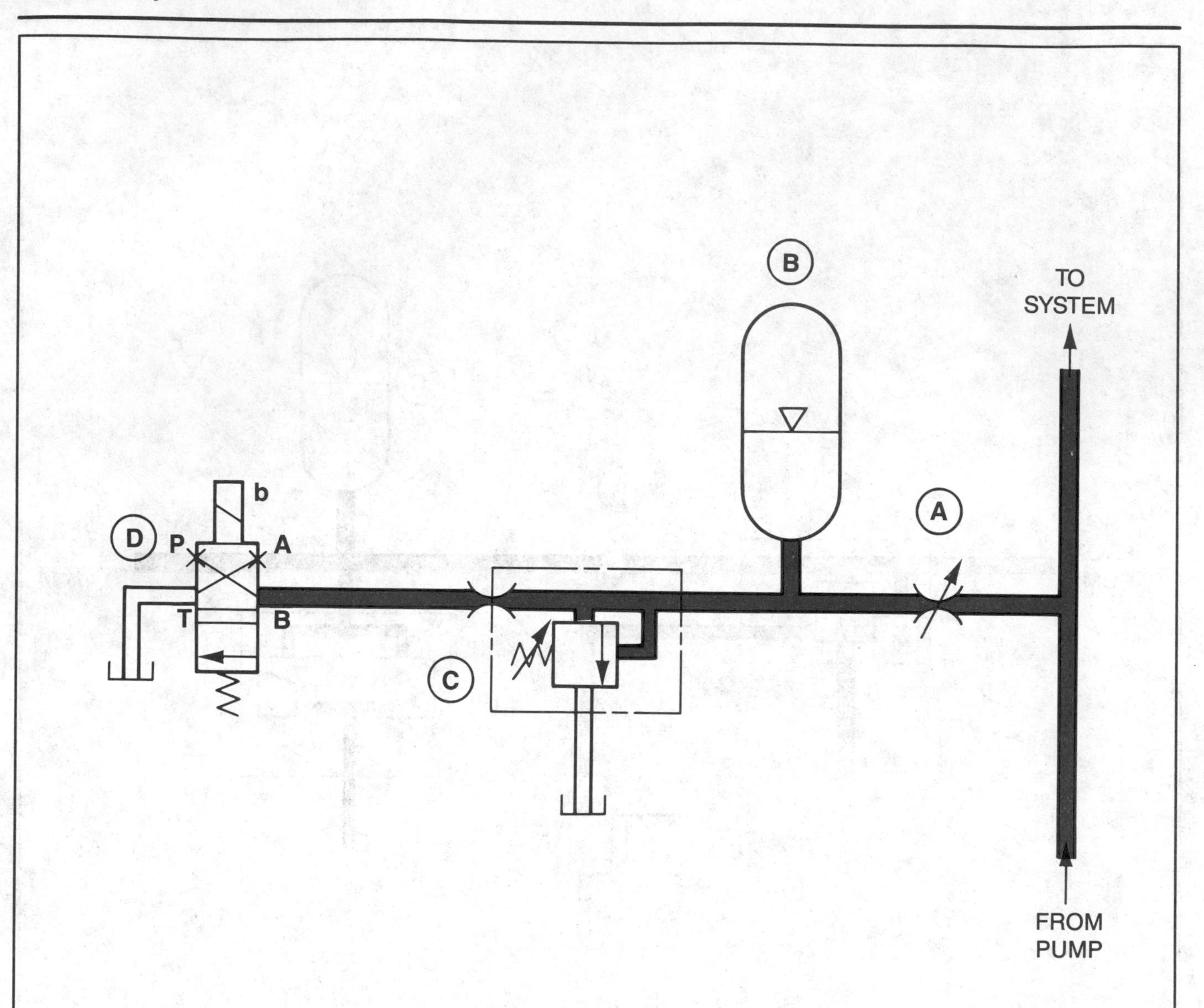

A. NORMAL OPERATION

The charge in accumulator (B) is automatically bled off to permit safe servicing of the system when pump motor is stopped. The circuit consists of needle valve (A), accumulator (B), relief valve (C) and directional valve (D). An electrical control circuit holds solenoid (Db) energized when the pump motor is running and de-energizes it when the motor is stopped.

View A shows circuit condition during normal operation of the system when the pump motor is running. Solenoid (Db) is energized to shift valve (D) and block flow to tank from accumulator (B). Accumulator is charged or discharged through valve (A) as dictated by requirements of the system. Needle valve (A) is often used to control rate of accumulator discharge to the system.

Figure 19-4. Accumulator bleed-off circuit. (1 of 2)

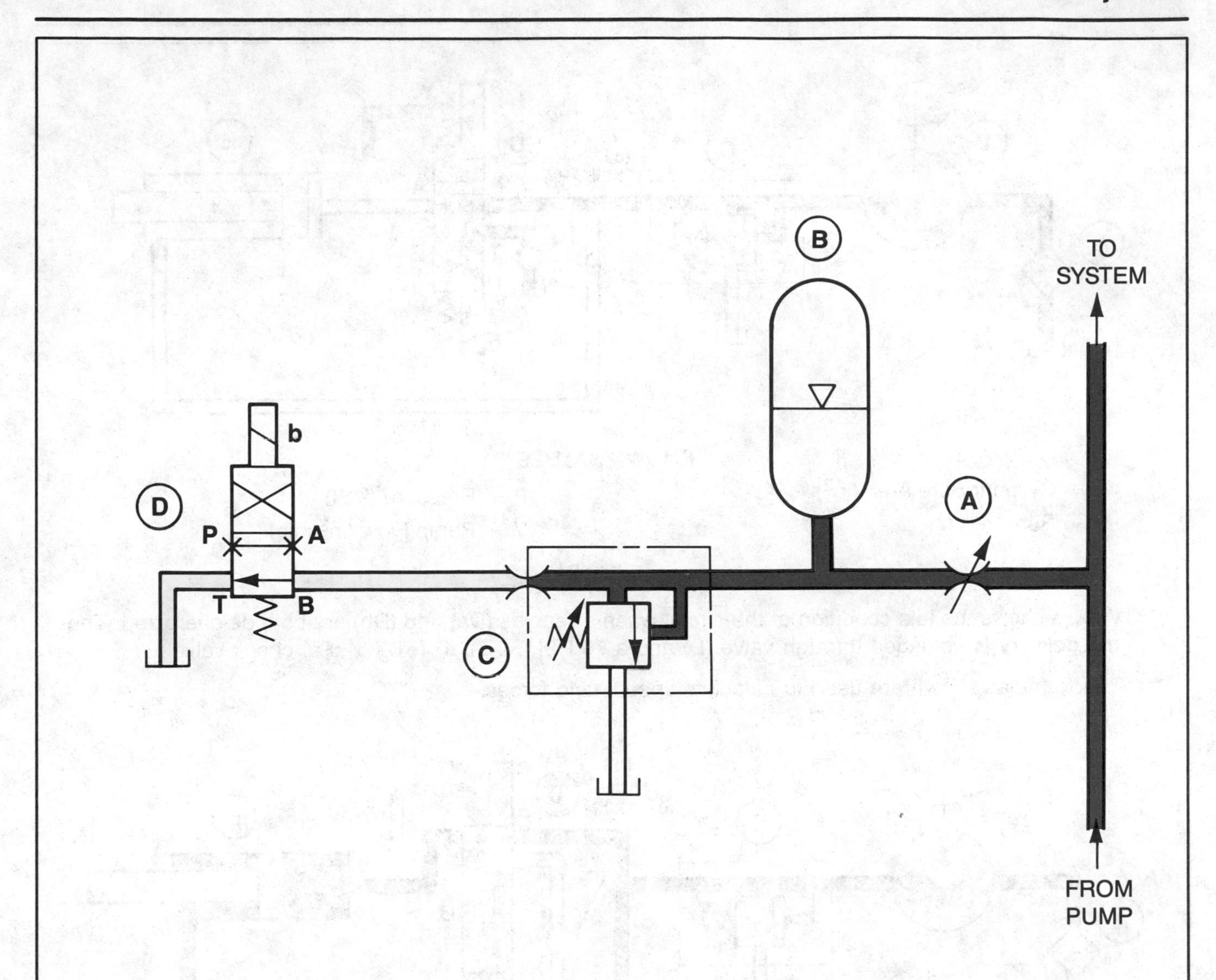

B. BLEED OFF

View B shows circuit condition when the pump motor is stopped. Solenoid (Db) is de-energized and the charge in accumulator (B) is bled off to tank through valve (D). Rate of bleed-off is controlled by a fixed restriction at valve (C).

Valve (C) is set slightly higher than the maximum pressure control and provides protection against excessive pressures due to thermal expansion.

Figure 19-4. Accumulator bleed-off circuit. (2 of 2)

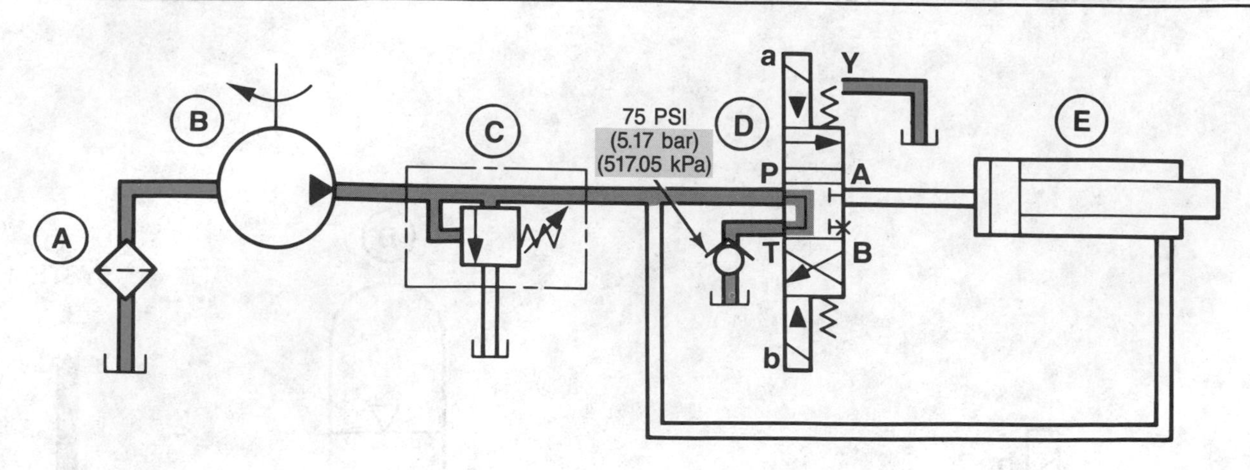

KNOWN VALUES

A = Cylinder Bore Area (in^2)
a = Rod Area (in^2)

$$K = \frac{A}{a}$$

P = Pressure (PSI)
V = Pump Flow (in^3/min)

A. IDLE

View A shows the idle condition of the circuit when solenoids (Da) and (Db) are both de-energized. The pump delivery is unloaded through valve (D) and a 75 PSI (5.17 bar) (517.05 kPa) check valve.

The formulas shown are used to calculate speeds and forces.

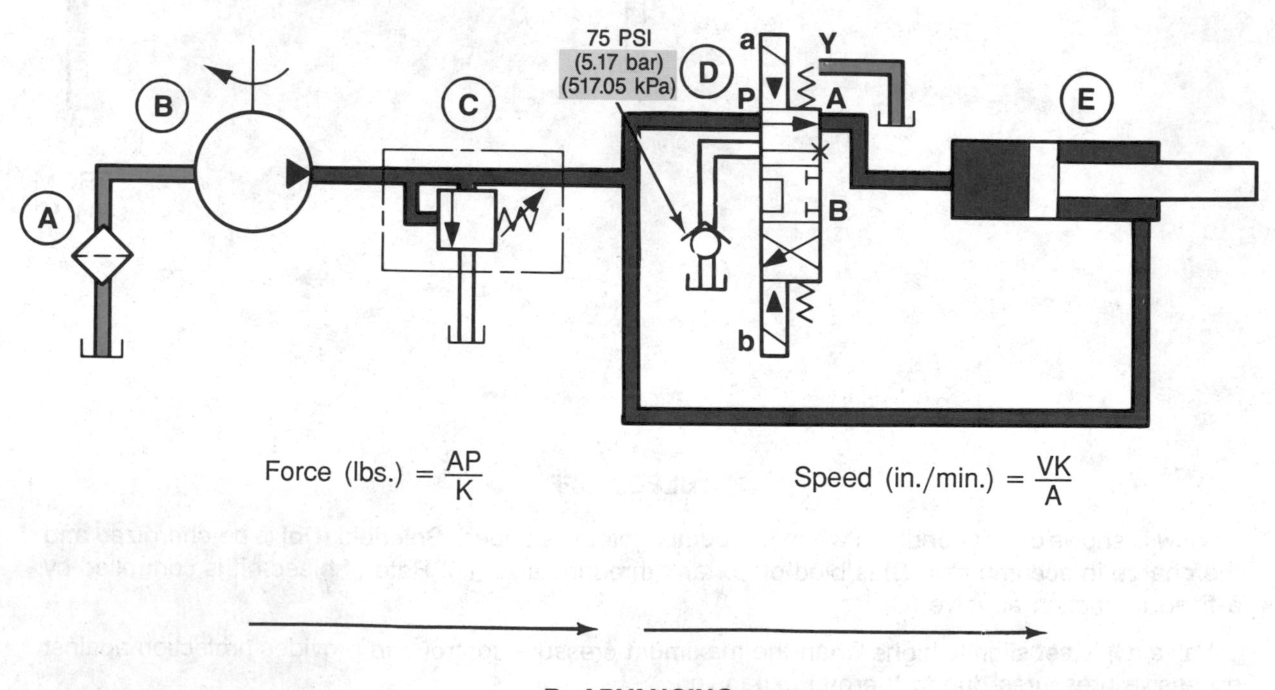

$$\text{Force (lbs.)} = \frac{AP}{K}$$

$$\text{Speed (in./min.)} = \frac{VK}{A}$$

B. ADVANCING

View B shows the flow and force conditions when solenoid (Da) is energized for regenerative advance. Discharge from the rod end of the cylinder joins the pump delivery at the "P" port of valve (D) to increase the piston speed. However, note that system pressure also acts in the rod end of the cylinder to reduce the output force capability. Formulas shown are used to calculate speeds, forces and flow rates. They also show that, during regenerative advance, speed increases and force decreases proportionately as the ratio of areas increases.

Figure 19-5. Regenerative circuit. (1 of 2)

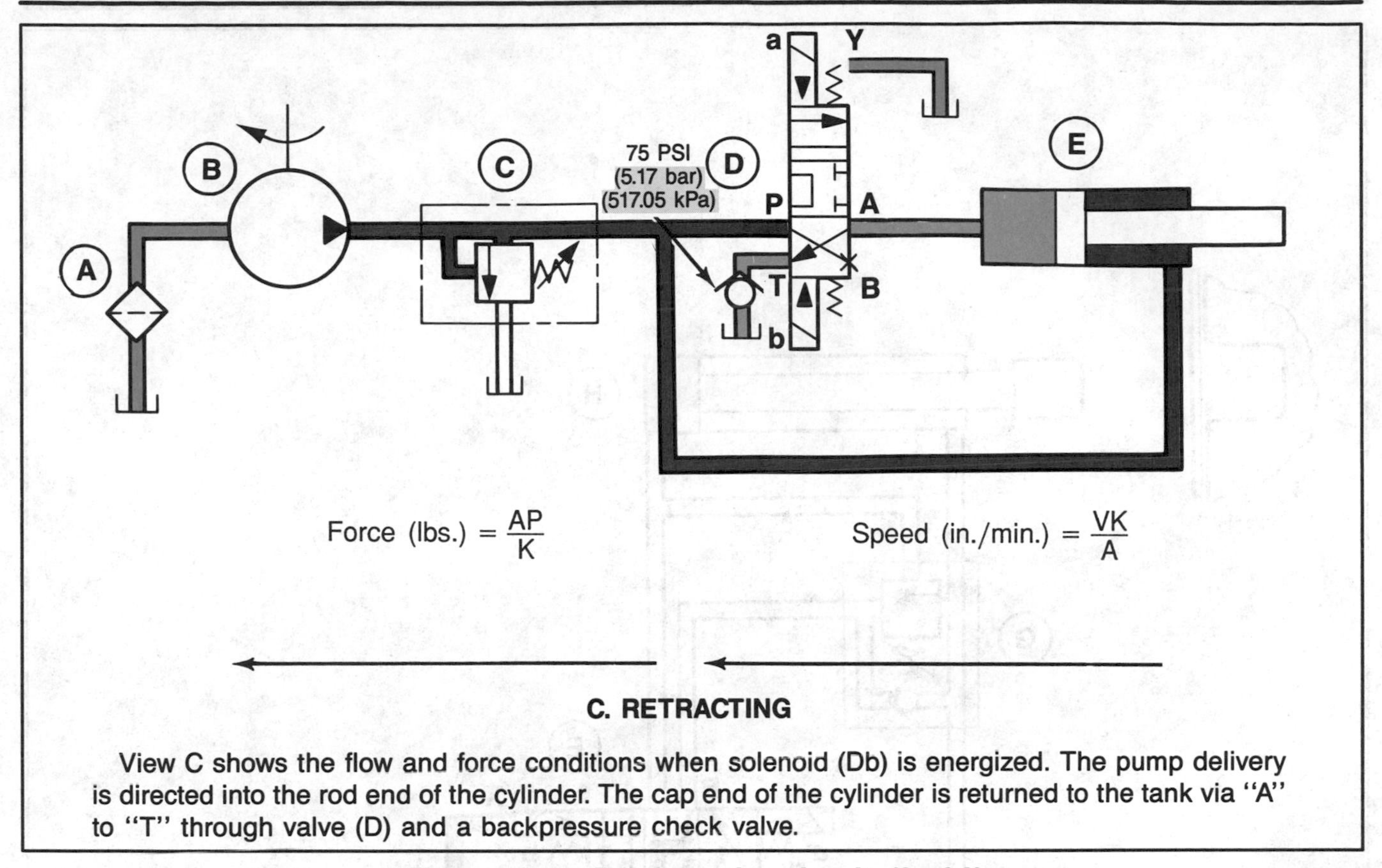

C. RETRACTING

View C shows the flow and force conditions when solenoid (Db) is energized. The pump delivery is directed into the rod end of the cylinder. The cap end of the cylinder is returned to the tank via "A" to "T" through valve (D) and a backpressure check valve.

Figure 19-5. Regenerative circuit. (2 of 2)

to the tank during conventional advance. When the directional valve shifts to retract the cylinder, the pump outputs to the rod-end through the check valve found in the pressure control valve.

Clamping and Sequence Circuits

In the application of clamping and machining a workpiece, it is necessary to have operations occur in a definite order. The pressure at the first operation must be maintained while the second occurs. One such circuit is shown in Figure 19-7.

Controlled Pressure Clamping Circuit. This circuit provides sequencing plus a controlled clamping pressure, which can be held while the work cylinder is feeding and retracting. Pressing the START button shifts a directional valve and causes the clamp cylinder to extend. Upon contact with the workpiece, a limit switch actuates the solenoid of another directional valve, initiating the work stroke. A sequence valve assures that during the work stroke, the clamp pressure is maintained at a predetermined minimum. When higher pressure is required for the work stroke, a pressure reducing valve limits the clamp pressure to a safe maximum.

Additional electric controls can reverse the work cylinder directional valve while pressure is maintained on the clamp. After the work cylinder is fully retracted, the clamp opens.

Brake Circuit

Figure 19-8 shows an application of a brake valve that holds a backpressure in a rotary motor when needed and brakes the motor when the open center direction valve is shifted to neutral.

Figure 19-8A shows the motor accelerating with the load pressure in the auxiliary remote control connection, holding the brake valve open. Figure 19-8B shows the motor in the RUN mode. Figure 19-8C shows the opcration when the motor tries to overrun the pump, creating a lower pressure in the drive line. Neutral braking through backpressure is shown in Figure 19-8D.

INJECTION MOLDING SYSTEM

Injection molding is one of the major processes by which thermoplastics are transformed into usable products. The two basic components of an injection molding machine are the clamp unit and the injection or plasticating unit. (See Figure 19-9.)

The development of cartridge valve and manifold technology has greatly affected the design of hydraulic

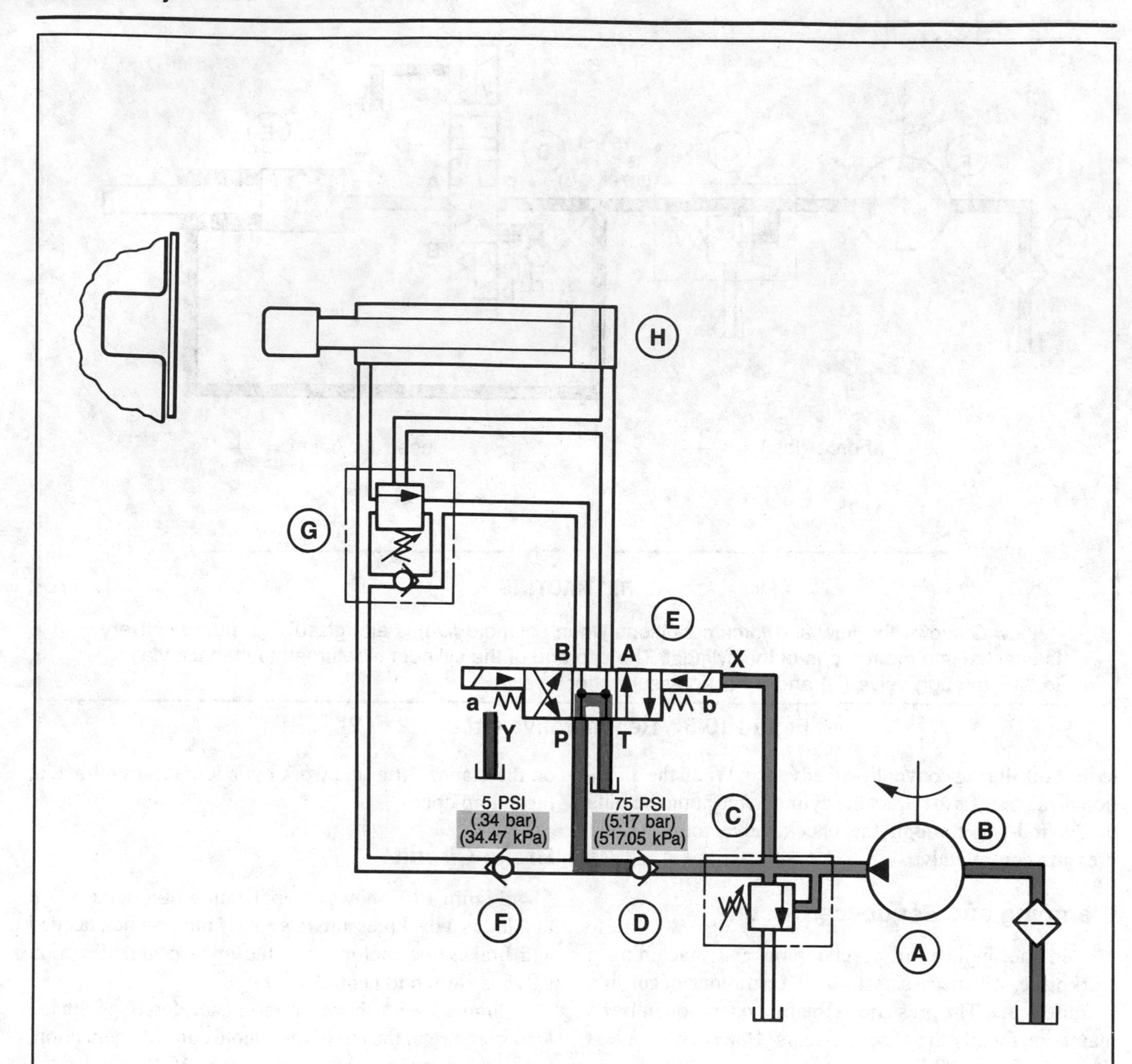

A. IDLE CONDITION

A differential cylinder with an area ratio of approximately 2:1 is used.

In the idle condition, solenoids (Ea) and (Eb) are both de-energized and delivery of the pump (B) is unloaded to the tank through valves (C), (D), and (E). Valve (D) provides pilot pressure for pilot operation of valve (E).

Figure 19-6. Regenerative advance with pressure changeover to conventional advance. (1 of 4)

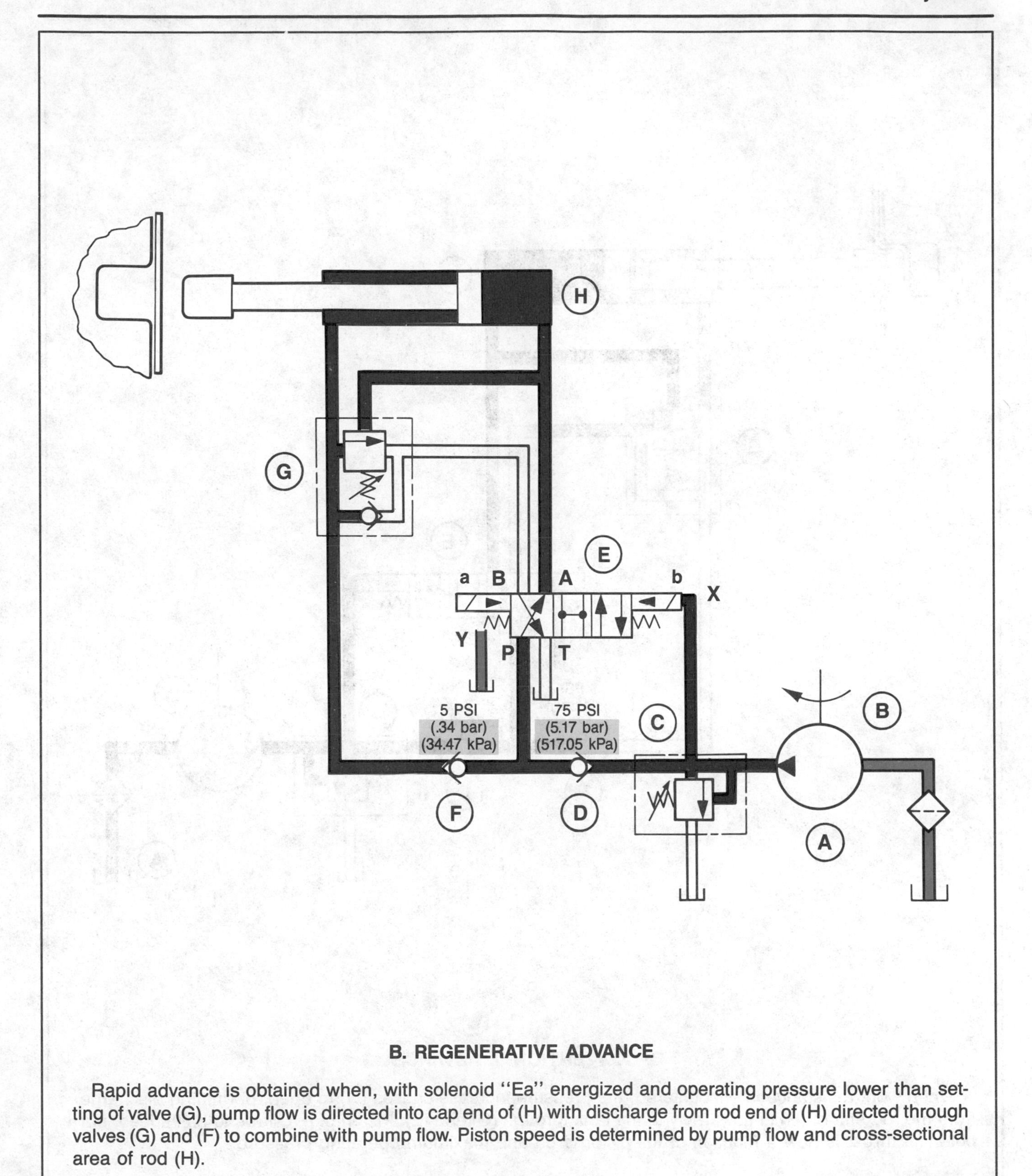

B. REGENERATIVE ADVANCE

Rapid advance is obtained when, with solenoid "Ea" energized and operating pressure lower than setting of valve (G), pump flow is directed into cap end of (H) with discharge from rod end of (H) directed through valves (G) and (F) to combine with pump flow. Piston speed is determined by pump flow and cross-sectional area of rod (H).

Figure 19-6. Regenerative advance with pressure changeover to conventional advance. (2 of 4)

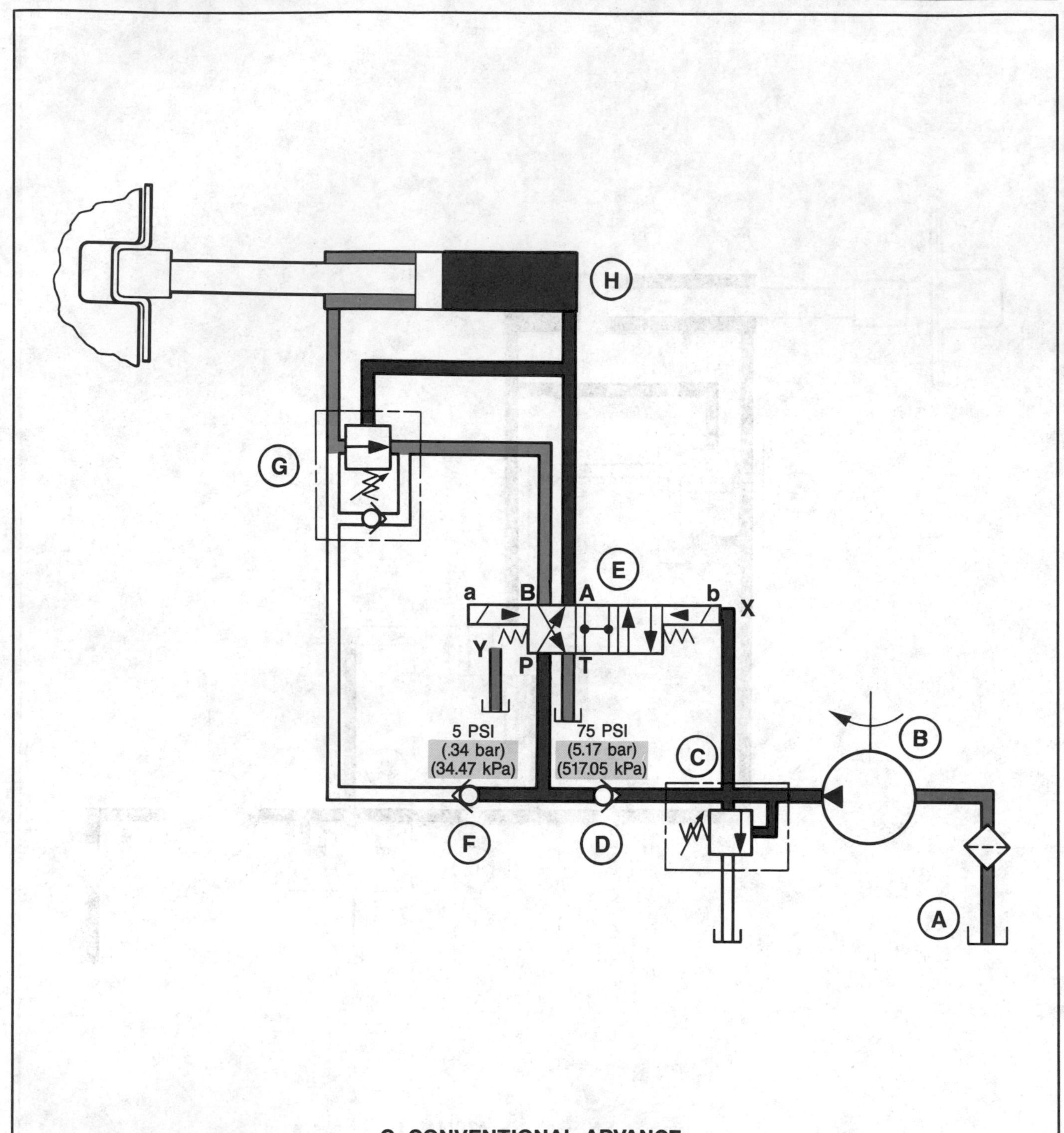

C. CONVENTIONAL ADVANCE

When work resistance is encountered, the pressure increase causes (G) to open, permitting discharge from the rod end of (H) to flow freely to the tank through (G) and (E). Piston of (H) slows to half speed, but potential thrust is now a function of full piston area and maximum operating pressure.

Figure 19-6. Regenerative advance with pressure changeover to conventional advance. (3 of 4)

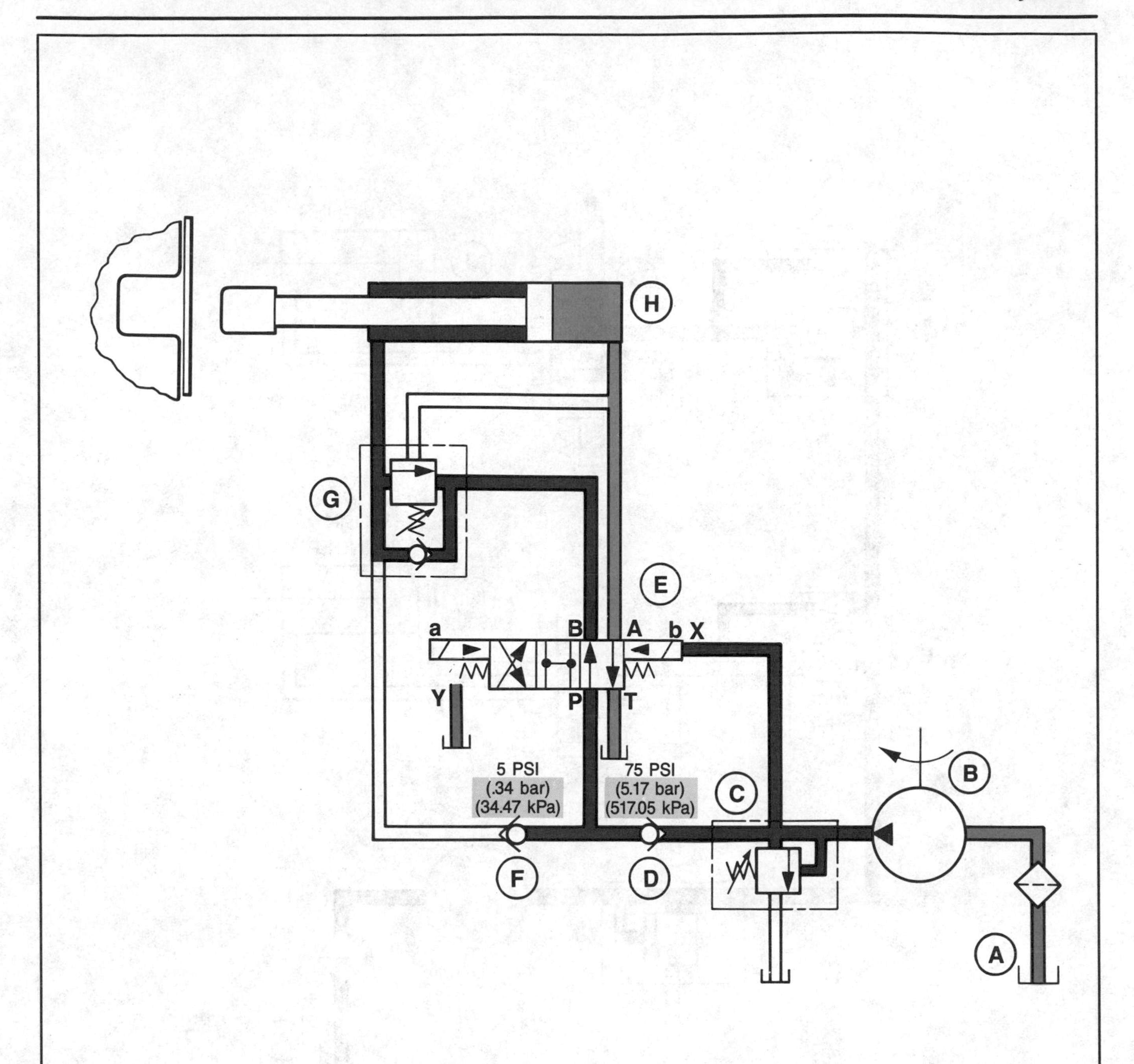

D. RAPID RETURN

Rapid return is obtained with solenoid (Eb) energized. Pump flow is directed through (E), the integral check valve in (G), and into the rod end of (H). Discharge from the cap end of (H) flows freely to the tank through (E). Piston speed is determined by pump flow and the annular area of (H), and is the same as advance speed.

Figure 19-6. Regenerative advance with pressure changeover to conventional advance. (4 of 4)

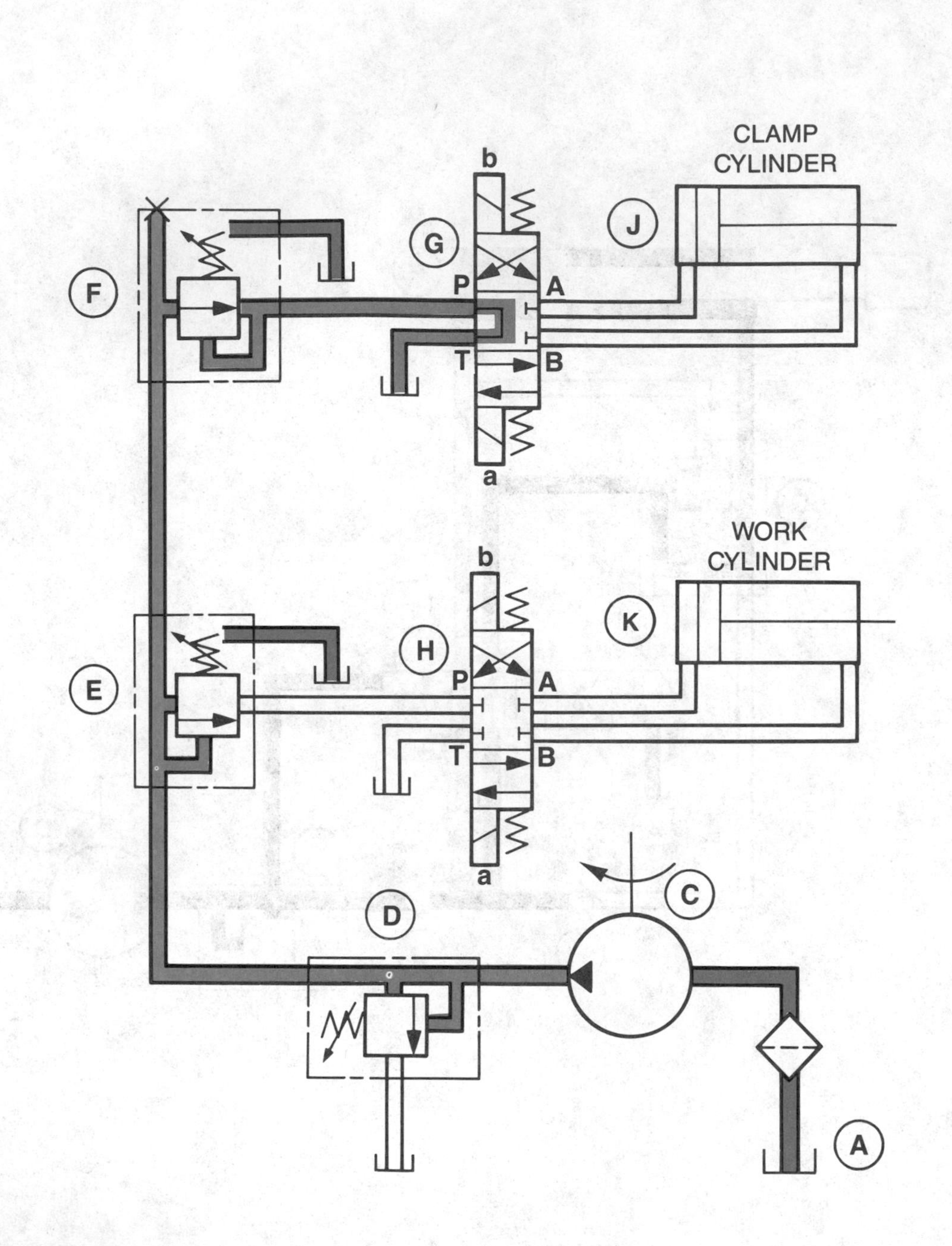

A. IDLE CONDITION

When all solenoids are de-energized, directional valves (G) and (H) are spring-centered and the delivery of pump (C) is unloaded to the tank through valves (D), (E), (F), and (G).

Figure 19-7. Controlled pressure clamping circuit. (1 of 6)

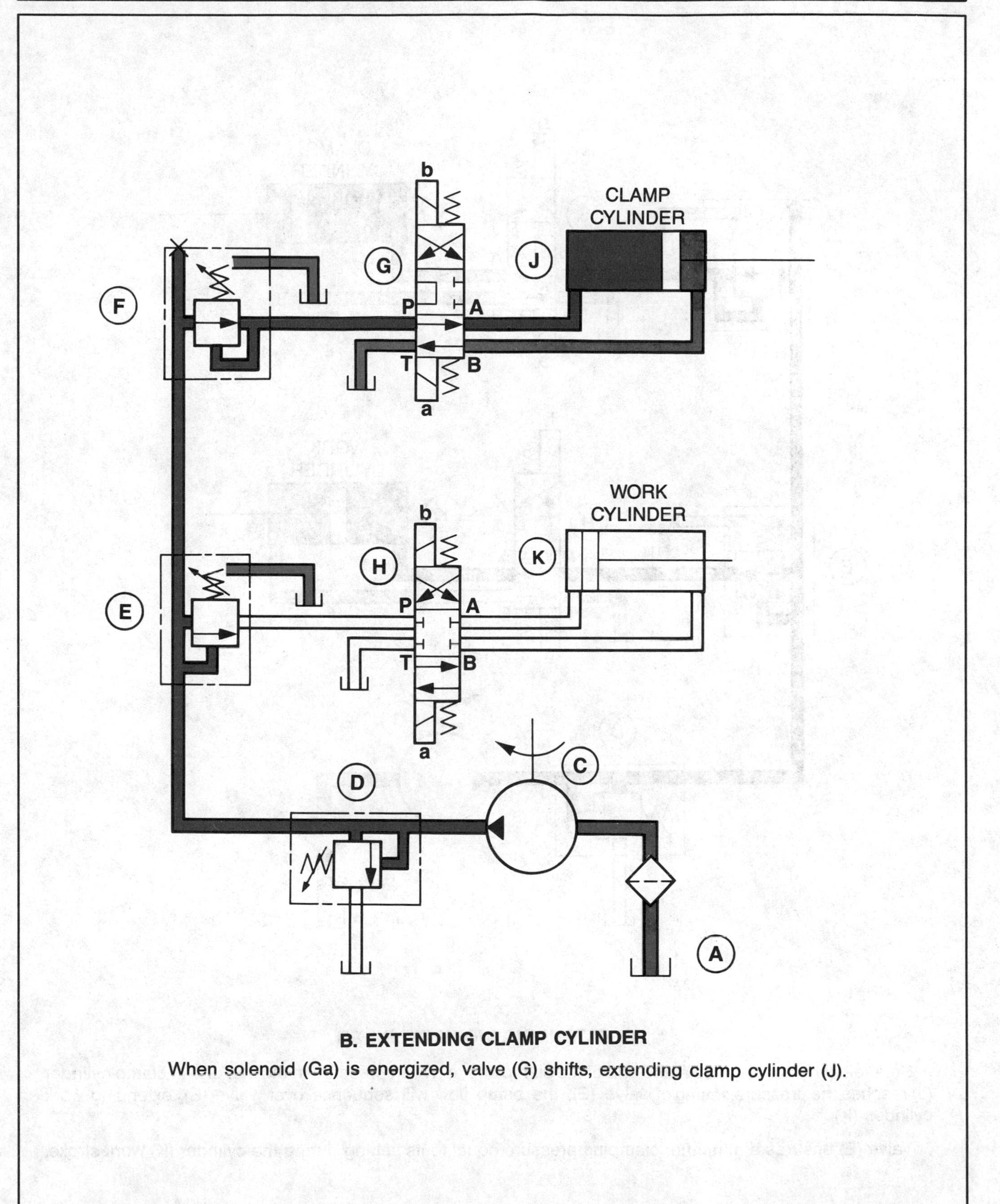

B. EXTENDING CLAMP CYLINDER

When solenoid (Ga) is energized, valve (G) shifts, extending clamp cylinder (J).

Figure 19-7. Controlled pressure clamping circuit. (2 of 6)

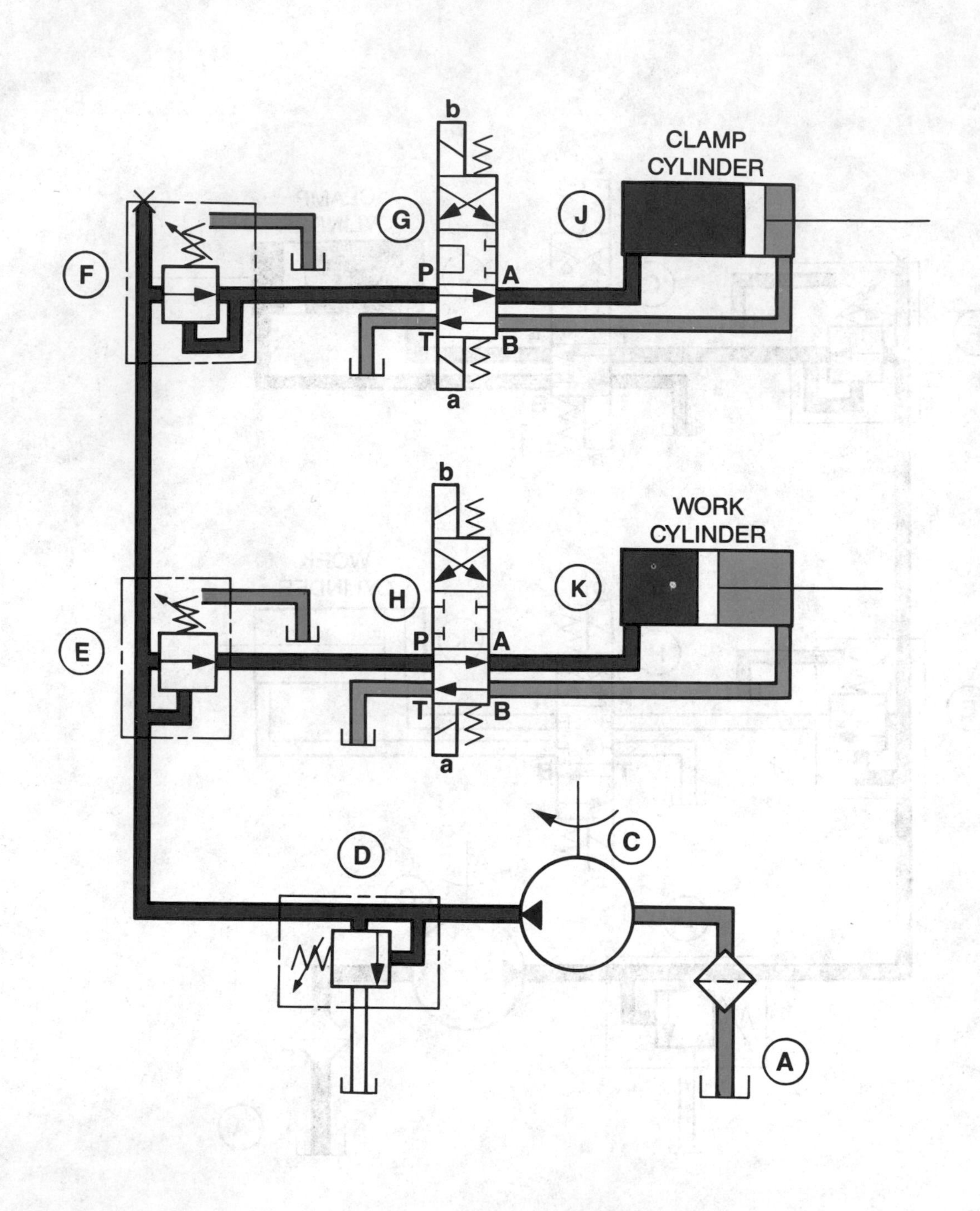

C. EXTENDING WORK CYLINDER

Solenoid (Ha) is energized at the end of the clamp stroke. When pressure in the cap end of clamp cylinder (J) reaches the pressure setting of valve (E), the pump flow will sequence over valve (E), extending work cylinder (K).

Valve (E) ensures a minimum clamping pressure equal to its setting during the cylinder (K) work-stroke.

Figure 19-7. Controlled pressure clamping circuit. (3 of 6)

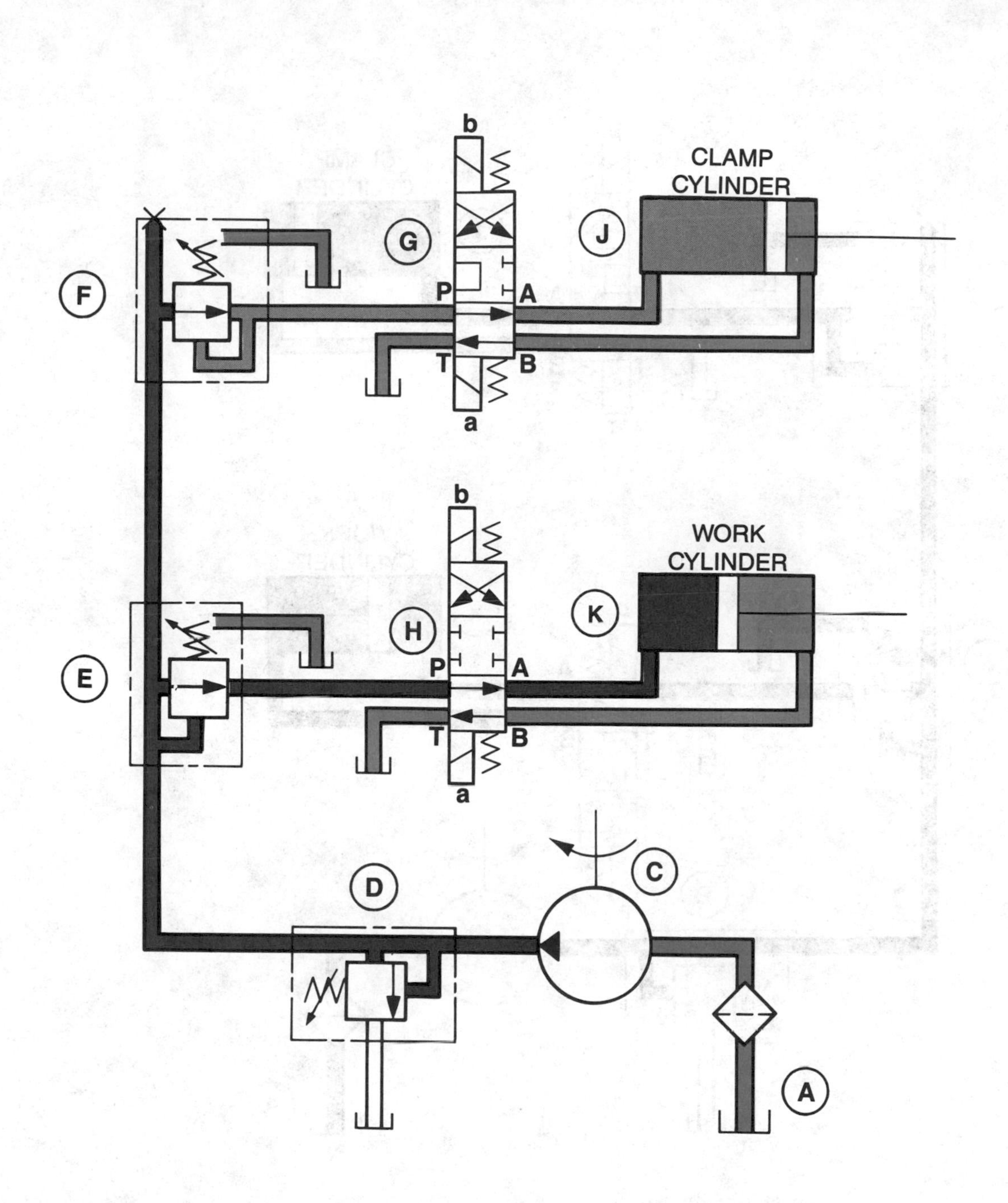

D. LIMITING MAXIMUM CLAMPING PRESSURE

When system pressure exceeds the allowed maximum clamping pressure during the working stroke, pressure-reducing valve (F) moves toward the closed position, keeping the clamping pressure at its pressure setting. System relief valve (D) limits the maximum working pressure.

Figure 19-7. Controlled pressure clamping circuit. (4 of 6)

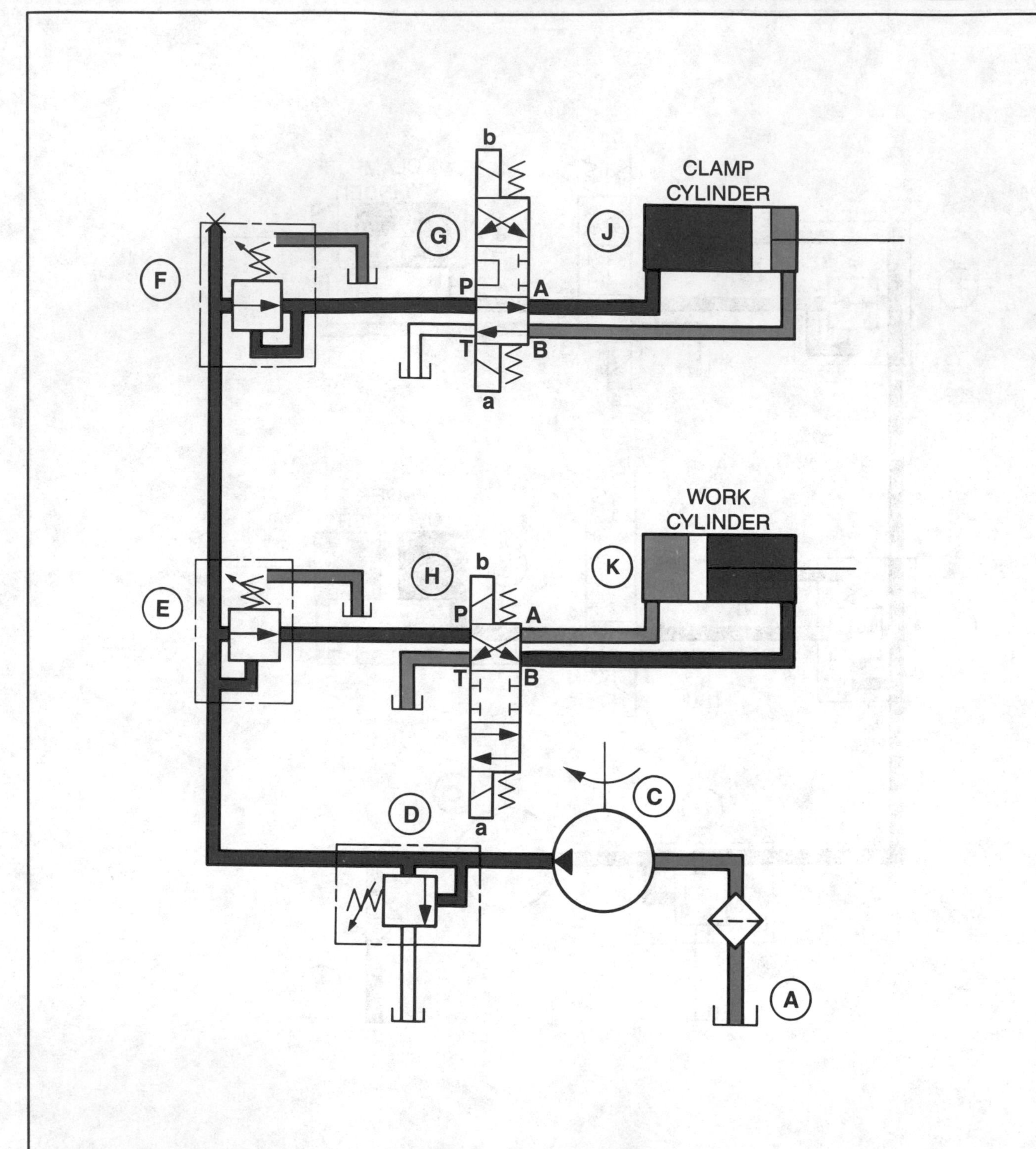

E. RETRACTING WORK CYLINDER

When solenoid (Ha) is de-energized and solenoid (Hb) is energized, work cylinder (K) retracts and sequence valve (E) maintains a minimum clamping pressure. Reducing valve (F) limits maximum clamping pressure.

Figure 19-7. Controlled pressure clamping circuit. (5 of 6)

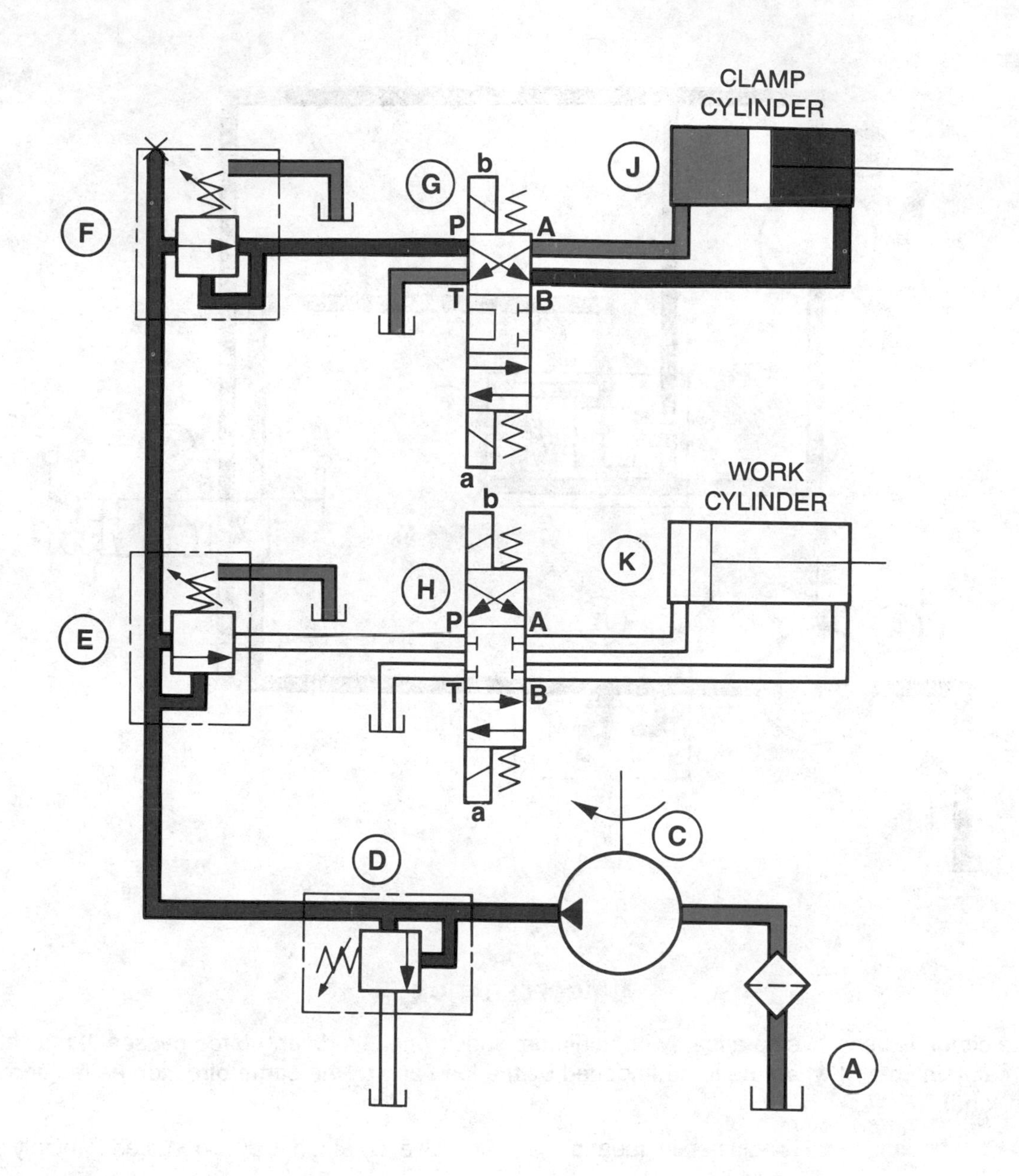

F. RETRACTING CLAMP CYLINDER

When solenoid (Hb) is de-energized, valve (H) moves to the spring-centered position and stops work cylinder (K).

When solenoid (Ga) is de-energized and solenoid (Gb) is energized, clamp cylinder (K) retracts. Solenoid (Gb) is de-energized at the end of the retraction. Valve (G) moves to the spring-centered position and the system returns to the idle condition.

Figure 19-7. Controlled pressure clamping circuit. (6 of 6)

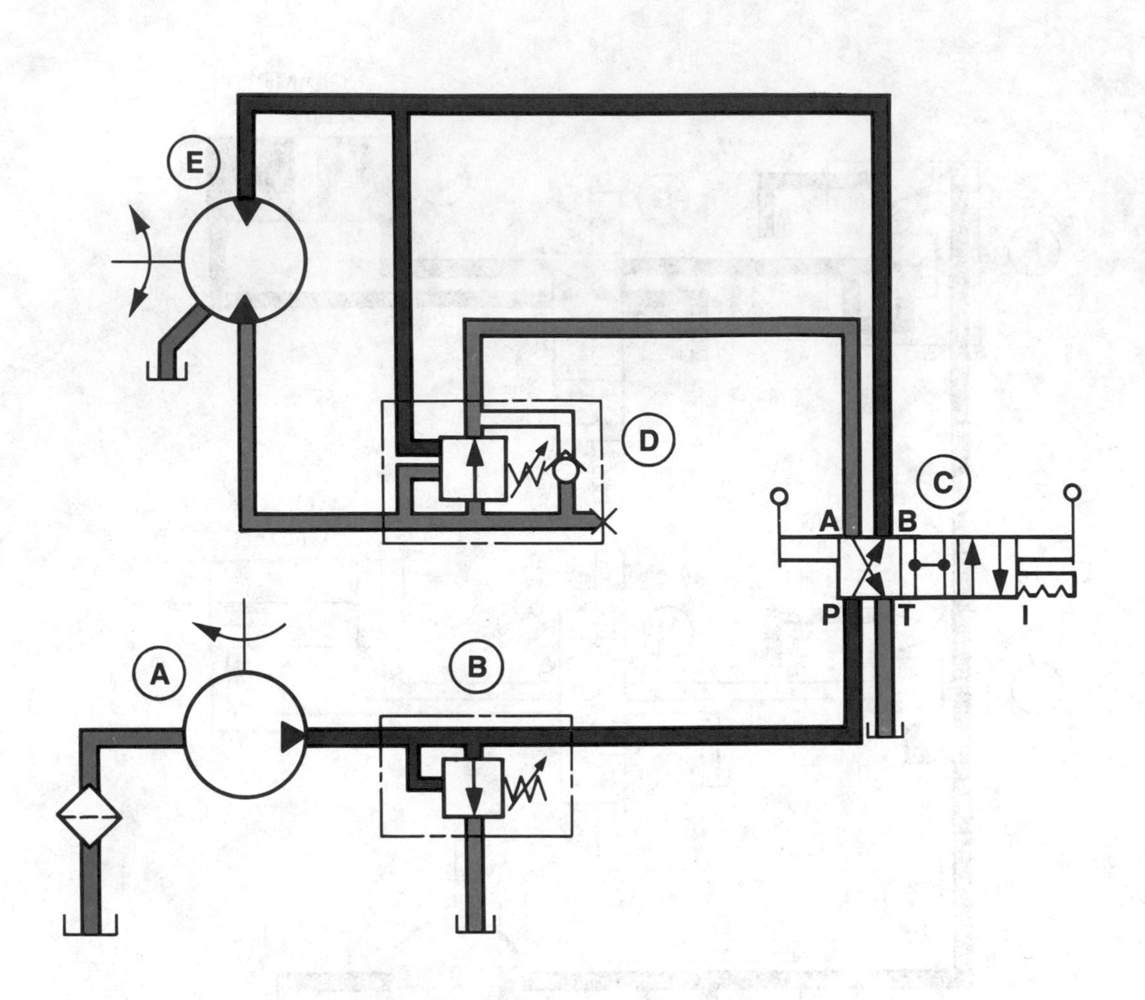

A. ACCELERATION

A brake circuit is used to stop a load with minimum shock when its driving force ceases. It may also be used to maintain control when the force imposed by the load acts in the same direction as motor rotation (negative load).

The desired braking force is adjusted by means of a brake valve (D) which is pilot-operated remotely and/or internally.

Remote control pressure is sampled from the input motor line and acts under the full area of the valve spool. Motor outlet pressure acts under the small piston of (D) through an internal passage.

Valve (D) is normally closed. It is opened by either or both of these pilot forces acting against an adjustable spring load.

As motor (E) accelerates, the inlet pressure at (E) and the remote pilot connection of valve (D) are equal to the pressure setting of a relief valve (B). Therefore, valve (D) is held wide open, causing no backpressure at the discharge of motor (E).

Figure 19-8. Brake circuit. (1 of 4)

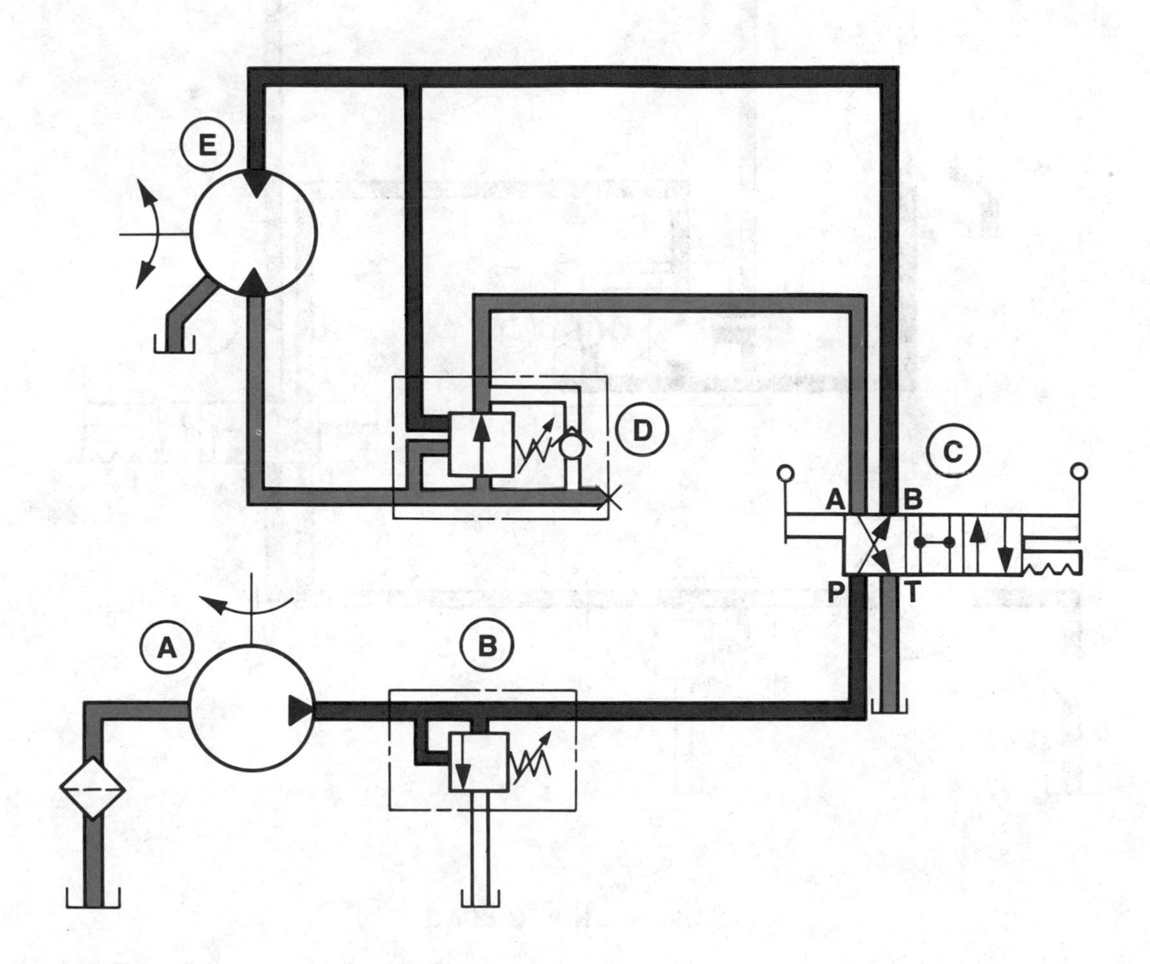

B. FULL SPEED RUNNING WITH AN OPPOSING LOAD

The load opposes the direction of rotation of motor (E) during "run." Working pressure required to drive this load acts under the large spool area of (D) to hold it fully open. Discharge from (E) returns freely to tank through (D) and (C). Delivery rate of pump (A) determines speed of (E).

Figure 19-8. Brake circuit. (2 of 4)

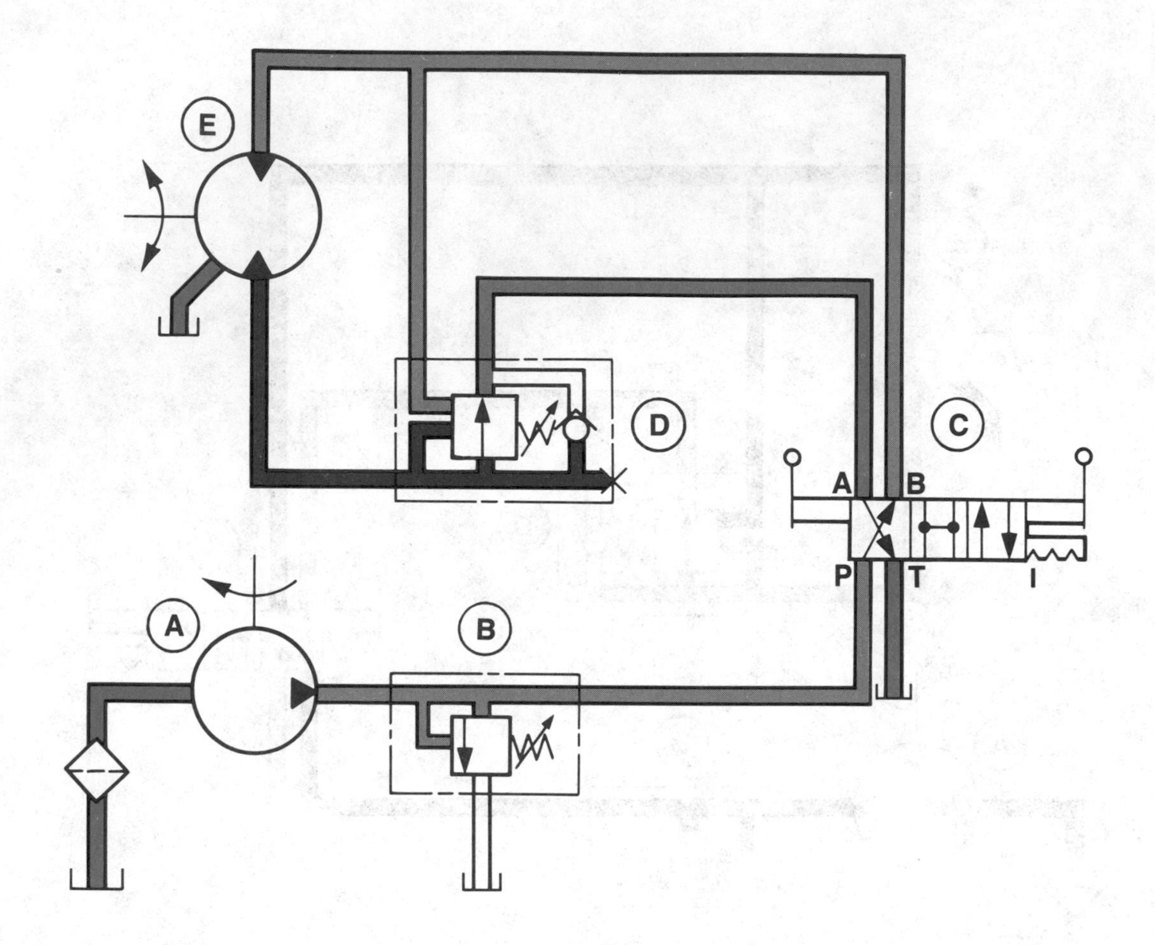

C. OVERRUNNING LOAD

The load may act in the same direction as rotation of motor (E) in certain applications. This "negative load" assumes a portion of the driving force on motor (E) which reduces pressure at the motor inlet.

Reduced pressure at motor inlet, effective under the valve spool of (D), permits the spool to move toward its closed position, thus restricting the discharge from (E).

Restricted flow through (D) creates backpressure in the outlet of (E). This backpressure acts under the small piston of (D).

The sum of the pressures acting under the valve spool and small piston of (D) holds the valve spool at the restricting position required for sufficient backpressure to maintain control of the load on (E).

The extent of negative loading determines the amount of backpressure on (E).

Figure 19-8. Brake circuit. (3 of 4)

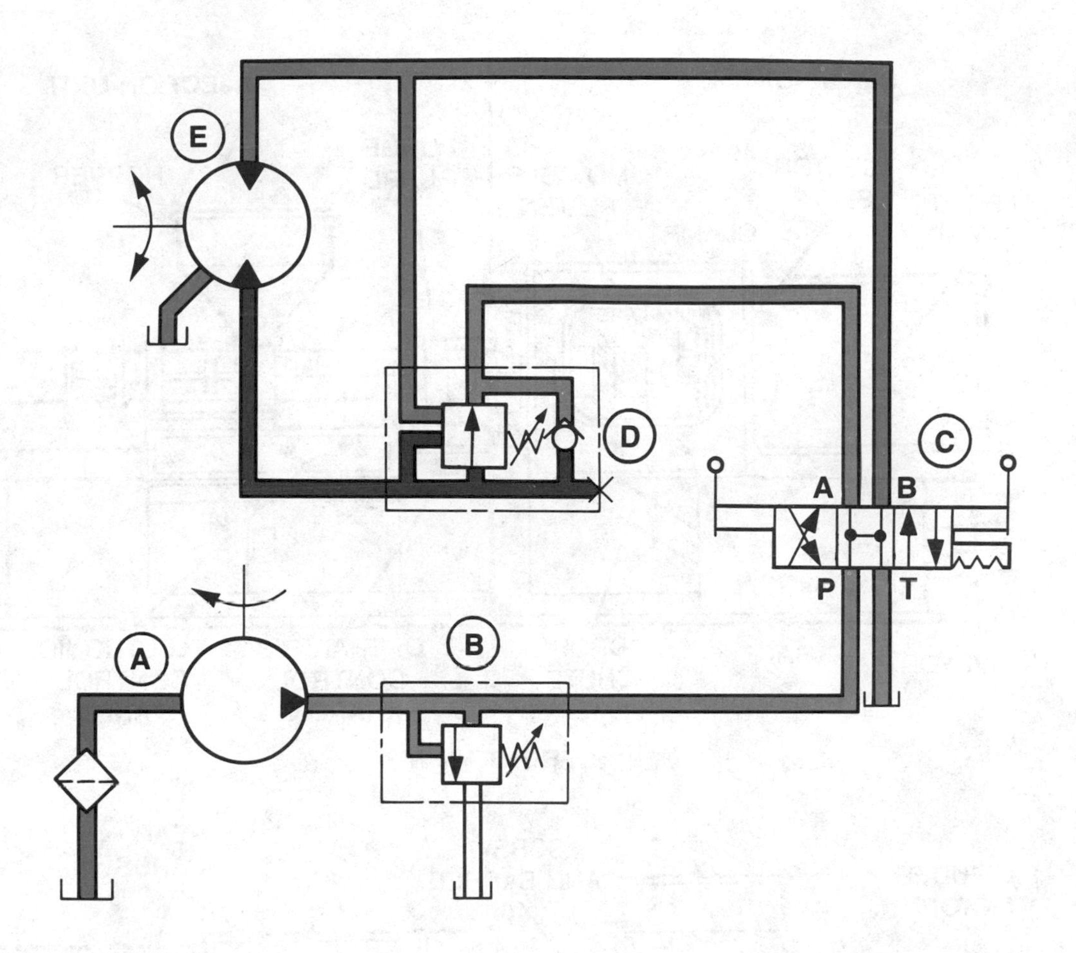

D. BRAKING

Valve (C) is shifted to the "Neutral" position to brake the load on motor (E). Pump (A) delivery is open to tank through valve (C).

Load inertia continues to drive (E) causing it to act as a pump. Inlet fluid to (E) is supplied through (C).

With the inlet of (E) open to tank, pilot pressure under the valve spool of (D) becomes zero, permitting it to move toward the closed position. This restricts discharge from (E) creating backpressure at its outlet.

Backpressure at outlet of (E) acts under the small piston of (D) opposing the spring force. These two opposing forces hold the valve spool at the restricting position. Adjusted setting of (D) therefore determines braking pressure and rate of deceleration.

Figure 19-8. Brake circuit. (4 of 4)

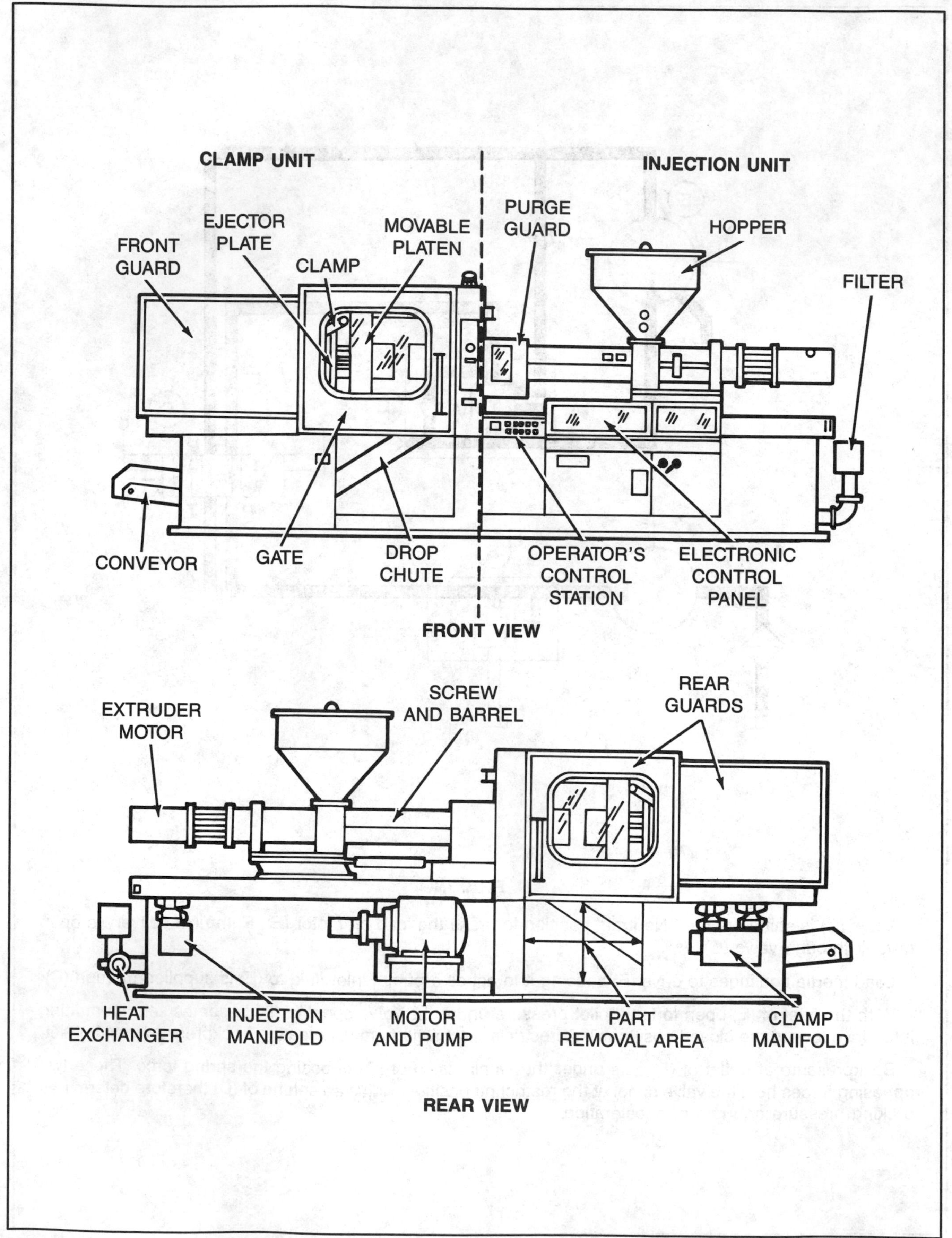

Figure 19-9. Horizontal injection molding machine.

controls for the injection molding machine industry. Cartridge valve circuits have replaced conventional spool valve circuits because they provide faster response times, less complicated circuitry, less leakage, and longer life.

Overall Machine Circuit

The injection molding process consists of several sequential stages of operation which are explained below in simplified form:

- Powdered or granular thermoplastic material is fed into the injection section of the machine.
- The main clamp cylinder moves forward, closing the mold halves.
- The full clamp area is pressurized to provide the force necessary to hold the two mold halves tightly together during injection of the molten thermoplastic material and also during the cooling down period.
- The material is transported toward the mold and at the same time melted and mixed by the rotating injection screw. The injection screw is turned by the extruder motor. As the screw turns, the plastic pellets are sheared by the flights on the screw and then melted by heaters as they move toward the front end of the screw chamber.
- The molten thermoplastic material is forced into the clamped and pressurized mold during the injection stroke. This step must be well controlled to ensure that all sections of the mold are filled at the proper speed. As the screw is retracted, new thermoplastic is drawn into the screw chamber from the hopper.
- The injected material solidifies in the shape of the mold cavity.
- The clamped mold opens and ejector cylinders push the completed pieces out of the mold.

It is very important to provide good pressure and flow control throughout the entire process. This ensures the repeatable cycle essential for keeping scrap losses to a minimum.

Figure 19-10 shows a block diagram of the overall injection molding machine circuitry. The system has three main manifolds that control the injection molding process. These manifolds and the functions they control are:

- Pressure/Flow Control manifold (P/Q manifold) controlling system pressure and flow.
- Clamp manifold controlling clamp close, mold protect, clamp pressurization and prefill shift, decompression, and clamp open functions.
- Injection manifold controlling injection forward, injection return, extruder run, and backpressure control functions.

Pressure/Flow (P/Q) Manifold

The P/Q manifold is a flow-demand-dependent pump control package designed for use with two or more fixed displacement pumps. On/off loading of the larger pump is regulated by hydraulic flow demand; no additional electronic signals are required.

Remotely generated electronic analog command signals provide programmable control of system pressure and flow. The system can easily be adapted for plant computer control of production.

Basic Functions of P/Q Manifold. Figure 19-11 shows components of the P/Q manifold, along with a large pump (Q2) and a small pump (Q1). The ratio of pump flow (Q2:Q1) depends on the machine requirements. Pump flow ratios between 60:40 percent and 85:15 percent may be used. In practice, a ratio of 80:20 percent is a good compromise between energy efficiency and application flexibility. Q2 may represent one or more pumps depending upon the amount of flow required.

The P/Q manifold operates in either a pressure or flow control mode. If the pressure (M1) at the inlet of proportional throttle valve 4.0 is less than the commanded pressure setting of proportional pressure control 2.3, the P/Q manifold operates in a load sensing, pressure compensated, flow control mode.

If flow to the load is less than commanded flow, for example when a cylinder has reached the end of its stroke, pressure (M1) rises to the commanded pressure setting. In this case, the P/Q manifold operates in a pressure control mode.

Small pump Q1 supplies lower flow functions such as slow movements and pressure holding. If system flow demand is less than the output of Q1, large pump Q2 is unloaded to the tank. If the system requires more flow, Q2 will automatically be loaded. On- and off-load response times are very short, with minimum pressure overshoot and virtually no shock.

Figure 19-12 shows pressure step response characteristics for a P/Q control manifold with a deadhead load, for example, a cylinder fully extended. Starting from the idle condition with zero command to the pilot pressure valve 2.3, relief valves 1.0 and 2.0 are open, unloading both pumps to tank at about 200 psi (13.78 bar / 1378.80

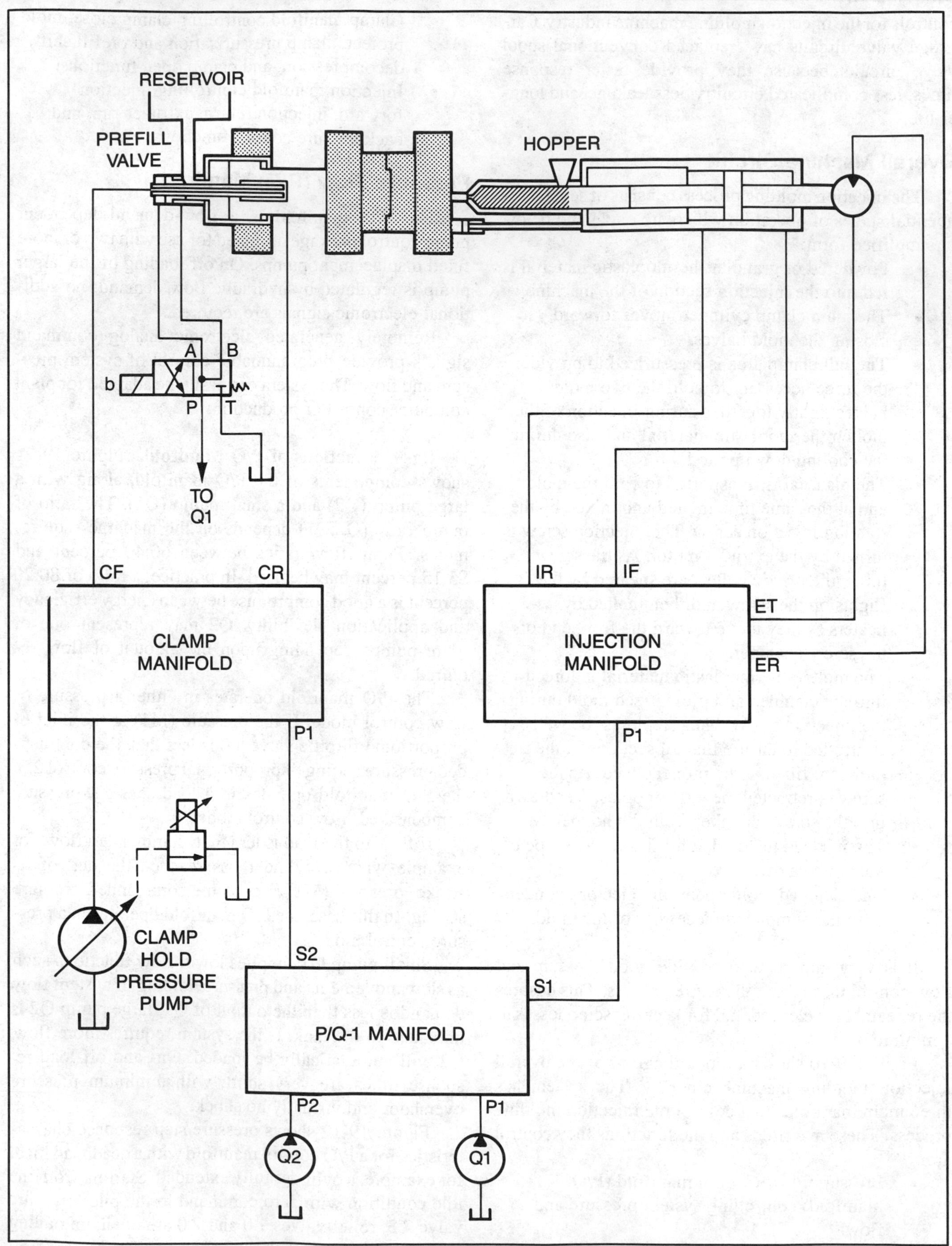

Figure 19-10. Overall machine block diagram.

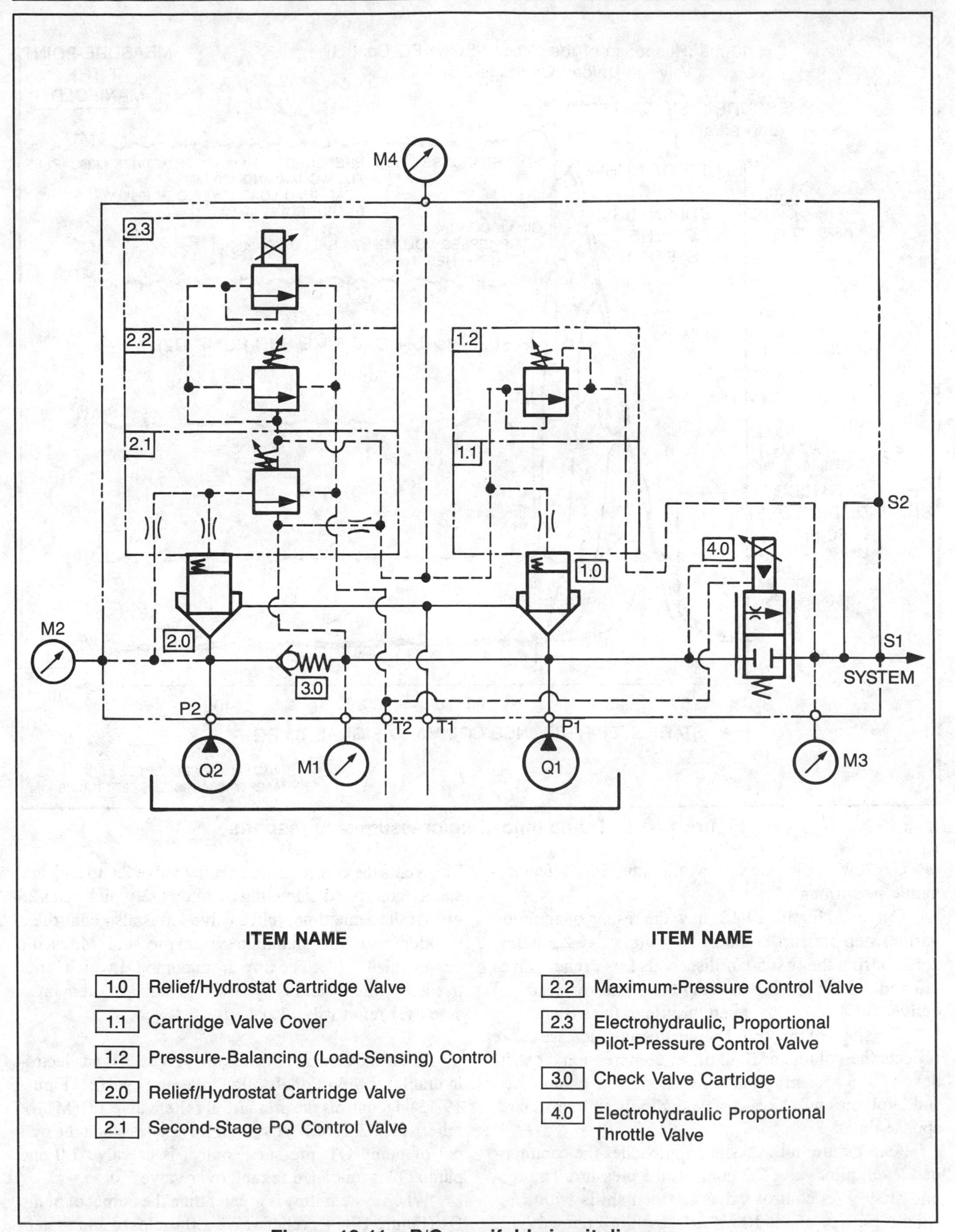
M4
2.3
2.2
1.2
2.1
1.1
S2
4.0
1.0
M2
2.0
S1
3.0
SYSTEM
P2
T2
T1
P1
Q2
M1
Q1
M3
ITEM NAME
1.0 Relief/Hydrostat Cartridge Valve
1.1 Cartridge Valve Cover
1.2 Pressure-Balancing (Load-Sensing) Control
2.0 Relief/Hydrostat Cartridge Valve
2.1 Second-Stage PQ Control Valve
ITEM NAME
2.2 Maximum-Pressure Control Valve
2.3 Electrohydraulic, Proportional Pilot-Pressure Control Valve
3.0 Check Valve Cartridge
4.0 Electrohydraulic Proportional Throttle Valve

Figure 19-11. P/Q manifold circuit diagram.

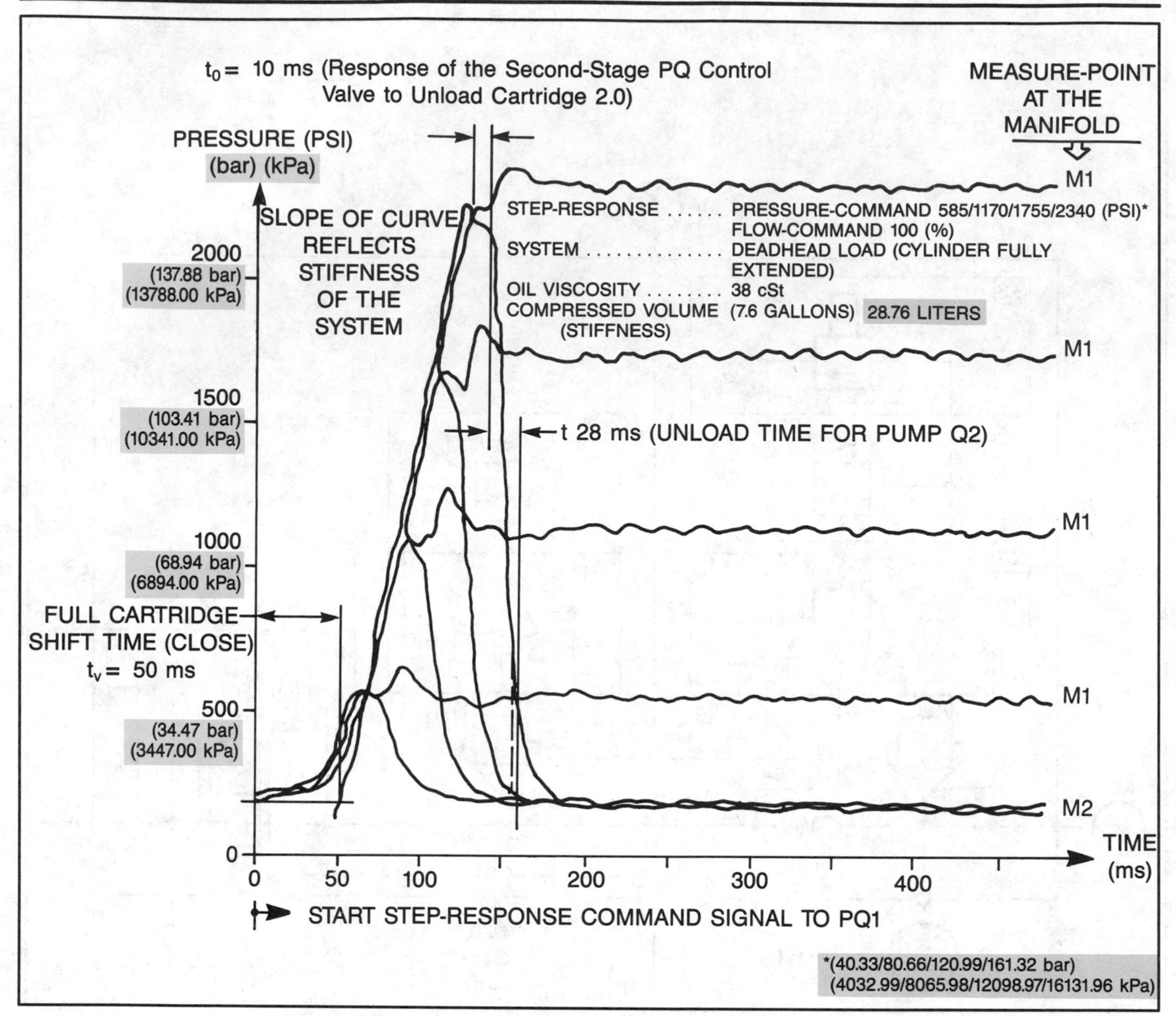

Figure 19-12. Pump onloading/pressure step response.

kPa). A lower idle pressure would adversely affect dynamic performance.

Curves in Figure 19-12 show the results of applying various step pressure commands to pilot valve 2.3 at time zero. During the first 50 milliseconds (ms), relief valves 1.0 and 2.0 are closing to bring the pumps on-load and deliver fluid across the open throttle to the system.

The slope of the straight-line pressure increase curve reflects the volume of fluid under compression, or stiffness of the system. Systems with smaller compressed fluid volumes will have correspondingly shorter step response times.

As pressure in the system approaches the commanded level, pilot valve 2.3 controls the pressure. The second stage P/Q control valve 2.1 then shifts with a response time of about 10 ms as shown in Figure 19-12. This vents the control fluid of relief valve 2.0 to tank and starts a controlled unloading of pump Q2, which takes 28 ms. At the same time, relief valve 1.0 is also controlled by pilot valve 2.3 to maintain system pressure (M3) at the commanded value. The flow from pump Q1 which is not needed by the system is bypassed to tank at system pressure over relief valve 1.0.

Pressure Control Function. Valve 2.3 is an electrohydraulic proportional pilot pressure valve (Figure 19-13). It controls the pressure at relief valve 1.0 (M1) or relief valve 2.0 (M2). If system flow is less than the output of pump Q1, pressure control is at valve 1.0 and pump Q2 is unloaded to tank over valve 2.0.

When system flow is greater than the output of pump Q1 (Figure 19-14), relief valve 1.0 is closed and system

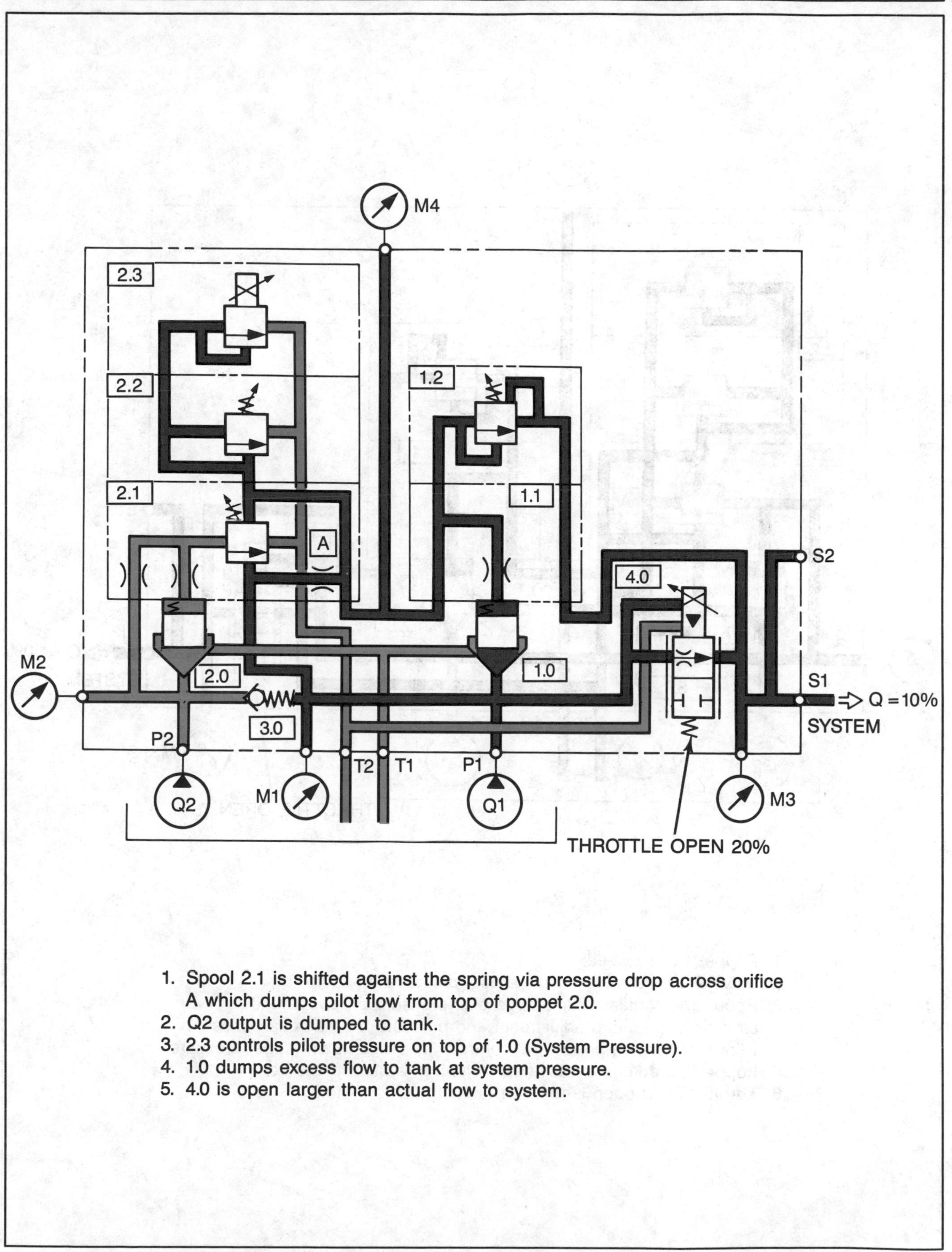

1. Spool 2.1 is shifted against the spring via pressure drop across orifice A which dumps pilot flow from top of poppet 2.0.
2. Q2 output is dumped to tank.
3. 2.3 controls pilot pressure on top of 1.0 (System Pressure).
4. 1.0 dumps excess flow to tank at system pressure.
5. 4.0 is open larger than actual flow to system.

Figure 19-13. Pressure control function System flow less than pump Q1 output.

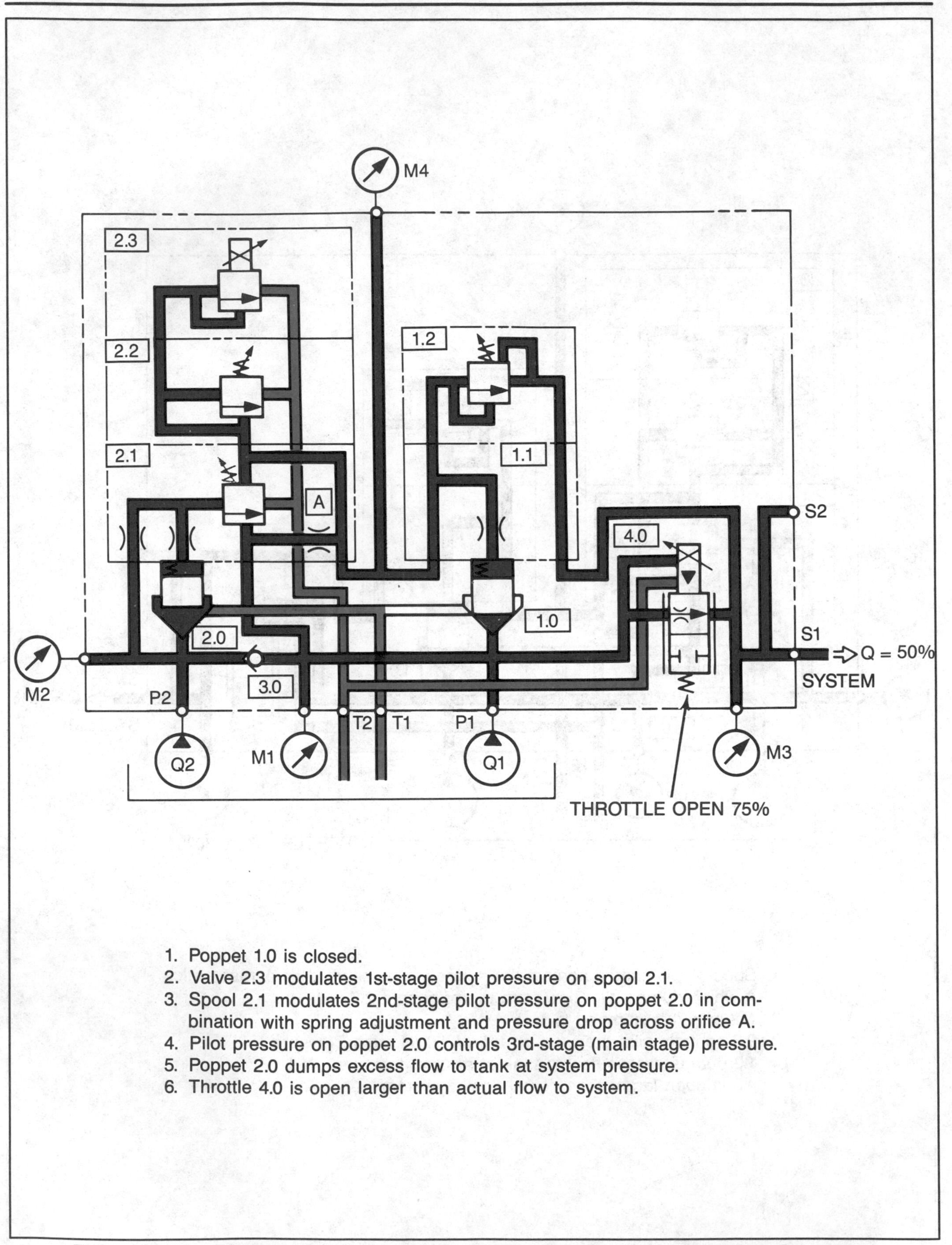

1. Poppet 1.0 is closed.
2. Valve 2.3 modulates 1st-stage pilot pressure on spool 2.1.
3. Spool 2.1 modulates 2nd-stage pilot pressure on poppet 2.0 in combination with spring adjustment and pressure drop across orifice A.
4. Pilot pressure on poppet 2.0 controls 3rd-stage (main stage) pressure.
5. Poppet 2.0 dumps excess flow to tank at system pressure.
6. Throttle 4.0 is open larger than actual flow to system.

Figure 19-14. Pressure control function System flow greater than pump Q1 output.

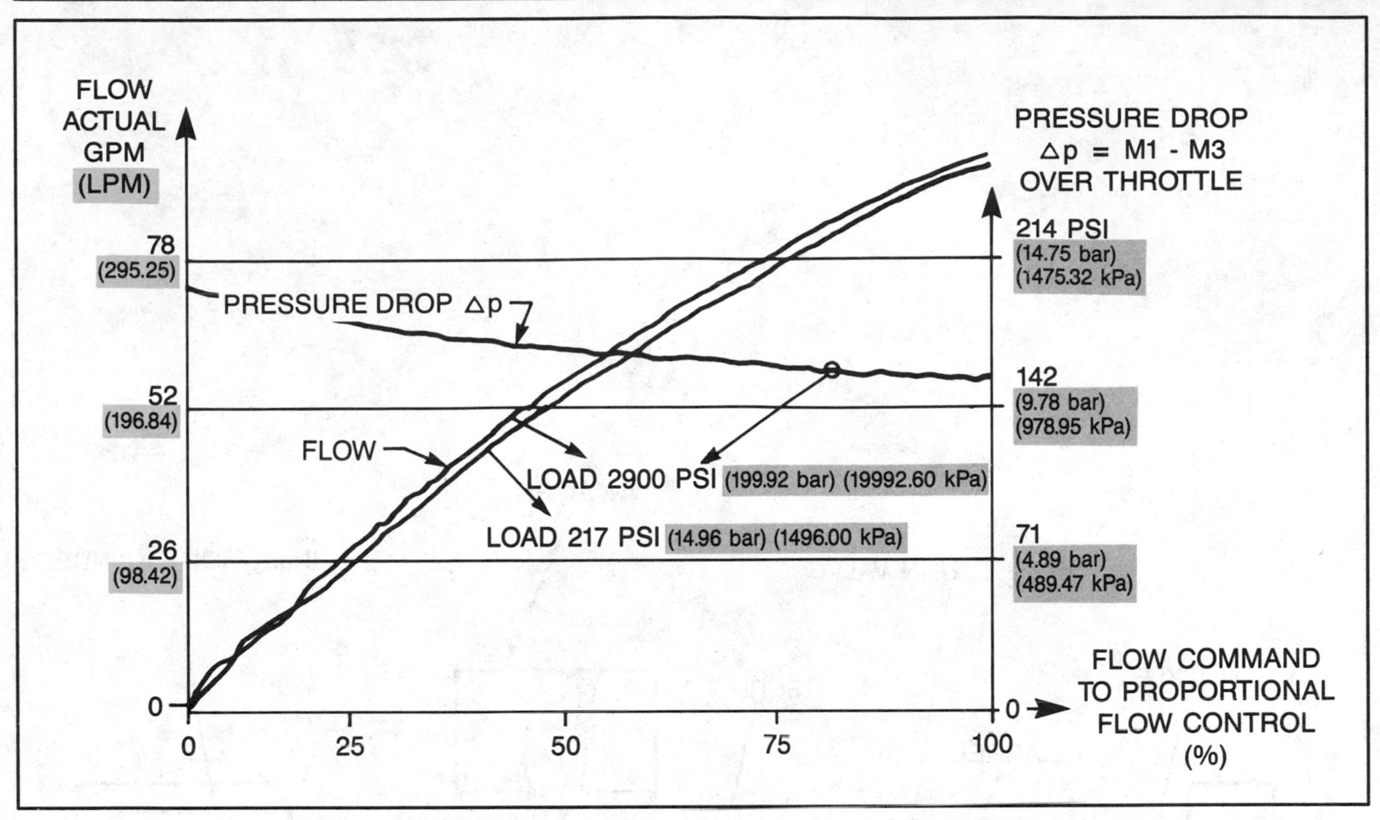

Figure 19-15. Flow and pressure drop vs. flow command.

pressure is controlled by relief valve 2.0.

Flow Control Function (Load Sensing). Load sensing is achieved by a cartridge valve hydrostat combined with a proportional throttle valve that keeps the pressure difference between M1 and M3 almost constant (Figure 19-15). Consequently, there is almost a linear flow/demand signal characteristic within the whole range of load pressure and flow variation. The proportional throttle with feedback position control assures high repeatability.

A good match between maximum flow and maximum demand signal can be achieved by selecting the appropriate available flow range for the proportional throttle valve 4.0.

Valve 4.0 (Figure 19-14) is an electrically modulated proportional throttle valve. Combined with cartridge valve 1.0 or 2.0 (acting as a hydrostat), the throttle provides a three-way (bypass type) pressure compensated flow control or load sensing function.

If the flow command to the throttle 4.0 is less than the output of pump Q1, excess flow from pump Q1 goes over cartridge valve 1.0 to tank. Valve 1.0 acts as the pressure compensating hydrostat to maintain a near constant pressure drop across throttle 4.0 of about 215 psi (14.83 bar / 1483.21 kPa) at zero flow. Cartridge valve 2.0 opens to unload pump Q2 to tank.

If the flow command to throttle 4.0 is greater than the output flow of pump Q1, cartridge valve 1.0 is closed and excess flow from pump Q2 goes over cartridge valve 2.0 to tank. Valve 2.0 acts as a pressure compensating hydrostat in this case.

The hydrostatic pressure drop (M3-M1) equals the 100 psi (6.89 bar / 689.40 kPa) cracking pressure of valve 1.0 plus the 115 psi (7.93 bar / 792.81 kpa) setting of valve 1.2.

The flow vs. command curves of Figure 19-15 show that the P/Q manifold provides excellent pressure compensation. Only a small change in flow results from a large change in load pressure.

Injection Manifold

The injection cycle consists of the following phases:

- Extruder Run/Backpressure Control
- Melt Decompression (Injection Return)
- Injection Forward

As shown in Figure 19-16, the injection unit consists of a screw and barrel, hopper, extruder motor, and injection cylinder which are all mounted on a moveable carriage. The screw, extruder motor and injection cylinder form a single unit that is capable of moving while the carriage remains stationary.

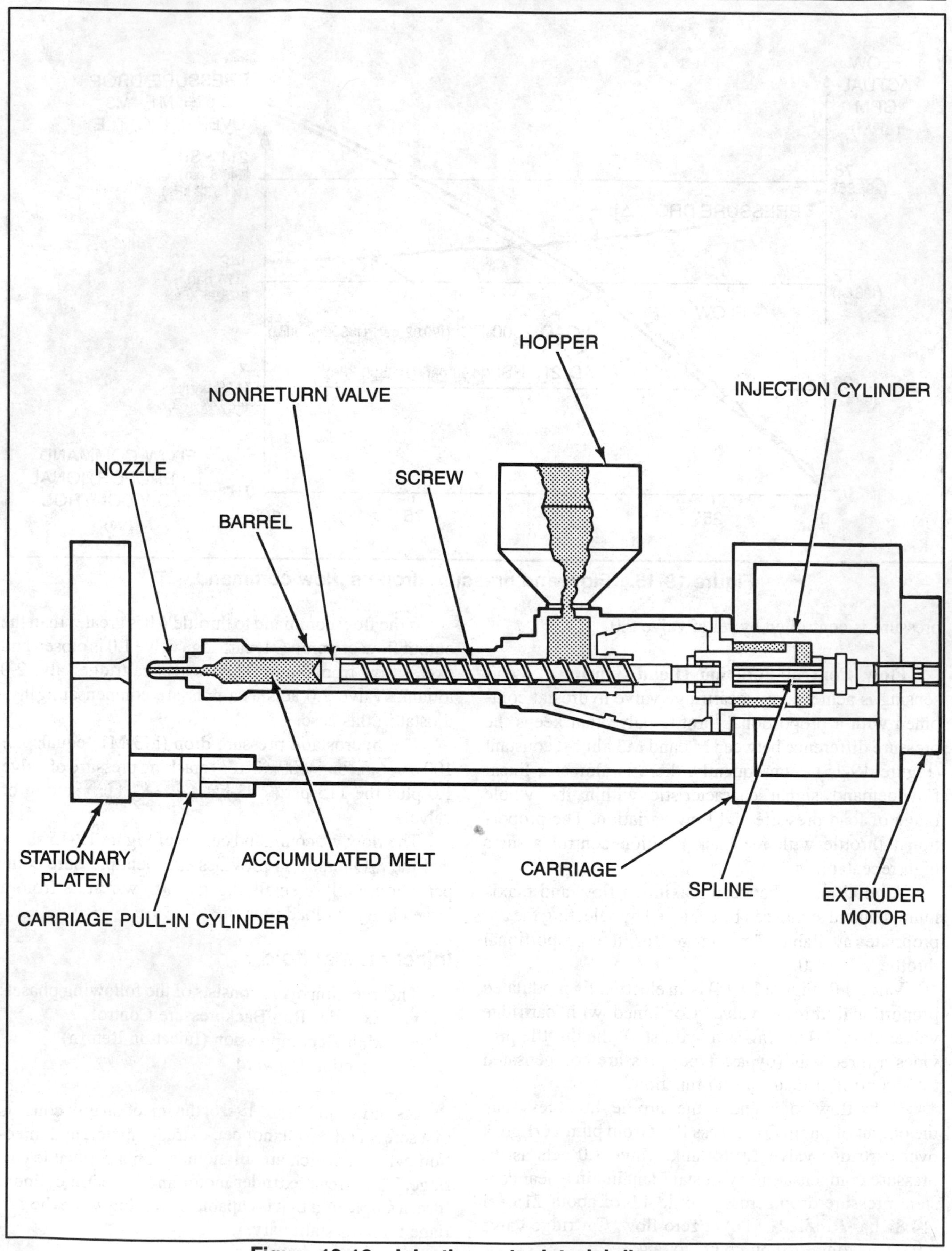

Figure 19-16. Injection unit pictorial diagram.

SOLENOID	Y1	Y2	Y3	Y4	Y5
EXTRUDER-RUN/BACKPRESSURE	X				X
INJECTION RETRACT		X			X
INJECTION FORWARD			X		X
CARRIAGE RETURN				X	
CARRIAGE FORWARD & HOLD					X

Solenoid Y1 (Valve 1) = Extruder-Run Cartridge Valve
Solenoid Y2 (Valve 2) = Injection-Back Cartridge Valve
Solenoid Y3 (Valve 3) = Injection-Forward Cartridge Valve
Solenoid Y4 (Valve 7) = Carriage Pull-In Directional Valve
Solenoid Y5 (Valve 7) = Cartridge Pull-In Directional Valve
Valve 1 = Extruder-Run Cartridge Valve (Solenoid Y1)
Valve 2 = Injection-Back Cartridge Valve (Solenoid Y2)
Valve 3 = Injection-Forward Cartridge Valve (Solenoid Y3)
Valve 4 = Pilot Shuttle Valve
Valve 5 = 4:1 Backpressure Cartridge Valve
Valve 6 = Tank Cartridge Valve
Valve 7 = Carriage Pull-In Directional Valve
Valve 8 = Manual Flow Control

Figure 19-17. Injection cycle solenoid chart.

Good control of the injection unit is essential for maintaining a precisely repeatable cycle that ensures the production of quality parts. Because molds are becoming more intricate, surface finish tolerances are more exacting, and the materials being used are more exotic, plastic processing machines in use today require very sophisticated controls.

The injection manifold must be able to control the following injection unit variables:

- Speed of the extruder motor during the melting of the plastic
- The backpressure on the plastic during the melting process
- Barrel heater temperatures
- Speed of injection (injection profiling)
- Pressure control of the mold fill and hold sequences

Figure 19-17 is a solenoid chart that indicates the status of valve solenoids during the various injection phases. Figures 19-18 through 19-20 show the injection unit circuitry during each portion of the cycle. Refer to these diagrams while proceeding through the injection cycle descriptions presented below.

In normal operation, the carriage is held in the forward position against the mold by the carriage pull-in cylinder. (See Figure 19-18.) This prevents any backward movement of the carriage that might be caused by the force of injection during the injection forward portion of the cycle. Solenoid Y5 on the carriage pull-in directional valve (7) remains energized throughout the entire injection process.

In the event that material must be cleaned from around the nozzle of the injection barrel, or if maintenance is required on the injection unit, solenoid Y5 is de-energized and solenoid Y4 is energized to produce movement of the carriage away from the mold.

Extruder Run/Backpressure Control. During this initial phase of the injection cycle, plastic pellets are transformed into a viscous material that can be injected into a mold. The extruder motor is a hydraulic low-speed, high-torque motor that is coupled to the injection screw. As the motor and screw turn, plastic pellets are gravity fed from the hopper, pushed forward by the motion of the screw across the flights of the screw and directed to the front of the barrel. As the pellets pass between the flights of the screw and the inside of the barrel, shear energy is exerted on the plastic. At the same time, heat energy is added to the polymer from the electrically controlled barrel heaters. The continuously rotating screw provides a mixing action which blends in any colorants and helps maintain a homogeneous melt. Some materials require more heat energy than others for optimum plastification and consistency.

The melting and mixing action of the injection unit can be controlled by regulating screw speed and by exerting a constant backpressure on the screw and melt as the screw turns. In this way, the energy exerted into the material is held constant, resulting in a uniform density of the material before it is injected into the mold.

The screw rotates and pushes the viscous plastic material forward to the front of the barrel where it accumulates. This action produces a pressure on the screw and injection cylinder causing the unit to back away from the nozzle end of the barrel. This movement is slowed and controlled by the opposition of the backpressure on the injection cylinder set up by the injection manifold. Movement continues until the injection unit has moved far enough to reach a preset shot size. The distance that the screw moves backward is usually monitored by some form of stroke transducer or limit switch. Feedback from this device is directed to the electronic controller that sends the run commands to the extruder motor and screw. When the appropriate shot size is reached, the controller turns off the extruder motor.

As shown in Figure 19-18, the extruder motor is activated by controller selection of solenoid Y1. At the same time, the motor speed is controlled by an analog command sent to the flow control circuitry on the P/Q manifold. Extruder motor speed can be controlled either open or closed loop depending on the sophistication needed.

In many cases, a very low backpressure control is needed to provide proper consistency of the melt. Backpressure is regulated by controlling the pilot pressure of cartridge valve 5 on the injection manifold. The controlling pressure is provided by the Auxiliary Pressure Control block through connection S1 on the injection manifold. All pressure valves have a minimum pressure they can control. The backpressure valve shown in Figure 19-18 has an area ratio of 4:1 between the bottom and top of the cartridge. By using this ratio, the minimum controllable pressure can be lowered and a pressure control of 1/4 of the pilot pressure can be achieved. Fluid from the injection cylinder is forced by the backward movement of the screw through valve 5 to tank at a pressure controlled by the valve. The other port (Z2) that closes the valve is vented to tank through a directional valve on cartridge 3 during this cycle.

Melt Decompression. While the extruder motor and screw are loading the barrel with molten material, the plastic in the mold from the previous shot has hardened. This part must be removed before a new injection forward cycle can be initiated. If the mold opens while backpressure is on the melt in the barrel, plastic material would be forced out of the nozzle end of the injection unit. For this reason, the melt in the barrel must be decompressed before the mold can be opened. This is accomplished by energizing solenoid Y2 on the injection return cartridge valve (2). (See Figure 19-19.) This vents pilot pressure on the AP area of the valve to tank, allowing the valve to open. Flow through the valve is applied to the screw end of the injection cylinder and moves the injection unit return for a very short stroke while the mold is opened for extraction of the previous part. Backpressure control from the P/Q manifold is removed by the controller and fluid from the other side of the injection cylinder returns to tank through valve 5. The mold is then closed for the injection forward portion of the cycle.

Injection Forward. The next step is to inject the polymer into the mold at a controlled speed and pressure. (See Figure 19-20.) By energizing solenoid Y3, valve 3 opens and diverts flow from the P/Q block into the injection cylinder. An analog flow command is provided to control the rate at which the plastic fills the various parts of the mold. As the injection cylinder is pushed forward, the oil in the opposite side of the cylinder is diverted to tank across cartridge valve 6. Pilot pressure is applied to both of the closing ports on valve 5 keeping it closed during injection forward. Valves 1 and 2 are also closed while the injection cylinder moves forward.

As the viscous plastic is forced into the mold, a resistance is created which tends to push the injection unit away from the mold, so the injection unit must be held

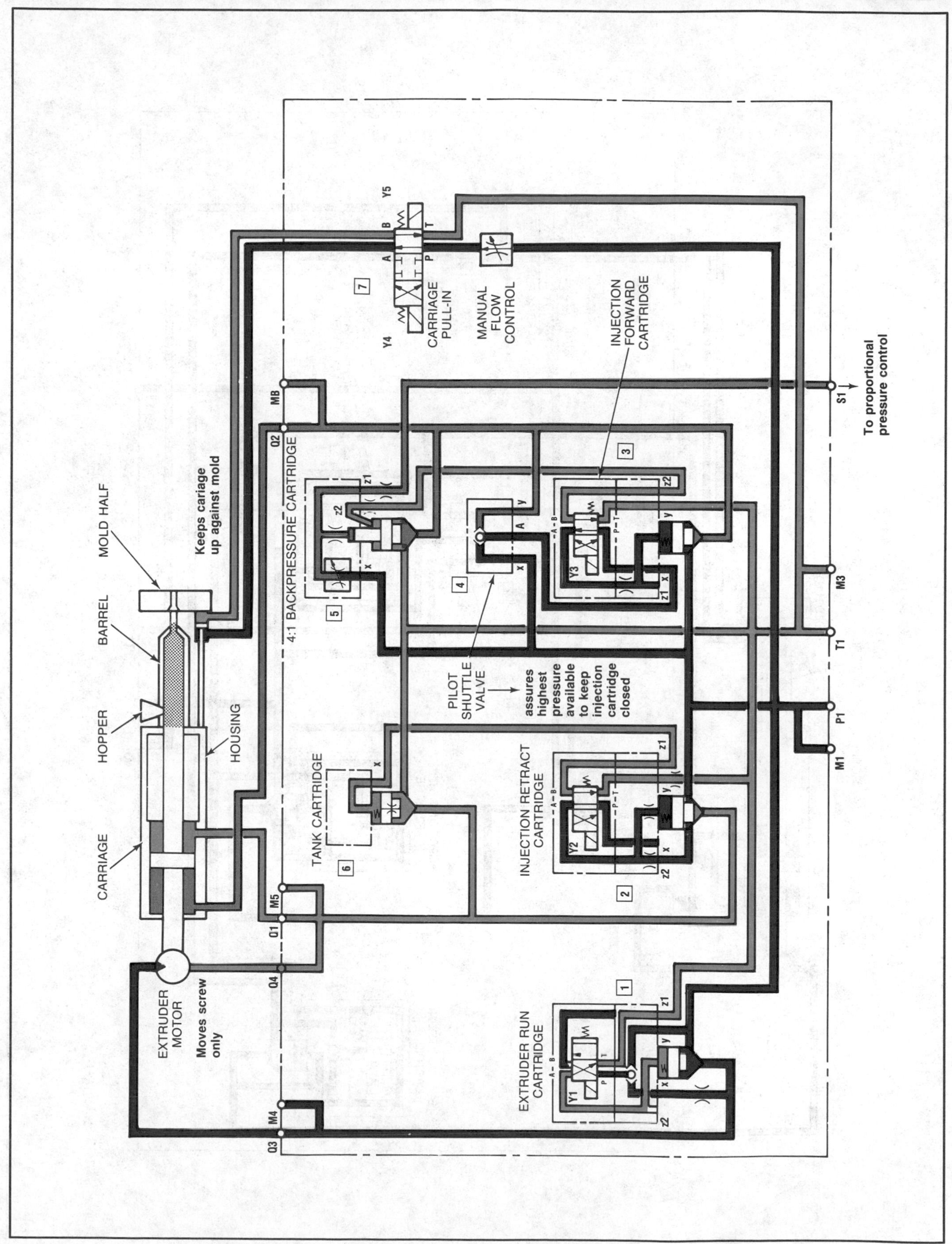

Figure 19-18. Extruder run/backpressure control circuit.

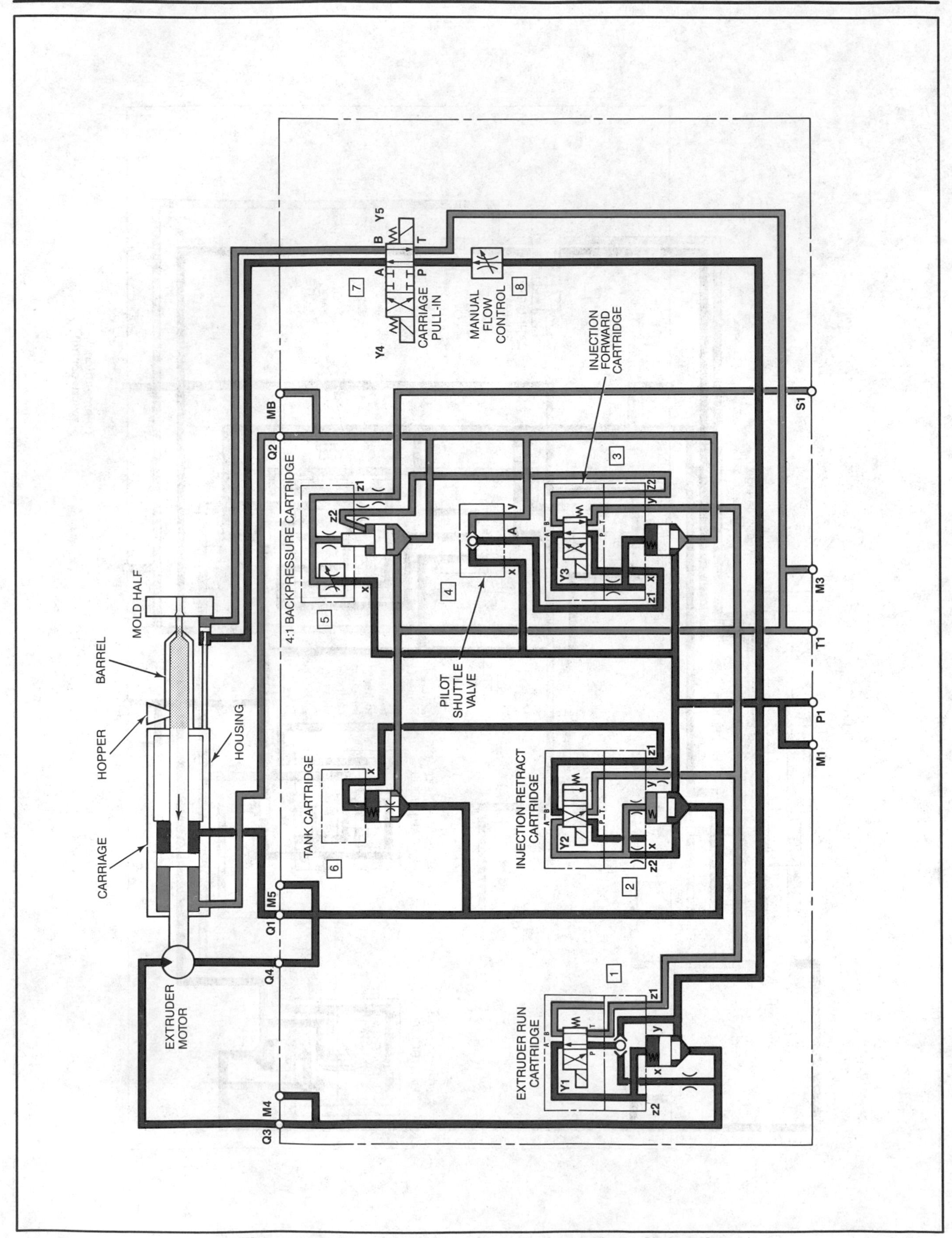

Figure 19-19. Melt decompression control circuit.

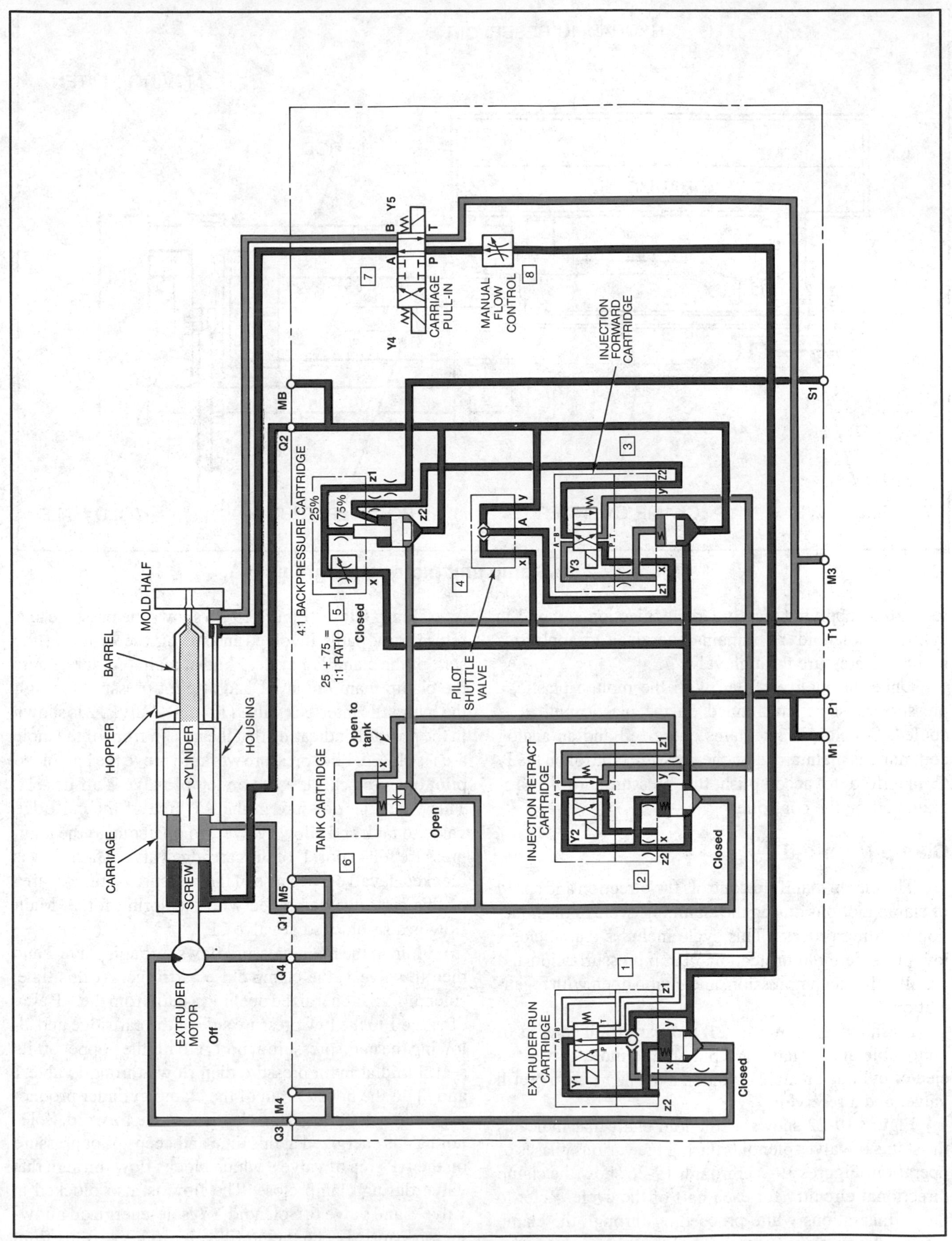

Figure 19-20. Injection forward control circuit.

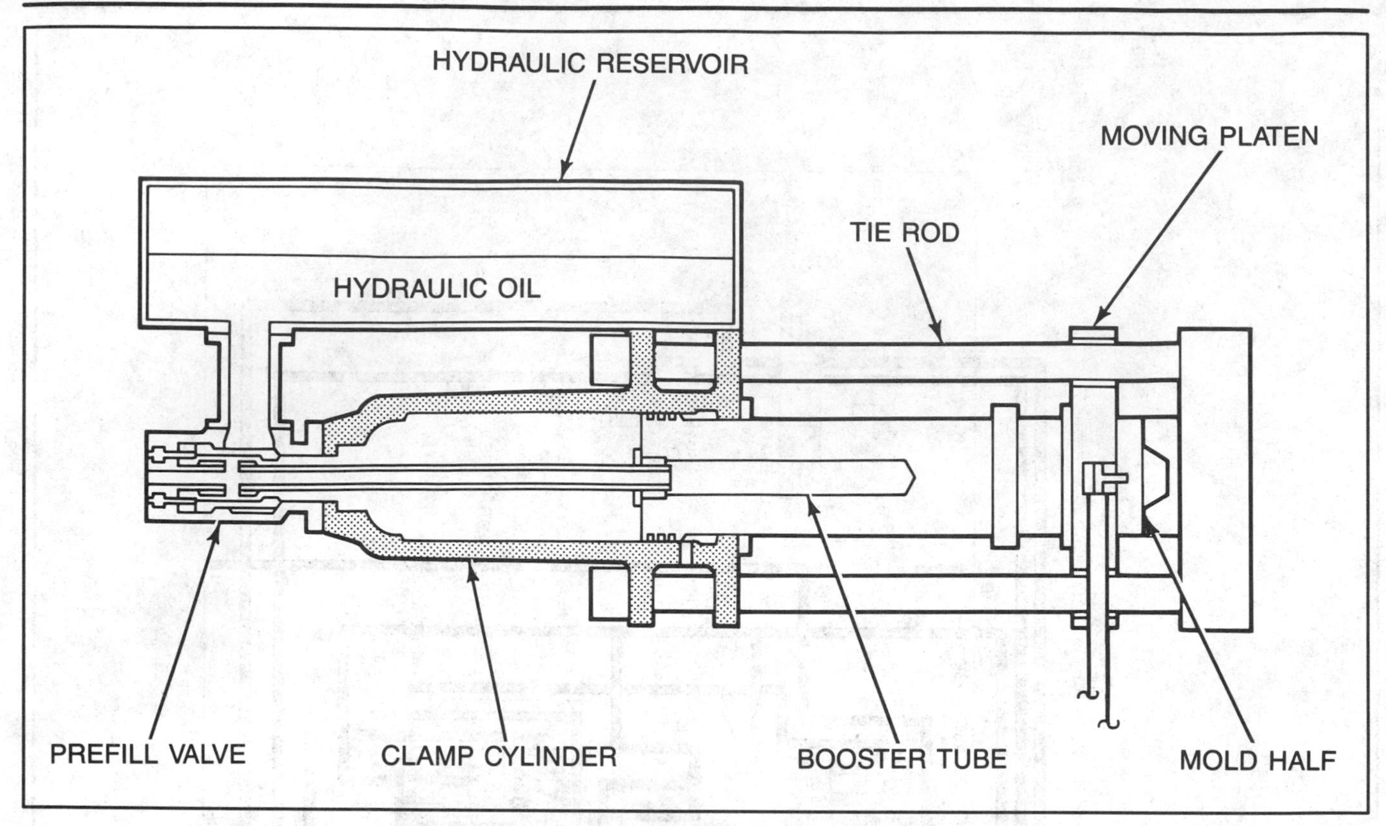

Figure 19-21. Clamp unit pictorial diagram.

forward by applying pressure to the injection unit pull-in cylinder. Solenoid Y5 remains energized to apply this hydraulic pressure through valve 7.

Once the mold is filled with the molten plastic, a pressure must be maintained on the plastic until it is cooled. Solenoid Y3 remains energized and an analog command is maintained on the pressure control on the P/Q manifold to accomplish the pressure profile programmed on the controller.

Clamp Manifold

The clamp manifold controls the direction and speed of clamp motions during the clamping cycle of the injection molding process. This cycle includes clamp close, low pressure mold protection, clamp pressurization and prefill shift, decompression, and clamp open with regeneration.

As shown in Figure 19-21, the clamp unit consists of a movable and a stationary platen, the mold halves, an ejector cylinder, a main cylinder, a booster ram, a prefill valve, and a reservoir.

Figure 19-22 shows a solenoid chart that indicates the status of valve solenoids during the various phases of operation. Figures 19-23 through 19-26 show the clamp directional circuitry for each part of the cycle. Refer to these illustrations while proceeding through the clamp motion descriptions presented below.

Clamp Close. Figure 19-23 shows the pressure and return flow paths involved in closing the clamp at the start of the clamping cycle. Fluid at main pressure enters the clamp manifold at P1 and at pilot pressure through the hydraulic interlock valve (12) at R. Valve 12 is shown in the position indicating that the gate in front of the mold area is closed. Pilot pressure works to move the ball in the pilot operated check valve on top of valve 2 off its seat. This vents the pressure in the AP area of the cartridge valve to tank and allows valve 2 to open due to the main pressure felt at port B of the cartridge valve. Main flow is blocked at valve 3 because of the pressure on the AP area of this cartridge combined with the spring force. Main flow is also blocked at valve 11.

Fluid at the main pressure flows through valve 2 and then to valve 1, the clamp close cartridge. At this time, solenoid Y5 is energized opening a path from the AP area of valve 1 to the low pressure side of the cartridge and allowing the main pressure at port A to lift the poppet off its seat. Fluid at main pressure then flows through valve 1 and on to the booster ram of the clamp cylinder producing a force that causes the clamp to move forward. Solenoid Y4 is energized at this time and keeps pilot pressure on the AP area of valve 9 which blocks flow through that valve during clamp close. The flow is also blocked at valve 8 and valve 6. Solenoid Y7 is de-energized allowing pressure to open the prefill valve. This permits fluid

SOLENOID	Y1	Y2	Y3	Y4	Y5	Y6	Y7
CLAMP CLOSE		X		X	X		
LPMP		X		X	X	X	
PRESSURIZATION		X		X	X		X
DECOMPRESSION							X
CLAMP OPEN/REGEN	X		X	X			

Solenoid Y1 (Valve 6) = Clamp Open Regenerative Pilot Valve
Solenoid Y2 (Valve 4) = Counterpressure Selection Directional Valve
Solenoid Y3 (Valve 3) = Clamp Open Cartridge
Solenoid Y4 (Valve 9) = Clamp Open to Tank Cartridge
Solenoid Y5 (Valve 1) = Clamp Close Cartridge
Solenoid Y6 (Valve 11) = Low-Pressure Mold-Protection Valve
Solenoid Y7 (Valve 10) = Prefill Shift Valve

Valve 1 = Clamp Close Cartridge (Solenoid Y5)
Valve 2 = Main-Stage Interlock Cartridge
Valve 3 = Clamp Open Cartridge (Solenoid Y3)
Valve 4 = Counterpressure Selection Directional Valve (Solenoid Y2)
Valve 5 = Intensification Relief Valve
Valve 6 = Regenerative Pilot Valve (Solenoid Y1)
Valve 7 = Clamp Close to Tank Cartridge
Valve 8 = Regenerative Cartridge
Valve 9 = Clamp Open to Tank Cartridge (Solenoid Y4)
Valve 10 = Prefill Shift Valve (Solenoid Y7)
Valve 11 = Low-Pressure Mold-Protection Valve (Solenoid Y6)
Valve 12 = Gate Valve

Figure 19-22. Clamp motion solenoid chart.

from the reservoir to fill the main cylinder area while the clamp is moving forward.

Fluid from the pullback area of the main clamp cylinder flows through valve 7, the clamp close tank cartridge, and then to tank. The pressure of the return oil permitted through valve 7 is determined by the pilot pressure on the AP area. The path for pilot flow is through orifice 18 in the insert, through the cover and across orifice 27 to tank through valve 4 (Y2 is energized). Orifice 18 determines the pilot flow, while orifice 27 dictates the pilot pressure on top of cartridge 7. These conditions, in turn, determine the counterpressure present during clamp close. This pressure should be just high enough to prevent cavitation from a runaway load as the clamp is closing. Return flow is blocked at both valve 3 and valve 11.

Low Pressure Mold Protection. Low pressure mold protection (LPMP) is a provision in the hydraulic circuit that helps to protect the mold from damage that might be caused by an obstruction between the mold halves or by excessive speed commands. During clamp close fast (before the mold protection stroke) the booster ram moves the clamp mass forward at high speed with a low counterpressure on the return line side.

When LPMP is selected (by setting a switch on the control panel) the LPMP circuitry is activated near the end of the clamp close stroke at a point that is determined by the LPMP stroke parameter (e.g., 2 inches (51 mm)). This parameter is inserted in a microprocessor control program during the machine mold setup procedure.

Figure 19-24 shows the flow paths in the clamp directional circuit during the LPMP stroke, when the overall force on the mold must be limited. When the LPMP

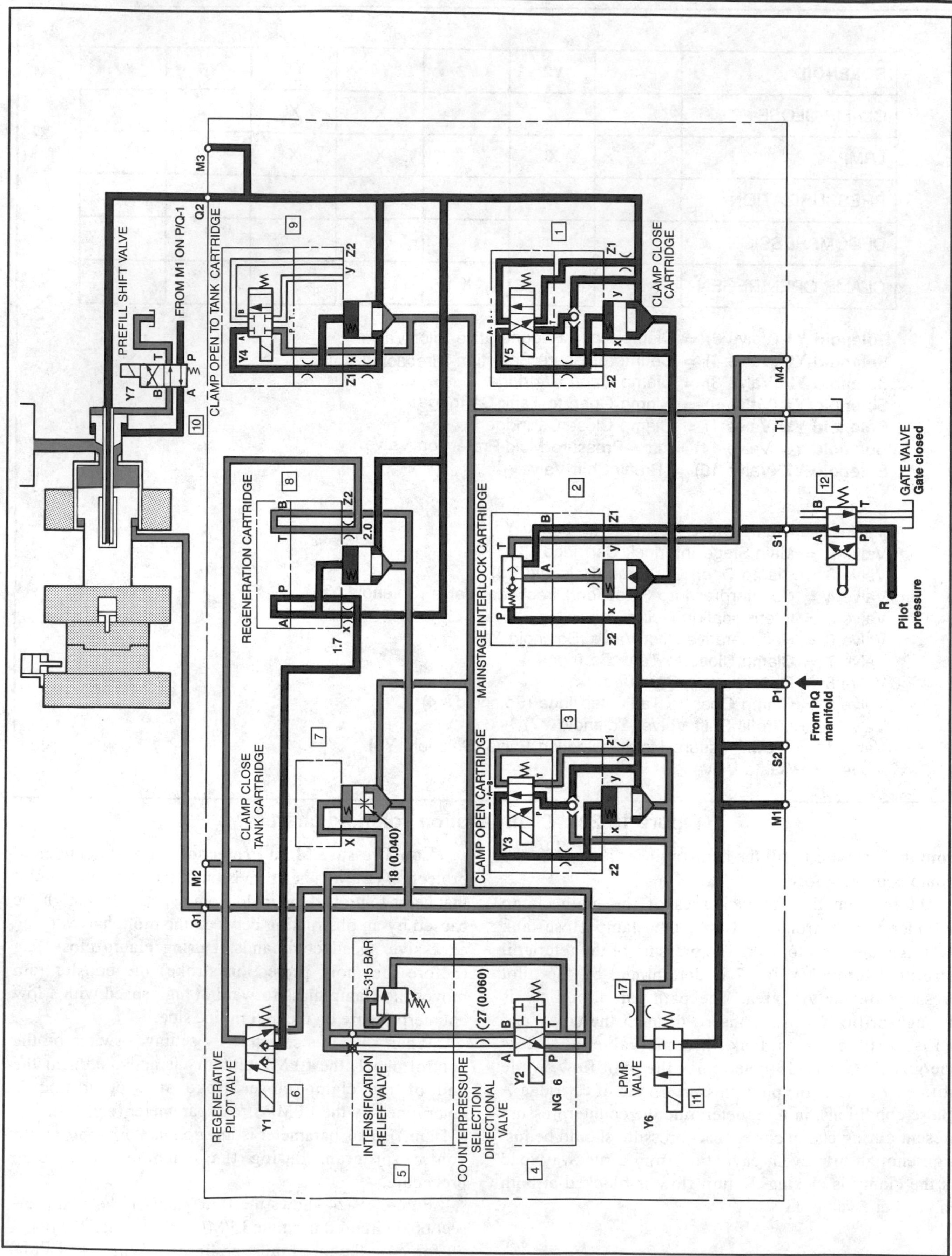

Figure 19-23. Clamp close (without LPMP) control circuit.

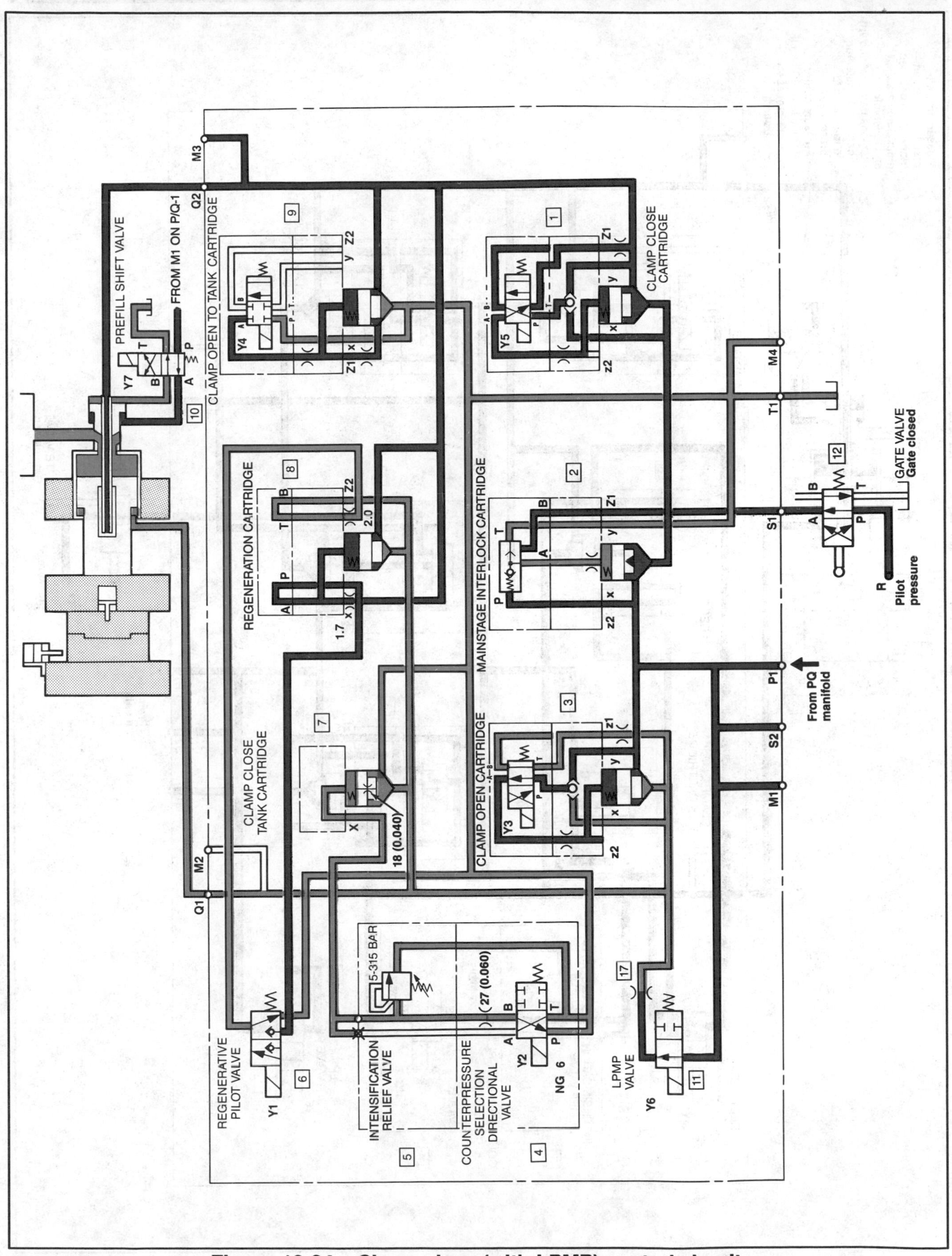

Figure 19-24. Clamp close (with LPMP) control circuit.

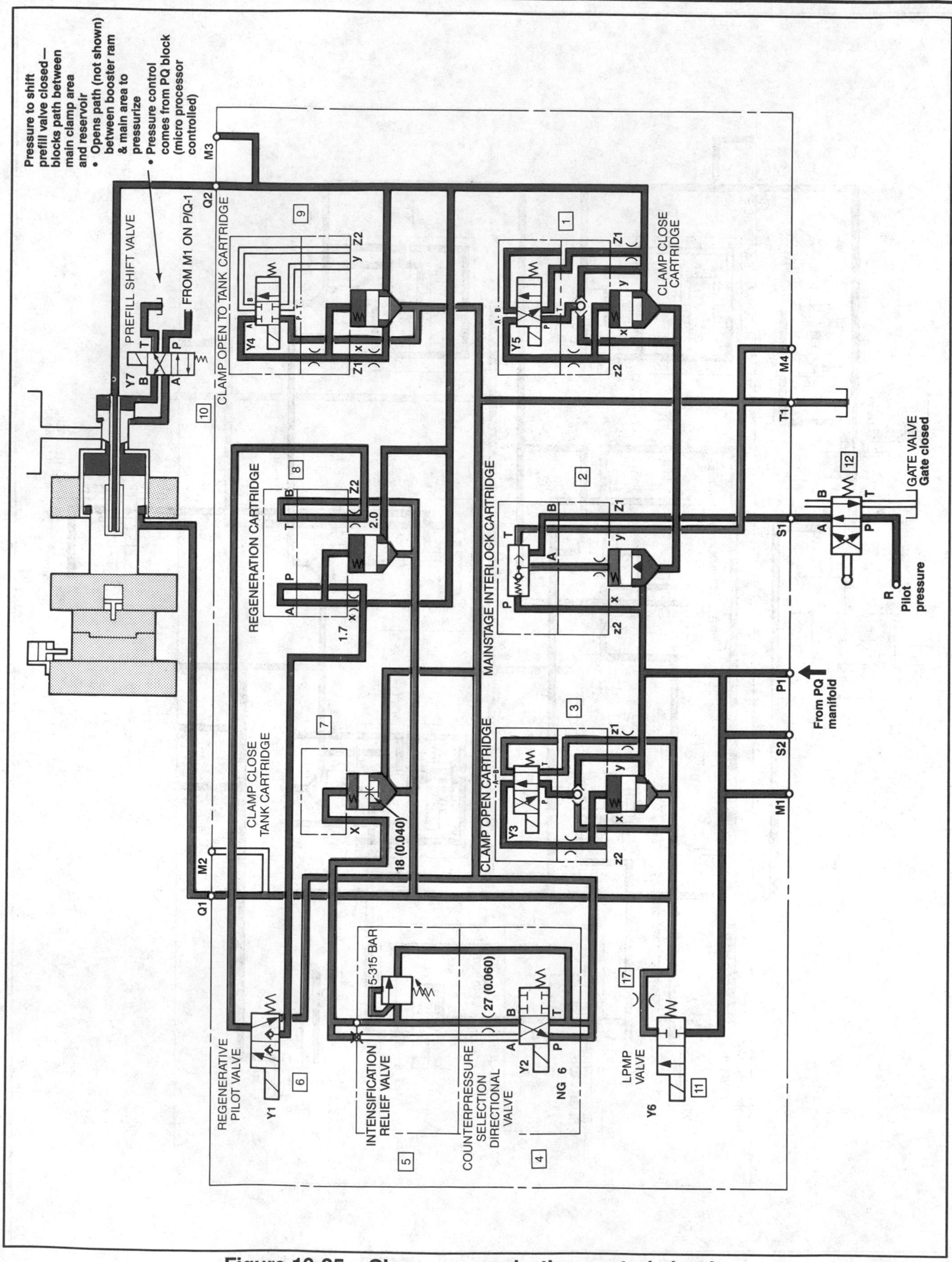

Figure 19-25. Clamp pressurization control circuit.

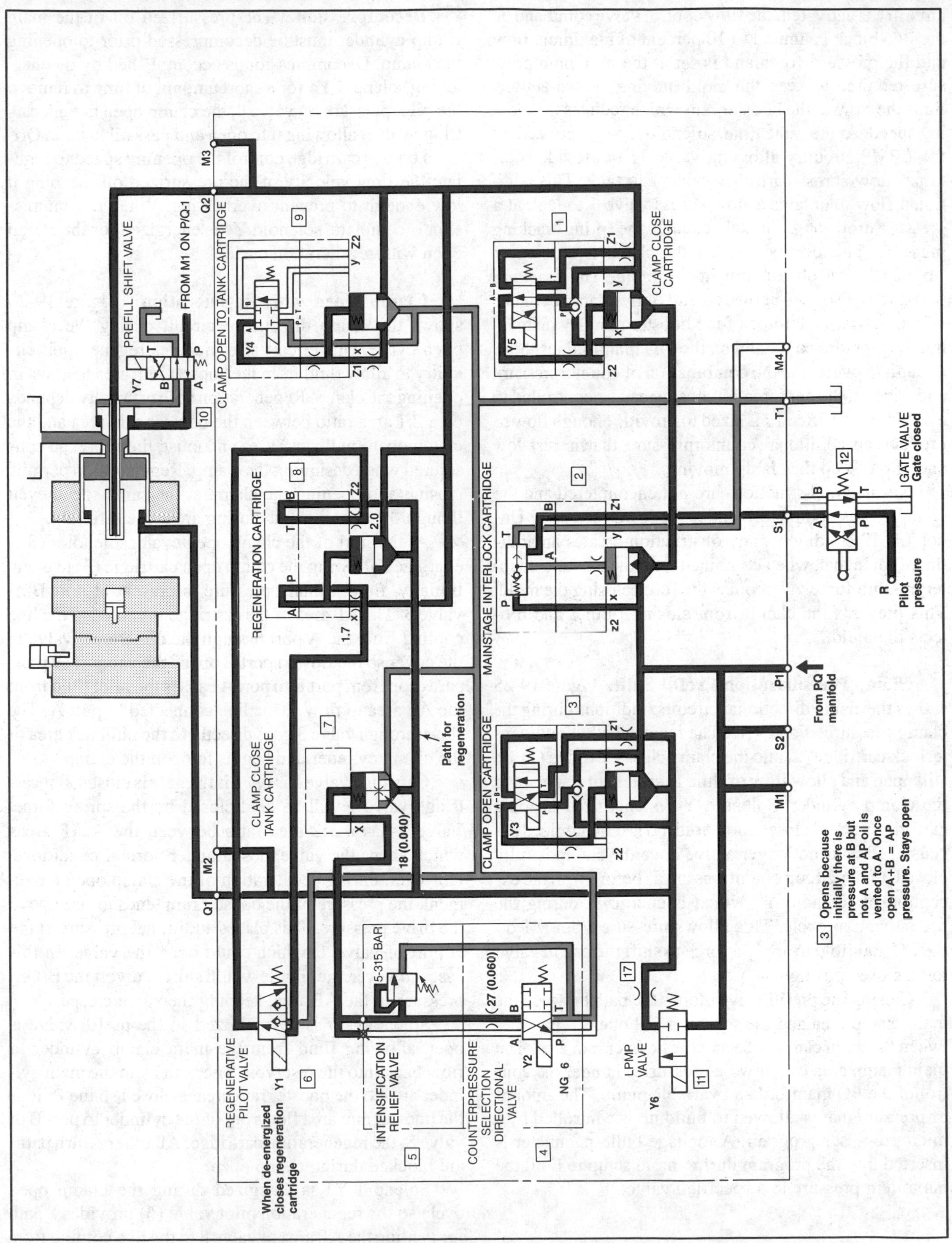

Figure 19-26. Clamp open with regeneration control circuit.

circuitry is activated, the flow control valve command on the P/Q block is limited to 10 percent of maximum flow and the pressure command is set at the minimum pressure required to keep the mold moving. These actions limit the flow to the booster ram which reduces the closing speed. At the same time, solenoid Y6 is energized by the LPMP circuitry allowing valve 11 to provide additional flow across cartridge valve 7 to tank. This additional flow maintains a flow across valve 7 to tank at a pressure drop approximately equal to the spring cracking pressure. This occurs whether the clamp is moving or stopped by an obstruction in the mold. The cracking pressure is also seen at the rod end of the clamp cylinder which opposes the force of the booster ram. By increasing the pressure command on the P/Q manifold just high enough to overcome the combination of counterpressure and friction, the net force for closing can be controlled to a minimum. Orifice 17 is sized to provide enough flow to produce an additional counterpressure that is just low enough to keep the clamp moving.

Normally, obstructions are not encountered and the clamp closes completely under the lower pressure. Under LPMP conditions, any obstruction creates an additional force that, when combined with the existing counterpressure force, overcomes the force closing the mold. This prevents the clamp from closing further and protects the mold.

Clamp Pressurization/Prefill Shift. Figure 19-25 shows the clamp directional circuit conditions during the clamp pressurization cycle. The mainstage hydraulic interlock cartridge (2) and the clamp close cartridge (1) are still open and allow flow of fluid at system pressure up to the clamp cylinder. Solenoid Y4 is energized keeping cartridge valve 9 closed. Solenoid Y6 is de-energized because LPMP is no longer active once the clamp is fully closed. Other circuit conditions are as before with the exception of solenoid Y7 which is energized during the pressurization cycle. This allows pressure from M1 on the PQ manifold to apply a force to shift the prefill valve to its closed position.

Closing the prefill valve blocks the path between the main clamp area and the reservoir but opens a path between the main clamp area and the booster ram area. The main clamp area can now be pressurized under the control of the PQ manifold and variable pump. The amount of pressure that is allowed to build up is controlled by a microprocessor program. A tonnage build parameter is inserted into the program during mold setup to limit the maximum pressure to a specified value.

Decompression. The pressurized oil in the main clamp cylinder must be decompressed prior to opening the clamp. Decompression is accomplished by de-energizing solenoid Y4 for a short amount of time to remove the pilot pressure on valve 9, the clamp open to tank cartridge valve, allowing it to open and pass oil to tank. Orifices on this cartridge control the opening speed to regulate the flow velocity of the pressurized oil, keeping it low enough to prevent overheating. When decompression is complete, solenoid Y4 is energized for the clamp open with regeneration cycle.

Clamp Open with Regeneration. Figure 19-26 shows the clamp directional circuit during the clamp open cycle. This circuit also includes regeneration circuitry to allow fluid from the booster ram area to assist in opening the clamp. Regeneration ability usually depends on a 2:1 area ratio between the booster ram area and the clamp open (pullback) area and must, therefore, be considered when designing the clamp. Regeneration permits closing and opening the clamp at the same speed even though the opening and closing areas are different.

At the start of the clamp open cycle, solenoid Y3 is energized allowing the clamp open cartridge (3) to open. Initially, fluid at main pressure is present at port B of valve 3. The AP area of the cartridge is vented to the line coming from the A port through the directional valve in the cover so the flow at port B opens the valve. The pressure drop from port B to port A causes the pilot fluid from the AP area to flow to the line connected to port A. The flow through valve 3 goes directly to the pullback area of the clamp cylinder and begins to open the clamp.

Cartridge valve 2 is open if the gate is closed, but cartridge valve 1 will be held closed by the spring force. Valve 1 has a 1:2 area ratio between the A:AP areas which keeps the valve closed under normal conditions. However, during deceleration of the clamp open movement, the pressure of the booster ram tends to rise above the drive pressure. This places additional pressure at the B port of valve 1, which could open the valve. In this case, however, the shuttle would shift to divert the B-port pressure to the AP area, keeping the valve closed.

Solenoid Y7 is de-energized so the prefill valve is open allowing fluid from the main clamp cylinder to flow back into the reservoir. The path from the main cylinder area to the booster ram area is closed so the fluid in the booster ram area flows out of the cylinder to port B of valve 8, the regeneration cartridge. All other return paths are blocked during clamp open.

Solenoid Y1 is energized during the clamp open cycle so the regeneration pilot valve (6) provides a path for venting the AP area of valve 8 to the line coming from port A of valve 8. This allows the force at port B, which is produced by the booster ram flow, to open the valve. Because of the pressure drop across the valve, the pressure at port A is lower than the pressure at port B so the pilot flow will join with the main flow through the valve. The flow through the regeneration cartridge is also directed to the pullback area of the main clamp and assists the main flow from valve 3 in opening the clamp.

Clamp Motion Safety Circuitry. Valve 5, the intensification safety valve, prevents damage to the hydraulic components that might be caused by pressure intensification. If valve 4 fails or gets plugged and the prefill is commanded to close before the mold halves touch, the main clamp area would be allowed to pressurize. The only opposition to this force would be the fluid in the clamp pullback area. This area is commonly one-twentieth (1/20) of the main clamp area, so a 2500 psi (172.35 bar / 17,235.00 kPa) clamp pressure setting would cause a 50,000 psi (3447.00 bar / 344,700.00 kPa) pressure on the smaller area. This would definitely cause component failure and, very likely, injury to personnel. Valve 5 is an adjustable pressure relief valve that can be set anywhere between 72 and 4569 psi (4.96 bar / 496.37 kPa and 314.99 bar / 31,498.69 kPa). Should pressure intensification occur, the valve opens at the preset pressure level providing a path to tank that prevents any further pressure buildup.

Another safety feature on the clamp manifold is provided by valve 2, the mainstage interlock cartridge. Industry safety standards require that both a hydraulic and an electrical interlock must be provided in the circuitry controlling the flow of oil to the clamp close cylinder. When the gate is closed, it creates both a hydraulic and an electrical signal allowing oil to be delivered to the clamp close side of the main clamp cylinder. The electrical signal is provided by a cam on the gate that closes an electrical limit switch.

The hydraulic signal is provided by pilot pressure from gate valve 30 to the pilot operated check valve on top of the mainstage interlock cartridge valve. When the gate is closed, a cam on the gate shifts the directional valve (12) to port the oil to the pilot valve. If the gate is opened or there is a pilot line break, the pilot operated check valve will close and main pressure will be present on the AP area of valve 2. This pressure along with the spring force closes valve 2, shutting off the main flow of oil to the clamp close cylinder.

Some circuits also include a proximity switch which senses the spool position to ensure that the poppet on valve 2 does not stick open. If the proximity switch sends a signal to the controller saying that the poppet is open when it should be closed, the software in the controller activates circuitry to shut off the pumps. System operation cannot resume until the error is corrected.

QUESTIONS

1. Name two reasons for using an unloading circuit.
2. Why is the high volume pump of a two pump system often unloaded during high pressure operation?
3. When is it appropriate to use automatic venting?
4. In an accumulator/pump unloading circuit, when is the pump unloaded?
5. If a cylinder in a regenerative circuit has a ratio of cap end area to rod end annular area of 2:1, will the cylinder advance and retract at the same speed? Will a higher area ratio cause the extending speed to proportionately increase or decrease?
6. How can pressure be held on a clamp while the work cylinder is extending?
7. To what degree does the brake valve restrict flow when a motor is accelerating?
8. If the pressure in the injection cylinder rises to the electrically modulated pressure valve setting because the mold is filled, what will P2 pump pressure be?
9. As the extruder motor turns the screw and pellets are being plasticized, how is the density of the melted plastic maintained?
10. What happens to the oil in the booster tube during regenerative clamp open?

APPENDIX

A

DEFINITION OF TECHNICAL TERMS

NOTE: The hydraulic definitions listed relate to the context in which these terms are used in this manual. A more general definition of terms is given in the "Glossary of Terms for Fluid Power NFPA Recommended Standard T2.70.1."

ABSOLUTE—A measure having as its zero point or base the complete absence of the entity being measured.

ABSOLUTE PRESSURE—The pressure above zero absolute, i.e., the sum of atmospheric and gauge pressure. In vacuum related work it is usually expressed in millimeters of mercury (mm Hg).

ACCUMULATOR—A container in which fluid is stored under pressure as a source of fluid power.

ACTUATOR—A device for converting hydraulic energy into mechanical energy. A motor or cylinder.

AERATION—Air in the hydraulic fluid. Excessive aeration causes the fluid to appear milky and components to operate erratically because of the compressibility of the air trapped in the fluid.

ALTERNATING CURRENT (AC)—A continuously changing magnitude of current, produced by a power source (e.g., generator, alternator).

AMERICAN WIRE GAGE—A device used to determine the standard sizes of wire conductors.

AMPLIFIER—A device for amplifying either voltage or current.

AMPLITUDE OF SOUND—The loudness of a sound.

ANALOG SIGNAL—An AC or DC voltage or current signal that represents continuously variable physical quantities (e.g. voltage, current, pressure, temperature, or speed).

AND LOGIC GATE—A digital circuit in which the output state equals a binary 1 if and only if all of its inputs are a binary 1 at the same time.

ANNULAR AREA—A ring shaped area—often refers to the net effective area of the rod-side of a cylinder piston, i.e. the piston area minus the cross-sectional area of the rod.

ATMOSPHERE (ONE)—A pressure measure equal to 14.7 psi (1.01 bar) (101.34 kPa).

ATMOSPHERIC PRESSURE—Pressure exerted by the atmosphere at any specific location. (Sea level pressure is approximately 14.7 pounds per square inch (1.01 bar) (101.34 kPa) absolute.)

ATTENUATOR—A variable resistive device used to reduce the value of current or voltage.

BACK CONNECTED—A condition where pipe connections are on normally unexposed surfaces of hydraulic equipment. (Gasket mounted units are back connected.)

BACKPRESSURE—Usually refers to pressure existing on the discharge side of a load. It adds to the pressure required to move the load.

BAFFLE—A device, usually a plate, installed in a reservoir to separate the pump inlet from return lines.

BAR—An international standard unit of pressure equal to 100,000 Pascals, approximately 14.5 psi.

BINARY NUMBER SYSTEM—A number system used in digital equipment that has a base of 2. The digits 0 and 1 are the only digits used in forming numbers in this system.

BLEED-OFF—To divert a specific controllable portion of pump delivery directly to reservoir.

BRAKE VALVE—A valve used in the exhaust line of a hydraulic motor to (1) prevent overspeeding when an overrunning load is applied to the motor shaft and (2) prevent excessive pressure buildup when decelerating or stopping a load.

BREATHER—A device which permits air to move in and out of a container or component to maintain atmospheric pressure.

BYPASS—A secondary passage for fluid flow.

CARTRIDGE—

1. The replaceable element of a fluid filter.
2. The pumping unit from a vane pump, composed of rotor, ring, vanes and one or both side plates.

CARTRIDGE VALVE—A valve that is inserted into a standard size cavity in a manifold block and is held in place with either self-contained screw threads or a cover secured with bolts. May be slip-in or screw-in types. Perform directional, pressure or flow control functions.

CAVITATION—A localized gaseous condition within a liquid stream which occurs where the pressure is reduced to the vapor pressure.

CHAMBER—A compartment within a hydraulic unit. May contain elements to aid in operation or control of a unit. Examples: spring chamber, drain chamber, etc.

CHARGE (supercharge)—

1. To replenish a hydraulic system above atmospheric pressure.
2. To fill an accumulator with fluid under pressure (see precharge pressure).

CHARGE PRESSURE—The pressure at which replenishing fluid is forced into the hydraulic system (above atmospheric pressure).

CHECK VALVE—A valve which permits flow of fluid in one direction only.

CHOKE—A restriction, the length of which is large with respect to its cross-sectional dimension.

CIRCUIT—An arrangement of components interconnected to perform a specific function within a system.

CLOSED CENTER VALVE—One in which all ports are blocked in the center or neutral position.

CLOSED CENTER CIRCUIT—One in which flow through the system is blocked in neutral and pressure at the pump outlet is maintained at the maximum pressure control setting.

CLOSED CIRCUIT—A circuit in which a complete path exists for the flow of electrical current.

CLOSED LOOP—A system in which the feedback from one or more elements is compared to a command signal, providing an error signal to control the output of the loop.

COMMAND SIGNAL (or input signal)—An external signal representing a new position or velocity to which the servo must respond.

COMPENSATOR CONTROL—A displacement control for variable pumps and motors which alters displacement when system pressure exceeds its adjusted pressure setting.

COMPONENT—A single hydraulic or electrical unit.

COMPRESSIBILITY—The change in volume of a unit volume of fluid when subjected to a unit change in pressure.

CONDUCTOR AMPACITY—The maximum number of amperes that an electrical conductor can safely carry continuously.

CONTAMINATION—Any material foreign to a hydraulic fluid that has a harmful effect on its performance in a system. Contaminants may be solid particles, liquids or gases.

CONTROL—A device used to regulate the function of a unit (see Hydraulic Control, Manual Control, Mechanical Control, and Compensator Control).

COOLER—A heat exchanger used to remove heat from the hydraulic fluid.

COUNTERBALANCE VALVE—A pressure control valve which maintains backpressure to prevent a load from falling.

CRACKING PRESSURE—The pressure at which a pressure actuated valve begins to pass fluid.

CURRENT—The directed flow of electrical charges from one point to another around a closed electrical circuit. Current is measured in units called amperes or amps.

CUSHION—A device sometimes built into the ends of a hydraulic cylinder which restricts the flow of fluid at the outlet port, thereby arresting the motion of the piston rod.

CYLINDER—A device which converts fluid power into linear mechanical force and motion. It usually consists of a movable element such as a piston and piston rod, plunger rod, plunger or ram, operating within a cylindrical bore.

DEADBAND—The region or band of no response where an error signal will not cause a corresponding actuation of the controlled variable.

DECOMPRESSION—The slow release of confined fluid to gradually reduce pressure on the fluid.

DELIVERY—The volume of fluid discharged by a pump in a given time, usually expressed in gallons per minute (gpm (LPM)).

DE-VENT—To close the vent connection of a pressure control valve permitting the valve to function at its adjusted pressure setting.

DIFFERENTIAL CYLINDER—Any cylinder in which the two opposed piston areas are not equal.

DIGITAL AND CLOSED LOOP (DCL)—A system in which microprocessor based circuitry is implanted within a servo valve to allow it to operate as an intelligent device by responding to addressable, high speed, serially transmitted digital command signals.

DIGITAL SIGNAL—A voltage or current that varies between two distinct and fixed levels.

DIODE RECTIFIER—A device used to convert AC voltage into DC pulses.

DIRECT CURRENT (DC)—A steady level of electrical current, produced by a power source (e.g. battery, thermocouple, etc.), that flows in only one direction in a circuit.

DIRECT DIGITAL LINK (DDL)—A digital feedback system used with servo valves to allow the valve to communicate with a digital host device such as a PLC or personal computer. The host device serially transmits command signals to the valve, and position information is sent back to the host which responds with an error correction signal.

DIRECTIONAL VALVE—A valve which selectively directs or prevents fluid flow to desired channels.

DISPLACEMENT—The quantity of fluid which can pass through a pump, motor or cylinder in a single revolution or stroke.

DITHER—A low amplitude, AC signal superimposed on the torque motor or solenoid input to improve system resolution. Dither is expressed by the dither frequency (H_z) and the peak-to-peak dither voltage.

DOUBLE ACTING CYLINDER—A cylinder in which fluid force can be applied to the movable element in either direction.

DRAIN—A passage in, or a line from, a hydraulic component which returns leakage fluid independently to reservoir or to a vented manifold.

EFFICIENCY—The ratio of output to input. Volumetric efficiency of a pump is the actual output in gpm (LPM) divided by the theoretical or design output. The overall efficiency of a hydraulic system is the output power divided by the input power. Efficiency is usually expressed as a percent.

ELECTROHYDRAULIC SERVO VALVE—A directional type valve which receives a variable or controlled electrical signal and which controls or meters hydraulic flow.

ELECTROMAGNETIC INTERFERENCE (EMI)—A modification of signals contained in circuit wiring (or stored in a microprocessor) caused by a strong magnetic field radiated from some other electrical or electronic device or conductor.

ELECTROMOTIVE FORCE (EMF)—The force produced by a difference of electrical potential that causes current to flow in a circuit. EMF is measured in units called volts.

ENERGY—The ability or capacity to do work. Measured in units of work.

ENCLOSURE—A rectangle drawn around a graphical component or components to indicate the limits of an assembly.

ERROR (signal)—The signal which is the algebraic summation of a command signal and a feedback signal.

FEEDBACK (or feedback signal)—The output signal from a feedback element.

FEEDBACK LOOP—Any closed circuit consisting of one or more feedback elements which transmit a signal to a summation point.

FILTER—A device used to separate and retain insoluble contaminants from a fluid.

FLOODED—A condition where the pump inlet is charged by placing the reservoir oil level above the pump inlet port.

FLOW CONTROL VALVE—A valve which controls the rate of flow.

FLOW RATE—The volume, mass, or weight of a fluid passing through any conductor per unit of time.

FLUID—
1. A liquid or gas.
2. A liquid that is specially compounded for use as a power-transmitting medium in a hydraulic system.

FOLLOW VALVE—A control valve which ports oil to an actuator so the resulting output motion is proportional to the input motion to the valve.

FORCE—Any push or pull measured in units of weight. In hydraulics, total force is expressed by the product of pressure (force per unit area) and the area of the surface on which the pressure acts (F = P x A).

FOUR-WAY VALVE—A directional valve having four flow paths.

FREQUENCY—The number of times an action occurs in a unit of time.

FRONT CONNECTED—A condition wherein piping connections are on normally exposed surfaces of hydraulic components.

FULL FLOW—In a filter, the condition where all the fluid must pass through the filter element or media.

GAUGE PRESSURE—A pressure scale which ignores atmospheric pressure. Its zero point is atmospheric pressure.

GROUND—A point of zero reference in electrical circuits to which all circuit voltages are compared. Also, to ground a device means to make connections to an earth ground for safety purposes.

HEAD—The height of a column or body of fluid above a given point expressed in linear units. Head is often used to indicate gauge pressure. Pressure is equal to the height times the density of the fluid.

HEAT—The form of energy that has the capacity to create warmth or to increase the temperature of a substance. Any energy that is wasted or used to overcome friction is converted to heat. Heat is measured in calories or British Thermal Units (BTUs). One BTU is the amount of heat required to raise the temperature of one pound of water one degree Fahrenheit. (One calorie is the amount of heat required to raise one gram of water one degree centigrade.)

HEAT EXCHANGER—A device which transfers heat through a conducting wall from one fluid to another.

HORSEPOWER (HP)—The power required to lift 550 pounds one foot in one second or 33,000 pounds one foot in one minute. A horsepower is equal to 746 watts or to 42.4 British Thermal Units per minute. (One watt is the power required to lift one newton one meter in one second).

HYDRAULIC BALANCE—A condition of equal opposed hydraulic forces acting within a hydraulic component.

HYDRAULIC CONTROL—A control which is actuated by hydraulically induced forces.

HYDRAULICS—Engineering science pertaining to liquid pressure and flow.

HYDRODYNAMICS—Engineering science pertaining to the energy of liquid flow and pressure.

HYDROSTATICS—Engineering science pertaining to the energy of confined liquids.

IMPEDANCE—The combination of AC and DC resistance in a circuit; measured in ohms.

INSULATOR—A material that blocks the flow of current which is used for short circuit and shock prevention.

KINETIC ENERGY—Energy that a substance or body has by virtue of its mass (weight) and velocity.

LAMINAR (FLOW)—A condition where the fluid particles move in continuous parallel paths; streamline flow.

LEVERAGE—A gain in output force over input force by sacrificing the distance moved at the output. Mechanical advantage or force multiplication.

LIFT—The height a body or column of fluid is raised; for instance, from the reservoir to the pump inlet. Lift is sometimes used to express a vacuum. The opposite of head.

LINE—A tube, pipe or hose which acts as a conductor of hydraulic fluid.

LINEAR ACTUATOR—A device for converting hydraulic energy into linear motion—a cylinder or ram.

LOGIC CIRCUIT—A digital circuit (also called a gate) that has binary inputs and outputs and is capable of performing a decision making function.

MANIFOLD—A fluid conductor which provides multiple connection ports.

MANUAL CONTROL—A control actuated by the operator, regardless of the means of actuation. Example: lever or foot pedal control for directional valves.

MANUAL OVERRIDE—A means of manually actuating an automatically-controlled device.

MAXIMUM PRESSURE VALVE—See relief valve.

MECHANICAL CONTROL—Any control actuated by linkages, gears, screws, cams or other mechanical elements.

METER—To regulate the rate of fluid flow.

METER-IN—To regulate the rate of fluid flow into an actuator or system.

METER-OUT—To regulate the rate of fluid flow from an actuator or system.

MICRON—One-millionth of a meter or about 0.00004 inch.

MICRON RATING—The smallest size particles a filter will remove.

MICROPROCESSOR—A digital computer based on a single miniature integrated circuit that offers programmability and computational ability for use in electrohydraulic control applications.

MODEM—Modulator/Demodulator. A device used to provide digital data communications between two separate microprocessing units. The modulator converts digital information into tones that are transmitted over telephone lines and the demodulator changes the received tones back to digital data.

MOTOR—A device which converts hydraulic fluid power into mechanical force and motion. It usually provides rotary mechanical motion.

MULTIMETER—A device (VOM or DMM) used to measure electrical quantities.

NATIONAL ELECTRICAL CODE (NEC)—Standards, recommended practices, specifications and directions for electrical wiring of residential, commercial, industrial and farm buildings. Published by the National Fire Prevention Association.

NOISE (ELECTRICAL)—Interference-type problems within electrical circuits and associated wiring that can cause erratic and improper operation of equipment.

OHM'S LAW—States that the current in a circuit is directly proportional to the voltage and indirectly proportional to the resistance. Expressed as a formula: $I = E/R$.

OPEN CENTER CIRCUIT—One in which pump delivery flows freely through the system and back to the reservoir in neutral.

OPEN CENTER VALVE—One in which all ports are interconnected and open to each other in the center or neutral position.

OPEN CIRCUIT—A circuit in which a complete path for electrical current flow does not exist.

OPERATIONAL AMPLIFIER (OP AMP)—An integrated circuit amplifier with special characteristics (high gain, high input impedance, low output impedance, differential amplification) that make it especially suitable for electrohydraulic control systems.

OR LOGIC GATE—A digital circuit in which the output state equals a binary 1 if any or all of its inputs are a binary 1.

ORIFICE—A restriction, the length of which is small in respect to its cross-sectional dimensions.

PARALLEL DATA TRANSMISSION—A form of digital data transmission in which all of the bits of a binary word are processed or transmitted at the same time. Faster than serial transmission but requires separate circuitry for each bit.

PASSAGE—A machined or cored fluid conducting path which lies within or passes through a component.

PILOT PRESSURE—Auxiliary pressure used to actuate or control hydraulic components.

PILOT VALVE—An auxiliary valve used to control the operation of another valve. The controlling stage of a two-stage valve.

PISTON—A cylindrically shaped part which fits within a cylinder and transmits or receives motion by means of a connecting rod.

PLUNGER—A cylindrically shaped part which has only one diameter and is used to transmit thrust; a ram.

POPPET—A valve that moves perpendicular to or from its seat.

PORT—An internal or external terminus of a passage in a component.

POSITIVE DISPLACEMENT—A characteristic of a pump or motor which has the inlet positively sealed from the outlet so that fluid cannot recirculate in the component.

POTENTIOMETER—A control element in the servo system which controls electrical potential.

POWER—Work per unit of time; measured in horsepower (hp) or watts (W).

POWER PACK—An integral power supply unit usually containing a pump, reservoir, relief valve and directional control.

PRECHARGE PRESSURE—The pressure of compressed gas in an accumulator prior to the admission of liquid.

PRESSURE—Force per unit area; usually expressed in pounds per square inch (psi), bar or kilopascals (kPa).

PRESSURE DROP—The difference in pressure between any two points of a system or a component.

PRESSURE LINE—The line carrying the fluid from the pump outlet to the pressurized port of the actuator.

PRESSURE OVERRIDE—The difference between the cracking pressure of a valve and the pressure reached when the valve is passing full flow.

PRESSURE PLATE—A side plate in a vane pump or motor cartridge on the pressure port side.

PRESSURE REDUCING VALVE—A valve which limits the maximum pressure at its outlet regardless of the inlet pressure.

PRESSURE SWITCH—An electric switch operated by fluid pressure.

PROGRAMMABLE LOGIC CONTROLLER (PLC)—A programmable solid-state device that provides the control of a machine or process on the basis of input and output conditions.

PROPORTIONAL VALVE—A valve which controls and varies pressure, flow, direction, acceleration and deceleration from a remote position. They are adjusted electrically and are actuated by proportional solenoids rather than by a force or torque motor. The output flow is proportional to the input signal. They provide moderately accurate control of hydraulic fluid.

PULSE WIDTH MODULATION—An electronic signal of constant frequency and amplitude that has varying pulse width to control the level of power to the solenoid.

PUMP—A device which converts mechanical force and motion into hydraulic fluid power.

RAM—A single-acting cylinder with a single diameter plunger rather than a piston and rod.

RECIPROCATION—Back-and-forth straight line motion or oscillation.

REGENERATIVE CIRCUIT—A piping arrangement for a differential type cylinder in which discharge fluid from the rod-end combines with pump delivery to be directed into the head-end.

RELAY—An electromagnetic device that allows one circuit to control another without a direct electrical connection between the two circuits.

RELAY LOGIC—A system for controlling a machine or process based on the status of various interconnected relays.

RELIEF VALVE—A pressure operated valve which by- passes pump delivery to the reservoir, limiting system pressure to a predetermined maximum value.

REPLENISH—To add fluid to maintain a full hydraulic system.

RESERVOIR—A container for storage of liquid in a fluid power system.

RESISTANCE—The opposition to current flow offered by the components of an electrical circuit.

RESTRICTION—A reduced cross-sectional area in a line or passage which produces a pressure drop.

RETURN LINE—A line used to carry exhaust fluid from the actuator back to sump.

ROTARY ACTUATOR—A device for converting hydraulic energy into a rotary motion hydraulic motor.

SEQUENCE—

1. The order of a series of operations or movements.
2. To divert flow to accomplish a subsequent operation or movement.

SEQUENCE VALVE—A pressure operated valve which, at its setting, diverts flow to a secondary line while holding a predetermined minimum pressure in the primary line.

SERIAL DATA TRANSMISSION—A form of digital data transmission in which the data bits of a binary word are processed or transmitted one at a time. Slower than parallel data transmission, but does not require separate circuitry for each bit.

SERVO MECHANISM (servo)—A mechanism subjected to the action of a controlling device which will operate as if it were directly actuated by the controlling device, but capable of supplying power output many times that of the controlling device, this power being derived from an external and independent source.

SERVO VALVE—
1. A valve which modulates output as a function of an input command.
2. A follow valve.

SINGLE ACTING CYLINDER—A cylinder in which hydraulic energy can produce thrust or motion in only one direction. (May be mechanically or gravity returned.)

SLIP—Internal leakage of hydraulic fluid.

SOLENOID—An electromechanical device that converts electrical energy into linear mechanical motion. Used to actuate directional valves. May be air gap or wet armature type.

SPOOL—A term loosely applied to almost any moving cylindrically shaped part of a hydraulic component which moves to direct flow through the component.

STRAINER—A coarse filter.

STREAMLINE FLOW—(See laminar flow.)

STROKE—
1. The length of travel of a piston or plunger.
2. To change the displacement of a variable displacement pump or motor.

SUBPLATE—An auxiliary mounting for a hydraulic component providing a means of connecting piping to the component.

SUCTION LINE—The hydraulic line connecting the pump inlet port to the reservoir or sump.

SUMP—A reservoir.

SUPERCHARGE—(See charge.)

SURGE—A transient rise of pressure or flow.

SWASH PLATE—A stationary canted plate in an axial type piston pump which causes the pistons to reciprocate as the cylinder barrel rotates.

SYNCHRO—A rotary electromagnetic device generally used as an AC feedback signal generator which indicates position. It can also be used as a reference signal generator.

TACHOMETER—(AC) (DC)—A device which generates an AC or DC signal proportional to the speed at which it is rotated and the polarity of which is dependent on the direction of rotation of the rotor.

TANK—The reservoir or sump.

THROTTLE—To permit passing of a restricted flow. May control flow rate or create a deliberate pressure drop.

TORQUE—A rotary thrust. The turning effort of a fluid motor usually expressed in inch-pounds or newton-meters.

TORQUE CONVERTER—A rotary fluid coupling that is capable of multiplying torque.

TORQUE MOTOR—A type of electromechanical transducer having rotary motion used to actuate servo valves.

TRANSDUCER (or feedback transducer)—A device that converts one type of energy to another. An example would be the transducer sensing pressure and generating an electrical signal in proportion to the pressure.

TRANSFORMER—A device that transfers AC energy from one circuit to another without electrical contact between the two circuits.

TURBULENT FLOW (TURBULENCE)—A condition where the fluid particles move in random paths rather than in continuous parallel paths.

TURBINE—A rotary device that is actuated by the impact of a moving fluid against blades or vanes.

TWO-WAY VALVE—A directional control valve with two flow paths.

UNLOAD—To conserve energy by releasing flow (usually directly to the reservoir).

UNLOADING VALVE—A valve which bypasses flow to tank when a set pressure is maintained on its pilot port.

VACUUM—Pressure less than atmospheric pressure. It is usually expressed in inches of mercury (in. Hg) or millimeters of mercury (mm Hg) as referred to the existing atmospheric pressure.

VALVE—A device which controls fluid flow direction, pressure, or flow rate.

VELOCITY—
1. The speed of flow through a hydraulic line. Expressed in feet per second (fps), inches per second (ips), meters per second (m/s) or centimeters per second (cm/s).
2. The speed of a rotating component measured in revolutions per minute (rpm) or radians per second (rad/s).

VENT—

1. To permit opening of a pressure control valve by opening its pilot port (vent connection) to atmospheric pressure.
2. An air breathing device on a fluid reservoir.
3. To remove trapped air from a component.

VISCOSITY—A measure of the internal friction or the resistance of a fluid to flow.

VISCOSITY INDEX—An arbitrary measure of a fluid's resistance to viscosity change with temperature changes.

VOLTAGE—The amount of electromotive force (emf) produced by a power source or available at points in a circuit; expressed in units called volts.

VOLUME—

1. The size of a space or chamber in cubic units.
2. Loosely applied to the output of a pump in gallons per minute (gpm) or liters per minute (LPM)

WATT'S LAW—States that when one amp of current flows through a device with a one volt voltage drop, one watt of power is dissipated in the form of heat (P = I x E).

WOBBLE PLATE—A rotating canted plate in an axial type piston pump which pushes the pistons into their bores as it "wobbles."

WORK—Exerting a force through a definite distance. Work is measured in units of force multiplied by distance; for example, pound-foot or newton-meter.

APPENDIX

B

SYMBOLS FOR HYDRAULIC SYSTEMS

This section of the manual provides a guide to basic symbols which are the building blocks of any hydraulic circuit. The composite diagrams shown here should be considered only as examples since many modifications are possible by changing elements such as flow paths and control devices.

LINES

LINE, WORKING (MAIN)		LINES CROSSING	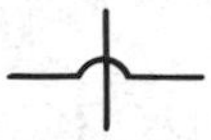 OR 
LINE, PILOT (FOR CONTROL)		LINES JOINING	
LINE, LIQUID DRAIN		LINE WITH FIXED RESTRICTION	
COMPONENT ENCLOSURE		FLEXIBLE	
HYDRAULIC FLOW		STATION, TESTING, MEASUREMENT, POWER TAKE-OFF, OR PLUGGED PORT	
PNEUMATIC FLOW			

FLUID STORAGE

RESERVOIR, VENTED		LINE, TO RESERVOIR • ABOVE FLUID LEVEL • BELOW FLUID LEVEL	
RESERVOIR, PRESSURIZED		VENTED MANIFOLD	

METHODS OF OPERATION

SPRING		DETENT	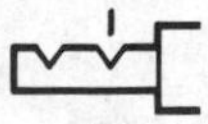
MANUAL	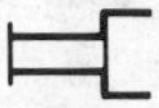	PRESSURE COMPENSATED	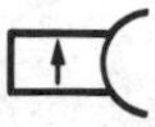
PUSH BUTTON	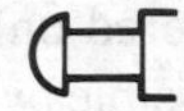	SOLENOID, SINGLE WINDING	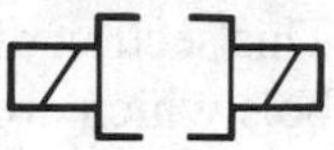
PUSH-PULL LEVER		SERVO MOTOR	
PEDAL OR TREADLE		PILOT PRESSURE • REMOTE SUPPLY	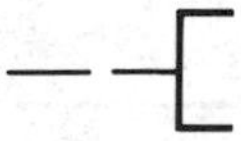
MECHANICAL	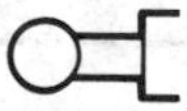	• INTERNAL SUPPLY	

FIXED DISPLACEMENT PUMPS

SINGLE, VANE, PISTON & GEAR TYPE	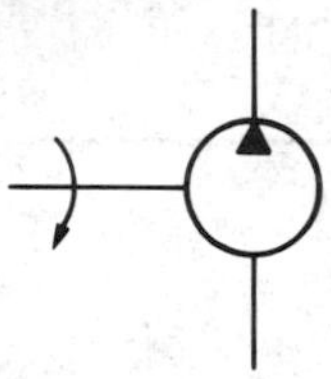	SINGLE, WITH INTEGRAL PRIORITY VALVE	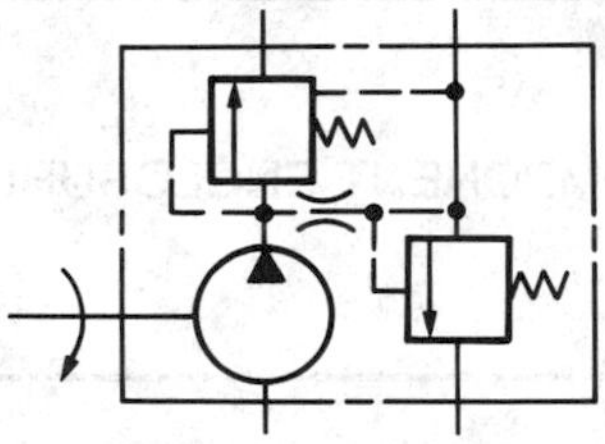
SINGLE, POWER STEERING PUMP WITH INTEGRAL FLOW CONTROL AND RELIEF VALVES	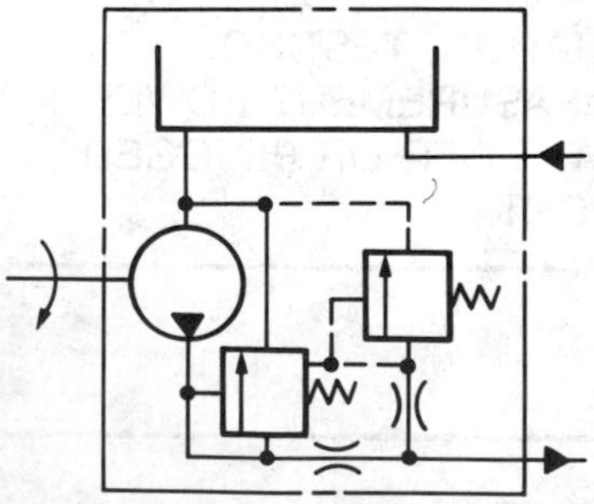	SINGLE, WITH INTEGRAL FLOW CONTROL VALVE	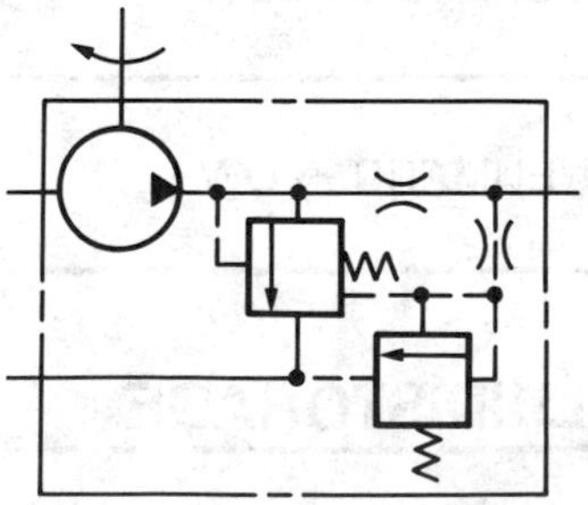
SINGLE, PISTON TYPE WITH DRAIN	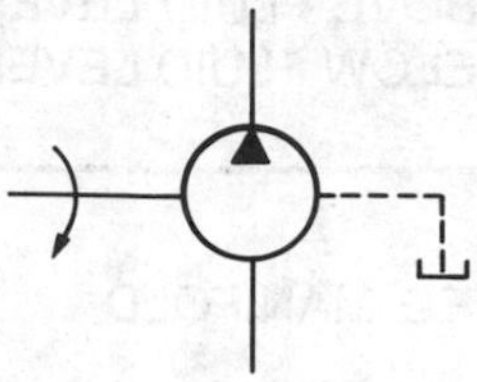	DOUBLE, VANE AND GEAR TYPE	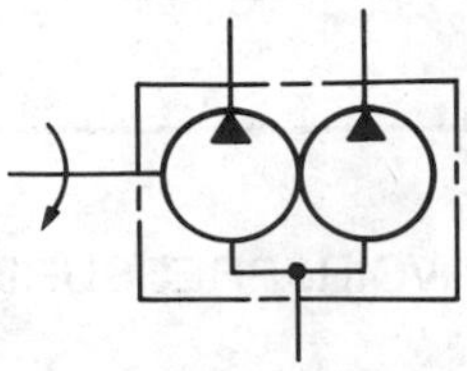

VARIABLE DISPLACEMENT PUMPS

MANUAL, HANDWHEEL CONTROL	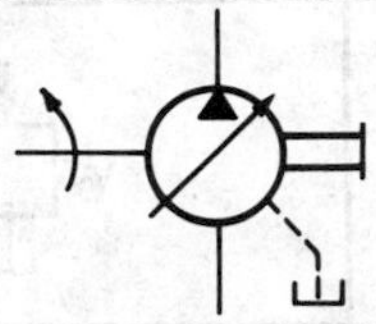
PRESSURE-COMPENSATOR CONTROL	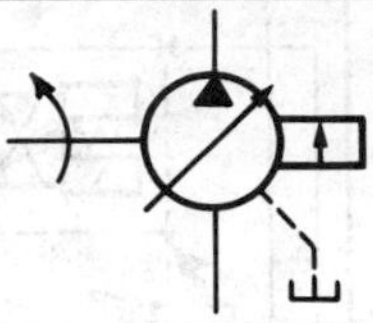
PRESSURE COMPENSATOR AND TORQUE LIMITER	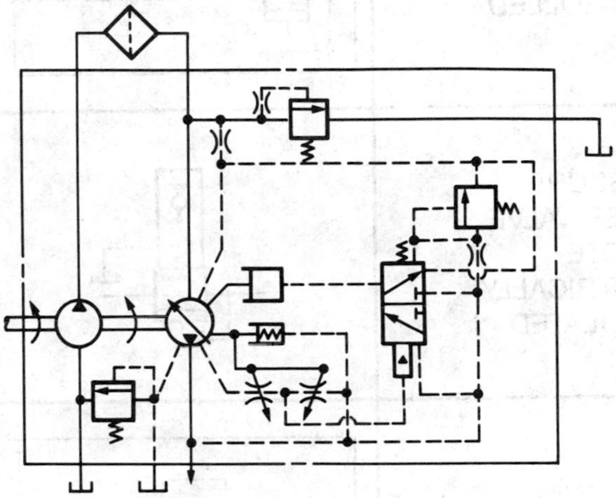
LOAD-SENSING CONTROL	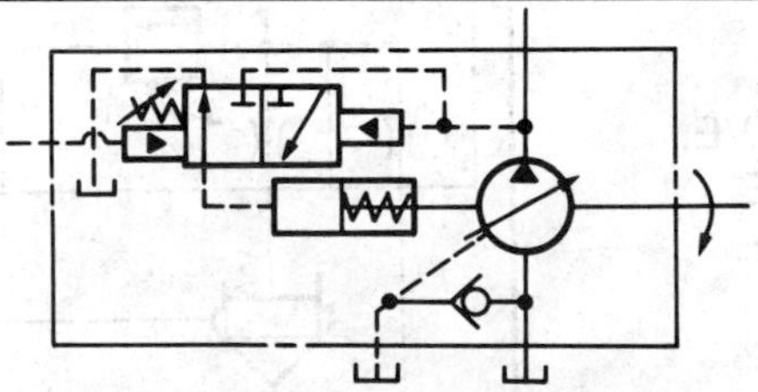
LOAD-SENSING AND PRESSURE-LIMITER CONTROL	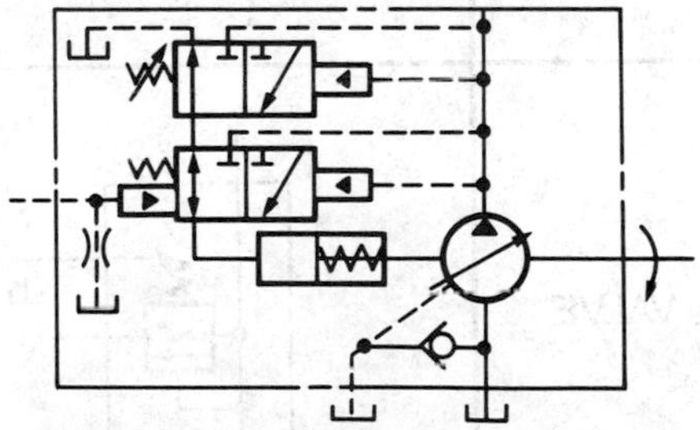
ELECTRO-HYDRAULIC CONTROL	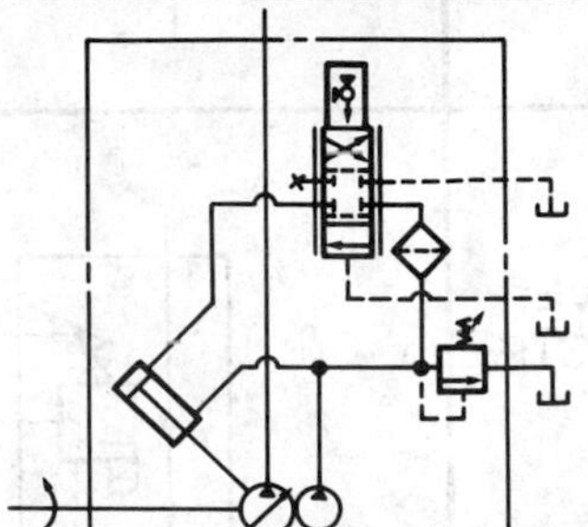

PRESSURE CONTROL VALVES

PRESSURE RELIEF	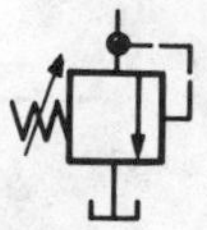	COUNTERBALANCE VALVE WITH INTEGRAL CHECK	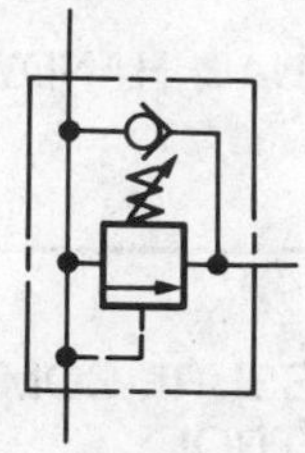
PRESSURE RELIEF VALVE, SOLENOID CONTROLLED	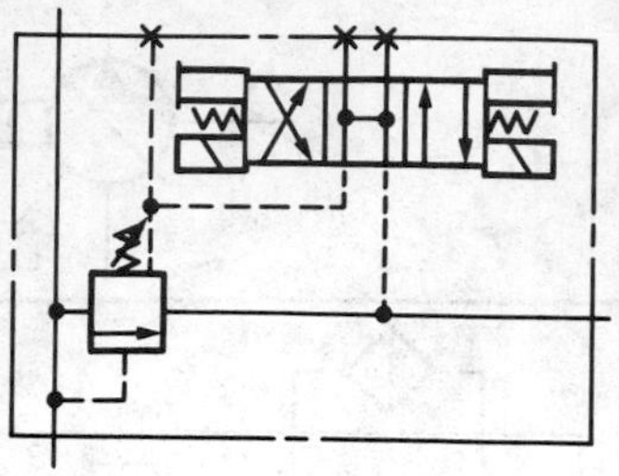	UNLOADING RELIEF VALVE WITH INTEGRAL CHECK	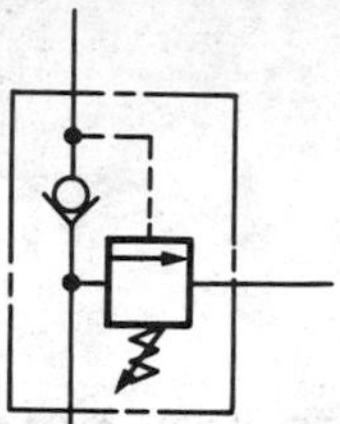
PRESSURE RELIEF VALVE, REMOTE ELECTRICALLY MODULATED	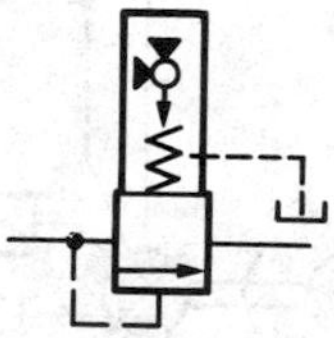	PRESSURE REDUCING	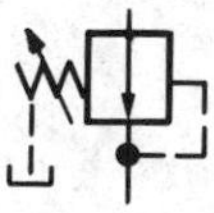
PRESSURE RELIEF VALVE, CARTRIDGE VALVE TYPE	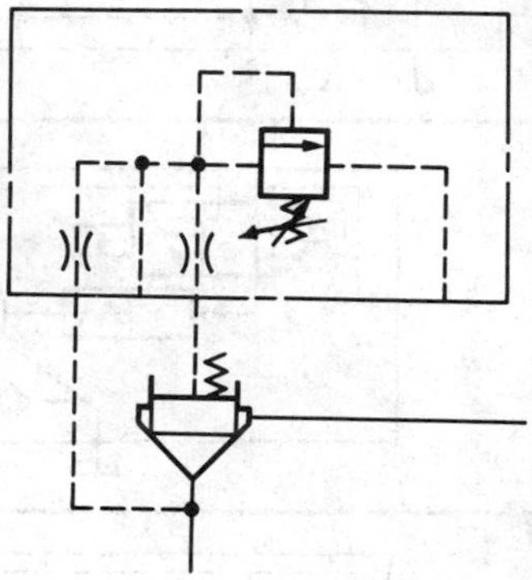	PRESSURE-REDUCING VALVE WITH INTEGRAL CHECK	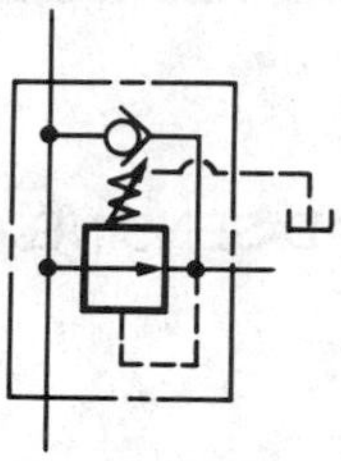
SEQUENCE VALVE	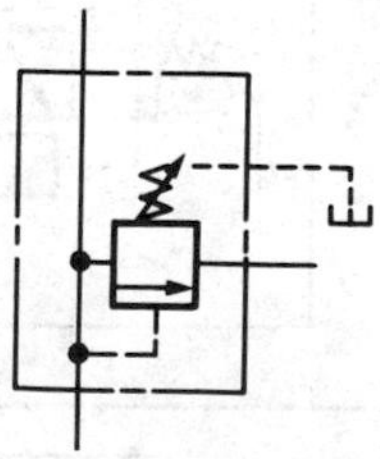	PRESSURE-REDUCING VALVE, CARTRIDGE VALVE-TYPE	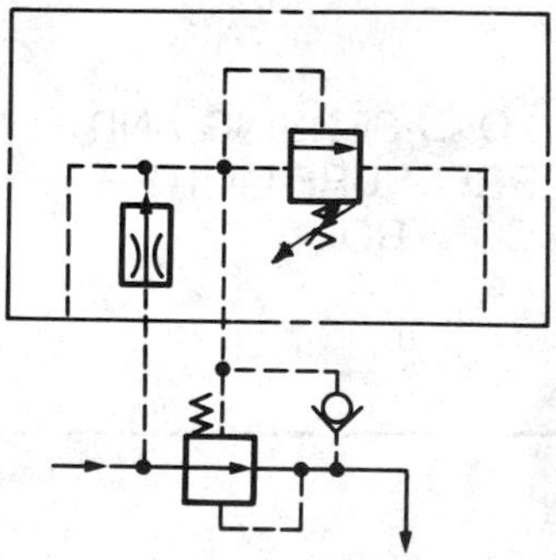
SEQUENCE VALVE, AUXILIARY REMOTE CONTROL OPERATION	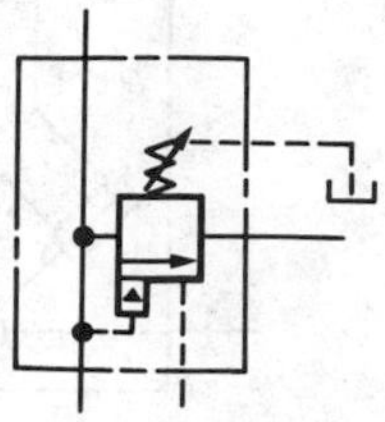		

FLOW CONTROL VALVES

FLOW CONTROL, ADJUSTABLE—NONCOMPENSATED	
FLOW REGULATOR WITH REVERSE FREE FLOW	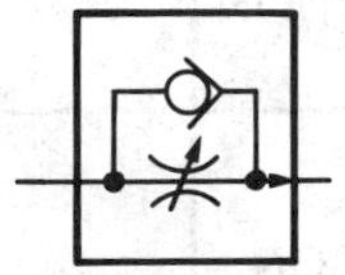
FLOW CONTROL VALVE	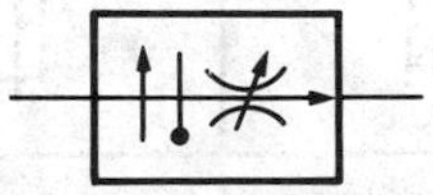
FLOW CONTROL VALVE WITH CHECK (SIMPLIFIED SYMBOL)	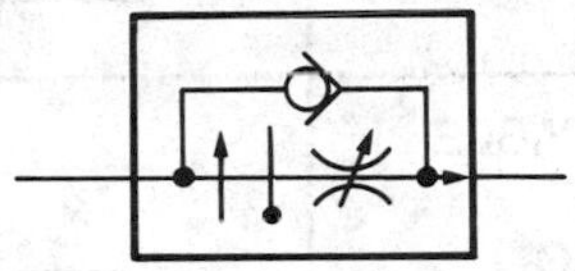
REMOTE CONTROLLED, ELECTRICALLY MODULATED WITH CHECK VALVE	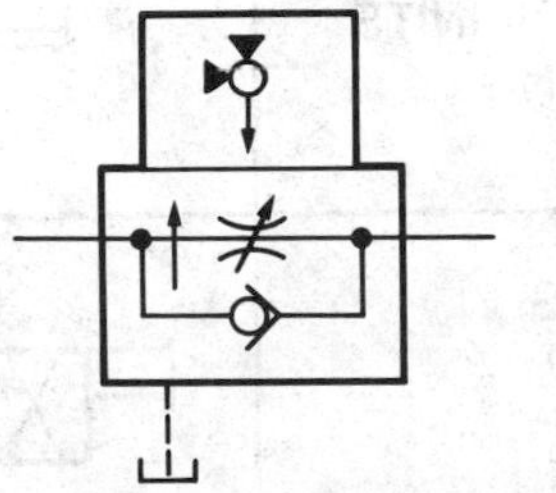
FLOW CONTROL AND OVERLOAD RELIEF VALVE	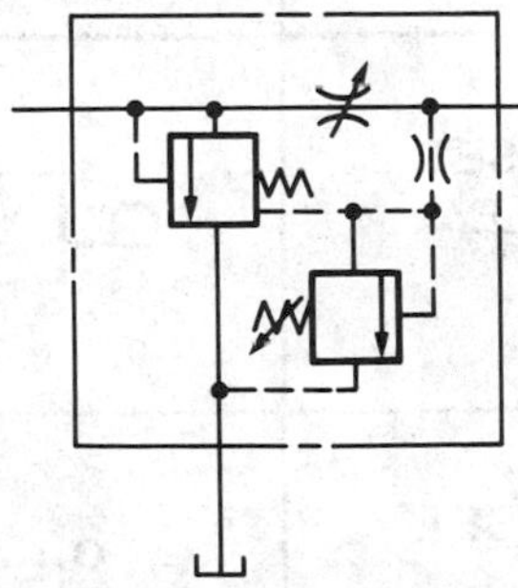
PRESSURE-COMPENSATED CARTRIDGE VALVE TYPE	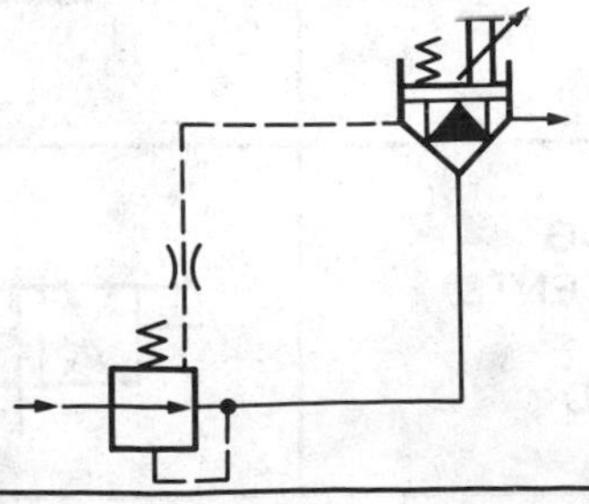

DIRECTIONAL CONTROL VALVES

TWO POSITION, TWO CONNECTION

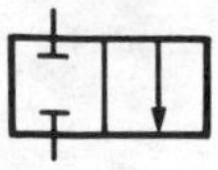

TWO POSITION, THREE CONNECTION

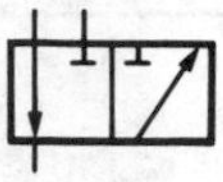

TWO POSITION, FOUR CONNECTION

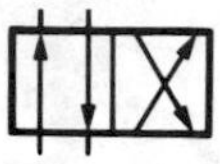

TWO POSITION IN TRANSITION

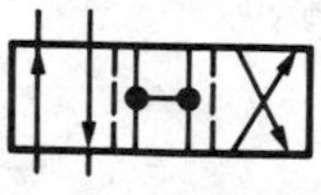

THREE POSITION, FOUR CONNECTION

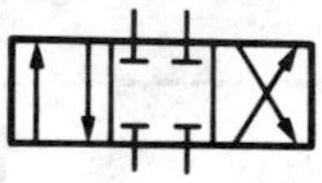

VALVES CAPABLE OF INFINITE POSITIONING (HORIZONTAL BARS INDICATE INFINITE POSITIONING ABILITY)

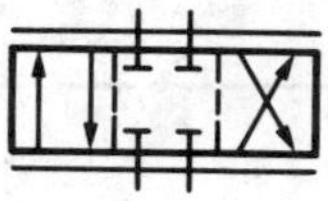

NO SPRING DETENTED, MANUALLY OPERATED

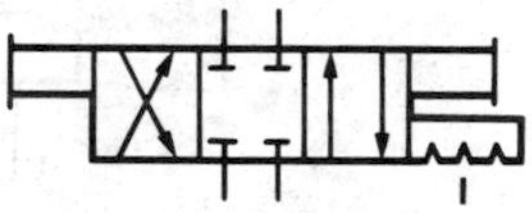

SPRING OFFSET, MECHANICALLY OPERATED

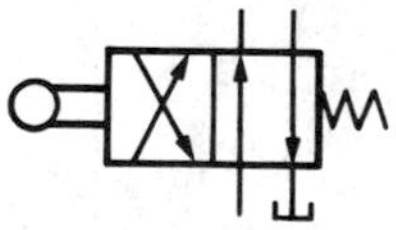

SPRING OFFSET, LEVER OPERATED—TWO WAY

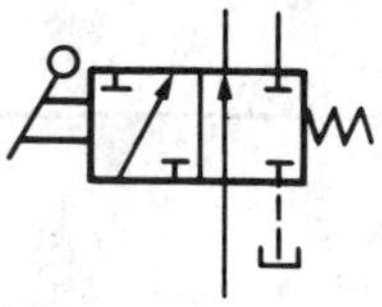

NO SPRING WITH DETENTS, LEVER OPERATED

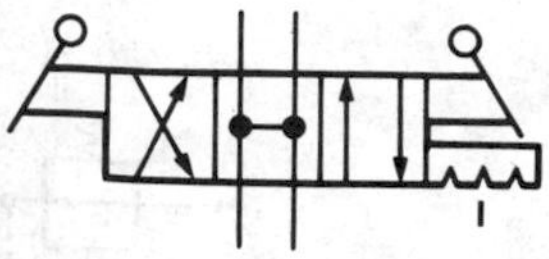

SPRING CENTERED, AIR OPERATED

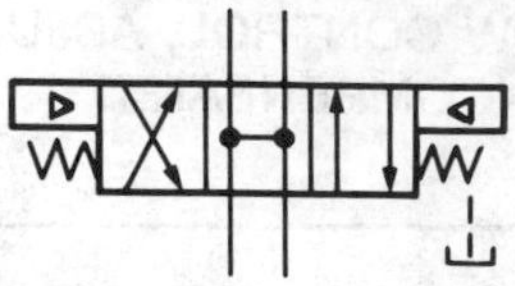

SPRING OR PRESSURE CENTERED, PILOT OPERATED

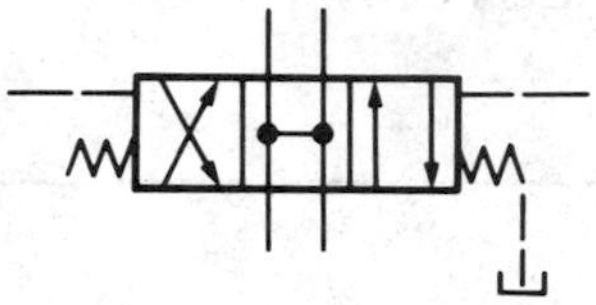

SPRING CENTERED, SOLENOID OPERATED

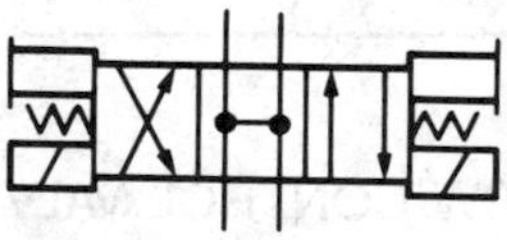

SOLENOID CONTROLLED, PILOT OPERATED

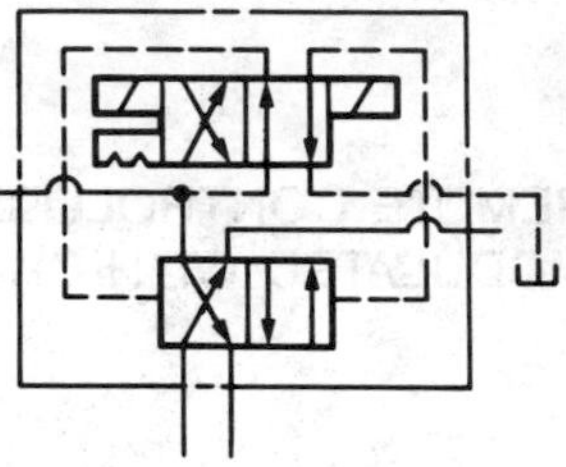

SPRING CENTERED, SOLENOID CONTROLLED, PILOT OPERATED

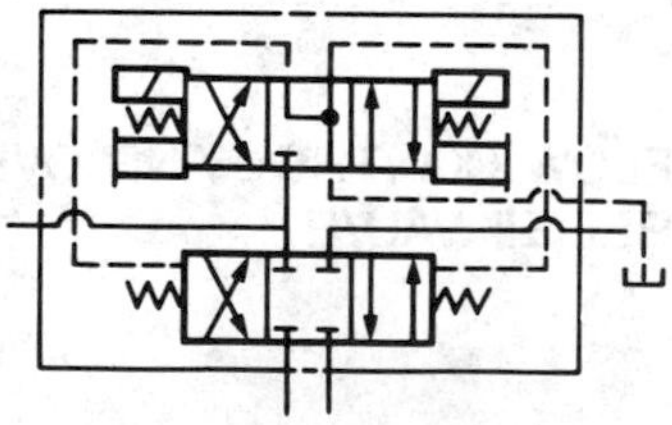

PRESSURE CENTERED, SOLENOID CONTROLLED, PILOT OPERATED (SIMIPLIFIED SYMBOL)

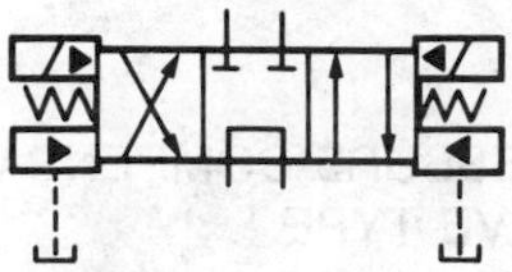

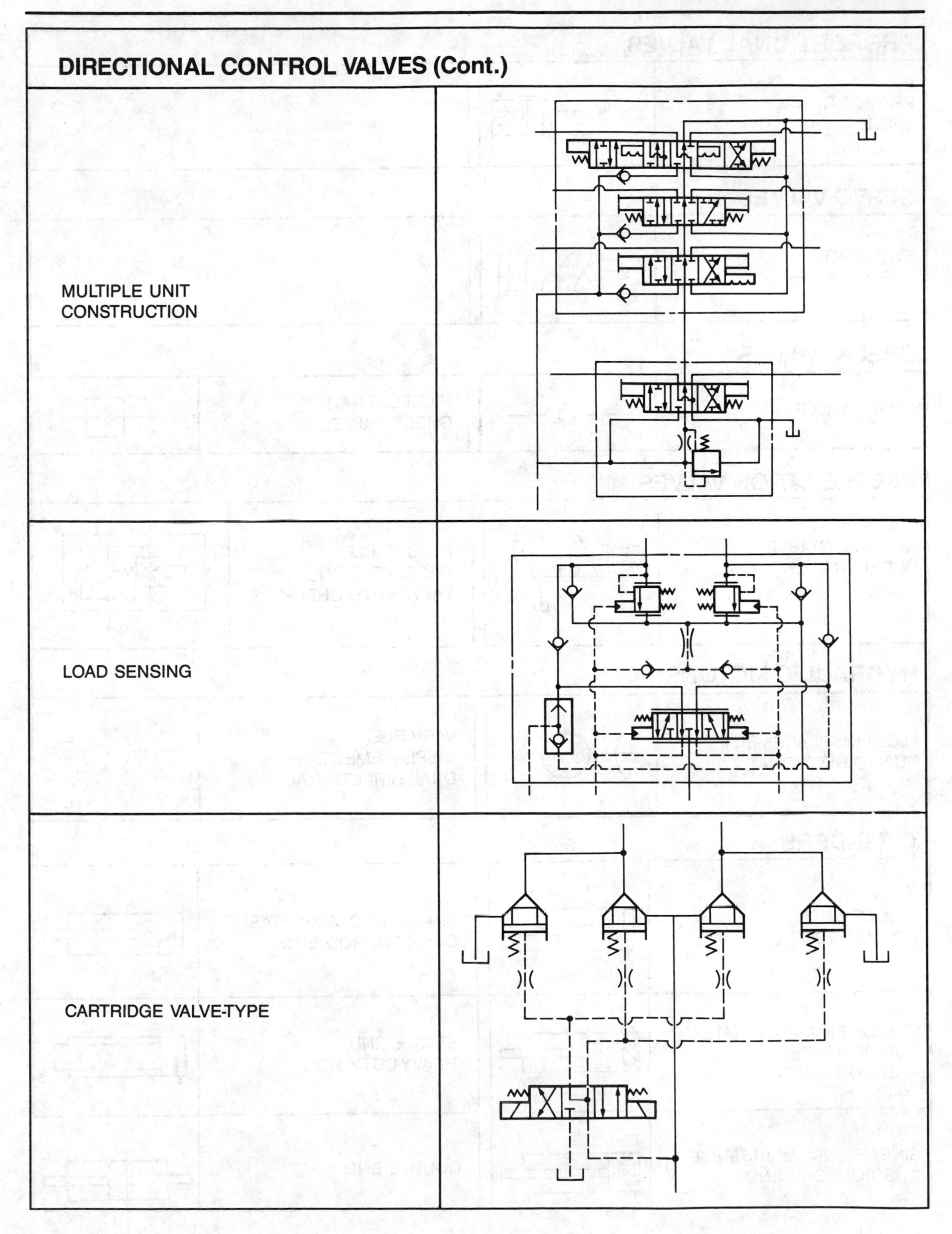
DIRECTIONAL CONTROL VALVES (Cont.)
MULTIPLE UNIT CONSTRUCTION
LOAD SENSING
CARTRIDGE VALVE-TYPE

PROPORTIONAL VALVES

PROPORTIONAL VALVE			

SERVO VALVES

TWO-STAGE SERVO VALVE	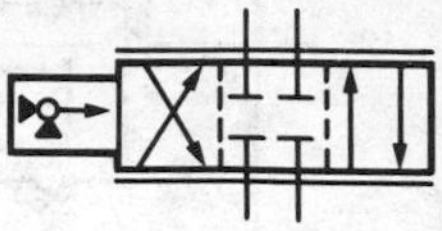		

CHECK VALVES

CHECK VALVE		PILOT-OPERATED CHECK VALVE	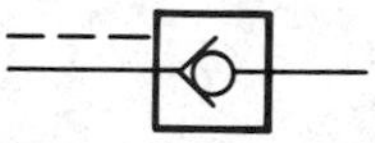

DECELERATION VALVES

NONADJUSTABLE DECELERATION VALVE	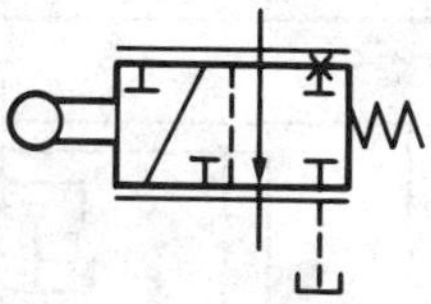	ADJUSTABLE DECELERATION VALVE WITH CHECK	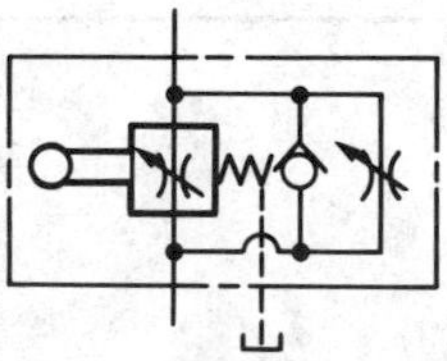

HYDRAULIC MOTORS

FIXED DISPLACEMENT, DUAL DIRECTIONAL	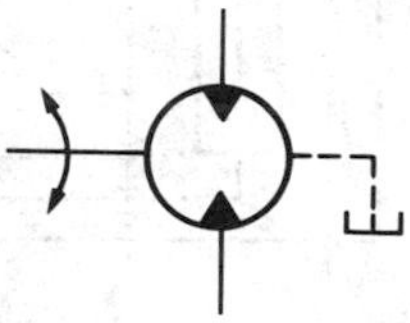	VARIABLE DISPLACEMENT, DUAL DIRECTIONAL	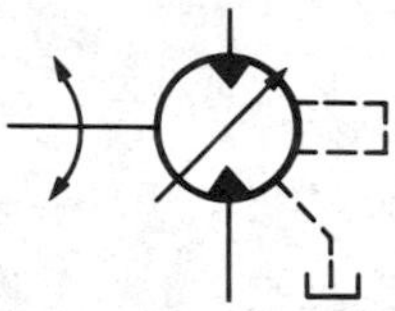

CYLINDERS

SINGLE END, NO CUSHIONS	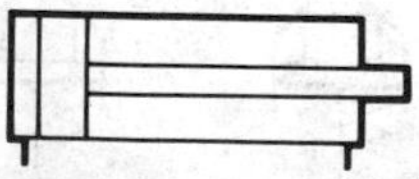	SINGLE END, ADJUSTABLE CUSHION, ROD END	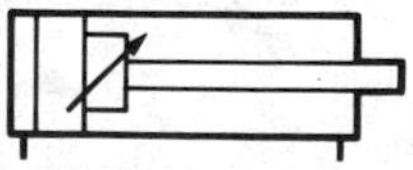
SINGLE END, ADJUSTABLE CUSHIONS, CAP AND ROD ENDS		SINGLE END, HEAVY-DUTY ROD	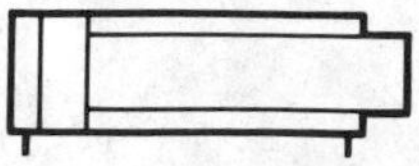
SINGLE END, ADJUSTABLE CUSHION, CAP END	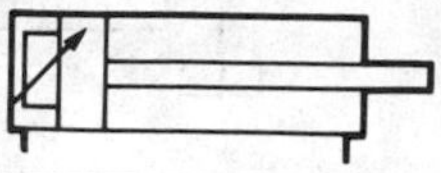	DOUBLE END	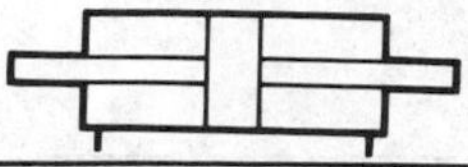

MISCELLANEOUS

Item	Symbol
ELECTRIC MOTOR	
INTERNAL COMBUSTION ENGINE  	
ACCUMULATOR, SPRING LOADED  	
ACCUMULATOR, GAS CHARGED  	
HEATER	
COOLER	
TEMPERATURE CONTROLLER	
TEMPERATURE CAUSE OR EFFECT	
FILTER, STRAINER	
PRESSURE SWITCH	
PRESSURE INDICATOR	
TEMPERATURE INDICATOR	
AIR BLEED	
STEERING BOOSTER	

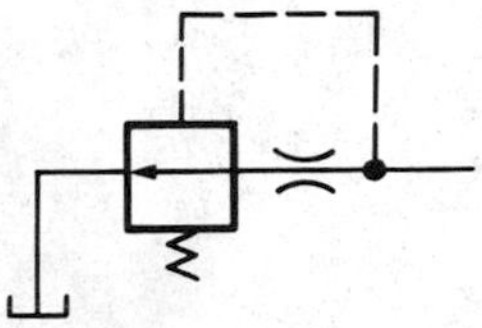

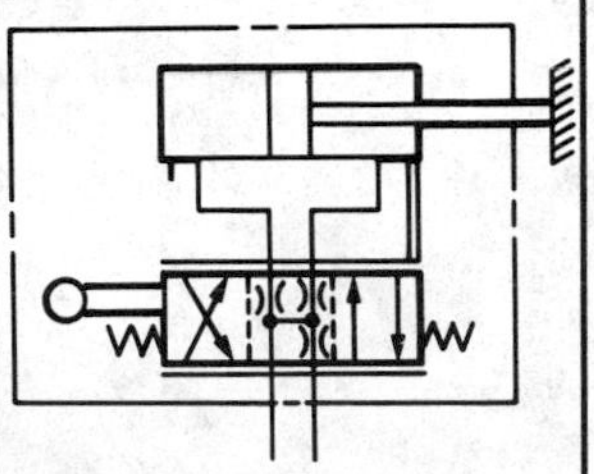

APPENDIX C
FLUID POWER DATA

FORMULAS AND UNITS (ENGLISH ONLY)

Pascal's Law and the Conservation of Energy

1. Hydraulics is a means of transmitting power. It may be used to multiply force or modify motions.
2. PASCAL'S LAW: Pressure exerted on a confined fluid is transmitted undiminished in all directions,acts with equal force on all equal areas, and acts at right angles to those areas.
3. To find the area of a round piston, square the diameter and multiply by .7854.

   $$Diameter_{(in)} = \sqrt{\frac{Area_{(in^2)}}{.7854}} \qquad Area_{(in^2)} = Diameter_{(in)} \times Diameter_{(in)} \times .7854$$

4. The force exerted by a piston can be determined by multiplying the piston area by the pressure applied. $Force_{(lb)} = Pressure_{(psi)} \times Area_{(in^2)}$
5. To determine the volume of fluid required to move a piston a given distance, multiply the piston area by the stroke required. $Volume_{(in^3)} = Area_{(in^2)} \times Length_{(in)}$ (231 Cubic inch $_{(in^3)}$ = One U.S. Gallon).
6. Work is force acting through a distance. $Work_{(in\text{-}lb)} = Force_{(lb)} \times Distance_{(in)}$
7. Power is the rate of doing work. $Power_{\left(\frac{in\text{-}lb}{sec}\right)} = \frac{Work_{(in\text{-}lb)}}{Time_{(sec)}} = \frac{Force_{(lb)} \times Distance_{(in)}}{Time_{(sec)}}$

Oil and Atmospheric Pressure

8. Hydraulic oil serves as a lubricant and is practically noncompressible. It will compress approximately 0.4 of 1% at 1000 psi and 1.1% at 3000 psi at 120 degrees Farenheit.
9. The weight of hydraulic oil may vary with a change in viscosity. However, 55-58 lbs. per cubic foot covers the viscosity range from 150 to 900 SSU at 100 degrees Farenheit.
10. Pressure at the bottom of a one foot column of oil will be approximately 0.4 psi. To find the approximate pressure at the bottom of any column of oil, multiply the height in feet by 0.4.
11. Atmospheric pressure equals 14.7 psia at sea level.
12. Gauge readings do not include atmospheric pressure unless marked psia.

Pumping Principles

13. There must be a pressure drop (pressure difference) across an orifice or other restriction to cause flow through it. Conversely, if there is no flow, there will be no pressure drop.
14. A fluid is pushed, not drawn, into a pump.
15. A pump does not pump pressure; its purpose is to create flow. Pumps used to transmit power are usually a positive displacement type.
16. Pressure is caused by resistance to flow. A pressure gauge indicates the workload at any given moment.

Series and Parallel Circuits

17. Fluids take the path of least resistance. Sum of the pressure drops is equal to total resistance in a series circuit.

Helpful information (1 of 2).
Revised 11/23/94

Calculating Velocity in a Cylinder

18. Speed of a cylinder piston is dependent upon its size (piston area) and the rate of flow into it.

$$Velocity_{(\frac{in}{min})} = \frac{Flow_{(\frac{in^3}{min})}}{Area_{(in^2)}} \qquad Flow_{(\frac{in^3}{min})} = Velocity_{(\frac{in}{min})} \times Area_{(in^2)}$$

19. To determine the pump capacity needed to extend a cylinder piston of a given area through a given distance in a specific time:

$$GPM = \frac{Area_{(in^2)} \times Stroke\ Length_{(in)} \times 60_{(seconds)}}{Time_{(seconds)} \times 231_{(\frac{in^3}{gal})}}$$

$$Time_{(seconds)} = \frac{Piston\ Area_{(in^2)} \times Stroke\ Length_{(in)} \times 60_{(seconds)}}{GPM \times 231_{(\frac{in^3}{gal})}}$$

Determining Pipe, Tubing and Hose Sizes

20. Flow velocity through a pipe varies inversely as the square of the inside diameter. Doubling the inside diameter increases the area four times.
21. Friction losses (pressure drop) of a liquid in a pipe vary with velocity.
22. To find the actual area of a pipe needed to handle a given flow, use the formula:

$$Area_{(in^2)} = \frac{GPM \times .3208}{Velocity_{(\frac{ft}{sec})}} \qquad Velocity_{(\frac{ft}{sec})} = \frac{GPM}{Area_{(in^2)} \times 3.117}$$

23. The actual inside diameter of standard pipe is usually larger than the nominal size quoted. A conversion chart should be consulted when selecting pipe.
24. Steel and copper tubing size indicates the outside diameter. To find the actual inside diameter, subtract two times the wall thickness from the tube size quoted. **ID = OD - 2 x Wall thickness.**
25. Hydraulic hose sizes are usually designated by their nominal inside diameter,. With some exceptions, this is indicated by a dash number representing the number of sixteenth inch increments in their inside diameter.

Horsepower and Torque

26. One H.P. = 33,000 ft.lbs./min.or 33,000 lbs. raised one foot in one minute or 550 lbs. raised one foot in one second. One H.P.= 746 Watts. One H.P. = 42.4 BTUs per minute.
27. To find the H.P. required for a given flow rate at a known pressure, use the formula:

$$Pump\ Output\ HP = GPM \times PSI \times .000583 = \frac{GPM \times PSI}{1714}$$

To find the H.P. required to drive a hydraulic pump of a given volume at a known pressure, use the formula:

$$Pump\ Input\ HP = \frac{GPM \times PSI \times .000583}{Pump\ Efficiency} = \frac{GPM \times PSI}{1714 \times Pump\ Efficiency}$$

If actual poump efficiency is not known, use the following rule of thumb formula for input horsepower:

$$Pump\ Input\ HP = GPM \times PSI \times .0007 \text{ (assumes 83.3\% efficiency)}$$

28. The relationship between Torque and Horsepower is:

$$Torque_{(lb\text{-}in)} = \frac{HP \times 63025}{RPM} = \frac{PSI \times CIR}{2\pi} \qquad HP = \frac{Torque_{(lb\text{-}in)} \times RPM}{63025}$$

Helpful information (2 of 2).

	PSIA (POUNDS PER SQUARE INCH ABSOLUTE) (bar) (kPa)	PSI (POUNDS PER SQUARE INCH GAUGE) GAUGE SCALE (bar) (kPa)	IN. HG ABS. (INCHES OF MERCURY ABSOLUTE) BAROMETER SCALE (mbar)	IN. HG (INCHES OF MERCURY) VACUUM SCALE (mbar)	FEET OF OIL ABSOLUTE (meters)	FEET OF WATER ABSOLUTE (bar)
3 ATMOSPHERES ABSOLUTE 2 ATMOSPHERES GAUGE	44.1 (3.04) (304.06)	29.4 (2.02) (202.68)	(90) (3047.75)		111 (33.83)	102 (3.04)
2 ATMOSPHERES ABSOLUTE 1 ATMOSPHERE GAUGE	29.4 (2.02) (202.68)	14.7 (1.01) (101.34)	(60) (2031.83)		74 (22.55)	68 (2.03)
1 ATMOSPHERE ABSOLUTE (ATMOSPHERIC PRESSURE)	14.7 (1.01) (101.34)	0 (0.00) (0.00)	29.92 (30) (1013.21)(1015.92)	0 (0.00)	37 (11.28)	34 (1.01)
	10 (0.68) (68.94)	−5 (−0.34) (−34.47)	20 (677.29)	10 (338.64)	24 (7.31)	$22\frac{2}{3}$ (0.67)
	5 (0.34) (34.47)	−10 (−0.68) (−68.94)	10 (338.64)	20 (677.29)	12 (3.65)	$11\frac{1}{2}$ (0.34)
PERFECT VACUUM	0 (0.00) (0.00)	−15 (−1.03) (−103.41)	0 (0.00)	29.92 (1013.21)	0 (0.00)	0 (0.00)

============== Indicates that the scale is not used in this range. Values are shown for comparison only.

Pressure and vacuum scale comparison.

QUANTITY NAME		METRIC UNITS	METRIC SYMBOLS	U.S. CUSTOMARY SYMBOLS
Acceleration		meter per second squared	m/s^2	ft/sec^2
Angle, Plane		degree	°	°
		minute	′	′
		second	″	″
Area		square millimeter	mm^2	in^2
Bulk Modulus (See Modulus)				
Conductivity, Thermal		watt per meter kelvin	W/m•K	Btu/hr•ft°F
Cubic Expansion, Coefficient (Note 8)		per degree Celsius	1/°C	1/°F
Current, Electric		ampere	A	A
Density	hydraulic fluids & other liquids	kilogram per liter	kg/L (Note 1)	lb/gal
	gases	kilogram per cubic meter	kg/m^3	lb/ft^3
	solids	gram per cubic centimeter	g/cm^3	lb/ft^3
Displacement (Unit discharge) (Note 2)	pneumatic	cubic centimeter	cm^3	in^3
		liter (Note 1)	L	gal
	hydraulic	milliliter (Note 1)	mL	in^3
Efficiency		percent	%	%
Energy, Heat		kilojoule	kJ	Btu
Flow Rate, Heat		watt	W	Btu/min
Flow Rate, Mass		gram per second	g/s	lb/min
		kilogram per second	kg/s	lb/s
Flow Rate, Volume (Note 3)	pneumatic	cubic decimeter per second	dm^3/s	ft^3/min (cfm)
		cubic centimeter per second	cm^3/s	in^3/min (cim)
	hydraulic	liter per minute	L/min (Note 1)	gal/min
		milliliter per minute	mL/min (Note 1)	in^3/min
Force		Newton	N	lb
Force per Length		Newton per millimeter	N/mm	lb/in
Frequency (Cycle)		hertz	Hz	Hz (cps)
		reciprocal minute	1/min (Note 4)	cpm
Frequency (Rotational)		reciprocal minute	1/min (Note 4)	rpm
Heat		kilojoule	kJ	Btu
Heat Capacity, Specific		kilojoule per kilogram kelvin	kJ/kg•K	Btu/lb•°F
Heat Transfer, Coefficient		watt per square meter kelvin	W/m^2•k (Note 9)	Btu/hr•ft^2•°F
Inertia, Moment of		kilogram meter squared	kg•m^2	lb•ft^2 lb•in^2

Metric units for fluid power applications (1 of 3).

QUANTITY NAME	METRIC		U.S. CUSTOMARY SYMBOLS
	UNITS	SYMBOLS	
Length	millimeter	mm	in
	meter	m	ft
	micrometer	µm	(micron)
			in
Linear Expansion, Coefficient (Note 7)	per degree	1/°C	1/°F
Mass	kilogram	kg	lb
Modulus, Bulk	megapascal	MPa	lb/in^2 (psi)
Momentum	kilogram meter per second	kg•m/s	lb•ft/sec
Potential, Electric	volt	V	V
Power	kilowatt	kW	hp
	watt	W	Btu/min
Pressure*	kilopascal (Note 5)	kPa	lb/in^2 (psi)
	bar (Note 5)	bar	lb/in^2 (psi)
Rotational Speed (Shaft Speed) (See Frequency, Rotational)			
Stress—Normal, Shear, Strength of Materials	megapascal	MPa	lb/in^2 (psi)
Surface Roughness	micrometer	µm	in
Temperature, Customary	degree Celsius	°C	°F
Temperature, Thermodynamic (Absolute)	kelvin	K	°R
Time	second, minute, hour	s, min, hr	s, min, hr
Torque (Moment of Force)	Newton meter	N•m	lb•ft
			lb•in
Volume — pneumatic	cubic decimeter	dm^3	ft^3
	cubic centimeter	cm^3	cm^3
Volume — hydraulic	liter (Note 1)	L	gal
	milliliter (Note 1)	mL	oz
Volumetric Flow (See Flow Rate)			
Velocity, Linear	meter per second	m/s	ft/s
	millimeter per second	mm/s	in/s
Velocity, Angular	radian per second	rad/s	rad/s
Viscosity, Dynamic	millipascal second	mPa•s	cP
Viscosity, Kinematic	square millimeter per second (Note 6)	mm^2/s	cSt
Work	joule	J	ft•lb

* Measurement of pressure:
above atmospheric, use --- kPa g or bar g ------ psi g
below atmospheric, use --- kPa vacuum or bar vacuum ------ in Hg
or kPa absolute or bar absolute --- psi abs

Metric units for fluid power applications (2 of 3).

NOTES

NOTE 1. The international symbol for liter is lower case "l" which can easily be confused with the numeral "1." Accordingly, the symbol "L" is to be used for U.S. fluid power.

NOTE 2. Indicate displacement of a rotary device as "per revolution" and of a nonrotary device as "per cycle."

NOTE 3. For gases, this quantity is frequently expressed as free gas at Standard Reference Atmosphere, as defined in reference 4.5 and specified in reference 4.6. In such cases, the abbreviation "ANR" is to follow the expression of the quantity of unit (m^3/min [ANR]). See reference 4.1 for more detailed information of the attachment of letters to a symbol.

NOTE 4. Mechanical oscillations are normally expressed in cycles per unit time and rotational frequency in revolutions per unit time. Since "cycle" and "revolution" are not units, they do not have internationally recognized symbols. Therefore, they are normally expressed by abbreviations which are different in various languages. In English, the symbology for mechanical oscillations is c/min and for rotation frequency, r/min.

NOTE 5. The bar and kilopascal are given equal status as pressure units. At this time, the domestic fluid power industry does not agree on one preferred unit. The pascal is the SI unit for pressure and a major segment of U.S. industry has accepted the multiple, kilopascal (kPa), as the preferred unit. On the other hand, the majority of the international fluid power industry has accepted the bar as the metric pressure unit. The bar is recognized by the EEC as an acceptable metric unit and is shown in ISO 1000 for use in specialized fields. Conversely, the bar is considered by both the International Committee on Weights and Measures and the NBS as a unit to be used for a limited time only. Further, the bar has been deprecated by Canada, ANMC, and some U.S. standards organizations and is illegal in some countries.

NOTE 6. Viscosity is frequently expressed in SUS (Saybolt Universal Seconds). SUS is the time in seconds for 60 mL of fluid to flow through a standard orifice at a specified temperature. Conversion between kinematic viscosity, mm^2/s (centistokes) and SUS can be made by reference to Tables in references 4.2 and 4.7.

NOTE 7. The linear expansion coefficient is a ratio, not a unit, and is expressed in customary U.S. units such as in/in and in metric units such as mm/mm per unit temperature change.

NOTE 8. The cubic expansion coefficient is a ratio, not a unit and is expressed in customary U.S. units such as in in^3/in^3 and in metric units such as cm^3/cm^3 per unit temperature change.

NOTE 9. In these expressions, "K" indicates temperature interval. Therefore, "K" may be replaced with °C if desired without changing the value or affecting the conversion factor.

Metric units for fluid power applications (3 of 3).

To convert →		Into →		Multiply by
Into ←		**To convert** →		**Divide by**
Unit	**Symbol**	**Unit**	**Symbol**	**Factor**
Atmospheres	Atm	bar	bar	1.013250
BTU/hour	Btu/h	kilowatts	kW	0.293071×10^{-3}
Cubic centimeters	cm^3	liters	l	0.001
Cubic centimeters	cm^3	milliliters	ml	1.0
Cubic feet	ft^3	cubic meters	m^3	0.0283168
Cubic feet	ft^3	liters	l	28.3161
Cubic inches	in^3	cubic centimeters	cm^3	16.3871
Cubic inches	in^3	liters	l	0.0163866
Degrees (angle)	°	radians	rad	0.0174533
Fahrenheit	°F	Celsius (centigrade)	°C	▲
Feet	ft	meters	m	0.3048
Feet of water	ft H_2O	bar	bar	0.0298907
Fluid ounces, US	US fl oz	cubic centimeters	cm^3	29.5735
Foot pounds f.	ft lbf	joules	J	1.35582
Foot pounds/minute	ft lbf/min	watts	W	81.3492
Gallons, US	US gal	liters	l	3.78531
Horsepower	hp	kilowatts	kW	0.7457
Inches of mercury	in Hg	millibar	mbar	33.8639
Inches of water	in H_2O	millibar	mbar	2.49089
Inches	in	centimeters	cm	2.54
Inches	in	millimeters	mm	25.4
Kilogram force	kgf	Newtons	N	9.80665
Kilogram f. meter	kgt m	Newton meters	Nm	9.80665
Kilogram f./sq centimeter	kgf/cm^2	bar	bar	0.980665
Kilopascals	kPa	bar	bar	0.01
Kiloponds	kp	Newtons	N	9.80665
Kilopond meters	kp m	Newton meters	Nm	9.80665
Kiloponds/square centimeter	kp/cm^2	bar	bar	0.980665
Microinches	μin	microns	μm	0.0254
Millimeters of mercury	mm Hg	millibar	mbar	1.33322
Millimeters of water	mm H_2O	millibar	mbar	0.09806
Newtons/square centimeter	N/cm^2	bar	bar	0.1
Newtons/square meter	N/m^2	bar	bar	10^{-5}
Pascals (Newton/sq meter)	Pa	bar	bar	10-5
Pints, US	US liq pt	liters	l	0.473163
Pounds (mass)	lb	kilograms	kg	0.4536
Pounds/cubic foot	lb/ft^3	kilograms/cubic meter	kg/m^3	16.0185
Pounds/cubic inch	lb/in^3	kilograms/cubic centimeter	kg/cm^3	0.0276799
Pounds force	lbf	Newtons	N	4.44822
Pounds f. feet	lbf ft	Newton meters	Nm	1.35582
Pounds f. inches	lbf in	Newton meters	Nm	0.112985
Pounds f./square inch	lbf/in^2	bar	bar	0.06894
Revolutions/minute	r/min	radians/second	rad/s	0.104720
Square feet	ft^2	square meters	m^2	0.092903
Square inches	in^2	square meters	m^2	6.4516×10^{-4}
Square inches	in^2	square centimeters	cm^2	6.4516

▲ °C = 5(°F−32)/9

Fluid power equivalents

1 bar = 10^5 N/m^2
1 bar = 10 N/cm^2 = 1 dN/mm^2
1 pascal = 1 N/m^2
1 liter = 1000.028 cm^3
1 centistoke (cSt) = 1 mm^2/s
1 joule = 1 wattsecond (Ws)
Hertz (Hz) = cycles/second

Prefixes denoting decimal multiples or sub-multiples

For multiples

Factor	Prefix	Symbol
$x10^{12}$	tera	T
$x10^9$	giga	G
$x10^6$	mega	M
$x10^3$	kilo	k
$x10^2$	hecto	h
x10	deka	da

For sub-multiples

Factor	Prefix	Symbol
$x10^{-1}$	deci	d
$x10^{-2}$	centi	c
$x10^{-3}$	milli	m
$x10^{-6}$	micro	μ
$x10^{-9}$	nano	n
$x10^{-12}$	pico	p
$x10^{-15}$	femto	f
$x10^{-18}$	atto	a

English—metric conversion factors.

SAE VISCOSITY GRADES FOR ENGINE OILS

SAE Viscosity Grade	Viscosity (cP) at Temperature °C (°F) Max	Viscosity (cSt) at 100°C (212°F)	
		Min	Max
0W	3250 at −30 (−22)	3.8	—
5W	3500 at −25 (−13)	3.8	—
10W	3500 at −20 (−4)	4.1	—
15W	3500 at −15 (5)	5.6	—
20W	4500 at −10 (14)	5.6	—
25W	6000 at −5 (23)	9.3	—
20	—	5.6	Less than 9.3
30	—	9.3	Less than 12.5
40	—	12.5	Less than 16.3
50	—	16.3	Less than 21.9

NOTE: 1cP = 1mPa·s; 1 cSt = 1mm²/s

SAE viscosity grades for engine oils.

ISO Viscosity Grade	Midpoint Kinematic Viscosity cSt at 40°C (104°F)	Kinematic Viscosity Limits cSt at 40°C (104°F)	
		Minimum	Maximum
ISOVG2	2.2	1.98	2.42
ISOVG3	3.2	2.88	3.52
ISOVG5	4.6	4.14	5.06
ISOVG7	6.8	6.12	7.48
ISOVG10	10	9.00	11.0
ISOVG15	15	13.5	16.5
ISOVG22	22	19.8	24.2
ISOVG32	32	28.8	35.2
ISOVG46	46	41.4	50.6
ISOVG68	68	61.2	74.8
ISOVG100	100	90.0	110
ISOVG150	150	135	165
ISOVG220	220	198	242
ISOVG320	320	288	352
ISOVG460	460	414	506
ISOVG680	680	612	748
ISOVG1000	1000	900	1100
ISOVG1500	1500	1350	1650

ISO viscosity grades for industrial oils.

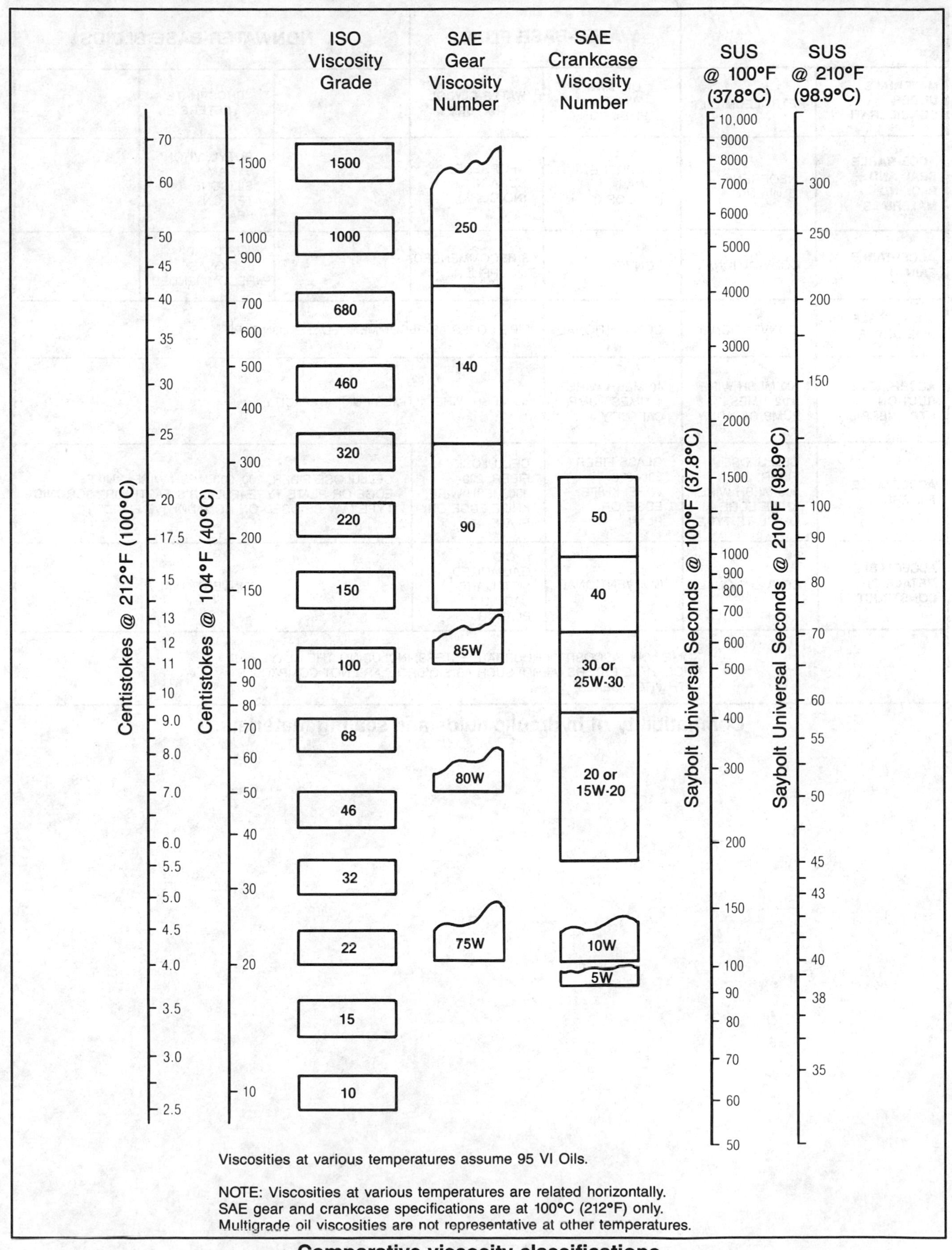

Comparative viscosity classifications.

		WATER-BASE FLUIDS		NONWATER-BASE BLUIDS		
MATERIALS UNDER CONSIDERATION	PETROLEUM OILS	OIL AND WATER EMULSION	WATER-GLYCOL MIXTURE		PHOSPHATE ESTERS	
ACCEPTABLE SEAL AND PACKING MATERIALS	NEOPRENE, BUNA N	NEOPRENE, BUNA N, (NO CORK)	NEOPRENE, BUNA N, (NO CORK)		BUTYL, VITON*, VYRAM, SILICONE, TEFLON FBA	
ACCEPTABLE PAINTS	CONVENTIONAL	CONVENTIONAL	AS RECOMMENDED BY SUPPLIER		"AIR CURE" EPOXY AS RECOMMENDED	
ACCEPTABLE PIPE DOPES	CONVENTIONAL	CONVENTIONAL	PIPE DOPES AS RECOMMENDED, TEFLON TAPE			
ACCEPTABLE SUCTION STRAINERS	100 MESH WIRE 1-1/2 TIMES PUMP CAPACITY	40 MESH WIRE 4 TIMES PUMP CAPACITY	50 MESH WIRE, 4 TIMES PUMP CAPACITY			
ACCEPTABLE FILTERS	CELLULOSE FIBER, 200–300 MESH WIRE, KNIFE EDGE OR PLATE TYPE	GLASS FIBER, 200–300 WIRE, KNIFE EDGE OR PLATE	CELLULOSE FIBER, 200–300 MESH WIRE, KNIFE EDGE OR PLATE	CELLULOSE FIBER, 200–300 MESH WIRE, KNIFE EDGE OR PLATE TYPE (FULLER'S EARTH OR MICRONIC TYPE MAY BE USED ON NONADDITIVE FLUIDS.)		
ACCEPTABLE METALS OF CONSTRUCTION	CONVENTIONAL	CONVENTIONAL	AVOID GALVANIZED METAL AND CADMIUM PLATING		CONVENTIONAL	

* SOME LOW VISCOSITY PHOSPHATE ESTERS, INCLUDING THE ALKYL PHOSPHATE ESTERS (AERO) SUCH AS SKYDROL, ARE NOT COMPATIBLE WITH VITON SEALS.

Compatibility of hydraulic fluids and sealing materials.

CONTAMINATION

Contaminant	Character	Source and Remarks
Acidic by-products	Corrosive	Breakdown of oil. May also arise from water-contamination of phosphate-ester fluids.
Sludge	Blocking	Breakdown of oil.
Water	Emulsion	Already in fluid or introduced by system fault or breakdown of oxidation-inhibitors.
Air	Soluble	Effect can be controlled by anti-foam additives.
	Insoluble	Excess air due to improper bleeding, poor system design or air leaks.
Other oils	Miscible but may react	Use of wrong fluid for topping up, etc.
Grease	May or may not be miscible	From lubrication points.
Scale	Insoluble	From pipes not properly cleaned before assembly.
Metallic particles	Insoluble with catalytic action	May be caused by water contamination, controllable with anti-rust additives.
Paint flakes	Insoluble, blocking	Paint on inside of tank old or not compatible with fluid.
Abrasive particles	Abrasive and blocking	Airborne particles (remove with air filter).
Elastomeric particles	Blocking	Seal breakdown. Check fluid, compatibility of seal design.
Sealing compound particles	Blocking	Sealing compounds should not be used on pipe joints.
Sand	Abrasive and blocking	Sand should not be used as a filler for manipulating pipe bends.
Adhesive particles	Blocking	Adhesives or jointing compounds should not be used on gaskets.
Lint or fabric threads	Blocking	Only lint-free cloths or rags should be used for cleaning or plugging dismantled components.

Contaminants in hydraulic systems.

Code	Number of particles per 100 milliliters			
	Over 5 μm		Over 15 μm	
	More than	& up to	More than	& up to
20/17	500k	1M	64k	130k
20/16	500k	1M	32k	64k
20/15	500k	1M	16k	32k
20/14	500k	1M	8k	16k
19/16	250k	500k	32k	64k
19/15	250k	500k	16k	32k
19/14	250k	500k	8k	16k
19/13	250k	500k	4k	8k
18/15	130k	250k	16k	32k
18/14	130k	250k	8k	16k
18/13	130k	250k	4k	8k
18/12	130k	250k	2k	4k
17/14	64k	130k	8k	16k
17/13	64k	130k	4k	8k
17/12	64k	130k	2k	4k
17/11	64k	130k	1k	2k
16/13	32k	64k	4k	8k
16/12	32k	64k	2k	4k
16/11	32k	64k	1k	2k
16/10	32k	64k	500	1k
15/12	16k	32k	2k	4k
15/11	16k	32k	1k	2k
15/10	16k	32k	500	1k
15/9	16k	32k	250	500
14/11	8k	16k	1k	2k
14/10	8k	16k	500	1k
14/9	8k	16k	250	500
14/8	8k	16k	130	250
13/10	4k	8k	500	1k
13/9	4k	8k	250	500
13/8	4k	8k	130	250
12/9	2k	4k	250	500
12/8	2k	4k	130	250
11/8	1k	2k	130	250

Table of ISO codes and corresponding contamination levels.

Target Contamination class to ISO Code		Suggested maximum Particle level		Sensitivity	Type of system	Suggested filtration rating
5 μm	15 μm	5 μm	15 μm			$\beta_\chi > 75$
13	9	4,000	250	Super critical	Silt sensitive control system with very high reliability. Laboratory or aerospace.	1–2
15	11	16,000	1,000	Critical	High performance servo and high pressure long life systems, i.e., aircraft, machine tools, etc.	3–5
16	13	32,000	4,000	Very important	High quality reliable systems. General machine requirements.	10–12
18	14	130,000	8,000	Important	General machinery and mobile systems. Medium pressure, medium capacity.	12–15
19	15	250,000	16,000	Average	Low pressure heavy industrial systems, or applications where long life is not critical.	15–25
21	17	1,000,000	64,000	Main protection	Low pressure systems with large clearance.	25–40

Suggested acceptable contamination levels.

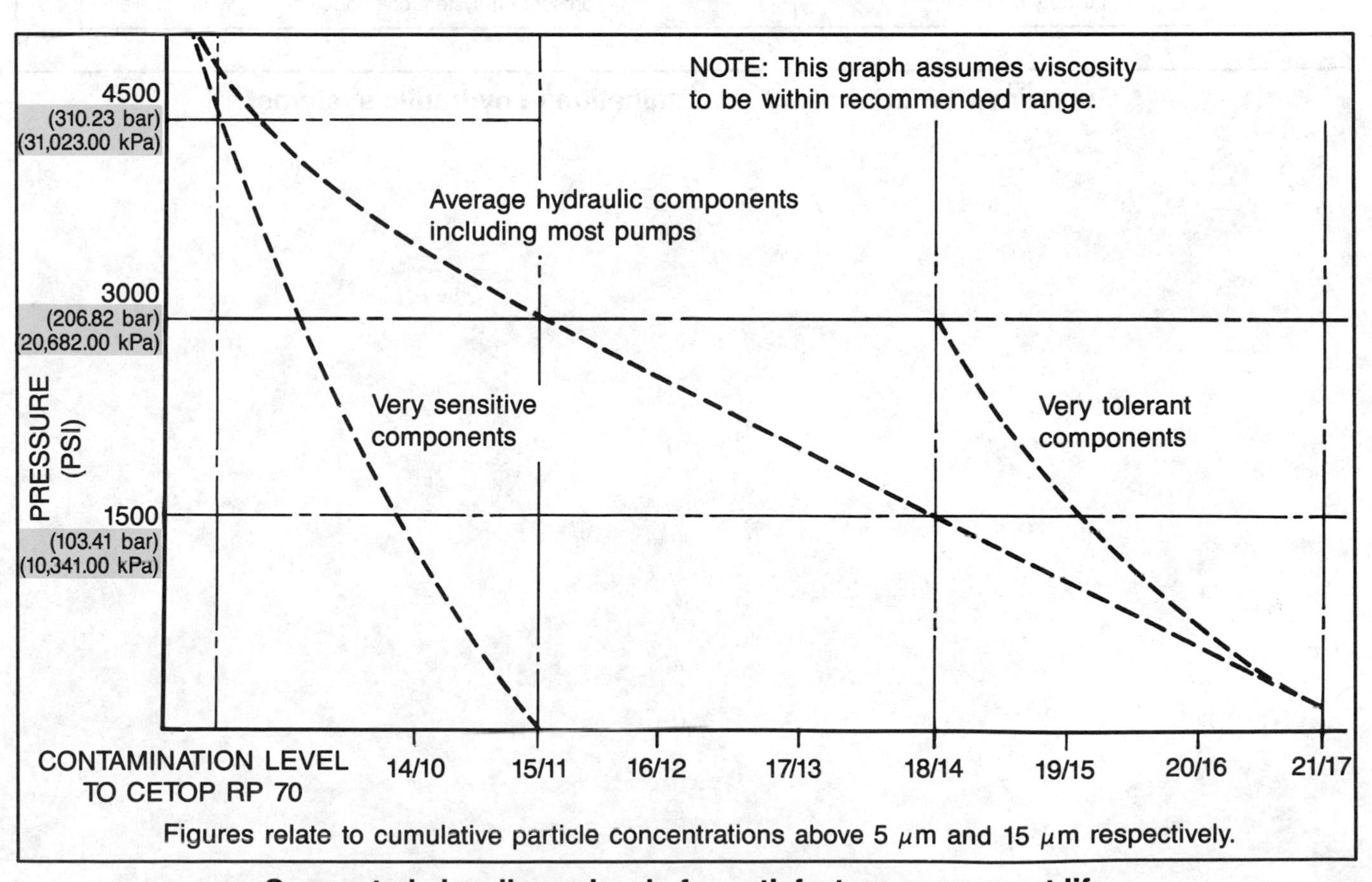

Suggested cleanliness levels for satisfactory component life.

Contamination Source	Controller
Inbuilt in components, pipes, manifolds, etc.	Good flushing procedures, system not operated on load until acceptable contamination level obtained.
plus Present in initial charge of fluid	Integrity of supplier. Fluid stored under correct conditions (exclusion of dirt, condensation, etc.). Fluid filtered during filling.
plus Ingressed through air breather	An effective air breather with rating compatible with degree of fluid filtration.
plus Ingressed during fluid replenishment	Suitable filling points which ensure some filtration of fluid before entering reservoir.
plus Ingressed during maintenance	This task undertaken by responsible personnel. Design should minimize the effects.
plus Ingressed through cylinder rod seals	Effective wiper seals or, if airborne contamination, rods protected by suitable gaiters.
plus Further generated contamination produced as a result of the above and the severity of the duty cycle.	Correct fluid selection and properties (viscosity and additives) maintained. Good system design minimizing effects of contamination present on system components.

Guidelines for controlling contamination in hydraulic systems.

PIPE, TUBING, AND HOSE

General requirements. When installing the various iron and steel pipes, tubes, and fittings of a hydraulic system, it is necessary that they be absolutely clean, free from scale, and all kinds of foreign matter. To attain this end, the following steps should be taken.

1. Tubing, pipes, and fittings should be brushed with boiler tube wire brush or cleaned with commercial pipe-cleaning apparatus. The inside edge of tubing and pipe should be reamed after cutting to remove burrs.
2. Short pieces of pipe and tubing and steel fittings are sandblasted to remove rust and scale. Sandblasting is a sure and efficient method for short straight pieces and fittings. Sandblasting is not used, however, if there is the slightest possibility that particles of sand will remain in blind holes or pockets in the work after flushing.
3. In the case of longer pieces of pipe or short pieces bent to complex shapes where it is not practical to sandblast, the parts are pickled in a suitable solution until all rust and scale is removed. Preparation for pickling requires thorough degreasing in TRICHLOROETHYLENE or other commercial degreasing solution.
4. Neutralize pickling solution.
5. Rinse parts and prepare for storage.
6. Tubing must not be welded, brazed, or silver soldered after assembly as proper cleaning is impossible in such cases. It must be accurately bent and fitted so that it will not be necessary to spring it into place.
7. If flange connections are used, flanges must fit squarely on the mounting faces and be secured with screws of the correct length. Screws or studnuts must be drawn up evenly to avoid distortion in the valve or pump body.
8. Be sure that all openings into the hydraulic system are properly covered to keep out dirt and metal slivers when work such as drilling, tapping, welding, or brazing is being done on or near the unit.
9. Threaded fittings should be inspected to prevent metal slivers from the threads getting into the hydraulic system.
10. Before filling the system with hydraulic oil, be sure that the hydraulic fluid is as specified and that it is clean. Do not use cloth strainers or fluid that has been stored in contaminated containers.
11. Use a #120 mesh screen when filling the reservoir. Operate the system for a short time to eliminate air in the lines. Add hydraulic fluid if necessary.
12. Safety precautions. Dangerous chemicals are used in the cleaning and pickling operations to be described. They should be kept only in the proper containers and handled with extreme care.

PICKLING PROCESS

1. Thoroughly degrease parts in degreaser, using trichloroethylene or other commercial degreasing solution.
2. **Tank No. 1**
   Solution. Use a commercially available derusting compound in solution as recommended by the manufacturer. The solution should not be used at a temperature exceeding that recommended by the manufacturer, otherwise, the inhibitor will evaporate and leave a straight acid solution. The length of time the part will be immersed in this solution will depend upon the temperature of the solution and the amount of rust or scale which must be removed. The operator must use his judgment on this point.
3. After pickling, rinse parts in cold running water and immerse in tank No. 2. The solution in this tank should be a neutralizer mixed with water in a proportion recommended by the manufacturer. This solution should be used at recommended temperatures and the parts should remain immersed in the solution for the period of time recommended by the manufacturer.
4. Rinse parts in hot water.
5. Place in tank No. 3. The solution in this tank should contain antirust compounds as recommended by the manufacturer. Usually the parts being treated should be left to dry with antirust solution remaining on them.

   If pieces are stored for any period of time, ends of the pipes should be plugged to prevent the entrance of foreign matter. Do not use rags or waste as they will deposit lint on the inside of the tube or pipe. Immediately before using pipes, tubes and fittings, they should be thoroughly flushed with suitable degreasing solution.

Preparation of pipes, tubes, and fittings before installation in a hydraulic system.

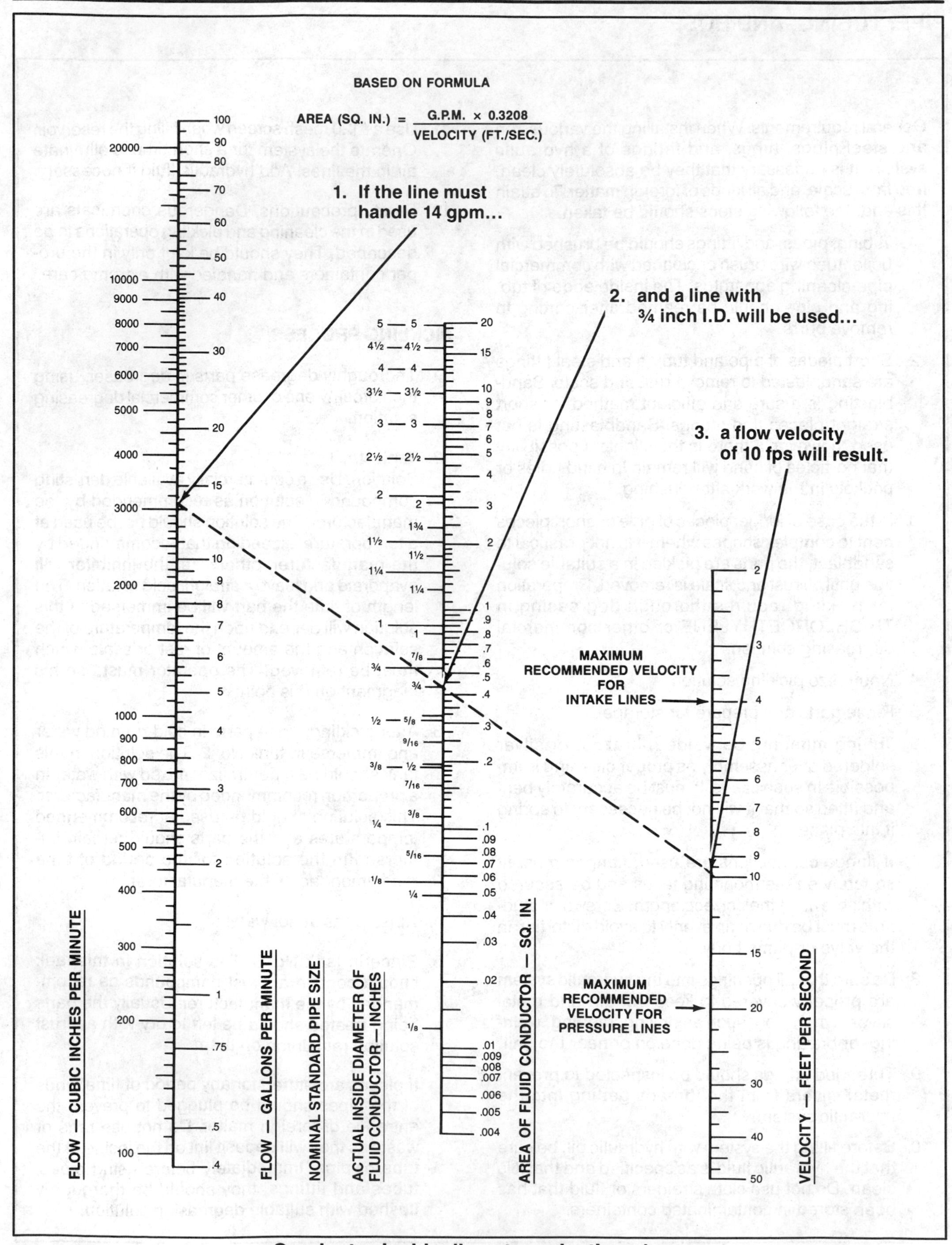
BASED ON FORMULA
AREA (SQ. IN.) = G.P.M. × 0.3208 / VELOCITY (FT./SEC.)
1. If the line must handle 14 gpm...
2. and a line with ¾ inch I.D. will be used...
3. a flow velocity of 10 fps will result.
MAXIMUM RECOMMENDED VELOCITY FOR INTAKE LINES
MAXIMUM RECOMMENDED VELOCITY FOR PRESSURE LINES
FLOW — CUBIC INCHES PER MINUTE
FLOW — GALLONS PER MINUTE
NOMINAL STANDARD PIPE SIZE
ACTUAL INSIDE DIAMETER OF FLUID CONDUCTOR — INCHES
AREA OF FLUID CONDUCTOR — SQ. IN.
VELOCITY — FEET PER SECOND

Conductor inside diameter selection chart.

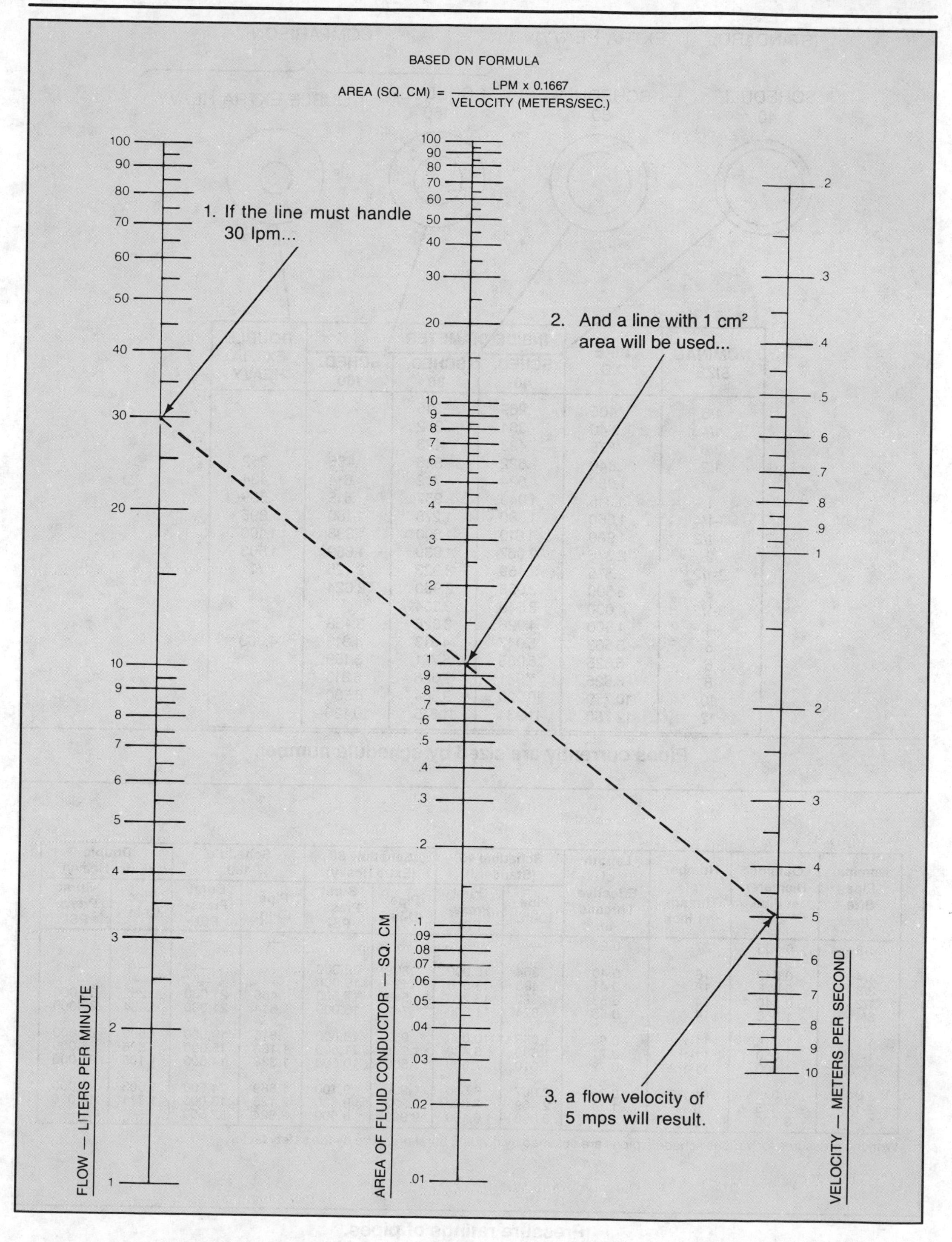
BASED ON FORMULA
AREA (SQ. CM) = LPM x 0.1667 / VELOCITY (METERS/SEC.)
1. If the line must handle 30 lpm...
2. And a line with 1 cm² area will be used...
3. a flow velocity of 5 mps will result.
FLOW — LITERS PER MINUTE
AREA OF FLUID CONDUCTOR — SQ. CM
VELOCITY — METERS PER SECOND
100
90
80
70
60
50
40
30
20
10
9
8
7
6
5
4
3
2
1
.9
.8
.7
.6
.5
.4
.3
.2
.1
.09
.08
.07
.06
.05
.04
.03
.02
.01

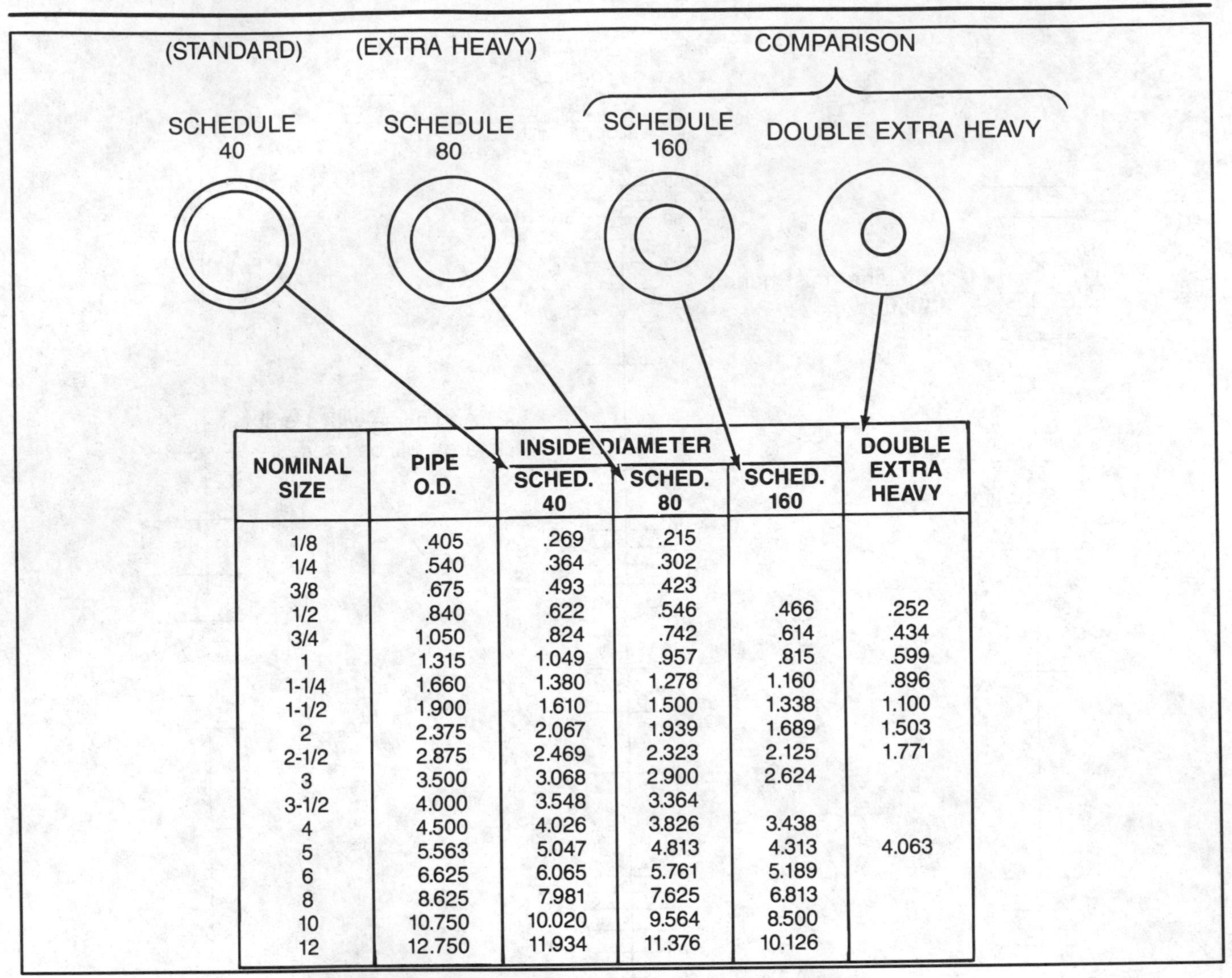

NOMINAL SIZE	PIPE O.D.	INSIDE DIAMETER SCHED. 40	SCHED. 80	SCHED. 160	DOUBLE EXTRA HEAVY
1/8	.405	.269	.215		
1/4	.540	.364	.302		
3/8	.675	.493	.423		
1/2	.840	.622	.546	.466	.252
3/4	1.050	.824	.742	.614	.434
1	1.315	1.049	.957	.815	.599
1-1/4	1.660	1.380	1.278	1.160	.896
1-1/2	1.900	1.610	1.500	1.338	1.100
2	2.375	2.067	1.939	1.689	1.503
2-1/2	2.875	2.469	2.323	2.125	1.771
3	3.500	3.068	2.900	2.624	
3-1/2	4.000	3.548	3.364		
4	4.500	4.026	3.826	3.438	
5	5.563	5.047	4.813	4.313	4.063
6	6.625	6.065	5.761	5.189	
8	8.625	7.981	7.625	6.813	
10	10.750	10.020	9.564	8.500	
12	12.750	11.934	11.376	10.126	

Pipes currently are sized by schedule number.

Nominal Pipe Size in.	Outside Diameter of Pipe -in.	Number of Threads Per Inch	Length of Effective Threads -in.	Schedule 40 (Standard)		Schedule 80 (Extra Heavy)		Schedule 160		Double (Extra Heavy)	
				Pipe ID-in.	Burst Press-PSI	Pipe ID-in.	Burst Press-PSI	Pipe ID-in.	Burst Press-PSI	Pipe ID-in.	Burst Press-PSI
1/8	0.405	27	0.26	—	—	—	—	—	—	—	—
1/4	0.540	18	0.40	.364	16,000	.302	22,000	—	—	—	—
3/8	0.675	18	0.41	.493	13,500	.423	19,000	—	—	—	—
1/2	0.840	14	0.53	.622	13,200	.546	17,500	.466	21,000	.252	35,000
3/4	1.050	14	0.55	.824	11,000	.742	15,000	.614	21,000	.434	30,000
1	1.315	11-1/2	0.68	1.049	10,000	.957	13,600	.815	19,000	.599	27,000
1-1/4	1.660	11-1/2	0.71	1.380	8,400	1.278	11,500	1.160	15,000	.896	23,000
1-1/2	1.900	11-1/2	0.72	1.610	7,600	1.500	10,500	1.338	14,800	1.100	21,000
2	2.375	11-1/2	0.76	2.067	6,500	1.939	9,100	1.689	14,500	1.503	19,000
2-1/2	2.875	8	1.14	2.469	7,000	2.323	9,600	2.125	13,000	1.771	18,000
3	3.500	8	1.20	3.068	6,100	2.900	8,500	2.624	12,500	—	—

Working pressures for various schedule pipes are obtained by dividing burst pressure by the safety factor.

Pressure ratings of pipes.

OIL FLOW CAPACITY OF TUBING

Figures in the chart are GPM flow capacities of tubing, and were calculated from the formula: GPM = V × A ÷ .3208, in which V = velocity of flow in feet per second, and A is inside square inch area of tube.

Figures in Body of Chart are GPM Flows							
Tube O.D.	**Wall Thick.**	**2 Ft/Sec**	**4 Ft/Sec**	**10 Ft/Sec**	**15 Ft/Sec**	**20 Ft/Sec**	**30 Ft/Sec**
1/2	.035	.905 GPM	1.81 GPM	4.52 GPM	6.79 GPM	9.05 GPM	13.6
	.042	.847	1.63	4.23	6.35	6.47	12.7
	.049	.791	1.58	3.95	5.93	7.91	11.9
	.058	.722	1.44	3.61	5.41	7.22	10.8
	.065	.670	1.34	3.35	5.03	6.70	10.1
	.072	.620	1.24	3.10	4.65	6.20	9.30
	.083	.546	1.09	2.73	4.09	5.46	8.16
5/8	.035	1.51	3.01	7.54	11.3	15.1	22.6
	.42	1.43	2.85	7.16	10.7	14.3	21.4
	.049	1.36	2.72	6.80	10.2	13.6	20.4
	.058	1.27	2.54	6.34	9.51	12.7	19.0
	.065	1.20	2.40	6.00	9.00	12.0	18.0
	.072	1.13	2.26	5.66	8.49	11.3	17.0
	.083	1.03	2.06	5.16	7.73	10.3	15.5
	.095	.926	1.85	4.63	6.95	9.26	13.9
3/4	.049	2.08	4.17	10.4	15.6	20.8	31.2
	.058	1.97	3.93	9.84	14.8	19.7	29.6
	.065	1.88	3.76	9.41	14.1	18.8	28.2
	.072	1.75	3.51	8.77	13.2	17.5	26.4
	.083	1.67	3.34	8.35	12.5	16.7	25.0
	.095	1.53	3.07	7.67	11.5	15.3	23.0
	.109	1.39	2.77	6.93	10.4	13.9	20.8
7/8	.049	2.95	5.91	14.8	22.2	29.5	44.3
	.058	2.82	5.64	14.1	21.1	28.2	42.3
	.065	2.72	5.43	13.6	20.4	27.2	40.7
	.072	2.62	5.23	13.1	19.6	26.2	39.2
	.083	2.46	4.92	12.3	18.5	24.6	36.9
	.095	2.30	4.60	11.5	17.2	23.0	34.4
	.109	2.11	4.22	10.6	15.8	21.1	31.7
1	.049	3.98	7.96	19.9	29.9	39.8	59.7
	.058	3.82	7.65	19.1	28.7	38.2	57.4
	.065	3.70	7.41	18.5	27.8	37.0	55.6
	.072	3.59	7.17	17.9	26.9	35.9	53.8
	.083	3.40	6.81	17.0	25.5	34.0	51.1
	.095	3.21	6.42	16.1	24.1	32.1	48.2
	.109	3.00	6.00	15.0	22.4	29.9	44.9
	.120	2.83	5.65	14.1	21.2	28.3	42.4
1-1/4	.049	6.50	13.0	32.5	48.7	64.9	97.4
	.058	6.29	12.6	31.5	47.2	62.9	94.4
	.065	6.14	12.3	30.7	46.0	61.4	92.1
	.072	6.00	12.0	30.0	44.9	59.9	89.8
	.083	5.75	11.5	28.8	43.1	57.5	86.3
	.095	5.50	11.0	27.5	41.2	55.0	82.5
	.109	5.21	10.4	26.1	39.1	52.1	78.2
	.120	5.00	10.0	25.0	37.4	50.0	74.9
1-1/2	.065	9.19	18.4	45.9	68.9	91.9	138
	.072	9.00	18.0	45.0	67.5	90.0	135
	.083	8.71	17.4	43.5	65.3	87.1	131
	.095	8.40	16.8	42.0	63.0	84.0	126
	.109	8.4	16.1	40.2	60.3	80.4	121
	.120	7.77	15.5	38.8	58.3	77.7	117
1-3/4	.065	12.8	25.7	64.2	96.3	128	193
	.072	12.6	25.2	63.1	94.7	126	189
	.083	12.3	24.6	61.4	92.1	123	184
	.095	11.9	23.8	59.6	89.3	119	179
	.109	11.5	23.0	57.4	86.1	115	172
	.120	11.2	22.3	55.8	83.7	112	167
	.134	10.7	21.5	53.7	80.6	107	161
2	.065	17.1	34.2	85.6	128	171	257
	.072	16.9	33.7	84.3	126	169	253
	.083	16.5	32.9	82.3	123	165	247
	.095	16.0	32.1	80.2	120	160	240
	.109	15.5	31.1	77.7	117	155	233
	.120	15.2	30.3	75.8	114	152	227
	.134	14.7	29.4	73.4	110	147	220

Oil flow velocity in tubing/pipe sizes and pressure ratings.

9.D. INCHES	DASH NO. REF	SAE NO. & TYPE SPEC.	O.D. MAX. INCHES	MIN. BEND (INTERNAL) RAD. IN. AT MAX. OPERATING PRESSURE	*MAX. OPERATING PRESSURE PSI
3/16	-3	100R1-A	0.531	3 1/2	3,000
3/16	-3	100R1-AT	0.494	3 1/2	3,000
3/16	-3	100R2-A&B	0.656	3 1/2	5,000
3/16	-3	100R2-AT&BT	0.567	3 1/2	5,000
3/16	-3	100R3	0.531	3	1,500
3/16	-4	100R5	0.539	3	3,000
3/16	-3	100R6	0.469	2	500
3/16	-3	100R7	0.423	3 1/2	3,000
3/16	-3	100R8	0.575	3 1/2	5,000
3/16	-3	100R10-A	0.781	4	10,000
3/16	-3	100R10-AT	-	4	10,000
3/16	-3	100R11	0.906	4	12,500
1/4	-4	100R1-A	0.656	4	2,750
1/4	-4	100R1-AT	0.557	4	2,750
1/4	-4	100R2-A&B	0.719	4	5,000
1/4	-4	100R2-AT&BT	0.619	4	5,000
1/4	-4	100R3	0.594	3	1,250
1/4	-5	100R5	0.601	3 3/8	3,000
1/4	-4	100R6	0.531	2 1/2	400
1/4	-4	100R7	0.513	4	2,750
1/4	-4	100R8	0.660	4	5,000
1/4	-4	100R10-A	0.844	5	8,750
1/4	-4	100R10-AT	-	5	8,750
1/4	-4	10R11	0.969	5	11,250
5/16	-5	100R1-A	0.719	4 1/2	2,500
5/16	-5	100R1-AT	0.619	4 1/2	2,500
5/16	-5	100R2-A&B	0.781	4 1/2	4,250
5/16	-5	100R2-AT&BT	0.682	4 1/2	4,250
5/16	-5	100R3	0.719	4	1,200
5/16	-6	100R5	0.695	4	2,250
5/16	-5	100R6	0.594	3	400
5/16	-5	100R7	0.590	4 1/2	2,500
3/8	-6	100R1-A	0.812	5	2,250
3/8	-6	100R1-AT	0.713	5	2,250
3/8	-6	100R2-A&B	0.875	5	4,000
3/8	-6	100R2-AT&BT	0.777	5	4,000
3/8	-6	100R3	0.781	4	1,125
3/8	-6	100R6	0.656	3	400
3/8	-6	100R7	0.700	5	2,250
3/8	-6	100R8	0.800	5	4,000
3/8	-6	100R9-A	0.875	5	4,500
3/8	-6	100R9-AT	0.831	5	4,500
3/8	-6	100R10-A	0.969	6	7,500
3/8	-6	100R10-AT	-	6	7,500
3/8	-6	100R11	1.094	6	10,000
3/8	-6	100R12	0.828	5	4,000
13/32	-6.5	100R1-A	0.844	5 1/2	2,250
13/32	-6.5	100R1-AT	0.744	5 1/2	2,250
13/32	-8	100R5	0.789	4 5/8	2,000
1/2	-8	100R1-A	0.938	7	2,000
1/2	-8	100R1-AT	0.846	7	2,000
1/2	-8	100R2-A&B	1.000	7	3,500
1/2	-8	100R2-AT&BT	0.908	7	3,500
1/2	-8	100R3	0.969	5	1,000
1/2	-10	100R5	0.945	5 1/2	1,750
1/2	-8	100R6	0.812	4	400
1/2	-8	100R7	0.860	7	2,000
1/2	-8	100R8	0.970	7	3,500
1/2	-8	100R9-A	1.000	7	4,000
1/2	-8	100R9-AT	0.958	7	4,000
1/2	-8	100R10-A	1.125	8	6,250
1/2	-8	100R10-AT	-	8	6,250
1/2	-8	100R11	1.250	8	7,500
1/2	-8	100R12	0.968	7	4,000
5/8	-10	100R1-A	1.062	8	1,500
5/8	-10	100R1-AT	0.971	8	1,500
5/8	-10	100R2-A&B	1.125	8	2,750
5/8	-10	100R2-AT&BT	1.034	8	2,750
5/8	-10	100R3	1.094	5 1/2	875
5/8	-12	100R5	1.101	8 1/2	1,500
5/8	-10	100R6	0.938	5	350
5/8	-10	100R7	0.990	8	1,500
5/8	-10	100R8	1.175	8	2,750

9.D. INCHES	DASH NO. REF.	SAE NO. & TYPE SPEC.	O.D. MAX. INCHES	MIN. BEND (INTERNAL) RAD. IN. AT MAX. OPERATING PRESSURE	*MAX. OPERATING PRESSURE PSI
3/4	-12	100R1-A	1.219	9 1/2	1,250
3/4	-12	100R1-AT	1.127	9 1/2	1,250
3/4	-12	100R2-A&B	1.281	9 1/2	2,250
3/4	-12	100R2-AT&BT	1.190	9 1/2	2,250
3/4	-12	100R3	1.281	6	750
3/4	-12	100R4	1.375	5	300
3/4	-12	100R7	1.100	9 1/2	1,250
3/4	-12	100R8	1.300	9 1/2	2,250
3/4	-12	100R9-A	1.266	9 1/2	3,000
3/4	-12	100R9-AT	1.255	9 1/2	3,000
3/4	-12	100R10-A	1.469	11	5,000
3/4	-12	100R10-AT	1.450	11	5,000
3/4	-12	100R11	1.594	11	6,250
3/4	-12	100R12	1.241	9 1/2	4,000
7/8	-14	100R1-A	1.344	11	1,125
7/8	-14	100R1-AT	1.252	11	1,125
7/8	-14	100R2-A&B	1.406	11	2,000
7/8	-14	100R2-AT&BT	1.315	11	2,000
7/8	-16	100R5	1.266	7 3/8	800
1	-16	100R1-A	1.547	12	1,000
1	-16	100R1-AT	1.440	12	1,000
1	-16	100R2-A&B	1.609	12	2,000
1	-16	100R2-AT&BT	1.531	12	2,000
1	-16	100R3	1.547	8	565
1	-16	100R4	1.825	6	250
1	-16	100R7	1.420	12	1,000
1	-16	100R8	1.520	12	2,000
1	-16	100R9-A	1.809	12	3,000
1	-16	100R9-AT	1.594	12	3,000
1	-16	100R10-A	1.797	14	4,000
1	-16	100R10-AT	1.790	14	4,000
1	-16	100R11	1.963	14	5,000
1	-16	100R12	1.542	12	4,000
1 1/8	-20	100R5	1.531	9	625
1 1/4	-20	100R1-A	1.875	16 1/2	825
1 1/4	-20	100R1-AT	1.768	16 1/2	625
1 1/4	-20	100R2-A&B	2.062	16 1/2	1,625
1 1/4	-20	100R2-AT&BT	1.053	16 1/2	1,625
1 1/4	-20	100R3	1.812	10	375
1 1/4	-20	100R4	2.000	8	200
1 1/4	-20	100R9-A	2.062	16 1/2	2,500
1 1/4	-20	100R9-AT	1.997	16 1/2	2,500
1 1/4	-20	100R10-A	2.062	18	3,000
1 1/4	-20	100R10-AT	2.060	18	3,000
1 1/4	-20	100R11	2.219	18	3,500
1 1/4	-20	100R12	1.912	16 1/2	3,000
1 3/8	-24	100R5	1.781	10 1/2	500
1 1/2	-24	100R1-A	2.125	20	500
1 1/2	-24	100R1-AT	2.047	20	500
1 1/2	-24	100R2-A&B	2.312	20	1,250
1 1/2	-24	100R2-AT&BT	2.203	20	1,250
1 1/2	-24	100R4	2.250	10	150
1 1/2	-24	100R9-A	2.312	20	2,000
1 1/2	-24	100R9-AT	-	20	2,000
1 1/2	-24	100R10-A	2.312	22	2,500
1 1/2	24	100R10-AT	2310	22	2,500
1 1/2	-24	100R11	2.469	22	3,000
1 1/2	-24	100R12	2.167	20	2,500
1 13/16	-32	100R5	2.266	13 1/4	350
2	-32	100R1-A	2.688	25	375
2	-32	100R2-AT	2.594	25	375
2	-32	100R2-A&B	2.812	25	1,125
2	-32	100R2-AT&BT	2.703	25	1,125
2	-32	100R4	2.750	12	100
2	-32	100R9-A	2.875	26	2,000
2	-32	100R9-AT	-	26	2,000
2	-32	100R10-A	2.844	28	2,500
2	-32	100R10-AT	2.840	28	2,500
2	-32	100R11	3.031	28	3,000
2	-32	100R12	2.688	25	2,500
2 1/2	-40	100R4	3.250	14	82
3	-48	100R4	3.750	18	56
3 1/2	-56	100R4	4.250	21	45
4	-64	100R4	4.750	24	35

SAE Hydraulic Hose Data

This chart lists the theoretical push and pull forces that cylinders will exert when supplied with various working pressures, plus theoretical piston velocities when supplied with 15 Ft./Sec. fluid velocity through SCH 80 size pipe.

Cyl. Bore Dia.	Piston Rod Dia.	Work Area Sq. In.	HYDRAULIC WORKING PRESSURE P.S.I.						Fluid Required Per In. Of Stroke		Port Size	Fluid Velocity @ 15 Ft./Sec.	
			500	750	1000	1500	2000	3000	Gal.	Cu. In.		Flow GPM	Piston Vel. In./Sec.
1-1/2	— —	1.767	883	1325	1767	2651	3534	5301	.00765	1.767	1/2	11.0	24.0
	5/8	1.460	730	1095	1460	2190	2920	4380	.00632	1.460			29.0
	1	.982	491	736	982	1473	1964	2946	.00425	.982			43.1
2	— —	3.141	1571	2356	3141	4711	6283	9423	.01360	3.141	1/2	11.0	13.5
	1	2.356	1178	1767	2356	3534	4712	7068	.01020	2.356			18.0
	1-3/8	1.656	828	1242	1656	2484	3312	4968	.00717	1.656			25.6
2-1/2	— —	4.909	2454	3682	4909	7363	9818	14727	.02125	4.909	1/2	11.0	8.6
	1	4.124	2062	3093	4124	6186	8248	12372	.01785	4.124			10.3
	1-3/8	3.424	1712	2568	3424	5136	6848	10272	.01482	3.424			12.4
	1-3/4	2.504	1252	1878	2504	3756	5008	7512	.01084	2.504			16.9
3-1/4	— —	8.296	4148	6222	8296	12444	16592	24888	.0359	8.296	3/4	20.3	9.4
	1-3/8	6.811	3405	5108	6811	10216	13622	20433	.0295	6.811			11.5
	1-3/4	5.891	2945	4418	5891	8836	11782	17673	.0255	5.891			13.3
	2	5.154	2577	3865	5154	7731	10308	15462	.0223	5.154			15.2
4	— —	12.566	6283	9425	12566	18849	25132	37698	.0544	12.566	3/4	20.3	6.2
	1-3/4	10.161	5080	7621	10161	15241	20322	30483	.0440	10.161			7.7
	2	9.424	4712	7068	9424	14136	18848	28272	.0408	9.424			8.3
	2-1/2	7.657	3828	5743	7657	11485	15314	22971	.0331	7.657			10.2
5	— —	19.635	9818	14726	19635	29453	39270	58905	.0850	19.635	3/4	20.3	4.0
	2	16.492	8246	12369	16492	24738	32984	49476	.0714	16492			4.7
	2-1/2	14.726	7363	11044	14726	22089	29542	44178	.0637	14.726			5.3
	3	12.566	6283	9424	12566	18849	25132	37698	.0544	12.566			6.2
	3-1/2	10.014	5007	7510	10014	15021	20028	30042	.0433	10.014			7.8
6	— —	28.274	14137	21205	28274	42411	56548	84822	.1224	28.274	1	33.8	4.6
	2-1/2	23.365	11682	17524	23365	35047	46730	70095	.1011	23.365			5.6
	3	21.205	10602	15904	21205	31807	42410	63615	.0918	21.205			6.1
	4	15.708	7854	11781	15708	23562	31416	47124	.0680	15.708			8.3
7	— —	38.485	19242	28864	38485	57728	76970	115455	.1666	38.485	1-1/4	60.2	6.0
	3	31.416	15708	23562	31416	47124	62832	94248	.1360	31.416			7.4
	4	25.919	12960	19439	25919	38878	51838	77757	.1122	25.919			8.9
	5	18.850	9425	14137	18850	28275	37700	56550	.0816	18.850			12.3
8	— —	50.265	25133	37699	50265	75398	100530	150795	.2176	50.265	1-1/2	83.0	6.4
	3-1/2	40.644	20322	30483	40644	60966	81288	121932	.1759	40.644			7.9
	4	37.699	18850	28274	37699	56548	75398	113097	.1632	37.699			8.5
	5-1/2	26.507	13253	19880	26507	39760	53014	79521	.1147	26.507			12.0
10	— —	78.540	39270	58905	78540	117810	157080	235620	.3400	78.540	2	139	6.8
	4-1/2	62.636	31318	46977	62636	93954	125272	187908	.2711	62.636			8.5
	5-1/2	54.782	27391	41086	54782	82173	109564	164346	.2371	54.782			9.8
	7	40.055	20027	30041	40055	60082	80110	120165	.1734	40.055			13.4
12	— —	113.10	56550	84825	113100	169650	226200	339300	.4896	113.10	2-1/2	199	6.8
	5-1/2	89.34	44670	67005	89340	134010	178680	268020	.3867	89.34			8.6
	7	74.62	37310	55965	74620	111930	149240	223860	.3230	74.62			10.3
	8	62.84	31420	47130	62840	94260	125680	188520	.2720	62.84			12.2
14	— —	153.94	76970	115455	153940	230910	307880	461820	.6664	153.94	2-1/2	199	5.0
	7	115.46	57730	86595	115460	173190	230920	346380	.4998	115.46			6.6
	8	103.68	51840	77760	103680	155520	207360	311040	.4488	103.68			7.4
	10	75.40	37700	56550	75400	113100	150000	226200	.3264	75.40			10.2

Oil consumption in gallons per minute = Gallons per inch × inches per minute of piston travel.
1 gallon = 231 cubic inches. Cylinder bore diameters and piston rod diameters are in inches.

Cylinder size selection.

This information is provided courtesy of American National Standards Institute "Hydraulic Fluid Power – Systems standard for stationary industrial machinery." (NFPA/JIC T2.24.1 – 1991)

12 FILTRATION AND FLUID CONDITIONING

12.1 Filtration

Filtration shall be provided to limit the in-service particulate contamination level to the values listed in table 1 for the given system operating pressure, fluid type and components used.

12.2 Location and Sizing of Filters and Fluid Conditioners

12.2.1 Filters shall be located in pressure, return or auxiliary circulation lines as necessary to achieve the cleanliness levels of filtration during normal equipment operation.

12.2.2 Unless specified in Part B of the Hydraulic equipment data form (or similar document), filtration on pump suction lines shall not be used. Inlet screens or strainers are acceptable.

12.2.3 If used, suction filtration devices shall be equipped with an integral bypass valve to limit the maximum pressure drop at rated system flow to a value that insures the requirements of 10.3.5 are satisfied.

12.2.4 Filters shall be sized to provide a minimum of 800 hours operation under normal system conditions.

12.2.5 All filter assemblies shall be equipped with some device which indicates when the filter requires servicing. This device shall be readily visible to the operator or maintenance personnel.

12.2.6 Filter assemblies whose elements cannot withstand full system differential pressure without damage shall be equipped with bypass valves. The bypass valve opening differential pressure shall be at least 20% higher than the differential pressure required to actuate the aforementioned indicator.

12.2.7 If specified in Part B of the Hydraulic equipment data form, the fluid system fill ports shall be routed through an auxiliary filter or one of the existing system filters.

12.2.8 Conduit connections to the fluid power filter shall be such as to eliminate external leakage. Taper pipe threads or connection mechanisms which require the use of a nonintegral sealing compound shall not be used, except as noted in 17.3.1.

REPORTING CLEANLINESS LEVELS OF HYDRAULIC FLUIDS
SAE J1165 OCT80

TABLE 1 — CLEANLINESS LEVEL CORRELATION TABLE

ISO CODE	PARTICLES PER MILLILITER >10 μm	ACFTD* GRAVIMETRIC LEVEL — mg/L	MIL STD 1246A (1967)	NAS 1638 (1964)	DISAVOWED "SAE" LEVEL (1963)
26/23	140,000	1,000			
25/23	85,000		1,000		
23/20	14,000	100	700		
21/18	4,500			12	
20/18	2,400		500		
20/17	2,300			11	
20/16	1,400	10			
19/16	1,200			10	
18/15	580			9	6
17/14	280		300	8	5
16/13	140	1		7	4
15/12	70			6	3
14/12	40		200		
14/11	35			5	2
13/10	14	0/1		4	1
12/9	9			3	0
11/8	5			2	
10/8	3		100		
10/7	2.3			1	
10/6	1.4	0.01			
9/6	1.2			0	
8/5	0.6			00	
7/5	0.3		50		
6/3	0.14	0.001			
5/2	0.04		25		
2/0.8	0.01		10		

*ACFTD (Air Cleaner Fine Test Dust) — ISO-Approved Test and Calibration Contaminant.

REPORTING CLEANLINESS LEVELS OF HYDRAULIC FLUIDS
SAE J1165 OCT80

TABLE 2 — ISO RANGE NUMBER TABLE

PARTICLE CONCENTRATION (PARTICLES PER MILLILITER)	RANGE NUMBER
10,000,000	30
5,000,000	29
2,500,000	28
1,300,000	27
640,000	26
320,000	25
160,000	24
80,000	23
40,000	22
20,000	21
10,000	20
5,000	19
2,500	18
1,300	17
640	16
320	15
160	14
80	13

PARTICLE CONCENTRATION (PARTICLES PER MILLILITER)	RANGE NUMBER
40	12
20	11
10	10
5.0	9
2.5	8
1.3	7
0.64	6
0.32	5
0.16	4
0.08	3
0.04	2
0.02	1
0.01	0.9
0.005	0.8
0.0025	0.7

NOTE: If cumulative particle count falls between two adjacent particle concentrations, the proper Range Number needed to formulate that portion of the ISO Contaminant Code is found opposite the higher particle concentration.

APPENDIX

D

BASIC ELECTRICAL AND ELECTRONIC DATA

This portion of the manual explains the rules, laws, formulas, and symbols of basic electrical and electronic data. Refer to Chapter 9 for a complete explanation of electrical symbols.

OHM'S LAW COMBINED WITH JOULE'S LAW

WHERE:

I = Current in Amperes
E = Potential in Volts
R = Resistance in Ohms
P = Power in Watts

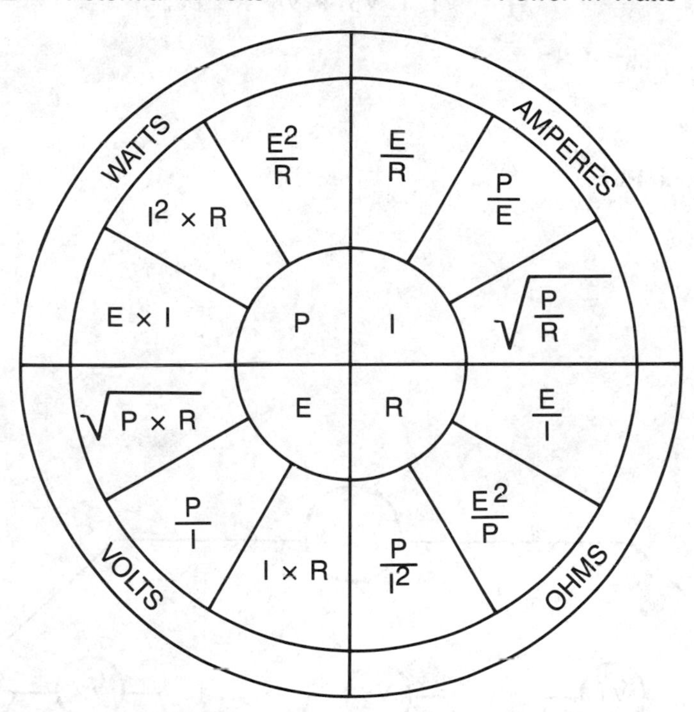

EXAMPLES OF CHART USE:

$$I = \frac{E}{R} = \frac{P}{E} = \sqrt{\frac{P}{R}}$$

$$E = \sqrt{P \times R} = \frac{P}{I} = I \times R$$

$$P = I^2 \times R = \frac{E^2}{R} = E \times I$$

$$R = \frac{P}{I^2} = \frac{E^2}{P} = \frac{E}{I}$$

RULES FOR USING A SERIES CIRCUIT

- The total resistance in a series circuit is the sum of the individual resistances.

$$R_t = R_1 + R_2 + R_3 + \text{...etc.}$$

- The current in a series circuit is the same through all components.

$$I_1 = I_2 = I_3 = I$$

- The sum of the individual voltage drops will add up to the source voltage.

$$V_t = V_1 + V_2 + V_3 + \text{...etc.}$$

EXAMPLE:

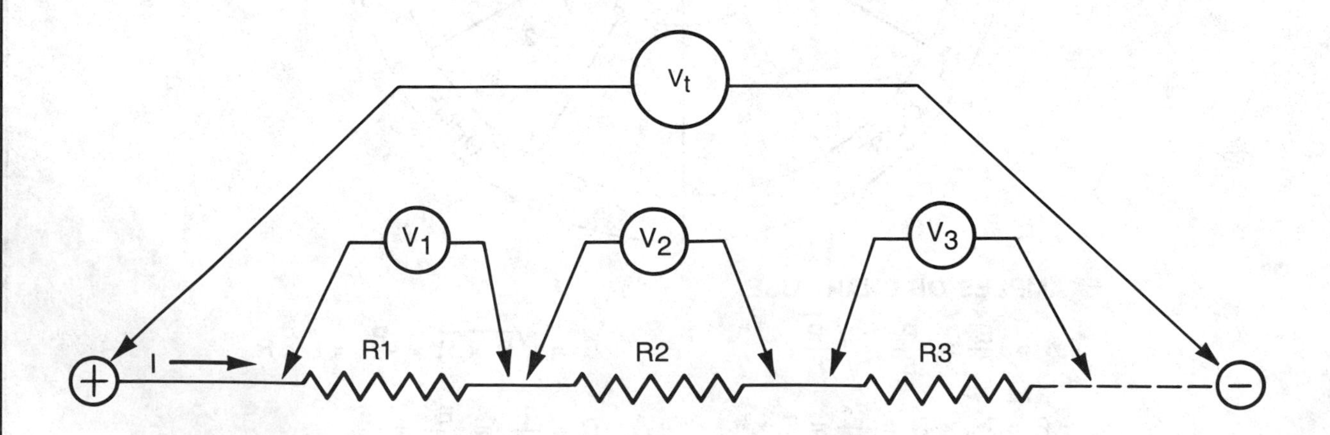

RULES FOR USING A PARALLEL CIRCUIT

The formulas listed below are used to find the total resistance of a parallel circuit.

- With two parallel resistors:

$$R_t = \frac{R_1 R_2}{R_1 + R_2}$$

- With more than two parallel resistors:

$$\frac{1}{R_t} = \frac{1}{R_1} + \frac{1}{R_2} + \frac{1}{R_3} + \text{etc.}$$

- With identical parallel resistors, divide the value of one resistor by the total number of identical resistors.

Example: For ten, 500 Ohm parallel resistors:

500 ÷ 10 = 50 Ohms

- The voltage drop across each branch (resistance) is the same.
- The total current in a parallel circuit is the sum of the currents in the individual branches.

$$I_T = I_1 + I_2 + I_3 + \text{etc.}$$

EXAMPLE:

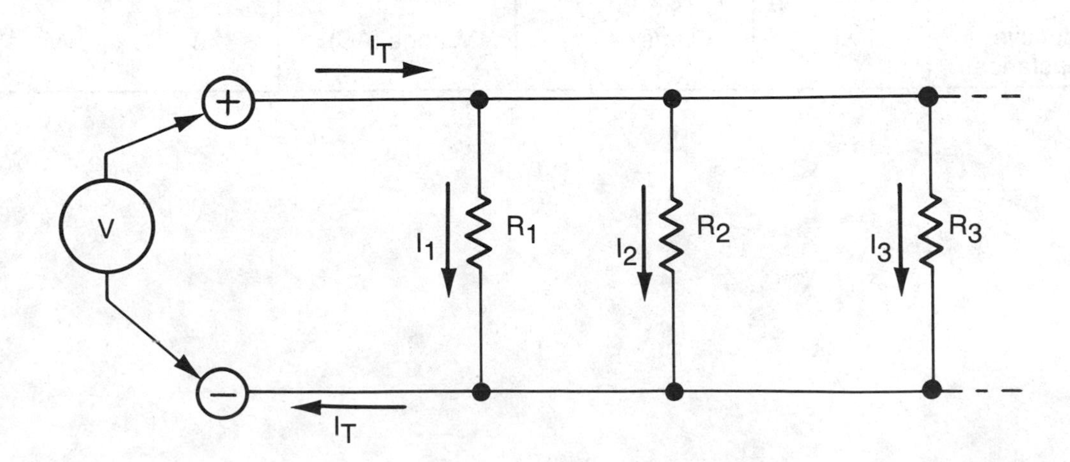

TERMS FOR ELECTRICAL CIRCUITS

TERM	SYMBOL	UNIT OF MEASURE	TERM	SYMBOL	UNIT OF MEASURE
Capacitance	C	Farads	Impedance	Z	Ohms
Capacitive Reactance	X_C	Ohms	Power	P	Watts
Charge	q	Coulombs	Resistance	R	Ohms
Current (DC or AC)	I	Amperes	Time Constant	Tc	Seconds
Inductance	L	Henries	Voltage or Potential (DC)	V or E	Volts
Inductive Reactance	X_L	Ohms	Voltage (AC)	V	Volts

POWER OF TEN PREFIXES

PREFIX	SYMBOL	MULTIPLIER
atto	a	10^{-18}
femto	f	10^{-15}
pico	p	10^{-12}
nano	n	10^{-9}
micro	u	10^{-6}
milli	m	10^{-3}
centi	c	10^{-2}
deci	d	10^{-1}

PREFIX	SYMBOL	MULTIPLIER
deka	da	10^{1}
hecto	h	10^{2}
kilo	k	10^{3}
mega	M	10^{6}
giga	G	10^{9}
tera	T	10^{12}

NUMBER CONVERSION CHART

DECIMAL NOTATION	POWER OF TEN NOTATION
0.000,000,001	10^{-9}
0.000,000,01	10^{-8}
0.000,000,1	10^{-7}
0.000,001	10^{-6}
0.000,01	10^{-5}
0.0001	10^{-4}
0.001	10^{-3}
0.01	10^{-2}
0.1	10^{-1}

DECIMAL NOTATION	POWER OF TEN NOTATION
1	10^{0}
10	10^{1}
100	10^{2}
1,000	10^{3}
10,000	10^{4}
100,000	10^{5}
1,000,000	10^{6}
10,000,000	10^{7}
100,000,000	10^{8}
1,000,000,000	10^{9}

DETERMINING POWER IN AC AND DC CIRCUITS

TO FIND	ALTERNATING CURRENT		DIRECT CURRENT
	THREE PHASE	SINGLE PHASE	
Amperes when hp (Watts) is known	$I = \frac{746 \times hp}{1.73 \times E \times Eff \times PF}$	$I = \frac{746 \times hp}{E \times Eff \times Pf}$	$I = \frac{746 \times hp}{E \times Eff}$
Amperes when kW is known	$I = \frac{1000 \times kW}{1.73 \times E \times PF}$	$I = \frac{1000 \times kW}{E \times PF}$	$I = \frac{1000 \times kW}{E}$
Amperes when kVA is known	$I = \frac{1000 \times kVA}{1.73 \times E}$	$I = \frac{1000 \times kVA}{E}$	— —
Kilowatts input	$kW = \frac{1.73 \times E \times I \times PF}{1000}$	$kW = \frac{E \times I \times PF}{1000}$	$kW = \frac{E \times I}{1000}$
Kilovolt-Amperes	$KVA = \frac{1.73 \times E \times I}{1000}$	$kVA = \frac{E \times I}{1000}$	— —
Horsepower Output .	$hp = \frac{1.73 \times E \times I \times Eff \times PF}{746}$	$hp = \frac{E \times I \times Eff \times PF}{746}$	$hp = \frac{E \times I \times Eff}{746}$

I = Amperes
E = Volts
Eff = Efficiency (in Decimals)

kW = Kilowatts
kVA = Kilovolt-amperes
hp = Horsepower Output
PF = Power Factor (in Decimals)

NOTE: Efficiency and Power Factor expressed as values between 0 and 1

Horsepower Required to Drive a Hydraulic Pump:

$$hp = \frac{GPM \times PSI}{1714 \times \text{pump efficiency}}$$

Torque of an Electrical Induction Motor:

$$\text{Torque (lb-ft)} = \frac{hp \times 5250}{RPM}$$

Speed of an Electrical Induction Motor:

$$RPM = \frac{120 \times \text{frequency}}{\text{No. of Poles}}$$

INDICATING LIGHT COLOR CODES

COLOR	CONDITION	SITUATION(S)
Red	Danger, Abnormal, Fault	Fault(s) in air, water, lubricating or filtering system(s). Excess pressure or temperature. Machine has been stopped by a protective device.
Amber or Yellow	Attention, Caution, Marginal	Automatic cycle engaged. Levels are nearing limits. Ground fault indicated.
Green	Machine Ready	All machine functions and auxiliary functions are operating at specified levels. Machine cycle completed and ready for restart.
White or Clear	Normal	Normal air, water, and lubrication pressures.
Blue	Any condition not covered by above colors	

PUSH-BUTTON COLOR CODE

COLOR	CONDITION	SITUATION(S)
Red	Stop, Emergency stop, Off	Motor(s) stopped, Master stop, or Emergency stop.
Yellow	Return, or Emergency return	Machine elements are returned to a safe position.
Green or Black	Start, On	Cycle or partial sequence is started. Motor(s) started.
White, Gray, or Blue	Any condition not covered by the above colors	

RESISTOR COLOR CODES

COLOR	FIRST DIGIT (A)	SECOND DIGIT (B)	MULTIPLIER (C)	TOLERANCE (D)
Black	0	0	1	—
Brown	1	1	10	—
Red	2	2	100	—
Orange	3	3	1,000	—
Yellow	4	4	10,000	—
Green	5	5	100,000	—
Blue	6	6	1,000,000	—
Violet	7	7	10,000,000	—
Gray	8	8	100,000,000	—
White	9	9	—	—
Brown	—	—	—	+/– 1%
Gold	—	—	—	+/– 5%
Silver	—	—	—	+/– 10%
No Color	—	—	—	+/– 20%

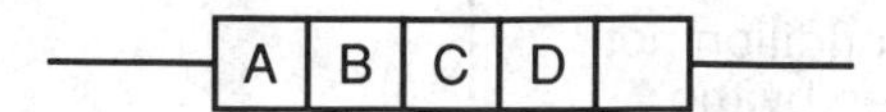

ELECTRICAL FORMULAS

RL Circuit Time Constant

$$\frac{\text{L (in henrys)}}{\text{R (in ohms)}} = \text{t (in seconds), or}$$

$$\frac{\text{L (in microhenrys)}}{\text{R (in ohms)}} = \text{t (in microseconds)}$$

RC Circuit (Series) Impedance

$$Z = \sqrt{R^2 + (X_C)^2}$$

RC Circuit Time Constant

R (ohms) × C (farads) = t (seconds)

R (megohms) × C (microfarads) = t (seconds)

R (ohms) × C (microfarads) = t (microseconds)

R (megohms) × C (picofarads) = t (microseconds)

Inductors in Series

$$L_T = L_1 + L_2 + \ldots$$

(No coupling between coils)

Two Capacitors (Series)

$$C_T = \frac{C_1 C_2}{C_1 + C_2}$$

Two Inductors in Parallel

$$L_T = \frac{L_1 L_2}{L_1 + L}$$

(No coupling between coils)

More Than Two Capacitors (Series)

$$\frac{1}{C_T} = \frac{1}{C_1} + \frac{1}{C_2} + \frac{1}{C_3} + \ldots$$

More Than Two Inductors in Parallel

$$\frac{1}{L_T} = \frac{1}{L_1} + \frac{1}{L_2} + \frac{1}{L_3} + \ldots$$

(No coupling between coils)

Capacitors in Parallel

$$C_T = C_1 + C_2 + \ldots$$

Inductive Reactance

$$X_L = 2\pi fL$$

Capacitive Reactance

$$X_C = \frac{1}{2\pi fC}$$

Q of a Coil

$$Q = \frac{X_L}{R}$$

ELECTRICAL FORMULAS

RL Circuit (Series) Impedance

$$Z = \sqrt{R^2 + (X_L)^2}$$

R, C, and L Circuit (Series) Impedance

$$Z = \sqrt{R^2 + (X_L - X_C)^2}$$

SINE-WAVE VOLTAGE RELATIONSHIPS

Effective or r.m.s. Value

$$E_{eff} = \frac{E_{max}}{\sqrt{2}} = \frac{E_{max}}{1.414} = 0.707E_{max}$$

Maximum Value

$$E_{max} = \sqrt{2}\ (E_{eff}) = 1.414E_{eff}$$

AC Circuit Voltage

$$E = IZ = \frac{P}{I \times P.F.}$$

AC Circuit Current

$$I = \frac{E}{Z} = \frac{P}{E \times P.F.}$$

AC CIRCUIT POWER

Apparent Power

$$P = EI$$

True Power

$$P = EI\cos\theta = EI \times P.F.$$

Power Factor

$$P.F. = \frac{P}{EI} = \cos\theta$$

$$\cos\theta = \frac{\text{true power}}{\text{apparent power}}$$

TRANSFORMERS

Voltage Relationship

$$\frac{E_p}{E_s} = \frac{N_p}{N_s} \text{ or } E_s = E_p \times \frac{N_s}{N_p}$$

Current Relationship

s = Secondary

p = Primary

N = Number of turns

$$\frac{I_p}{I_s} = \frac{N_s}{N_p}$$

Parallel Circuit Impedance

$$Z = \frac{Z_1 Z_2}{Z_1 + Z_2}$$

EQUIVALENT LOGIC/LADDER DIAGRAMS

Logic Diagrams | **Ladder Diagrams**

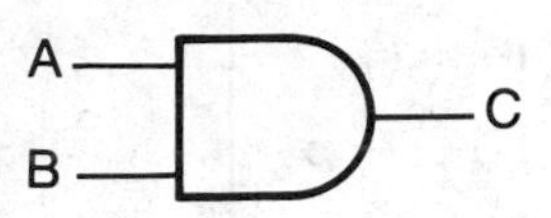

AND Gate

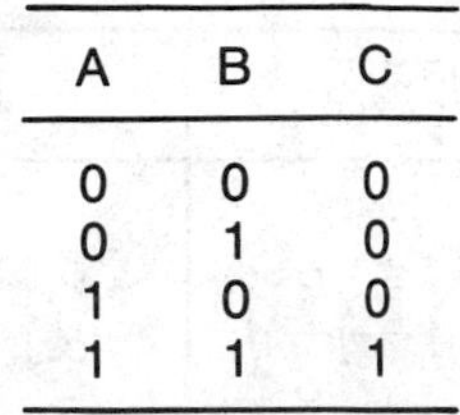

A	B	C
0	0	0
0	1	0
1	0	0
1	1	1

AND Truth Table

A B C

Equivalent Circuit

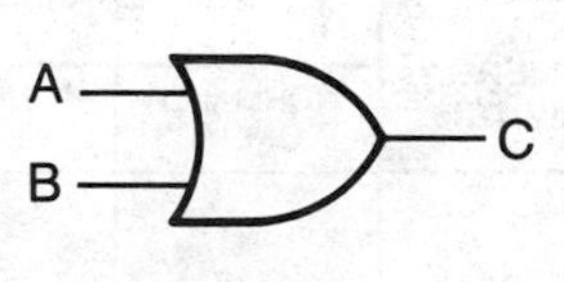

OR Gate

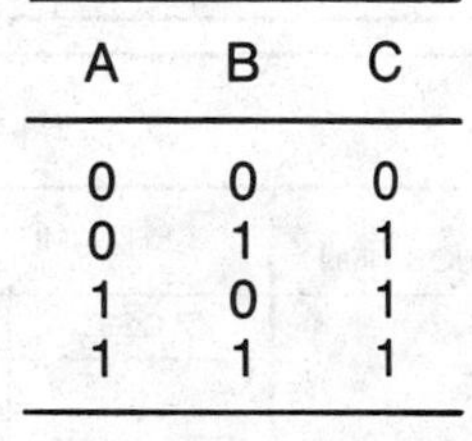

A	B	C
0	0	0
0	1	1
1	0	1
1	1	1

OR Truth Table

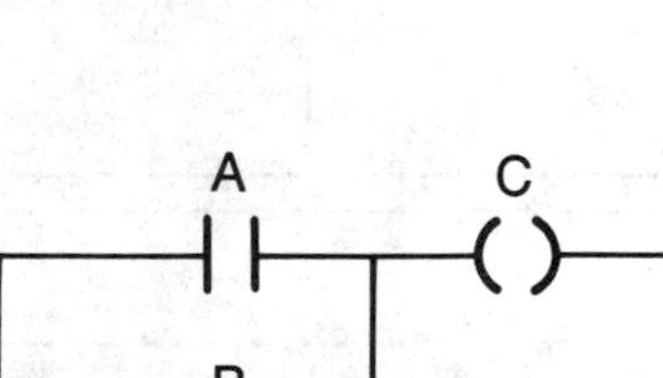

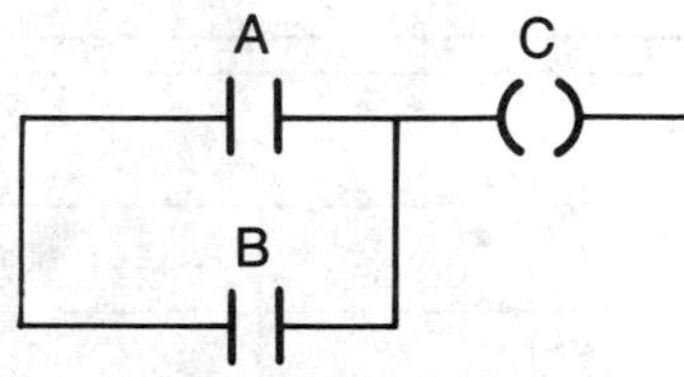

Equivalent Circuit

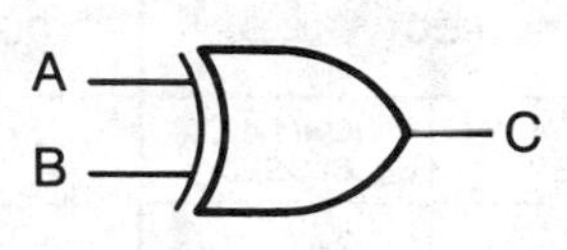

Exclusive - OR Gate

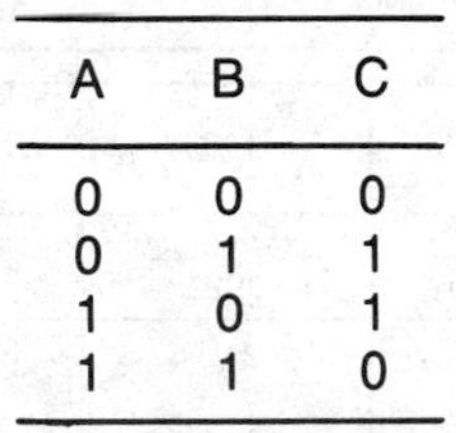

A	B	C
0	0	0
0	1	1
1	0	1
1	1	0

Exclusive - OR Truth Table

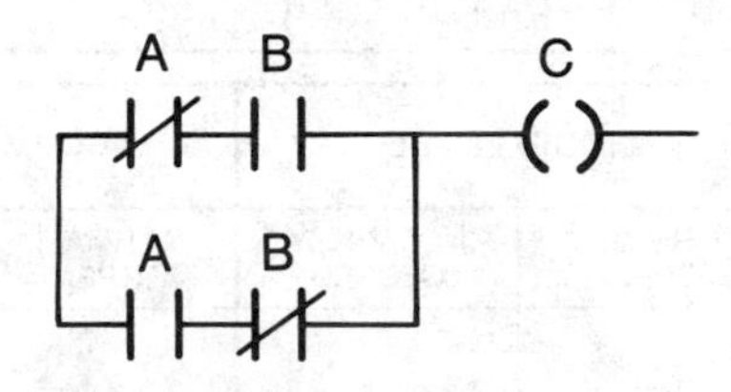

Equivalent Circuit

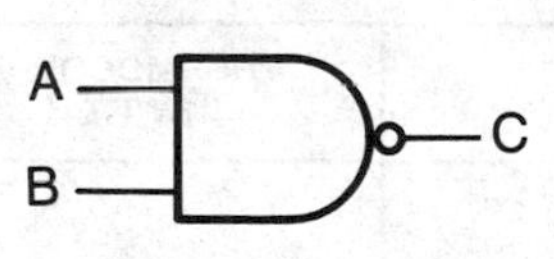

NAND Gate

A	B	C
0	0	1
0	1	1
1	0	1
1	1	0

NAND Truth Table

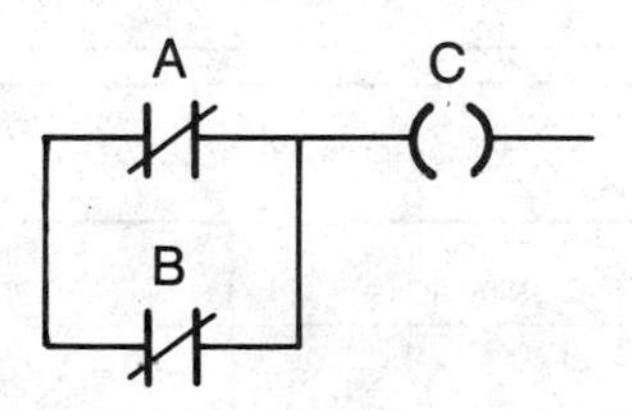

Equivalent Circuit

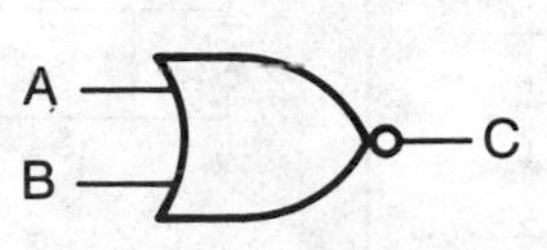

NOR Gate

A	B	C
0	0	1
0	1	0
1	0	0
1	1	0

NOR Truth Table

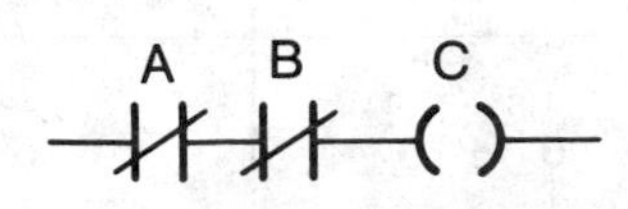

Equivalent Circuit

ELECTRICAL RELAY DIAGRAM SYMBOLS

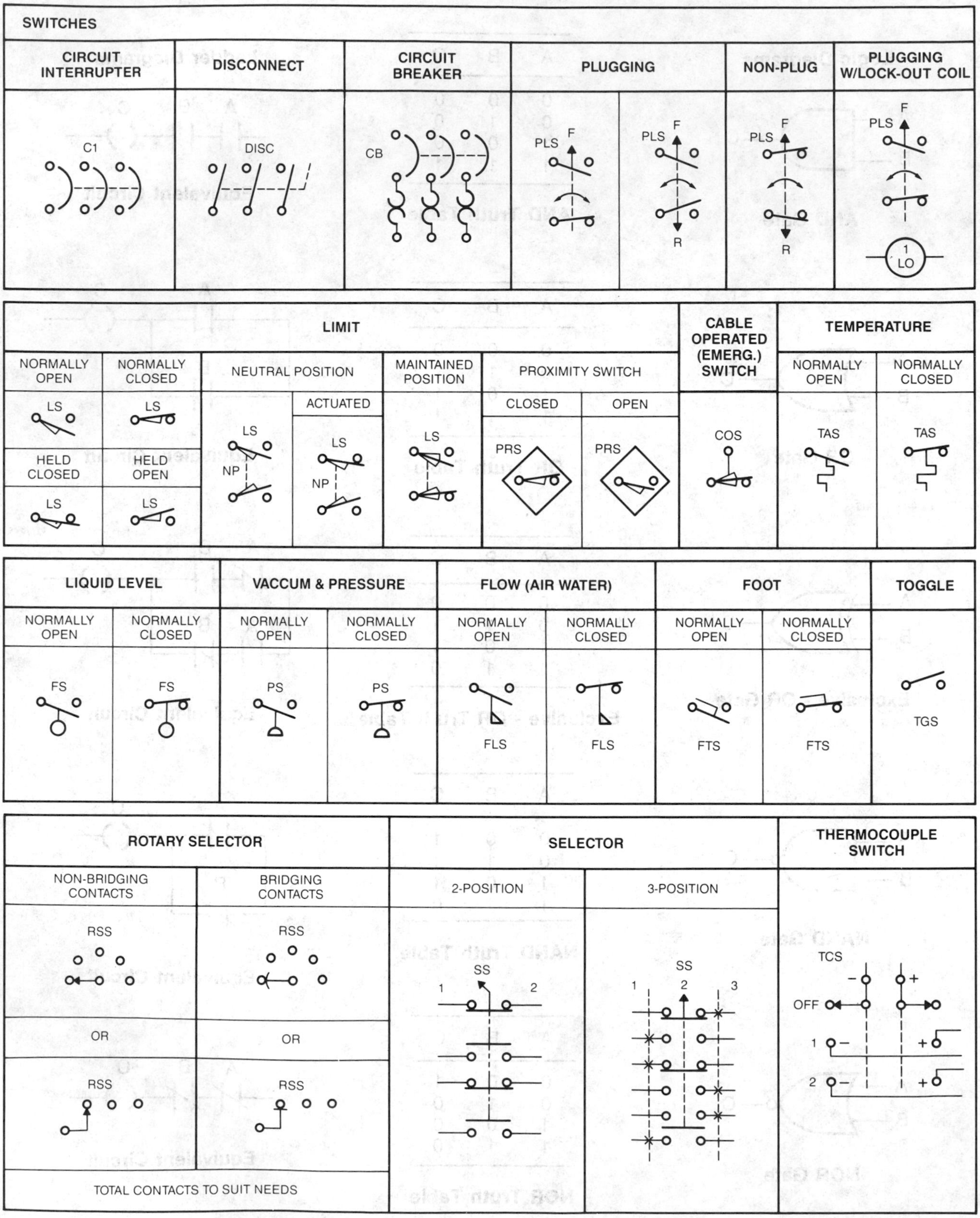

SWITCHES

CIRCUIT INTERRUPTER	DISCONNECT	CIRCUIT BREAKER	PLUGGING		NON-PLUG	PLUGGING W/LOCK-OUT COIL
C1	DISC	CB	PLS F	PLS F R	PLS F R	F PLS 1 LO

LIMIT

NORMALLY OPEN	NORMALLY CLOSED	NEUTRAL POSITION	ACTUATED	MAINTAINED POSITION	PROXIMITY SWITCH CLOSED	PROXIMITY SWITCH OPEN
LS	LS	LS NP	LS NP	LS	PRS	PRS
HELD CLOSED	HELD OPEN					
LS	LS					

CABLE OPERATED (EMERG.) SWITCH	TEMPERATURE NORMALLY OPEN	TEMPERATURE NORMALLY CLOSED
COS	TAS	TAS

LIQUID LEVEL / VACCUM & PRESSURE / FLOW (AIR WATER) / FOOT / TOGGLE

LIQUID LEVEL NORMALLY OPEN	LIQUID LEVEL NORMALLY CLOSED	VACCUM & PRESSURE NORMALLY OPEN	VACCUM & PRESSURE NORMALLY CLOSED	FLOW (AIR WATER) NORMALLY OPEN	FLOW (AIR WATER) NORMALLY CLOSED	FOOT NORMALLY OPEN	FOOT NORMALLY CLOSED	TOGGLE
FS	FS	PS	PS	FLS	FLS	FTS	FTS	TGS

ROTARY SELECTOR / SELECTOR / THERMOCOUPLE SWITCH

ROTARY SELECTOR NON-BRIDGING CONTACTS	ROTARY SELECTOR BRIDGING CONTACTS	SELECTOR 2-POSITION	SELECTOR 3-POSITION	THERMOCOUPLE SWITCH
RSS	RSS	SS 1 2	SS 1 2 3	TCS − + OFF 1 + 2 +
OR	OR			
RSS	RSS			
TOTAL CONTACTS TO SUIT NEEDS				

ELECTRICAL RELAY DIAGRAM SYMBOLS

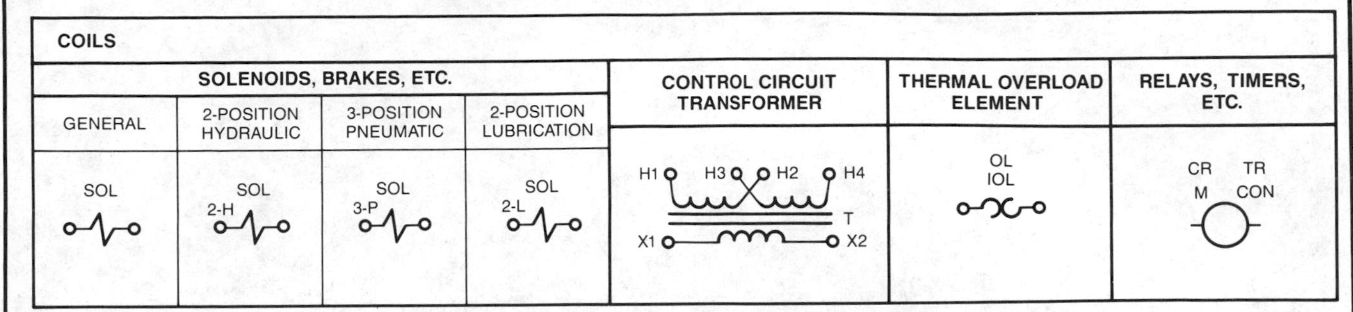

ADJUSTABLE IRON CORE

X

AIR CORE

X

MAGNETIC AMPLIFIER WINDING

MAX

CONNECTIONS, ETC.

GROUND

GRD

CHASSIS OR FRAME

NOT NECESSARILY GROUNDED

CH

PLUG AND RECP.

PL

RECP

CONDUCTORS

NOT CONNECTED

CONNECTED

CONTACTS

TIME DELAY AFTER COIL

NORMALLY OPEN — TR — ENERGIZED

NORMALLY CLOSED — TR — ENERGIZED

NORMALLY OPEN — TR — DEENERGIZED

NORMALLY CLOSED — TR — DEENERGIZED

RELAY, ETC.

NORMALLY OPEN — CR M CON

NORMALLY CLOSED — CR M CON

THERMAL OVERLOAD

OL IOL

PILOT LIGHTS

LT R

PUSH TO TEST

LT R

LETTER DENOTES COLOR

PUSHBUTTONS

SINGLE CIRCUIT

NORMALLY OPEN

PB

NORMALLY CLOSED

PB

DOUBLE CIRCUIT

PB

MUSHROOM HEAD

PB

MAINTAINED CONTACT

PB PB

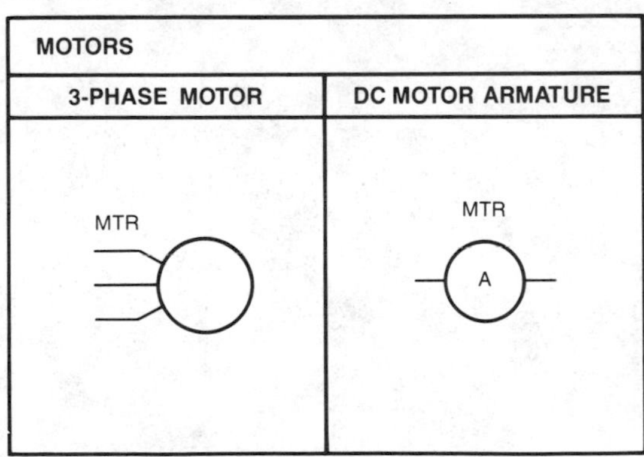

HORN, SIREN, ETC.

AH

BUZZER

ABU

BELL

ABE

APPENDIX E

LOGICAL TROUBLESHOOTING IN HYDRAULIC SYSTEMS

The object of this appendix is to provide a logical approach to hydraulic system troubleshooting which can be extended to cover specific machines in all areas of industry. The fundamentals used to develop this approach (control of flow, pressure, and direction) apply equally as well to a rolling mill in a steelworks or a winch drive on a trawler.

Probably the greatest aid to troubleshooting is the confidence that comes with knowing the system. Since every component has a purpose in the system, the construction and operating characteristics of each one should be clearly understood. For example, knowing that a solenoid controlled directional valve can be manually actuated will save considerable time in isolating a defective solenoid.

It is also important to know the capabilities of the system. Each component in the system has a maximum rated speed, torque, or pressure. If the system is loaded beyond the specifications, the possibility of failure is greatly increased.

The correct operating pressures of a system must be known and always checked and set with a pressure gauge. The hydraulic schematic should have the correct pressures. If not, assume that the correct operating pressure is the lowest pressure that will allow adequate performance of the system function and still remain below the maximum rating of the components and machine. Once the correct pressures have been established, note them on the hydraulic schematic for future reference.

Understanding the system also includes knowing the proper signal and feedback levels, as well as dither and gain settings in servo control systems.

Occasionally, a seemingly uncomplicated procedure such as relocating a system or changing a component part can cause problems. The points discussed below will aid in avoiding unnecessary complications.

- Each component in the system must be compatible with, and form an integral part of, the system. As an example, placing the wrong size filter on the inlet of a pump can cause cavitation and subsequent damage to the pump.
- All lines should be the correct size and free of restrictive bends. An undersized or restricted line results in a pressure drop in that line.
- Some components are meant to be mounted in a specific position, relative to the other components or lines. The housing of an in-line pump, for example, must remain filled with fluid to provide lubrication.
- Although they are not essential for system operation, having adequate test points for pressure readings will also expedite troubleshooting.

The ability to recognize trouble indicators in a specific system is usually acquired with experience. To help this process, analyze the system and develop a logical sequence for setting valves, mechanical stops, interlocks, and electrical controls. Tracing the flow paths can often be accomplished by listening for flow in the lines or feeling them for excessive warmth.

By working regularly with the system, a cause and effect troubleshooting guide, similar to the charts found in this appendix, can be developed. The initial time spent on such a project could save many hours of system down time later on.

Although troubleshooting and repair are a normal part of operating a system, down time can be minimized by keeping up on simple system maintenance. Regularly performing the three tasks noted here, the performance, efficiency, and life of the system will be greatly improved.

- Maintain a sufficient quantity of hydraulic fluid that is clean and is the correct type and viscosity.

- Change the filters and clean strainers often.
- Keep all connections tight enough so that air is kept out of the system but no distortion is present.

Whenever system troubleshooting is being carried out, the most important consideration is always safety. Although much of practicing good safety habits is common sense, the stress of a breakdown situation may mean that a potential hazard gets overlooked. For this reason, it is a good idea to establish a regular shutdown procedure, like the one shown in Figure E-1, which is carried out before beginning any work on a system.

These same safety considerations call for a restart procedure (Figure E-7) which is followed when the repairs have been made and the system is ready to run again.

TROUBLESHOOTING CHARTS

The following troubleshooting charts (Figures E-2—E-6) are arranged in five main categories. The heading of each is a symptom that indicates some malfunction in the system. For example, if a pump is excessively noisy, refer to Chart I (Excessive Noise). The noisy pump is referenced in Column A below the main heading. Underneath, listed in order of likelihood of occurrence or ease of checking, there are four probable causes for a noisy pump, Each cause references a remedy that can be found at the bottom of the page. If cavitation is occurring in the system, perform the procedure(s) listed under Remedy A. If not, move to the next cause, continuing until the cause of the problem is determined and eliminated.

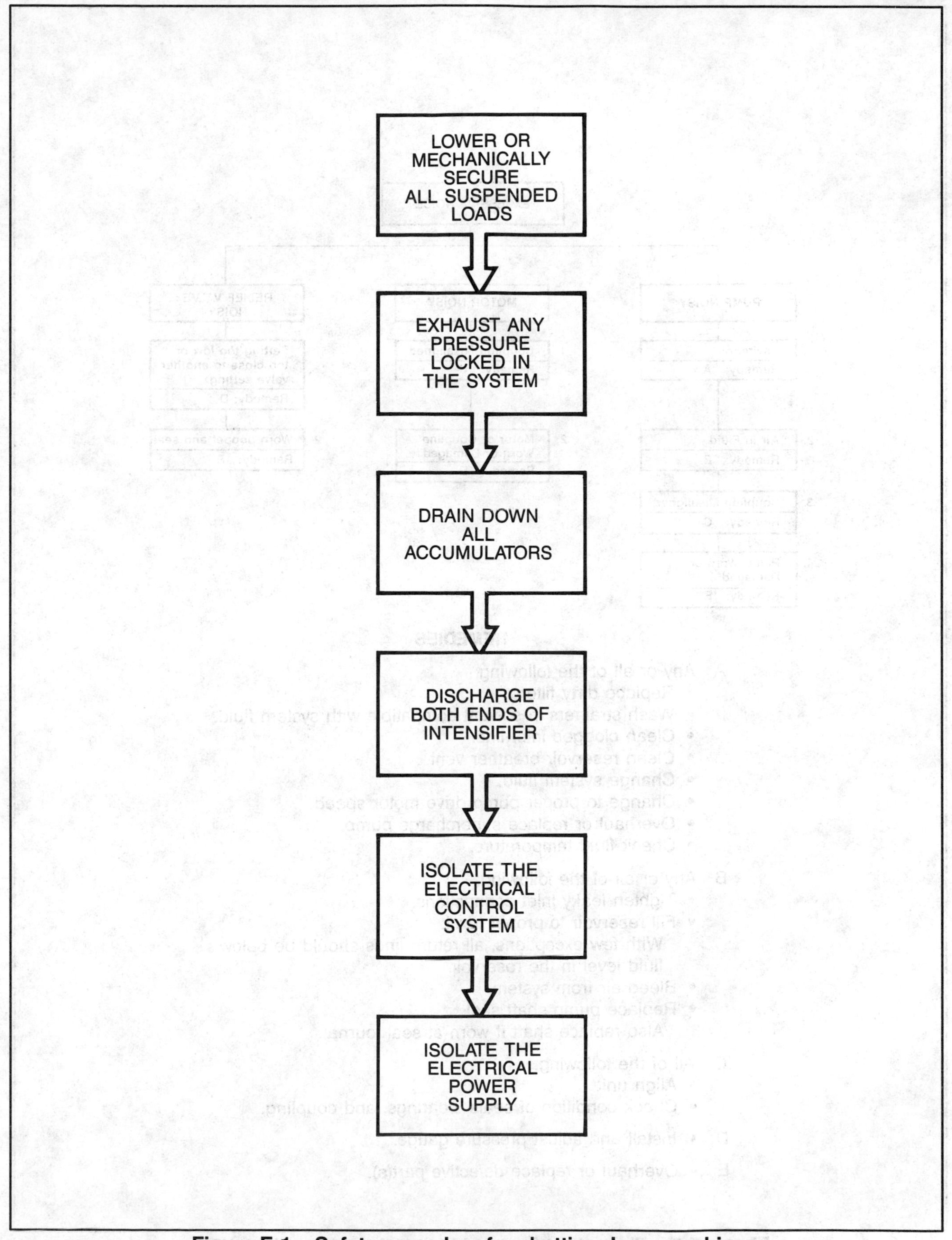

Figure E-1. Safety procedure for shutting down machines.

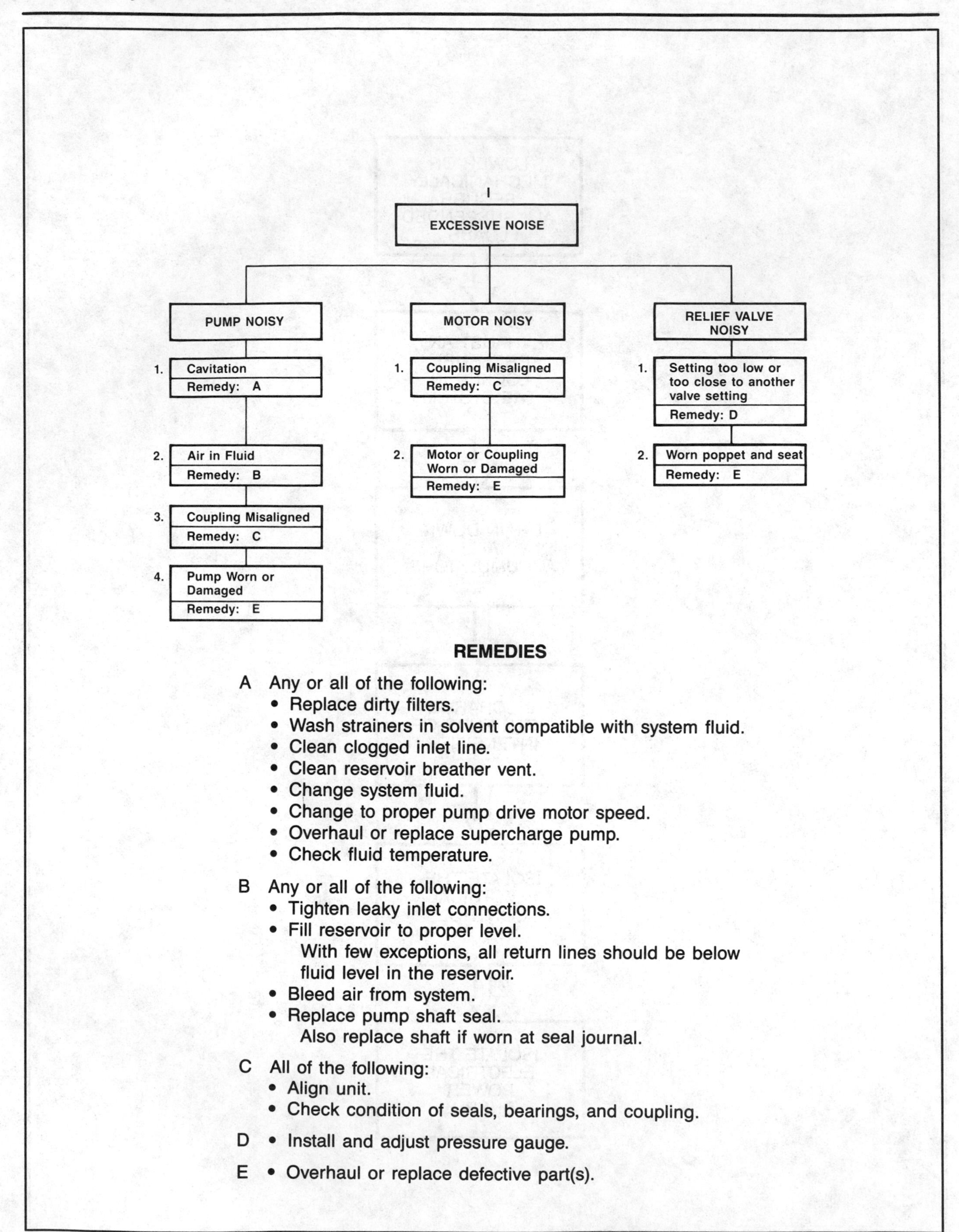
I
EXCESSIVE NOISE
PUMP NOISY
1. Cavitation
Remedy: A
2. Air in Fluid
Remedy: B
3. Coupling Misaligned
Remedy: C
4. Pump Worn or Damaged
Remedy: E
MOTOR NOISY
1. Coupling Misaligned
Remedy: C
2. Motor or Coupling Worn or Damaged
Remedy: E
RELIEF VALVE NOISY
1. Setting too low or too close to another valve setting
Remedy: D
2. Worn poppet and seat
Remedy: E
REMEDIES
A Any or all of the following:
• Replace dirty filters.
• Wash strainers in solvent compatible with system fluid.
• Clean clogged inlet line.
• Clean reservoir breather vent.
• Change system fluid.
• Change to proper pump drive motor speed.
• Overhaul or replace supercharge pump.
• Check fluid temperature.
B Any or all of the following:
• Tighten leaky inlet connections.
• Fill reservoir to proper level.
With few exceptions, all return lines should be below fluid level in the reservoir.
• Bleed air from system.
• Replace pump shaft seal.
Also replace shaft if worn at seal journal.
C All of the following:
• Align unit.
• Check condition of seals, bearings, and coupling.
D • Install and adjust pressure gauge.
E • Overhaul or replace defective part(s).

Figure E-2. Flow chart for troubleshooting excessive noise.

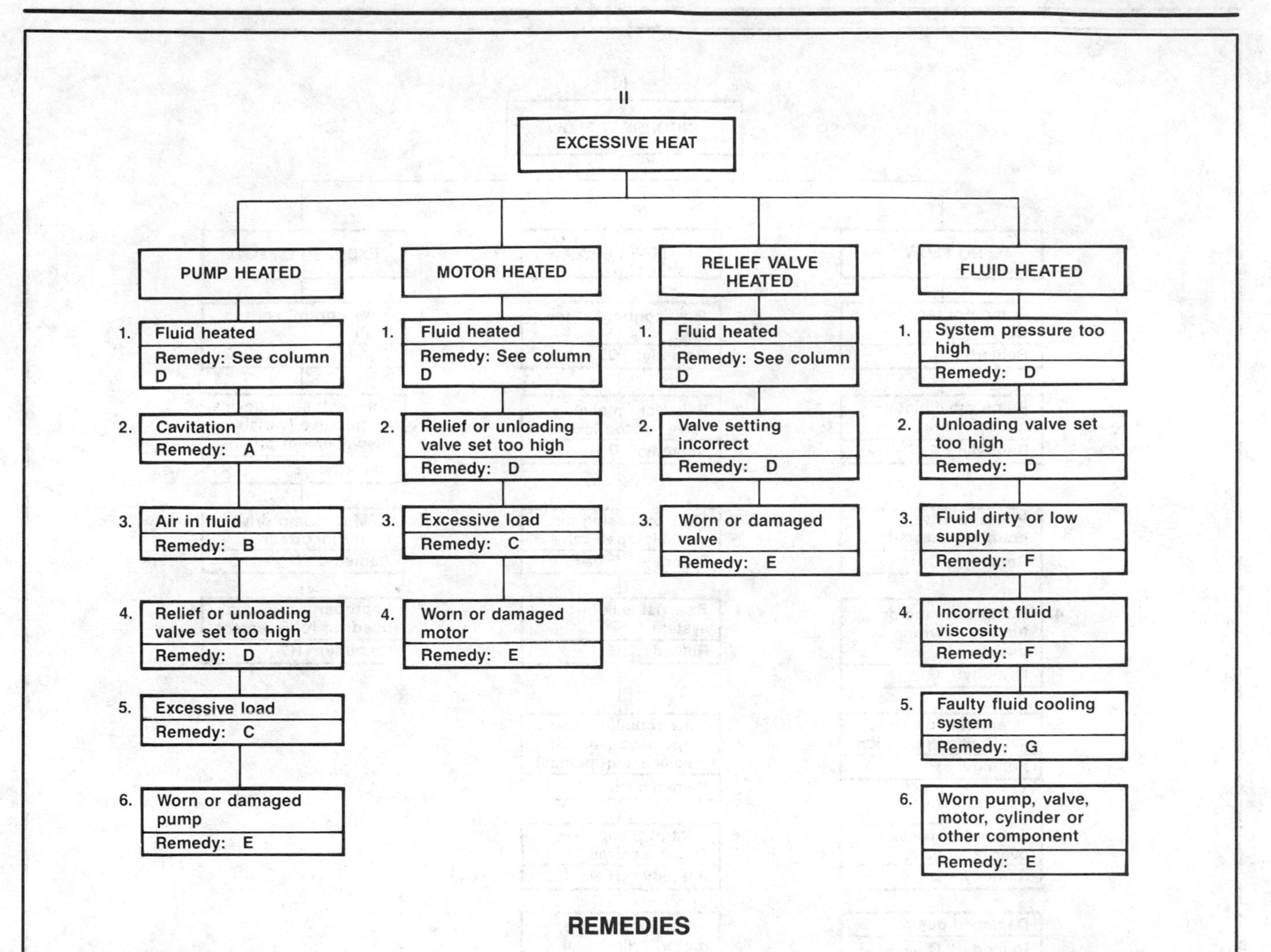

II
EXCESSIVE HEAT
PUMP HEATED
MOTOR HEATED
RELIEF VALVE HEATED
FLUID HEATED
1. Fluid heated
Remedy: See column D
2. Cavitation
Remedy: A
3. Air in fluid
Remedy: B
4. Relief or unloading valve set too high
Remedy: D
5. Excessive load
Remedy: C
6. Worn or damaged pump
Remedy: E
1. Fluid heated
Remedy: See column D
2. Relief or unloading valve set too high
Remedy: D
3. Excessive load
Remedy: C
4. Worn or damaged motor
Remedy: E
1. Fluid heated
Remedy: See column D
2. Valve setting incorrect
Remedy: D
3. Worn or damaged valve
Remedy: E
1. System pressure too high
Remedy: D
2. Unloading valve set too high
Remedy: D
3. Fluid dirty or low supply
Remedy: F
4. Incorrect fluid viscosity
Remedy: F
5. Faulty fluid cooling system
Remedy: G
6. Worn pump, valve, motor, cylinder or other component
Remedy: E

REMEDIES

A Any or all of the following:
- Replace dirty filters.
- Clean clogged inlet line.
- Clean reservoir breather vent.
- Change system fluid.
- Change to proper pump drive motor speed.
- Overhaul or replace supercharge pump.

B Any or all of the following:
- Tighten leaky inlet connections.
- Fill reservoir to proper level.
  With few exceptions, all return lines should be below fluid level in the reservoir.
- Bleed air from system.
- Replace pump shaft seal.
  Also replace shaft if worn at seal journal.

C All of the following:
- Align unit.
- Check condition of seals, bearings, and coupling.
- Locate and correct mechanical binding.
- Check for workload in excess of circuit design.

D
- Install and adjust pressure gauge.
  Keep at least 125 PSI (8.617 bar) (861.7 kPa) difference between valve settings.

E
- Overhaul or replace defective part(s).

F
- Change filters.
- Check system fluid viscosity. Change if necessary.
- Fill reservoir to proper level.

G
- Clean cooler and/or strainer.
- Replace cooler control valve.
- Repair or replace cooler.

Figure E-3. Flow chart for troubleshooting excessive heat.

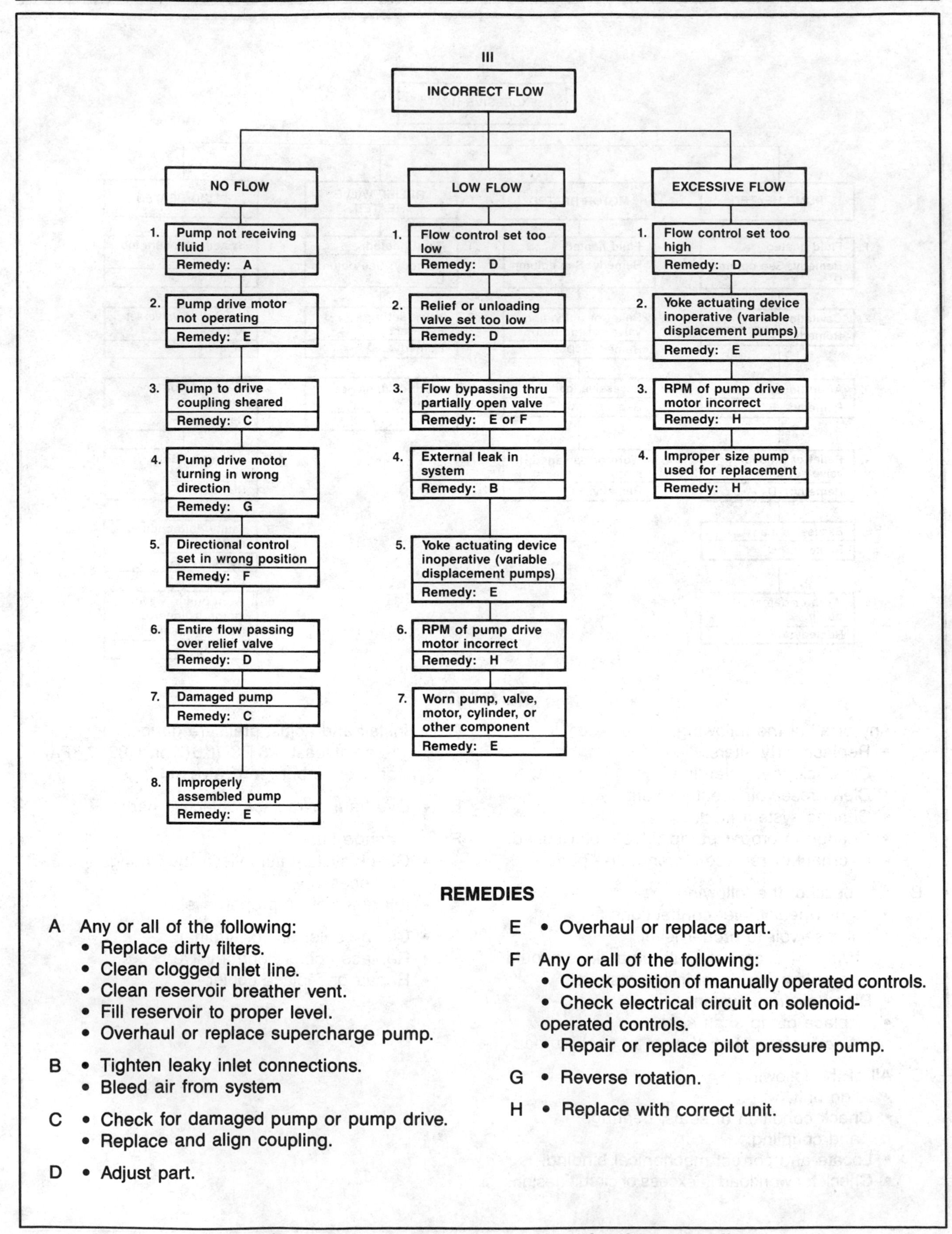
III
INCORRECT FLOW
NO FLOW
1. Pump not receiving fluid
Remedy: A
2. Pump drive motor not operating
Remedy: E
3. Pump to drive coupling sheared
Remedy: C
4. Pump drive motor turning in wrong direction
Remedy: G
5. Directional control set in wrong position
Remedy: F
6. Entire flow passing over relief valve
Remedy: D
7. Damaged pump
Remedy: C
8. Improperly assembled pump
Remedy: E
LOW FLOW
1. Flow control set too low
Remedy: D
2. Relief or unloading valve set too low
Remedy: D
3. Flow bypassing thru partially open valve
Remedy: E or F
4. External leak in system
Remedy: B
5. Yoke actuating device inoperative (variable displacement pumps)
Remedy: E
6. RPM of pump drive motor incorrect
Remedy: H
7. Worn pump, valve, motor, cylinder, or other component
Remedy: E
EXCESSIVE FLOW
1. Flow control set too high
Remedy: D
2. Yoke actuating device inoperative (variable displacement pumps)
Remedy: E
3. RPM of pump drive motor incorrect
Remedy: H
4. Improper size pump used for replacement
Remedy: H
REMEDIES
A Any or all of the following:
• Replace dirty filters.
• Clean clogged inlet line.
• Clean reservoir breather vent.
• Fill reservoir to proper level.
• Overhaul or replace supercharge pump.
B • Tighten leaky inlet connections.
• Bleed air from system
C • Check for damaged pump or pump drive.
• Replace and align coupling.
D • Adjust part.
E • Overhaul or replace part.
F Any or all of the following:
• Check position of manually operated controls.
• Check electrical circuit on solenoid-operated controls.
• Repair or replace pilot pressure pump.
G • Reverse rotation.
H • Replace with correct unit.

Figure E-4. Flow chart for troubleshooting incorrect flow.

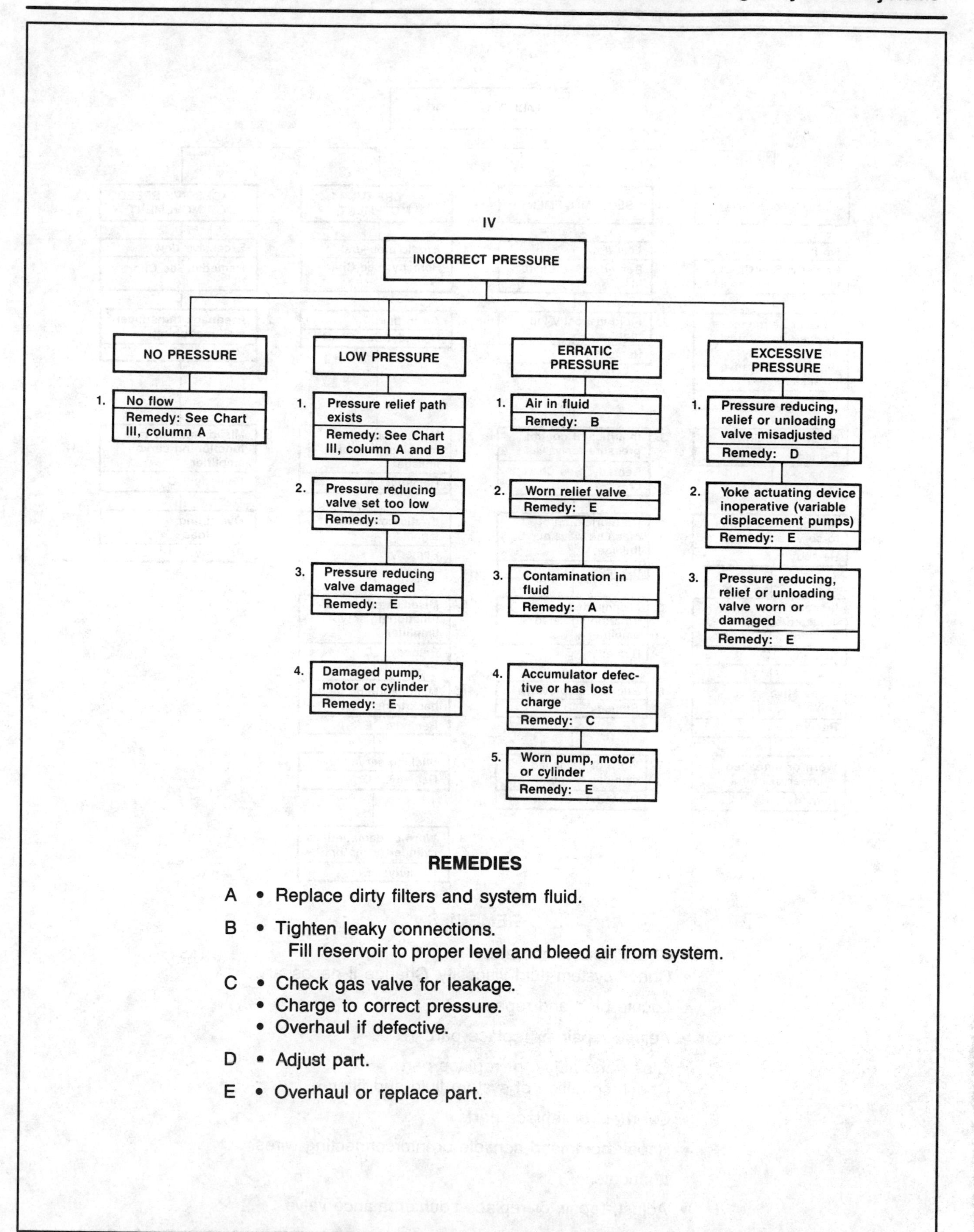
IV
INCORRECT PRESSURE
NO PRESSURE
1. No flow
Remedy: See Chart III, column A
LOW PRESSURE
1. Pressure relief path exists
Remedy: See Chart III, column A and B
2. Pressure reducing valve set too low
Remedy: D
3. Pressure reducing valve damaged
Remedy: E
4. Damaged pump, motor or cylinder
Remedy: E
ERRATIC PRESSURE
1. Air in fluid
Remedy: B
2. Worn relief valve
Remedy: E
3. Contamination in fluid
Remedy: A
4. Accumulator defective or has lost charge
Remedy: C
5. Worn pump, motor or cylinder
Remedy: E
EXCESSIVE PRESSURE
1. Pressure reducing, relief or unloading valve misadjusted
Remedy: D
2. Yoke actuating device inoperative (variable displacement pumps)
Remedy: E
3. Pressure reducing, relief or unloading valve worn or damaged
Remedy: E
REMEDIES
A • Replace dirty filters and system fluid.
B • Tighten leaky connections.
Fill reservoir to proper level and bleed air from system.
C • Check gas valve for leakage.
• Charge to correct pressure.
• Overhaul if defective.
D • Adjust part.
E • Overhaul or replace part.

Figure E-5. Flow chart for troubleshooting incorrect pressure.

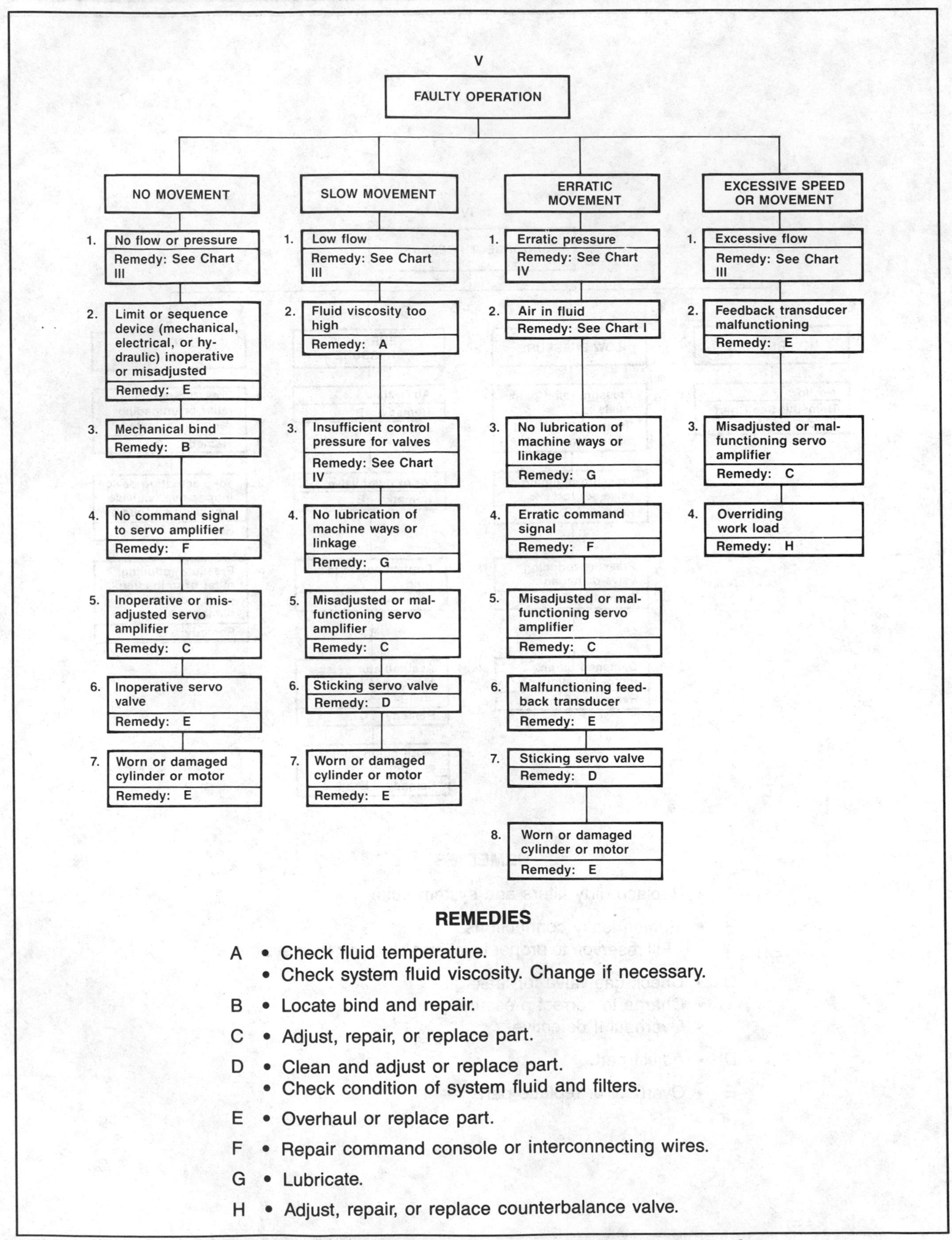

Figure E-6. Flow chart for troubleshooting faulty operation.

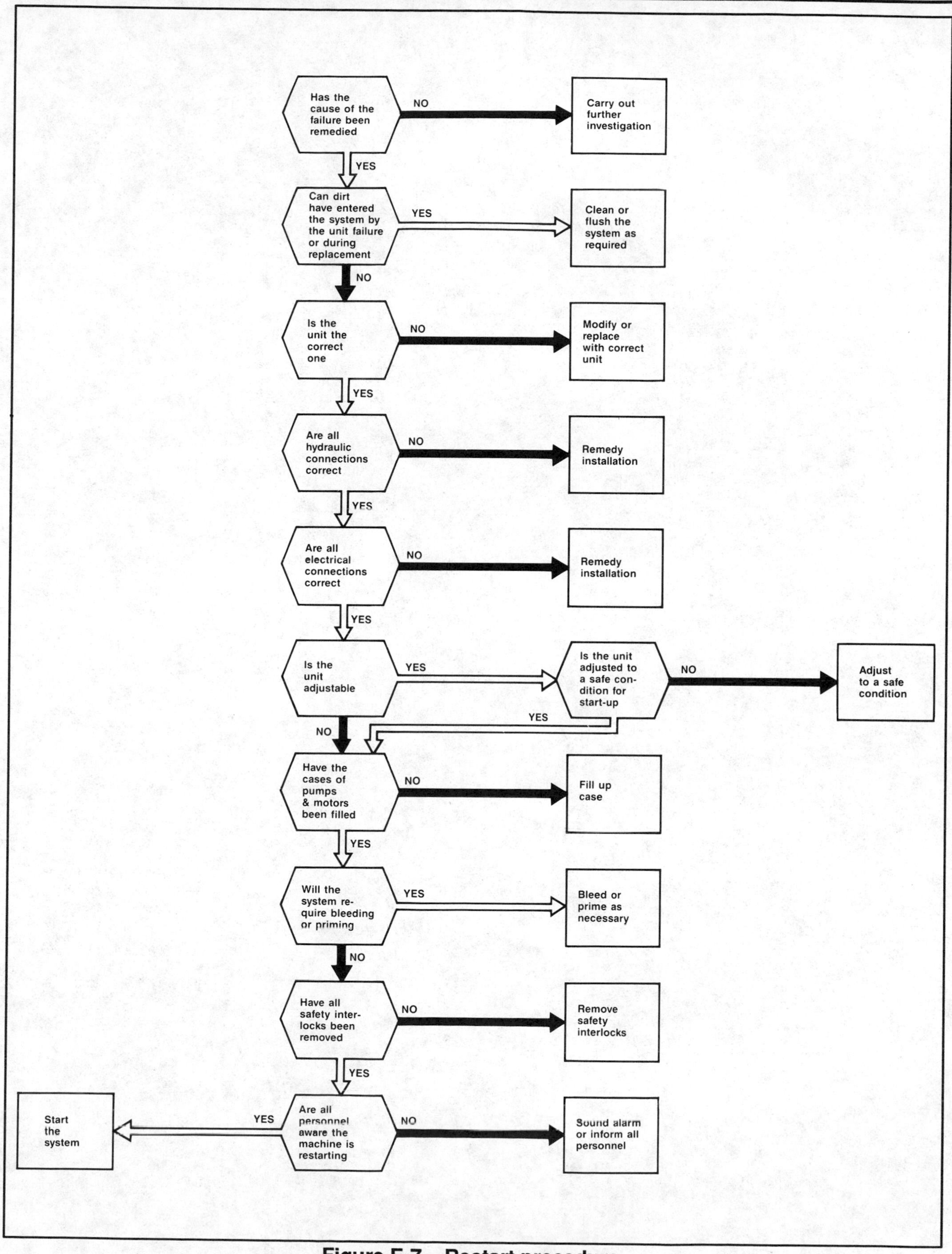

Figure E-7. Restart procedure.

Figure E-7. Restart procedure.

APPENDIX

F

NOISE CONTROL IN HYDRAULIC SYSTEMS

INTRODUCTION

Although a certain amount of noise control is required in hydraulic systems just to conform to government regulations, a conscientious noise control program actually provides a competitive edge. The combination of a quiet pump, well–engineered vibration and pulsation controls, and good, economical installation practices will result in a product with a distinct advantage in the marketplace.

The Vickers publication "More Sound Advice," issued in the 1970's, illustrated a variety of machine noise control methods. Because of the nearly infinite variety of hydraulic applications, it's not possible here to discuss the individual features of particular systems. There are, however, a number of installation techniques that can be applied to almost *all* hydraulic systems. When used correctly, these techniques can yield significant reductions in noise.

This appendix describes the following:

1. Noise generation and noise control techniques.
2. Noise terms, definition and the use of the decibel.
3. Noise measurement procedures.

SYSTEM NOISE CONTROL

Quiet Hydraulics – A Team Effort

A successful noise control program requires a team effort by individuals in several areas of expertise. A quiet hydraulic pump does not guarantee a quiet system. The choice of a quiet pump should be only one part of a multi-faceted program that calls upon the talents of the system designer, fabricator, installer, and maintenance technicians. And if any member of the team fails to do their job, it can mean failure of the entire noise control program.

The pump and system designers play a key role in achieving successful noise control. They must evaluate every noise control technique available from the standpoints of both cost and practicality. Three of the basic approaches used in quieting hydraulic power systems are:

1. Internal and external pump pulsation control
2. Pump and structure isolation
3. Damping and/or stiffening

Noise Transmission and Generation

Noise is defined as the unwanted by–product of fluctuating forces in a component or system. In a hydraulic system, this noise can be transmitted in three ways: through the air, through the fluid, and through the system's physical structure. We generally think of noise as travelling only through the medium of the air, going directly from its source to some receiver (our ear). This is called *airborne noise*. That airborne noise, however, must have a source within some component of the hydraulic system. That component is normally the pump. Whether it's a piston, vane, or gear pump, the internal pumping and porting design can never be perfect. As a result, uneven flow characteristics and pressure waves are created and transmitted through the fluid. This is known as *fluidborne noise*. The pressure wave fluctuations of fluidborne noise in turn create corresponding force fluctuations. These result in vibration, also known as *structureborne noise*. This structureborne noise is transmitted not only through the pump body, but through attached structures as well. These structures then emit an audible sound.

The surrounding structures and surface areas in a hydraulic system tend to be much larger than the pump itself, and therefore radiate noise more efficiently. For this reason, while the pump design should minimize internal pulsations, it's also important to use proper isolation techniques to keep the remaining vibrations from reaching adjoining structures.

Design for Low Noise

An intelligent program of noise control should start at the source: the pump. A quiet pump is the responsibility of the pump manufacturer. The problem for the designer is that although a hydraulic pump is required to perform over a wide range of speeds and pressures, noise control can only be optimized for a relatively narrow portion of that range. The most common strategy is to use

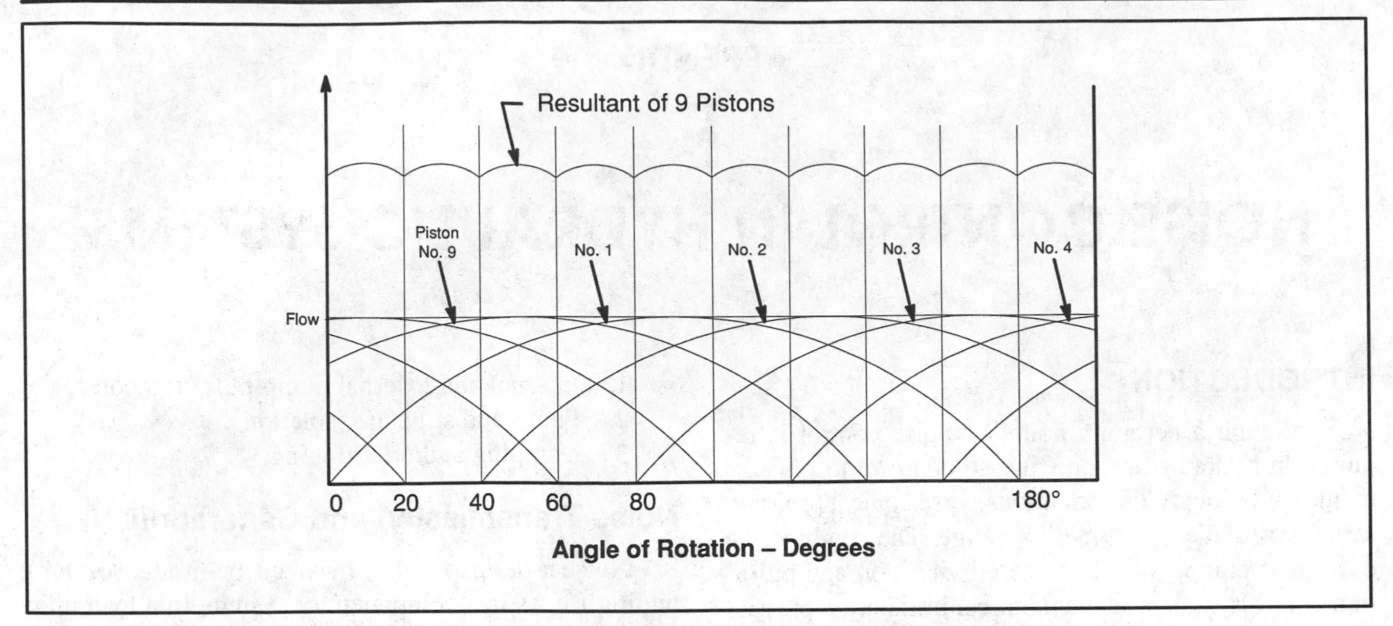

Figure F-1. Inherent Piston Pump Discharge Ripple

porting design to limit the pressure pulsations at the pump's rated speed and pressure. The pulsations are reduced as much as possible without creating a large amount of noise at lower speeds and pressures. Piston, vane, and gear pumps are similar in that their total output flow is the sum of the flows from the individual pumping elements or chambers. Fluid fills the chambers at the pump inlet, is compressed mechanically and/or hydraulically through orifices, and is then combined into a single discharge flow. Each pumping element in a piston pump delivers its fluid to the discharge port in a half–sine profile. The pump discharge is the total of the equally spaced half–sines added in phase. The result is an inherent flow ripple, as shown in Figure F-1 for a nine–piston pump. This ripple is independent of any fluid compression, either through piston motion or any type of internal hydraulic metering. Vane pump flow ripple is more controllable. Cam contours can be designed to reduce mechanical compression effects. This is done by making pressure transitions in the dwell section, where there is controlled change in vane chamber volume. For this reason, vane pumps will normally generate less noise over a wider range of speeds and pressures than piston pumps.

Noise Frequencies

Pump noise energies are generated in several ways. Vibrational energy is created by an imbalance in the pump, drive motor, or couplings. It can also be produced by some undesired interaction in the assembly, but it's rare for any significant audible noise to be generated by these interactions. Nonetheless, care should be taken to minimize its effects on pump or motor life. Figure F-2 shows the frequency spectrum of a ten–vane pump operating at 1800 rpm with shaft rotation frequency of 30 Hz. Any misalignments in the power train will produce noise components at twice and four times this frequency.

The strongest energy components occur at pumping frequency. This frequency equals the number of pumping chambers times the shaft frequency (300 Hz in Figure F-2). Noise energy is also produced at multiples, or harmonics, of this frequency. 600 Hz and 900 Hz are the second and third harmonics seen in Figure F-2. These harmonics have enough amplitude to produce significant noise. This noise comes not only from the pump itself, but from attached structures which are often more efficient at radiating the noise transmitted from the pump.

Vibration Noise Control

Vibration control is used to prevent pulsation energy from the pump from being transmitted to machine structures. Most pump and drive motor assemblies are attached through a flexible coupling and mounted on a common base to maintain alignment. The common base is resiliently mounted to the support structure, as seen in Figure F-3. An isolator should be selected that has a natural frequency approximately $^1/_2$ or less the pump's rotational frequency. For example, an isolator with a natural frequency of 10 Hz or less would be appropriate for a pump with a rotational or forcing frequency of 20 Hz at 1200 rpm, and would work even better for a pump with a frequency of 30 Hz at 1800 rpm. The higher the ratio between the forcing frequency and the natural frequency of the isolator system, the greater the amount of isolation (see Figure F-4). A typical commercial isolator (which

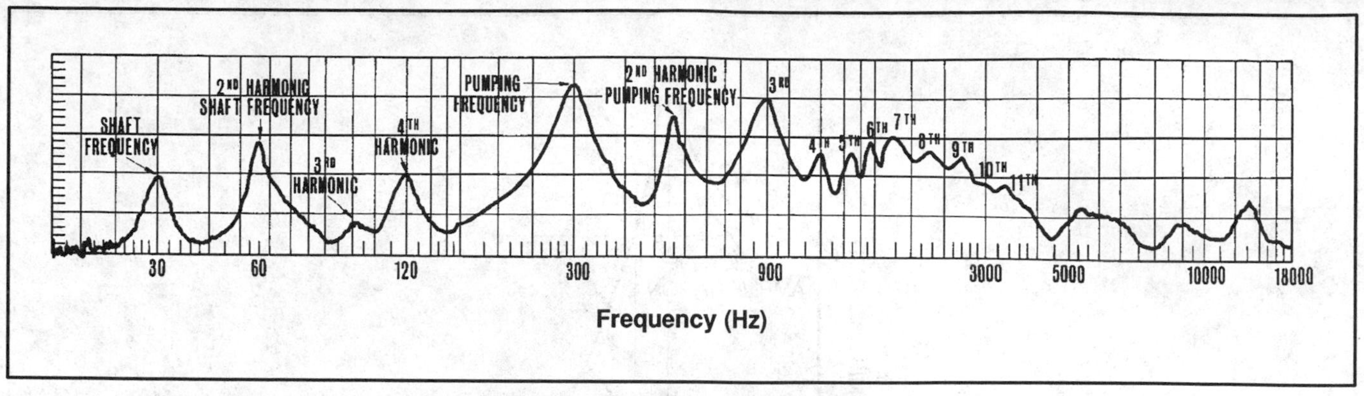

Figure F-2. Structureborne or Fluidborne Spectrum Identifying Shaft and Pumping Frequency and Harmonics

costs about $15 in moderate quantities) can reduce transmitted vibration energy by 10 dB at 1200 rpm and 15 dB at 1800 rpm (Figure F-5).

Isolators are classified by their load carrying capacity and related natural frequency. When pump, motor, and subplate assembly weights are known, the amount of evenly distributed weight on each of the isolators can be calculated. Isolators should be selected that will not be loaded above 60% to 70% of their capacity. This will allow a sufficient safety margin in the event of shock loading.

The same type of isolators can also be used on power units with overhead reservoirs. Eight isolators can be installed either under the reservoir feet (Figure F-6), or under the upright leg structures supporting the reservoir. The isolators shown would be very effective because the ratio of forcing frequency to natural frequency is very high. For example, a nine-piston pump operating at 1200 rpm would have a pumping frequency of 9 x 20 rev/sec or 180 Hz. If a 10 Hz isolator system were used, there would be a very low level of vibration transmission, because the ratio between the two frequencies would be 18:1.

The chart below lists load ratings of typical isolators with carrying capacities of 60 to 4400 lbs. per isolator.

Load Series	Isolator Number	Max. Static Load Per Isolator (lbs.)
Light	L1 L2 L3 L4 L5	60 100 130 200 260
Medium	M1 M2 M3 M4	300 450 550 700
Heavy	H1 H2 H3	700 1000 1500
Extra Heavy	EH1 EH2 EH3 EH4	1500 2000 3000 4400

Figure F-7 provides two examples of proper isolator selection.

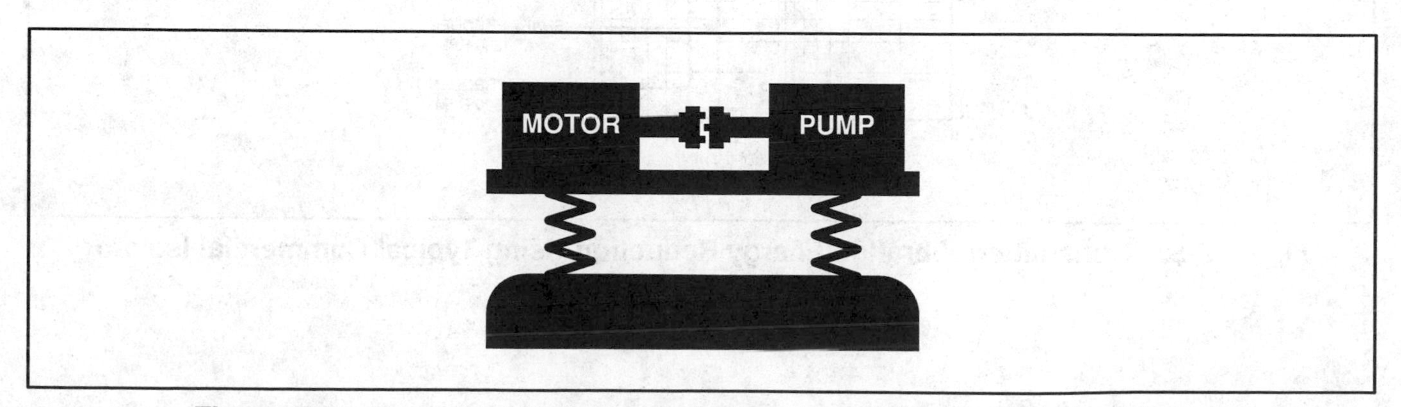

Figure F-3. Pump and Motor on Subplate, Isolated from Stiff Foundation

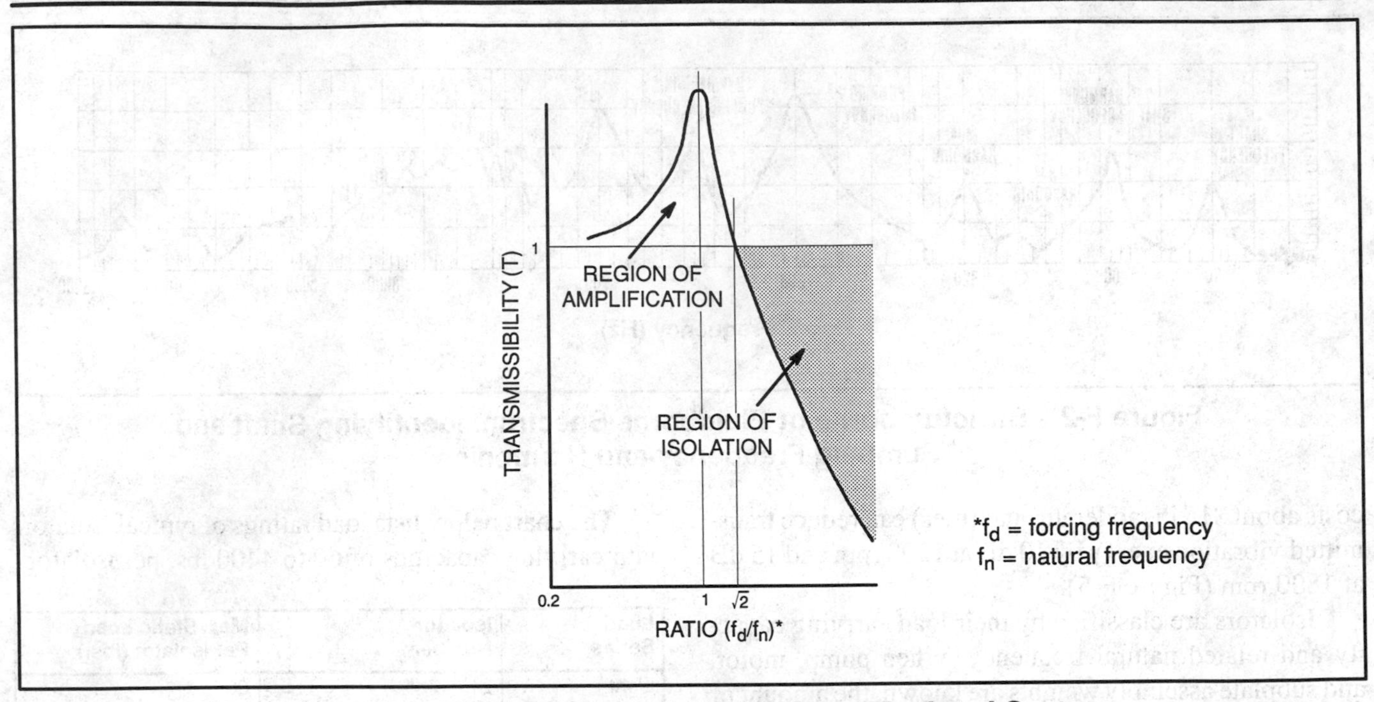

Figure F-4. Typical Transmissibility for an Isolated System

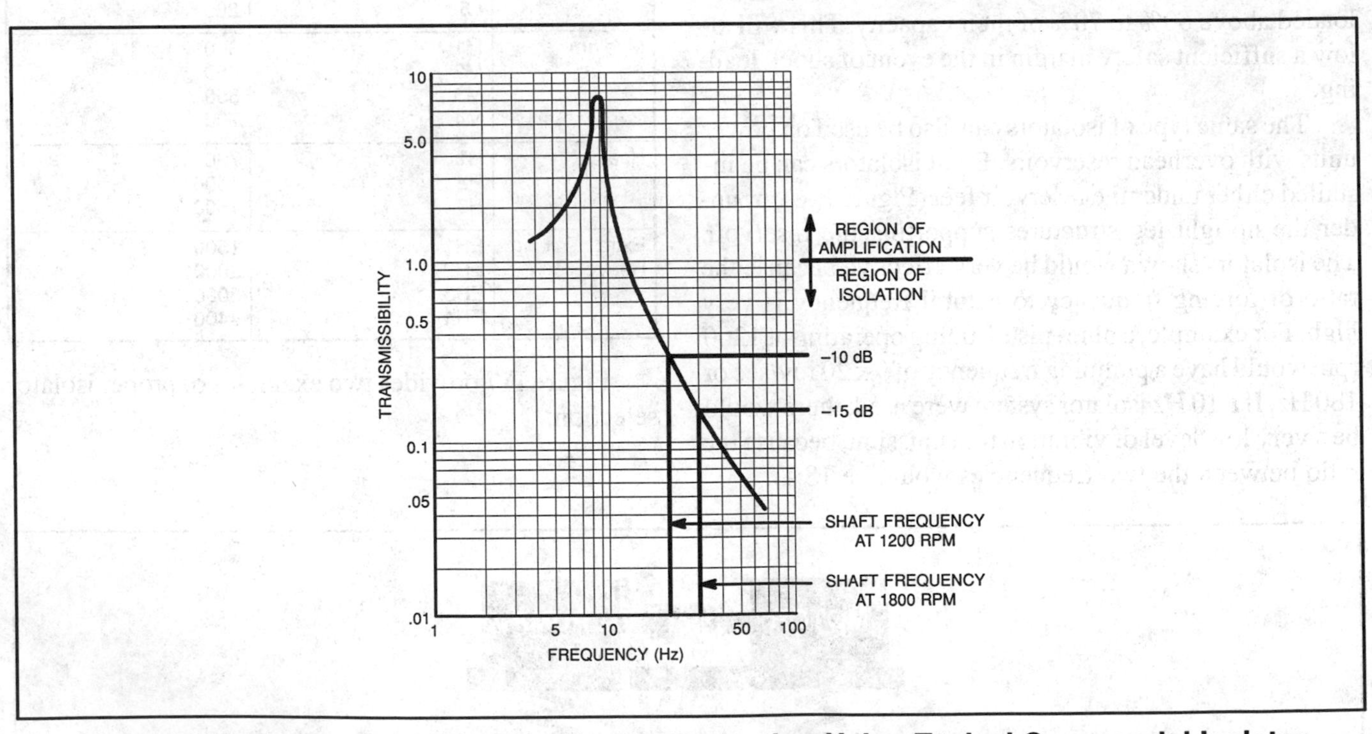

Figure F-5. Transmitted Vibration Energy Reduction Using Typical Commercial Isolator

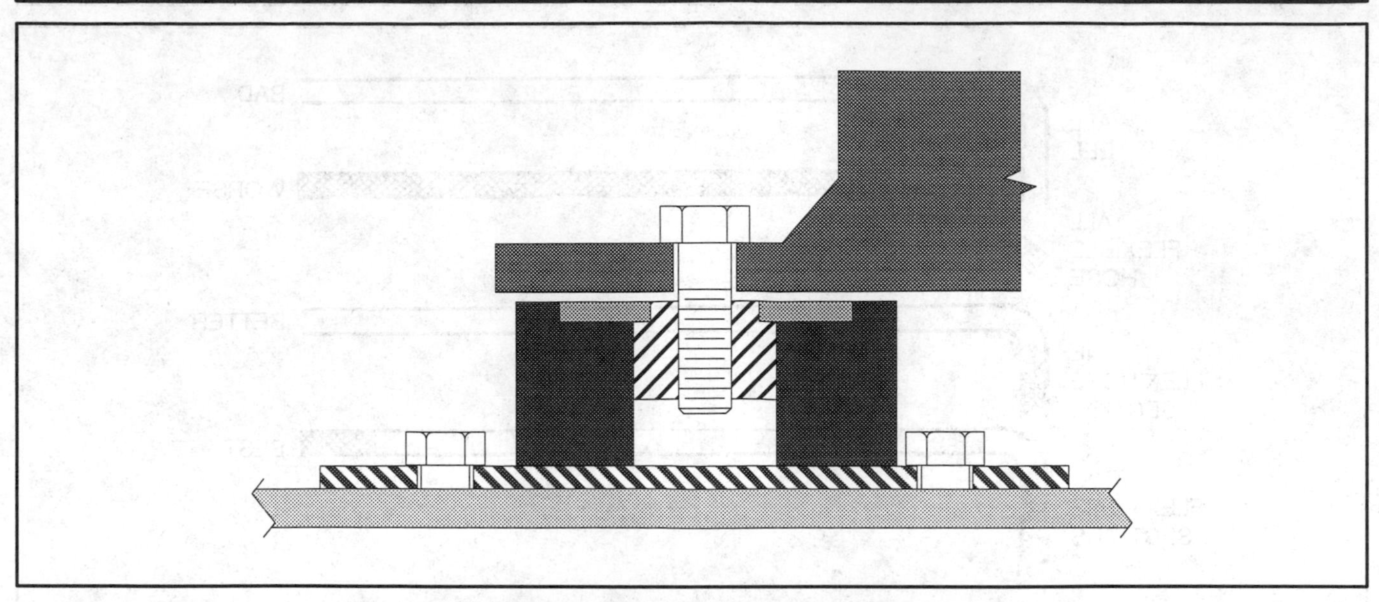

Figure F-6. Reservoir Foot on Isolator

A)	Pump, motor, and subplate weight	=	800 lbs.
	Load on each of 6 isolators	=	133 lbs.
	Maximum static load when loaded to 60 to 70% of capacity	=	190 to 222 lbs.
	Selection:		**# L4 (from chart on page 5)**
B)	Reservoir weight	=	550 lbs.
	Attached accessories weight	=	150 lbs.
	100 gallons oil weight (7 lbs./gal.)	=	700 lbs.
	TOTAL	=	1400 lbs.
	Load on each of 4 isolators	=	350 lbs.
	Maximum static load when loaded to 60 to 70% of capacity	=	500 to 583 lbs.
	Selection:		**# M3 (from chart on page 5)**

Figure F-7. Isolator Selection

Isolation With Hose

Rubber hose must be used to maintain the vibration isolation afforded when the pump and motor assembly are mounted on isolators. The isolation capabilities of hose will reduce the amount of vibration energy entering the system. Unfortunately, improper use of hose is probably the main cause of noise in many systems.

Although structureborne noise can be reduced by using long lengths of hose, the pressure pulsations from the pump will cause the hose to undergo cyclic radial expansion. One of the drawbacks of hose is that it acts as an efficient radiator in the frequency range where most of the energy is generated: the first few harmonics. Because of these two factors, long lines of hose are less effective for noise reduction than the use hose at either end of a solid line (Figure F-8).

Two other shortcomings of hose are that its length changes with pressure, and that when it's bent through a radius it acts like a Bourdon tube, trying to straighten out with increasing pressure. Both produce forces that act on connecting structures. The best way to maintain noise control while making bends with hose is to use a solid elbow with a hose section on each end. This eliminates problems caused by the Bourdon tube effect. Any change in the length of one hose is accommodated by bending in the other hose. Figure F-9 illustrates the preferred configurations for 90° and 180° bends.

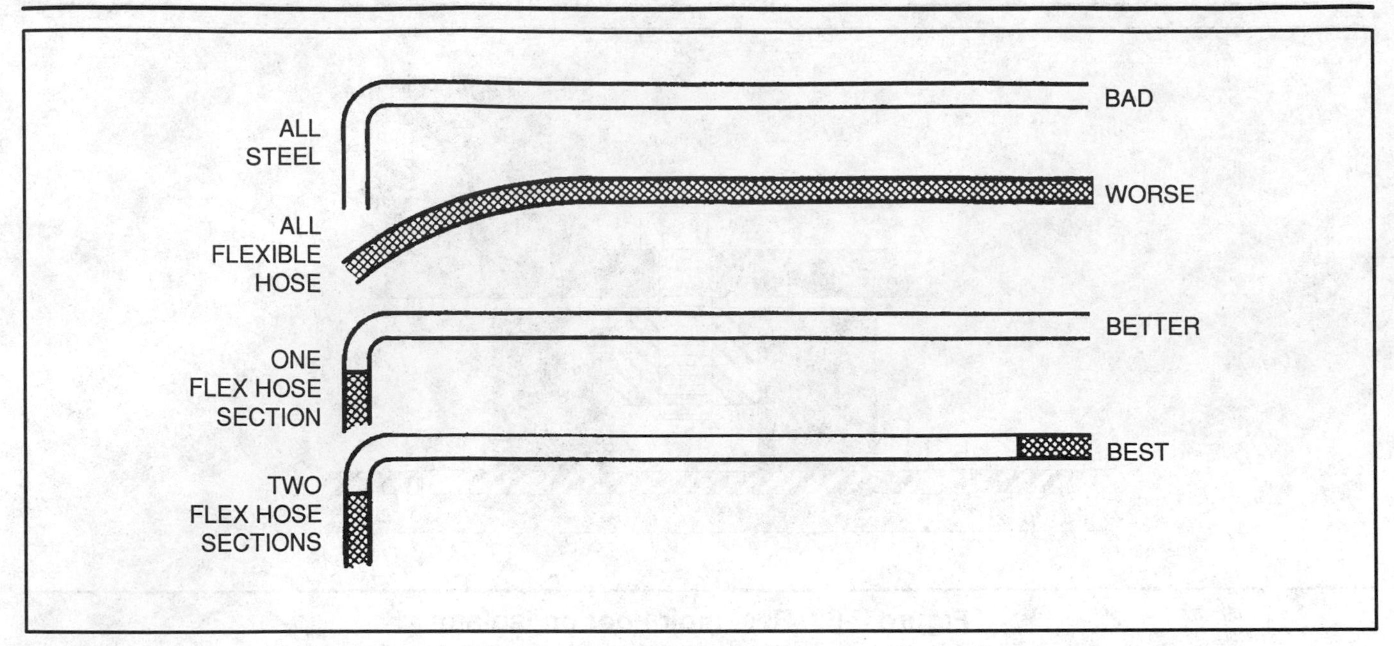

Figure F-8. Long Hydraulic Line Configurations

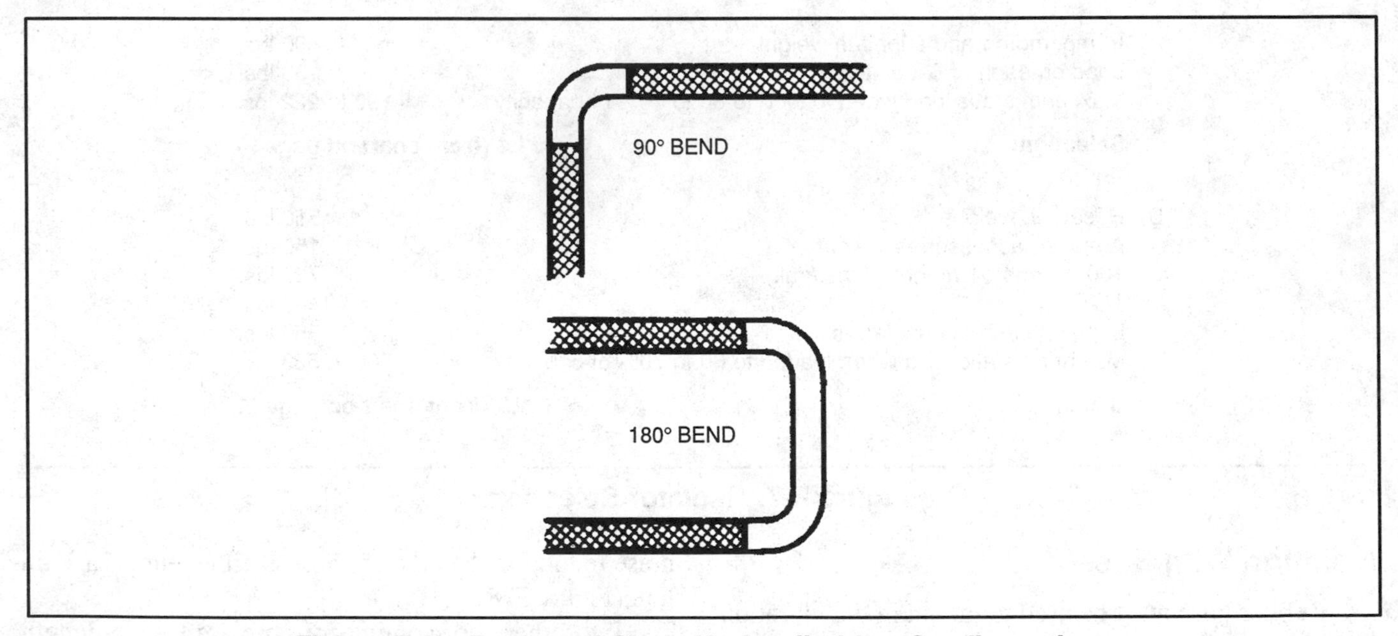

Figure F-9. Preferred Short Hydraulic Line Configurations

Fluidborne Noise Control

Fluidborne noise control begins with a pump's internal design. The ports should be configured so that the lowest practical pressure pulsations are generated. Additional external controls can be added to prevent as much pulsation energy as possible from being communicated to the system. This is usually done by adding expansion volume at the outlet of the pump. These acoustic filters, as they are called, can take many forms. The two most common are gas charged side branch accumulators and flow–through type pulsation filters. Each has its advantages and disadvantages. The side branch type is cheaper, but limited to attenuating pulsations in only a narrow range of frequencies. As a result, it isn't totally effective throughout the first four pump harmonics (where most of the pulsation energy is generated). The flow–through device, although generally larger and more expensive, has a distinct performance advantage: pulsations throughout the spectrum are reduced, including those at the most significant harmonics. The filters (shown in Figure F-10) have optimum effectiveness when gas charged to $^1/_3$ the maximum operating pressure of the hydraulic system.

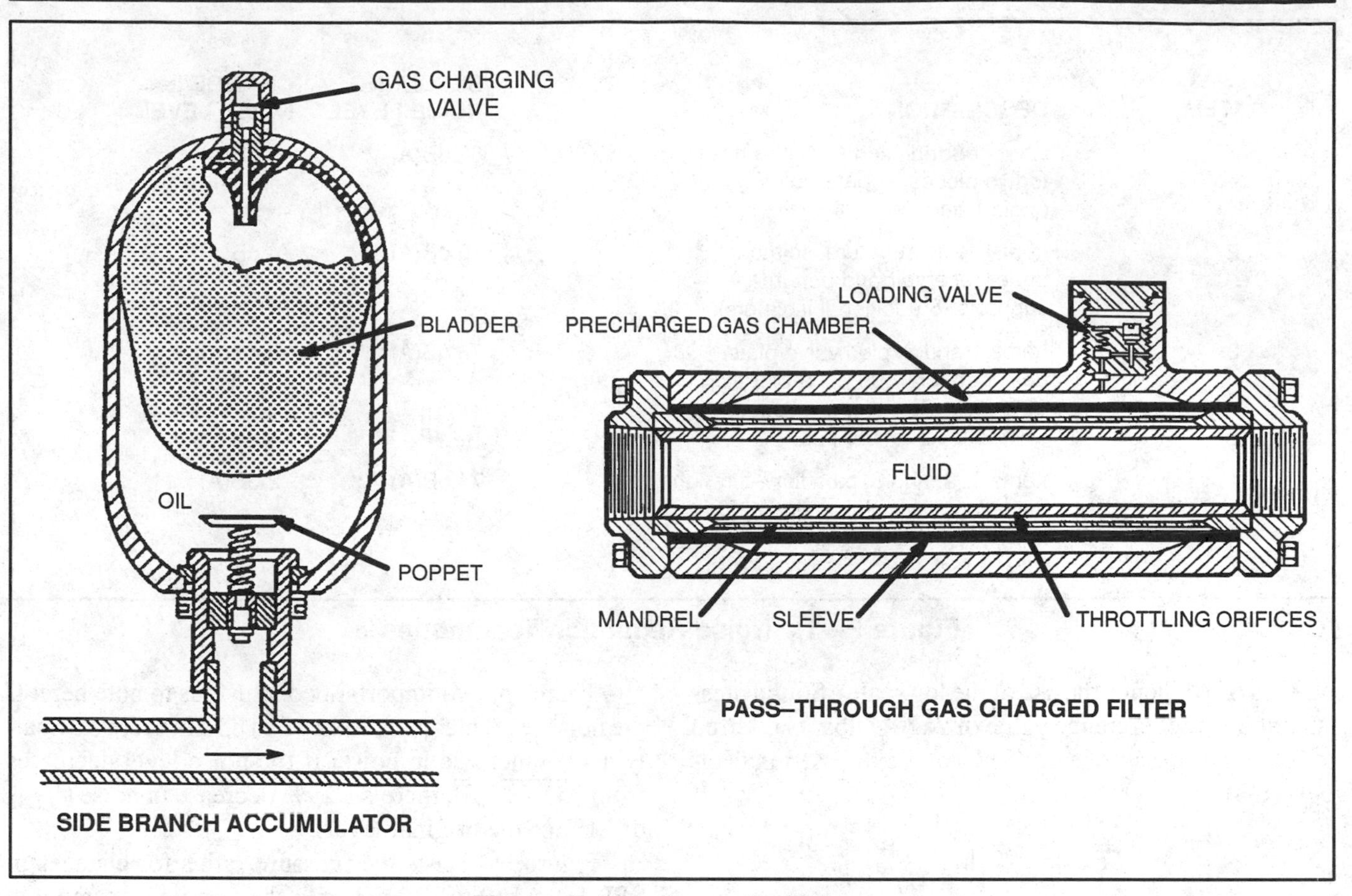

Figure F-10. Acoustic Filters

Application of Techniques

Experimental evaluations were made on a typical automotive application: a power unit with a 150 gallon overhead reservoir supplying oil to a piston pump. The pump delivered 40 gpm at 1200 rpm at a pressure of 750 psi. The noise level, as received, was 88 dB(A). (Standard accepted noise measurement procedures are explained in the subsection entitled System Noise Evaluation.) Four different noise reduction techniques were applied to the system, starting with those that would have the greatest effect on noise levels. The results are shown in Figure F-11.

It's important to note that if the changes outlined above had been made in some other sequence, the amount of noise reduction at each step would have been different from that shown – particularly if items 3 and 4 had been evaluated before items 1 and 2. The final noise level of 74 dB(A) would be the same, but the first item tried wouldn't have yielded such a significant reduction. In the example shown, the areas of highest noise radiation were addressed first. This is essential because no appreciable noise reduction can be achieved unless the most significant noise source is identified and its level reduced first.

NOISE TERMINOLOGY

Noise Terms and Equations

Sound at a particular point in air is defined as the rapid variation in air pressure around a steady state value. Sound pressure is measured in the same units as atmospheric pressure. Since it's an alternating quantity, the term "sound pressure" is usually referred to by its root mean square (rms) value. At a frequency of 1000 Hz, a sound with an rms pressure of 2×10^{-4} microbars (ubar), or about 2×10^{-10} atmospheres, is just below the hearing threshold of someone with good ears. Expressed in more familiar terms, that level of sound pressure would be 2.9×10^{-9} psi. The fact that slightly greater pressures become audible shows the amazing sensitivity of the human ear. It can detect variations in atmospheric pressure as small as a few parts in 20,000,000,000.

In addition to this sensitivity, the human ear has an enormous dynamic range. Not only can it detect sounds as small as 2×10^{-4} ubar, it can accommodate sound pressures as high as 2000 ubar without being overloaded, i.e. causing pain. That's a dynamic range ratio from threshold to pain of 10,000,000:1 (Figure F-13). Because this range is so large, it's more convenient to express ratios in

ITEM	DESCRIPTION	RESULTING NOISE LEVEL	CHANGE IN NOISE LEVEL
1	Changed radiused pressure hose to two pieces, separated by right angle fitting.	83 dB(A)	–5 dB(A)
2	Item 1, plus structure isolators under reservoir and upright supports (8 additional isolators).	79 dB(A)	–4 dB(A)
3	Items 1 and 2, plus valve plate designed for 1000 psi rather than 3600 psi. (Pulsations reduced from 200 psi to 140 psi.)	76 dB(A)	–3 dB(A)
4	Items 1, 2, and 3, plus flow–through pulsation filter. (Pulsations reduced from 140 psi to 35 psi.)	74 dB(A)	–2 dB(A)

Figure F-11. Noise Reduction Techniques

powers of 10 (hence the use of the log scale). Sound pressure above the reference value of $2x10^{-4}$ ubar is referred to as sound pressure level (SPL) and expressed in decibels (dB).

$$SPL = 20 \log \frac{P}{P_0}$$

OR

$$SPL = 10 \log \left(\frac{P}{P_0} \right)^2$$

Where:
SPL = Sound pressure level (dB)
P = Sound pressure (bar)
P_o = Reference pressure (.0002 ubar)

From this equation, a pressure ratio of 10,000,000 (10^7) would result in the following noise level:
$SPL = 20 \log 10^7$
= 7 (20) log 10 = 7 (20) (1) = 140 dB (ie., painful)

Pressure ratios can also be calculated based on the changes in sound pressure levels (ΔSPL). For example, what is the pressure ratio (R_p) if the noise level changes by 3 dB?

$$R_P = 10^{\Delta SPL/20}$$

Where:
R_p = pressure ratio
= $10^{3/20}$ = 1.41

There are two important conclusions to note here: If the noise level increases from 82 to 85 dB, there's actually a 41% increase in noise; if the noise level decreases from 85 to 82 dB, there's a 29% decrease in noise ($^1/_{1.41}$ or 71% of the original level).

Figure F-12 lists the pressure ratios for changes in SPL from +10 to –10 dB with the previous example in bold:

Change In SPL	Pressure Ratio	Change In SPL	Pressure Ratio
1	1.12	-1	.89
2	1.26	-2	.79
3	**1.41**	**-3**	**.71**
4	1.59	-4	.63
5	1.78	-5	.56
6	2	-6	.5
7	2.24	-7	.45
8	2.51	-8	.4
9	2.82	-9	.35
10	3.16	-10	.32

Figure F-12. Pressure Ratios

(A chart of typical noise levels is shown in Figure F-13.)

Human Response to Noise – The "A" Scale

A microphone measures actual sound pressures emitted from a noise source, but the human ear doesn't treat equal levels with equal tolerance over the audible frequency range of up to 12,000 Hz. The ear is more sensitive to noise above 1000 Hz. This sensitivity is simulated by using the "A" scale filtering system in the signal processing of measured noise. This internationally standardized system gauges the ear's response to noise

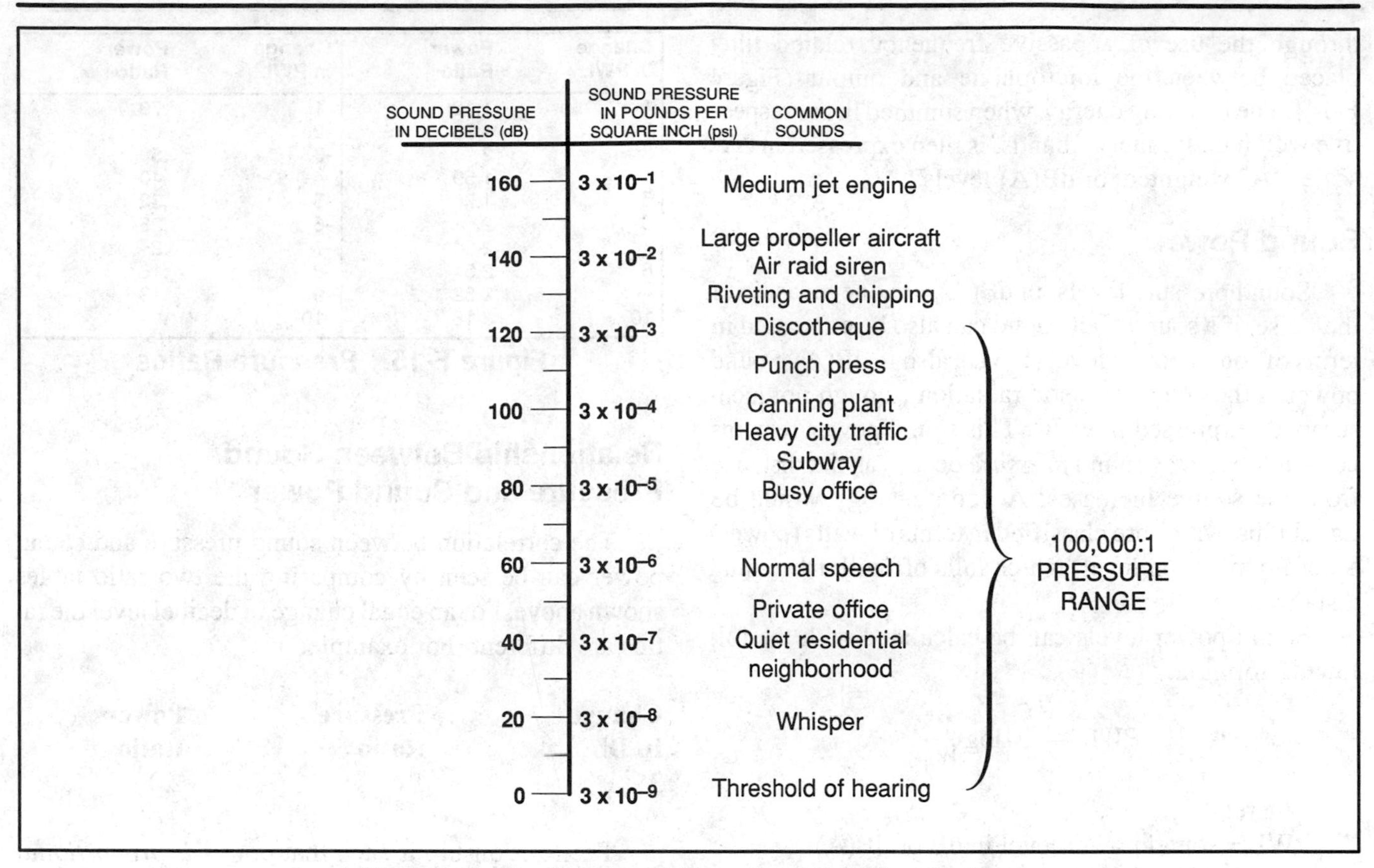

Figure F-13. Dynamic Range of the Human Ear

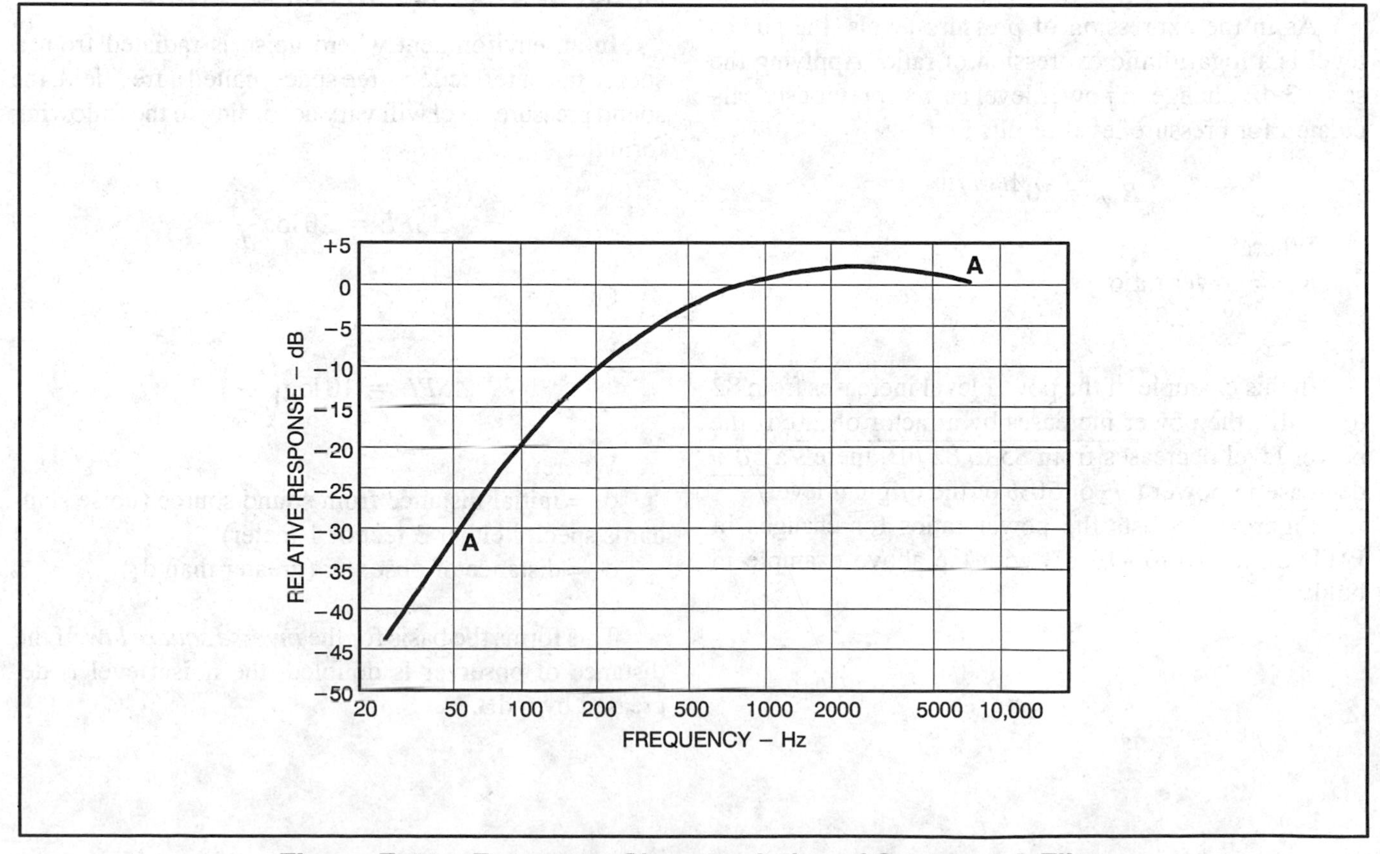

Figure F-14. Response Characteristics of Standard A Filter

through the use of a passive frequency related filter placed between the microphone and output (Figure F-14). The resulting energy, when summed in the respective weighted frequency bands, is then expressed in "A" scale, "A" weighted, or dB(A) levels.

Sound Power

Sound pressure levels, in dB(A), are one measure of the noise of a source, but sound can also be expressed in terms of sound power level (PWL), also in dB(A). Sound power is the actual acoustic radiation (though not measurable), expressed in watts. This sound power remains constant, whereas sound *pressure* decays as the distance from the source increases. A good analogy would be light bulbs, which are classified in terms of watts (power) – not illumination level, which falls off with increasing distance.

Sound power levels can be calculated with the following formula:

$$PWL = 10 \log \frac{W}{W_0}$$

Where:
PWL = sound power level in dB or dB(A)
W = acoustic radiation of source in watts
W_o = reference radiation in watts (10^{-12})

As in the expression of pressure levels, the power level is a logarithmic expression of ratio. Applying the same 3 dB change in power level as was previously calculated for pressure level results in:

$$R_W = 10^{\Delta PWL/10}$$

Where:
R_W = power ratio
$= 10^{3/10} = 2.0$

In this example, if the power level increases from 82 to 85 dB, the power increases by a factor of 2.0; if the power level decreases from 85 to 82 dB, there's a 50% decrease in power ($^1/_2$ or 50% of the original level).

Figure F-15 lists the power ratios for changes in PWL from +10 to –10 dB with the above example in bold:

Change In PWL	Power Ratio	Change In PWL	Power Ratio
1	1.12	-1	.79
2	1.26	-2	.63
3	**2**	**-3**	**.5**
4	1.59	-4	.40
5	1.78	-5	.32
6	2	-6	.25
7	2.24	-7	.2
8	2.51	-8	.16
9	2.82	-9	.13
10	3.16	-10	.1

Figure F-15. Pressure Ratios

Relationship Between Sound Pressure and Sound Power

The correlation between sound pressure and sound power can be seen by comparing the two ratio tables shown above. For an equal change in decibel level the ratios are different. For example:

Change In Db	Pressure Ratio	Power Ratio
3	1.41	2.0

The relationship is such that ***power is proportional to the pressure squared.***

Effect of Distance on Noise Levels

In an environment where noise is radiated from a source into a reflection–free space, called a free field, the sound pressure level will vary according to the following formula:

$$\Delta SPL = 20 \log \frac{d_1}{d_2}$$

OR

$$\Delta SPL = 10 \log \left(\frac{d_1}{d_2}\right)^2$$

Where:
d_1 = initial distance from sound source (noise standards specify either 3 feet or 1 meter)
d_2 = distance of observer (greater than d_1)

This forms the basis for the *inverse square law.* If the distance of observer is doubled, the noise level is decreased by 6 dB.

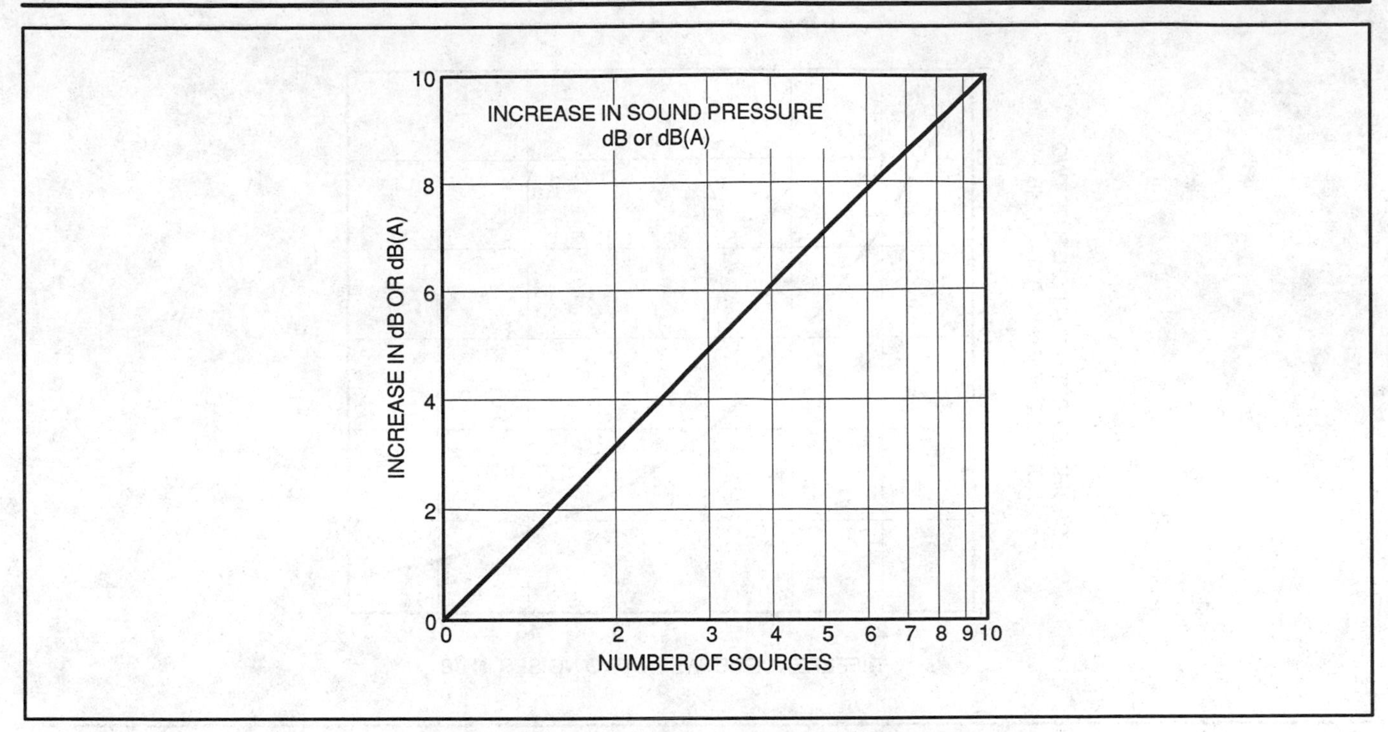

Figure F-16. Addition of Equal Sounds

For example:
d_1 = 3 feet
d_2 = 6 feet

Then:
$\Delta SPL = 20 \log .5 = -6$ dB
OR
$\Delta SPL = 10 \log (.5)^2 = -6$ dB

Noise Addition

Noises can only be added or subtracted on the basis of acoustic **power** – *not pressure*. Noise sources of equal sound pressure levels are combined as follows:

$$SPL_2 = SPL_1 + 10 \log X$$

SPL_2 = noise level of all sources
SPL_1 = noise level of 1 source
X = number of sources

For example:
SPL_1 = 80
X = 3 sources
$SPL_2 = 80 + 10 \log 3 = 80 + 4.8 = 84.8$ dB

A chart showing this relationship appears in Figure F-16.

Unequal sound pressure levels can also be added (See Figure F-17). The difference in the pressure levels of two sounds is used to determine how much their combined level will exceed the higher of the two. To add a third level, use the same process to combine it with the total from the first two levels.

The following example shows the two steps needed to find the total noise from sources of 70, 74, and 77 dB:

1. 74 – 70 = 4 dB (X Axis)
   dB to add to larger = 1.5 (Y Axis)
   TOTAL = 75.5 dB
2. 77 – 75.5 = 1.5 dB (X Axis)
   dB to add to larger = 2.3 (Y Axis)
   TOTAL = 77 + 2.3 = 79.3 dB

The total can also be calculated from the equation:

$$SPL = 10 \log \Sigma 10^{SPL/10}$$

$$= 10 \log (10^{7.4} + 10^{7.0} + 10^{7.7}) = 79.3 \text{ dB}$$

NOISE MEASUREMENTS

Component Evaluation and Rating

To assist machine tool builders in their selection of components on the basis of noise, the National Fluid Power Association (NFPA) developed a standard that assures uniformity in the measurement and reporting of sound levels. This standard, T3.9.12, contains guidelines for obtaining standardized sound ratings. It is concerned only with the radiated noise of components, primarily pumps.

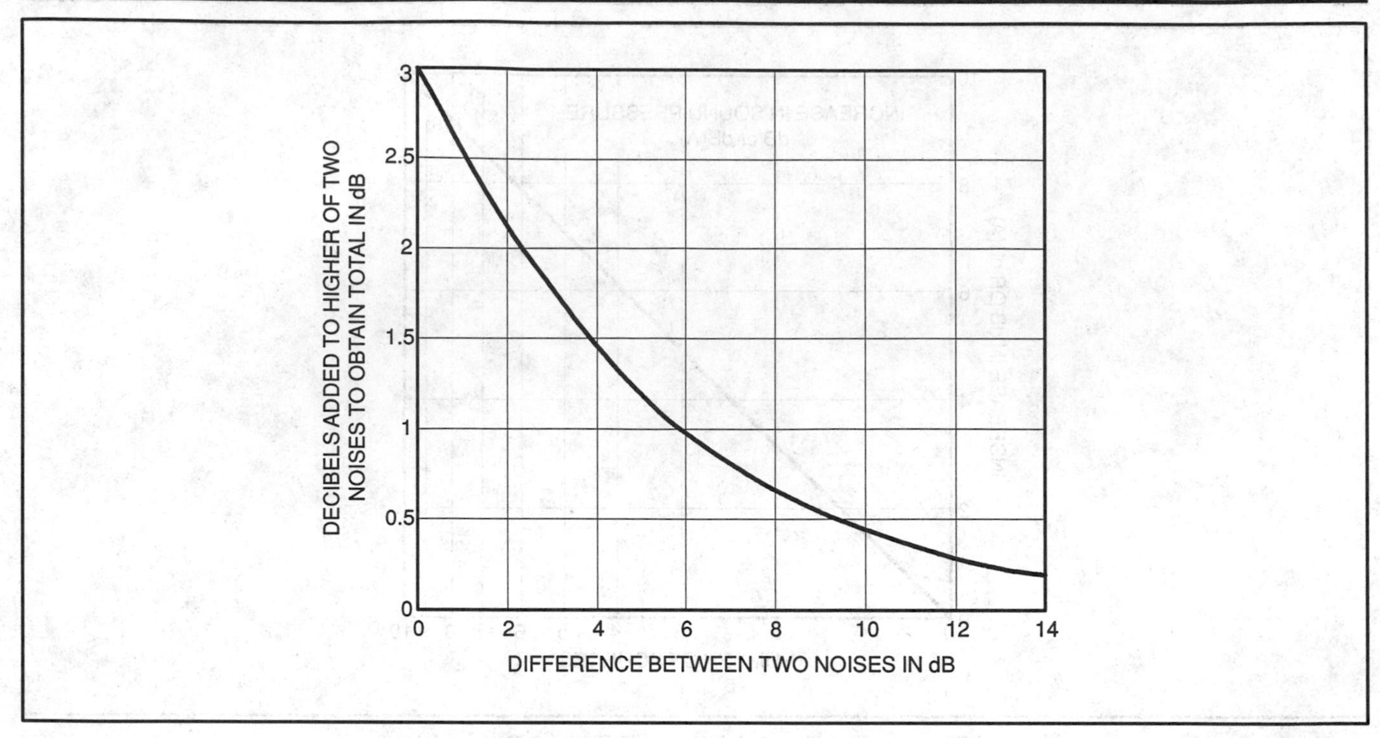

Figure F-17. Addition of Unequal Sounds

The rating is usually expressed in dB(A) at a distance three feet in a free field above a reflecting plane (semi anechoic). This is a computed figure derived from a mathematical model. This model assumes that all the sound power from a pump is radiated from a single point located in the center of a hypothetical test hemisphere (Figure F-18). The standard allows for masking all parts of the circuit that might contribute to noise. This includes wrapping hydraulic lines and enclosing any load valves. ***All radiated noise must be attributable to the pump, with no corrections for background.***

Microphones are positioned on spatial coordinates, each of which is located at the centroid of equal areas of the hemisphere surface. The rule of thumb is to use one microphone for each square meter of area. With the area of the hemisphere at $2\pi r^2$, and r equal to approximately 1 meter, six microphones are sufficient.

Power is defined as follows:

$$F = P \times A$$

Where:
F = power
P = sound pressure
A = area acted on by P ($2\pi r^2$)

In terms of decibels:

$$PWL = \overline{SPL} + 10 \log 2\pi r^2$$

Where:
PWL = sound power level
$\overline{SPL}$ = average sound pressure level of 6 microphones

r = radius in meters (3 feet = .914 meters)

Therefore:
PWL = $\overline{SPL}$ + 10 log 2π + 20 log .914
= $\overline{SPL}$ + 8 − .8 = $\overline{SPL}$ + 7.2 (for 3 feet)
OR
PWL = $\overline{SPL}$ + 8.0 (for 1 meter)

System Noise Evaluation

The procedure for measuring system noise is different from the one used for components. Power unit systems are normally located in areas where background acoustics cannot be controlled. Guidelines for measurement in such environments are included in the National Machine Tool Builders Association's (NMTBA) "Noise Measurement Techniques" booklet.

Microphones are positioned ***1 meter from the perimeter of the machine and 1.5 meters above floor level***, as shown in Figure F-19. It's extremely important that these distances be measured accurately. This insures uniformity of measurement and comparison, and compliance to customer noise level specifications. A 4-inch position error at a nominal 1 meter distance can result in a 1 dB error in measurement accuracy.

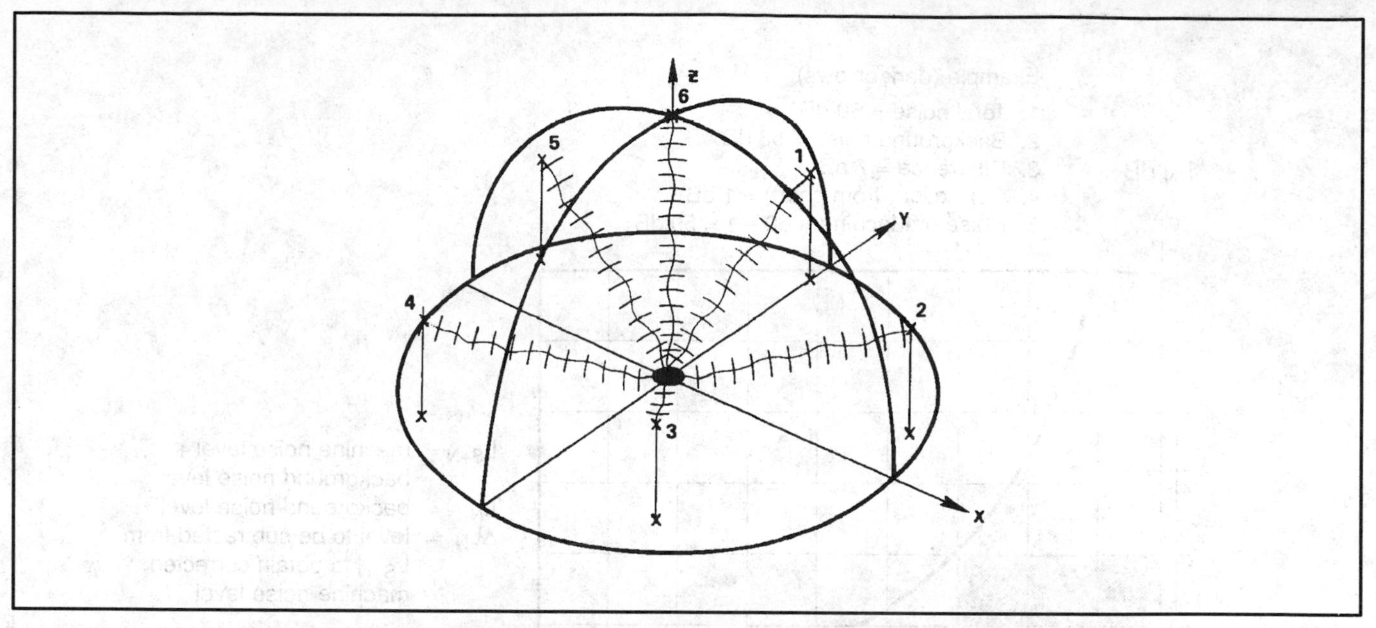

Figure F-18. Microphone Positions for Measuring Pump Noise

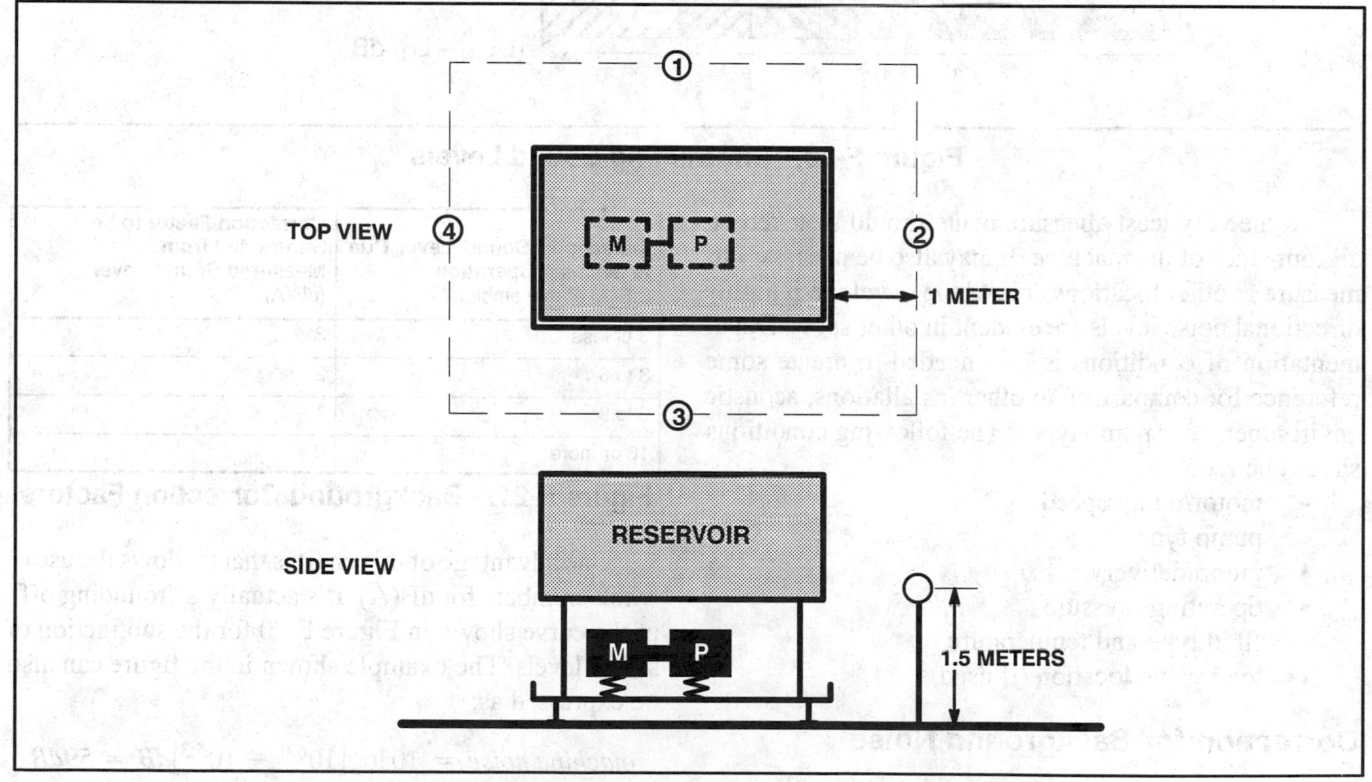

Figure F-19. Microphone Positions for Measuring System Noise

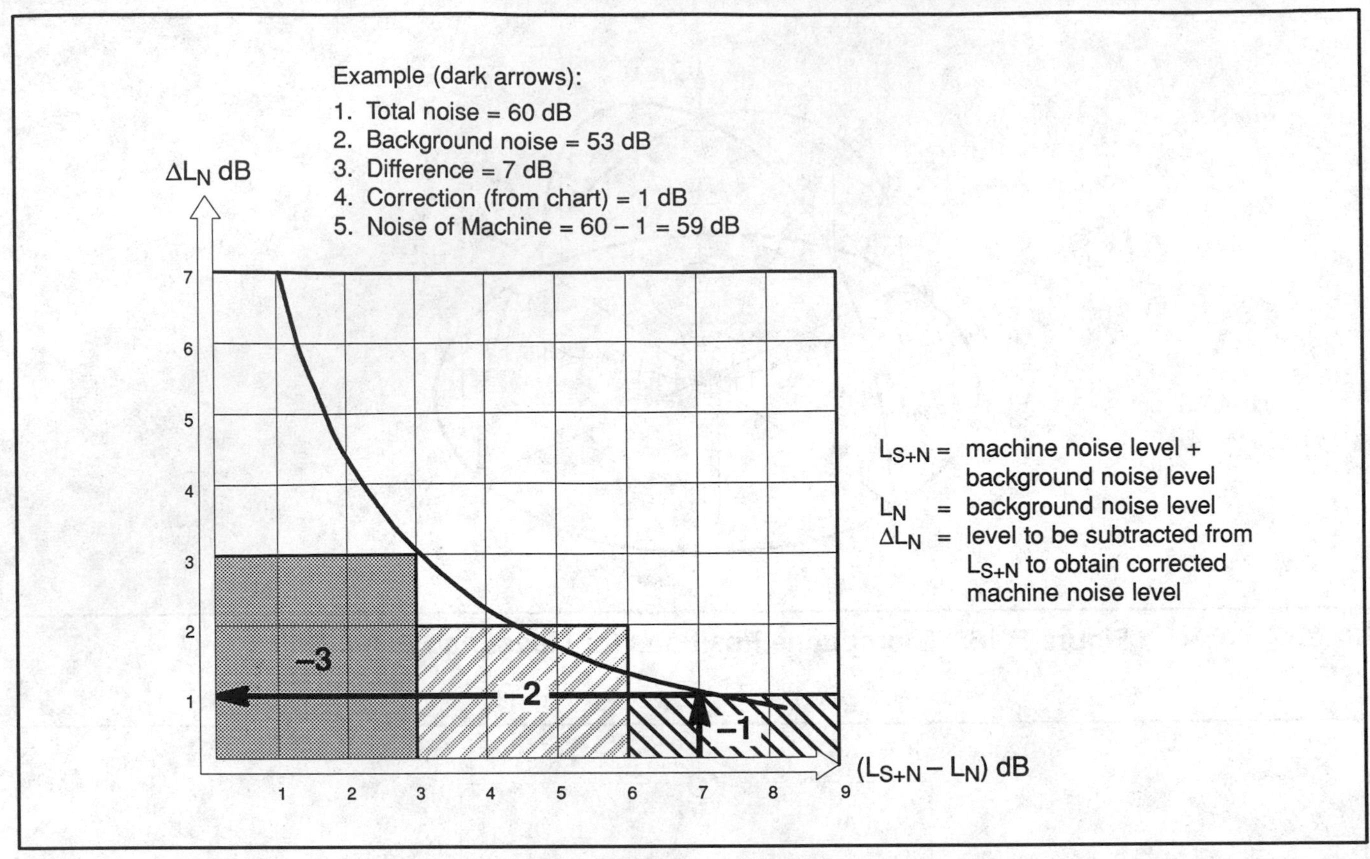
Example (dark arrows):
1. Total noise = 60 dB
2. Background noise = 53 dB
3. Difference = 7 dB
4. Correction (from chart) = 1 dB
5. Noise of Machine = 60 – 1 = 59 dB
ΔLN dB
7
6
5
4
3
2
1
–3
–2
–1
1 2 3 4 5 6 7 8 9
(LS+N – LN) dB
LS+N = machine noise level + background noise level
LN = background noise level
ΔLN = level to be subtracted from LS+N to obtain corrected machine noise level

Figure F-20. Subtracting Sound Levels

At the very least, measurements should be taken on all four sides of the machine. It may also be necessary to measure at other locations around the envelope if highly directional noise levels are evident in other spots. Documentation of conditions is also needed to create some reference for comparison to other installations, acoustic environments, or pump types. The following conditions should be recorded:

- motor/pump speed
- pump type
- pump delivery
- operating pressure
- fluid type and temperature
- load valve location (if used)

Correction for Background Noise

Accurate system noise measurement may involve correcting for the noise of the surrounding area. When ambient sound levels are within 10 dB(A) of the levels when the machine is operating, correction factors may be applied. This is done in accordance with Figure F-21, derived from the NMTBA booklet.

Increase in Sound Level Due to Machine Operation (dB(A) above ambient)	**Correction Factor to be Subtracted from Measured Sound Level** (dB(A))
3 or less	3
3 to 6	2
6 to 9	1
10 or more	0

Figure F-21. Background Correction Factors

One advantage of this chart is that it allows the use of whole numbers for dB(A). It's actually a "rounding off" of the curve shown in Figure F-20 for the subtraction of sound levels. The example shown in the figure can also be expressed as:

$$machine\ noise = 10\log[10^{6.0} - 10^{5.3}]dB = 59dB$$

Documentation of noise measurements made using four microphone positions might look like that shown in Figure F-22.

Noise level should be expressed as the maximum level measured (in this example, 78 dB in position 3) or possibly as the average of the four levels:

$$SPL = 10\log\left[\frac{10^{7.4} + 10^{7.0} + 10^{7.8} + 10^{7.5}}{4}\right]$$

$$= 75 \text{ dB(A)}$$

Measuring Position	1	2	3	4
	db(A)			
Total noise	76	73	79	75
Background noise	71	70	70	65
Background correction	-2	-3	-1	0
Machine noise	74	70	78	75

Figure F-22. Noise Measurement Documentation

INDEX OF TERMS

C

D

E

F

G

H

I

J

K

L

M

N

O

P

R

S

T

U